ex libris

45.-

Geo.

Rhizosphere: Achievements and Challenges

Yves Dessaux · Philippe Hinsinger · Philippe Lemanceau
Editors

Plants and Soil

Reprinted from *Plant and Soil*, Vol 321, nos 1-2.

Editors
Yves Dessaux
CNRS, Inst. Sciences due Vegetal (ISV)
avenue de la Terrasse,
91198 Gif-sur-Yvette, France

Philippe Hinsinger
INRA – SupAgro, UMR 1222 Biogeochimie du,
Sol et de la, Rhizosphere, place Viala,
34060 Montpellier, France

Philippe Lemanceau
Université Bourgogne, INRA-CMSE,
UMR 1229 Microbiologie du, Sol et de,
21065 Dijon, France

Cover caption:

Background photograph: Fababean (*Vicia faba* L.) grown in the long-term P-fertilizer field trial at Auzeville (INRA Toulouse), exhibiting roots with N_2-fixing nodules, abundant roots hairs and adhering soil, i.e. key players and features in the rhizosphere of legumes (photograph by P. Hinsinger).

Left insert photograph: *In situ* detection of *gfp*-tagged *Pseudomonas* sp. DSMZ 13134 cells on root surface of barley (*Hordeum vulgare* L.) using the CLSM (confocal laser scanning microscope LSM510, Carl Zeiss, Jena, Germany). Two-day old seedlings were inoculated with a bacterial suspension (10^8 cells per seedling). Plants were grown for two weeks in agricultural soil in pots in a greenhouse before analysis of the root colonization. Autofluorescent soil particles can be seen in the upper right corner (courtesy of K. Buddrus-Schiemann, Helmholtz Zentrum München, Neuherberg, Germany).

Right insert photograph: *In situ* detection of bacterial cells on the root surface of potato (*Solanum tuberosum* L.) grown under field conditions four weeks after planting. Fluorescence *in situ* hybridization (FISH) was performed using the oligonucleotide probe EUB-338-mix labeled with Fluos. Bacterial cells appear with the CLSM as green fluorescent signals and a clay particle can be seen as redish autofluorescence (courtesy of K. Buddrus-Schiemann, Helmholtz Zentrum München, Neuherberg, Germany).

ISBN: 978-90-481-2855-6   eISBN: 978-90-481-2857-0

DOI: 10.1007/978-90-481-2857-0

Library of Congress Control Number: 9789048128556

© Springer Science + Business Media B.V., 2010
No part of this work may be reproduced, stored in a retrieval system, or transmitted
in any form or by any means, electronic, mechanical, photocopying, microfilming, recording
or otherwise, without written permission from the Publisher, with the exception
of any material supplied specifically for the purpose of being entered
and executed on a computer system, for exclusive use by the purchaser of the work.

Printed on acid-free paper

springer.com

Contents

Editorial

Rhizosphere: so many achievements and even more challenges
Y. Dessaux · P. Hinsinger · P. Lemanceau   1–3

Review Articles

Carbon flow in the rhizosphere: carbon trading at the soil–root interface
D.L. Jones · C. Nguyen · R.D. Finlay   5–33

Nitrogen-fixing bacteria associated with leguminous and non-leguminous plants
C. Franche · K. Lindström · C. Elmerich   35–59

Biochemical cycling in the rhizosphere having an impact on global change
L. Philippot · S. Hallin · G. Börjesson · E.M. Baggs   61–81

Plant-microbe-soil interactions in the rhizosphere: an evolutionary perspective
H. Lambers · C. Mougel · B. Jaillard · P. Hinsinger   83–115

Rhizosphere: biophysics, biogeochemistry and ecological relevance
P. Hinsinger · A.G. Bengough · D. Vetterlein · I.M. Young   117–152

Plant root growth, architecture and function
A. Hodge · G. Berta · C. Doussan · F. Merchan · M. Crespi   153–187

The rhizosphere zoo: An overview of plant-associated communities of microorganisms, including phages, bacteria, archaea, and fungi, and of some of their structuring factors
M. Buée · W. De Boer · F. Martin · L. van Overbeek · E. Jurkevitch   189–212

Rhizosphere fauna: the functional and structural diversity of intimate interactions of soil fauna with plant roots
M. Bonkowski · C. Villenave · B. Griffiths   213–233

Plant-driven selection of microbes
A. Hartmann · M. Schmid · D. van Tuinen · G. Berg   235–257

Rhizosphere microbiota interfers with plant-plant interactions
A. Sanon · Z.N. Andrianjaka · Y. Prin · R. Bally · J. Thioulouse · G. Comte · R. Duponnois   259–278

Molecular communication in the rhizosphere
D. Faure · D. Vereecke · J.H.J. Leveau   279–303

Acquisition of phosphorus and nitrogen in the rhizosphere and plant growth promotion by microorganisms
A.E. Richardson · J.-M. Barea · A.M. McNeill · C. Prigent-Combaret   305–339

The rhizosphere: a playground and battlefield for soilborne pathogens and beneficial microorganisms
J.M. Raaijmakers · T.C. Paulitz · C. Steinberg · C. Alabouvette · Y. Moënne-Loccoz   341–361

Rhizosphere engineering and management for sustainable agriculture
P.R. Ryan · Y. Dessaux · L.S. Thomashow · D.M. Weller   363–383

Rhizosphere processes and management in plant-assisted bioremediation (phytoremediation) of soils
W.W. Wenzel   385–408

Novel approaches in plant breeding for rhizosphere-related traits
M. Wissuwa · M. Mazzola · C. Picard   **409–430**

Strategies and methods for studying the rhizosphere—the plant science toolbox
G. Neumann · T.S. George · C. Plassard   **431–456**

Sampling, defining, characterising and modeling the rhizosphere—the soil science tool box
J. Luster · A. Göttlein · B. Nowack · G. Sarret   **457–482**

Molecular tools in rhizosphere microbiology—from single-cell to whole-community analysis
J. Sørensen · M. Haubjerg Nicolaisen · E. Ron · P. Simonet   **483–512**

Iron dynamics in the rhizosphere as a case study for analyzing interactions between soils, plants and microbes
P. Lemanceau · P. Bauer · S. Kraemer · J.-F. Briat   **513–535**

EDITORIAL

Rhizosphere: so many achievements and even more challenges

Yves Dessaux · Philippe Hinsinger ·
Philippe Lemanceau

Received: 15 April 2009 / Accepted: 2 June 2009 / Published online: 11 June 2009
© Springer Science + Business Media B.V. 2009

The story of this "Rhizosphere book project" started about 3 years ago, as the three of us were discussing the organization of the "International Rhizosphere 2 Conference" held in Montpellier in 2007 (Hartmann et al. 2008a; Jones and Hinsinger 2008). At one point of this conversation, we noticed that previous text books related to rhizosphere—though remarkable—only dealt with partial aspects of rhizosphere knowledge, e.g. rhizosphere and plant nutrition (Marschner 1995), rhizosphere and plant health (Parker et al. 1985), rhizosphere microbiology (Mukerji et al. 2006; Varma et al. 2004), biochemistry (Pinton et al. 2001 and 2007) and ecology (Cardon and Whitbeck 2007), while the more general ones needed significant update, having been published prior to the era of modern molecular biology and molecular microbial ecology (Curl and Truelove 1986; Kesters and Cregan 1989; Lynch 1990).

With the above in mind, we decided to contact a number of colleagues all over the world to assess first whether they shared the need for such an update, and second, whether they would contribute to this venture. Their enthusiastic answers definitively prompted us to initiate the project.

The aim of this book was to provide a **holistic view of the rhizosphere,** keeping in mind the pioneer rhizosphere concept of Hiltner (Hartmann et al. 2008b) and its unique functioning that implies numerous, strong and complex interactions between plant roots, soil constituents and microorganisms. Furthermore, this book not only aimed at addressing **current knowledge** and achievements but also at outlining the **future challenges** that stand in front of rhizosphere sciences. We, as editors, therefore rapidly faced the difficulty to tackle, as much as possible, **all these facets of rhizosphere knowledge in a single book**. A meeting of the three editors at a café in Paris Quartier Latin led to the definition of the current structure of the book, with five sections covering soil sciences to microbial ecology, plant sciences to biotechnology.

The first section of this book describes **rhizosphere as a central component of ecosystems and**

Responsible Editor: Hans Lambers.

Y. Dessaux (✉)
UPR2355, Institut des Sciences du Végétal,
CNRS, Avenue de la terrasse,
91198 Gif-sur-Yvette CEDEX, France
e-mail: dessaux@isv.cnrs-gif.fr

P. Hinsinger
UMR 1222 Eco&Sols Ecologie Fonctionnelle et
Biogéochimie des Sols (INRA—IRD—SupAgro),
INRA, Place Viala,
34060 Montpellier, France

P. Lemanceau
UMR 1229 Microbiologie du Sol et de l'Environnement,
INRA, Université de Bourgogne,
17 Rue Sully, BV 86510,
21034 Dijon, France

biogeochemical cycles, especially with respect to carbon and nitrogen.. The hot topic of global change is addressed in a separate chapter, given the environmental relevance of this issue at the beginning of the 21st Century. The last chapter opens an even broader perspective of the evolution of the rhizosphere and co-evolution of its key components (soil, plants and soil (micro)biota) over long terms, including geological time scales.

The second section takes into account the simple observation that rhizosphere is a **multiple interface between soils, plant roots, microbes and fauna**. These "key players" are not presented in a descriptive approach as usual, but rather with a functional approach of their complex roles in the numerous processes that occur in the rhizosphere, and make it so unique as a soil habitat.

As Lorenz Hiltner reported as early as 1904 (Hiltner 1904, see Hartmann et al. 2008b)—a visionary notion at that time—**the rhizosphere is a place where the above mentioned biological components strongly interact**. These interactions occur not only between soils and plant roots, or plant roots and microbes, but also between plants themselves, and microbes themselves, through numerous signaling molecules and complex pathways. Such complex interactions have major implications for **plant nutrition and health**. These various aspects are presented in the third section of this book. They provide the bases for potential applications in the context of the current challenge to significantly reduce the use of fertilizers and pesticides in agroecosystems, and to further increase their productivity at the global level to meet the food demand of a growing world population (Griffon 2006).

The **need for the development of sustainable agricultural practices and a "cleaner" Earth environment** is supported by an increasing social demand. These features justify the three chapters gathered as section 4 under the title "**ecological engineering**". They address the agronomic issues either through plant breeding or rhizosphere engineering. They also address bioremediation, a stimulating and crucial topic for preserving soil and water quality, in relation with pesticide residues, resulting from agricultural practices, and contaminants, such as polycyclic aromatic hydrocarbons or heavy metals, resulting from industrial activities. As such, these chapters describe some of the outcomes of several years of investigations on rhizosphere interactions, as innovative management techniques of agroecosystems.

Last, section 5 is a more methodological one as it was designed to gather **technique-oriented contributions**. Three articles introduce the currently available and **state-of-the-art tool boxes** that allow the study of rhizosphere processes and properties, from the perspectives of major rhizosphere components, namely plants, soils, and microbes. The last chapter aggregates these research strategies and tools, and presents how they can be best combined through a case study dealing with iron (bio)availability, sequestration and metabolism in the rhizosphere.

In order to achieve the ambitious objectives of a comprehensive, holistic update of rhizosphere knowledge, we proposed an **unusual work organization** to the authors. Rather than asking a single expert research team—the "usual suspects"—to contribute to one chapter, we contacted several authors whose expertise best complemented each other's and asked them to work together to produce a given chapter. We hoped that each chapter would strongly benefit from the complimentary points of view of the authors, in terms of **quality, originality and comprehensiveness**. This unusual organization worked quite well in several instances, and appeared less successful in others. Whatever the difficulties faced by the authors or by us as editors, we wish to thank very warmly all contributors for having played with these atypical rules to produce the excellent chapters that will most likely become highly cited, classic references. We also warmly thank the Springer editing group, Maryse Walsh in the first place, whose constant support contributed noticeably to the completion of this book. Thanks to Hans Lambers who made it possible to have this opus published both as a special issue of the refereed journal *Plant and Soil* and a book, which shall greatly enhance the dissemination of rhizosphere knowledge to both rhizosphere and non-rhizosphere specialists.

In the last years, rhizosphere scientists have progressively built up a vast, interacting and active community, via the organization of two International Rhizosphere conferences. The first of these, "Rhizosphere 2004 Conference—Perspectives and Challenges—A Tribute to Lorenz Hiltner", was held in Munich, Germany in September 2004 and gathered 480 scientists from worldwide (Hartmann et al. 2004; Hinsinger and

Marschner 2006; Jones 2006; Jones et al. 2006; Sen 2005; Smalla et al. 2006). It was hosted by Anton (Toni) Hartmann, and organized largely with the support of a European Science Foundation COST Action "Understanding and Modeling Plant-Soil Interactions in the Rhizosphere Environment" chaired from 2002 to 2004 by Walter Wenzel and from 2004 to 2006 by Philippe Hinsinger. The idea to organize such an event originated from the success of a series of Rhizosphere meetings held in France in 1989, 1997 and 2001 and from Tara Singh Gahoonia's idea to make something very special the year of the centenary of the rhizosphere concept. This Rhizosphere 2004 Conference considerably helped structuring the vast and so far fragmented (along disciplines) community of rhizosphere scientists, who strongly supported the idea to make such events regular. The Rhizosphere 2 Conference was thus held in Montpellier, France, in August 2007, hosted by Philippe Hinsinger, gathered 570 participants from 48 countries, while Rhizosphere 3 will be organized by Hans Lambers in Perth, Australia in September 2011. This young and successful series of International Conferences largely made it possible for us all, editors and co-authors of this book, to meet, exchange ideas and open our respective views of the rhizosphere as a unique micro-environment. They have definitely been the ferment of the various chapters of this book.

In a foreword of the proceedings of the "Conférence Rhizosphère" held in Dijon in 2001, the "ancestor" of the International Rhizosphere Conferences, we quoted Leonardo Da Vinci who stated in the Renaissance era *"We know better the mechanics of celestial bodies than the functioning of the soil below our feet"* (Dessaux et al. 2003). Though there is still some truth in this assertion, we strongly hope that this rhizosphere book will reveal to the reader how much progress have been made in understanding the various aspects of rhizosphere composition, structure, function and applications, as well as the analytical and methodological issues and the rapid evolution of the corresponding techniques to study the rhizosphere, especially over the last 20 years.

Acknowledgements We thank all of our many authors for their commitment to this project and for the patience of those who stuck to the initial, rather short deadlines. We also thank the numerous reviewers for having contributed to improving the quality of this special issue of Plant and Soil and book on Rhizosphere achievements and challenges.

References

Cardon Z, Whitbeck J (2007) The Rhizosphere. Elsevier, An Ecological Perspective

Curl EA, Truelove B (1986) The rhizosphere. Springer-Verlag, Berlin, New York, Heidelberg, Tokyo

Dessaux Y, Hinsinger P, Lemanceau P (2003) Foreword of the special issue "Third Rhizosphere Conference". Agronomie 23:373

Griffon M (2006) Nourrir la planète. Odile Jacob Sciences, Paris. 455p.

Hartmann A, Schmid M, Wenzel W, Hinsinger P. (2004) Rhizosphere 2004—Perspectives and Challenges—A Tribute to Lorenz Hiltner. Book of Abstracts, GSF-Bericht, GSF-National Research Center for Environment and Health, Neuherberg. 333 p.

Hartmann A, Lemanceau P, Prosser JI (2008a) Multitrophic interactions in the rhizosphere—Rhizosphere microbiology: at the interface of many disciplines and expertises. FEMS Microbiol Ecol 65:179

Hartmann A, Rothballer M, Schmid M (2008b) Lorenz Hiltner, a pioneer in rhizosphere microbial ecology and soil bacteriology research. Plant Soil 312:7–14

Hiltner L (1904) Über neuere Erfahrungen und Probleme auf dem Gebiete der Bodenbakteriologie unter besonderer Berücksichtigung der Gründüngung und Brache. Arbeiten der Deutschen Landwirtschaftlichen Gesellschaft 98:59–78

Hinsinger P, Marschner P (2006) Rhizosphere 2004 Conference—Perspectives and Challenges—A Tribute to Lorenz Hiltner. Plant Soil 283:VII–VIII

Jones DL (2006) Rhizosphere Congress 2004—Perspectives and Challenges, Munich, September 2004. Soil Biol Biochem 38:1177

Jones DL, Hinsinger P (2008) The rhizosphere: complex by design. Plant Soil 312:1–6

Jones DL, Kirk GJD, Staunton S (2006) Foreword to the "Rhizosphere 2004" papers. Eur J Soil Sci 57:1–1

Kesters DL, Cregan PB (1989) The rhizosphere and plant growth. Kluwer Academic Publishers, Dordrecht

Lynch JM (1990) The Rhizosphere, Wiley Interscience, John Wiley & Sons Ltd., Chichester. 581 p.

Marschner H (1995) Mineral Nutrition of Higher Plants. 2nd ed. Academic Press, London. 899 p.

Mukerji KG, Manoharachary C, Singh J (2006) Microbial Activity in the Rhizosphere, Springer

Parker CA, Rovira AD, Moore KJ, Wong PTW, Kollmorgen JF (1985) Ecology and management of soil borne pathogens, APS Press. St-Paul, XXX p

Pinton R, Varanini Z, Nannipieri P (2001) The rhizosphere. Biochemistry and Organic Substances at the Soil-Plant Interface, CRC Press

Pinton R, Varanini Z, Nannipieri P (2007) The rhizosphere. Biochemistry and Organic Substances at the Soil-Plant Interface, 2nd Edition. CRC Press

Sen R (2005) Towards a multifunctional rhizosphere concept: back to the future ? New Phytol 168:266–268

Smalla K, Sessitsch A, Hartmann A (2006) The Rhizosphere: "soil compartment influenced by the root". FEMS Microb Ecol 56:165

Varma A, Abbott L, Werner D, Hampp R (2004) Plant Surface Microbiology, Springer-Verlag, Heidelberg. 616 p.

REVIEW ARTICLE

Carbon flow in the rhizosphere: carbon trading at the soil–root interface

D. L. Jones · C. Nguyen · R. D. Finlay

Received: 30 July 2008 / Accepted: 4 February 2009 / Published online: 25 February 2009
© Springer Science + Business Media B.V. 2009

Abstract The loss of organic and inorganic carbon from roots into soil underpins nearly all the major changes that occur in the rhizosphere. In this review we explore the mechanistic basis of organic carbon and nitrogen flow in the rhizosphere. It is clear that C and N flow in the rhizosphere is extremely complex, being highly plant and environment dependent and varying both spatially and temporally along the root. Consequently, the amount and type of rhizodeposits (e.g. exudates, border cells, mucilage) remains highly context specific. This has severely limited our capacity to quantify and model the amount of rhizodeposition in ecosystem processes such as C sequestration and nutrient acquisition. It is now evident that C and N flow at the soil–root interface is bidirectional with C and N being lost from roots and taken up from the soil simultaneously. Here we present four alternative hypotheses to explain why high and low molecular weight organic compounds are actively cycled in the rhizosphere. These include: (1) indirect, fortuitous root exudate recapture as part of the root's C and N distribution network, (2) direct re-uptake to enhance the plant's C efficiency and to reduce rhizosphere microbial growth and pathogen attack, (3) direct uptake to recapture organic nutrients released from soil organic matter, and (4) for inter-root and root–microbial signal exchange. Due to severe flaws in the interpretation of commonly used isotopic labelling techniques, there is still great uncertainty surrounding the importance of these individual fluxes in the rhizosphere. Due to the importance of rhizodeposition in regulating ecosystem functioning, it is critical that future research focuses on resolving the quantitative importance of the different C and N fluxes operating in the rhizosphere and the ways in which these vary spatially and temporally.

Keywords Carbon cycling · Nitrogen cycling · Mycorrhizas · Organic matter · Review · Rhizodeposition · Root processes · Signal transduction

Responsible Editor: Philippe Hinsinger.

D. L. Jones (✉)
School of the Environment & Natural Resources,
Bangor University,
Bangor, Gwynedd LL57 2UW, UK
e-mail: afs080@bangor.ac.uk

C. Nguyen
INRA, UMR1220 TCEM,
71 Avenue Edouard Bourlaux, BP 81,
33883 Villenave d'Ornon, France

R. D. Finlay
Uppsala BioCenter,
Department of Forest Mycology and Pathology, SLU,
Box 7026, SE-750 07, Uppsala, Sweden

Introduction

For over a century it has been established that plants can dramatically modify their soil environment giving rise to the so called rhizosphere effect (Clark 1949;

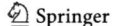

Rovira 1965; Whipps 2001). Although the initial trigger of this rhizosphere effect was not identified, subsequent research has shown that it is largely induced by the release of carbon (C) from roots into the surrounding soil. Although roots can release large amounts of inorganic C which may directly affect the biogeochemistry of the soil (Cheng et al. 1993; Hinsinger 2001; Hinsinger et al. 2009), it is the release of organic carbon that produces the most dramatic changes in the physical, biological and chemical nature of the soil. In its broadest sense, this release of organic C is often termed rhizodeposition (Jones et al. 2004). The term rhizodeposition includes a wide range of processes by which C enters the soil including: (1) root cap and border cell loss, (2) death and lysis of root cells (cortex, root hairs etc), (3) flow of C to root-associated symbionts living in the soil (e.g. mycorrhizas), (4) gaseous losses, (5) leakage of solutes from living cells (root exudates), and (5) insoluble polymer secretion from living cells (mucilage; Fig. 1). Although these loss pathways can be clearly differentiated between at a conceptual level it is often extremely difficult at the experimental level to discriminate between them in both space and time. Consequently, while individual studies have shown that these can all occur, probably simultaneously in the same plant root system, it is almost impossible to rank the relative importance of each process. Further, as we understand more about the mechanisms of C flow in both soil and roots we find that many of the published results are severely biased by the experimental system in which individual factors or processes were examined (Jones and Darrah 1993; Meharg 1994; Kuzyakov 2006). This has left the literature on rhizosphere C flow awash with studies which may bear no relationship to real world events, particularly those performed in the absence of soil. Despite this, however, it is clearly apparent that our incremental approach to understanding C flow is paying dividends from both a commercial and environmental perspective. Firstly, from a commercial perspective it is clear that root C excretions can be useful for the non-destructive production of high value pharmaceuticals, pigments and flavours for use in the medical and cosmetic industries (Oksman-Caldentey and Inze 2004). In these applications, roots are typically transformed with *Agrobacterium rhizogenes* which induces hairy root disease. The neoplastic (cancerous) transformed roots are genetically stable and can grow

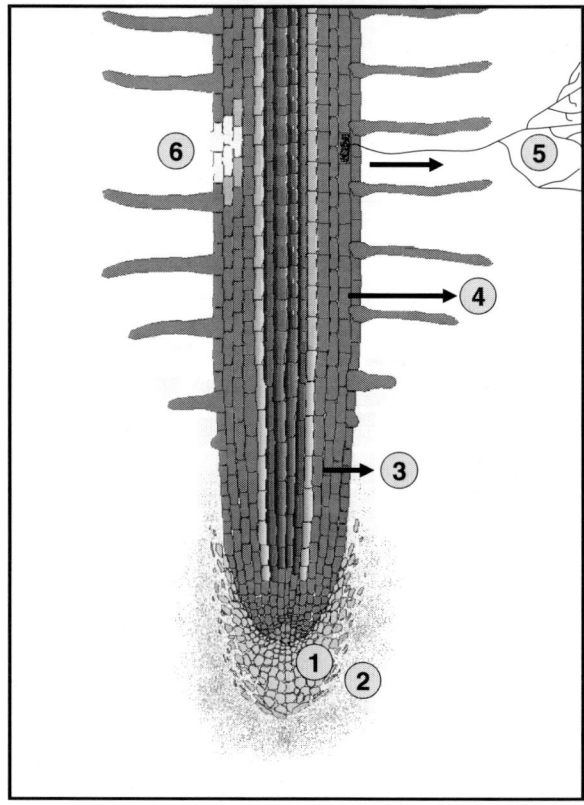

Fig. 1 Schematic representation of a longitudinal section of a growing root showing the six major sites of rhizodeposition: *1* loss of root cap and border cells, *2* loss of insoluble mucilage, *3* loss of soluble root exudates, *4* loss of volatile organic C, *5* loss of C to symbionts (e.g. arbuscular mycorrhizas), and *6* loss of C due to death and lysis of root epidermal and cortical cells

rapidly in the absence of shoots in a hormone free medium making them suitable for the controlled excretion and collection of secondary metabolites (Srivastava and Srivastava 2007). In our efforts to create a more sustainable environment, it is also clear that rates of release of C compounds from roots can be manipulated to increase food production, enhance water conservation, speed up the remediation of contaminated sites, and reduce the need for artificial fertilizers and pesticides (Lasat 2002; Vessey 2003; Welbaum et al. 2007). Thus while the intricate complexity of the rhizosphere continues to amaze us and presents a real challenge to scientists trying to unravel its diverse web of interactions, it also has the potential to offer great benefits to society. As C flow from roots is essentially the starting point from which the rhizosphere develops it is important that we improve our understanding of this process.

This review aims to critically assess our current understanding of rhizosphere C flow and to highlight areas for further research. Due to the large number of publications in this research area (>5000 individual publications) it is not our aim to cover the entire literature but to highlight examples to support themes. Readers requiring more comprehensive historical reviews of the literature should consult Rovira (1969), Curl and Truelove (1986), Lynch (1990) and Pinton et al. (2001).

Roots release a great variety of compounds by different mechanisms

Virtually, all compounds contained in root tissues can be released into the soil. In hydroponic culture, carbohydrates, organic and amino acids, phenolics, fatty acids, sterols, enzymes, vitamins, hormones, nucleosides have been found in the root bathing solution (Rovira 1969; Grayston et al. 1996; Dakora and Phillips 2002; Read et al. 2003; Leinweber et al. 2008). These compounds are released by various mechanisms including secretion, diffusion or cell lysis. Depending upon the question being addressed, different nomenclatures for rhizodeposits have been proposed based for instance on the mechanisms of release, on the biochemical nature of rhizodeposits or on their functions in the rhizosphere. The classification first proposed by Rovira et al. (1979) is generic and has been extensively used.

Mucilage

Root mucilage forms a gelatinous layer surrounding the root tip and is one of the few clearly visible signs of organic C excretion from roots (Fig. 2). It is mainly composed of polysaccharides of 10^6–10^8 Da in size (Paull et al. 1975) and is actively secreted by exocytosis from root cap cells (Morre et al. 1967; Paull and Jones 1975a, b, 1976a, b). Alongside polysaccharides, it also contains proteins (ca. 6% of dry weight; Bacic et al. 1987) and some phospholipids (Read et al. 2003). In most situations mucilage released into the soil confers a wide range of benefits to the plant. For example, the carboxylic groups of mucilage can complex potentially toxic metals (e.g. Al, Cd, Zn, Cu), protecting the root meristem (Morel et al. 1986; Mench et al. 1987). In addition, mucilage enhances soil aggregate stability which in the long-term promotes soil aeration, root growth and reduces

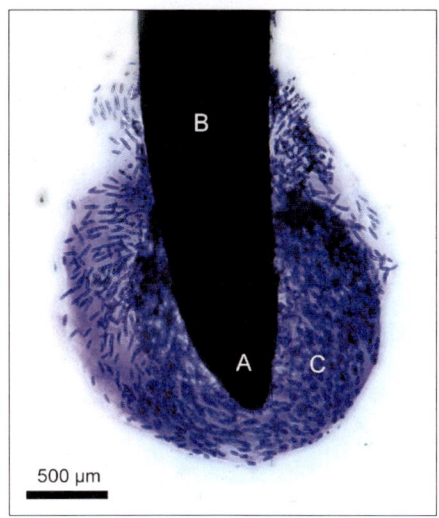

Fig. 2 Light microscope image showing the large amount of mucilage (*blue* halo surrounding the root) and border cells production in a *Zea mays* L. root tip. *Labels* indicate the root quiescent centre (*A*), the main root elongation zone (*B*), and the mucilage halo in which the border cell are embedded (*C*). The mucilage is stained with *aniline blue*

soil erosion (Guckert et al. 1975; Morel et al. 1990; Czarnes et al. 2000). Mucilage also possesses a high intrinsic affinity for water and when fully hydrated, has a water content 100,000 times greater than its dry weight (McCully and Boyer 1997) and expands to form a viscous droplet covering the root tip. Such properties play a role in maintaining the continuity of water flow towards the rhizoplane (Read and Gregory 1997; Read et al. 2003) and in reducing the frictional resistance as the root tip moves through the soil (Iijima et al. 2004). Recent work has also suggested that specific mucilage components (e.g. prenylated stilbenes) possess antimicrobial properties and may be important in preventing pathogen attack (Sobolev et al. 2006). Apart from the C employed to synthesize and secrete mucilage, its loss into the soil appears to have no known negative effects on soil and plant health. Of most concern is that mucilage represents a source of labile C in the soil and is consequently rapidly consumed by soil microorganisms (typical half-life of 3 days). In some instances this can induce the proliferation of root rot disease-causing organisms in the rhizosphere (e.g. *Pythium aphanidermatum*; Zheng et al. 2000) while in other situations due to its high C:N ratio (ca. 65), its biodegradation induces a transient net immobilisation of N in the rhizosphere (Mary et al. 1993; Nguyen et al. 2008). While increasing the rate of polysaccharide mucilage release from roots is possible and would only constitute a minor C drain (Darrah 1991a), it is unlikely to yield major benefits in comparison to alteration in other rhizodeposition processes.

Border cells

Also called sloughed-off cells, border cells are the cells that detach from the external layers of the root cap, which is continuously renewed (Barlow 1975; Fig. 2). The controlled separation of the border cells reduces the frictional force experienced by the root tip (Bengough and McKenzie 1997; Bengough and Kirby 1999; Iijima et al. 2004). The daily rate of border cell production is highly variable among plant species, from none to tens of thousands with release rates highly dependent upon the prevailing environmental conditions (Hawes et al. 1998; Iijima et al. 2000; Zhao et al. 2000). Once detached from the cap, border cells remain alive in the soil for several days (Stubbs et al. 2004). They are surrounded by the mucilage they secrete, which binds heavy metals away from the root meristem (Miyasaka and Hawes 2001). Border cells also produce signal compounds involved in the protection of meristem against pathogens (Hawes et al. 2000) and in the promotion of symbiosis (Brigham et al. 1995; Hawes et al. 1998). Recent work has also suggested that border cells can act as a decoy luring pathogenic nematodes and fungi away from the main root axis (Gunawardena and Hawes 2002; Rodger et al. 2003). However, contradictory results have also been found highlighting the difficulties of manipulating border cell release and physiology for disease control (Wuyts et al. 2006; Knox et al. 2007). While border cells may provide a convenient mechanism for compound delivery to soil, further fundamental work is required to characterise the metabolomic and proteomic expression patterns in comparison to other root cells to understand and capitalize on their unique attributes (Jiang et al. 2006). In the total rhizodeposition C budget, however, border cells only constitute a small proportion of the C entering the soil (Iijima et al. 2000; Farrar et al. 2003).

Exudates

Exudates are defined as diffusible compounds which are lost passively by the root and over which the root exerts little direct control. Rates of loss of individual compounds depend upon three critical factors, namely (1) the root-soil concentration gradient, (2) the permeability of the plasma membrane, and (3) the spatial location of the solutes in the root tissue (e.g. epidermis versus stele). The dominant organic compounds in roots reflect those compounds central to cell metabolism and include free sugars (e.g. glucose, sucrose), amino acids (e.g. glycine, glutamate) and organic acids (e.g. citrate, malate, oxalate; Kraffczyk et al. 1984). Their concentration inside the root is typically orders of magnitude greater than that in the surrounding soil solution due to continual removal from the soil by the soil microbial community and replenishment of internal pools by the root. Although we know a great deal about the concentrations of solutes in whole roots our understanding of the spatial and temporal dynamics of organic solutes in roots is severely limited. As there is significant internal partitioning of solutes in root cells (e.g. cytoplasm versus vacuole; Gout et al. 1993; Ciereszko et al. 1999), of critical importance is the actual concentration gradient that exists between the cytoplasm and the cell wall space rather than that between the whole root and bulk soil. In addition, at a tissue level, the role of the cortex in root exudation versus that of the epidermis remains unknown. Although apoplastic loss may represent a slow diffusion pathway in comparison to direct loss from the epidermis (Canny 1995; Fleischer and Ehwald 1995), evidence suggests that gaps between epidermal cells (where apoplastic loss ultimately manifests itself) are strong regions of microbial colonization and therefore C availability (Quadt-Hallmann et al. 1997; Watt et al. 2006). Therefore more work is required to characterise apoplastic loss pathways from the root cortex and its contribution to maintaining the endo- and ectorhizosphere microbial community. Further work is also required to determine the temporal dynamics of solutes in root tissues (e.g. diurnal versus ontogenetic) and the relationship with exudation.

The cytoplasmic pH of most root cells ranges from 7.2–7.5. Within this range most organic acids are negatively charged while most amino acids and sugars carry no net charge. Due to plasma membrane H^+-ATPases pumping H^+ out of the cells, the outside of the plasma membrane carries more positive charge than the inside (Fig. 3). Consequently, there is a greater tendency for anionic organic solutes to be

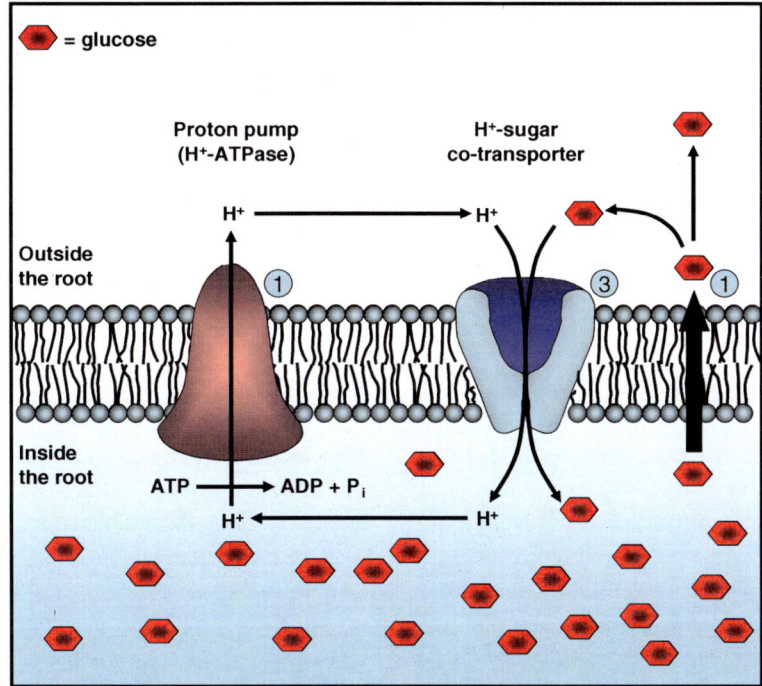

Fig. 3 Schematic representation of the three main processes involved in the bi-directional flux of low molecular weight organic solutes (e.g. glucose) across the soil root interface. Flux (*1*) denotes the passive transport of glucose across the plasma membrane in response to the large cytoplasm (20 mM) to soil solution (10 μM) concentration gradient. Flux (*2*) denotes the active energization of the plasma membrane by the H^+-ATPase which pumps H^+ out of the cell using ATP as the energy source. Flux (*3*) denotes the active re-uptake of sugars from the soil solution back into the cytoplasm using a H^+-cotransport protein. The cell wall is not drawn for clarity

drawn across the membrane at faster rates than non-charged solutes (Ryan et al. 2001). Studies in non-plant systems suggest that although solutes can diffuse through the lipid bilayer, faster rates of diffusion occur at the lipid–protein boundary. Further, organic solute loss may be accelerated at sites where active growth is occurring as membrane vesicle contents are released during fusion with the plasma membrane. Rates of exudation can also be greatly speeded up by the opening of solute specific channels in the membrane. Probably the best known example of this is the release of organic acids when roots experience either P deficiency or high external concentrations of free toxic Al^{3+} (Zhang et al. 2004; Ligaba et al. 2006). The release of organic acids such as citrate, malate and oxalate can complex the Al^{3+} rendering it non-toxic. Detailed reviews of the role of organic acid channels in metal detoxification and nutrient uptake (e.g. P) can be found in Ryan et al. (2001), Jones et al. (2004) and Roberts (2006).

Generally, rates of exudate loss are greater at root tips in comparison to mature root regions (McDougal and Rovira 1970; Hoffland et al. 1989). Potential reasons to explain this enhanced C loss from tips include: (1) higher solute concentrations in root tip regions thereby creating a larger diffusion gradient (Jones and Darrah 1996; Jones et al. 1996), (2) small vacuolar volume of root tip cells inducing higher cytoplasmic concentrations (Patel et al. 1990), greater surface area-to volume ratio of tip cells, (3) the lack of an endodermal layer to minimize cortical loss (Schraut et al. 2004), (4) increased rates of apoplastic solute unloading from the vascular tissue leading to greater apoplastic loss (Bockenhoff et al. 1996), (5) greater apoplastic volume inducing higher rates of solute diffusion (Kramer et al. 2007), (6) higher rates of growth in tip regions and therefore solute loss during vesicle fusion and signalling events (Beemster and Baskin 1998; Roux and Steinebrunner 2007), and (7) localized loss of root border and cap cells which may undergo apoptosis releasing solutes (Shishkova and Dubrovsky 2005). Like many other aspects of rhizodeposition our conceptual understanding is good, however, our detailed mechanistic understanding of the relative importance of the individual flux pathways remains poor and this must remain as a priority for future research.

Secretions (excluding mucilage)

Plant roots actively secrete various compounds in response to a range of environmental conditions and our understanding of the role of these compounds in rhizosphere processes often remains poor (Wen et al. 2007). One exception is the characterisation of phytosiderophore release by grasses under conditions of low Fe availability (Negishi et al. 2002). In this situation, Fe-phytosiderophore complexes are also actively taken back into the plant (Haydon and Cobbett 2007). Phenolics are also secreted from roots and have been implicated in the mobilization of nutrients such as Fe and P, however, their quantitative importance remains unknown (Dakora and Phillips 2002). Enzymes (e.g. phosphatase) and many other compounds such as secondary metabolites may also be secreted into the rhizosphere and participate in the interactions between the roots and their environment (Bais et al. 2004). High molecular weight compounds or toxic molecules are likely to be released by exocytosis (Verpoorte et al. 2000). However, the mode of release is not always clearly established and much further work is required to elucidate mechanisms of release and their quantitative significance in the soil.

Senescence-derived compounds

Depending upon the conditions experienced by the root, a variable part of the epidermis including root hairs and of the cortical cells can degenerate and release their content into the rhizosphere (Fusseder 1987; McCully 1999). As roots rarely senesce in hydroponic culture, this process is largely thought to occur in soil where pathogens and mineral abrasion can induce cell death. Little is known about the magnitude of this flux pathway as it is almost impossible to study in soil. Consequently, most measurements typically rely on quantifying the amount of epidermal and cortical cell loss rather than the amount of C transferred to the soil. The amount entering the soil can be expected to depend upon whether the roots undergo programmed (apoptosis) or spontaneous cell death, however, little is known about the relative importance of these two processes. However, we do know that plant roots contain a significant amount of soluble and insoluble C and that their death will results in a significant C and N input to the soil and an elevation of microbial populations in their necrosphere (McClaugherty et al. 1982; Nadelhoffer and Raich 1992; Stewart and Frank 2008). Typically, there is a positive correlation between root diameter and lifespan (Gill and Jackson 2000). Consequently, in temperate agricultural grasslands containing an abundance of fine roots, we can calculate the magnitude of the C input to soil from root turnover. The soil organic C content of a temperate, grazed grassland soil typically ranges from 10 to 50 g C kg soil^{-1} while the standing root biomass typically ranges from 5 to 15 g root-C kg soil^{-1} and the microbial biomass from 0.5 to 1 g C kg soil^{-1} of which we assume only 10% is active (Jones, unpublished). It has been estimated that in the growing season approximately 25% of the roots turn over each month equating to approximately 2 to 10 g C kg soil^{-1} month^{-1} (i.e. enough C to generate 50 to 100 times the size of the active microbial biomass in soil). This can be compared to the rates of C exudation from grass roots which typically range from 1 to 10 mg C g root-C^{-1} day^{-1} (Hodge et al. 1997; Paterson and Sim 1999; Paterson et al. 2003). Consequently, we can estimate the amount of C entering grassland soils from root exudation to be in the range 0.1 to 5 g C kg soil^{-1} month^{-1} similar to that derived from root turnover.

Carbon flow to mycorrhizal and bacterial symbionts

Apart from the bacterial-legume symbiosis which has been reviewed extensively, little is known about the flow of C to other bacterial symbionts in the rhizosphere (Dilworth et al. 2008; Ohyama et al. 2009). Consequently, here we will focus on mycorrhizas. Most plants in natural and semi-natural vegetation systems form symbiotic associations with mycorrhizal fungi and there is increasing evidence to suggest that the flow of C to and through this symbiotic interface may be of significance in many plant–soil interactions, playing an important role in different biogeochemical processes (Finlay and Rosling 2006; Finlay 2008). Mycorrhizal symbionts contribute to carbon flow in the rhizosphere in three main ways. Firstly, the investment of C in production of biomass of intra- and extraradical mycelial structures is, in itself, substantial (Leake et al. 2004). Secondly, there is a flow of C through these structures, resulting in release of a range of exudates into the mycorrhizosphere, and thirdly, these compounds, and the mycorrhizal mycelium itself, can be used as energy rich substrates by other organisms, resulting in respiratory loss of carbon as CO_2. As with studies of other components of rhizodeposition, considerable effort has been directed at quantifying the contribution of these processes in relation to total rhizosphere C flow, whilst fewer studies have focused on their potential functional roles.

Because of their fine dimensions and fragility, mycorrhizal hyphae are even more difficult to study than fine roots. The mycelium is easily damaged when excavating roots, it consists of viable and non-viable fractions and must be distinguished from the mycelia of saprotrophic and pathogenic fungi. Despite these difficulties, much knowledge has been gained about the structure, biology and impact of mycorrhizal mycelia (see Leake et al. 2004 for an extensive review). Over 50 estimates of mycelial production by arbuscular mycorrhizal (AM) fungi or ectomycorrhizal (EM) fungi are cited from a range of pot and field studies. Estimates of hyphal length for AM fungi typically range from 3–30 m g^{-1} soil but 68–101 m g^{-1} soil have been recorded in undisturbed grasslands with permanent plant cover. EM hyphae are more difficult to distinguish morphologically from saprotrophic fungi and hyphal length estimates are less reliable but available data suggest hyphal length densities of between 3 to 600 m g^{-1} soil. Wallander et al. (2001) used a combination of techniques such as in-growth mesh bags, measurements of fungal markers such as phospholipid fatty acids and ergosterol, $\delta^{13}C$ values and trenching to distinguish mycorrhizal fungi from soil dwelling saprotrophs. The total amount of EM mycelium colonising the mesh bags was calculated to be 125–200 kg ha^{-1} and the total amount of EM mycelium, including EM mantles was estimated to be 700–900 kg ha^{-1}.

Clearly the investment of C in mycelial structures is considerable and many attempts have been made to estimate C allocation to mycorrhizal mycelium. Many of these involve labelling studies with radioactive or stable isotopes and are subject to different sources of error. Microcosm studies may result in unnaturally high mycelial biomass and/or exclude soil biota which may graze fungal mycelia. However, short pulse-labelling experiments may underestimate C allocation to mycorrhizal mycelia since they only measure cytoplasmic allocation and exclude C allocation to previously formed fungal cell walls. Many experiments fail to measure respiratory losses of labelled CO_2 which complicates the construction of complete C budgets. C flow through arbuscular mycorrhizal (AM) mycelia has been measured in grassland ecosystems dominated by AM mycelia and found to be at least as large as that of fine roots, with at least 5.4–7.7% of the C lost by plants being respired from AM fungal mycelium and 3.9–6.2% being fixed in mycorrhizal mycelium within 21 h of labelling (Johnson et al. 2002). These figures are comparable with those for fine roots and suggest that there is a very rapid flux of C through mycorrhizal hyphae. A particular strength of these data is that they were obtained under field conditions. Additional studies using similar methods have investigated the effects of soil invertebrates and shown that they can disrupt C transport through hyphal networks but that there is still a significant, rapid flow (Johnson et al. 2005). Analyses of ^{14}C content of AM hyphae by accelerator mass spectrometry (Staddon et al. 2003) suggest that most hyphae live for 5–6 days, again suggesting that there is a large and rapid pathway of C flow through the AM extraradical mycelium.

Measurements of C flow to ectomycorrhizal mycelium colonising forest trees are more difficult to obtain due to the size of the plant hosts, but data from smaller plants in microcosm systems (Leake et al. 2001) showed that the extraradical mycelium of the ectomycorrhizal fungus *Suillus bovinus* colonising *Pinus sylvestris* seedlings contained 9% of the ^{14}C contained in the plants 56 h after labelling. Over 60% of the C allocated to the extraradical mycelium was allocated to mycelium colonising patches of litter, which only represented 12% of the available area for colonisation, suggesting that this C allocation was associated with nutrient acquisition. Data from a range of microcosm-based labelling studies (see Leake et al. 2004 for details) suggest that 7–30% of net C fixation is allocated to ectomycorrhizal mycelium and that 16–71% of this C is lost as respiration. These data are likely to be underestimates of C transfer to the mycelium since short term pulse-labelling experiments do not measure the carbon in the fungal cell walls. Although microcosm experiments may not accurately reflect field conditions, manipulation of the ectomycorrhizal extraradical mycelium in forest ecosystems using the methods employed by Johnson et al. (2002) is not possible due to the large size of the plants. Tree girdling experiments in a 45–55 year old pine forest by Högberg et al. (2001), however, suggest that soil respiration is directly coupled to the flux of current assimilate to mycorrhizal roots and fungi. Decreases of 37% were recorded within 1–2 days, however, this method does not allow separate determination of the root and fungal components. Further observations following a large-scale girdling experiment suggest that ectomycorrhizas may contribute at least 32% of soil microbial biomass and as much as half the dissolved organic carbon in forest soil (Högberg and Högberg 2002). The below-ground flux of recent photosynthate has been followed with high temporal resolution using ^{13}C labelling of 4-m-tall *Pinus sylvestris* trees (Högberg et al. 2008). C in the active pools in needles, soluble carbohydrates in phloem and in soil respiratory efflux had half-lives of 22, 17 and 35 h, respectively. C in soil microbial cytoplasm had a half-life of 280 h, while the C in ectomycorrhizal root tips turned over much more slowly. Simultaneous labelling of the soil with $^{15}NH_4^+$ showed that the ectomycorrhizal roots, which were the strongest sinks for photosynthate, were also the largest sinks for N. Tracer levels peaked after 24 h in the phloem, after 2–4 days in the soil respiratory efflux and soil microbial cytoplasm and 4–7 days in the ectomycorrhizal roots. The results indicate close temporal coupling between tree canopy photosynthate and soil biological activity. Other recent studies using free air carbon dioxide enrichment (FACE) experiments as a means of ^{13}C labelling (Körner et al. 2005) and bomb ^{14}C estimates of root age (Gaudinski et al. 2001) suggest that fine roots of trees may turn over much more slowly than previously assumed. This suggests that more of the below-ground C flux may take place through mycorrhizal fungi and other soil biota associated with roots (Högberg and Read 2006). A recent FACE study of a *Populus* plantation supports this idea, suggesting that extraradical mycorrhizal mycelium is the dominant pathway (62%) through which C enters the soil organic matter pool (Godbold et al. 2006).

Both arbuscular mycorrhizal and ectomycorrhizal plants can regulate their C allocation to roots. *Trifolium repens* plants have been shown to increase their rates of photosynthesis in response to increased sink strength of mycorrhizal roots and to increase activities of cell wall and cytoplasmic invertases and sucrose synthase (Wright et al. 1998). In ectomycorrhizal plants the symbiotic partners receive up to 19 times more carbohydrates from their roots than normal leakage would cause, resulting in a strong C sink. To avoid parasitism the plants appear to have developed mechanisms to regulate the C drain to the fungal symbiont in relation to the supply of fungus-derived supply of nutrients (Nehls 2008). Increased expression of plant and fungal hexose transporter genes has been detected at the plant fungus interface in ectomycorrhizas, but it appears there may also be mechanisms to restrict carbohydrate loss to the fungus. Hexoses generated from sucrose hydrolysis by plant-derived acid invertases could be taken up by plant or fungal cells through monosaccharide transporters. One Poplar sugar transporter gene (*PttMST3.1*) is expressed at least 10 times more highly than other hexose transporter genes and it is postulated that this may be regulated at the post-transcriptional level by phosphorylation which would allow activation of the transporter as a reaction to the amount of nutrients delivered by the fungus. If the fungus provided sufficient nutrients the activity of the transporter would be shut off, while the protein would be activated as soon as the nutrient transfer is insufficient (Nehls 2008). Unpublished data support this hypothesis but further studies of the genetic basis of regulation of carbon flow at the symbiotic interface are still needed in a range of different mycorrhizal associations.

One disadvantage of simple labelling experiments showing transport of a labelled element from a source to a sink is that they provide no information about *net* movement of the element in question, since there may be an equal (or greater) movement of the same (unlabelled) element in the reverse direction. The issue of C transport between plants connected by a common mycorrhizal mycelium has been controversial. Experiments by Francis and Read (1984) demonstrated the potential for transfer of C along concentration gradients from sources to sinks induced by shading, however, these studies were criticised for the above reasons. Experiments by Simard et al. (1997) using reciprocal labelling with ^{14}C and ^{13}C demonstrated net transfer of C from *Betula papyrifera* to *Pseudotsuga menziesii* but the overall ecological significance of inter-plant C transfer has been questioned by Robinson and Fitter (1999). NMR studies of common AM mycelial networks by Pfeffer et al. (2004) revealed that, although significant amounts of C were transferred between different roots connected by a common fungal mycelium, the labelled C remained within fungal compounds and no transfer of C from fungus to plant took place. As pointed out by Pfeffer et al. (2004) and earlier by Finlay and Söderström (1992) such distribution of C within mycelial networks may be of significance even in the absence of net transfer of C from fungus to plants since it would reduce the C demand of the fungal mycelium colonising newly connected host plants and enable them to gain access to nutrients taken up by the mycelium. Although the predominant movement of C in fully autotrophic mycorrhizal hosts is likely to be from plant to fungus, over 400 plant species are achlorophyllous and described as '*myco-heterotrophic*', obtaining their C from fungi. DNA-based studies of these fungi have revealed most of them to be mycorrhizal species colonising other autotrophic plants. The mycoheterotrophic species are thus effectively 'cheaters' or epiparasites obtaining their C and nutrients through mycorrhizal connections with neighbouring autotrophic plants (Bidartondo 2005; Bidartondo et al. 2002; Leake 2004). In orchids the direction of C transfer is often reversed since about 100 species are completely achlorophyllous and all others pass through a germination and early developmental phase in which they are dependent on an external supply of nutrients and C since they have minute, dust-like seeds with no reserves. Survival of germinating seedlings is thus dependent upon rapid integration into fungal mycelial networks. Although this pathway of C transfer is sometimes dismissed as a 'special case' in discussions concerning the overall significance of C transfer via mycorrhizal hyphal connections, the Orchidaceae is the largest family in the plant kingdom with over 30,000 species so the habit is arguably widespread and of evolutionary significance.

Acquisition of N (Bending and Read 1995) and P (Lindahl et al. 2001) by ectomycorrhizal fungi colonising organic substrates is dependent on resources allocated to the mycelium. Ectomycorrhizal and ericoid mycorrhizal fungi play a pivotal role in the mobilisation of N and P from organic polymers (Read and Perez-Moreno 2003) and their enzymatic capacities have been reviewed by Lindahl et al. (2005).

Increased ectomycorrhizal mycelial growth and biomass production, resulting in selective spatial allocation of C to nutrient rich substrates has been demonstrated in a range of studies (see Read and Perez-Moreno 2003) and been shown to be associated with mobilisation of N and P. Energy is undoubtedly required for the synthesis of enzymes involved in the mobilisation of nutrients but the partitioning of C between fungal biomass production and hydrolytic activity is not yet fully understood. Experiments by Lindahl et al. (2007) suggest that decomposition of litter by saprotrophs and mobilisation of N from well-decomposed organic matter may be spatially and temporally separated in boreal forests. Many of the organic N compounds taken up by ectomycorrhizal mycelium contain C derived from photosynthetic products originally translocated to the soil via the same mycelium. This may reduce the C drain imposed upon the host plant by ectomycorrhizal symbionts. In axenically grown *Betula pendula* plants supplied with ^{14}C labelled protein as the sole exogenous N source, only ectomycorrhizal plants were able to exploit this N source. Heterotrophic uptake of C associated with utilisation of this organic N source was estimated to be up to 9% of plant C over a 55 day period (Abuzinadah and Read 1989). Simple amino acid sources are taken up intact by a range of mycorrhizal plants as demonstrated in field experiments by (Näsholm et al. 1998) and this also contributes to the reverse flow of C through the rhizosphere to plant roots. Utilisation of organic N sources by arbuscular mycorrhizal plants is less well understood but Hodge et al. (2001) demonstrated enhanced decomposition and capture of N from decaying grass leaves in the presence of AM fungi. Further experiments are needed to distinguish between direct capture and uptake of organic N by the hyphae and indirect uptake of inorganic N through enhanced decomposition. It is possible that mycorrhizal hyphae contribute to rhizosphere priming via a release of energy rich C which is utilised by microbial saprotrophs. The mycorrhizal mycelium provides a vastly increased surface area (compared with roots alone) for interactions with other microorganisms and an important pathway for translocation into the soil of energy-rich compounds derived from plant assimilates. Soluble C compounds released by the extraradical mycelium of arbuscular fungi have been shown to influence the activity of both fungi and bacteria associated with the mycorrhizosphere (Filion et al. 1999; Toljander et al. 2007). Both stimulatory and inhibitory interactions are possible and these have been reviewed with respect to their relevance in sustainable agriculture by Johansson et al. (2004). Production of mycorrhizal mycelial exudates has been shown to influence bacterial species composition and vitality (Toljander et al. 2007) and vitality of mycorrhizal hyphae in turn has been shown to influence attachment of different bacteria to AM hyphae (Toljander et al. 2006). Other recent experiments indicate that AM fungi may influence bacterial assemblages in roots but that the effect is not reciprocal (Singh et al. 2008). AM fungi also produce a glycoprotein, glomalin, which is deposited in soil as hyphae senesce and has been estimated to constitute as much as 5% of soil C (see Treseder and Turner 2007). As well as playing a role in soil aggregation glomalin production is thought to sequester significant amounts of C on a global scale (Treseder and Turner 2007).

Exudation and reabsorption of some C compounds from fluid droplets produced at ectomycorrhizal hyphal tips has been demonstrated by Sun et al. (1999) who concluded that it might represent an important mechanism for conditioning the hyphal environment in the vicinity of tips, creating an interface for the exchange of nutrients and C compounds with the adjacent soil environment and its other micro-organisms. Ectomycorrhizal fungi produce significant amounts of organic acids (Sun et al. 1999; Ahonen-Jonnarth et al. 2000) which may play a role in weathering of minerals, complexation of toxic Al^{3+} or in antibiosis. The microbial decomposition of these organic acids could also contribute significantly to soil respiration (van Hees et al. 2005). Experiments by Rosling et al. (2004a, b) suggest that mycorrhizal and other fungi differ in their ability to allocate C to different mineral substrates and that more labelled C is allocated to easily weatherable minerals such as potassium feldspar than to quartz.

Despite the fact that the rhizosphere is defined in terms of its elevated levels of soil microbiological activity, we still know surprisingly little about the role of rhizosphere communities in C flow, and little is known about the roles of different members of the community in assimilating plant exudates. Experiments by Ostle et al. (2003) and Rangel-Castro et al. (2005a) demonstrated rapid allocation and incorporation of recently photosynthesized ^{13}C into soil microbial biomass. Labelled C is incorporated within hours and the half life of microbial pools of ^{13}C was calculated to be 4.7 days. RNA-based stable isotope probing experiments by Rangel-Castro et al. (2005b) using DGGE analysis of bacterial, fungal and archaea, showed that active communities in limed soils were more complex than those in unlimed soils and were more active in utilization of recently exuded ^{13}C compounds. This suggests that in unlimed soils the active microbial community may have been utilizing other sources of C but the results may also reflect differences in the amount of root exudation in limed and unlimed grasslands. Another approach which has been used to study bacterial communities associated with mycorrhizal and non-mycorrhizal root systems is the use of symbiosis-defective plant mutants. In experiments by Offre et al. (2008), Oxalobacteraceae isolates were more abundant in mycorrhizal roots of *Medicago truncatula* than in non-mycorrhizal roots of symbiosis-defective plants, whereas *Comamonadaceae* isolates were more abundant in non-mycorrhizal roots.

New approaches based on stable isotope probing, RNA analysis, and metagenomics (Vandenkoornhuyse et al. 2007) indicate that there are many hitherto unidentified root symbionts and that bacteria and AM fungi occupying roots show differential activity in C consumption with much higher C flow to some fungi than others. Therefore, while it is clear that symbionts are important determinants of rhizodeposition, our understanding remains poor in many respects. While this article is about C flow in the rhizosphere, and there has been a general tendency in rhizosphere research to concentrate on "quantitatively significant" C fluxes, it should be remembered that plants produce a wide spectrum of chemicals which are usually called secondary metabolites because of their presumed secondary role in plant growth. Chemicals released in the rhizosphere play vital roles in signalling between plant roots and different microorganisms. Although these chemicals may only constitute a small proportion of the total photosynthetically derived C flow from roots they can play a key role in plant survival through defence against pathogens or in attracting beneficial symbionts. One example of this is the strigolactones, that are produced in the root exudates of many monocot and dicot species (Bouwmeester et al. 2007). These compounds induce branching of arbuscular mycorrhizal fungi but also stimulate the germination of seeds of parasitic plants (*Striga* and *Orobanche* spp.). However, infection by *Striga* is reduced in plants colonised by AM fungi through down-regulating the production of the germination stimulant. Phosphate starvation is known to induce strigolactone production, and also to favour AM colonisation, while AM fungi are known to improve the P status of their hosts, which in turn would repress strigolactone production. The effects of environmental factors on numerous other signalling molecules are still entirely unknown, although their effects on plant growth and survival may be of paramount importance. Therefore, although more quantitative studies of C and N flux in the rhizosphere are still needed, these should also be complemented by further qualitative studies of the role of different signalling molecules, the roles these play in plant–soil–microbe interactions and the way in which they are influenced by different environmental conditions.

Carbon flow in the rhizosphere is bi-directional

Prior to 1990, the general consensus was that rhizodeposition was a unidirectional flux whereby plant C was lost from roots into the soil (Curl and Truelove 1986). Once in the soil it was assumed to undergo a number of fates including movement away from the root in the soil solution due to diffusion and mass flow, capture by soil microorganisms, and sorption to the solid (Martin 1975; Newman and Watson 1977). However, experiments undertaken in hydroponic culture and subsequently soil revealed that plant roots can also take up a range of organic compounds from the soil into the roots with subsequent transfer to the shoots (Jones and Darrah 1992, 1993, 1994). Of the compounds investigated so far, roots from a range of species have been shown to take up predominantly low molecular weight solutes such as organic acids, sugars and amino acids (Jones and Darrah 1995; Sacchi et al. 2000; Thornton 2001). In addition, roots may also take up inorganic C from outside the root when present in a dissolved form (e.g. HCO_3^-; Cram 1974; Amiro and Ewing 1992; Ford et al. 2007). Although HCO_3^- can be readily converted to organic acids inside the root, the contribution of this inwardly directed inorganic C flux to the overall C economy of the plants is small especially in view of the large amount of HCO_3^- generated in respiratory processes (Ford et al. 2007). One potential exception occurs within proteoid roots of lupin roots where significant uptake and assimilation of HCO_3^- into malate and citrate occurs (Johnson et al. 1996). These HCO_3^- derived organic acids are then exuded back into the soil to aid in P mobilization in the rhizosphere.

Discrimination also needs to be made between organic C that is taken up and assimilated in a controlled (i.e. active transport) way and that which is inadvertently taken up as a consequence of its physicochemical properties (i.e. passive transport). In the case of compounds with a high octanol–water partition coefficient (K_{OW}) value, these can simply become sorbed to cell membranes and subsequently metabolised (e.g. pesticides, chlorinated hydrocarbons; Scheunert et al. 1994). This passive process can be expected to have no positive benefit to the plant. Similarly, positively charged organic compounds can become sorbed to cell walls with no subsequent assimilation. Some neutrally charged compounds (e.g. acetic acid) can also passively enter the cell if the concentration outside is greater than that inside. While this has been used as an experimental tool to understand membrane function its significance in soil remains unknown (Herrmann and Felle 1995).

Of greatest ecological significance is the active root uptake of sugars and organic nitrogen compounds (e.g. amino acids, polyamines etc) from soil. Typically, these compounds are taken into the plant by co-transporters which are constitutively expressed and located throughout the root system (Jones and Darrah 1994, 1996; Fig. 3). These co-transporters are powered by the plasma membrane H^+-ATPases which are predominantly located in the epidermis rather than in the root cortex although levels of H^+-ATPases are also high in the stellar regions (Samuels et al. 1992; Jahn et al. 1998). The transport proteins simultaneously transport H^+ across the plasma membrane together with individual organic solutes. The transporters are also relatively solute specific with transport families for amino acids and sugars being well characterised at both the physiological and molecular level (Fischer et al. 1998; Williams et al. 2000; Hirner et al. 2006). In addition, membrane transporters also exist for other solutes such as peptides, flavonoids and polyamines although these protein families remain less well characterised (DiTomaso et al.

1992; Hart et al. 1992; Buer et al. 2007; Jones et al. 2005a, b). There is also strong evidence to suggest that plant roots can take up larger molecular weight solutes by endocytosis (Samaj et al. 2005). Current evidence suggest that this process is important for auxin-mediated cell–cell communication, polar growth, gravitropic responses, cytokinesis and cell wall morphogenesis (Ovecka et al. 2005).

As the plant expends energy in the uptake of these compounds from soil we assume that the process must confer some benefit to the plant. At present there are four principal hypotheses to explain why plants might take up organic solutes from soil (Fig. 4). Although there is no reason to suggest that these are mutually exclusive it is likely that their importance varies in space and time within a root system and between plant species.

Hypothesis 1: direct root exudate recapture

The first explanation is that the root is simply recapturing C back from the soil that it previously lost in response to passive root exudation, the latter being a process over which it exerts little direct control (Jones et al. 1996). This recapture of exudate C not only enhances C use efficiency in the plant but also prevents C accumulation in the rhizosphere thereby reducing the growth of the soil microbial community. This may serve three purposes by (1) reducing microbial competition for poorly available nutrients required by the root (e.g. N and P), (2) reducing the growth of potentially pathogenic organisms, and (3) minimizing chemotactic gradients for pathogenic organisms. When chemotaxis of beneficial organisms is required, current evidence suggests that more chemically specific signals at low concentrations are released in root exudates in a spatially and temporally controlled manner (e.g. flavonoids; Antunes et al. 2006; Sugiyama et al. 2007).

Hypothesis 2: indirect, fortuitous root exudate recapture

The second explanation is that re-uptake of C from soil might simply be indirectly related to normal source-sink C delivery mechanisms in plants. In most roots, solutes arriving from the shoots are unloaded symplastically from the phloem, however, some subsequently leak into the apoplast where retrieval by active transporters can occur (Eleftheriou and Lazarou 1997; Patrick 1997). This is unlikely to be of significance in areas with a well developed exodermis and tissues with high symplastic connectivity, however, it may be important in root caps and cells where plasmodesmata have been blocked (Zhu and Rost 2000; Hukin et al. 2002). These re-uptake processes may also be indirectly linked to cell wall bound invertases (Huang et al. 2007). These enzymes convert apoplastic sucrose to glucose and fructose which are then taken into the cell by co-localized sugar transporters (Dimou et al. 2005). Import of extracellular hexose sugars has been linked to a range of sensing and signalling pathways in addition to their potential role in supplying sugars for cellular expansion (Sherson et al. 2003).

Hypothesis 3: nutrient capture from soil

The third explanation is that the uptake of organic compounds from the soil may be a mechanism to supply organic nutrients in addition to traditional inorganic uptake routes (i.e. NO_3^-, NH_4^+, $H_2PO_4^-$ etc). This may be particularly relevant in situations where the supply of inorganic nutrients is limiting due to either their low intrinsic solubility (e.g. P), low rate

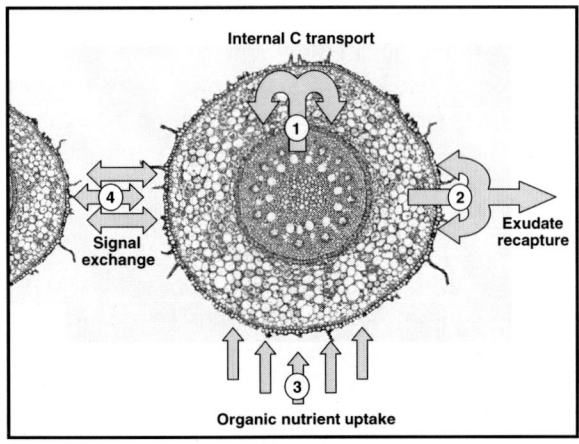

Fig. 4 Schematic representation of a transverse root section illustrating the four principal hypotheses explaining the uptake of organic C from soil: *1* indirect, fortuitous root exudate recapture in the root's internal apoplastic transport and signalling pathways, *2* direct recapture of root exudates from the soil with the aim of reducing microbial growth and pathogen chemotaxis, *3* uptake of organic nutrients (e.g. amino acids) released during the mineralization of soil organic matter in the rhizosphere, and *4* transfer of chemical signals involved in inter-root and root–microbial communication pathways

of ecosystem addition or a block in organic matter mineralization preventing their release back into the soil (e.g. N). It may also be particularly relevant to non-mycorrhizal plants which lack the capability to directly mineralize organic matter. Addition of isotopically labeled organic compounds to soil (^{15}N, ^{13}C, ^{14}C) has shown that roots have the potential to take up and assimilate a wide range of compounds. In agricultural soils, however, it has been shown that plants are poor competitors for amino acids and sugars in comparison to the soil microbial community (Owen and Jones 2001; Bardgett et al. 2003; Kuzyakov and Jones 2006). Consequently, unless concentrations of organic solutes in the soil are very high the uptake of exogenous organic N is likely to be of minimal significance (Jones et al. 2005a, b). In contrast, work in predominantly arctic and alpine soils has suggested that organic N taken up from the soil in the form of amino acids may contribute significantly to a plant's N budget (Chapin et al. 1993). In this case the direct uptake of organic N circumnavigates the need for the soil microbial community to mineralize soil organic matter (Lipson and Nasholm 2001). The uptake of organic N by roots is often viewed in the literature as being unidirectional. The lack of consideration for an outward flux (i.e. root exudation) therefore brings into question many of the rates of flux reported in the literature. In most experiments, rates of organic N uptake into roots are measured with dual ^{15}N-^{13}C labeled compounds. As exudation of organic N is derived predominantly from non-isotopically labeled organic N, due to the large internal organic N reservoir relative to the amount added, most isotopic tracer experiments will greatly overestimate the rates of uptake. What we really require are not measurements of gross rates of uptake, but moreover net rates of uptake (i.e. influx minus efflux; Philips et al. 2006). In the case of amino acids and sugars, measurements in sterile hydroponic culture have shown that the point of net zero uptake (i.e. influx = efflux) occurs when the external concentration is between 0.5 and 10 μM (Jones and Darrah 1994, 1996; Phillips et al. 2004). In most situations this is extremely similar to the concentrations which exist naturally in soil solution (Andersson and Berggren 2005; Jones et al. 2005a, b; Boddy et al. 2007) suggesting that the contribution of organic N uptake from soil may be less important than as a mechanism for retaining the resources it already has (i.e. recapture of exudates). The interpretation of isotopic flux measurements is also complicated by the knowledge that some organic N compounds can be firstly broken down in the soil and the ^{15}N released taken up as $^{15}NH_4^+$ or $^{15}NO_3^-$. Measurements of the relative enrichment of ^{15}N and ^{13}C in the roots can potentially be used to discriminate between ^{15}N taken up in an intact form versus that previously mineralized in the soil. However, after amino acids enter the root they can undergo a number of metabolic reactions that can ultimately lead to approximately 40–60% of the ^{13}C being released as $^{13}CO_2$ (e.g. transamination and deamination; Owen and Jones 2001). Similarly, the loss of organic N derived CO_2 can also occur after uptake by mycorrhizas again leading to an underestimation of the organic N flux. Consequently, isotopic flux measurements are fraught with potential pitfalls that make interpretation of organic C and N fluxes at the root–soil interface extremely difficult (Jones et al. 2005a, b).

Hypothesis 4: rhizosphere signalling

The fourth explanation for plants actively taking up organic compounds from soil is for inter and intra-root signalling and for root–microbial signal exchange. In comparison to the other three potential explanations, this is a poorly explored aspect of rhizosphere ecology (Bais et al. 2004). Although sugars are important in plant signalling, there is currently no evidence to suggest that they would provide specific signals to enable effective communication between roots and other organisms in the rhizosphere. More likely is that root transporters would be involved in the uptake of highly specific signalling molecules (e.g. peptides). In other cases, compounds released from microorganisms can have a direct effect on plant growth and metabolism (Brown 1972), however, the mode of transport of these signalling molecules into the root remains unknown (e.g. lumichrome; Phillips et al. 2004; Matiru and Dakora 2005). As our understanding of the diversity and control of signalling processes in plants increases it is likely that some of these will have functional significance in the rhizosphere (Bais et al. 2004; Bahyrycz and Konopinska 2007; Jun et al. 2008). Further discussion of this issue can be found in Hartmann et al. (2009), Lambers et al. (2009) and Faure et al. (2009).

Rhizodeposition in the plant C and N budget

Methods of investigation

While rhizodeposition can be quantified relatively easily in the absence of soil by growing roots in sterile hydroponic culture and collecting the C accumulating in the external media, this method lacks ecological relevance (Ryan et al. 2001). Quantifying rhizosphere C-flow in relation to soil environments, however, has proved extremely difficult. The amount of rhizodeposition entering soil during a growing season typically represents only a small amount of C and N in comparison to that already present in the soil organic matter (SOM) and therefore measuring changes in soil C in response to rhizodeposition remains virtually impossible. This is also highly pertinent to the uncertainties surrounding the effects of environmental change on the soil C-balance where, although significant effects on soil C-sequestration are predicted, changes in soil C are difficult to detect due to the large SOM-C background and high degree of spatial variability in SOM. Consequently, tracing root-derived C and N by isotopic techniques is a prerequisite for the quantification of rhizodeposition in soil (Warembourg and Kummerow 1991). For C, a widely used technique involves the exposure of shoots to a ^{13}C or ^{14}C enriched atmosphere to label the photoassimilates. Subsequently, the fixed isotope tracer becomes partitioned to range of operationally defined below-ground compartments (roots, soil residues including microbial biomass, and root-derived CO_2). Changes in isotope abundance in these pools is typically followed over time to estimate rhizodeposition as part of the plant or ecosystem C budget. The experimental conditions have to be carefully considered when interpreting the partitioning of photoassimilates. For example, the length of the isotopic labeling and subsequent chase period is a major determinant of the amount of C delivered into the soil (Meharg 1994). Short-term pulse labeling (minutes to hours) traces rhizodeposits derived predominantly from recent photoassimilates (i.e. root exudates, mucilage and border cells). Accordingly, pulse-labeling by this method tends to underestimate total rhizodeposition but remains useful in the investigation of assimilate partitioning in relation to plant metabolism (Phillips and Fahey 2005; Allard et al. 2006; Hill et al. 2007). Longer isotopic labeling periods (weeks to months) trace not only the senescence and turnover of roots but also the fraction of root exudates that may not be derived from recent C (Swinnen et al. 1995). In some cases this may be a significant part of total root exudation (Thornton et al. 2004).

Most experiments in soil have tended to focus on the total amount of C lost in rhizodeposition, while hydroponic studies carried out in the laboratory have tended to focus on the tracking of individual compounds lost from roots. The isotopic tracking of individual root-derived compounds into soil, however, has only recently become routinely possible (Paterson et al. 2008). The use of gas chromatography coupled to isotope ratio mass spectrometry allows the dynamic tracking of specific rhizodeposits such as sugars from roots and associated symbionts into soil (Derrien et al. 2004; Paterson et al. 2007). Another recent breakthrough for tracing C flow in the rhizosphere is stable isotope probing (SIP). In this technique, isotopically labeled plant assimilates are released into the soil and then subsequently taken up and incorporated into the soil microbial community. The isotopically labeled microbial DNA, RNA or phospholipids can then be extracted and their isotope ratio determined whilst genetic material can be sequenced to identify members of the soil microbial community consuming the rhizodeposits (Singh et al. 2004, Rangel-Castro et al. 2005a, b; Shrestha et al. 2008). However, labeling of photoassimilates often requires a sophisticated experimental set-up particularly for large plants (Warembourg and Kummerow 1991). In the case of ^{14}C, its use may be problematic, particularly in the field, due to safety and environmental concerns. This severely limits isotopic investigations on field-grown plants over their entire life cycle (unless ^{13}C pulse-labelling is used, Högberg et al. 2008). Ultimately, this represents a serious concern when calculating agro- or natural-ecosystem C budgets and the potential contribution of rhizodeposition to C sequestration. The use of natural abundance of the stable isotope ^{13}C can provide an elegant alternative (i.e. $\delta^{13}C$; Ekblad and Hogberg 2001). Growing a C_4 plant on a C_3 history soil (and vice versa) allows the tracing of new plant-derived C from the C_4-plant in soil because of the difference in ^{13}C natural abundance in plant material between C_3 and C_4 vegetation (Boutton 1996; Rochette et al. 1999). However, this approach is restricted to limited contexts, typically to maize grown on a C_3 soil because of the difficulty to find a soil with a known C_3 or C_4 history (Balesdent and Balabane 1992; Qian et al. 1997). Non-isotopic approaches for quantifying rhizodeposition are available using a range of microbial biosensors. These have been employed as semi-quantitative measures of total root C flow and more recently for spatially localising the release of specific exudate components (Paterson et al. 2006).

One major difficulty when attempting to quantify the transfer of labeled C below ground is the mineralization of rhizodeposits by rhizosphere microorganisms. Typically, low molecular weight (MW) root exudates are believed to only have a residence time of a few hours in soil solution as they are rapidly consumed by the C-limited rhizosphere microbial community (Nguyen and Guckert 2001; van Hees et al. 2005). However, as it was observed for glucose, the microbial uptake of substrate and its subsequent mineralization may be decoupled in time. Therefore, the turnover of low molecular weight exudates in soil solution as determined from the kinetics of mineralization are likely to be underestimated by an order of magnitude, indicating turnover times of minutes rather than hours (Hill et al. 2008). Although higher MW rhizodeposits have a slightly longer persistence time in soil, they are still mineralized within a few days (Mary et al. 1992, 1993; Nguyen et al. 2008). This rapid biodegradation of rhizodeposits means that a significant proportion of the rhizodeposits are quickly lost from the soil as labeled CO_2 (rhizomicrobial respiration). The longer the labeling and the chase period, the greater the amounts of rhizodeposits are lost in this way. Because a large proportion of the rhizodeposits are low MW and labile, rhizomicrobial respiration overlaps directly in time and space with root/symbiont respiration of the same labeled photoassimilates (Dilkes et al. 2004). Experimentally, the flux of labeled CO_2 derived directly from roots and indirectly from rhizodeposits are therefore determined together (rhizosphere respiration; Todorovic et al. 2001). Knowledge of the partitioning of rhizosphere respiration into root, symbiont and rhizomicrobial components, however, is crucial if we are to gain a deeper understanding of rhizosphere C flow (Paterson 2003; Paterson et al. 2005). Many attempts have been made to partition rhizosphere respiration, from the simple use of antibiotics to more sophisticated models based on isotopic methods, however, none has proved satisfactory from a quantitative perspective (Cheng et al. 1993; Kuzyakov 2006; Sapronov and Kuzyakov 2007). Current estimates suggest that approximately 50% of rhizosphere respiration is due to the turnover of rhizodeposits and 50% to direct root (and mycorrhizal) respiration, however, it is clear that this needs to be major focus for research in the future (Kuzyakov 2002).

Estimating N rhizodeposition in soil is commonly undertaken with a ^{15}N isotopic tracer (Hertenberger and Wanek 2004). The tracer is supplied by a foliar application as a solution or spray, by stem feeding, by a pre-culture on a labeled substrate or by using split-root systems, one compartment for being used for the labeling and the other for determining the release of ^{15}N from roots (Jensen 1996; Hogh-Jensen and Schjoerring 2001; Mayer et al. 2003). All methods currently assume, (1) a homogenous mixing of the tracer within the plant N pool, and (2) that the isotopic signature of N-rhizodeposits is the same as that of the roots. Further rigorous validation of these assumptions is required. Depending on the soil conditions, a fraction of the rhizodeposited ^{15}N also may be unrecovered due to denitrification leading to an underestimation of N rhizodeposition (de Graaff et al. 2007). Recent advances have also been made in the dual ^{13}C-^{15}N isotopic labeling of plants in situ (Wichern et al. 2007).

To overcome the difficulties related to non-sterile soil conditions, many studies were and are still conducted in sterile hydroponic culture. Under these conditions both the amount and the nature of compounds released from roots can be determined. However, one has to be aware of the limitations of such experimental conditions. For instance, both the nature and quantity of compounds released from roots depends on the plant/root physiology, which greatly differs between a simple sterile nutrient solution and a complex soil environment (Neumann and Römheld 2001). Furthermore, exudation has been quantified for decades in nutrient solution which are not regularly renewed, a system that exacerbates the re-uptake of exudates by roots and that leads to a large underestimation in exudation rates (Jones and Darrah 1993). Consequently, it is necessary to adapt the experimental set-up used to study exudation so that it account for the re-uptake of exudates. This can be done by using microcosms percolated by nutrient solution (Hodge et al. 1996) or by modelling the kinetics of exudate accumulation in the root bathing solution as the net output between the gross efflux and the re-uptake of exudates (Personeni et al. 2007). The use of bioreporter microorganisms is also an interesting approach to spatially localize the release of some specific compounds or class of compounds, however, quantitative information about the rhizodeposition flux are difficult to achieve (Yeomans et al. 1999; Darwent et al. 2003).

Investigations of rhizodeposition are hampered by many technical difficulties and sometimes by unresolved methodological problems arising from the numerous interactions between roots, the soil matrix and microorganisms (van Hees et al. 2005). Availability of robust methodologies for the qualitative and quantitative determination of rhizodeposition in soil clearly remains an unsolved issue. Current methods are often incomplete or biased and consequently, estimates of the flux of C (and to a lesser extend of N) to the rhizosphere are associated with significant uncertainty.

How much C is lost via rhizodeposition?

In the last few decades, hundreds of attempts have been made to quantify the amount of photoassimilate C partitioned below ground (Nguyen 2003). Most of the initial studies used ^{14}C although ^{13}C is now increasingly being used for tracing purposes. Results are commonly expressed as partition coefficients describing that amount of net fixed C allocated between shoots, roots, rhizosphere respiration (root and symbiont respiration + respiration of rhizodeposits) and soil residues. Soil residues include rhizodeposits, microbial biomass-C and metabolites derived from rhizodeposits (including mycorrhizal hyphae) but also fine roots debris that cannot be effectively separated from the soil (e.g. root hairs, epidermal cells etc). Figure 5 summarizes a review of whole plant C partitioning averaged across a wide range of published studies and updates previous reviews (Bidel et al. 2000; Nguyen 2003). Overall, it is clear that most isotopic labeling studies have focussed on young plants at a vegetative stage (typically <1 month old). The focus on young plants is due to methodological difficulties in growing and labeling plants to maturity in controlled conditions. However, plant age has a strong effect of the partitioning of photoassimilate to the rhizosphere. For example, in annual plants pulse-labelled with ^{14}C, when comparing two plant ages ranging within the range 28–600 days of culture, the partitioning of C to rhizosphere clearly decreases with plant age of −43%, −28%, −20% for roots, rhizosphere respiration and soil residues, respectively (median values; Nguyen 2003). Although C partitioning has been investigated in a wide range of plant species, almost half of the published data are for wheat and rye-grass and 76% of the studies are related to five crop/grassland species. Hence, we currently have a very incomplete picture of C rhizodeposition particularly in mixed plant communities. In particular, dicotyledonous plants have received little attention and the amount of rhizodeposition by trees in natural ecosystems remains virtually unknown with the exception of recent studies by (Högberg et al. 2008). Furthermore, when very young trees have been used to study rhizodeposition, the experimental conditions employed often bear little relevance to natural forest stands. Consequently, our poor knowledge of rhizodeposition in trees is problematic, particularly when quantifying C sequestration in forests.

Studies indicate that roughly 40% of net fixed C is allocated belowground. For cereals and grasses, this approximates to around 1.5–2.2 t C ha^{-1} for the vegetation period (Kuzyakov and Domanski 2000). Of the C partitioned below ground about 50% of it is retained in root biomass (19% of net fixed C), 33% is returned to the atmosphere as rhizosphere respiration (12% of net fixed C), 12% can be recovered as soil residues (5% of net fixed C) and a small amount is lost by leaching and surface runoff. Assuming that roots and microorganisms contribute equally to rhizosphere respiration (Kuzyakov 2006), an assumption that must be treated with caution, then a rough estimate of rhizodeposition would be around 11% of the net fixed C or 27% of C allocated to roots. This would correspond to 400–600 kg C ha^{-1} for the vegetation period of grasses and cereals. These values only provide a rough estimate, however, due to the uncertainty surrounding the partitioning of rhizosphere respiration and because soil residues often include small roots and living mycorrhizal mycelium that cannot be realistically separated from soil by current protocols. This probably explains the skewed distribution of the soil residue partitioning coefficients (Fig. 5).

Studies quantifying the amount of N rhizodeposition are much less numerous and a survey of the

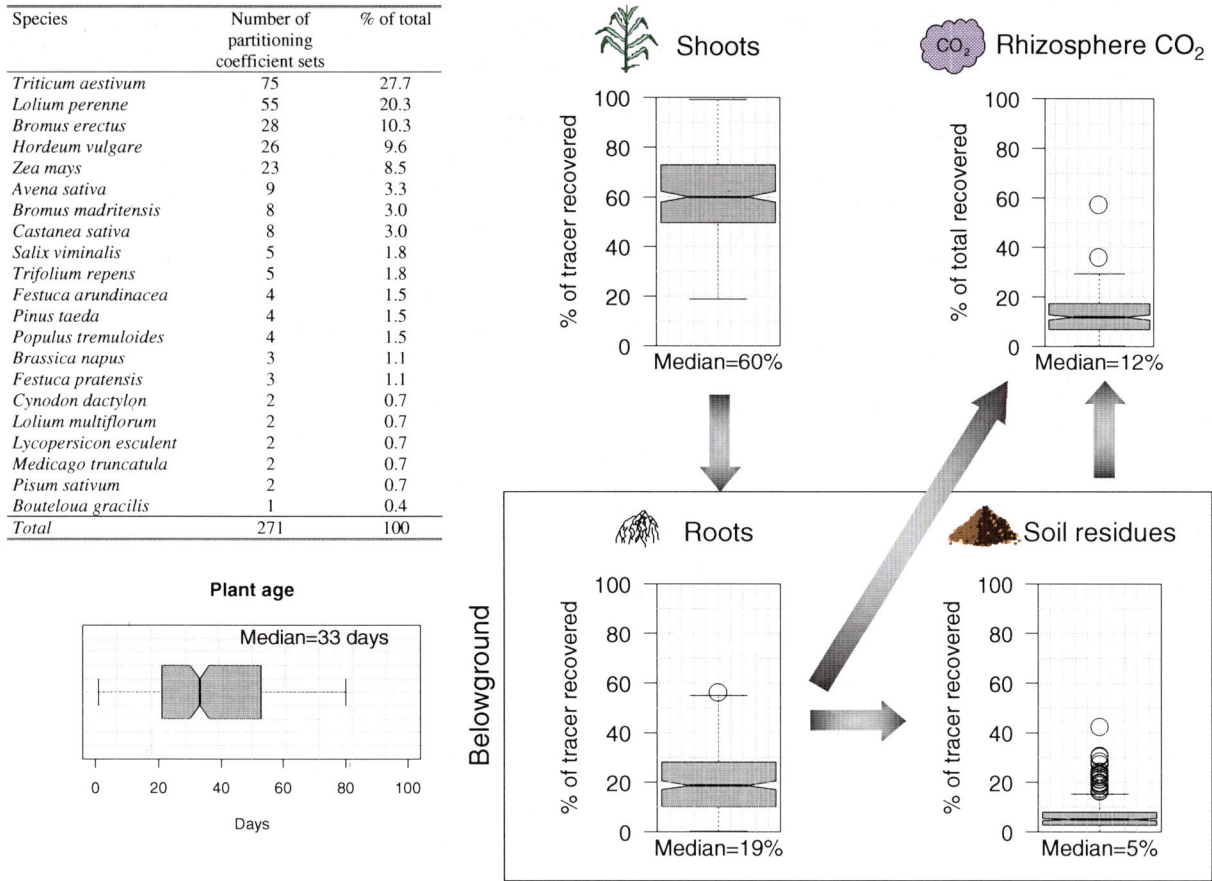

Fig. 5 Partitioning of labeled net fixed C after a pulse or continuous exposure of shoots to a $^{14}CO_2$ enriched atmosphere. For each compartment, *boxplots* show the distribution of 271 individual partition coefficients drawn from a review of the literature by Nguyen (2003) and updated to 2007. The plant species and the distribution of plant ages are provided on the *left*. The *box* represents the second and third quartiles separated by the median. The *whiskers* extend to 1.5 times the interquartile range. The *circles* denote outliers

literature shows that our knowledge of this phenomenon is very incomplete (Fig. 6). More than 60% of the available data pertain to pea and wheat. In these studies, N rhizodeposition accounts for 10–16% of total plant N with losses higher in legumes in comparison to non-legume species. This observation may be biased, however, as the legume based studies have tended to use older plants (Fig. 6) where a larger part of rhizodeposited N may be attributed to root turnover. As free amino acids and proteins represent only a minor component of root exudates (typically 1–2% of exudate-C; Kraffczyk et al. 1984; Jones and Darrah 1993) we assume that they contribute little to plant N rhizodeposition. We conclude that N rhizodeposition must be largely due to root turnover or possibly to an efflux of labeled ammonium and/or nitrate (Feng et al. 1994; Scheurwater et al. 1999).

Small root debris that cannot be separated by common sampling protocols may also lead to an overestimation of N rhizodeposition.

Published reports also show that the partition coefficients of C and N both below ground and to rhizodeposition are highly variable. This illustrates that plant species, plant ecotype/cultivar, age and environmental conditions all exert a strong impact on rhizodeposition. From conventional tracer experiments it is often difficult to conclude about how rhizodeposition is affected by environmental conditions as the partitioning of rhizosphere respiration between the root and the microbial components may also be altered. However, when the partitioning of C to root, to rhizosphere respiration and to soil residues changes in the same way, some conclusions may be drawn. Hence, it can be assumed that the percentage

Species	Number of results	% of total
Avena sativa	3	4.0
Cajanus cajan	1	1.3
Cicer arietinum	1	1.3
Glycine max	1	1.3
Hordeum vulgare	2	2.7
Lathyrus sativus	1	1.3
Lolium perenne	1	1.3
Lupinus albus	1	1.3
Lupinus angustifolius	1	1.3
Medicago sativa	1	1.3
Ornithopus compressus	1	1.3
Pisum sativum	27	36.0
Trifolium pratense	1	1.3
Trifolium pratense	1	1.3
Trifolium repens	2	2.7
Trifolium subterraneum	1	1.3
Triticum aestivum	20	26.7
Triticum turgidum	6	8.0
Vicia faba	2	2.7
Vigna radiata	1	1.3
Total	75	100

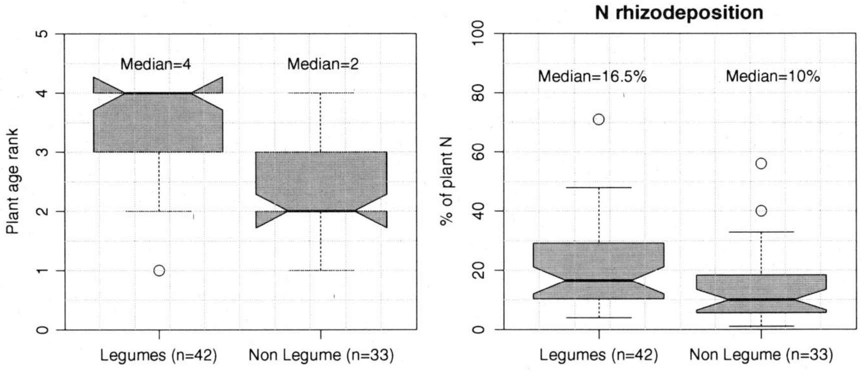

Fig. 6 Summary of published studies on N rhizodeposition expressed as a percentage of total plant N. Rhizosphere N derived from roots was determined by labeling of plant N with ^{15}N supplied as $^{15}NH_3$, $^{15}NO_3$ or ^{15}N-urea. The technique used was one of the following: split-root cultures, stem/petiole infiltration/injection, leaf dipping, $^{15}NH_3$-enriched atmosphere or preculture on a ^{15}N-labelled substrate. The plant species and the ranked distribution of plant age are given on the *left*. Ranks for plant ages are defined as follows: *1* early vegetative stage, *2* end of vegetative stage, *3* flowering/grain filling, *4* maturity. The *box* represents the second and third quartiles separated by the median. The *whiskers* extend to 1.5 times the interquartile range. *Circles* represent outliers

of assimilates ending up as rhizodeposition generally decreases with plant age and is increased by the presence of microorganisms and by elevated atmospheric CO_2.

We now have almost 30 years of knowledge from C rhizodeposition research. From tracer experiments, we can reasonably predict the order of magnitude of this C flux for agroecosystems. These studies all attest to rhizodeposition being a major C flux. In hindsight, however, it is also evident that a quantitative approach to assessing the functional role of rhizodeposition in soil is strongly limited by technical difficulties arising from the complex interactions occurring in the rhizosphere and the tight link between rhizodeposition and the plant's physiological status. Accordingly, there is an urgent need to develop new approaches and methods for probing rhizodeposition. The coupling of plant labeling with molecular tools is promising for understanding the link between the plant-derived C and microbial processes in the rhizosphere but the current information remains more qualitative than quantitative. Considering the need to have a quantitative understanding of C and N fluxes in the rhizosphere to predict ecosystem behaviour, modelling approaches should be considered to be of major importance. For example, integrated modelling of rhizosphere functioning could help to assess previous estimates of rhizodeposition by cross validation of rhizodeposition models with other models, for which the output variables are tightly connected to rhizodeposition and are more accessible (e.g. microbial growth, N dynamics). This could help to integrate our knowledge, to link rhizodeposition with plant functioning and to upscale case studies to the ecosystem level.

Modelling approaches

Mathematical modelling has the potential to predict C flows at spatial and temporal scales that are beyond the capability of current experimental techniques (Darrah et al. 2006). The construction and use of these models, however, are only as good as the knowledge of the individual processes and the values they are parameterized with. We know that the rhizosphere is inherently complex and that by default, current mathematical models are highly simplistic from a mechanistic standpoint. Despite this, however, there is also no doubt that they have greatly improved our understanding of rhizosphere processes (Barber 1995; Nye and Tinker 2000). In addition, it is also clear that major advances in mathematically describing the complexity of the rhizosphere have been made in recent years (Roose and Fowler 2004; Schnepf and Roose 2006). These advances have been only become possible through interdisciplinary interaction between applied mathematicians and rhizosphere biologists.

In a rhizodeposition context, one of the first quantitative modelling approaches was that taken by Newman and Watson (1977) where rhizosphere C flow was used to drive a soil microbial growth model. This model was subsequently refined by Darrah (1991a, b) with microbial growth placed in a growing root context. In terms of whole plant modelling, a photosynthesis model was used to calculate the flux of C entering the soil using photoassimilate partition coefficients (Swinnen 1994). However, due to the tight relationship between rhizodeposition and plant physiology, the input of C into the soil is not a constant part of the net fixed C or even of the C allocated to roots (see above). Therefore, it is necessary to have a more mechanistic approach, by modelling rhizodeposition along with plant physiology and more particularly with root system functioning. Indeed, exudation, which is a major component of rhizodeposition, is dependent upon root surface area and on the C concentration in root tissue relative to that in the soil solution. Subsequently, exudation can be simplistically modelled by a diffusion equation placed in a vegetation model that simulates plant phenology, canopy assimilation and carbohydrate partitioning above and belowground (Grant 1993). Due to the higher rates of exudation at root apices (McCully and Canny 1985; Darwent et al. 2003), the number and type of lateral branching is an important characteristic to be considered (Henry et al. 2005). Figure 7 shows an example of how this can be done. Exudation of an individual root was modelled from the root surface area (given by the root length and diameter) and by including the longitudinal variability of the C efflux (Personeni et al. 2007). Upscaling this model to the whole root system was achieved by coupling the exudation model to a root architecture model that simulates root emergence, their length and diameter as a function of thermal time (Pages and Pellerin 1996). When this was done the simulated cumulative exudation was 4.9 g C plant^{-1} or 390 kg C ha^{-1} (eight plant m^{-2}) at 860 growing degree days (flowering). This estimate accounts for the longitudinal variability of C efflux along individual roots, for the number of branches and for the root surface area of a model maize root system. This value is consistent with rhizodeposition estimates from tracer experiments, provided that rhizodeposition includes not only exudation but also mucilage, border cells and lysed cells.

There is great interest in coupling the modelling of rhizodeposition with root architecture models as it allows users to simulate changes in rhizodeposition in response to environmental conditions or photoassimilate availability through modifications of the characteristics of the root system. For example, N availability commonly increases root branching and consequently the number of root apices with higher rates of exudation. Similarly, limitation in C allocation to roots induces a reduction in root branching and in root diameter (Thaler and Pages 1998; Bidel et al. 2000)

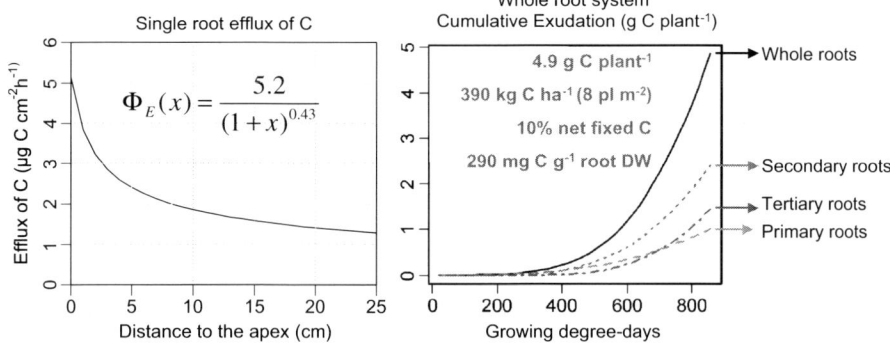

Fig. 7 Modelling of root exudation in maize. *Left*: Experimentally parameterized efflux profile of C from a single root (from Personeni et al. 2007). *Right*: Mathematical simulation of whole root system exudation in maize from germination until flowering (860 growing degree days, base 6°C). The simulation was performed by coupling the single root efflux of C model to the root architecture model of Pages and Pellerin (1996)

and consequently exudation. Furthermore, recent root architecture models have also included C availability in root tissue (Thaler and Pages 1998; Bidel et al. 2000), which is potentially important for modelling diffusive losses. Therefore, further investigations are needed to elucidate if a change in rhizodeposition occurring in response to a modification in photo-assimilate availability (Dilkes et al. 2004) is related to changes in root architecture and/or to changes in C availability within root tissues, which would change the rhizodeposition by individual roots. Much work has yet to be done to understand the mechanisms of N release from roots but a similar approach to that presented for C can be considered to model rhizodeposition of N in relation to the root system structure and functioning. Hence, modelling rhizodeposition with root architecture models that integrate C and N availability in root tissue is undoubtedly a promising perspective for predicting the release of C and N by roots under various environmental conditions.

Rhizodeposition—future outlook

It is clear from the previous discussion that we have made great progress in highlighting the importance of rhizosphere C flow in numerous aspects of ecosystem functioning. However, it is also apparent that we have a very long way to go before we can realistically harness the full extent of this knowledge for landscape level management (e.g. forest sustainability, biodiversity enhancement etc). This is exemplified by our process level understanding of C and nutrient flow at the single root level, however, how this scales up in landscapes which contain a mosaic of hydrologically interconnected vegetation and soil types remains unknown. This is certainly a goal which will only occur through an integration and enhancement of mathematical model scaling techniques. Indeed, the rhizosphere can be expected to play a major role in all the major challenges facing the planet including greenhouse gas mitigation, sustainable food production and food security, bioenergy production, preservation of water quality, accelerated restoration of post-industrial sites etc. One of the major obstacles to achieving this is the shear complexity of the rhizosphere and the lack of experimental techniques for teasing apart the myriad of interactions between roots and their biological, chemical and physical environment. While our knowledge of rhizodeposition has focused on crop plants, for both practical and economic reasons, there is a critical need to assess rhizosphere C flow in complex plant communities. A broader understanding of rhizosphere responses throughout the plant world can yield great insights into plant–soil functioning which cannot be provided by working on crop plants alone (Lambers et al. 2008). Similarly, there is also a need to look at rhizosphere processes in mature trees and particularly in mixed plantations where many synergistic relationships have been reported to occur (e.g. in litter decomposition, mycorrhizal interactions etc; Rothe and Binkley 2001). With the ongoing advances in our experimental and theoretical understanding of plant and microbial genomics, proteomics and metabolomics and the current focus on systems biology (Meldrum 2000), it is evident that rhizodeposition will remain a major focus of research for the foreseeable future. This increase in technology inevitably brings new challenges. In particular, finding robust statistical approaches to disentangle massive datasets produced from temporal and spatial sampling will be a necessity to maximise the potential of the technology (e.g. those produced by biodiversity measures such as pyrosequencing; Emerson et al. 2008; Fulthorpe et al. 2008). Consequently, rhizosphere bioinformatics is likely to grow in importance in the next decade. One of the most promising areas for further development is manipulating rhizosphere C flow to produce sustainable agricultural production systems. If we can interpret the C signals in the rhizosphere and then manipulate their flow, there is a potential to influence rhizosphere development. This could help to reduce our reliance on pesticides if we can stimulate and preserve the activity of biocontrol agents in the rhizosphere. However, a note of caution must also be made in our attempts to manipulate rhizosphere C flow. Although there are many researchers around the world attempting to alter rhizodeposition to help reduce our over-reliance on chemical fertilizers and pesticides, we must be careful not to over-mine or exploit the natural resources to a point at which the soils are left in a highly degraded state (i.e. unsuitable for colonization by native plants by excessively stripping the soil P pool). Consequently, there is also a pressing need for a debate on the ethics of manipulating the rhizosphere if we want to preserve public support for our research.

Acknowledgements The authors would like to address special thanks to L. Pagès (INRA, Avignon) for providing simulations from root architecture models.

References

Abuzinadah RA, Read DJ (1989) Carbon transfer associated with assimilation of organic nitrogen sources by silver birch (*Betula pendula* Roth.). Trees (Berl) 3:17–23 doi:10.1007/BF00202396

Ahonen-Jonnarth U, Van Hees PAW, Lundström US, Finlay RD (2000) Production of organic acids by mycorrhizal and non-mycorrhizal *Pinus sylvestris* L. seedlings exposed to elevated concentrations of aluminium and heavy metals. New Phytol 146:557–567 doi:10.1046/j.1469-8137.2000.00653.x

Allard V, Robin C, Newton PCD, Lieffering M, Soussana JF (2006) Short and long-term effects of elevated CO_2 on *Lolium perenne* rhizodeposition and its consequences on soil organic matter turnover and plant N yield. Soil Biol Biochem 38:1178–1187 doi:10.1016/j.soilbio.2005.10.002

Amiro BD, Ewing LL (1992) Physiological conditions and uptake of inorganic ^{14}C by plant–roots. Environ Exp Bot 32:203–211 doi:10.1016/0098-8472(92)90003-K

Andersson P, Berggren D (2005) Amino acids, total organic and inorganic nitrogen in forest floor soil solution at low and high nitrogen input. Water Air Soil Pollut 162:369–384 doi:10.1007/s11270-005-7372-y

Antunes PM, Rajcan I, Goss MJ (2006) Specific flavonoids as interconnecting signals in the tripartite symbiosis formed by arbuscular mycorrhizal fungi, *Bradyrhizobium japonicum* (Kirchner) Jordan and soybean (*Glycine max* (L.) Merr.). Soil Biol Biochem 38:533–543 doi:10.1016/j.soilbio.2005.06.008

Bacic A, Moody SF, McComb JA, Hinch JM, Clarke AE (1987) Extracellular polysaccharides from shaken liquid cultures of *Zea mays*. Aust J Plant Physiol 14:633–641

Bahyrycz A, Konopinska D (2007) Plant signalling peptides: some recent developments. J Pept Sci 13:787–797 doi:10.1002/psc.915

Bais HP, Park SW, Weir TL, Callaway RM, Vivanco JM (2004) How plants communicate using the underground information superhighway. Trends Plant Sci 9:26–32 doi:10.1016/j.tplants.2003.11.008

Balesdent J, Balabane M (1992) Maize root-derived soil organic-carbon estimated by natural ^{13}C abundance. Soil Biol Biochem 24:97–101 doi:10.1016/0038-0717(92)90264-X

Barber SA (1995) Soil nutrient bioavailability. Wiley, New York

Bardgett RD, Streeter TC, Bol R (2003) Soil microbes compete effectively with plants for organic-nitrogen inputs to temperate grasslands. Ecology 84:1277–1287 doi:10.1890/0012-9658(2003)084[1277:SMCEWP]2.0.CO;2

Barlow PW (1975) The root cap. In: Torrey JG, Clarkson DT (ed) The development and function of roots (Third Cabot Symposium). Academic, London, pp 21–54

Beemster GTS, Baskin TI (1998) Analysis of cell division and elongation underlying the developmental acceleration of root growth in *Arabidopsis thaliana*. Plant Physiol 116:1515–1526 doi:10.1104/pp.116.4.1515

Bending GD, Read DJ (1995) The structure and function of the vegetative mycelium of ectomycorrhizal plants. V. The foraging behaviour of ectomycorrhizal mycelium and the translocation of nutrients from exploited organic matter. New Phytol 130:401–409 doi:10.1111/j.1469-8137.1995.tb01834.x

Bengough AG, Kirby JM (1999) Tribology of the root cap in maize (*Zea mays*) and peas (*Pisum sativum*). New Phytol 142:421–425 doi:10.1046/j.1469-8137.1999.00406.x

Bengough AG, McKenzie BM (1997) Sloughing of root cap cells decreases the frictional resistance to maize (*Zea mays* L.) root growth. J Exp Bot 48:885–893 doi:10.1093/jxb/48.4.885

Bidartondo MI (2005) The evolutionary ecology of mycoheterotrophy. New Phytol 167:335–352 doi:10.1111/j.1469-8137.2005.01429.x

Bidartondo MI, Redecker D, Hijri I, Wiemken A, Bruns TD, Domínguez L, Sérsic A, Leake JR, Read DJ (2002) Epiparasitic plants specialized on arbuscular mycorrhizal fungi. Nature 419:389–392 doi:10.1038/nature01054

Bidel LPR, Pages L, Riviere LM, Pelloux G, Lorendeau JY (2000) MassFlowDyn I: A carbon transport and partitioning model for root system architecture. Ann Bot (Lond) 85:869–886 doi:10.1006/anbo.2000.1149

Bockenhoff A, Prior DAM, Grundler FMW, Oparka KJ (1996) Induction of phloem unloading in *Arabidopsis thaliana* roots by the parasitic nematode *Heterodera schachtii*. Plant Physiol 112:1421–1427 doi:10.1104/pp.112.4.1421

Boddy E, Hill PW, Farrar J, Jones DL (2007) Fast turnover of low molecular weight components of the dissolved organic carbon pool of temperate grassland field soils. Soil Biol Biochem 39:827–835 doi:10.1016/j.soilbio.2006.09.030

Boutton TW (1996) Stable carbon isotope ratios of soil organic matter and their use as indicators of vegetation and climate change. In: Boutton TW, Yamasaki S (eds) Mass spectrometry of soils. Marcel Dekker, New York, pp 47–82

Bouwmeester HJ, Roux C, Lopez-Raez JA, Bécard G (2007) Rhizosphere communication of plants, parasitic plants and AM fungi. Trends Plant Sci 12:224–230 doi:10.1016/j.tplants.2007.03.009

Brigham LA, Woo H, Nicoll SM, Hawes MC (1995) Differential expression of proteins and mRNAs from border cells and root tips of pea. Plant Physiol 109:457–463

Brown ME (1972) Plant-growth substances produced by microorganisms of soil and rhizosphere. J Appl Bacteriol 35:443–451

Buer CS, Muday GK, Djordjevic MA (2007) Flavonoids are differentially taken up and transported long distances in Arabidopsis. Plant Physiol 145:478–490 doi:10.1104/pp.107.101824

Canny MJ (1995) Apoplastic water and solute movement—new rules for an old space. Annu Rev Plant Physiol Plant Mol Biol 46:215–236 doi:10.1146/annurev.pp.46.060195.001243

Chapin FS, Moilanen L, Kielland K (1993) Preferential use of organic nitrogen for growth by a nonmycorrhizal arctic sedge. Nature 361:150–153 doi:10.1038/361150a0

Cheng WX, Coleman DC, Carroll CR, Hoffman C (1993) In-situ measurement of root respiration and soluble C-concentrations in the rhizosphere. Soil Biol Biochem 25:1189–1196 doi:10.1016/0038-0717(93)90251-6

Ciereszko I, Farrar JF, Rychter AM (1999) Compartmentation and fluxes of sugars in roots of *Phaseolus vulgaris* under phosphate deficiency. Biol Plant 42:223–231 doi:10.1023/A:1002108601862

Clark FE (1949) Soil microorganisms and plant roots. Adv Agron 1:241–288 doi:10.1016/S0065-2113(08)60750-6

Cram WJ (1974) Effects of Cl^- on HCO_3^- and malate fluxes and CO_2 fixation in carrot and barley root cells. J Exp Bot 25:253–268 doi:10.1093/jxb/25.2.253

Curl EA, Truelove (1986) The rhizosphere. Advanced series in agricultural science 15. Springer, Berlin

Czarnes S, Hallett PD, Bengough AG, Young IM (2000) Root- and microbial-derived mucilages affect soil structure and water transport. Eur J Soil Sci 51:435–443 doi:10.1046/j.1365-2389.2000.00327.x

Dakora FD, Phillips DA (2002) Root exudates as mediators of mineral acquisition in low-nutrient environments. Plant Soil 245:35–47 doi:10.1023/A:1020809400075

Darrah PR (1991a) Models of the rhizosphere. 1. Microbial-population dynamics around a root releasing soluble and insoluble carbon. Plant Soil 133:187–199 doi:10.1007/BF00009191

Darrah PR (1991b) Models of the rhizosphere. 2. A quasi 3-dimensional simulation of the microbial-population dynamics around a growing root releasing soluble exudates. Plant Soil 138:147–158 doi:10.1007/BF00012241

Darrah PR, Jones DL, Kirk GJD, Roose T (2006) Modelling the rhizosphere: a review of methods for 'upscaling' to the whole-plant scale. Eur J Soil Sci 57:13–25 doi:10.1111/j.1365-2389.2006.00786.x

Darwent MJ, Paterson E, McDonald AJS, Tomos AD (2003) Biosensor reporting of root exudation from *Hordeum vulgare* in relation to shoot nitrate concentration. J Exp Bot 54:325–334 doi:10.1093/jxb/54.381.325

de Graaff MA, Six J, van Kessel C (2007) Elevated CO_2 increases nitrogen rhizodeposition and microbial immobilization of root-derived nitrogen. New Phytol 173:778–786 doi:10.1111/j.1469-8137.2006.01974.x

Derrien D, Marol C, Balesdent J (2004) The dynamics of neutral sugars in the rhizosphere of wheat. An approach by ^{13}C pulse-labelling and GC/C/IRMS. Plant Soil 267:243–253 doi:10.1007/s11104-005-5348-8

Dilkes NB, Jones DL, Farrar J (2004) Temporal dynamics of carbon partitioning and rhizodeposition in wheat. Plant Physiol 134:706–715 doi:10.1104/pp.103.032045

Dilworth MJ, James EK, Sprent JI, Newton WE (2008) Nitrogen-fixing leguminous symbioses. Springer, New York

Dimou M, Flemetakis E, Delis C, Aivalakis G, Spyropoulos KG, Katinakis P (2005) Genes coding for a putative cell-wall invertase and two putative monosaccharide/H^+ transporters are expressed in roots of etiolated *Glycine max* seedlings. Plant Sci 169:798–804 doi:10.1016/j.plantsci.2005.05.037

DiTomaso JM, Hart JJ, Kochian LV (1992) Transport kinetics and metabolism of exogenously applied putrescine in roots of intact maize seedlings. Plant Physiol 98:611–620 doi:10.1104/pp.98.2.611

Ekblad A, Hogberg P (2001) Natural abundance of ^{13}C in CO_2 respired from forest soils reveals speed of link between tree photosynthesis and root respiration. Oecologia 127:305–308 doi:10.1007/s004420100667

Eleftheriou EP, Lazarou DS (1997) Cytochemical localization of ATPase activity in roots of wheat (*Triticum aestivum*). Biologia 52:573–583

Emerson D, Agulto L, Liu H, Liu LP (2008) Identifying and characterizing bacteria in an era of genomics and proteomics. Bioscience 58:925–936 doi:10.1641/B581006

Farrar J, Hawes M, Jones D, Lindow S (2003) How roots control the flux of carbon to the rhizosphere. Ecology 84:827–837 doi:10.1890/0012-9658(2003)084[0827:HRCTFO]2.0.CO;2

Faure D, Bloemberg G, Leveau J, Veereecke D (2009) Molecular communications in the rhizosphere. Plant Soil (this volume)

Feng JN, Volk RJ, Jackson WA (1994) Inward and outward transport of ammonium in roots of maize and sorghum—contrasting effects of methionine sulfoximine. J Exp Bot 45:429–439 doi:10.1093/jxb/45.4.429

Filion M, St-Arnaud M, Fortin JA (1999) Direct interaction between the arbuscular mycorrhizal fungus *Glomus intraradices* and different rhizosphere micro-organisms. New Phytol 141:525–533 doi:10.1046/j.1469-8137.1999. 00366.x

Finlay RD (2008) Ecological aspects of mycorrhizal symbiosis with special emphasis on the functional diversity of interactions involving the extraradical mycelium. J Exp Bot 59:1115–1126 doi:10.1093/jxb/ern059

Finlay RD, Rosling A (2006) Integrated nutrient cycles in forest ecosystems, the role of ectomycorrhizal fungi. In: Gadd GM (ed) Fungi in biogeochemical cycles. Cambridge University Press, Cambridge, pp 28–50

Finlay R, Söderström B (1992) Mycorrhiza and carbon flow to the soil. In: Allen MJ (ed) Mycorrhizal functioning. Chapman & Hall, New York, pp 134–160

Fischer WN, Andre B, Rentsch D, Krolkiewicz S, Tegeder M, Breitkreuz K, Frommer WB (1998) Amino acid transport in plants. Trends Plant Sci 3:188–195 doi:10.1016/S1360-1385(98)01231-X

Fleischer A, Ehwald R (1995) The free-space of sugars in plant-tissues—external film and apoplastic volume. J Exp Bot 46:647–654 doi:10.1093/jxb/46.6.647

Ford CR, Wurzburger N, Hendrick RL, Teskey RO (2007) Soil DIC uptake and fixation in *Pinus taeda* seedlings and its C contribution to plant tissues and ectomycorrhizal fungi. Tree Physiol 27:375–383

Francis R, Read DJ (1984) Direct transfer of carbon between plants connected by vesicular arbuscular mycorrhizal mycelium. Nature 307:53–56 doi:10.1038/307053a0

Fulthorpe RR, Roesch LFW, Riva A, Triplett EW (2008) Distantly sampled soils carry few species in common. ISME J 2:901–910 doi:10.1038/ismej.2008.55

Fusseder A (1987) The longevity and activity of the primary root of maize. Plant Soil 101:257–265 doi:10.1007/BF02370653

Gaudinski JB, Trumbore SE, Davidson EA, Cook AC, Markewitz D, Richter DD (2001) The age of fine-root carbon in three forests of the eastern United States measured by radiocarbon. Oecologia 129:420–429

Gill RA, Jackson RB (2000) Global patterns of root turnover for terrestrial ecosystems. New Phytol 147:13–31 doi:10.1046/j.1469-8137.2000.00681.x

Godbold DL, Hoosbeek MR, Lukac M, Cotrufo MF, Janssens IA, Ceulemans R, Polle A, Velthorst EJ, Scarascia-

Mugnozza G, DeAngelis P, Miglietta F, Peressotti A (2006) Mycorrhizal hyphal turnover as a dominant process for carbon input into soil organic matter. Plant Soil 281:15–24 doi:10.1007/s11104-005-3701-6

Gout E, Bligny R, Pascal N, Douce R (1993) ^{13}C nuclear-magnetic-resonance studies of malate and citrate synthesis and compartmentation in higher-plant cells. J Biol Chem 268:3986–3992

Grant RF (1993) Rhizodeposition by crop plants and its relationship to microbial activity and nitrogen distribution. Model Geo-Biosph Process 2:193–209

Grayston SJ, Vaughan D, Jones D (1996) Rhizosphere carbon flow in trees, in comparison with annual plants: the importance of root exudation and its impact on microbial activity and nutrient availability. Appl Soil Ecol 5:29–56 doi:10.1016/S0929-1393(96)00126-6

Guckert A, Breisch H, Reisinger O (1975) Soil/root interface. 1. Electron microscope study of mucigel/clay microorganism relations. Soil Biol Biochem 7:241–250 doi:10.1016/0038-0717(75)90061-9

Gunawardena U, Hawes MC (2002) Tissue specific localization of root infection by fungal pathogens: role of root border cells. Mol Plant Microbe Interact 15:1128–1136 doi:10.1094/MPMI.2002.15.11.1128

Hart JJ, DiTomaso JM, Linscott DL, Kochian LV (1992) Transport interactions between paraquat and polyamines in roots of intact maize seedlings. Plant Physiol 99:1400–1405 doi:10.1104/pp.99.4.1400

Hartmann A, Berg G, van Tuinen D (2009) Plant-driven selection of microbes. Plant Soil (this volume)

Hawes MC, Brigham LA, Wen F, Woo HH, Zhu Z (1998) Function of root border cells in plant health: pioneers in the rhizosphere. Annu Rev Phytopathol 36:311–327 doi:10.1146/annurev.phyto.36.1.311

Hawes MC, Gunawardena U, Miyasaka S, Zhao XW (2000) The role of root border cells in plant defence. Trends Plant Sci 5:128–133 doi:10.1016/S1360-1385(00)01556-9

Haydon MJ, Cobbett CS (2007) Transporters of ligands for essential metal ions in plants. New Phytol 174:499–506 doi:10.1111/j.1469-8137.2007.02051.x

Henry F, Nguyen C, Paterson E, Sim A, Robin C (2005) How does nitrogen availability alter rhizodeposition in *Lolium multiflorum* Lam. during vegetative growth? Plant Soil 269:181–191 doi:10.1007/s11104-004-0490-2

Herrmann A, Felle HH (1995) Tip growth in root hair-cells of *Sinapis-alba*. l—significance of internal and external Ca^{2+} and pH. New Phytol 129:523–533 doi:10.1111/j.1469-8137.1995.tb04323.x

Hertenberger G, Wanek W (2004) Evaluation of methods to measure differential ^{15}N labeling of soil and root N pools for studies of root exudation. Rapid Commun Mass Spectrom 18:2415–2425 doi:10.1002/rcm.1615

Hill PW, Marshall C, Williams GG, Blum H, Harmens H, Jones DL, Farrar JF (2007) The fate of photosynthetically-fixed carbon in *Lolium perenne* grassland as modified by elevated CO_2 and sward management. New Phytol 173:766–777 doi:10.1111/j.1469-8137.2007.01966.x

Hill PW, Farrar JF, Jones DL (2008) Decoupling of microbial glucose uptake and mineralization in soil. Soil Biol Biochem 40:616–624 doi:10.1016/j.soilbio.2007.09.008

Hinsinger P (2001) Bioavailability of soil inorganic P in the rhizosphere as affected by root-induced chemical changes: a review. Plant Soil 237:173–195 doi:10.1023/A:1013351617532

Hinsinger P, Bengough AG, Vetterlein D, Young IM (2009) Rhizosphere: biophysics, biogeochemistry and ecological relevance. Plant Soil (this volume)

Hirner A, Ladwig F, Stransky H, Okumoto S, Keinath M, Harms A, Frommer WB, Koch W (2006) Arabidopsis LHT1 is a high-affinity transporter for cellular amino acid uptake in both root epidermis and leaf mesophyll. Plant Cell 18:1931–1946 doi:10.1105/tpc.106.041012

Hodge A, Grayston SJ, Ord BG (1996) A novel method for characterisation and quantification of plant root exudates. Plant Soil 184:97–104 doi:10.1007/BF00029278

Hodge A, Paterson E, Thornton B, Millard P, Killham K (1997) Effects of photon flux density on carbon partitioning and rhizosphere carbon flow of *Lolium perenne*. J Exp Bot 48:1797–1805

Hodge A, Campbell CD, Fitter AH (2001) An arbuscular mycorrhizal fungus accelerates decomposition and acquires nitrogen directly from organic material. Nature 413:297–299 doi:10.1038/35095041

Hoffland E, Findenegg GR, Nelemans JA (1989) Solubilization of rock phosphate by rape. 2. Local root exudation of organic-acids as a response to P-starvation. Plant Soil 113:161–165 doi:10.1007/BF02280176

Högberg MN, Högberg P (2002) Extramatrical ectomycorrhizal mycelium contributes one-third of microbial biomass and produces, together with associated roots, half the dissolved organic carbon in a forest soil. New Phytol 154:791–795 doi:10.1046/j.1469-8137.2002.00417.x

Högberg P, Read DJ (2006) Towards a more plant physiological perspective on soil ecology. Trends Ecol Evol 21:548–554 doi:10.1016/j.tree.2006.06.004

Högberg P, Nordgren A, Buchmann N, Taylor AFS, Ekblad A, Högberg MN, Nyberg G, Ottosson-Löfvenius M, Read DJ (2001) Large-scale forest girdling shows that current photosynthesis drives soil respiration. Nature 411:789–792 doi:10.1038/35081058

Högberg P, Högberg MN, Göttlicher SG, Betson NR, Keel SG, Metcalfe DB, Campbell C, Schindlbacher A, Hurry V, Lundmark T, Linder S, Näsholm T (2008) High temporal resolution tracing of photosynthate carbon from the tree canopy to forest soil microorganisms. New Phytol 177:220–228

Hogh-Jensen H, Schjoerring JK (2001) Rhizodeposition of nitrogen by red clover, white clover and ryegrass leys. Soil Biol Biochem 33:439–448 doi:10.1016/S0038-0717(00)00183-8

Huang LF, Bocock PN, Davis JM, Koch KE (2007) Regulation of invertase: a suite of transcriptional and post-transcriptional mechanisms. Funct Plant Biol 34:499–507 doi:10.1071/FP06227

Hukin D, Doering-Saad C, Thomas CR, Pritchard J (2002) Sensitivity of cell hydraulic conductivity to mercury is coincident with symplasmic isolation and expression of plasmalemma aquaporin genes in growing maize roots. Planta 215:1047–1056 doi:10.1007/s00425-002-0841-2

Iijima M, Griffiths B, Bengough AG (2000) Sloughing of cap cells and carbon exudation from maize seedling roots in compacted sand. New Phytol 145:477–482 doi:10.1046/j.1469-8137.2000.00595.x

Iijima M, Higuchi T, Barlow PW (2004) Contribution of root cap mucilage and presence of an intact root cap in maize (*Zea mays*) to the reduction of soil mechanical impedance. Ann Bot (Lond) 94:473–477 doi:10.1093/aob/mch166

Jahn T, Baluska F, Michalke W, Harper JF, Volkmann D (1998) Plasma membrane H^+-ATPase in the root apex: evidence for strong expression in xylem parenchyma and asymmetric localization within cortical and epidermal cells. Physiol Plant 104:311–316 doi:10.1034/j.1399-3054.1998.1040304.x

Jensen ES (1996) Rhizodeposition of N by pea and barley and its effect on soil N dynamics. Soil Biol Biochem 28:65–71 doi:10.1016/0038-0717(95)00116-6

Jiang K, Zhang SB, Lee S, Tsai G, Kim K, Huang HY, Chilcott C, Zhu T, Feldman LJ (2006) Transcription profile analyses identify genes and pathways central to root cap functions in maize. Plant Mol Biol 60:343–363 doi:10.1007/s11103-005-4209-4

Johansson JF, Paul LR, Finlay RD (2004) Microbial interactions in the mycorrhizosphere and their significance for sustainable agriculture. FEMS Microbiol Ecol 48:1–13 doi:10.1016/j.femsec.2003.11.012

Johnson JF, Allan DL, Vance CP, Weiblen G (1996) Root carbon dioxide fixation by phosphorus-deficient *Lupinus albus*—contribution to organic acid exudation by proteoid roots. Plant Physiol 112:19–30 doi:10.1104/pp.112.1.31

Johnson D, Leake JR, Ostle N, Ineson P, Read DJ (2002) In situ $^{13}CO_2$ pulse-labelling of upland grassland demonstrates that a rapid pathway of carbon flux from arbuscular mycorrhizal mycelia to the soil. New Phytol 153:327–334 doi:10.1046/j.0028-646X.2001.00316.x

Johnson D, Krsek M, Wellington EMH, Stott AW, Cole L, Bardgett RD, Read DJ, Leake JR (2005) Soil invertebrates disrupt carbon flow through fungal networks. Science 309:1047 doi:10.1126/science.1114769

Jones DL, Darrah PR (1992) Resorption of organic-components by roots of *Zea mays* L. and its consequences in the rhizosphere. 1. Resorption of ^{14}C labelled glucose, mannose and citric-acid. Plant Soil 143:259–266 doi:10.1007/BF00007881

Jones DL, Darrah PR (1993) Re-sorption of organic-compounds by roots of *Zea mays* L. and its consequences in the rhizosphere. 2. Experimental and model evidence for simultaneous exudation and re-sorption of soluble C compounds. Plant Soil 153:47–59 doi:10.1007/BF00010543

Jones DL, Darrah PR (1994) Amino-acid influx at the soil–root interface of *Zea mays* L. and its implications in the rhizosphere. Plant Soil 163:1–12

Jones DL, Darrah PR (1995) Influx and efflux of organic-acids across the soil–root interface of *Zea mays* L. and its implications in rhizosphere C flow. Plant Soil 173:103–109 doi:10.1007/BF00155523

Jones DL, Darrah PR (1996) Re-sorption of organic compounds by roots of *Zea mays* L. and its consequences in the rhizosphere. 3. Characteristics of sugar influx and efflux. Plant Soil 178:153–160 doi:10.1007/BF00011173

Jones DL, Darrah PR, Kochian LV (1996) Critical evaluation of organic acid mediated iron dissolution in the rhizosphere and its potential role in root iron uptake. Plant Soil 180:57–66

Jones DL, Hodge A, Kuzyakov Y (2004) Plant and mycorrhizal regulation of rhizodeposition. New Phytol 163:459–480 doi:10.1111/j.1469-8137.2004.01130.x

Jones DL, Healey JR, Willett VB, Farrar JF (2005a) Dissolved organic nitrogen uptake by plants—an important N uptake pathway? Soil Biol Biochem 37:413–423 doi:10.1016/j.soilbio.2004.08.008

Jones DL, Shannon D, Junvee-Fortune T, Farrar JF (2005b) Plant capture of free amino acids is maximized under high soil amino acid concentrations. Soil Biol Biochem 37:179–181 doi:10.1016/j.soilbio.2004.07.021

Jun JH, Fiume E, Fletcher JC (2008) The CLE family of plant polypeptide signalling molecules. Cell Mol Life Sci 65:743–755 doi:10.1007/s00018-007-7411-5

Knox OGG, Gupta VVSR, Nehl DB, Stiller WN (2007) Constitutive expression of Cry proteins in roots and border cells of transgenic cotton. Euphytica 154:83–90 doi:10.1007/s10681-006-9272-7

Körner C, Asshoff R, Bignucolo O, Hättenschwiler R, Keel SG, Peláez-Riedl S, Pepin S, Siegwolf RTW, Zotz G (2005) Carbon flux and growth in mature deciduous forest trees exposed to elevated CO_2. Science 309:1360–1362

Kraffczyk I, Trolldenier G, Beringer H (1984) Soluble root exudates of maize: influence of potassium supply and rhizosphere microorganisms. Soil Biol Biochem 16:315–322 doi:10.1016/0038-0717(84)90025-7

Kramer EM, Frazer NL, Baskin TI (2007) Measurement of diffusion within the cell wall in living roots of *Arabidopsis thaliana*. J Exp Bot 58:3005–3015 doi:10.1093/jxb/erm155

Kuzyakov Y (2002) Separating microbial respiration of exudates from root respiration in non-sterile soils: a comparison of four methods. Soil Biol Biochem 34:1621–1631 doi:10.1016/S0038-0717(02)00146-3

Kuzyakov Y (2006) Sources of CO_2 efflux from soil and review of partitioning methods. Soil Biol Biochem 38:425–448 doi:10.1016/j.soilbio.2005.08.020

Kuzyakov Y, Domanski G (2000) Carbon input by plants into the soil. Review. J Plant Nutr Soil Sci 163:421–431 doi:10.1002/1522-2624(200008)163:4<421::AID-JPLN421>3.0.CO;2-R

Kuzyakov Y, Jones DL (2006) Glucose uptake by maize roots and its transformation in the rhizosphere. Soil Biol Biochem 38:851–860 doi:10.1016/j.soilbio.2005.07.012

Lambers H, Raven JA, Shaver GR, Smith SE (2008) Plant nutrient-acquisition strategies change with soil age. Trends Ecol Evol 23:95–103 doi:10.1016/j.tree.2007.10.008

Lambers H, Mougel C, Jaillard B, Hinsinger P (2009) Plant–microbe–soil interactions in the rhizosphere: an evolutionary perspective. Plant Soil (this volume)

Lasat MM (2002) Phytoextraction of toxic metals: a review of biological mechanisms. J Environ Qual 31:109–120

Leake JR (2004) Myco-heterotroph/epiparasitic plant interactions with ectomycorrhizal and arbuscular mycorrhizal fungi. Curr Opin Plant Biol 7:422–428 doi:10.1016/j.pbi.2004.04.004

Leake JR, Donnelly DP, Saunders EM, Boddy L, Read DJ (2001) Rates and quantities of carbon flux to ectomycorrhizal mycelium following ^{14}C pulse labeling of *Pinus sylvestris* seedlings: effects of litter patches and interaction with a wood decomposer fungus. Tree Physiol 21:71–82

Leake JR, Johnson D, Donnelly D, Muckle G, Boddy L, Read DJ (2004) Networks of power and influence: the role of mycorrhizal mycelium in controlling plant communities and agroecosystem functioning. Can J Bot 82:1016–1045 doi:10.1139/b04-060

Leinweber P, Eckhardt KU, Fischer H, Kuzyakov Y (2008) A new rapid micro-method for the molecular–chemical characterization of rhizodeposits by field-ionization mass spectrometry. Rapid Commun Mass Spectrom 22:1230–1234 doi:10.1002/rcm.3463

Ligaba A, Katsuhara M, Ryan PR, Shibasaka M, Matsumoto H (2006) The BnALMT1 and BnALMT2 genes from rape encode aluminum-activated malate transporters that enhance the aluminum resistance of plant cells. Plant Physiol 142:1294–1303 doi:10.1104/pp.106.085233

Lindahl B, Olsson S, Stenlid J, Finlay RD (2001) Effects of resource availability on mycelial interactions and ^{32}P-transfer between a saprotrophic and an ectomycorrhizal fungus in soil microcosms. FEMS Microbiol Ecol 38:43–52 doi:10.1111/j.1574-6941.2001.tb00880.x

Lindahl BD, Finlay RD, Cairney JWG (2005) Enzymatic activities of mycelia in mycorrhizal fungal communities. In: Dighton J, Oudemans P, White J (eds) The fungal community: its organization and role in the ecosystem. Marcel Dekker, New York, pp 331–348

Lindahl BD, Ihrmark K, Boberg J, Trumbore S, Högberg P, Stenlid J, Finlay RD (2007) Spatial separation of litter decomposition and mycorrhizal nitrogen uptake in boreal forests. New Phytol 173:611–620 doi:10.1111/j.1469-8137.2006.01936.x

Lipson D, Nasholm T (2001) The unexpected versatility of plants: organic nitrogen use and availability in terrestrial ecosystems. Oecologia 128:305–316 doi:10.1007/s004420100693

Lynch JM (1990) The rhizosphere. Wiley, London

Martin JK (1975) ^{14}C-labeled material leached from rhizosphere of plants supplied continuously with ^{14}CO$_2$. Soil Biol Biochem 7:395–399 doi:10.1016/0038-0717(75)90056-5

Mary B, Mariotti A, Morel JL (1992) Use of ^{13}C variations at natural abundance for studying the biodegradation of root mucilage, roots and glucose in soil. Soil Biol Biochem 24:1065–1072 doi:10.1016/0038-0717(92)90037-X

Mary B, Fresneau C, Morel JL, Mariotti A (1993) C-cycling and N-cycling during decomposition of root mucilage, roots and glucose in soil. Soil Biol Biochem 25:1005–1014 doi:10.1016/0038-0717(93)90147-4

Matiru VN, Dakora FD (2005) The rhizosphere signal molecule lumichrome alters seedling development in both legumes and cereals. New Phytol 166:439–444 doi:10.1111/j.1469-8137.2005.01344.x

Mayer J, Buegger F, Jensen ES, Schloter M, Hess J (2003) Estimating N rhizodeposition of grain legumes using a ^{15}N in situ stem labelling method. Soil Biol Biochem 35:21–28 doi:10.1016/S0038-0717(02)00212-2

McClaugherty CA, Aber JD, Melillo JM (1982) The role of fine roots in the organic-matter and nitrogen budgets of 2 forested ecosystems. Ecology 63:1481–1490 doi:10.2307/1938874

McCully ME (1999) Roots in soil: unearthing the complexities of roots and their rhizospheres. Annu Rev Plant Physiol Plant Mol Biol 50:695–718 doi:10.1146/annurev.arplant.50.1.695

McCully ME, Boyer JS (1997) The expansion of maize root-cap mucilage during hydration. 3. Changes in water potential and water content. Physiol Plant 99:169–177 doi:10.1111/j.1399-3054.1997.tb03445.x

McCully ME, Canny MJ (1985) Localisation of translocated ^{14}C in roots and root exudates of field-grown maize. Physiol Plant 65:380–392 doi:10.1111/j.1399-3054.1985.tb08661.x

McDougal BM, Rovira AD (1970) Sites of exudation of ^{14}C-labelled compounds from wheat roots. New Phytol 69:999–1002 doi:10.1111/j.1469-8137.1970.tb02479.x

Meharg AA (1994) A critical-review of labeling techniques used to quantify rhizosphere carbon-flow. Plant Soil 166:55–62 doi:10.1007/BF02185481

Meldrum D (2000) Automation for genomics, part two: sequencers, microarrays, and future trends. Genome Res 10:1288–1303 doi:10.1101/gr.157400

Mench M, Morel JL, Guckert A (1987) Metal binding properties of high molecular weight soluble exudates from maize (Zea mays L.) roots. Biol Fertil Soils 3:165–169 doi:10.1007/BF00255778

Miyasaka SC, Hawes MC (2001) Possible role of root border cells in detection and avoidance of aluminum toxicity. Plant Physiol 125:1978–1987 doi:10.1104/pp.125.4.1978

Morel JL, Mench M, Guckert A (1986) Measurement of Pb^{2+}, Cu^{2+} and Cd^{2+} binding with mucilage exudates from maize (Zea mays L.) roots. Biol Fertil Soils 2:29–34 doi:10.1007/BF00638958

Morel HJL, Guckert A, Plantureux S, Chenu C (1990) Influence of root exudates on soil aggregation. Symbiosis 9:87–91

Morre DJ, Jones DD, Mollenhauer HH (1967) Golgi apparatus mediated polysaccharide secretion by outer root cap cells of Zea mays. 1. Kinetics and secretory pathway. Planta 74:286–301 doi:10.1007/BF00384849

Nadelhoffer KJ, Raich JW (1992) Fine root production estimates and belowground carbon allocation in forest ecosystems. Ecology 73:1139–1147 doi:10.2307/1940664

Näsholm T, Ekblad A, Nordin A, Giesler R, Högberg M, Högberg P (1998) Boreal forest plants take up organic nitrogen. Nature 392:914–916 doi:10.1038/31921

Negishi T, Nakanishi H, Yazaki J, Kishimoto N, Fujii F, Shimbo K, Yamamoto K, Sakata K, Sasaki T, Kikuchi S, Mori S, Nishizawa NK (2002) cDNA microarray analysis of gene expression during Fe-deficiency stress in barley suggests that polar transport of vesicles is implicated in phytosiderophore secretion in Fe-deficient barley roots. Plant J 30:83–94 doi:10.1046/j.1365-313X.2002.01270.x

Nehls U (2008) Mastering ectomycorrhizal symbiosis: the impact of carbohydrates. J Exp Bot 59:1097–1108 doi:10.1093/jxb/erm334

Neumann G, Römheld V (2001) The release of root exudates as affected by the plant's physiological status. In: Pinton R, Varini Z, Nannipieri P (eds) The rhizosphere. Biochemistry and organic substances at the soil–plant interface. Marcel Dekker, New York, pp 41–93

Newman EI, Watson A (1977) Microbial abundance in rhizosphere—computer-model. Plant Soil 48:17–56 doi:10.1007/BF00015157

Nguyen C (2003) Rhizodeposition of organic C by plants: mechanisms and controls. Agronomie 23:375–396 doi:10.1051/agro:2003011

Nguyen C, Guckert A (2001) Short-term utilisation of C-14-[U] glucose by soil microorganisms in relation to carbon availability. Soil Biol Biochem 33:53–60 doi:10.1016/S0038-0717(00)00114-0

Nguyen C, Froux F, Recous S, Morvan T, Robin C (2008) Net N immobilisation during the biodegradation of mucilage in soil as affected by repeated mineral and organic fertilization. Nutr Cycl Agroecosyst 80:39–47 doi:10.1007/s10705-007-9119-1

Nye PH, Tinker PB (2000) Solute movement in the rhizosphere. Oxford University Press, Oxford

Offre P, Pivato B, Mazurier S, Siblot S, Berta G, Lemanceau P, Mougel C (2008) Microdiversity of Burkholderiales associated with mycorrhizal and nonmycorrhizal roots of *Medicago truncatula*. FEMS Microbiol Ecol 65:180–192 doi:10.1111/j.1574-6941.2008.00504.x

Ohyama T, Ohtake T, Sueyoshi K, Tewari K, Takahashi Y, Ito S, Nishiwaki T, Nagumo Y, Ishii S, Sato T (2009) Nitrogen fixation and metabolism in soybean plants. Nova, Hauppauge

Oksman-Caldentey KM, Inze D (2004) Plant cell factories in the post-genomic era: new ways to produce designer secondary metabolites. Trends Plant Sci 9:433–440 doi:10.1016/j.tplants.2004.07.006

Ostle N, Whiteley AS, Bailey MJ, Sleep D, Ineson P, Manefield M (2003) Active microbial RNA turnover in a grassland soil estimated using a $^{13}CO_2$ spike. Soil Biol Biochem 35:877–885 doi:10.1016/S0038-0717(03)00117-2

Ovecka M, Lang I, Baluska F, Ismail A, Illes P, Lichtscheidl IK (2005) Endocytosis and vesicle trafficking during tip growth of root hairs. Protoplasma 226:39–54 doi:10.1007/s00709-005-0103-9

Owen AG, Jones DL (2001) Competition for amino acids between wheat roots and rhizosphere microorganisms and the role of amino acids in plant N acquisition. Soil Biol Biochem 33:651–657 doi:10.1016/S0038-0717(00)00209-1

Pages L, Pellerin S (1996) Study of differences between vertical root maps observed in a maize crop and simulated maps obtained using a model for the three-dimensional architecture of the root system. Plant Soil 182:329–337

Patel DD, Barlow PW, Lee RB (1990) Development of vacuolar volume in the root-tips of pea. Ann Bot (Lond) 65:159–169

Paterson E (2003) Importance of rhizodeposition in the coupling of plant and microbial productivity. Eur J Soil Sci 54:741–750 doi:10.1046/j.1351-0754.2003.0557.x

Paterson E, Sim A (1999) Rhizodeposition and C-partitioning of *Lolium perenne* in axenic culture affected by nitrogen supply and defoliation. Plant Soil 216:155–164 doi:10.1023/A:1004789407065

Paterson E, Thornton B, Sim A, Pratt S (2003) Effects of defoliation and atmospheric CO_2 depletion on nitrate acquisition, and exudation of organic compounds by roots of *Festuca rubra*. Plant Soil 250:293–305 doi:10.1023/A:1022819219947

Paterson E, Thornton B, Midwood AJ, Sim A (2005) Defoliation alters the relative contributions of recent and non-recent assimilate to root exudation from *Festuca rubra*. Plant Cell Environ 28:1525–1533 doi:10.1111/j.1365-3040.2005.01389.x

Paterson E, Sim A, Standing D, Dorward M, McDonald AJS (2006) Root exudation from *Hordeum vulgare* in response to localized nitrate supply. J Exp Bot 57:2413–2420 doi:10.1093/jxb/erj214

Paterson E, Gebbing T, Abel C, Sim A, Telfer G (2007) Rhizodeposition shapes rhizosphere microbial community structure in organic soil. New Phytol 173:600–610 doi:10.1111/j.1469-8137.2006.01931.x

Paterson E, Osler G, Dawson LA, Gebbing T, Sim A, Ord B (2008) Labile and recalcitrant plant fractions are utilised by distinct microbial communities in soil: independent of the presence of roots and mycorrhizal fungi. Soil Biol Biochem 40:1103–1113 doi:10.1016/j.soilbio.2007.12.003

Patrick JW (1997) Phloem unloading: sieve element unloading and post-sieve element transport Ann Rev Plant Physiol. Plant Mol Biol 48:191–222 doi:10.1146/annurev.arplant.48.1.191

Paull RE, Jones RL (1975a) Studies on the secretion of maize root cap slime. 2. Localization of slime production. Plant Physiol 56:307–312 doi:10.1104/pp.56.2.307

Paull RE, Jones RL (1975b) Studies on the secretion of maize root-cap slime. 3. Histochemical and autoradiographic localization of incorporated fucose. Planta 127:97–110 doi:10.1007/BF00388371

Paull RE, Jones RL (1976a) Studies on the secretion of maize root cap slime. 4. Evidence for the involvement of dictyosomes. Plant Physiol 57:249–256 doi:10.1104/pp.57.2.249

Paull RE, Jones RL (1976b) Studies on the secretion of maize root cap slime. 5. The cell wall as a barrier to secretion. Zeit Pflanzenphysiol 79:154–164

Paull RE, Johnson CM, Jones RL (1975) Studies on the secretion of maize root cap slime. 1. Some properties of the secreted polymer. Plant Physiol 56:300–306 doi:10.1104/pp.56.2.300

Personeni E, Nguyen C, Marchal P, Pagès L (2007) Experimental evaluation of an efflux–influx model of C exudation by individual apical root segments. J Exp Bot 58:2091–2099 doi:10.1093/jxb/erm065

Pfeffer PE, Douds DD, Bücking H, Schwartz DP, Shachar-Hill Y (2004) The fungus does not transfer carbon to or between roots in an arbuscular mycorrhizal symbiosis. New Phytol 163:617–627

Philips DA, Fox TC, Six J (2006) Root exudation (net efflux of amino acids) may increase rhizodeposition under elevated CO_2. Glob Change Biol 12:561–567 doi:10.1111/j.1365-2486.2006.01100.x

Phillips RP, Fahey TJ (2005) Patterns of rhizosphere carbon flux in sugar maple (*Acer saccharum*) and yellow birch (*Betula alleghaniensis*) saplings. Glob Change Biol 11:983–995 doi:10.1111/j.1365-2486.2005.00959.x

Phillips DA, Fox TC, King MD, Bhuvaneswari TV, Teuber LR (2004) Microbial products trigger amino acid exudation from plant roots. Plant Physiol 136:2887–2894 doi:10.1104/pp.104.044222

Pinton R, Varanini Z, Nannipieri P (2001) The rhizosphere. Biochemistry and organic substances at the soil–plant interface. CRC, Boca Raton

Qian JH, Doran JW, Walters DT (1997) Maize plant contributions to root zone available carbon and microbial transformations of nitrogen. Soil Biol Biochem 29:1451–1462 doi:10.1016/S0038-0717(97)00043-6

Quadt-Hallmann A, Hallmann J, Kloepper JW (1997) Bacterial endophytes in cotton: location and interaction with other plant associated bacteria. Can J Microbiol 43:254–259

Rangel-Castro JI, Killham K, Ostle N, Nicol GW, Anderson IC, Scrimgeour CM, Ineson P, Meharg A, Prosser JI (2005a) Stable isotope probing analysis of the influence of liming on root exudate utilization by soil microorganisms. Environ Microbiol 7:828–838 doi:10.1111/j.1462-2920. 2005.00756.x

Rangel-Castro JI, Prosser JI, Ostle N, Scrimgeour CM, Killham K, Meharg A (2005b) Flux and turnover of fixed carbon in soil microbial biomass of limed and unlimed plots of an upland grassland ecosystem. Environ Microbiol 7:544–552 doi:10.1111/j.1462-2920.2005.00722.x

Read DB, Gregory PJ (1997) Surface tension and viscosity of axenic maize and lupin root mucilages. New Phytol 137:623–628 doi:10.1046/j.1469-8137.1997.00859.x

Read DJ, Perez-Moreno J (2003) Mycorrhizas and nutrient cycling in ecosystems—a journey towards relevance? New Phytol 157:475–492 doi:10.1046/j.1469-8137.2003. 00704.x

Read DB, Bengough AG, Gregory PJ, Crawford JW, Robinson D, Scrimgeour CM, Young IM, Zhang K, Zhang X (2003) Plant roots release phospholipid surfactants that modify the physical and chemical properties of soil. New Phytol 157:315–326 doi:10.1046/j.1469-8137.2003.00665.x

Roberts SK (2006) Plasma membrane anion channels in higher plants and their putative functions in roots. New Phytol 169:647–666 doi:10.1111/j.1469-8137.2006.01639.x

Robinson D, Fitter AH (1999) The magnitude and control of carbon transfer between plants linked by a common mycorrhizal network. J Exp Bot 50:9–13 doi:10.1093/jexbot/50.330.9

Rochette P, Flanagan LB, Gregorich EG (1999) Separating soil respiration into plant and soil components using analyses of the natural abundance of ^{13}C. Soil Sci Soc Am J 63:1207–1213

Rodger S, Bengough AG, Griffiths BS, Stubbs V, Young IM (2003) Does the presence of detached root border cells of *Zea mays* alter the activity of the pathogenic nematode *Meloidogyne incognita?*. Phytopath 93:1111–1114 doi:10.1094/PHYTO.2003.93.9.1111

Roose T, Fowler AC (2004) A mathematical model for water and nutrient uptake by plant root systems. J Theor Biol 228:173–184 doi:10.1016/j.jtbi.2003.12.013

Rosling A, Lindahl BD, Finlay RD (2004a) Carbon allocation in intact mycorrhizal systems of *Pinus sylvestris* L. seedlings colonizing different mineral substrates. New Phytol 162:795–802 doi:10.1111/j.1469-8137.2004. 01080.x

Rosling A, Lindahl BD, Taylor AFS, Finlay RD (2004b) Mycelial growth and substrate acidification of ectomycorrhizal fungi in response to different minerals. FEMS Microbiol Ecol 47:31–37 doi:10.1016/S0168-6496(03) 00222-8

Rothe A, Binkley D (2001) Nutritional interactions in mixed species forests: a synthesis. Can J Res 31:1855–1870 doi:10.1139/cjfr-31-11-1855

Roux SJ, Steinebrunner I (2007) Extracellular ATP: an unexpected role as a signaller in plants. Trends Plant Sci 11:522–527 doi:10.1016/j.tplants.2007.09.003

Rovira AD (1965) Interactions between plant roots and soil microorganisms. Annu Rev Microbiol 19:241–266 doi:10.1146/annurev.mi.19.100165.001325

Rovira AD (1969) Plant root exudates. Bot Rev 35:35–59 doi:10.1007/BF02859887

Rovira AD, Foster RC, Martin JK (1979) Note on terminology: origin, nature and nomenclature of the organic materials in the rhizosphere. In: Harley JL, Scott Russell R (eds) The soil–root interface. Academic, London, pp 1–4

Ryan PR, Delhaize E, Jones DL (2001) Function and mechanism of organic anion exudation from plant roots. Ann Rev Plant Physiol Plant Mol Biol 52:527–560

Sacchi GA, Abruzzese A, Lucchini G, Fiorani F, Cocucci S (2000) Efflux and active re-absorption of glucose in roots of cotton plants grown under saline conditions. Plant Soil 220:1–11 doi:10.1023/A:1004701912815

Samaj J, Read ND, Volkmann D, Menzel D, Baluska F (2005) The endocytic network in plants. Trends Cell Biol 15:425–433 doi:10.1016/j.tcb.2005.06.006

Samuels AL, Fernando M, Glass ADM (1992) Immunofluorescent localization of plasma-membrane H^+-ATPase in barley roots and effects of K-nutrition. Plant Physiol 99:1509–1514 doi:10.1104/pp.99.4.1509

Sapronov DV, Kuzyakov YV (2007) Separation of root and microbial respiration: comparison of three methods. Eurasian Soil Sci 40:775–784 doi:10.1134/S1064229307070101

Scheunert I, Topp E, Attar A, Korte F (1994) Uptake pathways of chlorobenzenes in plants and their correlation with n-octanol/water partition-coefficients. Ecotoxicol Environ Saf 27:90–104 doi:10.1006/eesa.1994.1009

Scheurwater I, Clarkson DT, Purves JV, van Rijt G, Saker LR, Welschen R, Lambers H (1999) Relatively large nitrate efflux can account for the high specific respiratory costs for nitrate transport in slow-growing grass species. Plant Soil 215:123–134 doi:10.1023/A:1004559628401

Schnepf A, Roose T (2006) Modelling the contribution of arbuscular mycorrhizal fungi to plant phosphate uptake. New Phytol 171:669–682

Schraut D, Ullrich CI, Hartung W (2004) Lateral ABA transport in maize roots (*Zea mays*): visualization by immunolocalization. J Exp Bot 55:1635–1641 doi:10.1093/jxb/erh193

Sherson SM, Alford HL, Forbes SM, Wallace G, Smith SM (2003) Roles of cell-wall invertases and monosaccharide transporters in the growth and development of Arabidopsis. J Exp Bot 54:525–531 doi:10.1093/jxb/erg055

Shishkova S, Dubrovsky JG (2005) Developmental programmed cell death in primary roots of Sonoran Desert Cactaceae. Am J Bot 92:1590–1594 doi:10.3732/ajb.92.9.1590

Shrestha M, Abraham WR, Shrestha PM, Noll M, Conrad R (2008) Activity and composition of methanotrophic bacterial communities in planted rice soil studied by flux measurements, analyses of pmoA gene and stable isotope probing of phospholipid fatty acids. Environ Microbiol 10:400–412 doi:10.1111/j.1462-2920.2007.01462.x

Simard SW, Perry DA, Jones MD, Myrold DD, Durall DM, Molina R (1997) Net transfer of carbon between ectomycorrhizal tree species in the field. Nature 388:579–582 doi:10.1038/41557

Singh BK, Millard P, Whiteley AS, Murrell JC (2004) Unravelling rhizosphere–microbial interactions: opportunities and limitations. Trends Microbiol 12:386–393 doi:10.1016/j.tim.2004.06.008

Singh BK, Naoise N, Ridgway KP, McNicol J, Young JPW, Daniell TJ, Prosser JI, Millard P (2008) Relationship between assemblages of mycorrhizal fungi and bacteria on grass roots. Environ Microbiol 10:534–541 doi:10.1111/j.1462-2920.2007.01474.x

Sobolev VS, Potter TL, Horn BW (2006) Prenylated stilbenes from peanut root mucilage. Phytochem Anal 17:312–322 doi:10.1002/pca.920

Srivastava S, Srivastava AK (2007) Hairy root culture for mass-production of high-value secondary metabolites. Crit Rev Biotechnol 27:29–43 doi:10.1080/07388550601173918

Staddon PL, Bronk Ramsey C, Ostle N, Ineson P, Fitter AH (2003) Rapid turnover of hyphae of mycorrhizal fungi determined by AMS microanalysis of ^{14}C. Science 300:1138–1140 doi:10.1126/science.1084269

Stewart AM, Frank DA (2008) Short sampling intervals reveal very rapid root turnover in a temperate grassland. Oecologia 157:453–458 doi:10.1007/s00442-008-1088-9

Stubbs VEC, Standing D, Knox OGG, Killham K, Bengough AG, Griffiths B (2004) Root border cells take up and release glucose-C. Ann Bot (Lond) 93:221–224 doi:10.1093/aob/mch019

Sugiyama A, Shitan N, Yazaki K (2007) Involvement of a soybean ATP-binding cassette—type transporter in the secretion of genistein, a signal flavonoid in legume–rhizobium symbiosis. Plant Physiol 144:2000–2008 doi:10.1104/pp.107.096727

Sun Y-P, Unestam T, Lucas SD, Johanson KJ, Kenne L, Finlay RD (1999) Exudation–reabsorption in mycorrhizal fungi, the dynamic interface for interaction with soil and other microorganisms. Mycorrhiza 9:137–144 doi:10.1007/s005720050298

Swinnen J (1994) Rhizodeposition and turnover of root-derived organic material in barley and wheat under conventional and integrated management. Agric Ecosyst Environ 51:115–128 doi:10.1016/0167-8809(94)90038-8

Swinnen J, van Veen JA, Merckx R (1995) Root decay and turnover of rhizodeposits in field-grown winter-wheat and spring barley estimated by ^{14}C pulse-labeling. Soil Biol Biochem 27:211–217 doi:10.1016/0038-0717(94)00161-S

Thaler P, Pages L (1998) Modelling the influence of assimilate availability on root growth and architecture. Plant Soil 201:307–320 doi:10.1023/A:1004380021699

Thornton B (2001) Uptake of glycine by non-mycorrhizal *Lolium perenne*. J Exp Bot 52:1315–1322 doi:10.1093/jexbot/52.359.1315

Thornton B, Paterson E, Midwood AJ, Sim A, Pratt SM (2004) Contribution of current carbon assimilation in supplying root exudates of *Lolium perenne* measured using steady-state ^{13}C labelling. Physiol Plant 120:434–441 doi:10.1111/j.0031-9317.2004.00250.x

Todorovic C, Nguyen C, Robin C, Guckert A (2001) Root and microbial involvement in the kinetics of C-14-partitioning to rhizosphere respiration after a pulse labelling of maize assimilates. Plant Soil 228:179–189 doi:10.1023/A:1004830011382

Toljander JF, Artursson V, Paul LR, Jansson JK, Finlay RD (2006) Attachment of different soil bacteria to arbuscular mycorrhizal fungi is determined by hyphal vitality and fungal species. FEMS Microbiol Lett 254:34–40 doi:10.1111/j.1574-6968.2005.00003.x

Toljander JF, Paul L, Lindahl BD, Elfstrand M, Finlay RD (2007) Influence of AM fungal exudates on bacterial community structure. FEMS Microbiol Ecol 61:295–304 doi:10.1111/j.1574-6941.2007.00337.x

Treseder KK, Turner KM (2007) Glomalin in ecosystems. Soil Sci Soc Am J 71:1257–1266 doi:10.2136/sssaj2006.0377

van Hees PAW, Jones DL, Finlay R, Godbold DL, Lundström U (2005) The carbon we do not see—the impact of low molecular weight compounds on carbon dynamics and respiration in forest soils: a review. Soil Biol Biochem 37:1–13 doi:10.1016/j.soilbio.2004.06.010

Vandenkoornhuyse P, Mahé S, Ineson P, Staddon P, Ostle N, Cliquet J-B, Francez A-J, Fitter AH, Young JPW (2007) Active root-inhabiting microbes identified by rapid incorporation of plant-derived carbon into RNA. Proc Natl Acad Sci U S A 104:16970–16975 doi:10.1073/pnas.0705902104

Verpoorte R, van der Heijden R, Memelink J (2000) Engineering the plant cell factory for secondary metabolite production. Transgenic Res 9:323–343 doi:10.1023/A:1008966404981

Vessey JK (2003) Plant growth promoting rhizobacteria as biofertilizers. Plant Soil 255:571–586 doi:10.1023/A:1026037216893

Wallander H, Nilsson LO, Hagerberg D, Bååth E (2001) Estimation of the biomass and seasonal growth of external mycelium of ectomycorrhizal fungi in the field. New Phytol 151:752–760 doi:10.1046/j.0028-646x.2001.00199.x

Warembourg FR, Kummerow J (1991) Photosynthesis/translocation studies in terrestrial ecosystems. In: Coleman DC, Fry B (eds) Carbon isotope techniques. Academic, London, pp 11–37

Watt M, Hugenholtz P, White R, Vinall K (2006) Numbers and locations of native bacteria on field-grown wheat roots quantified by fluorescence in situ hybridization (Fish). Environ Microbiol 8:871–884 doi:10.1111/j.1462-2920.2005.00973.x

Welbaum GE, Sturz AV, Dong ZM, Nowak J (2007) Managing soil microorganisms to improve productivity of agro-ecosystems. Crit Rev Plant Sci 23:175–193 doi:10.1080/07352680490433295

Wen FS, VanEtten HD, Tsaprailis G, Hawes MC (2007) Extracellular proteins in pea root tip and border cell exudates. Plant Physiol 143:773–783 doi:10.1104/pp.106.091637

Whipps JM (2001) Microbial interactions and biocontrol in the rhizosphere. J Exp Bot 52:487–511

Wichern F, Mayer J, Joergensen RG, Muller T (2007) Rhizodeposition of C and N in peas and oats after ^{13}C-^{15}N double labelling under field conditions. Soil Biol Biochem 39:2527–2537 doi:10.1016/j.soilbio.2007.04.022

Williams LE, Lemoine R, Sauer N (2000) Sugar transporters in higher plants—a diversity of roles and complex regulation. Trends Plant Sci 5:283–290 doi:10.1016/S1360-1385(00)01681-2

Wright DP, Read DJ, Scholes JD (1998) Mycorrhizal sink strength influences whole plant carbon balance of *Trifoli-*

um repens L. Plant Cell Environ 21:881–891 doi:10.1046/j.1365-3040.1998.00351.x

Wuyts N, Maung ZTZ, Swennen R, De Waele D (2006) Banana rhizodeposition: characterization of root border cell production and effects on chemotaxis and motility of the parasitic nematode *Radopholus similis*. Plant Soil 283:217–228 doi:10.1007/s11104-006-0013-4

Yeomans CV, Porteous F, Paterson E, Meharg AA, Killham K (1999) Assessment of lux-marked *Pseudomonas fluorescens* for reporting on organic carbon compounds. FEMS Microbiol Lett 176:79–83 doi:10.1111/j.1574-6968.1999.tb13645.x

Zhang WH, Ryan PR, Tyerman SD (2004) Citrate-permeable channels in the plasma membrane of cluster roots from white lupin. Plant Physiol 136:3771–3783 doi:10.1104/pp.104.046201

Zhao XW, Schmitt M, Hawes MC (2000) Species-dependent effects of border cell and root tip exudates on nematode behaviour. Phytopath 90:1239–1245 doi:10.1094/PHYTO.2000.90.11.1239

Zheng J, Sutton JC, Yu H (2000) Interactions among *Pythium aphanidermatum*, roots, root mucilage, and microbial agents in hydroponic cucumbers. Can J Plant Pathol 22:368–379

Zhu T, Rost TL (2000) Directional cell-to-cell communication in the Arabidopsis root apical meristem. III. Plasmodesmata turnover and apoptosis in meristem and root cap cells during four weeks after germination. Protoplasma 213:99–107 doi:10.1007/BF01280510

REVIEW ARTICLE

Nitrogen-fixing bacteria associated with leguminous and non-leguminous plants

Claudine Franche · Kristina Lindström · Claudine Elmerich

Received: 8 July 2008 / Accepted: 10 November 2008 / Published online: 3 December 2008
© Springer Science + Business Media B.V. 2008

Abstract Nitrogen is generally considered one of the major limiting nutrients in plant growth. The biological process responsible for reduction of molecular nitrogen into ammonia is referred to as nitrogen fixation. A wide diversity of nitrogen-fixing bacterial species belonging to most phyla of the *Bacteria* domain have the capacity to colonize the rhizosphere and to interact with plants. Leguminous and actinorhizal plants can obtain their nitrogen by association with rhizobia or *Frankia* via differentiation on their respective host plants of a specialized organ, the root nodule. Other symbiotic associations involve heterocystous cyanobacteria, while increasing numbers of nitrogen-fixing species have been identified as colonizing the root surface and, in some cases, the root interior of a variety of cereal crops and pasture grasses. Basic and advanced aspects of these associations are covered in this review.

Keywords Nitrogen fixation · Symbiosis · Rhizobia · *Frankia* · Cyanobacteria · *Azospirillum*

Introduction

Fixed nitrogen is a limiting nutrient in most environments, with the main reserve of nitrogen in the biosphere being molecular nitrogen from the atmosphere. Molecular nitrogen cannot be directly assimilated by plants, but it becomes available through the biological nitrogen fixation process that only prokaryotic cells have developed. Proliferation of bacteria in soil adhering to the root surface was discovered toward the end of the nineteenth century, at the same time as the discovery of nitrogen fixation. The term "rhizosphere" was then coined by Hiltner in 1901 to designate soil immediately surrounding roots under the influence of the plant (see Rovira 1991).

For many years, a limited number of bacterial species were believed to be nitrogen fixers (Postgate 1981), but in the last 30 years nitrogen fixation has been shown to be a property with representatives in most of the phyla of *Bacteria* and also in methanogenic *Archaea* (Young 1992). The property of symbiotically fixing nitrogen within nodules of vascular plants is found in two major groups of bacteria

Responsible Editor: Robert Reid.

C. Franche
Equipe Rhizogénèse, UMR DIAPC,
Institut de Recherche pour le Développement,
B. P. 64501, 34394 Montpellier Cedex 5, France

K. Lindström
Department of Applied Chemistry and Microbiology,
Viikki Biocenter, University of Helsinki,
P. O. Box 56, 00014 Helsinki, Finland

C. Elmerich (✉)
Biologie Moléculaire du Gène chez les Extrêmophiles,
Département de Microbiologie, Institut Pasteur,
25 rue du Dr. Roux,
75724 Paris Cedex 15, France
e-mail: elmerich@pasteur.fr

not phylogenetically related: rhizobia (*Alpha-proteobacteria*) that associate essentially with leguminous plants belonging to one superfamily of angiosperms (Fabaceae), (Sprent 2001) and *Frankia* (in *Actinobacteria*) that associate with a broader spectrum of plants from eight families (Huss-Danell 1997; Vessey et al. 2004). Another important group of nitrogen-fixing bacteria is that of the cyanobacteria, found in association with a large variety of higher and lower plants, fungi and algae (Meeks and Elhai 2002). Associative symbiosis refers to a wide variety of nitrogen-fixing species that colonize the root surface of non-leguminous plants, without formation of differentiated structures (Elmerich and Newton 2007). Among these, the frequent isolation of bacteria from surface-sterilized root led to identification of a new category, nitrogen-fixing endophytes (Döbereiner 1992; Reinhold-Hurek and Hurek 1998; Baldani and Baldani 2005).

The present paper presents an overview of the general biology of the different types of nitrogen-fixing associations and of basic mechanisms by which bacteria can interact with the root system of their host plant. It also reports the phylogeny and particular physiology of bacteria involved in these associations, as well as the relative agronomic importance of the different systems.

The nitrogen fixation process

Enzymatic conversion of molecular nitrogen to ammonia is catalyzed by nitrogenase, an oxygen-labile enzyme complex highly conserved in free-living and symbiotic diazotrophs. The most common form of nitrogenase, referred to as Mo-nitrogenase or conventional nitrogenase, contains a prosthetic group with molybdenum, FeMoCo. Some bacteria, such as *Azotobacter* and several photosynthetic nitrogen fixers (including some cyanobacteria), carry additional forms of nitrogenase whose cofactor contains vanadium (V-nitrogenase) or only iron (Fe-nitrogenase) (Rubio and Ludden 2005; Newton 2007). We shall limit this chapter to the conventional enzyme.

The nitrogenase enzyme, which has been purified from different sources, is composed of two metalloproteins. Component 1, also designated MoFe protein, is a tetramer of 220,000 Da composed of two non-identical subunits α and β, while component 2, also designated Fe protein, is a dimer of 68,000 Da formed by identical subunits. Two FeMoCo are bound to the α subunits of the MoFe protein. In addition, there are two other prosthetic groups containing 4Fe-4S clusters. 'P-clusters' are covalently bound to cysteine residues of MoFe protein bridging α and β subunits. The third type of Fe-S group is linked to the Fe protein (Zheng et al. 1998; Hu et al. 2007; Newton 2007; Rubio and Ludden 2008).

Nitrogen reduction is a very complex mechanism not as yet fully elucidated. The result of net reduction of molecular nitrogen to ammonia is generally accounted for by the following equation:

$$N_2 + 16\,MgATP + 8\,e^- + 8\,H^+$$
$$\rightarrow 2\,NH_3 + H_2 + 16\,MgADP + 16\,Pi$$

The genetics of nitrogen fixation was initially elucidated in *Klebsiella oxytoca* strain M5a1 (first identified as *K. pneumoniae*). In that strain, *nif* genes necessary for synthesis of a functional nitrogenase are clustered in a 24 kb region (Arnold et al. 1988). This is the most compact organization of *nif* genes ever described. The three structural genes coding for Mo nitrogenase polypeptides are *nifD* and *nifK* for Mo protein subunits and *nifH* for Fe protein. Full assembly of nitrogenase requires products of other *nif* genes involved in synthesis of FeMoCo (*nifB, nifQ, nifE, nifN, nifX, nifU, nifS, nifV, nifY* also *nifH*) and in assembly of iron sulfur clusters (*nifS* and *nifU*) and maturation of the nitrogenase components (*nifW* and *nifZ*) (Zheng et al., 1998; Rubio and Ludden, 2005; Hu et al. 2007; Rubio and Ludden, 2008). In addition, *Klebsiella* contains genes required for electron transport to nitrogenase (*nifF* and *nifJ*), as well as the regulatory *nifLA* genes controlling expression of the *nif* cluster (Merrick and Edwards 1995; Dixon and Kahn 2004). It is now established that a core of *nif* genes (*nifH, nifD nifK, nifY, nifB, nifQ, nifE, nifN, nifX, nifU, nifS, nifV, nifW, nifZ*) required for nitrogenase synthesis and catalysis is conserved in all diazotrophs. Depending on the system, other genes are required for in vivo nitrogenase activity, such as those coding for components of physiological electron transport chains (flavodoxin, ferredoxin and the NADH-ubiquinone oxidoreductase (NQR) encoded by the *rnfABCDGEF* cluster) to nitrogenase, molybdenum uptake and homeostasis, and oxygen protection and regulation, including respiratory chains adapted to oxygen conditions at which the nitrogen fixation process can operate (for details see Fischer 1994; Dixon and Kahn 2004; Pedrosa and Elmerich, 2007).

Soil and rhizosphere microbiology

Discovery of nitrogen fixers

The discovery of nitrogen fixation was attributed to the German scientists Hellriegel and Wilfarth, who in 1886 reported that legumes bearing root nodules could use gaseous (molecular) nitrogen. Shortly afterwards, in 1888, Beijerinck, a Dutch microbiologist, succeeded in isolating a bacterial strain from root nodules. This isolate happened to be a *Rhizobium leguminosarum* strain. Beijerinck (in 1901) and Lipman (in 1903) were responsible for isolation of *Azotobacter* spp., while Winodgradsky (in 1901) isolated the first strain of *Clostridium pasteurianum* (see Stewart 1969). Discovery of nitrogen fixation in blue-green algae (now classified as cyanobacteria) was established much later (Stewart 1969). By 1960, the nitrogen fixation capacities of free-living soil bacteria had been established for only a dozen genera. This was a long way from our present knowledge of the distribution of nitrogen fixation ability in most phyla of the *Bacteria* domain (Postgate 1981; Balandreau 1983; Young 1992; Henson et al. 2004; Lindström and Martínez-Romero 2007; Schmid and Hartmann 2007) (Fig. 1). Indeed, identification of new nitrogen-fixing genera and species has long been hampered by technical limitations due to both the unavailability of proper tools for taxonomy and phylogeny and the difficulty in proving nitrogen-fixing capacity.

Measurement of nitrogen fixation

Measurement of the net gain in fixed nitrogen by nitrogen balance was greatly facilitated, at the end of the 19th century, when Kjeldahl, a Danish scientist, introduced an analytical method for determination of total nitrogen (Bergersen 1980). The technique proved to be useful for quantifying nitrogen in different ecosystems. Development of ^{15}N isotopic tracer techniques (in the 1940's) enabled scientists to more effectively demonstrate that a gain in nitrogen resulted from nitrogen fixation (Burris and Miller 1941). Under field conditions, the use of ^{15}N isotopic gas is impracticable. Instead, a common method is based on differences in the natural abundance of ^{15}N contained in mineral sources of nitrogen compared to atmospheric nitrogen (Boddey et al. 2000; 2001). It enables determining the proportion of plant N derived from atmospheric nitrogen (%Ndfa). A major achievement for rapid determination of nitrogenase activity was the development of an assay using the acetylene reduction technique (Hardy et al. 1973), which could be applied not only to pure cultures and cellular extracts, but also to excised roots, soil cores and greenhouse experiments (Bergersen 1980). This technique makes use of the ability of nitrogenase to reduce alternative substrates such as acetylene into ethylene, which can be determined by gas chromatography. Nonetheless, the nitrogen fixation capacity of isolated bacteria is often difficult to prove, e. g. a majority of root nodule bacteria (*Frankia* and rhizobia) are unable to grow at the expense of molecular nitrogen ex planta, and a large number of bacterial isolates from non-symbiotic systems do not express nitrogenase at a high rate in the free-living state. Thus, detection of nitrogenase activity by

Fig. 1 Phylogenetic 16S tree with prokaryotes carrying *nif* genes (by courtesy of German Jurgens)

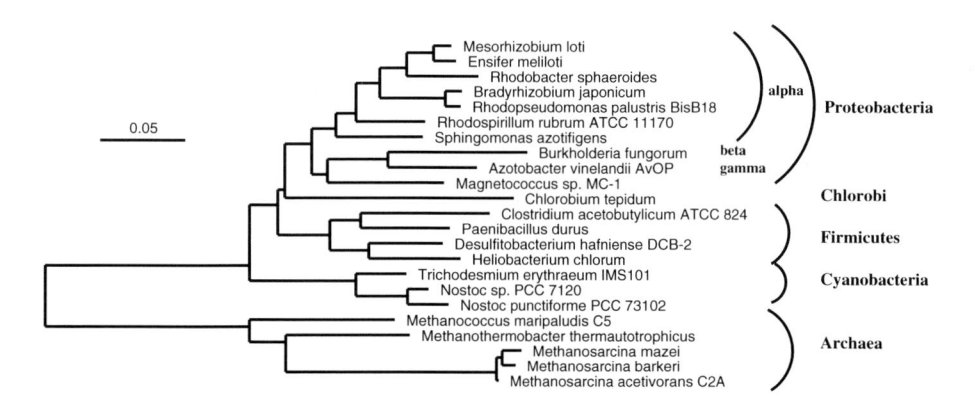

acetylene reduction test is sometimes ineffectual and inconclusive.

Amplification of *nif* DNA from environmental samples and community analyses

An alternative means of identifying nitrogen fixers became popular with the development of *nif* gene cloning and sequencing and of DNA amplification by polymerase chain reaction (PCR). This led to the demonstration of the presence of *nif* DNA in putative nitrogen-fixing isolates by PCR amplification, followed by nucleotide sequencing of the amplicon. A series of oligonucleotides such as universal *nifH* primers initially developed by Zehr and McReynolds (1989) were designed to amplify *nifH* fragments from environmental samples. This is of importance in ecological surveys of nitrogen fixers in the soil of rhizospheres. Another advantage is that it enables assessment of the biodiversity of bacteria without strain isolation, taking into account the non-culturable population (Ueda et al. 1995; Elbeltagy et al. 2001; Hamelin et al 2002; Roesch et al. 2008). Strain-specific probes based on conserved sequences of 16S rRNA genes were also developed (see e. g. Stoffels et al. 2001) to follow populations of a particular nitrogen-fixing species in environmental samples.

Rhizobium-legume symbiosis

Host plants

Many leguminous plant species can enter into a symbiotic relationship with root-nodule bacteria, collectively referred to as rhizobia. The legumes belong to the order Fabales, family Leguminosae (alternatively Fabaceae), in eurosid clade I (Doyle and Luckow 2003). Traditionally, three main subfamilies are distinguished: Caesalpinoidae, Mimosoidae and Papilionidae. The Caesalpinoidae has very few nodulating members, whereas most of the important agricultural crops are members of the Papilionidae. Mimosoidae has recently received attention, since, in many cases, bacteria recovered from their nodules belong to the beta subclass of *Proteobacteria*, while Papilionidae symbionts belongs to the alpha subclass (see below; Chen et al. 2003, 2005). Only one non-legume, the woody plant *Parasponia* sp., can be nodulated by rhizobia and utilize nitrogen fixed by the bacteria. Legumes are thought to have been around for 59 Ma, and all leguminous subfamilies evolved 56–50 Ma ago (Sprent and James 2007). In comparison with rhizobia, they are young; the *Sinorhizobium–Bradyrhizobium* split was pinpointed to about 500 Ma ago (Turner and Young 2000), implying that rhizobia were around before there were any legumes to nodulate.

Alpha and beta rhizobia

The rhizobia are Gram-negative and belong to the large and important *Proteobacteria* division (Fig. 2). The alpha-proteobacterial genera *Agrobacterium, Allorhizobium, Azorhizobium, Bradyrhizobium, Mesorhizobium, Rhizobium, Sinorhizobium, Devosia, Methylobacterium, Ochrobactrum* and *Phyllobacterium* all harbor nodule-forming bacteria, and so do the beta-proteobacterial *Burkholderia* and *Cupriavidus* (Lindström and Martínez-Romero 2007; http://edzna.ccg.unam.mx/rhizobial-taxonomy). The taxonomic classification of rhizobia follows standard procedures and is based on the phylogeny of housekeeping genes and whole-genome similarities (Lindström et al. 2006). Since nodulation functions did not evolve until long after bacterial housekeeping properties, it is thus not always possible to distinguish nodule formers by their names. Only genera in which nodulating bacteria were first discovered have "rhizobium" in their names, whereas e.g. *Burkholderia* (former *Pseudomonas*) species were first recognized through other properties.

Nodulation genes and *nod* factors

A common genetic determinant for rhizobia is the presence of genes encoding nodulation and nitrogen fixation functions (*nod, nol, noe, nif* and *fix* genes). These genes are often carried on plasmids or other accessory elements, such as symbiotic islands, and properties encoded by them can be easily lost or gained (for a recent review, see MacLean et al. 2007). The *nod, nol* and *noe* gene products are involved in production of a nodulation signal, the Nod factor, which is a lipo-chitooligosaccharide. Initiation of nodule formation on compatible host plants results from a molecular dialogue between the host and the bacteria (Fig. 3) (Dénarié et al. 1993; Schultze and

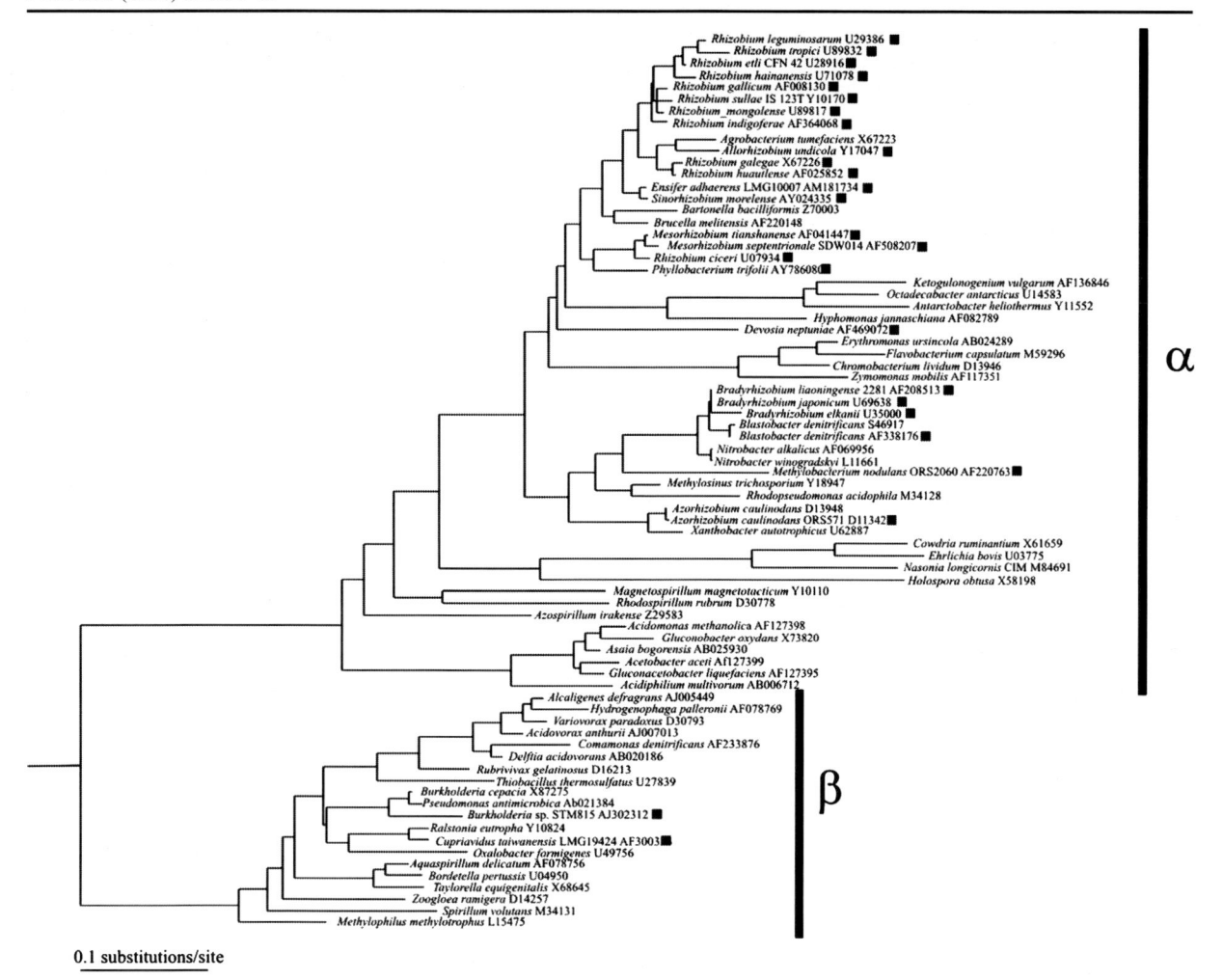

Fig. 2 Phylogeny of rhizobia. A maximum likelihood tree based on *rrs* genes from 75 taxa from alpha- and beta-subdivisions of *Proteobacteria*. Representatives of species capable of forming nodules are marked with a *black box*. (Reprinted from Dresler-Nurmi et al. 2007, with permission)

Fig. 3 Signal exchange in Rhizobium-plant symbiosis. Flavonoids produced by the host plant induce rhizobial *nod* genes. This leads to production of Nod factors. The insert shows an infection thread passing the root cortex toward a cluster of dividing cells that will become a root promordium. (Reprinted from Schultze and Kondorosi 1998, with permission)

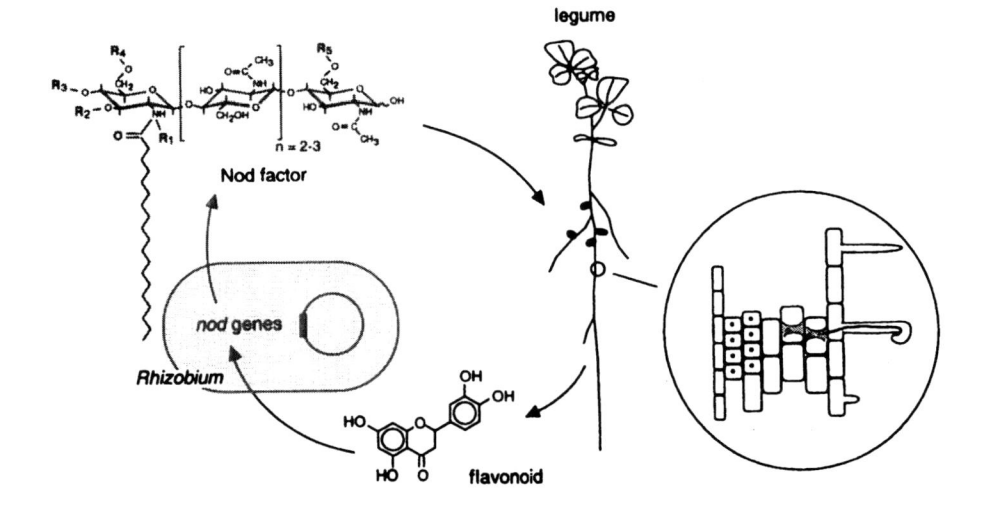

Kondorosi 1998; Perret et al. 2000; Spaink 2000). The host plants produce flavonoids (and related secondary metabolites) in the rhizosphere. These signals can be perceived by a specific bacterial receptor, NodD, which acts as a transcriptional activator of other nodulation genes. The core of the Nod factor molecule is encoded by canonical *nodA, nodB* and *nodC* whereas, for example, *nodFE* are involved in polyunsaturation of the fatty acyl group attached to the core molecule (Yang et al. 1999). Other nodulation genes encode enzymes which add a variety of substituents to the core, as in the case of Nod factors produced by *Azorhizobium caulinodans* (Mergaert et al. 1993). The Nod factor acts as an elicitor of root nodule formation by the plant by triggering a developmental program leading to construction of the root nodule and entry of rhizobia into the nodule (Long 2001; Geurts and Bisseling 2002; Gage 2004). It is an important host specificity determinant (Spaink 2000).

Recently, the Nod factor paradigm was challenged by Giraud et al. (2007), who discovered that certain photosynthetic, stem- and root-nodulating bradyrhizobia do not possess canonical *nodABC* genes but use other mechanisms for signalling to the plant. Their experiments led them to hypothesize that a purine derivative might play a role in triggering nodule formation instead of the Nod factor. This points to the complexity of the symbiotic system and shows that bacteria have employed diverse strategies to gain entry into the roots.

The infection process and nodule organogenesis

The textbook example of rhizobial infection is via plant root hairs which, prior to the infection process, respond to the presence of compatible rhizobia by deformation (shepherd's crooks, cauliflower structures, etc.). At the deformation stage, the plant perceives the rhizobial signal and initiates a developmental program aimed at formation of symbiotically nitrogen-fixing nodules (Dénarié et al. 1996). A set of plant genes, initially called nodulins, is specifically activated in response to nodulation factor perception (Geurts and Bisseling 2002).

The first receptors of Nod Factors are LysM-type receptor kinases named NFR1 and NFR5 in the model legume *Lotus japonicus* and LYK3 and NFP in *Medicago truncatula*. Different legumes mutants bocked in early step of nodulation, referred as *DMI* (does not make infection) were found to be also blocked in colonization with arbuscular mycorrhiza fungi (AM) suggesting common signalling pathways. The LRR receptor like kinase SYMRK or NORK is central for signal transduction and conserved among both nitrogen fixing and AM symbioses. Calcium spiking is observed in the early steps of root hair infection suggesting that calcium plays a role of secondary messenger in the infection process. Further steps in the signalling cascade lead to induction of cortical cell division (Limpens and Bisseling 2003; Kinkema et al. 2006; Oldroyd and Downie 2008).

A nodule meristem is thus formed within the root while the rhizobia enter through a plant-derived infection thread—a tube formed to facilitate rhizobia entry to the deeper layers. The infection threads grow transcellularly and finally, rhizobia wrapped into a plant-derived membrane, now called symbiosome membrane, are delivered into plant cells (Fig. 3). Nodules are either of an indeterminate type with an apical meristem, or they are determinate, meaning that the peripherally located meristem stops functioning after nodule completion (Foucher and Kondorosi 2000). In some nodules, all plant cells are infected with rhizobia, whereas in other nodule types, there are interstitial cells without symbiosomes. All mimosoid legumes and over half of the papilionid legumes represent this latter infection and nodulation type (Sprent and James 2007).

In their interesting and speculative review, however, Sprent and James (2007) present a tentative scheme for evolution of different nodule structures in which other infection and nodulation types are presented. Caesalpinoid legumes, which are seldom nodulated, display symbiosis in which the bacteria are retained within infection threads throughout symbiosis. Another evolutionarily interesting nodulation mode is present in many agronomically important crop plants as well, namely, infection through cracks in the root ("crack entry"). This type is represented in e.g. *Lupinus* and *Arachis*. Interestingly, in the case of the aquatic legumes (e.g. *Sesbania rostrata*) infection occurred through root hair curling except under flooding conditions where the mode of infection was by crack entry (Goormachtig et al. 2004).

Nodule physiology

During nodule formation, host tissues develop to form a specialized tissue that maintains an environment in which nitrogen fixation can occur. Functioning of the nodule was recently reviewed by White et al. (2007). In the nodule, specialized organelle-like forms of bacteria called bacteroids are engulfed in plant-derived membranes, forming symbiosomes. The reduction in dinitrogen inside the nodule requires energy, which is provided by the plant. Photosynthate in the form of sucrose is transported to the nodule, whereas dicarboxylic acids further provide the bacteroids with carbon and energy through the symbiosome membrane. For generation of energy through respiration, a high flux but a low internal concentration of oxygen is achieved with the aid of leghemoglobin.

Ammonia produced in the bacteroid needs to be transported to the plant through the symbiosome membrane. In addition to ammonia, alanine is transported. An amino acid flux back through the symbiosome membrane has also been proposed to be involved in the transport mechanism (Prell and Poole 2006). Ammonia is further assimilated into glutamine or asparagine in the plant cytosol. In determinate nodules, these are further converted into ureides in uninfected cells adjacent to the infected ones. In indeterminate nodules, this does not occur, and all plant cells are normally infected. The study of symbiosome biochemistry is impaired by technical difficulties involved when intact but isolated symbiosomes are used.

Host specificity and effectiveness

The origin of nodulation genes in bacteria is still unknown, but by studying infection and nodulation modes in extant legumes, information about the evolution of nodulation can be deduced. Some plant species can be infected by rhizobia representing different bacterial genera, whereas other species are extremely restrictive, accepting only a very narrow range of symbionts. The composition of the root exudates, on one hand, and the structure of the Nod factor, on the other, confer host specificity to the symbiotic interaction (Perret et al. 2000). The presence of type III secretion systems was discovered in several rhizobial genomes (Marie et al. 2001). Nodulation outer proteins (Nops) were identified as protein secreted by this apparatus and Nops were reported to play a role in nodulation efficiency and in some cases host specificity (Marie et al. 2001; Ausmees et al 2004; Cooper 2007).

The plant genus *Galega*, with representatives *G. orientalis* and *G. officinalis* (in the Hologalegina group of the Papilionidae; Doyle and Luckow 2003) are nodulated only by the rhizobial species *Rhizobium galegae* (Lipsanen and Lindström 1988). Host specificity of the symbiotic interaction is manifested at two levels. *Galega* plant root exudate induces *R. galegae* strains in a very specific interaction with NodD1 (Räsänen et al. 1991, Suominen et al. 2003). The Nod factor produced by the bacteria in response to exudate induction carries a unique mixture of molecules with polyunsaturated fatty acyl chains and an acyl group on the penultimate chito-oligosaccharide residue (Yang et al. 1999). These features result in very specific symbiosis. Bacteria producing polyunsaturated acyl chain substituents are taxonomically diverse, but all nodulate closely related legumes in the Hologalegina group (Suominen et al. 2001; Yang et al. 1999).

Host specificity is also expressed at the level of nitrogen fixation. Strains effective (fixing nitrogen) on *G. orientalis* will be ineffective (non-fixing) on *G. officinalis* and *vice versa*. It is interesting, though, that the plant cannot distinguish between effective and ineffective bacteria, but accepts both types in mixed inoculation experiments (Tas et al. 1996). Based on these phenotypic features, which are also reflected in the genomic makeup (Kaijalainen and Lindström 1989), two biovars orientalis and officinalis we distinguished (Radeva et al. 2001).

Rhizobium leguminosarum is a classical example of a rhizobial species having biovars (Laguerre et al. 1996; Mutch and Young 2004). The biovars confer host specificity for nodulation: *R. leguminosarum* biovar viciae typically nodulates peas (*Pisum* sp.) and vetches (*Vicia* sp.), biovar phaseoli beans (*Phaseolus* sp.) and biovar trifolii clovers (*Trifolium* sp.). In these biovars, the core genomes are similar (one species), whereas symbiotic genes are carried on plasmids which can be interchanged (Johnston et al. 1978; Laguerre et al. 1992; Young et al. 2006).

Sinorhizobium sp. strain NGR 234 represents another extreme. This strain was found to form nodules on 112 plant genera. This is due to a very versatile NodD protein induced by a variety of root extracts, to the capacity to produce a mixture of diverse Nod factors and to synthesis of Nops proteins by a type III secretion system (Freiberg et al. 1997; Pueppke and Broughton 1999; Ausmees et al 2004).

Our own studies (Dresler-Nurmi et al. unpublished) of 293 strains of bacteria isolated from the plant species *Calliandra calothyrsus* growing in different parts of the world revealed that several rhizobial species and even genera could nodulate this plant. Isolates from the Central American gene center of the plant were more diverse than those from countries into which the plant had been introduced. However, when the *nodA* gene of 70 isolates was sequenced and compared, they all grouped together and were closely related to the *nodA* gene of *R. tropici*. Thus, *Calliandra* is a promiscuous plant which, when introduced into a new site, seems to nodulate with local indigenous rhizobia with this *nodA* type. Other studies have shown that plants from the Mimosoidae (tribe Ingae, species *Calliandra calothyrsus* and tribe Mimosae, species *Mimosa diplotricha, Prosopsis* sp., *Leucaena leucocephala*) as well as the Papilionidae (tribe *Robinieae,* species *Gliricidia sepium*) form one cross-inoculation group (rhizobial isolates from one plant can nodulate all other plants in the group).

In the case of plant-rhizobium combinations which result in nodules with a proper structure, but without nitrogen-fixation activity, the bacteria act as parasites, probably attracted by the carbon and energy provided by the plant. In the case of *Rhizobium galegae,* the plant seemed not to be able to discriminate between effective and ineffective symbiotic partners, but both were equally competitive and the plant starved to death because of the absence of combined nitrogen (Tas et al. 1996). From a practical point of view, competition and effectiveness are, at the moment, among the most challenging questions for the scientific community.

Actinorhizal symbiosis

Host plants

Actinorhizal plants represent about 200 species distributed among 24 genera in eight angiosperm families (Huss-Danell 1997). Almost all genera are nodulated by Frankia in the Casuarinaceae, Coriariaceae, Eleagnaceae, Datisticaceae and Myricaceae families, whereas nodulation occurs occasionally in Betulaceae, Rhamnaceae and Rosaceae (Benson and Clawson 2000). This wide distribution contrasts with rhizobial symbiosis that, with the exception of *Parasponia* in the Ulmaceae, is limited to the Fabaceae family. All actinorhizal plants are woody trees or shrubs except for *Datisca*, a genus of flowering plants. These perennial dicotyledon angiosperms are distributed worldwide, from cold high latitudes with strong seasonal influences to warm tropical regions with no pronounced difference between seasons. Examples of well-known genera include *Alnus* (alder), *Eleagnus* (autumn olive), *Hippophae* (sea buckthorn) and *Casuarina* (beef wood).

Frankia

The genus *Frankia* comprises high mol% G+C Gram-positive genera belonging to the family *Frankiaceae* in the order *Actinomycetales* (Normand et al. 1996). The closest relative to *Frankia* among actinomycetes is the non-sporulating, rod-shaped cellulolytic thermophile *Acidothermus cellulolyticus*. Unlike rhizobia obtained in pure culture toward the end of the 19th century, the first successful isolation of an effective *Frankia* strain was only achieved from nodules of *Comptonia peregrina* in 1978 (Callaham et al. 1978). *Frankia* is a filamentous bacterium forming hyphal colonies without an aerial mycelium and characterized by a slow growth rate. One striking feature is its ability to differentiate two unique developmental structures that are critical to its survival: vesicles and spores (Fig. 4a) (Lechevalier 1994). Vesicles are the site for actinorhizal nitrogen fixation, while spores contained in multilocular sporangia are the reproductive structures of *Frankia*.

Over 200 strains of *Frankia* have been isolated from 20 plant genera. They are closely phylogenetically related and there is no evidence of the presence of nodulating ability in related actinobacteria (Clawson et al. 2004). Trees generated from sequence alignments of 16S rDNA, nitrogen fixation and glutamine synthetase genes generally yield three major closely related clades with respect to the nodulated host plants (Benson and Clawson 2007). A fourth contains related "*Frankia*-like" actinomycetes (Nod⁻/Fix⁻), unable to fix nitrogen or to induce nodules (Benson and Clawson 2000). *Frankia* from Clade I include strains nodulating members of the hamamelid families *Betulaceae, Casuarinaceae* and *Myricaceae*. *Frankia* from Clade II are typically associated with members of the *Coriariaceae, Datiscaceae, Rosaceae* and

Fig. 4 *Frankia* and actinorhizal nodules **a** *Frankia* in pure culture; nitrogen-fixing vesicles (v) and sporangia (s) can be observed. **b** Actinorhizal multilobed nodules on the root system of the actinorhizal plant *Allocasuarina verticillata*. **c** Pseudolongitudinal section of a nodular lobe from *A. verticillata*; the nitrogen-fixing zone contains large cells filled with *Frankia* (f), and the infection zone (iz) is located in the apex of the nodular lobe. Bars: A= 10 μm; B=5 mm; C=200 μm

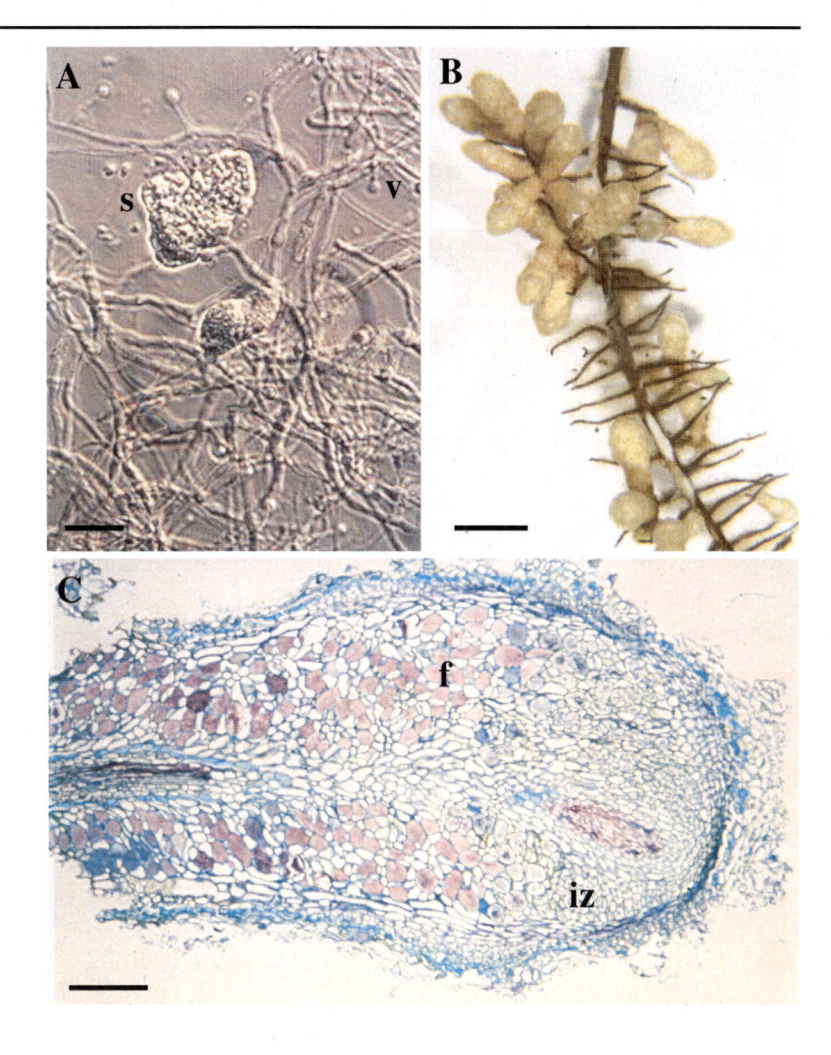

Ceanothus of the *Rhamnaceae*. This Clade II is characterized by low diversity, supporting a recent origin for symbiosis in this lineage. Clade III is still poorly known; strains appear to nodulate most *Ceanothus* sp. and also appear in *Myricaceae*, *Eleagnaceae*, *Rosaceae*, *Betulaceae* and *Gymnostoma* of the *Casuarinaceae* family (Benson and Clawson 2007).

Compared to that in rhizobia, the development of molecular genetic tools in *Frankia* has been difficult to implement mainly due to the relatively slow growth rate of filamentous hyphae; in most cases, genetic transformation, mutagenesis and functional complementation failed to provide conclusive results (Lavire and Cournoyer 2003; Normand and Mullin 2008). It was hypothesized that the absence of DNA-mediated transformation could result either from lack of gene expression, DNA restriction or the use of an inappropriate replicon. Hence, for some time, genetic analysis of *Frankia* has been mainly based on gene cloning via hybridization to genes from other organisms, phylogenetic analyses of selected gene sequences and isolation and characterization of plasmids (Simonet et al. 1990; Wall, 2000). Several groups have focused their research on the development of appropriate *Frankia* cloning vectors from native plasmids (John et al. 2001; Lavire et al. 2001; Xu et al. 2002). Studies on codon usage and *Frankia* promoter recognition have also been initiated (Cournoyer and Normand 1994; McEwan and Gatherer 1999; Bock et al. 2001). The genomics era together with the use of molecular biology methods have led to significant progress, including mRNA transcript analyses and overexpression of *Frankia* proteins in *Escherichia coli*. In 2007, the complete genome sequence of three *Frankia* strains was established (Normand et al. 2007). Their sizes varied from 5.4 Mbp to 9 Mpb. Efforts to detect genes homologous to the *nod* genes of rhizobia in *Frankia* had failed (Cérémonie et al. 1998) until preliminary analysis of the *Frankia* genome revealed disperse putative *nod*-like genes. However, these do not appear to be organized in clusters as in rhizobia, and the key *nodA* gene is absent.

Infection process

Two modes of infection of actinorhizal plants by *Frankia* have been described: intracellular root hair infection and intercellular root invasion (Wall and Berry 2008). Similarly to the situation encountered in rhizobia, the mode of infection depends on the host plant. Intracellular infection via root hairs (e.g. of *Casuarina*, *Alnus*, *Myrica*) starts with root hair curling following signal exchange between *Frankia* and the host plant. The signalling molecule pathway has not yet been identified, despite investigations in several laboratories (Prin and Rougier 1987; van Ghelue et al. 1997). However, preliminary characterization of a *Frankia* molecule capable of inducing root hair curling in host plants indicates that it differs from Nod factors in rhizobia (Cérémonie et al. 1999), consistent with the absence of *nod* genes in *Frankia* genome. After invagination of growing filaments of *Frankia* in the curled root hairs, infection proceeds intracellularly in the root cortex. *Frankia* hyphae become encapsulated by a cell wall deposit that is believed to consist of xylans, cellulose and pectins of host origin (Berg 1990, 1999). At the same time, limited cell divisions occur in the cortex near the invading root hair, leading to formation of a small external protuberance called the prenodule (Berry and Sunell 1990). Infection threads consist of lines of encapsulated *Frankia* hyphae progressing intracellularly toward the mitotically active zone and finally invading most cells of the prenodule. As the prenodule develops, cell divisions are induced in the pericycle located opposite the protoxylem pole, giving rise to another nodule primordium. In fact, actinorhizal prenodules do not evolve into nodules and the distantly induced primordium constitutes the nodule. The actual function of the prenodule was investigated in *Casuarina glauca*. A study of symbiosis-related gene expression coupled with cellular modification (cell wall lignification) indicated that prenodules displayed the same characteristics as nodules and hence could be considered very simple symbiotic organs (Laplaze et al. 2000, 2008). Thus, sequential differentiation of prenodules and then nodules constitutes a major difference from the situation in legumes, where cortical cell divisions lead to formation of a unique nodule primordium evolving into a mature nodule. The prenodule might thus be a parallel symbiotic organ of its own or the remaining form of a common nodule ancestor for legumes and actinorhizal plants.

Prenodule formation does not occur in the intercellular root invasion process (e.g. *Discaria*, *Ceanothus*, *Elaeagnus*, *Hyppophae*). *Frankia* hyphae penetrate between two adjacent rhizoderm cells and progress apoplastically through cortical cells within an electron-dense matrix secreted into the intercellular spaces (Miller and Baker 1985; Liu and Berry 1991; Wall and Berry 2008). Once the nodule primordium has developed from the pericycle, intracellular penetration by *Frankia* and formation of infection threads is initiated acropetally in developing cortical cells of the nodule lobe primordium, following a pattern similar to that described in plant species invaded through root-hairs.

Nodule development and functioning

For both intracellular and intercellular modes of infection by *Frankia*, an apical meristem is responsible for primordium growth towards the root surface in regions not infected by *Frankia*. As previously indicated, the nodule primordium does not incorporate the prenodule, but becomes infected by hyphae arising from the prenodule. Further development of the primordium gives rise to an indeterminate actinorhizal nodule lobe with a central vascular bundle surrounded by an endoderm, an expanded cortex containing *Frankia*-infected cells and a periderm (Fig. 4b,c). New lobes arise continuously to form a coralloid nodule. Some species like *Casuarina* or *Myrica* develop a so-called root nodule at the apex of each lobe (Duhoux et al. 1996). This root nodule lacks root hairs, has a reduced root cap and displays negative geotropism. It might be involved in diffusion of gas, especially oxygen, in and out of the nodule lobe.

Based on cytological and gene expression studies, four zones are recognized in mature actinorhizal nodules (Duhoux et al. 1996; Laplaze et al. 2008). Zone I is the apical meristem free of *Frankia*. Adjacent to the meristem is zone II, an infection zone in which some of the young cortical cells resulting from the meristem activity are infected by *Frankia*. The encapsulated bacterium starts to proliferate and the plant cells enlarge. Zone III is the fixation zone containing both infected and uninfected cortical cells. Infected cells are hypertrophied and are filled with *Frankia* filaments that differentiate vesicles where nitrogen fixation takes place. The appearance and shape of these vesicles is controlled by the plant. In some species like *Casuarina*, infected cells have a lignified cell wall and there is no vesicle differentiation. Uninfected cells are smaller and, in some species, contain amyloplast and phenolic compounds and might be involved in nitrogen and carbon metabolism. Finally, a basal senescence zone (zone IV) is observed in old nodules; plant cells and bacteria degenerate and the nitrogenase is switched off. More recently, a second level of compartmentalization was described in *Casuarina glauca* nodules based on accumulation of flavans, which occurs in uninfected cells in the endodermis and cortex. These cells form layers that delimit *Frankia*-infected compartments in the nodule lobe and may play a role in restricting bacterial infection to certain zones of the nodule (Laplaze et al. 1999).

Molecular biology and actinorhizal nodule and plant gene expression

During differentiation of the actinorhizal nodule, a set of genes called actinorhizal genes is activated in the developing nodule. Heterologous probing and differential screening of nodule cDNA libraries with root and nodule-specific cDNA resulted in isolation and characterization of more than 25 nodule-specific or nodule-enhanced plant genes in several actinorhizal plants, including *Alnus, Datisca, Eleagnus* and *Casuarina* (Pawlowski and Bisseling 1996; Laplaze et al. 2008; Pawlowski and Sprent, 2008). One of the earliest symbiotic genes characterized thus far is *cg12*, which encodes a subtilisin-like protease expressed in *Frankia*-infected root hairs of *C. glauca* (Svistoonoff et al. 2003). A homologue of the receptor-like kinase gene SymRK found in legumes was also recently shown to be necessary for actinorhizal nodule formation in the tree *Casuarina glauca* (Gherbi et al. 2008a).

Recently emerging tools should contribute to increasing our knowledge of the molecular mechanisms of actinorhizal symbiosis over the next few years. One is the development of the first genomic platform for the study of plant gene expression in actinorhizal symbiosis (Hocher et al. 2006). This was recently applied to a *C. glauca* gene nodule library and it revealed increased expression of genes involved in primary metabolism, protein synthesis, cell division and defense. The second tool is the use of hairpin RNA to achieve post-transcriptional gene silencing in *C. glauca*, providing a versatile approach to assessing gene function during the nodulation process induced by *Frankia* (Gherbi et al. 2008b). Expressed sequence tags (ESTs) that exhibit homology with the early symbiotic genes *DMI2* and *DMI3* (Catoira et al. 2000) from legumes involved in the Nod factor transduction pathway are currently being characterized using this RNA interference approach.

Associations with cyanobacteria

Cyanobacteria and symbiosis

Cyanobacteria are widely distributed in aquatic and terrestrial environments. Long regarded as algae because they performed oxygenic photosynthesis, they are now classified into the domain of *Bacteria*, in five Sections based mostly on morphological criteria (Rippka et al. 1979). Indeed, cyanobacteria constitute the largest and most diverse group of Gram-negative prokaryotes. While nitrogen fixation is found both in unicellular and filamentous species, associations with plants are essentially limited to heterocystous cyanobacteria *Nostocales*, primarily of the genus *Nostoc* and *Anabaena*. Besides vascular plants, there exist a wide variety of non-vascular lower plant belonging to bryophytes, including liverworts and hornworts, algae and fungi, that develop associations with cyanobacteria, as well as many marine eukaryotes (Rai 1990; Bergman et al. 1996; Rai et al. 2000, 2002). We shall limit this review to associations with gymnosperms (Cycads), angiosperms (*Gunnera*) and pteridophytes (*Azolla*).

Differentiation of heterocysts, nitrogen-fixing specialized cells

Some filamentous cyanobacteria from Sections IV and V are able to differentiate specialized cells called heterocysts under nitrogen limitation conditions

(Fig. 5a) (Rippka et al. 1979) (Fig. 5a). An anaerobic environment compatible with the functioning of nitrogenase in heterocysts is linked to formation of multilayered envelopes external to the outer membrane, elimination of a functional oxygen-producing photosystem II, and additional changes in their physiology not detailed here (Buikema and Haselkorn 1993). Nitrogen fixed by heterocysts is exported to vegetative cells of the filaments; in return, vegetative cells provide heterocysts with carbohydrates derived from their photosynthetic activity. This interdependence ensures filament growth under conditions of nitrogen limitation.

Initial information on heterocyst differentiation came essentially from the study of *Anabaena* sp. strain PCC 7120, a strain not known to associate with plants (Buikema and Haselkorn 1993). Heterocysts develop within about 24 h from vegetative cells located at semi-regular intervals in the filaments (Fig. 5a). The signalling pathway that leads to initiation of heterocyst differentiation at a particular location in the filament is complex, and the number of genes identified as being involved in the developmental process is regularly increasing (Ehira et al. 2003; Golden and Yoon 2003; Zhang et al. 2006; Xu et al. 2008). Two of them, *hetR* and *ntcA*, have a critical function in initial steps of heterocyst differentiation (Buikema and Haselkorn 1991; Wei et al. 1994; Shi et al. 2006), while *patS* is involved in heterocyst spacing (Yoon and Golden 1998). HetR is a transcriptional regulator with autoprotease activity, which functions as a master switch in heterocyst differentiation. NtcA is a global nitrogen regulator that belongs to the CRP (cAMP receptor proteins) superfamily and which acts as a sensor of nitrogen deprivation in response to internal concentrations of 2-oxoglutarate. NtcA plays a role in control of *hetR* expression under N deprivation consistent with the

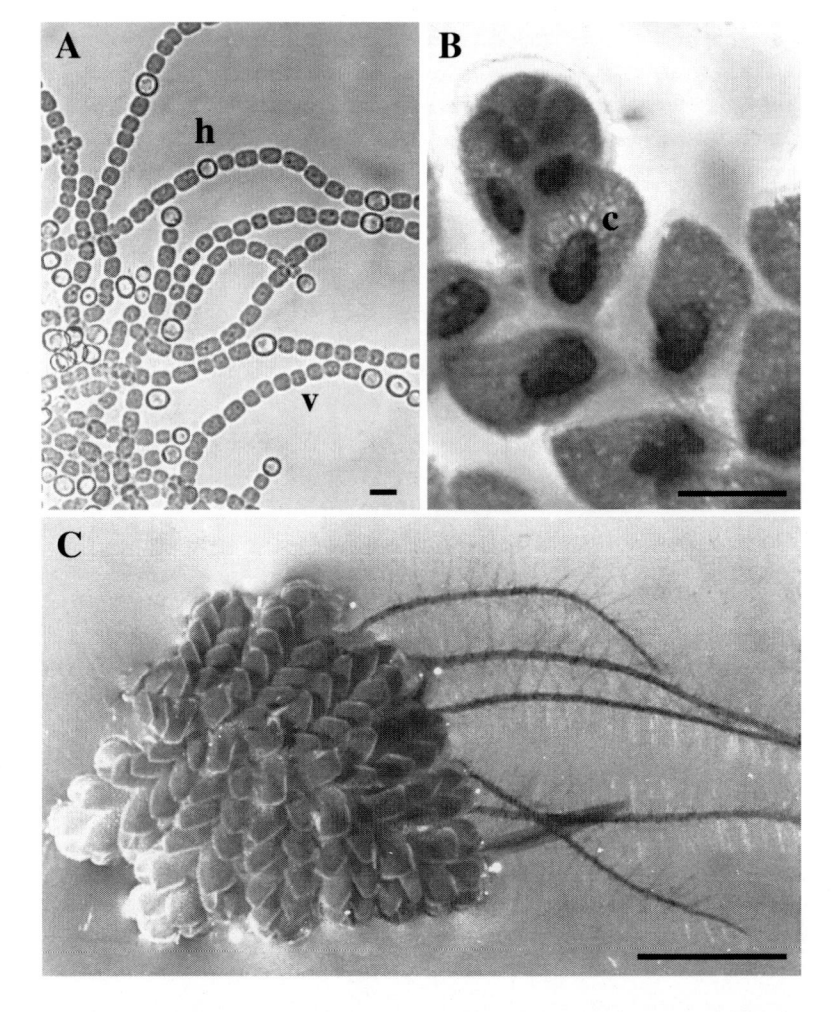

Fig. 5 Free-living *Anabaena* and *Azolla* **a** Free-living *Anabaena* strain cultured in medium deprived of nitrogen; heterocysts (h) can be observed among vegetative cells (v). **b** Frond of *Azolla pinnata* digested by cellulase and pectinase; cavities (c) filled with symbiotic *Anabaena azollae* are visible. **c**: Frond of *A. pinnata*. Bars: A=10 μm; B=50 μm; C=500 μm

fact that *ntcA* mutants cannot form heterocysts. PatS is a small diffusible peptide inhibitor of heterocyst differentiation, probably by inhibiting the HetR transcription activation function (Huang et al. 2004). Thus, the semiregular pattern of heterocyst formation may derive from the autoregulatory activity of HetR and diffusion of the PatS peptide (Xu et al. 2008). HetN, a protein similar to ketoacyl reductase, is also thought to downregulate expression of *hetR* (Callahan and Buikema 2001; Borthakur et al. 2005).

The percentage of heterocysts in filaments grown in the free-living state is in the range of 5 to 10% of cells, whereas it reaches 30 to 40% of cells within the filaments hosted by the plant and, in the particular case of *Gunnera,* up to 60–80% (Meeks and Elhai 2002; Wang et al. 2004). This reflects a direct correlation between the efficiency of nitrogen fixation and heterocyst frequency. Whereas *Anabaena azollae* appears to be an obligate symbiont, other symbiotic cyanobacteria can be grown in free-living culture and retain their ability to infect their host plant. Thus, properties of mutants impaired in heterocyst differentiation can be assayed in the host plant. Mutants of *ntcA*, *hetR* and *hetF* have been obtained in *N. punctiforme* strain PCC 92293, an isolate from *Gunnera,* which can also infect *Anthoceros punctatus*. None could differentiate heterocysts, similarly to what was found with corresponding mutants of non-symbiotic strain PCC 7120 (Wong and Meeks 2001, 2002). Both *hetF* and *hetR* mutants can infect *A. punctatus* with a frequency similar to that of the wild type, but are unable to support growth of the plant because of their inability to develop heterocysts and fix dinitrogen. The *ntcA* mutant failed to infect *Anthoceros* because it is also impaired in formation of hormogonia, which is the "infection unit" (see below).

Differentiation of hormogonia, the "infection unit"

Hormogonia are filaments, motile by gliding, formed by cells with reduced metabolic activity, of smaller size than vegetative cells produced by the genus *Nostoc* (Rippka et al. 1979). Because of their motility, they provide a means of dispersing cyanobacteria and play a critical role in establishment of symbiosis with host plants by enabling access to plant structures that will house symbiotic colonies. Hormogonia differentiation results from division of vegetative filaments without growth. Their formation is induced in response to different stress conditions at the free-living state and in response to host plant factors called HIF (for hormogonia-inducing factor) in symbiosis. After 48 h, hormogonia can regenerate vegetative filaments and differentiate heterocysts (Campbell and Meeks 1989).

Inactivation of the *N. punctiforme* ATCC 29133 *hrmA* gene by transposon mutagenesis results in a higher frequency of hormogonia formation in response to HIF of *A. punctatus* (Cohen and Meeks 1997). The *hrmA* gene is part of an *hrmRIUA* operon that has high sequence similarity to sugar uronate metabolism operons of other bacteria. *Nostoc hrmA* mutants are unable to survive in long-term coculture with *Anthoceros* due to their continued formation of hormogonia. In contrast, wild-type filaments of *Nostoc* spp., after an initial burst of HIF-induced hormogonia formation, show a period of immunity to HIF enabling growth and nitrogen fixation.

Symbiotic associations with vascular plants

Gunnera, a genus of about 40 species, is the only angiosperm with which the cyanobacterium referred to as *Nostoc punctiforme* is associated (Rasmussen and Svenning 2001). Specialized stem glands, susceptible to infection by the cyanobacterium *Nostoc*, are formed at the base of petioles (Bergman 2002). Acidic mucilage, containing HIF (Rasmussen et al. 1994), secreted from the stem glands, induces differentiation of vegetative *Nostoc* filaments into motile hormogonia, which move into channels present in the glands. Then, compatible *Nostoc* strains induce divisions in the host cells lining the channel and *Nostoc* cells are subsequently taken up into plant cells. Once intracellular, a high frequency of differentiation of vegetative cells into heterocysts occurs and nitrogen is fixed at a high rate (Rasmussen et al. 1996).

Cycads are the only gymnosperms that fix dinitrogen, including 90 species in 9 genera (Costa and Lindblad 2002). They develop specialized lateral roots, so-called coralloid roots, in which a cell layer has been prepared for infecting cyanobacteria. Following an unknown infection process, the endophyte invades a mucilage-filled space in the outer cortex of the root nodule. Morphological features of the cyanobacterium change as the root nodules mature, with heterocyst frequency increasing with age, while vegetative cell enlargement also occurs with age. Large amounts of phenolic compounds are found in the cortical cells surrounding the zone of cyanobacteria. It has been suggested that these compounds may inhibit growth of other organisms and contribute to containment of the symbiont.

Azolla species are native to Asia, Africa and the Americas, and have been dispersed by man and by natural pathways to various parts of the world. Some are strictly tropical or subtropical, while others grow and thrive in lakes, swamps and streams, and other small bodies of water under either temperate or tropical climates. The genus *Azolla* includes seven species that have been grouped into two sections, *Euazolla* and *Rhizosperma*, based on the structure of their sporocarps. A view of *Azolla pinnata* is shown in Fig. 5c. Symbiosis between the aquatic fern *Azolla* and *A. azollae* is of particular interest because it is the only plant-prokaryote symbiosis known to persist throughout the reproductive cycle of the host plant (Lumpkin and Plucknett 1980; Nierzwicki-Bauer 1990; Lechno-Yossef and Nierzwicki-Bauer 2002). During vegetative growth, the symbiont is located in a distinct leaf cavity at the base of the dorsal lobe of the leaves (Fig. 5b). Vegetative maintenance of the association depends on retention of *A. azollae* filaments, morphologically similar to hormogonia, at the apical meristem of fronds. The directed movement of the cyanobacterium within the host is accomplished by specialized plant epidermal trichomes. It has been hypothesized that the specific surface properties of hormogonia, which differ from those of vegetative filaments, enable recognition by trichomes, so that only generative hormogonium cells serve as inocula for new cavities developing at the apex of the frond. The leaf cavity of *Azolla* can also host other bacteria together with the symbiotic *Anabaena;* the function of these bacteria remains unknown. The cavity is surrounded by mucilage and completely lined by an envelope. As leaf maturation occurs, the non-heterocystous hormogonia of the youngest leaves develop into heterocystous filaments located at the periphery of the cavity. Finally, the symbiotic cavities respond to the presence of cyanobacteria by elaborating long, finger-like cells that may serve to increase the surface area for nutrient exchange. Nitrogen is released from the cyanobiont almost exclusively as NH_4^+. During sexual reproduction, the cyanobiont colony survives in the indusium cap of the megaspore. In *Azolla*, deoxyanthocyanins produced by the aquatic fern also contribute to induction of *hrmA* expression (Cohen et al. 2002). This result suggests that in *Azolla,* appropriate localization of phenolic compounds could function in plant-mediated mechanisms for repressing hormogonium formation after penetration of cyanobacteria into the host plant.

Associative and endophytic nitrogen fixers

Nitrogen-fixing plant-growth-promoting rhizobacteria

Nitrogen-fixing bacteria that contribute to plant growth stimulation or to disease prevention and suppression are referred to as plant growth-promoting rhizobacteria (PGPR). Their isolation from the roots of forage grasses and cereal crops and many other plants, in both natural and cultivated ecosystems, has been extensively performed (Döbereiner and Pedrosa 1987; Okon and Labandera-Gonzales 1994; Baldani and Baldani 2005). This led to identifying two groups with respect to the degree of association with the host plant: rhizospheric and endophytic colonizers. Classical microbiological techniques involving cultivation of bacteria identified soil bacteria belonging to genera such as *Azospirillum, Azotobacter, Alcaligenes, Bacillus, Beijerinckia, Campylobacter, Derxia,* several members of *Enterobacteriaceae* (*Klebsiella, Pantoae*) and *Pseudomonas stutzeri* (Rennie 1980; Balandreau 1983; Elmerich et al. 1992; Yan et al. 2008). As most of these strains were isolated from surface-sterilized root samples, this suggests that proportions of these cells are protected from sterilizing agents, but it may also reflect some colonization of the root tissues. Other isolates, such as those of *Azoarcus* (Hurek and Reinhold-Hurek 2003), *Burkholderia* (Caballero-Mellado et al. 2004), *Herbaspirillum* and *Gluconacetobacter* (Baldani and Baldani 2005), or *K. pneumoniae* strain 342 (Chelius and Triplett 2000), happened to belong to endophytes. The complete genome of *Azoarcus* and *P. stutzeri* A1501 was established (Krause et al. 2006; Yan et al. 2008) genomes projects of *Azospirillum, Azotobacter, Herbaspirillum* and *Gluconacetobacter* are under completion.

Colonization of the root system: *Azospirillum* as a model bacterium

Colonization of the root surface has been best studied in *Azospirillum. Spirillum*-like bacteria were first isolated by Beijerinck in 1923; they were rediscovered by Becking in 1963 and by Johanna Döbereiner's team in Brazil in the 1970's (Von Bülow and Döbereiner 1975). The *Azospirillum* genus was described by Tarrand et al. (1978); it belongs to the *Alphaproteobacteria* phylum and 7 species were recognized in 2007 (Schmid and Hartmann 2007): *A. brasilense, A. lipoferum, A. amazonense, A. irakense, A. halopraeferens, A. largimobile* and *A. doebereinerae. Azospirillum* species display an extremely wide ecological distribution and are associated in nature with a wide diversity of plants, including those of agronomic importance such as wheat, rice, sorghum and maize and several non-gramineous species (Döbereiner and Pedrosa 1987). These bacteria are aerobic non-fermentative chemoorganotrophs, vibrioid to S-shaped, containing polyhydroxyalkanoate granules (PHA). In liquid medium motility is ensured by a polar flagellum. In some species (e. g. *A. brasilense, A. lipoferum, A. amazonense*), lateral flagellation enables swarming on a solid surface. Another important property of azospirilla is the ability to differentiate resistant forms which are non-motile ovoid cyst-like cells, much larger than vegetative cells and surrounded by a thick capsule (Lamm and Neyra 1981). Cyst formation is concomitant with cell aggregation into macroscopic flocs occurring in some culture conditions. Encapsulated azospirilla display greater resistance to desiccation and heat.

Cyst formation and production of siderophores and bacteriocins (spirilobactin) (Tapia-Hernández et al. 1990) are likely to play a role in survival of these bacteria under unfavorable conditions and in competition with other members of the soil microflora. Bacterial motility as well as chemotactic responses towards root exudates are involved in the initial step of the root colonization process (Vande Broek et al. 1998). Attachment to the root system is mediated by the polar flagellum and is followed by irreversible anchoring of the bacteria (Steenhoudt and Vanderleyden 2000). The polar flagellum is glycosylated and binds wheat root, whereas the lateral flagella are not essential during the adsorption phase. As *rpoN* controls flagellar biogenesis, *rpoN* mutants are impaired in colonization. An operon carrying chemotaxis genes (*che*) was identified in *A. brasilense* (Hauwaerts et al. 2002). However, *cheB* and *cheR* mutants were only partially impaired in their chemotactic response, suggesting multiple chemotaxis systems in this bacteria (Stephens et al. 2006). Several genes governing motility in *A. brasilense* Sp7 have been mapped onto a 90 MDa plasmid, while other genes are located on the chromosome. The complete nucleotide sequencing of this plasmid was established (Vanbleu et al. 2004). The plasmid also carries several genes governing surface polysaccharides that might play some role in the colonization process. The structural gene for major outer membrane protein, *omaA,* was characterized and found to encode an adhesin with high affinity to roots (Burdman et al. 2001). A transcriptional regulator of the LuxR-UhpA family, *flcA,* controlling differentiation into cysts and flocculation, was also described as playing a role in surface colonization (Pereg-Gerk et al. 1998).

Electron micrographs of colonized roots revealed that azospirilla were anchored to roots by fibrillar material, probably similar to the fibrillar material produced during flocculation. Bacteria colonize the rhizoplane and are found in high numbers upon emergence of lateral roots and also near the root cap (De Oliveira Pinheiro et al. 2002). The degree of invasion of plant tissues differs between strains. In an early report, using fluorescent antibody staining techniques, it was found that *Azospirillum* colonizes the intercellular spaces between the epidermis and the cortex of the root (Schank et al. 1979). Other techniques involving bacteria carrying *gus* and *lacZ* fusions were helpful not only at locating the bacteria on the root surface, but also for assaying *nif* gene expression during the colonization process (Vande Broek et al. 1993; Arsène et al. 1994). The fluorescent *in situ* hybridization (FISH) technique developed for phylogenetic purposes, coupled with the use of confocal laser scanning microscopy (CLSM), enabled better *in situ* localization of bacteria on the root (Aßmuss et al. 1995). In particular, it established that endophytic colonization of some *Azospirillum* strain such as *A. brasilense* Sp245 was found in the intercellular spaces of the root epidermis (Rothballer et al. 2003). Attempts to identify genes expressed at early stages of the interaction with the host plants are in progress (Pothier et al. 2007).

Nitrogen-fixing endophytes

An increasing number of reports describe the occurrence of nitrogen-fixing bacteria within plant tissues of a host plant that does not show diseases symptoms, with the most studied genera being *Azoarcus* sp, *Gluconacetobacter* and *Herbaspirillum* (Hurek and Reinhold-Hurek, 2003, Lery et al. 2008). Endophytes multiply and spread within plant tissues without causing damage. Early steps in infection may be similar to those reported with rhizospheric bacteria, initially involving surface colonization at the site of emergence of root hairs (Hurek and Reinhold-Hurek, 2003). In the case of *Azoarcus*, type IV pili were found to be essential for that process and hydrolytic enzymes, or endoglucanases, are involved in tissue penetration (Dörr et al. 1998; Krause et al. 2006). The concentration of bacteria recovered after sterilization of the root system can reach up to 10^8 CFU per g of dry weight. Another characteristic is systemic spreading of bacteria, which can be found in plant xylem vessels and in shoots, as described in the case of sugar cane infection with *G. diazotrophicus* (James and Olivares 1998) and in the case of infection of the C4-gramineous plant *Miscanthus sinensis* by *H. frisingense* (Rothballer et al. 2008). Bacteria are located mostly in intercellular spaces, but intracellular location is also seen in dead cells (Hurek and Reinhold-Hurek 2003). The main difference from rhizospheric bacteria (that can also be found, in some cases, in the first layers of the root cortex), is the fact that endophytes do not persist in the soil. Therefore, the frontier between rhizospheric and endophytic systems is not truly strict.

Factors involved in plant growth promotion and crop protection

Colonization of the plant root affects both the morphology and physiology of the host plant. A typical response after inoculation with *Azospirillum* is enhanced proliferation of lateral roots and root hairs. In general, this is accompanied by changes in root physiology, such as increased mineral and water uptake, increased root respiration, delay in leaf senescence and increased dry weight (Okon 1985; Dobbelaere and Okon 2007). The plant growth promotion effect was tentatively attributed to production of auxin-like compounds such as indole-3-acetic acid (IAA), commonly produced by soil bacteria. In general, biosynthesis of IAA uses tryptophan (Trp) as a precursor, and several pathways for conversion of Trp into IAA have been described (Costacurta and Vanderleyden 1995; Baca and Elmerich 2007). The indole acetamide route common in phytopathogenic bacteria is not present in nitrogen-fixing root colonizers, including *Azospirillum*. Instead, many soil bacteria possess the indole pyruvate route, involving an indole-3-pyruvate decarboxylase encoded by *ipdC* (Koga et al. 1991). *Azospirillum* insertion mutants in *ipdC* still produce IAA, and strain Sp7 contains two alternative pathways (Carreño-Lopez et al. 2000). Indeed, it is difficult to distinguish between hormones produced by the plant and bacteria. Elevated IAA pools in plants colonized with *Azospirillum* was reported, but it could not be concluded as to whether this increase was linked to hydrolysis of IAA-conjugates produced by the plant, to bacterial production, or to both (Fallik et al. 1989). Other plant hormones, such as gibberellins and cytokinins, are also produced by many soil bacteria, but as yet, little is known concerning the involvement of these compounds in the plant response to inoculation (Baca and Elmerich 2007). In contrast, the role of bacteria in preventing ethylene synthesis has been better studied. Ethylene is a plant hormone that prevents root elongation. Many soil bacteria encode a deaminase (*acdS* gene) which degrades the direct precursor of ethylene, i.e. 1-aminocyclopropane-1-carboxylic acid (ACC) (Glick 2005), including strains of *A. lipoferum* (Prigent-Combaret et al. 2008), *P. stutzeri* A1501 (Yan et al. 2008) and *H. frisingense* (Rothballer et al. 2008). Introduction of *acds* gene into *A. brasilense* resulted in increased root elongation in some plants (Holguin and Glick 2001). Production of N-acylhomoserine lactone that may be involved in plant growth promotion was also observed in some cases (Boyer et al. 2008; Rothballer et al. 2008).

To date, there is still limited information related to the role of nitrogen-fixing bacteria as biocontrol agents. Production of antimicrobial metabolites as well as hydrolytic enzymes has been reported, but their role in crop protection remains to be established. A particular strain of *A. brasilense* was reported to be antagonistic of the parasitic weed *Striga*, that affect many tropical cereals by preventing *Striga* seed germination (Miché et al. 2000).

Nitrogen fixation and crop productivity

Biological nitrogen fixation represents, annually, up to 100 million tons of N for terrestrial ecosystems, and from 30 to 300 million tons for marine ecosystems. In addition, 20 million tons result from chemical fixation due to atmospheric phenomena (Mosier 2002). The first industrial production of rhizobium inoculant began by the end of the 19th century. However, to sustain production of cereal crops, legumes and other plants of agricultural importance, the supply of nitrogenous chemical fertilizers has been regularly increasing since the Second World War. According to an FAO report, production of N fertilizer for 2007 was 130 million tons of N, and this should further increase in the coming years (FAO 2008). This extensive use has certain drawbacks. A proportion of added fertilizer is lost as a result of denitrification and leaching of soil by rainfall and irrigation. In addition, leaching leads to water pollution caused by eutrophication. As a consequence, extending application of biological nitrogen fixation by any means is an important issue.

Benefits from legumes and actinorhizal plants

Because the concentration of fixed nitrogen is a limiting factor for growth, nitrogen fixers have a selective advantage that enables them to adapt to the most extreme conditions and to colonize diverse ecological niches. Indeed, nitrogen-fixing symbiotic microorganisms play an important role in the life of plants, ensuring not only their nutrition, but also their defense against pathogens and pests, and adaptation to various environmental stress.

Legumes are often considered to be the major nitrogen-fixing systems, as they may derive up to 90% of their nitrogen from N_2 (e.g. faba bean, lupin, soybean, groundnut). The rate of fixation, in the range of 200 to 300 kg N/ha/crop, can be attained for most species, but this largely depends on cultivars and culture conditions (Peoples et al 1995). Among grain legumes, soybean represents more than 50% of the world oilseed production.

Many actinorhizal plants, but also legumes, are capable of sustaining a mycorrhizal association as well, thus forming tripartite symbiosis and enhancing the success of these plants under poor soil conditions. Due to these properties, actinorhizal species can grow and improve soil fertility in disturbed sites and are used in recolonization and reclaiming of eroded areas, sand dunes, moraines and areas of industrial waste and road cuts, and are planted following fires, volcanic eruptions and logging (Wheeler and Miller 1990). In addition, some actinorhizal species can grow well under a range of environmental stresses such as high salinity, heavy metal and extreme pH (Dawson 1990). Actinorhizal plant nitrogen fixation rates are comparable to those found in legumes (Torrey and Tjepkema 1979; Dawson 1983). Alders, in particular, are known to be beneficial at improving nutrient-poor soils (Wheeler and Miller 1990; Myrold and Huss-Danell 2003). The annual input of N from N_2 fixation in alder stands ranges from 20 to 300 kg N/ha, depending on stand age, stand density and site conditions (Wheeler and Miller 1990). As much as 85–100% of foliar N in a speckled alder stand is estimated to have been derived from the atmosphere (Hurd et al. 2001).

Benefits of association with cyanobacteria

The rate of dinitrogen fixation in plant-associated cyanobacteria is much greater than that of the same free-living strains, and it correlates broadly with the increased heterocyst frequency observed in symbiosis. In the *Anthoceros* association, dinitrogen fixation is 4- to 35-fold higher than that of the free-living *Nostoc*. In *Anthoceros* and in *Blasia*, fixed dinitrogen is released to the plant as ammonia and it has been shown that as little as 20% of dinitrogen is retained by the cyanobiont (Adams 2002). Thus far, nitrogen-fixing *Azolla*-cyanobacterial symbiosis is the only one of economic importance to farming systems. Symbiosis has for centuries been used to produce green manure plants, especially in China, Vietnam, and Southeast Asia (Lumpkin and Plucknett 1982), and there is no doubt of the beneficial effect of *Azolla* in terms of increasing rice grain yield. The entire nitrogen requirement of *Azolla* is fulfilled by the cyanobacteria and *Azolla* can accumulate 2-4 or more kg N/ha/day. The fern grows rapidly and can double its biomass in 3 days under optimal conditions. These unique properties have made it possible for rice production to replace part of the chemical fertilizer with *Azolla*-cyanobacteria symbiosis (Peters and Meeks 1989; Liu and Zheng 1992). The nitrogen fertilizer fixed by *Azolla* becomes available to rice after the *Azolla* mat is incorporated into soil and its nitrogen begins to be released through decomposition. Furthermore, it has been shown that the presence of an *Azolla* mat on the surface of water significantly reduces volatilization of nitrogen fertilizers (Vlek et al. 1995). Though *Azolla* use as a rice fertilizer is currently declining, it is still used at the farm level in China, India, Senegal, the Philippines, Colombia, Bolivia and Brazil. However, the widespread use of *Azolla* is limited by environmental factors such as high temperature, insects and disease. Furthermore, genetic improvement in symbiosis remains difficult (van Hove and Lejeune 2002).

Benefits of association with grasses

Taken together, legumes occupy about 10% of cultivated land, while cereals represent almost 50% (Peoples et al. 1995). Nitrogen fixation with cereal crops is often considered negligible in comparison to rates of symbiotic nitrogen fixation attained by root nodulated plants. Hence, the possibility of increasing nitrogen fixation with cereal crops by inoculation with wild type or engineered nitrogen-fixing bacteria is an extremely challenging project. Inoculation of cereals and vegetables with *Azotobacter* and bacilli, called "bacterization", was in common use in Russia in the 1950s and beneficial effects were reported (Macura 1966). Whether or not the benefit is due to nitrogen fixation depends on the systems, plant cultivars, soil and many other parameters. It was long known that some crops like sugar cane could grow with little addition of nitrogen fertilizer in Brazil, and many other examples, including rice and wheat, have been reported (van Berkum and Bohlool 1980; Boddey and Döbereiner 1982). In particular, *Paspalum notatum* cv. batatais that specifically associates with *Azotobacter paspali* was reported to fix from 15 to 90 kg N/ha/year (Döbereiner et al. 1972). At present, sugar cane is a good example of a crop that can benefit from nitrogen fixation, since certain cultivars can derive more than 150 kg N/ha/year from BNF (Boddey et al. 2001). Thus, some crops naturally benefit from the nitrogen fixation capacity of associated bacteria. In their survey of 20 years of field inoculation worldwide, Okon and Labandera-Gonzales (1994) concluded that significant increases in yields, from 5 to 30%, could be achieved by inoculation with *Azospirillum*, in particular when the use of chemical N fertilizer was low. However, they believed that the growth-promoting effect was probably linked to phytohormone production rather than nitrogen fixation. Recently, Rodrigues et al. (2008) reported significant nitrogen fixation, determined by the ^{15}N isotope dilution method, after inoculation of rice with certain strains of *A. amazonense*. Plant growth promotion due to nitrogen fixation of endophytes was also demonstrated (Sevilla et al. 2001; Hurek et al. 2002). This reinforces the view that research on a rhizosphere association might be of great value for agriculture.

Acknowledgments The authors are indebted to Dr. German Jurgens, Department of Applied Chemistry and Microbiology, Helsinki University, Finland, for drawing a phylogenetic tree of nitrogen fixers, and to Ms. Jerri Bram for improving the language.

References

Aßmus B, Hutzler P, Kirchhof G, Amann R, Lawrence JR, Hartmann A (1995) In situ localization of *Azospirillum brasilense* in the rhizosphere of wheat with fluorescently labeled, rRNA-targeted oligonucleotide probes and scanning confocal laser microscopy. Appl Environ Microbiol 61:1013–1019

Adams DG (2002) Symbioses with hornworts and liverworts. In: Rai AN, Bergman B, Rasmussen U (eds) Cyanobacteria in symbiosis. Kluwer academic, Dordrecht, pp 117–136

Arnold W, Rump A, Klipp W, Priefer UB, Pühler A (1988) Nucleotide sequence of a 24,206-base-pair DNA fragment carrying the entire nitrogen fixation gene cluster of *Klebsiella pneumoniae*. J Mol Biol 203:715–738 doi:10.1016/0022-2836(88)90205-7

Arsène F, Katupitiya S, Kennedy IR, Elmerich C (1994) Use of *lacZ* fusions to study the expression of *nif* genes of *Azospirillum brasilense* in association with plants. Mol Plant Microbe Interact 7:748–757

Ausmees N, Kobayashi H, Deakin WJ, Marie C, Krishnan HB, Broughton WJ, Perret X (2004) Characterization of NopP, a type III secreted effector of *Rhizobium* sp. strain NGR234. J Bacteriol 18:4774–4780 doi:10.1128/JB.186.14.4774-4780.2004

Baca BE, Elmerich C (2007) Microbial production of plant hormones. In: Elmerich C, Newton WE (eds) Associative and endophytic nitrogen-fixing bacteria and cyanobacterial associations. Springer, Dordrecht, pp 113–143

Balandreau J (1983) Microbiology of the association. Can J Microbiol 29:851–859

Baldani JI, Baldani VLD (2005) History on the biological nitrogen fixation research in gramineaceous plants: special emphasis on the Brazilian experience. An Acad Bras Cienc 77:549–579 doi:10.1590/S0001-37652005000300014

Benson DR, Clawson ML (2007) Recent advances in the biogeography and genecology of symbiotic *Frankia* and its host plants. Physiol Plant 130:318–330 doi:10.1111/j.1399-3054.2007.00934.x

Benson DR, Clawson ML (2000) Evolution of the actinorhizal plant symbioses. In: Triplett EW (ed) Prokaryotic nitrogen fixation: A model system for analysis of biological process. Horizon Scientific Press, Wymondham, UK, pp 207–224

Berg RH (1990) Cellulose and xylans in the interface capsule in symbiotic cells of actinorhizae. Protoplasma 159:35–43 doi:10.1007/BF01326633

Berg RH (1999) *Frankia* forms infection threads. Can J Bot 77:1327–1333 doi:10.1139/cjb-77-9-1327

Bergersen F (1980) Methods for evaluating biological nitrogen fixation. Willey and Sons, Chichester

Bergman B (2002) *Nostoc-Gunnera* symbiosis. In: Rai AN, Bergman B, Rasmussen U (eds) Cyanobacteria in symbiosis. Kluwer academic, Dordrecht, pp 207–232

Bergman B, Matveyev A, Rasmussen U (1996) Chemical signalling in cyanobacterial-plant symbioses. Trends Plant Sci 1:191–197 doi:10.1016/1360-1385(96)10021-2

Berry MA, Sunell AL (1990) The infection process and nodule development. In: Schwintzer RC, Tjepkema JD (eds) The Biological of *Frankia* and actinorhizal plants. Academic press Inc, San Diego, pp 61–88

Bock JV, Battershell T, Wiggington J, John TR, Johnson JD (2001) *Frankia* sequences exhibiting RNA polymerase promoter activity. Microbiology 147:499–506

Boddey RM, Döbereiner J (1982) Association of *Azospirillum* and other diazotrophs with tropical gramineae. In: Non symbiotic nitrogen fixation and organic matter in the tropics. Indian Society of Soil Science, New Delhi, pp 28–47

Boddey RM, Peoples MB, Palmer B, Dart PJ (2000) The use of ^{15}N natural abundance technique to quantify biological nitrogen fixation by woody perennials. Nutr Cycl Agroecosyst 57:235–270 doi:10.1023/A:1009890514844

Boddey RM, Polidoro JC, Resende AS, Alves BJR, Urquiaga S (2001) Use of the ^{15}N natural abundance technique for the quantification of the contribution of N_2 fixation to sugar cane and other grasses. Aust J Plant Physiol 28:889–895

Borthakur PB, Orozco CC, Young-Robbins SS, Haselkorn R, Callahan SM (2005) Inactivation of *patS* and *hetN* causes lethal levels of heterocyst differentiation in the filamentous cyanobacterium *Anabaena* sp PCC 7120. Mol Microbiol 57:111–113 doi:10.1111/j.1365-2958.2005.04678.x

Boyer M, Bally R, Perrotto S, Chaintreuil C, Wisniewski-Dyé F (2008) A quorum quenching approach to identify quorum-sensing-regulated functions in *Azospirillum lipoferum*. Res Microbiol •••:159 doi:10.1016/j.resmic.2008.08.003

Buikema WJ, Haselkorn R (1991) Characterization of a gene controlling heterocyst differentiation in the cyanobacterium *Anabaena* sp PCC 7120. Genes Dev 5:321–330 doi:10.1101/gad.5.2.321

Buikema WJ, Haselkorn R (1993) Molecular genetics of cyanobacterial development. Annu Rev Plant Physiol 44:33–52 doi:10.1146/annurev.pp.44.060193.000341

Burdman S, Dulguerova G, Okon Y, Jurkevitch E (2001) Purification of the major outer membrane protein of *Azospirillum brasilense*, its affinity to plant roots and its involvement in cell aggregation. Mol Plant Microbe Interact 14:555–561 doi:10.1094/MPMI.2001.14.4.555

Burris RH, Miller CE (1941) Application of ^{15}N to the study of biological nitrogen fixation. Science 93:114–115 doi:10.1126/science.93.2405.114

Caballero-Mellado J, Martínez-Aguilar L, Paredes-Valdez G, Estrada-de los Santos P (2004) *Burkholderia unamae* sp nov, an N2-fixing rhizospheric and endophytic species. Int J Syst Evol Microbiol 5:1165–1172 doi:10.1099/ijs.0.02951-0

Callaham D, DelTredici P, Torrey JG (1978) Isolation and cultivation in vitro of the actinomycete causing root nodulation in *Comptonia*. Science 199:899–902 doi:10.1126/science.199.4331.899

Callahan SM, Buikema WJ (2001) The role of HetN in maintenance of the heterocyst pattern in *Anabaena* sp PCC 7120. Mol Microbiol 40:941–950 doi:10.1046/j.1365-2958.2001.02437.x

Campbell EL, Meeks JC (1989) Characteristics of hormogonia formation by symbiotic *Nostoc* spp in response to the presence of *Anthoceros punctatus* or its extracellular products. Appl Environ Microbiol 55:125–131

Carreño-Lopez R, Campos-Reales NB, Elmerich C, Baca BE (2000) Physiological evidence for differently regulated tryptophan-dependent pathways for indole-3-acetic acid synthesis in *Azospirillum brasilense*. Mol Gen Genet 264:521–530 doi:10.1007/s004380000340

Catoira R, Galera C, de Billy F, Penmetsa RV, Journet EP, Maillet F, Rosenberg C, Cook D, Dénarié J (2000) Four genes of *Medicago truncatula* controlling components of a Nod factor transduction pathway. Plant Cell 12:1647–1666

Cérémonie H, Cournoyer B, Maillet F, Normand P, Fernandez MP (1998) Genetic complementation of rhizobial nod mutants with *Frankia* DNA: artefact or realty? Mol Gen Genet 260:115–119 doi:10.1007/s004380050877

Cérémonie H, Debellé F, Fernandez MP (1999) Structural and functional comparison of Frankia root hair deforming factor and rhizobia Nod factor. Can J Bot 77:1293–1301 doi:10.1139/cjb-77-9-1293

Chelius MK, Triplett EW (2000) Immunolocalization of dinitrogenase reductase produced by *Klebsiella pneumoniae* in association with *Zea mays* L. Appl Environ Microbiol 66:783–787 doi:10.1128/AEM.66.2.783-787.2000

Chen WM, James EK, Chou JH, Sheu SY, Yang SZ, Sprent JI (2005) Beta-rhizobia from *Mimosa pigra*, a newly discovered invasive plant in Taiwan. New Phytol 168:661–675 doi:10.1111/j.1469-8137.2005.01533.x

Chen WM, Moulin L, Bontemps C, Vandamme P, Béna G, Boivin-Masson C (2003) Legumes symbiotic nitrogen fixation by beta-proteobacteria is widespread in nature. J Bacteriol 185:7266–7272 doi:10.1128/JB.185.24.7266-7272.2003

Clawson ML, Bourret A, Benson DR (2004) Assessing the phylogeny of *Frankia*-related plant nitrogen-fixing root nodule symbioses with *Frankia* 16SRNA and glutamine synthetase gene sequences. Mol Phylogenet Evol 31:131–138 doi:10.1016/j.ympev.2003.08.001

Cohen MF, Meeks JC (1997) A hormogonium regulating locus, *hrmUA*, of the cyanobacterium *Nostoc punctiforme* strain ATCC29133 and its response to an extract of a symbiotic plant partner *Anthoceros punctatus*. Mol Plant Microbe Interact 10:280–289 doi:10.1094/MPMI.1997.10.2.280

Cohen MF, Sakihama Y, Takagi YC, Ichiba T, Yamasaki H (2002) Synergistic effect of deoxyanthocyanins from symbiotic fern *Azolla* spp on *hrmA* gene induction in the cyanobacterium *Nostoc punctiforme*. Mol Plant Microbe Interact 9:875–882 doi:10.1094/MPMI.2002.15.9.875

Cooper JE (2007) Early interactions between legumes and rhizobia: disclosing complexity in a molecular dialogue. J

Appl Microbiol 103:1355–1365 doi:10.1111/j.1365-2672. 2007.03366.x

Costa JL, Lindblad P (2002) Cyanobacteria in symbiosis in cycads. In: Rai AN, Bergman B, Rasmussen U (eds) Cyanobacteria in symbiosis. Kluwer academic, Dordrecht, pp 195–206

Costacurta A, Vanderleyden J (1995) Synthesis of phytohormones by plant-associated bacteria. Crit Rev Microbiol 21:1–18 doi:10.3109/10408419509113531

Cournoyer B, Normand P (1994) Gene expression in *Frankia*: characterization of promoters. Microbiology 78:229–236

Dawson JO (1983) Dinitrogen fixation in forest ecosystems. Can J Microbiol 29:979–992

Dawson JO (1990) Interaction among actinorhizal and associated plant species. In: Schwintzer RC, Tjepkema JD (eds) The Biological of *Frankia* and actinorhizal plants. Academic press Inc, San Diego, pp 299–316

De Oliveira Pinheiro R, Boddey LH, James EK, Sprent JI, Boddey RM (2002) Adsorption and anchoring of *Azospirillum* strains to roots of wheat. Plant Soil 246:151–166 doi:10.1023/A:1020645203084

Dénarié J, Debellé F, Promé JC (1996) *Rhizobium* lipochitooligosaccharide nodulation factors: signalling moleculaes mediating regcognition and morphogenosis. Annu Rev Biochem 65:503–535 doi:10.1146/annurev.bi.65.070196.002443

Dénarié J, Debellé F, Truchet G, Promé JC (1993) *Rhizobium* and legume nodulation: A molecular dialogue. In: Palacios R, Mora J, Newton WE (eds) New horizons in nitrogen fixation. Kluwer, Dordrecht, pp 19–30

Dixon R, Kahn D (2004) Genetic regulation of biological nitrogen fixation. Nat Rev Microbiol 2:621–631 doi:10.1038/nrmicro954

Dobbelaere S, Okon Y (2007) The plant growth promoting effect and plant response. In: Elmerich C, Newton WE (eds) Associative and endophytic nitrogen-fixing bacteria and cyanobacterial associations. Springer, Dordrecht, pp 145–170

Döbereiner J, Day JM, Dart PJ (1972) Nitrogenase activity and oxygen sensitivity of the *Paspalum notatum-Azotobacter paspali* association. J Gen Microbiol 71:103–116

Döbereiner J, Pedrosa FO (1987) Nitrogen-fixing bacteria in non-leguminous crop plants. Science Tech, Madison and Springer Verlag, Berlin

Döbereiner J (1992) History and new perspectives of diazotrophs in association with non-leguminous plants. Symbiosis 13:1–13

Dörr J, Hurek T, Reinhold-Hurek B (1998) Type IV pili are involved in plant-microbe and fungus-microbe interactions. Mol Microbiol 30:7–17 doi:10.1046/j.1365-2958.1998.01010.x

Doyle JJ, Luckow MA (2003) The rest of the iceberg. Legume diversity and evolution in a phylogenetic context. Plant Physiol 1331:900–910 doi:10.1104/pp.102.018150

Dresler-Nurmi A, Fewer D, Räsänen LA, Lindström K (2007) The diversity and evolution of rhizobia. In: Pawlowski K (ed) Prokaryotic endosymbionts in plants. Springer Verlag doi:101007/7171

Duhoux E, Diouf D, Gherbi H, Franche C, Ahée J, Bogusz D (1996) Le nodule actinorhizien. Acta Bot Gallica 143:593–608

Ehira S, Ohmori M, Sato N (2003) Genome-wide expression analysis of the responses to nitrogen deprivation in the heterocyst-forming cyanobacterium *Anabaena* sp strain PCC 7120. DNA Res 10:97–113 doi:10.1093/dnares/10.3.97

Elbeltagy A, Nishioka K, Sato T, Suzuki H, Ye B, Hamada T, Isawa T, Mitsui H, Minamisawa K (2001) Endophytic colonization and in planta nitrogen fixation by a *Herbaspirillum* sp isolated from wild rice species. Appl Environ Microbiol 67:5285–5293 doi:10.1128/AEM.67.11.5285-5293.2001

Elmerich C, Newton WE (2007) Associative and Endophytic Nitrogen-fixing Bacteria and Cyanobacterial Associations. Springer, The Netherlands

Elmerich C, Zimmer W, Vieille C (1992) Associative nitrogen-fixing bacteria. In: Stacey G, Burris RH, Evans HJ (eds) Biological nitrogen fixation. Chapman and Hall Inc, New York, pp 212–258

Fallik E, Okon Y, Epstein E, Goldman A, Fischer M (1989) Identification and quantification of IAA and IBA in *Azospirillum brasilense*-inoculated maize roots. Soil Biol Biochem 21:147–153 doi:10.1016/0038-0717(89)90024-2

FAO (2008) Current world fertilizer trends and outlook 2011/2012. Food and agricultural organization of the United Nations, Rome

Fischer HM (1994) Genetic regulation of nitrogen fixation in rhizobia. Microbiol Rev 58:352–386

Freiberg C, Fellay R, Bairoch A, Broughton WJ, Rosenthal A, Perret X (1997) Molecular basis of symbiosis between *Rhizobium* and legumes. Nature 387:394–401 doi:10.1038/387394a0

Foucher C, Kondorosi E (2000) Cell cycle regulation in the course of nodule organogenesis in Medicago. Plant Mol Biol 43:773–786 doi:10.1023/A:1006405029600

Gage DJ (2004) Infection and invasion of roots by symbiotic, nitrogen-fixing rhizobia during nodulation of temperate legumes. Microbiol Mol Biol Rev 68:280–300 doi:10.1128/MMBR.68.2.280-300.2004

Geurts R, Bisseling T (2002) *Rhizobium* nod factor perception and signalling. Plant Cell 14(Suppl):S239–S249

Gherbi H, Markmann K, Svistoonoff S, Estevan J, Autran D, Giczey G, Auguy F, Péret B, Laplaze L, Franche C, Parniske M, Bogusz D (2008a) SymRK defines a common genetic basis for plant root endosymbioses with arbuscular mycorrhiza fungi, rhizobia, and Frankiabacteria. Proc Natl Acad Sci USA 105:4928–4932 doi:10.1073/pnas.0710618105

Gherbi H, Nambiar-Veetil M, Zhong C, Félix J, Autran D, Girardin R, Auguy F, Bogusz D, Franche C (2008b) Post-transcriptional gene silencing in the root system of the actinorhizal tree *Allocasuarina verticillata*. Mol Plant Microbe Interact 21:518–524 doi:10.1094/MPMI-21-5-0518

Giraud E, Moulin L, Vallenet D, Barbe V, Cytryn E, Avarre JC, Jaubert M, Simon D, Cartieaux F, Prin Y, Bena G, Hannibal L, Fardoux J, Kojadinovic M, Vuillet L, Lajus A, Cruveiller S, Rouy Z, Mangenot S, Segurens B, Dossat C, Franck WL, Chang WS, Saunders E, Bruce D, Richardson P, Normand P, Dreyfus B, Pignol D, Stacey G, Emerich D, Verméglio A, Médigue C, Sadowsky M (2007) Legumes symbioses: absence of *nod* genes in photosynthetic bradyrhizobia. Science 316:1307–1312 doi:10.1126/science.1139548

Glick BR (2005) Modulation of plant ethylene levels by the bacterial enzyme ACC deaminase. FEMS Microbiol Lett 251:1–7 doi:10.1016/j.femsle.2005.07.030

Golden JW, Yoon HS (2003) Heterocyst development in *Anabaena*. Curr Opin Microbiol 6:557–563 doi:10.1016/j.mib.2003.10.004

Goormachtig S, Capoen W, Holsters M (2004) Rhizobium infection : lessons from the versatile nodulation behaviours of water tolerant legumes. Trends Plant Sci 9:518–522 doi:10.1016/j.tplants.2004.09.005

Hamelin J, Fromin N, Tarnawski S, Teyssier-Cuvelle S, Aragno M (2002) *nifH* gene diversity in the bacterial community associated with the rhizosphere of *Molinia coerulea*, an oligonitrophilic perennial grass. Environ Microbiol 4:477–481 doi:10.1046/j.1462-2920.2002.00319.x

Hardy RWF, Burns RC, Holsten RD (1973) Application of the acetylene-ethylene reduction assay for measurement of nitrogen fixation. Soil Biol Biochem 5:47–81 doi:10.1016/0038-0717(73)90093-X

Hauwaerts D, Alexandre G, Das SK, Vanderleyden J, Zhulin IB (2002) A major chemotaxis gene cluster in *Azospirillum brasilense* and relationships between chemotaxis operons in α-proteobacteria. FEMS Microbiol Lett 208:61–67

Henson BJ, Watson LE, Barnum SR (2004) The evolutionary history of nitrogen fixation, as assessed by *nifD*. J Mol Evol 58:390–399 doi:10.1007/s00239-003-2560-0

Hocher V, Auguy F, Argout X, Laplaze L, Franche C, Bogusz D (2006) Expressed sequence-tag analysis in *Casuarina glauca* actinorhizal nodule and root. New Phytol 169:681–688 doi:10.1111/j.1469-8137.2006.01644.x

Holguin G, Glick BR (2001) Expression of the ACC deaminase gene from *Enterobacter cloacae* UW4 in *Azospirillum brasilense*. Microb Ecol 41:281–288

Hu Y, Fay AW, Lee CC, Ribbe MW (2007) P-cluster maturation on nitrogenase MoFe protein. Proc Natl Acad Sci USA 104:10424–10429 doi:10.1073/pnas.0704297104

Huang X, Dong Y, Zhao J (2004) HetR homodimer is a DNA binding protein required for heterocyst differentiation, and the DNA-binding activity is inhibited by PatS. Proc Natl Acad Sci USA 101:4848–4853 doi:10.1073/pnas.0400429101

Hurd TM, Raynal DJ, Schwintzer CR (2001) Symbiotic N2 fixation of *Alnus incana* spp *rugosa* in shrub wetlands of the Adirondack mountains, New York, USA. Oecologia 126:94–103 doi:10.1007/s004420000500

Hurek T, Reinhold-Hurek B (2003) *Azoarcus* sp strain BH72 as a model for nitrogen-fixing grass endophytes. J Biotechnol 106:169–178 doi:10.1016/j.jbiotec.2003.07.010

Hurek T, Handlley LL, Reinhold-Hurek B, Piché Y (2002) *Azoarcus* grass endophytes contribute fixed nitrogen to the plant in an unculturable state. Mol Plant Microbe Interact 15:233–242 doi:10.1094/MPMI.2002.15.3.233

Huss-Danell K (1997) Actinorhizal symbioses and their N2 fixation. New Phytol 136:375–405 doi:10.1046/j.1469-8137.1997.00755.x

James EK, Olivares FL (1998) Infection and colonisation of sugarcane and other graminaceous plants by endophytic diazotrophs. Crit Rev Plant Sci 17:77–119 doi:10.1016/S0735-2689(98)00357-8

John TR, Rice JM, Johnson JD (2001) Analysis of pFQ12, a 224-kb *Frankia* plasmid. Can J Microbiol 47:608–617 doi:10.1139/cjm-47-7-608

Johnston AWB, Beynon JL, Buchanan-Wollaston AV, Setchell SM, Hirsch PR, Beringer JE (1978) High frequency transfer of nodulating ability between strains and species of *Rhizobium*. Nature 276:634–636 doi:10.1038/276634a0

Kaijalainen S, Lindström K (1989) Restriction fragment length polymorphism analysis of *Rhizobium galegae* strains. J Bacteriol 171:5561–5566

Kinkema M, Scott PT, Gresshoff PM (2006) Legume nodulation: successful symbiosis through short and long-distance signalling. Funct Plant Biol 33:1–15 doi:10.1071/FP06056

Koga J, Adachi T, Hidaka H (1991) Molecular cloning of the gene for indolepyruvate decarboxylase from *Enterobacter cloacae*. Mol Gen Genet 226:10–16 doi:10.1007/BF00273581

Krause A, Ramakumar A, Bartels D, Battistoni F, Bekel T, Boch J, Böhm M, Friedrich F, Hurek T, Krause L, Linke B, McHardy AC, Sarkar A, Schneiker S, Syed AA, Thauer R, Vorhölter FJ, Weidner S, Pühler A, Reinhold-Hurek B, Kaiser O, Goesmann A (2006) Complete genome of the mutualistic, N(2)-fixing grass endophyte *Azoarcus* sp. strain BH72. Nat Biotechnol 24:1385–1391 doi:10.1038/nbt1243

Laguerre G, Géniaux E, Mazurier SI, Rodriguez Casartelli R, Amarger N (1992) Conformity and diversity among field isolates of *Rhizobium leguminosarum* bv. *viciae*, bv. *trifolii*, and bv. *phaseoli* revealed by DNA hybridization using chromosome and plasmid probes. Can J Microbiol 39:412–419

Laguerre G, Mavingui P, Allard MR, Charnay MP, Louvrier P, Mazurier SI, Rigottier-Gois L, Amarger N (1996) Typing of rhizobia by PCR DNA fingerprinting and PCR restriction fragment length polymorphism analysis of chromosomal and symbiotic gene regions: application to *Rizobium leguminosarum* and its different biovars. Appl Environ Microbiol 62:2029–2036

Lamm RB, Neyra CA (1981) Characterization and cyst production of *Azospirilla* isolated from selected grasses growing in New Jersey and New York. Can J Microbiol 27:1320–1325

Laplaze L, Duhoux E, Franche C, Frutz T, Svistoonoff S, Bogusz D (2000) *Casuarina glauca* prenodule cells display the same differentiation as the corresponding nodule cells. Mol Plant Microbe Interact 13:107–112 doi:10.1094/MPMI.2000.13.1.107

Laplaze L, Gherbi H, Frutz T, Pawlowski K, Franche C, Macheix J-J, Auguy F, Bogusz D, Duhoux E (1999) Flavan-containing cells delimit *Frankia*-infected compartments in *Casuarina glauca* nodules. Plant Physiol 121:113–122 doi:10.1104/pp.121.1.113

Laplaze L, Svistoonoff S, Santi C, Auguy F, Franche C, Bogusz D (2008) Molecular biology of actinorhizal symbioses. In: Pawlowski K, Newton WE (eds) Nitrogen-fixing actinorhizal symbioses. Springer, Dordrecht, pp 235–259

Lavire C, Cournoyer B (2003) Progress on the genetics of the N2-fixing actinorhizal symbiont *Frankia*. Plant Soil 254:125–137 doi:10.1023/A:1024915300030

Lavire C, Louis D, Perriere G, Briolay J, Normand P, Cournoyer B (2001) Analysis of pFQ31, a 8551-bp cryptic plasmid from three symbiotic nitrogen-fixing actinomycete. *Frankia*. FEMS Microbiol Lett 197:111–116 doi:10.1111/j.1574-6968.2001.tb10591.x

Lechevalier MP (1994) Taxonomy of the genus *Frankia* (Actinomycetales). Int J Syst Bacteriol 44:1–8

Lechno-Yossef S, Nierzwicki-Bauer SA (2002) *Azolla-Anabaena azollae* symbiosis. In: Rai AN, Bergman B, Rasmussen U (eds) Cyanobacteria in symbiosis. Kluwer academic publishers, Dordrecht, pp 153–178

Lery LMS, von Krüger WMA, Viana FC, Teixeira KRS, Bisch PM (2008) a comparative proteomic analysis of *Gluconacetobacter diazotrophicus* PAL5 at exponential and stationary phases of cultures in the presence of high and low levels of inorganic nitrogen compound. Biochim Biophys Acta

Limpens E, Bisseling T (2003) Signaling in symbiosis. Curr Opin Plant Biol 6:343–350 doi:10.1016/S1369-5266(03)00068-2

Lindström K, Martínez-Romero E (2007) International committee on systematics of prokaryotes subcommittee on the taxonomy of *Agrobacterium* and *Rhizobium*: minutes of the meeting, 23-24 July 2006, Århus, Denmark. Int J Syst Evol Microbiol 57:1365–1366 doi:10.1099/ijs.0.65255-0

Lindström K, Kokko-Gonzales P, Terefework Z, Räsänen LA (2006) Differentiation of nitrogen-fixing legume root nodule bacteria. In: Cooper JE, Rao JR (eds) Molecular techniques for soil and rhizosphere microorganisms. CABI Publishing, Wallingford, pp 236–258

Lipsanen P, Lindström K (1988) Infection and root nodule structure in the *Rhizobium galegae* sp nov—*Galega* sp symbiosis. Symbiosis 6:81–96

Liu CC, Zheng WW (1992) Nitrogen fixation of *Azolla* and its utilization in agriculture in China. In: Hong GF (ed) Nitrogen fixation and its research in China. Springer-Verlag, Berlin, pp 526–537

Liu Q, Berry AM (1991) The infection process and nodule initiation in the *Frankia-Ceanothus* root nodule symbiosis. Protoplasma 163:82–92 doi:10.1007/BF01323332

Long SR (2001) Gene and signals in the *Rhizobium*-legume symbiosis. Plant Physiol 125:69–72 doi:10.1104/pp.125.1.69

Lumpkin TA, Plucknett DL (1980) *Azolla*: botany, physiology, and use as a green manure. Econ Bot 34:111–153

Lumpkin TA, Plucknett DL (1982) *Azolla* as a green manure: use and management in crop production. Westview, Boulder, Colorado

MacLean AM, Finan TM, Sadowsky MJ (2007) Genomes of the symbiotic nitrogen-fixing bacteria of legumes. Plant Physiol 144:615–622 doi:10.1104/pp.107.101634

Macura J (1966) Rapport général. Ann Inst Pasteur (Paris) 111 (suppl 3):9–38

Marie C, Broughton WJ, Deakin WJ (2001) *Rhizobium* type III secretion systems: legume charmers or alarmers? Curr Opin Plant Biol 4:336–342 doi:10.1016/S1369-5266(00)00182-5

McEwan NR, Gatherer D (1999) Codon indices as a predictor of gene functionality in a *Frankia* operon. Can J Bot 77:1287–1292 doi:10.1139/cjb-77-9-1287

Meeks JC, Elhai J (2002) Regulation of cellular differentiation in filamentous cyanobacteria in free-living and plant-associated symbiotic growth states. Microbiol Mol Biol Rev 66:94–121 doi:10.1128/MMBR.66.1.94-121.2002

Mergaert P, Van Montagu M, Promé JC, Holsters M (1993) Three unusual modifications, a D-arabinosyl, an N-methyl, and a carbamoyl group, are present on the Nod factors of *Azorhizobium caulinodans* strain ORS571. Proc Natl Acad Sci USA 90:1551–1555 doi:10.1073/pnas.90.4.1551

Merrick MJ, Edwards RA (1995) Nitrogen control in bacteria. Microbiol Rev 59:604–622

Miché L, Bouillant ML, Rohr R, Salle G, Bally R (2000) Physiological and cytological studies on the inhibition of Striga seed germination by the plant growth-promoting bacterium *Azospirillum brasilense*. Eur J Plant Pathol 106:347–351 doi:10.1023/A:1008734609069

Miller IM, Baker DD (1985) The initiation, development and structure of root nodules in *Eleagnus angustifolia* L (*Eleagnaceae*). Protoplasma 128:107–119 doi:10.1007/BF01276333

Mosier AR (2002) Environmental challenges associated with needed increases in global nitrogen fixation. Nutr Cycl Agroecosyst 63:101–116 doi:10.1023/A:1021101423341

Mutch LA, Young JPW (2004) Diversity and specificity of *Rhizobium leguminosarum* biovar *viciae* on wild and cultivated legumes. Mol Ecol 13:2435–2444 doi:10.1111/j.1365-294X.2004.02259.x

Myrold DD, Huss-Danell K (2003) Alder and lupine enhance nitrogen cycling in a degraded forest soil in Northern Sweden. Plant Soil 254:47–56 doi:10.1023/A:1024951115548

Newton WE (2007) Physiology, biochemistry and molecular biology of nitrogen fixation. In: Bothe H, Ferguson SJ, Newton WE (eds) Biology of the nitrogen cycle. Elsevier, Amsterdam, pp 109–130

Nierzwicki-Bauer SA (1990) *Azolla-Anabaena* symbiosis. In: Rai AN (ed) Handbook of symbiotic cyanobacteria. CRC, Boca Raton, pp 119–136

Normand P, Mullin BC (2008) Prospects for the study of a ubiquitous actinomycete, *Frankia*, and its host plants. In: Pawlowski K, Newton WE (eds) Nitrogen-fixing actinorhizal symbioses. Springer, Dordrecht, pp 289–303

Normand P, Lapierre P, Tisa LS, Gogarten JP, Alloisio N, Bagnarol E, Bassi CA, Berry AM, Bickhart DM, Choisne N, Couloux A, Cournoyer B, Cruveiller S, Daubin V, Demange N, Francino MP, Goltsman E, Huang Y, Martinez M, Mastronunzio JE, Mullin BC, Nieman J, Pujic P, Rawnsley T, Rouy Z, Schenowitz C, Sellstedt A, Tvares F, Tomkins JP, Vallenet D, Valverde C, Wall L, Wang Y, Médigue C, Benson DR (2007) Genome characteristics of facultatively symbiotic *Frankia* sp strains reflect host range and host plant biogeography. Genome Res 17:7–15 doi:10.1101/gr.5798407

Normand P, Orso S, Cournoyer B, Jeannin P, Chapelon C, Dawson J, Evtushenko L, Misra AK (1996) Molecular phylogeny of the genus *Frankia* and related genera and emendation of family *Frankiaceae*. Int J Syst Bacteriol 46:1–9

Okon Y (1985) *Azospirillum* as a potential inoculant for agriculture. Trends Biotechnol 3:223–228 doi:10.1016/0167-7799(85)90012-5

Okon Y, Labandera-Gonzales CA (1994) Agronomic applications of *Azospirillum*: an evaluation of 20 years worldwide field inoculation. Soil Biol Biochem 26:1591–1601 doi:10.1016/0038-0717(94)90311-5

Oldroyd GED, Downie JA (2008) Coordinating nodule morphogenesis with rhizobial infection in Legumes. Annu Rev Plant Biol 59:519–546 doi:10.1146/annurev.arplant.59.032607.092839

Pawlowski K, Bisseling T (1996) Rhizobial and actinorhizal symbioses: what are the shared features? Plant Cell 6:1899–1913

Pawlowski K, Sprent JI (2008) Comparison between actinorhizal symbiosis and legume symbiosis. In: Pawlowski K, Newton WE (eds) Nitrogen-fixing actinorhizal symbioses. Springer, Dordrecht, pp 261–288

Pedrosa FO, Elmerich C (2007) Regulation of nitrogen fixation and ammonium assimilatiuon in associated and endophytic nitrogen fixing bacteria. In: Elmerich C, Newton WE (eds) Associative and endophytic nitrogen-fixing bacteria and cyanobacterial associations. Springer, The Netherlands, pp 41–71

Peoples MB, Herridge DF, Ladha JK (1995) Biological nitrogen fixation: an efficient source of nitrogen for sustainable agricultural production? Plant Soil 174:2–28

Pereg-Gerk L, Paquelin A, Gounon A, Kennedy IR, Elmerich C (1998) A transcriptional regulator of the LuxR-UhpA family, FlcA, controls flocculation and wheat root surface colonization by *Azospirillum brasilense* Sp7. Mol Plant Microbe Interact 11:177–187 doi:10.1094/MPMI.1998.11.3.177

Perret X, Staehelin C, Broughton WJ (2000) Molecular basis of symbiotic promiscuity. Microbiol Mol Biol Rev 64:180–201 doi:10.1128/MMBR.64.1.180-201.2000

Peters GA, Meeks JC (1989) The *Azolla-Anabaena* symbiosis: basic biology. Annu Rev Plant Physiol Plant Mol Biol 40:193–210 doi:10.1146/annurev.pp.40.060189.001205

Postgate J (1981) Microbiology of the free-living nitrogen-fixing bacteria, excluding cyanobacteria. In: Gibson AH, Newton WE (eds) Current perspectives in nitrogen fixation. Elsevier/North-Holland Biomedical, Amsterdam, pp 217–228

Pothier JF, Wisniewski-Dyé F, Weiss-Gayet M, Moënne-Loccoz Y, Prigent-Combaret C (2007) Promoter-trap identification of wheat seed extract-induced genes in the plant-growth-promoting rhizobacterium *Azospirillum brasilense* Sp245. Microbiology 153:3608–3622 doi:10.1099/mic.0.2007/009381-0

Prell J, Poole P (2006) Metabolic changes of rhizobia in legume nodules. Trends Microbiol 14:161–168 doi:10.1016/j.tim.2006.02.005

Prigent-Combaret C, Blaha D, Pothier F, Vial L, Poirier M-A, Wisniewski-Dyé F, Moënne-Loccoz Y (2008) FEMS Microbiol Ecol 65:220-219 doi:10.1111/j.1574-6941.2008.00545.x

Prin Y, Rougier M (1987) Preinfection events in the establishment of *Alnus-Frankia* symbiosis: study of the root hair deformation step. Plant Physiol 6:99–106

Pueppke SG, Broughton WJ (1999) *Rhizobium* sp strain NGR234 and *R fredii* USDA257 share exceptionally broad, nested host ranges. Mol Plant Microbe Interact 12:293–318 doi:10.1094/MPMI.1999.12.4.293

Radeva G, Jurgens G, Niemi M, Nick G, Suominen L, Lindström K (2001) Description of two biovars in the *Rhizobium galegae* species: biovar orientalis and biovar officinalis. Syst Appl Microbiol 24:195–205 doi:10.1078/0723-2020-00029

Rai AN (1990) Handbook of symbiotic cyanobacteria. CRC Press, Boca Raton, Florida, USA

Rai AN, Bergman B, Rasmussen U (2002) Cyanobacteria in symbiosis. Kluwer Academic, Dordrecht

Rai AN, Söderbäck E, Bergman B (2000) Cyanobacterium-plant symbiosis. New Phytol 147:449–481 doi:10.1046/j.1469-8137.2000.00720.x

Räsänen LA, Heikkilä-Kallio U, Suominen L, Lipsanen P, Lindström K (1991) Expression of common nodulation genes of *Rhizobium galegae* in various backgrounds. Mol Plant Microbe Interact 4:535–544

Rasmussen U, Svenning MM (2001) Characterization of genotypic methods of symbiotic *Nostoc* strains isolated from five species of *Gunnera*. Arch Microbiol 176:204–210 doi:10.1007/s002030100313

Rasmussen U, Johansson C, Bergman B (1994) Early communication in the *Gunnera-Nostoc* symbiosis: plant-induced cell differentiation and protein synthesis in the cyanobacterium. Mol Plant Microbe Interact 6:696–702

Rasmussen U, Johansson C, Renglin A, Petersson C, Bergman B (1996) A molecular characterization of the *Gunnera-Nostoc* symbiosis: comparison with *Rhizobium*- and *Agrobacterium*-plant interactions. New Phytol 133:391–398 doi:10.1111/j.1469-8137.1996.tb01906.x

Reinhold-Hurek B, Hurek T (1998) Life in grasses, diazotrophic endophytes. Trends Microbiol 6:139–144 doi:10.1016/S0966-842X(98)01229-3

Rennie RJ (1980) Dinitrogen-fixing bacteria: computer-assisted identification of soil isolates. Can J Microbiol 26:1275–1283

Rippka R, Deruelles J, Waterbury JB, Herdman M, Stanier RY (1979) Genetic assignments, strain histories and properties of pure cultures of cyanobacteria. J Gen Microbiol 111:1–61

Rodrigues EP, Rodrigues LS, Martinez de Oliveira AL, Baldani VLD, dos Santos Teixeira KR, Urquiaga S, Reis VM (2008) *Azospirillum amazonense* inoculation: effect on growth yield and N_2 fixation of rice (*Oryza sativa* L.). Plant Soil doi:10.1007/s11104-00079476-1

Roesch LFW, Camargo FAO, Bento FM, Triplett EW (2008) Biodiversity of diazotrophs within the soil, root and stem of field grown maize. Plant Soil 302:91–104 doi:10.1007/s11104-007-9458-3

Rothballer M, Eckert B, Schmid M, Fekete A, Scholter M, Lehner A, Pollmann S, Hartmann A (2008) Endophytic root colonization of gramineous plants by *Herbaspirillum frisingense*. FEMS Microbiol Ecol 66:85–95 doi:10.1111/j.1574-6941.2008.00582.x

Rothballer M, Schmid M, Hartmann A (2003) *In situ* localization and PGPR-effect of *Azospirillum brasilense* colonizing roots of different wheat varieties. Symbiosis 34:261–279

Rovira AD (1991) Rhizosphere research, 85 years of progress and frustration. In: Keister DL, Cregan PB (eds) The rhizosphere and plant growth. Kluwer Academic, The Netherlands, pp 3–13

Rubio LM, Ludden PW (2005) Maturation of nitrogenase: a biochemical puzzle. J Bacteriol 187:405–414 doi:10.1128/JB.187.2.405-414.2005

Rubio LM, Ludden PW (2008) Biosynthesis of the iron-molybdenum cofactor of nitrogenase. Annu Rev Microbiol 62:93–111 doi:10.1146/annurev.micro.62.081307.162737

Schank SC, Smith RL, Weiser GC, Zuberer DA, Bouton JH, Quesenberry KH, Tyler ME, Milam JR, Littel RC (1979) Fluorescent antibody technique to identify *Azospirillum brasilense* associated with roots of grasses. Soil Biol Biochem 11:287–295 doi:10.1016/0038-0717(79)90074-9

Schmid M, Hartmann A (2007) Molecular phylogeny and ecology of root associated diazotrophic α and β Proteobacteria. In: Elmerich C, Newton WE (eds) Associative and endophytic nitrogen-fixing bacteria and cyanobacterial associations. Springer, Dordrecht, pp 41–71

Schultze M, Kondorosi A (1998) Regulation of symbiotic root nodule development. Annu Rev Genet 32:33–57 doi:10.1146/annurev.genet.32.1.33

Sevilla M, Burris RH, Gunapala N, Kennedy C (2001) Comparison of benefit to sugar cane plant growth and $^{15}N_2$ incorporation following inoculation of sterile plants with Acetobacter diazotrophicus wild-type and Nif⁻-mutant strains. Mol Plant Microbe Interact 14:359–366 doi:10.1094/MPMI.2001.14.3.358

Shi Y, Zhao W, Zhang W, Ye Z, Zhao J (2006) Regulation of intracellular free calcium concentration during heterocyst differentiation by HetR and NtcA in Anabaena sp Pcc7120. Proc Natl Acad Sci USA 103:11334–11339 doi:10.1073/pnas.0602839103

Simonet P, Normand P, Hirch M, Akkermans ADL (1990) The genetics of the Frankia-actinorhizal symbiosis. In: Gresshoff PM (ed) Molecular biology of symbiotic nitrogen fixation. CRC, Bocaraton, USA, pp 77–109

Spaink HP (2000) Root nodulation and infection factors produced by rhizobial bacteria. Annu Rev Microbiol 54:257–288 doi:10.1146/annurev.micro.54.1.257

Sprent JI (2001) Nodulation in legumes. Royal Botanic Gardens, Kew, UK

Sprent JI, James EK (2007) Legume evolution: where do nodules and mycorrhizas fit in? Plant Physiol 144:575–581 doi:10.1104/pp.107.096156

Steenhoudt O, Vanderleyden J (2000) Azospirillum, a free-living nitrogen-fixing bacterium closely associated with grasses: genetic, biochemical and ecological aspects. FEMS Microbiol Rev 24:487–506 doi:10.1111/j.1574-6976.2000.tb00552.x

Stephens BB, Loar SN, Alexandre G (2006) Role of CheB and CheR in the complex chemotactic and aerotactic pathway of Azospirillum brasilense. J Bacteriol 188:4759–4768 doi:10.1128/JB.00267-06

Stewart WDP (1969) Biological and ecological aspects of nitrogen fixation by free-living microorganisms. Proc Roy Soc B (London) 172:367–388

Stoffels M, Castellanos T, Hartmann A (2001) Design and application of new 16S-rRNA-targeted oligonucleotide probes for the Azospirillum-Skermanella-Rhodocista-cluster. Syst Appl Microbiol 24:83–97 doi:10.1078/0723-2020-00011

Suominen L, Lortet G, Roos C, Paulin L, Lindström K (2001) Identification and structure of the Rhizobium galegae common nodulation genes: evidence for horizontal gene transfer. Mol Biol Evol 18:906–916

Suominen L, Luukkainen R, Lindström K (2003) Activation of the nodA promoter by the nodD genes of Rhizobium galegae induced by synthetic flavonoids or Galega orientalis root exudates. FEMS Microbiol Lett 19:225–232 doi:10.1016/S0378-1097(02)01206-5

Svistoonoff S, Laplaze L, Auguy F, Runions J, Duponnois R, Haseloff J, Franche C, Bogusz D (2003) cg12 expression is specifically linked to infection of root hairs and cortical cells during Casuarina glauca and Allocasuarina verticillata actinorhizal nodule development. Mol Plant Microbe Interact 16:600–607 doi:10.1094/MPMI.2003.16.7.600

Tapia-Hernández A, Mascarúa-Esperza MA, Caballero-Mellado J (1990) Production of bacteriocins and siderophore-like activity by Azospirillum brasilense. Microbios 64:73–83

Tarrand JJ, Krieg NR, Döbereiner J (1978) A taxonomic study of the Spirillum lipoferum group with description a new genus, Azospirillum gen nov and two species, Azospirillum lipoferum (Beijerinck) comb nov and Azospirillum brasilense sp nov. Can J Microbiol 24:967–980

Tas E, Leinonen P, Saano A, Piippola S, Kaijalainen S, Räsänen LA, Hakola S, Lindström K (1996) Assessment of the competitiveness of rhizobia infecting Galega orientalis using plant yield, nodulation, and strain identification by PCR and antibiotic resistance. Appl Environ Microbiol 62:529–535

Torrey JG, Tjepkema JD (1979) Symbiotic nitrogen fixation in actinomycete-nodulated plants. Bot Gaz 140(Suppl):i–ii doi:10.1086/337026

Turner SL, Young JPW (2000) The glutamine synthetases of rhizobia: phylogenetics and evolutionary implications. Mol Biol Evol 17:309–319

Ueda T, Suga Y, Yahiro N, Matsuguchi T (1995) Remarkable N_2-fixing bacterial diversity detected in rice roots by molecular evolutionary analysis of nifH gene sequences. J Bacteriol 177:1414–1417

van Berkum P, Bohlool BB (1980) Evaluation of nitrogen fixation by bacteria in association with roots of tropical grasses. Microbiol Rev 44:491–517

van Ghelue M, Lovaas E, Ringo E, Solheim B (1997) Early interaction between Alnus glutinosa and Frankia strain Arl3. Production and specificity of root hair deformation factors. Physiol Plant 9:579–587 doi:10.1111/j.1399-3054.1997.tb05360.x

Van Hove C, Lejeune A (2002) Applied aspects of Azolla-Anabaena symbiosis. In: Rai AN, Bergman B, Rasmussen U (eds) Cyanobacteria in symbiosis. Kluwer academic, Dordrecht, pp 179–194

Vanbleu E, Marchal K, Lambrecht M, Mathys J, Vanderleyden J (2004) Annotation of the pRhico plasmid of Azospirillum brasilense reveals its role in determining the outer surface composition. FEMS Microbiol Lett 232:165–172 doi:10.1016/S0378-1097(04)00046-1

Vande Broek A, Lambrecht M, Vanderleyden J (1998) Bacterial chemotactic motility is important for the initiation of wheat root colonization by Azospirillum brasilense. Microbiology 144:2599–2606

Vande Broek A, Michiels J, Van Gool AP, Vanderleyden J (1993) Spatial-temporal colonization patterns of Azospirillum brasilense on the wheat root surface and expression of the bacterial nifH gene during association. Mol Plant Microbe Interact 6:592–600

Vessey JK, Pawlowski K, Bergman B (2004) Root-based N_2-fixing symbioses: legumes, actinorhizal plants, Parasponia and cycads. Plant Soil 266:205–230 doi:10.1007/s11104-005-0871-1

Vlek PLG, Diakite MY, Mueller H (1995) The role of Azolla in curbing ammonia volatilization from flooded rice systems. Fert Res 42:165–174 doi:10.1007/BF00750511

Von Bülow JFW, Döbereiner J (1975) Potential for nitrogen fixation in maize genotypes in Brazil. Proc Natl Acad Sci USA 72:2389–2393 doi:10.1073/pnas.72.6.2389

Wall LG (2000) The actinorhizal symbiosis. J Plant Growth Regul 19:167–182

Wall LG, Berry AM (2008) Early interactions, infection and nodulation in actinorhizal symbiosis. In: Pawlowski K, Newton WE (eds) Nitrogen-fixing actinorhizal symbioses. Springer, Dordrecht, pp 147–166

Wang CM, Ekman M, Bergman B (2004) Expression of cyanobacterial genes involved in heterocyst differentiation and dinitrogen fixation along a plant symbiosis development profile. Mol Plant Microbe Interact 17:436–443 doi:10.1094/MPMI.2004.17.4.436

Wei TF, Ramasubramanian TS, Golden JW (1994) *Anabaena* sp strain PCC 7120 *ntcA* gene required for growth on nitrate and heterocyst development. J Bacteriol 176:4473–4482

Wheeler CT, Miller IM (1990) Current potential uses of actinorhizal plants in Europe. In: Schwintzer RC, Tjepkema JD (eds) The biological of *Frankia* and actinorhizal plants. Academic press Inc, San Diego, pp 365–389

White J, Prell J, James EK, Poole P (2007) Nutrient sharing between symbionts. Plant Physiol 144:604–614 doi:10.1104/pp.107.097741

Wong FCY, Meeks JC (2001) The *hetF* gene product is essential to heterocyst differentiation and affects HetR function in the cyanobacterium *Nostoc punctiforme*. J Bacteriol 183:2654–2661 doi:10.1128/JB.183.8.2654-2661.2001

Wong FCY, Meeks JC (2002) Establishment of a functional symbiosis between the cyanobacterium *Nostoc punctiforme* and the bryophyte *Anthoceros punctatus* requires genes involved in nitrogen control and initiation of heterocyst differentiation. Microbiology 148:315–323

Xu X, Elhai J, Wolk CP (2008) Transcriptional and developmental responses by *Anabaena* to deprivation of fixed nitrogen. In: Herrero A, Flores E (eds) Cyanobacteria: Molecular biology, genomics and evolution. Horizon Scientific, Norwich, pp 383–422

Xu XD, Kong RQ, de Bruijn FJ, He SY, Murry MA, Newman T, Wolk P (2002) DNA sequence and genetic characterization of plasmid pFQ11 from *Frankia alni* strain CpI1. FEMS Microbiol Lett 207:103–107 doi:10.1111/j.1574-6968.2002.tb11036.x

Yan Y, Yang J, Dou Y, Chen M, Ping S, Peng J, Lu W, Zhang W, Yao Z, Li H, Liu W, He S, Geng L, Zhang X, Yang F, Yu H, Zhan Y, Li D, Lin Z, Wang Y, Elmerich C, Lin M, Jin Q (2008) Nitrogen fixation island and rhizosphere competence traits in the genome of root-associated *Pseudomonas stutzeri* A1501. Proc Natl Acad Sci USA 105:7564–7569 doi:10.1073/pnas.0801093105

Yang GP, Debellé F, Savagnac A, Ferro M, Schiltz O, Maillet F, Promé D, Treilhou M, Vialas C, Lindström K, Dénarié J, Promé JC (1999) Structure of the *Mesorhizobium huakuii* and *Rhizobium galegae* Nod factors: a cluster of phylogenetically related legumes are nodulated by rhizobia producing Nod factors with alpha,beta-unsaturated N-acyl substitutions. Mol Microbiol 34:227–237 doi:10.1046/j.1365-2958.1999.01582.x

Yoon HS, Golden JW (1998) Heterocyst pattern formation controlled by a diffusible peptide. Science 282:935–938 doi:10.1126/science.282.5390.935

Young P (1992) Phylogenetic classification of nitrogen-fixing organisms. In: Stacey G, Burris RH, Evans HJ (eds) Biological nitrogen fixation. Chapman and Hall Inc, New York, pp 43–86

Young JPW, Crossman LC, Johnston AWB, Thomson NR, Ghazoui ZF, Hull KH, Wexler M, Curson ARJ, Todd JD, Poole PS, Mauchline TH, East AK, Quail MA, Churcher C, Arrowsmith C, Cherevach I, Chillingworth T, Clarke K, Cronin A, Davis P, Fraser A, Hance Z, Hauser H, Jagels K, Moule S, Mungall K, Norbertczak H, Rabbinowitsch E, Sanders M, Simmonds M, Whitehead S, Parkhill J (2006) The genome of *Rhizobium leguminosarum* has recognizable core and accessory components. Genome Biol 7:R34 doi:10.1186/gb-2006-7-4-r34

Zehr JP, McReynolds LA (1989) Use of degenerate oligonucleotides for amplification of the *nifH* gene from the marine cyanobacterium *Trichodesmium* spp. Appl Environ Microbiol 55:2522–2526

Zhang CC, Laurent S, Sakr S, Peng L, Bedu S (2006) Heterocyst differentiation and pattern formation in cyanobacteria: a chorus of signals. Mol Microbiol 59:367–375 doi:10.1111/j.1365-2958.2005.04979.x

Zheng L, Cash DL, Flint DH, Dean DR (1998) Assembly of iron-sulfur clusters. Identification of an iscSUA-hscBA-fdx gene cluster from *Azotobacter vinelandii*. J Biol Chem 273:13264–1327272 doi:10.1074/jbc.273.21.13264

REVIEW ARTICLE

Biochemical cycling in the rhizosphere having an impact on global change

L. Philippot · S. Hallin · G. Börjesson ·
E. M. Baggs

Received: 5 March 2008 / Accepted: 30 September 2008 / Published online: 21 October 2008
© Springer Science + Business Media B.V. 2008

Abstract Changes in chemical properties in soil around plant roots influence many microbial processes, including those having an impact on greenhouse gas emissions. To potentially mitigate these emissions according to the Kyoto protocol, knowledge about how and where these gases are produced and consumed in soils is required. In this review, we focus on the greenhouse gases nitrous oxide and methane, which are produced by nitrifying and denitrifying prokaryotes and methanogenic archaea, respectively. After describing the microbial processes involved in production and consumption of nitrous oxide and methane and how they can be affected in the rhizosphere, we give an overview of nitrous oxide and methane emissions from the rhizosphere and soils and sediments with plants. We also discuss strategies to mitigate emissions from the rhizosphere and consider possibilities for carbon sequestration.

Keywords Nitrifiers · Denitrifiers · Methanogens · Methanotrophs · Greenhouse gas · Rhizosphere

Responsible Editor: Philippe Lemanceau.

L. Philippot
INRA,
UMR 1229,
21000 Dijon, France

L. Philippot
University of Burgundy,
UMR 1229,
21000 Dijon, France

S. Hallin · G. Börjesson
Department of Microbiology,
Swedish University of Agricultural Sciences,
Box 7025, 750 07 Uppsala, Sweden

E. M. Baggs
Institute of Biological and Environmental Sciences,
University of Aberdeen,
Cruickshank Building, St Machar Drive,
Aberdeen AB24 3UU, UK

L. Philippot (✉)
Soil and Environmental Microbiology,
UMR 1229, 17 rue Sully,
21065 Dijon Cedex, France
e-mail: Laurent.Philippot@dijon.inra.fr

Introduction

Plants affect local conditions in the rhizosphere soil in many ways that influence microbial activity, abundance and community composition (Lynch 1990; Sørensen 1997). Several of these factors have a direct impact on microbial communities emitting greenhouse gases (GHG), which are of major concern for global change (Molina and Rovira 1964a; Tiedje 1988). The three main terrestrial GHG subject to the Kyoto protocol are carbon dioxide (CO_2), methane (CH_4), and nitrous oxide (N_2O). While CO_2 is produced by all living organisms, N_2O and CH_4 are both produced and reduced by microbial guilds

(Conrad 1996). These gases are of major concern since they have global warming potentials about 298 and 25 times, respectively, that of carbon dioxide over a 100 years period (IPCC 2007). Nitrous oxide is a side product of the aerobic nitrification process and an obligate intermediate in the denitrification pathway (Conrad 1996), and can therefore be emitted by both nitrifiers and denitrifiers. However, only the latter are also a sink of N_2O. Whether soils are a net source or sink of atmospheric N_2O depends on the environmental factors regulating consumption and production, but most soils are a net source (Conrad 1996). Methane is produced by methanogenic archaea in anaerobic soil and consumed by CH_4 oxidizing bacteria in aerobic soil. The main terrestrial CH_4 sources are wetland ecosystems, where both methanogens and methanotrophs are present and active. Methane oxidation occurs in most soils, and upland soils are mostly sinks (LeMer and Roger 2001).

Nitrous oxide production and consumption are regulated by oxygen partial pressure, and nitrification is additionally controlled by the concentration of ammonia and pH, while denitrification is also controlled by availability of carbon and nitrate (Tiedje 1988). Methanogenesis is dependent on strict anaerobic and low Redox conditions as well as on the fermentative production of precursors for the methanogens, whereas CH_4 oxidation is mainly dependent on oxygen and CH_4 availability (LeMer and Roger 2001). All the above mentioned factors are affected by the presence of plant roots. The oxygen partial pressure can be altered in the rhizosphere because of respiration by roots and root-associated microorganisms, root consumption of water, and root penetration into the soil, which decreases soil compaction and creates channels for gas transfer. In contrast, wetland plants can alter the oxygen partial pressure by diffusion of the oxygen through aerenchyma to the roots and the surrounding soil (Armstrong 1971). Plants also release readily available organic compounds in soil solution through rhizodeposition, of which root exudation is the largest component (Nguyen 2003). These root-derived organic compounds are considered as a major driving force for many microbial processes in the rhizosphere (Lynch 1990). Finally concentrations of nitrate and ammonium also fluctuate in the rhizosphere due to root uptake.

With better understanding of the controls on GHG production and reduction in arable soil, it will be possible to develop appropriate management strategies for mitigation. The IPCC (2007) report comprehensively covers options for mitigation of N_2O and CH_4, in addition to CO_2 from agricultural systems. However, few strategies really fully utilize the unique nature of the rhizosphere, and with greater understanding of controls on rhizosphere biogeochemistry, we will be better placed to mitigate GHG emissions at the site of production within the rhizosphere soil, in addition to indirectly through agricultural management. In this review, we describe the microbial processes involved in production and consumption of N_2O and CH_4 and how they can be regulated in the rhizosphere. We then give an overview of GHG emissions from the rhizosphere and cropped soils and discuss strategies to mitigate emissions and possibilities for carbon sequestration.

Microbial processes producing and reducing nitrous oxide and methane in rhizosphere soil

Nitrification

Nitrification is a two-step process, consisting of the conversion of ammonia (NH_3) to nitrite (NO_2^-) and its subsequent conversion to nitrate (NO_3^-). The pioneering work of Winogradsky established that this process is performed by chemolithotrophic bacteria that respire with oxygen and assimilate CO_2. These chemolithotrophic bacteria are classified into two groups, based on their ability to oxidize ammonia to nitrite (ammonia-oxidizing bacteria) or nitrite to nitrate (nitrite-oxidizing bacteria) (Kowalchuk and Stephen 2001). The nitrifying bacteria are phylogenetically affiliated to the β- and γ-Proteobacteria, but recent discoveries have demonstrated Crenarchaea to also be important ammonia oxidizers in soil (Leininger et al. 2006). In addition to chemolithotrophic nitrification, some bacteria and fungi possess the potential for heterotrophic nitrification, oxidizing both organic and inorganic nitrogen compounds, and this process is believed to play a role mainly in forest soils (Killham 1986). However, many of the approaches to study heterotrophic nitrification have been performed in pure culture systems and the significance of heterotrophic nitrification in soils still needs to be determined (DeBoer and Kowalchuk 2001; Stams et al. 1990). During nitrification, the conversion of ammonia to the highly mobile nitrate ion minimizes emissions of ammonia, but provides opportunities for nitrogen losses by leaching or denitrification from soil and the root zone (Giles 2005). The loss of nitrogen from the root zone is an economic drain due to fertilizer loss, but also has environmental implications, such as nitrate pollution of ground water, eutrophication of surface waters and emissions of the greenhouse gas N_2O.

Nitrification in the rhizosphere of upland soils

Plants affect several factors that influence nitrification. A long-term field trial comparing unfertilized cropped soil and unfertilized bare fallow showed that plants stimulate nitrification (Enwall et al. 2007). This could be due to the increased organic matter that in turn enhances nitrogen turnover in the soil, in combination with increased aeration. Nevertheless, several studies have reported nitrification to be negatively affected in the rhizosphere (Lensi et al. 1992; Molina and Rovira 1964b; Norton and Firestone 1996; Priha et al. 1999; Robinson 1972). As an example, Wheatley et al. (1990) showed that *Pisum sativum*, *Hordeum vulgare*, *Brassica campestris rapifera* and *Lolium perenne* depressed potential nitrification at a certain plant development stage. A recent study demonstrated that this negative rhizosphere effect on gross nitrification rates was variable along the plant root (Herman et al. 2006). Thus, gross nitrification rates in soil near the root tip of *Avena barbata* were the same as those in bulk soil, whereas nitrification was lower in soil near the older root sections. This was due to rapid uptake of NH_4^+ by the older parts of the root, which limited nitrification rates. Not only plants, but also plant species specific effects on nitrification have been reported. During the growing season, nitrification rates were four times greater in *Deschampsia* patches than in *Acomastylis* patches (Steltzer and Bowman 1998), and when comparing potential nitrification in the rhizosphere of *Pisum sativum*, *Hordeum vulgare*, *Brassica campestris rapifera* and *Lolium perenne*, differences up to 10 fold between the plants were shown (Wheatley et al. 1990). Abundance of ammonia oxidizing bacteria, nitrification rates and nitrate concentrations were also significantly lower in the rhizosphere of *Brachiaria humidicola* compared to other pasture species (Ishikawa et al. 2003; Sylvester-Bradley et al. 1988). The observed plant species effects were attributed to large nitrogen inputs by non-symbiotic nitrogen fixation in the rhizosphere of some plants (Brejda et al. 1994) or differential nitrogen uptake or root respiration by the various species. In general, the lower activity of nitrifiers in the rhizosphere can be explained by a decrease in ammonium concentration due to plant uptake or by the heteretrophic microbes being more competitive compared to autotrophic nitrifiers in this carbon rich environment.

Some studies have suggested that the negative effects on nitrification could be due to an inhibition phenomenon, since the existence of plant-derived nitrification inhibitors is well known. However, the hypothesis that the plant itself is capable of releasing inhibitors of nitrification into soil has been at the centre of a controversy for many years because of the absence of direct evidence (Lata et al. 2000; Munro 1966; Rice and Pancholy 1972). The first indirect evidence of an inhibition phenomenon was provided by Moore and Waid (1971), who showed that addition of root washings from different plants reduced the rate of nitrification up to 84% in proportion to the added amount (Moore and Waid 1971). More recently, using transplantation of *Hyparrhenia diplandra* grass originating from high- or low-nitrifying soils, Lata et al. (2004) showed that there was a significant individual plant effect on nitrification. Thus, plants that originated from the low-nitrifying soil decreased nitrification activity in the high-nitrifying soil, and vice-versa.

A direct demonstration of plants decreasing ammonia oxidation activity in soil was obtained by Subbarao et al. (2007), who used a bioluminescence assay based on a recombinant *Nitrosomonas europaea* (Iizumi et al. 1998) to detect ammonia oxidation inhibitors in root exudates of 18 plant species. Inhibition of nitrification varied widely among the different plant species, and the authors concluded that nitrification inhibition was probably a widespread phenomenon in tropical pasture grass (Subbarao et al. 2006, 2007). Inhibition of nitrification has also been observed when cultivating oil seed rape. The tissues of *Brassica* contain many secondary compounds, including glucosinolates, which, upon disruption of tissues, are hydrolyzed to form iso-thiocyanates (ITCs) and other toxic volatile sulphuric compounds (Bending and Lincoln 1999). ITCs can inhibit nitrification by either reducing the abundance of nitrifying bacteria or lowering nitrification rates (Bending and Lincoln 2000). However, the identification of the chemical mediator(s) in root exudates responsible for inhibition of nitrification in the rhizosphere is still missing. Plant control of nitrification could provide an advantage in competition for nitrogen, and since nitrification is the prior step to processes that can reduce the plant available pool of nitrogen, this ability of the plant to inhibit nitrification is a sophisticated way to reduce nitrogen losses through nitrate leaching or denitrification (Fillery 2007). Thus, inhibition of nitrification can induce environmentally significant changes in the ecosystem nitrogen balance (Lata et al. 2004).

Plant effects on nitrification in rice paddies and wetland sediments

Under flooded conditions soils become anoxic almost immediately beneath the soil–water interface. As a result, nitrification is restricted to a millimetre-thick surface layer. However, wetland plants have developed several strategies to transport oxygen to the root-zone, where it can radially diffuse to the rhizosphere (Armstrong 1971; Frenzel et al. 1992, Colmer 2003, Voesnek et al. 2006), thus establishing an aerobic habitat for nitrification (Fig. 1).

Contrasting effects of wetland plants on nitrifiers have been described. When comparing rice-planted and unplanted pots, Chen et al. (1998) could not detect any difference in nitrification rates. On the other hand, a study conducted in irrigated rice fields planted with three different rice cultivars revealed significant differences in both size of the nitrifier community and nitrification rates between the cultivars (Gosh and Kashyap 2003), which were attributed to variation in root porosity among the cultivars. As in upland soils, the effect of plants on nitrification in water saturated systems is most likely dependent on nitrogen concentration, since ammonia oxidizers are competing for nitrogen with plants and heterotrophic bacteria (Verhagen et al. 1994). Thus, Arth and Frenzel (2000) observed that in unfertilized rice paddy, assimilation by the rice roots lowered the available ammonium to a level where nitrification virtually could not occur.

Studies have also been performed to spatially locate root-associated nitrification (Arth and Frenzel 2000; Li et al. 2004). In a fertilized rice paddy, nitrification was detected by multi-channel micro-

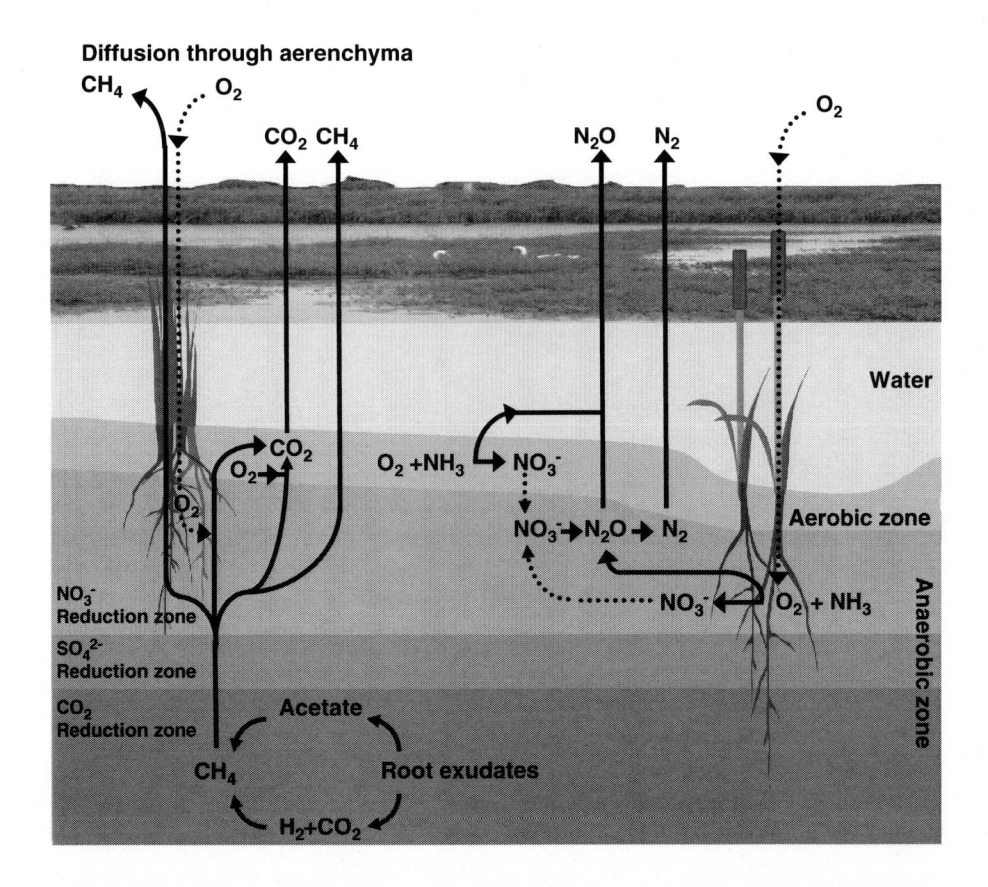

Fig. 1 Production and consumption of N_2O and CH_4 in the rhizosphere of wetland plants

electrodes at a distance of 0–2 mm from the surface of the rice roots, demonstrating that the effect of rice on nitrification is limited to the root surface (Arth and Frenzel 2000). Accordingly, Briones et al. (2002) showed enrichment of ammonia oxidizing bacteria on rice root surfaces, which suggests that root surface populations of ammonia oxidizing bacteria play a major role in determining nitrification rates in the rice rhizosphere. Stimulation of numbers and activity of nitrifying bacteria has also been described for other oxygen-releasing plants (Bodelier et al. 1996; Engelaar et al. 1995), indicating that nitrification in the rhizosphere of aquatic plants could be a common phenomenon in conditions where nitrogen is not limiting. In constructed wetlands, ammonium removal has been shown to be higher in *Phragmites* sp. stands than in those planted with *Typha* sp. (Gersberg et al. 1986), but it remains unclear whether this is due mainly to higher nitrification rates in the *Phragmites* sp. rhizosphere, or due to a more efficient denitrification in sediments covered with this species. Despite the importance of wetland vegetation for nitrogen removal, studies of plant and plant-species effects on nitrification rates and ecology of nitrifiers are scarce.

Nitrate reduction and denitrification

Dissimilatory nitrate reduction into nitrite can be performed by microorganisms that, in contrast to nitrifiers, belong to most of the prokaryotic families (Philippot 1999). The produced nitrite can be either reduced into ammonia by dissimilatory nitrate reduction to ammonium (DNRA, also termed nitrate ammonification) or into nitric oxide (NO), nitrous oxide (N_2O) or dinitrogen gas (N_2) during denitrification. In both processes, nitrogen oxides are used as terminal electron acceptors instead of oxygen for generation of a transmembrane proton electrochemical potential across the cytoplasmic membrane. Denitrification is the main biological process responsible for returning fixed nitrogen to the atmosphere, thus closing the nitrogen cycle. This reduction of soluble nitrogen to gaseous nitrogen is negative for agriculture, since it can deplete the soil of nitrate, an essential plant nutrient. The denitrification N_2O/N_2 product ratio is variable, and N_2O may even be the dominant end product (Chèneby et al. 1998). However, denitrification also provides a valuable ecosystem service by mediating nitrogen removal from nitrate-polluted waters in sediments and other water saturated soils. The ecology of denitrifiers in agricultural soils has recently been reviewed in detail (Philippot et al. 2007).

Nitrate reduction and denitrification in the rhizosphere of upland soils

Several studies have reported that plants can influence the activity, diversity and abundance of nitrate reducers and denitrifiers. Woldendorp (1962) was the first to show that the living root system stimulated denitrification. This early study was followed by several more quantitative measurements of nitrate reduction or denitrification activities. Rate increases ranging from two to 22 times were observed in rhizosphere soil compared to bulk soil (Bakken 1988; Hojberg et al. 1996; Klemedtsson et al. 1987; Philippot et al. 2006; Smith and Tiedje 1979). The stimulation of denitrification in the rhizosphere is positively correlated with soil nitrate concentration. At low NO_3^- concentrations, denitrification rates can even be lower in the rhizosphere compared to the bulk soil (Qian et al. 1997; Smith and Tiedje 1979). It has also been reported that the rhizosphere effect on denitrification was associated with air-filled pore space (Wollersheim et al. 1987). Thus, denitrification rate increased ten-fold at a low moisture tension, while at medium, or high moisture tension, plants had no, or even a negative, effect on denitrification (Bakken 1988). Accordingly, Prade and Trolldenier (1988) showed that the rhizosphere effect on denitrification was confined to air-filled porosity below 10–12% (v/v).

The primary driver of rhizosphere microbial community development is the release of plant-derived low molecular weight organic compounds into the soil, and thus denitrification rates are often positively correlated with total C or soluble organic C (Baggs and Blum 2004; Bijay-Singh et al. 1988; Paul and Beauchamp 1989). However, contradictory results have been published concerning the influence of the organic compounds released by roots on denitrification. On one hand, it has been reported that root exudates could not provide metabolizable organic compounds to the denitrification process (Haider et al. 1987), or that root-derived organic compounds were rapidly immobilized or mineralized by microorganisms in the rhizosphere, and thus had little influence on denitrification (McCarty and Bremner 1993). On the other hand, Qian et al. (1997) argued that labile organic compounds from roots influence denitrification losses of nitrogen. In two recent studies, Mounier et al. (2004) and Henry et al. (2008) demonstrated that addition of root exudates or mucilage to soil without plants could stimulate nitrate reduction or denitrification activity with increases in the range of those observed in planted soil. This suggests that the higher denitrification activity in soil surrounding the plant roots is mainly due to rhizodeposition. However, factors regulating denitrification in the rhizosphere are strongly interwoven and the stimulating effect of root-derived organic compounds on denitrification can only be observed under non-limiting concentrations of nitrate and oxygen.

While effects of plants on the activity of the nitrate reducer or denitrifier communities have been widely investigated, there are fewer studies on how plants affect the composition of these functional communities. The distribution of denitrifying isolates from soil with or without maize, differed, and *Agrobacterium*-related denitrifiers were enriched in the planted soil (Chèneby et al. 2004). The nitrate reducer community structure was also significantly different in the maize rhizosphere compared to bulk soil (Chèneby et al. 2003; Philippot et al. 2002). Analysis of the effect of root-derived organic compounds on the structure and density of nitrate reducing and denitrifying communities revealed minor or no changes after addition of mucilage or artificial root exudates, even though nitrate reduction and denitrification activity were strongly stimulated (Henry et al. 2008; Mounier et al. 2004). Therefore, even though root-derived organic compounds can stimulate denitrification activity, it does not seem to be a strong driver of the denitrifier community structure in soil (Philippot et al. 2007).

Effects of plant species have mainly been studied on denitrifier activity rather than denitrifier community structure and are attributed to differences in quality and quantity of organic compound flow from roots. Higher denitrification rates in the rhizosphere of legumes compared to other plants were observed in several studies (Kilian and Werner 1996; Scaglia et al. 1985; Svensson et al. 1991). Significant differences in denitrification activity below grass tufts among three species were also reported by Patra et al. (2006). Some studies have shown plant species to have a significant influence on the composition of the denitrifier community (Bremer et al. 2007; Patra et al. 2006). Nevertheless, comparison of the composition of the nitrate reducer community under *Lolium perenne* and *Trifolium repens* did not reveal any species effect (Deiglmayr et al. 2004). Analysis of the denitrification gene transcripts in the rhizosphere of three plant species revealed that the active denitrifiers differed, even though the denitrifier community structure based on the total gene pool was similar for all plant species investigated (Sharma et al. 2005).

Plant effects on nitrate reduction and denitrification in rice paddies and wetland sediments

The release of oxygen by the roots of wetland plants can stimulate nitrification and subsequently denitrification after diffusion of nitrate into the reduced zone of the sediment (Fig. 1). Thus, it is generally agreed that denitrification rates in the rhizosphere of aerenchymatous plants are regulated by the rate of nitrification (Arth et al. 1998; Reedy et al. 1989). Furthermore, aerenchymatous plants could also affect the nitrate reducers and denitrifiers by nitrate uptake and exudation of organic compounds. Arth and Frenzel (2000) showed that while nitrification occurred at a distance of 0–2 mm from the surface around individual rice roots, denitrification occurred at 1.5–5.0 mm. There is a large body of literature estimating denitrification rates from paddy rice (e.g. Arth et al. 1998; Buresh and DeDatta 1990; Xing et al. 2002a; Zhu et al. 2003) and denitrification has been recognized as one of the major ways of nitrogen loss in this agroecosystem, thus contributing to the low nitrogen fertilizer efficiency (Cassman et al. 1993; Reddy and Patrick 1986).

In wetlands, vegetation coverage is an important supplier of organic compounds, fueling denitrification (Kallner-Bastviken et al. 2005). In addition, organic compounds can also indirectly enhance denitrification by increasing aerobic respiration, which lowers oxygen levels in the sediment (Nielsen et al. 1990). Thus, an increase in both size and activity of the nitrate reducers was observed in the *Glyceria maxima* rhizosphere (Nijburg et al. 1997). The composition of the nitrate-reducer community was shown to be driven by the presence of *G. maxima* when nitrate was limiting, but when input levels of nitrate were high, nitrate availability determined the community composition. It is not known whether or not the observed positive effects of wetland plants depend on plant species. Kallner-Bastviken et al. (2003) did not find any difference in potential denitrification activities in intact cores with *Phragmites* sp. or *Typha* sp. shoots, although others have shown that samples from *Typha latifolia* and *Phragmites australis* rhizospheres exhibited significantly different nitrate reduction and denitrification rates (Ruiz-Rueda et al. 2008). These differences were connected to typical *Typha* sp. and *Phragmites* sp. associated denitrifying communities. Not sure what this sentence means Accordingly, in another wetland, the denitrifying community structure differed in sediment with an invasive cattail hybrid *Typha* x *glauca* compared to sediment with the native plant species, *Scirpus* sp. (Angeloni et al. 2006).

Methanogenesis and methane oxidation

Methanogenesis is the microbiological production of CH_4 using small organic compounds as a terminal electron acceptor. Methanogenic organisms belong to the phylum Euryarcheota within the domain Archaea, and produce CH_4 either by converting acetic acid to CH_4 and CO_2, or by converting CO_2 with H_2 to CH_4 (Conrad 2007). These simple substrates are provided by other organisms through fermentation. Other forms of carbon, such as formate or methylated compounds, can also be used by methanogens. Methanogenesis requires strict anaerobiosis and low Redox potential. Thus, CH_4 is produced only after depletion of other electron acceptors; nitrate, sulphate, Mn(IV) and Fe (III) (Conrad 2007), which should occur after the Redox potential has dropped to $Eh \approx -300$ mV (Kludze et al. 1993). However, CH_4 emissions have been observed from irrigated rice fields already at $Eh > 300$ mV (Jiao et al. 2006).

A significant proportion of the CH_4 produced in anaerobic layers is oxidized before it reaches the atmosphere. Therefore, net CH_4 emissions are the results of two opposite processes: CH_4 production by methanogenic archaea and CH_4 oxidation by methanotrophic bacteria. Methane oxidizers use CH_4 as their sole carbon and energy source and have an obligatory aerobic metabolism, thereby depending on access to oxygen. Diffusion rates of methane and oxygen are key factors controlling the activity of methanotrophs. They are divided into two families: the *Methylococcaeae*, belonging to *Gammaproteobacteria*, and the *Methylocystaceae*, belonging to the *Alphaproteobacteria*, also known as type I and type II (Bowman 1999; Hanson and Hanson 1996).

Due to homology between the enzymes catalyzing the first steps in methane oxidation and ammonia oxidation, ammonia oxidizing bacteria may also hold the potential to co-oxidize CH_4. However, several studies have excluded a significant role of ammonia oxidizers in CH_4 oxidation (Bodelier and Frenzel 1999; Klemedtsson et al. 1999). The ecology of both methanogens and methanotrophs has recently been reviewed by Conrad (2007).

Plant effects on methane production and consumption in rice paddies and other soils

Wetland plants regulate the CH_4 budget in several ways (Fig. 1). First, exudation by plant roots provides carbon compound precursors to methanogenic archaea (Aulakh et al. 2001a, b; Frenzel and Bosse 1996; Kankaala and Bergström 2004; van Veen et al. 1989). Pulse labelling of rice plants with $^{13}C-CO_2$ or $^{14}C-CO_2$ showed that plant photosynthates excreted from the roots are converted to CH_4 after being fermented to acetate and H_2, which indicates that plant photosynthates are a major source of CH_4 in the rhizosphere (Dannenberg and Conrad 1999; Minoda and Kimura 1994; Minoda et al. 1996). Watanabe et al. (1999) estimated that the supply of organic compounds from rice plants in the form of exudates and sloughed tissues could represent between 37% and 40% of the carbon sources for CH_4 emission. Stimulation of methanogenesis by exudation has also been shown in the rhizosphere of natural wetland plants (Kludze and DeLaune 1994; Saarnio et al. 2004).

A second effect of wetland plants is the passive transport of CH_4 from the anoxic soil to the atmosphere through the plant aerenchyma. Transport of CH_4 from plant roots to the shoots and release into the atmosphere can represent up to 90% of the total CH_4 flux (Butterbach-Bahl et al. 1997; Cicerone and Shetter 1981; Holtzapfel-Pschorn et al. 1986; Nouchi et al. 1990; Schültz et al. 1989). The aerenchyma of wetland plants is also a conduit pipe for oxygen, allowing oxygen diffusion into the rhizosphere and the adjacent sediment, which can stimulate methane oxidizing bacteria. The transport of oxygen by rice roots was illustrated by the work of Frenzel et al. (1992), who detected oxygen down to the depth of 40 mm in a flooded soil planted with rice, whereas it was confined to a thin surface layer of 3.5 mm in the unplanted soil. Another important consequence of the increased oxygen concentration in the soil is that the Redox potential will change, and reductants such as Fe^{2+}, Mn^{2+} and H_2S will be re-oxidized (Kludze et al. 1993). The increased Redox potential will severely hamper CH_4 production and so lower emissions.

A stimulatory effect of rice plants on CH_4 production has been reported in several studies (Dannenberg and Conrad 1999; Holtzapfel-Pschorn et al. 1986) with decreasing rates with depth and distance from the plant (Sass et al. 1991). Whether the net increase of CH_4 production observed in these studies results from stimulation of methanogenesis, CH_4 plant-mediated transport or an inhibition of CH_4 oxidation is difficult to know. However, in rice fields, variations in CH_4 emission were mostly attributed to variations in methanotrophic activity (Schütz et al. 1989).

In contrast to CH_4 production, which shows pronounced variation during the year, the composition of the methanogenic community in rice fields seems to be rather stable (Krüger et al. 2005). The methanogenic community structure was also very similar between rice root and soil samples, with a relatively lower abundance of *Methanosaetacae* on the roots as the only observed difference. Recently, Lu and Conrad (2005) demonstrated that rice cluster I methanogens, an uncultured lineage forming a distinct clade within the phylogenetic radation of *Methanosarcinales* and *Methanomicrobiales*, were the key players in CH_4 production from plant-derived organic compounds in rice microcosms. In addition to rice cluster I, *Methanosarcinae*, *Methanosaetaceae*, and *Methanosarcinaceae*, were shown to be present on rice roots (Chin et al. 2004).

The alteration of CH_4 oxidation rates by plants have been observed in several studies showing higher potential rates in the root compartments than in root-free compartments of rice microcosms (Bodelier and Frenzel 1999; Gilbert and Frenzel 1998). However, the extent to which root-associated methane oxidation varies among plant taxa and among wetland ecosystems is unknown (King 1996). Similarly to CH_4 production, temporal variation of CH_4 oxidation was observed in rice paddies. Thus, Eller and Frenzel (2001) found that in situ CH_4 oxidation was important only during the vegetative growth phase of the plants and then later became negligible. In contrast, Bosse and Frenzel (1998) observed that CH_4 oxidation occurred during the whole growth period of rice. The fact that methanotrophs are able to profit from the oxygen release from the rice plants is reflected not only by increase of their potential activities, but also by their increase in numbers in the rhizosphere. Thus, MPN counts of methanotrophs were 15 times higher in the root compartment compared to in the non-root compartment (Bodelier and Frenzel 1999). A similar increase was reported by Gilbert and Frenzel (1998), who observed one order of magnitude higher numbers of methane-oxidizing bacteria in the rhizosphere than in the bulk soil. Methanotrophs are also found in surface-sterilised roots and basal culms, which indicates their ability to colonise the interior of roots and culms (Bosse and Frenzel 1997). Investigation of the methanotroph community structure in rice paddies revealed the presence of both the *Methylococcaceae* and *Methylocystaceae* families in soil and root compartments over the whole season (Eller and Frenzel 2001). A recent study demonstrated that *Methylococcaceae* and *Methylocystaceae* populations in the rhizospheric soil and on the rice roots changed differently over time with respect to activity and population size, and that *Methylococcaceae* methanotrophs played a particularly important role in the rice field ecosystem (Shrestha et al. 2008).

Emissions of greenhouse gases from the rhizosphere

Nitrous oxide emissions from rhizosphere soil

Evidence from different cropping systems

Emissions of N_2O are typically greater in the presence of growing plants, particularly legumes, than from bare soil (e.g. Kilian and Werner 1996; Klemedtsson et al. 1987). Emission factors vary from 0.1% to 7% of nitrogen applied in different agricultural systems (Skiba and Smith 2000), reflecting differences in vegetation type, crop management and climate. Measured emissions can vary significantly with crop type, for example ranging from 0.2 to 0.7 kg N_2O–N 100 kg^{-1} N applied for small grain cereals, 0.3–5.8 kg N_2O–N 100 kg^{-1} N applied from cut grassland (Dobbie et al. 1999), and 3.9–8.7 kg N_2O–N ha^{-1} $year^{-1}$ from maize fields (Sehy et al. 2003). In legume fields, emissions range from 0.34 to 4.6 kg N_2O–N ha^{-1} $year^{-1}$ (Eichner 1990), including natural emissions, those associated with cultivation, and those derived from nitrogen fixed by the legume. Yang and Cai (2006) demonstrated the effect of soybean growth on N_2O emission to vary with plant growth stage, primarily being controlled by available nitrogen and mineralization during the early growth stage, but in later growth by quantity of root exudates, itself being closely related to plant photosynthesis.

There are several reports of low N_2O emissions from rice paddy fields were low (Buresh and Austin 1988; Lindau et al. 1990; Smith et al. 1982), with less than 0.1% of the applied nitrogen emitted as N_2O in temperate and tropical rice fields when soils are flooded (Freney 1997). However, it has since been found that N_2O is mainly emitted during the non-flooding periods (Xing 1998). For example, the annual N_2O emission from a rice-flooding fallow system, which received 300 kg N fertilizer, and a rice–wheat cropping system receiving 680 kg N fertilizer were 1.4 and 4.3 kg N_2O–N ha^{-1} year^{-1}, respectively (Xing et al. 2002b). Similar emission rates (1.3 and 3.6 kg N_2O–N ha^{-1} year^{-1}) were reported in other nitrogen fertilized rice cropping systems with crop rotations including fallow or green manure (Xiong et al. 2002).

Despite the plethora of data on emissions from different cropping systems, few attempts have been made to attribute N_2O emission to rhizosphere soil per se, where, for example comparisons are made within and between crop rows. These have demonstrated a strong influence of plant roots, with decreasing emissions measured with distance away from the root (Smith and Tiedje 1979). We also lack long-term studies encompassing several cropping seasons or crop rotations, so that any gradual loss of residual fertilizer- or residue nitrogen remains unquantified, despite it being recognized that in a variable climate, several years' data is required to obtain a robust estimate of emissions (Dobbie et al. 1999).

Primary drivers of nitrous oxide production in the rhizosphere and their effects on the N_2O-to-N_2 ratio

Nitrogen application, oxygen partial pressure, carbon availability and pH are considered the primary determinants of rates of ammonia oxidation and denitrification in the rhizosphere. Nitrogen fertilizer application results in short-term increased N_2O emissions (Bouwman 1996; Mosier 1994) that last between several days and up to a few weeks (VanCleemput et al. 1994). This increase in N_2O emissions can be exacerbated by a raised denitrifier N_2O-to-N_2 product ratio following nitrogen fertilizer application since nitrate is preferred over N_2O as an electron acceptor for denitrifiers at concentrations of >10 μg g^{-1} (Baggs et al. 2003; Blackmer and Bremner 1978). Inubishi et al. (1996) found that denitrifier N_2O production rapidly responded to nitrate application, whereas there was a lag in the response of nitrifiers, even when a large quantity of ammonium was added to soils. In contrast, Baggs et al. (2003) observed that nitrification was the predominant N_2O producing process over denitrification in the rhizosphere of *Lolium perenne* during the first seven days after application of NH_4NO_3. In conditions where nitrification and denitrification are limited by ammonium and nitrate, respectively, roots compete with the microorganisms for nitrogen and may lower emissions. This means that sometimes greater emissions are reported for fallow than for cropped systems (Duxbury et al. 1982).

The anaerobic volume of soil is a key factor affecting both nitrification and denitrification. Tillage has an important role to play in altering the aeration status of soil through modifying the soil structure, with typically higher N_2O emission from no-till soils compared to tilled soils (Baggs et al. 2003, 2006; Linn and Doran 1984). In terms of potential for denitrification, this is exacerbated by the often higher soil organic matter availability in the upper topsoil of no-till soils (Nieder et al. 1989). Denitrification is the predominant N_2O producing process above 70–80% water-filled pore space (WFPS; Davidson 1991), or at oxygen partial pressures below 0.5% (Parkin and Tiedje 1984), with ammonia oxidation demonstrated to be predominant at lower WFPS (Bateman and Baggs 2005). The N_2O-to-N_2 ratio falls approaching 100% WFPS, but the nitrous oxide reductase is thought to lag behind the nitrate reductase in time following anoxic conditions (Letey et al. 1980).

The role of organic carbon in the regulation of N_2O-to N_2 ratios is still poorly understood and the importance of root-derived organic compounds flow in the rhizosphere is unknown against that of soil organic matter. Haller and Stolp (1985) provided evidence for rhizosphere stimulation of denitrification, with *Pseudomonas aeruginosa* producing 1.8 ml N_2O–

N day^{-1}, which was equivalent to consumption of 72 mg glucose-C day^{-1} from root exudation. Emissions of N$_2$O have also been reported to be raised, by a factor of 10 or more, following cutting or damage of plants (Beck and Christensen 1987). This is most likely in response to organic compounds released from the roots stimulating denitrification. Recently, an effect of root exudate composition on the denitrifier N$_2$O-to-N$_2$ ratio was reported by Henry et al. (2008), where, N$_2$O-to-N$_2$ ratios of 0.3 and 1 were observed in microcosms amended with artificial root exudates containing 80% and 40% of sugar, respectively. Increased belowground organic compounds allocation by *Lolium perenne* swards under elevated partial pressure of carbon dioxide has also been demonstrated to stimulate denitrifier N$_2$O production (Baggs and Blum 2004; Baggs et al. 2003).

Another key factor that can affect N$_2$O production in soil is pH, but plant-mediated pH effects on N$_2$O production during nitrification and denitrification have yet to be directly determined. Activity of ammonia oxidizing bacteria may be expected to be reduced at low pH, due to a decline in NH$_3$ availability. However, the pH effect on N$_2$O emissions by nitrification is not clear and both greater (Martikainen and DeBoer 1993) and less (Goodroad and Keeney 1984) nitrifier-N$_2$O production has been reported at soil pH 4 than at pH 6. Production of N$_2$O by denitrification can also be influenced by decreased pH in the rhizosphere. The nitrous oxide reductase is known to be sensitive to low pH (Firestone et al. 1980) and Thomsen et al. (1994) showed that reduction of N$_2$O to N$_2$ was inhibited at low pH values in *Paracoccus*. Accordingly, several studies reported that decreasing soil pH increases N$_2$O production by denitrification (Nägele and Conrad 1990; Šimek and Cooper 2002).

Where is nitrous oxide produced in the rhizosphere?

Most of the above studies have measured net emissions from cropped soil, but it is unknown spatially where this N$_2$O is produced in the rhizosphere. However, it is generally accepted that denitrifier activity decreases with distance from roots (Smith and Tiedje 1979). The different drivers of nitrification and denitrification support the idea that these processes are spatially distinct within the rhizosphere, with denitrification being more dependent on root-derived organic compounds and lowered oxygen availability, but nitrification sensitive to pH effects, and competition with plants for available ammonium (Fig. 2). Spatial location of N$_2$O production may also diverge over time in conjunction with altered root exudation or root respiration during plant growth and development. The significance and location of any 'hotspots and hot moments' have yet to be verified, and we currently still rely on theoretical models (Arah and Smith 1989; Smith 1980). Poor characterization at the micro scale raises the question of whether the key process drivers, for denitrification and nitrification are the same, and of the same ranked significance, with differing scale? Only if this is so, can known responses at the plot scale be used to understand interactions within the rhizo-

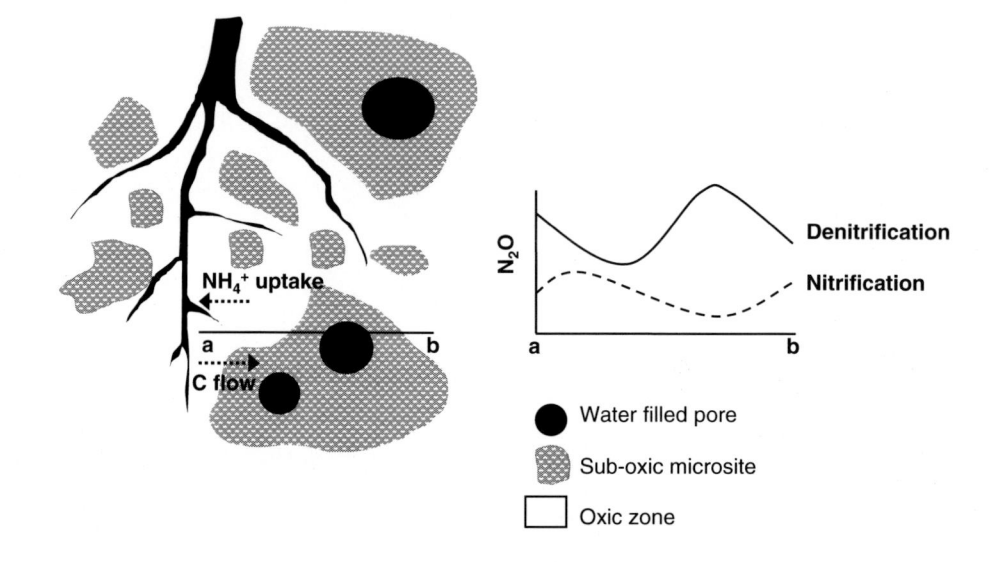

Fig. 2 Conceptual representation of the spatial arrangement of microsites in the rhizosphere, and hypothesized N$_2$O production by nitrification and denitrification with distance from a plant root as influenced by carbon, oxygen and [NH$_4^+$] gradients. N$_2$O production is based on putative nitrification and denitrification rates

sphere. However, at the plot or field level soil hydrology and frequency and duration of rainfall events often become the primary drivers of denitrifier-N_2O production (Dobbie and Smith 2006; Sextone et al. 1985).

Methane emissions from rhizosphere soil

Rice paddies are one of the most important sources of atmospheric CH_4, with a global emission ranging from 30 to 50 Tg CH_4 year^{-1}, which account for about 10% to 20% of the global CH_4 budget. Rice photosynthates can comprise up to more than 50% of total CH_4 emissions (Watanabe et al. 1999) and transport of CH_4 up to 90%. There is an impressive literature on CH_4 emissions from rice paddies. A seasonal pattern of CH_4 emissions is commonly reported with two or three maxima observed in irrigated rice fields during the cropping season. A diurnal pattern is also observed with maximum rates in the afternoon (Schültz et al. 1989). Mean CH_4 emission rates observed in rice fields during the growing season in China, India or the Philippines ranged from 0.02 to 1.3 g CH_4 m^{-2} day^{-1}, depending mainly on the irrigation and fertilization regimes (Jing et al. 2002; Gosh et al. 2003; Wassmann et al. 1999). As examples of the great variability of CH_4 emissions, rates ranging from 0.0035 to 0.180 g CH_4 m^{-2} day^{-1} were reported from flooded rice paddies in California (Cicerone and Shetter 1981).

Methane emissions from cultivated or natural wetlands are usually lower than 0.2 g CH_4 m^{-2} h^{-1} (Le Mer and Roger 2001). In wetlands, aquatic plants generally transport ten times the amount of CH_4 relative to non-vegetated areas (Chanton 2005) and this differ between plant species (Ding et al. 2005). CH_4 transport through the aerenchyma are estimated to account for 50% to 90% of total emissions (Cicerone and Shetter 1981). However, ebullition is thought to account for 18–50% of total CH_4 emissions from Swedish wetlands (Christensen et al. 2003).

Fifteen percent of the net carbon fixed by wetlands may be released to the atmosphere as CH_4 (Brix et al. 2001). Thus, most of the CH_4 flux in a northern Minnesota peatland was derived from recently fixed carbon in living vegetation, and not much from decomposition of old peat (Chanton et al. 1995). In peat-forming wetlands, bryophytes (liverworts, hornworts and mosses) are more sensitive to water table position than vascular plants, and may therefore be used as predictors of CH_4 emission (Joabsson et al. 1999). Nilsson and Bohlin (1993) found that both CH_4 and CO_2 concentrations in Swedish mires were positively correlated with *Sphagnum* remains and negatively correlated with *Carex* remains in peat. This difference was attributed to less easily degradable carbon in *Carex* compared to *Sphagnum*. Importance of vegetation type was confirmed in a large national inventory of Swedish mires comprising 3,157 measured chamber flux rates, where it was estimated that sedge mires accounted for 96% of the CH_4 emitted from natural wetlands in Sweden (Nilsson et al. 2001). Vegetation composition was also found to be an important factor controlling CH_4 emission from an ombrotrophic peatland, with greater CH_4 emissions observed from *Eriophorum* sp. areas than from *Sphagnum* areas (Frenzel and Rudolph 1998). Accordingly, CH_4 emissions of about 72 mg CH_4 m^{-2} day^{-1} were observed in areas containing both *E. vaginatum* L. and *Sphagnum*, which was more than six times higher than areas without *E. vaginatum* (Greenup et al. 2000). Similar results were reported by Minkkinen and Laine (2006), who observed the highest emissions of 29 mg CH_4 m^{-2} day^{-1} from *E. vaginatum* L., with a decreasing trend to *Sphagna* (10.0 mg CH_4 m^{-2} day^{-1}) and forest moss (2.6 mg CH_4 m^{-2} day^{-1}). An effect of plant cover was also reported in freshwater marshes in China with higher CH_4 fluxes during the summer season of 168 to 744 mg CH_4 m^{-2} day^{-1} in the rhizosphere of a *Carex* marsh than in the *Deyeuxia angustifolia* marsh (Ding et al. 2004).

Potential feedback controls of greenhouse gas production in the rhizosphere

Carbon sequestration in the rhizosphere

Carbon sequestration in soil is described as a promising way for reducing the increasing atmospheric carbon dioxide concentration (3.2 Pg C y^{-1}) and carbon storage in agricultural soils is mentioned under Article 3.4 of the Kyoto Protocol. The terrestrial carbon reservoir is 1,500~1,600 Pg of organic-C in the first meter depth (Eswaran et al. 1995), which is more than twice that in the vegetation or the atmospheric pools (Lal 2004). Terrestrial carbon sequestration is controlled by the balance of carbon inputs from primary production and subsequent storage in soil, and outputs through degradation of organic compounds, for which both plants and microbes are accountable. Carbon input from plants mainly includes transfer of the carbon stored in dead plant biomass into the soil by decomposition, and accumulation of soil organic matter due to the humification process after plant death is well documented. The humification rate varies depending on climatic conditions, and plant biochemical composition. Perennial and early successional systems increase storage of soil carbon (Robertson et al. 2000), but in agricultural soils, carbon losses exceed the gains. Conversion of natural ecosystems to agricultural land resulted in the loss of 30% to 75% of their antecedent soil organic carbon pool, which is estimated at 50 to 100 Pg of C (0.8 Pg per year) (Jarecki and Lal 2003; McLauchlan et al. 2006; Schlesinger 1984). The rhizosphere could hypothetically make a significant contribution to carbon input, since about 17% of the plant-fixed carbon is transferred to the rhizosphere soil through root exudates, which corresponds to up to 50% of the plant biomass (Nguyen 2003). However, only a few studies on soil carbon input from rhizodeposits exist and hitherto, it has been shown that most of the root exudates are oxidized to carbon dioxide within a few hours (Jones and Hodge 1999; Kuzyakov and Demin 1998; Verburg et al. 1998), and less than 5% of the carbon transferred to the rhizosphere through root exudates is incorporated into soil organic matter (Kumar et al. 2006; Nguyen 2003). Another source of C input in the rhizosphere is Mycorrhiza which act as a plant C-sink. Thus, a recent study suggested that turnover of mycorrhizal external mycelium may be of importance for the transfer of root derived C to soil organic matter (Godbold et al. 2006). On the other hand, the rhizosphere can also be source of carbon dioxide through the decay of soil organic matter, which can be stimulated by plant-derived carbon and is referred to as the 'rhizosphere priming effect' (Kuzyakov and Demin 1998). However, the supply of labile carbon, such as soluble sugars, amino acids, root mucilage or rhizosphere extract, induces no or little affect on decomposition of soil organic matter, compared to more recalcitrant plant derived carbon, such as ryegrass, cellulose or wheat straw (Fontaine et al. 2007; Mary et al. 1992, 1993).

Whilst the rhizosphere may have a potential contribution to carbon sequestration, the amount of carbon stored in soil mainly depends on land-management practices, edaphic factors and climate. There is a large body of literature on management strategies to increase the net carbon storage in agricultural soils (Post et al. 2004). Such practices include reduced tillage, increasing residue inputs, crop rotation with cover crops, green manures, or perennial crops. Most of the increases in soil carbon associated with these practices result from reversing processes by which traditional management has depleted the soil carbon stocks that accumulated under native perennial vegetation (Cole et al. 1997). Thus, assuming a recovery of 50% of carbon losses in agricultural soils, the global potential for C sequestration over the next 50–100 years would be approximately 25–50 Pg C. In Europe, estimates of the carbon sequestration capacity of agricultural soils are up to 16–23 Tg C year^{-1} (Freibauer et al. 2004; Smith et al. 1998).

Opportunities for mitigation of nitrous oxide emissions in the rhizosphere

Most pertinent to mitigation of rhizosphere N_2O emissions in arable soil are synchronization of nitrogen application to crop demand, precision farming strategies with use of slow- or controlled-release fertilizers (McTaggart and Tsuruta 2003), application of nitrification inhibitors such as dicyandiamide (Di et al. 2007; Hoogendorn et al. 2008) and drainage and aeration of soil (Monteny et al. 2006). It has been estimated that a better synchronization of nitrogen application to crop demand and more closely integrating animal waste and crop residue application with crop production, could decrease N_2O emissions by about 0.38 Tg N_2O–N (Cole et al. 1997). Controlled-release fertilizer, nitrification inhibitors and water management could further lower these emissions by about 0.3 Tg N_2O–N, resulting in a total potential reduction of 0.7 (0.36 to 1.1) Tg N_2O–N, representing 9% to 26% of current emissions from agricultural soil (Cole et al. 1997). Another option to mitigate N_2O emissions in arable soil is no-tillage farming (Li et al. 2005), but its benefits may only be realized in the long-term (Six et al. 2004). One emerging and potentially promising option that did not appear in the IPCC report is the combined application of lime and zeolite. This has recently been demonstrated to lower N_2O emission and increase reduction to dinitrogen in urea amended soil (Zaman et al. 2007), and its potential to be used in the rhizosphere warrants further investigation. A more speculative mitigation possibility is in manipulating exudated carbon compounds and their flow into the rhizosphere, through plant breeding. However, we still do not know if there is any active selection for organic compounds within the denitrifier community, and any carbon preference, and the impact of this on the selection of denitrifiers that have a nitrous oxide reductase or the impact on the regulation of the nitrous oxide reductase itself. Until these relationships are verified, it may be preferable to manage agro-ecosystems to encourage temporary nitrogen immobilization with re-mineralization in the spring in timing with crop demand.

Opportunities for mitigation of CH_4 in the rhizosphere

Mitigation of CH_4 emission can be achieved by lowering methanogenesis and/or CH_4 transport to the atmosphere, or by stimulating methanotrophy. In upland soils, such as forest and agricultural soils, no or little CH_4 is produced, instead atmospheric CH_4 can be oxidized. Forest soils tend to be good sinks for methane, because the trees help to keep the water table well below the surface, which allows for methanotrophy. Thus, well drained non-agricultural soils contribute 5% to 10% of the global CH_4 sink (Cicerone and Oremland 1988; Crutzen 1991). However, the low concentration of CH_4 in upland soils, is likely to be the limiting factor for CH_4 oxidation. Increasing the CH_4 sink can be done through selection of tree species, since forest composition affects CH_4 uptake rates (Menyailo and Hungate 2003). Conversion of natural ecosystems to agricultural land usually lowers CH_4 oxidation, but mitigation of CH_4 emissions can be achieved by limiting cultural practices affecting CH_4 and O_2 availability. For example, soil compaction by tractors may reduce CH_4 oxidation by 50% (Hansen et al. 1993), and drainage can also be crucial in determining the size of the soil methane sink.

Mitigation of methane production in peatland ecosystems

Plant species differ in properties that constrain microbial respiration as well as properties that promote CH_4 oxidation, and this plant associated CH_4 oxidation has been reported from a wide range of wetland species (Sorrell et al. 2002). Further examples include CH_4 oxidation associated with roots and rhizomes of *Sparganium eurycarpum*, where 1% to 58% (mean 27%) of the total CH_4 flux was oxidized (King 1996). In the rhizosphere of *Carex lasiocarpa* and *C. meyeriana*, CH_4 oxidation lowered potential CH_4 emissions by 3.2–35.9% and 4.3–38.5%, respectively (Ding et al. 2004). Similarly, Popp et al. (2000), found that rhizospheric CH_4 oxidation in a *Carex*-dominated fen in Canada lowered net CH_4 emissions by around 20%. Lower CH_4 oxidation was observed in *Carex*-dominated wetlands compared to other types of sedge vegetation (*Eriophorum* and *Juncus*) in southern Sweden (Ström et al. 2005). In agreement, it was shown that CH_4 oxidation was not associated with *Eriophorum* (Frenzel and Rudolph 1998).

The importance of plant cover in influencing net CH_4 emissions is also related to their capacity to transport CH_4 through their aerenchyma. Transport is linked to root porosity (intercellular gas spaces and aerenchyma), which differ substantially among plants (Colmer 2003) but also within genus (e.g. 5% to 30% in *Rumex*, Laan et al. 1989) or between cultivars (Huang et al. 1994). Thus, the choice of plant species in, for example, constructed wetlands could be a way to contribute to CH_4 mitigation. Plant cover can also have implications for the management of peatlands as sources or sinks for CH_4. In addition, lowering of CH_4 emissions could be accomplished through drainage, independent of any ecological or financial considerations. On the other hand, drainage of peat increases the emissions of CO_2 and N_2O. Thus it has been shown that drainage for forestry stimulates N_2O emission on fertile and fertilized sites and that agricultural use of peatland induces considerable and long-lasting emissions of CO_2 and N_2O (Alm et al. 2007).

Mitigation of methane production in rice paddies

The complexity of the role of the rice plant for regulating CH_4 production has been well investigated and reviewed (Aulakh et al. 2001b; Conrad 2002; Frenzel 2000; Wassmann and Aulakh 2000). Up to 90% of the CH_4 emitted in rice paddies is released through rice transport (Cicerone and Shetter 1981; Conrad 2007), while between 19% and 90% of the

CH_4 produced is oxidized (Bosse and Frenzel 1997; Conrad 1996; Gilbert and Frenzel 1995; Holtzapfel-Pschorn et al. 1985), with up to 75% of the CH_4 oxidation taking place in the rhizosphere (Frenzel 2000). Accordingly, strategies to lower net CH_4 emission from rice fields include reduction of CH_4 production, increasing CH_4 oxidation, and lowering CH_4 transport through the plant. Among the CH_4 emission mitigation strategies that do not compromise rice productivity, introduction of drainage periods during the crop cycle appears to be the most efficient (Neue 1993). Thus, it has been estimated that intermittent drainage periods in one third of the poorly drained rice fields in China could reduce 10% the agricultural CH_4 emissions (9.9 Tg, Kern et al. 1997). However, a major drawback of this strategy is that it consumes two to three times more water than continuous flooding and can increase the N_2O emissions (Hou et al. 2000). Another strategy to lower emissions may be oriented toward rice cultivar selection and use of rice cultivar characterized by a low root exudation and porosity to limit production and transport of CH_4. There are 90,000 known rice cultivars with large variations in genotype and phenotype that can affect CH_4 production, rhizospheric CH_4 oxidation and plant-mediated CH_4 transport efficiency. Accordingly, several studies reported large differences in CH_4 emissions between cultivars that can reach up to 500% (Jia et al. 2002; LeMer and Roger 2001; Wassmann and Aulakh 2000). Mitigation of CH_4 emissions in rice paddies also includes amendment with compost residues instead of uncomposted material (Conrad 2007; LeMer and Roger 2001) and direct seeding instead of transplanting.

Concluding remarks

Emissions of CH_4 and N_2O from soils and how they are affected by the presence of plants have now been investigated for several decades and most studies have shown a strong influence of plant roots. The soil oxygen partial pressure is a major factor regulating nitrification, denitrification, methanogenesis and methanotrophy, which is clearly reflected in studies reported in this review indicating that the rhizosphere effect on these underpinning processes controlling greenhouse gas emissions vary widely from upland to wetlands soils. Thus, in uplands soils, inhibition of nitrification by plants is commonly reported whereas denitrification is most often stimulated in the rhizosphere. Also in upland soils, methanotrophy largely dominates over methanogenesis. By contrast, nitrification in wetlands is stimulated next to the roots where radial oxygen losses occurred. Wetland plants also stimulate methanogenesis through root exudation and can facilitate CH_4 transportation to the atmosphere in their tissues. The development of molecular approaches allowed significant progress in the knowledge of the ecology of the microbial guilds involved in greenhouse gas emissions and now we know that the size and/or the diversity of the nitrifier, denitrifier, methanogen and methanotroph communities are also influenced by the presence of plant roots. However, it remains unclear whether these changes in microbial communities in the rhizosphere affect greenhouse gas emissions. Because the rhizosphere effect is complex and results from the action of several strongly interwoven factors such as organic carbon and oxygen availability, further research is required in order to reconcile apparently conflicting results. In addition, prediction of a general rhizosphere effect is difficult since there is evidence that it is both plant species and soil-type dependent. Thus, our incomplete knowledge of the complex rhizosphere effect on microbial guilds controlling greenhouse gas emissions is still limiting the development of plant-based strategies to mitigate emissions of both CH_4 and N_2O.

Acknowledgements The Environment and Agronomy research division and the International Relations Department of the INRA and the Department of Microbiology of the Swedish University of Agricultural Sciences in Uppsala are gratefully acknowledged for supporting and hosting L. Philippot's long-term mission in Sweden. E.M. Baggs is supported by an Advanced Research Fellowship awarded by the Natural Environment Research Council (NERC), U.K. S. Halin is supported by the Formas financed Uppsala Microbiomics Center.

References

Alm J, Shrupali NJ, Minkinen K, Aro L, Hytönen J, Laurila T, Lohila A, Maljanen M, Martikainen PJ, Mäkiranta P, Penttilä T, Saarnio S, Silvan N, Tuittila ES, Laine J (2007) Emission factors and their uncertainty for the exchange of CO_2, CH_4 and N_2O in Finnish managed peatlands. Boreal Environ Res 12:191–209

Angeloni LA, Jankowski KJ, Tuchmann NC, Kelly JJ (2006) Effects of an invasive cattail species (*Typha × glauca*) on sediment nitrogen and microbial community structure composition in a freshwater wetland. FEMS Microbiol Lett 263:86–92 doi:10.1111/j.1574-6968.2006.00409.x

Arah JRM, Smith KA (1989) Steady-state denitrification in aggregated soils—a mathematical model. J Soil Sci 40:139–149 doi:10.1111/j.1365-2389.1989.tb01262.x

Armstrong W (1971) Radial oxygen losses from intact rice roots as affected by distance from the apex, respiration and waterlogging. Physiol Plant 25:192–197 doi:10.1111/j.1399-3054.1971.tb01427.x

Arth I, Frenzel P (2000) Nitrification and denitrification in the rhizosphere of rice: the detection of processes by a new multi-channel electrode. Biol Fertil Soils 31:427–435 doi:10.1007/s003749900190

Arth I, Frenzel P, Conrad R (1998) Denitrification coupled to nitrification in the rhizosphere of rice. Soil Biol Biochem 4:509–515 doi:10.1016/S0038-0717(97)00143-0

Aulakh MS, Wassmann R, Bueno C, Rennenberg H (2001a) Impact of root exudates of different cultivars and plant development stages of rice (*Oryza sativa* L.) on methane production in a paddy soil. Plant Soil 230:77–86 doi:10.1023/A:1004817212321

Aulakh MS, Wassmann R, Rennenberg H (2001b) Methane emission from rice fields—quantification, mechanisms, role of management, and mitigation options. Adv Agron 91:193–260 doi:10.1016/S0065-2113(01)70006-5

Baggs EM, Blum H (2004) CH_4 oxidation and CH_4 and N_2O emissions from *Lolium perenne* swards under elevated atmospheric CO_2. Soil Biol Biochem 36:713–723 doi:10.1016/j.soilbio.2004.01.008

Baggs EM, Richter M, Cadish G, Hartwig UA (2003) Denitrification in grass swards is increased under elevated atmospheric CO_2. Soil Biol Biochem 35:729–732 doi:10.1016/S0038-0717(03)00083-X

Baggs EM, Chebii J, Ndufa JK (2006) A short-term investigation of trace gas emissions following tillage and no-tillage of agroforestry residues in western Kenya. Soil Tillage Res 90:69–76 doi:10.1016/j.still.2005.08.006

Bakken LR (1988) Denitrification under different cultivated plants: effects of soil moisture tension, nitrate concentration, and photosynthetic activity. Biol Fertil Soils 6:271–278 doi:10.1007/BF00261011

Bateman EJ, Baggs EM (2005) Contributions of nitrification and denitrification to N_2O emissions from soils at different water-filled pore space. Biol Fertil Soils 41:379–388 doi:10.1007/s00374-005-0858-3

Beck H, Christensen S (1987) The effect of grass maturing and root decay on nitrous oxide production in soil. Plant Soil 103:269–273 doi:10.1007/BF02370399

Bending GD, Lincoln SD (1999) Characterization of volatile sulphur compounds from soils treated with sulfur-containing organic materials. Soil Biol Biochem 31:695–703 doi:10.1016/S0038-0717(98)00163-1

Bending GD, Lincoln SD (2000) Inhibition of soil nitrifying bacteria communities and their activities by glucosinolate hydrolysis products. Soil Biol Biochem 32:1261–1269 doi:10.1016/S0038-0717(00)00043-2

Bijay-Singh JC, Ryden A, Whitchhead DC (1988) Some relationships between denitrification potential and fractions of organic carbon in air-dried and field-moist soils. Soil Biol Biochem 20:737–741 doi:10.1016/0038-0717(88)90160-5

Blackmer AM, Bremner JM (1978) Inhibitory effect of nitrate on reduction of N_2O to N_2 by soil microorganisms. Soil Biol Biochem 10:187–191 doi:10.1016/0038-0717(78)90095-0

Bodelier PLE, Frenzel P (1999) Contribution of methanotrophic and nitrifying bacteria to CH_4 and NH_4^+ oxidation in the rhizosphere of rice plants as determined by new methods of discrimination. Appl Environ Microbiol 65:1826–1833

Bodelier PLE, Libochant J, Blom C, Laanbroek H (1996) Dynamics of nitrification and denitrification in root-oxygenated sediments and adaptation of ammonia-oxidizing bacteria to low oxygen or anoxic habitats. Appl Environ Microbiol 62:4100–4107

Bosse U, Frenzel P (1997) Activity and distribution of methane-oxidizing bacteria in flooded rice soil microcosms and in rice plants (*Oryza sativa*). Appl Environ Microbiol 63:1199–1207

Bosse U, Frenzel P (1998) Methane emissions from rice microcosms: the balance of production, accumulation and oxidation. Biogeochem 41:199–214 doi:10.1023/A:1005909313026

Bouwman AF (1996) Direct emission of nitrous oxide from agricultural soils. Nutr Cycl Agroecosyst 45:53–70 doi:10.1007/BF00210224

Bowman JP (1999) The methanotrophs—the families Methylococcaceae and Methylocystaceae. In: Dorwin M (ed) The prokaryotes. Springer, New York, pp 266–289

Brejda JJ, Kremer RJ, Brown JR (1994) Indication of associative nitrogen fixation in eastern grama grass. J Range Manage 47:192–196 doi:10.2307/4003014

Bremer C, Braker G, Matthies D, Reuter A, Engels C, Conrad R (2007) Impact of plant functional group, plant species, and sampling time on the composition of *nirK*-type denitrifier communities in soil. Appl Environ Microbiol 73:6876–6884 doi:10.1128/AEM.01536-07

Briones AM, Okabe S, Umemiya Y, Ramsing N-B, Reichardt W, Okuyama H (2002) Influence of different cultivars on populations of ammonia-oxidizing bacteria in the root environment of rice. Appl Environ Microbiol 68:3067–3075 doi:10.1128/AEM.68.6.3067-3075.2002

Brix H, Sorrell BK, Lorenzen B (2001) Are *Phragmites*-dominated wetlands a net source or net sink of greenhouse gases. Aquat Bot 69:313–324 doi:10.1016/S0304-3770(01)00145-0

Buresh RJ, Austin RA (1988) Direct measurement of dinitrogen and nitrous oxide flux in flooded rice fields. Soil Sci Soc Am J 52:681–687

Buresh RJ, DeDatta SK (1990) Denitrification losses from puddled rice soils in the tropics. Biol Fertil Soils 9:1–13 doi:10.1007/BF00335854

Butterbach-Bahl K, Papen H, Rennenberg H (1997) Impact of gas transport through rice cultivars on methane emission from paddy fields. Plant Cell Environ 20:1175–1183 doi:10.1046/j.1365-3040.1997.d01-142.x

Cassman KG, Kropf MJ, Gaunt J, Peng S (1993) Nitrogen use efficiency of rice reconsidered: what are the key constraints. Plant Soil 155–156:359–362 doi:10.1007/BF0 0025057

Chanton JP (2005) The effect of gas transport on the isotope signature of methane in wetlands. Org Geochem 36:753–768 doi:10.1016/j.orggeochem.2004.10.007

Chanton JP, Bauer JE, Glaser PA, Siegel DI, Kelley CA, Tyler SC, Romanowicz EH, Lazrus A (1995) Radiocarbon evidence for the substrates supporting methane formation within northern Minnesota peatlands. Geochim Cosmochim Acta 59:3663–3668 doi:10.1016/0016-7037(95)00240-Z

Chen DL, Chalk PM, Freney JR, Luo QX (1998) Nitrogen transformations in a flooded soil int the presence and absence of rice plants: 1. Nitrification. Nutr Cycl Agroecosyst 51:259–267 doi:10.1023/A:1009736729518

Chèneby D, Hartmann A, Hénault C, Topp E, Germon JC (1998) Diversity of denitrifying microflora and ability to reduce N_2O in two soils. Biol Fertil Soils 28:19–26 doi:10.1007/s003740050458

Chèneby D, Hallet S, Mondon M, Martin-Laurent F, Germon JC, Philippot L (2003) Genetic characterization of the nitrate reducing community based on narG nucleotide sequence analysis. Microb Ecol 46:113–121 doi:10.1007/s00248-002-2042-8

Chèneby D, Perrez S, Devroe C, Hallet S, Couton Y, Bizouard F, Iuretig G, Germon JC, Philippot L (2004) Denitrifying bacteria in bulk and maize-rhizospheric soil: diversity and N_2O-reducing abilities. Can J Microbiol 50:469–474 doi:10.1139/w04-037

Chin K-J, Lueders T, Friedrich MW, Klose M, Conrad R (2004) Archaeal community structure and pathway of methane formation on rice roots. Microb Ecol 47:59–67 doi:10.1007/s00248-003-2014-7

Christensen TR, Panikov N, Mastepanov M, Joabsson A, Stewart A, Öquist M, Sommerkorn M, Reynaud S, Svensson B (2003) Biotic controls on CO_2 and CH_4 exchange in wetlands—a closed environment study. Biogeochem 64:337–354 doi:10.1023/A:1024913730848

Cicerone RJ, Oremland RS (1988) Biochemical aspects of atmospheric methane. Global Biogeochem Cycles 2:299–327 doi:10.1029/GB002i004p00299

Cicerone RJ, Shetter JD (1981) Sources of atmospheric methane: measurements in rice paddies and a discussion. J Geophys Res 86:7203–7209 doi:10.1029/JC086iC08p07203

Cole C, Duxbury J, Freney J, Heinemeyer O, Minami K, Mosier A, Paustian K, Rosenberg N, Sampson N, Sauerbeck D, Zhao Q (1997) Global estimates of potential mitigation of greenhouse gas emissions by agriculture. Nutr Cycl Agroecosyst 49:221–228 doi:10.1023/A:1009 731711346

Colmer TD (2003) Long-distance transport of gases in plants: a perspective on internal aeration and radial oxygen loss from roots. Plant Cell Environ 26:17–36 doi:10.1046/j.1365-3040.2003.00846.x

Conrad R (1996) Soil microorganisms as controllers of atmospheric trace gases (H_2, CO, CH_4, OCS, N_2O, and NO). Microbiol Rev 60:609–640

Conrad R (2002) Control of microbial methane production in wetalnd rice fields. Nutr Cycl Agroecosyst 64:59–69 doi:10.1023/A:1021178713988

Conrad R (2007) Microbial ecology of methanogens and methanotrophs. Adv Agron 96:1–63 doi:10.1016/S0065-2113(07)96005-8

Crutzen P (1991) Methane's sinks and sources. Nature 350:380–381 doi:10.1038/350380a0

Dannenberg S, Conrad R (1999) Effect of rice plants on methane production and rhizospheric metabolism in paddy soil. Biochem 45:53–71

Davidson EA (1991) Fluxes of nitrous oxide and nitric oxide from terrestrial ecosystems. A global inventory of nitric oxide emissions from soils. In: Rogers J, Whitman W (eds) Microbial production and consumption of greenhouse gases: methane, nitrogen oxides, and halomethanes. American Society for Microbiology, Washington DC, pp 219–235

DeBoer W, Kowalchuk GA (2001) Nitrification in acid soils: micro-organisms and mechanisms. Soil Biol Biochem 33:853–866 doi:10.1016/S0038-0717(00)00247-9

Deiglmayr K, Philippot L, Hartwig UA, Kandeler E (2004) Structure and activity of the nitrate-reducing community in the rhizosphere of Lolium perenne and Trifolium repens under long-term elevated atmospheric pCO_2. FEMS Microbiol Ecol 49:445–454 doi:10.1016/j.femsec.2004. 04.017

Di HJ, Cameron KC, Sherlock RR (2007) Comparison of the effectiveness of a nitrification inhibitor, dicyandiamide, in reducing nitrous oxide emissions in four different soils under different climatic and management conditions. Soil Use Manage 23:1–9 doi:10.1111/j.1475-2743.2006.00057.x

Ding W, Cai Z, Tsuruta H (2004) Summertime variation of methane oxidation in the rhizosphere of a Carex dominated freshwater marsh. Atmos Environ 38:4165–4173 doi:10.1016/j.atmosenv.2004.04.022

Ding W, Cai Z, Tsurutua H (2005) Plant species effects on methane emissions from freshwater marshes. Atmos Environ 39:3199–3207 doi:10.1016/j.atmosenv.2005.02.022

Dobbie KE, Smith KA (2006) The effect of water table depth on emissions of N_2O from a grassland soil. Soil Use Manage 22:22–28 doi:10.1111/j.1475-2743.2006.00002.x

Dobbie K, Taggart I, Smith K (1999) Nitrous oxide emissions from intensive agricultural systems: Variations between crops and seasons, key driving variables, and mean emission factors. J Geo Res 104:891–899

Duxbury JM, Bouldin DR, Terry RE, Tate RL (1982) Emissions of nitrous oxide from soils. Nature 298:462–464 doi:10.1038/298462a0

Eichner MJ (1990) Nitrous oxide emission from fertilized soils: Summary of available data. J Environ Qual 19:272–280

Eller G, Frenzel P (2001) Changes in activity and community structure of methane-oxidizing bacteria over the growth period of rice. Appl Environ Microbiol 67:2395–2403 doi:10.1128/AEM.67.6.2395-2403.2001

Engelaar WHMG, Symens JC, Lanbroek HJ, Blom CWPM (1995) Preservation of nitrifying capacity and nitrate availability in waterlogged soils by radial oxygen loss from roots of wetland plants. Biol Fertil Soils 20:243–248 doi:10.1007/BF00336084

Enwall K, Nyberg K, Bertilsson S, Cederlund H, Stenström J, Hallin S (2007) Long-term impact of fertilization on activity and composition of bacterial communities and metabolic guilds in agricultural soil. Soil Biol Biochem 39:106–115 doi:10.1016/j.soilbio.2006.06.015

Eswaran H, Van den Berg E, Reich P, Kimble JM (1995) Global soil C ressources. In: Lal R, Kimble J, Levine E (eds) Soils and global change. Lewis, Boca Raton, FL, pp 27–43

Fillery IRP (2007) Plant-based manipulation of nitrification in soil: a new approach to managing N loss. Plant Soil 294:1–4 doi:10.1007/s11104-007-9263-z

Firestone MK, Firestone RB, Tiedje JM (1980) Nitrous oxide from soil denitrification: factors controlling its biological production. Science 208:749–751 doi:10.1126/science. 208.4445.749

Fontaine S, Barot S, Barré P, Bdioui N, Mary B, Rumpel C (2007) Stability of organic carbon in deep soil layers controlled by fresh carbon supply. Nature 450:277–281 doi:10.1038/nature06275

Freibauer A, Rounsevell MDA, Smith P, Verhagen J (2004) Carbon sequestration in the agricultural soils of Europe. Geoderma 122:1–23 doi:10.1016/j.geoderma.2004.01.021

Freney JR (1997) Emission of nitrous oxide from soils used for agriculture. Nutr Cycl Agroecosyst 49:1–6 doi:10.1023/A:1009702832489

Frenzel P (2000) Plant-associated methane oxidation in rice fields and wetlands. Adv Microb Ecol 16:85–114

Frenzel, P, Bosse, U (1996) Methyl fluoride, an inhibitor of methane oxidation and methane production. FEMS Microbiol Ecol 21:25–36

Frenzel P, Rudolph J (1998) Methane emission from a wetland plant: the role of CH_4 oxidation in *Eriophorum*. Plant Soil 202:27–32 doi:10.1023/A:1004348929219

Frenzel P, Rothfuss F, Conrad R (1992) Oxygen profiles and methane turnover in a flooded rice microcosm. Biol Fertil Soils 14:84–89 doi:10.1007/BF00336255

Gersberg RM, Elkins BV, Goldman CR (1986) Role of aquatic plants in wastewater treatment by artifical wetlands. Water Res 20:363–368 doi:10.1016/0043-1354(86)90085-0

Gilbert B, Frenzel P (1995) Methanotrophic bacteria in the rhizosphere of rice microcosms and their effect on porewater methane concentration and methane emission. Biol Fertil Soils 20:93–100 doi:10.1007/BF00336586

Gilbert B, Frenzel P (1998) Rice roots and CH_4 oxidation: the activity of bacteria, their distribution and the microenvironment. Soil Biol Biochem 30:1903–1916 doi:10.1016/S0038-0717(98)00061-3

Giles J (2005) Nitrogen study fertilizes fears of pollution. Nature 433:791 doi:10.1038/433791a

Godbold DL, Hoosebeek MR, Lukac M, Cotrufo MF, Janssens IA, Ceulemans R, Polle A, Velthorst EJ, Scarascia-Mugnozza G, DeAngelis P, Miglietta F, Peressottu A (2006) Mycorrhizal hyphal turnover as a dominant process for carbon input into soil organic matter. Plant Soil 281:15–24 doi:10.1007/s11104-005-3701-6

Goodroad LL, Keeney DR (1984) Nitrous oxide production in aerobic soils under varying pH, temperature and water content. Soil Biol Biochem 16:39–43 doi:10.1016/0038-0717(84)90123-8

Gosh P, Kashyap AK (2003) Effect of rice on rate of N-mineralization, nitrification and nitrifier population size in an irrigated rice ecosystem. Appl Soil Ecol 24:27–41 doi:10.1016/S0929-1393(03)00068-4

Gosh S, Majumdar D, Jain MC (2003) Methane and nitrous oxide emissions from an irrigated rice of North India. Chemsphere 51:181–195 doi:10.1016/S0045-6535(02)00822-6

Greenup AL, Bradford MA, McMamara NP, Ineson P, Lee JA (2000) The role of *Eriophorum vaginatum* in CH_4 flux from ombotrophic peatland. Plant Soil 227:265–272 doi:10.1023/A:1026573727311

Haider K, Mosier AR, Heinemeyer O (1987) Effect of growing plants on denitrification at high nitrate concentrations. Soil Sci Soc Am J 51:97–102

Haller T, Stolp H (1985) Quantitative estimation of root exudation of maize plant. Plant Soil 86:207–216 doi:10.1007/BF02182895

Hansen S, Maehlum JE, Bakken LR (1993) N_2O and CH_4 fluxes in soil influenced by fertilization and tractor traffic. Soil Biol Biochem 25:621–630 doi:10.1016/0038-0717(93)90202-M

Hanson RS, Hanson TE (1996) Methanogenic bacteria. Microbiol Rev 60:439–471

Henry S, Texier S, Hallet S, Bru D, Dambreville C, Chèneby D, Bizouard F, Germon JC, Philippot L (2008) Disentangling the rhizosphere effect on nitrate reducers and denitrifiers: insight into the role of root exudates. Environ Microbiol 10(11):3082–3092

Herman DJ, Johnson KK, Jaeger CH III, Schwartz E, Firestone MK (2006) Root influence on nitrogen mineralization and nitrification in *Avena barbata* rhizosphere soil. Soil Sci Soc Am J 70:1504–1511 doi:10.2136/sssaj2005.0113

Hojberg O, Binnerup SJ, Sorensen J (1996) Potential rates of ammonium oxidation, nitrate reduction and denitrification in the young barley rhizosphere. Soil Biol Biochem 28:47–54 doi:10.1016/0038-0717(95)00119-0

Holtzapfel-Pschorn A, Conrad R, Seiler W (1985) Production, oxidation and emission of methane in rice paddies. FEMS Microbiol Ecol 31:343–351 doi:10.1111/j.1574-6968.1985.tb01170.x

Holtzapfel-Pschorn A, Conrad R, Seiler W (1986) Effects of vegetation of the emission of methane from submerged paddy soil. Plant Soil 92:223–231 doi:10.1007/BF02372636

Hoogendorn CJ, Klein CAMd, Rutheford AJ, Letica S, Devantier BP (2008) The effect of increasing rates of nitrogen fertilizer and a nitrification inhibitor on nitrous oxide emissions from urine patches on sheep grazed hill country pasture. Aust J Exp Agric 48:147–151 doi:10.1071/EA07238

Hou AX, Chen GX, Wang ZP, Van Cleemput O, Partick WH Jr (2000) Methane and nitrous oxide emissions from a rice field in relation to soil redox and microbiological processes. Soil Sci Am J 64:2180–2186

Huang B, Jonhson JW, Nesmith S, Bridges DC (1994) Growth, physiological and anatomical responses of two wheat genotypes to waterlogging and nutrient supply. J Exp Bot 45:193–202 doi:10.1093/jxb/45.2.193

Iizumi T, Mizumoto M, Nakamura K (1998) A bioluminescence assay using *Nitrosomonas europaea* for rapid and sensitive detection of nitrification inhibitors. Appl Environ Microbiol 64:3656–3662

Inubishi K, Naganuma H, Kitahara S (1996) Contribution of denitrification and autotrophic and heterotrophic nitrification to nitrous oxide production in andosols. Biol Fertil Soils 23:292–298 doi:10.1007/BF00335957

IPCC (2007) Climate Change 2007: Synthesis Report. Contribution of Working Groups I, II and III to the Fourth Assessment Report of the Intergovernmental Panel on Climate Change [Core Writing Team, Pachauri RK, Reisinger A (eds.)]. IPCC, Geneva, Switzerland, p 104

Ishikawa I, Subbarao GV, Ito O, Okado K (2003) Suppresion of nitrification and nitrous oxide emission by the tropical grass *Brachiaria humidicola*. Plant Soil 255:413–419 doi:10.1023/A:1026156924755

Jarecki MK, Lal R (2003) Crop management for soil carbon sequestration. Crit Rev Plant Sci 22:471–502 doi:10.1080/713608318

Jia ZJ, Cai ZC, Xu H, Tsuruta H (2002) Effect of rice cultivars on methane fluxes in a paddy soil. Nutr Cycl Agroecosyst 64:87–94 doi:10.1023/A:1021102915805

Jiao ZH, Hou AX, Shi Y, Huang GH, Wang YH, Chen X (2006) Water management influencing methane and nitrous oxide emissions from rice field in relation to soil redox and microbial community. Commun Soil Sci Plant Anal 37:1889–1903 doi:10.1080/00103620600767124

Jing L, Mingxing W, Yao H, Yuesci W (2002) New estimates of methane emissions from Chinese rice paddies. Nutr Cycl Agroecosyst 64:33–42 doi:10.1023/A:1021184314338

Joabsson A, Christensen TR, Wallén B (1999) Vascular plant controls on methane emissions from northern peatforming wetlands. Trends Ecol Evol 14:385–388 doi:10.1016/S0169-5347(99)01649-3

Jones DL, Hodge A (1999) Biodegradation kinetics ans sorption reactions of three differently charged amino acids in soil and their effect on plant organic nitrogen availability. Soil Biol Biochem 31:1331–1342 doi:10.1016/S0038-0717(99)00056-5

Kallner-Bastviken S, Eriksson PG, Martins I, Neto JM, Leonardson L, Tonderski K (2003) Potential nitrification and denitrification on different surfaces in a constructed treatment wetland. J Environ Qual 32:2414–2420

Kallner-Bastviken S, Eriksson PG, Premrov A, Tonderski K (2005) Potential denitrification in wetland sediments with different plant species detritus. Ecol Eng 25:183–190 doi:10.1016/j.ecoleng.2005.04.013

Kankaala P, Bergström I (2004) Emission and oxidation of methane in *Equisetum fluviatile* stands growing on organic sediment and sand bottoms. Biogeochemistry 67:21–37 doi:10.1023/B:BIOG.0000015277.17288.7a

Kern JS, Gong ZT, Zhang GL, Zhuo HZ, Luo GB (1997) Spatial analysis of methane emission from paddy soil in China and the potential for emission reduction. Nutr Cycl Agroecosyst 49:181–195 doi:10.1023/A:1009710425295

Kilian S, Werner D (1996) Enhanced denitrification in plots of N_2-fixing faba beans compared to plots of a non-fixing legume and non-legumes. Biol Fertil Soils 21:77–83 doi:10.1007/BF00335996

Killham K (1986) Heterotrophic nitrification. In: Prosser J (ed) Nitrification. IRL Press, Oxford

King GM (1996) In situ analyses of methane oxidation associated with the roots and rhizomes of a bur reed, *Sparganium eurycarpum*, in a marine wetland. Appl Environ Microbiol 62:4548–4555

Klemedtsson L, Svensson BH, Rosswall T (1987) Dinitrogen and nitrous oxide produced by denitrification and nitrification in soil with and without barley plants. Plant Soil 99:303–310 doi:10.1007/BF02370877

Klemedtsson L, Jiang Q, Klemedtsson AK, Bakken L (1999) Autotrophic ammonium-oxidising bacteria in Swedish mor humus. Soil Biol Biochem 31:839–847 doi:10.1016/S0038-0717(98)00183-7

Kludze HK, DeLaune RD (1994) Methane emissions and growth of *Spartina patens* in response to soil redox intensity. Soil Sci Soc Am J 58:1838–1845

Kludze HK, Delaune RD, Patrick WH Jr (1993) Aerenchyma formation and methane and oxygen-exchange in rice. Soil Sci Soc Am J 57:386–391

Kowalchuk GA, Stephen JR (2001) Ammonia-oxidizing bacteria: a model for molecular biology. Annu Rev Microbiol 55:485–529 doi:10.1146/annurev.micro.55.1.485

Krüger M, Frenzel P, Kemnitz D, Conrad R (2005) Activity, structure and dynamics of the methanogenic archaeal community in a flooded Italian rice field. FEMS Microbiol Ecol 51:323–331 doi:10.1016/j.femsec.2004.09.004

Kumar R, Pandey S, Pendey A (2006) Plant roots and carbon sequestration. Curr Sci 91:885–890

Kuzyakov Y, Demin V (1998) CO_2 efflux by rapid decomposition of low molecular organic substances in soils. Sci Soils 3:1–12 doi:10.1007/s10112-998-0002-2

Laan P, Berrevoets MJ, Lythe S, Armstrong W, Blom CWPM (1989) Root morphology and aerenchyma formation as indicators of the flooded-tolerance of *Rumex* species. J Ecol 77:693–703 doi:10.2307/2260979

Lal R (2004) Agricultural activities and the global carbon cycle. Nutr Cycl Agroecosyst 70:103–116 doi:10.1023/B:FRES.0000048480.24274.0f

Lata JC, Degrange V, Abbadie L, Lensi R (2000) Relationships between root density of the African grass *Hyparrhenia diplandra* and nitrification at the decimetric scale: an inhibition–stimulation balance hypothesis. Proc R Soc Lond B. Biol Sci 276:1–6

Lata JC, Degrange V, Raynaud X, Maron P, Lensi R, Abbadie L (2004) Grass population control nitrification in savanna soils. Funct Ecol 18:605–611 doi:10.1111/j.0269-8463.2004.00880.x

Leininger S, Urich T, Schloter M, Schwark L, Qi J, Nicol GW, Prosser JI, Schuster SC, Schleper C (2006) Archaea predominate among ammonia-oxidizing prokaryotes in soils. Nature 442:806–809 doi:10.1038/nature04983

LeMer J, Roger P (2001) Production, oxidation and consumption of methane by soils: a review. Eur J Soil Biol 37:25–50 doi:10.1016/S1164-5563(01)01067-6

Lensi R, Domenach AM, Abbadie L (1992) Field study of nitrification and denitrification in a wet savanna of west africa (Lamto, Côte d'ivoire). Plant Soil 147:107–113 doi:10.1007/BF00009376

Letey J, Valoras N, Hadas A, Focht DD (1980) Effect of air-filled porosity, nitrate concentration and time on the ratio of N_2O/N_2 evolution during denitrification. J Environ Qual 9:227–231

Li YL, Zhang YL, Hu J, Shen QR (2004) Contribution of nitrification happened in rhizospheric soil growing with different rice cultivars to N nutrition. Biol Fertil Soils 43:417–425 doi:10.1007/s00374-006-0119-0

Li CS, Frolking S, Butterbach-Bahl K (2005) Carbon sequestration in arable soils is likely to increase nitrous oxide emissions, offsetting reductions in climate radiative forcing. Clim Change 72:321–338 doi:10.1007/s10584-005-6791-5

Lindau CW, DeLaune RD, Patrick WHJ, Bollich PK (1990) Fertilizer effects on dinitrogen, nitrous oxide, and methane emissions from lowland rice. Soil Sci Soc Am J 54:1789–1794

Linn DM, Doran JW (1984) Effect of water-filled pore space on carbon dioxide and nitrous oxide production in tilled and nontilled soils. Soil Sci Soc Am J 48:1267–1272

Lu Y, Conrad R (2005) In situ stable isotope probing of methanogenic Archaea in the rice rhizosphere. Science 309:1088–1090 doi:10.1126/science.1113435

Lynch JM (1990) The rhizosphere. Wiley, Chichester

Martikainen PJ, DeBoer W (1993) Nitrous oxide production and nitrification in acidic soil from a Dutch coniferous forest. Soil Biol Biochem 25:343–347 doi:10.1016/0038-0717(93)90133-V

Mary B, Mariotti A, Morel JL (1992) Use of $_{13}$C variations at natural abundance for studying the biodegradation of root mucilage, roots and glucose in soil. Soil Biol Biochem 24:1065–1072 doi:10.1016/0038-0717(92)90037-X

Mary B, Fresneau C, Morel JL, Mariotti A (1993) C and N cycling during decomposition of root mucilage, roots and glucose in soil. Soil Biol Biochem 25:1005–1014 doi:10.1016/0038-0717(93)90147-4

McCarty GW, Bremner JM (1993) Factors affecting the availability of organic-carbon for denitrification of nitrate in subsoils. Biol Fertil Soils 15:132–136 doi:10.1007/BF00336431

McLauchlan KL, Hobbie SE, Post WM (2006) Conversion from agriculture to grassland buils soil organic matter on decadal timescales. Ecol Appl 16:143–153 doi:10.1890/04-1650

McTaggart IP, Tsuruta H (2003) The influence of controlled release fertilisers and the form of applied nitrogen on nitrous oxide emissions from an andosol. Nutr Cycl Agroecosyst 67:47–54 doi:10.1023/A:1025108911676

Menyailo M, Hungate BA (2003) Interactive effects of tree species and soil moisture on methane consumption. Soil Biol Biochem 35:73–79 doi:10.1016/S0038-0717(03)00018-X

Minkkinen K, Laine J (2006) Vegetation heterogeneity and ditches create spatial variability in methane fluxes from peatlands drained for forestry. Plant Soil 285:289–304 doi:10.1007/s11104-006-9016-4

Minoda T, Kimura M (1994) Contribution of photosynthesized carbon to methane emitted from paddy fields. Geophys Res Lett 21:2007–2010 doi:10.1029/94GL01595

Minoda T, Kimura M, Wada E (1996) Photosynthates as dominant source of CH_4 and CO_2 in soil water and CH_4 emitted to the atmosphere from paddy fields. J Geo Res 101:21091–21097 doi:10.1029/96JD01710

Molina JAE, Rovira AD (1964a) Influence of plant roots on autotrophic nitrifying bacteria. Can J Microbiol 10:249–255

Molina JAE, Rovira AD (1964b) The influence of plant roots on autotrophic nitrifying bacteria. Can J Microbiol 10:249–247

Monteny GJ, Bannink A, Chadwick D (2006) Greenhouse gas abatement strategies for animal husbandry. Agric Ecosyst Environ 112:163–170 doi:10.1016/j.agee.2005.08.015

Moore DRE, Waid JS (1971) The influence of washings of living roots on nitrification. Soil Biol Biochem 3:69–83 doi:10.1016/0038-0717(71)90032-0

Mosier AR (1994) Nitrous oxide emissions from agricultural soils. Fertil Res 37:191–200 doi:10.1007/BF00748937

Mounier E, Hallet S, Chèneby D, Benizri E, Gruet Y, Nguyen C, Piutti S, Robin C, Slezack-Deschaumes S, Martin-Laurent F, Germon JC, Philippot L (2004) Influence of maize mucilage on the diversity and activity of the denitrifying community. Environ Microbiol 6:301–312 doi:10.1111/j.1462-2920.2004.00571.x

Munro PE (1966) Inhibition of nitrite-oxidizers by roots of grass. J Appl Ecol 3:227–229 doi:10.2307/2401247

Nägele W, Conrad R (1990) Influence of pH on the release of NO and N2O from fertilized and unfertilized soil. Biol Fertil Soils 10:139–144

Neue HU (1993) Methane emission from rice fields. Bioscience 43:466–474 doi:10.2307/1311906

Nguyen C (2003) Rhizodeposition of organic C by plants: mechanisms and controls. Agronomie 23:375–396 doi:10.1051/agro:2003011

Nieder R, Schollmayer G, Richter J (1989) Denitrification in the rooting zone of cropped soils with regard to methodology and climate: A review. Biol Fertil Soils 8:219–226 doi:10.1007/BF00266482

Nielsen LP, Christensen PB, Revsbech NP, Sorensen J (1990) Denitrification and oxygen respiration in biofilms studied with microsensor for nitrous oxide and oxygen. Microb Ecol 19(1):63–72

Nijburg JW, Coolen MJL, Gerards S, Klein Gunnewiek PJA, Laanbroek HJ (1997) Effect of nitrate availability and the presence of *Glyceria maxima* on the composition and activity of the dissimilatory nitrate-reducing bacterial community. Appl Environ Microbiol 63:931–937

Nilsson M, Bohlin E (1993) Methane and carbon dioxide concentrations in bogs and fens—with special reference to the effects of the botanical composition of the peat. J Ecol 81:615–625 doi:10.2307/2261660

Nilsson M, Mikkelä C, Sundh I, Granberg G, Svensson BH, Ranneby B (2001) Methane emission from Swedish mires: National and regional budgets and dependence on mire vegetation. J Geophys Res 106:20,847–820,860

Norton J, Firestone M (1996) N dynamics in the rhizosphere of *Pinus ponderosa* seedlings. Soil Biol Biochem 28:351–362 doi:10.1016/0038-0717(95)00155-7

Nouchi I, Mariko S, Aoki K (1990) Mechanism of methane transport from the rhizosphere to the atmosphere through rice plants. Plant Physiol 94:59–66

Parkin TB, Tiedje JM (1984) Application of a soil core method to investigate the effect of oxygen concentration on denitrification. Soil Biol Biochem 16:331–334 doi:10.1016/0038-0717(84)90027-0

Patra A, Abbadie L, Clays-Josserand A, Degrange V, Grayston SJ, Guillaumaud N, Loiseau P, Louault F, Mahmood S, Nazaret S, Philippot L, Poly F, Prosser JI, Le Roux X (2006) Effects of management regime and plant species on the enzyme activity and genetic structure of N-fixing, denitrifying and nitrifying bacterial communities in grassland soils. Environ Microbiol 8:1005–1016 doi:10.1111/j.1462-2920.2006.00992.x

Paul JW, Beauchamp EG (1989) Denitrification and fermentation in plant-residue-amended soil. Biol Fertil Soils 7:303–309 doi:10.1007/BF00257824

Philippot L (1999) Dissimilatory nitrate reductases in bacteria. Biochim Biophys Acta 1446:1–23

Philippot L, Piutti S, Martin-Laurent F, Hallet S, Germon JC (2002) Molecular analysis of the nitrate-reducing community from unplanted and maize-planted soil. Appl Environ Microbiol 68:6121–6128 doi:10.1128/AEM.68.12.6121-6128.2002

Philippot L, Kufner M, Chèneby D, Depret G, Laguerre G, Martin-Laurent F (2006) Genetic structure and activity of the nitrate-reducers community in the rhizosphere of different cultivars of maize. Plant Soil 287:177–186 doi:10.1007/s11104-006-9063-x

Philippot L, Hallin S, Schloter M (2007) Ecology of denitrifying prokaryotes in agricultural soil. Adv Agron 96:135–190

Popp TJ, Chanton JP, Whiting GJ, Grant N (2000) Evaluation of methane oxidation in the rhizosphere of a Carex dominated fen in north central Alberta. Biogeochemistry 51:259–281 doi:10.1023/A:1006452609284

Post W, Izzaurralde R, Jastrow J, McCarl B, Amonette J, Bailey V, Jardine P, West T, Zhou J (2004) Enhencement of carbon sequestration in US soils. Bioscience 54:895–908 doi:10.1641/0006-3568(2004)054[0895:EOCSIU]2.0.CO;2

Prade K, Trolldenier G (1988) Effect of wheat roots on denitrification at varying soil air-filled porosity and organic-carbon content. Biol Fertil Soils 7:1–6 doi:10.1007/BF00260723

Priha O, Grayston SJ, Pennanen T, Smolander A (1999) Microbial activities related to C and N cycling and microbial community structure in the rhizospheres of *Pinus sylvestris*, *Picea abies* and *Betula pendula* seedlings in an organic and mineral soil. FEMS Microbiol Ecol 30:187–199 doi:10.1111/j.1574-6941.1999.tb00647.x

Qian JH, Doran JW, Walters DT (1997) Maize plant contributions to root zone available carbon and nitrogen transformations of nitrogen. Soil Biol Biochem 29:1451–1462 doi:10.1016/S0038-0717(97)00043-6

Reddy K, Patrick WJ (1986) Denitrification losses in flooded rice fields. Fertil Res 9:99–116 doi:10.1007/BF01048697

Reedy K, Patrick WJ, Lindau C (1989) Nitrification-denitrification at the plant root-sediment interface in wetlands. Limnol Oceanogr 34:241–302

Rice C, Pancholy S (1972) Inhibition of nitrification by climax ecosystem. Am J Bot 59:1033–1040 doi:10.2307/2441488

Robertson G, Paul E, Harwood R (2000) Greenhouse gases in intensive agriculture: contributions of individual gases to the radiative forcing of the atmosphere. Science 289:1922–1925 doi:10.1126/science.289.5486.1922

Robinson J (1972) Nitrification in a New Zealand grassland soil. Plant Soil 14:173–183

Ruiz-Rueda O, Hallin S, Bañeras L (2008) Structure and function of denitrifying and nitrifying bacterial communities in relation to the plant species in a constructed wetland. FEMS Microbiol Ecol (in press)

Saarnio S, Wittenmayer L, Merbach W (2004) Rhizospheric exudation of *Eriophorum vaginatum* L.—potential link to methanogenesis. Plant Soil 267:343–355 doi:10.1007/s11104-005-0140-3

Sass RL, Fisher FM, Harcombe PA, Turner FT (1991) Methane emission from rice fields as influenced by solar radiation, temperature and straw incorporation. Glob Change Cycles 5:335–350 doi:10.1029/91GB02586

Scaglia J, Lensi R, Chalamet A (1985) Relationship between photosynthesis and denitrification in planted soil. Plant Soil 84:37–43 doi:10.1007/BF02197865

Schlesinger W (1984) Soil organic matter: a source of atmospheric CO_2. In: Woodwell G (ed) The role of terrestrial vegetation in the global carbon cycle. Wiley, New York, pp 111–127

Schütz H, Holzapfel-Pschorn A, Conrad R, Rennenberg H, Seiler W (1989) A three years continuous record on the influence of daytime, season and fertilizer treatment on methane emission rate from an Italian rice paddy field. J Geophys Res 94:16405–16416 doi:10.1029/JD094iD13p16405

Sehy U, Ruser R, Munch J (2003) Nitrous oxide fluxes from maize fields: relationship to yield, site-specific fertilization, and soil conditions. Agric Ecosyst Environ 99:97–111 doi:10.1016/S0167-8809(03)00139-7

Sextone A, Revsbech N, Parkin T, Tiedje J (1985) Direct measurement of oxygen profiles and denitrification rates in soil aggregates. Soil Sci Soc Am J 49:645–651

Sharma S, Aneja MK, Mayer J, Munch JC, Schloter M (2005) Diversity of transcripts of nitrite reductase genes (*nirK* and *nirS*) in rhizospheres of grain legumes. Appl Environ Microbiol 71:2001–2007 doi:10.1128/AEM.71.4.2001-2007.2005

Shrestha M, Abraham W-R, Shrestha PM, Noll M, Conrad R (2008) Activity and composition of methanotrophic bacterial communities in planted rice soil studied by flux measurements, analyses of *pmoA* gene and stable isotope probing of phospholipid fatty acids. Environ Microbiol 10:400–412 doi:10.1111/j.1462-2920.2007.01462.x

Šimek M, Cooper JE (2002) The influence of soil pH on denitrification: Progress towards the understanding of this interaction over the last 50 years. Eur J Soil Sci 53:345–354 doi:10.1046/j.1365-2389.2002.00461.x

Six J, Ogle S, Breidt F, Conant R, Mosier A, Paustian K (2004) The potential to mitigate global warming with no-tillage management is only realized when practiced in the long term. Glob Change Biol 10:155–160 doi:10.1111/j.1529-8817.2003.00730.x

Skiba U, Smith K (2000) The control of nitrous oxide emissions from agricultural and natural soils. Chemosphere Glob Chang Sci 2:379–386

Smith KA (1980) A model of the extent of anaerobic zones in aggregated soil and its potential application to estimates of denitrification. J Soil Sci 31:263–277 doi:10.1111/j.1365-2389.1980.tb02080.x

Smith MS, Tiedje JM (1979) The effect of roots on soil denitrification. Soil Sci Soc Am J 43:951–955

Smith CJ, Brandon M, Patrick WJ (1982) Nitrous oxide emission following urea-N fertilization of wetland rice. Soil Sci Plant Nutr 28:161–171

Smith P, Powlson D, Glendining M, Smith J (1998) Preliminary estimates of the potential for carbon mitigation in European soils through no-till farming. Glob Change Biol 4:679–685 doi:10.1046/j.1365-2486.1998.00185.x

Sørensen J (1997) The rhizosphere as a habitat for soil microorganisms. In: Elsas JDV, Trevors JT, Wellington EMK (eds) Modern soil microbiology. Marcel Dekker, New York

Sorrell BK, Downes MT, Stanger CL (2002) Methanotrophic bacteria and their activity on submerged aquatic macrophytes. Aquat Bot 72:107–119 doi:10.1016/S0304-3770(01)00215-7

Stams A, Flameling E, Marnette E (1990) The importance of autotrophic versus heterotrophic oxidation of atmospheric ammonium in forest ecosystems with acid soil. FEMS Microbiol Ecol 74:337–344 doi:10.1111/j.1574-6968.1990.tb04080.x

Steltzer H, Bowman W (1998) Differential influence of plant species on soil nitrogen transformation within moist meadow alpine toundra. Ecosystems (N Y, Print) 1:464–474 doi:10.1007/s100219900042

Ström L, Mastepanov M, Christensen TR (2005) Species-specific effects of vascular plants on carbon turnover and methane emissions from wetlands. Biogeochem 75:65–82 doi:10.1007/s10533-004-6124-1

Subbarao GV, Ishikawa I, Ito O, Nakahara K, Wang H, Berry WL (2006) A bioluminescence assay to detect nitrification inhibitors released from plant roots: a case study with *Brachiaria humidicola*. Plant Soil 288:101–112 doi:10.1007/s11104-006-9094-3

Subbarao GV, Rondon M, Ito O, Ishikawa I, Rao IM, Nakahara K, Lascano C, Berry WL (2007) Biological nitrification inhibition (BNI)—is it a widespread phenomenon. Plant Soil 294:5–18 doi:10.1007/s11104-006-9159-3

Svensson BH, Klemedtsson L, Simkins S, Paustin K, Rosswall T (1991) Soil denitrification in three cropping systems characterized by differences in nitrogen and carbon supply. I. Rate-distribution frequencies, comparison between systems and seasonal N losses. Plant Soil 138:257–271 doi:10.1007/BF00012253

Sylvester-Bradley R, Mosquera D, Mendez J (1988) Inhibition of nitrate accumulation in tropical grassland soils: effect of nitrogen fertilization and soil disturbance. J Soil Sci 39:407–416 doi:10.1111/j.1365-2389.1988.tb01226.x

Thomsen J, Geest T, Cox R (1994) Mass spectrometric studies of the effect of pH on the accumulation of intermediates in denitrification in *Paracoccus denitrificans*. Appl Environ Microbiol 60:536–541

Tiedje JM (1988) Ecology of denitrification and dissimilatory nitrate reduction to ammonium. In: Zehnder AJB (ed) Biology of anaerobic microorganisms. Wiley, New York, pp 179–244

van Veen JA, Merckx R, van de Geijn SC (1989) Plant- and soil-related controls of the flow of carbon from roots through soil microbial biomass. Plant Soil 115:179–188 doi:10.1007/BF02202586

VanCleemput O, Vermoesen A, DeGroot C, VanRyckeghem K (1994) Nitrous oxide emission out of grassland. Environ Monit Assess 31:145–152 doi:10.1007/BF00547190

Verburg P, Gorissen A, Arp W (1998) Carbon allocation and decomposition of root-derived organic matter in a plant–soil system of *Calluna vulgaris* as affected by elevated CO_2. Soil Biol Biochem 30:1251–1258 doi:10.1016/S0038-0717(98)00055-8

Verhagen F, Hageman P, Woldendorp JM, Laanbroek H (1994) Competition for ammonium between nitrifying bacteria and plant roots in soil in pots—effects of grazing by flagellates and fertilization. Soil Biol Biochem 26:89–96 doi:10.1016/0038-0717(94)90199-6

Voesenek LACJ, Colmer TD, Pieril R, Milenaar FF, Peeters AJM (2006) How plants cope with complete submergence. New Phytol 170:213–226 doi:10.1111/j.1469-8137.2006.01692.x

Wassmann R, Aulakh MS (2000) The role of rice plants in regulating mechanisms of methane emissions. Biol Fertil Soils 31:20–29 doi:10.1007/s003740050619

Wassmann R, Neue HU, Lantin RS, Aduna JB, Alberto MCR, Andales MJ, Tan MJ, Vandergon HACD, Hoffmann H, Papen H, Rennberg H, Seiler W (1999) Temporal patterns of methane emissions from wetland ricefields treated by different modes of N-application. J Geophys Res Atmo 99:16457–16462 doi:10.1029/94JD00017

Watanabe A, Takeda T, Kimura M (1999) Evaluation of origins of CH_4 carbon emitted rice paddies. J Geo Res 104:23623–23629 doi:10.1029/1999JD900467

Wheatley R, Ritz K, Griffiths B (1990) Microbial biomass and mineral N transformations in soil planted with barley, ryegrass, pea or turnip. Plant Soil 127:157–167 doi:10.1007/BF00014422

Woldendorp JM (1962) The quantitative influence of the rhizosphere on denitrification. Plant Soil 17:267–270 doi:10.1007/BF01376229

Wollersheim R, Trolldenier G, Beringer H (1987) Effect of bulk density and soil water tension on denitrification in the rhizosphere of spring wheat (*Triticum vulgare*). Biol Fertil Soils 5:181–187 doi:10.1007/BF00256898

Xing GX (1998) N_2O emission from ropland in China. Nutr Cycl Agroecosyst 52:249–254 doi:10.1023/A:1009776008840

Xing GX, Cao YC, Shi SL, Sun GQ, Du LJ, Zhu JG (2002a) Denitrification in underground saturated soil in a rice paddy region. Soil Biol Biochem 34:1593–1598 doi:10.1016/S0038-0717(02)00143-8

Xing GX, Shi SL, Shen GY, Du LJ, Xiong ZQ (2002b) Nitrous oxide emissions from paddy soil in three rice-based cropping system in china. Nutr Cycl Agroecosyst 64:135–143 doi:10.1023/A:1021131722165

Xiong ZQ, Xing GX, Tsuruta H, Shen GY, Shi SL, Du LJ (2002) Measurement of nitrous oxide emissions from two rice-based cropping systems in china. Nutr Cycl Agroecosyst 64:125–133 doi:10.1023/A:1021179605327

Yang L, Cai Z (2006) Effects of shading soybean plants on N_2O emission from soil. Plant Soil 283:265–274 doi:10.1007/s11104-006-0017-0

Zaman M, Nguyen M, Matheson F, Blennerhassett J, Quin B (2007) Can soil amendments (zeolite or lime) shift the balance between nitrous oxide and dinitrogen emissions from pasture and wetland soils receiving urine or urea-N. Aust J Soil Res 45:543–553 doi:10.1071/SR07034

Zhu J, Liu G, Han Y, Zhang Y, Xing G (2003) Nitrate distribution and denitrification in the saturated zone of a paddy field under rice/wheat rotation. Chemosphere 50:725–732 doi:10.1016/S0045-6535(02)00212-6

REVIEW ARTICLE

Plant-microbe-soil interactions in the rhizosphere: an evolutionary perspective

Hans Lambers · Christophe Mougel · Benoît Jaillard · Philippe Hinsinger

Received: 29 December 2008 / Accepted: 18 May 2009 / Published online: 20 June 2009
© Springer Science + Business Media B.V. 2009

Abstract Soils are the product of the activities of plants, which supply organic matter and play a pivotal role in weathering rocks and minerals. Many plant species have a distinct ecological amplitude that shows restriction to specific soil types. In the numerous interactions between plants and soil, microorganisms also play a key role. Here we review the existing literature on interactions between plants, microorganisms and soils, and include considerations of evolutionary time scales, where possible. Some of these interactions involve intricate systems of communication, which in the case of symbioses such as the arbuscular mycorrhizal symbiosis are several hundreds of millions years old; others involve the release of exudates from roots, and other products of rhizodeposition that are used as substrates for soil microorganisms. The possible reasons for the survival value of this loss of carbon over tens or hundreds of millions of years of evolution of higher plants are discussed, taking a cost-benefit approach. Co-evolution of plants and rhizosphere microorganisms is discussed, in the light of known ecological interactions between various partners in terrestrial ecosystems. Finally, the role of higher plants, especially deep-rooted plants and associated microorganisms in the weathering of rocks and minerals, ultimately contributing to pedogenesis, is addressed. We show that rhizosphere processes in the long run are central to biogeochemical cycles, soil formation and Earth history. Major anticipated discoveries will enhance our basic understanding and allow applications of new knowledge to deal with nutrient deficiencies, pests and diseases, and the challenges of increasing global food production and agroecosystem productivity in an environmentally responsible manner.

Keywords Biogeochemistry · Evolution · Nutrient acquisition · Pedogenesis · Rhizodeposition · Rhizosphere · Symbiosis · Root exudation · Weathering

Responsible Editor: Yves Dessaux.

H. Lambers (✉)
School of Plant Biology,
The University of Western Australia,
Crawley, WA 6009, Australia
e-mail: hans.lambers@uwa.edu.au

C. Mougel
INRA, UMR 1229 Microbiologie du Sol et de l'Environnement, CMSE,
17 Rue Sully, BV 86510,
21034 Dijon, France
e-mail: mougel@dijon.inra.fr

B. Jaillard · P. Hinsinger
INRA, UMR 1222 Eco&Sols Ecologie Fonctionnelle
& Biogéochimie des Sols (INRA – IRD – SupAgro),
1 Place Pierre Viala,
34060 Montpellier, France

B. Jaillard
e-mail: jaillard@montpellier.inra.fr

P. Hinsinger
e-mail: philippe.hinsinger@supagro.inra.fr

Introduction

Most plants depend on soil, but plants and their associated microorganisms also play a crucial role in the formation or modification of soil (Pate et al. 2001; Pate and Verboom 2009; Taylor et al. 2009). Soil results from the weathering of rocks and minerals, and can have various properties according to the origin of the parent material, climate and vegetation. Soil C is predominantly derived from plants, directly or indirectly, and whilst weathering may be due to physical and chemical influences, most weathering processes involve plants, primarily roots, or microbial activities that depend on root-derived C (Raven and Edwards 2001; Beerling and Berner 2005; Taylor et al. 2009).

The focus of this review is on the long-term, evolutionary dimension of the interactions between plant roots, microorganisms and soil in the rhizosphere. We discuss their consequences for the development of terrestrial ecosystems, the functional diversity of the plant and microbial communities, biogeochemical cycles and soil formation. These interactions require communication between numerous organisms involved in rhizosphere processes. In recent years, much has been learned about these interactions, and how organisms have coevolved. We also discuss how different processes involved in plant-soil interactions dominate at different stages of primary or secondary succession.

We present a selection of major ecosystem processes that occur in the rhizosphere, but then focus on a limited number of examples in an evolutionary context, given the scarcity of information that is available in the literature. Inevitably, a substantial part of our knowledge on long-term evolutionary processes is based on information on contemporary processes, and hence the discussion on the evolutionary context remains rather speculative.

Nutrient acquisition

Next to water and temperature, nutrients are the environmental factor that most strongly constrains terrestrial plant growth. The productivity of virtually all natural ecosystems, even arid ecosystems, responds to addition of one or more nutrients, indicating widespread nutrient limitation. Mineral nutrients such as P or Fe are very reactive and strongly bound to soil particles. Their availability is generally low, especially in calcareous soils. Plant species differ greatly in their capacity to acquire nutrients from soil. Some plants are capable of acquiring Fe, P or other ions from calcareous soils, whereas others cannot extract enough nutrients to persist on such soils (Lambers et al. 2008b). Nutrient acquisition from calcareous soils involves rhizosphere processes, such as the exudation of phosphate-mobilising carboxylates (Hinsinger 2001) or the release of Fe-chelating phytosiderophores (Römheld 1987; Ma et al. 2003; Robin et al. 2008). Phytosiderophores also mobilise other micronutrients whose availability at high pH is low, e.g., Zn (Römheld 1991; Cakmak et al. 1996, and Cu (Chaignon et al. 2002; Michaud et al. 2008).

Phosphate acquisition from soils with low P concentrations in solution as well as plant growth can be enhanced by mycorrhizal symbioses (Bolan 1991; Richardson et al. 2009). However, even when P acquisition or plant growth are not enhanced in the presence of mycorrhizal fungi, the P taken up by the fungus may represent a major fraction of the total amount of P acquired by the mycorrhizal plant (Smith et al. 2003). Approximately 80% of all higher plant species can form a mycorrhizal symbiosis; of these, the arbuscular mycorrhizal (AM) association is the most common (Brundrett 2009) (Fig. 1), especially on relatively young soils (Lambers et al. 2008a). AM is also the most ancient among mycorrhizal symbioses (Table 1), the first evidence dating back to more than 400 million years ago (Remy et al. 19994; Brundrett 2002). On somewhat older soils, AM are partly replaced by ectomycorrhizas and ericoid mycorrhizas, which are considered more advanced and diverse mycorrhizal symbioses (Brundrett 2002); the latter symbioses are capable of accessing forms of both P and N (Leake and Read 1989; Cairney and Burke 1998) that are not available for AM fungi (Bolan et al. 1987; Yao et al. 2001). Mycorrhizal associations are frequently beneficial for both symbiotic partners (Table 1). Plants benefit from the fungi because these acquire nutrients, which are inaccessible for the plant because of distance from the roots, location in pores that are too small for roots to access, or, occasionally, occurrence as forms that are unavailable to plants. Conversely, fungi ensure a supply of C derived from photosynthesis by the plant (Smith and Read 2008). On extremely poor soils, when virtually all P is

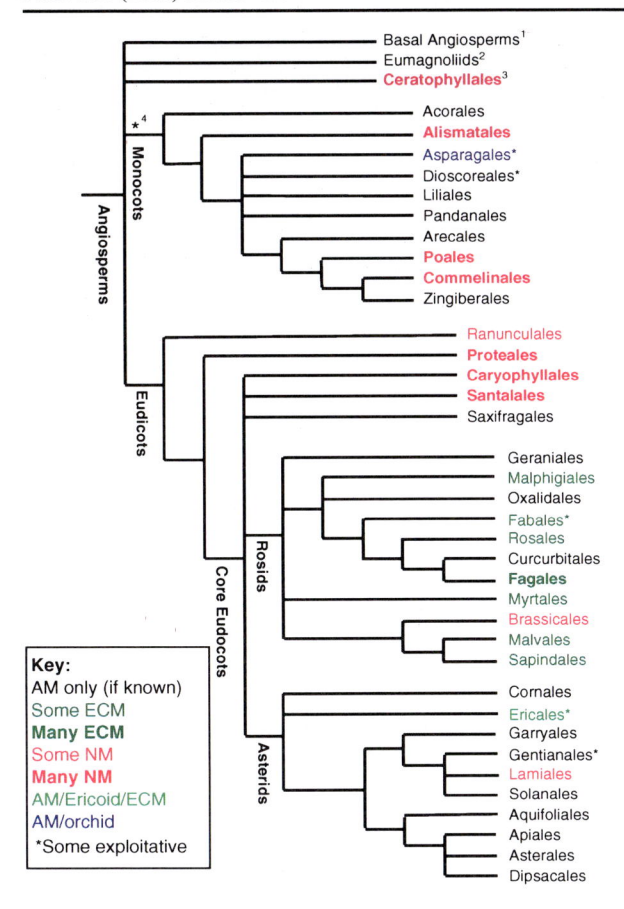

Fig. 1 Phylogenetic position of mycorrhizal lineages in a simplified Angiosperm family tree, with indications of the extent of the mycorrhizal status within each plant family (modified after Brundrett 2002, with kind permission of Blackwell Science Ltd.)

role of root clusters in these environments, and further research is warranted.

The non-mycorrhizal habit of many cluster-bearing plant species (Shane and Lambers 2005b) presents an intriguing situation from an evolutionary perspective, because ancestors of these non-mycorrhizal species were most likely all arbuscular mycorrhizal (Brundrett 2002). We know that some of the non-mycorrhizal families with root clusters, e.g., Proteaceae, are as old as early- to mid-Tertiary (Hopper and Gioia 2004), but there is no information about the time these lineages became non-mycorrhizal. Brundrett (2002) provided evidence for the view that the evolution of specialised strategies of nutrient acquisition, such as cluster roots and also new types of mycorrhizas, coincided with the origin of numerous plant families, which thereby became more competitive, especially so in certain nutrient-limited habitats. Such nutrient-acquisition mechanisms may have provided a selective advantage to those plant lineages in which these new strategies evolved, resulting in increased nutrient acquisition, albeit presumably at increased C costs.

Brundrett (2009) pointed out that cost/benefit analyses are rather complex to make, given that mycorrhizal plants remain dominant in most habitats, while a major group of non-mycorrhizal plant species is found in marginal environments, especially extremely infertile soils in the case of cluster-bearing species (Lambers et al. 2008a). Non-mycorrhizal species also occur in waterlogged, saline, dry, metal-contaminated, or cold habitats where plant productivity is low and inoculum of mycorrhizal fungi could be scarce (Brundrett 2002). Interestingly, at least one species in the Proteaceae is mycorrhizal as well as cluster-bearing, i.e. *Hakea verrucosa*, which is endemic on ultramaphic soils, which have high Ni concentrations (Boulet and Lambers 2005). The AM habit of a species endemic on soil rich in Ni and belonging to a typically non-mycorrhizal family is not unique for Proteaceae. The grassland community of a California serpentine soil includes two species from non-mycorrhizal families, *Arenaria douglasii* (Caryophyllaceae) and *Streptanthus glandulosus* (Brassicaceae) (Hopkins 1987). Boulet and Lambers (2005) speculate that massive release of carboxylic acids would mobilise not only phosphate, but also Ni, rendering this toxic for root growth. Indeed, root growth, in particular cluster-root growth, of *H. verrucosa* is at least as sensitive to Ni in nutrient

strongly sorbed onto soil particles, the 'scavenging' strategy of mycorrhizas is not effective (Parfitt 1979). On such soils, which are common in old landscapes, species with root clusters that release a range of exudates that effectively 'mine' P are prominent (Lambers et al. 2006, 2008a) (Fig. 2). Many species that produce root clusters (Fig. 3) are non-mycorrhizal, but some are capable of associating with mycorrhizal fungi as well as making clusters (Reddell et al. 1997; Lambers et al. 2006). On strongly acidic or alkaline soils, where P is bound to Al and Fe or Ca complexes, respectively, the mining strategy would also be effective. Indeed, many actinorhizal species and Cyperaceae with root clusters are common in acidic bogs (e.g., Crocker and Schwintzer 1993; Skene et al. 2000) or on calcareous dunes (e.g., Oremus and Otten 1981; Bakker et al. 2005). However, no systematic studies have focused on the

Table 1 Estimated age, evolutionary and functional categories of plant-fungus associations (after Brundrett 2002)

	Pathogen	Endophyte	AM	ECM	Ericoid	Orchid	Exploited
Association:							
Estimated age of association (million years)	> 1000	> 400	> 400	> 100	< 100	+/−100	recent
Plant provides a key habitat for fungus	+	+	+	+ / −	− ?	?	− ?
Fungus efficient at mineral nutrient acquisition from soil	−	−	+	+	+	+	− or +
Interface hyphae highly specialised	+	−	+	+	−	−	+ or −
Plant-fungus coevolution	−	−	+	+	?	−?	−
Host-fungus specificity	varies	high?	low	medium	medium	high, medium	extremely high
Ecological role:							
Mineral nutrient transfer to plant	−	−	+	+	+ ?	+	+
Energy transfer to fungus	+	+ / −	+	+	+ ?	− or + ?	−
Energy transfer to plant	−	−	− (+)	− (+)	−	+ or −	+
Plant:							
Switched to new fungus lineages			−	+	+	+	+
Recruitment of new plant lineages continues	+	+	?	+	−	−	+
Photosynthesis	+	+	+ (−)	+ (−)	+	+ (−)	−
Obligate requirement for association	−	−	+ or +/−	+	+?	+ or −	+
Fungus:							
Multiple lineages of fungi	+	+	−	+	+	+	+
Recruitment of new fungal lineages continues	+	+	−	+	+?	+	+
Obligate (host required for growth)	+	+ or − ?	+	+	?	−	−
Capable of independent growth (in axenic culture)	+ or −	+	−	+ or −	+	+	+ or −

Question marks indicate uncertain roles where further investigation is required, while brackets indicate unusual states that occasionally occur

solution as is the root growth of the congeneric *H. prostrata* (F.M. Boulet and H. Lambers, unpubl.). Given that Proteaceae, Caryophyllaceae and Brassicaceae are typically non-mycorrhizal, the AM habit of *H. verrucosa*, *A. douglasii* and *S. glandulosus* has probably evolved fairly recently, from non-mycorrhizal ancestors, but nothing is known about the time or nature of such events.

There is at least one species in the Proteaceae that is mycorrhizal without cluster roots, i.e. *Placospermum coriaceum* from the rainforest in north-eastern Australia (P. Reddell, V. Gordon and M. Webb, pers. comm.). In comparison with Proteaceae from the same rainforests (*Darlingtonia darlingiana*, *Carnarvonia* sp. and *Musgravea heterophylla*), *Placospermum coriaceum* produces relatively little dry weight at very low total soil P concentrations (< 25 mg P kg^{-1} soil), but performs similarly or better at higher P concentrations (>50 mg P kg^{-1} soil), when *Musgravea heterophylla* suffers from P toxicity (P. Reddell, V. Gordon and M. Webb, pers. comm.). Species in the genus *Persoonia* (Proteaceae) produce neither root clusters (Purnell 1960) nor mycorrhizas (Brundrett and Abbott 1991). This is puzzling, as *Persoonia* species occur on severely P-impoverished soils, suggesting an alternative root adaptation, which has yet to be discovered. Again, we can only speculate about the mycorrhizal status of the ancestors of *Persoonia* species.

Nitrogen acquisition can be enhanced greatly by symbiotic N_2 fixation, which is common in legumes, with the exception of species belonging to the less-specialised subfamily Caesalpinioideae (Vessey et al. 2005). The fact that nodulation is uncommon in this subfamily, more common in Mimosoideae and very common in Papilionoideae is in line with the idea that the three legume subfamilies evolved in the order Caesalpinioideae, Mimosoideae and Papilionoideae.

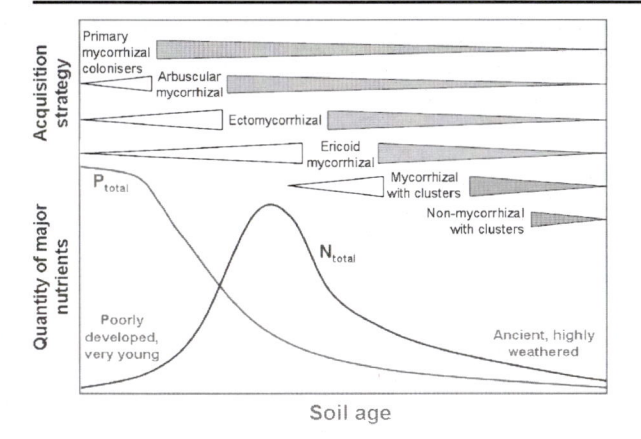

Fig. 2 Changes in total soil P and total N as a function of soil age and in plant nutrient-acquisition strategies (modified after Lambers et al. 2008a). The soil age scale spans from 'poorly developed, very young soils' (which stands for soils resulting from recent volcanic eruptions) to 'ancient, weathered soils' (i.e. soils that have been above sea level and have not been rejuvenated by glaciations over several millions of years). Some mycorrhizal species may co-occur with non-mycorrhizal cluster-bearing species in severely P-impoverished soils, but they never become dominant. The width of the triangles referring to the different ecological strategies of nutrient acquisition provides a (relative) measure of the abundance of these strategies as dependent on soil age. The total P levels in soils range from 30 to 800 mg kg^{-1}, while N levels range from <5 to 8000 mg kg^{-1}

However, as pointed out by Sprent (2007), recent results suggest that Caesalpinioideae and Papilionoideae appeared about 60 million years ago, early in the Tertiary, while Mimosoideae evolved 20 million years later. It has been hypothesised that nodulation evolved about 55 million years ago, i.e. at a time when the atmospheric CO_2 concentration had increased substantially. Nitrogen thus likely became limiting for plant growth, favouring the process of N_2 fixation, which occurs at the expense of the total C fixed by the host plant (Sprent 2007; Sprent and James 2007). Legumes form root nodules with rhizobia (Franche et al. 2009). Actinorhizal species (e.g., *Alnus*, *Casuarina*, *Myrica*) form symbiotic N_2-fixing nodules (rhizothamnia) with Actinobacteria, and cycads (e.g., *Ceratozamia*, *Macrozamia*) form N_2-fixing structures (coralloid roots) with cyanobacteria (Vessey et al. 2005; Franche et al. 2009). Outside the plant families mentioned above, symbiotic N_2 fixation is extremely rare. Symbiotic N_2 fixation is particularly important during primary succession on young soils with very low soil N levels (Walker and Syers 1976; Crews et al. 1995) as well as during secondary succession after

a fire, when a large fraction of N in biomass has been volatilised (Halliday and Pate 1976). During primary or secondary succession, soil N levels steadily increase, and the ecosystem changes from predominantly N limited to being limited increasingly by P (Parfitt et al. 2005; Lambers et al. 2008a).

Some ericoid mycorrhizas and ectomycorrhizas (ECM) belonging to groups that are incapable of symbiotic N_2 fixation may access complex organic N, including peptides and proteins (Högberg 1990). Some may even access very poorly accessible N that is locked up in protein-tannin complexes (Read 1996; Read and Perez-Moreno 2003). This capacity to access protein-tannin complexes is not universal, however (Bending and Read 1996). For the ECM

Fig. 3 Root morphology of Australian Proteaceae. Clusters of *Hakea ceratophylla*, freshly dug up in Alyson Baird reserve, Perth, Western Australia (**a**); clusters of *Hakea prostrata* plants grown in nutrient solution (**b**). Photos courtesy of Michael W. Shane (School of Plant Biology, the University of Western Australia)

association of red pine (*Pinus resinosa*) and *Pisolithus tinctorius,* N uptake into red pine occurs only in the presence of saprotrophic microorganisms (Wu et al. 2003).

Symbiotic microorganisms can obviously play a key role in accessing complex organic N, bypassing saprotrophs involved in mineralisation. However, in some mycorrhizal systems these saprotrophs play a pivotal role in making N available to the plants. The AM fungi also increase N nutrition by extending the absorptive zone due to hyphal extensions (Jonsson et al. 2001; Lerat et al. 2003) described as the 'mycorrhizosphere'. This increase in N uptake is related to the stimulation of bacteria grazing in the rhizosphere.

Predator-prey interactions in the rhizosphere are an important aspect to stimulate plant growth according to the 'microbial loop in soil'; this mechanism involves protozoa and nematodes (Clarholm 1985; Bonkowski 2004; Bonkowski et al. 2009). These interactions have recently been modelled by Raynaud et al. (2006). An increase in microbial density in the rhizosphere stimulates bacterial grazers such as protozoa. Grazing of bacteria contributes to excretion of one third of the ingested N as ammonia (Zwart et al. 1994). Ammonia contributes to plant growth directly or after its transformation by nitrifying soil microorganisms, which provide nitrate to the plant (Bonkowski et al. 2001).

The presence of protozoa decreases the length of fungal hyphae by 18%, while the presence of ectomycorrhizas leads to a reduction in numbers of bacterial and protozoan grazers (38% and 34%, respectively) (Bonkowski et al. 2001). These results point to a significant trade-off in C allocation between bacterial and fungal communities in the rhizosphere due to mutualistic root symbionts that act as a sink for C (Vance and Heichel 1991; Bago et al. 2000; Johnson et al. 2002). Despite this conflict in C allocation, a synergistic microbial effect enhances uptake of N (17%) and P (55%) in the combined association of mycorrhizas and protozoa where hyphal networks increase N uptake due to protozoa-mobilised N (Bonkowski et al. 2001). Symbiotic N_2 fixation is typically very important in young landscapes and during secondary succession, as discussed above; conversely, the role of mycorrhizas capable of accessing complex nitrogenous compounds, including protein-tannin complexes tends to become more important on older soils (Lambers et al. 2008a).

Molecular communication

Communication in the rhizosphere is pivotal for many interactions and their corresponding evolution. Several molecular signalling pathways, particularly in bacteria, have been well characterised, highlighting the ability of bacteria to 'sense' plant-derived compounds. Bacteria are also able to modulate plant functioning by producing phytohormone-like molecules. In addition, severe competition between microorganisms in the rhizosphere involves specific communication between microorganisms, including quorum-sensing and complex mechanisms that modulate it (Faure et al. 2009). Symbiotic associations as discussed above require intricate signalling between the macrosymbiont (host plant) and the microsymbionts (guest microorganisms). This signalling is best understood for the legume-rhizobium symbiosis, where flavonoids and betaines are released by legume roots (Vessey et al. 2005; Franche et al. 2009). However, other signalling molecules are also involved in the legume-rhizobium interactions that lead to a symbiosis (Faure et al. 2009). The signalling molecules released from roots bind with a bacterial gene product, and then interact with a specific promoter in the genome of rhizobium. This promoter is associated with the genes responsible for inducing nodulation (the nodulation, or *nod* genes). The products of these genes (Nod factors) induce root-hair curling on the plant and cortical cell divisions, which are among the earliest, microscopically observable events in the nodulation of most legume species. When rhizobia adhere to root hairs, the cell wall of the affected root hair is partly hydrolysed at the tip, allowing the bacteria to enter. An infection thread is formed by invagination of the cell wall (Gage and Margolin 2000). Alternatively, rhizobia may enter through cracks in the epidermis, associated with lateral-root formation, or wounds (Sprent 2007). However, also in that event, effective signalling must occur for a nodule to emerge.

The actinorhizal symbiosis between plant species like *Alnus glutinosa* and the Actinobacteria *Frankia* also involves the release of specific compounds (flavonols) that enhance the level of nodulation, but their exact role in the process is yet unknown (Van Ghelue et al. 1997, Vessey et al. 2005). Very little is known about the chemical nature of attractants from hosts to cyanobacteria (Vessey et al. 2005).

Major progress is being made on signalling between roots and AM fungi, highlighting the importance of strigolactones (Akiyama et al. 2005, Paszkowski 2006). Strigolactones are a group of sesquiterpene lactones, previously isolated as seed-germination stimulants for some parasitic weeds (Yoneyama et al. 2008). Strigolactones induce extensive hyphal branching in germinating spores of the AM fungus *Gigaspora margarita* at very low concentrations (Bouwmeester et al. 2007). In AM plants, strigolactone release is greatly enhanced under phosphate starvation (López-Ráez et al. 2008; Yoneyama et al. 2008). Isolation and identification of plant symbiotic signals open up new ways for studying the molecular basis of plant-AM fungus interactions. This discovery also provides a clear answer to a long-standing question on the evolutionary origin of the release from host roots of signalling molecules that stimulate seed germination in parasitic plants (Akiyama and Hayashi 2006). Signalling between host and fungus also plays a role in mycorrhizal associations other than AM, but we know much less about this (Martin et al. 2001).

Plants not only communicate via the release of root exudates with symbiotic microorganisms and parasitic higher plants, but also with roots of neighbouring plants (Bais et al. 2004). Some of the released exudates have positive effects, e.g., Fe-mobilising phytosiderophores and phosphate-mobilising carboxylates, as discussed above. These positive effects may lead to *facilitation*, i.e. amelioration of the environment of neighbouring plants (Lambers et al. 2008b), as recently modelled by Raynaud et al. (2008). Other exudates have a negative effect on growth of neighbours. If the roots of the neighbouring plants are of a different species, these effects are called *allelopathic*; if the roots are of plants of the same species or even the same plant, the effects are called *autotoxic* (Lambers et al. 2008b).

Allelopathy or interference competition is hard to demonstrate in nature. One of the more convincing studies was done on *Centaurea maculosa*, a Eurasian species that has become an invasive weed in North America (Ridenour and Callaway 2001). Using activated carbon, which adsorbs allelochemicals, it can be demonstrated that the allelochemical released from the roots of *C. maculosa*, (±)-catechin, inhibits root growth of neighbouring grasses such as *Festuca idahoensis*. Somewhat surprisingly, (±)-catechin concentrations in bulk soil are rarely very high, suggesting that the effect of (±)-catechin is exhibited only when roots are in close proximity and casting doubt on the claim that this allelochemical accounts for the invasive nature of *C. maculosa* (Perry et al. 2007). Some species, e.g., *Lupinus sericeus* and *Gaillardia grandiflora*, are resistant to (±)-catechin, because they exude increased amounts of oxalate upon exposure to (±)-catechin. Oxalate blocks (±)-catechin-triggered generation of reactive oxygen species and thus reduces oxidative damage (Weir et al. 2006). Another allelochemical (3,4,5-trihydroxybenzoic acid or gallic acid) that gives rise to the formation of elevated levels of reactive oxygen species in roots of neighbouring plants is released by another invasive weed, *Phragmites australis*. Gallic acid leads to acute rhizotoxicity. Unlike (±)-catechin, gallic acid is persistent in *P. australis*'s rhizosphere (Rudrappa et al. 2007).

Autotoxicity can lead to soil sickness, i.e. gradual declines in crop yield in the absence of crop rotations (Lambers et al. 2008b). It is quite common in cucurbit crops, e.g., *Citrullus lanatus*, *Cucumis melo* and *C. sativus* (Yu et al. 2000). Cinnamic acid is one of the autotoxic compounds involved; again, it induces formation of reactive oxygen species (Ding et al. 2007). Autotoxicity may also be involved in responding to obstructions in soil. Semchenko et al. (2008) found that grass species characteristic of nutrient-poor habitats restrict placement of their root mass in substrate containing obstructions. This response disappears in the presence of activated carbon, indicating the effect of inhibitory chemicals.

Upon attack of roots of *Thuja occidentalis* by larvae of *Otiorhynchus sulcatus* (a weevil), these roots release chemicals and thus attract *Heterorhabditis megidis*, a parasitic nematode, which preys on the weevil larvae (Van Tol et al. 2001). Similar below-ground tritrophic interactions occur in *Zea mays*. Upon attack by beetle larvae, their roots release a sesquiterpene, (E)-β-caryophyllene, which attracts entomopathogenic nematodes and increases the effectiveness of these nematodes in finding and killing herbivore larvae (Rasmann et al. 2005; Turlings and Ton 2006). Our rapidly increasing knowledge of below-ground tritrophic interactions is expected to provide opportunities for applications in plant management systems, similar to those existing for above-ground tritrophic interactions (Turlings and Wäckers 2004). From an evolutionary point of view, little is known, however, about when such complex interactions first appeared.

Rhizodeposition

Rhizodeposition is the release of C compounds from living plant roots into the surrounding soil; it is a ubiquitous phenomenon (Jones et al. 2004, 2009). The loss of C from root epidermal and cortical cells leads to a proliferation of microorganisms inside (endophytes), on the surface and outside the root. Rhizodeposition results in different chemical, physical and biological characteristics in the rhizosphere compared with those of the bulk soil. The magnitude of these changes is determined by the amount and type of C released from the root, as well as intrinsic soil characteristics. Rhizodeposition basically results from two different processes: (1) leakage of compounds over which the plant exerts little control; (2) exudation of specific compounds with a specific function and over which the plant exerts control (Jones 1998). Leakage of compounds as defined here represents a minor component of a plant's C budget, less than 5% of all C daily fixed in photosynthesis (Lambers et al. 2008b). Higher values cited in the literature probably include C released by either root respiration or from dying root cells (Lambers 1987). Jones et al. (2009) provide quantitative data derived from meta-analyses of published datasets in their recent review.

Plant roots may release massive amounts of organic compounds via rhizodeposition

Rates of exudation *sensu stricto* vary widely among species and environmental conditions. At neutral pH and optimum P supply, rates tend to be low. In Al-resistant species, carboxylate-exudation rates tend to increase at low pH, in the presence of Al (Kochian et al. 2005). Dependent on species, the predominant carboxylates are malate (Delhaize et al. 1993) or citrate (Miyasaka et al. 1991), and in a few plant species, oxalate (Zheng et al. 1998). Cultivars of rice (*Oryza sativa*) that are resistant to Pb release oxalate when exposed to Pb (Yang et al. 2000). Similarly, roots of *Triticum aestivum* release malate and citrate when exposed to Cu (Clemens et al. 2002). The role of the released carboxylates is to chelate Al and other metals, and thus prevent harmful effects of the toxic metal, when occurring as free metal cationic species.

Iron-efficient grasses release phytosiderophores when their growth is limited by the availability of Fe (Ma 2005; Robin et al. 2008). The phytosiderophores are released from the root tips only (Marschner et al. 1987), predominantly during the early morning (Takagi et al. 1984; Ma and Nomoto 1994). Robin et al. (2008) stress that this is an efficient strategy to maximise the positive impact on Fe acquisition by minimising the breakdown of phytosiderophores by rhizosphere microorganisms, in agreement with model calculations of Darrah (1991). Neighbouring Fe-inefficient plants may benefit from these released phytosiderophores (facilitation), and this knowledge can be applied in intercropping Fe-efficient crops with calcifuge ones, e.g., maize (*Zea mays*) with peanuts (*Arachis hypogaea*) (Zuo et al. 2000) or red fescue (*Festuca rubra*) with fruit trees (Ma et al. 2003).

Many plants enhance their rate of carboxylate exudation when their P supply is severely limiting (Vance et al. 2003). Massive exudation rates are exhibited by species that produce root clusters at very low P supply (Watt and Evans 1999; Shane and Lambers 2005b). Root-cluster-bearing species, e.g., *Lupinus albus* (Fabaceae; Watt and Evans 1999), *Hakea prostrata* (Proteaceae; Shane et al. 2004) and *Schoenus unispiculatus* (Cyperaceae; Shane et al. 2006) release carboxylates in an exudative burst. This ensures mobilisation of P before microorganisms have an opportunity to decompose the released exudates, according to model calculations of Darrah (1991). Moreover, root clusters of *Lupinus albus* drastically reduce the cluster-root rhizosphere pH, thus inhibiting bacterial activity; they also release phenolics, which induce fungal sporulation, as well as chitinases and glucanases, which degrade fungal cell walls, prior to the exudative burst (Weisskopf et al. 2006). This complex strategy ensures minimal degradation and maximum efficiency of exuded carboxylates to mobilise scarcely available P and micronutrients.

Whilst rhizodeposition incurs a loss of C to the plant, there are obviously also major benefits. First, there is the signalling to microsymbionts, as discussed above; this only incurs a minor C cost. Second, exudation *sensu stricto* can have a major impact on nutrient acquisition. Phytosiderophores and other chelating agents play a pivotal role in acquiring Fe and other micronutrients, especially from calcareous soil (Römheld 1987; Ma 2005; Robin et al. 2008). They incur a relatively small C cost, because release rates are relatively low and restricted in space (from root tips only) and time (in the early morning only) (Cakmak et al. 1996). Carbon costs associated with carboxylate release to chelate and detoxify Al or other metals are also relatively small, as only root tips appear to be involved in this process (Delhaize and Ryan 1995).

Carbon costs associated with carboxylate exudation to enhance P acquisition are substantial, since the carboxylates must be exuded in quantities sufficient to solubilise sorbed P (Lambers et al. 2006). Given that root clusters are short-lived, i.e. approx. three weeks in *Lupinus albus* (Watt and Evans 1999) and *Hakea prostrata* (Shane et al. 2004), and less than two weeks in *Schoenus unispiculatus* (Shane et al. 2006), this strategy is probably very costly in terms of C requirements. Solid data on the C costs of root clusters, including the costs of the carboxylates they release, are unavailable. Estimates of the costs of the production, respiration and carboxylate release of cluster roots of *Hakea prostrata* are as high as 50–100% of daily rates of photosynthesis. However, it should be borne in mind that *Hakea prostrata*, like other cluster-bearing plants in Mediterranean environments, produces root clusters only during the wet winter months, so that the C costs on an annual basis are probably four times less (Lambers et al. 2006).

The cost of rhizodeposition is balanced by benefits provided by microorganisms

Plants live in association with a rich diversity of microorganisms during their entire development. These associations correspond to a gradient of plant-microbe interactions from mutualism to parasitism, even in the case of symbionts such as mycorrhizal fungi (Johnson et al. 1997) (Fig. 4, Table 1). These interactions involve complex trophic relations between plants and soil microorganisms. These interactions, called the rhizosphere effect (Hartmann et al. 2008), correspond to the stimulation of microbial community activities in the rhizosphere compared with those in the bulk soil due to C released by roots, i.e. rhizodeposition.

Whilst some plants release exudates in such a way as to minimise rapid breakdown by microorganisms, as discussed above for root-cluster-bearing species and grasses releasing phytosiderophores, plants may

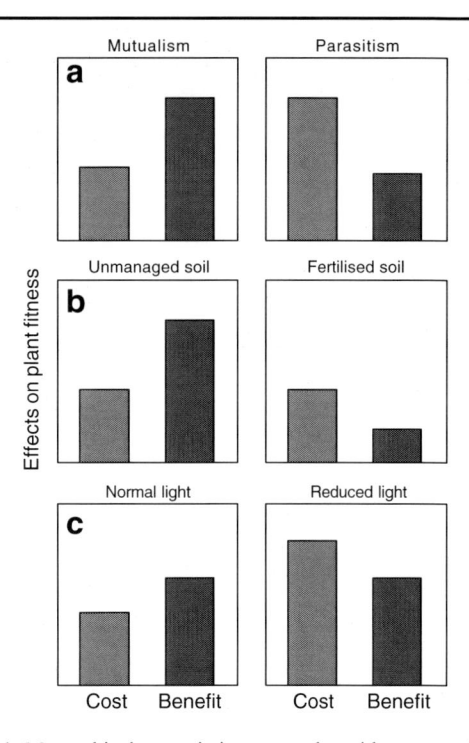

Fig. 4 Mycorrhizal associations can be either mutualistic, when the benefit of the association for the plant exceeds the cost, or parasitic when costs exceed benefits (**a**) (reproduced after Johnson et al. 1997, with kind permission from Blackwell Science Ltd.). The cost-benefit relationship varies with environmental conditions where the symbiosis occurs; upon fertilisation, the growth benefit of the mycorrhizal association decreases, and the cost may outweigh the benefit, ultimately leading to a parasitic association (**b**). Similarly, a low light intensity may increase the cost of photosynthate production above the benefit from the symbiotic association, thus also resulting in a parasitic association (**c**)

also stimulate the growth of rhizosphere organisms. Some of these microbial-growth promoting effects are clearly beneficial for plant growth, e.g., by acting as antagonists for deleterious microorganisms (Schippers et al. 1987; Raaijmakers et al. 2009), by releasing plant growth substances (Dobbelaere et al. 1999; Faure et al. 2009) or promoting non-symbiotic N_2-fixing microorganisms (Döbereiner and Pedrosa 1987; Saubidet et al. 2002; Richardson et al. 2009). However, the potential of associative N_2-fixing bacteria to promote the growth of cereals and grasses by enhancing the availability of combined N is generally limited. Positive effects of inoculation with these bacteria on plant growth are due to, e.g., synthesis of phytohormones and vitamins, and inhibition of plant ethylene synthesis (Richardson et al. 2009). These bacteria may also be able to decrease the deleterious effects of pathogenic microorganisms, mostly through the synthesis of antibiotics and/or fungicidal compounds (Dobbelaere et al. 2003).

From a plant perspective, C flow into the rhizosphere can be considered as a cost for the plant, because a significant amount of C does not contribute to dry matter production. Thus, maintaining rhizodeposition during plant evolution should be balanced by expected benefits.

In natural terrestrial ecosystems (i.e. without major human impact, such as agroecosystems), mineral N can be a limiting factor for plant growth (Vitousek and Howarth 1991) and plants maintain interactions with soil microorganisms to enhance N nutrition. Non-specific interactions include the stimulation of microbial activities in the rhizosphere providing the essential energy to degrade soil organic matter and to remobilise some of the N immobilised in soil organic matter (Hodge et al. 2000). This stimulation of the mineralisation of soil organic matter following fresh organic matter inputs corresponds to the so-called 'priming effect' (Kuzyakov et al. 2000; Fontaine et al. 2003; Paterson 2003). The role of microorganisms in such priming effect is considered to be due to two separate mechanisms, depending on the species strategy (r or K) (Fontaine et al. 2003). (1) An increase in the release of extracellular enzymes by r-strategists; these enzymes are directed to fresh organic matter depolymerisation, but they can also degrade soil organic matter (mechanism 1). (2) The stimulation of K-strategists by metabolites released during the partial degradation of either fresh or soil organic matter by r-strategists, ultimately resulting in an enhanced mineralisation of the soil organic matter by enzymes produced by K-strategists (mechanism 2, implying co-metabolism between r-strategists and K-strategists). The priming effect in the rhizosphere is a direct consequence of rhizodeposition (Kuzyakov et al. 2000).

Exudates may enhance the activity of phosphate-solubilising bacteria, and hence increase the plant's P supply (Richardson 2001). Although the ability to solubilise phosphate is common in isolates from the rhizosphere, the application of these organisms as plant inoculants varies in its effectiveness (Kucey et al. 1989; Richardson et al. 2009). Inoculation of seeds of chickpea (*Cicer arietinum*) with phosphate-solubilising bacteria on agar plates, followed by growth in pots with sterilised vermiculite significantly enhances plant growth and shoot P concentrations (Gull et al. 2004). Trials with wheat (*Triticum aestivum*) in pots with unsterilised soil showed that bacterial strains that are good at solubilising phosphate in laboratory tests result in increased grain yield and grain P content in the presence or absence of applied dicalcium phosphate (Harris et al. 2006). We are still a long way off application of phosphate-solubilising bacteria under real-world field conditions.

In addition to microorganisms that depend on the release of C in the soil surrounding the roots, a number of microorganisms occur in what is considered the inner rhizosphere, e.g., endophytes living in the root apoplast. These endophytes get first served upon the exudation of C from the surrounding root cells. Recent results by Deshmukh et al. (2006) show that some fungal endophytes, e.g., *Piriformospora indica* are beneficial to host crop plants (e.g., barley, *Hordeum vulgare*) via induced local and systemic resistance to fungal diseases and by increasing resistance to abiotic stress. These endophytes require host-cell death for proliferation in rhizodermal and cortical cells. Deshmukh et al. (2006) suggest that this endophyte interferes with the host cell-death program to form a mutualistic interaction with the plant host. The prevailing view of endophytic, symbiotic fungi is that selection pressures have imposed a directionality on the evolution of symbiotic lifestyles (Saikkonen et al. 1998). Accordingly, mutualistic endophytes would be expected to have evolved from parasitic or pathogenic fungi. However, some pathogenic *Colletotrichum* spp. have the ability to express different symbiotic lifestyles based on host genotypes in a single geographical location. A single fungal isolate may be pathogenic in some plant species, but provide mutualistic benefits such as disease resistance, drought tolerance, and growth enhancement to other plant species (Redman et al. 2001; Rodriguez and Redman 2008). According to these authors, either *Colletotrichum* spp. have evolved to possess maximum symbiotic flexibility or directional evolution has occurred in a host genotype-specific manner. Such observations warrant a re-evaluation of hypotheses describing the evolution, ecology, and dynamics of fungal and plant communities.

The best-documented case of endophytic fungi for which cost-benefit approaches have been developed is that of mycorrhizal fungi. The cost of such symbioses is especially well documented, the host plant delivering from 4 to 20% of the total C assimilated by photosynthesis. This means a significant cost in all cases (Peng et al. 1993; Rygiewicz and Andersen 1994; Tinker et al. 1994; Watkins et al. 1996). It is slightly more difficult to estimate the benefits for the plant which include increased access to poorly mobile soil resources such as P or micronutrients, if these nutrients are limiting growth. Resource limitation is indeed a key component of cost-benefit analyses of the mycorrhizal symbiosis (Johnson et al. 1997). When P is limiting, increasing the volume of soil from which P can be depleted (i.e. the volume of the rhizosphere) can be achieved at minimal extra C cost via root hairs. Conversely, mycorrhizal hyphae are likely more efficient, but also less cost-effective (Lynch and Ho 2005). Peng et al. (1993) showed that at high, non-limiting P supply, the growth of *Citrus* seedlings is significantly decreased as a result of mycorrhizal infection. Although many mycorrhiza researchers dislike describing some mycorrhizal interactions as parasitic, it is technically an accurate description of cases where the fungus is detrimental to the plant (Johnson et al. 1997).

From a microbial perspective, the rhizosphere is a niche of high microbial activity, due to the release of rhizodeposits, compared with the bulk soil, which is considered an oligotrophic environment limited by C. This leads to important modifications in the genetic structure and diversity of bacterial and fungal communities resulting from the competition between soil microorganisms for nutrients provided by plants (e.g., Mougel et al. 2006). Using different *Arabidopsis thaliana* accessions, Micallef et al. (2009) recently demonstrated that modifications of soil microbial communities are plant genotype-specific. This may offer the possibility to manage microbial communities in the rhizosphere.

As the flux of rhizodeposits corresponds to efflux and re-uptake of C, the plant plays an important role in the control of the soil food web, and also in the molecular control points for the co-evolution of plants and rhizosphere organisms. When considering the effect of C allocation to microbial populations, it is worth separating the effect on specific mutualistic and parasitic populations (endophytic life style) and on other components of microbial communities (rhizospheric life style). This separation is especially relevant when considering co-evolution (see section below), which correspond to evolutionary forces that happen when different species have close ecological interactions. Additional research at the microbial community level is required to further our understanding of why high rates of rhizodeposition have been maintained during plant evolution.

Mutualistic and parasitic interactions

The term coevolution is used to describe cases where two (or more) species reciprocally affect each other's evolution. Coevolution is likely to happen when different species have close ecological interactions with one another. These ecological relationships include a number of interactions that occur in the rhizosphere, involving parasite/host, competitive species or mutualistic species. The most obvious and best-studied of these plant-microbe interactions are those with symbiotic mutualists and parasites. The ecological relationships, including mutualistic symbionts, can be classified into two broad types: protectors against pathogens and providers of nutrients, e.g., mycorrhizal fungi and N_2-fixing microorganisms (Reynolds et al. 2003). Spatial variation in the interactions between parasites (and symbiotic mutualists) and their hosts is thought to be a major force in coevolution and in generating biodiversity (Thompson 1999). Due to their faster generation time, microorganisms evolve faster than their hosts. A greater evolutionary potential of parasites and symbiotic mutualists leads to the general prediction that they should be locally adapted to their hosts, and that their fitness should decrease with distance of the host population (Ebert 1994, 1998).

Plants interact with two major types of mutualistic soil microorganisms: mycorrhizal fungi and N_2-fixing microorganisms. The AM association represents an ancient symbiosis (Table 1), with fossil evidence dating back 400 million years (Remy et al. 1994; Brundrett 2002) which has led to the idea that AM have contributed to the colonisation of terrestrial ecosystems by early land plants (Redecker et al. 2000) (Fig. 5). The coevolution of mycorrhizal fungi and roots is now well established in the light of evidence from palaeobotanical, morphological studies and DNA-based phylogenies (Brundrett 2002). Different types of mycorrhizas are recognised, the most common being arbuscular mycorrhizas, in which the fungi belong to the Glomeromycota (Schüßler et al.

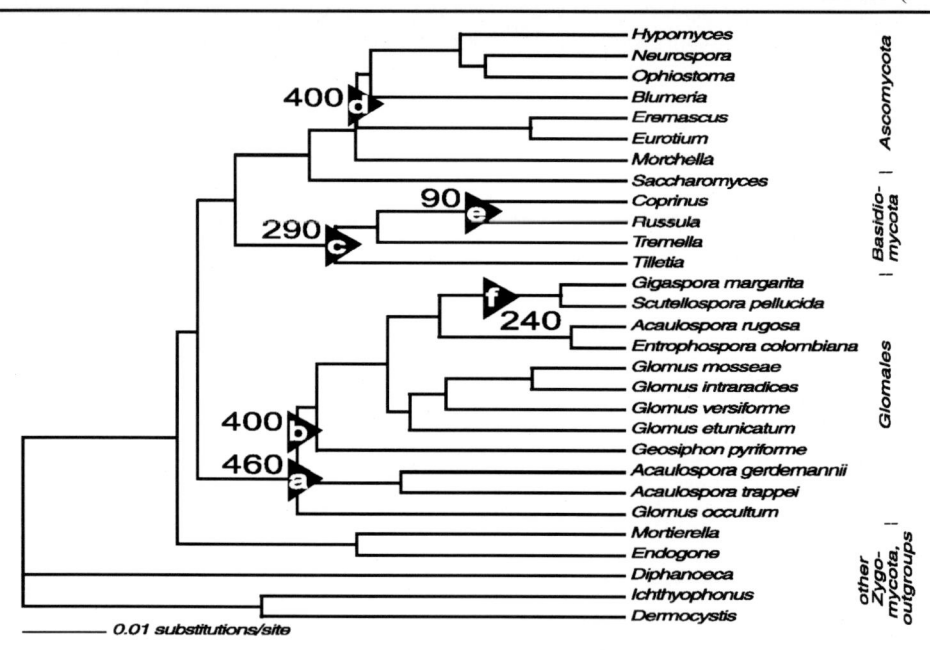

Fig. 5 Phylogenetic tree of fungi as constructed based on small subunit ribosomal sequences, with tree branching points referring to fungal fossils (letters *a* to *e*) and respective geological times indicated as numbers on the tree. From Redecker et al. (2000), Glomalean fungi from the Ordovician. Science 289:1920–1921. Reprinted with permission from AAAS. The *triangles* indicate that all fossils could also have been deposited later in the history of each clade, allowing the origins of the clades to be shifted back in time. Point *a* stands for the glomalean fossils reported in Redecker et al. (2000); point *c* stands for the fossil arbuscular mycorrhizas from the Rhynie Chert; point *c* indicates a fossil clamp connection (earliest evidence of Basidiomycota); point *d* indicates the Ascomycota from the Rhynie Chert; point *e* stands for the gilled mushroom in amber; point *f* indicates the arbuscular mycorrhizas of the Gigasporaceae type from the Triassic

2001), occurring in 92% of land plant families (Wang and Qiu 2006; Brundrett 2009), and the ectomycorrhizas, which occur in certain families of gymnosperms, dicotyledons and one monocotyledon genus (Brundrett 2002). There is a strong relationship between the age of plant-fungus associations and the degree of dependence of mycorrhizal fungi on their hosts, as all AM (about 400 million years old) and some ECM (about 100 million years old) are incapable of independent growth (Table 1).

Brundrett (2002) proposed the following order of evolution:

(1) Endophytic associations that could be considered as the source of new plant-fungus associations. The first stage of evolution from endophytic to mycorrhizal fungus would be specialisation of both partners; at this stage benefits to the plant would be limited.

(2) Mutualistic associations, where the exchange processes evolved to access limiting resources (mineral nutrients for the plant and C compounds for the mycorrhizal fungi). It is possible that the first mycorrhizas were formed by a Geosiphon-like fungus with an abundant supply of N obtained from associated cyanobacteria (Schüßler and Kluge 2000). In this context, both AM and ECM evolved as a complex interface with active exchange of limited duration. Increasing control of associations by the host along with increasing interface complexity are the strongest evolutionary trends.

(3) The third proposed stage of evolution consists of myco-heterotrophic plants where the fungi apparently do not benefit from the association with myco-heterotrophic plants, but where plants get C from their mycorrhizal associate, as occurs in orchids.

The Glomeromycota are a single unique ancient lineage, in contrast with other fungi that have multiple origins and coevolved with plants much more rapidly (Redecker et al. 2000; Schüßler et al. 2001). Some plants may continue to acquire new fungal lineages as for ECM where rapid diversification of these fungi continues to this day (Hibbert et al. 2000). Most lineages of mycorrhizal fungi have descended from saprophytes with enzymes that can penetrate plant cell walls that presumably first became endophytes after attraction to roots by exudates (Kohzu et al. 1999). New types of mycorrhizas do not always result from the adoption of new lineages of fungi, as some mycoheterotrophs exploit ECM fungi or saprophytes. Orchid mycorrhizal fungi may not benefit from associations with orchids and thus would not coevolve with plants, or form separate lineages from their saprophytic or parasitic relatives. This is in line with the prevailing view of a directionality of the evolution of symbiotic lifestyles, in which mutualistic endophytes are expected to have evolved from parasitic or pathogenic fungi (Saikkonen et al. 1998). However, as discussed above, this view is currently challenged by the discovery of fungal endophytes that exhibit a whole array of lifestyles in a given environment, as a function of the host plant species that is colonised, the latter being responsible for the control of the fungal lifestyle (Redman et al. 2001).

N_2-fixing symbiotic associations include symbioses between Fabaceae and rhizobia, actinorhizal plants and *Frankia*, *Parasponia* sp. and rhizobia, and cycads and cyanobacteria. This cyanobacterial association is possibly the most ancient and could have evolved 250 million years ago (Raven 2002). Cycads represent an ancient life form and are the most primitive seed-plants, dating back to the mid-late Devonian, 380-360 million years ago (Brenner et al. 2003). They develop symbiotic associations with cyanobacteria, which evolved 3 billion years ago (Schopf et al. 2002), suggesting that competent cyanobacteria may have evolved well before the cycads did. However, compared with other symbioses, we know very little about this primitive form and how it has evolved. There is virtually no direct evidence for the origin and evolution of land plant/cyanobacteria from fossil records. Krings et al. (2009) recently presented evidence for a filamentous cyanobacterium associated with mycorrhizal axes of an Early Devonian land plant, *Aglaophyton major*. The cyanobacteria enter the plant through the stomata and colonise the substomatal cavity and intercellular spaces in the outer cortex. This is the earliest direct evidence for cyanobacterial associations with land plants.

In other symbiotic associations, the result of the plant-microbe interactions involves the formation of specific organs, called nodules (in legumes) or rhizothamnia (in actinorhizal species). Studies based on the chloroplast gene *rbcL* suggest that the flowering plant families involved in rhizobial or actinorhizal symbioses belong to the same large lineage, suggesting that a predisposition for forming nodules with rhizobia or rhizothamnia with actinorhizal bacteria has originated only once during the evolution flowering plant (Soltis et al. 1995). Nevertheless, phylogenetic studies based on *rbcL* sequences indicate that the capacity to nodulate appeared at different times during plant evolution (Doyle et al. 1997). Genes involved in nodulation belong to multigene families with a complex evolutionary pattern as illustrated by leghaemoglobin genes (Clegg et al. 1997). In *Vicia*, the most distantly related member of the globin gene is also induced in a mycorrhizal symbiosis, suggesting a connection between the two symbioses (Frühling et al. 1997). Origins of the legume, *Parasponia* and actinorhizal symbioses are considerably more recent than the cycad-cyanobacteria symbiosis, given that angiosperms did not evolve until 250 to 150 million years ago (Sprent and Raven 1992). Comparison of phylogenetic trees using representative *Frankia* (based on 16 S rDNA and *nifH* sequences) and actinorhizal plants (based on *rbcL* sequences) suggest some degree of coevolution (Jeong et al. 1997). Concerning the Fabaceae and rhizobia association, the coevolution process is less clear, as shown in the *Medicago-Sinorhizobium* association, despite the fact that cospeciation in *Sinorhizobium* and *Medicago* occurred (Béna et al. 2005). This result is not surprising, because coevolution is rare in associations in which the microsymbiont has a free-living stage (Ronquist 1998). Aguilar et al. (2004) suggested coevolution of *Phaseolus vulgaris* and *Rhizobium etli* in the centres of host plant genetic diversification, stressing the necessity to take into account the geographical pattern of local co-adaptation (Thompson 1997). More recently, in *Medicago truncatula* the ability to develop symbioses with both AM fungi and rhizobia involves some common genes (e.g., *DMI1*), pointing towards a common evolutionary pathway in the ability in *Medicago truncatula* to develop symbiotic associations in the early stages of the interaction (Ané et al. 2004). Cospeciation of symbiotic partners involves stabilising mechanism (sanctions from plant partner) to maintain such association during evolution (Sachs et al. 2004). Theory predicts that mutualisms may be evolutionary unstable because of higher fitness of 'cheater' genotypes (receiving greater fitness benefit than they confer) (Denison 2000). The presence of cheaters in rhizobium genotypes appears to be due to the symbiotic and saprophytic lifestyle of rhizobium genotype corresponding to horizontal transmission in which evolution of rhizobia could be influenced by plant functioning (in nodules) or the soil environment. Some evidence for adaptive partner choice, which might stabilise mutualims against the invasion of cheaters was reported in *Medicago truncatula*-rhizobium mutualism. These results stress the role of plants in the rhizobium genotype partner choice from soil diversity to reward more cooperative rhizobium strains (Heath and Tiffin 2007, 2009).

Parasitic associations between higher plants and microorganisms involve two independent sequences of infection processes corresponding to infectivity (infection success) and virulence (host damage). In plant pathology, virulence is synonymous with infectivity in accordance with classic gene-for-gene interactions where specific pathogen gene products trigger resistance reactions in the plant. Gene-for-gene is an important genetic system of interaction in coevolution between airborne parasites and plants (Thompson and Burdon 1992). In soil-borne parasites, mechanism of plant resistance or tolerance of parasites often involves complex genetic determinants, i.e. polygenic rather than monogenic resistance as for air-borne parasites.

Less is known about the role of plant genes involved in resistance or tolerance. In some cases resistance to soil-borne diseases is monogenic and dominant, as well as conferred through polygenic effects depending upon the physiological race of the microbial pathogen and the host plant genotype examined (Wissuwa et al. 2009). An interesting specificity in microbe-microbe, antagonistic interaction in the rhizosphere is the interactions of pathogens with beneficial microorganisms able to suppress disease with mechanisms like antibiosis, i.e. involving the production of antibiotics targeted against microbial pathogens, or competition for nutrients (Raaijmakers et al. 2009). Microbes that produce antibiotics are readily isolated from natural disease-suppressive soils, e.g., antibiotics against take-all of wheat (Weller 1988), black root rot (Keel et al. 1996), or *Fusarium* wilt (Tamietti et al. 1993). In addition, competition for C and Fe are also responsible for tolerance to *Fusarium* wilt in disease-suppressive soils (Lemanceau and Alabouvette 1993). Microbial antagonism against parasites decreases the frequency of root infections and the severity of the disease in plants, whereas antagonism against deleterious microorganisms enhances plant growth (Schippers et al. 1987). Plant growth promotion can also be due to metabolites affecting the plant's physiology, e.g., growth substances (Glick 1995; Richardson et al. 2009). Specific bacterial metabolites may induce systemic resistance by eliciting defence reactions of the host plant (Van Loon et al. 1998). Much less is known about these beneficial plant-microbe interactions and in particular the plant genetic determinants to support or enhance them. Using a quantitative genetic analysis, some quantitative trait loci of tomato were identified in the interaction of a biocontrol strain (*Bacillus cereus*) and a parasite (*Pythium torulosum*). The corresponding traits identified were associated with the resistance to *Pythium torulosum*, growth parameters of the biocontrol strain, seedling emergence and microbial biocontrol (Smith et al. 1999). There is some evidence that modern breeding efforts in crop plants have inadvertently selected against hosting such beneficial microorganisms as documented for the AM symbiosis (Hetrick et al. 1995; Wissuwa et al. 2009). Further research is needed to explore possible coevolution between host plants and particular microbial populations or communities that are mutualistic but not symbiotic.

Soil formation, biogeochemical cycling and Earth history

Given the many ways roots and rhizosphere microorganisms interact with soil properties, one may ask: what would a soil be like in the absence of plants? More to the point: would there be any soil in the absence of plants? Studying the first steps of colonisation by plants of bare rocks (such as glacial moraines or basalt flows) provides some answers to these questions (Cochran and Berner 1996). While rock-lichen interactions have received more interest as a model of biological weathering (Drever 1994; Banfield et al. 1999), Cochran and Berner (1996) and Moulton et al. (2000) stress that much higher weathering rates are achieved once the basalt rock had been colonised by higher plants, compared with that of bare rock or rock covered with lichens. There have also been experimental approaches to show that roots and their associated microorganisms can considerably increase the rate of weathering of rocks (Bormann et al. 1998; Hinsinger et al. 2001) or minerals (Hinsinger et al. 1993; Bakker et al. 2004; Calvaruso et al. 2006). Based on such observations and measurements, it is concluded that higher plants play a pivotal role in soil formation in the root/rock interface, i.e. the very first stage of a rhizosphere can be considered as the site where soil formation actually starts.

From an evolutionary perspective, the emergence of vascular plants, and, amongst these, of higher plants with deep and mycorrhizal root systems, is considered a major process in the Earth history and the biogeochemical cycles of CO_2 (and O_2) via their impact on rock weathering and soil formation in terrestrial ecosystems (Berner 1992, 1997; Kenrick and Crane 1997; Retallack 1997; Beerling et al. 2001; Beerling and Berner 2005; Taylor et al. 2009). Whilst the considerable decrease in atmospheric CO_2 concentration during the Devonian period (410 to 360 million years ago) was due to higher plants locking C up in their large trunks and leaves, even more significant was the role of roots (Kenrick 2001).

The removal of enormous quantities of CO_2 from the Devonian atmosphere is now thought to be the consequence of the development of deep-rooted plants that have contributed to achieving: (1) the formation of cracks and mechanical breakdown of rock particles as a consequence of root growth, with potential consequences for the circulation of water throughout the profile, its residence time and the overall water cycle; (2) an input of C deep in the regolith, thereafter creating the soil and subsoil profile, allowing microorganisms to grow at greater depth, along with root growth; and (3) the chemical changes of their rhizosphere, as a consequence of water and nutrient uptake, respiration, exudation of acidic and chelating compounds (Hinsinger et al. 2006, 2009). These various processes are not all equally well documented, especially on long (evolutionary) time scales.

Root-induced physical breakdown of rocks and consequences for deep soil formation and the water cycle

The colonisation of the terrestrial environment by plants profoundly modified the landscape, the soils and terrestrial ecosystems, and at an even broader scale the composition of the atmosphere. Phylogenetic studies favour a single origin of land plants from Charophyceae, which are predominantly freshwater green algae from the mid-Paleozoic era (Kenrick and Crane 1997). During the early Devonian period, the terrestrial vegetation would have looked like a more or less continuous short chlorophyllous carpet, largely composed of bryophytes, and other plants that did not have roots, but rhizomes instead. Plant cover consisted of rather short shoots erected along those running stems.

Plants with roots appeared in the late Devonian and more so during the Carboniferous period (starting 360 million years ago), with the most primitive being pteridophytes (ferns), horsetails and cycads (Kenrick and Crane 1997; Brundrett 2002). The rooting depth and root diameters significantly increased over the late-Devonian to early-Carboniferous transition period (Retallack 1997) (Fig. 6). The differentiation of roots presumably resulted from an adaptation to the new organo-mineral medium, porous, wet and relatively nutrient-rich. Major functions of roots are to acquire water and mineral nutrients for the plant, and to exude organic compounds, allowing the growth of significant microbial communities in the rhizosphere. Even the oldest fossils of roots show traces of close associations with fungi resembling AM (Brundrett 2002).

Evidence for endophytic associations is available for the first bryophyte-like land plants, which may have been the precursors of the first mycorrhizas (hyphae, vesicles and arbuscules) reported to occur in fossil rhizomes (Brundrett 2002) in the early Devonian, i.e. in land plants that had no roots, and thus no rhizosphere. From this time onwards, the interactions between plants, soil and microorganisms had the same significance ascribed to them today. Each component interacted with the others, plant-soil and microorganism-soil, while plants and microorganisms coevolved; these types of interactions have persisted ever since. The evolution of strong, deep, mycorrhizal root systems of vascular plants started at this period and occurred to a large extent over the late Devonian-early Carboniferous transition period (Kenrick and Crane 1997; Retallack 1997; Brundrett 2002; Taylor et al. 2009) (Fig. 6); their adaptation to most climatic and pedological conditions found on Earth contributed to the colonisation of continental surfaces (Kenrick and Crane 1997).

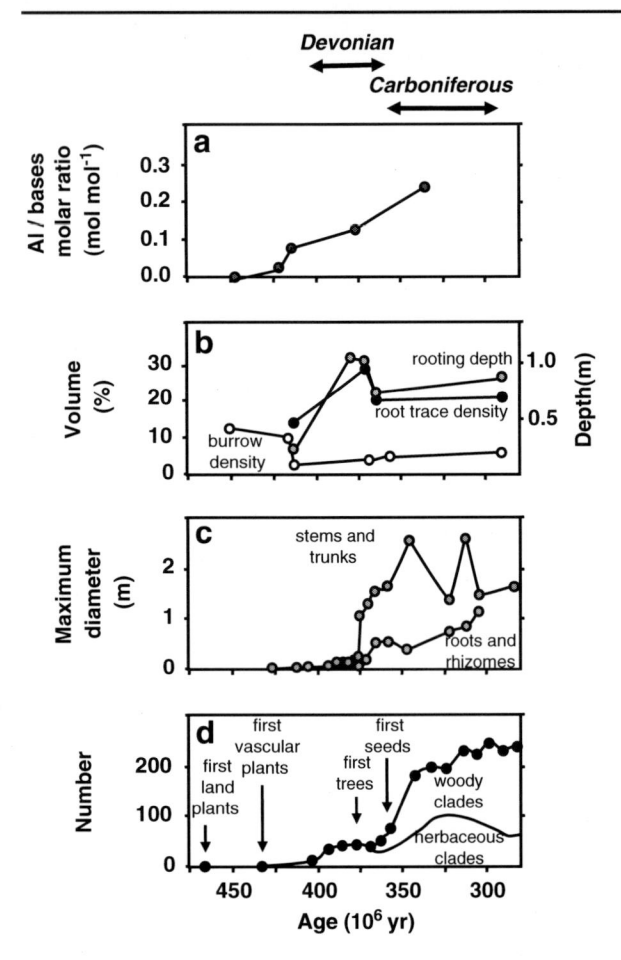

Fig. 6 Evolution of paleosol features in the mid-Paleozoic period (modified after Retallack (1997) Early forest soils and their role in Devonian global change. Science 276:583–585. Reprinted with permission from AAAS). Index of subsurface silicate weathering as assessed by the molar ratios of aluminium/bases in subsurface minus surface horizons (**a**), indices of bioturbation, as assessed along a line transect of hand specimens occupied either by burrows or by roots (expressed as volume percent), and corresponding root depth (**b**), as related to the maximum diameter of fossil stems and trunks, or of roots and rhizomes (**c**) and diversity of fossil plant species records, as expressed as numbers of species (**d**)

Plant cover plays several major roles on the surface of the continents (Berner 1997). Most significant is to limit erosion via retaining and fixing the fine, organic and mineral particles. These fine particles, clay- and silt-sized minerals and organic matter, have a high specific surface area and exchange capacity. They are organised in aggregates, which are physically stabilised, contributing to soil structure. A well structured soil supports deep rooting of plants which itself contributes to stabilise and structure the soil. The retention of fine particles by plant cover thus results in a perpetuating cycle, which finally benefits both plant cover and soil formation (Van Breemen 1993). The soil structure is important for biological activity by providing a stable habitat to all soil-borne organisms, and by allowing the transfer of water and exchange of gases involved in respiratory processes. It also allows infiltration of water. This soil water-retention capacity contributes locally to the maintenance of conditions favourable for the development of roots and other biological activities. At a larger, regional scale, the development of higher plants has played a key role in regulating water flows and limiting erosion processes (Berner 1997).

The property of plants to improve soil structure, or to increase the porosity of degraded horizons, is widely used in agronomy and ecological engineering. Blanchart et al. (2004) studied the effect of various cultural practices on the change of the physical properties of vertisols degraded by intensive gardening. They showed that a culture of the grass *Digitaria decubens* during 4 years made it possible to double the porosity of the top 40 cm of the soil. At the same time, the proportion of stable aggregates increased from less than 30 to nearly 70% which resulted in a 50% reduction of surface erosion. Consequently, the content of soil organic C increased from 15 to 25 g kg^{-1} soil. Akhter et al. (2003, 2004) used different methods to improve the physical properties of degraded sodic soils. They showed that a culture of kallar grass (*Leptochloa fusca*) significantly increased porosity, water retention, permeability and structural stability of soils. The effects were significant as of the third year of culture, and increased the following years. In this case, again, the improvements coincided with a significant increase in the soil organic matter content. In such soils, which often exhibit physical constraints at depth which limit the infiltration of water and root growth, young roots tend to re-use former root chanels (Creswell and Kirkegaard 1995), resulting in cumulative rhizosphere effects (Pankhurst et al. 2002; Hinsinger et al. 2005). Jassogne (2008) has recently shown that in such environments some plant species play a key, pioneer role in creating new biopores in the heavily textured soil horizons that limit the growth of other plant species; these biopores can subsequently be re-used by roots of primer plant as well as other plant species. The concept of primer plants, as defined by Yunusa and Newton (2003) for such environments, would be worth being extrapolated to other physically contrained environments, in line with the concept of ecosystem engineers, *sensu* Jones et al. (1994).

The effect of plant growth on soil structure occurs at the same time at different scales. At the scale of a single root, the growing root leads to aggregation of the surrounding soil particles (Hinsinger et al. 2009). In 1862, Sachs (quoted by Dehérain 1873) showed that soil binds to the youngest parts of the root system of wheat (*Triticum aestivum*), forming a rhizosheath. This rhizosheath results initially from physical interactions between the root and the soil, via root hairs or mycorrhizal hyphae, which emerge from roots (Dorioz et al. 1993; Czarnes et al. 1999; McCully 1999). It also results from the exudation of polysaccharides by roots or associated microorganisms. Exudation, root-hair growth and mycorrhizal infection occur especially in the youngest region of the roots which explains Sachs's observation and those of many authors since. Note that this first effect of binding of rhizosphere particles may be followed by a secondary effect; namely, that organic matter may decrease the wettability of aggregates which slows their rate of wetting and thus decreases the intensity of constraints that act inside the aggregates and tend to disrupt them (Chenu et al. 2000).

Root-induced and microbially mediated sequestration of C as organic matter and carbonates and consequences for the global C cycle

The input of C into the soil at greater depths than achieved by the pre-Devonian vegetation comprising non-rooted plants is a major root-induced process, which dramatically influenced the Earth history via its impact on global pedogenesis. The consequence of the advent of higher plants in the Paleozoic (542–251 million years ago) on the biogeochemistry of C has become a hot topic, given its relevance for better understanding the biogeochemical cycle of C and its link with elevated CO_2 concentrations in the atmosphere (Berner 1992, 1997; Mora et al. 1996; Beerling et al. 2001; Beerling and Berner 2005). Indeed, the atmospheric CO_2 concentration dropped from 3000–4500 μmol mol^{-1} in the early-mid Devonian (410–380 million years ago) to about 1000 μmol mol^{-1} towards the end of the Devonian (360 million years ago) (Fig. 7a). That coincided with the advent of deep-rooted vascular plants, which were mainly responsible for this dramatic global change (Berner 1992, 1997; Retallack 1997; Beerling and Berner 2005) (Figs. 6b and 7b).

Beerling and Berner (2005) concluded that the decrease in atmospheric CO_2 concentration was due to enhanced rates of photosynthesis, and subsequent sequestration of C into the soil (Fig. 6c), burial of C into sediments, and enhanced weathering of Ca- and

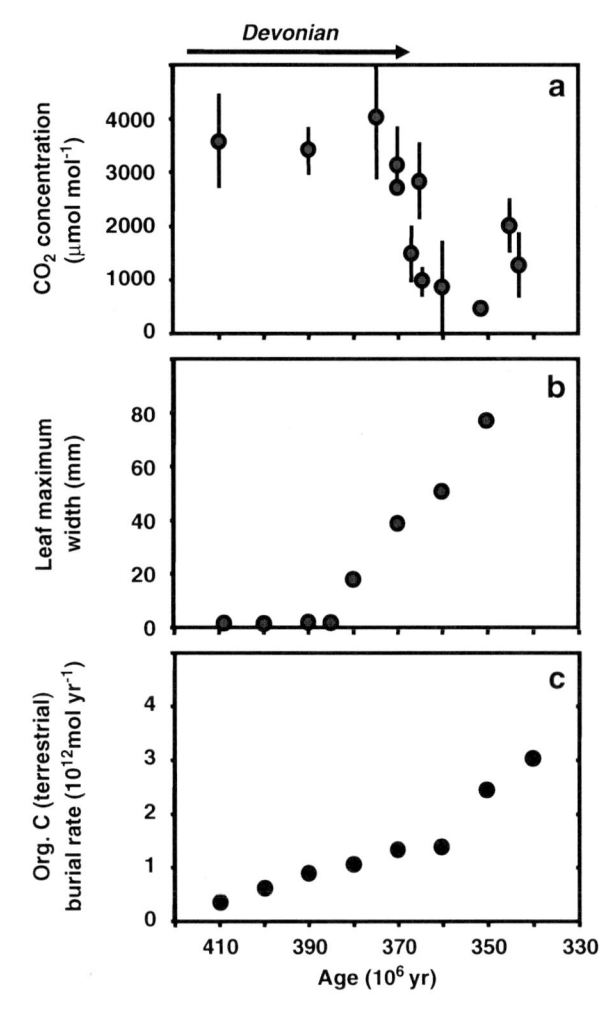

Fig. 7 Changes in the atmospheric CO_2 concentration during the Late Palaeozoic (**a**), concurrent increase in maximum width of megaphyll leaves calculated from fossils for 10 million year intervals (**b**), and changes in terrestrial carbon burial, calculated as the difference between global and marine burial (**c**), the latter being estimated from pyrite burial rates and a Cambrian–Silurian mean molar organic carbon/pyrite sulfur burial ratio of 1.5 (adapted from Beerling and Berner 2005, with kind permission from the The National Academy of Science of the USA)

Mg-bearing silicates, leading ultimately to the formation of Ca- and Mg-carbonates in the oceans. This input of C is linked with the decay of deep roots, as accounted for by models such as those developed by Berner and Kothavala (2001) and Beerling and Berner (2005). The role of root and microbial respiration in the rhizosphere should also be acknowledged, given its expected impact on the rates of dissolution (weathering) of minerals such as silicates and carbonates (Retallack 1997; Kuzyakov et al. 2006; Taylor et al. 2009). The additional input of C linked with rhizodeposition and the subsequent microbial loop needs to be accounted for in biogeochemical models used for reconstructing the C history of the Earth, as current knowledge of rhizodeposition shows it is of similar magnitude as C investment in root architecture (Nguyen 2003; Jones et al. 2004, 2009).

Via their direct role in the formation of biogenic carbonates, i.e. the precipitation of $CaCO_3$, predominantly calcite, which ultimately forms calcretes (Freytet et al. 1997; Verboom and Pate 2006a, b), roots of higher plants can leave long-lasting morphological features in soil profiles and be a major contributor to pedogenesis. Calcareous rhizoconcretions have been observed in some of the oldest paleosoils as described by Retallack (1997), who stressed "the early Paleozoic greenhouse may have been curbed by the evolution of rhizospheres with an increased ratio of primary to secondary production and by more effective silicate weathering". A first type of biogenic precipitation of calcite in the rhizosphere is related to the build-up of Ca which results from the large amount of Ca transferred towards the root surface as a consequence of mass flow (Jaillard 1982; Callot et al. 1983; Hinsinger 1998). In calcareous soils, this process can lead to the precipitation of calcite crystals around the roots which remains clearly visible around root pores once the root has died (Fig. 8a). Ultimately, the coalescence of the calcite precipitates around neighbouring roots can result in continuous calcrete formation when root density is large and ambient conditions are prone to biogenic precipitation of Ca carbonate in the rhizosphere (Jaillard 1982; Callot et al. 1983). This process is very clearly illustrated in 'The Pinnacles', 200 km north of Perth in Western Australia, where the soil was blown away following a bushfire, exposing the calcrete formations formed around roots over many years before (Fig. 8b).

Fig. 8 Root-induced calcrete formation as a result of calcium carbonate (calcite) precipitation around a peach (*Prunus persica*) root biopore (reproduced from Callot et al. 1983, with kind permission of INRA Publishers) as a consequence of mass-flow and increased concentration of Ca ions in the rhizosphere (**a**) and remnant calcrete formations presumably formed around tree roots after the soil was blown away following a bushfire, exposing the so-called 'Pinnacles' in Nambung National Park (**b**), 200 km north of Perth in Western Australia (photo by Philippe Hinsinger)

Although it is not as strictly related to rhizosphere processes as the example illustrated by 'The Pinnacles', Jaillard (1982, 1983, 1984, 1985) and Jaillard et al. (1991) have described another fascinating case of biogenic precipitation of calcite induced by root activity, leading to the formation of calcified roots in calcareous soils. These calcified roots, when fully preserved in the soil, occur and appear as calcite crystals shaping the root cortex cells, while the stele remains free of precipitates (Fig. 9). This process occurs in living plants, and is therefore not a fossilisation process (Jaillard 1987a, b). It is actually promoted by root activity, which can dissolve

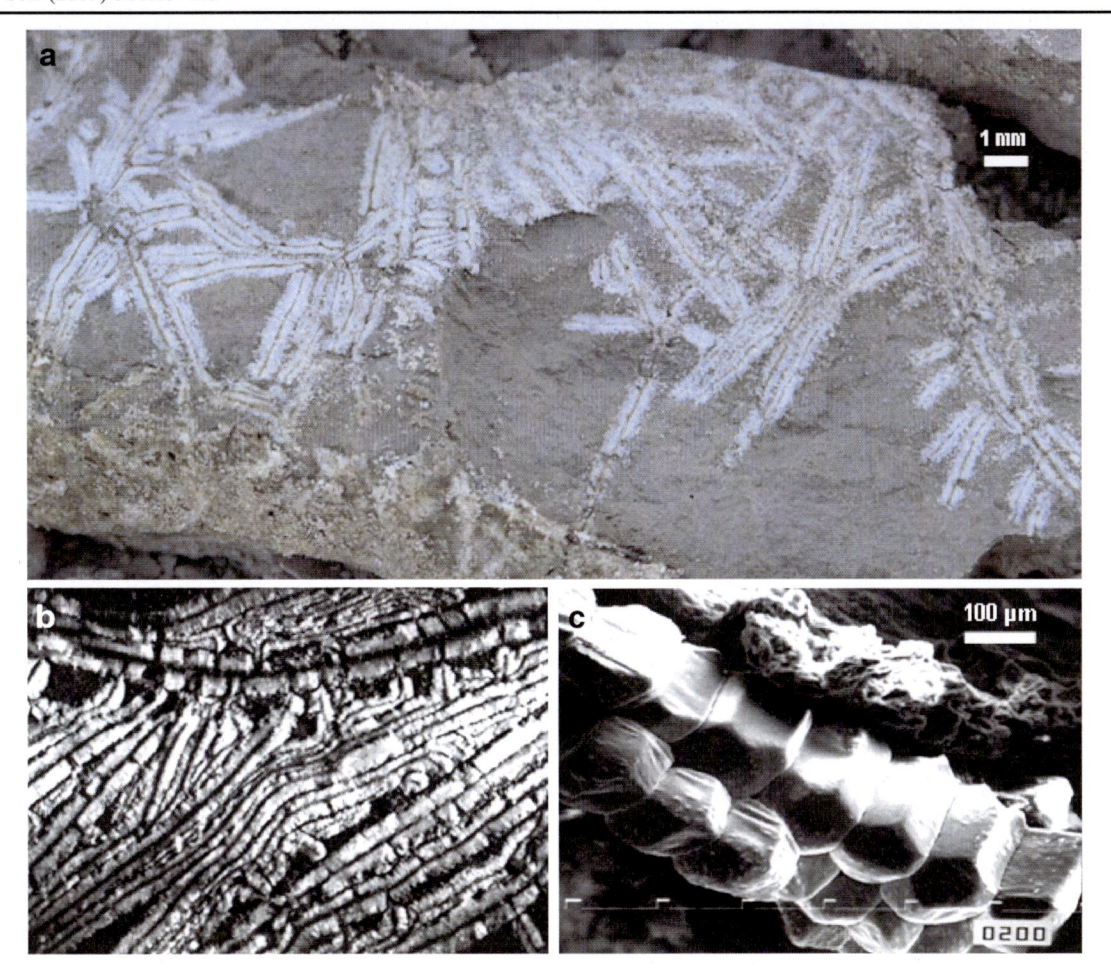

Fig. 9 Root-induced calcrete appearing as calcified roots forming as a result of calcium carbonate (calcite) precipitation in the cortical cells of perennial grasses growing in the 'garrigues' Mediterranean bush North of Montpellier in calcareous soils developed on a calcareous marl parent material (photo by Philippe Hinsinger) (**a**), closer view under light microscope showing distinctly that the stele is not calcified (**b**), and scanning electron micrograph of two rows of calcified cortical cells (**c**) (taken from Jaillard et al. 1991, with kind permission of Elsevier Ltd.)

abundant amounts of Ca carbonates in the outer rhizosphere as a consequence of respiration and proton release, and alleviate Ca toxicity by precipitating it as calcite in the vacuoles of root cortical cells (Jaillard 1987a, b). Although root-induced, this biogenic process is not just a rhizosphere process as it occurs inside root cells, similarly as the well documented precipitation of Ca oxalate (Horner and Wagner 1995). However, Ca oxalate formation seldom occurs to the same extent as found for calcite, which ultimately can completely fill the vacuole and the cell, so that the whole cortex tissue is full of large crystals shaping each individual cell. Remarkably, Jaillard (1984) showed that this biogenic process is responsible for a considerable shift of the particle size of calcareous soils, because of the rather isometric size of the calcified cells (around 80 μm diameter). Jaillard (1984) calculated that within about 1000 years, a third of all the Ca carbonate contained in calcareous soils can be dissolved and re-precipitate due to root activity. This process is thus one of the most dramatic cases of biogenic mineral formation in terrestrial ecosystems.

It should be stressed that the process of re-precipitation of Ca carbonate is a redistribution of previously existing Ca carbonate, and thus it does not involve any net C sequestration in the soil. That is, the dissolution of Ca carbonate in the rhizosphere is compensated for by re-precipitation inside the root cells. Based on the measurement of ^{14}C incorporation in Ca carbonate derived from ^{14}C-labelled root exudates, Kuzyakov et al. (2006) calculated that full re-crystallisation of loess carbonate would take 400 to 2000 years, which is fairly close to the estimated 1000 years provided by Jaillard (1984). This root-borne phenomenon may also account for the formation of widespread fossils that are especially abundant at the Cretaceous-Tertiary junction, the so-called microcodium, as suggested by Jaillard et al. (1991) and subsequently substantiated by Kosir (2004). Using morphological and isotopic geochemical approaches, these authors demonstrated that microcodium has many similarities with present-day rhizomorphic calcretes (Jaillard 1987a, Jaillard et al. 1991; Morin 1993), confirming early views of Klappa (1980). Such features, called rhizoliths, can be well preserved in paleosols and ancient calcretes (e.g., Retallack 1997). They have recently been used by Wang and Greenberg (2007), because they provide a high-resolution and continuous record of paleoclimates and past environments, based on isotopic geochemistry.

The formation of Ca oxalate crystals is well documented for ectomycorrhizal and saprophytic or pathogenic fungi, amongst which some species produce massive amounts of oxalic acid (Cromack et al. 1979; Dutton and Evans 1996; Connolly et al. 1999; Wallander 2000; Casarin et al. 2004), with a potential impact on the weathering of silicate minerals (Paris et al. 1996; Wallander 2000). This occurs in the rhizosphere as related to the activity of ectomycorrhizal fungi (Lapeyrie 1988; Wallander 2000; Casarin et al. 2003, 2004), especially in calcareous soils (Casarin et al. 2003), where the crystals of Ca oxalate may ultimately develop into needle-shaped crystals of Ca carbonate (Callot et al. 1985a, b; Verrecchia 1990; Verrecchia and Dumont 1996). The latter process would hardly change soil mineralogical composition or contribute any net sequestration of atmospheric C when the predominant source of Ca is the dissolution of soil Ca carbonate as would be the case for calcretes formed in calcareous soils. However, it may well be the case when the source of Ca is the weathering of other Ca-bearing minerals, e.g., plagioclase feldspars (Cromack et al. 1979; Landeweert et al. 2001). If so, the role of Ca oxalate in biogeochemical cycles as described by Graustein et al. (1977) should be reconsidered, as pointed out by Cailleau et al. (2004).

Root-induced and microbially mediated weathering of rocks and minerals and consequences for the cycle of nutrients and soil formation

The contributions of higher plants and soil microorganisms to the weathering of primary silicate minerals and rocks are manifold, and largely based on rhizosphere processes. Weathering models now account for the additional effect of plants due to the uptake of nutrients forming silicate minerals and rocks (e.g., Taylor and Velbel 1991; Bormann et al. 1998; Moulton et al. 2000; Velbel and Price 2007). However, they do not fully account for all other rhizosphere processes likely to dramatically influence the weathering rates. While the role of roots and rhizosphere microorganisms on elevated soil CO_2 concentrations via respiration and exudation of organic acids is well acknowledged, the potential impact of protons and siderophores released by roots and rhizosphere microorganisms on the weathering of silicate rocks is poorly accounted for.

Hinsinger and Jaillard (1993) showed that the uptake of K by roots of ryegrass (*Lolium multiflorum*) and the subsequent depletion of K in the rhizosphere (decrease in solution K concentration) is the driving force for the weathering of a Mg-bearing silicate such as phlogopite mica and concurrent formation of a clay mineral (vermiculite) in the absence of any pH decrease. Hinsinger et al. (2006) showed that the sink effect of plant roots for K leads to an approximately 5- and 6-fold increase of the rate of release of interlayer K, and of the concurrent weathering of micas such as phlogopite and biotite, respectively, compared with unplanted systems where only leaching is responsible for the observed weathering. This agrees with weathering rates of biotite computed by Taylor and Velbel (1991) and Velbel and Price (2007), who showed that taking account of the uptake of K by the vegetation of forested watersheds leads to rates that are up to 3.5 times greater than when ignoring this term in the K budget. This rhizosphere process, which is simply the consequence of the uptake activity of plant roots is also the only one reconciling K budgets of long-term field trials conducted by agronomists, and current knowledge of soil mineralogy and chemistry. Indeed, fertiliser trials conducted in Europe have often shown that, in the absence of fertiliser K application, the cumulated removal of K by successive crops usually does not match with an equivalent decrease in exchangeable K, even when this pool is considered as the bioavailable K for crops. Such field trials suggest that the release of non-exchangeable K amounts to 10 up to 100 kg K ha^{-1} yr^{-1}, as an average value computed over the whole duration of these long-term (several decades) fertiliser trials (Hinsinger 2002). Holmqvist et al. (2003) and Simonsson et al. (2007) found similar rates of weathering in long-term K-fertiliser trials of Northern Europe, pointing to an important contribution of the concurrent release of non-exchangeable K to the overall soil K budget. This contradicts current opinions of soil scientists that in such arable soils, the release of non-exchangeable K is negligible, because of the bulk soil solution K concentration typically ranging from 100 to 1000 μM.

The release of non-exchangeable K, which is actually largely based on the release of interlayer K and concomitant weathering of micas and micaceous clay (illite-like) minerals is governed by soil solution K concentration. Springob and Richter (1998) have clearly shown that the rate of this process is dramatically enhanced at soil solution K concentrations below 3 μM, and almost nil at those concentrations common in bulk soils of arable land in Europe. Based on such findings, virtually no release of non-exchangeable K is expected, although short-term pot experiments (Kuchenbuch and Jungk 1982; Niebes et al. 1993; Coroneos et al. 1996) as well as long-term fertiliser trials (Hinsinger 2002) have shown that this process may contribute a major proportion of K uptake by crops. The explanation lies in the specific chemical conditions of the rhizosphere. Claassen and Jungk (1982) showed, indeed, that while bulk soil solution K concentrations are in the order of several hundreds of μM, in the immediate vicinity (about 1 mm) of maize (*Zea mays*) roots, such concentrations could be as low as 2–3 μM. The latter are compatible with large rates of release of non-exchangeable K, as shown by Springob and Richter (1998). Therefore, as shown for single minerals (Hinsinger and Jaillard 1993), the sink effect of plant roots and subsequent depletion of soil solution K in the rhizosphere is the driving process that leads to the release of interlayer K and concomitant weathering of micaceous minerals in soils (Hinsinger 2002).

A recent study by Barré et al. (2007a) has further shown the causal link between these two processes, with a quantitative assessment of X-ray diffraction data. These authors showed that the uptake of K by ryegrass (*Lolium multiflorum*) in a rhizobox experiment quantitatively matches the increased amount of interstratified illite-smectite minerals in the rhizosphere which formed at the expense of the illite-like clay minerals. These findings confirmed previous findings of Hinsinger and Jaillard (1993), with a similar approach applied to a single micaceous mineral instead of a whole soil clay fraction, as well as the earlier report of an increased amount of interstratified illite-vermiculite minerals in the rhizosphere of field-grown maize plants (Kodama et al. 1994). Applying the same method to soil clay fractions collected in a long-term fertiliser trial, Barré et al. (2008) confirmed on a longer time scale that plant uptake of K matches the formation of interstratified illite-smectite minerals at the expense of the illite-like clay minerals, as a consequence of the release of interlayer, non-exchangeable K from the latter. According to these authors, plants thus play a major role in the biogeochemical cycle of K and in the formation and dynamics of expandable clay minerals (vermiculites or smectites) in top soils, with illite-like clay minerals playing the role of a huge K reservoir (Barré et al. 2007b, 2008). Indeed, no release of interlayer K contained in micaceous minerals would be expected to occur in the absence of plants and the efficient sink effect of their roots depleting soil solution K in the rhizosphere. This shows that through rhizosphere processes, higher plants are major drivers of weathering of rocks and formation of soils.

As pointed out by Taylor et al. (2009) and Hinsinger et al. (2006), rhizosphere acidification should be taken into account when considering the relevance of rhizosphere processes in bulk soil formation (pedogenesis), as expected given that pH has a dramatic impact on the rate of weathering of most minerals, especially in the acidic range (Banfield et al. 1999). The role of pH in the biologically mediated dissolution of minerals as related to the production of organic acids by roots and rhizosphere microorganisms has been studied in detail for forest tree seedlings for various silicate minerals (Wallander and Wickman 1999; Bakker et al. 2004). Wallander and Wickman (1999) showed that ectomycorrhizal fungi are largely responsible for the observed elevated concentrations of citric and oxalic acids, which account for the release of K and Mg from biotite.

These studies did not distinguish the direct pH effect from that of the chelating carboxylates, which also directly impact dissolution/weathering reactions (Huang and Keller 1970; Jones et al. 1996; Jones 1998). Several studies show rhizosphere acidification in forest stands measured in situ, with potential consequences for accelerated weathering rates (Courchesne and Gobran 1997; Turpault et al. 2005), although such studies do not demonstrate the causal relationship between these two processes. The impact of pH changes on the weathering of micas and feldspars in the soil around ectomycorrhizal fungi was, however, clearly shown by Arocena et al. (1999) and Arocena and Glowa (2000). Some ectomycorrhizal fungi supposedly play a major role in these processes, rather than the host plant itself (Landeweert et al. 2001; Hoffland et al. 2004). Fungal hyphae are able to dissolve silicate minerals, thereby forming tunnel-like pores inside mineral grains (Jongmans et al. 1997).

When considering the potential involvement of ectomycorrhizal fungi in soil formation over longer time scales, we have to be aware that ECM occur only in a small proportion of all species among angiosperms and gymnosperms. They have evolved rather recently, compared with AM (Table 1); they first appeared around 100 million years ago (Brundrett 2002), and thus are not old enough to have contributed to a major extent to the enhanced weathering that occurred in the late Devonian with the advent of rooted vascular plants (Landeweert et al. 2001). On the other hand, AM fungi (Glomeromycota), which belong to a less diverse group than ectomycorrhizal fungi, evolved prior to the advent of deep-rooted vascular plants, 380 to 360 million years ago (Brundrett 2002; Beerling and Berner 2005; Taylor et al. 2009). However, their direct involvement in the weathering of rocks and minerals is questionable. This means that the enhanced weathering that occurred prior to the evolution of ectomycorrhizal fungi resulted likely from root-borne processes or non-symbiotic rhizosphere microorganisms, rather than mycorrhizas. Thus, the involvement of mycorrhizal fungi in the weathering of rocks and pedogenesis, and especially that of ectomycorrhizal species associated mostly with temperate, forest tree species, is of major importance in forest ecosystems, but restricted to the last 50–100 million years (Landeweert et al. 2001).

Another point when considering the implication of mycorrhizas in biological weathering is the potential role of bacteria associated with ectomycorrhizal roots and hyphae (Calvaruso et al. 2007; Uroz et al. 2007). Frey-Klett et al. (2005) reported selection of so-called mycorrhiza-helper bacteria belonging to fluorescent pseudomonads in ECM. Calvaruso et al. (2007) have shown that several ectomycorrhizal symbioses select for different bacterial communities that are potentially involved in the production of acidic or chelating metabolites, ultimately affecting the weathering of minerals. They also found that more bacteria exhibiting a large weathering potential were found in the ectomycorrhizosphere than in the bulk soil. This was confirmed by Uroz et al. (2007), who assessed the weathering capacity as the ability to acidify the growing medium and to release Fe from biotite particles in the growing medium. Among 61 isolates collected from *Scleroderma citrinum* mycorrhizas, mycorrhizosphere and adjacent bulk soil in an oak (*Quercus petraea*) forest stand, they identified bacteria of the *Burkholderia, Collimonas, Pseudomonas* and *Sphingomonas* genera, with the largest weathering capacity shown by the *Burkholderia* and *Collimonas* genera. Calvaruso et al. (2006) inoculated pine (*Pinus sylvestris*) seedlings with three strains of *Burkholderia glathei* amongst the most efficient isolates identified by Uroz et al. (2007) and measured the weathering of biotite in microcosms, based on Mg and K release rates. The weathering rate was increased 1.4-fold for Mg and 1.5-fold for K, when compared with that of pine alone, indicating a congruent dissolution, likely mediated by a proton- or organic acid-promoted dissolution of the mineral.

As in the recent study of Uroz et al. (2007), Leyval and Berthelin (1989, 1991) had shown earlier with axenic pine seedlings inoculated with an ectomycorrhizal fungus and rhizobacteria that the weathering of biotite is largely due to the effect of the host plant itself, with a substantial additional effect of rhizosphere microorganisms. This was even more so for a similar microcosm study conducted with maize and an AM fungus with or without additional effect of inoculated bacteria (Berthelin and Leyval 1982; Leyval et al. 1990); the host plant is responsible for the largest portion of the biologically mediated weathering of biotite, as shown by the release of K and Fe.

Although previous studies assessing the effect of plants and rhizosphere microorganisms on the biologically mediated weathering of minerals have often stressed the microbial contribution, many have actually shown that the higher plants *per se* induces a major shift in the rates of weathering, relative to abiotic weathering. This might explain why the advent of vascular plants had such a dramatic effect on the weathering of silicates (Fig. 6a) as indicated by the decrease in atmospheric CO_2 concentration in the late Devonian (Fig. 7a) (Berner 1997; Retallack 1997; Beerling and Berner 2005). Biogeochemical models, however, attribute the effect of those plants to: (1) acidification of the rhizosphere as a consequence of respiration and the production of organic acids (assuming that these are largely due to rhizosphere microbial activities); and (2) the higher rate of uptake of nutrients from nutrient-bearing minerals (Beerling and Berner 2005), but the latter effect is seldom stressed.

In addition to its potential direct effect on shifting the dissolution equilibrium as clearly shown for K in micas (Hinsinger and Jaillard 1993), the uptake of nutrients by higher plants has a major effect on rhizosphere pH, excess cation over anion uptake resulting in proton efflux from roots and thus in rhizosphere acidification (Hinsinger et al. 2003). When one considers proton sinks and sources in terrestrial ecosystems, the balance of cations over anions taken up by the vegetation is, indeed, a major contributor (cation uptake being a source, while anion uptake is a sink for protons), sometimes greater than respiration and the resulting build-up of soil CO_2 concentrations (Van Breemen et al. 1984; Bourrié and Lelong 1994; Frey et al. 2004). This is especially so in acidic environments where respiration has little impact on pH, as opposed to the situation in neutral or alkaline environments. It should also be borne in mind that the proton-promoted dissolution of minerals is a major proton sink in terrestrial ecosystems (Van Breemen et al. 1984) which means that the weathering of rocks may mask the actual pH changes occurring in the rhizosphere, thereby hiding the actual importance of such a phenomenon (Hinsinger et al. 2003). This was also stressed by Hinsinger et al. (2001), who studied the weathering of a basalt rock in a microcosm study. They showed that, while a planted system leads to 2- to 10-fold increased dissolution rates for most elements (Ca, Mg, Si), compared with unplanted systems, where only leaching is responsible for the observed weathering, the impact of the plants on Fe-dissolution rates is much greater, up to 500-fold. A significant root-induced acidification was observed which possibly explains the kinetics of release of Ca and Mg from the basalt, but obviously not that of Fe. Although Hinsinger et al. (2001) did not demonstrate the underlying processes, it is likely that the much greater effect for Fe than for other elements is due to more specific rhizosphere processes, rather than shifting the dissolution equilibrium via plant uptake and root-mediated acidification, e.g., redox changes or Fe chelation by root or microbial siderophores. This is also suggested by results obtained for the rhizosphere of various crop species supplied with goethite (a poorly soluble Fe oxyhydroxide) as sole source of Fe (Bertrand et al. 1999; Bertrand and Hinsinger 2000). Recently, Reichard et al. (2005) provided direct evidence for the involvement of phytosiderophores secreted by wheat (*Triticum aestivum*) in the ligand-promoted dissolution of goethite. A similar mechanism operates for microbial siderophores (Reichard et al. 2007).

Pate et al. (2001) noted a close association of species belonging to the Proteaceae and lateritic soils. As discussed above, almost without exception, species of the Proteaceae produce proteoid root clusters, which release vast amounts of carboxylates (Shane and Lambers 2005b). These carboxylates chelate Fe and other micronutrients which explains why Mn concentrations in cluster-bearing species tend to be high (Gardner and Boundy 1983; Bolland et al. 2000) and correlated with the fraction of the total root mass invested in cluster roots (Shane and Lambers 2005a). Pate and coworkers proposed that Fe is solubilised during the wet season, and then moves down the profile, where it is precipitated following microbial breakdown of the chelating carboxylates, giving rise to a lateritic or podzolic subsoil. This clearly shows that plants, through their exudation of specific compounds, can give rise to a subsoil that is very hard to penetrate, except through cracks and biopores, this hardpan being composed in this case of a ferricrete (Verboom and Pate 2003).

In later work, Verboom and Pate (2006a, b) propose that precipitates of silcrete, calcrete and ferricrete are formed in analogous manners, but further work is required to find out which exudation or other root-induced processes, if any, are involved in these pedogenic processes. This has been discussed above for the case of calcretes. Verboom and Pate (2006b) recently showed that in a similar environment and with similar parent material, the occurrence of calcretes spatially coincide with myrtaceous woodland, while ferricretes are predominantly found under proteaceous shrub-heathland. This led Verboom and Pate (2006a) to further develop their 'phytotarium' concept, to emphasise the major role of the vegetation cover on soil profile development. Particularly remarkable evidence of bioengineered ferricretes are Fe-coated root channels that are recorded in lateritic or deep podzolic soil profiles (Verboom and Pate 2006a), as evidenced by the pale brown to red colour around root channels or around roots growing both vertically and sub-vertically (Fig. 10). The hypothetical processes leading to the formation of such rhizosphere traits and ultimately of the bioengineered ferricretes, as suggested by Verboom and Pate (2003, 2006a) have not been demonstrated so far. The exact sequence of events under such aerobic conditions still needs to be elucidated.

Taking account of all of the above-mentioned rhizosphere processes, there is an obvious need to re-examine the impact of higher plants and rhizosphere microbes on the weathering of rocks and minerals for: (1) understanding current rates of weathering and corresponding fluxes in biogeochemical cycles of nutrients (Hobbie 1992) and other elements (e.g., Si, see Lucas et al. 1993; Gérard et al. 2008); and (2) elucidating their role in the biogeochemical history of the Earth, early formation of deep soils in terrestrial ecosystems (pedogenesis) and composition of the atmosphere (CO_2 and O_2 concentrations). These are highly relevant for another major ecosystem service, which is the provision of habitats, because soils are the greatest reservoir of biodiversity on Earth, and many of the above-mentioned processes have triggered the emergence of this unique diversity (Crawford et al. 2005; Hinsinger et al. 2009). Therefore, higher plants play a central role as drivers of the habitability of the Earth's terrestrial ecosystems (Schwartzman and Volk 1989). *Sensu* Jones et al. (1994), they definitively need to be recognised as key 'ecosystem engineers'.

Concluding remarks

Lovelock (1965) compared the atmosphere of the Earth and other planets, and concluded that living organisms can affect atmospheric composition, finally to their own global benefit. This first idea was then extended to a set of functions that are essential for the development of life at the surface of the Earth. Lovelock considered Earth as a super-organism capable of regulating itself in its own interest. The novelist William Golding poetically called this super-organism Gaia, after the goddess Earth-Mother (Lovelock 2003). Van Breemen (1993) took inspiration from this idea to examine the soils as living constructions (Lovelock 1993). Indeed, soils have been co-constructed by plants and rhizosphere microorganisms. These have contributed to improving the physical and chemical properties of soils to allow enhanced growth of plants and associated microorganisms. This has led to increased depth and loosening, structuring, better water retention, better aeration and gas exchange, increased availability of inorganic and organic compounds, enhanced rock weathering and C sequestration. All these properties make the soil a more favourable habitat to sustain the development of living organisms in terrestrial ecosystems. This is certainly linked with the soil being the greatest reservoir of biodiversity on the planet (Crawford et al. 2005). The central and positive role that higher plants have played to improve and maintain the Earth's terrestrial ecosystems in this favourable state, including via a number of rhizosphere processes, needs to be recognised, as stressed by Pate et al. (2001) and Verboom and Pate (2006a). From a global perspective, Schwartzman and Volk (1989) emphasised the role of higher plants in the habitability of Earth. In the current context of global change, it is worth noting that, according to Berner (1992) and Berner and Kothalava (2001), a key

Fig. 10 Biogenic ferricrete around a root of *Eucalyptus* sp. growing vertically, several meters below the soil surface in a deep podzolic soil developed in a sandy parent material at Jandakot, Western Australia. Note iron oxide precipitation in the rhizosphere (photo by Philippe Hinsinger)

process is the plant-promoted rock weathering, which ultimately allows long-term sequestration of C as carbonate precipitates in the oceans, and thus determines the atmospheric CO_2 concentrations. In that sense, plants and associated microorganisms should be recognised as key 'ecosystem engineers', *sensu* Jones et al. (1994), their major common construction being the rhizosphere.

Major challenges lie ahead of us, to make new discoveries on the signalling processes between the various organisms that play a role in rhizosphere processes. These discoveries will not only enhance our basic understanding, but also allow exciting applications of this new knowledge to deal with pests and diseases in an environmentally responsible manner. There are also major challenges to work towards new crops and cropping systems that are better able to acquire nutrients from soil, in particular P, given that our non-renewable P resources are rapidly running out, with phosphate-fertiliser prices rapidly increasing. Rhizosphere ecology is now a firmly established research field, with many exciting challenges, both from a fundamental and a strategic-applied perspective.

Major developments in our understanding of rhizosphere biogeochemistry and ecology are expected, including greater consideration of long-term temporal scales as well as global scales. Its links with evolutionary processes need to be further understood which is a challenge for the future. While much progress has been made in the case of symbiotic associations, there are considerable knowledge gaps for the many other biological interactions that play a role in the rhizosphere with respect to when and how they have evolved. Answering the puzzling question of why higher plants maintained such a large loss of C via rhizodeposition over tens or hundreds of million years of evolution will be a major challenge. Yet, rhizodeposition is central to many interactions in the soil and to processes beyond rhizosphere ecology. In fact, they are crucial for soil and terrestrial ecosystem functioning and biodiversity.

References

Aguilar OM, Riva O, Peltzer E (2004) Analysis of *rhizobium elti* and of its symbiosis wild *Phaseolus vulgaris* support coevolution in centers of host diversification. Proc Natl Acad Sci USA 101:13548–13553

Akhter J, Mahmood K, Malik KA, Ahmed S, Murray R (2003) Amelioration of a saline sodic soil through cultivation of a salt-tolerant grass *Leptochloa fusca*. Environ Conserv 30:168–174

Akhter J, Murray R, Mahmood K, Malik KA, Ahmed S (2004) Improvement of degraded physical properties of a saline-sodic soil by reclamation with kallar grass (*Leptochloa fusca*). Plant Soil 258:207–216

Akiyama K, Hayashi H (2006) Strigolactones: chemicals signals for fungal symbionts and parasitic weeds in plant roots. Ann Bot 97:925–931

Akiyama K, Matsuzaki K, Hayashi H (2005) Plant sesquiterpenes induce hyphal branching in arbuscular mycorrhizal fungi. Nature 435:824–827

Ané JM, Kiss GB, Riely BK, Penmetsa RV, Oldroyd GED, Ayax C, Lévy J, Debellé F, Baek J-M, Kalo P, Rosenberg C, Roe BA, Long SR, Dénarié J, Cook DR (2004) Medicago truncatula DMI1 required for bacterial and fungal symbioses in legumes. Science 303:1364–1367

Arocena JM, Glowa KR (2000) Mineral weathering in ectomycorrhizosphere of subalpine fir (*Abies lasiocarpa* (Hook.) Nutt.) as revealed by soil solution composition. For Ecol Manage 133:61–70

Arocena JM, Glowa KR, Massicotte HB, Lavkulich L (1999) Chemical and mineral composition of ectomycorrhizosphere soils of subalpine fir (*Abies lasiocarpa* (Hook.) Nutt.) in the Ae horizon of a luvisol. Can J Soil Sci 79:25–35

Bago B, Pfeffer PE, Shachar-Hill Y (2000) Carbon metabolism and transport in arbuscular mycorrhizas. Plant Physiol 124:949–957

Bais HP, Park S-W, Weir TL, Callaway RM, Vivanco JM (2004) How plants communicate using the underground information superhighway. Trends Plant Sci 9:26–32

Bakker MR, George E, Turpault MP, Zhang JL, Zeller B (2004) Impact of Douglas-fir and Scots pine seedlings on plagioclase weathering under acidic conditions. Plant Soil 266:247–259

Bakker C, Rodenburg J, Van Bodegom PM (2005) Effects of Ca- and Fe-rich seepage on P availability and plant performance in calcareous dune soils. Plant Soil 275:111–122

Banfield JF, Barker WW, Welch SA, Taunton A (1999) Biological impact on mineral dissolution: application of the lichen model to understanding mineral weathering in the rhizosphere. Proc Natl Acad Sci 96:3404–3411

Barré P, Velde B, Catel N, Abbadie L (2007a) Quantification of potassium addition or removal through plant activity on clay minerals by X-ray diffraction. Plant Soil 292:137–146

Barré P, Velde B, Abbadie L (2007b) Dynamic role of 'illite-like' clay minerals in temperate soils: facts and hypotheses. Biogeochemistry 82:77–88

Barré P, Montagnier C, Chenu C, Abbadie L, Velde B (2008) Clay minerals as a soil potassium reservoir: observation and quantification through X-ray diffraction. Plant Soil 302:213–220

Beerling DJ, Berner RA (2005) Feedbacks and the coevolution of plants and atmospheric CO_2. Proc Natl Acad Sci USA 102:1302–1305

Beerling DJ, Osborne CP, Chaloner WG (2001) Evolution of leaf-form in land plants linked to atmospheric CO2 decline in the Late Palaeozoic era. Nature 410:352–354

Béna G, Lyet A, Huguet T, Olivieri I (2005) Medicago–Sinorhizobium symbiotic specificity evolution and the geographic expansion of Medicago. J Evol Biol 18:1547–1558

Bending GD, Read DJ (1996) Nitrogen mobilization from protein-polyphenol complex by ericoid and ectomycorrhizal fungi. Soil Biol Biochem 28:1603–1612

Berner RA (1992) Weathering, plants, and the long-term carbon cycle. Geochim Cosmochim Acta 56:3225–3231

Berner RA (1997) Paleoclimate–The rise of plants and their effect on weathering and atmospheric CO_2. Science 276:544–546

Berner RA, Kothavala Z (2001) GEOCARB III: a revised model of atmospheric CO_2 over Phanerozoic time. Am J Sci 30:182–204

Berthelin J, Leyval C (1982) Ability of symbiotic and non-symbiotic rhizospheric microflora of maize (Zea mays) to weather micas and to promote plant growth and plant nutrition. Plant Soil 68:369–377

Bertrand I, Hinsinger P (2000) Dissolution of iron oxyhydroxide in the rhizosphere of various crop species. J Plant Nutr 23:1559–1577

Bertrand I, Hinsinger P, Jaillard B, Arvieu JC (1999) Dynamics of phosphorus in the rhizosphere of maize and rape grown on synthetic, phosphated calcite and goethite. Plant Soil 211:111–119

Blanchart E, Albrecht A, Chevallier T, Hartmann C (2004) The respective roles of roots and earthworms in restoring physical properties of Vertisol under a Digitaria decumbens pasture (Martinique, WI). Agric Ecosyst Environ 103:343–355

Bolan NS (1991) A critical review on the role of mycorrhizal fungi in the uptake of phosphorus by plants. Plant Soil 134:189–207

Bolan NS, Robson AD, Barrow NJ (1987) Effect of vesicular-arbuscular mycorrhiza on the availability of iron phosphates to plants. Plant Soil 99:401–410

Bolland MDA, Sweetingham MW, Jarvis RJ (2000) Effect of applied phosphorus on the growth of Lupinus luteus, L. angustifolius and L. albus in acidic soils in the south-west of Western Australia. Aust J Exp Agric 40:79–92

Bonkowski M (2004) Protozoa and plant growth: the microbial loop in soil revisited. New Phytol 162:617–631

Bonkowski M, Jentschke G, Scheu S (2001) Contrasting effects of microbial partners in the rhizosphere: interactions between Norway Spruce seedlings (Picea abies Karst.), mycorrhiza (Paxillus involutus (Batsch) Fr.) and naked amoebae (protozoa). Appl Soil Ecol 18:193–204

Bonkowski M, Villenave C, Griffiths B, (2009) Rhizosphere fauna: functional and structural diversity of intimate interactions of soil fauna with plant roots. Plant Soil 321:213–232. doi:10.1007/s11104-009-0013-2

Bormann BT, Wang D, Bormann FH, Benoit R, April D, Snyder MC (1998) Rapid plant induced weathering in an aggrading experimental ecosystem. Biogeochemistry 43:129–155

Boulet FM, Lambers H (2005) Characterisation of arbuscular mycorrhizal colonisation in the cluster roots of Hakea verrucosa F. Muell (Proteaceae) and its effect on growth and nutrient acquisition in ultramafic soil. Plant Soil 269:357–367

Bourrié G, Lelong F (1994) Les solutions du sol: du profil au bassin versant. In: Bonneau M, Souchier B (eds) Pédologie 2: constituants et Propriétés du Sol. Masson, Paris, pp 239–273

Bouwmeester HJ, Roux C, Lopez-Raez JA, Becard G (2007) Rhizosphere communication of plants, parasitic plants and AM fungi. Trends Plant Sci 12:224–230

Brenner ED, Stevenson DW, Twigg RW (2003) Cycads: evolutionary innovations and the role of plant-derived neurotoxins. Trends Plant Sci 8:446–452

Brundrett MC (2002) Coevolution of roots and mycorrhizas of land plants. New Phytol 154:275–304

Brundrett MC (2009) Mycorrhizal associations and other means of nutrition of vascular plants: understanding the global diversity of host plants by resolving conflicting information and developing reliable means of diagnosis. Plant Soil 320:37–77. doi:10.1007/s11104-008-9877-9

Brundrett MC, Abbott LK (1991) Roots of jarrah forest plants. I. Mycorrhizal associations of shrubs and herbaceous plants. Aust J Bot 39:445–457

Cailleau G, Braissant O, Verrecchia EP (2004) Biomineralization in plants as a long-term carbon sink. Naturwissenschaften 91:191–194

Cairney JWG, Burke RM (1998) Extracellular enzyme activities of the ericoid mycorrhizal endophyte Hymenoscyphus ericae (Read) Korf & Kernan: their likely roles in decomposition of dead plant tissue in soil. Plant Soil 205:181–192

Cakmak I, Sari N, Marschner H, Ekiz H, Kalayci M, Yilmaz A, Braun HJ (1996) Phytosiderophore release in bread and durum wheat genotypes differing in zinc efficiency. Plant Soil 180:183–189

Callot G, Chamayou H, Maertens C, Salsac L (1983) Mieux comprendre les interactions sol-racine. Incidence sur la nutrition minérale. INRA, Paris, p 326

Callot G, Guyon A, Mousain D (1985a) Inter-relation entre les aiguilles de calcite et hyphes mycéliens. Agronomie 5:209–216

Callot G, Mousain D, Plassard C (1985b) Concentrations de carbonate de calcium sur les parois des hyphes mycéliens. Agronomie 5:143–150

Calvaruso C, Turpault MP, Frey-Klett P (2006) Root-associated bacteria contribute to mineral weathering and to mineral nutrition in trees: a budgeting analysis. Appl Environ Microbiol 72:1258–1266

Calvaruso C, Turpault MP, Leclerc E, Frey-Klett P (2007) Impact of ectomycorrhizosphere on the functional diversity of soil bacterial and fungal communities from a forest stand in relation to nutrient mobilization processes. Microb Ecol 54:567–577

Casarin V, Plassard C, Souche G, Arvieu JC (2003) Quantification of oxalate ions and protons released by ectomycorrhizal fungi in rhizosphere soil. Agronomie 23:461–469

Casarin V, Plassard C, Hinsinger P, Arvieu JC (2004) Quantification of ectomycorrhizal fungal effects on the bioavailability and mobilisation of soil P in the rhizosphere of Pinus pinaster. New Phytol 163:177–185

Chaignon V, Di Malta D, Hinsinger P (2002) Fe-deficiency increases Cu acquisition by wheat cropped in a Cu-contaminated vineyard soil. New Phytol 154:121–130

Chenu C, Le Bissonnais Y, Arrouays D (2000) Organic matter influence on clay wettability and soils aggregate stability. Soil Sci Soc Am J 64:1479–1486

Claassen N, Jungk A (1982) Kaliumdynamik im wurzelnahen Boden in Beziehung zur Kaliumaufnahme von Maispflanzen. Z Pflanzenern Bodenkd 145:513–525

Clarholm M (1985) Interactions of bacteria, protozoa and plants leading to mineralization of soil nitrogen. Soil Biol Biochem 17:181–187

Clegg MT, Cummings MP, Durbin ML (1997) The evolution of plant nuclear genes. Proc Natl Acad Sci USA 94:7791–7798

Clemens S, Palmgren MG, Kramer U (2002) A long way ahead: understanding and engineering plant metal accumulation. Trends Plant Sci 7:309–315

Cochran MF, Berner RA (1996) Promotion of chemical weathering by higher plants: field observations on Hawaiian basalts. Chem Geol 132:71–77

Connolly JH, Shortle WC, Jellison J (1999) Translocation and incorporation of strontium carbonate derived strontium into calcium oxalate crystals by the wood decay fungus *Resinicium bicolor*. Can J Bot 77:179–187

Coroneos C, Hinsinger P, Gilkes RJ (1996) Granite powder as a source of potassium for plants: a glasshouse bioassay comparing two pasture species. Fert Res 45:143–152

Courchesne F, Gobran GR (1997) Mineralogical variation of bulk and rhizosphere soils from a Norway spruce stand. Soil Sci Soc Am J 61:1245–1249

Crawford JW, Harris JA, Ritz K, Young IM (2005) Towards an evolutionary ecology of life in soil. Trends Ecol Evol 20:81–86

Creswell HP, Kirkegaard JA (1995) Subsoil amelioration by plant roots – the process and the evidence. Austr J Soil Res 33:221–239

Crews TE, Kitayama K, Fownes JH, Riley RH, Herbert DA, Muellerdombois D, Vitousek PM (1995) Changes in soil-phosphorus fractions and ecosystem dynamics across a long chronosequence in Hawaii. Ecology 76:1407–1424

Crocker LJ, Schwintzer CR (1993) Factors affecting formation of cluster roots in Myrica gale seedlings in water culture. Plant Soil 152:287–298

Cromack K, Sollins P, Graustein WC, Speidel K, Todd AW, Spycher G, Li CY, Todd RL (1979) Calcium oxalate accumulation and soil weathering in mats of hypogeous fungus. *Hysterangium crassum*. Soil Biol Biochem 11:463–468

Czarnes S, Hiller S, Dexter AR, Hallett PD, Bartoli F (1999) Root : soil adhesion in the maize rhizosphere: the rheological approach. Plant Soil 211:69–86

Darrah PR (1991) Models of the rhizosphere. I. Microbial population dynamics around a root releasing soluble and insoluble carbon. Plant Soil 133:187–199

Dehérain PP (1873) Cours de Chimie agricole. Hachette, Paris

Delhaize E, Ryan PR (1995) Aluminum toxicity and tolerance in plants. Plant Physiol 107:315–321

Delhaize E, Ryan PR, Randall PJ (1993) Aluminum tolerance in wheat (*Triticum aestivum* L.). II. Aluminum-stimulated excretion of malic acid from root apices. Plant Physiol 103:695–702

Denison RF (2000) Legume sanctions and the evolution of symbiotic cooperation by rhizobia. Am Nat 156:567–576

Deshmukh S, Hückelhoven R, Schäfer P, Imani J, Sharma M, Weiss M, Waller F, Kogel K-H (2006) The root endophytic fungus Piriformospora indica requires host cell death for proliferation during mutualistic symbiosis with barley. Proc Natl Acad Sci 103:18450–18457

Ding J, Sun Y, Xiao CL, Shi K, Zhou YH, Yu JQ (2007) Physiological basis of different allelopathic reactions of cucumber and figleaf gourd plants to cinnamic acid. J Exp Bot 58:3765–3773

Dobbelaere S, Croonenborghs A, Thys A, Vande Broek A, Vanderleyden J (1999) Phytostimulatory effect of *Azospirillum brasilense* wild type and mutant strains altered in IAA production on wheat. Plant Soil 212:155–164

Dobbelaere S, Vanderleyden J, Okon Y (2003) Plant growth-promoting effects of diazotrophs in the rhizosphere. Crit Rev Plant Sci 22:107–149

Döbereiner J, Pedrosa FO (1987) Nitrogen-fixing bacteria in nonleguminous crop plants. Science Tech, Inc., Madison

Dorioz JM, Roberts M, Chenu C (1993) The role of roots, fungi and bacteria, on clay particle organization. An experimental approach. Geoderma 56:179–194

Doyle JJ, Doyle JL, Ballenger JA, Dickson EE, Kajita T, Ohashi H (1997) A phylogeny of the chloroplast gene rbcL in the Leguminosae: taxonomic correlations and insights into the evolution of modulation. Am J Bot 84:541–554

Drever JI (1994) The effect of land plants on weathering rates of silicate minerals. Geochim Cosmochim Acta 58:2325–2332

Dutton MV, Evans CS (1996) Oxalate production by fungi: its role in pathogenicity and ecology in the soil environment. Can J Microbiol 42:881–895

Ebert D (1994) Virulence and local adaptation of a horizontally transmitted parasite. Science 265:1084–1086

Ebert D (1998) Experimental evolution of parasites. Science 282:1432–1435

Faure D, Vereecke D, Leveau JHJ (2009) Molecular communication in the rhizosphere. Plant Soil 321:279–303. doi:10.1007/s1104-008-9839-2

Fontaine S, Mariotti A, Abbadie L (2003) The priming effect of organic matter: a question of microbial competition? Soil Biol Biochem 35:837–843

Franche C, Lindström K, Elmerich C (2009) Nitrogen-fixing bacteria associated with leguminous and non-leguminous plants. Plant Soil 321:35–59. doi:10.1007/s11104-008-9833-8

Frey J, Frey T, Pajuste K (2004) Input-output analysis of macroelements in ICP-IM catchment area, Estonia. Landsc Urban Plan 67:217–223

Freytet P, Plaziat JC, Verrecchia EP (1997) A classification of rhizogenic (root-formed) calcretes, with examples from the upper Jurassic lower Cretaceous of Spain and upper Cretaceous of southern France. Sediment Geol 110:299–303

Frey-Klett P, Chavatte M, Clausse ML, Courrier S, Le Roux C, Raaijmakers J, Martinotti MG, Pierrat JC, Garbaye J (2005) Ectomycorrhizal symbiosis affects functional diversity of rhizosphere fluorescent pseudomonads. New Phytol 165:317–328

Frühling M, Roussel H, Gianinazzi-Pearson V, Pühler A, Perlick AM (1997) The *Vicia faba* leghemoglobin gene *VfLb29* is induced in root nodules and in roots colonized by the arbuscular mycorrhizal fungus *Glomus fasciculatum* Mol. Plant-Microbe Interact 10:124–131

Gage DJ, Margolin W (2000) Hanging by a thread: invasion of legume plants by rhizobia. Curr Opin Microbiol 3:613–617

Gardner WK, Boundy KA (1983) The acquisition of phosphate by *Lupinus albus* L. IV. The effect of interplanting wheat and white lupin on the growth and mineral composition of the two species. Plant Soil 70:391–402

Gérard F, Mayer KU, Hodson MJ, Ranger J (2008) Modelling the biogeochemical cycle of silicon in soils: application to a temperate forest ecosystem. Geochim Cosmochim Acta 72:741–758

Glick BR (1995) The enhancement of plant-growth by free-living bacteria. Can J Microbiol 41:109–117

Graustein WC, Cromack K, Sollins P (1977) Calcium oxalate: occurrence in soils and effect on nutrient and geochemical cycles. Science 198:1252–1254

Gull M, Hafeez FY, Saleem M, Malik KA (2004) Phosphorus uptake and growth promotion of chickpea by co-inoculation of mineral phosphate solubilising bacteria and a mixed rhizobial culture. Aust J Exp Agric 38:1521–1526

Halliday J, Pate JS (1976) Symbiotic nitrogen fixation by blue green algae in the cycad Macrozamia riedlei: physiological characteristics and ecological significance. Aust J Plant Physiol 3:349–358

Harris JN, New PB, Martin PM (2006) Laboratory tests can predict beneficial effects of phosphate-solubilising bacteria on plants. Soil Biol Biochem 38:1521–1526

Hartmann A, Rothballer M, Schmid M (2008) Lorenz Hiltner, a pioneer in rhizosphere microbial ecology and soil bacteriology research. Plant Soil 312:7–14

Heath KD, Tiffin P (2007) Context dependence in the coevolution of plant and rhizobial mutualists. Proc Roy Soc B 274:1905–1912

Heath KD, Tiffin P (2009) Stabilizing mechanisms in a legume-rhizobium mutualism. Evolution 63:652–662

Hetrick BAD, Wilson GWT, Gill BS, Cox TS (1995) Chromosome location of mycorrhizal responsive genes in wheat. Can J Bot 73:891–897

Hibbert DS, Gilbert LB, Donoghue M (2000) Evolutionary instability of ectomycorrhizal symbiosis in basidiomycetes. Nature 407:506–508

Hinsinger P (1998) How do plant roots acquire mineral nutrients? Chemical processes involved in the rhizosphere. Adv Agron 64:225–265

Hinsinger P (2001) Bioavailability of soil inorganic P in the rhizosphere as affected by root-induced chemical changes: a review. Plant Soil 237:173–195

Hinsinger P (2002) Potassium. In: Lal R (ed) Encyclopedia of soil science. Marcel Dekker, Inc., New York

Hinsinger P, Jaillard B (1993) Root-induced release of interlayer potassium and vermiculitization of phlogopite as related to potassium depletion in the rhizosphere of ryegrass. J Soil Sci 44:525–534

Hinsinger P, Elsass F, Jaillard B, Robert M (1993) Root-induced irreversible transformation of a trioctahedral mica in the rhizosphere of rape. J Soil Sci 44:535–545

Hinsinger P, Fernandes Barros ON, Benedetti MF, Noack Y, Callot G (2001) Plant-induced weathering of a basaltic rock: experimental evidence. Geochim Cosmochim Acta 65:137–152

Hinsinger P, Plassard C, Tang C, Jaillard B (2003) Origins of root-induced pH changes in the rhizosphere and their responses to environmental constraints: a review. Plant Soil 248:43–59

Hinsinger P, Gobran GR, Gregory PJ, Wenzel WW (2005) Rhizosphere geometry and heterogeneity arising from root-mediated physical and chemical processes. New Phytol 168:293–303

Hinsinger P, Plassard C, Jaillard B (2006) The rhizosphere: a new frontier in soil biogeochemistry. J Geochem Explor 88:210–213

Hinsinger P, Bengough AG, Vetterlein D, Young IM (2009) Rhizosphere: biophysics, biogeochemistry and ecological relevance. Plant Soil 321:117–152. doi:10.1007/s11104-008-9885-9

Hobbie SE (1992) Effects of plant-species on nutrient cycling. Trends Ecol Evol 7:336–339

Hodge A, Robinson D, Fitter AH (2000) Are microorganisms more effective than plants at competing for nitrogen ? Trends Plant Sci 5:304–308

Hoffland E, Kuyper TW, Wallander H, Plassard C, Gorbushina A, Haselwandter K, Holmström S, Landeweert R, Lundström US, Rosling A, Sen R, Smits MM, Van Hees PAW, Van Breemen N (2004) The role of fungi in weathering. Frontiers Ecol Environ 2:258–264

Högberg P (1990) ^{15}N natural abundance as a possible marker of the ectomycorrhizal habit of trees in mixed African woodlands. New Phytol 115:483–486

Holmqvist J, Øgaard AF, Öborn I, Edwards AC, Mattsson L, Sverdrup H (2003) Application of the PROFILE model to estimate potassium release from mineral weathering in Northern European agricultural soils. Eur J Agron 20:149–163

Hopkins NA (1987) Mycorrhizae in a California serpentine grassland community. Can J Bot 65:484–487

Hopper SD, Gioia P (2004) The southwest Australian floristic region: evolution and conservation of a global hot spot of biodiversity. Annu Rev Ecol Evol Syst 35:623–650

Horner HT, Wagner BL (1995) Calcium oxalate formation in higher plants. In: Khan SR (ed) Calcium oxalate in biological systems. CRC, Boca Raton, pp 53–72

Huang WH, Keller WD (1970) Dissolution of rock-forming silicate minerals in organic acids: simulated fist-stage weathering of fresh mineral surfaces. Am Mineral 57:2076–2094

Jaillard B (1982) Relation entre dynamique de l'eau et organisation morphologique d'un sol calcaire. Science du Sol 20:31–52

Jaillard B (1983) Mise en évidence de la calcitisation des cellules corticales de racines de Graminées en milieu carbonaté. Compte Rendu de l'Académie des Sciences de Paris, t 297, série II, 293–296

Jaillard B (1984) Mise en évidence de la néogenèse de sables calcaires sous l'influence des racines: incidence sur la granulométrie du sol. Agronomie 4:91–100

Jaillard B (1985) Activité racinaire et rhizostructures en milieu carbonaté. Pédologie 35:297–313

Jaillard B (1987a) Les structures rhizomorphes calcaires: modèle de réorganisation des minéraux du sol par les racines. Thèse d'Etat, USTL, Montpellier, 228p

Jaillard B (1987b) Techniques for studying the ionic environment at the soil-root interface. In: Methodology in soil-K research (I.P.I. Ed.). International Potassium Institute, Bâle, pp 231-245

Jaillard B, Guyon A, Maurin AF (1991) Structure and composition of calcified roots, and their identification in calcareous soils. Geoderma 50:197–210

Jassogne L (2008) Characterisation of porosity and root growth in a sodic texture-contrast soil. PhD Thesis, The University of Western Australia

Jeong SC, Liston L, Myrold DD (1997) Molecular phylogeny of the genus Ceanothus using ndhF and rbcL sequences. Theor Appl Genet 94:825–857

Johnson NC, Graham JH, Smith FA (1997) Functioning of mycorrhizal associations along the mutualism-parasitism continuum. New Phytol 135:575–585

Jonsson LM, Nilsson LC, Wardle DA, Zackrisson O (2001) Context dependent effects of ectomycorrhizal species richness on tree seedling productivity. Oikos 93:353–364

Johnson D, Leake JR, Ostle N, Ineson P, Read DJ (2002) In situ (CO_2)-^{13}C pulse labelling of upland grassland demonstrates a rapid pathway of carbon flux from arbuscular mycorrhizal mycelia to the soil. New Phytol 153:327–334

Jones DL (1998) Organic acids in the rhizosphere–a critical review. Plant Soil 205:25–44

Jones CG, Lawton JH, Shachak M (1994) Organisms as ecosystem engineers. Oikos 69:373–386

Jones DL, Darrah PR, Kochian LV (1996) Critical evaluation of organic acid mediated dissolution in the rhizosphere and its potential role in root iron uptake. Plant Soil 180:57–66

Jones DL, Hodge A, Kuzyakov Y (2004) Plant and mycorrhizal regulation of rhizodeposition. New Phytol 163:459–480

Jones DL, Nguyen C, Finlay RD (2009) Carbon flow in the rhizosphere: carbon trading at the soil–root interface. Plant Soil 321:5–33. doi:10.1007/s11104-009-9925-0

Jongmans AG, Van Breemen N, Lundström U, Van Hees PAW, Finlay RD, Srinivasan M, Unestam T, Giesler R, Melkerud P-A, Olsson M (1997) Rock-eating fungi. Nature 389:682–683

Keel C, Weller DM, Nastch A, Défago G, Cook RJ, Thomashow LS (1996) Conservation of the 2, 4-diacetylphloroglucinol biosynthesis locus among fluorescent Pseudomonas strains from diverse geographic locations. Appl Environ Microbiol 62:552–563

Kenrick P (2001) Turning over a new leaf. Nature 410:309–310

Kenrick P, Crane PR (1997) The origin and early evolution of plants on land. Nature 389:33–39

Klappa CF (1980) Rhizoliths in terrestrial carbonates: classification, recognition, genesis and significance. Sedimentology 27:613–629

Kochian LV, Piñeros MA, Hoekenga OA (2005) The physiology, genetics and molecular biology of plant aluminum resistance and toxicity. Plant Soil 274:175–195

Kodama H, Nelson S, Yang F, Kohyama N (1994) Mineralogy of rhizospheric and non-rhizospheric soils in corn fields. Clays Clay Min 42:755–763

Kohzu A, Yoshioka T, Ando T, Takahashi M, Koba K, Wada E (1999) Natural ^{13}C and ^{15}N abundance of field-collected fungi and their ecological implications. New Phytol 144:323–330

Kosir A (2004) Microcodium revisited: root calcification products of terrestrial plants on carbonate-rich substrates. J Sediment Res 74:845–857

Krings M, Hass H, Kerp H, Taylor TN, Agerer R, Dotzler N (2009) Endophytic cyanobacteria in a 400-million-yr-old land plant:a scenario for the origin of a symbiosis? Rev Palaeobot Palynol 153:62–69

Kucey RMN, Janzen HH, Leggett ME (1989) Microbially mediated increases in plant-available phosphorus. Adv Agron 42:199–228

Kuchenbuch R, Jungk A (1982) A method for determining concentration profiles at the soil-root interface by thin slicing rhizospheric soil. Plant Soil 68:391–394

Kuzyakov Y, Friedel JK, Stahr K (2000) Review of mechanisms and quantification of priming effects. Soil Biol Biochem 32:1485–1498

Kuzyakov Y, Shevtzova TE, Pustovoytov K (2006) Carbonate re-crystallization in soil revealed by ^{14}C labeling: experiment, model and significance for paleo-environmental reconstructions. Geoderma 131:45–58

Lambers H (1987) Growth, respiration, exudation and symbiotic associations: the fate of carbon translocated to the roots. In: Gregory PJ, Lake JV, Rose DA (eds) Root development and function - effects of the physical environment. Cambridge University Press, Cambridge, pp 125–145

Lambers H, Shane MW, Cramer M, Pearse SJ, Veneklaas EJ (2006) Root structure and functioning for efficient acquisition of phosphorus: matching morphological and physiological traits. Ann Bot 98:693–713

Lambers H, Shaver G, Raven JA, Smith SE (2008a) N and P-acquisition change as soils age. Trends Ecol Evol 23:95–103

Lambers H, Chapin FS III, Pons TL (2008b) Plant physiological ecology, 2nd edn. Springer, New York

Landeweert R, Hoffland E, Finlay RD, Kuyper TW, Van Breemen N (2001) Linking plants to rocks: ectomycorrhizal fungi mobilize nutrients from minerals. Trends Ecol Evol 16:248–254

Lapeyrie F (1988) Oxalate synthesis from soil bicarbonate by the mycorrhizal fungus *Paxillus involutus*. Plant Soil 110:3–8

Leake JR, Read DJ (1989) The biology of mycorrhiza in the Ericaceae. New Phytol 112:69–76

Lemanceau P, Alabouvette C (1993) Suppression of fusarium-wilts by fluorescent psedomonads: mechanisms and applications. Biocontrol Sci Technol 3:219–234

Lerat S, Lapointe L, Gutjahr S, Piché Y, Vierheilig H (2003) Carbon partitioning in a split-root system of arbuscular mycorrhizal plants is fungal and plant species dependent. New Phytol 157:589–595

Leyval C, Berthelin J (1989) Interactions between *Laccaria laccata*, *Agrobacterium radiobacter* and beech roots: influence on P, K, Mg and Fe mobilization from minerals and plant growth. Plant Soil 117:103–110

Leyval C, Laheurte F, Belgy G, Berthelin J (1990) Weathering of micas in the rhizospheres of maize, pine and beech seedlings influenced by mycorrhizal and bacterial inoculation. Symbiosis 9:105–109

López-Ráez JA, Charnikhova T, Gómez-Roldán V, Matusova R, Kohlen W, De Vos R, Verstappen F, Puech-Pages V, Bécard G, Mulder P, Bouwmeester H (2008) Tomato strigolactones are derived from carotenoids and their biosynthesis is promoted by phosphate starvation. New Phytol 178:863–874

Lovelock JE (1965) A physical basis for life detection experiments. Nature 207:568–570

Lovelock JE (1993) The soil as a model for the Earth. Geoderma 57:213–215

Lovelock JE (2003) The living Earth. Nature 426:769–770

Lucas Y, Luizão FJ, Chauvel A, Rouiller J, Nahon D (1993) The relation between biological activity of the rain forest and mineral composition of soils. Science 260:521–523

Lynch JP, Ho MD (2005) Rhizoeconomics: carbon costs of phosphorus acquisition. Plant Soil 269:45–56

Ma JF (2005) Plant root responses to three abundant soil minerals: silicon, aluminum and iron. Crit Rev Plant Sci 24:267–281

Ma JF, Nomoto K (1994) Biosynthetic pathway of 3-epihydroxymugineic acid and 3-hydroxymugineic acid in gramineous plants. Soil Sci Plant Nutri 40:311–317

Ma JF, Ueno H, Ueno D, Rombolà AD, Iwashita T (2003) Characterization of phytosiderophore secretion under Fe deficiency stress in Festuca rubra. Plant Soil 256:131–137

Marschner H, Römheld V, Kissel M (1987) Localization of phytosiderophores release and of iron uptake along intact barley roots. Physiol Plant 71:157–162

Martin F, Duplessis S, Ditengou F, Lagrange H, Voiblet C, Lapeyrie F (2001) Developmental cross talking in the ectomycorrhizal symbiosis: signals and communication genes. New Phytol 152:145–154

McCully ME (1999) Roots in soil: unearthing the complexities of roots and their rhizospheres. Ann Rev Plant Physiol Plant Mol Biol 50:695–718

Michaud AM, Chappellaz C, Hinsinger P (2008) Copper phytotoxicity affects root elongation and iron nutrition in durum wheat (*Triticum turgidum durum* L.). Plant Soil 310:151–165

Micallef SA, Shiaris MP, Colon-Carmona A (2009) Influence of *Arabidopsis thaliana* accessions on rhizobacterial communities and natural variation in root exudates. J Exp Bot. doi:10.1093/jxb/erp053

Miyasaka SC, Buta JG, Howell RK, Foy CD (1991) Mechanism of aluminum tolerance in snapbeans: root exudation of citric acid. Plant Physiol 96:737–743

Mora CI, Driese SG, Colarusso LA (1996) Middle to late Paleozoic atmospheric CO_2 levels from soil carbonate and organic matter. Science 271:1105–1107

Morin N (1993) Microcodium: architecture, structure et composition. Comparaison avec les racines calcifiées. Thèse, USTL, Montpellier, 137 p

Mougel C, Offre P, Ranjard L, Corberand T, Gamalero E, Robin C, Lemanceau P (2006) Dynamic of the genetic structure of bacterial and fungal communities at different development stages of *Medicago truncatula* Jemalong J5. New Phytol 170:165–175

Moulton KL, West J, Berner RA (2000) Solute flux and mineral mass balance approaches to the quantification of plant effects on silicate weathering. Am J Sci 300:539–570

Nguyen C (2003) Rhizodeposition of organic C by plants: mechanisms and controls. Agronomie 23:375–396

Niebes JF, Hinsinger P, Jaillard B, Dufey JE (1993) Release of non exchangeable potassium from different size fractions of two highly K-fertilized soils in the rhizosphere of rape (*Brassica napus* cv Drakkar). Plant Soil 155(156):403–406

Oremus PAI, Otten H (1981) Factors affecting growth and nodulation of *Hippophae rhamnoides* L. ssp. *rhamnoides* in soils from two successional stages of dune formation. Plant Soil 63:317–331

Pankhurst CE, Pierret A, Hawke BG, Kirby JM (2002) Microbiological and chemical properties of soil associated with macropores at different depths in a red-duplex soil in NSW Australia. Plant Soil 238:11–20

Parfitt RL (1979) The availability of P from phosphate-goethite bridging complexes. Desorption and uptake by ryegrass. Plant Soil 53:55–65

Parfitt RL, Ross DJ, Coomes DA, Richardson SJ, Smale MC, Dahlgren RA (2005) N and P in New Zealand soil chronosequences and relationships with foliar N and P. Biogeochemistry 75:305–328

Paris F, Botton B, Lapeyrie F (1996) In vitro weathering of phlogopite by ectomycorrhizal fungi. II. Effect of K^+ and Mg^{2+} deficiency and N sources on accumulation of oxalate and H^+. Plant Soil 179:141–150

Paszkowski U (2006) Mutualism and parasitism: the yin and yang of plant symbioses. Curr Opin Plant Biol 9:364–370

Pate JS, Verboom WH (2009) Contemporary biogenic formation of clay pavements by eucalypts: further support for the phytotarium concept. Ann Bot 103:673–685

Pate JS, Verboom WH, Galloway PD (2001) Co-occurrence of Proteaceae, laterite and related oligotrophic soils: coincidental associations or causative inter-relationships? Aust J Bot 49:529–560

Paterson E (2003) Importance of rhizodeposition in the coupling of plant and microbial productivity. Eur J Soil Sci 54:741–750

Peng S, Eissenstat DM, Graham JH, Williams K, Hodge NC (1993) Growth depression in mycorrhizal citrus at high phosphorus supply: analysis of carbon costs. Plant Physiol 101:1063–1071

Perry LG, Thelen GC, Ridenour WM, Callaway RM, Paschke MW, Vivanco JM (2007) Concentrations of the Allelochemical (±)-catechin in *Centaurea maculosa* soils. J Chem Ecol 33:2337–2344

Purnell HM (1960) Studies of the family Proteaceae. I. Anatomy and morphology of the roots of some Victorian species. Austr J Bot 8:38–50

Raaijmakers JM, Paulitz TC, Steinberg C, Alabouvette C, Moënne-Loccoz Y (2009) The rhizosphere: a playground and battlefield for soilborne pathogens and beneficial microorganisms. Plant Soil 321:341–361. doi:10.1007/s11104-008-9568-6

Rasmann S, Köllner TG, Degenhardt J, Hiltpold I, Toepfer S, Kuhlmann U, Gershenzon J, Turlings TCJ (2005) Recruitment of entomopathogenic nematodes by insect-damaged maize roots. Nature 434:732–737

Raven JA (2002) The evolution of cyanobacterial symbioses. Proc R Irish Acad 102B:3–6

Raven JA, Edwards D (2001) Roots: evolutionary origins and biogeochemical significance. J Exp Bot 52:381–401

Raynaud X, Lata J-C, Leadley PW (2006) Soil microbial loop and nutrient uptake by plants: a test using a coupled C:N model of plant–microbial interactions. Plant Soil 287:95–116

Raynaud X, Jaillard B, Leadley PW (2008) Plants may alter competition by modifying nutrient bioavailability in rhizosphere: a modeling approach. Amer Nat 171:44–58

Read DJ (1996) The structure and function of the ericoid mycorrhizal root. Ann Bot 77:365–374

Read DJ, Perez-Moreno J (2003) Mycorrhizas and nutrient cycling in ecosystems–a journey towards relevance? New Phytol 157:475–492

Reddell P, Yun Y, Shipton WA (1997) Cluster roots and mycorrhizae in Casuarina cunninghamiana: their occurrence and formation in relation to phosphorus supply. Aust J Bot 45:41–51

Redecker D, Kodner R, Graham LE (2000) Glomalean fungi from the Ordovician. Science 289:1920–1921

Redman RS, Dunigan DD, Rodriguez RJ (2001) Fungal symbiosis from mutualism to parasitism: who controls the outcome, host or invader? New Phytol 151:705–716

Reichard PU, Kraemer SM, Frazier SW, Kretzschmar R (2005) Goethite dissolution in the presence of phytosiderophores: rates, mechanisms, and the synergistic effect of oxalate. Plant Soil 276:115–132

Reichard PU, Kretzschmar R, Kraemer SM (2007) Dissolution mechanisms of goethite in the presence of siderophores and organic acids. Geochim Cosmochim Acta 71:5635–5650

Remy W, Taylor TN, Hass H, Kerp H (1994) Four hundred-million-year-old vesicular arbuscular mycorrhizae. Proc Natl Acad Sci USA 91:11841–11843

Retallack GJ (1997) Early forest soils and their role in Devonian global change. Science 276:583–585

Reynolds HL, Packer A, Bever JD, Clay K (2003) Grassroots ecology: plant-microbe-soil interactions as drivers of plant community structure and dynamics. Ecology 84:2281–2291

Richardson AE (2001) Prospects for using soil microorganisms to improve the acquisition of phosphorus by plants. Aust J Plant Physiol 28:897–906

Richardson AE, Barea J-M, McNeill AM, Prigent-Combaret C (2009) Acquisition of phosphorus and nitrogen in the rhizosphere and plant growth promotion by microorganisms. Plant Soil 321:305–339. doi:10.1007/s11104-009-9895-2

Ridenour WM, Callaway RM (2001) The relative importance of allelopathy in interference: the effects of an invasive weed on a native bunchgrass. Oecologia 126:444–450

Robin A, Vansuyt G, Hinsinger P, Meyer JM, Briat JF, Lemanceau P (2008) Iron dynamics in the rhizosphere: consequences for plant health and nutrition. Adv Agron 99:183–225

Rodriguez R, Redman R (2008) More than 400 million years of evolution and some plants still can't make it on their own: plant stress tolerance via fungal symbiosis. J Exp Bot 59:1109–1114

Römheld V (1987) Different strategies for iron acquisition in higher plants. Physiol Plant 70:231–234

Römheld V (1991) The role of phytosiderophores in acquisition of iron and other micronutrients in graminaceous species: an ecological approach. Plant Soil 130:127–134

Ronquist F (1998) Phylogenetic approaches in coevolution and biogeography. Zool Scripta 26:313–322

Rudrappa T, Bonsall J, Gallagher JL, Seliskar DM, Bais HP (2007) Root-secreted allelochemical in the noxious weed *Phragmites australis* deploys a reactive oxygen species response and microtubule assembly disruption to execute rhizotoxicity. J Chem Ecol 33:1898–1918

Rygiewicz PT, Andersen CP (1994) Mycorrhizae alter quality and quantity of carbon allocated below ground. Nature 369:58–60

Sachs JL, Mueller UG, Wilcox TP, Bull JJ (2004) The evolution of cooperation. Q Rev Biol 79:135–160

Saikkonen K, Faeth SH, Helander M, Sullivan TJ (1998) Fungal endophytes: a continuum of interactions with host plants. Ann Rev Ecol Syst 29:319–343

Saubidet MI, Fatta N, Barneix AJ (2002) The effect of inoculation with *Azospirillum brasilense* on growth and nitrogen utilization by wheat plants. Plant Soil 245:215–222

Schippers B, Bakker AW, Bakker P (1987) Interactions of deleterious and beneficial rhizosphere microorganisms and the effect of cropping practices. Annu Rev Phytopathol 25:339–358

Schopf JW, Kudryavtsev AB, Agresti DG, Wdowlak TJ, Czaja AD (2002) Laser-Raman imagery of Earth's earliest fossils. Nature 416:73–76

Schüßler A, Kluge M (2000) *Geosiphon pyriforme*, an endocytosymbiosis between fungus and cyanobacteria, and its meaning as a model system for Arbuscular mycorrhizal research. In: Hock B (ed) The mycota IX fungal associations. Springer-Verlag, Berlin, pp 151–161

Schüßler A, Gehrig H, Schwarzott D, Walker C (2001) Analysis of partial Glomales SSU rRNA gene sequences: implications for primer design and phylogeny. Mycol Res 105:5–15

Semchenko M, Zobel K, Heinemeyer A, Hutchings MJ (2008) Foraging for space and avoidance of physical obstructions by plant roots: a comparative study of grasses from contrasting habitats. New Phytol 179:1162–1170

Schwartzman D, Volk T (1989) Biotic enhancement of weathering and the habitability of Earth. Nature 340:457–460

Shane MW, Lambers H (2005a) Manganese accumulation in leaves of *Hakea prostrata* (Proteaceae) and the significance of cluster roots for micronutrient uptake as dependent on phosphorus supply. Physiol Plant 124:441–450

Shane MW, Lambers H (2005b) Cluster roots: a curiosity in context. Plant Soil 274:99–123

Shane MW, Cramer MD, Funayama-Noguchi S, Cawthray GR, Millar AH, Day DA, Lambers H (2004) Developmental physiology of cluster-root carboxylate synthesis and exudation in harsh hakea. Expression of phosphoenolpyruvate carboxylase and the alternative oxidase. Plant Physiol 135:549–560

Shane MW, Cawthray GR, Cramer MD, Kuo J, Lambers H (2006) Specialized 'dauciform' roots of Cyperaceae are structurally distinct, but functionally analogous with 'cluster' roots. Plant Cell Environ 29:1989–1999

Simonsson M, Andersson S, Andrist-Rangel Y, Hillier S, Mattsson L, Öborn I (2007) Potassium release and fixation as a function of fertilizer application rate and soil parent material. Geoderma 140:188–198

Skene KR, Sprent JI, Raven JA, Herdman L (2000) *Myrica gale* L. J Ecol 88:1079–1094

Smith SE, Read DJ (2008) Mycorrhizal symbiosis, 3rd edn. Elsevier, City

Smith KP, Handelsman J, Goodman RM (1999) Genetic basis in plants for interactions with disease-suppressive bacteria. Proc Natl Acad Sci USA 96:4786–4790

Smith SE, Smith FA, Jakobsen I (2003) Mycorrhizal fungi can dominate phosphate supply to plants irrespective of growth responses. Plant Physiol 133:16–20

Soltis DE, Soltis PS, Morgan DR, Swensen SM, Mullin BC, Dowd JM, Martin PG (1995) Chloroplast gene sequence data suggest a single origin of the predisposition for symbiotic nitrogen fixation in angiosperms. Proc Natl Acad Sci USA 92:2647–2651

Sprent JI (2007) Evolving ideas of legume evolution and diversity: a taxonomic perspective on the occurrence of nodulation. New Phytol 174:11–25

Sprent JI, James EK (2007) Legume evolution: where do nodules and mycorrhizas fit in? Plant Physiol 144:575–581

Sprent JI, Raven JA (1992) Evolution of nitrogen-fixing root nodules symbioses. In: Stacey G, Burris RH, Evans HJ (eds) Biological nitrogen fixation: achievements and objectives. Chapman & Hall, New York, pp 461–496

Springob G, Richter J (1998) Measuring interlayer potassium release rates from soil materials. II. A percolation procedure to study the influence of the variable 'solute K' in the < 1…10 μM range. Z Pflanzenern Bodenkd 161:323–329

Takagi S, Nomoto K, Takemoto T (1984) Physiological aspect of mugineic acid, a possible phytosiderophore of graminaceous plant. J Plant Nutr 7:469–477

Tamietti G, Ferraris L, Matta A (1993) Physiological-responses of tomato plants grown in Fusarium suppressive soil. J Phytopathol - Phytopathologische Zeitschrift 138:66–76

Taylor AB, Velbel MA (1991) Geochemical mass balances and weathering rates in forested watersheds of the Southern Blue Ridge II. Effects of botanical uptake terms. Geoderma 51:29–50

Taylor LL, Leake JR, Quirk J, Hardy K, Banwarts SA, Beerling DJ (2009) Biological weathering and the long-term carbon cycle: integrating mycorrhizal evolution and function into the current paradigm. Geobiology 7:171–191

Thompson JN (1997) Evaluating the dynamics of coevolution among geographically structured populations. Ecology 78:1619–1623

Thompson JN (1999) The evolution of species interaction. Science 284:2116–2118

Thompson JN, Burdon JJ (1992) Gene-for-gene coevolution between plants and parasites. Nature 360:121–125

Tinker PB, Durall DM, Jones MD (1994) Carbon use efficiency in mycorrhizas–Theory and sample calculations. New Phytol 128:115–122

Turlings TCJ, Ton J (2006) Exploiting scents of distress: the prospect of manipulating herbivore-induced plant odours to enhance the control of agricultural pests. Curr Opin Plant Biol 9:421–427

Turlings TCJ, Wäckers FL (2004) Recruitment of predators and parasitoids by herbivore-damaged plants. In: Cardé RT, Millar J (eds) Advances in insect chemical ecology. Cambridge University Press, Cambridge, pp 21–75

Turpault MP, Utérano C, Boudot JP, Ranger J (2005) Influence of mature Douglas fir roots on the solid soil phase of the rhizosphere and its solution chemistry. Plant Soil 275:327–336

Uroz S, Calvaruso C, Turpault MP, Pierrat JC, Mustin C, Frey-Klett P (2007) Effect of mycorrhizosphere on the genotypic and metabolic diversity of the bacterial communities involved in mineral weathering in a forest soil. Appl Environ Microbiol 79:3019–3027

Van Breemen N (1993) Soils as biotic constructs favouring net primary productivity. Geoderma 57:183–211

Van Breemen N, Driscoll CT, Mulder J (1984) Acidic deposition and internal proton sources in acidification of soils and waters N. Nature 307:599–604

Van Ghelue M, Løvaas E, Ringø E, Solheim B (1997) Early interactions between *Alnus glutinosa* (L.) Gaertn. And *Frankia* strain ArI3. Production and specificity of root hair deformation factor(s). Physiol Plant 99:579–587

Van Loon LC, Bakker PAHM, Pieterse CMJ (1998) Systemic resistance induced by rhizosphere bacteria. Annu Rev Phytopathol 36:453–483

Van Tol RWHM, Van der Sommen ATC, Boff MIC, Van Bezooijen J, Sabelis MW, Smits PH (2001) Plants protect their roots by alerting the enemies of grubs. Ecol Lett 4:292–294

Vance CP, Heichel GH (1991) Carbon in N2 fixation: limitation or exquisite adaptation. Annu Rev Plant Physiol Plant Mol Biol 42:373–392

Vance CP, Uhde-Stone C, Allen DL (2003) Phosphorus acquisition and use: critical adaptations by plants for securing a non-renewable source. New Phytol 157:423–447

Velbel MA, Price JR (2007) Solute geochemical mass-balances and mineral weathering rates in small watersheds: Methodology, recent advances, and future directions. Appl Geochem 22:1682–1700

Verboom WH, Pate JS (2003) Relationships between cluster root-bearing taxa and laterite across landscapes in south-west Western Australia: an approach using airborne radiometric and digital elevation models. Plant Soil 248:321–333

Verboom WH, Pate JS (2006a) Bioengineering of soil profiles in semiarid ecosystems: the 'phytotarium' concept. A review. Plant Soil 289:71–102

Verboom WH, Pate JS (2006b) Evidence of active biotic influences in pedogenetic processes. Case studies from semiarid ecosystems of south-west Western Australia. Plant Soil 289:103–121

Verrecchia EP (1990) Litho-diagenetic implications of the calcium oxalate-carbonate biogeochemical cycle in semi-arid calcretes, Nazareth, Israel. Geomicrobiol J 8:87–99

Verrecchia EP, Dumont JL (1996) A biogeochemical model for chalk alteration by fungi in semiarid environments. Biogeochemistry 35:447–470

Vessey JK, Pawlowski K, Bergman B (2005) N2-fixing symbiosis: legumes, actinorhizal plants, and cycads. Plant Soil 274:51–78

Vitousek PM, Howarth RW (1991) Nitrogen limitation on land and in the sea: how can it occur? Biogeochemistry 13:87–115

Walker TW, Syers JK (1976) Fate of phosphorus during pedogenesis. Geoderma 15:1–19

Wallander H (2000) Uptake of P from apatite by *Pinus sylvestris* seedlings colonised by different ectomycorrhizal fungi. Plant Soil 218:249–256

Wallander H, Wickman T (1999) Biotite and microcline as potassium sources in ectomycorrhizal and non-mycorrhizal *Pinus sylvestris* seedlings. Mycorrhiza 9:25–32

Wang H, Greenberg SE (2007) Reconstructing the response of C3 and C4 plants to decadal-scale climate change during the late Pleistocene in southern Illinois using isotopic analyses of calcified rootlets. Quaternary Res 67:136–142

Wang B, Qiu Y-L (2006) Phylogenetic distribution and evolution of mycorrhizas in land plants. Mycorrhiza 16:299–363

Watkins NK, Fitter AH, Graves JD, Robinson D (1996) Carbon transfer between C-3 and C-4 plants linked by a common mycorrhizal network, quantified using stable carbon isotopes. Soil Biol Biochem 28:471–477

Watt M, Evans JR (1999) Linking development and determinacy with organic acid efflux from proteoid roots of white lupin grown with low phosphorus and ambient or elevated atmospheric CO_2 concentration. Plant Physiol 120:705–716

Weir T, Bais H, Stull V, Callaway R, Thelen G, Ridenour W, Bhamidi S, Stermitz F, Vivanco J (2006) Oxalate contributes to the resistance of *Gaillardia grandiflora* and *Lupinus sericeus* to a phytotoxin produced by *Centaurea maculosa*. Planta 223:785–795

Weisskopf L, Abou-Mansour E, Fromin N, Tomasi N, Santelia D, Edelkott I, Neumann G, Aragno M, Tabacchi R, Martinoia E (2006) White lupin has developed a complex strategy to limit microbial degradation of the secreted citrate required for phosphate nutrition. Plant Cell Environ 29:919–927

Weller DM (1988) Biological control of soilborne plant pathogens in the rhizosphere with bacteria. Annu Rev Phytopathol 26:379–407

Wissuwa M, Mazzola M, Picard C (2009) Novel approaches in plant breeding for rhizosphere–related traits. Plant Soil 321:409–430. doi:10.1007/s11104-008-9693-2

Wu T, Sharda JN, Koide RT (2003) Exploring interactions between saprotrophic microbes and ectomycorrhizal fungi using a protein–tannin complex as an N source by red pine (*Pinus resinosa*). New Phytol 159:131–139

Yao Q, Li X, Feng G, Christie P (2001) Mobilization of sparingly soluble inorganic phosphates by the external mycelium of an arbuscular mycorrhizal fungus. Plant Soil 230:279–285

Yang Y-Y, Jung J-Y, Suh SW-Y, H-S LY (2000) Identification of rice varieties with high tolerance or sensitivity to lead and characterization of the mechanism of tolerance. Plant Physiol 124:1019–1026

Yoneyama K, Xie X, Sekimoto H, Takeuchi Y, Ogasawara S, Akiyama K, Hayashi H, Yoneyama K (2008) Strigolactones, host recognition signals for root parasitic plants and arbuscular mycorrhizal fungi, from Fabaceae plants. New Phytol 179:484–494

Yu JQ, Shou SY, Qian YR, Zhu ZJ, Hu WH (2000) Autotoxic potential of cucurbit crops. Plant Soil 223:147–151

Yunusa IAM, Newton PJ (2003) Plants for amelioration of subsoil constraints and hydrological control: the primer-plant concept. Plant Soil 257:261–281

Zheng SJ, Ma JF, Matsumoto H (1998) High aluminum resistance in buckwheat. I. Al-induced specific secretion of oxalic acid from root tips. Plant Physiol 117:745–751

Zuo Y, Zhang F, Li X, Cao Y (2000) Studies on the improvement in iron nutrition of peanut by intercropping with maize on a calcareous soil. Plant Soil 220:13–25

Zwart KB, Kuikman PJ, van Veen JA (1994) Rhizosphere protozoa: their significance in nutrient dynamics. In: Darbyshire JF (ed) Soil protozoa. CAB international, Wallingford, pp 93–122

REVIEW ARTICLE

Rhizosphere: biophysics, biogeochemistry and ecological relevance

Philippe Hinsinger · A. Glyn Bengough ·
Doris Vetterlein · Iain M. Young

Received: 27 May 2008 / Accepted: 29 December 2008 / Published online: 21 January 2009
© Springer Science + Business Media B.V. 2009

Abstract Life on Earth is sustained by a small volume of soil surrounding roots, called the rhizosphere. The soil is where most of the biodiversity on Earth exists, and the rhizosphere probably represents the most dynamic habitat on Earth; and certainly is the most important zone in terms of defining the quality and quantity of the Human terrestrial food resource. Despite its central importance to all life, we know very little about rhizosphere functioning, and have an extraordinary ignorance about how best we can manipulate it to our advantage. A major issue in research on rhizosphere processes is the intimate connection between the biology, physics and chemistry of the system which exhibits astonishing spatial and temporal heterogeneities. This review considers the unique biophysical and biogeochemical properties of the rhizosphere and draws some connections between them. Particular emphasis is put on how underlying processes affect rhizosphere ecology, to generate highly heterogeneous microenvironments. Rhizosphere ecology is driven by a combination of the physical architecture of the soil matrix, coupled with the spatial and temporal distribution of rhizodeposits, protons, gases, and the role of roots as sinks for water and nutrients. Consequences for plant growth and whole-system ecology are considered. The first sections address the physical architecture and soil strength of the rhizosphere, drawing their relationship with key functions such as the movement and storage of elements and water as well as the ability of roots to explore the soil and the definition of diverse habitats for soil microorganisms. The distribution of water and its accessibility in the rhizosphere is considered in detail, with a special emphasis on spatial and temporal dynamics and heterogeneities. The physical architecture and water content play a key role in determining the biogeochemical ambience of the rhizosphere, via their effect on partial pressures of O_2 and CO_2, and thereby on redox potential and pH of the rhizosphere, respectively. We address the

Responsible Editor: Philippe Lemanceau.

P. Hinsinger (✉)
UMR 1222 Eco&Sols Ecologie Fonctionnelle
& Biogéochimie des Sols (INRA–IRD–SupAgro), INRA,
Place Viala,
34060 Montpellier, France
e-mail: philippe.hinsinger@supagro.inra.fr

A. G. Bengough
Scottish Crop Research Institute,
Dundee DD2 5DA, United Kingdom

D. Vetterlein
Department Soil Physics,
Helmholtz Centre for Environmental Research-UFZ,
Theodor-Lieser-Str. 4,
06120 Halle/Saale, Germany

I. M. Young
School of Environmental and Rural Sciences,
University of New England,
Armidale, NSW 2351, Australia

various mechanisms by which roots and associated microorganisms alter these major drivers of soil biogeochemistry. Finally, we consider the distribution of nutrients, their accessibility in the rhizosphere, and their functional relevance for plant and microbial ecology. Gradients of nutrients in the rhizosphere, and their spatial patterns or temporal dynamics are discussed in the light of current knowledge of rhizosphere biophysics and biogeochemistry. Priorities for future research are identified as well as new methodological developments which might help to advance a comprehensive understanding of the co-occurring processes in the rhizosphere.

Keywords Soil strength · Soil structure · Water potential · pH · Redox potential · Nutrient availability

Introduction

Soils are the largest reservoir of biodiversity on Earth. They are important habitats for Prokaryotes and a diversity of Eukaryotes, which comprise fungi among soil microorganisms, as well as large variety of invertebrates (from protozoa and nematodes to mites, collembola, insects and earthworms). The diversity of Prokaryotes in soil has been estimated to be about three orders of magnitude larger than in all other environmental compartments of the Earth's ecosystems combined (Curtis et al. 2002; Crawford et al. 2005; Curtis and Sloan 2005). Roots of higher plants anchor the above-ground diversity of terrestrial ecosystems, and provide much of the carbon to power the soil ecosystem. Besides their role in biodiversity, soils are even more remarkable from a functional perspective, in sustaining all other forms of terrestrial diversity and providing many ecosystem services.

A major feature of soils is their temporal and spatial heterogeneities from the nm to the km scales (Young and Ritz 2000; Pierret et al. 2007). Soils are complex assemblages of extremely diverse habitats, which certainly explain why they harbour such a diversity of organisms. For instance, Ramette and Tiedje (2007) have shown that the interactions of environmental heterogeneities and spatial distance are central determinants of the relatedness and abundance of rhizosphere bacteria of the *Burkholderia cepacia* complex. Besides species richness, species abundances are also remarkable in soils (e.g. in Watt et al. 2006a). Even though a single gram of soil may contain about 10^7–10^{12} bacteria, 10^4 protozoa, 10^4 nematodes, 5–25 km of fungal hyphae, given an average specific surface area of about 20 $m^2 \ g^{-1}$ and the very small size of most of these microorganisms, their surface coverage amounts in total to only 10^{-5}–10^{-6}% of the total soil surface area (Young and Crawford 2004). The soil can be considered a huge desert, where life is discretely distributed, even more so when one accounts for the tendency of many of these soil microorganisms to form colonies and to aggregate, forming hot spots of activity (Ranjard and Richaume 2001; Nunan et al. 2003; Watt et al. 2006b). One of the most fascinating hot spots of activity and diversity in soils is the rhizosphere (Jones and Hinsinger 2008).

The rhizosphere is best defined as the volume of soil around living roots, which is influenced by root activity (the "Einflusssphäre der Wurzel" according to Hiltner (1904) in Hartmann et al. 2008). As stressed by Hinsinger et al. (2005) and Gregory (2006) this means that, depending on the activity that one considers (exudation of reactive compounds, respiration, uptake of more or less mobile nutrients and water), the radial extension of the rhizosphere can range from sub-µm to supra-cm scales. As stressed by Darrah (1993), the inner boundary of the rhizosphere is not better defined. When one considers the movement of water, nutrients or endophytic microorganisms through the apoplasm, the inner boundary is inadequately represented by the outer surface of the root, as depicted in most rhizosphere models (Watt et al. 2006c). The temporal development of the rhizosphere is equally relevant to consider (Jones et al. 2004; Watt et al. 2006a and 2006c), although relatively poorly documented. Spatial and temporal components of the rhizosphere will thus especially be addressed in this review.

Soil is a physical environment where it is often difficult for roots, microorganisms and soil fauna to move, and where resources (water, air, nutrients) are frequently scarce and patchy, with considerable vertical variation down the soil profile. Even when abundant, soil resources are often poorly available to organisms due to the capacity of soil matrix to bind water and nutrients, so that roots have evolved to adapt and to influence their environment (Lambers et al. 1998; Raven and Edwards 2001; Hinsinger et al. 2005), optimizing their functional architecture to explore and make use of resources in heterogeneous soils (Leyser and Fitter 1998; Pierret et al. 2007). Roots of higher plants (and their associated microbes) have coevolved with soils as they play a major role in soil formation processes, via a range of physical, chemical and biological processes (Verboom and Pate 2006; Lambers et al. 2009). The aim of this paper is to take the reader through a journey in the biophysical and biogeochemical environment of plant roots. In each topic we consider the ecological relevance, underlying processes and spatial/temporal heterogeneity operating in this crucial micro-environment.

Physical architecture of the rhizosphere

Ecological relevance

There are two fundamental reasons to attempt to understand the physical architecture of the volume of soil immediately surrounding the root. Firstly, the stability of this inner physical structure is a key determinant of a root's ability to explore and exploit the soil resource. Secondly, the geometry of the pore space (Fig. 1) defines the allocation of resources to soil biota, the permeability of gases and solutes to and from the root, and the diversity of microbial habitats in the area of highest carbon resource.

Underlying processes

For decades we have had at least a qualitative understanding of the impact of roots on the stability of soil. Simply put, the rhizosphere volume exhibits a greater resistance to an external, mechanical stress than soil not associated with roots, thus soil in the presence of roots generally exhibits greater stability. Typically this is assessed using some form of aggregate stability test (see Young et al., 2001) or rheological tests (Czarnes et al. 1999). Examining the influence of six crop species on aggregate stability Haynes and Beare (1997) found the presence of roots significantly increased stability (50–100% increase compared to a non-planted control), but only after the soil was air-dried. This result is directly related to the lower resistance of dry soil to slaking and highlights the importance of understanding the stability of soil systems across a range of environments. Additionally, legume crops were shown to have a greater influence on aggregation compared with non-legume crops. Overall, the authors attributed an important, yet undefined role, of the microbial community in increasing stability within plant species. In particular the role of saprophytic fungi in association with legume crops was identified as important. Caravaca et al. (2005) showed that a combination of plant type and rhizosphere microbial community affected aggregate stability. In their study arbuscular mycorrhizal fungi was implicated in increasing stability of soil associated with roots, which concurs with the work of Kabir and Koide (2000). In a recent study, Moreno-Espindola et al. (2007) found that root hairs were more important in the adhesion and stability of soil (predominantly sand –70%) than fungal hyphae. They reported a ratio of 40:1 for maize and 100:1 for Bermuda grass (*Cynodon dactylon* L.), even in the presence of arbuscular mycorrhizal fungi. The predominance of root hairs over hyphae may be due to the nature of the soil. Predominantly sand, this generates temporal changes in moisture (wet~dry cycles) occurring at the root-soil interface, which in addition to dense root hairs, increases stability and adhesion, as compared with soil more associated with hyphae. Soil-water is known to dramatically increase the cohesion and strength of sand – e.g. sand on a beach – and soil, through a mechanism known as effective mechanical stress (Mullins and Panayiotopoulos 1984).

Through many studies it is clear that biological activity may increase the stability of soil within and outwith the rhizosphere. What is evident is the importance of wet~dry cycles to 'lock-in' that stability. Clearly a combination of biophysical factors, the exact nature dependent on a wide range of conditions, impacts on the stability of soil at the root-soil interface. Another important factor relates to the chemical make-up of the carbon involved. Martens (2000) conclusively demonstrated the importance of phenolic acids, predominantly plant-derived, in soil aggregation. His works supports the conceptual framework of Tisdall and Oades (1992) that states *"residues with slower decomposition rates resulted in persistent soil aggregation."* Kaci et al. (2005), focussing on the production of exopolysaccharides by bacterial populations found exopolysaccharides exuded by *Rhizobium* to be composed of a tetrasaccharide repeating unit. This was considered as a thickening agent with polyelectrolyte properties which provided significant increases in soil aggrega-

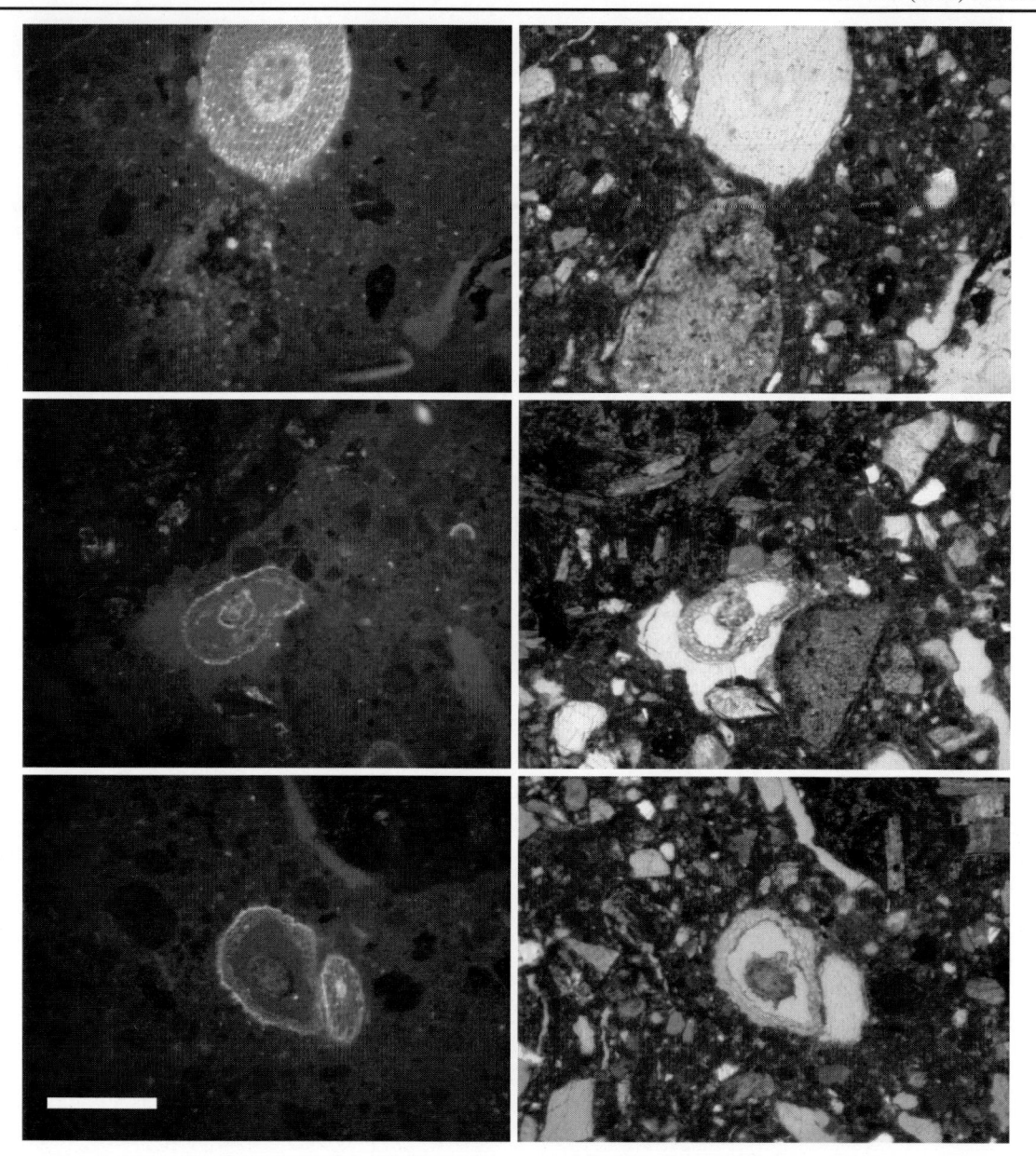

Fig. 1 Micrographs of soil thin sections showing barley roots growing in a sandy loam soil. Fluorescence images on left, transmission images on right. Images show a root with intact cortex in soil with few macropores present (top), a root with disintegrated cortex growing in a macropore (middle), and a main root axis with one of its lateral branches, distorted by the pressure of soil particles (bottom–only the lateral has intact cortex). Scale bar, bottom left, is 0.5 mm (Reproduced by kind permission of A. Glyn Bengough)

tion. Besides microbial exopolysaccharides, roots are also directly responsible for production of mucilages that altogether considerably affect soil structure in the rhizosphere and, ultimately structure-dependent processes such as water transport (Czarnes et al. 2000).

The combined effect of root hairs and mucilage either produced by the root itself or by rhizosphere microorganisms (Watt et al. 1993) can lead to the formation of specific structures called rhizosheaths (Fig. 2) which have been evidenced for a wide range of plant species and especially in grasses (Watt et al. 1994; Young 1995; North and Nobel 1997; McCully 1999; Moreno-Espindola et al. 2007). These structures are remarkably stable and play a dual role in soil-root water transfers, whilst their formation is definitely linked with water dynamics (Watt et al. 1994; Young 1995). How stable a physical structure is, is obviously important in terms of the impact of external and internal perturbations on that structure. However, the geometry of the structure has a vital role in its functioning (Crawford et al. 2005). In the context of soil, the nature of the spatial and temporal geometry of the porosity is a key factor.

Gradients, spatial heterogeneity and temporal development

Young (1998) provides a review of research in the variations of certain aspects of the geometry of soil from rhizosphere to bulk soil (i.e. soil without roots). A summary of his review shows a small but significant body of work that observed increases in bulk density close to the root-soil interface. This presents a picture of a root punching through soil, deforming and packing relatively wet soil to form the start of a new rhizosphere. Tighter packing around the root increases root-soil contact and thus, in theory increases hydraulic contact, and thus the probability of resource exchange from soil to root and vice versa. However, in front of the root tip any increase in density would be counterproductive. Slightly counter to this work is the research by Martens and Frankenberger (1992) on the impact of bacterial polymers which showed significant increases (20%) in porosity in pre-packed soils. In a similar vein Alami et al (2000) examined the role of exopolysaccharide-producing *Rhizobium* on the structure of rhizosphere soil, finding significant increases in soil porosity (12–60 μm), irrespective of initial soil water regime. An interesting secondary observation was the potential role of exopolysaccharides in reducing the impact of water deficit on plants. This neatly ties into the role of exopolysaccharides in minimizing the effects of desiccation on bacterial populations within biofilms.

In a recent study Feeney et al. (2006) carried out an extensive analysis of the impact of plant roots and microorganisms on the structure of the rhizosphere. Using a combination of high resolution x-rays and 3D geostatistical analysis they analysed micropore properties (>4.4 μm) from rhizosphere and non-rhizosphere soil. These results showed, for the first time in 3D, large and significant increases in micropore porosity associated with root+microbe, and microbe only soil, 12% and 8%, respectively compared with 4% in control. Such increased rhizosphere porosity does however not contradict increased bulk density, as reported by other authors when using techniques that do not resolve the pore size distribution with high resolution. Indeed root growth may result in larger pores being squashed to become smaller pores. Hence whilst the total porosity decreases (and thus the bulk density increases), the volume of particular size ranges of pores may actually increase, especially in the micropore range as reported within the rhizosphere aggregates as revealed by the use of high resolution x-rays and 3D geostatistical analysis.

Importantly, Feeney et al. (2006) also measured the spatial correlation that exists between pore volume

Fig. 2 Rhizosheaths formed around roots of *Lyginia barbata* R.Br. (top photograph) and barley (*Hordeum vulgare* L., bottom micrograph). Micrograph obtained by cryoscanning electron microscopy of the rhizosheath sampled in situ in field-grown barley plants. Development of long root hairs and their role in aggregating the soil thereby forming the rhizosheath is clearly visible (Reproduced by kind permission of Philippe Hinsinger (top photograph) and Margaret E. McCully)

neighbours. This is an important measure as Crawford et al. (2005) and Young and Crawford (2004) suggest that as the spatial correlation increases this signifies a move from random to a correlated structure. The latter relates directly to an increase in local diffusion rates and thus resource allocation to the microsites where many of the microbial populations reside. This work provided the first substantive proof that soil-plant-microbe systems operate as a self-organised unit, with the microporosity as a driving force.

A key issue for future research related to the physical structure of soil, is to see a move away from relatively descriptive work (the physics of numbers and differences) to a more functional approach. It is less what is different and more what is the functional relevance of any changes in structure. Linking this into predictive models will provide a much needed input on the spatio-temporal dynamics of the soil system for all processes.

Soil strength in the rhizosphere

Ecological consequences of rhizosphere strength

The soil strength around the root apex greatly influences the pressure that a root must exert to penetrate the soil. If a pre-existing channel does not exist, a root must exert sufficient pressure to rearrange the soil particles and either push them aside, or ahead of the root apex. The soil within a radius of up to 20 times the radius of a penetrating probe can exert a mechanical influence on the probe (Greacen et al. 1969), and it is likely that this is also the case for a root. This is the zone where, depending on the soil mechanical properties, plastic (irreversible) and elastic (reversible) deformation occurs.

Mechanical impedance to root growth decreases the root elongation rate and increases root diameter (Taylor and Ratliff 1969; Bengough and Mullins 1990). Plants with shorter root axes explore a smaller volume of soil, and are therefore more likely to suffer nutrient and water shortage if these resources are scarce, limiting shoot growth. Mechanical impedance also restricts shoot growth directly, even when water and nutrient supply are non-limiting. Leaf expansion in young wheat seedlings decreased by two thirds as penetrometer resistance increased from 1.5 MPa to 5.5 MPa, and was unresponsive to increasing nutrient supply (Masle and Passioura 1987). A rapid shoot response to mechanically stressing the root system was shown clearly by applying an external confining stress to roots of wheat (*Triticum aestivum* L.) and barley (*Hordeum vulgare* L.) growing in sand, causing decreased shoot elongation rates within 10 min (Young et al. 1997). This decrease in shoot growth associated with increasing soil strength is caused by an unidentified, and possibly complex, system of root-shoot signalling (Passioura 1988; Passioura 2002).

Plants growing in the field experience a wide range of soil physical conditions throughout a growing season (Bengough et al. 2006). Analysis of the strength of soil as a function of its matric potential indicates that mechanical impedance will often limit root elongation severely at matric potentials in the range −0.10 MPa to −0.25 MPa (Whalley et al. 2005a). Matric potentials in this range are not normally major limitations to root elongation in the absence of mechanical impedance (Sharp et al. 2004).

Soil strength also influences the colonisation of root tips by soil bacteria. Slower root elongation rates associated with compacted soil increased the numbers of bacteria in the rhizosphere around root tips of wheat grown in lab and field experiments (Watt et al. 2003). Watt et al. (2003) showed that root axes extending at one third of the elongation rate in loose soil had eight times as many bacteria and 20 times as many *Pseudomonas* spp. per unit length of root, with the biggest differences being in the apical 10 mm. The strength of the soil surrounding the root therefore has both direct and indirect effects on plant growth and rhizosphere ecology and, in the next section we consider some of the mechanisms influencing rhizosphere strength.

Underlying processes influencing rhizosphere strength

The mechanical properties of the rhizosphere depend on the local soil density, its matric potential, and the introduction of any materials that influence physical interactions between neighbouring soil particles.

The soil density in the rhizosphere will depend largely on the root diameter, and the path followed by the growing root tip. The total decrease in pore space around the root must be at least as big as the volume occupied by the root. The seminal root axes of most cereal crops rapidly approach their maximum diameter within a few mm of the root apex. Increasing soil strength decreases the length of the elongation zone of the root, and can as much as double the root diameter (Bengough and Mullins 1990; Watt et al. 2005). The trajectory followed by a root tip will depend on the spatial variation in soil strength. For instance, roots often locate and occupy large channels in the soil more frequently than expected by chance alone (Stirzaker et al. 1996). Indeed, 80% of wheat roots were found within 2.2 mm of soil macropores, in two Australian vertisols (Stewart et al. 1999). The mechanism whereby roots locate such pores may well be linked to circumnutation (the spiralling pattern of root growth), that has been observed since Darwin in humid air and in relatively soft media such as agar. Circumnutation has been found to interact with gravitropism and the mechanical properties of the growth medium, to produce a waving pattern of root growth in *Arabidopsis*, even on a simple gel surface (Thompson and Holbrook 2004). The stress regime around a growing root tip will be complex, but cracks and fissures in the soil may represent low resistance pathways for root growth, that are detected when they come within a few root diameters of the root tip.

The matric potential of the rhizosphere becomes more negative as roots extract water from the soil. As soil strength may increase by >100 fold as soil dries from −1 kPa to −1.5 MPa, the pattern of water extraction in the rhizosphere will greatly influence soil conditions (Whalley et al. 2005a). It is also reasonable to assume that water, transported via hydraulic lift through the root system from deeper roots growing in wet soil, and released into drier surface soil at night, when the transpiration demand is minimal, will decrease rhizosphere strength in the upper portion of the soil profile. Similarly, if the water potential of mucilage released by the root cap is more positive than the matric potential of rhizosphere soil, this may soften the way for a penetrating root (Passioura 2002). Feedback between rhizosphere strength and exudation results in greater release of root exudates by mechanically impeded roots (Barber and Gunn 1974; Boeuf-Tremblay et al. 1995). Barley plants grown in glass beads released approximately twice the dry mass of exudates as compared with plants in liquid culture, with maize (*Zea mays* L.) exuding a nine-fold greater mass of carbohydrate (Barber and Gunn 1974). Boeuf-Tremblay et al. (1995) demonstrated that this was due to root exudates rather than rhizosphere microbial products as they observed an increased exudation with increasing strength in axenically-grown maize seedlings. Such a large release of exudates into the rhizosphere may well give rise to the greater numbers of bacteria observed around the tips of mechanically impeded roots in soil (Watt et al. 2003).

The increase in exudation with an increase in soil strength is also accompanied by an increase in the release of root border cells per mm of root elongation (Iijima et al. 2000). A 12-fold increase in border cells/mm root elongation was measured for maize radicles in compacted, as compared with loose sand. The number of border cells released was sufficient to cover completely the surface of the root cap in compacted sand, decreasing the frictional resistance to root penetration and acting as a lubricating sleeve around the root tip. This disposable sleeve of cells may decrease the colonisation of the root tip by bacteria. Indeed both frictional resistance and bacterial colonisation of the root tip increase markedly if the root cap is removed from maize (Iijima et al. 2003; Humphris et al. 2005). Relatively little is known about border cell release in older plants, and there is interesting new evidence that border cell release may decrease substantially with increasing root age (Odell et al. 2008). Both border cells (Fig. 3) and root cap mucilage may lubricate root penetration (Iijima et al. 2004), and there is evidence that enhanced root exudation in mechanically impeded roots persists and even increases with plant development stage (Boeuf-Tremblay et al. 1995).

Gradients, spatial heterogeneity and temporal development

Gradients in rhizosphere strength depend on the associated gradients in soil density, matric potential, and root deposition of organic material. The mechanical stress in the soil around the root tip is likely to be greatest immediately in front of the root apex; the distribution of stress simulated using a Finite Element Model was changed by thickening of the root tip such that a 60% increase in the root diameter decreased the peak axial stress in front of the root tip by a quarter, facilitating root penetration (Kirby and Bengough 2002). Deformation of the rhizosphere will be greatest where roots have historically exerted most mechanical

1.54 g cm^{-3} in the bulk soil (Bruand et al. 1996; Young 1998). The variation in soil porosity has been modelled as decreasing exponentially away from the root surface (Dexter 1987), although this form of model is derived mainly from empirical fits to the data available, rather than to a mechanically based model.

The local physical environment of a root can be influenced greatly by the presence of root channels from preceding crops, especially in hard soils (e.g. maize following the tropical legume *Stylosanthes hamata*, Lesturgez et al. 2004). This is illustrated clearly by minirhizotron images showing soybean roots growing along channels in compact soil left by decomposing roots of canola just 2 to 3 months earlier (Williams and Weil 2004). In similarly dense soils that were unploughed, more than half of the entire length of wheat root systems contacted the decaying skeletons of roots (Watt et al. 2005). Such a large local input of root material is likely to greatly alter the local environment within the macropore sheath – the zone of soil typically within 1 to 3 mm of the macropore wall (Stewart et al. 1999). Indeed, the fungal and bacterial populations within the macropore sheath can be greatly enhanced in comparison to those of the bulk soil (Pierret et al. 1999), even more so than these authors found in the rhizosphere of the same soil after repacking. Bundt et al. (2001) also showed that preferential flow paths in soils were to be considered as biological hot spots, possibly because of being sites of deposition of organic matter and nutrients. This is especially the case for biopores such as earthworms galleries and root channels, i.e. either present day or relic rhizospheres (Pierret et al. 2007). While penetration of root systems to depth will enable roots to access valuable reserves of water deep in the subsoil, clustering of roots within relatively sparse macropore channels means that the extraction of water from large blocks of subsoil may be relatively much slower than if the roots were uniformly distributed throughout the subsoil volume (Passioura 1991).

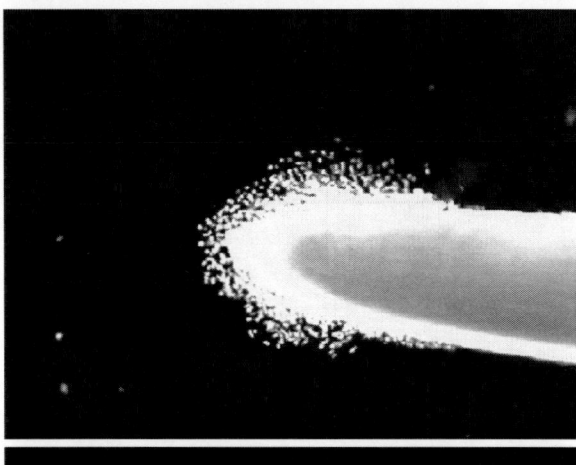

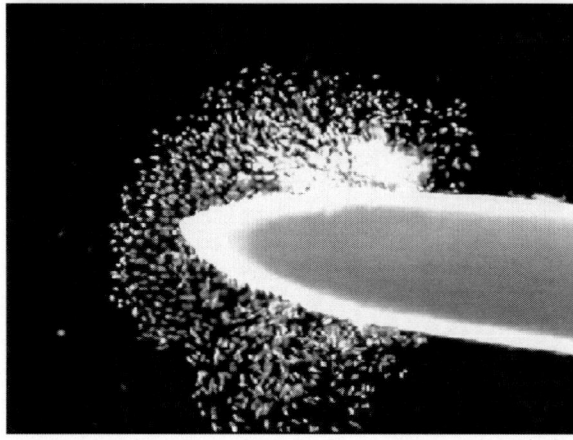

Fig. 3 Timelapse sequence of border cell release during mucilage hydration following immersion of a maize (*Zea mays* L.) primary root tip in water, after 1, 3, and 9 min. Maize root diameter is approximately 1 mm (Reproduced by kind permission of A. Glyn Bengough)

stress on the surrounding soil. Such deformations can be large and estimates of local density around maize roots indicate dry bulk densities as great as 1.8 g cm^{-3} next to the root surface, as compared with

Water distribution and accessibility in the rhizosphere

Ecological relevance

Water is the solvent and transport medium in natural systems. Water is a major constituent of plants and microbes and a reactant or substrate in many important processes crucial for metabolic activity. Another role of water is the maintenance of turgor which is essential for cell enlargement and growth (Kramer and Boyer 1995). However, most of the water taken up by plants is transpired to the atmosphere in their attempt to assimilate CO_2 (Jackson et al. 2000).

Roots provide the hydraulic continuity between soil and atmosphere and thereby play a key role in the global water cycle. Water lost through stomata during photosynthesis has to be replaced by uptake from the soil. As discussed above, with decreasing water content in soil its mechanical resistance increases and penetration of roots is hindered (Pardales and Kono 1990, Sharp and Davies 1985). In dry soils roots (fine roots in particular) may desiccate and loose their function. Nutrient uptake is reduced due to decreasing nutrient mobility and vanishing uptake capacity (root activity). Resources in dry soils are not available to the plants and the associated microorganisms, thus competition for the limiting resources increases if parts of the soil dry. The ability of some roots to continue elongation at water potentials that are low enough to inhibit shoot growth completely is an important species specific response to soil drying (Sharp et al. 2004). Plant species able to maintain root activity and growth in drying soil or to compensate the uptake of nutrients and water by other parts of the root system or by an association with microorganisms which help to overcome the negative effects of soil drying may have a competitive advantage under water limiting conditions. Such differences between plant species in certain traits may determine community composition in natural ecosystems but also the efficiency of intercropping system in agriculture or agro forestry (Callaway et al. 2003).

Soil micro-organisms are predominantly aquatic in nature, i.e. are living in the liquid phase and not in the air-filled pores. Key aspects of moisture in the rhizosphere include the matric potential which determines the distribution of water-filled pores (providing hydraulic connectivity) which in turn act as valves in soils altering the diffusion rates of gases to and from microbial populations (Focht 1992). This regulates the activity of aerobic against anaerobic organisms (Young and Ritz 2000). It is important here to draw a distinction between an anaerobic environment which exists due to say a pore being filled with water, and a pore separated from other pores due to say annular water rings held by capillary action onto organo-mineral surfaces surrounding it.

Additionally, the matric potential regulates the thickness of the water films adhering to organo-mineral surfaces. This is again linked to the hydraulic connectivity of soil pores, and directly impacts the movement of bacteria and protozoa. At a higher order, nematode movement has been shown to be intimately linked to water-film thickness (Wallace 1958), which has been recently shown to be implicated in potential gene flow and the creation of biodiversity.

An important aspect of microbial activity in the rhizosphere is the ability of microbes to adapt to a highly varied moisture regime so close to a major sink of water; and of course under some circumstances, for instance hydraulic lift, an important source of water.

Underlying processes

Root water uptake in the soil-plant-atmosphere continuum is a passive water flux driven by the water potential gradient between the soil and the atmosphere. It can be regulated by stomatal movement, but for open stomata it is a function of the potential gradient and the conductivity/resistance in the system.

In older textbooks, roots are regarded as nearly prefect osmometers. Clarkson et al. (1971) and Sanderson (1983) showed higher uptake fluxes in the younger regions of the root of barley and related it to the endodermis development in older regions of the root. Although water flow through the root cortex can occur in parallel pathways, through the apoplast (cell walls), through the symplast (plasmodesmata) or transcellular (aquaporins), it was indeed assumed for a long time that the endodermis with its casparian bands stops apoplastic flow completely and thus acts as the 'root membrane' (Passioura and Munns 1984, Steudle and Peterson 1998). This classical view was challenged in recent years as there are numerous indications for an apoplastic bypass through the endodermis. Casparian bands are not yet developed in some areas like root primordia and root tips and, in addition, casparian bands are not absolutely impermeable to water as analysis of the isolated material has shown (Schreiber et al. 2005, Zimmermann et al. 2000). This, results of puncturing experiments in which the effects of small holes in the endodermis on hydraulic conductivity were measured (Steudle 1993), and detailed measurements with root and cell pressure probes of root and cell hydraulic conductivities and reflection coefficients, led to the development of the 'composite transport model', i.e. parallel pathways for water uptake with different relevant driving forces occurring also across the endodermis (Steudle 2000, 2001).

Along the apoplastic path, water movement will be hydraulic in nature, i.e. driven by gradients in hydrostatic potential. Cell walls have no selective properties for solutes (reflection coefficient σ is close to zero), thus osmotic gradients are not relevant for this path.

Along the cell-to-cell path (plasmodesmata and aquaporins) the reflection coefficient σ, a measure for the semi-permeability of a membrane, is close to one and hence osmotic and hydrostatic potential gradients act together in an additive manner.

The relative contribution of hydraulic (predominantly apoplastic pathway) and osmotic (only cell-to-cell path) flow to total water flux changes with root development (Frensch et al. 1996) and environmental conditions (Vandeleur et al. 2005).

The hydraulic conductivity (or its reciprocal, the resistance) differs between the two pathways. It has been shown in numerous measurements, especially for tree roots, that the conductivity for the hydraulic flow can be up to three orders of magnitude greater than for the osmotic flow (Steudle and Peterson 1998). The hydraulic conductivity in either pathway is not constant but changes with maturation of root tissue and number and function of aquaporins (Frensch and Steudle 1989; Frensch et al. 1996; Barrowclough et al. 2000; Tyerman et al. 2002; Vandeleur et al. 2005). Changes based on aquaporin expression can occur within hours and result in diurnal fluctuation of hydraulic conductivity (Henzler et al. 1999). Similarly, diurnal fluctuations in root diameter have been observed in rhizotron studies, and may give rise to large changes in root-soil contact for roots located in macropores (Huck et al. 1970). Any decrease in root-soil contact will decrease the hydraulic conductivity of the root-soil interface, although there is evidence that frequent exposure to water deficit may harden roots, decreasing such fluctuations in diameter during drying cycles (Lemcoff et al. 2006).

A important highlight in any discourse on soil water is an explicit recognition that, generally the root and microbes are living in and adapting to a complex mix of solutes that make up soil moisture, and our general notions of the simplicity of soil 'water' fail. A growth area in research is directly related to microbes' ability to alter the surface tension and contact angle of soil moisture (Urbanek et al. 2007). This will have important implications for the shape of the moisture characteristic and the sorptivity of soil, which of course feeds forward to a wide range of important processes.

The presence of mucilage in the rhizosphere has given rise to studies on its ability to retain water. The first of these studies used freezing point depression (Guinel and McCully 1986), and may well have been flawed due to the difficulty of using this technique with a gel (McCully and Boyer 1997). Later studies using thermocouple psychrometry suggested that maize mucilage may contain 99.9% water at potentials of −50 kPa (Read et al. 1999), although under wetter conditions the mucilage contained even greater quantities of water (McCully and Boyer 1997).

Surfactants, such as lecithin, contained in mucilage may change the water-release properties of the rhizosphere, such that more water is released, especially from relatively wet coarse textured soils (Read et al. 2003). Measurements of the water-release characteristic of the rhizosphere, as compared with bulk soil, showed that rhizospheres of barley and maize were drier than bulk soil at the same matric potential, partly as a result of changes in rhizosphere pore-size-distribution and angle of wetting (Whalley et al. 2005b). Besides hydrophilic substances, hydrophobic compounds can also be produced by either roots or microbes, resulting in increased water repellency in the rhizosphere relative to bulk soil (Czarnes et al. 2000; Hallett et al. 2003), which further complicates the biophysics of the rhizosphere.

Gradients and spatial heterogeneities and their temporal development

The site of maximum water uptake along a root is determined by the interplay between radial root resistance (see above) and axial root resistance and can be compared with hydraulics in porous pipes (Zwieniecki et al. 2003). Axial resistance is a function of xylem maturation and xylem vessel diameter as well as number and organisation of xylem vessels (Shane et al. 2000, Steudle and Peterson 1998). These traits are highly species specific, but are altered by environmental conditions (Tyree and Sperry 1989). While axial conductivity increases with increasing distance from the tip as xylem vessels are formed and mature, radial conductivity decreases due to formation of endodermis and exodermis with casparian bands and suberin lamellae. For fully differentiated root tissue axial conductivity is larger by orders of magnitude than radial conductivity.

For barley (*Hordeum vulgare* L.) and pumpkin (*Curcurbita pepo* L.) the maximum water uptake was reported to occur 3–8 mm behind the root tip (Clarkson et al. 1971; Sanderson 1983; Kramer and Boyer 1995). For tree roots maximum uptake was observed close to the root tips or where lateral roots are emerging (Häussling et al. 1988). Clearly, non-invasive techniques such as magnetic resonance imaging, X-ray computer tomography or neutron radiography which enable visualizing roots and water in soil simultaneously (Fig. 4), bear a great potential for investigating water uptake profiles along single roots (MacFall et al. 1990; Pierret et al. 2005; Menon et al. 2006). However, despite recent technological advances, non-invasive observations of plant roots and their environment still face a trade-off between spatial resolution, field-of-view and three-dimensionality (Pierret et al. 2003; Garrigues et al. 2006).

Gradients in soil water content may not only develop along a root as a result of different uptake rates but may also develop radially around a root if soil hydraulic conductivity becomes limiting for uptake (Fig. 5). This is likely to occur under high evaporative demand, a small root-shoot ratio and a coarse soil texture (Gardner 1960; Sperry et al. 2002) and was demonstrated experimentally (Hainsworth and Aylmore 1989; MacFall et al. 1990). Such gradients in soil water content or soil matric potential are taken into account in microscopic approaches of modelling plant water uptake and can there function as a threshold. This is in contrast to macroscopic models on plant water uptake in which roots are only regarded as a diffuse sink which varies in size as a function of soil depth (Feddes et al. 2001). Recently Doussan et al. (2006) have combined models describing explicitly root architecture (root system growth and deployment in space) with the microscopic approach of describing water uptake based on potential gradients in soil-plant-atmosphere continuum. With this approach they were able to reproduce the water distribution patterns obtained in experiments.

Gradients in soil water content develop around roots, along developing roots but also on a larger scale between different soil horizons. Root distribution throughout the soil profile varies with plant development (annual crops) season and plant species and shows an extremely high plasticity, i.e. is not only genetically controlled but adapts to environmental conditions (Callaway et al. 2003). Likewise water is not distributed evenly throughout the soil profile even under equilibrium conditions, due to the shape of the water retention curve and the effect of gravity. As a consequence the contribution of different soil horizons to cover evaporative demand shows a large fluctuation and changes with time. In regions with seasonal precipitation patterns topsoil, where the highest rooting density prevails, usually dries up first, but deeper roots may tap ground water or soil layers close to the water table showing little fluctuation in water content.

Roots may partly buffer such large scale gradients by redistribution of water from wetter soil regions to drier regions of the soil profile. Water is released from roots during periods when transpiration ceases (usually at night) and soil water potential in the dry soil region becomes more negative than plant water potential. This phenomenon, also known as 'hydraulic lift' has been shown to exist in about 30 different plant species and the reported amounts of water transferred per night range from 14 to 30% of daily evapotranspiration. However, despite its potential importance for water use efficiency, facilitation or water parasitism, nutrient uptake from dry topsoil or maintenance of root function, the magnitude, path-

Fig. 4 Volumetric water content [%] distribution along 3 week old tap root of lupin (*Lupinus albus* L.) growing in a sandy soil obtained by neutron radiography (performed at the facilities of Paul Scherer Institute, Zürich). The root, due to its high water content, shows up as a longitudinal body in the centre. The field of view of the picture is ten by 40 mm, it is a detail of a neutron radiogram with an original field of view of 150×150 mm, pixel size is 0.272 mm. The scale from red to blue corresponds to volumetric water content ranging from 40 to 28% (Reproduced by kind permission of Andrea Carminati et al.)

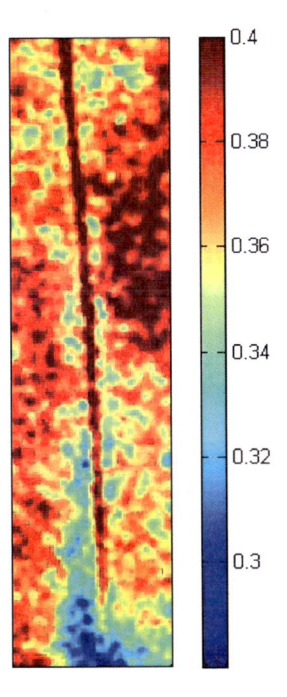

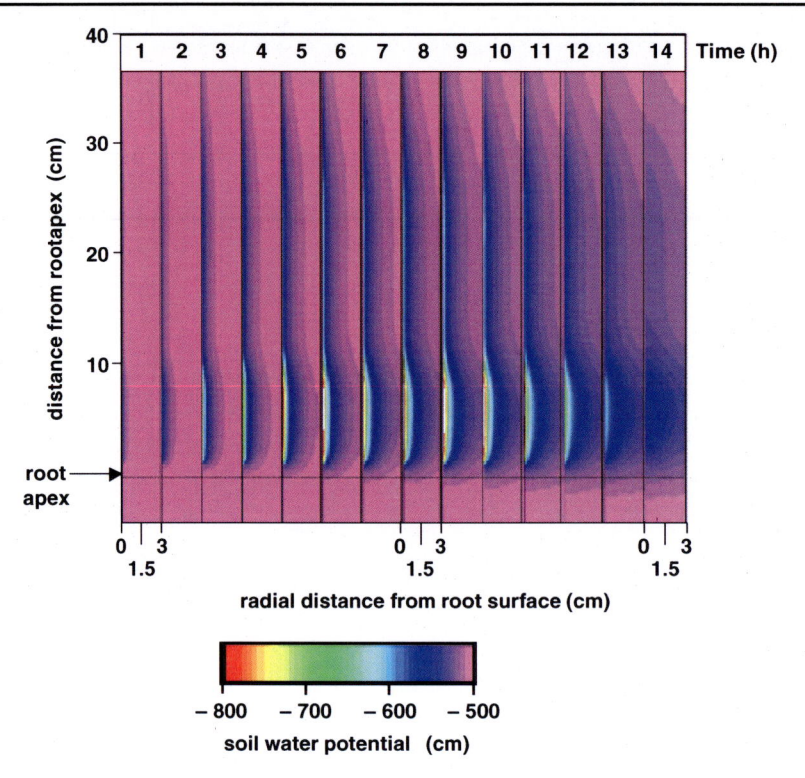

Fig. 5 Temporal development of radial and longitudinal gradients of water potential as a consequence of water uptake by a root of maize (*Zea mays* L.) growing in a clay loam at an initial soil water potential of −0.05 MPa (equivalent to 500 cm water column), as obtained by modelling. The root axis is located along the left axis of each box. In order to account for the day/night cycle of transpiration, a sinusoidal variation (between −0.1 and −1.2 MPa) of the xylem water potential was imposed. Axial and radial variations of the hydraulic conductance along the root were included in the simulation and generated an heterogeneous pattern of uptake and water potential in the soil along the root, with greater variations near the root tip (Reproduced from Doussan et al. (2003) by kind permission of EDP Sciences)

ways and resistances of these redistribution processes are still poorly understood (Caldwell et al. 1998; Callaway et al. 2003; Newman et al. 2006).

It is well proven by direct measurements that water can move in either direction within the root system depending on the direction of water potential gradients (Burgess et al. 2000) and there is no indication of a general "rectifier like behaviour of roots", i.e. a higher resistance to water efflux compared with influx, from anatomical or physiological features. However, root radial resistance can increase by formation of suberin lamellae in the tangential walls in the exodermis or dehydration of root tissue, resulting in a decrease of root radius and increased formation of an air gap between the root surface and the soil, i.e. a loss of root-soil contact (Nobel and Cui 1992). For a long time it is known that variation in root diameter occurs reversibly on a diurnal basis (Huck et al. 1970). Consequences of changes in root diameter for root-soil contact are described in detail by Veen et al. (1992).

Whether hydraulic redistribution is observed in measurable quantities not only depends on root resistance but also on hydraulic conductivity of the soil (Vetterlein and Marschner 1993). Hydraulic redistribution was observed less frequently in coarse textured soils than in fine textured ones (Yoder and Nowak 1999). Another important factor for hydraulic redistribution, apart from the size of the water potential gradient between the wet and dry soil regions, is the ratio between the uptake capacity in the moist region and the density of functional roots in the dry region (Ryel et al. 2002).

Last but not least the presence of salt, decreasing osmotic potential, may affect water redistribution. Salts can accumulate around roots and thus, as for soil matric potential, steep gradients for osmotic potential can be formed around roots (Stirzaker and Passioura

1996; Vetterlein and Jahn 2004a). Whether and to which degree salts accumulate depends on initial salt concentration in soil solution, soil texture and the ratio between evaporative demand and plant requirement for the ion in question (Vetterlein et al. 2004, 2007a). The extent to which water uptake is reduced due to such gradients in osmotic potential depends on the relative contribution of osmotic and hydraulic flow to total water uptake and thus varies with environmental conditions like evaporative demand. Likewise plants growing in saline soils can adjust osmotically to salinity and this may constrain water release to drying soil.

Chemical ambience (pH and redox potential) of the rhizosphere

Ecological relevance

The soil pH has dramatic importance for below-ground life. One of the most striking pieces of evidence is shown by recent biogeographical studies, e.g. Fierer and Jackson (2006) study which investigated a data set of 98 soils sampled across the Americas. This study showed that temperature, rainfall and latitude had virtually no effect on the diversity and richness of soil microbial communities, whilst soil pH had a major effect, by far the largest amongst the investigated parameters. Bacterial diversity was highest in neutral soils and minimal in acidic soils. Extremes of pH are also well documented to impose major constraints on root growth due primarily to the toxicities of ions such as Al^{3+}, Mn^{2+} and H^+ in the acidic range (Marschner 1995; Kinraide and Yermiyahu 2007), or HCO_3^- in the alkaline range (Tang et al. 1993). Additional effects are related to nutrient deficiencies such as that of iron in calcareous soils (Lemanceau et al. 2009) or phosphate in both acidic and alkaline soils (Richardson et al. 2009). Interestingly, through their physiological functions, plant roots and soil microbes are however capable of considerably altering soil pH relative to the bulk soil. Rhizosphere pH has been reported to be up to 1–2 pH units below or above bulk soil pH as shown in microcosms (e.g. Riley and Barber 1971; Gahoonia et al. 1992) or less frequently in situ (e.g. Yang et al. 1996; Michaud et al. 2007). This may have a dramatic effect on soil biogeochemistry, microbial communities (including at the microsite scale, e.g. Strong et al. 1997) and may ultimately feed back on plant physiology or symbiosis (e.g. Cheng et al. 2004).

The pO_2 and hence the redox potential is highly variable in soils, with values which range from atmospheric pO_2 in the most aerobic conditions down to zero in strictly anaerobic conditions. These changes sometimes occur locally along short distances (Rappoldt and Crawford 1999), e.g. within small soil aggregates (Renault and Stengel 1994) as an effect of water content which affects the gas exchanges in the air-filled porosity, and largely as a result of biologically-mediated processes of O_2 consumption (Brune et al. 2000; Khalil et al. 2004; Pidello and Jocteur Monrozier 2006). As stressed by Brune et al (2000), the availability of O_2 has a major impact not only on the redox potential of the environment and many biogeochemical cycles, but also on the energetic situation of microorganisms. This is illustrated for nitrification/denitrification processes that rely on different bacterial communities which function either at high/low soil pO_2 (Focht 1992; Khalil et al. 2004). In addition, most plant species are highly sensitive to hypoxia/anoxia, only a few of them being able to cope with prolonged periods of low pO_2 as occur in submerged soil conditions (Perata and Alpi 1993). This is the case for wetland plants especially which have evolved specific strategies to cope with hypoxic conditions that prevail in the environments in which they grow.

Underlying processes

A primary function of below-ground organisms which can substantially impact soil pH is respiration and the subsequent increase in pCO_2. Because of respiration, bulk soil pCO_2 is well-known to be much (ten to hundred-fold) higher than that of the atmosphere (360 cm^3 m^{-3}). Karberg et al. (2005) reported values ranging from 7,000 to 24,000 (up to 32,000 under elevated atmospheric pCO_2) cm^3 m^{-3} in a forest soil. Given that roots and microbes are major contributors to soil respiration, it is expected that rhizosphere especially in the region behind the root tip should be a hot spot of elevated pCO_2 and decreased pO_2 as shown by Bidel et al. (2000). This is however little documented in soils except the few data published by Gollany et al. (1993), who measured pCO_2 values in the order of about 100,000 cm^3 m^{-3} at 1–3mm from roots. The same holds for pO_2 values, although it is widely accepted that respiration should result in a decrease in pO_2 in the rhizosphere, with a notable exception for wetland plants (see below). Bidel et al. (2000) measured pO_2 values along the roots of *Prunus persica* (L.) Batsch seedlings and clearly showed that these were much smaller in the meristematic region of the root as a consequence of intense metabolic activity and respiration. They showed indeed that O_2 consumption in this region of the root was positively correlated with root growth. Only about 5×10^{-14} mol O_2 s^{-1} were consumed when no growth of the root tip occurred, whereas the respiration rate reached values greater than 35×10^{-14} mol O_2 s^{-1} for active meristems. Bidel et al. (2000) also estimated the relative contribution of microbial respiration to the observed decrease in pO_2. The corresponding flux decreased abruptly from 10 to 1 nmol O_2 m^{-3} gel s^{-1} within the 300 μm surrounding the root surface near the apex. These figures were however obtained in agar media, which is known to be rather hypoxic, and thus mimic the situation of a poorly aerated soil. Fischer et al. (1989) reported a decrease in redox potential (shift of Eh from about 700 to less than 380 mV) when the root tip of soil-grown faba bean (*Vicia faba* L.) reached the microelectrode. This phenomenon was reverted (Eh went back to initial value) about one-day later, once the root tip had moved away from the microelectrode, confirming that respiration was especially large near the meristematic zone of the root (apex). In contrast with aerobic conditions which are little documented for changes in pO_2 values in the rhizosphere, the case of wetland plants growing in hypoxic (submerged) soils has been extensively studied. To ensure the respiration of their root cells, those plants have evolved aerenchyma which conducts O_2 from the shoots to the roots (Armstrong 1979). Leakage of O_2 from roots can result in a local build-up of pO_2 in the rhizosphere of wetlands plants (Flessa and Fischer 1992; Revsbech et al. 1999; Armstrong et al. 2000; Blossfeld and Gansert 2007), which has been especially studied for rice (*Oryza sativa* L.). Revsbech et al. (1999) reported for instance that pO_2 increased up to a fifth of atmospheric pO_2 at rice root surface, while being almost nil at distances greater than 0.4 mm from the root surface.

Given that CO_2 rapidly forms H_2CO_3 which is a weak acid (pK = 6.36), increased pCO_2 thus results in a decreased pH, in all but the most acidic soils (where H_2CO_3 remains essentially undissociated). This actually means that in situ values for rhizosphere pH of calcareous soils are expected to be close to 7, rather than 8.3 as dictated by the dissolution/precipitation equilibrium of $CaCO_3$ at ambient pCO_2. For instance, based on the values measured by Gollany et al. (1993), Hinsinger et al. (2003) computed rhizosphere pH values of about 6.7–6.8. In spite of the current attention on aboveground pCO_2 it is rather astonishing that so little data is available about rhizosphere pCO_2 and its impact on belowground organisms and biogeochemical cycles. Greenway et al. (2006) have recently addressed this issue, especially the feedback effect of high pCO_2 on root growth, in the specific context of waterlogged soils where the excess of water impedes gas exchange and leads to elevated pCO_2 and low levels of pO_2.

In contrast, the major implications of proton influx/efflux from roots in rhizosphere pH changes have been studied in detail (see reviews by Nye 1981 and, more recently Hinsinger et al 2003). As elegantly shown by Marschner and co-workers with the use of dye indicators (Römheld and Marschner 1981; Marschner and Römheld 1983; Luster et al. 2009, Fig. 6), this process occurs in order to balance cation/anion net uptake (Raven 1986), and actually one should account for all charged compounds (ions) crossing the root cell membranes, e.g. organic anions (carboxylates) exuded by roots (Hinsinger et al. 2003). Net influx of excess cations results in a net efflux of protons and thus rhizosphere acidification, while alkalisation occurs for a net influx of excess anions over cations. Nitrogen which is in high demand by plants, has a major impact on this process as it is predominantly used as either an anion (nitrate) or a cation (ammonium), while it can be used as the uncharged species N_2 in symbiosis, such as in legumes. The former is expected to make the rhizosphere more alkaline, while the two latter forms of N acidify the rhizosphere (Riley and Barber 1971; Marschner and Römheld 1983; Le Bot et al. 1990; Gahoonia et al. 1992; Plassard et al. 1999; Tang et al. 2004). Rhizosphere alkalisation as related to proton influx confers an adaptative advantage for plant roots growing in acid soils by alleviating aluminium and other metal toxicities (Degenhardt et al. 1998; Pineros et al. 2005; Michaud et al. 2007). Rhizosphere acidification as related to proton efflux from roots is well known as an adaptative response of many plant

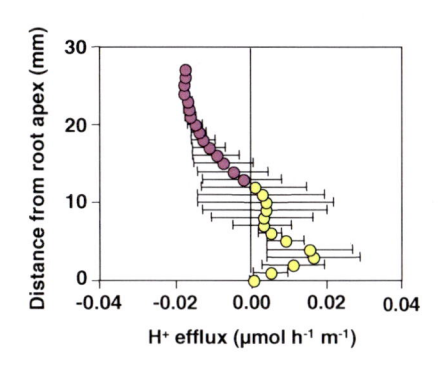

Fig. 6 Root-induced pH changes in Fe-deficient tobacco (*Nicotiana tabacum* L.). Roots were embedded in an agarose gel containing bromocresol purple as dye indicator to reveal the actual pH. The fluxes of protons as computed from the temporal change of pH are indicated to show that significant alkalisation (negative values of proton efflux) occurred along the basal portions of the roots, in contrast with the distinct apical acidification which was visible by the yellow color of the dye on the photograph (using the method described by Vansuyt et al. (2003). Reproduced by kind permission of Gérard Vansuyt, Gérard Souche and Benoît Jaillard)

species to iron and phosphorus deficiencies (Römheld and Marschner 1981; Tang et al. 2004; Lemanceau et al. 2009; Richardson et al. 2009).

Although many authors have been refering to the exudation of the so-called "organic acids" by roots (as reviewed by Jones 1998 and Ryan et al. 2001), it has been shown that carboxylates are dissociated at the cytosolic pH of root cells and thus exuded as anions. Their contribution to rhizosphere acidification thus largely depends on the previously-mentioned process, as stressed by Hinsinger et al. (2003). Beside roots, many soil microbes can produce organic acids and thus contribute to rhizosphere acidification as documented for ectomycorrhizal and saprophytic or pathogenic fungi massively producing oxalic acid (Dutton and Evans 1996; Wallander 2000; Casarin et al. 2004). Dramatic changes of pH can also occur as a consequence of the microbially-mediated oxidation of nitrogen (nitrification, e.g. Strong et al. 1997) or sulfur. Enhanced nitrification, which produces protons and nitrate ions with a stoechiometric 1:1 molar ratio, has been reported to occur in the rhizosphere relative to the bulk soil (Binnerup and Sorensen 1992; Hojberg et al. 1996; Herman et al. 2006) and in wetland plants such as rice (Kirk and Kronzucker 2005). In such plants, the intimate coupling of rhizosphere pH changes with redox processes has also been documented for the case of iron oxidation and subsequent iron oxide precipitation. Begg et al (1994) and Kirk and Le Van Du (1997) calculated that this process resulted in a major proportion of the acidification measured in the rhizosphere of lowland rice, the other contributor being the use of ammonium as an important source of nitrogen under ambient reduced conditions, thus resulting in net proton efflux from rice roots. Neubauer et al. (2007) showed for *Juncus effusus* L. that this oxidation process was partly mediated by lithotrophic bacteria in the rhizosphere, besides oxygen leakage from root aerenchyma. Denitrification is another major process in the nitrogen cycle which is largely controlled by redox conditions. Pidello et al. (1993) showed that rhizosphere bacteria such as *Azospirillum brasilense* increased the soil redox potential, compared with the control (not inoculated) soil and consequently decreased by several-fold the denitrifying activity of the soil. Pidello (2003) showed for another rhizosphere bacterium, *Pseudomonas fluorescens*, that strains varying in pyoverdine production affected the soil redox potential differently. Pyoverdines are strong, electro-active Fe chelators. Both strains of *P. fluorescens* decreased the soil redox potential, but the mutant strain that did not produce pyoverdine had a greater effect. The coupling of these processes implied in the biogeochemical cycling of Fe in the rhizosphere is further considered by Lemanceau et al. (2009).

Gradient, spatial and temporal heterogeneities

Soil pH, as many other chemical and physical properties, can substantially vary in space and time, as evidenced in both agricultural and forest ecosystems. Spatial variation is especially documented with changes over 0.5–3 units that are frequently reported within a small plot, a soil horizon and down to millimetric scales (Yang et al. 1995; Göttlein et al. 1996; Göttlein and Matzner 1997; Yanai et al. 2003; El Sebai et al. 2007). The spatial variability of pH buffering capacity is much less documented than pH heterogeneity, in spite of its functional relevance. Localised patches of organic matter or discrete distribution of $CaCO_3$ grains are likely to have a large influence on this parameter and on the subsequent changes of pH over time and space. One has to take into account this pre-existing heterogeneity of soil properties when investigating changes of pH in the rhizosphere in situ (in field-grown plants). This can be shown from the work of Schöttelndreier and Falkengren-Grerup (1999) who stressed that they could hardly distinguish between root-induced alteration of pH and utilisation of soil heterogeneity. Redox potential can also be subject to considerable variations in space and time, which are largely related to changes in soil water content, water saturation leading to decreases in pO_2 depending on the biological activities responsible for O_2 consumption and gas diffusivity. Besides the vertical gradient which is expected to occur in soils from the upper, aerated horizons down to deeper, water-saturated horizons, considerable heterogeneities can be observed at a much smaller scale within the soil matrix as shown when studying changes in redox potentials across small soil aggregates (Renault and Stengel 1994; Pidello and Jocteur Monrozier 2006).

Rhizosphere processes are obviously an additional source of heterogeneity (Hinsinger et al. 2005) and especially so for pH. These processes are driving forces for the formation of radial pH gradients around living roots. Evidence for such gradients which can extend up to several mm from roots has been reported by many authors (e.g. Schaller 1987; Gahoonia et al. 1992; Begg et al. 1994; Hinsinger and Gilkes 1996; Nichol and Silk 2001; Kopittke and Menzies 2004; Vetterlein and Jahn 2004b; Vetterlein et al. 2007b) since the early work of Farr et al. (1969) who showed a marked decrease in pH close to a root mat of onion (*Allium cepa* L.). With a refined root mat approach, Gahoonia et al. (1992) showed for ryegrass (*Lolium perenne* L.) that when fed with nitrate, the increase in pH reached about 1 pH unit while the decrease in pH amounted to 1.5 units when ammonium was supplied, the spatial extension of this phenomenon was about 2 mm in a luvisol and 4 mm in an oxisol. Recently Vetterlein and Jahn (2004b), Vetterlein et al. (2007b) and Bravin et al. (2008) have been studying the temporal development of similar gradients over time scales of several weeks, with a temporal resolution of several days, while Cornu et al. (2007) monitored pH changes in the rhizosphere with a daily resolution (but without studying the radial gradient in this case). Most of these studies have shown a more or less steady build-up of rhizosphere acidification or alkalisation over time. A finer temporal resolution would have been needed to account for the diurnal rhythm that one may expect, given that the uptake of ions and thus the resulting production/consumption of protons is known to follow diurnal rhythms as previously shown in hydroponically grown plants (Le Bot and Kirkby 1992; Rao et al. 2002; Tang et al. 2004), as well as the photosynthesis-driven diurnal patterns of exudation and thus of rhizosphere respiration. Most of the published work on pH gradients in the rhizosphere as described so far have however been obtained with root mat techniques which only give access to the average effect of many roots. This lead to an overestimation of the extent of the process compared with normal rhizosphere geometry and does not account for heterogeneities of pH along the root system of a single plant (Jaillard et al. 2003; Hinsinger et al. 2005). Limited studies, based on the use of microelectrodes have shown local gradients of pH (e.g. Schaller 1987; Nichol and Silk 2001) that vary according to the location along the root axis. Häussling et al. (1985)

showed in situ that rhizosphere acidification (0.3 pH unit change) occurred only behind the root tip of 60-yr old spruce (*Picea abies* (L.) Karst.) tree roots, while alkalisation occurred at the apex and more basal parts of the roots (reaching up to 0.8 pH unit change at the apex) in an acidic soil (bulk pH 4.2).

Remarkable studies of the spatial heterogeneities of pH changes in the rhizosphere have been published based on the use of dye indicators as initially developed by Weisenseel et al. (1979) and made popular by Marschner/Römheld and co-workers. Römheld and Marschner (1981), Marschner and Römheld (1983) showed for instance the typical localised acidification that occurs behind the root tips as a response to Fe deficiency in plant species belonging to the Strategy I of Fe acquisition. Such strategy has been shown to occur in all plant species but grasses (i.e. graminaceous plant species, which belong to Strategy II, see below) and is defined by enhanced proton efflux combined with enhanced Fe-reductase activity occurring behind root tips when such plants are exposed to low Fe availability which typically occurs in calcareous soils (Marschner 1995; Robin et al. 2008). Using a pH dye indicator combined with image analysis according to the method developed by Jaillard et al. (1996), Vansuyt et al. (2003) measured the variation of proton efflux along the axis of roots of Fe-deficient tobacco (Fig. 6). They showed that slight alkalisation (proton influx) was occurring along the basal part of the root while acidification (proton efflux) occurred at the root apex (up to 5–15 mm from the tip, Fig. 6). Similarly, a few works have been showing that P deficiency was also resulting in enhanced acidification of the rhizosphere, although this was not always localised at the root tip as for Fe (Gregory and Hinsinger 1999; Hinsinger et al. 2003; Tang et al. 2004). Plassard et al. (1999) comparing this technique with the use of microelectrodes in hydroponically-grown plants to derive proton effluxes confirmed the heterogeneity of such fluxes along root axes, even for plants which were not exposed to Fe or P deficiency. The combined use of proton and ion-selective electrodes (for measuring e.g. ammonium, nitrate and potassium concentration gradients and the corresponding fluxes) confirmed whether these patterns of proton efflux along roots were largely related to patterns of influx of major nutrients such as N and K which also substantially vary along roots and according to the mycorhizal status of roots in ectomycorrhizae (Plassard et al. 2002; Hawkins et al. 2008). Plassard et al. (2002) showed that nitrate and potassium influxes in long roots of mycorrhizal pine (*Pinus pinaster* Soland *in* Ait.) were not significantly different from those of non mycorrhizal roots, suggesting that proton efflux would be unaffected as well. In contrast they showed that influxes of nitrate and potassium were mainly affected by the mycorhizal status and species in short mycorrhizal roots. Contrary to the long roots, much higher nitrate than potassium influx was found to occur in those roots (Plassard et al. 2002), suggesting that alkalisation should occur in the rhizosphere of short, mycorrhizal roots while slight acidification was expected in the rhizosphere of non infected, long roots. This suggests that the rhizosphere pH can be extremely heterogeneous along the root system of ectomycorrhizal plants, as also shown at a broader scale for seedlings growing in soil-filled rhizotrons by Casarin et al. (2003). These authors showed that acidification occurred in the rhizosphere of ectomycorrhizal roots, compared with the non mycorrhizal plant. Rigou et al. (1995) had formerly shown with a dye indicator that for pine seedlings grown in agar gels the roots of mycorrhizal plants were exhibiting larger proton effluxes and thus rhizosphere acidification than roots of non mycorrhizal plants. However, most of the literature on localised pH changes in the rhizosphere has been using techniques where plants are grown in rather artificial conditions such as hydroponic solutions or agar gels. Such simplified media which are transparent and thus allow visual observation of heterogeneities in the rhizosphere have been also successfully used to visualize redox changes occurring in rice (Trolldenier 1988). Using an oxygen depleted fluid agar medium, combined with redox microelectrodes, Armstrong et al. (2000) measured pO_2 gradients with high spatial resolution (10 µm) across roots and the rhizosphere of a wetland plant (*Phragmites australis* L.). They showed that at the apex, pO_2 increased from 5,000 cm^3 m^{-3} at 2 mm from the root surface to about 100,000 cm^3 m^{-3} at the root suface (and slightly more in the root cortex). Interestingly they showed that a much sharper gradient of pO_2 or almost no leakage of O_2 occurred at more basal parts of the root, suggesting that root-induced rhizosphere oxidation was rather confined to the apical region of the roots. The use of microsensors of O_2 was successfully used in both agar media and water-saturated soils to study the gradient of pO$_2$ in the rhizosphere of rice (Revsbech et al. 1999).

Only few attempts have been made to assess pH and redox potential changes in soil grown plants. Recently Blossfeld and Gansert (2007) made a major step forward by using a foliar optical pH sensor which provided access to spatial heterogeneities of pH over several tens of cm^2 around roots of plants grown in rhizotrons. In addition their technique was sensitive enough to show a distinct diurnal rhythm with a larger acidification occurring in day-time than at night (Fig. 7). Although applied so far only to a wetland plant species (*Juncus effusus* L.) growing under reduced soil conditions, this technique and the future development of other similar sensors are promising tools to further our understanding of actual pH changes occurring in situ. Previously, assessment of the temporal and spatial heterogeneities of pH changes had been achieved with the use of arrays of microelectrodes (Fischer et al. 1989) or soil solution samplers designed by Göttlein et al. (1996). Other solution samplers such as rhizons also provided valuable, discrete information on pH changes in the rhizosphere but they proved unable to monitor short-term (less than daily) or short-distance changes along a root or radially from the root because of sampling a quite large volume of soil solution (Cornu et al. 2007; Bravin et al. 2008). The design of non-invasive techniques as the one developed by Blossfeld and Gansert (2007) is in this respect much more helpful and promising (Luster et al. 2009).

Fig. 7 Temporal development of root-induced pH changes in the rhizosphere of growing roots of *Juncus effusus* L. Snapshots were obtained at different times from day 1 (D1) to day 2 (D2), with illumination starting at 0800 (8 am) and ending at 2200 (10 pm) each day. The colours indicate different pH values (see legend at the bottom) as measured non-invasively by the planar pH optode. The crossing points of the grid (top left) indicate the positions of fibre-optic pH measurements. The digital photograph (bottom right) shows the investigated section of the planar optode at the end of the time series with two roots (labelled I and II) growing across the pH optode (taken from Blossfeld and Gansert (2007) with kind permission by Blackwell Publish. Ltd.)

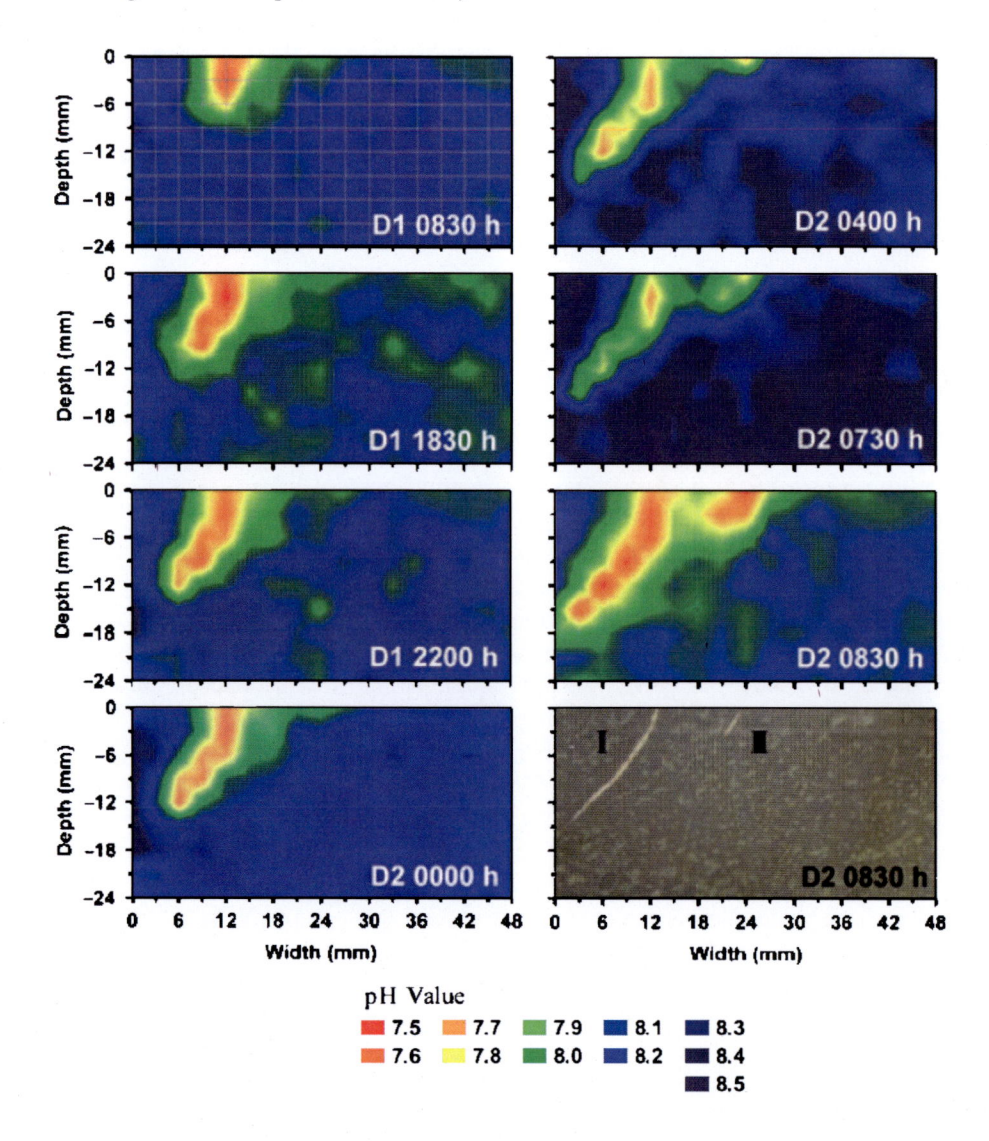

The pH buffering capacity of the soil is subject to considerable variations between soils and also possibly within a given soil as a function of the heterogeneities of distribution of constituents that play a key role in buffering the pH (e.g. particles or patches of organic matter or $CaCO_3$). As stressed by Hinsinger et al. (2003) based on the earlier work of Schubert et al. (1990) and Hinsinger and Gilkes (1996), little or no significant pH change may not mean the absence of proton fluxes in the rhizosphere, as these protons may be implied in a range of reactions that result in proton consumption (which make up the buffering capacity of the soil). These fluxes may have an important functional impact (e.g. on the subsequent mobilisation/immobilisation of nutrients) while pH changes would remain unaffected if the buffering capacity of the soil is large. This is illustrated in the work of Göttlein et al. (1999) who reported an increase in Al concentration close to growing roots (Fig. 8), in spite of an absence of significant pH change. Presumably, protons released by roots were consumed in reactions with the soil solid phase, e.g. exchange with Al-ions, which ultimately resulted in the observed increase in Al concentration. Beside its effect on the change in pH, the pH buffering capacity was found linearly correlated with the radial extension of pH gradients in the rhizosphere, as measured with microelectrodes around single roots (Schaller 1987).

The same holds for redox potential and may explain the reason why rhizosphere oxidation is often found to be confined to very short distances in water-saturated soils, typically less than a few hundred μm (Flessa and Fischer 1992; Revsbech et al. 1999; Bravin et al. 2008), possibly up to a few mm from the root surface (Begg et al. 1994; Armstrong et al. 2000). O_2 leaking from the roots is rapidly consumed in redox reactions such as the precipitation of iron oxides and a range of microbially-mediated processes (Revsbech et al. 1999; Bravin et al. 2008). Callaway and King (1996) have shown that, in addition to this, O_2 leaking from the roots of wetland plants such as *Typha latipholia* L. can be used by neighbouring plants (for respiration purposes) that would otherwise not withstand the low ambient pO_2. This was the evidence for O_2-mediated facilitation-competition taking place within this plant community.

In contrast with the spatial component of pH and Eh changes in the rhizosphere, the temporal component has been poorly documented, especially so over rather long terms (more than a few hours or days). Turpault et al. (2007) have been comparing in situ pH changes at two distinct seasons and have basically related the seasonal effect to the nitrification process and its link with cation/anion uptake balance. Investigating such long time scales requires in situ monitoring of pH, thus raising many methodological challenges, and the question of the location of the probes/sampling devices given the dynamic nature of root growth and development. Mathematical models are additional tools to solve these interdependent, complex temporal/spatial patterns as nicely shown over shorter scales by Kim et al. (1999) and Peters (2004). Kim et al. (1999) modeled the temporal evolution over a few hours up to 2 days of the axial and radial patterns of pH around a growing root tip, taking account of the growth of the root during this period of time. It should be stressed that efflux of protons also feeds back on root elongation (via cell-wall loosening) according to the acid growth theory, which further justify the necessity for coupling these two processes (Peters 2004). For redox potential, a few studies have been monitoring temporal changes of Eh over time. Cornu et al. (2007) and Bravin et al. (2008) have been monitoring redox potential close to a root mat (rhizobox approaches) with a daily resolution, in aerobic and anaerobic conditions, respectively. Continuous monitoring was achieved by Fischer et al. (1989) who could thereby relate the observed change of redox potential to the growth of the root of faba bean (*Vicia faba* L.). They showed that the decrease in Eh was fast (300 mV within about 1h) when the root apex approached the microelectrode. Then it slowly increased to reach its initial value after about 24 h (at a stage where the apex had grown a few mm further away), suggesting that such reduction was probably related to intense respiration in the meristematic zone of the root (apex). The combined pH-O_2 sensor foils developed recently by Blossfeld and Gansert (2007) shall provide a unique opportunity to map and monitor pH and redox potential changes in the rhizosphere of soil-grown plants over fine spatial and temporal resolutions, as shown for pH in the wetland plant *Juncus effusus* L. (Fig. 7). So far this has only been applied to ambient, reducing conditions and it would thus be very helpful to apply a similar technique for studying the spatial and temporal changes of redox conditions in plants growing in aerobic soil conditions.

Fig. 8 Temporal changes of Ca and Al concentrations, and Ca/Al ratios in the rhizosphere of growing roots of an oak (*Quercus ruber* L.) seedling. Soil solution was sampled at 3-week intervals with an array of microsuction cups positioned along a grid in a rhizotron, and analyses were performed with capillary electrophoresis (taken from Göttlein et al. (1999) with kind permission by Springer)

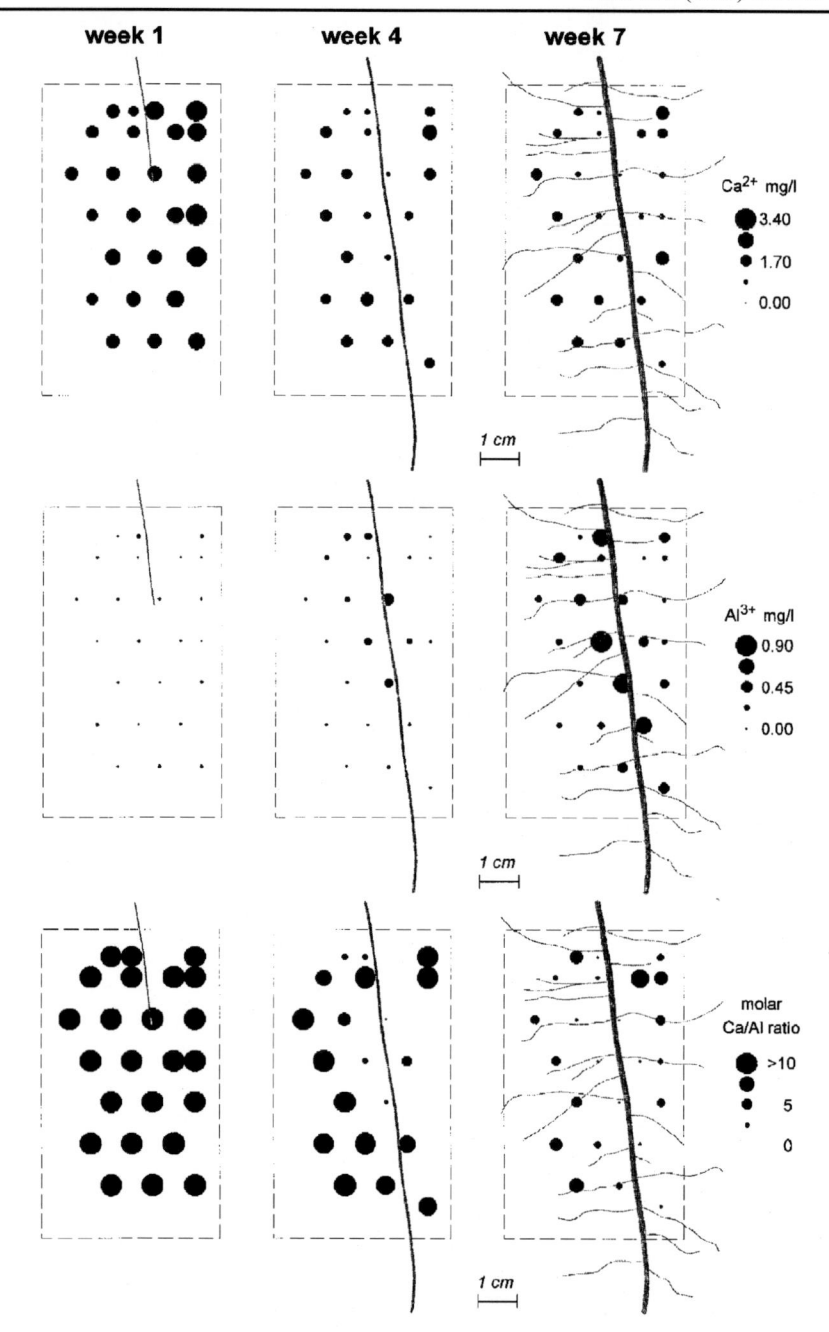

Nutrient distribution and accessibility in the rhizosphere

Ecological relevance

Compared with aquatic environments, soil environments are harsher because many nutrients are present in only small concentrations in the soil solution, being largely bound to the solid phase constituents. Their spatial distribution is far more heterogeneous than in aquatic environments and thus the spatial component of the bioavailability (accessibility) of nutrients is crucial. Plants and soil microorganisms have only limited ability to move towards nutrient-enriched zones, compared with animals. To cope with those conditions they have therefore evolved a whole range of strategies for accessing nutrient resources in the soil (Marschner 1995; Lambers et al. 1998; Richardson et al 2009). Many of the underlying processes play a key role in nutrient cycling in soils (Kandeler et al. 2002; Philippot et al. 2009). Plants may create positive feedbacks to nutrient cycling because of species' differences in carbon deposition and competition with microbes for nutrients in the rhizosphere. Plant species' effects can be as or more important than abiotic factors, such as climate, in controlling ecosystem fertility (Hobbie 1992). The availability of organic matter is a major driver for heterotrophic microorganisms, and plants play a considerable role through either litter deposition or rhizodeposition in the rhizosphere (Jones et al 2004, 2009; Wardle et al. 2004).

The scarcity of many nutrients in soils also means that harsh competition occurs for these nutrients and ultimately determines the structure of microbial communities, as evidenced for iron and phosphorus. The most nutrient-efficient microorganisms are thereby efficiently selected in the rhizosphere (Marschner et al. 2006; Calvaruso et al. 2007; Robin et al. 2008; Lemanceau et al 2009). Such competition also occurs between plant roots of neighbouring species and may ultimately drive the composition of plant communities (e.g. Callaway et al. 2002; Raynaud et al. 2008). In alpine and arctic soils where the availability of inorganic nitrogen is especially low due to slow rates of organic matter mineralisation, plant communities are dominated by those species that can make use of amino-acids or ammonium rather than nitrate (Chapin et al 1993; Raab et al. 1996; Lambers et al. 1998). Modelling diffusion of nutrients in the rhizosphere of neighbouring roots proved an efficient approach to further understanding some of the key processes involved in plant-plant interactions. Craine et al. (2005) showed for instance that higher nitrate uptake capacity (I_{max}) of one plant species led to larger depletion zones around its roots and thus played a central role in pre-empting nitrate and out-competing less efficient roots of neighbour plants. Raynaud et al. (2008) further refined this model by accounting also for the diffusion of exudates, showing that large diffusion rates of root exudates may result in increased bioavailability of nutrients such as phosphate for neighbouring plant roots which do not exhibit same exudation potential, thereby contributing to facilitation rather than competition.

The occurrence of nitrogen-fixing legumes also plays a major role in structuring plant and microbial communities as a result of the build-up of nitrogen availability in soils (Spehn et al. 2000). This means that besides competition for nutrients, facilitation and complementarity are other major driving forces of ecosystem productivity (Callaway and Walker 1997; Callaway et al. 2002; Gross et al. 2007). This is obvious for mutualistic relationships between the host plant and the microsymbiont in mycorrhizal or rhizobial associations. However, there is an increasing interest for understanding how it operates between plant species living in communities (Callaway and Walker 1997; Callaway et al. 2002), either in grassland or agroforestry and other intercropping systems (Paynel and Cliquet 2003; Gross et al. 2007; Li et al. 2007; Li et al. 2008).

Underlying processes

Rhizodeposition is a primary process altering nutrient abundance in the rhizosphere. The corresponding flow of carbon results in a local build-up in C-rich substrates that fuel the growth of heterotrophic microbial communities (Lynch and Whipps 1990; Nguyen 2003; Jones et al. 2004, 2009). The rhizosphere thus appears as enriched soil microsites, along with other hotspots of microbial activities such as plant debris and dead bodies (detritusphere). While the bulk soil is C-limiting for microbial growth (Wardle 1992), the rhizosphere is rather N-limiting than C-limiting. A priming effect can occur in the rhizosphere leading to enhanced decomposition of soil organic matter, as reviewed by Kuzyakov (2002) who stressed that reduced decomposition could sometimes occur as well. Such priming effect is the consequence of rhizodeposition and especially so root exudation releasing simple substrates with low C:N ratios, stimulating rhizosphere microorganisms to decompose soil organic matter to meet their N requirements. Rhizosphere enrichment in inorganic N is however unlikely to occur due to microbes and plants competing for the acquisition of the released N, as recently modelled by Raynaud et al. (2006) who also accounted for the exudation of N-rich compounds such as amino-acids. The fate of C and N is further complicated in the rhizosphere if one accounts for the whole microbial loop including predators such as protozoae and nematodes (Bonkowski 2004; Bonkowski et al. 2009).

Nutrient transfer via mass-flow and diffusion occurs in the rhizosphere because of water and nutrient uptake by roots, and is another major driving force of nutrient redistribution in the rhizosphere. Nutrients that occur as solutes in the soil solution are transferred via mass-flow towards the root surface when the plant is transpiring. Nutrients that are abundant in the soil solution may thereby accumulate in the rhizosphere whenever their flow exceeds plant's demand. This typically occurs for Ca and Mg, and has been also reported to occur for K (Lorenz et al. 1994; Barber 1995; Clegg and Gobran 1997; Vetterlein and Jahn 2004b; Turpault et al. 2005). Precipitation of gypsum (Ca sulfate, Fig. 9) and calcite (Ca carbonate) have been reported to occur as a consequence of the build-up of Ca concentration in the rhizosphere (Hinsinger 1998). In contrast, when the flow of nutrient transferred via mass-flow is less than plant's requirement, their concentration decreases in the rhizosphere (Lorenz et al. 1994; Barber 1995; Jungk 2002; Hinsinger 2004). Such depletion typically occurs for P and micronutrients, as well as for N and K in many cases, and generates a diffusion gradient towards the root surface (Hendriks et al. 1981; Kuchenbuch and Jungk 1982; Gahoonia et al. 1992; Ge et al. 2000; Jungk 2002; Hinsinger et al. 2005). The rhizosphere can thereby appear as a nutrient-enriched or -impoverished zone.

Besides the above-mentioned physical processes, roots and rhizosphere microorganisms can also alter the nutrient concentration via a range of chemical and biochemical processes (Darrah 1993; Hinsinger 1998; Hinsinger et al. 2005). Changes of pH in the rhizosphere can for instance dramatically influence the availability of nutrients via competition of protons for metal cations on cation exchange sites (e.g. for Cu and Zn; Loosemore et al. 2004; Michaud et al. 2007) or via shifting the dissolution/precipitation equilibria of nutrient-bearing minerals as shown for e.g. Mg (Hinsinger et al. 1993; Calvaruso et al. 2006) and P (Gahoonia et al., 1992; Begg et al. 1994; Hinsinger and Gilkes 1996; Bertrand et al. 1999; George et al. 2002a). In addition, exudation of organic ligands by roots and rhizosphere microorganisms may further increase the availability of nutrients via either ligand exchange-promoted desorption of anions (e.g. phosphate-ions desorbed by mucilage or carboxylates: Geelhoed et al. 1999; Hinsinger 2001; Ryan et al. 2001; Read et al. 2003; Dunbabin et al. 2006) and/or complexation of metal cations, as documented for carboxylates (citrate or oxalate) and siderophores of microbial or plant origin. These can thereby promote the dissolution of metal-bearing minerals as shown for Ca and Fe (Jones and Darrah 1994; Wallander 2000; Casarin et al. 2004; Reichard et al. 2005, 2007) and/or increase the total concentration of metal cations in the soil solution (Gerke et al. 2000; Reichard et al. 2005, 2007). In submerged soils where ambient anoxic conditions prevail and lead to excess (toxicity) of metal nutrients such as Fe and Mn (occurring as reduced species), their concentration can be decreased in the rhizosphere of wetland plants and rice as a

Gypsum crystals

Ca=blue, S=green, Al=red

Fig. 9 Crystals of gypsum (Ca sulfate) precipitating close to a root of pine (*Pinus sylvestris* L.) growing along a 30 μm nylon mesh (on the other side of the mesh) in a rhizobox experiment. The micrograph (left box) was obtained with a scanning electron microscope, while the corresponding elemental maps (right box) were obtained by EDX analyses (Reproduced by kind permission of Doris Vetterlein)

consequence of redox reactions and Fe plaque formation (Begg et al. 1994; Bravin et al. 2008). Besides playing a major role in root- and microbe-mediated Fe cycling in the rhizosphere (Neubauer et al. 2007), this Fe plaque can sequester other nutrients such as phosphate and zinc (Kirk and Bajita 1995; Saleque and Kirk 1995). Another major process that can concur to altering the availability of nutrients in the rhizosphere is the release by roots and microorganisms of enzymes implied in the hydrolysis of organic forms of nutrients such as N, P and S, as documented extensively for phosphatases (George et al. 2002b, 2004 and 2006; Kandeler et al. 2002; Li et al. 2004; Denton et al. 2006; Richardson et al. 2009).

The concurrence of the various above-mentioned processes can lead to both enrichment and impoverishment of nutrients in the rhizosphere as shown for instance for P (Fig. 10). A given fraction of soil P can be depleted close to roots and accumulate at further distance from roots as shown by Hübel and Beck (1993) or Hinsinger and Gilkes (1996). It has been more frequently reported that some P fractions are depleted while others are increased in the rhizosphere, with neat differences between plant species (Geelhoed et al. 1999; George et al. 2002a; Li et al. 2008). What is little known is the quantitative contribution of the various above-mentioned processes to the actual pattern of nutrient distribution observed in the rhizosphere, even more so when different species live in a community. Modelling can help unravel these processes (Raynaud et al. 2008). Most of the few investigations on the rhizosphere of soil-grown plants have reported greater nutrient concentrations than in bulk soil (Courchesne and Gobran 1997; Schöttelndreier and Falkengren-Grerup 1999; Pankhurst et al. 2002; Turpault et al. 2005). However, as stressed by Schöttelndreier and Falkengren-Grerup (1999) and Hinsinger et al. (2005), it is difficult to distinguish between actual rhizosphere enrichment or preferential colonization of nutrient-rich patches or root macropores in such studies. This is a clear limitation of in situ experiments as the nutrient foraging behavior has been shown as an important strategy for nutrient acquisition in higher plants (Leyser and Fitter 1998; Robinson et al. 1999; Hodge 2004; Hodge et al. 2009) including ectomycorrhizal (roots and hyphae) plants (Read and Perez-Moreno 2003; Rosling et al. 2004). Root architecture matters more for the more poorly mobile nutrient P than for relatively mobile N, but conversely the plasticity of root systems has been shown to be greater for patchy distribution of N than P (Fitter et al. 2002; Linkohr et al. 2002).

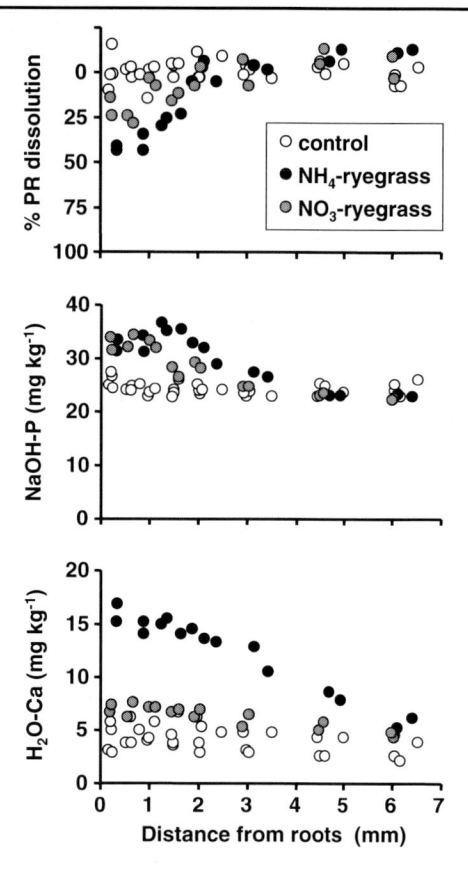

Fig. 10 Gradients of P and Ca in the rhizosphere of ryegrass (*Lolium rigidum* L.) supplied with a phosphate rock (PR) as sole source of P and Ca, and with N as either nitrate only (NO_3) or nitrate and ammonium (NH_4), relative to control treatment without plant. The extent of the dissolution of PR was deduced from the measurement of total Ca and was greater when ammonium was supplied as a consequence of greater root-induced acidification. Part of the dissolved P and Ca (resulting from PR dissolution) which were not taken up by ryegrass accumulated in the rhizosphere and were recovered in the alkali-soluble and water-soluble fractions respectively, over varying distances and extents, depending on N source (taken from Hinsinger and Gilkes (1996) with kind permission by Blackwell Publish. Ltd.)

Gradients, spatial and temporal heterogeneities

Distribution of nutrients is far from uniform in soils, and roots are an additional source of heterogeneity, which builds up over time in the rhizosphere, along the course of root growth. This was shown for instance by Göttlein et al. (1999) for Ca in an acidic forest soil (Fig. 8). As for water, nutrient uptake is not uniform along a growing root (Clarkson 1991). Even if environmental conditions like water content, temperature, and root soil contact are homogeneous, there is a distinct pattern of uptake for specific nutrients as shown in nutrient solution by Harrison-Murray and Clarkson (1973) and Ferguson and Clarkson (1975). They clearly showed that Ca uptake was much larger at the root tip that in older regions of the root because of the suberised endodermis restricting the radial flow of those nutrients which preferentially use the apoplastic pathway. In their studies, other nutrients which are known to use the symplasmic pathway of radial flow towards the root stele, such as P and K, were evenly taken up along the root length. For nutrients other than Ca and Mg, contradictory results can be found in the literature. For maize roots highest phosphate uptake was found for 1-day old root zones and uptake declines to 25 to 30% in 26-day old zones (Ernst et al. 1989). This uptake pattern was likely related to the presence and life span of root hairs. A similar pattern of uptake was observed for nitrate by Reidenbach and Horst (1997) who suggested that such differences in uptake should be reflected in uptake models by forming root age classes with different uptake rates. Reidenbach and Horst (1997) also observed a marked diurnal variation in uptake, which was larger at lower light intensity. Diurnal patterns of exudation which derive from the change of availability of photosynthates in the plant as shown for instance for citrate release by proteoid roots (Keerthisinghe et al. 1998, Dessureault-Rompré et al. 2007) are also possibly implied in a diurnal pattern of P acquisition. In this respect, the most remarkable circadian pattern is that described for phytosiderophore release in iron deficient graminaceous plant species (Walter et al. 1995; Reichman and Parker 2007). A distinct peak of secretion occurs between 3 and 6 h after the onset of light, which is quite a more efficient strategy for efficient Fe acquisition than evenly releasing such ligands at a lower flux all day long (Robin et al. 2008). Focussing their secretion in a narrow temporal and spatial (it is occurring at greater flux behind the apex) window minimises the risk of being metabolised by rhizosphere microorganisms prior to chelating Fe (Darrah 1991). Specialized organs such as ectomycorrhizal roots and cluster roots are additional sources of heterogeneities in exudation, uptake and acquisition of nutrients along root systems (Plassard et al. 2002; Casarin et al. 2003; Shane and Lambers 2005; Lambers et al. 2006).

Radial gradients of nutrients around roots due to depletion or accumulation can extend over less than 1 mm up to several cm depending on their mobility in the soil (Fig. 10). Depletion of P, K and nitrate typically occurs over about 1, 5 and more than 10 mm for these ions, respectively (e.g. Jungk and Claassen 1986; Darrah 1993; Jungk 2002; Hinsinger 2004). For phosphate, the least mobile of major nutrients, the radial gradient which normally does not extend more than about 1 mm can be increased up to about 2 mm by root hairs (Gahoonia et al. 2001; Ma et al. 2001). If plants are mycorrhizal, the hyphae can extend even further into the soil and resulting in a depletion of P over several mm (Li et al. 1991). Although, little documented, reverse gradients can occur for nutrients such as Ca which tend to accumulate in the rhizosphere as a consequence of mass-flow exceeding the actual uptake flux into the root (Fig. 10).

Water uptake, nutrient uptake, exudation of protons and release of organic acids or enzymes, are processes that are likely to have a different temporal and spatial pattern. Nutrient gradients in the rhizosphere, which results from the interplay of all these processes and transport processes (mass flow and diffusion), can thus considerably vary in both space and time. At a spatial point of view, in a few cases, complex concentration profiles have been shown to occur for nutrients such as P (Fig. 10), as a consequence of the combination of an uptake-driven depletion gradient and a soil P mobilisation-driven accumulation gradient, the latter being the consequence of the exudation of protons (Hinsinger and Gilkes 1996), carboxylates (Kirk 1999) or possibly phosphatase enzymes (Hübel and Beck 1993). This means that nutrient uptake and transport processes (of the considered nutrient only) as accounted for in most models derived from that of Nye and Marriott (1969) only explain a portion of the observed gradients for nutrients such as P. Indeed, one should also account for the gradients of protons (Gahoonia et al. 1992; Hinsinger and Gilkes 1996) and exudates (George et al. 2002b) that may contribute to further mobilisation/immobilisation of P in the rhizosphere, as well as phosphatases of both root and microbial origins (Tarafdar and Jungk 1987; Chen et al. 2002; George et al. 2002b; Kandeler et al. 2002). Those exudates that are more mobile (e.g. protons rather than enzymes) are thus likely more efficient at providing extra nutrients to the root (Hinsinger et al. 2005). For instance, the poor mobility of enzymes in the soil (due to strong adsorption onto soil particles) is likely to explain the poor performance for P acquisition of transgenic plants transformed to secrete a phytase of fungal origin relative to control plants once grown in soils, while the difference was considerable in soil-free media (George et al. 2004 and 2005).

Such complex interplay of processes is dynamic. For example as K requirement is high for young plants but decreases with age, K depletion at the root surface can be observed in young plants (12 days old) while accumulation was found for older plants (29 days old) by Vetterlein and Jahn (2004b). Accumulation of easily available K in the rhizosphere as recorded in situ for field-grown plants such as forest trees can also result from the combination of uptake-driven depletion and weathering of K-bearing minerals or organic matter which over longer time scales may contribute a net increase in soil solution and exchangeable K in the rhizosphere (Clegg and Gobran 1997; Turpault et al. 2005). If studies are conducted in soil and not in hydroponics, care should be taken in interpreting soil solution concentrations. Soil solution concentrations may increase/decrease due to transport exceeding/falling short of plant uptake or mobilisation/immobilisation reactions. However, changes in soil solution concentrations can also be a consequence of changes in the amount of solute, i.e. soil water content (Vetterlein and Jahn 2004b). Typically, soil water content/soil matric potential rapidly decreases at the root surface compared with the bulk soil (see above). Thus increased soil solution concentrations at the root surface may just reflect decreases in soil water content. The same is true for chemical substances released from the roots like organic anions, protons or enzymes. The amount released dissolves in a smaller volume of water and their diffusion into the bulk soil is hindered as diffusion coefficients decrease with decreasing water content.

Temporal dynamics of nutrient gradients, neglecting additional water content gradients, are illustrated by Kirk (2002) and Geelhoed et al. (1999) for P. In their model calculations these authors predicted soil solution P concentration profiles with increasing distance from the root surface. In both models, the release of citrate by roots and citrate degradation with time and interaction with P adsorbed on surfaces or P in mineral phases were taken into account as well as citrate diffusion away from the root surface. Both models predict, for certain time intervals, a strong increase in soil solution P concentration close to the root surface and a decrease within the first millimetre from the root surface. With time the steepness of gradients decreases. These predictions are in close agreement with experimental data based on soil solution sampling over time (Vetterlein et al. 2007b).

Kirk's and Geelhoed's models, in contrast to earlier models of Nye and Marriott (1969) or Claassen et al. (1986), already take into account the interaction of a number of chemical species. Kirk's approach (Kirk 2002) is basically an extension of that proposed by Nye (1983) to account for the interaction of two solutes diffusing in opposite directions (e.g. phosphate and a root exudate) via an interaction term. It can hardly be generalised as it cannot account for the more complex situation of P biogeochemistry and biochemistry in the rhizosphere, where solution P concentration results from many more than two counteracting processes (mass-flow and diffusion of P-ions and exudation of diverse exudates being possibly interacting). The approach developed by Geelhoed et al. (1999) is more mechanistic but applies to a simplistic situation where only one type of solid constituent (goethite) interacts via a single mechanism (adsorption/desorption) while in practice P concentrations also depend on a range of dissolution/precipitation equilibria and organic P hydrolysis in the rhizosphere (Hinsinger 2001; Hinsinger et al. 2005). Only sophisticated geochemical models or ideally models that couple numerous, biogeochemical reactions and transport processes can account for such complex rhizosphere chemistry (Anoua et al. 1997; Nowack et al. 2006). This type of approach which takes into account the chemical speciation in soil solution was recently extended by Szegedi et al. (2008) by the development of the RhizoMath model. It is based on coupling the mathematical package MATLAB with the geochemical code PHREEQC. RhizoMath's greatest advantage is that different geochemical models (with and without charge balance) and geometries (planar and radial) are already included. Moreover, due to its graphical user interface the tool can be applied without changing the source code or a complex input file. RhizoMath can be used as a tool to integrate processes occurring simultaneously in the rhizosphere and it has the potential to function as a submodule for larger scale plant growth models to build in rhizosphere processes.

Conclusions

Life on planet Earth is sustained by a small volume of soil surrounding roots and influenced by living roots, called the rhizosphere. This micro-environment concentrates much of Earth's biodiversity, being probably the most dynamic habitat on Earth (Jones and Hinsinger 2008). As it represents the interface between soils and plants, the rhizosphere plays a key role in defining the quality and quantity of the Human terrestrial food resource on the one hand, and on soil formation on the other hand, via the input of organic carbon deep in the substratum. Despite its central importance to all life below and aboveground, we know very little about the intimate functioning of the rhizosphere, and even less about how best we can manipulate it to our advantage. We especially lack a holistic perception of the rhizosphere and have an extraordinary ignorance of the intimate connections between the biology, physics and chemistry of the system, which thereby exhibits astonishing spatial and temporal heterogeneities. Whilst important insights have been made by individual disciplines, we postulate that the next great advances will be through interdisciplinary approaches that implicitly account for biophysical and biogeochemical processes, driven by a range of theoretical frameworks designed to predict important functions such as water uptake, food production and soil sustainability.

The rhizosphere exhibits a unique biophysics and biogeochemistry which contrasts from that of the bulk soil, thereby explaining the diversity of habitats it harbours for microorganisms. Besides determining those microbial habitats, the inner physical structure is the basis for describing movement and storage of elements and water as well as the ability of roots to explore the soil. This ultimately determines the accessibility of water and nutrients to the roots and associated microorganisms, as well as the biogeochemical environment, via the partial pressures of O_2 and CO_2, and thereby the redox potential and pH of the rhizosphere, respectively. We urgently need to improve our methodology to describe this physical architecture at the micro scale and to express these findings not only in a descriptive manner but to interpret the data in respect to their functional relevance.

Noninvasive methods may be the tool to improve knowledge on the physical architecture and may in addition provide new insights into spatial and temporal dynamics of porosity, soil strength and water distribution around single roots. Unfortunately these methods still face a trade-off between spatial resolution and field of view and different methods are optimum for obtaining detailed information on structure (e.g. X-ray), water (e.g. neutron radiography) and roots (e.g. microscopy). Non-invasive methods for visualising the distribution of elements in situ with a similar spatial resolution are unfortunately not available at present. Two-dimensional developments as exemplified by the optode foil for the measurement of temporal changes of pH, redox potential and pCO_2 are however opening up new opportunities. While much progress has been made recently for exploring the spatial dimension of rhizosphere processes, our knowledge of their temporal dimension is still scarce. In addition, as stressed by Watt et al. (2006a), the paucity of reliable data obtained under field conditions underlines the rudimentary state of our knowledge of root-microorganism interactions in real world conditions.

With improvements in spatial and temporal resolution of measurements the need for models to integrate this information increases as the different parameters interact with each other and show different dynamics in space and time. There are promising approaches under way that allow new modules to be added as knowledge increases. There remains the challenge of integrating such detailed models into root growth and root architecture models for upscaling of rhizosphere processes (Standing et al. 2007). This last step is crucial to reflect rhizosphere processes in field, regional or global scale models. Our knowledge of the rhizosphere biophysics and biogeochemistry has been increasing tremendously during the last few decades, opening possible new applications to a range of major ecosystem services at a much broader scale than the soil/root/microorganism interface. Major progress will be achieved if scientists are able to bridge the micro- and macro-scales, from the rhizosphere of individual roots to soils at field scales and larger.

Acknowledgements The Scottish Crop Research Institute receives grant-in-aid support from the Scottish Government Rural and Environment Research and Analysis Directorate.

References

Alami Y, Achouak W, Marol C, Heulin T (2000) Rhizophere soil aggregation and plant growth promotion of sunflowers by an exopolysaccharide-producing Rhizobium sp. Isolated from sunflower roots. Appl Environ Microbiol 66:3393–3398 doi:10.1128/AEM.66.8.3393-3398.2000

Anoua M, Jaillard B, Ruiz T, Bénet JC (1997) Couplages entre transferts de matière et réactions chimiques dans un sol. II. Application au transfert de matière dans la rhizosphère. Entropie 207:13–24

Armstrong W (1979) Aeration in higher plants. Adv Bot Res 7:225–232 doi:10.1016/S0065-2296(08)60089-0

Armstrong W, Cousins D, Armstrong J, Turner DW, Becket PM (2000) Oxygen distribution in wetland plant roots and permeability barriers to gas-exchange with the rhizosphere: a microelectrode and modelling study with *Phragmites australis*. Ann Bot (Lond) 86:687–703 doi:10.1006/anbo.2000.1236

Barber SA (1995) Soil nutrient bioavailability: a mechanistic approach, 2nd edn. John Wiley, New York, USA, p 414

Barber DA, Gunn KB (1974) The effect of mechanical forces on the exudation of organic substances by the roots of cereal plants. New Phytol 73:39–45 doi:10.1111/j.1469-8137.1974.tb04604.x

Barrowclough DE, Peterson CA, Steudle E (2000) Radial hydraulic conductivity along developing onion roots. J Exp Bot 51:547–557 doi:10.1093/jexbot/51.344.547

Begg CBM, Kirk GJD, MacKenzie AF, Neue H-U (1994) Root-induced iron oxidation and pH changes in the lowland rice rhizosphere. New Phytol 128:469–477 doi:10.1111/j.1469-8137.1994.tb02993.x

Bengough AG, Mullins CE (1990) Mechanical impedance to root growth - a review of experimental techniques and root growth responses. J Soil Sci 41:341–358

Bengough AG, Bransby MF, Hans J, McKenna SJ, Roberts TJ, Valentine TA (2006) Root responses to soil physical conditions; growth dynamics from field to cell. J Exp Bot 57:437–447 doi:10.1093/jxb/erj003

Bertrand I, Hinsinger P, Jaillard B, Arvieu JC (1999) Dynamics of phosphorus in the rhizosphere of maize and rape grown on synthetic, phosphated calcite and goethite. Plant Soil 211:111–119 doi:10.1023/A:1004328815280

Bidel LPR, Renault P, Pages L, Riviere LM (2000) Mapping meristem respiration of *Prunus persica* (L.) Batsch seedlings: potential respiration of the meristems, O_2 diffusional constraints and combined effects on root growth. J Exp Bot 51:755–768 doi:10.1093/jexbot/51.345.755

Binnerup SJ, Sorensen J (1992) Nitrate and nitrite microgradients in barley rhizosphere as detected by a highly sensitive denitrification bioassay. Appl Environ Microbiol 58:2375–2380

Blossfeld S, Gansert D (2007) A novel non-invasive optical method for quantitative visualization of pH dynamics in the rhizosphere of plants. Plant Cell Environ 30:176–186 doi:10.1111/j.1365-3040.2006.01616.x

Boeuf-Tremblay V, Plantureux S, Guckert A (1995) Influence of mechanical impedance on root exudation of maize seedlings at two development stages. Plant Soil 172:279–287 doi:10.1007/BF00011330

Bonkowski M (2004) Protozoa and plant growth: the microbial loop in soil revisited. New Phytol 162:617–631 doi:10.1111/j.1469-8137.2004.01066.x

Bonkowski M, Griffiths B, Villenave C (2009) Rhizosphere fauna: functional and structural diversity. Plant Soil (this volume, in preparation, provisional title)

Bravin MN, Travassac F, Le Floch M, Hinsinger P, Garnier JM (2008) Oxygen input controls the spatial and temporal dynamics of arsenic at the surface of a flooded paddy soil and in the rhizosphere of lowland rice (*Oryza sativa* L.): a microcosm study. Plant Soil 312:207–218 doi:10.1007/s11104-007-9532-x

Bruand A, Cousin I, Nicoullaud B, Duval O, Bégon JC (1996) Backscattered electron scanning images of soil porosity for analysing soil compaction around roots. Soil Sci Soc Am J 60:895–901

Brune A, Frenzel P, Cypionka H (2000) Life at the oxic-anoxic interface: microbial activities and adaptations. FEMS Microbiol Rev 24:691–710

Bundt M, Widmer F, Pesaro M, Zeyer J, Blaser P (2001) Preferential flow paths: biological 'hot spots' in soils. Soil Biol Biochem 33:729–738 doi:10.1016/S0038-0717(00)00218-2

Burgess SSO, Pate JS, Adams MA, Dawson TE (2000) Seasonal water acquisition and redistribution in the Australian woody phreatophyte, *Banksia prionotes*. Ann Bot (Lond) 85:215–224 doi:10.1006/anbo.1999.1019

Caldwell MM, Dawson TE, Richards JH (1998) Hydraulic lift: consequences of water efflux from the roots of plants. Oecologia 113:151–161 doi:10.1007/s004420050363

Callaway RM, King L (1996) Temperature-driven variation in substrate oxygenation and the balance of competition and facilitation. Ecology 77:1189–1195 doi:10.2307/2265588

Callaway RM, Walker LR (1997) Competition and facilitation: a synthetic approach to interactions in plant communities. Ecology 78:1958–1965

Callaway RM, Brooker RW, Choler P, Kikvidze Z, Lortie CJ, Michalet R, Paolini L, Pugnaire FL, Newingham B, Ascheoug ET, Armas C, Kikodze D, Cook BJ (2002) Positive interactions among alpine plants increase with stress. Nature 417:844–848 doi:10.1038/nature00812

Callaway RM, Pennings SC, Richards CL (2003) Phenotypic plasticity and interactions among plants. Ecology 84:1115–1128 doi:10.1890/0012-9658(2003)084[1115:PPAIAP]2.0.CO;2

Calvaruso C, Turpault MP, Frey-Klett P (2006) Root-associated bacteria contribute to mineral weathering and to mineral nutrition in trees: a budgeting analysis. Appl Environ Microbiol 72:1258–1266 doi:10.1128/AEM.72.2.1258-1266.2006

Calvaruso C, Turpault MP, Leclerc E, Frey-Klett P (2007) Impact of ectomycorrhizosphere on the functional diversity of soil bacterial and fungal communities from a forest stand in relation to nutrient mobilization processes. Microb Ecol 54:567–577 doi:10.1007/s00248-007-9260-z

Caravaca F, Alguacil MM, Torres P, Roldan A (2005) Plant type mediates rhizospheric microbial activities and soil aggregation in a semiarid Mediterranean salt marsh. Geoderma 124:375–382 doi:10.1016/j.geoderma.2004.05.010

Casarin V, Plassard C, Souche G, Arvieu JC (2003) Quantification of oxalate ions and protons released by ectomycorrhizal fungi in rhizosphere soil. Agronomie 23:461–469 doi:10.1051/agro:2003020

Casarin V, Plassard C, Hinsinger P, Arvieu JC (2004) Quantification of ectomycorrhizal fungal effects on the bioavailability and mobilisation of soil P in the rhizosphere of *Pinus pinaster*. New Phytol 163:177–185 doi:10.1111/j.1469-8137.2004.01093.x

Chapin FS, Moilanen L, Kielland K (1993) Preferential use of organic nitrogen for growth by a non-mycorrhizal arctic sedge. Nature 361:1550–1553 doi:10.1038/361150a0

Chen CR, Condron LM, Davis MR, Sherlock RR (2002) Phosphorus dynamics in the rhizosphere of perennial ryegrass (*Lolium perenne* L.) and radiata pine (*Pinus radiata* d. don.). Soil Biol Biochem 34:487–499 doi:10.1016/S0038-0717(01)00207-3

Cheng Y, Howieson JG, O'Hara GW, Watkin ELJ, Souche G, Jaillard B, Hinsinger P (2004) Proton release by roots of *Medicago murex* and *Medicago sativa* growing in acidic conditions, and implications for rhizosphere pH changes and nodulation at low pH. Soil Biol Biochem 36:1357–1365 doi:10.1016/j.soilbio.2004.04.017

Claassen N, Syring KM, Jungk A (1986) Verification of a mathematical model by simulating potassium uptake from soil. Plant Soil 95:209–220 doi:10.1007/BF02375073

Clarkson DT (1991) Root structure and sites of ion uptake. In: Waisel Y, Eshel A, Kafkafi U (eds) Plant Roots – the Hidden Half. Marcel Dekker, Inc., New York, USA, pp 417–453

Clarkson DT, Robards AW, Sanderson J (1971) The tertiary endodermis in barley roots: fine structure in relation to radial transport of ions and water. Planta 96:292–305 doi:10.1007/BF00386944

Clegg S, Gobran GR (1997) Rhizospheric P and K in forest soil manipulated with ammonium sulfate and water. Can J Soil Sci 77:525–533

Cornu JY, Staunton S, Hinsinger P (2007) Copper concentration in plants and in the rhizosphere as influenced by the iron status of tomato (*Lycopersicum esculentum* L.). Plant Soil 292:63–77 doi:10.1007/s11104-007-9202-z

Courchesne F, Gobran GR (1997) Mineralogical variation of bulk and rhizosphere soils from a Norway spruce stand. Soil Sci Soc Am J 61:1245–1249

Craine JM, Fargione J, Sugita S (2005) Supply pre-emption, not concentration reduction, is the mechanism of competition for nutrients. New Phytol 166:933–940 doi:10.1111/j.1469-8137.2005.01386.x

Crawford JW, Harris JA, Ritz K, Young IM (2005) Towards an evolutionary ecology of life in soil. Trends Ecol Evol 20:81–86 doi:10.1016/j.tree.2004.11.014

Curtis TP, Sloan WT (2005) Exploring microbial diversity-A vast below. Science 309:1331–1333 doi:10.1126/science.1118176

Curtis TP, Sloan WT, Scannell JW (2002) Estimating prokaryotic diversity and its limits. Proc Natl Acad Sci USA 99:10494–10499 doi:10.1073/pnas.142680199

Czarnes S, Hiller S, Dexter AR, Hallett PD, Bartoli F (1999) Root:soil adhesion in the maize rhizosphere: the rheological approach. Plant Soil 211:69–86 doi:10.1023/A:1004656510344

Czarnes S, Hallett PD, Bengough AG, Young IM (2000) Root- and microbial-derived mucilages affect soil structure and water transport. Eur J Soil Sci 51:435–443 doi:10.1046/j.1365-2389.2000.00327.x

Darrah PR (1991) Models of the rhizosphere. I. Microbial population dynamics around a root releasing soluble and insoluble carbon. Plant Soil 133:187–199 doi:10.1007/BF00009191

Darrah PR (1993) The rhizosphere and plant nutrition: a quantitative approach. Plant Soil 155/156:1–20 doi:10.1007/BF00024980

Degenhardt J, Larsen PB, Howell SH, Kochian LV (1998) Aluminum resistance in the *Arabidopsis* mutant alr-104 is caused by an aluminum-induced increase in rhizosphere pH. Plant Physiol 117:19–27 doi:10.1104/pp.117.1.19

Denton MD, Sasse C, Tibbett M, Ryan MH (2006) Root distributions of Australian herbaceous perennial legumes in response to phosphorus placement. Funct Plant Biol 33:1091–1102 doi:10.1071/FP06176

Dessureault-Rompré J, Nowack B, Schulin R, Luster J (2007) Spatial and temporal variation in organic acid anion exudation and nutrient anion uptake in the rhizosphere of *Lupinus albus* L. Plant Soil 301:123–134 doi:10.1007/s11104-007-9427-x

Dexter AR (1987) Compression of soil around roots. Plant Soil 97:401–406 doi:10.1007/BF02383230

Doussan C, Pagès L, Pierret A (2003) Soil exploration and resource acquisition by plant roots: an architectural and modelling point of view. Agronomie 23:419–431 doi:10.1051/agro:2003027

Doussan C, Pierret A, Garrigues E, Pages L (2006) Water uptake by plant roots: II-Modelling of water transfer in the soil root-system with explicit account of flow within the root system-Comparison with experiments. Plant Soil 283:99–117 doi:10.1007/s11104-004-7904-z

Dunbabin VM, McDermott S, Bengough AG (2006) Upscaling from rhizosphere to whole root system: modelling the effects of phospholipid surfactants on water and nutrient uptake. Plant Soil 283:57–72 doi:10.1007/s11104-005-0866-y

Dutton MV, Evans CS (1996) Oxalate production by fungi: its role in pathogenicity and ecology in the soil environment. Can J Microbiol 42:881–895 doi:10.1139/m96-114

El Sebai T, Lagacherie B, Soulas G, Martin-Laurent F (2007) Spatial variability of isoproturon mineralizing activity within an agricultural field: geostatistical analysis of simple physicochemical and microbiological soil parameters. Environ Pollut 145:680–690 doi:10.1016/j.envpol.2006.05.034

Ernst M, Römheld V, Marschner H (1989) Estimation of phosphorus uptake capacity by different zones of the primary root of soil-grown maize (*Zea mays* L.). J Plant Nutr Soil Sci 152:21–25 doi:10.1002/jpln.19891520105

Farr E, Vaidyanathan LV, Nye PH (1969) Measurement of ionic concentration gradients in soil near roots. Soil Sci 107:385–391 doi:10.1097/00010694-196905000-00012

Feddes RA, Hoff H, Bruen M, Dawson T, de Rosnay P, Dirmeyer O, Jackson RB, Kabat P, Kleidon A, Lilly A, Pitman AJ (2001) Modeling root water uptake in hydrological and climate models. Bull Am Meteorol Soc 82:2797–2809 doi:10.1175/1520-0477(2001)082<2797:MRWUIH>2.3.CO;2

Feeney DS, Crawford JW, Daniell T, Hallett PD, Nunan N, Ritz K, Rivers M, Young IM (2006) Three-dimensional microorganisation of the soil-root-microbe system. Microb Ecol 52:151–158 doi:10.1007/s00248-006-9062-8

Ferguson IB, Clarkson DT (1975) Ion transport and endodermal suberization in the roots of *Zea mays*. New Phytol 75:69–79 doi:10.1111/j.1469-8137.1975.tb01372.x

Fierer N, Jackson RB (2006) The diversity and biogeography of soil bacterial communities. Proc Natl Acad Sci USA 103:626–631 doi:10.1073/pnas.0507535103

Fischer WR, Flessa H, Schaller G (1989) pH values and redox potentials in microsites of the rhizosphere. Z Pflanzenern Bodenkd 152:191–195 doi:10.1002/jpln.19891520209

Fitter AH, Williamson L, Linkohr B, Leyser O (2002) Root system architecture determines fitness in an *Arabidopsis* mutant in competition for immobile phosphate ions but not for nitrate ions. Proc R Soc Lond B Biol Sci 269:2017–2022 doi:10.1098/rspb.2002.2120

Flessa H, Fischer WR (1992) Plant-induced changes in the redox potentials of rice rhizospheres. Plant Soil 143:55–60 doi:10.1007/BF00009128

Focht DD (1992) Diffusional constraints on microbial processes in soil. Soil Sci 154:300–307 doi:10.1097/00010694-199210000-00006

Frensch J, Steudle E (1989) Axial and radial hydraulic resistance to roots of maize (*Zea mays* L.). Plant Physiol 91:719–726

Frensch J, Hsiao TC, Steudle E (1996) Water and solute transport along developing maize roots. Planta 198:348–355 doi:10.1007/BF00620050

Gahoonia TS, Claassen N, Jungk A (1992) Mobilization of phosphate in different soils by ryegrass supplied with ammonium or nitrate. Plant Soil 140:241–248 doi:10.1007/BF00010600

Gahoonia TS, Nielsen NE, Joshi PA, Jahoor A (2001) A root hairless barley mutant for elucidating genetic of root hairs and phosphorus uptake. Plant Soil 235:211–219 doi:10.1023/A:1011993322286

Gardner WR (1960) Dynamic aspects of water availability to plants. Soil Sci 89:63–73 doi:10.1097/00010694-196002000-00001

Garrigues E, Doussan C, Pierret A (2006) Water uptake by plant roots: I. Formation and propagation of a water extraction front in mature root systems as evidenced by 2D light transmission imaging. Plant Soil 283:83–98 doi:10.1007/s11104-004-7903-0

Ge Z, Rubio G, Lynch JP (2000) The importance of root gravitropism for inter-root competition and phosphorus acquisition efficiency: results from a geometric simulation model. Plant Soil 218:159–171 doi:10.1023/A:1014987710937

Geelhoed JS, van Riemsdijk WH, Findenegg GR (1999) Simulation of the effect of citrate exudation from roots on the plant availability of phosphate adsorbed on goethite. Eur J Soil Sci 50:379–390 doi:10.1046/j.1365-2389.1999.00251.x

George TS, Gregory PJ, Robinson JS, Buresh RJ (2002a) Changes in phosphorus concentrations and pH in the rhizosphere of some agroforestry and crop species. Plant Soil 246:65–73 doi:10.1023/A:1021523515707

George TS, Gregory PJ, Wood M, Read DJ, Buresh RJ (2002b) Phosphatase activity and organic acids in the rhizosphere of potential agroforestry species and maize. Soil Biol Biochem 34:1487–1494 doi:10.1016/S0038-0717(02)00093-7

George TS, Richardson AE, Hadobas PA, Simpson RJ (2004) Characterization of transgenic *Trifolium subterraneum* L. which expresses *phyA* and releases extracellular phytase: growth and P nutrition in laboratory media and soil. Plant Cell Environ 27:1351–1361 doi:10.1111/j.1365-3040.2004.01225.x

George TS, Richardson AE, Simpson RJ (2005) Behaviour of plant-derived extracellular phytase upon addition to soil. Soil Biol Biochem 37:977–978 doi:10.1016/j.soilbio.2004.10.016

George TS, Turner BL, Gregory PJ, Richardson AE (2006) Depletion of organic phosphorus from oxisols in relation to phosphatase activities in the rhizosphere. Eur J Soil Sci 57:47–57 doi:10.1111/j.1365-2389.2006.00767.x

Gerke J, Beissner L, Römer W (2000) The quantitative effect of chemical phosphate mobilization by carboxylate anions on P uptake by a single root. I. The basic concept and determination of soil parameters. J Plant Nutr Soil Sci 163:207–212 doi:<207::AID-JPLN207>3.0.CO;2-P

Gollany HT, Schumacher TE, Rue RR, Liu S-Y (1993) A carbon dioxide microelectrode for in situ pCO_2 measurement. Microchem J 48:42–49 doi:10.1006/mchj.1993.1069

Göttlein A, Matzner E (1997) Microscale heterogeneity of acidity related stress-parameters in the soil solution of a forested cambic podzol. Plant Soil 192:95–105 doi:10.1023/A:1004260006503

Göttlein A, Hell U, Blasek R (1996) A system for microscale tensiometry and lysimetry. Geoderma 69:147–156 doi:10.1016/0016-7061(95)00059-3

Göttlein A, Heim A, Matzner E (1999) Mobilization of aluminium in the rhizosphere soil solution of growing tree roots in an acidic soil. Plant Soil 211:41–49 doi:10.1023/A:1004332916188

Greacen EL, Barley KP, Farrell DA (1969) The mechanics of root growth in soils with particular reference to the implications for root distribution. In: Whitington WJ (ed) Root growth. Butterworths, London, UK, pp 256–268

Greenway H, Armstrong W, Colmer TD (2006) Conditions leading to high CO_2 (>5 kPa) in waterlogged-flooded soils and possible effects on root growth and metabolism. Ann Bot (Lond) 98:9–32 doi:10.1093/aob/mcl076

Gregory PJ (2006) Roots, rhizosphere, and soil: the route to a better understanding of soil science? Eur J Soil Sci 57:2–12 doi:10.1111/j.1365-2389.2005.00778.x

Gregory PJ, Hinsinger P (1999) New approaches to studying chemical and physical changes in the rhizosphere: an overview. Plant Soil 211:1–9 doi:10.1023/A:1004547401951

Gross N, Suding KN, Lavorel S, Roumet C (2007) Complementarity as a mechanism of coexistence between functional groups of grasses. J Ecol 95:1296–1305 doi:10.1111/j.1365-2745.2007.01303.x

Guinel FC, McCully ME (1986) Some water-related physical properties of maize root-cap mucilage. Plant Cell Environ 9:657–666 doi:10.1111/j.1365-3040.1986.tb01624.x

Hainsworth JM, Aylmore LAG (1989) Non-uniform soil water extraction by plant roots. Plant Soil 113:121–124 doi:10.1007/BF02181929

Hallett PD, Gordon DC, Bengough AG (2003) Plant influence on rhizosphere hydraulic properties: direct measurements using a miniaturized infiltrometer. New Phytol 157:597–603 doi:10.1046/j.1469-8137.2003.00690.x

Harrison-Murray RS, Clarkson DT (1973) Relationship between structural development and the absorption of ions by the root system of Cucurbita pepo. Planta 114:1–16 doi:10.1007/BF00390280

Hartmann A, Rothballer M, Schmid M (2008) Lorenz Hiltner, a pioneer in rhizosphere microbial ecology and soil bacteriology research. Plant Soil 312:7–14 doi:10.1007/s11104-007-9514-z

Häussling M, Leisen E, Marschner H, Römheld V (1985) An improved method for non-destructive measurements of the pH at the root–soil interface (rhizosphere). J Plant Physiol 117:371–375

Häussling M, Jorns CA, Lehmbecker G, Hecht-Buchholz C, Marschner H (1988) Ion and water uptake in relation to root development in Norway spruce (Picea abies (L.) Karst.). J Plant Physiol 133:486–491

Hawkins BJ, Boukcim H, Plassard C (2008) A comparison of ammonium, nitrate and proton net fluxes along seedling roots of Douglas-fir and lodgepole pine grown and measured with different inorganic nitrogen sources. Plant Cell Environ 31:278–287 doi:10.1111/j.1365-3040.2007.01760.x

Haynes RJ, Beare MH (1997) Influence of six crop species on aggregate stability and some labile organic matter fractions. Soil Biol Biochem 29:1647–1653 doi:10.1016/S0038-0717(97)00078-3

Hendriks L, Claassen N, Jungk A (1981) Phosphatverarmung des wurzelnahen Bodens und Phosphataufnahme von Mais und Raps. Z Pflanzenern Bodenkd 144:486–499 doi:10.1002/jpln.19811440507

Henzler T, Waterhouse RN, Smyth AJ, Carvajal M, Cooke DT, Schäffner AR, Steudle E, Clarkson DT (1999) Diurnal variations in hydraulic conductivity and root pressure can be correlated with the expression of putative aquaporins in the roots of Lotus japonicus. Planta 210:50–60 doi:10.1007/s004250050653

Herman DJ, Johnson KK, Jaeger CH III, Schwartz E, Firestone MK (2006) Root influence on nitrogen mineralization and nitrification in Avena barbata rhizosphere soil. Soil Sci Soc Am J 70:1504–1511 doi:10.2136/sssaj2005.0113

Hiltner L (1904) Über neuere Erfahrungen und Probleme auf dem Gebiete der Bodenbakteriologie unter besonderer Berücksichtigung der Gründüngung und Brache. Arb DLG 98:59–78

Hinsinger P (1998) How do plant roots acquire mineral nutrients? Chemical processes involved in the rhizosphere. Adv Agron 64:225–265 doi:10.1016/S0065-2113(08)60506-4

Hinsinger P (2001) Bioavailability of soil inorganic P in the rhizosphere as affected by root-induced chemical changes: a review. Plant Soil 237:173–195 doi:10.1023/A:1013351617532

Hinsinger P (2004) Nutrient availability and transport in the rhizosphere. In: Goodman RM (ed) Encyclopedia of Plant and Crop Science. Marcel Dekker, Inc., New York, USA, pp 1094–1097

Hinsinger P, Gilkes RJ (1996) Mobilization of phosphate from phosphate rock and alumina-sorbed phosphate by the roots of ryegrass and clover as related to rhizosphere pH. Eur J Soil Sci 47:533–544 doi:10.1111/j.1365-2389.1996.tb01853.x

Hinsinger P, Elsass F, Jaillard B, Robert M (1993) Root-induced irreversible transformation of a trioctahedral mica in the rhizosphere of rape. Eur J Soil Sci 44:535–545 doi:10.1111/j.1365-2389.1993.tb00475.x

Hinsinger P, Plassard C, Tang C, Jaillard B (2003) Origins of root-induced pH changes in the rhizosphere and their responses to environmental constraints: a review. Plant Soil 248:43–59 doi:10.1023/A:1022371130939

Hinsinger P, Gobran GR, Gregory PJ, Wenzel WW (2005) Rhizosphere geometry and heterogeneity arising from root-mediated physical and chemical processes. New Phytol 168:293–303 doi:10.1111/j.1469-8137.2005.01512.x

Hobbie SE (1992) Effects of plant-species on nutrient cycling. Trends Ecol Evol 7:336–339 doi:10.1016/0169-5347(92)90126-V

Hodge A (2004) The plastic plant: root responses to heterogenous supplies of nutrients. New Phytol 162:9–24 doi:10.1111/j.1469-8137.2004.01015.x

Hodge A, Berta G, Crespi M, Doussan C (2009) Plant roots: growth and architecture. Plant Soil (this volume, in preparation, provisional title)

Hojberg O, Binnerup SJ, Sorensen J (1996) Potential rates of ammonium oxidation, nitrite oxidation, nitrate reduction and denitrification in the young barley rhizosphere. Soil Biol Biochem 28:47–54 doi:10.1016/0038-0717(95)00119-0

Hübel F, Beck E (1993) In-situ determination of the P-relations around the primary root of maize with respect to inorganic and phytate-P. Plant Soil 157:1–9

Huck MG, Klepper B, Taylor HM (1970) Diurnal variation in root diameter. Plant Physiol 45:529–530

Humphris SN, Bengough AG, Griffiths BS, Kilham K, Rodger S, Stubbs V, Valentine TA, Young IM (2005) Root cap influences root colonisation by Pseudomonas fluorescens SBW25 on maize. FEMS Microbiol Ecol 54:123–130 doi:10.1016/j.femsec.2005.03.005

Iijima M, Griffiths B, Bengough AG (2000) Sloughing of cap cells and carbon exudation from maize seedling roots in compacted sand. New Phytol 145:477–482 doi:10.1046/j.1469-8137.2000.00595.x

Iijima M, Higuchi T, Barlow PW, Bengough AG (2003) Root cap removal increases root penetration resistance in maize (Zea mays L.). J Exp Bot 54:2105–2109 doi:10.1093/jxb/erg226

Iijima M, Higuchi T, Barlow PW (2004) Contribution of root cap mucilage and presence of an intact root cap in maize (Zea mays) to the reduction of soil mechanical impedance. Ann Bot (Lond) 94:473–477 doi:10.1093/aob/mch166

Jackson RB, Sperry JS, Dawson TE (2000) Root water uptake and transport: using physiological processes in global predictions. Trends Plant Sci 5:482–488 doi:10.1016/S1360-1385(00)01766-0

Jaillard B, Ruiz L, Arvieu JC (1996) pH mapping in transparent gel using color indicator videodensitometry. Plant Soil 183:85–95 doi:10.1007/BF02185568

Jaillard B, Plassard C, Hinsinger P (2003) Measurements of H^+ fluxes and concentrations in the rhizosphere. In: Rengel Z (ed) Handbook of Soil Acidity. Marcel Dekker, Inc., New York, USA, pp 231–266

Jones DL (1998) Organic acids in the rhizosphere-a critical review. Plant Soil 205:25–44 doi:10.1023/A:1004356007312

Jones DL, Darrah PR (1994) Role of root derived organic acids in the mobilization of nutrients from the rhizosphere. Plant Soil 166:247–257 doi:10.1007/BF00008338

Jones DL, Hinsinger P (2008) The rhizosphere: complex by design. Plant Soil 312:1–6 doi:10.1007/s11104-008-9774-2

Jones DL, Hodge A, Kuzyakov Y (2004) Plant and mycorrhizal regulation of rhizodeposition. New Phytol 163:459–480 doi:10.1111/j.1469-8137.2004.01130.x

Jones DL, Nguyen C, Finlay RD (2009) Carbon flow in the rhizosphere: carbon trading at the soil-root interface. Plant Soil (this volume, in preparation)

Jungk A (2002) Dynamics of nutrient movement at the soil-root interface. In: Waisel Y, Eshel A Kafkafi U (eds) Plant Roots: The Hidden Half, 3rd Ed. Marcel Dekker, Inc., New York, USA, pp 587–616

Jungk A, Claassen N (1986) Availability of phosphate and potassium as the result of interactions between root and soil in the rhizosphere. J Plant Nutr Soil Sci 149:411–427 doi:10.1002/jpln.19861490406

Kabir Z, Koide RT (2000) The effect of dandelion or a cover crop on mycorrhiza inoculum potential, soil aggregation and yield of maize. Agric Ecosyst Environ 78:167–174 doi:10.1016/S0167-8809(99)00121-8

Kaci Ym Heyraund A, Barakat M, Heulin T (2005) Isolation and identification of an EPS-producing *Rhizobium* strain from arid soil (Algeria): characterisation of its EPS and the effect of inoculation on wheat Rhizosphere soil structure. Microbiol 156:522–531

Kandeler E, Marschner P, Tscherko D, Gahoonia TS, Nielsen NE (2002) Microbial community composition and functional diversity in the rhizosphere of maize. Plant Soil 238:301–312 doi:10.1023/A:1014479220689

Karberg NJ, Pregitzer KS, King JS, Friend AL, Wood JR (2005) Soil carbon dioxide partial pressure and dissolved inorganic carbonate chemistry under elevated carbon dioxide and ozone. Oecologia 142:296–306 doi:10.1007/s00442-004-1665-5

Keerthisinghe G, Hocking PJ, Ryan PR, Delhaize E (1998) Effect of phosphorus supply on the formation and function of proteoid roots of white lupin (*Lupinus albus* L.). Plant Cell Environ 21:467–478 doi:10.1046/j.1365-3040.1998.00300.x

Khalil K, Mary B, Renault P (2004) Nitrous oxide production by nitrification and denitrification in soil aggregates as affected by O_2 concentration. Soil Biol Biochem 36:687–699 doi:10.1016/j.soilbio.2004.01.004

Kim TK, Silk WK, Cheer AY (1999) A mathematical model for pH patterns in the rhizospheres of growth zones. Plant Cell Environ 22:1527–1538 doi:10.1046/j.1365-3040.1999.00512.x

Kinraide TB, Yermiyahu U (2007) A scale of metal ion binding strengths correlating with ionic charge, Pauling electronegativity, toxicity, and other physiological effects. J Inorg Biochem 101:1201–1213 doi:10.1016/j.jinorgbio.2007.06.003

Kirby JM, Bengough AG (2002) Influence of soil strength on root growth: experiments and analysis using a critical-state model. Eur J Soil Sci 53:119–127 doi:10.1046/j.1365-2389.2002.00429.x

Kirk GJD (1999) A model of phosphate solubilization by organic anion excretion from plant roots. Eur J Soil Sci 50:369–378 doi:10.1046/j.1365-2389.1999.00239.x

Kirk GJD (2002) Modelling root-induced solubilization of nutrients. Plant Soil 255:49–57 doi:10.1023/A:1020667416624

Kirk GJD, Bajita JB (1995) Root-induced iron oxidation, pH changes and zinc solubilization in the rhizosphere of lowland rice. New Phytol 131:129–137 doi:10.1111/j.1469-8137.1995.tb03062.x

Kirk GJD, Le Van Du (1997) Changes in rice root architecture, porosity, and oxygen and proton release under phosphorus deficiency. New Phytol 135:191–200 doi:10.1046/j.1469-8137.1997.00640.x

Kirk GJD, Kronzucker HJ (2005) The potential for nitrification and nitrate uptake in the rhizosphere of wetland plants: a modelling study. Ann Bot (Lond) 96:639–646 doi:10.1093/aob/mci216

Kopittke PM, Menzies NW (2004) Effect of Mn deficiency and legume inoculation on rhizosphere pH in highly alkaline soils. Plant Soil 262:13–21 doi:10.1023/B:PLSO.0000037023.18127.7a

Kramer PJ, Boyer JS (1995) Water Relations of Plants and Soils. Academic Press, San Diego, USA, p 495

Kuchenbuch R, Jungk A (1982) A method for determining concentration profiles at the soil-root interface by thin slicing rhizospheric soil. Plant Soil 68:391–394 doi:10.1007/BF02197944

Kuzyakov Y (2002) Review: factors affecting rhizosphere priming effects. J Plant Nutr Soil Sci 165:382–396 doi:<382::AID-JPLN382>3.0.CO;2-#

Lambers H, Chapin SF, Pons T (1998) Plant physiological ecology. Springer, New York

Lambers H, Shane MW, Cramer MD, Pearse SJ, Veneklaas EJ (2006) Root structure and functioning for efficient acquisition of phosphorus: matching morphological and physiological traits. Ann Bot (Lond) 98:693–713 doi:10.1093/aob/mcl114

Lambers H, Mougel C, Jaillard B, Hinsinger P (2009) Plant-microbe-soil interactions in the rhizosphere: an evolutionary perspective. Plant Soil (this volume, in preparation, provisional title)

Le Bot J, Kirkby EA (1992) Diurnal uptake of nitrate and potassium during the vegetative growth of tomato. J Plant Nutr 15:247–264 doi:10.1080/01904169209364316

Le Bot J, Alloush GA, Kirkby EA, Sanders FE (1990) Mineral nutrition of chickpea plants supplied with NO_3 or NH_4–N. II. Ionic balance in relation to phosphorus stress. J Plant Nutr 13:1591–1605 doi:10.1080/01904169009364177

Lemcoff JH, Ling F, Neumann PM (2006) Short episodes of water stress increase barley root resistance to radial shrinkage in a dehydrating environment. Physiol Plant 127:603–611 doi:10.1111/j.1399-3054.2006.00688.x

Lemanceau P, Kraemer SM, Briat JF (2009) Iron in the rhizosphere: a case study. Plant Soil (this volume, in preparation, provisional title)

Lesturgez G, Poss R, Hartmann C, Bourdon E, Noble A, Ratana-Anupap S (2004) Roots of *Stylosanthes hamata* create macropores in the compact layer of a sandy soil. Plant Soil 260:101–109 doi:10.1023/B:PLSO.0000030184.24866.aa

Leyser O, Fitter A (1998) Roots are branching out in patches. Trends Plant Sci 3:203–204 doi:10.1016/S1360-1385(98)01253-9

Li X-L, George E, Marschner H (1991) Extension of the phosphorus depletion zone in VA-mycorrhizal white clover in a calcareous soil. Plant Soil 136:41–48 doi:10.1007/BF02465218

Li SM, Li L, Zhang FS, Tang C (2004) Acid phosphatase role in chickpea/maize intercropping. Ann Bot (Lond) 94:297–303 doi:10.1093/aob/mch140

Li L, Li SM, Sun JH, Zhou LL, Bao XG, Zhang HG, Zhang FS (2007) Diversity enhances agricultural productivity via rhizosphere phosphorus facilitation on phosphorus-deficient soils. Proc Natl Acad Sci USA 104:11192–11196 doi:10.1073/pnas.0704591104

Li H, Shen J, Zhang F, Clairotte M, Drevon JJ, Le Cadre E, Hinsinger P (2008) Dynamics of phosphorus fractions in the rhizosphere of common bean (*Phaseolus vulgaris* L.) and durum wheat (*Triticum turgidum durum* L.) grown in monocropping and intercropping systems. Plant Soil 312:139–150 doi:10.1007/s11104-007-9512-1

Linkohr BI, Williamson LC, Fitter AH, Leyser HMO (2002) Nitrate and phosphate availability and distribution have different effects on root system architecture of *Arabidopsis*. Plant J 29:751–760 doi:10.1046/j.1365-313X.2002.01251.x

Loosemore N, Straczek A, Hinsinger P, Jaillard B (2004) Zinc mobilization from a contaminated soil by three genotypes of tobacco as affected by soil and rhizosphere pH. Plant Soil 260:19–32 doi:10.1023/B:PLSO.0000030173.71500.e1

Lorenz SE, Hamon RE, McGrath SP (1994) Differences between soil solutions obtained from rhizosphere and non-rhizosphere soils by water displacement and soil centrifugation. Eur J Soil Sci 45:431–438 doi:10.1111/j.1365-2389.1994.tb00528.x

Luster J, Göttlein A, Nowack B, Sarret G (2009) Sampling, defining, characterising and modeling the rhizosphere – The soil science tool box. Plant Soil (this volume, in preparation)

Lynch JM, Whipps JM (1990) Substrate flow in the rhizosphere. Plant Soil 129:1–10 doi:10.1007/BF00011685

Ma Z, Walk TC, Marcus A, Lynch JP (2001) Morphological synergism in root hair length, density, initiation and geometry for phosphorus acquisition in *Arabidopsis thaliana*: a modeling approach. Plant Soil 236:221–235 doi:10.1023/A:1012728819326

MacFall JS, Johnson GA, Kramer PJ (1990) Observation of a water-depletion region surrounding loblolly pine roots by magnetic resonance imaging. Proc Natl Acad Sci USA 87:1203–1207 doi:10.1073/pnas.87.3.1203

Marschner H (1995) Mineral Nutrition of Higher Plants, 2nd edn. Academic Press, London, p 889

Marschner H, Römheld V (1983) In vivo measurement of root-induced pH changes at the soil-root interface: effect of plant species and nitrogen source. Z Pflanzenernaehr Bodenkd 111:241–251

Marschner P, Solaiman Z, Rengel Z (2006) Rhizosphere properties of Poaceae genotypes under P-limiting conditions. Plant Soil 283:11–24 doi:10.1007/s11104-005-8295-5

Martens DA (2000) Relationship between plant phenolic acids released during soil mineralisation and aggregate stabilisation. Soil Sci Soc Am J 66:1857–1867

Martens DA, Frankenberger WT (1992) Decomposition of bacteria polymers in soil and their influence on soil structure. Biol Fertil Soils 13:65–73 doi:10.1007/BF00337337

Masle J, Passioura JB (1987) The effect of soil strength on the growth of young wheat plants. Aust J Plant Physiol 14:643–656

McCully ME (1999) Roots in soil: unearthing the complexities of roots and their rhizospheres. Ann. Rev. Plant Physiol Plant Mol Biol 50:695–718 doi:10.1146/annurev.arplant.50.1.695

McCully ME, Boyer JS (1997) The expansion of maize root-cap mucilage during hydration. 3. Changes in water potential and water content. Physiol Plant 99:169–177 doi:10.1111/j.1399-3054.1997.tb03445.x

Menon M, Robinson B, Oswald SE, Kaestner A, Abbaspour KC, Lehmann E, Schulin R (2006) Visualization of root growth in heterogeneously contaminated soil using neutron radiography. Eur J Soil Sci 58:802–810 doi:10.1111/j.1365-2389.2006.00870.x

Michaud AM, Bravin MN, Galleguillos M, Hinsinger P (2007) Copper uptake and phytotoxicity as assessed in situ for durum wheat (*Triticum turgidum durum* L.) cultivated in Cu-contaminated, former vineyard soils. Plant Soil 298:99–111 doi:10.1007/s11104-007-9343-0

Moreno-Espindola IP, Rivera-Becerril F, de Jesus Ferrara-Guerrero M, De Leon-Gonzalez F (2007) Role of root-hairs and hyphae in adhesion of sand particles. Soil Biol Biochem 39:2520–2526 doi:10.1016/j.soilbio.2007.04.021

Mullins CE, Panayiotopoulos KP (1984) The strength of unsaturated mixtures of sand and kaolin and the concept of effective stress. Eur J Soil Sci 35:459–468 doi:10.1111/j.1365-2389.1984.tb00303.x

Neubauer SC, Toledo-Durán GE, Emerson D, Megonigal JP (2007) Returning to their roots: iron-oxidizing bacteria enhance short-term plaque formation in the wetland-plant rhizosphere. Geomicrobiol J 24:65–73 doi:10.1080/01490450601134309

Newman BD, Wilcox BP, Archer SR, Breshears DD, Dahm CN, Duffy CJ, McDowell NG, Phillips FM, Scanlon BR, Vivoni ER (2006) Ecohydrology of water-limited environments: a scientific vision. Water Resour Res 42:1–15 doi:10.1029/2005WR004141

Nguyen C (2003) Rhizodeposition of organic C by plants: mechanisms and controls. Agronomie 23:375–396 doi:10.1051/agro:2003011

Nichol SA, Silk WK (2001) Empirical evidence of a convection-diffusion model for pH patterns in the rhizospheres of root tips. Plant Cell Environ 24:967–974 doi:10.1046/j.0016-8025.2001.00739.x

Nobel PS, Cui M (1992) Hydraulic conductances of the soil, the root-soil air gap, and the root: Changes for desert succulents in drying soil. J Exp Bot 43:319–326 doi:10.1093/jxb/43.3.319

North GB, Nobel PS (1997) Drought-induced changes in soil contact and hydraulic conductivity for roots of *Opuntia ficus-indica* with and without rhizosheaths. Plant Soil 191:249–258 doi:10.1023/A:1004213728734

Nowack B, Mayer KU, Oswald SE, van Beinum W, Appelo CAJ, Jacques D, Seuntjens P, Gérard F, Jaillard B, Schnepf A, Roose T (2006) Verification and intercomparison of reactive transport codes to describe root-uptake. Plant Soil 285:305–321 doi:10.1007/s11104-006-9017-3

Nunan N, Wu K, Young IM, Crawford JW, Ritz K (2003) Spatial distribution of bacterial communities and their relationships with the micro-architecture of soil. FEMS Microbiol Ecol 44:203–215 doi:10.1016/S0168-6496(03)00027-8

Nye PH (1981) Changes of pH across the rhizosphere induced by roots. Plant Soil 61:7–26 doi:10.1007/BF02277359

Nye PH (1983) The diffusion of two interacting solutes in soil. J Soil Sci 34:677–691

Nye PH, Marriott FHC (1969) A theoretical study of the distribution of substances around roots resulting from simultaneous diffusion and mass flow. Plant Soil 30:459–472 doi:10.1007/BF01881971

Odell RE, Dumlao MR, Samar D, Silk WK (2008) Stage-dependent border cell and carbon flow from roots to rhizosphere. Am J Bot 95:441–446 doi:10.3732/ajb.95.4.441

Pankhurst CE, Pierret A, Hawke BG, Kirby JM (2002) Microbiological and chemical properties of soil associated with macropores at different depths in a red-duplex soil in NSW Australia. Plant Soil 238:11–20 doi:10.1023/A:1014289632453

Pardales JR, Kono Y (1990) Development of sorghum root system under increasing drought stress. Jpn J Crop Sci 59:752–761

Passioura JB (1988) Root signals control leaf expansion in wheat seedlings growing in drying soil. Aust J Plant Physiol 15:687–693

Passioura JB (1991) Soil structure and plant-growth. Aust J Soil Res 29:717–728 doi:10.1071/SR9910717

Passioura JB (2002) Soil conditions and plant growth. Plant Cell Environ 25:311–318 doi:10.1046/j.0016-8025.2001.00802.x

Passioura JB, Munns R (1984) Hydraulic resistance of plants. II. Effects of rooting medium, and time of day, in barley and lupin. Aust J Plant Physiol 11:341–350

Paynel F, Cliquet JB (2003) N transfer from white clover to perennial ryegrass, via exudation of nitrogenous compounds. Agronomie 23:503–510 doi:10.1051/agro:2003022

Perata P, Alpi A (1993) Plant-responses to anaerobiosis. Plant Sci 93:1–17 doi:10.1016/0168-9452(93)90029-Y

Peters WS (2004) Growth rate gradients and extracellular pH in roots: how to control an explosion. New Phytol 162:571–574 doi:10.1111/j.1469-8137.2004.01085.x

Philippot L, Hallin S, Börjesson G, Baggs EM (2009) Biogeochemical cycling in the rhizosphere having an impact on global change. Plant Soil (this volume, in preparation)

Pidello A (2003) The effect of *Pseudomonas fluorescens* strains varying in pyoverdine production on the soil redox status. Plant Soil 253:373–379 doi:10.1023/A:1024875824350

Pidello A, Jocteur Monrozier L (2006) Inoculation of the redox effector *Pseudomonas fluorescens* C7R12 strain affects soil redox status at the aggregate scale. Soil Biol Biochem 38:1396–1402 doi:10.1016/j.soilbio.2005.10.010

Pidello A, Menendez L, Lensi R (1993) Azospirillum affects Eh and potential denitrification in a soil. Plant Soil 157:31–34

Pineros MA, Shaff JE, Manslank HS, Alves VMC, Kochian LV (2005) Aluminum resistance in maize cannot be solely explained by root organic acid exudation. A comparative physiological study. Plant Physiol 137:231–241 doi:10.1104/pp.104.047357

Pierret A, Moran CJ, Pankhurst CE (1999) Differentiation of soil properties related to the spatial association of wheat roots and soil macropores. Plant Soil 211:51–58 doi:10.1023/A:1004490800536

Pierret A, Doussan C, Garrigues E, Mc Kirby J (2003) Observing plant roots in their environment: current imaging options and specific contribution of two-dimensional approaches. Agronomie 23:471–479 doi:10.1051/agro:2003019

Pierret A, Moran C, Doussan C (2005) Conventional detection methodology is limiting our ability to understand the roles and functions of fine roots. New Phytol 166:967–980 doi:10.1111/j.1469-8137.2005.01389.x

Pierret A, Doussan C, Capowiez Y, Bastardie F, Pagès L (2007) Root functional architecture: a framework for modeling the interplay between roots and soil. Vadose Zone J 6:269–281 doi:10.2136/vzj2006.0067

Plassard C, Meslem M, Souche G, Jaillard B (1999) Localization and quantification of net fluxes of H^+ along roots of maize by combined use of videodensitometry of indicator dye and ion-selective microelectrodes. Plant Soil 211:29–39 doi:10.1023/A:1004560208777

Plassard C, Guerin-Laguette A, Véry AA, Casarin V, Thibauld JB (2002) Local measurements of nitrate and potassium fluxes along roots of maritime pine. Effetcs of ectomycorrhizal symbiosis. Plant Cell Environ 25:75–84 doi:10.1046/j.0016-8025.2001.00810.x

Raab TK, Lipson DA, Monson RK (1996) Non-mycorrhizal uptake of amino acids by roots of the alpine sedge *Kobresia myosuroides*: implications for the alpine nitrogen cycle. Oecologia 108:488–494 doi:10.1007/BF00333725

Ranjard L, Richaume AS (2001) Quantitative and qualitative microscale distribution of bacteria in soil. Res Microbiol 152:707–716 doi:10.1016/S0923-2508(01)01251-7

Rao TR, Yano K, Iijima M, Yamauchi A, Tatsumi J (2002) Regulation of rhizosphere acidification by photosynthetic activity in cowpea (*Vigna unguiculata* L.Walp.) seedlings. Ann Bot (Lond) 89:213–220 doi:10.1093/aob/mcf030

Ramette A, Tiedje JM (2007) Multiscale responses of microbial life to spatial distance and environmental heterogeneity in a patchy ecosystem. Proc Natl Acad Sci USA 104:2761–2766 doi:10.1073/pnas.0610671104

Rappoldt C, Crawford JW (1999) The distribution of anoxic volume in a fractal model of soil. Geoderma 88:329–347 doi:10.1016/S0016-7061(98)00112-8

Raven JA (1986) Biochemical disposal of excess H^+ in growing plants? New Phytol 104:175–206 doi:10.1111/j.1469-8137.1986.tb00644.x

Raven JA, Edwards D (2001) Roots: evolutionary origin and biogeochemical significance. J Exp Bot 52:381–408

Raynaud X, Lata JC, Leadley PW (2006) Soil microbial loop and nutrient uptake by plants: a test using a coupled C : N model of plant-microbial interactions. Plant Soil 287:95–116 doi:10.1007/s11104-006-9003-9

Raynaud X, Jaillard B, Leadley PW (2008) Plants may alter competition by modifying nutrient bioavailability in rhizosphere: a modeling approach. Am Nat 171:44–58 doi:10.1086/523951

Read DJ, Perez-Moreno J (2003) Mycorrhizas and nutrient cycling in ecosystems – a journey towards relevance? New Phytol 157:475–492 doi:10.1046/j.1469-8137.2003.00704.x

Read DB, Gregory PJ, Bell AE (1999) Physical properties of axenic maize root mucilage. Plant Soil 211:87–91 doi:10.1023/A:1004403812307

Read DB, Bengough AG, Gregory PJ, Crawford JW, Robinson D, Scrimgeour CM, Young IM, Zhang K, Zhang X (2003) Plant roots release phospholipid surfactants that modify the physical and chemical properties of soil. New Phytol 157:315–326 doi:10.1046/j.1469-8137.2003.00665.x

Reichard PU, Kraemer SM, Frazier SW, Kretzschmar R (2005) Goethite dissolution in the presence of phytosiderophores: rates, mechanisms, and the synergistic effect of oxalate. Plant Soil 276:115–132 doi:10.1007/s11104-005-3504-9

Reichard PU, Kretzschmar R, Kraemer SM (2007) Dissolution mechanisms of goethite in the presence of siderophores and organic acids. Geochim Cosmochim Acta 71:5635–5650 doi:10.1016/j.gca.2006.12.022

Reichman SM, Parker DR (2007) Probing the effects of light and temperature on diurnal rhythms of phytosiderophore release in wheat. New Phytol 174:101–108 doi:10.1111/j.1469-8137.2007.01990.x

Reidenbach G, Horst W (1997) Nitrate-uptake capacity of different root zones of Zea mays (L.) in vitro and in situ. Plant Soil 196:295–300 doi:10.1023/A:1004280225323

Renault P, Stengel P (1994) Modeling oxygen diffusion in aggregated soils. I. Anaerobiosis inside the aggregates. Soil Sci Soc Am J 58:1017–1023

Revsbech NP, Pedersen O, Reichardt W, Briones A (1999) Microsensor analysis of oxygen and pH in the rice rhizosphere under field and laboratory conditions. Biol Fertil Soils 29:379–385 doi:10.1007/s003740050568

Richardson AE, Barea JM, McNeil AM, Prigent-Combaret C (2009) Acquisition of phosphorus and nitrogen in the rhizosphere and plant growth promotion by microorganisms. Plant Soil (this volume, in preparation)

Rigou L, Mignard E, Plassard C, Arvieu JC, Rémy JC (1995) Influence of ectomycorrhizal infection on the rhizosphere pH around roots of maritime pine (*Pinus pinaster* Soland in Ait.). New Phytol 130:141–147 doi:10.1111/j.1469-8137.1995.tb01824.x

Riley D, Barber SA (1971) Effect of ammonium and nitrate fertilization on phosphorus uptake as related to root-induced pH changes at the root-soil interface. Soil Sci Soc Am Proc 35:301–306

Robin A, Vansuyt G, Hinsinger P, Meyer JM, Briat JF, Lemanceau P (2008) Iron dynamics in the rhizosphere: consequences for plant health and nutrition. Adv Agron 99:183–225 doi:10.1016/S0065-2113(08)00404-5

Robinson D, Hodge A, Griffiths BS, Fitter AH (1999) Plant root proliferation in nitrogen-rich patches confers competitive advantage. Proc R Soc Lond B Biol Sci 265:431–435

Römheld V, Marschner H (1981) Iron deficiency stress induced morphological and physiological changes in root tips on sunflower. Physiol Plant 53:354–360 doi:10.1111/j.1399-3054.1981.tb04512.x

Rosling A, Lindahl BD, Finlay RD (2004) Carbon allocation to ectomycorrhizal roots and mycelium colonising different mineral substrates. New Phytol 162:795–802 doi:10.1111/j.1469-8137.2004.01080.x

Ryan PR, Delhaize E, Jones DL (2001) Function and mechanism of organic anion exudation from plant roots. Annu Rev Plant Physiol Plant Mol Biol 52:527–560 doi:10.1146/annurev.arplant.52.1.527

Ryel RJ, Caldwell MM, Yoder CK, Or D, Leffler AJ (2002) Hydraulic redistribution in a stand of *Artemisia tridentata*: evaluation of benefits to transpiration assessed with a simulation model. Oecologia 130:173–184

Saleque MA, Kirk GJD (1995) Root-induced solubilization of phosphate in the rhizosphere of lowland rice. New Phytol 129:325–336 doi:10.1111/j.1469-8137.1995.tb04303.x

Sanderson J (1983) Water uptake by different regions of the barley root. Pathways of radial flow in relation to development of the endodermis. J Exp Bot 34:240–253 doi:10.1093/jxb/34.3.240

Schaller G (1987) pH changes in the rhizosphere in relation to the pH-buffering of soils. Plant Soil 97:444–449 doi:10.1007/BF02383234

Schöttelndreier M, Falkengren-Grerup U (1999) Plant induced alteration in the rhizosphere and the utilisation of soil heterogeneity. Plant Soil 209:297–309 doi:10.1023/A:1004681229442

Schreiber L, Franke R, Hartmann K-D, Ranathunge K, Steudle E (2005) The chemical composition of suberin in apoplastic barriers affects radial hydraulic conductivity differently in the roots of rice (*Oryza sativa* L. cv. IR64) and corn (*Zea mays* L. cv. Helix). J Exp Bot 56:1427–1436 doi:10.1093/jxb/eri144

Schubert S, Schubert E, Mengel K (1990) Effect of low pH of the root medium on proton release, growth, and nutrient uptake of field beans (*Vicia faba*). Plant Soil 124:239–244 doi:10.1007/BF00009266

Shane MW, Lambers H (2005) Cluster roots: a curiosity in context. Plant Soil 274:101–125 doi:10.1007/s11104-004-2725-7

Shane MW, McCully ME, Canny MJ (2000) Architecture of branch-root junctions in maize: structure of connecting xylem and the porosity of pit membranes. Ann Bot (Lond) 85:613–624 doi:10.1006/anbo.2000.1113

Sharp RE, Davies WJ (1985) Root growth and water uptake by maize plants in drying soil. J Exp Bot 36:1441–1456 doi:10.1093/jxb/36.9.1441

Sharp RE, Poroyko V, Hejlek LG, Spollen WG, Springer GK, Bohnert HJ, Nguyen HT (2004) Root growth maintenance during water deficits: physiology to functional genomics. J Exp Bot 55:2343–2351 doi:10.1093/jxb/erh276

Spehn EM, Joshi J, Schmid B, Alphei J, Körner C (2000) Plant diversity effects on soil heterotrophic activity in experimental grassland ecosystems. Plant Soil 224:217–230 doi:10.1023/A:1004891807664

Sperry JS, Hacke UG, Oren R, Comstock JP (2002) Water deficits and hydraulic limits to leaf water supply. Plant Cell Environ 25:251–263 doi:10.1046/j.0016-8025.2001.00799.x

Standing D, Baggs EM, Wattenbach M, Smith P, Killham K (2007) Meeting the challenge of scaling up processes in the plant–soil–microbe system. Biol Fertil Soils 44:245–257 doi:10.1007/s00374-007-0249-z

Steudle E (1993) Pressure probe techniques: Basic principles and application to studies of water and solute relations at the cell, tissue, and organ level. In: Smith JAC, Griffiths H (eds) Water Deficits: Plant Responses from Cell to Community. Bios Scientific Publishers, Oxford, UK, pp 5–36

Steudle E (2000) Water uptake by plant roots: an integration of views. Plant Soil 226:45–56 doi:10.1023/A:1026439226716

Steudle E (2001) The cohesion-tension mechanism and the acquisition of water by plant roots. Annu Rev Plant

Physiol Plant Mol Biol 52:847–875 doi:10.1146/annurev.arplant.52.1.847

Steudle E, Peterson CA (1998) How does water get through roots? J Exp Bot 49:775–788 doi:10.1093/jexbot/49.322.775

Stewart JB, Moran CJ, Wood JT (1999) Macropore sheat: quantification of plant root and soil macropore association. Plant Soil 211:59–67 doi:10.1023/A:1004405422847

Stirzaker RJ, Passioura JB (1996) The water relations of the root-soil interface. Plant Cell Environ 19:201–208 doi:10.1111/j.1365-3040.1996.tb00241.x

Stirzaker RJ, Passioura JB, Wilms Y (1996) Soil structure and plant growth: impact of bulk density and biopores. Plant Soil 185:151–162 doi:10.1007/BF02257571

Strong DT, Sale PWG, Helyar KR (1997) Initial soil pH affects the pH at which nitrification ceases due to self-induced acidification of microbial microsites. Aust J Soil Res 35:565–570 doi:10.1071/S96055

Szegedi K, Vetterlein D, Nietfeld H, Neue H-U, Jahn R (2008) New tool RhizoMath for modeling coupled transport and speciation in the rhizosphere. Vadose Zone J 7:712–720 doi:10.2136/vzj2007.0064

Tang C, Kuo J, Longnecker NE, Thomson CJ, Robson AD (1993) High pH causes disintegration of the root surface in *Lupinus angustifolius* L. Ann Bot (Lond) 71:201–207 doi:10.1006/anbo.1993.1025

Tang C, Drevon JJ, Jaillard B, Souche G, Hinsinger P (2004) Proton release of two genotypes of bean (*Phaseolus vulgaris* L.) as affected by N nutrition and P deficiency. Plant Soil 260:59–68 doi:10.1023/B:PLSO.0000030174.09138.76

Tarafdar JC, Jungk A (1987) Phosphatase activity in the rhizosphere and its relation to the depletion of soil organic phosphorus. Biol Fertil Soils 3:199–204 doi:10.1007/BF00640630

Taylor HM, Ratliff LH (1969) Root elongation rates of cotton and peanuts as a function of soil strength and water content. Soil Sci 108:113–119 doi:10.1097/00010694-196908000-00006

Thompson MV, Holbrook NM (2004) Root-gel interactions and the root waving behavior of *Arabidopsis*. Plant Physiol 135:1822–1837 doi:10.1104/pp.104.040881

Tisdall JM, Oades JM (1992) Organic matter and water stable aggregates in soils. J Soil Sci 33:141–163 doi:10.1111/j.1365-2389.1982.tb01755.x

Trolldenier G (1988) Visualization of oxidizing power of rice roots and of possible participation of bacteria in iron deposition. Z Pflanzenernaehr Bodenkd 151:117–121 doi:10.1002/jpln.19881510209

Turpault MP, Utérano C, Boudot JP, Ranger J (2005) Influence of mature Douglas fir roots on the solid soil phase of the rhizosphere and its solution chemistry. Plant Soil 275:327–336 doi:10.1007/s11104-005-2584-x

Turpault MP, Gobran GR, Bonnaud P (2007) Temporal variations of rhizosphere and bulk soil chemistry in a Douglas fir stand. Geoderma 137:490–496 doi:10.1016/j.geoderma.2006.10.005

Tyerman SD, Niemietz CM, Bramley H (2002) Plant aquaporins: multifunctional water and solute channels with expanding roles. Plant Cell Environ 25:173–194 doi:10.1046/j.0016-8025.2001.00791.x

Tyree MT, Sperry JS (1989) Vulnerability of xylem to cavitation and embolism. Annu Rev Plant Physiol Plant Mol Biol 40:19–38 doi:10.1146/annurev.pp.40.060189.000315

Urbanek E, Hallett P, Feeney D, Horn R (2007) Water repellency and distribution of hydrophilic and hydrophobic compounds in aggregates from different tillage systems. Geoderma 140:147–155 doi:10.1016/j.geoderma.2007.04.001

Vandeleur R, Niemietz C, Tilbrook J, Tyerman SD (2005) Roles of aquaporins in root responses to irrigation. Plant Soil 274:141–161 doi:10.1007/s11104-004-8070-z

Vansuyt G, Souche G, Straczek A, Briat JF, Jaillard B (2003) Flux of protons released by wild type and ferritin over-expressor tobacco plants: effect of phosphorus and iron nutrition. Plant Physiol Biochem 41:27–33 doi:10.1016/S0981-9428(02)00005-0

Veen BW, Van Noordwijk M, De Willigen P, Boone FR, Kooistra MJ (1992) Root-soil contact of maize, as measured by a thin-section technique. III. Effects on shoot growth, nitrate and water uptake efficiency. Plant Soil 139:131–138 doi:10.1007/BF00012850

Verboom WH, Pate JS (2006) Bioengineering of soil profiles in semiarid ecosystems: the 'phytotarium' concept. A review. Plant Soil 289:71–102 doi:10.1007/s11104-006-9073-8

Vetterlein D, Jahn R (2004a) Combination of micro suction cups and time-domain reflectometry to measure osmotic potential gradients between bulk soil and rhizosphere at high resolution in time and space. Eur J Soil Sci 55:497–504 doi:10.1111/j.1365-2389.2004.00612.x

Vetterlein D, Jahn R (2004b) Gradients in soil solution composition between bulk soil and rhizosphere–In situ measurement with changing soil water content. Plant Soil 258:307–317 doi:10.1023/B:PLSO.0000016560.84772.d1

Vetterlein D, Marschner H (1993) Use of microtensiometer technique for studies of hydraulic lift in a sandy soil planted with pearl millet (*Pennisetum americanum* [L.] Leeke). Plant Soil 149:275–282 doi:10.1007/BF00016618

Vetterlein D, Kuhn K, Schubert S, Jahn R (2004) Consequences of sodium exclusion for the osmotic potential in the rhizosphere - Comparison of two maize cultivars differing in Na^+ uptake. J Plant Nutr Soil Sci 167:337–344 doi:10.1002/jpln.200420407

Vetterlein D, Szegedi K, Stange F, Jahn R (2007a) Impact of soil texture on temporal and spatial development of osmotic-potential gradients between bulk soil and rhizosphere. J Plant Nutr Soil Sci 170:347–356 doi:10.1002/jpln.200521952

Vetterlein D, Szegedi K, Ackermann J, Mattusch J, Neue H-U, Tanneberg H, Jahn R (2007b) Competitive mobilisation of phosphate and arsenate associated with goethite by root activity. J Environ Qual 36:1811–1820 doi:10.2134/jeq2006.0369

Wallace HR (1958) Movement of eelworms. Ann Appl Biol 46:74–85 doi:10.1111/j.1744-7348.1958.tb02179.x

Wallander H (2000) Uptake of P from apatite by *Pinus sylvestris* seedlings colonised by different ectomycorrhizal fungi. Plant Soil 218:249–256 doi:10.1023/A:1014936217105

Walter A, Pich A, Scholz G, Marschner H, Römheld V (1995) Diurnal variations in release of phytosiderophores and in concentrations of phytosiderophores and nicotianamine in roots and shoots of barley. J Plant Physiol 147:191–196

Wardle DA (1992) A comparative asssessment of factors which influence microbial growth carbon and nitrogen levels in soil. Biol Rev Camb Philos Soc 67:321–358 doi:10.1111/j.1469-185X.1992.tb00728.x

Wardle DA, Bardgett RD, Klironomos JN, Setala H, van der Putten WH, Wall DH (2004) Ecological linkages between aboveground and belowground biota. Science 304:1629–1633 doi:10.1126/science.1094875

Watt M, McCully ME, Jeffree CE (1993) Plant and bacterial mucilages of the maize rhizosphere - comparison of their soil-binding properties and histochemistry in a model system. Plant Soil 151:151–165 doi:10.1007/BF00016280

Watt M, McCully ME, Canny MJ (1994) Formation and stabilisation of maize rhizosheaths: Effect of soil water content. Plant Physiol 106:179–186

Watt M, McCully ME, Kirkegaard JA (2003) Soil strength and rate of root elongation alter the accumulation of *Pseudomonas* spp. and other bacteria in the rhizosphere of wheat. Funct Plant Biol 30:483–491 doi:10.1071/FP03045

Watt M, Kirkegaard JA, Rebetzke GJ (2005) A wheat genotype developed for rapid leaf growth copes well with the physical and biological constraints of unploughed soil. Funct Plant Biol 32:695–706 doi:10.1071/FP05026

Watt M, Silk WK, Passioura JB (2006a) Rates of root and organism growth, soil conditions, and temporal and spatial development of the rhizosphere. Ann Bot (Lond) 97:839–855 doi:10.1093/aob/mcl028

Watt M, Hugenholtz P, White R, Vinall K (2006b) Numbers and locations of native bacteria on field-grown wheat roots quantified by fluorescence in situ hybridization (FISH). Environ Microbiol 8:871–884 doi:10.1111/j.1462-2920.2005.00973.x

Watt M, Kirkegaard JA, Passioura JB (2006c) Rhizosphere biology and crop productivity-a review. Aust J Soil Res 44:299–317 doi:10.1071/SR05142

Weisenseel MH, Dorn A, Jaffe FJ (1979) Natural H^+ currents traverse growing roots and root hairs of barley (*Hordeum vulgare* L.). Plant Physiol 64:512–518

Whalley WR, Leeds-Harrison PB, Clark LJ, Gowing DJG (2005a) Use of effective stress to predict the penetrometer resistance of unsaturated agricultural soils. Soil Tillage Res 84:18–27 doi:10.1016/j.still.2004.08.003

Whalley WR, Riseley B, Leeds-Harrison PB, Bird NRA, Leech PK, Adderley WP (2005b) Structural differences between bulk and rhizosphere soil. Eur J Soil Sci 56:353–360 doi:10.1111/j.1365-2389.2004.00670.x

Williams SM, Weil RR (2004) Crop cover root channels may alleviate soil compaction effects on soybean crop. Soil Sci Soc Am J 68:1403–1409

Yanai J, Sawamoto T, Oe T, Kusa K, Yamakawa K, Sakamoto K, Naganawa T, Inubushi K, Hatano R, Kosaki T (2003) Spatial variability of nitrous oxide emissions and their soil-related determining factors in an agricultural field. J Environ Qual 32:1965–1977

Yang J, Hammer RD, Blanchar RW (1995) Microscale pH spatial-distribution in the Ap horizon of Mexico silt loam. Soil Sci 160:371–375 doi:10.1097/00010694-199511000-00006

Yang J, Blanchar RW, Hammer RD, Thompson AL (1996) Soybean growth and rhizosphere pH as influenced by horizon thickness. Soil Sci Soc Am J 60:1901–1907

Yoder CK, Nowak RS (1999) Hydraulic lift among native plant species in the Mojave Desert. Plant Soil 215:93–102 doi:10.1023/A:1004729232466

Young IM (1995) Variations in moisture contents between bulk soil and the rhizosheath of wheat (*Triticum aestivum* L. cv Wembley). New Phytol 130:135–139 doi:10.1111/j.1469-8137.1995.tb01823.x

Young IM (1998) Biophysical interactions at the root-soil interface: a review. J Agric Sci 130:1–7 doi:10.1017/S002185969700498X

Young IM, Ritz K (2000) Tillage, habitat space and function of soil microbes. Soil Tillage Res 53:201–213 doi:10.1016/S0167-1987(99)00106-3

Young IM, Crawford JW (2004) Interactions and self-organisation in the soil-microbe complex. Science 304:1634–1637 doi:10.1126/science.1097394

Young IM, Montagu K, Conroy J, Bengough AG (1997) Mechanical impedance of root growth directly reduces leaf elongation rates of cereals. New Phytol 135:613–619 doi:10.1046/j.1469-8137.1997.00693.x

Young IM, Crawford JW, Rappoldt C (2001) New methods and models for characterising structural heterogeneity of soil. Soil Tillage Res 61:33–45 doi:10.1016/S0167-1987(01)00188-X

Zimmermann HM, Hartmann K, Schreiber L, Steudle E (2000) Chemical composition of apoplastic transport barriers in relation to radial hydraulic conductivity of corn roots (*Zea mays* L.). Planta 210:302–311 doi:10.1007/PL00008138

Zwieniecki MA, Thompson MV, Holbrook NM (2003) Understanding the hydraulics of porous pipes: tradeoffs between water uptake and root length utilization. J Plant Growth Regul 21:315–323 doi:10.1007/s00344-003-0008-9

REVIEW ARTICLE

Plant root growth, architecture and function

Angela Hodge · Graziella Berta ·
Claude Doussan · Francisco Merchan ·
Martin Crespi

Received: 3 December 2008 / Accepted: 6 February 2009 / Published online: 5 March 2009
© Springer Science + Business Media B.V. 2009

Abstract Without roots there would be no rhizosphere and no rhizodeposition to fuel microbial activity. Although micro-organisms may view roots merely as a source of carbon supply this belies the fascinating complexity and diversity of root systems that occurs despite their common function. Here, we examine the physiological and genetic determinants of root growth and the complex, yet varied and flexible, root architecture that results. The main functions of root systems are also explored including how roots cope with nutrient acquisition from the heterogeneous soil environment and their ability to form mutualistic associations with key soil microorganisms (such as nitrogen fixing bacteria and mycorrhizal fungi) to aid them in their quest for nutrients. Finally, some key biotic and abiotic constraints on root development and function in the soil environment are examined and some of the adaptations roots have evolved to counter such stresses discussed.

Keywords Root systems · Auxin · Root architecture · Soil heterogeneity · Abiotic and biotic stresses · Soil micro-organisms (including nitrogen-fixing bacteria and mycorrhizal fungi)

Responsible Editor: Yves Dessaux.

A. Hodge (✉)
Department of Biology,
Area 14, P.O. Box 373, University of York,
York YO10 5YW, UK
e-mail: ah29@york.ac.uk

G. Berta
Dipartimento di Scienze dell'Ambiente e della Vita,
Università del Piemonte Orientale,
via Bellini 25/G,
Alessandria 15100, Italy

C. Doussan
UMR 1114 EMMAH INRA/UAPV Domaine Saint-Paul,
Site Agroparc,
Avignon Cedex 9 84914, France

F. Merchan
Departamento de Microbiología y Parasitología,
Facultad de Farmacia, Universidad de Sevilla,
c/ Profesor García González,
Sevilla 41012, España

M. Crespi
Institut des Sciences du Végétal (ISV), CNRS,
1 Avenue de la Terrasse,
Gif-sur-Yvette 91198, France

Physiological and genetic determinants of root growth and achitecture

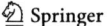

A major difference between plant and animal development is that positional information rather than cell lineage determines cell fate in plants (Singh and Bhalla 2006). Post-embryonically, plant development is essentially driven by stem cells localized in apical regions of shoots and roots, and referred to as apical meristems. This particular characteristic allows plants, which are sessile organisms, to adapt their morphology and organ development to the encountered environmental conditions. The spatial configuration of the root system (number and length of lateral organs), so-called root architecture, vary greatly depending on the plant species, soil composition, and particularly on water and mineral nutrients availability (Malamy 2005). Plants can optimize their root architecture by initiating lateral root primordia and influencing growth of primary or lateral roots. The root system results from the coordinated control of both genetic endogenous programs (regulating growth and organogenesis) and the action of abiotic and biotic environmental stimuli (Malamy 2005). The interactions between these extrinsic and intrinsic signals however complicate the dissection of specific transduction pathways. Such complex traits likely depending on multiple genes may be analyzed through quantitative genetics via the identification of quantitative trait loci (QTL) linked to root architecture (e.g. Mouchel et al. 2004, Fitz Gerald et al. 2006). Understanding the molecular mechanisms governing such developmental plasticity is therefore likely to be crucial for crop improvement in sustainable agriculture. In this section, we focus on hormonal and genetic determinants of root architecture.

The embryonic root apical meristem (RAM) specification occurs very early in embryo development (Benfey and Scheres 2000). The RAM constitutes the stem cell niche that eventually produces all below-ground organs, including lateral roots (Sabatini et al. 2003). The cellular organization of *Arabidopsis* RAMs is very regular, with the initials of the various tissue types surrounding a quiescent center (QC). An asymmetric division of the hypophysis generates a large basal daughter and a small apical daughter, called the lens-shaped cell, which is the progenitor of the QC. During root initiation, the QC appears to act like an organizing center that can direct surrounding cells to produce a set of initials (Aida et al. 2004, Jenik et al. 2007). Different mechanisms specify the initials above the QC that give rise to the ground tissue and the stele (proximal initials) as well as the initials below the QC that produce the root cap (distal initials). This root cap and the underlying root apical meristem form the root zone called the distal root apex, where cells divide actively. At certain distance from the RAM, the anticlinal and asymmetric divisions of a group of pericycle cells are the initial event of the formation of a lateral root. Several consequent divisions result in the formation of a dome-shape primordium that grows through outer cell layers and a meristem is established. Upon activation of this meristem, the lateral root emerges from the parental root and continue to grow in the soil (Malamy 1995 and references therein).

Phytohormonal regulation of the root system: Auxin as a major player

The different stages of root development are controlled and regulated by various phytohormones with auxin playing a major role (reviewed by Leyser 2006). In roots, auxin is involved in lateral root formation, maintenance of apical dominance and adventitious root formation. All these developmental events require correct auxin transport and signaling.

During embryonic RAM formation many auxin-related mutants, such as those involved in its biosynthesis *yuc1 yuc4 yuc10 yuc11* quadruple mutant (Cheng et al. 2007), or mutants affected in transport or signaling (*monopteros, bodenlos, auxin transport inhibitor resistant 1* (*tir1*) and related *tir1/afb1–3* quadruple mutant (Dharmasiri et al. 2005, Jenik et al. 2007) are unable to form the embryonic RAM. The auxin flux coming from the apical region of the embryo into the hypophysis leads to TIR1 (and related redundant AFBs) pathway activation and induction of auxin-response genes such as *PIN* genes (coding for auxin efflux carriers), whose products will increase auxin transport and accumulation to further differentiate the hypothesis (Benkova et al. 2003).

Auxin also plays a major role in lateral root initiation and development. Lateral root development can be divided in different steps: primordium initiation and development, emergence, and meristem activation. Auxin local accumulation in *Arabidopsis* root pericycle cells adjacent to xylem vessels, triggers lateral root initiation by re-specifying these cells into lateral root founder cells (Dubrovsky et al. 2008). Furthermore, it is also involved in the growth and organization of lateral root primordia and emergence from the parent root (Laskowski et al. 2006). Indeed, mutants or transgenic lines with elevated auxin biosynthesis and endogenous levels of IAA display significant increased root branching (Seo et al. 1998, King et al. 1995). In addition, overexpression of the

DFL1/GH3-6 or the *IAMT1* genes, which encode enzymes modulating free IAA levels, results in a reduction in lateral root formation (Staswick et al. 2005; Qin et al. 2005). Hence, normal auxin biosynthesis and homeostasis are necessary for lateral root initiation in *Arabidopsis*. However, overall auxin content is not always positively correlated with lateral root number (Ivanchenko et al. 2006).

Auxin transport into the regions where lateral root initiate also seems crucial for the regulation of root branching (Casimiro et al. 2001). Functional analyses of mutants impaired in auxin transport, such as the *aux1, axr4,* and *pin* multiple mutants (Benkova et al. 2003; Marchant et al. 2002), have demonstrated the crucial role of auxin transport in lateral root formation. In the root, auxin is transported in both an acropetal (base-to-apex; Ljung et al. 2005) and basipetal (apex-to-base) direction within inner and outer tissues of the root apex (Swarup and Bennett 2003). Studies using auxin transport inhibitors have shown that the polarity of auxin movement provides an important developmental signal during lateral root initiation (Casimiro et al. 2001) and root gravitropism (Parry et al. 2001). The strong bias in the direction of auxin transport within a tissue results from the asymmetrical cellular localization of an influx system involving the AUX1 protein (Swarup and Bennett 2003) and an efflux apparatus with PIN-type efflux facilitators (reviewed by Paponov et al. 2005).

The auxin efflux system regulated by PIN protein family members is also crucial for lateral root development (Benkova et al. 2003). Although not all *pin* mutations affect lateral root formation, multiple *pin* mutants have dramatic defects in root patterning, including lateral root development (Benkova et al. 2003; Blilou et al. 2005). Cellular localisation of PIN proteins is also important as a *gnom* mutant allele is defective in PIN-dependent developmental processes, including lateral root primordium development, presumably due to disorganized PIN1 localization (Geldner et al. 2004).

When auxin reaches the target tissue it induces a transcriptional response. Several auxin-responsive genes and gene families involved in auxin signalling have been identified, of which the *Aux/IAAs* and *ARFs* are the best studied. Auxin induces expression of Aux/IAA proteins, which in many cases reduces the sensitivity of cells toward auxin. This induction is mediated by ARF proteins that bind to the auxin-responsive elements (AuxREs) in the promoters of auxin-responsive genes, and activate or repress transcription through interaction with specific Aux/IAA proteins (Liscum and Reed 2002). These loops represent extensive feedback as well as feed-forward regulation (for further information on Aux/IAA proteins, their degradation and their interaction with ARF transcription factors see the review by Benjamins and Scheres 2008). Alterations in these control systems affect several auxin-dependent plant pathways including root development. For example, normal differentiation of a root cap from the distal initials depends on a pair of redundantly acting auxin response factors, *ARF10* and *ARF16* (Wang et al. 2005). In double mutants, the root cap initials overproliferate, thereby generating a mass of undifferentiated progeny. Another example are the gain-of-function *solitary root* (*slr*) mutants which carries a stabilizing mutation in the IAA14 negative regulator of auxin signaling and cannot form lateral root primordia. Detailed analyses of the *slr* mutant demonstrated that the *slr* mutation blocks early cell divisions during initiation (Fukaki et al. 2002)

Other hormones involved in the control of root development

Although auxin plays a fundamental role in root growth and development, several other phytohormones modulate auxin action and consequently affect root development and architecture. In contrast to auxin, exogenous cytokinin application suppresses lateral root formation, and transgenic *Arabidopsis* plants with decreased cytokinin levels display increased root branching and enhanced primary root growth (Werner et al. 2003). The notion that cytokinin negatively regulates root growth has also been verified by studies of cytokinin perception and signalling. For instance, double mutants for the redundant *Arabidopsis* cytokinin receptors AHK2 and AHK3 display a faster growing primary root and greatly increased root branching (Riefler et al. 2006). To identify the stage of lateral root development sensitive to cytokinin, Laplaze et al. (2007) performed transactivation experiments that revealed that xylem-pole pericycle cells, but not lateral root primordial, are sensitive to cytokinin. Cytokinins perturb the expression of PIN genes in lateral root founder cells and prevent the formation of the auxin gradient required to pattern the primordia. Recently cytokinin have also been shown to induce differentiation of cells situated between the root meristem and the elongation zone (Dello Ioio et al. 2007), providing new insights on the molecular mechanisms involved in root meristem maintenance.

The actively growing primary root of dicotyledon plants may exhibit apical dominance forcing their lateral roots to develop further away from the root tip (Lloret and Casero 2002). Root apical dominance may be triggered by the root-cap-synthesized cytokinin and the balance between auxin and cytokinin regulates root gravitropism, a major determinant of root architecture (Aloni et al. 2006). The control of lateral root initiation depends on the primary root vascular system which conveys spatial information as well as photosynthetic carbon to the newly formed primordial (Parizot et al. 2008). Three types of vascular differentiation can be distinguished: (1) primary differentiation, which occurs in cells originating from the primary vascular meristem, the procambium; (2) secondary differentiation, where the derivates originate from the vascular cambium; and (3) regenerative differentiation, in which the vascular elements re-differentiate from parenchyma cells at lateral root junctions. The vascular tissues are induced by polar auxin movement along the plant body, from the hydathodes of young leaves downward to the root tips. Cytokinin by itself does not induce vascular tissues, but in the presence of auxin, it promotes vascular differentiation and regeneration, a critical process for the establishment of a root system (Aloni et al. 2006).

Another hormone involved in lateral root formation is ethylene. Treatments that induce C_2H_4 production (i.e. flooding) promote adventitious and lateral root formation (Mergemann and Sauter 2000). Ethylene stimulates auxin biosynthesis and basipetal auxin transport toward the elongation zone, where it activates a local auxin response leading to inhibition of cell elongation. Stepanova et al. (2005) showed that ethylene-triggered inhibition of root growth is mediated by the action of the weak ethylene insensitive2/anthranilate synthase α1 (*WEI2/ASA1*) and wei7/anthranilate synthase β1 (*ASB1*) genes that encode rate-limiting tryptophan biosynthesis enzymes In addition, ethylene modulates the transcription of several components of the auxin transport machinery and induces a local activation of the auxin signaling pathway to regulate root growth (Ruzicka et al. 2007).

Abscisic acid (ABA), a universal stress hormone, plays a significant role in stimulating elongation of the main root and emergence of lateral roots in response to drought (De Smet et al. 2006b). As with other hormones, the complexity of the plant responses induced by this hormone makes it difficult to distinguish primary from secondary effects (e.g. stomata closure affects water movement in the plant). In general, ABA has an antagonistic effect on lateral root primordial formation and emergence to auxin, which initiates lateral roots (Malamy 2005). This role of ABA in the control of the activity of the lateral root meristems has a significant impact on the final size and architecture of the root system (De Smet et al. 2006b). Indeed, responsiveness to certain abiotic factors is lost in certain ABA signaling mutants (Signora et al. 2001; He et al. 2005; Fujii et al. 2007) such as the systemic inhibition of lateral root formation in high N conditions in the ABA-insensitive mutant *abi4* (Signora et al. 2001).

Genetic and genomic approaches to analyze root growth and architecture

Forward and reverse genetic approaches have been undertaken to elucidate the genetic control of lateral root formation. As expected, most of the lateral root mutants (approximately 40%) identified in *A. thaliana* are affected in a specific component of the auxin pathways. Their phenotypes can usually be rescued or mimicked through auxin application. Several mutants having lateral root phenotypes (De Smet et al. 2006a) have been linked to auxin signaling (*axr* mutants, *tir1*, *msg2-1*, *shy2*), auxin transport (*aux1* and *pin* mutants), and auxin homeostasis (*alf1*, *ydk1*, *df11*). Analysis of these mutants has led to the identification of several genes that regulate lateral root formation and/or coordinate this process in response to environmental cues (reviewed in Osmont et al. 2007). Few *Arabidopsis* mutants such as the previously mentioned *slr-1/AUXIAA14* (Fukaki et al. 2002) and *alf4-1* (Celenza et al. 1995), are not capable of initiating any lateral roots. The encoded ALF4 protein is conserved among plants and has no similarities to proteins from other kingdoms. The *Arabidopsis gene* is involved in lateral root initiation but not in primary root formation as the alf4-1 mutant forms a primary root. DiDonato et al. (2004) have shown that ALF4 functions are required to maintain the pericycle in a mitosis-competent state needed for lateral root formation. Other mutants are affected in vascular tissues and consequently cannot produce lateral roots (e.g. Ohashi-Ito and Bergmann 2007).

In contrast, ectopic expression of *PLT* genes, coding for AP2-type transcription factors, trigger *de novo* formation of roots from shoot structures. *PLT1* was isolated by reverse genetics technologies based on the identification of a QC– and stem-cell–specific enhancer trap line. Indeed, *PLT* genes are pivotal determinants of QC and stem cell identity (Aida et al. 2004). Although *plt1* mutants only display mild root growth defects, the establishment and maintenance of the QC and stem cells in the root meristem, are severely impaired in the *plt1 plt2* double mutant. NAC1 is another transcription factor implicated in lateral root development since its overexpression enhances lateral root initiation (Xie et al. 2002). A genetic framework for the control of cell division and differentiation in root meristematic cells have been recently proposed (Dello Ioio et al. 2008).

A second strategy used to identify non-essential modulators of root development, is exploiting natural genetic variation by analysis of quantitative trait loci (QTL). QTL mapping has the advantage of identifying genomic regions containing genes with subtle effects masked in a particular genetic background. For instance, in *A. thaliana* and maize QTLs linked to root architecture have been identified (Mouchel et al. 2004; Tuberosa et al. 2002; Loudet et al. 2005). Exploring the polymorphisms underlying natural variation can identify the DNA sequence changes that lead to a modified root system architecture in nature. In recent years, several new genomics resources and tools (e.g. genome sequences, tens of thousands of molecular markers, microarrays, and knock-out collections) have become available to assist in QTL mapping and cloning, even if these genes have subtle effects on phenotype (see Paran and Zamir 2003). For example, quantitative morphological and physiological variation analysis of a cross between isogenized *Arabidopsis* accessions revealed that the BREVIS RADIX (BRX) transcription factor controls the extent of cell proliferation and elongation in the growth zone of the root tip (Mouchel et al. 2004).

Genomics have provided new tools to explore downstream and upstream genes of broad regulatory signals such as hormone-responsive factors. Transcriptional changes during root branching in *Arabidopsis* have been monitored to identify key regulators of root architecture (Himanen et al. 2004), including several core cell cycle genes. In addition, Vanneste et al. (2005) compared the early transcriptome of lateral root formation using synchronised induction of these organs in wild type and *slr* mutants. This revealed negative and positive feedback mechanisms that regulate auxin homeostasis and signalling in the pericycle initial cells.

The advent of genomic resources have allowed the combination of global profiling with novel biological approaches such as enhancer trap lines driving green fluorescent protein (GFP) expression in specific root cell types (Brady et al. 2007). These authors used transgenic lines that expressed GFP in specific root tissues to isolate corresponding fluorescent protoplast populations derived from them and characterise their expression patterns using microarrays. Thus were able to determine which genes were expressed and at what level in a particular root cell type. Recently, these transgenic plants were used to characterise the transcriptional response to high salinity of different cell layers and developmental stages of the *Arabidopsis* root (Dinneny et al. 2008). Transcriptional responses are highly constrained by developmental parameters further revealing interactions between developmental fates and environmental stresses.

Genomic approaches mainly rely on the expression patterns of protein-coding genes. However, the discovery of small RNAs (microRNAs, miRNA; and small interfering RNAs, siRNA) in the last decade altered the paradigm that protein coding genes are the only significant components of gene regulatory networks (Chapman and Carrington 2007). Small RNAs are involved in a variety of phenomena that are essential for genome stability, establishment or maintenance of organ identity, adaptive responses to biotic and abiotic stresses. In plants, MIRNAs are genes generally encoded in intergenic regions whose maturation requires a particular type III RNase named DICER-LIKE 1 (DCL1) (Chapman and Carrington 2007). The small mature miRNA is then incorporated in a protein complex, the so-called RISC (RNA-Induced Silencing Complex) that can recognize mRNAs partially complementary to the miRNA nucleotide sequence. This recognition event mediated by the RISC-loaded miRNA leads to cleavage or translational inhibition of the target mRNA. Mutants affected in miRNA metabolism (biosynthesis, action and transport as *dcl1, ago1, hen1, hyl1, hst1, se*) show pleiotropic phenotypes confirming the role of miRNAs in diverse developmental processes (Chapman and Carrington 2007).

Several MIRNAs have been involved in root development. For example, *Arabidopsis mir164a* and *mir164b* mutant plants produced more lateral roots than wild-type plants (Guo et al. 2005). These phenotypes are similar to those obtained by NAC1 overexpression, a transcription factor targetted by miR164 (Xie et al. 2002). Overexpression of miRNA-resistant *NAC1* mRNA results in slight increases in lateral root numbers. In addition, the auxin-related transcription factors ARF10 and ARF16 are also targets of three related microRNA (miRNA) genes, *miR160a, miR160b* and *miR160c*. These miRNAs limit the expression domain of *ARF10* and *ARF16* to the columella, and have a complementary pattern. The over-expression of a miRNA-resistant *ARF16* mRNA under its own promoter results in pleiotropic developmental defects. *MIR160* overexpressing plants, in which the expression of ARF10 and ARF16 is repressed, and the *arf10-2 arf16-2* double mutants display the same root tip defect, with uncontrolled cell division and blocked cell differentiation in the root distal region showing a tumor-like root apex and loss of gravity-sensing (Wang et al. 2005). Apart from these examples, due to the large diversity of these novel regulatory RNAs, it is likely that many other miRNAs or other small RNAs participate in the regulation of root development and architecture.

Root meristematic cells integrate signals from the environment to regulate specific developmental responses and cope with external constraints. This post-embryonic growth and development requires the activation of hormone homeostasis and signalling pathways, transcriptional regulation by specific transcription factors and post-transcriptional regulation of developmental regulators by non-coding RNAs. These regulatory mechanisms may be particularly relevant to adjust differentiation processes to the environmental conditions encountered during growth, notably on primary and lateral root developmental programs.

Root system achitecture

Root systems and architecture definitions

Soil is a complex medium with high spatial and temporal environmental variability at a wide range of scales, including those relevant to plant roots. The root system, with its extensive but structured development, can be considered as an evolutionary response to such spatio-temporal variability in resource supply and associated constraints upon growth (Harper et al. 1991). As a consequence, the extension in space and time of the root system is governed by genetically driven developmental rules which are modulated by environmental conditions. As demonstrated by QTL mapping which recently revealed that root morphology is in most cases regulated by a suite of small-effect loci that interact with the environment (de Dorlodot et al. 2007).

The architecture of a root system is the result of developmental processes and is a dynamic notion. Root architecture addresses two important concepts: the shape of the root system and its structure. The shape defines the location of roots in space and the way the root system occupies the soil. Its quantification is generally achieved by measuring variables such as root depth, lateral root expansion and root length densities. In contrast, root structure describes the variety of the components constituting the root system (roots and root segments) and their relationship (e.g. topology: connexion between roots; root gradients). While, root differentiation has important impacts upon structure—function relations (Clarkson 1996).

The rhizosphere (i.e. the volume of soil around living plant roots that is influenced by root activity; Hinsinger et al. 2005), is often simply thought of as a cylindrical shape around the root. However, this oversimplification does not account for integration at the root system level or for the inherent complexity of root systems which arise from geometry, temporal dynamics and the heterogeneous aspects of roots. These complexities are incorporated into the concept of root architecture. Root geometry is complex because of the specific motion in space of each root, the relative locations between roots and the possible overlapping of their zones of influence. The temporal dynamic comes both from the growth of the different root axes and from physiological processes associated with root segments (i.e. tissue differentiation) resulting in temporal and spatial variability of function along the root axes. The diversity among roots within the root system and soil heterogeneity further increase this variability.

Root classification and elaboration of the root system

It has long been recognized that the *in situ* morphology of a root system can be complex and may vary greatly, even within a species (Weaver 1926; Cannon 1949; Kutschera 1960), reflecting the interplay between developmental processes and environmental constraints (Fig. 1). Consequently, the complexity of root systems has led to a number of different classification systems. These classifications can be based on: branch structure (topology) (Fitter 1987; 2002), root activity (Wahid 2000) or development (Cannon 1949). The latter is the more classical approach and is useful in understanding growth as well as obtaining a more global view of root architecture, for example, in relation to plant habitat. Thus, developmental classification has been widely used in modelling approaches to simulate architecture (cf section below).

From a developmental stand point roots are classified according to their ontogenesis into three main categories: primary, nodal and lateral roots (Cannon 1949; Harper et al. 1991; Klepper 1992), the formation of which has been shown to be genetically separable (Hochholdinger et al. 2004). This classification also reflects differences between monocot– and dicotyledon species: dicots root system is derived from primary roots and lateral branching (primary root system), with roots that may exhibit radial growth. Depending on the extension of laterals relative to the primary axis, the morphology of the root system will vary between taprooted and diffuse or fasciculate (Fig. 1). In monocots, root systems derive not only from branching of primary roots but also from emmited nodal roots (adventitious root system). Monocot roots do not undergo secondary radial growth. The primary root differentiates from the seed's radicle already present in the seed embryo. This generally gives rise to a single-axis root system, or taproot system, with dominant vertical growth (gravitropism). Adventitious (or nodal) roots differentiate from organs other than roots (e.g. rhizomes, stems etc) and are initiated at precise locations (near stem nodes for example) with a defined temporal pattern, in coordination with the development of the shoot. They are often abundant and give rise to a fasciculated root system. Adventitious roots are much less sensitive to gravitropism than primary roots (Klepper 1992). Lateral roots originate from the branching of a parent axis, often at near right angles,

Fig. 1 Diversity of root systems. On the top row, root systems are little branched, while more profusely branched on the bottom row. Dominance of a single main axis increases from left to right (from Kutschera 1960)

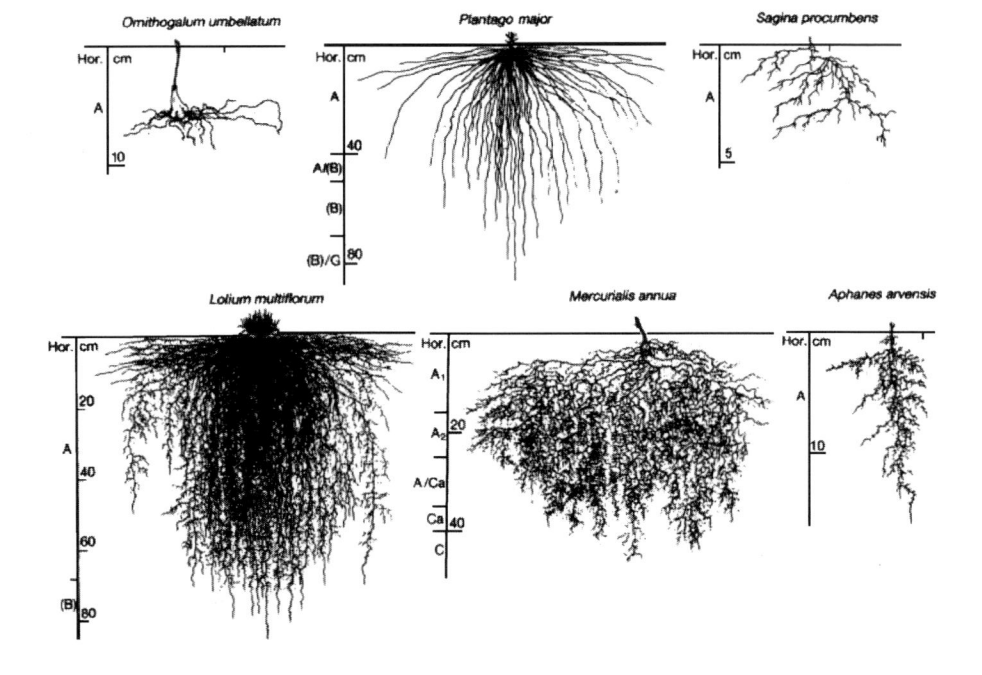

and differentiate from parent roots by initiation of a primordium in the pericycle cells located adjacent to protoxylem poles (although with some exceptions e.g. wheat, see Demchenko and Demchenko 2001) at some distance from the apex. This process results in a branching front which follows the parent root's apex (acropetal branching). However, lateral roots may also appear out of this main sequence, differentiating from older tissues generally near the main root base. The maximum number of branching orders seems to be a genotypic feature. Branching can also develop by reiteration. Here the parent axis duplicates into two (or three) axes of the same morphological type and creates forks in the root system which can profoundly impact upon tree root system architecture (Vercambre et al. 2003).

In a number of plant species (e.g. cereals), the primary root system will dominate the early growth stages while the adventitious root system will dominate in older plants. In the latter case, the location and dynamic of the nodal root emissions are of prime importance for the general structure of the root system as has been demonstrated using cereals (Klepper 1992; Pellerin 1993; Yamazaki and Nakamoto 1983).

The respective importance of the primary and adventitious root systems (i.e. the relative growth rates of main axes and laterals, and the number of branching orders), varies among plant species and can be further modulated by environmental conditions. Hence, different plant species develop different soil exploration strategies. For example, maize (a monocotyledon) emits nodal roots throughout its vegetative phase allowing shallow soil horizons to be repeatedly explored. In contrast, narrow leaf lupine (a dicotyledon) explores shallow soil horizons by means of a single generation of branch roots and much less new roots are emitted (see Fig. 2).

Developmental processes contributing to root system architecture

Common basic developmental processes act simultaneously upon root system dynamics as follows:

(i) *Emission of new main axes* (primary or adventitious), which is highly coordinated with shoot system growth (leaf and tiller development) as discussed above;

(ii) *Branching*, which is the development of lateral roots along a bearing root of lower order. This is the main process for multiplying the number of roots. Root elongation and branching are coordinated processes (Lecompte and Pagès 2007) and an equilibrium is maintained between root number and length. Formation of lateral root primordia (in terms of locations and cell division) are now well described (see *Physiological and genetic determinants* section and Lloret and Casero 2002). In the acropetal branching sequence under homogeneous conditions, lateral emergence occurs after a defined thermal time, allowing the length of the unbranched apical bearing axis to vary with its growth rate (Pellerin and Tabourel 1995). However, under heterogeneous soil conditions root branching is highly sensitive to environmental stimuli, particularly nutrient availability (Drew and Saker 1975; see also *Root Function* section below). It has been shown that not all lateral root primordia develop into lateral roots and that hormones influence primordia development (Malamy 2005). The rapid activation of primordia and the consecutive growth of lateral roots may be how plants respond to nutrient perception. For example, in Arabidopsis four effects of N supply on root development have been identified (Zhang et al. 2007) namely: (1) a localized stimulatory effect of external nitrate on lateral root elongation; (2) a systemic inhibitory effect of high tissue nitrate concentrations on the activation of lateral root meristems; (3) a suppression of lateral root initiation by a high C:N ratio, and (4) an inhibition of primary root growth and stimulation of root branching by external L-glutamate (organic N).

(iii) *Axial growth*, defines the length and trajectory of roots and contributes to the dynamic colonization of new soil zones. From an architectural point both the elongation and the direction have to be considered as elongation will define the dynamics of soil colonization, while direction will define its shape. Moreover, both are highly variable between roots and are sensitive to external conditions. Axial root growth occurs from the distal end (root tip) as a result of both cell division in the meristem and elongation within the elongation zone. The optimal soil

Fig. 2 The difference in the rooting pattern of **a** a mature maize plant, with numerous adventitious roots repetitively exploiting the soil (from Weaver 1926) and **b** a narrow leaf lupine plant with a typical herring-bone root system and with less roots emitted with time for exploiting the soil surface (from Doussan et al. 2006)

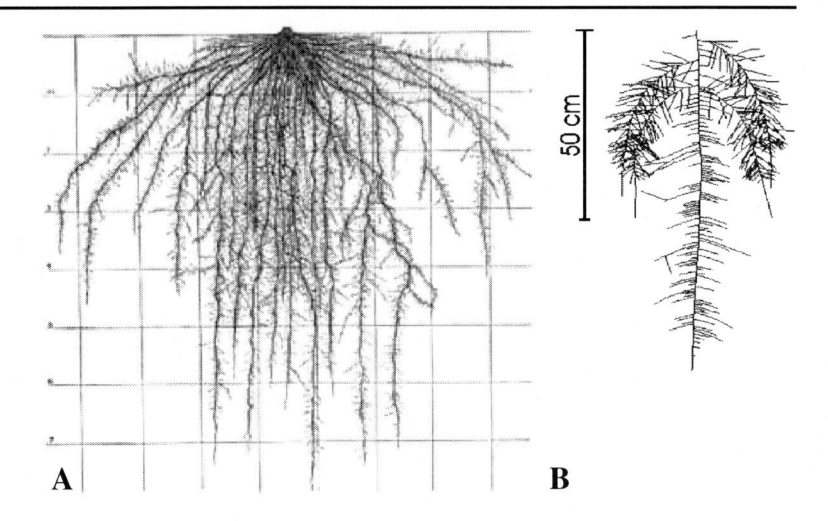

temperature range for root growth varies depending on the species and their origin (17–35°C; McMichael and Burke 2002). Trajectory of the root is influenced by different tropisms and the root cap seems to be involved in sensing these (e.g. gravitropism, hydrotropism, P tropism etc; Kaneyasu et al. 2007; Svistoonoff et al. 2007). The influence of tropisms varies among root branching order or adventitious/seminal roots (Lynch and Brown 2001). During elongation, the root tip is pushed forward into the soil, the root part behind the elongation zone being anchored in soil. However, during movement, the root tip has to overcome the mechanical resistance of the soil (see section on *Root response to abiotic stress* below). Presence of zones of high mechanical impedance is one of the most common physical limitations to soil root exploration (Hoad et al. 1992) as it impacts not only on root growth rate and morphology (Dexter 1987; Masle 2002) but also growth direction with tortuous trajectories (Pagès 2002), soil colonization (Tardieu and Katerji 1991), and the local soil-root environment (Pierret et al. 2007). Consequently, in natural soil roots follow the path of least mechanical resistance such as through cracks and biopores (Rasse and Smucker 1998). Root colonization of soil macropores leads to a specific biological and chemical environment from that of the bulk soil (Pierret et al. 1999), but this environment may also vary with root type. Indeed, Pierret et al. (2005) demonstrated that the thinner the roots, the denser the soil in which they were located, indicating that soil pores of varying radius were accessed by different root types from main axes to laterals of different order.

(iv) *Radial growth*: some, but not all, roots of dicotyledonous plants show a radial growth that results from the activity of laterally situated secondary meristems (cambium). Aside from morphological variations, this process is important for a range of root functions including increased axial transport properties (particularly axial hydraulic conductivity), increased mechanical strength and anchorage, storage capacity and protection against predation, drought or pathogens.

(v) *Root senescence and decay:* root mortality and turnover are important processes in the development and function of root systems, particularly in perennial plants. From a plant resource allocation view, if root systems exhibit plasticity in producing new roots in favorable soil zones, they should also be able to shed these roots when resource uptake is no longer efficient. Replacement of decaying roots by more efficient new ones (i.e. root turnover) may account for *c*. 30% of global terrestrial net primary production (Jackson et al. 1997). Considerable variations in root longevity are found (Fig. 3), which may be due to the methodology used: ≤ 1 year from minirhizotron techniques, to > 4 years from data derived from carbon isotopes techniques (Pritchard and Strand 2008). In addition, the influence of environmental conditions (e.g. temperature, nutrients) on

Fig. 3 The relationship between root longevity (yr) and root diameter for roots produced in shallow (0–15 cm) and deep soil (15–30) at the Duke FACE forest site from Fall 1998 through January 2007. Longevity was determined based on visual observations of roots using minirhizotrons. Root longevity was not influenced by diameter in shallow soil, but increased with diameter in deeper soil. The data highlight the heterogeneity in longevity of fine roots (< 2.0 mm in diameter). From Pritchard and Strand (2008)

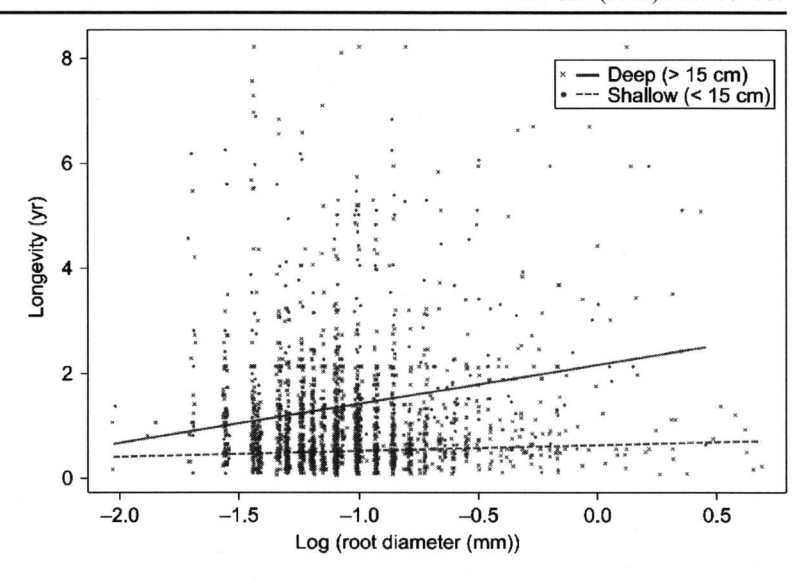

root life span is contradictory (see Eissentat and Yanai 2002). Roots originating from the seed (i.e. radicle, seminal) or primary adventitious roots are generally perennial, determining root system shape. Higher orders lateral roots may live longer when they undergo secondary growth or development of a periderm, often signaled by browning of the root (Hishi and Takeda 2005; Eissentat and Yanai 2002). Fine roots lacking secondary growth, wall thickening or with a low C:N ratio are shorter lived. Morphological markers, such as root tip diameter, have been shown to correlate with root longevity (Thaler and Pagès 1999). Decaying roots enter a necrosis process starting from the distal end and superficial tissues which progressively extends to the whole root. However, a decaying root may still be functional (Robinson 2001).

Diversity of roots within the root system

The diversity of root function is illustrated by water and nitrate uptake. If all roots behave identically, ex-situ measured rates of uptake based on root length result in a large over-estimation of uptake rates (Robinson 1991). In contrast, actual *measured* uptake rates suggest that only 10% and 30% of the total root system length is involved in nitrate and water uptake respectively. This raises the question of which parts of the root system are active and does this value vary with nutrient availability? The answer relies on the assessment of the heterogeneity of root functioning *in-situ*. This anatomical, physiological, and morphological diversity has been called "heterorhizy" (Noelle 1910; Waisel and Eshel 1992; Hishi 2007). Variations in morphology or function can occur among roots of different origin (ontogenesis) or along the root (differentiation).

Morphologically, large differences among roots occur due to root diameter: for example in the monocotyledon, *Musa acuminate* (Banana tree) variations in root diameter can reach 2 orders of magnitude (i.e. from 0.06 to 6 mm; Lecompte et al. 2001). Such variations are common in many species, although generally at lower orders of magnitude (*c.* 1) and are linked to variation in root structure (Fitter 2002, Varney and McCully 1991a). In dicotyledons species such variations are reinforced by secondary growth. This has large consequences on root transport properties: roots showing secondary growth may exhibit a 100–1,000 increase in axial hydraulic conductivity compared to roots without secondary growth (Vercambre et al. 2002). Morphologically different, specialized roots may also be present in different species such as contractile roots, cluster roots, dauciform roots and so forth (see *Specialized root morphologies* section below).

Differences in functioning between roots of different origins have been demonstrated. For example, in maize, seminal roots, even if relatively few in number, play a dominant role in water supply during a significant part of the plants life span (Navara 1987), but are able to acquire less P than nodal roots (Mistrik and Mistrikova 1995). Variations between the taproot and laterals in Cl or K uptake have been demonstrated in pea (Waisel and Eshel 1992), while in maize, nitrate uptake rate was 1.5 times greater in branch roots compared to the main axis (Lazof et al. 1992).

Large changes in physiological properties also occur along roots as a result of tissue aging and differentiation. Consequently, variation in root function including nutrient uptake rates are observed at increasing distance from the root tip (Clarkson 1996). These include gradients in:

- respiration : high respiration variation was found in the vicinity of the apex and also up to 20 cm from the root tip in *Prunus persica* (Bidel et al. 2000);
- ion uptake : NO_3 and NH_4 uptake and P, K, Ca uptake and translocation, varies along roots with different zones of active (mainly apical) and passive uptake (Cruz et al. 1995; Lazof et al. 1992; Clarkson 1996);
- proton/hydroxyl excretion: parts of roots may excrete H^+, while other release OH^- depending on the N source (Jaillard et al. 2000);
- carbon exudation, which is mainly concentrated in the apical zone (Personeni et al. 2007);
- axial and radial root water transport properties which is linked to tissue maturation (i.e. suberization and opening of xylem vessels). In maize this resulted in a tenfold decrease of radial root hydraulic conductivity while the root axial conductance increased by *c.* three orders of magnitude (Doussan et al. 1998). Radial growth in perennial roots of dicotyledons also results in a drastic increase of root axial conductance (Vercambre et al. 2002).

Thus, the global functioning of a root system is the reflection of more or less independent physiological processes coordinated at the root system level but varying among different roots and even along a single root according to its age. This has a profound influence upon root-soil interactions and the dynamics of the rhizosphere as shown by water flux along a maize and peach tree root system, estimated from a coupled hydraulic-root architecture model (Fig. 4).

Modeling of the root system architecture and function

Modeling of root architecture and function has been the subject of many recent reviews (see Pagès 2002, Fourcaud et al. 2008; Pierret et al. 2007), thus is only briefly considered here. Such models are helpful in linking knowledge gained at the level of the individual root to that of the entire root system. Moreover, because they include explicit quantitative information about the soil volume occupied as well as the location, number and, possibly, functionality of roots, modeling approaches provide opportunities to understand root exploration of soil in combination with soil bio-physico-chemical processes (Pagès et al. 2000, Wang et al. 2006). Modeling approaches also open new avenues for genetic improvement of crops by estimating the importance of different root traits/functions in low-input or low-fertility soils (Lynch 2007; de Dorlodot et al. 2007, Wu et al. 2005) or in plant competition (Dunbabin 2007).

Conceptually, root system architecture can be modeled in different ways, depending on the goals and actual knowledge and parameterization of the different processes available. Root architecture can be described by synthetic descriptors such as fractal dimensions, which represent the space filled by the roots (Tatsumi et al. 1989), or the relationship between the basal root and total root system size (van Noordwijk and Mulia 2002). Topological modeling is another synthetic way of describing root system architecture. Here, the root system is represented by a set of links which can connect two branching points (interior link) or an apex to a branching point (external link) allowing derived topological parameters such as the magnitude (number of external links), altitude (largest number of links between an external link and the base link) and the ratio of magnitude to expected altitude, to be defined (see Fitter 2002). These attributes have been related to the carbon cost, the transport properties and the exploitation efficiency of topologically different architectures (Fitter 2002, Bernston 1994) and has proved fruitful in describing the global implications of contrasted branching patterns (e.g. herringbone vs dichotomous).

Another approach to modeling root system architecture and growth is to represent the developmental processes of the root system. This results in a three-dimensional set of connected axes or segments, defining the roots and their apex, each characterized by properties such as age, root type, diameter, length and possibly water or nutrient-uptake ability. Such models have been developed with variable degrees of

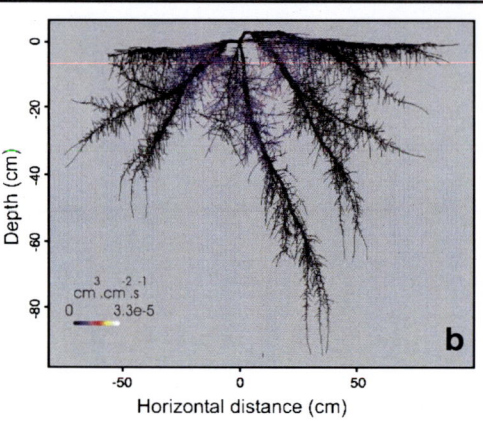

Fig. 4 The heterogeneity of functioning of root systems in the case of water uptake. Graphs show the uptake rates of water along roots derived from modelling of **a** a maize root system and **b** a young peach tree root system, both taking up water from a homogeneous medium of freely available water. The pattern of uptake is heterogeneous for both plants, but show a distinct distal-proximal gradient for the peach tree, while uptake is more distributed along the root system and patchy for the maize. This intra- and interspecific variations of fluxes linked to root system architecture may have significant impacts upon rhizospheric processes (redrawn from Doussan et al. 1999)

dynamic complexity: for example, Ge et al. (2000) showed the effect of altered gravitropism of the basal roots of bean plants in P acquisition efficiency; Somma et al. (1998) and Dunbabin et al. (2002) incorporated aspects of the effect of nitrate availability upon root system development in an attempt to simulate root plasticity and efficiency. Doussan et al. (1998, 2006) introduced root functional heterogeneity into root architectural models to address the interaction between water uptake by the root system and soil water transfer and hydrodynamic properties. A similar approach was recently conducted by Javaux et al. (2008) also for water (Fig. 5) and is important as it tries to define which parts of the root system are effectively active.

However, very few, if any, of these models have thus far coupled interactions among the spatial pattern of resource availability (e.g. P availability may be greater at the surface while water may be available at depth), heterogeneity of roots in physiological capacity and interactions between the root and above ground parts. In relation to rhizosphere interactions integration of detailed geochemical soil processes as relates to, for example, nutrient mobility, have only been vaguely incorporated into such models and presently lack rigor. Further, the interactions between exudates, nutrient uptake and the root system are still in their infancy (Dunbabin et al. 2006). However, progress in reactive transport codes and root models (Nowack et al. 2006), along with enhanced computing power, will likely allow interactions between root and soil biogeochemistry to be invesigated in the near future. Coupling between mycorrhiza processes and root systems also has yet to be fully achieved but may, in conjunction with experimental data, provide clues to the respective importance of these key soil micro-organisms upon plant nutrition (Smith 2000). To this end, a first attempt in modeling mycorrhizal growth, uptake and translocation of P has been done by Schnepf and Roose (2006). Finally, although not all the interactions and processes of the soil-plant system need, or can, be simulated with the same degree of accuracy, progress will be made by focusing models on some key, specific processes while approximating others.

Root functions

General root functions in the heterogeneous soil environment

As indicated above, the main functions of a root system are anchorage and uptake of water and nutrients. In trees and other woody species, extensive belowground structures whose main role is to provide support rather than nutrient acquisition are required but in smaller plant species anchorage occurs largely as a secondary function of root growth and development in soil. The overall form or 'architecture' of the

Fig. 5 Functional root architecture models interacting with their environment using water as an example. **a** Cross-sections of the three-dimensional soil water potential distribution after 7 days with constant water flux at the root collar for three soil types. High gradients are found in the vicinity of roots, depending on the soil type. Barley root architecture is shown in white (from Javaux et al. 2008). **b** Propagation with time of a water extraction front along the root system during daily transpiration for a (modified) Lupine root system in a sandy soil. Water flux density is in $cm^3\ cm^{-2}\ s^{-1}$ (from Doussan et al. 2006)

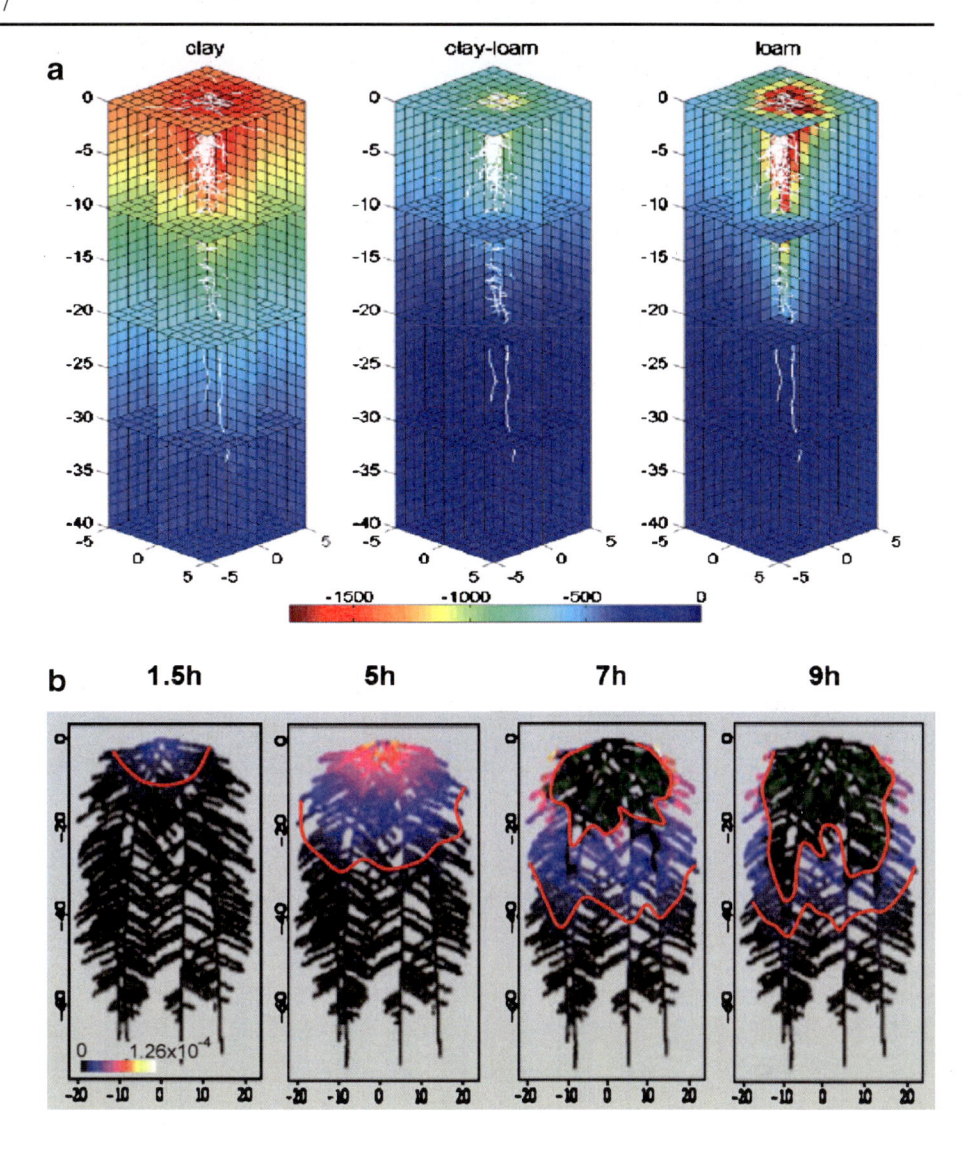

root system is also important for anchorage and water-nutrient uptake. Root system architecture is very varied among different plant species but within species architecture is flexible (see above) and can alter as a result of prevailing soil conditions. This flexibility arises due to the modular structure of roots which enables root deployment in zones or patches rich in moisture or nutrients. The genetic controls on this root deployment are still largely unknown although the gene *ANR1* is involved in the first stages of the nitrate (NO_3^-) signalling system when NO_3^- levels are locally enhanced (Zhang and Forde 1998) and a novel gene, *miz1* (from *mizu-kusseil* meaning 'water' and 'tropism' respectively in Japanese) has recently been identified which is essential for hydrotropism (growth towards water) in *Arabidopsis* (Kobayashi et al. 2007). While the genetic controls remain largely undiscovered, it is well established that roots exhibit a high degree of physiological and morphological plasticity in order to capture nutrients from transient patches in the soil (reviewed by Hodge 2004, 2006). Physiological responses, which tend to occur first, can be both large and rapid: roots of perennial species common to the Great Basin region of North America showed as much as an 80% increase in phosphate (P) uptake rates in patches enriched with P compared to those roots present in control (distilled water) patches after only a few days (Jackson et al. 1990). Morphological responses where roots proliferate in response to nutrient patches tend to take longer as, strictly speaking, root proliferation involves the initiation of new laterals (rather than simply root elongation).

It is often stated that fast-growing species from fertile habitats are capable of greater morphological plasticity than slow-growing plants from infertile habitats, as the former have more resources to allocate to new root construction (Chapin 1980; Grime et al. 1986). However, in a survey of the literature Reynolds and d'Antonio (1996) found that while more biomass was allocated below-ground to roots under low nitrogen (N) conditions, this was a general plant response and did not differ among plants from habitats of contrasting fertility. While Reynolds and d'Antonio (1996) considered gross partitioning above- and below-ground to high or low N, Campbell et al. (1991) investigated biomass allocation *within* the root system of competitively dominant and subordinate plant species in a nutrient patchy environment. Campbell et al. (1991) created nutrient patches by dripping nutrient solution continuously onto the surface of sand resulting in two-nutrient rich and two nutrient poor quadrants in the microcosm units without the use of barriers thus allowing free root growth within the microcosm. When the increment in root biomass in nutrient-rich sections was examined competitively dominant plants had the greatest amount of roots in the nutrient rich quadrants but the subordinate plants allocated more of their *new* root growth into the nutrient-rich zone: i.e. subordinates placed their new roots with more precision. Campbell et al. (1991) referred to this as a trade-off between the scale and precision of the response. Although this is an intriguing idea it has proved to be controversial, with the results of some studies suggesting that larger root systems actually show greater precision than smaller (subordinate) ones (e.g. Einsmann et al. 1999; Farley & Fitter 1999). Kembel and Cahill (2005) recently re-examined support for the scale/precision hypothesis by using published data for 121 plant species in total. To allow for differences between experimental conditions, in particular if root biomass or surface area was measured, Kembel and Cahill (2005) divided the data into three subsets: British flora, Great plains flora and combined biomass. In the first two subsets data was derived from a single study in each case whereas in the third subset numerous studies were combined. Root surface area was only used as a parameter for the Great plains flora. Kembel and Cahill (2005) found no relationship between root foraging precision and scale in any of the three data sets examined. There was however, weak support for a relationship between relative growth rate (RGR) and precision, but only in the Great plains flora subset, which was the only data set where root surface area (rather than biomass) was used as the parameter for root foraging response, and the only case were all species came from a single plant community. Root biomass although commonly reported (and easily measured) is not necessarily the best measurement to determine foraging ability as changes in the architecture of the root system can occur without a change in biomass (see Hodge 2004). Biomass measurements also do not take into account root longevity which can increase (Pregitzer et al. 1993), decrease (Hodge et al. 1999c) or remain unchanged (Hodge et al. 2000d) in nutrient patches. The nature of the patch itself (i.e. duration, size and concentration) may influence how long roots persist (see Hodge et al. 1999b, 2000d) but to fully understand root demography in relation to soil heterogeneity, we need to understand the controls on root production and death. Although there is information on the former, we know little about what causes roots to die. Some are inevitably eaten, but the plant controls on root death remain elusive.

Root *length* (a measure of root surface area) has been shown to be a good predictor of nutrient capture from patches (see Hodge et al. 1999a, 2000c; Robinson et al. 1999) however the context in which the proliferation response is produced also needs to be considered. For example, although plants grown individually respond to complex N-rich patches by root proliferation, an actual benefit of this proliferation in relation to plant N capture could not be shown (van Vuuren et al. 1996; Hodge et al. 1998). However, when plants were grown in interspecific competition for these N-sources, then the root length produced in the patch was directly related to the amount of patch N each plant captured (Hodge et al. 1999a). Thus, in the absence of interspecific plant competition the importance of root proliferation for N capture was obscure. The activity of soil microorganisms and fauna in releasing nutrients from complex patches for plant uptake is also critical (Bradford et al. 2007; Hodge et al. 2000a, d, e), thus studies need to be conducted under more ecologically relevant conditions for the root response to be correctly interpreted.

Root-mycorrhiza enhanced nutrient capture

Most plants have an additional mechanism of nutrient acquisition, namely that of mycorrhizal (literally meaning 'fungus-root') symbiosis, and in the natural environment being mycorrhizal is the norm rather than the exception (for the impact of mycorrhizal colonisation upon actual root structure see 'Root responses to micro-organisms' section below). Of the seven types of mycorrhizal associations that can form depending on the fungal-plant combination, the three most important and globally widespread are the ericoid (ERM), ecto- (ECM) and arbuscular mycorrhizal (AM) associations. Of these three, the AM is the ancestral condition and likely aided the first land plants to colonise the terrestrial environment (Smith and Read 2008). The AM association is also the most widespread type of mycorrhizal symbiosis forming on two-thirds of all plant species (Smith and Read 2008).

ERM and ECM fungi are known to have saprophytic capabilities that can enhance nutrient acquisition (particularly N) for their host plant (Read and Perez-Moreno 2003). Arbuscular mycorrhizal fungi (AMF) were traditionally thought to be important only for P acquisition (see Smith and Read 2008). Phosphate ions form insoluble complexes with most of the abundant cations (e.g. Al^{3+}, Fe^{3+} and Ca^{2+}) in soil. Consequently phosphate diffuses very slowly and depletion zones rapidly develop around roots. AM fungi extend out from these depletion zones and thus can explore a larger soil volume for P acquisition. However, there is now evidence that AMF can also capture and transfer N to their associated host plant (Govindarajulu et al. 2005) including from complex organic sources in the soil (Hodge et al. 2001). The exact mechanism by which AMF can capture N from these complex patches still needs to be defined, as these fungi have no known saprophytic capabilities (Smith and Read 2008). Although there are vast amounts of literature on the nutritional benefits of mycorrhizal associations much less is known about how mycorrhizal fungi alter the responses of roots in heterogeneous nutrient environments, and data from those studies that have investigated this tend to be on the AM association.

When both AMF and colonised roots are present in organic patches the root responses appear to be more important than that of the actual fungus, at least in terms of N capture from the patch (Hodge 2001a, b). However, when only the fungus was permitted access to the patch, the amount of patch N in the plant increased and was directly related to the extent which the AMF proliferated hyphae in the patch zone (Hodge et al. 2001). Often in controlled pot experiments, and from which much of this data is derived, conditions for root growth are more favourable than they would be in the field. Thus the benefit of AMF under field conditions will likely be greater than the studies with both roots and AMF experiencing the patch suggest. In addition, because of its extensive external mycelium the AMF hyphae should be able to access patches the plant cannot, either because of distance from the root or because they are too small or transient to evoke a root response. Moreover, colonisation by AMF can enhance the host root proliferation response specifically in a patch zone (Hodge et al. 2000b), although this did not result in a net benefit in terms of N capture, when plants were grown in monoculture. However, as stated previously, root proliferation responses only conferred a net benefit in terms of N capture when plants were grown in interspecific competition (Hodge et al. 1999a). Thus, when the effects of AM colonisation on plants grown in interspecific competition were specifically tested, colonisation by AMF did enhance N capture for their associated host plant (Hodge 2003a) but this affect may depend upon the plant species present (see Hodge 2003b). Although AMF are well known to enhance P capture for their associated host plant few studies have actually considered heterogeneous (i.e. patchy) supplies of P. Cui and Caldwell (1996a, b) studied the impact of AMF colonisation when both roots and AMF were present in a P-rich solution patch and when the fungus only was allowed access to the enriched P zone. Colonisation by the AMF enhanced P capture under both conditions. However, large differences among different AMF may exist in their ability to enhance P (Cavagnaro et al. 2005) and N (Leigh et al. 2009) capture from a patchy environment.

Specialised root morphologies

Along with mycorrhizal and nitrogen-fixing symbiosis (see Root response to micro-organisms section below), cluster roots represent the third major strategy of nutrient acquisition to have evolved among land plants (Skene 1998). Unlike the other two however, cluster root formation does not involve a tight root-microbial interaction. Although some cluster rooted species, such as lupins, can also form nitrogen-fixing symbiosis, they tend to be non- or very weakly mycorrhizal (but see Boulet and Lambers 2005). Thus, cluster roots represent an alternative mechanism to capture soil P, although Fe uptake may also be important (Zaïd et al. 2003). Originally thought to be confined to the Proteaceae, these roots were previously termed 'proteoid' (Purnell 1960), however, root clusters have now been shown to develop on a range of different plant species belonging to a number of different families including Betulaceae, Fabaceae (notably lupins), Casuarinaceae, Myricaceae, Mimosaceae, Eleagnaceae and Moreacea (Skene 1998), thus, the more inclusive 'cluster roots' is now used.

There are two main types of cluster roots: 'simple', bottle-brush-like cluster roots and 'compound' cluster roots made up of a collection of many 'simple' cluster roots together and which can form dense mats of roots in the upper soil-organic horizons (Shane and Lambers 2005). The rootlets that make up a cluster root grow only 0.6–35 mm long but can be densely covered in root hairs, thus further increasing the root's surface area. According to Lamont (2003) individual 'clusters' are generally 2–75 mm in length and 1–34 mm in width, although some can be much larger (i.e. up to 200 mm long and 70 mm wide in some *Hakea* species). The development of cluster roots is generally followed by a short, marked exudative burst of predominantly carboxylates (notably citrate and malate) and other organic chelators such as phenolics as well as enzymes such as acid phosphatases (Gardner et al. 1983; Gilbert et al. 1999; Roelofs et al. 2001; Weisskopf et al. 2006a). The tri-carboxylate citrate is particularly effective at replacing P held on exchange surfaces (such as crystalline $Al(OH)_3$ or $Fe(OH)_3$) and complexing with metal ions in the solid phase which constitute the exchange matrix on which the P is held (e.g. Ca in rock phosphate, Fe^{3+} in $Fe(OH)_3$; Jones 1998). In many soils P is only released in appreciable amounts following these very high levels of carboxylate addition. The exudative burst can last a few days and in lupins at least, follows a diurnal pattern (Watt and Evans 1999).

Both formation of cluster roots and the exudative burst are regulated by P concentrations in the shoot rather than in the soil (Shane et al. 2003), although the actual exudative burst is the more tightly regulated of the two (Keerthisinghe et al. 1998). This regulation makes economic sense for the plant given that cluster-roots under P stress can exude 3–23% (depending on plant age) of their dry weight as citrate (Johnson et al. 1996; Gardner et al. 1983; Dinkelaker et al. 1989). This concentrated release of carboxylates allows much more P to be extracted from smaller volumes of soil than a mycorrhizal fungus would be able to capture from the same soil volume, thus in these nutrient deficient soils this is a much more effective P capture mechanism, albeit at a higher carbon cost to the plant. In some species such as lupins the release of protons (H^+) is also thought to be important in acidifying the rhizosphere to further enhance P capture (Dinkelaker et al. 1989), although H^+ release may simply be related to carboxylate export (Neumann et al. 1999).

Exudation of acid phosphatases (APases) by cluster roots may further enhance P acquisition by allowing *organic* forms of P to be utilised (Adams and Pate 1992; Gilbert et al. 1999; Neumann and Martinoia 2002). In many soils the availability of organic P is limited by the low solubility of recalcitrant organic P forms (such as Ca/Mg-phytates and Fe/Al-phytates). Solubilization of these organic P forms by the exuded carboxylates may allow these P forms to be broken down by the APases, thus releasing P that would otherwise be unavailable to the plant. Braum and Helmke (1995) showed that in lupins isotopically exchangeable P values were significantly larger (by 2–5 times) and ^{32}P specific activities were lower than those of soybean indicating lupin was accessing a chemically different (and larger) P pool than soybean. Cluster roots can also release isoflavonoids at higher rates than non-cluster roots (Weisskopf et al. 2006a). Bacterial populations in the rhizosphere of cluster roots are reduced at the time of citrate release (Weisskopf et al. 2006b). As isoflavonoids potentially have anti-microbial properties, their function may be to inhibit microbial activity in the rhizosphere prior to the release of citrate to reduce the possibility of microbial degradation (Weisskopf et al. 2006a, b).

Many cluster root studies have used lupins as the model plant (e.g. Dinkelaker et al. 1989; Gardner et al. 1983; Watt and Evans 1999) probably due to their economic value. Therefore it remained to be determined if different species varied in the compounds exuded from their cluster roots under comparable experimental conditions. Roelofs et al. (2001) addressed this question by studying a range of Australian *Banksia, Hakea* and *Dryandra* species (Proteaceae) for the carboxylates released. The relative contributions of different carboxylates did vary between species, although certain compounds (e.g. malate, malonate, lactate, citrate and *trans*-aconitate) made up a large proportion of the total carboxylates released. Unlike lupins however, no proton release accompanied the release of the carboxylates in these species (Roelofs et al. 2001). Similarly, Denton et al. (2007) found that widespread *Banksia* species did not exhibit greater plasticity in the composition of exuded carboxylates compared to rare *Banksia* species. Thus, rarity could not be explained by limited phenotypic adjustment to carboxylate exudation. Moreover, when the widespread species were grown on the soil in which the rare species occurred, the widespread species did not exude the same composition of carboxylates as that by the rare species. This is important as it argues against a particular carboxylate composition being optimal for maximising phosphorus acquisition from a particular soil.

In sedges (Cyperaceae) another specialised root morphology has been recognised: dauciform roots (after the Latin *daucus* for carrot; Lamont 1974). Dauciform roots can vary in size and shape among plant genera but tend to be short lateral roots with dense clusters of long root hairs, cone or 'carrot-like' shaped and are short-lived (Shane et al. 2005). Although Lamont (1974) noted up to forty differences in the morphological and anatomical characters of dauciform and cluster roots, and while cluster roots are associated with dense clusters of *roots* while dauciform roots are associated with dense mats of root *hairs*, these two root forms are actually functionally similar. Both release large amounts of carboxylates (notably citrate) and at similar rates from mature roots, in both the composition of carboxylates released alters with developmental stage, both release APases, both are short lived and both are induced at low plant P and suppressed at high plant P levels (Playsted et al. 2006; Shane et al. 2006). The distribution of the Cyperaceae is much more widespread than that of cluster rooted species but the extent to which Cyperaceae rely on dauciform roots in different habitats remains to be addressed.

Root response to biotic and abiotic stress

Root response to micro-organisms

Though root systems are genetically determined (see 'Physiological and genetic determinants of root growth and architecture' above) they can be strongly influenced by a wide range of abiotic and biotic factors, including the presence of soil micro-organisms. In some cases the effect upon root growth and morphogenesis are clearly evident with the formation of visible novel organs (e.g. root nodules in the *Rhizobium*-legume symbioses), while in others the impact upon roots are much less evident (e.g. following colonisation by AMF or rhizobacteria)..

Pathogenic fungi

Pathogenic fungi reduce plant growth (see other chapters in this volume), and affect root architecture. In tomato plants infected by *Rhizoctonia solani*, the root system is characterized by the scarcity of short adventitious roots and the emergence of many short laterals, leading to a more branched root system (Berta et al. 2005; Sampò et al. 2007) and a reduction of mitotically active apices (Fusconi et al. 1999). A higher degree of branching has also been reported for tomato roots infected with *Phytophthora nicotianae* var. *parasitica* (Trotta el al. 1996). *R. solani* and *Pithium sp.* have been shown to induce a more monopodial type of branching pattern, with fewer orders of branching than uninfected controls in tomato and *Medicago sativa*, respectively (Larkin 1995; Sampò et al. 2007). These modifications, which result in a root system that is inefficient both in nutrient capture and transport (Fitter and Stickland 1991), may be explained by hormonal imbalance.

Cyanobacteria-invaded coralloid roots

These roots are typical of Cycads, a very ancient group of Gymnosperms naturally distributed over four of the five continents with the exception of Europe, although here they are cultivated for ornamental purposes. Coralloid roots are characterized by dichotomous branching, forming coral-like shapes. Their development begins with the formation of young roots named precoralloids that, when mature, are invaded by cyanobacteria located between the cells of the root cortex (Douglas 2002). New precoralloid roots originate at the basis of preexisting ones (Ahern and Staff 1994).

Coralloid roots infected by cyanobacteria are surrounded by a pronounced layer of mucilaginous material, where cyanobacteria occur as short hormogonial filaments, the infective units of the *Nostoc* species involved in the symbiotic association (Grilli Caiola 2002). Hormogonia penetrate the roots through breaks in the dermal layer and reach, through a cortical channel, the cyanobacterial zone, which is the structural and physiological site of the Cycas-Cyanobacteria symbiosis. In this zone, even though the cyanobionts contain photosynthetic pigments, they behave as mixotrophic in terms of their carbon source, while they perform nitrogen fixation in specialized cells (i.e. heterocysts), where the nitrogenase enzyme is synthesized and protected (Grilli Caiola 2002).

Legume nodules

Nodules are specialized root organs in which symbiotic bacteria (Rhizobia) are able to convert atmospheric nitrogen into ammonia as a nitrogen source (see also Biological nitrogen fixation section in this volume). Establishment of the *Rhizobium*-legume symbiosis depends on a molecular dialogue: flavonoids excreted by host plant roots induce the expression of bacterial nod genes, which encode protein involved in the synthesis and excretion of specific lipochitooligosaccharide signalling molecules called Nod factors, that in turn are recognised by host legumes (Amor et al. 2003; Sadowsky 2005). Responses to Nod factors by root hairs include: altered ion fluxes and plasma membrane depolarisation, calcium spiking, root hair deformation (due to changes in the actin cytoskeleton), and early nodulin gene expression. In cortical cells, Nod factors induce nodulin gene expression and cell division leading to nodule primordium formation (Amor et al. 2003).

In legumes Rhizobia induce two types of nodules: determinate and indeterminate (Douglas, 2002). The latter are the most commonly formed on temperate legumes by *Rhizobium* species, while the former are induced by *Bradyrhizobium* (the name is associated with the slow growth of these bacteria) on tropical legumes. Nodule formation is a multistep process. Rhizobia move to roots by positive chemotaxis in response to root exudates. The bacteria then infect the host roots via root hairs, or via wounds and lesions, or through spaces occurring around root primordia or adventitious roots. In the case of root-hair infection, the attachment of Rhizobia leads to root hair curling, Rhizobia then enter the root by invagination of the plasma membrane, and induce formation of an infection thread, a growing tube with cell wall material filled with growing bacteria (Sadowsky 2005). Rhizobia move down the infection thread towards the root cortex, where cell division leads to the production of nodule primordia, functioning as a meristem (Amor et al. 2003). The primordium gives rise to the mature nodule, in which meristem, invasion, infected and degenerative zones occur. The bacteria are released (but are always surrounded by the host plasma membrane) in the invasion zone, in cells which elongate and increase in size while bacteria proliferate and differentiate into the active form (bacterioids) (Nardon and Charles 2002). In pea, the genes controlling nodule morphogenesis have been identified for plant tissue colonization and differentiation of bacteria into bacterioids and for the development of nodule tissue (Borisov et al. 2007). The genetic system controlling nitrogen-fixing symbiosis development in plants likely evolved from the system controlling AM symbiosis (Borisov et al. 2007).

Rhizobia obtain carbon compounds, especially malate, from their host and in turn perform nitrogen fixation at the centre of the nodule in a microaerobic zone surrounded by a layer of very closely packed cells with few air spaces. In this zone, the nitrogenase enzyme can catalyze the reaction of nitrogen fixation without being inactivated by oxygen (Douglas 2002). Rhizobia respire aerobically which is possible due to a plant pigment, leghemoglobin, that binds oxygen with high affinity, so carrying it to bacterial cells. When symbiotic interactions are efficient, the result is a pink root system, due to the presence of this pigment (Andersson et al. 1996; Kawashima et al. 2001).

Ectomycorrhizal (ECM) roots

ECM are formed by mutualistic interactions between soil fungi and the roots of woody plants (see also the section on Rhizosphere micro-organisms, this volume). In ECM plants, roots usually have very few, or no, root hairs and tend to be short and thick. The root apices are generally covered with a fungal mantle (sheath), from which hyphae extend into the soil, and become mitotically inactive (Clowes 1981). Within the root, hyphae always remain apoplastic and can colonize the epidermal (angiosperms) and the cortical cell (gymnosperms) layers, forming the Hartig net, a complex branched structure, which mediates nutrient transfer between fungus and plant (Smith and Read 2008). The most evident root modifications are the early formation of lateral roots and a dichotomy of the apical meristems in a number of species. External hyphae extend out of the depletion root zones, to explore the soil substrate and are responsible for the nutrient capture and water uptake of the symbiotic tissues. ECM fungi provide phosphate, the most immobile ion in the soil, and ammonium and in return they receive plant carbonic compounds (Harley and Smith 1983). The fungi may also acquire nutrients from more complex organic substrates, but large differences between different species and even among strains of the same fungus exist in accessing these complex substrates (Read and Perez-Moreno 2003; Hodge et al. 1995). The structure of ECM is essentially determined by the fungal species rather than the host plant, however there is considerable variation in the degree of host specificity among species and even among strains of ECM fungi (Le Quéré et al. 2004).

Arbuscular mycorrhizal (AM) associations

AM are symbioses between soil fungi (order Glomeromycota) and the roots of most terrestrial plants (see also Rhizosphere micro-organisms section, this volume). Colonization of the root system by the fungal partner is restricted to parenchymatous cells of the cortex, where it grows both intra- and inter-cellularly. Intracellular haustoria invaginate the host protoplast and form structures known as arbuscules. These finely branched hyphae are surrounded by a metabolically active plant cytoplasm, bounded by a plant membrane; reciprocal exchanges of metabolites are thought to occur at this interface (Smith and Read 2008).

It was previously believed that AM fungi did not influence the development of the root system and that growth of the root apices was normal (Harley and Smith 1983). However, developmental and topological approaches have now revealed that AMF do influence the root system structure of their host and that these effects occur in both monocotyledon and dicotyledon plants. Developmental approaches describe the root system based on chronology of development, where higher order laterals develop from a lateral of lower order whereas topological analysis is based on graph theory, a mathematical approach that describes branch structure (see *Root architecture* section above). The latter, applied to root systems by Fitter (1985), allows a branching analysis not confined to just the degree of branching, but extended to the root arrangement or pattern: i.e. two root systems with the same degree of branching can differ in their structure (from herringbone-like to dichotomous). Berta et al. (1990) demonstrated increased branching of secondary roots of leek, and an overall longer root system when colonized by the AMF *Glomus* sp. (strain E3), however, although single roots were shorter than those from uncolonised plants, no evidence was found for an altered topology. Similarly, Schellenbaum et al. (1991) also reported increased branching of *Vitis vinifera* roots when colonised by the AMF *G. fasciculatum*, with greater effects on higher order laterals, whose length, contrary to the resuts from leek, also increased. Effects on number and length of laterals, rather than of primary roots, have been also shown in another woody plant, *Platanus acerifolia* (Tisserant et al. 1996). In contrast to the data on leek, both these latter studies showed evidence of altered topology, with AM *V. vinifera* root systems exhibiting a more random pattern and *P. acerifolia* a more dichotomous one, thus suggesting a higher capability for nutrient adsorption (Fitter 1985; Tisserant et al. 1992). Other studies also support the finding that AMF colonization effects root system development generally by an increase in root branching (Berta et al. 1995, 2005; Citernesi et al. 1998) (Fig. 1 b,c-d) which in turn will increase the surface area for nutrient absorption although this response may not be universal (Forbes et al. 1996; Vigo et al. 2000).

The root system depends on variations in the structure and activity of its apices. All the factors that directly or indirectly affect root morphogenesis, affect apical meristem activity. Mycorrhizal colonisation is one of these factors, but its impact upon root meristem structure and activity has received little attention (see Berta et al. 2002). Studies on leek show that adventitious roots when colonized by AMF have a determined pattern of growth. In the colonised tissue vacuolation and differentiation of stele and cortex tissues occur close to the apical tissues, thus root apices display a pattern of senescence and/or necrosis more rapidly than in non-colonized roots, (Berta et al. 1990; Varney and McCully 1991b), and the mitotic cycle of meristematic cells become longer with increasing colonization (Berta et al. 1991). Another effect of AM colonization is an increase of the diameter of root apices, leading to thicker roots (Berta et al. 2002). Root growth inhibition observed in *A. porrum* and in maize, however, is not a general characteristic of mycorrhizal roots and it does not occur in *M. truncatula* (Oláh et al. 2005), hence, it may be related to a specific strategy of growth (or of resource acquisition).

The modifications described, especially an increase in root branching, lead to an increased volume of tissue available for AMF colonisation (Fusconi et al. 1999), but could also potentially result in higher susceptibility to pathogens. However, in strawberry plants infected by *Phytophthora fragariae*, disease was found to decrease as the degree of root branching increased (Norman et al. 1996), as a consequence of different root exudates from AM colonised roots that stimulated the production of sporangia by the pathogen less than exudates from non mycorrhizal strawberry plants (Norman and Hooker, 2000). Studies with *Lycopersicon esculentum* have suggested a similar involvement of AM root exudates in the biocontrol of the pathogen *P. nicotianae* (Vigo et al. 2000).

One of the main benefit of AMF on plants is the increased uptake of nutrients, particularly phosphate, and, as a consequence, the P status of the plant tissues is higher than in non-AM plants grown on the same medium (Fusconi et al. 2005). The tissue P content may be, by itself, a cause of growth modification (Fusconi et al. 1994, 2000). Morphogenetic effects of AMF on the root system, dependent or not by P, are probably mediated by changes in hormone synthesis, transport or sensitivity. Altered levels of auxin (Kaldorf and Ludwig-Müller 2000; Torelli et al. 2000), cytokinin (Barker and Tagu 2000; Torelli et al. 2000), ethylene (Vierheilig et al. 1994) and gibberellins (Shaul-Keinan et al. 2002) have all been reported in mycorrhizal plants. However, other more specific aspects of the mycorrhizal interaction may be influenced by hormones. For example, the lengthening of the mitotic cycle reported for mycorrhizal root apices of *A. porrum*, may be related to an increase of abscisic acid, known to have an inhibitory effect on the progression of the cell cycle, and which was found at high levels in AM roots of *Zea mays* (Danneberg et al. 1992) and also in extraradical hyphae of two *Glomus* sp. isolates (Esch et al. 1994). Other morphogenetic factors that have been observed in AM roots include hyperpolarization of the root cortical cells (Fieschi et al. 1992).

More recently, studies using the model plant *Medicago truncatula* have demonstrated that AMF effect root morphogenesis through an increase in lateral root number at a very early stage of the root-fungus interaction before hyphae have established contact with roots, suggesting that secretion of diffusible factors may be involved (Oláh et al. 2005). Lateral root initiation is triggered by the plant hormone auxin (De Smet et al. 2006b), but in *M. truncatula* the diffusible AMF factors are unlikely to be an auxin-like substance because of the different morphogenetic effect following auxin administration. Moreover, Nod factors stimulate lateral root formation in a way similar to that of AMF, suggesting that these molecular signals may have a general role as plant growth regulators, in addition to their role in the nodulation process (Oláh et al. 2005).

The form of a root system is a consequence not only of root development but also of root mortality. How long a root lives (i.e. root longevity) is important both in driving the final form of the root system, and impacting upon nutrient and carbon fluxes between plants and the soil (Berta et al. 2002). The precocious inactivation and loss of the meristem of the AM root apices of *A. porrum* does not lead to the senescence of the differentiated, colonized root tissues. On the contrary, in these tissues senescence is delayed (Lingua et al. 1999). In contrast, in tree roots AMF promote senescence of the fine rootlets and, consequently, their turnover, so making a significant contribution to biogeochemical cycling (Atkinson et al. 1994). The different fungal effects on root longevity could be related to root function, as *A. porrum* adventitious roots are primarily engaged in exploration, whilst higher order roots of trees are primarily engaged in absorption. These two strategies (delayed senescence and root turnover) enable AM fungi to accomplish the same result, namely a more vital and efficient root system.

Plant growth promoting rhizobacteria (PGPR)

Root system structure is also influenced by other beneficial soil micro-organisms such as the root colonizing "Plant Growth Promoting Rhizobacteria (PGPR)" (see sections on Plant growth promotion and beneficial bacteria, elsewhere in this volume). PGPR's affect root architecture by increasing total root length and branching as a consequence of hormone production and improved plant mineral nutrition (Gamalero et al. 2002, 2004; Lemanceau et al. 2005). In turn, these changes to the root system architecture likely impact upon microbial dynamics in the rhizosphere through altered rhizodeposition including changes in signalling molecules released (Bais et al. 2006; Jones et al. 2004)

To summarise, the information presented in this section supports the idea of root systems as very plastic structures, influenced by the soil microorganisms, yet also able to influence them. For example, in coralloid roots, Cycas are able to regulate the growth of cyanobacteria outside the root by inducing hormogonia, and also inside the root to promote conditions for maximum nitrogen fixation (Grilli Caiola 2002); plants control the extent of AM colonization (Vierheilig 2004) while hyphal density and anastomosis rate also show a plant species effect (Giovannetti et al. 2004). On the other hand, plants with different root system structure and physiology, are differently dependent upon mycorrhization (Azcón and Ocampo 1981; Fitter 2004) while, in turn, colonisation by mycorrhizal fungi may impact upon root system structure (Berta et al. 2005).

Root response to abiotic stress

Soil is the most complex of all environments containing liquid, gaseous and solid phases, the ratios of which can change depending on the prevailing conditions. Thus, this complexity presents several challenges to the growing root and in this section some of the abiotic stresses that impact upon the root and the roots response to these stresses will be discussed. In particular, we will focus upon mechanical impedance, water stress and poor aeration which can all severely limit root growth in soil. The root response to poor aeration will be considered in detail and the impacts on global processes as well as microbial activity and nutrient cycling discussed.

Mechanical impedance

Roots are subject to mechanical impedance when the force required to displace soil particles as the root grows increases. As a result, root diameter behind the root tip increases and root elongation decreases with increasing soil strength. Materechera et al. (1991) reported that dicotyledons (with large diameters) penetrated strong soils more than graminaceous monocotyledons (with smaller diameters), and from this suggested that dicotyledonous species may exert greater axial growth pressures than monocotyledonous species. However, in studies where maximum axial root growth pressure (σ_{max}) has been directly measured, no relation between the σ_{max} and the ability to penetrate strong soil has been found (Whalley and Dexter 1993), including in studies where monocotyledons and dicotyledonous species have been directly compared (Clark and Barraclough 1999). Thus, the ability to penetrate strong soils may instead be directly related to root diameter (see Materechera et al. 1992; Clark et al. 2003). Changes in root elongation and diameter due to mechanical impedance are likely controlled by ethylene production (He et al. 1996; Sarquis et al. 1991). Sarquis et al. (1991) demonstrated that evolution of ethylene increased within 1 h following application of mechanical impedance to roots. Furthermore, addition of various inhibitors of ethylene action and synthesis helped restore root elongation rates in these mechanically impeded roots, highlighting the importance of ethylene in control of the root response to mechanical impedance (Sarquis et al. 1991).

Water stress

Generally, water is believed to be available for plant uptake at matric potentials greater than -1.5 MPa (the wilting point of many mesophytic plants), but root growth can be severely slowed by potentials greater than this value (Bengough 2003). Schenk and Jackson (2002a) surveyed the literature on root system sizes for individual plants from deserts, scrublands, grasslands and savannas all with ≤ 1,000 mm mean annual precipitation. They found that maximum rooting depth showed a strong positive relationship with mean annual precipitation for all plant growth forms, except shrubs and trees. Moreover, maximum rooting depth for all growth forms tended to be shallowest in arid regions and deepest in subhumid regions, which was thought to be the result of more restricted water infiltration depths in areas with lower precipitation. These results appear to contradict the widely held view that rooting depth increases in drier environments. However, as Schenk and Jackson (2002a) state, the distinction between *relative* and *absolute* rooting depths is critical. Thus, for a given canopy size herbaceous plants do have deeper maximum rooting depths in drier environments, but as canopy size increases so too does the absolute rooting depth. However, the relationship is not simply due to increased plant size as above– and below–ground allometrics also change with climate. Moreover, depths at which plants have 50% or 95% of their *total* root biomass are significantly deeper in drier than in humid environments (Schenk and Jackson 2002b). Water availability is not the only abiotic factor that influences rooting depth, soil texture, organic horizon size (Schenk and Jackson 2002b) and plant species composition will also dictate the rooting depth achieved.

Soil aeration and aerenchyma formation

Excess water leading to waterlogging is also problematic for the root as this can lead to hypoxia (low oxygen). In response to such stress many plants (but not all i.e. *Brassica napus*; Voesenek et al. 1999) can form aerenchyma, enlarged gas spaces in tissues which exceed those found as intracellular spaces. Aerenchyma form either by cell separation ('schizogenous' aerenchyma) or by cell death and removal ('lysigenous' aerenchyma); more is known about formation of the latter as it occurs in many crop species including barley, rice and wheat (Drew et al. 2000; Evans 2003; Justin and Armstrong 1991). Aerenchyma formation may be induced by elevated ethylene concentrations (Fig. 6; Colmer et al. 2006; Drew et al. 2000; but see also Visser and Bögemann 2006). As noted above ethylene also is important in the roots response to mechanical impedance. However different mechanisms may be involved in the promotive effects of these two stresses as while hypoxia and mechanical impedance had strong synergistic effects on ethylene production actual aerenchyma formation did not show a synergistic effect (He et al. 1996). Aerenchyma allow gas (notably oxygen) to be transported from the shoot to the root either via simple diffusion, or possibly under pressure flow (Jackson and Armstrong 1999). Aerenchyma are therefore essential for the internal ventilation of roots in an anaerobic environment. However, there may be a trade-off between root mechanical strength and aerenchyma formation particularly in plant species that lack a multiseriate ring of tissue for mechanical protection in the outer cortex (Striker et al. 2007).

The movement of oxygen from the shoots to the roots also has important consequences for rhizosphere micro-organisms, as some oxygen is released from the root (termed radical oxygen loss (ROL)) into the zone immediately surrounding the root. The resulting "oxidation zone" (Fig. 6) has been estimated to extend 1– 3 mm from the root (Van Bodegom et al. 2001) although its size will depend upon oxygen supply versus consumption and the redox buffering capacity of the soil. The plant species present is also a factor as many wetland plant species are capable of forming a barrier to ROL in root basal areas, thus ensuring oxygen reaches the root apex allowing continued root growth (Armstrong et al. 1991; McDonald et al. 2002). In contrast, in non-wetland species oxygen may be more readily lost from the root, as shown for wheat (Thomson et al. 1992) and barley (McDonald et al. 2001). Even in wetland species, however, this barrier to ROL may be more 'leaky' in some species than others, and monocotyledons may produce a tighter barrier to oxygen loss than dicotyledonous species (Laan et al. 1989; Visser et al. 2000). The type of aerenchyma that form may be important in determining how effective the barrier to ROL has to be in order to ensure the root tip receives adequate oxygen. Visser et al. (2000) suggested that the barrier to ROL needs to be more effective in species with lysigenous aerenchyma because porosity increases for a considerable length behind the tip (2–6 cm), whereas in species with schizogenous aerenchyma high porosity is achieved much nearer the tip (i.e. within 0.5–1 cm). In addition, ROL barrier formation may be related to sediment type, redox conditions and microbial oxygen demand in the plants habitat (Smits et al. 1990; Laskov et al.

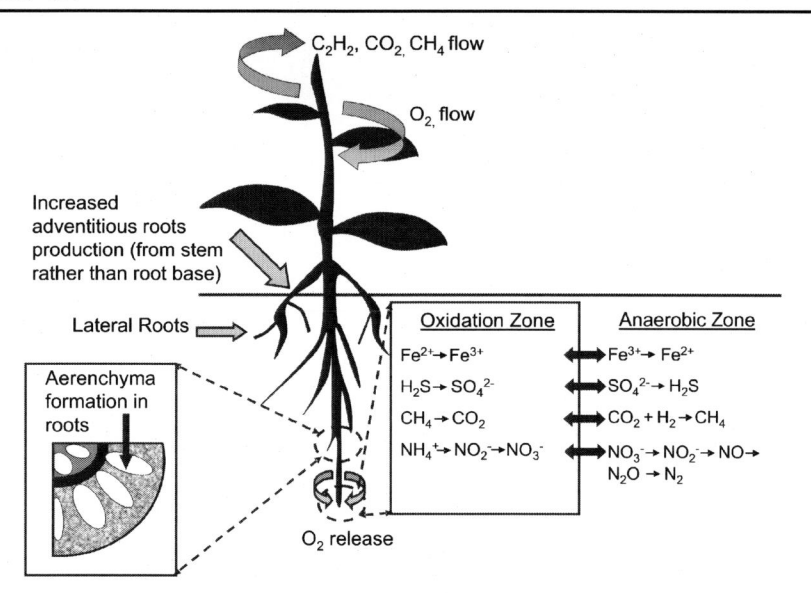

Fig. 6 Some responses of the plant to oxygen stress caused by waterlogging in the soil. Aerenchyma (enlarged gas spaces in plant tissues) are formed by cell separation or death and subsequent removal of cellular material. Oxygen (O_2) is transported from above ground organs to the roots while ethylene (C_2H_2), carbon dioxide (CO_2) and methane (CH_4) from the soil-root environment flow in the opposite direction. Plants can also increase production of adventitious roots near the surface of the soil where O_2 concentrations may be less limiting. Some O_2 will also leaked to the soil resulting in an oxidation zone around the root. The size of this oxidation zone depends on the extent and effectiveness of any radical oxygen barrier produced in the roots which may prevent oxygen loss in the root basal zones but not at the root tips

2006), and upon the type and age of the root; ROL in young adventitious and fine lateral roots can be rapid whereas in older parts of adventitious roots and rhizomes no ROL occurs (Armstrong and Armstrong 1988). The ROL barrier may be constitutively produced in some plant species (Visser et al. 2000) but induced under stagnant, deoxygenated conditions in others (Colmer 2003; Insalud et al. 2006; but see McDonald et al. 2002). How the barrier to ROL actually forms is not known but in rice plants a layer of sclerenchymatous fibres with thick secondary walls on the external side of the cortex forms (Clark and Harris 1981), which may, together with adjacent cell layers, become impregnated with phenolic compounds or lipid complexes to form a barrier to ROL (Insalud et al. 2006).

Consequences for nutrient uptake in oxygen-deficient environments

The formation of a barrier to ROL may also be expected to interfere with nutrient uptake and translocation of water and ions (Clark and Harris 1981). Evidence suggests however, that the barrier to ROL does not affect root hydraulic conductivity i.e. water uptake ability (Garthwaite et al. 2006; Ranathunge et al. 2003) except when plants are also under P stress (Fan et al. 2007). Thus, in O_2-deficient environments there may be a trade-off between root morphology and physiology to balance the need for nutrient and water capture and retain adequate oxygen supply to the root system for root function. Using modelling approaches Kirk (2003) showed that a system of coarse, aerenchmymatous primary roots with gas-impermeable walls allowing oxygen flow to short, fine, gas-permeable lateral roots was the best compromise between the need for internal aeration and maximum absorbing surface per unit root mass, which, in fact is, the basic architecture of many wetland root systems.

Some oxygen loss to the rhizosphere may actually be beneficial to the plant as it may stimulate oxygen requiring nitrifying bacteria and prevent reduction of toxic ions (e.g. Fe^{3+} to Fe^{2+}) in the immediate vicinity of the root. Kirk (2003) estimated that the oxygen lost to the rhizosphere via fine lateral roots was enough to allow sufficient nitrification to satisfy half the plant's N demand as absorbed NO_3^-. More studies have been conducted investigating the influence of low oxygen concentrations on P uptake but the results of these investigations are conflicting. Lorenzen et al. (2001) found that under low oxygen concentration plants adjusted root length less in response to low P than plants in aerated solution but that this did not adversely impact upon nutrient capture. Insalud et al. (2006) found although numbers of adventitious roots, root dry mass and root-to-shoot ratios increased in rice plants 4 d after transferral to stagnant conditions, maximum root length and shoot growth declined as did relative P uptake, at least initially, under low P conditions. In contrast, Rubio et al. (1997) found that while waterlogging (thus reduced O_2 supply) did reduce root-to-shoot ratios of *Paspalum dilatatum*, P concentrations were not reduced. Thus, P uptake per unit of root biomass must have increased under waterlogging conditions which was due to production of finer roots with higher physiological capacity to absorb P as well as an increase in the availability of P in the waterlogged soil (Rubio et al. 1997). The study by Rubio et al. (1997) was the only study of the three discussed above which used soil. Studies in stagnant solution culture enable the responses of roots to be easily studied but they do not reflect the complex nutrient dynamics that occur in the soil matrix. Saleque and Kirk (1995) proposed O_2 release from roots results in Fe^{2+} oxidation which, due to the release of H^+, acidifies the rhizosphere sufficiently to increase the pool of available P. Under P deficiency, *plant* release of H^+ also increased due to uptake of NH_4^+ rather than NO_3^- in O_2-stressed plants resulting in an additional mechanism to acidify the rhizosphere and potentially increase P solubilization (Kirk and Du Le Van 1997). Thus, studies which do not use soil as the growth medium will not have the same spatial-temporal P release dynamics and consequently may underestimate the capture of P under oxygen deficient conditions.

Aerenchyma and soil micro-organisms

Oxygen release by aerenchymatous plants also has important impacts on nutrient cycling in flooded soils (Fig. 6) via impacts on the bacterial community (Bodelier and Frenzel 1999; Reddy et al. 1989). The stimulation of nitrifying bacteria in the oxidation zone may positively impact on the plant's nutrition by releasing NO_3^- from the soil/sediment for root uptake, although some of the converted NO_3^- may diffuse into anaerobic areas and be lost via denitrification with the aerenchyma facilitating the release of the denitrified gases into the atmosphere (Reddy et al. 1989). Moreover, some wetland species such as rice have been shown to preferentially uptake NH_4^+ rather than NO_3^- (see Reddy et al. 1989), which may further exasperate the loss of N from the plant-soil system via the denitrification pathway. Many wetland plant species are also AM (Cooke and Lefor 1998; Bauer et al. 2003), although colonisation levels and arbuscule frequency may be more depressed than in non-waterlogged conditions (Mendoza et al. 2005). The benefit, in terms of enhanced P capture may also vary with plant-fungal combination (see previous sections, and elsewhere in this volume). In a study on AM colonisation of fens Cornwell et al. (2001) found that nine out of ten dicotyledonous species examined formed AMs while the six monocotyledonous species studied were generally non-mycorrhizal. Cornwell et al. (2001) attributed these differences to the extent of aerenchyma formation in these plants. In the monocotyledonous species well developed aerenchyma formed which would be able to oxidise the rhizosphere promoting phosphorus mineralisation and its subsequent availability to the plant, thus rendering these species less reliant on AM fungi to enhance P capture. However, Šraj-Kržič et al. (2006) found no relationship between AM colonisation and aerenchyma ratio in their study on various emerged and submerged plant species, but did find positive correlations between AM colonisation frequency and intensity and plant available phosphorous. As AM fungi are known to be multi-functional (Newsham et al. 1995), AM colonisation may confer benefits to their host other than simply enhanced P capture under waterlogged conditions.

Aerenchyma formation and methane emissions

In addition to allowing the flow of oxygen from shoots to roots and the surrounding soil, aerenchyma also act as important conduits of gases such as carbon dioxide, ethylene and methane in the opposite direction (Fig. 6). Methane (CH_4) emission in particular has important consequences for global carbon cycle and global warming. The International Panel on Climate Change (IPCC) report that CH_4 is *c.* 21 times more effective at trapping heat in the earth's atmosphere than CO_2 over a 100 year period. Natural and agricultural wetlands contribute *c.* 40–50% of the total methane emitted to the atmosphere on a global basis and a positive correlation between CH_4 emission and net ecosystem production for a range of global wetlands has been found (Whiting and Chanton 1993). Moreover, c. 3% of daily net ecosystem productivity can be emitted directly back to the atmosphere as CH_4 (Whiting and Chanton 1993). Thus, CH_4 efflux through wetland species can be considerable, with values of 64–90% of the total CH_4 efflux lost through this pathway (Butterbach-Bahl et al. 1997; Shannon et al. 1996). Higher efflux of CH_4 in the presence of aerenchymatous plants has been attributed to: (i) an increased supply of substrates through rhizodeposition and litter (Whiting and Chanton 1993), although this may vary with plant species (Shannon et al. 1996; Yavitt and Lang 1990) and variety (Wang and Adachi 2000) present; (ii) a reduction in CH_4 oxidation by the plant release pathway allowing the CH_4 to avoid aerobic zones containing methane oxidising bacteria (Shannon and White 1994) and (iii) ease of transport from the site of production to the atmosphere via the aerenchyma. Specifically, CH_4 dissolved in the water surrounding the root is thought to diffuse into the cell-wall water of the root cells, gasify in the root cortex and passes into the aerenchyma until it is released from the micropores in leaf sheaths (Nouchi et al. 1990). As global rice production is forecast to increase than so likely will CH_4 emissions but careful selection of cultivars can help reduce this environmental problem (see Butterbach-Bahl et al. 1997).

To conclude, growth in soil can be problematic for the root. Although traditionally scientists have tended to tackle the problems that roots experience one at a time, in reality the root experiences a wide range of stresses simultaneously. Moreover, different parts of the root system may well be experiencing different stresses. A better understanding of how roots cope with multiple stresses, though scientifically more challenging, is now required in order for plant responses and plasticity be fully exploited and manipulated to enhance, or even to try and sustain, crop production in the future given the challenges that lie ahead with decreased rock phosphate stocks and global climate change. The impacts of root 'stress' responses upon microbial community structure and function also have considerable potential for manipulation but here again a better understanding of soil biology is required first.

Acknowledgements AH thanks Alastair Fitter, Anna Armstrong and Deirdre Rooney for comments on the '*Root Function*' and '*Root Response to Abiotic Stress*' sections. We also thank three anonymous referees for their comments which helped greatly improve the manuscript.

References

Adams MA, Pate JS (1992) Availability of organic and inorganic forms of phosphorus to lupins (*Lupinus* spp.). Plant Soil 145:107–113 doi:10.1007/BF00009546

Ahern CP, Staff IA (1994) Symbiosis in cycads: the origin and development of coralloid roots in *Macrozamia communis* (Cycadaceae). Am J Bot 81:1559–1570 doi:10.2307/2445333

Aida M, Beis D, Heidstra R, Willemsen V, Blilou I, Galinha C, Nussaume L, Noh YS, Amasino R, Scheres B (2004) The PLETHORA genes mediate patterning of the *Arabidopsis* root stem cell niche. Cell 119:109–120 doi:10.1016/j.cell.2004.09.018

Aloni R, Aloni E, Langhans M, Ullrich CI (2006) Role of cytokinin and auxin in shaping root architecture: regulating vascular differentiation, lateral root initiation, root apical dominance and root gravitropism. Ann Bot (Lond) 97:883–893 doi:10.1093/aob/mcl027

Amor BB, Sidney L, Shaw SL, Oldroyd GED, Maillet F, Penmetsa RV, Cook D, Long SR, Dénarié J, Gough C (2003) The NFP locus of *Medicago truncatula* controls an early step of Nod factor signal transduction upstream of a rapid calcium flux and root hair deformation. Plant J 34:495–506 doi:10.1046/j.1365-313X.2003.01743.x

Andersson CR, Jensen EO, Llewellyn DJ, Dennis ES, Peacock WJ (1996) A new hemoglobin gene from soybean: a role for hemoglobin in all plants. Proc Natl Acad Sci USA 93:5682–5687 doi:10.1073/pnas.93.12.5682

Armstrong J, Armstrong W (1988) *Phragmites australis*–a preliminary study of soil-oxidizing sites and internal gas transport pathways. New Phytol 108:373–382 doi:10.1111/j.1469-8137.1988.tb04177.x

Armstrong W, Justin SHFW, Beckett PM, Lythe S (1991) Root adaptation to soil waterlogging. Aquat Bot 39:57–73 doi:10.1016/0304-3770(91)90022-W

Atkinson D, Berta G, Hooker JE (1994) Impact of root colonization in root architecture, root longevity and the formation of growth regulators. In: Gianinazzi S, Schüepp H (eds) Impact of arbuscular mycorrhizas on sustainable agriculture and natural ecosystem. Birkhäuser Verlag, Basel, pp 89–99

Azcón R, Ocampo JA (1981) Factors affecting the vesicular-arbuscular infection and mycorrhizal dependency of thirteen wheat cultivar. New Phytol 87:677–685 doi:10.1111/j.1469-8137.1981.tb01702.x

Bais HP, Weir TL, Perry LG, Gilroy S, Vivanco JM (2006) The role of root exudates in rhizosphere interactions with plants and other organisms. Annu Rev Plant Biol 57:233–266 doi:10.1146/annurev.arplant.57.032905.105159

Barker S, Tagu D (2000) The roles of auxins and cytokinins in mycorrhizal symbioses. J Plant Growth Regul 19:144–154

Bauer CR, Kellogg CH, Bridgham SD, Lamberti GA (2003) Mycorrhizal colonisation across hydrologic gradients in restored and reference fresh-water wetlands. Wetlands 23:961–968 doi:10.1672/0277-5212(2003)023[0961: MCAHGI]2.0.CO;2

Benfey PN, Scheres B (2000) Root development. Curr Biol 10: R813–R815 doi:10.1016/S0960-9822(00)00814-9

Benjamins R, Scheres B (2008) Auxin: the looping star in plant development. Annu Rev Plant Biol 59:443–465 doi:10.1146/annurev.arplant.58.032806.103805

Bengough AG (2003) Root growth and function in relation to soil structure, composition, and strength. In: de Kroon H, Visser EJW (eds) Ecological studies, vol. 168: root ecology. Springer-Verlag, Berlin-Heidelberg, pp 151–171

Benkova E, Michniewicz M, Sauer M, Teichmann T, Seifertova D, Jurgens G, Friml J (2003) Local, efflux-dependent auxin gradients as a common module for plant organ formation. Cell 115:591–602 doi:10.1016/S0092-8674(03) 00924-3

Bernston GM (1994) Modelling root architecture: are there tradeoffs between efficiency and potential of resource acquisition? New Phytol 127:483–493 doi:10.1111/j.1469-8137.1994.tb03966.x

Berta G, Fusconi A, Trotta A, Scannerini S (1990) Morphogenetic modifications induced by the mycorrhizal fungus *Glomus* strain E3 in the root system of *Allium porrum* L. New Phytol 114:207–215 doi:10.1111/j.1469-8137.1990. tb00392.x

Berta G, Tagliasacchi AM, Fusconi A, Gerlero D, Trotta A, Scannerini S (1991) The mitotic cycle in root apical meristems of *Allium porrum* L. is controlled by the endomycorrhizal fungus *Glomus* sp. strain E_3. Protoplasma 161:12–16 doi:10.1007/BF01328892

Berta G, Trotta A, Fusconi A, Hooker J, Munro M, Atkinson D, Giovannetti M, Marini S, Fortuna P, Tisserant B, Gianinazzi-Pearson V, Gianinazzi S (1995) Arbuscular mycorrhizal induced changes to plant growth and root system morphology in *Prunus cerasifera* L. Tree Physiol 15:281–293

Berta G, Fusconi A, Hooker J (2002) Arbuscular mycorrhizal modifications to plant root systems: scale, mechanisms and consequences. In: Gianinazzi S, Schuepp H, Barea JM, Haselwandter K (eds) Mycorrhizal technology in agriculture. Birkhäuser Verlag, Basel, pp 71–86

Berta G, Sampò S, Gamalero E, Massa N, Lemanceau P (2005) Suppression of *Rhizoctonia* root-rot of tomato by *Glomus mosseae* BEG12 and *Pseudomonas fluorescens* A6RI is associated with their effect on the pathogen growth and on root morphogenesis. Eur J Plant Pathol 111:279–288 doi:10.1007/s10658-004-4585-7

Bidel L, Renault P, Pagès L, Rivière LM (2000) An improved method to measure spatial variation in root respiration: application to the taproot of a young peach tree *Prunus persica*. Agronomie 21:179–192 doi:10.1051/agro:2001116

Blilou I, Xu J, Wildwater M, Willemsen V, Paponov I, Friml J, Heidstra R, Aida M, Palme K, Scheres B (2005) The PIN auxin efflux facilitator network controls growth and patterning in *Arabidopsis* roots. Nature 433:39–44 doi:10.1038/nature03184

Bodelier PLE, Frenzel P (1999) Contribution of methanotrophic and nitrifying bacteria to CH_4 and NH_4^+ oxidation in the rhizosphere of rice plants as determined by new methods of discrimination. Appl Environ Microbiol 65:1826–1833

Borisov AY, Danilova TN, Koroleva TA, Kuznetsova EV, Madsen L, Mofett M, Naumkina TS, Nemanki TA, Ovchinnikov ES, Pavlova ZB, Petrova NE, Pinaev AG, Radutoiu S, Rozov SM, Rychagova ST, Shtark OY, Solovov II, Stougaard JS, Tikhonovich IA, Topunov AF, Tsyganov VE, Vasil'chikov AG, Voroshilova VA, Weeden NF, Zhernakov AI, Zhukov VA (2007) Regulatory genes of garden pea (*Pisum sativum* L.) controlling the development of nitrogen-fixing nodules and arbuscular mycorrhiza: a review of basic and applied aspects. Appl Biochem Microbiol 43:237–243 doi:10.1134/S00036 83807030027

Boulet FM, Lambers H (2005) Characterisation of arbuscular mycorrhizal fungi colonisation in cluster roots of *Hakea verrucosa* F. Muell (*Proteaceae*), and its effect on growth and nutrient acquisition in ultramafic soil. Plant Soil 269:357–367 doi:10.1007/s11104-004-0908-x

Braum SM, Helmke PA (1995) White lupin utilizes soil phosphorus that is unavailable to soybean. Plant Soil 176:95–100 doi:10.1007/BF00017679

Bradford MA, Eggers T, Newington JE, Tordoff GM (2007) Soil faunal assemblage composition modifies root ingrowth to plant litter patches. Pedobiologia (Jena) 50:505–513 doi:10.1016/j.pedobi.2006.07.001

Brady SM, Orlando DA, Lee JY, Wang JY, Koch J, Dinneny JR, Mace D, Ohler U, Benfy PN (2007) A high-resolution root spatiotemporal map reveals dominant expression patterns. Science 318:801–806

Butterbach-Bahl K, Papen H, Rennenberg H (1997) Impact of gas transport through rice cultivars on methane emission from paddy fields. Plant Cell Environ 20:1175–1183 doi:10.1046/j.1365-3040.1997.d01-142.x

Campbell BD, Grime JP, Mackey JML (1991) A trade-off between scale and precision in resource foraging. Oecologia 87:532–538 doi:10.1007/BF00320417

Cannon WA (1949) A tentative classification of root sytems. Ecology 30:452–458 doi:10.2307/1932458

Casimiro I, Marchant A, Bhalerao RP, Beeckman T, Dhooge S, Swarup R, Graham N, Inzé D, Sandberg G, Casero PJ, Bennett P (2001) Auxin transport promotes *Arabidopsis* lateral root initiation. Plant Cell 13:843–852

Cavagnaro TR, Smith FA, Smith SE, Jakobsen I (2005) Functional diversity in arbuscular mycorrhizas: exploitation of soil patches with different phosphate enrichment differs among fungal species. Plant Cell Environ 28:642–650 doi:10.1111/j.1365-3040.2005.01310.x

Celenza JL Jr, Grisafi PL, Fink GR (1995) A pathway for lateral root formation in *Arabidopsis thaliana*. Genes Dev 9:2131–2142 doi:10.1101/gad.9.17.2131

Chapman EJ, Carrington JC (2007) Specialization and evolution of endogenous small RNA pathways. Nat Rev Genet 8:884–896 doi:10.1038/nrg2179

Chapin FS III (1980) The mineral nutrition of wild plants. Annu Rev Ecol Syst 11:233–260 doi:10.1146/annurev. es.11.110180.001313

Cheng Y, Dai X, Zhao Y (2007) Auxin synthesized by the YUCCA flavin monooxygenases is essential for embryogenesis and leaf formation in *Arabidopsis*. Plant Cell 19:2430–2439 doi:10.1105/tpc.107.053009

Citernesi AS, Vitagliano C, Giovannetti M (1998) Plant growth and root system morphology of *Olea europea* L. rooted cuttings as influenced by arbuscular mycorrhizas. J Hortic Sci Biotechnol 73:647–654

Clark LH, Harris WH (1981) Observations on the root anatomy of rice (*Oryza sativa* L.). Am J Bot 68:154–161 doi:10.2307/2442846

Clark LJ, Barraclough PB (1999) Do dicotyledons generate greater maximum axial root growth pressures than monocotyledons? J Exp Bot 50:1263–1266 doi:10.1093/jexbot/50.336.1263

Clark LJ, Whalley WR, Barraclough PB (2003) How do roots penetrate strong soil? Plant Soil 255:93–104 doi:10.1023/A:1026140122848

Clarkson DT (1996) Root structure and sites of ion uptake. In: Waisel Y, Eshel A, Kafkafi U (eds) Plant roots: the hidden half. Marcel Dekker Pub, New York, pp 483–510

Clowes FAL (1981) Cell proliferation in ectotrophic mycorrhizas of *Fagus sylvatica*. New Phytol 87:547–555 doi:10.1111/j.1469-8137.1981.tb03225.x

Colmer TD (2003) Aerenchyma and an inducible barrier to radial oxygen loss facilitate root aeration in upland, paddy and deep-water rice (*Oryza sativa* L.). Ann Bot (Lond) 91:301–309 doi:10.1093/aob/mcf114

Colmer TD, Cox MCH, Voesenek LACJ (2006) Root aeration in rice (*Oryza sativa*): evaluation of oxygen, carbon dioxide, and ethylene as possible regulators of root acclimatizations. New Phytol 170:767–778 doi:10.1111/j.1469-8137.2006.01725.x

Cooke JC, Lefor MW (1998) The mycorrhizal status of selected plant species from Connecticut wetlands and transition zones. Restor Ecol 6:214–222 doi:10.1111/j.1526-100X.1998.00628.x

Cornwell WK, Bedford BL, Chapin CT (2001) Occurrence of arbuscular mycorrhizal fungi in a phosphorus-poor wetland and mycorrhizal response to phosphorus fertilization. Am J Bot 88:1824–1829 doi:10.2307/3558359

Cruz C, Lips SH, Martins-Loucao MA (1995) Uptake of inorganic nitrogen in roots of carob seedlings. Physiol Plant 95:167–175 doi:10.1111/j.1399-3054.1995.tb00824.x

Cui M, Caldwell MM (1996a) Facilitation of plant phosphate acquisition by arbuscular mycorrhizas from enriched soil patches. I. Roots and hyphae exploiting the same soil volume. New Phytol 133:453–460 doi:10.1111/j.1469-8137.1996.tb01912.x

Cui M, Caldwell MM (1996b) Facilitation of plant phosphate acquisition by arbuscular mycorrhizas from enriched soil patches. II. Hyphae exploiting root-free soil. New Phytol 133:461–467 doi:10.1111/j.1469-8137.1996.tb01913.x

Danneberg G, Latus C, Zimmer W, Hundeshagen B, Schneider-Poetsc HJ, Bothe H (1992) Influence of vesicular-arbuscular mycorrhiza on phytohormone balance in maize (*Zea mays* L.). J Plant Physiol 141:33–39

de Dorlodot S, Forster B, Pagès L, Price A, Tuberosa R, Draye X (2007) Root system architecture: opportunities and constraints for genetic improvement of crops. Trends Plant Sci 12:474–481 doi:10.1016/j.tplants.2007.08.012

Dello Ioio R, Linhares FS, Scacchi E, Casamitjana-Martinez E, Heidstra R, Costantino P, Sabatini S (2007) Cytokinins determine *Arabidopsis* root-meristem size by controlling cell differentiation. Curr Biol 17:678–682 doi:10.1016/j.cub.2007.02.047

Dello Ioio R, Nakamura K, Moubayidin L, Perilli S, Taniguchi M, Morita MT, Aoyama T, Costantino P, Sabatini S (2008) A genetic framework for the control of cell division and differentiation in the root meristem. Science 322:1380–1384 doi:10.1126/science.1164147

Demchenko NP, Demchenko KN (2001) Resumption of DNA synthesis and cell division in wheat roots as related to lateral root initiation. Russ J Plant Physiol 48:755–763 doi:10.1023/A:1012552307270

Denton MD, Veneklaas EJ, Lambers H (2007) Does phenotypic plasticity in carboxylate exudation differ among rare and widespread *Banksia* species (*Proteaceae*)? New Phytol 173:592–599 doi:10.1111/j.1469-8137.2006.01956.x

De Smet I, Vanneste S, Inzé D, Beeckman T (2006a) Lateral root initiation or the birth of a new meristem. Plant Mol Biol 60:871–887 doi:10.1007/s11103-005-4547-2

De Smet I, Zhang H, Inzé D, Beeckman T (2006b) A novel role for abscisic acid emerges from underground. Trends Plant Sci 11:434–439 doi:10.1016/j.tplants.2006.07.003

Dexter AR (1987) Mechanics of root growth. Plant Soil 97:401–406 doi:10.1007/BF02383230

Dharmasiri N, Dharmasiri S, Estelle M (2005) The F-box protein TIR1 is an auxin receptor. Nature 435:441–445 doi:10.1038/nature03543

DiDonato RJ, Arbuckle E, Buker S, Sheets J, Tobar J, Totong R, Grisafi P, Fink GR, Celenza JL (2004) *Arabidopsis* ALF4 encodes a nuclear-localized protein required for lateral root formation. Plant J 37:340–353 doi:10.1046/j.1365-313X.2003.01964.x

Dinkelaker B, Römheld V, Marschner H (1989) Citric acid excretion and precipitation of calcium citrate in the rhizosphere of white lupin (*Lupinus albus* L.). Plant Cell Environ 12:285–292 doi:10.1111/j.1365-3040.1989.tb01942.x

Dinneny JR, Long TA, Wang JY, Jung JW, Mace D, Pointer S, Barron C, Brady SM, Schiefelbein J, Benfey PN (2008) Cell identity mediates the response of *Arabidopsis* roots to abiotic stress. Science 320:942–945 doi:10.1126/science.1153795

Doussan C, Vercambre G, Pagès L (1998) Modelling of the hydraulic architecture of root systems: An integrated approach to water absorption—Distribution of axial and radial conductances in maize. Ann Bot (Lond) 81:225–232 doi:10.1006/anbo.1997.0541

Doussan C, Vercambre G, Pages L (1999) Water uptake by two contrasting root systems (maize, peach tree): results from a model of hydraulic architecture. Agronomie 19:255–263 doi:10.1051/agro:19990306

Doussan C, Pierret A, Garrigues E, Pagès L (2006) Water uptake by plant roots: II–Modelling of water transfer in the soil root-system with explicit account of flow within the root system—Comparison with experiments. Plant Soil 283:101–119 doi:10.1007/s11104-004-7904-z

Drew MC, Saker LR (1975) Nutrient supply and the growth of the seminal root system in barley. II- Localised compensatory increases in lateral root growth and rates of nitrate uptake when nitrate supply is restricted to only one part of the root system. J Exp Bot 26:79–90 doi:10.1093/jxb/26.1.79

Dubrovsky JG, Sauer M, Napsucialy-Mendivil S, Ivanchenko MG, Friml J, Shishkova S, Celenza J, Benková E (2008) Auxin acts as a local morphogenetic trigger to specify lateral root founder cells. Proc Natl Acad Sci USA 105:8790–8794 doi:10.1073/pnas.0712307105

Dunbabin V (2007) Simulating the role of rooting traits in crop-weed competition. Field Crops Res 104:44–51 doi:10.1016/j.fcr.2007.03.014

Dunbabin V, Diggle AJ, Rengel Z, van Hungten R (2002) Modelling the interactions between water and nutrient uptake and root growth. Plant Soil 239:19–38 doi:10.1023/A:1014939512104

Dunbabin VM, McDermott S, Bengough AG (2006) Upscaling from rhizosphere to whole root system: modelling the effects of phospholipid surfactants on water and nutrient uptake. Plant Soil 283:57–72 doi:10.1007/s11104-005-0866-y

Douglas AE (2002) Symbiotic interactions. Oxford University Press, Oxford

Drew MC, He C-J, Morgan PW (2000) Programmed cell death and aerenchyma formation in roots. Trends Plant Sci 5:123–127 doi:10.1016/S1360-1385(00)01570-3

Einsmann JC, Jones RH, Pu M, Mitchell RJ (1999) Nutrient foraging traits in 10 co-occurring plant species of contrasting life forms. J Ecol 87:609–619 doi:10.1046/j.1365-2745.1999.00376.x

Eissenstat DM, Yanai RD (2002) Root life span, efficiency and turnover. In: Waisel Y, Eshel A, Kafkafi U (eds) Plant roots: the hidden half. Marcel Dekker Pub, New York, pp 221–238

Esch H, Hundeshagen B, Schneider-Poetsch H, Bothe H (1994) Demonstration of abscisic acid in spores and hyphae of the arbuscular-mycorrhizal fungus *Glomus* and the N_2-fixing cyanobacterium *Anabaena variabilis*. Plant Sci 99:9–16 doi:10.1016/0168-9452(94)90115-5

Evans DE (2003) Aerenchyma formation. New Phytol 161:35–49 doi:10.1046/j.1469-8137.2003.00907.x

Fan MS, Bai RQ, Zhao XF, Zhang JH (2007) Aerenchyma formed under phosphorus deficiency contributes to the reduced root hydraulic conductivity in maize roots. J Integr Plant Biol 49:598–604 doi:10.1111/j.1744-7909.2007.00450.x

Farley RA, Fitter AH (1999) The response of seven co-occurring woodland herbaceous perennials to localized nutrient-rich patches. J Ecol 87:849–859 doi:10.1046/j.1365-2745.1999.00396.x

Fieschi M, Alloatti G, Sacco S, Berta G (1992) Cell membrane potential hyperpolarisation in vesicular arbuscular mycorrhizae of *Allium porrum* L.: a non-nutritional long-distance effect of the fungus. Protoplasma 168:136–140 doi:10.1007/BF01666259

Fitz Gerald JN, Lehti-Shiu MD, Ingram PA, Deak KI, Biesiada T, Malamy JE (2006) Identification of quantitative trait loci that regulate *Arabidopsis* root system size and plasticity. Genetics 172:485–498 doi:10.1534/genetics.105.047555

Fitter AH (1985) Functional significance of root morphology and root system architecture. In: Fitter AH, Atkinson D, Read DJ, Usher MB (eds) Ecological interactions in soil. Blackwell Scientific Publications, Oxford, pp 37–42

Fitter A (1987) An architectural approach to the comparative ecology of plant root systems. New Phytol 106:61–77

Fitter A (2002) Characteristics and functions of roots systems. In: Waisel Y, Eshel A, Kafkafi U (eds) Plant roots: the hidden half. Marcel Dekker Pub, New York, pp 15–32

Fitter AH (2004) Magnolioid root-hairs, architecture and mycorrhizal dependency. New Phytol 164:15–16 doi:10.1111/j.1469-8137.2004.01193.x

Fitter AH, Stickland TR (1991) Architectural analysis of plant root systems. 2: Influence of nutrient supply on architecture in contrasting plant species. New Phytol 118:383–389 doi:10.1111/j.1469-8137.1991.tb00019.x

Forbes PJ, Ellison CH, Hooker JE (1996) The impact of arbuscular mycorrhizal fungi and temperature on root system development. Agronomie 16:617–620 doi:10.1051/agro:19961004

Fourcaud T, Zhang X, Stokes A, Lambers H, Korner C (2008) Plant growth modeling and applications: the increasing importance of plant architecture in growth models. Ann Bot (Lond) 101:1053–1063 doi:10.1093/aob/mcn050

Fujii H, Verslues PE, Zhu JK (2007) Identification of two protein kinases required for abscisic acid regulation of seed germination, root growth, and gene expression in *Arabidopsis*. Plant Cell 19:485–494 doi:10.1105/tpc.106.048538

Fukaki H, Tameda S, Masuda H, Tasaka M (2002) Lateral root formation is blocked by a gain-of-function mutation in the *SOLITARY-ROOT/IAA14* gene of *Arabidopsis*. Plant J 29:153–168 doi:10.1046/j.0960-7412.2001.01201.x

Fusconi A, Berta G, Tagliasacchi AM, Scannerini S, Trotta A, Gnavi E, De Padova S (1994) Root apical meristems of arbuscular mycorrhizae of *Allium porrum* L. Environ Exp Bot 43:181–193 doi:10.1016/0098-8472(94)90037-X

Fusconi A, Gnavi E, Trotta A, Berta G (1999) Apical meristems of tomato roots and their modifications induced by arbuscular mycorrhizal and soilborne pathogenic fungi. New Phytol 142:505–516 doi:10.1046/j.1469-8137.1999.00410.x

Fusconi A, Tagliasacchi AM, Berta G, Trotta A, Brazzaventre S, Ruberti F, Scannerini S (2000) Root apical meristems of *Allium porrum* L. as affected by arbuscular mycorrhizae and phosphorus. Protoplasma 214:219–226 doi:10.1007/BF01279066

Fusconi A, Lingua G, Trotta A, Berta G (2005) Effects of arbuscular mycorrhizal colonization and phosphorus application on nuclear ploidy in *Allium porrum* plants. Mycorrhiza 15:313–321 doi:10.1007/s00572-004-0338-x

Gamalero E, Martinotti MG, Trotta A, Lemanceau P, Berta G (2002) Morphogenetic modifications induced by *Pseudomonas fluorescens* A6RI and *Glomus mosseae* BEG12 in the root system of tomato differ according to plant growth conditions. New Phytol 155:293–300 doi:10.1046/j.1469-8137.2002.00460.x

Gamalero E, Trotta A, Massa N, Copetta A, Martinotti MG, Berta G (2004) Impact of two fluorescent pseudomonads and an arbuscular mycorrhizal fungus on tomato plant growth, root architecture and P acquisition. Mycorrhiza 14:185–192 doi:10.1007/s00572-003-0256-3

Gardner WK, Barber DA, Parbery DG (1983) The acquisition of phosphorus by *Lupinus albus* L. III. The probable mechanism by which the phosphorus movement in the soil/root interface is enhanced. Plant Soil 70:107–124 doi:10.1007/BF02374754

Garthwaite AJ, Steudle E, Colmer TD (2006) Water uptake by roots of *Hordeum marinum*: formation of a barrier to radial

O$_2$ loss does not affect root hydraulic conductivity. J Exp Bot 57:655–664 doi:10.1093/jxb/erj055

Ge Z, Rubio G, Lynch JP (2000) The importance of root gravitropism for inter-root competition and phosphorus acquisition efficiency: Results from a geometric simulation model. Plant Soil 218:159–171 doi:10.1023/A:1014987710937

Geldner N, Richter S, Vieten A, Marquardt S, Torres-Ruiz RA, Mayer U, Jürgens G (2004) Partial loss-of-function alleles reveal a role for GNOM in auxin transport-related, post-embryonic development of *Arabidopsis*. Development 131:389–400 doi:10.1242/dev.00926

Gilbert GA, Knight JD, Vance CP, Allan DL (1999) Acid phosphatase activity in phosphorus-deficient white lupin roots. Plant Cell Environ 22:801–810 doi:10.1046/j.1365-3040.1999.00441.x

Giovannetti M, Sbrana C, Avio L (2004) Patterns of belowground plant interconnections established by means of arbuscular mycorrhizal networks. New Phytol 164:175–181 doi:10.1111/j.1469-8137.2004.01145.x

Govindarajulu M, Pfeffer PE, Jin H, Abubaker J, Douds DD, Allen JW, Bücking H, Lammers PJ, Shachar-Hill Y (2005) Nitrogen transfer in the arbuscular mycorrhizal symbiosis. Nature 435:819–823 doi:10.1038/nature03610

Grilli-Caiola M (2002) Cycad coralloid roots housing cyanobacteria. In: Seckbach J (ed) Symbiosis: mechanisms and model systems. Kluwer Academic Publishers, Dordrecht, pp 399–409

Grime JP, Crick JC, Rincon JE (1986) The ecological significance of plasticity. In: Jennings DH, Trewavas AJ (eds) Plasticity in plants. Company of Biologists, Cambridge, pp 5–30

Guo HS, Xie Q, Fei JF, Chua NH (2005) MicroRNA directs mRNA cleavage of the transcription factor NAC1 to downregulate auxin signals for *Arabidopsis* lateral root development. Plant Cell 17:1376–1386 doi:10.1105/tpc.105.030841

Harley JL, Smith SE (1983) Mycorrhizal symbiosis. Academic Press, London

Harper JL, Jones M, Hamilton NR (1991) The evolution of roots and the problem of analysing their behaviour. In: Atkinson D (ed) Plant root growth: an ecological perspective. Blackwell Scientific Publications, Oxford, pp 3–24

He C-J, Finlayson SA, Drew MC, Jordan WR, Morgan PW (1996) Ethylene biosynthesis during aerenchyma formation in roots of maize subjected to mechanical impedance and hypoxia. Plant Physiol 112:1679–1685

He XJ, Mu RL, Cao WH, Zhang ZG, Zhang JS, Chen SY (2005) AtNAC2, a transcription factor downstream of ethylene and auxin signaling pathways, is involved in salt stress response and lateral root development. Plant J 44:903–916 doi:10.1111/j.1365-313X.2005.02575.x

Himanen K, Vuylsteke M, Vanneste S, Vercruysse S, Boucheron E, Alard P, Chriqui D, Van Montagu M, Inzé D, Beeckman T (2004) Transcript profiling of early lateral root initiation. Proc Natl Acad Sci USA 101:5146–5151 doi:10.1073/pnas.0308702101

Hinsinger P, Gobran GR, Gregory PJ, Wenzel WW (2005) Rhizosphere geometry and heterogeneity arising from root-mediated physical and chemical processes. New Phytol 168:293–303 doi:10.1111/j.1469-8137.2005.01512.x

Hishi T (2007) Heterogeneity of individual roots within the fine root architecture: causal links between physiological and ecosystem functions. J For Res 12:126–133 doi:10.1007/s10310-006-0260-5

Hishi T, Takeda H (2005) Life cycles of individual roots in fine root system of *Chamaecyparis obtusa* Sieb. et Zucc. J For Res 10:181–187 doi:10.1007/s10310-004-0120-0

Hoad SP, Russel G, Lucas ME, Bingham IJ (1992) The management of wheat, barley and oat root systems. Adv Agron 74:193–246 doi:10.1016/S0065-2113(01)74034-5

Hochholdinger F, Woll K, Sauer M, Dembinsky D (2004) Genetic dissection of root formation in maize (*Zea mays*) reveals root-type specific developmental programs. Ann Bot (Lond) 93:359–368 doi:10.1093/aob/mch056

Hodge A (2001a) Arbuscular mycorrhizal fungi influence decomposition of, but not plant nutrient capture from, glycine patches in soil. New Phytol 151:725–734 doi:10.1046/j.0028-646x.2001.00200.x

Hodge A (2001b) Foraging and the exploitation of soil nutrient patches: in defence of roots. Funct Ecol 15:416 doi:10.1046/j.1365-2435.2001.00519.x

Hodge A (2003a) Plant nitrogen capture from organic matter as affected by spatial dispersion, interspecific competition and mycorrhizal colonisation. New Phytol 157:303–314 doi:10.1046/j.1469-8137.2003.00662.x

Hodge A (2003b) N capture by *Plantago lanceolata* and *Brassica napus* from organic material: the influence of spatial dispersion, plant competition and an arbuscular mycorrhizal fungus. J Exp Bot 54:2331–2342 doi:10.1093/jxb/erg249

Hodge A (2004) The plastic plant: root responses to heterogeneous supplies of nutrients. New Phytol 162:9–24 doi:10.1111/j.1469-8137.2004.01015.x

Hodge A (2006) Plastic plants and patchy soils. J Exp Bot 57:401–411 doi:10.1093/jxb/eri280

Hodge A, Alexander IJ, Gooday GW (1995) Chitinolytic enzymes of pathogenic and ectomycorrhizal fungi. Mycol Res 99:934–941 doi:10.1016/S0953-7562(09)80752-1

Hodge A, Stewart J, Robinson D, Griffiths BS, Fitter AH (1998) Root proliferation, soil fauna and plant nitrogen capture from nutrient-rich patches in soil. New Phytol 139:479–494 doi:10.1046/j.1469-8137.1998.00216.x

Hodge A, Robinson D, Griffiths BS, Fitter AH (1999a) Why plants bother: root proliferation results in increased nitrogen capture from an organic patch when two grasses compete. Plant Cell Environ 22:811–820 doi:10.1046/j.1365-3040.1999.00454.x

Hodge A, Robinson D, Griffiths BS, Fitter AH (1999b) Nitrogen capture by plants grown in N-rich organic patches of contrasting size and strength. J Exp Bot 50:1243–1252 doi:10.1093/jexbot/50.336.1243

Hodge A, Stewart J, Robinson D, Griffiths BS, Fitter AH (1999c) Plant, soil fauna and microbial responses to N-rich organic patches of contrasting temporal availability. Soil Biol Biochem 31:1517–1530 doi:10.1016/S0038-0717(99)00070-X

Hodge A, Robinson D, Fitter AH (2000a) Are microorganisms more effective than plants at competing for nitrogen? Trends Plant Sci 5:304–308 doi:10.1016/S1360-1385(00)01656-3

Hodge A, Robinson D, Fitter AH (2000b) An arbuscular mycorrhizal inoculum enhances root proliferation in, but not nitrogen capture from, nutrient-rich patches in soil. New Phytol 145:575–584 doi:10.1046/j.1469-8137.2000.00602.x

Hodge A, Stewart J, Robinson D, Griffiths BS, Fitter AH (2000c) Spatial and physical heterogeneity of N supply from soil does not influence N capture by two grass species. Funct Ecol 14:575–584 doi:10.1046/j.1365-2435.2000.t01-1-00470.x

Hodge A, Stewart J, Robinson D, Griffiths BS, Fitter AH (2000d) Competition between roots and micro-organisms for nutrients from nitrogen-rich patches of varying complexity. J Ecol 88:150–164 doi:10.1046/j.1365-2745.2000.00434.x

Hodge A, Stewart J, Robinson D, Griffiths BS, Fitter AH (2000e) Plant N capture and microfaunal dynamics from decomposing grass and earthworm residues in soil. Soil Biol Biochem 32:1763–1772 doi:10.1016/S0038-0717(00)00095-X

Hodge A, Campbell CD, Fitter AH (2001) An arbuscular mycorrhizal fungus accelerates decomposition and acquires nitrogen directly from organic material. Nature 413:297–299 doi:10.1038/35095041

Insalud N, Bell RW, Colmer TC, Rerkasem B (2006) Morphological and physiological responses of rice (*Oryza sativa*) to limited phosphorus supply in aerated and stagnant solution culture. Ann Bot (Lond) 98:995–1004 doi:10.1093/aob/mcl194

Ivanchenko MG, Coffeen WC, Lomax TL, Dubrovsky JG (2006) Mutations in the *Diageotropica* (*Dgt*) gene uncouple patterned cell division during lateral root initiation from proliferative cell division in the pericycle. Plant J 46:436–447 doi:10.1111/j.1365-313X.2006.02702.x

Jackson M, Armstrong W (1999) Formation of aerenchyma and the processes of plant ventilation in relation to soil flooding and submergence. Plant Biol 1:274–287 doi:10.1111/j.1438-8677.1999.tb00253.x

Jackson RB, Manwaring JH, Caldwell MM (1990) Rapid physiological adjustment of roots to localized soil enrichment. Nature 344:58–60 doi:10.1038/344058a0

Jackson RB, Mooney HA, Schulze E-D (1997) A global budget for fine root biomass, surface area, and nutrient contents. Proc Natl Acad Sci USA 94:7362–7366 doi:10.1073/pnas.94.14.7362

Jaillard B, Schneider A, Mollier A, Pellerin S (2000) Modélisation du prélèvement minéral par les plantes fondée sur le fonctionnement bio-physico-chimique de la rhizosphère. In: Maillard P, Bonhomme R (eds) Fonctionnement des peuplements végétaux sous contraintes environnementales. Coll. INRA 93:253-287

Javaux M, Schroder T, Vanderborght J (2008) Use of a three-dimensional detailed modeling approach for predicting root water uptake. Vadose Zone J 7:1079–1088 doi:10.2136/vzj2007.0115

Jenik PD, Gillmor CS, Lukowitz W (2007) Embryonic patterning in *Arabidopsis thaliana*. Annu Rev Cell Dev Biol 23:207–236 doi:10.1146/annurev.cellbio.22.011105.102609

Johnson JF, Allan DL, Vance CP, Weiblen G (1996) Root carbon dioxide fixation by phosphorus-deficient *Lupinus albus*. Plant Physiol 112:19–30 doi:10.1104/pp.112.1.31

Jones DL (1998) Organic acids in the rhizosphere- a critical review. Plant Soil 205:25–44 doi:10.1023/A:1004356007312

Jones D, Hodge A, Kuzyakov Y (2004) Plant and mycorrhizal regulation of rhizodeposition. New Phytol 163:459–480 doi:10.1111/j.1469-8137.2004.01130.x

Justin SHFW, Armstrong W (1991) Evidence for the involvement of ethene in aerenchyma formation in adventitious roots of rice (*Oryza sativa*). New Phytol 118:49–62 doi:10.1111/j.1469-8137.1991.tb00564.x

Kaldorf M, Ludwig-Müller J (2000) AM fungi affect the root morphology of maize by increasing indole-3-butyric acid biosynthesis. Physiol Plant 109:58–67 doi:10.1034/j.1399-3054.2000.100109.x

Kaneyasu T, Kobayashi A, Nakayama M, Fujii N, Takahashi H, Miyazawa Y (2007) Auxin response, but not its polar transport, plays a role in hydrotropism of *Arabidopsis* roots. J Exp Bot 58:1143–1150 doi:10.1093/jxb/erl274

Kawashima K, Suganuma N, Tamaoki M, Kouchi H (2001) Two types of pea leghemoglobin genes showing different O_2-binding affinities and distinct patterns of spatial expression in nodules. Plant Physiol 125:641–651 doi:10.1104/pp.125.2.641

Keerthisinghe G, Hocking PJ, Ryan PR, Delhaize E (1998) Effect of phosphorus supply on the formation and function of proteoid roots of white lupin (*Lupinus albus* L.). Plant Cell Environ 21:467–478 doi:10.1046/j.1365-3040.1998.00300.x

Kembel SW, Cahill JF Jr (2005) Plant phenotypic plasticity belowground: a phylogenetic perspective on root foraging trade-offs. Am Nat 166:216–230 doi:10.1086/431287

King JJ, Stimart DP, Fisher RH, Bleecker AB (1995) A mutation altering auxin homeostasis and plant morphology in Arabidopsis. Plant Cell 7:2023–2037

Kirk GJD (2003) Rice root properties for internal aeration and efficient nutrient acquisition in submerged soils. New Phytol 159:185–194 doi:10.1046/j.1469-8137.2003.00793.x

Kirk GJD, Van DL (1997) Changes in rice root architecture, porosity, and oxygen and proton release under phosphorus deficiency. New Phytol 135:191–200 doi:10.1046/j.1469-8137.1997.00640.x

Klepper B (1992) Development and growth of crop root systems. Adv Soil Sci 19:1–25

Kobayashi A, Takahashi A, Kakimoto Y, Miyazawa Y, Fujii N, Higashitani A, Takahashi H (2007) A gene essential for hydrotropism in roots. Proc Natl Acad Sci USA 104:4724–4729 doi:10.1073/pnas.0609929104

Kutschera L (1960) Wurzelatlas mitleleuropaïsher Ackerunkräuter und Kulturpflanzen. Verlag Pub., Franckfurt am Main

Laan P, Smolders A, Blom CWPM, Armstrong W (1989) The relative roles of internal aeration, radical oxygen losses, iron exclusion and nutrient balances in flood-tolerance of *Rumex* species. Acta Bot Neer 38:131–145

Lamont B (1974) The biology of dauciform roots in the sedge *Cyathochaete avenacea*. New Phytol 73:985–996 doi:10.1111/j.1469-8137.1974.tb01327.x

Lamont B (2003) Structure, ecology and physiology of root clusters–a review. Plant Soil 248:1–19 doi:10.1023/A:1022314613217

Laplaze L, Benkova E, Casimiro I, Maes L, Vanneste S, Swarup R, Weijers D, Calvo V, Parizot B, Herrera-Rodriguez MB,

Offringa R, Graham N, Doumas P, Friml J, Bogusz D, Beeckman T, Bennett M (2007) Cytokinins act directly on lateral root founder cells to inhibit root initiation. Plant Cell 19:3889–3900 doi:10.1105/tpc.107.055863

Larkin RP (1995) Effects of infection by *Pythium* spp. on root system morphology of Alfalfa seedlings. Phytopathol 85:430–435 doi:10.1094/Phyto-85-430

Laskov C, Horn O, Hupfer M (2006) Environmental factors regulating the radial oxygen loss from roots of *Myriophyllum spicatum* and *Potamogeton crispus*. Aquat Bot 84:333–340 doi:10.1016/j.aquabot.2005.12.005

Laskowski M, Biller S, Stanley K, Kajstura T, Prusty R (2006) Expression profiling of auxin-treated *Arabidopsis* roots: toward a molecular analysis of lateral root emergence. Plant Cell Physiol 47:788–792 doi:10.1093/pcp/pcj043

Lazof D, Rufty TW, Redingbaugh MG (1992) Localization of nitrate absorption and translocation within morphological regions of corn roots. Plant Physiol 100:1251–1258 doi:10.1104/pp.100.3.1251

Lecompte F, Ozier-Lafontaine H, Pagès L (2001) The relationships between static and dynamic variables in the description of root growth. Consequences for field interpretation of rooting variability. Plant Soil 236:19–31 doi:10.1023/A:1011924529885

Lecompte F, Pagès L (2007) Apical diameter and branching density affect lateral root elongation rates in banana. Environ Exp Bot 59:243–251 doi:10.1016/j.envexpbot.2006.01.002

Lemanceau P, Offre P, Mougel C, Gamalero E, Dessaux Y, Moënne-Loccoz Y, Berta G (2005) Microbial ecology of the rhizosphere. In: Bloem J, Hopkins DW, Benedetti A (eds) Microbiological methods for assessing soil quality. CABI Publishing, Wallingford, pp 228–230

Leigh J, Hodge A, Fitter AH (2009) Arbuscular mycorrhizal fungi can transfer substantial amounts of nitrogen to their host plant from organic material. New Phytol 181:199–207 doi:10.1111/j.1469-8137.2008.02630.x

Le Quéré A, Schützendübel A, Rajashekar B, Canbäck B, Hedh J, Erland S, Johansson T, Anders Tunlid A (2004) Divergence in gene expression related to variation in host specificity of an ectomycorrhizal fungus. Mol Ecol 12:3809–3819 doi:10.1111/j.1365-294X.2004.02369.x

Leyser O (2006) Dynamic integration of auxin transport and signalling. Curr Biol 16:R424–R433 doi:10.1016/j.cub.2006.05.014

Lloret PG, Casero PJ (2002) Lateral root initiation. In: Waisel Y, Eshel A, Kafkafi U (eds) Plant roots: the hidden half. Marcel Dekker Pub, New York, pp 127–156

Lingua G, Sgorbati S, Citterio A, Fusconi A, Trotta A, Gnavi E, Berta G (1999) Arbuscular mycorrhizal colonization delays nucleus senescence in leek root cortical cells. New Phytol 141:161–169 doi:10.1046/j.1469-8137.1999.00328.x

Liscum E, Reed JW (2002) Genetics of Aux/IAA and ARF action in plant growth and development. Plant Mol Biol 49:387–400 doi:10.1023/A:1015255030047

Ljung K, Hull AK, Celenza J, Yamada M, Estelle M, Normanly J, Sandberg G (2005) Sites and regulation of auxin biosynthesis in *Arabidopsis* roots. Plant Cell 17:1090–1104 doi:10.1105/tpc.104.029272

Lorenzen B, Brix H, Mendelssohn IA, McKee KL, Miao SL (2001) Growth, biomass allocation and nutrient use efficiency in *Cladium jamaicense* and *Typha domingensis* as affected by phosphorus and oxygen availability. Aquat Bot 70:117–133 doi:10.1016/S0304-3770(01)00155-3

Loudet O, Gaudon V, Trubuil A, Daniel-Vedele F (2005) Quantitative trait loci controlling root growth and architecture in *Arabidopsis thaliana* confirmed by heterogeneous inbred family. Theor Appl Genet 110:742–753 doi:10.1007/s00122-004-1900-9

Lynch JP (2007) Roots of the second green revolution. Aust J Bot 55:493–512 doi:10.1071/BT06118

Lynch JP, Brown KM (2001) Topsoil foraging–an architectural adaptation of plants to low phosphorus avaibility. Plant Soil 237:225–237 doi:10.1023/A:1013324727040

Malamy JE (2005) Intrinsic and environmental response pathways that regulate root system architecture. Plant Cell Environ 28:67–77 doi:10.1111/j.1365-3040.2005.01306.x

Marchant A, Bhalerao R, Casimiro I, Eklof J, Casero PJ, Bennett M, Sandberg G (2002) AUX1 promotes lateral root formation by facilitating indole-3-acetic acid distribution between sink and source tissues in the *Arabidopsis* seedling. Plant Cell 14:589–597 doi:10.1105/tpc.010354

Masle J (2002) High soil strength: mechanical forces at play on root morphogenesis. In: Waisel Y, Eshel A, Kafkafi U (eds) Plant roots: the hidden half. Marcel Dekker Pub, New York, pp 807–820

Materechera SA, Dexter AR, Alston AM (1991) Penetration of very strong soils by seedling roots of different plant species. Plant Soil 135:31–41 doi:10.1007/BF00014776

Materechera SA, Alston AM, Kirby JM, Dexter AR (1992) Influence of root diameter on the penetration of seminal roots into a compacted subsoil. Plant Soil 144:297–303 doi:10.1007/BF00012888

McMichael BL, Burke JJ (2002) Temperature effects on root growth. In: Waisel Y, Eshel A, Kafkafi U (eds) Plant roots: the hidden half. Marcel Dekker Pub, New York, pp 717–728

McDonald MP, Galwey NW, Colmer TD (2001) Waterlogging tolerance in the tribe Triticeae: the adventitious roots of *Critesion marinum* have relatively high porosity and a barrier to radical oxygen loss. Plant Cell Environ 24:585–596 doi:10.1046/j.0016-8025.2001.00707.x

McDonald MP, Galwey NW, Colmer TD (2002) Similarity and diversity in adventitious root anatomy as related to root aeration among a range of wetland and dryland grass species. Plant Cell Environ 25:441–451 doi:10.1046/j.0016-8025.2001.00817.x

Mendoza R, Escudero V, Garcia I (2005) Plant growth, nutrient acquisition and mycorrhizal symbioses of a waterlogging tolerant legume (*Lotus glaber* Mill.) in a saline-sodic soil. Plant Soil 275:305–315 doi:10.1007/s11104-005-2501-3

Mergemann H, Sauter M (2000) Ethylene induces epidermal cell death at the site of adventitious root emergence in rice. Plant Physiol 124:609–614 doi:10.1104/pp.124.2.609

Mistrik I, Mistrikova I (1995) Uptake, transport and metabolim of phosphate by individual roots of *Zea mays* L. Biologia (Bratisl) 50:419–426

Mouchel CF, Briggs GC, Hardtke CS (2004) Natural genetic variation in *Arabidopsis* identifies BREVIS RADIX, a novel regulator of cell proliferation and elongation in the root. Genes Dev 18:700–714 doi:10.1101/gad.1187704

Nardon P, Charles H (2002) Morphological aspects of symbiosis. In: Seckbach J (ed) Symbiosis: Mechanisms and

Model Systems. Kluwer Academic Publishers, Dordrecht, pp 13–44

Navara J (1987) Participation of individual root types in water uptake by maize seedlings. Biologia (Bratisl) 42:17–26

Neumann G, Martinoia E (2002) Cluster roots–an underground adaptation for survival in extreme environments. Trends Plant Sci 7:162–167 doi:10.1016/S1360-1385(02)02241-0

Neumann G, Massonneau A, Martinoia E, Römheld V (1999) Physiological adaptations to phosphorus deficiency during proteoid root development in white lupin. Planta 208:373–382 doi:10.1007/s004250050572

Newsham KK, Fitter AH, Watkinson AR (1995) Multi-functionality and biodiversity in arbuscular mycorrhizas. Trends Ecol Evol 10:407–411 doi:10.1016/S0169-5347(00)89157-0

Noelle W (1910) Studin zur vergleighenden anatomie und morphologie der koniferenwuzeln mit rücksicht auf die systematik. Botanik. Zeit 68:169–266

Norman JR, Hooker JE (2000) Sporulation of *Phytophthora fragariae* shows greater stimulation by exudates of non-mycorrhizal than by mycorrhizal strawberry roots. Mycol Res 104:1069–1073 doi:10.1017/S0953756299002191

Norman JR, Atkinson D, Hooker JE (1996) Arbuscular mycorrhizal fungal-induced alteration to root architecture in strawberry and induced resistance to the root pathogen *Phytophthora fragariae*. Plant Soil 185:191–198 doi:10.1007/BF02257524

Nouchi I, Mariko S, Aoki K (1990) Mechanism of methane transport from the rhizosphere to the atmosphere through rice plants. Plant Physiol 94:59–66 doi:10.1104/pp.94.1.59

Nowack B, Mayer KU, Oswald SE, van Beinum W, Appelo CAJ, Jacques D, Seutjens P, Gérard F, Jaillard B, Schnepf A, Roose T (2006) Verification and intercomparison of reactive transport codes to describe root-uptake. Plant Soil 285:305–321 doi:10.1007/s11104-006-9017-3

Ohashi-Ito K, Bergmann DC (2007) Regulation of the Arabidopsis root vascular initial population by LONESOME HIGHWAY. Development 134:2959–2968 doi:10.1242/dev.006296

Oláh B, Brière C, Bécard G, Dénarié J, Gough C (2005) Nod factors and a diffusible factor from arbuscular mycorrhizal fungi stimulate lateral root formation in *Medicago truncatula* via the DMI1/DMI2 signalling pathway. Plant J 44:195–207 doi:10.1111/j.1365-313X.2005.02522.x

Osmont KS, Sibout R, Hardtke CS (2007) Hidden branches: developments in root ystem architecture. Annu Rev Plant Biol 58:93–113 doi:10.1146/annurev.arplant.58.032806.104006

Pagès L (2002) Modeling root system architecture. In: Waisel Y, Eshel A, Kafkafi U (eds) Plant roots: the hidden half. Marcel Dekker Pub, New York, pp 359–382

Pagès L, Doussan C, Vercambre G (2000) An introduction on below ground environment and ressource acquisition with special reference on trees—Simulation models should include plant structure and function. Ann Sci 57:513–520 doi:10.1051/forest:2000138

Paponov IA, Teale WD, Trebar M, Blilou I, Palme K (2005) The PIN auxin efflux facilitators: evolutionary and functional perspectives. Trends Plant Sci 10:170–177 doi:10.1016/j.tplants.2005.02.009

Paran I, Zamir D (2003) Quantitative traits in plants: beyond the QTL. Trends Genet 19:303–306 doi:10.1016/S0168-9525(03)00117-3

Parizot B, Laplaze L, Ricaud L, Boucheron-Dubuisson E, Bayle V, Bonke M, De Smet I, Poethig SR, Helariutta Y, Haseloff J, Chriqui D, Beeckman T, Nussaume L (2008) Diarch symmetry of the vascular bundle in Arabidopsis root encompasses the pericycle and is reflected in distich lateral root initiation. Plant Physiol 146:140–148 doi:10.1104/pp.107.107870

Parry G, Delbarre A, Marchant A, Swarup R, Napier R, Perrot-Rechenmann C, Bennett MJ (2001) Novel auxin transport inhibitors phenocopy the auxin influx carrier mutation aux1. Plant J 25:399–406 doi:10.1046/j.1365-313x.2001.00970.x

Pellerin S (1993) Rate of differentiation and emergence of nodal maize roots. Plant Soil 148:155–161 doi:10.1007/BF00012853

Pellerin S, Tabourel F (1995) Length of the apical unbranched zone of maize axile roots: its relationship to root elongation rate. Environ Exp Bot 35:193–200 doi:10.1016/0098-8472(94)00043-5

Personeni E, Nguyen C, Marchal P, Pagès L (2007) Experimental evaluation of an efflux-influx model of C exudation by individual apical root segments. J Exp Bot 58:2091–2099 doi:10.1093/jxb/erm065

Pierret A, Moran CJ, Pankurst CE (1999) Differentiation of soil properties related to the spatial association of wheat roots and soil macropores. Plant Soil 211:51–58 doi:10.1023/A:1004490800536

Pierret A, Moran CJ, Doussan C (2005) Conventional detection methodology is limiting our ability to understand the roles and functions of fine roots. New Phytol 166:967–980 doi:10.1111/j.1469-8137.2005.01389.x

Pierret A, Doussan C, Capowiez Y, Bastardie F, Pagès L (2007) Root functional architecture: a framework for modeling the interplay between roots and soil. Vadose Zone J 6:269–281 doi:10.2136/vzj2006.0067

Playsted CWS, Johnston M, Ramage CM, Edwards DG, Cawthray GR, Lambers H (2006) Functional significance of dauciform roots: exudation of carboxylates and acid phosphatase under phosphorus deficiency in *Caustis blakei* (Cyperaceae). New Phytol 170:491–500 doi:10.1111/j.1469-8137.2006.01697.x

Pregitzer KS, Hendrick RL, Fogel R (1993) The demography of fine roots in response to patches of water and nitrogen. New Phytol 125:575–580 doi:10.1111/j.1469-8137.1993.tb03905.x

Pritchard SG, Strand AE (2008) Can you believe what you see? Reconciling minirhizotron and isotopically derived estimates of fine root longevity. New Phytol 177:287–291

Purnell HM (1960) Studies of the family Proteaceae. I. Anatomy and morphology of the roots of some Victorian species. Aust J Bot 8:38–50 doi:10.1071/BT9600038

Qin G, Gu H, Zhao Y, Ma Z, Shi G, Yang Y, Pichersky E, Chen H, Liu M, Chen Z, Qu LJ (2005) An indole-3-acetic acid carboxyl methyltransferase regulates *Arabidopsis* leaf development. Plant Cell 17:2693–2704 doi:10.1105/tpc.105.034959

Ranathuge K, Steudle E, Lafitte R (2003) Control of water uptake by rice (*Oryza sativa* L.): role of the outer part of the root. Planta 217:193–205

Rasse DP, Smucker AJM (1998) Root recolonization of previous root channels in corn and alfalfa rotations. Plant Soil 204:203–212 doi:10.1023/A:1004343122448

Read DJ, Perez-Moreno J (2003) Mycorrhizas and nutrient cycling in ecosystems–a journey towards relevance. New Phytol 157:475–492 doi:10.1046/j.1469-8137.2003.00704.x

Reddy KR, Patrick WH Jr, Lindau CW (1989) Nitrification-denitrification at the plant root-sediment interface in wetlands. Limnol Oceanogr 34:1004–1013

Reynolds HL, d'Antonio C (1996) The ecological significance of plasticity in root weight ratio in response to nitrogen: opinion. Plant Soil 185:75–97 doi:10.1007/BF02257566

Riefler M, Novak O, Strnad M, Schmülling T (2006) *Arabidopsis* cytokinin receptor mutants reveal functions in shoot growth, leaf senescence, seed size, germination, root development, and cytokinin metabolism. Plant Cell 18:40–54 doi:10.1105/tpc.105.037796

Robinson D (1991) Roots and ressource fluxes in plants and communities. In: Atkinson D (ed) Plant root growth: an ecological perspective. Special publication of the British ecological society n°10. Blackwell Scientific Pub., London, pp 103–130

Robinson D (2001) Root proliferation, nitrate inflow and their carbon costs during nitrogen capture by competing plants in patchy soil. Plant Soil 232:41–50 doi:10.1023/A:1010377818094

Robinson D, Hodge A, Griffiths BS, Fitter AH (1999) Plant root proliferation in nitrogen-rich patches confers competitive advantage. Proc R Soc Lond B Biol Sci 265:431–435

Roelofs RFR, Rengel Z, Cawthray GR, Dixon KW, Lambers H (2001) Exudation of carboxylates in Australian *Proteaceae*: chemical composition. Plant Cell Environ 24:891–903 doi:10.1046/j.1365-3040.2001.00741.x

Rubio G, Oesterheld M, Alvarez CR, Lavado RS (1997) Mechanisms for the increase in phosphorus uptake of waterlogged plants: soil phosphorus availability, root morphology and uptake kinetics. Oecologia 112:150–155 doi:10.1007/s004420050294

Ruzicka K, Ljung K, Vanneste S, Podhorsk'a R, Beeckman T, Friml J, Benková E (2007) Ethylene regulates root growth through effects on auxin biosynthesis and transport-dependent auxin distribution. Plant Cell 19:2197–2212 doi:10.1105/tpc.107.052126

Sabatini S, Heidstra R, Wildwater M, Scheres B (2003) SCARECROW is involved in positioning the stem cell niche in the *Arabidopsis* root meristem. Genes Dev 17:354–358 doi:10.1101/gad.252503

Saleque MA, Kirk GJD (1995) Root-induced solubilization of phosphate in the rhizosphere of lowland rice. New Phytol 129:325–336 doi:10.1111/j.1469-8137.1995.tb04303.x

Sampò S, Avidano L, Berta G (2007) Morphogenetic effects induced by pathogenic and non pathogenic *Rhizoctonia solani* Kühn strains on tomato roots. Caryologia 60:1–20

Sadowsky MJ (2005) Soil stress factors influencing symbiotic nitrogen fixation. In: Wernerand D, Newton WE (eds) Nitrogen fixation in agriculture, forestry, ecology and the environment. Springer, The Netherlands, pp 89–10

Sarquis JI, Jordon WR, Morgan PW (1991) Ethylene evolution from maize (*Zea mays* L.) seedling roots and shoots in response to mechanical impedance. Plant Physiol 96:1171–1177 doi:10.1104/pp.96.4.1171

Schellenbaum L, Berta G, Ravolanirina F, Tisserant B, Gianinazzi S, Fitter AH (1991) Influence of endomycorrhizal infection on root morphology in a micropropagated woody plant species (*Vitis vinifera* L.). Ann Bot (Lond) 67:135–141

Schenk HJ, Jackson RB (2002a) Rooting depths, lateral root spreads and below-ground/above-ground allometries of plants in water-limited ecosystems. J Ecol 90:480–494 doi:10.1046/j.1365-2745.2002.00682.x

Schenk HJ, Jackson RB (2002b) The global biogeography of roots. Ecol Monogr 73:311–328

Schnepf A, Roose T (2006) Modelling the contribution of arbuscular mycorrhizal fungi to plant phosphate uptake. New Phytol 171:669–682

Seo M, Akaba S, Oritani T, Delarue M, Bellini C, Caboche M, Koshiba T (1998) Higher activity of an aldehyde oxidase in the auxin-overproducing superroot1 mutant of *Arabidopsis thaliana*. Plant Physiol 116:687–693 doi:10.1104/pp.116.2.687

Shane MW, Lambers H (2005) Cluster roots: a curiosity in context. Plant Soil 274:101–125 doi:10.1007/s11104-004-2725-7

Shane MW, De Vos M, De Roock S, Lambers H (2003) Shoot P status regulates cluster-root growth and citrate exudation in *Lupinus albus* grown with a divided root system. Plant Cell Environ 26:265–273 doi:10.1046/j.1365-3040.2003.00957.x

Shane MW, Dixon KW, Lambers H (2005) The occurrence of dauciform roots amongst Western Australian reeds, rushes and sedges, and the impact of phosphorus supply on dauciform-root development in *Schoenus unispiculatus* (Cyperaceae). New Phytol 165:887–898 doi:10.1111/j.1469-8137.2004.01283.x

Shane MW, Cawthray GR, Cramer MD, Kuo J, Lambers H (2006) Specialized "dauciform" roots of *Cyperaceae* are structurally distinct, but functionally analogous with "cluster" roots. Plant Cell Environ 29:1989–1999 doi:10.1111/j.1365-3040.2006.01574.x

Shannon RD, White JR (1994) A three-year study of controls on methane emissions from two Michigan peatlands. Biogeochemisty 27:35–60

Shannon RD, White JR, Lawson JE, Gilmour BS (1996) Methane efflux from emergent vegetation in peatlands. J Ecol 84:239–246 doi:10.2307/2261359

Shaul-Keinan O, Gadkal V, Ginzberg I, Grunzweig JM, Chet I, Elad Y, Wininger S, Belausov E, Eshed Y, Arzmon N, Ben-Tal Y, Kapulnik Y (2002) Hormone concentrations in tobacco roots change during arbuscular mycorrhizal colonization with *Glomus intraradices*. New Phytol 154:501–507 doi:10.1046/j.1469-8137.2002.00388.x

Signora L, De Smet I, Foyer CH, Zhang HM (2001) ABA plays a central role in mediating the regulatory effects of nitrate on root branching in *Arabidopsis*. Plant J 28:655–662 doi:10.1046/j.1365-313x.2001.01185.x

Singh MB, Bhalla PL (2006) Plant stem cells carve their own niche. Trends Plant Sci 11:241–246 doi:10.1016/j.tplants.2006.03.004

Skene KR (1998) Cluster roots: some ecological considerations. J Ecol 86:1060–1064 doi:10.1046/j.1365-2745.1998.00326.x

Smith FA (2000) Measuring the influence of mycorrhizas. New Phytol 148:4–6

Smith SE, Read DJ (2008) Mycorrhizal symbiosis, 3rd Edn. Academic Press, London

Smits AJM, Laan P, Thier RH, van der Velde G (1990) Root aerenchyma, oxygen leakage patterns and alcoholic fermentation ability of the roots of some nymphaeid and isoetid macrophytes in relation to sediment type of their habitat. Aquat Bot 38:3–17 doi:10.1016/0304-3770(90)90095-3

Somma F, Hopmans JW, Clausnitzer V (1998) Transient three-dimensional modeling of soil water and solute transport with simultaneous root growth, root water and nutrient uptake. Plant Soil 202:281–293 doi:10.1023/A:1004378602378

Šraj-Kržič N, Pongrac P, Klemenc M, Kladnik A, Regvar M, Gaberščik A (2006) Mycorrhizal colonisation in plants from intermittent aquatic habitats. Aquat Bot 85:331–336 doi:10.1016/j.aquabot.2006.07.001

Staswick PE, Serban B, Rowe M, Tiryaki I, Maldonado MT, Maldonado MC, Suza W (2005) Characterization of an *Arabidopsis* enzyme family that conjugates amino acids to indole-3-acetic acid. Plant Cell 17:616–627 doi:10.1105/tpc.104.026690

Stepanova AN, Hoyt JM, Hamilton AA, Alonso JM (2005) A link between ethylene and auxin uncovered by the characterization of two root-specific ethylene-insensitive mutants in *Arabidopsis*. Plant Cell 17:2230–2242 doi:10.1105/tpc.105.033365

Striker GG, Insausti P, Grimoldi AA, Vega AS (2007) Trade-off between root porosity and mechanical strength in species with different types of aerenchyma. Plant Cell Environ 30:580–589 doi:10.1111/j.1365-3040.2007.01639.x

Svistoonoff S, Creff A, Reymond M, Sigoillot-Claude C, Ricaud L, Blanchet A, Laurent Nussaume L, Desnos T (2007) Root tip contact with low-phosphate media reprograms plant root architecture. Nat Genet 39:792–796 doi:10.1038/ng2041

Swarup R, Bennett M (2003) Auxin transport: the fountain of life in plants? Dev Cell 5:824–826 doi:10.1016/S1534-5807(03)00370-8

Tardieu F, Katerji N (1991) Plant-response to the soil-water reserve: consequences of the root-system environment. Irrig Sci 12:145–152 doi:10.1007/BF00192286

Tatsumi J, Yamauchi A, Kono Y (1989) Fractal analysis of plant root systems. Ann Bot (Lond) 64:499–503

Thaler P, Pagès L (1999) Why are laterals less affected than main axes by homogeneous unfavourable physical conditions? A model-based hypothesis. Plant Soil 217:151–157 doi:10.1023/A:1004677128533

Thomson CJ, Colmer TD, Watkins I, Greenway H (1992) Tolerance of wheat (*Triticum aestivum* cvs. Gamenya and Kite) and triticale (*Triticosecale* cv. Muir) to waterlogging. New Phytol 120:335–344 doi:10.1111/j.1469-8137.1992.tb01073.x

Tisserant B, Schellenbaum L, Gianinazzi-Pearson V, Gianinazzi S, Berta G (1992) Influence of infection by an endomycorrhizal fungus on root development and architecture in *Platanus acerifolia*. Allionia 30:173–183

Tisserant B, Gianinazzi S, Gianinazzi-Pearson V (1996) Relationships between lateral root order, arbuscular mycorrhiza development, and the physiological state of the symbiotic fungus in *Platanus acerifolia*. Can J Bot 74:1947–1955 doi:10.1139/b96-233

Torelli A, Trotta A, Acerbi L, Arcidiacono G, Berta G, Branca C (2000) IAA and ZR content in leek (*Allium porrum* L.), as influenced by P nutrition and arbuscular mycorrhizae, in relation to plant development. Plant Soil 226:29–35 doi:10.1023/A:1026430019738

Trotta A, Varese GC, Gnavi E, Fusconi A, Sampo S, Berta G (1996) Interactions between the soilborne pathogen *Phytophthora nicotianae* var *parasitica* and the arbuscular mycorrhizal fungus *Glomus mosseae* in tomato plants. Plant Soil 185:199–209 doi:10.1007/BF02257525

Tuberosa R, Sanguineti MC, Landi P, Giuliani MM, Salvi S, Conti S (2002) Identification of QTLs for root characteristics in maize grown in hydroponics and analysis of their overlap with QTLs for grain yield in the field at two water regimes. Plant Mol Biol 48:697–712 doi:10.1023/A:1014897607670

Vanneste S, De Rybel B, Beemster GT, Ljung K, De Smet I, Van Isterdael G, Naudts M, Iida R, Gruissem W, Tasaka M, Inzé D, Fukaki H, Beeckman T (2005) Cell cycle progression in the pericycle is not sufficient for SOLITARY ROOT/IAA14-mediated lateral root initiation in *Arabidopsis thaliana*. Plant Cell 17:3035–3050 doi:10.1105/tpc.105.035493

Van Noordwijk M, Mulia R (2002) Functional branch analysis as tool for fractal scaling above and belowground trees for their additive and non-additive properties. Ecol Modell 149:41–51 doi:10.1016/S0304-3800(01)00513-0

van Vuuren MMI, Robinson D, Griffiths BS (1996) Nutrient inflow and root proliferation during the exploitation of a temporally and spatially discrete source of nitrogen in soil. Plant Soil 178:185–192 doi:10.1007/BF00011582

Van Bodegom P, Goudriaan J, Leffelaar P (2001) A mechanistic model on methane oxidation in a rice rhizosphere. Biogeochemistry 55:145–177 doi:10.1023/A:1010640515283

Varney GT, McCully ME (1991a) The branch roots of *Zea*: I–First order branches, their number, sizes and division into classes. Ann Bot (Lond) 67:357–364

Varney GT, McCully ME (1991b) The branch root of *Zea*. II. Developmental loss of the apical meristem in field grown roots. New Phytol 118:535–546 doi:10.1111/j.1469-8137.1991.tb00993.x

Vercambre G, Doussan C, Pagès L, Habib R, Pierret A (2002) Influence of xylem development on axial hydraulic conductance within *Prunus* root systems. Trees Struct Funct 16:479–487

Vercambre G, Pagès L, Doussan C, Habib R (2003) Architectural analysis and synthesis of the plum tree root system in orchard using a quantitative modelling approach. Plant Soil 251:1–11 doi:10.1023/A:1022961513239

Vierheilig H (2004) Further root colonization by arbuscular mycorrhizal fungi in already mycorrhizal plants is suppressed after a critical level of root colonization. J Plant Physiol 161:339–341 doi:10.1078/0176-1617-01097

Vierheilig H, Alt M, Mohr U, Boller T, Weimken A (1994) Ethylene biosynthesis and α-1,3-glucanase in the roots of host and non-host plants of vesicular-arbuscular mycorrhizal fungi after inoculation with *Glomus mosseae*. Plant Physiol 143:337–343

Vigo C, Norman JR, Hooker JE (2000) Biocontrol of the pathogen *Phytophthora parasitica* by arbuscular mycorrhizal fungi is a consequence of effects on infection loci. Plant Pathol 49:509–514 doi:10.1046/j.1365-3059.2000. 00473.x

Visser EJW, Bögemann GM (2006) Aerenchyma formation in the wetland plant *Juncus effusus* is independent of ethylene. New Phytol 171:305–314 doi:10.1111/j.1469-8137.2006.01764.x

Visser EJW, Colmer TD, Blom CWPM, Voesenek CJ (2000) Changes in growth, porosity, and radical oxygen loss from adventitious roots of selected mono- and dicotyledonous wetland species with contrasting types of aerenchyma. Plant Cell Environ 23:1237–1245 doi:10.1046/j.1365-3040.2000.00628.x

Voesenek LACJ, Armstrong W, Bogemann GM, McDonald MP, Colmer TD (1999) A lack of aerenchyma and high rates of radial oxygen loss from the root base contribute to the waterlogging intolerance of *Brassica napus*. Aust J Plant Physiol 26:87–93

Wahid PA (2000) A system of classification of woody perennials based on their root activity patterns. Agrofor Syst 49:123–130 doi:10.1023/A:1006309927504

Waisel Y, Eshel A (1992) Differences in ion uptake among roots of various types. J Plant Nutr 15:945–958 doi:10.1080/01904169209364373

Wang B, Adachi K (2000) Differences among rice cultivars in root exudation, methane oxidation, and populations of methanogenic and methanotrophic bacteria in relation to methane emission. Nutr Cycl Agroecosyst 58:349–356 doi:10.1023/A:1009879610785

Wang H, Inukai Y, Yamauchi A (2006) Root development and nutrient uptake. Crit Rev Plant Sci 25:279–301 doi:10.1080/07352680600709917

Wang JW, Wang LJ, Mao YB, Cai WJ, Xue HW, Chen XY (2005) Control of root cap formation by microRNA-targeted auxin response factors in *Arabidopsis*. Plant Cell 17:2204–2216 doi:10.1105/tpc.105.033076

Watt M, Evans JR (1999) Linking development and determinacy with organic acid efflux from proteoid roots of *Lupinus albus* L. grown with low phosphorus and ambient or elevated atmospheric CO_2 concentration. Plant Physiol 120:705–716 doi:10.1104/pp.120.3.705

Weaver JE (1926) Root development of field crops. McGraw-Hill, New York Pub

Weisskopf L, Tomasi N, Santelia D, Martinoia E, Langlade NB, Tabacchi R, Abou-Mansour E (2006a) Isoflavonoid exudation from white lupin is influenced by phosphate supply, root type and cluster-root stage. New Phytol 171:657–668

Weisskopf L, Abou-Mansour E, Fromin N, Tomasi N, Santelia D, Edelkott I, Neumann G, Aragno M, Tabacchi R, Martinoia E (2006b) White lupin has developed a complex strategy to limit microbial degradation of secreted citrate required for phosphate acquisition. Plant Cell Environ 29:919–927 doi:10.1111/j.1365-3040.2005.01473.x

Werner T, Motyka V, Laucou V, Smets R, Van Onckelen H, Schmülling T (2003) Cytokinin-deficient transgenic *Arabidopsis* plants show multiple developmental alterations indicating opposite functions of cytokinins in the regulation of shoot and root meristem activity. Plant Cell 15:2532–2550 doi:10.1105/tpc.014928

Whalley WR, Dexter AR (1993) The maximum axial growth pressure of roots of spring and autumn cultivars of lupin. Plant Soil 157:313–318 doi:10.1007/BF00011059

Whiting GJ, Chanton JP (1993) Primary production control of methane emission from wetlands. Nature 364:794–795 doi:10.1038/364794a0

Wu L, McGechan MB, Watson CA, Baddeley JA (2005) Developing existing plant root system architecture models to meet future, agricultural challenges. Adv Agron 85:181–219 doi:10.1016/S0065-2113(04)85004-1

Xie Q, Guo HS, Dallman G, Fang S, Weissman AM, Chua NH (2002) SINAT5 promotes ubiquitin-related degradation of NAC1 to attenuate auxin signals. Nature 419:167–170 doi:10.1038/nature00998

Yamazaki K, Nakamoto T (1983) Primary root formation in growing shoots of several cereal crops. Jpn J Crop Sci 52:342–348

Yavitt JB, Lang GE (1990) Methane production in contrasting wetland sites: response to organic-chemical components of peat and to sulfate reduction. Geomicrobiol J 8:27–46 doi:10.1080/01490459009377876

Zaïd EH, Arahou M, Diem HG, Morabet RE (2003) Is Fe deficiency rather than P deficiency the cause of cluster root formation in *Casuarina* species? Plant Soil 248:229–235 doi:10.1023/A:1022320227637

Zhang H, Forde BG (1998) An *Arabidopsis* MADS box gene that controls nutrient-induced changes in root architecture. Science 279:407–409 doi:10.1126/science.279. 5349.407

Zhang H, Rong H, Pilbeam D (2007) Signalling mechanisms underlying the morphological responses of the root system to nitrogen in *Arabidopsis thaliana*. J Exp Bot 58:2329–2338 doi:10.1093/jxb/erm114

REVIEW ARTICLE

The rhizosphere zoo: An overview of plant-associated communities of microorganisms, including phages, bacteria, archaea, and fungi, and of some of their structuring factors

M. Buée · W. De Boer · F. Martin ·
L. van Overbeek · E. Jurkevitch

Received: 5 December 2008 / Accepted: 1 April 2009 / Published online: 28 April 2009
© Springer Science + Business Media B.V. 2009

Introduction

Rhizosphere microorganisms have two faces, like Janus the Roman god of gates and doors who symbolizes changes and transitions, from one condition to another. One face looks at the plant root, the other sees the soil. The ears and the nose sense the other gods around and the mouths are wide open, swallowing as much as they can, and as described in Chapter 11, they also are busy talking. These faces may as well represent Hygieia (the Greek god of Health and Hygiene, the prevention of sickness and the continuation of good health) and Morta (the Roman god of death) for rhizosphere microbes can be beneficial, and promote plant growth and well being (Chapter 12) or detrimental, causing plant sickness and death (Chapter 13). It can be argued that many rhizosphere microbes are "neutral", faceless saprophytes that decompose organic materials, perform mineralization and turnover processes. While most may not directly interact with the plant, their effects on soil biotic and abiotic parameters certainly have an impact on plant growth. Maybe they are Janus' feet, the unsung heroes of the rhizosphere.

This chapter addresses some aspects of the taxonomical and functional microbial diversity of the rhizosphere. Bacteria, Archea, viruses and Fungi will be at the heart of our discussion, while other root-associated eukaryotes are the subjects of other chapters.

The vast amount of organic carbon secreted by plant roots is what forms, sustains and drives the rhizosphere web, a continuum of microbial populations colonizing niches from the plant's interior and into the bulk soil, responding to the plant, interacting with each other and impacting upon their environment. Within this continuum, the rhizosphere forms a transition zone between the bulk soil and the plant root surface. Plant roots exert strong effects on the rhizosphere through 'rhizodeposition' (root exudation, production of mucilages and release of shoughed-off root cells) and by providing suitable

Responsible Editor: Yves Dessaux.

M. Buée · F. Martin
UMR INRA-UHP, IFR 110, INRA-Nancy,
Champenoux, France

W. De Boer
NIOO-KNAW, Centre for Terrestrial Ecology,
Heteren, The Netherlands

L. van Overbeek
Plant Research International,
P.O. Box 16, Wageningen, The Netherlands

E. Jurkevitch (✉)
Faculty of Agriculture, Food and Environment,
The Hebrew University of Jerusalem,
Rehovot, Israel
e-mail: jurkevi@agri.huji.ac.il

ecological niches for microbial growth (Bais et al. 2006). Bacteria residing in the rhizosphere most likely originate from the surrounding bulk soil and will thrive under conditions prevailing in the neighbourhood of plant roots. It must therefore be assumed that bacterial communities in the rhizosphere form a subset of the total bacterial community present in bulk soils. Important parameters, like the quantity and the quality of available carbon compounds originating from plants, as well as novel sites for microbial attachment discriminate rhizosphere from bulk soil (Curl and Truelove 1986). Other parameters, intrinsic to the plant's physiology, genetic make up, life history and ecology, and the soil itself, certainly have major influences on the structure of rhizomicrobial communities, impacting on their spatial, temporal and functional components. While our knowledge of the microbial diversity in the rhizosphere is augmenting, understanding the principles underlying its structuring are today's and tomorrow's challenge.

Microorganisms present in the rhizosphere play important roles in the growth and in ecological fitness of their plant host. Important microbial processes that are expected to occur in the rhizosphere include pathogenesis and its counterpart, plant protection, as well as the production of antibiotics, geochemical cycling of minerals and plant colonization (Kent and Triplett 2002). Most of these processes are well studied in bacterial and fungal groups isolated from the rhizosphere of many plant species. However, microbes in the rhizosphere—Bacteria, Archaea and Fungi—do not escape the "big plate anomaly", by which a huge gap exists between the numbers of (viable) cells present in any sample and that of colonies retrieved on culture media. This leaves many species in the unknown: about 90% of the microbial cells present on plant roots and observed by microscopy are not recovered by cultivation in vitro (Goodman et al. 1998). Molecular, "culture independent" technologies are now doing much to close this gap. More so, the advent of parallel, very high throughput sequencing is promising to not only reveal "who's there" (richness) but also to answer questions such as "how many are you" (abundance) as well as "what are you doing there" (function). As an example, Fulthorpe et al. (2008) analyzed 139 819 bacterial sequences from the V9 region of the 16 S rRNA marker gene from four soils of different geographical regions. The most abundant taxa (>1%) were quite different from the "classical" dominant soil bacteria, with *Chitinobacteria* spp., *Acidobacterium* spp. and *Acidovorax* spp. constituting between 13% to 20% of the identified operational taxonomic units (OTU). Only three of the "classics" (*Bacillus*, *Flavobacterium* and *Pseudomonas*) were present in the list of the 10 most dominant taxa that included Proteobacteria, Bacteroidetes, Acidobacteria, Firmicutes and Gemmatimonades. Recent developments in rhizosphere soil metagenomics reveal that many hitherto unexplored bacterial groups (Kielak et al. 2009; Nunes de Rocha personal communication) are also present in the rhizosphere. Exploring these groups by unraveling their possible relationships with plants will start a new and fascinating area in rhizosphere research. It should also be noticed that in the Fulthorpe analysis, similarly to what is found in other metagenomic studies, about 40% of the reads were not attributed to any known OTU.

The analytical tool chosen for anyone's rhizosphere exploration may lead to different perceptions about bacterial species composition. By applying cultivation-based methods, particular groups will be strongly selected for, depending on the type of medium used. Like in all other terrestrial habitats, the cultured bacterial fraction in the rhizosphere must be considered as a small subset of the total, uncultured community (Staley and Konopka 1985; Hugenholtz et al. 1998). Therefore, the use of cultivation-independent techniques for recovery of bacterial groups from the rhizosphere must be considered as less biased. However, the main advantage of cultivation-based over cultivation-independent techniques is that bacterial isolates will be obtained that can be used for studying biotic interactions, e.g. with other microbial groups or with the plant host.

Bacterial communities are not uniformly distributed along root axes, but differ between root zones. Distinct bacterial community compositions are obtained by molecular fingerprints in different root zones, like those of emerging roots and root tips, elongating roots, sites of emergence of lateral roots, and older roots (Yang and Crowley 2000). It has been proposed that populations residing in the rhizosphere oscillate along root axes in a wave-like fashion (Semenov et al. 1999). Accordingly, bacterial communities temporarily profit from the nutrients released by younger roots in the root hair zones, and wave-like fluctuations in bacterial cell numbers can be explained by death and lysis of bacterial cells upon starvation when nutrients become depleted, followed by cell divisions in surviving and thus viable populations as promoted by the release of nutrients from dead and decaying cells (Semenov et al. 1999). Bacterial communities in rhizosphere soils are thus not static, but will fluctuate over time in different root zones, and bacterial compositions will differ between different soil types, plant species, plant growth seasons and local climates.

Changes induced in the soil by the growing root provide additional niches for soil microbes. Rhizosphere conditions sustain communities which differ from those found in bulk soil. Hence, these communities exhibit a "rhizosphere effect" (Curl and Truelove 1986; Lynch and Hobbie 1988).

Prokaryotes in the rhizosphere

Bacteria

Cultivation-based analyses of rhizosphere bacteria The structure of soil bacterial communities has long known to be strongly influenced by the presence of a root. Studies based on the use of growth media steadily showed that bacterial populations residing in the rhizosphere are one to two orders of magnitude larger than those residing in bulk soils and, are constituted by a larger fraction of gram negative cells, more symbionts and more r strategists (Curl and Truelove 1986). It was thought that fast-growing and opportunistic species, like pseudomonad species, would dominate in the rhizosphere because available nutrients are present in excess. This is in contrast to the situation in bulk soils in which the growth of such copiotrophes (*r* strategists) was considered to be limited due to low nutrient availability. Remarkable were the reports made by Miller and coworkers in 1989 and 1990 showing that not pseudomonads but rather Actinobacteria and 'coryneforms' dominate the rhizosphere of maize, grass and wheat. Nowadays, it is generally accepted that copiotrophs (*r* strategists) and also oligotrophs (*K* strategists) dominate in the rhizosphere but are spatially separated. In general, copiotrophs will thrive in places with high nutrient availability like root hair zones, while oligotrophs will gain hold in root zones exhausted in nutrients like older root parts (Semenov et al. 1999). Bacterial stress caused by nutrient limitations may be a common situation occurring in the rhizosphere and it can be held responsible for the presence of a wide variety of physiologically different bacterial groups in this habitat. Isolation of bacteria from grass rhizosphere soil on low nutrient agar media (ten times diluted TSB and soil extract medium) resulted in the recovery of a large diversity of Gram-positive and Gram-negative species, belonging to the Proteobacteria, Actinobacteria and Firmicutes (Nijhuis et al. 1993). Isolates were selected and tested for their capacity to colonize grass roots. One isolate in particular, identified as *Pseudomonas* (later redefined as *Burkholderia*) *cepacia* P2 was able to colonize the entire grass root system, including younger and older roots, at levels between 10^3 and 10^7 colony forming units per gram rhizosphere soil. Application of media low in nutrients thus resulted in the recovery of 'novel' species, naturally residing in the rhizosphere.

Cultivation-independent analyses of rhizosphere bacteria Novel cultivation-independent techniques are rapidly changing this 'panel' of rhizobacteria, to include uncultured microorganisms, providing a different representation of bacterial community composition. The precision of diversity analyses is totally dependent upon the taxonomic resolution achieved by any one study, and it should be remembered that closely related organisms, as inferred from their 16S rRNA sequences, could still exhibit very large physiological and ecological differences (Cohan and Perry 2007). Nevertheless, defined bacterial taxa appear to be commonly found in, or totally absent from the rhizosphere under many conditions, enabling comparisons between bulk soil and rhizosphere to be performed, thereby building an empirical basis for understanding rhizosphere processes.

In a study of rhizosphere soil of *Medicago sativa* and *Chenopodium album*, 16S rRNA gene sequences from isolates and library clones derived from rhizosphere DNA extracts were sequenced and compared with database sequences (Schwieger and Tebbe 2000). The most abundant isolates recovered from both plants showed closest similarities with *Variovorax* sp. strain WFF052 (β proteobacteria), *Acinetobacter calcoaceticus* (γ proteobacteria) and *Arthrobacter ramosus* (high GC Gram-positives), whereas via direct 16S rRNA gene PCR amplification and sequencing, the most abundant groups were Proteobacteria, members of the (former) *Cytophaga-Flavobacterium-Bacteroides* (CFB) group and high GC Gram-positives. This, as well as other studies (Heuer et al. 2002) clearly show that both approaches sometimes overlap; i.e. the same bacterial groups are detected by both techniques, but differences exist due to the absence of particular, supposedly uncultured groups of bacteria from collections of isolates.

Many studies indeed suggest that the Proteobacteria and the Actinobacteria form the most common of the dominant populations (>1%, usually much more) found in the rhizosphere of many different plant species (Singh et al. 2007; Green et al. 2006; Sharma et al. 2005; Chelius and Triplett 2001; Kaiser et al. 2001; Schallmach et al. 2000; Marilley and Aragno 1999). These groups contain many 'cultured' members. They are the most studied of the rhizobacteria, and as such, contain the majority of the organisms investigated, both as beneficial microbial inoculants and as pathogens.

Culture-independent analyses performed with various plants often reveal that members of the (former) *Cytophaga–Flavobacterium–Bacteroidetes* constitute a dominant fraction of the rhizobacteria, as with *Lolium perenne* (Marilley and Aragno 1999), maize (*Zea mays*; Chelius and Triplett 2001) or Lodgepole pine (*Pinus contorta*; Chow et al. 2002), and cucumber (*Cucumis sativus*; Green et al. 2006). In this latter publication, in which the rhizobacteria associated with the transition from seeds to root were analyzed, *Chryseobacterium* (Bacteroidetes) and Oxalobacteraceae (ß proteobacteria) were the most persistent populations. However, while Kaiser et al. (2001) also found dominant Bacteroidetes populations in canola (*Brassica napus*), Smalla et al. (2001) did not, although both studies appear to have been conducted at the same location. Other groups that may dominate the rhizosphere as shown by culture-independent analyses, include Acidobacterium, Verrucomicrobia, Gemmatimonadetes and Nitrospirae (Ludwig et al. 1997; Singh et al. 2007; Filion et al. 2004; Chow et al. 2002).

Acidobacteria and Verrucomicrobia can form highly dominant rhizosphere populations. Members of this group were found in the rhizosphere of two plant species endemic to South Africa, *Leucospermum truncatulum* and *Leucandendron xanthoconus* (Stafford et al. 2005). Nineteen percent of a 16S rRNA gene-based clone library from *Pinus contorta* (Lodgepole pine) was found to originate from seven different subgroups of Acidobacteria (Chow et al. 2002). In the rhizosphere of *Thlaspi caerulescens*, 14.5% of the 16S rRNA gene-based library clones were affiliated with acidobacterial sequences present in databases, belonging to five different subgroups (Gremion et al. 2003). In libraries of *Lolium perenne* (Perrenial ryegrass) (Singh et al. 2007) and *Castanea crenatam* (Chestnut) rhizosphere soils (Lee et al. 2008), 54.5% and 65% of the 16S rRNA gene clones were *Acidobacteria,* respectively. Verrucomicrobial species were found in the rhizosphere of *Pinus contorta* (Chow et al. 2002), *T. caerulescens* (Gremion et al. 2003), *L. truncatulum* and *L. xanthoconus* (Stafford et al. 2005), although at a lower abundance than that of Acidobacteria. Most likely, members of the Acidobacteria and Verrucomicrobia are common in the rhizosphere and occasionally, acidobacterial species are present at high abundance.

This indicates that both groups must be important members of rhizosphere communities. Their high diversities and abundances, especially those observed amongst acidobacterial species, might imply that these groups are important for the functioning of rhizosphere ecosystems and may be linked to plant health and fitness. In a study done Lee and coworkers (2008) it was shown that recovered Acidobacteria were metabolically active in the rhizosphere of *C. crenata,* suggesting that the rhizosphere is a natural habitat for this group of species. The difficulty of putting these bacteria in culture hampers the investigation of the roles they may play in the rhizosphere. In order to circumvent such problems, a partial genomic fragment of an acidobacterial group 1 species was obtained from an *R. solani* AG3-suppressive soil (Van Elsas et al. 2008a). Attempts to measure gene activities from this metagenomic clone remained unsuccessful (Van Elsas et al. 2008a, b). It may be a long lasting effort, if ever be successful, to establish the ecological relevance of these hard-to-culture species in the rhizosphere by making use of culture-independent approaches only. Therefore, recovering cultured members of the Acidobacteria and of the Verrucomicrobia from the rhizosphere should still be considered a priority in the study of the interactions of these bacteria with plants and of their contributions to rhizosphere processes. So far, only a few cultured Acidobacteria strains were recovered from bulk soils (Janssen et al. 2002). Recently, a few isolates belonging to Acidobacteria group 8 and to Verrucomicrobia were obtained from the rhizosphere of *Allium porrum* (Leek) and from potato (*Solanum tuberosum*; Nunes pers comm). All recovered isolates are to be soon tested in a soil-plant experimental set up. Based on the common occurrence and high population levels reached by these bacteria in the rhizosphere as well as in other natural habitats, it must be assumed that they fill important functions in numerous microbial ecological processes. Table 1 summarizes the abundance of the most dominant groups in the rhizosphere based on culture-independent approaches.

Proteobacteria It is clear that the internal diversity of high level taxa such as the Proteobacteria, Actinobacteria, Acidobacteria and else can be enormous and can fluctuate between studies and treatments within studies. For example, within the Proteobacteria, maybe the most dominant of the rhizobacteria, Chow et al. (2002) found that the alpha sub-class was the most abundant with, in decreasing order, the beta, gamma and delta sub-class following. The alpha-Proteobacteria contain the Rhizobiaceae, a large family that includes the important rhizobacteria nitrogen-fixing symbionts of legumes, as well as the plant pathogens *Agrobacterium* spp. In other studies, under other conditions the order can be quite different (Filion et al. 2004; McCaig et al. 1999). Clearly, the greatest interest in this type of diversity data lies in comparative analyses between soils, plants, plant age, and/or cultural conditions, providing some insight as to the main factors that structure these communities. One should also remember that as in bulk soil, a large fraction (up to 30%) of the sequence reads obtained in molecular studies does not match any identified taxon.

Since many of the cultured rhizobacteria are Proteobacteria, it only makes sense that these are the most studied and therefore the best known, providing a much more detailed picture of the genetic diversity of specific groups, and of the ecology of the groups' members.

The *Burkholderia* group of species, ß-Proteobacteria associated with roots of many different plants, is genetically diverse and is one of these well studied Proteobacteria. Members of this group can: positively

Table 1 Dominant bacterial groups found in the rhizosphere

Taxon	Abundance (% of total)	Quantification method	References[a]
Proteobacteria	27–66	Clone libraries	Chow et al. (2002); Filion et al. (2004); McCaig et al. (1999)
α	1–24	Clone libraries	
β	2–19	Clone libraries	
Burkholderia spp.	up to 93[b]	Clone libraries, isolates, colony hybridization	Di Cello et al.(1997); Ramette et al.(2005)
γ	0–9	Clone libraries	
Pseudomonas spp	0–19	Fluorescent in situ hybridization, isolates	Gochnauer et al. (1989); Watt et al. (2003), (2006)
δ [c1]	2–3	Clone libraries	
Acidobacteria	0–65	Clone libraries	Lee et al. (2008); Singh et al. (2007)
Actinobacteria	1–3	Clone libraries	Chelius and Triplett (2001); Singh et al. (2007), Stafford et al. (2005)
Cytophaga-Flavobacterium-Bacteroridetes[c]	0–3	Clone libraries	Chow et al. (2002), Marilley and Aragno (1999)
Verrucomicrobia	0–3	Clone libraries	Chow et al. (2002) Lee et al. (2008)
Unclassified	15–30	Clone libraries	

Only those most frequently found are included

[a] References relate to the appropriate phylum except for those in italics which are relevant to the genus.

[b] Population sizes and diversity vary between individual maize plants

[c] Although this taxon has been changed, it is used here for practical purposes

or negatively affect their hosts, are able to degrade toxic compounds and also act as opportunistic human pathogens (the *B. cepacia* complex) with this latter particular group being intensively studied (Coenye et al. 2001; Jacobs et al. 2008). *Burkholderia* spp. can represent more than 3.5% of the total cultured microbiota of the maize rhizosphere (Di Cello et al. 1997). However, abundance in individual plants may vary by more than a hundredfold, and above 90% of the colonies isolated from a maize plant were reported to belong to the *Burkholderia cepacia* complex (Ramette et al. 2005). In this latter study, up to seven isolates belonging to the *B. cepacia* complex species were recovered from the rhizosphere of individual plants, some exhibiting substantial allelic polymorphism, with *recA* used as a marker gene. Using new primers for the same gene, Payne et al. (2006) revealed an even higher diversity of uncultured *Burkholderia* species in the maize rhizosphere. Moreover, *Burkholderia* species composition and abundance greatly varied between individual plants and communities of the *B. cepacia* complex species and closely related strains of the same species were found to coexist at high population levels (Ramette et al. 2005). Temporal changes have also been noticed, the diversity of *Burkholderia* being markedly higher during the first stages of plant growth while decreasing over time (Di Cello et al. 1997). While time affected community structure, spatial distribution of *Burkholderia* strains isolated from individual plants, and from distinct root compartments was random, with rhizosphere and rhizoplane populations not significantly differing (Dalmastri et al. 1999). This is an interesting finding as the structure of rhizobacterial populations usually responds to this spatial gradient, with diversity decreasing from soil to root (Costa et al. 2006; Lerner et al. 2006; Marilley and Aragno 1999). In some studies, differences in bacterial community composition in bulk and rhizosphere soils were not always that clear but that may be due to limitations in the experimental set up where rhizosphere and bulk soils could not be distinguished from each other (Duineveld et al. 1998). Finally, it was also found that soil type and agricultural management regime were major factors affecting the genetic diversity of maize root–associated *B. cepacia* populations (Dalmastri et al. 1999; Salles et al. 2006). In the latter study it was found that more *Burkholderia* isolates antagonizing *Rhizoctonia solani* AG-3 were present in soil permanently covered with grass, demonstrating functional changes upon shifts in *Burkholderia* populations.

Plant and soil effects on bacterial community structure Based on such studies, some principles start to emerge. Among those, the impact of soil type on the structure of rhizobacterial communities appears to be general and dominant, and not restricted to one bacterial group (Latour et al. 1996; Marschner et al. 2004; Seldin et al. 1998).

The plant growth stage may be as important a factor in shaping rhizobacterial community structure (Herschkovitz et al. 2005; Lerner et al. 2006), and it may be the strongest factor affecting bacterial communities in the potato rhizosphere (Van Overbeek and Van Elsas 2008). Other factors shown to have an impact on rhizobacterial communities include seasonal changes (Dunfield and Germida 2003; van Overbeek and van Elsas 2008), plant species (Grayston et al. 1998; Smalla et al. 2001) and even cultivar (genotype) within the same plant species (Van Overbeek and Van Elsas 2008; Andreote et al. 2009). These data indicate that the rhizosphere is a highly dynamic environment for bacterial communities and that possibly more factors may affect rhizobacterial community composition, making the rhizosphere more variable and thus rather unpredictable for the presence or absence of particular bacterial species. While these studies show that such dynamics in bacterial community composition make it difficult to assign bacterial groups that are 'typical' to the rhizosphere, they also suggest that the driving forces shaping these communities can be elucidated.

Physical changes in soil structure after tillage were shown to reduce the diversity of bacteria, more so in bulk soil than in the rhizosphere (Lupwayi et al. 1998). A similar effect was detected when compost was applied to soil (Inbar et al. 2005). Finally, crop rotation increased bacterial diversity in the rhizosphere (Lupwayi et al. 1998; Alvey et al. 2003). As the plant species or its genotype (Dalmastri et al. 1999; Marschner et al. 2004) may alter rhizobacterial groups, crop rotation may enhance different populations. Such knowledge of the soil-plant interface is important for agricultural sustainability. It shows that agricultural practices that alter physical, chemical and biological soil properties can consequently affect rhizosphere processes. Biotic and abiotic factors influencing rhizosphere microbial communities are summarized in Table 2.

Microbial interactions in the rhizosphere Many types of interactions between bacterial populations take place within the rhizosphere. We shall only briefly mention two. One, antagonistic interactions between rhizobacteria and plant pathogens, is certainly the most relevant for agriculture and among the most studied. The other, predator-prey interactions between rhizobacteria is virtually virgin research territory. Interestingly, at least in some cases, the two may be linked.

Cultivation-based approaches are still the most commonly applied and useful methods for assessing possible interactions between antagonistic rhizobacterial isolates and plant pathogens. Dual plate tests make it possible to screen hundreds to thousands of isolates for antagonism against different phytopathogens. Many isolated antagonists are *Pseudomonas* spp., which as already mentioned are common rhizobacteria. As *Pseudomonas* spp. can lead to natural suppressiveness against take all disease (Chapter 13), this association has been extensively studied. Pseudomonads (*P. putida*) antagonistic to the soil-borne fungal pathogen *Verticillium dahliae* were isolated at high incidence from the rhizosphere of strawberry, potato and oilseed rape (Berg et al. 2002). Other important antagonists included other *Pseudomonas* species as well as various *Serratia* spp. Comparable observations were made in potato rhizosphere soils and showed that *Pseudomonas* (*P. chlororaphis*, *P. fluorescens*, *P. putida* and *P. syringae*) and *Serratia* (*S. grimesii*, *S. plymuthica*, *S. proteamaculans*), antagonized *Pectobacterium carotovorum* (formely *Erwinia carotovora*) and *Verticillium dahliae* (Lottmann et al. 1999). *P. putida* isolates also played an important role in the suppression of possible agents responsible for the apple replant disease, namely *Cylindrocarpon destructans*, *Pythium ultimum* and *Rhizoctonia solani* (Gu and Mazzola 2003). Most interestingly in this study was that disease suppressive groups were supported by wheat growth in a cultivar-dependent fashion. Also, addition of wheat root exudates to an apple orchard soil resulted in the stimulation of the disease-suppressive populations which demonstrates the importance of root exudation in the selection of antagonistic groups in the rhizosphere.

Cultivation-independent technologies are also usefully applied to study potential antagonist rhizobacteria. Bacterial group-specific primer systems,

Table 2 Biotic and abiotic factors affecting the structure of microbial rhizosphere communities

Biotic factor	Abiotic factor
Plant developmental stage	Soil type
B: Herschkovitz et al. (2005)	B: Latour et al. (1996); Marschner et al. (2004)
F: Mougel Gomes et al. (2006); Viebahn et al. (2005)	F: Mougel Gomes et al. (2006), Viebahn et al. (2005)
Plant species	Soil heterogeneity
B: Grayston et al. (1998), Smalla et al. (2001)	F: Viebahn et al. (2005)
F: Viebhan et al. (2005)	
Plant cultivar	Season
B: Van Overbeek and Van Elsas (2008a)	Smalla et al. (2001), Van Overbeek and Van Elsas (2008a)
Crop rotation	Tillage
B: Alvey et al. (2003), Lupwayi et al. (1998)	B: Lupwayi et al. (1998)
Viebahn et al. (2005)	
Proximity to root	Compost amendment
B: Costa et al. (2006), Lerner et al. (2006)	Inbar et al. (2005)
F: Mougel Gomes et al. (2006)	
Root zone	Wastewater irrigation
Semenov et al. (1999), Yang and Crowley (2000)	B: Oved et al. (2001)

The possibly most dominant parameters affecting community structure are written with bold characters

B bacteria, *F* fungi

developed and tested for specific detection of particular taxonomic groups, were applied to study the fingerprints of rhizosphere *Pseudomonas* (Garbeva et al. 2004) and Actinobacteria (Heuer et al. 2002) of field-grown potato. The rational of these studies was that these groups would cover the majority of the antagonists against major pathogens for potato (van Overbeek and van Elsas 2008). In accordance with other studies mentioned above, the effect of plant growth stage and genotype on the community composition of these rhizobacterial groups was demonstrated. Species belonging to both groups thus fluctuate depending on plant growth stage and applied cultivar, and most likely the antagonists belonging to both groups also do so. Accordingly, differences in levels of plant protection activity against phytopathogens by plant-associated communities can be expected, an important understanding for the development of agricultural management tools to control plant diseases.

Burkholderia, *Pseudomonas* and *Serratia* species can produce the antibiotic pyrrolnitrin (PRN; Kirner et al. 1998). The biosynthetic pathway of PRN is composed of four reactions encoded in the *prn*A, B, C and D operon, with *prn*D being highly conserved over all bacterial groups known to carry this operon (Garbeva et al. 2004; Costa et al. 2009). Primers targeting the *prn*D locus were applied to quantify the number of *prn* gene copies by real-time PCR in soils in crops with different histories, establishing a relationship between plant coverage and suppressiveness against *R. solani* AG3 (Garbeva et al. 2004), i.e. higher disease suppression as a result of cropping regime was related to *prn* locus copy number in soil.

Isolates from the potato rhizosphere belonging to the genera of *Lysobacter* appeared as a dominating group that were also able to antagonize the major potato pathogens *R. solani* AG3 and *Ralstonia solanacearum* biovar 2 (van Overbeek and van Elsas 2008). In another study, *Lysobacter* species (*L. antibioticus* and *L. gummosus*) were correlated with suppression of *R. solani* AG2.2IIIB in soils where sugar beets were grown (Postma et al. 2008). Interestingly, *Lysobacter* 16S rRNA gene clones dominated clone libraries obtained from the rhizosphere of *Calystegia soldanella* (beach morning glory) and *Elymus mollis* (wild rye) (Lee et al. 2006). The *Lysobacter* group may therefore be considered as an important group of rhizobacteria capable to suppress soil-borne phytopathogens.

It is noteworthy that many *Lysobacter* are predatory bacteria that can lyze and utilize other bacteria as nutrient sources (Reichenbach 2001). Other predatory bacteria that are found in rhizosphere soil are the *Bdellovibrio* and like organisms (BALOs). BALOs have been isolated from the rhizosphere of common bean (*Phaseolus vulgaris*), tomato (*Lycopersicon esculentum*; Jurkevitch 2007), Chinese cabbage (*Brassica campestris*; Elsherif and Grossmann 1996), strawberry (*Fragaria ananassa*), pepper (*Capsicum annum* L.), Lily of the valley (*Convallaria majalis* L.; Markelova and Kershentsev 1998) and soybean (*Glycine max*; Scherff 1973). BALOs are gram negative, obligate predators of other bacteria; they seem to be ubiquitous in soil and are commonly found in very different environments (Jurkevitch 2007). Rhizosphere BALO populations differ from those of bulk soil (Jurkevitch 2007) and are affected by plant age (Hershkovitz et al. 2005). Since BALO isolates can differ in prey ranges (Jurkevitch 2007), BALO populations may be altered by changes affecting the community structure of rhizobacterial communities at large. An increase in rhizosphere BALOs able to prey upon rhizobacterial fluorescent pseudomonads following an increase in population levels of the latter perhaps reflects such interactions (Elsherif and Grossmann 1996). Predator-prey interactions between rhizobacteria may thus be a significant—yet largely ignored—contributing force in the rhizosphere.

Mineral cycling Mineral cycling is a central process occurring in the rhizosphere. Growing roots are considered as a significant source of carbon and support a milieu of high microbial activity and high biomass where elemental turnover is enhanced. *Cytophaga*-like bacteria (former CFB) may be important contributors to this increased nutrient turnover in rhizospheric soils (Johansen and Binnerup 2002).

Key processes occurring in the rhizosphere are N fixation by diazotrophs (Nelson and Mele 2007) and ammonia oxidation (Nelson and Mele 2007; Chen et al. 2008). Phylogenetic groups involved in N fixation were found to be very broad, as was shown by DNA sequence analyses of a *nif*H-based clone library from the wheat rhizosphere (Nelson and Mele 2007). *Azospirillum*, *Rhodobacter* and *Beijerinckia* (all α Proteobacteria); *Azotobacter chroococcum*, *Vibrio diazotrophicus*, *Klebsiella pneumoniae* (all γ Proteo-

bacterial species) and cyanobacteria were all represented. In contrast, ammonia monooxygenase α subunit (*amo*A), a key enzyme involved in ammonia oxidation, is mainly limited to two genera: *Nitrosomonas* and *Nitrosospira*. Analyses of library clones based on *amo*A cDNA from the rice rhizosphere (Chen et al. 2008) and *amo*A PCR amplicons from wheat rhizosphere DNA (Nelson and Mele 2007) revealed the presence of *Nitrosospira* and some other bacterial groups, but not of *Nitrosomonas*. Supposedly, ammonia oxidizing bacteria belonging to *Nitrosospira* dominated over *Nitrosomonas* in both rhizosphere soils. Yet, under changing conditions these groups may react differently. The impact or wastewater irrigation, a growing water resource, on ammonia-oxidizing bacteria was examined in orchard soils. *Nitrosomonas*-like populations became dominant in effluent-irrigated soils in contrast to the *Nitrosospira*-like populations that dominated in soils irrigated with fertilizer-amended water (Oved et al. 2001). The ecological function is preserved but the service provider is changed.

Archaea

Much less is known on Archaea in soils, let alone in the rhizosphere, than on Bacteria. Archaea are "new comers" in biology (since Woese and Fox 1977). They were discovered in extreme environments and while much of the research on Archaea still concentrates on extreme biotops, they were shown to be essential actors in global processes, such as nitrification in the ocean (Wuchter et al. 2006), ammonification in soils (Leininger et al. 2006), and methane evolution (Erkel et al. 2006). For an enlighten narrative on the third domain of life, see Friend (2007).

Biotic and abiotic effects on dominant rhizosphere Archaea In earlier 16S rRNA gene amplification studies, both Crenarcheoata and Euryarcheota sequences were retrieved from soils (Bintrim et al. 1997; Borneman and Triplett 1997; Jurgens and Saano 1997). In a later analysis of seven diverse soils, one lineage of Crenarcheaota was consistently found and its abundance varied between 0.5% and 3% (relative to Bacteria) in bulk soil samples of a sandy habitat. However, only 0.16% archaeal sequences were retrieved from the rhizosphere of *Festuca ovina s.l* growing in this soil (Ochsenreiter et al. 2003). Different results were obtained by Simon et al. (2001), using fluorescent in-situ hybridization. They found that non-thermophylic Crenarcheota extensively colonized tomato roots, more so senescent roots, and constituted between 4% and 16% (relative to Bacteria) of the root colonizers. Bomberg et al. (2003) analyzed the extensive micorrhizosphere of pine boreal forest humus microcosms. Crenarcheaotal sequences were found in ectomycorrhizae-colonized roots but not in fungus-free roots. Also, the diversity of Crenarcheaota appeared to be larger in the rhizosphere than in adjacent bulk soil. These studies suggest that crenarchaeotes of the C1a, C1b, and C1c clades associate with a variety of terrestrial plant root systems. An in-depth study was conducted to determine to what extend the bulk soil and the rhizosphere selected for different crenarcheotes of the C1b clade (Sliwinski and Goodman 2004). Terrestrial plant roots were collected from 12 sampling locations harboring divergent flora. Significant differences were found between the PCR-SSCP profiles from the rhizosphere and bulk soils. In general, rhizosphere crenarchaeal richness was higher than that of bulk soil but appeared to be plant-lineage independent. Furthermore, some undefined effects were found to influence the way the rhizosphere differed from bulk soil. In another study by Nicol et al. (2004), amendments (nitrogen fertilization as ammonium or sheep urine, and pH change) to upland pasture rhizosphere soils had no effect on their crenarcheotal dominant and active 1.1b and 1.1c groups as determined by RT-PCR DGGE of the 16S rRNA gene. In contrast, the various amendments affected organisms belonging to a variety of major soil bacterial groups including alpha- and gamma-Proteobacteria, Actinobacteria and Acidobacteria. This led the authors to suggest that these factors are not major drivers of archaeal diversity in soil. Crenarchaeal associations with plants in native environments may therefore be dictated by environmental conditions, and not only result from the interactions between plant and microbe. This is markedly different from what is known with rhizosphere Bacteria, where the proximity to roots seems to reduce diversity, and soil type (and agricultural amendments) significantly affects community structure, suggesting that different factors affect the Archaea and the Bacteria that colonize plants.

Recently, David Valentine (2007) suggested that adaptation to energy stress dictates the ecology and evolution of Archaea, in contrast to many Bacteria that became adapted to maximize energy availability. If this is correct, these communities certainly largely diverge in their reactions to environmental factors, enabling them to co-exist as they fill different niches in a shared physical space.

Importance of rhizosphere Archaea in elemental cycling Rhizosphere Archaea are associated with the major crop, rice. Rice fields contribute 10% to 25% of global methane emissions, a potent greenhouse gas to the atmosphere (Houweling et al. 1999). The source of this methane is in large proportion methanogenic Archaea, and it is released through the gas vascular system of the rice plant (Nouchi et al. 1990). Although soil in rice paddies is anaerobic, oxygen diffuses into the roots, leading to transient oxic conditions on the root surface and in the adjacent rhizosphere soil. Differently to soils where crenarchaeotes are the largely dominant (if not exclusive) Archaea, members of the Euryarchaeota are the main archeal colonizers of rice roots (Conrad et al. 2008; Ramakrishnan et al. 2001). A large diversity of Archaea from the families *Methanosarcinaceae*, *Methanosaetaceae*, *Methanomicrobiaceae*, and *Methanobacteriaceae* and new euryarchaeotal sequence clusters named (RC) I, II, III and V, and crenarchaeotal clusters RC-IV and RC-VI were detected in rice soils (Ramakrishnan et al. 2001). Using pulse labeling of rice plants with $^{13}CO_2$, Lu and Conrad (2005) demonstrated the incorporation of ^{13}C into the ribosomal RNA of Rice Cluster I Archaea in rice rhizospheric soil. Thus, this archaeal group may play an important role in CH_4 production from plant-derived carbon. Although not isolated in pure culture, an RC-I genome was completed and revealed hitherto unknown functions in methanogens (Erkel et al. 2006). This includes aerotolerance and H_2/CO-dependence, the Embden–Meyerhof–Parnas pathway for carbohydrate metabolism, and assimilatory sulfate reduction. RC-I also possesses a unique set of antioxidant enzymes, DNA repair mechanisms, and oxygen-insensitive enzymes that provide it with the capacities to thrive in the relatively oxydized rice rhizosphere.

Herrmann et al. (2008) reported that archaeal ammonia monooxygenase genes were strongly enriched in the rhizosphere of the submersed macrophyte *Littorella uniflora* in comparison to the surrounding sediments. These sequences were 500 to 8,000 times more abundant than the cognate bacterial genes (as determined by q-PCR). This strongly suggests that rhizosphere Archaea play yet another important role in nutrient cycling and are major contributors to nitrification in freshwater sediments. Another recent study put forward data showing that ammonia oxidizing archaea predominate in paddy soil (Chen et al. 2008). Moreover, these Archaea were more abundant in the rhizosphere than in bulk soil. It was proposed that ammonium-oxidizing Archaea were more influenced by exudation from rice root (e.g. oxygen, carbon dioxide) than their bacterial counterparts.

To summarize, Archaea appear to be common but not major root inhabitants. Under specific conditions, such as reduced oxygen and/or high CO or CO_2 pressure, they may become very abundant and active contributors to rhizosphere processes. While few of their roles are now known, Archaea may largely be unsung heroes, participating in nutrient turnover and sustaining important ecological functions in plant roots.

Phages in the rhizophere

Viruses are the most abundant biological entities on Earth. With an estimated total population size of $>10^{30}$ in the ocean alone, phages contribute 20% to 40% of the bacterial mortality, and are therefore a driving force of biogeochemical nutrient cycles (Kimura et al. 2008; Suttle 2005). In addition to contributing to mortality, phages act as vectors of lateral gene transfer, thereby contributing to bacterial genome evolution (Canchaya et al. 2003). They also constitute the largest genetic reservoir on Earth (Kimura et al. 2008). However, few studies address phage diversity and impact in soils, and even less do so for the rhizosphere.

Bacteria and their phages exhibit predator-prey relationships. They co-exist in the same habitats through coevolution and "arms race" but trade-off between competitive ability and resistance to predators are at the basis of this coexistence. These trade-offs are modulated by genetic factors, environmental factors, and gene-by-environment interactions (Bohannan et al. 2002). Lysogeny may be a solution for the arms race in coevolution in various environments, and it seems to be extremely distributed and common (Kimura et al. 2008).

The effects of viruses on beneficial bacteria and on soil-borne plant pathogens have been extensively studied and as such stand out in comparison to the dearth of data on rhizosphere viruses (for a list of references, see Kimura et al. 2008). In general, phages are abundant in soils where bacterial hosts can be found. As an example, phages active against rhizobia can be detected in legume fields (Sharma et al. 2005). A similar trend is observed with predatory bacteria such as the *Bdellovibrio* and like organisms, which are more abundant in soils inoculated with their prey (Elsherif and Grossmann 1996, Jurkevitch, unpublished data).

Ashelford et al. (2003), using electron microscopy reported a mean of $1.5 \cdot 10^7$ virus-like particles (VLP) g^{-1} soil. These authors estimated that viruses in soils were equivalent to 4% of the total population of bacteria, much less than the viral-to-bacterial (VBR) ratios in the oceans of between three and 25 (Fuhrman 1999). In another study, up to 10^9 VLP were measured in different soils (Williamson et al. 2005). These particles exhibited a wide range of capsid diameters (20 to 160 nm) and morphologies, including filamentous and elongated capsid forms. Moreover, there was a significant link between land use (agricultural or forested), soil organic matter and water content to both bacterial and viral abundances. Viruses thus respond to physico-chemical parameters in the soil, yet the major structuring force may be the prey community. In contrast to the Ashelford et al. (2003) study, extremely high VBRs were found in agricultural soils (approximately 3,000) but they were much lower in forested soils (approximately ten). In rice paddies, 5.6×10^6 to 1.2×10^9 VLP mL^{-1} were detected (Nakayama et al. 2007). Another study examined the diversity of T4 phages, also in rice paddies (Jia et al. 2007): The sequences obtained from rice field were distinct from those obtained from marine environments. Phylogenetic analysis showed that most sequences belonged to two novel subgroups of T4-type bacteriophages, although some of them were related to already known subgroups of these T4-type viruses.

Archaeal viruses of Euryarchaeota and of Crenarchaeota have been described, and many exhibit exceptionally complex morphologies. They mostly have been isolated from extreme habitats, as specific screening for archaeal viruses has only been done in such extreme hydrothermal and hypersaline environments. However, some head-tail phages, with non-enveloped virions carrying icosahedral heads and helical tails that infect extreme halophiles or mesophilic or moderately thermophilic methanogens (Euryarchaeota) have been described (Prangishvili et al. 2006).

In summary, the soil and the rhizosphere's virosphere, are essentially uncharted. Undoubtedly, a lot of exciting biology lies ahead.

Fungal decomposers in the rhizosphere

Relative importance

Organotrophic fungi, better known as saprotrophic fungi, occur in the rhizosphere as has been shown by both cultivable—, biochemical—and PCR-based techniques (Smit et al. 1999; Viebahn et al. 2005; Berg et al. 2005; Baum and Hrynkiewicz 2006; De Boer et al. 2008; Zachow et al. 2008). The rhizosphere saprotrophic fungal community appears to consist of both yeasts and filamentous fungi with representatives of all major terrestrial phyla (Ascomycota and Basidiomycota) and sub-phyla (Mucoromycotina; Marcial Gomes et al. 2003; Renker et al. 2004; Berg et al. 2005; Vujanovic et al. 2007).

Plant species, plant developmental stage and soil type have been indicated as major factors determining the composition of rhizosphere fungal communities (Marcial Gomes et al. 2003: Viebahn et al. 2005; Mougel et al. 2006; Broz et al. 2007; Singh et al. 2007.; Broeckling et al. 2008). With respect to plant species, the composition of root exudates has been shown to be an important criterion for selection of rhizosphere fungi (Broeckling et al. 2008). The effect of plant developmental stage is probably caused by changes in the composition and quantity of rhizodeposits (exudates and sloughed-off root cells) during plant maturation (Marcial Gomes et al. 2003; Mougel et al. 2006; Houlden et al. 2008).

The phylogenetic composition of fungi in the rhizosphere does not give information on the abundance of saprotrophic fungi in the rhizosphere. Biomass measurements of fungal saprotrophs have been done using fungal specific biomarkers, in particular ergosterol. Ergosterol is a cell membrane sterol specific for fungi, but appears to be virtually absent in arbuscular mycorrhizal fungi (Weete 1989; Grandmougin-Ferjani et al. 1999; Olsson et al. 2003). AM fungi form associations with grasses and herbs. Hence, ergosterol-based fungal biomass in the rhizosphere of grasses and herbs will largely consist of saprotrophs. These are not only obligate saprotrophs but also plant-pathogens that are growing as saprotrophs in the rhizosphere. Since most tree species form mycorrhizal associations with basidiomycetes, ergosterol determinations in rhizosphere soil of trees cannot be used as an estimate for saprotrophic fungi.

Upon conversion of ergosterol to fungal biomass C (multiplying with 90) it was revealed that fungi can make up a significant amount (average 39%; range 20–66%) of the microbial biomass in the rhizosphere of grassland plants (Joergensen 2000). Interestingly, fungal biomass in the bulk soil was significantly correlated with that of rhizosphere soil. This may indicate that saprotrophic fungi involved in soil organic matter degradation do extend into the rhizosphere. This is supported by ergosterol-based measurements of fungal biomass in the rhizosphere of the non-mycorrhizal plant sand sedge (*Carex arenaria*) that was the main understory plant in some Dutch pine forests. The ergosterol concentration in the rhizosphere of sand sedge was always higher than in the bulk soil of these forests (Fig. 1). This suggests that saprotrophic fungi in the bulk soil do not only extend into rhizosphere but do also produce extra biomass in this habitat.

Plant species effects are not only apparent for the composition of saprotrophic fungi in the rhizosphere but also for the ergosterol-based fungal biomass (Appuhn and Joergensen 2006). This could be due to differences in the quantity of rhizodeposits between plant species. However, fungal community composition in the bulk soil has also been indicated as an important factor in determining differences in the amount of rhizosphere-associated fungal biomass between plant species (Broeckling et al. 2008).

Fungal decomposition niches

Based on several studies done on fungal- and bacterial activities in terrestrial ecosystems, a general picture has arisen of a major niche differentiation between organotrophic fungi and bacteria: Bacteria are mostly involved in the degradation of simple, soluble substrates whereas fungi are the main decomposers of solid, recalcitrant substrates (De Boer et al. 2005, 2006; Van der Wal et al. 2007; Paterson et al. 2006). This niche differentiation is, however, not strict. Certain bacteria, e.g. actinomycetes, are also specialized to degrade solid polymers like cellulose, whereas yeasts and sugar fungi are rapidly reacting when simple compounds like sugars are added to a soil (Kirby 2006; Van der Wal et al. 2006). In addition, fungi that are specialized to degrade the most abundant polymeric substances (cellulose, hemicellulose, lignin) do this by producing extracellular enzymes. The monomers and oligomers, e.g. sugars and disaccharides, which are released by the extracellular hydrolytic enzymes, are the actual substrates taken up by these fungi. Such compounds form also part of the exudates released by roots and could, therefore, be metabolized by fungal decomposers of soil organic matter (De Boer et al. 2008; Gramms and Bergmann 2008).

From the above it will be obvious that rhizosphere saprotrophic fungi may be involved in the degradation of both simple root exudates and the more complex compounds in sloughed off root cells. $^{13}CO_2$ pulse-labeling of plants indicated a predominant role of fungi in the decomposition of simple root exudates (Butler et al. 2003, Treonis et al. 2004). However, the fungal biomarker that was found to be rapidly enriched in ^{13}C was the phospholipid fatty acid

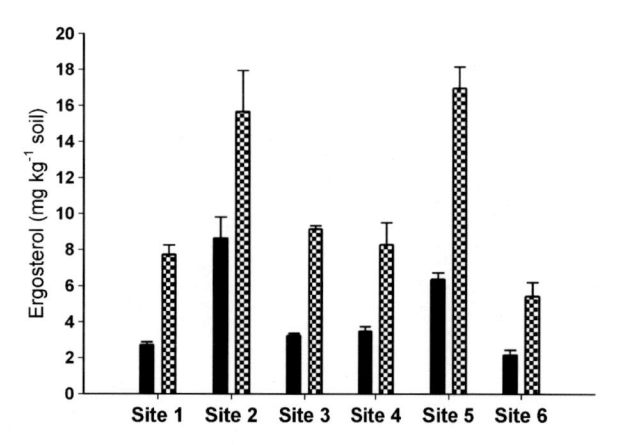

Fig. 1 Concentration of the fungal biomarker ergosterol in six Dutch sandy forest soils and in the rhizosphere soil of sand sedge plants (*Carex arenaria*) growing as understory plant in these forests. Data represent the mean and standard deviation of bulk soil (*black bars*) and rhizosphere soil (*checkered bars*) sampled from the upper 5 cm mineral soil layer at the different locations

(PLFA) 18:2ω6,9. This PLFA also occurs in the membranes of plant cells. Hence, overestimation of fungal ^{13}C enrichment cannot be excluded although low ^{13}C enrichment of other plant PLFAs shows that this is not likely (Treonis et al. 2004). Upon pulse-labeling of grassland plants with $^{13}CO_2$, Denef et al. (2007) showed rapid incorporation of ^{13}C in PLFAs that are characteristic for general fungi (18:2ω6,9) and arbuscular mycorrhizal fungi (16:1ω5). Only later in the experiment, bacterial PLFAs became enriched in ^{13}C. Similar observations had been made by Olsson and Johnson (2005). Denef et al. (2007) proposed that the first transfer of carbon from the roots to the soil is via arbuscular mycorrhizal fungi and other fungi that are closely associated with roots. The ^{13}C-enrichment of bacterial PLFAs in a later phase should then be considered as growth on fungal exudates or on dead/living hyphae. This hypothesis of bacteria growing on fungal exudates rather than direct on root exudates is in line with the concept of the mycorrhizosphere (Timonen and Marschner 2005).

By including saprotrophic fungi that have close associations with roots, the mycorrhizosphere concept is even extended. Indeed, it has been shown that *Trichoderma* spp. that are known as common soil saprotrophs and mycoparasites (growing on other fungi) can colonize the root epidermis and even a few cell layers below this level (Harman et al. 2004) Hence, direct uptake of root derived carbon by such root-colonizing *Trichoderma* spp. seems likely. Also for wood-decay fungi colonization of fine, living roots of trees has been shown (Vasiliauskas et al. 2007). If the colonization of roots by saprotrophic fungi is a general phenomenon, then apparently saprotrophic rhizosphere fungi may, at least partly, consist of fungi that have direct contacts with roots and obtain their energy from within the root rather than from consumption of root exudates.

Paterson et al. (2006) developed an artificial root from which simple root exudates were released into the soil at a rate similar to that done by young *Lolium perenne* plants. PLFA analysis showed that most of the root exudates were decomposed by gram-negative bacteria. However, fungal PLFAs did also increase in the rhizosphere and ^{13}C incorporation of selected root exudates (glucose and fumaric acid) into fungal PLFAs was observed. Hence, involvement of fungi in the degradation of simple root exudates seems likely. Yet, the relative contribution of fungi to decomposition of simple root exudates is not known for the time being. Fungal decomposition of simple root exudates in the rhizosphere may be related to the amount of exudates that are released per time unit. An increase in the loading of artificial root exudates to a pasture soil resulted in a shift from bacterial-dominated decomposition to fungal decomposition (Griffiths et al. 1999).

A predominant role of saprotrophic fungi in the degradation of more complex, solid rhizodeposits is likely. However, direct measurements of this have not been made and the evidence is circumstantial. For instance, Paterson et al. (2006)) showed that ^{13}C-labeled insoluble fractions of the grass *Lollium perenne* in a grassland soil were mainly decomposed by fungi. Mougel et al. (2006) observed strong shifts in fungal community composition in the rhizosphere of *Medicago truncata* during maturation of the plant whereas this was not seen for the rhizosphere bacterial community. They ascribed this to a response of fungal species to the release of more recalcitrant rhizodeposits. Interestingly, plants can also affect the fungal decomposition of soil organic matter (Subke et al. 2004). Hence, this may indicate that fungal degraders of recalcitrant compounds may be able to acquire root-derived compounds and that this extra source of energy stimulates their activities to degrade soil organic matter. In a similar way plants may stimulate the fungal degradation of organic pollutants in soil (Gramms and Bergmann 2008).

It is obvious that more studies are needed to evaluate the importance of decomposing activities of saprotrophic fungi in the rhizosphere. Based on the current knowledge, it is clear that contribution of fungi to decomposition of simple soluble root exudates cannot be neglected. In addition, direct consumption of root-derived compounds by hyphae colonizing and penetrating roots may be an important source of energy for saprotrophic fungi in the rhizosphere. However, the general idea that bacteria, in particular gram-negative bacteria, are the most important degraders of simple root or mycorrhizal exudates is not (yet) proven to be wrong. The same holds true for the idea that fungi are the most important degraders of more complex rhizodeposits.

Impact on plant performance

So far, most studies on the effects of saprotrophic fungi on plant performance are dealing with the suppression of root pathogens. Detailed studies have been done on *Fusarium oxysporum* of which pathogenic and non-pathogenic strains can co-occur in the rhizosphere (Nel et al. 2006). In fact, the pathogenic strains can be considered as saprotrophic fungi with the ability to infect plants, whereas the non-pathogenic strains only behave as saprotrophs. The non-pathogenic strains can suppress the pathogenic ones by competition for nutrients (root-exudates) and space (infection-sites on the roots) or by inducing resistance in the plant (Fravel et al. 2003).

Many opportunistic saprotrophic fungal strains of the genus *Trichoderma* strains are rhizosphere-competent and are often able to attack other fungi in the rhizosphere (Vinale et al. 2008). This mycoparasitic growth lies at the basis of control of plant-pathogenic fungi by *Trichoderma* spp. Indeed, the positive effect that *Trichoderma* spp. often has on plant performance is ascribed to direct control of plant-pathogens via mycoparasitism or antibiosis (Vinale et al. 2008). However, indirect effects by triggering the resistance of plants against plant pathogens appears also important, especially by *Trichoderma* strains that are able to infect the outer root layers (Harman et al. 2004; Van Wees et al. 2008). In addition, there is evidence that plant hormone-like compounds produced by *Trichoderma* spp. can stimulate plant growth (Vinale et al. 2008).

Saprotrophic fungi in the rhizosphere may also have an indirect effect on suppression of plant-pathogenic fungi. De Boer et al. (2008) showed that the frequency of potential antifungal properties of rhizosphere bacteria in fungal-rich soils is higher than in fungal-poor soils. This has been interpreted as the result of a selective competitive pressure of saprotrophic rhizosphere fungi on rhizosphere bacteria. The higher frequency of antifungal bacteria may help protecting plants against plant-pathogenic fungi.

Saprotrophic fungi in the rhizosphere could also contribute to the mineral nutrition of plants (Baum and Hrynkiewicz 2006). The aforementioned rhizosphere-induced stimulation of fungal soil organic matter degradation may release organically bound nutrients that can be taken up by roots directly or via mycorrhiza (Subke et al. 2004). However, for most plants it will be difficult to separate the actual impact of saprotrophic fungi on plant nutrition from that of mycorrhizal fungi, especially because effects of saprotrophic fungi on plant nutrition may operate via interactions with mycorrhizal fungi (Cairney and Meharg 2002; Werner and Zadworny 2003; Tiunov and Scheu 2005).

Perspectives

Whereas it has become evident that saprotrophic fungi are important inhabitants of the (mycor)rhizosphere, their actual functioning in the rhizosphere and the implication for the plant are poorly understood. This is an area of soil microbial ecology where much is to be gained not only for basic knowledge but also for applications, e.g. biocontrol, plant nutrition and bioremediation. The onset of such studies could be made with non-mycorrhizal plants to avoid interference with mycorrhizal fungi. Such studies could reveal the actual impact of saprotrophic fungi on plant nutrition and vice versa the impact of root-derived compounds on degradation of soil organic matter by fungi.

Analysis of gene expression profiles of saprotrophs has been done for soil samples (Yergeau et al. 2007). Using similar techniques will be most important to obtain better insight in the functioning of saprotrophic fungi in the rhizosphere and will also give a better insight of their interactions with mycorrhizal fungi. Niche differentiation between saprotrophic fungal species in the rhizosphere may be clarified by using RNA-stable isotope probing techniques in combination with $^{13}CO_2$-pulse labeling of plants or addition of ^{13}C labeled model root exudates (Vandenkoornhuyse et al. 2007).

Molecular ecology of ectomycorrhizal symbiosis

The mycorrhizal symbiosis

The rhizosphere hosts large and diverse communities of microorganisms that compete and interact with each other, and with plant roots. Within these communities, mycorrhizal fungi are almost ubiquitous. Their vegetative mycelium and host root tips differentiate a mutualistic symbiosis. The novel composite mycorrhizal root is the site of nutrient and carbon transfers between the symbiotic partners. Mycorrhizal associations allow most terrestrial plants to colonize and grow efficiently in suboptimal and marginal soil environments. Among the various types of mycorrhizal symbioses, arbuscular endomycorrhiza

(AM), ectomycorrhiza (ECM) or ericoid associations are found on most annual and perennial plants (probably>90%). About two-thirds of these plants are symbiotic with AM glomalean fungi. Ericoid mycorrhizas are ecologically important, but mainly restricted to heathlands. While a relatively small number of plants develop ECM, they dominate forest ecosystems in boreal, temperate and mediterranean regions. In the different mycorrhizal associations, extramatrical and intraradicular hyphal networks are active metabolic entities that provide essential scarce nutrient resources (e.g. phosphate and amino acids) to their host plants. These nutrient contributions are reciprocated by the provision of a stable carbohydrate-rich niche in roots for the fungal partner, making the relationship a true mutualistic symbiosis. The ecological performance of mycorrhizal fungi is a complex phenotype affected by many different genetic traits and by biotic and abiotic environmental factors. Anatomical features, such as the extension of extramatrical hyphae, resulting from symbiosis development are of paramount importance to the ecophysiological fitness of the mycorrhiza.

Molecular ecology of ectomycorrhizal fungi

Many PCR-based molecular tools have been developed to identify mycorrhizal fungi, and their potential for mycorrhizal ecology has been demonstrated in dozen of studies. The application of these molecular methods has provided detailed insights into the complexity of ECM fungal communities and offers exciting prospects to elucidate processes that structure these communities. They will improve our understanding of both fungal and plant ecology, such as plant-microbe interactions and ecosystem processes. Molecular studies have provided the following insights:

- Any single mycorrhiza can potentially be identified to the species level either by terminal RFLP or by DNA sequencing of PCR-amplified nuclear ribosomal DNA internal transcribed spacers.
- Fruiting body production is unlikely to reflect below-ground symbiont communities. Not all ectomycorrhizal fungi produce conspicuous epigeous fruiting bodies and of those fungi that do produce conspicuous fruiting bodies, a species' fruiting body production does not necessarily reflect its below-ground abundance.

- A few fungal ECM taxa account for most of the mycorrhizal abundance and are widely spread, whereas the majority of species are only rarely encountered.
- The spatial and temporal variation of ECM fungi is very high and most species show aggregated distributions with seasonal variations.
- The relative importance of vegetative spread and longevity of genotypes *vs* novel colonization from meiospores structures ECM communities/populations.

As stressed above a large proportion of ECM fungi do not produce conspicuous fruit bodies, but most importantly there were no techniques available to assess the extensive and highly active webs of extraradical hyphae permeating the soil. However, during the past decade, PCR-based molecular methods and DNA sequencing have been routinely used to identify mycorrhizal fungi, and the application of these molecular methods has provided detailed insights into the complexity of mycorrhizal fungal communities and populations, and offers exciting prospects for elucidation of the processes that structure ectomycorrhizal fungal communities. In the following section, we will detail some of these studies.

Diversity demography of ectomycorrhizal fungi

Molecular tools not only have managed to reveal the tremendous diversity of mycorrhizal fungi interacting with their hosts in space and time, but also how different environmental factors and forest land usage could alter the composition of these soil fungal communities. These molecular ecology studies will spur work on the dynamics and functions of mycorrhizal communities and populations, and generate hypotheses about their roles in changing forest ecosystems. For example, it appears that the formidable web of extramatrical hyphae of mycorrhizal fungi not only colonizes mineral soil horizons (Dickie et al. 2002; Genney et al. 2005), but it is also very abundant in litter and decaying wood debris (Tedersoo et al. 2003; Buée et al. 2007). Thanks to PCR-based methods, several studies have shown that the structure of ECM fungal populations and communities was affected by several biotic and abiotics factors, including soil and litter quality, climate, seasons, micro-sites heterogeneity, host tree or forest management (Peter et al. 2001; Rosling et al. 2003; Tedersoo et al. 2003, 2008; Buée et al. 2005; Ishida et al. 2007; Koide et al., 2007a). With on-going improvements in molecular techniques (Bruns et al. 2008) and dedicated DNA databases (Koljalg et al. 2005), identification of fungal taxa has expanded from fruit bodies to mycorrhizal roots to extraradical hyphae.

Genotyping of ectomycorrhizal mycobionts after DNA extraction from single mycorrhizal tips, followed by PCR amplification, DNA sequencing and comparison with sequences deposited in databases (NCBI: http://www.ncbi.nlm.nih.gov/; UNITE: http://unite.zbi.ee; Koljalg et al. 2005), confirmed the mycorrhizal status of different fungi, such as several tomentelloid fungi, *Clavulina* and *Ramaria* species and *Sebacinaceae* (Peter et al. 2001; Tedersoo et al. 2003; Buée et al. 2005; Nouhra et al. 2005).

Several studies have demonstrated that niche differentiation in soil horizons, host preference and soil nutrient gradients contribute to the high diversity of ECM fungi in boreal and temperate forests (Dickie et al. 2002; Lilleskov et al. 2002; Ishida et al. 2007). Moreover, in a Norway spruce plantation, Korkama et al. (2006) have also shown that the ECM community structure was related to the growth rate of the plant host and these authors have observed an influence of host genotype, suggesting that individual trees were partially responsible for this high diversity. Nevertheless, ECM fungal diversity and phylogenetic community structure are similar, in some degree, between the Southern Hemisphere and Holartic realm (Tedersoo et al. 2008). Combined with isotopic tracers, molecular approaches provide novel insights into soil fungal ecology. For instance, Lindahl et al. (2007) reported on the spatial patterns of ectomycorrhizal and saprotrophic fungi from soil profiles in a *Pinus sylvestris* forest in Sweden, and compared those patterns with profiles of bulk carbon/nitrogen (C/N) ratios, and ^{15}N and ^{14}C contents (as a proxy for age). Saprotrophic fungi were found to primarily colonize relatively recently shed litter components on the surface of the forest floor, where organic C was mineralized while N was retained. Mycorrhizal fungi were prominent in the underlying, more decayed litter and humus, where they apparently mobilized N and made it available to their host plants.

Mycorrhizas not only shape the plant communities, but they also affect the functional diversity of rhizospheric bacteria (Frey-Klett et al. 2005). In their seminal paper, Schrey et al. (2005) have shown that a molecular cross-talk is taking place between members of these multitrophic associations, i.e. bacteria, fungi and other microbes leaving on the root or in its vicinity. In forest soils, all mycorrhizal fungi interact with complex microbial communities, including bacteria and saprotrophic fungi (Wallander et al. 2006). The ectomycorhizosphere, which forms a very specific interface between the soil and the trees, hosts a large and diverse community of microorganisms. Some of these bacteria consistently promote mycorrhizal development, leading to the concept of "mycorrhization" helper bacteria (Garbaye 1994; Frey-Klett et al. 2005). This microbial complex likely plays a role in mineral weathering and solubilization processes (Calvaruso et al. 2006; Uroz et al. 2007). The ability of ectomycorrhizal roots to degrade organic matter may also reflect the action of the non-mycorrhizal microbiota, because the ECM extramatrical mycelium supports the activity of free-living decomposers.

Spatiotemporal dynamics of mycorrhizal species and communities

To study the impacts of ECM fungi on the plant communities, understanding spatial and temporal patterns present in ECM fungal community structure is critical. The spatial structure of these fungal communities has now been well documented, particularly in relation with soil microsites and dead woody debris (Tedersoo et al. 2003; Buée et al. 2007), and through the fine scale vertical distribution of ECM root tips (Dickie et al. 2002; Genney et al. 2005). Nevertheless, very little is known about these changes over time. Recently, few observations in temperate forests have reported temporal changes in the ECM communities and the potential role of ECM root tips turnover in response to climatic stress, especially during the dry period or in winter (Buée et al. 2005; Izzo et al. 2005). For instance *Laccaria amethystina*, *Clavulina cristata* and *Cenococcum geophilum* displayed very contrasted seasonal patterns: *L. amethystina* and *C. cristata* formed ECM mostly in winter, while *C. geophilum* populations built up during the summer dry period, suggesting for this last species a role in drought stress resistance (Jany et al. 2003; Buée et al. 2005; Courty et al. 2008). The specific spatiotemporal dynamics of mycorrhizal species and communities in the underground remain still elusive. The physical, chemical and biological complexity of the soil makes this kind of investigation a daunting project. The current situation could be eased by the development of high-throughput molecular diagnostic tools for cataloging soil microbes on the larger scale imposed by field studies of a very heterogeneous subterranean world. In addition, with improvements in molecular techniques, appropriate DNA databases (Koljalg et al. 2005) and powerful in silico capacities of data compilation and analysis (Nilsson et al. 2005, 2008), identification of taxa in fungal ecology could be easily expanded from traditional molecular methods to high-throughput molecular diagnostic tools, such as DNA oligoarrays (Bruns et al. 2008). Actually, DNA arrays have been used for genome-wide transcript profiling of ECM fungi (Martin et al. 2008) and, more recently, also for identification of microorganisms in complex environmental samples (Sessitsch et al. 2006). The development of phylochips dedicated to major fungal groups will facilitate the study of the dynamics of ECM communities.

Function of ECM communities

The ECM fungi assist the roots of forest trees in exploiting the soil solution, and the many fungal symbionts potentially associated with a single tree are very diverse in their ability to use water resources (Jany et al. 2003). ECM fungi are capable of nutrient uptake directly by solubilizing soil minerals (Gadd 2007). Among the mechanisms involved in this mineral weathering is the exudation of complexing compounds, especially low molecular weight organic acids and siderophores (Landeweert et al. 2001). These organic acids, and in particular oxalic acid, are considered to be the most important soil weathering agents produced by ECM fungi, because of their dual acidifying and complexing ability (Landeweert et al. 2001). ECM fungi may play a role in the turnover of carbon and nutrients. Many ECM species are able to produce a wide range of extra-cellular and cell wall-bound nutrient-mobilizing enzymes involved carbohydrate polymers (Koide et al. 2007b). Recently, studies based on the measurement of enzymatic activities of individual ECM tips confirmed the secretion of ß-glucosidases and cellobiohydrolase activities in several species (Courty et al. 2005; Buée et al. 2007). Nevertheless, even if the cellulolytic capacity is particularly well developed among some ECM fungi, genome analysis of the ECM fungus *Laccaria bicolor* showed an extreme reduction in the number of enzymes involved in the degradation of plant cell wall (Martin et al. 2008). ECM fungi are also largely involved in mobilizing organic forms of nitrogen and phosphorus from the organic polymers in which they are sequestered (Read and Perez-Moreno 2003; Hobbie and Horton 2007; Lindahl et al. 2007). Recent studies have focused on the production of extracellular proteases by ECM fungi under controlled conditions (Nehls et al. 2001; Nygren et al. 2007). The production of these hydrolytic enzymes and their impact on nutrient mobilisation are very variable at the intra- and inter-specific levels. The experimental demonstration of functional diversity or complementarity between ECM species will await additional field investigations. PCR-based molecular methods have shown that tree roots are colonized by a wide diversity of ectomycorrhizal fungal species, many of which develop extensive mycelia networks which scavenge nutrients and transport them over long distances. These extraradical mycelia can interconnect roots belonging to the same or to different plant species, and form common mycorrhizal networks that allow net transfer of carbon and nutrients between host plants (Simard and Durall 2004). Moreover, pulse-labelling studies have confirmed the carbon transfer from autotrophic to myco-heterotrophic plants via shared ECM fungi (McKendrick et al. 2000) and recent studies, with stable isotopes enrichments, have shown that several green forest orchids displayed ^{13}C level intermediate between fully autotrophic plants and myco-heterotrophic plants which received their total carbon from their associated fungi (Trudell et al. 2003; Julou et al. 2005).

The next challenge for future research is to identify the functions played by the assemblages of mycorrhizal fungi in situ (Read and Perez-Moreno 2003). Despite the fact that mycorrhizal fungi play an important role in N, P and C cycling in ecosystems in decomposing organic materials, the detailed function of fungi in nutrient dynamics is still unknown. Mycorrhizal fungi differ in their functional abilities and the different mycorrhizae they establish thus offer distinct benefits to the host plant. Some fungi may be particularly effective in scavenging organic N, and may associate with plants for which acquisition of N is crucial (Peter et al. 2001); others may be more effective at P uptake and transport. An important goal is therefore to develop approaches by which the functional abilities of the symbiotic guilds are assessed in the field. In any case, it is necessary to study the subterranean microbial networks to identify below ground activities. Analyses of ^{13}C and ^{15}N isotopic signatures have a significant potential to provide clues in this area (Hobbie and Horton 2007), although further investigation is required to understand the isotopic enrichment phenomenon (Henn and Chapela 2001). Combined community/population structure and function studies applying genomics may, in the future, significantly promote our understanding of the interactions between mycorrhizal fungal species with their hosts and with their biotic and abiotic environment.

What can we expect to dig up next?

As a prerequisite for large-scale functional ecology studies, we now need to characterize genes controlling the functioning of mycorrhizal symbioses. Critical in this endeavor will be the use of genomic information on the recently sequenced *Populus trichocarpa* (Tuskan et al. 2006) and its mycorrhizal symbionts. The completion of the genome sequences of the ectomycorrhizal *Laccaria bicolor* (Martin et al. 2008) and *Tuber melanosporum* provides an unprecedented opportunity to identify the key components of interspecific and organism–environment interactions (Whitham et al. 2006). By examining, modelling and manipulating patterns of gene expression, we can identify the genetic control points regulating the mycorrhizal response to changing host physiology, and better understand how these interactions control ecosystem function. Moreover, stable isotope probing has proved useful in separating different functional groups of soil bacteria (Radajewski et al. 2000). Significantly, it has also recently been applied to soil fungi (Lueders et al. 2004) and these authors demonstrated that this approach can be used for the analysis of fungal DNA in soil. Quantitative real-time PCR has also been used for the quantification of fungal DNA and RNA in soil (Landeweert et al. 2003) and offers unprecedented perspectives for the future, as do advances being made in the development and application of metagenomics approaches and microarray technology for studying the diversity and functioning of microbial communities. These methods, used in conjunction with the development of high-quality databases, have the potential to enhance further our understanding of fungal communities and their functional roles in soil ecological processes.

Final comments

In this chapter, we presented an overview on the ecology of different micro-organisms found in the rhizosphere. The rhizospheric community is complex, and made out of a myriad of organisms interconnecting in numerous ways, acting upon each other and reacting to their environment. In fact, rhizosphere microbes provide integrated solutions, which are more than the sum of their individual functions. The study of the genetic and functional diversity and of the ecological properties of rhizosphere microbes is already yielding tangible results: we start to understand the selective forces at play, the mechanisms of action of particular microbial groups and their relationships with their biotic and abiotic environment, and as a result of this knowledge new agricultural practices and environmental consciousness are emerging. Yet, much remains to be learned on ecological mechanisms in the rhizosphere. To understand them, in-situ functional analyses have to be consulted. Furthermore, while our knowledge of the diversity of the Bacteria and Fungi has increased tremendously and is growing rapidly, we still know very little on particular bacterial goups like Acidobacterium and Verrucomicrobia and also on Archaea and phages in the rhizosphere. We are however confident that novel technologies, along with proven analytical methods will, in the near future, solve these pressing queries, and summon, like the second face of knowledge, still unthought of new questions.

References

Alvey S, Yang CH, Buerkert A, Drowley DE (2003) Cereal/legume rotation effects on rhizosphere bacterial community structure in west African soils. Biol Fertil Soil 37:73–82

Andreote FD, Araújo WL, Azevedo JL, Van Elsas JD, Van Overbeek L (2009) Endophytic colonization of potato (*Solanum tuberosum* L.) by a novel competent bacterial endophyte, *Pseudomonas putida* strain P9, and the effect on associated bacterial communities. Appl Environ Microbiol. doi:10.1128/AEM.00491-09

Appuhn A, Joergensen RG (2006) Microbial colonization of roots as a function of plant species. Soil Biol Biochem 38:1040–1051

Ashelford KE, Day MJ, Fry JC (2003) Elevated abundance of bacteriophage infecting bacteria in soil. Appl Environ Microbiol 69:285–289

Bais HP, Weir TL, Perry LG, Gilroy S, Vivanco JM (2006) The role of root exudates in rhizosphere interactions with plants and other organisms. Ann Rev Plant Biol 57:233–266

Baum C, Hrynkiewicz K (2006) Clonal and seasonal shifts in communities of saprotrophic microfungi and soil enzyme activities in the mycorrhizopshere of *Salix* spp. J Plant Nutr 169:481–487

Berg G, Roskot N, Steidle A, Eberl L, Zock A, Smalla K (2002) Plant-dependent genotypic and phenotypic diversity of antagonistic rhizobacteria isolated from different *Verticillium* host plants. Appl Environ Microbiol 68:3328–3338

Berg G, Zachow C, Lottmann J, Götz M, Costa R, Smalla K (2005) Impact of plant species and site on rhizosphere-associated fungi antagonistic to Verticillium dahliae Kleb. Appl Environ Microbiol 71:4203–4213

Bintrim SB, Donohue TJ, Handelsman J, Roberts GP, Goodman RM (1997) Molecular phylogeny of Archaea from soil. PNAS 94:277–282

Bohannan BJM, Kerr B, Jessup CM, Hughes JB, Sandvik G (2002) Trade-offs and coexistence in microbial microcosms. Antonie van Leeuwenhoek 81:107–115

Bomberg M, Jurgens G, Saan A, Sen R, Timonen S (2003) Nested PCR detection of Archaea in defined compartments of pine mycorrhizospheres developed in boreal forest humus microcosms. FEMS Microbiol Ecol 43:163–171

Borneman J, Triplett EW (1997) Molecular microbial diversity in soils from Eastern Amazonia: Evidence for unusual microorganism and microbial population shifts associated with deforestation. Appl Environ Microbiol 63:2647–2653

Broeckling CD, Broz AK, Bergelson J, Manter DK, Vivanco JM (2008) Root exudates regulate soil fungal community composition and diversity. Appl Environ Microbiol 74:738–744

Broz AK, Manter DK, Vivanco JM (2007) Soil fungal abundance and diversity: another victim of the invasive plant *Centaurea maculosa*. ISME J 1:763–765

Bruns T, Arnold AE, Hughes K (2008) Fungal networks made of humans: UNITE, FESIN and frontiers in fungal ecology. New Phytol 177:586–588

Buée M, Vairelles D, Garbaye J (2005) Year-round monitoring of diversity and potential metabolic activity of the ectomycorrhizal community in a beech *Fagus silvatica* forest subjected to two thinning regimes. Mycorrhiza 15:235–245

Buée M, Courty PE, Mignot D, Garbaye J (2007) Soil niche effect on species diversity and catabolic activities in an ectomycorrhizal fungal community. Soil Biol Biochem 39:1947–1955

Butler JL, Williams MA, Bottomley PJ, Myrold DD (2003) Microbial community dynamics associated with rhizosphere carbon flow. Appl Environ Microbiol 69:6793–6800

Cairney JWG, Meharg AA (2002) Interactions between ectomycorrhizal fungi and soil saprotrophs: implications for decomposition of organic matter in soils and degradation of organic pollutants in the rhizosphere. Can J Bot 80:803–809

Canchaya C, Fournous G, Chibani-Chennoufi S, Dillmann M-L, Brussow H (2003) Phage as agents of lateral gene transfer. Curr Opin Microbiol 6:417–424

Calvaruso C, Turpault MP, Frey-Klett P (2006) Root-associated bacteria contribute to mineral weathering and to mineral nutrition in trees: A budgeting analysis. Appl Environ Microbiol 72:1258–1266

Courty PE, Pritsch K, Schloter M, Hartmann A, Garbaye J (2005) Activity profiling of ectomycorrhiza communities in two forest soils using multiple enzymatic tests. New Phytol 167:309–319

Courty PE, Franc A, Pierrat J-C, Garbaye J (2008) Temporal changes in the ectomycorrhizal community on two soil horizons of a temperate oak forest. Appl Environ Microbiol 74:5792–5801

Chelius MK, Triplett EW (2001) The diversity of Archaea and Bacteria in association with the roots of *Zea mays* L. Microb Ecol 41:252–163

Chen X, Zhu YG, Xia Y, Shen J-P, He J-Z (2008) Ammonia-oxidizing archaea: important players in paddy rhizosphere soil? Environ Microbiol 10:1978–1987

Chow M, Radomski CC, McDermott JM, Davies J, Axelrood PE (2002) Molecular characterization of bacterial diversity in Lodgepole pine (*Pinus contorta*) rhizosphere soils from British Columbia forest soils differing in disturbance and geographic source. FEMS Microbiol Ecol 42:347–357

Coenye T, Vandamme P, Govan JRW, LiPuma JJ (2001) Taxonomy and identification of the Burkholderia cepacia complex. J Clin Microbiol 39:3427–3436

Cohan FM, Perry EB (2007) Systematics for discovering the fundamental units of bacterial diversity. Curr Biol 17: R373–R386

Conrad R, Klose M, Noll M, Kemnitz D, Bodelier PLE (2008) Soil type links microbial colonization of rice roots to methane emission. Glob Chan Biol 14:657–669

Costa R, Götz M, Mrotzek N, Lottmann J, Berg G, Smalla K (2006) Effects of site and plant species on rhizosphere community structure as revealed by molecular analysis of microbial guilds. FEMS Microbiol Ecol 56:236–249

Costa R, Van Aarle IM, Mendes R, Van Elsas JD (2009) Genomics of pyrrolnitrin biosynthetic loci: evidence for conservation and whole-operon mobility within Gram-negative bacteria. Environ Microbiol 11(1):159–175

Curl EA, Truelove B (1986) The rhizosphere. Springer, New York

Dalmastri C, Chiarini L, Cantale C, Bevivino A, Tabacchioni S (1999) Soil type and maize cultivar affect the genetic diversity of maize root-associated *Burkholderia cepacia* populations. Microb Ecol 38:273–284

De Boer W, Folman LB, Summerbell RC, Boddy L (2005) Living in a fungal world: impact of fungi on soil bacterial niche development. FEMS Microbiol Ecol 29:795–811

De Boer W, Kowalchuk GA, van Veen JA (2006) 'Root-food' and the rhizosphere microbial community composition. New Phytol 170:3–6

De Boer W, de Ridder-Duine AS, Klein Gunnewiek PJA, Smant W, van Veen JA (2008) Rhizosphere bacteria from sites with higher fungal densities exhibit greater levels of potential antifungal properties. Soil Biol Biochem 40:1542–1544

Denef K, Bubenheim H, Lenhart K, Vermeulen J, van Cleemput O, Boeckx P, Müller C (2007) Community shifts and carbon translocation within metabolically-active rhizosphere microorganisms in grasslands under elevated CO_2. Biogeosciences 4:769–779

Di Cello F, Bevivino A, Chiarini L, Fani R, Paffetti D, Tabacchioni S, Dalmastri C (1997) Biodiversity of a *Burkholderia cepecia* population isolated from the maize rhizosphere at different plant growth stages. Appl Environ Microbiol 63:4485–4493

Dickie IA, Xu B, Koide RT (2002) Vertical niche differentiation of ectomycorrhizal hyphae in soil as shown by T-RFLP analysis. New Phytol 156:527–535

Duineveld MD, Rosado AS, Van Elsas JD, Van Veen JA (1998) Analysis of the dynamics of bacterial communities in the rhizosphere of chrysanthemum via denaturing gradient gel electrophoresis and substrate utilization patterns. Appl Environ Microbiol 64:4950–4957

Dunfield KE, Germida JJ (2003) Seasonal changes in the rhizosphere microbial communities associated with field-grown genetically modified Canola (*Brassica napus*). Appl Environ Microbiol 69:7310–7318

Elsherif M, Grossmann F (1996) Role of biotic factors in the control of soil-borne fungi by fluorescent pseudomonads. Microbiol Res 151:351–357

Erkel C, Kube M, Reinhardt R, Liesack W (2006) Genome of rice Cluster I Archaea–the key methane producers in the rice rhizosphere. Science 313:370–372

Filion M, Hamelin RC, Bernier L, St-Arnaud M (2004) Molecular profiling of rhizosphere microbial communities associated with healthy and diseased black spruce (*Picea mariana*) seedlings grown in a nursery. Appl Environ Microbiol 70:3541–3551

Fravel D, Olivain C, Alabouvette C (2003) *Fusarium oxysporum* and its biocontrol. New Phytol 157:493–502

Frey-Klett P, Chavatte M, Clausse ML, Courrier S, Le Roux C, Raaijmakers J, Martinotti MG, Pierrat JC, Garbaye J (2005) Ectomycorrhizal symbiosis affects functional diversity of rhizosphere fluorescent pseudomonads. New Phytol 165:317–328

Friend T (2007) The Third Domain. Joseph Henry Press, Washington, DC

Fuhrman JA (1999) Marine viruses and their biogeochemical and ecological effects. Nature 399:541–548

Fulthorpe RR, Roesch LFW, Riva A, Triplett EW (2008) Distantly sampled soils carry few species in common. ISME J 2:901–910

Gadd GM (2007) Geomycology: biogeochemical transformatins of rocks, minerals, metals and radionuclides by fungi, bioweathering and bioremediation. Mycol Res 111:3–49

Gochnauer MB, McCully ME, Labbé H (1989) Different populations of bacteria associated with sheathed and bare regions of roots of field-grown maize. Plant Soil 114:107–120

Garbaye J (1994) Helper bacteria: a new dimension to the mycorrhizal symbiosis. New Phytol 128:197–210

Garbeva P, Voesenek K, Van Elsas JD (2004) Quantitative detection and diversity of the pyrrolnitrin biosynthetic locus in soil under different treatments. Soil Biol Biochem 36:1453–1463

Genney DR, Anderson IC, Alexander IJ (2005) Fine-scale distribution of pine extomycorrhizas and their extrametrical mycelium. New Phytol 170:381–390

Goodman RM, Bintrim SB, Handelsman J, Quirino BF, Rosas JC, Simon HM, Smith KP (1998) A dirty look: soil microflora and rhizosphere microbiology. In: Flores HE, Lynch JP, Eissenstat D (eds) Radical biology: advances and perspectives on the function of plant roots. American Society of Plant Physiologists, Rockville, pp 219–231

Gramms G, Bergmann H (2008) Role of plants in the vegetative and reproductive growth of saprobic basidiomycetous ground fungi. Microbiol Ecol 56(4):660–670

Grandmougin-Ferjani A, Delpé Y, Hartmann M-A, Laruelle F, Sancholle M (1999) Strerol distribution in arbuscular mycorrhizal fungi. Phytochem 50:1027–1031

Grayston SJ, Wang S, Campbell CD, Edwards AC (1998) Selective influence of plant species on microbial diversity in the rhizosphere. Soil Biol Biochem. 30:369–378

Green SJ, Inbar E, Michel FC, Jr HY, Minz D (2006) Succession of bacterial communities during early plant development: Transition from seed to root and effect of compost amendment. Appl Environ Microbiol 72:3975–3983

Gremion F, Chatzinotas A, Harms H (2003) Comparative 16 S rDNA and 16 S rRNA sequence analysis indicates that Actinobacteria might be a dominant part of the metabolically active bacteria in heavy metal-contaminated bulk and rhizosphere soil. Environ Microbiol 5:896–907

Griffiths BS, Ritz K, Ebblewhite N, Dobson G (1999) Soil microbial community structure: effects of substrate loading rates. Soil Biol Biochem 31:145–153

Gu Y-H, Mazzola M (2003) Modification of fluorescent pseudomonad community and control of apple replant disease induced in a wheat cultivar-specific manner. Appl Soil Ecol 24:57–72

Harman GE, Howell CR, Viterbo A, Chet I, Lorito M (2004) Trichoderma species–opportunistic, avirulent plant symbionts. Nature Rev Microbiol 2:43–56

Henn MR, Chapela IH (2001) Ecophysiology of C-13 and N-15 isotopic fractionation in forest fungi and the roots of the saprotrophic-mycorrhizal divide. Oecologia 128:480–487

Herrmann M, Saunders AM, Schramm A (2008) Archaea dominate the ammonia-oxidizing community in the rhizosphere of the freshwater macrophyte *Littorella uniflora*. Appl Environ Microbiol 74:3279–3283

Herschkovitz Y, Lerner A, Davidov Y, Rothballer M, Hartmann A, Okon Y, Jurkevitch E (2005) Inoculation with the plant growth promoting rhizobacterium *Azospirillum brasilense* causes little disturbance in the rhizosphere and rhizoplane of maize (*Zea mays*). Microb Ecol 50:277–288

Hobbie EA, Horton TR (2007) Evidence that saprotrophic fungi mobilise carbon and mycorrhizal fungi mobilise nitrogen during litter decomposition. New Phytol 173:447–449

Houlden A, Timms-Wilson TM, Day MJ, Bailey MJ (2008) Influence of plant developmental stage on microbial community structure and activity in the rhizosphere of three field crops. FEMS Microbiol Ecol 65:193–201

Hugenholtz P, Goebel BM, Pace NR (1998) Impact of culture-independent studies on the emerging phylogenetic view of bacterial diversity. J Bacteriol 180:4765–4774

Heuer H, Kroppenstedt RM, Lottmann J, Berg G, Smalla K (2002) Effects of T4 lysozyme release from transgenic potato roots on bacterial rhizosphere communities are negligible relative to natural factors. Appl Environ Microbiol 68:1325–1335

Houweling S, Kaminski T, Dentener F, Lelieveld J, Heimann M (1999) Inverse modeling of methane sources and sinks using the adjoint of a global transport model. J Geophys Res 104:26137–26160

Inbar E, Green SJ, Hadar Y, Minz D (2005) Competing factors of compost concentration and proximity to root affect the distribution of *Streptomycetes*. Microb Ecol 50:73–81

Ishida TA, Nara K, Hogetsu T (2007) Host effects on ectomycorrhizal fungal communities: insight from eight host species in mixed conifer–broadleaf forests. New Phytol 174:430–440

Izzo A, Agbowo J, Bruns TD (2005) Detection of plot-level changes in ectomycorrhizal communities across years in an old-growth mixed-conifer forest. New Phytol 166:619–630

Jacobs JL, Fasi AC, Ramette A, Smith JJ, Hammerschmidt R, Sundin GW (2008) Identification and onion pathogenicity of *Burkholderia cepacia* complex isolates from the onion rhizosphere and onion field soil. Appl Environ Microbiol 74:3121–3129

Janssen PH, Yates PS, Grinton BE, Taylor PM, Sait M (2002) Improved culturability of soil bacteria and isolation in pure culture of novel members of the divisions Acidobacteria, Actinobacteria, Proteobacteria, and Verrucomicrobia. Appl Environ Microbiol 68:2391–2396

Jany JL, Martin F, Garbaye J (2003) Respiration activity of ectomycorrhizas from *Cenococcum geophilum* and *Lactarius* sp in relation to soil water potential in five beech forests. Plant Soil 255:487–494

Jia Z, Ishihara R, Nakajima Y, Asakawa S, Kimura M (2007) Molecular characterization of T4-type bacteriophages in a rice field. Environ Microbiol 9:1091–1096

Joergensen RG (2000) Ergosterol and microbial biomass in the rhizosphere of grassland soils. Soil Biol Biochem 32:647–652

Johansen JE, Binnerup SJ (2002) Contribution of *Cytophaga*-like bacteria to the potential of turnover of carbon, nitrogen, and phosphorus by bacteria in the rhizosphere of barley (*Hordeum vulgare* L). Microb Ecol 43:298–306

Julou T, Burhardt B, Gebauer G, Berviller D, Damesin C, Selosse M-A (2005) Mixotrophy in orchids: insights from a comparative study of green individuals and non-photosynthetic mutants of *Cephalanthera damasonium*. New Phytol 166:639–653

Jurgens GLK, Saano A (1997) Novel group within the kingdom *Crenarcheota* from boreal forest soil. Appl Environ Microbiol 63:803–805

Jurkevitch E (2007) A brief history of short bacteria: a chronicle of *Bdellovibrio* (and like organisms) research. In: Jurkevitch E (ed) Predatory Prokaryotes - Biology, ecology, and evolution. 2007. Springer-Verlag, Heidelberg

Kaiser O, Puhler A, Selbitschka W (2001) Phylogenetic analysis of microbial diversity in the rhizoplane of oilseed rape (*Brassica napus* cv Westar) employing cultivation-dependent and cultivation-independent approaches. Microb Ecol 42:136–149

Kent AD, Triplett EW (2002) Microbial communities and their interactions in soil and rhizosphere ecosystems. Annu Rev Microbiol 56:211–236

Kimura M, Jia Z-J, Nakaya N, Asakawa S (2008) Ecology of viruses in soils: Past, present and future perspectives. Soil Sci Plant Nut 54:1–32

Kirby R (2006) Actinomycetes and lignin degradation. Adv Appl Microbiol 58:125–168

Kirner S, Hammer PE, Hill DS, Altmann A, Fischer I, Weislo LJ, Lanahan M, Van Pée K-H, Ligon JM (1998) Functions encoded by pyrrolnitrin biosynthetic genes from Pseudomonas fluorescens. J Bacteriol 180:1939–1943

Ludwig W, Bauer SH, Bauer M, Held I, Kirchhoh G, Schulze R, Huber I, Spring S, Hartmann A, Schleifer KH (1997) Detection and in situ identification of representatives of a widely distributed new bacterial phylum. FEMS Microbiol Lett 153:181–190

Kielak A, Pijl AS, van Veen JA, Kowalchuk GA (2009) Phylogenetic diversity of Acidobacteria in a former agricultural soil. ISME J 3:378–382

Koide RT, Shumway DL, Xu B, Sharda JN (2007a) On temporal partinioning of a community of ectomycorrhizal fungi. New Phytol 74:420–429

Koide R, Courty P-E, Garbaye J (2007b) Research perspectives on functional diversity in ectomycorrhizal fungi. New Phytol 174:240–243

Koljalg U, Larsson KH, Abarenkov K et al (2005) UNITE: a database providing web-based methods for the molecular identification of ectomycorrhizal fungi. New Phytol 166:1063–1068

Korkama T, Pakkanen A, Pennanen T (2006) Ectomycorrhizal community structure varies among Norway spruce (*Picea abies*) clones. New Phytol 171:815–824

Landeweert R, Hoffland E, Finlay RD, Kuyper T, van Breemen N (2001) Linking plants to rocks ectomycorrhizal fungi mobilize nutrients from minerals. Trends Ecol Evol 16:248–254

Landeweert R, Veeman C, Kuyper T, Fritze H, Wernars K, Smit E (2003) Quantification of ectomycorrhizal mycelium in soil by real-time PCR compared to conventional quantification techniques. FEMS Microbiol Ecol 45:283–292

Latour X, Corberand T, Laguerre G, Allard F, Lemanceau P (1996) The composition of fluorescent pseudomonad populations associated with roots is influenced by plant and soil type. Appl Environ Microbiol 62:2449–2456

Lee MS, Do JO, Park MS, Jung S, Lee KH, Bae KS, Park SJ, Kim SB (2006) Dominance of *Lysobacter* sp. in the rhizosphere of two coastal sand dune plant species. Calystegia soldanella and Elymus mollis Antonie van Leeuwenhoek 90:19–27

Lee S-H, Ka J-O, Cho J-E (2008) Members of the phylum Acidobacteria are dominant and metabolically active in rhizosphere soil. FEMS Microbiol Lett 285:263–269

Leininger S, Urich T, Schloter M, Schwark L, Qi J, Nicol GW, Prosser JI, Schuster SC, Schleper C (2006) Archaea predominate among ammonia-oxidizing prokaryotes in soils. Nature 442:806–809

Lerner A, Herschkovitz Y, Baudoin E, Nazaret S, Moënne-Loccoz Y, Okon Y, Jurkevitch E (2006) Effect of Azospirillum brasilense on rhizobacterial communities analyzed by denaturing gradient gel electrophoresis and automated intergenic spacer analysis. Soil Biol Biochem 38:1212–1218

Lilleskov EA, Hobbie EA, Fahey TJ (2002) Ectomycorrhizal fungal taxa differing in response to nitrogen deposition also differ in pure culture organic nitrogen use and natural abundance of nitrogen isotopes. New Phytol 154:219–231

Lindahl BD, Ihrmark K, Boberg J, Trumbore SE, Högberg P, Stenlid J, Finlay RD (2007) Spatial separation of litter decomposition and mycorrhizal nitrogen uptake in a boreal forest. New Phytol 173:611–620

Lottmann J, Heuer H, Smalla K, Berg G (1999) Influence of transgenic T4-lysozyme-producing potato plants on potentially beneficial plant-associated bacteria. FEMS Microbiol Ecol 29:365–377

Lu Y, Conrad R (2005) In situ stable isotope probing of methanogenic Archaea in the rice rhizosphere. Science 309:1088–1090

Lueders T, Wagner B, Claus P, Friedrich MW (2004) Stable isotope probing of rRNA and DNA reveals a dynamic methylotroph community and trophic interactions with fungi and protozoa in oxic rice field soil. Environ Microbiol 6:60–72

Lupwayi NZ, Rice WA, Clayton GW (1998) Soil microbial diversity and community structure under wheat as influenced by tillage and crop rotation. Soil Biol Biochem 30:1733–1741

Lynch JM, Hobbie JE (1988) The terrestrial environment. In: Lynch JM, Hobbie JE (eds) Microorganisms in action: concepts and application in microbial ecology. Blackwell Scientific Publications, Oxford, GB, pp 103–131

Marcial Gomes NC, Fagbola O, Costa R, Rumjanek NG, Buchner A, Mendona-Hagler L, Smalla K (2003) Dynamics of fungal communities in bulk and maize rhizosphere soil in the tropics. Appl Environ Microbiol 69:3758–3766

Markelova NY, Kershentsev AS (1998) Isolation of a new strain of the genus *Bdellovibrio* from plant rhizosphere and its lytic spectrum. Microbiologya 67:837–841

Marilley L, Aragno M (1999) Phylogenetic diversity of bacterial communities differing in degree of proximity of *Lolium perenne* and *Trifolium repens* roots. Appl Soil Ecol 13:127–136

Marschner P, Crowley DE, Yang CH (2004) Development of specific rhizosphere bacterial communities in relation to plant species, nutrition and soil type. Plant Soil 261:199–208

Martin F, Aerts A, Ahrén D, Brun A, Duchaussoy F, Kohler A et al (2008) The genome sequence of the basidiomycete fungus *Laccaria bicolor* provides insights into the mycorrhizal symbiosis. Nature 452:88–92

McCaig AE, Glover LA, Prosser JI (1999) Molecular analysis of bacterial community Structure and diversity in unimproved and improved upland grass pastures. Appl Environ Microbiol 65:1721–1730

McKendrick SL, Leake JR, Read DJ (2000) Symbiotic germination and development of myco-heterotrophic plants in nature: transfer of carbon from ectomycorrhizal *Salix repens* and *Betula pendula* to the orchid *Corallorhiza trifida* through shared hyphal connections. New Phytol 145:539–548

Miller HJ, Henken G, Van Veen JA (1989) Variation and composition of bacterial populations in the rhizospheres of maize, wheat, and grass cultivars. Can J Microbiol 35:656–660

Miller HJ, Liljeroth E, Henken G, Van Veen JA (1990) Fluctuations in the fluorescent pseudomonad and actinomycete populations of rhizosphere and rhizoplane during the growth of spring wheat. Can J Microbiol 36:254–258

Mougel C, Offre P, Ranjard L, Corberand T, Gamalero E, Robin C, Lemanceau P (2006) Dynamic of the genetic structure of bacterial and fungal communities at different development stages of *Medicago truncata* Geartn. Cv. Jemalong line J5. New Phytol 170:165–175

Nakayama N, Okumura M, Inoue K, Asakawa S, Kimura M (2007) Seasonal variations in the abundance of virus-like particles and bacteria in the floodwater of a Japanese paddy field. Soil Sci Plant Nut 53:420–429

Nehls U, Bock A, Einig W, Hampp R (2001) Excretion of two proteases by ectomycorrhizal fungus *Amanita muscaria*. Plant Cell Environ 24:741–747

Nel B, Steinberg C, Labuschagne N, Viljoen A (2006) Isolation and characterization of nonpathogenic *Fusarium oxysporum* isolates from the rhizosphere of healthy banana plants. Plant Pathol 55:207–216

Nelson DR, Mele PM (2007) Subtle changes in rhizosphere microbial community structure in response to increased boron and sodium chloride concentrations. Soil Biol Biochem 39:340–351

Nicol G, Webster G, Glover AL, Prosser JI (2004) Differential response of archaeal and bacterial communities to nitrogen inputs and pH changes in upland pasture rhizosphere soil. Environ Microbiol 6:861–867

Nijhuis EH, Maat MJ, Zeegers IWE, Waalwijk C, Van Veen JA (1993) Selection of bacteria suitable for introduction into the rhizosphere of grass. Soil Biol Biochem 25:885–895

Nilsson H, Kristiansson E, Ryberg M, Larsson H (2005) Approaching the taxonomic affiliation of unidentified sequences in public databases – an example from the mycorrhizal fungi. BMC Bioinf 6:178–185

Nilsson H, Kristiansson E, Ryberg M, Hallenberg N, Larsson H (2008) Intraspecific ITS variability in the kingdom fungi as expressed in the internal sequence databases and its implications for molecular species identification. Evol Bioinfo 4:193–201

Nouchi I, Mariko S, Aoki K (1990) Mechanism of methane transport from the rhizosphere to the atmosphere through rice plants. Plant Physiol 94:59–66

Nouhra E, Horton T, Cazares E, Castellano M (2005) Morphological and molecular characterization of selected *Ramaria* mycorrhizae. Mycorrhiza 15:55–59

Nygren CM, Edqvist J, Elfstrand M, Heller G, Taylor AFS (2007) Detection of extracellular protease activity in different species and genera of ectomycorrhizal fungi. Mycorrhiza 17:241–248

Olsson PA, Johnson NC (2005) Tracking carbon from the atmosphere to the rhizosphere. Ecol Lett 8:1264–1270

Olsson PA, Larsson L, Bago B, Wallander H, van Aarle IM (2003) Ergosterol and fatty acids for biomass estaimation of mycorrhizal fungi. New Phytol 159:7–10

Ochsenreiter T, Seleki D, Quaiser A, Bonch-Osmolovskaya L, Schleper C (2003) Diversity and abundance of Crenarchaeota in terrestrial habitats studied by 16 S RNA surveys and real time PCR. Environ Microbiol 5:787–797

Oved T, Shaviv A, Goldrath T, Mandelbaum R, Minz D (2001) Influence of effluent irrigation on community composition and function of ammonia-oxidizing bacteria in soil. Appl Environ Microbiol 67:3426–3433

Paterson E, Gebbing T, Abel C, Sim A, Telfer G (2006) Rhizodeposition shapes rhizosphere microbial community structure in organic soil. New Phytol 173:600–610

Payne G, Ramette A, Rose HL, Weightman AJ Hefin T, James J, Tiedje M, Mahenthiralingam E (2006) Application of a *recA* gene-based identification approach to the maize rhizosphere reveals novel diversity in *Burkholderia* species. FEMS Microbiol Let 259:126–132

Peter M, Ayer F, Egli S, Honegger R (2001) Above- and belowground community structure of ectomycorrhizal fungi in three Norway spruce (*Picea abies*) stands in Switzerland. Can J Bot 79:1134–1151

Postma J, Schilder MT, Bloem J, Van Leeuwen-Haagsma WK (2008) Soil suppressiveness and functional diversity of the soil microflora in organic farming systems. Soil Biol Biochem 40:2394–2406

Prangishvili D, Forterre P, Garrett RA (2006) Viruses of the Archaea: a unifying view. Nat Rev Micro 4:837–848

Radajewski S, Philip I, Nisha P, Colin MJ (2000) Stable-isotope probing as a tool in microbial ecology. Nature 403:646–649

Ramakrishnan B, Lueders T, Dunfield PF, Conrad R, Friedrich MW (2001) Archaeal community structures in rice soils from different geographical regions before and after initiation of methane production. FEMS Microbiol Ecol 37:175–186

Ramette A, LiPuma JJ, Tiedje JM (2005) Species abundance and diversity of *Burkholderia cepacia* complex in the environment. Appl Environ Microbiol 71:1193–1201

Read DJ, Perez-Moreno J (2003) Mycorrhizas and nutrient cycling in ecosystems–a journey towards relevance? New Phytol 157:475–492

Reichenbach H (2001) The genus *Lysobacter*. In: Dworkin M et al (eds) The prokaryotes: An evolving electronic resource for the microbiological community. Springer-Verlag, New York

Renker C, Blanke V, Börstler B, Heinrichs J, Buscot F (2004) Diversity of *Cryptococcus* and *Dioszegia* yeasts (Basidiomycota) inhabiting arbuscular mycorrhizal roots or spores. FEMS Yeast Res 4:597–603

Rosling A, Landerweert R, Lindahl BD, Larsson K-H, Kuyper TW, Taylor AFS, Finlay RD (2003) Vertical distribution of ectomycorrhizal fungal taxa in a podzol soil profile. New Phytol 159:775–783

Schallmach E, Minz D, Jurkevitch E (2000) Culture-independent detection of shifts occurring in the structure of root-associated bacterial populations of common bean (*Phaseolus vulgaris* L) following nitrogen depletion. Microb Ecol 40:309–316

Salles JF, Van Elsas JD, Van Veen JA (2006) Effect of Agricultural management regime on *Burkholderia* community structure in soil. Microbiol Ecol 52:267–279

Scherff RH (1973) Control of bacterial blight of soybean by *Bdellovibrio bacteriovorus*. Phytopathol 63:400–402

Schrey D, Schellhammer M, Ecke M, Hampp R, Tarkka M (2005) Mycorrhizal helper bacterium *Streptomyces* AcH505 induces differential gene expression in the ectomycorrhizal fungus *Amanita muscaria*. New Phytol 168:205–216

Schwieger F, Tebbe CC (2000) Effect of field inoculation with *Sinorhizobium meliloti* L33 on the composition of bacterial communities in rhizospheres of a target plant (*Medicago sativa*) and a non-target plant (*Chenopodium album*)—linking of 16 S rRNA gene-based single-strand conformation polymorphism community profiles to the diversity of cultivated bacteria. Appl Enviro Microbiol 66:3556–3565

Seldin L, Rosado AS, da Cruz DW, Nobrega A, van Elsas JD, Paiva E (1998) Comparison of *Paenibacillus azotofixans* strains isolated from rhizoplane, rhizosphere, and non-root-associated soil from maize planted in two different Brazilian soils. Appl Environ Microbiol 64:3860–3868

Semenov AM, van Bruggen AHC, Zelenev VV (1999) Moving waves of bacterial populations and total organic carbon along roots of wheat. Microbiol Ecol 37:116–128

Sessitsch A, Hackl E, Wenzl P, Kilian A, Kostic T, Stralis-Pavese N, Tankouo, Sandjong B, Bodrossy L (2006) Diagnostic microbial microarrays in soil ecology. New Phytologist 171:719–736

Sharma S, Aneja MK, Mayer J, Munch JC, Schloter M (2005) Characterization of bacterial community structure in rhizosphere soil of grain legumes. Microb Ecol 49:407–415

Simard SW, Durall DM (2004) Mycorrhizal networks: a review of their extent function and importance. Can J Bot 82:1140–1165

Simon HM, Smith KP, Dodsworth JA, Guenthner B, Handelsman J, Goodman RM (2001) Influence of tomato genotype on growth of inoculated and indigenous bacteria in the spermosphere. Appl Environ Microbiol 67:514–520

Singh BK, Munro S, Potts JM, Millard P (2007) Influence of grass species and soil type on rhizosphere microbial community structure in grassland soils. Appl Soil Ecol 36:147–155

Sliwinski MK, Goodman RM (2004) Comparison of crenarchaeal consortia inhabiting the rhizosphere of diverse terrestrial plants with those in bulk soil in native environments. Appl Environ Microbiol 70:1821–1826

Smalla K, Wieland G, Buchner A, Zock A, Parzy J, Kaiser S, Roskot N, Heuer H, Berg G (2001) Bulk and rhizosphere soil bacterial communities studied by denaturing gradient gel electrophoresis: plant-dependent enrichment and seasonal shifts revealed. Appl Environ Microbiol 67:4742–4751

Smit E, Leeflang P, Glandorf B, van Elsas JD, Wernars K (1999) Analysis of fungal diversity in the wheat rhizosphere by sequencing of cloned PCR-amplified genes encoding 18 S rRNA and temperature gradient gel electrophoresis. Appl Environ Microbiol 65:2614–2621

Stafford WHL, Baker GC, Brown SA, Burton SG, Cowan DA (2005) Bacterial diversity in the rhizosphere of Proteaceae species. Environ Microbiol 7:1755–1768

Staley JT, Konopka A (1985) Measurement of in situ activities of nonphototrophic microorganisms in aquatic and terrestrial habitats. Annu Rev Microbiol 39:321–346

Subke J-A, Hahn V, Battipaglia G, Linder S, Buchman N, Cotrufo MF (2004) Feedback interactions between needle litter decomposition and rhizosphere activity. Oecologia 139:551–559

Suttle CA (2005) Viruses in the sea. Nature 437:356–361

Tedersoo L, Kõljalg U, Hallenberg N, Larsson KH (2003) Fine scale distribution of ectomycorrhizal fungi and roots across substrate layers including coarse woody debris in a mixed forest. New Phytol 159:153–165

Tedersoo L, Jairus T, Horton BM, Abarenkov K, Suvi I, Saar I, Koljalg U (2008) Strong host preference of ectomycorrhizal fungi in a Tasmania wet sclerophyll forest as revealed by DNA barcoding and taxon-specific primers. New Phytol 180:479–490

Timonen S, Marschner P (2005) Mycorrhizosphere concept. In: Mukerji KG, Manoharachary C, Singh J (eds) Microbial activity in the rhizosphere. Springer-Verlag, Berlin, pp 155–172

Tiunov AV, Scheu S (2005) Arbuscular mycorrhiza and Collembola interact in affecting community composition-jof saprotrophic microfungi. Oecologia 142:636–642

Treonis AM, Ostleb NJ, Stotth AW, Primrosea R, Graystona SJ, Ineson P (2004) Identification of groups of metabolically-active rhizosphere microorganisms by stable isotope probing of PLFAs. Soil Biol Biochem 36:533–537

Trudell SA, Rygiewicz PT, Edmonds RL (2003) Nitrogen and carbon stable isotope abundances support the mycoheterotrophic nature and host-specificity of certain achlorophyllous plants. New Phytol 160:391–401

Tuskan GA, DiFazio S, Jansson S, Bohlmann J, Grigoriev I, Hellsten U et al (2006) The genome of black cottonwood, *Populus trichocarpa*. Science 313:1596

Uroz S, Calvaruso C, Turpault MP, Pierrat JC, Mustin C, Frey-Klett P (2007) Effect of the mycorrhizosphere on the genotypic and metabolic diversity of the bacterial communities involved in mineral weathering in a forest soil. Appl Environ Microbiol 73:3019–3027

Valentine DL (2007) Adaptations to energy stress dictate the ecology and evolution of the Archaea. Nat Rev Micro 5:316–323

Vandenkoornhuyse P, Mahé S, Ineson P, Staddon P, Ostle N, Cliquet J-B, Francez A-J, Fitter AH, Young PW (2007) Active root-inhabiting microbes identified by rapid incorporation of plant-derived carbon into RNA. Proc Nat Acad Sci 104:16970–16975

Van der Wal A, van Veen JA, Pijl AS, Summerbell RC, de Boer W (2006) Constraints on development of fungal biomass and decomposition processes during restoration of arable sandy soils. Soil Biol Biochem 38:2890–2902

Van der Wal A, de Boer W, Smant W, van Veen JA (2007) Initial decay of woody fragments in soil is influenced by size, vertical position, nitrogen availability and soil origin. Plant Soil 301:189–201

Van Overbeek L, Van Elsas JD (2008) Effects of plant genotype and growth stage on the structure of bacterial communities associated with potato (*Solanum tuberosum* L.). FEMS Microbiol Ecol 64:283–296

Van Elsas JD, Jansson J, Sjöling S, Bailey M, Nalin R, Vogel T, Costa R, Van Overbeek L (2008a) The metagenomics of disease-suppressive soils - Experiences from the METACONTROL project. Trend Biotechnol 26(11):591–601

Van Elsas JD, Speksnijder AJ, Van Overbeek LS (2008b) A novel procedure for the metagenomics exploration of disease-suppressive soils. J Microbiol Meth 75(3):515–522

Van Wees SCM, van der Ent S, Pieterse CJM (2008) Plant immune responses triggered by beneficial microbes. Curr Opin Plant Biol 11:443–448

Vasiliauskas R, Menkis A, Finlay RD, Stenlid J (2007) Wood-decay fungi in the fine living roots of conifer seedlings. New Phytol 174:441–446

Viebahn M, Veenman C, Wernars K, van Loon LC, Smit E, Bakker PAHM (2005) Assessment of differences in ascomycete communities in the rhizosphere of field-grown wheat and potato. FEMS Microbiol Ecol 53:245–253

Vinale F, Sivasithamparam K, Ghisalberti EL, Marra R, Woo SL, Lorito M (2008) Trichoderma–plant–pathogen interactions. Soil Biol Biochem 40:1–10

Vujanovic V, Hamelin RC, Bernier L, Vujanovic G, St-Arnaud M (2007) Fungal diversity, dominance, and community structure in the rhizosphere of clonal *Picea marina* plants throughout nursery production chronosequences. Microbiol Ecol 54:672–684

Wallander H, Lindahl BD, Nilsson LO (2006) Limited transfer of nitrogen between wood decomposing and ectomycorrhizal mycelia when studied in the field. Mycorrhiza 16:213–217

Watt M, McCully ME, Kirkegaard JA (2003) Soil strength and rate of root elongation alter the accumulation of spp. and other bacteria in the rhizosphere of wheat. Funct Plant Biol 30:483–491

Watt M, Hugenholtz P, White R, Vinall K (2006) Numbers and locations of native bacteria on field-grown wheat roots quantified by fluorescence in situ hybridization (FISH). Env Microbiol 8:871–884

Whitham TG, Bailey JK, Schweitzer JA, Shuster SM, Bangert RK, LeRoy CJ et al (2006) A framework for community and ecosystem genetics: from genes to ecosystems. Nat Rev Genet 7:510–523

Weete JD (1989) Structure and function of sterols in fungi. Adv Lipid Res 23:115–167

Werner A, Zadworny M (2003) In vitro evidence of mycoparasitism of the ectomycorrhizal fungus *Laccaria laccata* against *Mucor hiemalis* in the rhizosphere of *Pinus sylvestris*. Mycologia 13:41–47

Williamson KE, Radosevich M, Wommack E (2005) Abundance and diversity of viruses in six Delaware soils. Appl Environ Microbiol 71:3119–3125

Woese C, Fox GE (1977) Phylogenetic structure of the prokaryotic domain: The primary kingdoms. PNAS 74:5088–5090

Wuchter C, Abbas B, Coolen MJL, Herfort L, van Bleijswijk J, Timmers P, Strous M, Teira E, Herndl GJ, Middelburg JJ, Schouten S, Sinninghe Damste J (2006) Archaeal nitrification in the ocean. PNAS 103:12317–12322

Yang C-H, Crowley DE (2000) Rhizosphere microbial community structure in relation to root location and plant iron nutritional status. Appl Environ Microbiol 66:345–351

Yergeau E, Kang S, He Z, Zhou J, Kowalchuk GA (2007) Functional microarray analysis of nitrogen and carbon cycling along an Antarctic latitudinal gradient. ISME J 1:163–179

Zachow C, Tilcher R, Berg G (2008) Sugar beet-associated bacterial and fungal communities show a high indigenous antagonistic potential against plant pathogens. Microb Ecol 55:119–129

REVIEW ARTICLE

Rhizosphere fauna: the functional and structural diversity of intimate interactions of soil fauna with plant roots

Michael Bonkowski · Cécile Villenave · Bryan Griffiths

Received: 28 January 2009 / Accepted: 24 April 2009 / Published online: 20 June 2009
© Springer Science + Business Media B.V. 2009

Abstract For decades, the term "rhizosphere fauna" has been used as a synonym to denote agricultural pests among root herbivores, mainly nematodes and insect larvae. We want to break with this constrictive view, since the connection between plants and rhizosphere fauna is far more complex than simply that of resource and consumer. For example, plant roots have been shown to be neither defenceless victims of root feeders, nor passive recipients of nutrients, but instead play a much more active role in defending themselves and in attracting beneficial soil microorganisms and soil fauna. Most importantly, significant indirect feed-backs exist between consumers of rhizosphere microorganisms and plant roots. In fact, the majority of soil invertebrates have been shown to rely profoundly on the carbon inputs from roots, breaking with the dogma of soil food webs being mainly fueled by plant litter input from aboveground. In this review we will highlight areas of recent exciting progress and point out the black boxes that still need to be illuminated by rhizosphere zoologists and ecologists.

Keywords Rhizosphere food web · Root herbivores · Signalling · Microbial vectors · Root growth · Energy channel

Introduction

For decades, the term "rhizosphere fauna" has been used as a synonym to denote agricultural pests among root herbivores, mainly nematodes and insect larvae. We want to break with this constrictive view, since the connection between plants and rhizosphere fauna is far more complex than simply that of resource and consumer. For example, plant roots have been shown to be neither defenceless victims of root feeders, nor passive recipients of nutrients, but instead play a much more active role in defending themselves and in attracting beneficial soil microorganisms and soil fauna. Most importantly, significant indirect feed-backs exist between consumers of rhizosphere microorganisms and plant roots. In fact, the majority of soil invertebrates have been shown to rely profoundly on the carbon inputs from roots, breaking with the dogma of soil food webs being mainly fueled by

Responsible Editor: Phillipe Lemanceau.

M. Bonkowski (✉)
Department of Terrestrial Ecology,
University of Cologne, Zoological Institute,
Weyertal 119, 50931 Cologne, Germany
e-mail: bonkowski@rhizosphere.de

C. Villenave
Research Institute for Development,
IRD-SeqBio/SupAgro, 2 place Viala, Bât. 12,
34060 Montpellier cedex 1, France

B. Griffiths
Teagasc, Environment Research Centre,
Johnstown Castle, Wexford, Co.,
Wexford, Ireland

plant litter input from aboveground. In this review we will highlight areas of recent exciting progress and point out the black boxes that still need to be illuminated by rhizosphere zoologists and ecologists.

The consumers of plant roots: direct impacts, re-programming of plant cells and indirect plant feed-back

Roots anchor the plant in soil, but most importantly, the whole nutrient transfer from soil to aboveground plant parts is channeled through the roots, and roots are important storage organs in perennial plants. This makes roots attractive to herbivores. However, plant roots are no passive victims of attacking herbivores and microorganisms. In fact, they have evolved a whole arsenal of direct defence compounds, such as terpenoids, and also indirect defenses, involving communication strategies to interact with soil fauna, soil microorganisms and other plant roots to ward off attack (Huber-Sannwald et al. 1997; Boff et al. 2001; van Tol et al. 2001; Mathesius et al. 2003; Bais et al. 2004, 2006; Rasmann et al. 2005; Dudley and File 2007). Therefore, successful root feeders are expected to employ highly coevolved strategies to counter root defence systems. The most important root feeders are considered to be root-feeding nematodes and insect larvae.

Root-feeding nematodes

All plant-parasitic nematodes have a stylet; a strong, hollow, needle-like structure that is used to pierce plant cells, inject nematode secretions and to feed on plant cell contents. Stylets vary in shape and size according to the feeding strategy of the nematode; for example, nematodes such as *Trichodorus* that feed on epidermal cells have short stylets, whereas those such as *Xiphinema* or *Longidorus* have considerably longer stylets and can feed on cells deeper within the plant (Gheysen and Jones 2006) (Fig. 1). To find the root, invade the root and induce a feeding site nematodes rely on an arsenal of secreted molecules and signaling pathways (see reviews by: Williamson and Gleason 2003; McKenzie Bird 2004; Gheysen and Jones 2006).

Migrating nematodes can locate their target by sensing chemical gradients (Robinson 2003), plant cell—specific surface determinants or electrical sig-

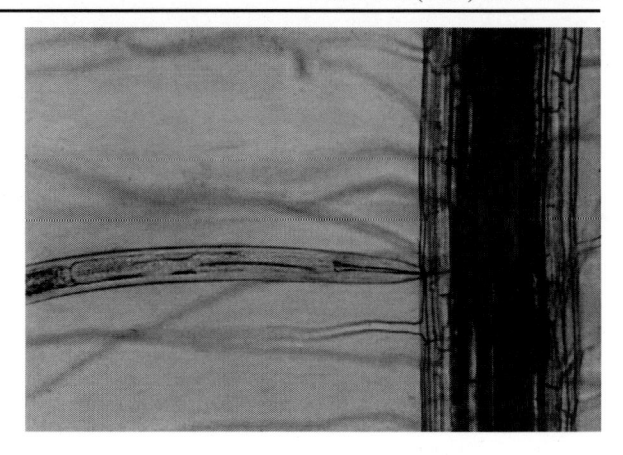

Fig. 1 A plant-parasitic nematode (*Xiphinema* sp.) piercing root cells

nals (Riga 2004). However, only a narrow part of roots, between the tip and the root hair zone is vulnerable to nematode attack. Detached root border cells, secreted from root tips are suggested to play a significant role to misdirect plant-parasitic nematodes until the vulnerable part of the root has outgrown its attackers (Rodger et al. 2003) (Fig. 2). If a suitable cell is located, endoparasitic nematodes then enter into the root (as opposed to ectoparasites that simply use their stylet to feed on cells without entering the root themselves), secreting a wide range of enzymes (including cellulases, chitinases and extensins) that are specifically targeted to degrade or modify host tissues, during their migration through the root (Davis et al. 2000). Many nematodes induce the plant to make specialized cells, or feeding sites, which are metabolically active and provide a source of sustained nutrition for the nematode (Fig. 2). Nematode feeding sites tend to have structural characteristics of metabolically active tissues including: cytoplasm highly enriched in sub-cellular organelles; signs of DNA replication and enlarged or multiple nuclei (Wyss 2002). It has been hypothesized that some of the nematode genes encoding these enzymes were acquired from soil bacteria via horizontal gene transfer (Popeijus et al. 2000; Veronico et al. 2001). Comparisons of host transcription patterns using a variety of techniques have indicated that nematode infection initiates complex changes in plant gene expression (Gheysen and Jones 2006). Genes that are induced in defence responses against other pathogens are up-regulated by nematode infection. Thus the plants are not passive in the face of nematode invasion but have

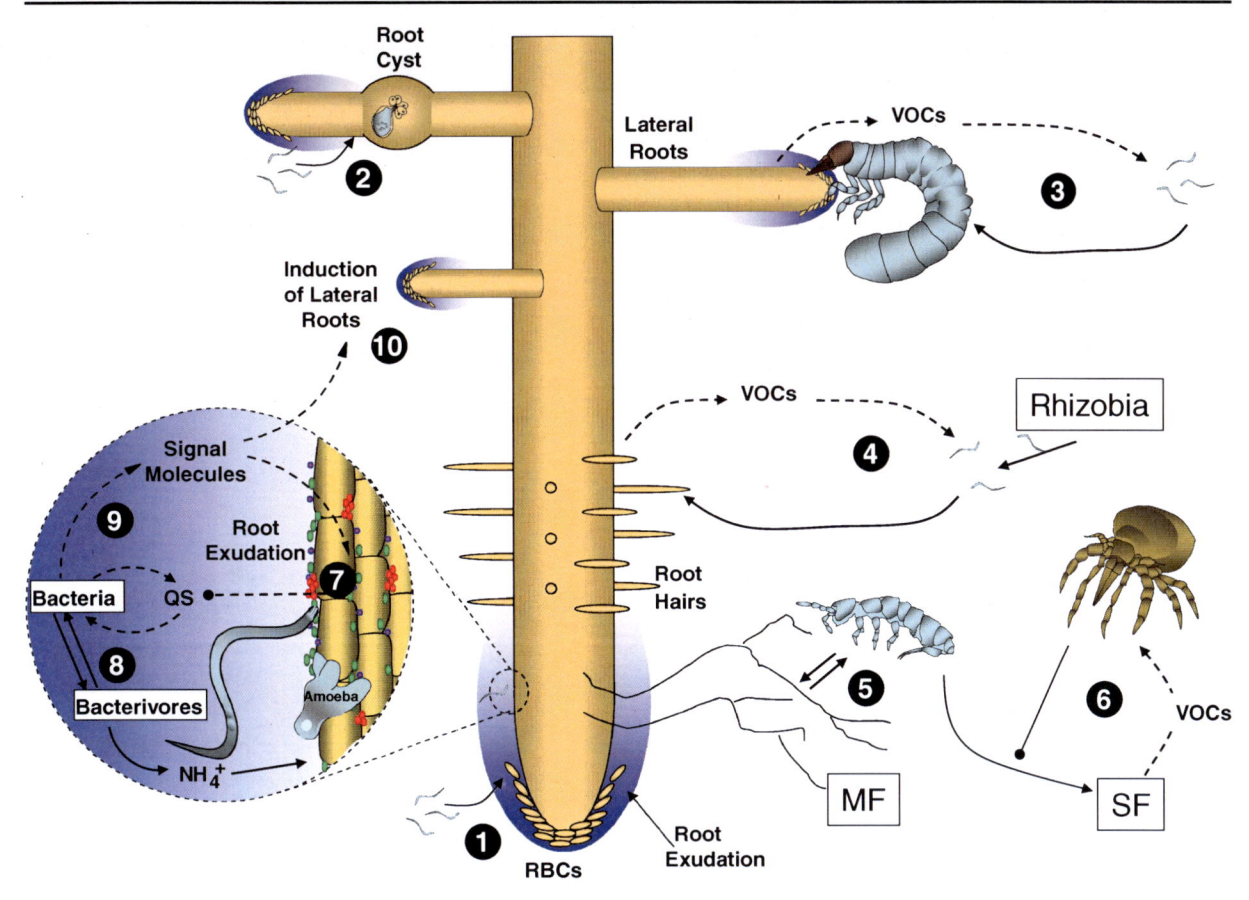

Fig. 2 What is known on signalling between rhizosphere fauna, microorganisms and plant roots? (*1*) Migrating root-feeding nematodes locate roots by chemical gradients of root-specific signals (Williamson and Gleason 2003). Root border cells (RBCs) can misdirect root feeding nematodes until the vulnerable part of the root tip has outgrown its attackers (Rodger et al. 2003). (*2*) Once targeted, root invading nematodes secrete a wide array of enzymes and signal molecules specifically targeted to downregulate host defense responses and to modify host tissues (McKenzie Bird 2004), e.g. root knot nematodes secrete signal molecules to induce cell growth of specific feeding sites. (*3*) Upon attack by root-feeding insect larvae, roots emit specific volatiles attracting entomopathogenic nematodes to kill the herbivore (Rasmann et al. 2005). (*4*) Legume roots emit specific volatiles to attract bacterivorous nematodes that carry symbiotic rhizobia on their cuticula to inoculate the roots (Horiuchi et al. 2005). (*5*) Fungivores affect the balance between mycorrhizal (MF) and saprophytic (SF) fungi in a density dependant manner. Some ectomycorrhiza species have been shown to use collembola as nutrient source (Klironomos and Hart 2001). (*6*) Predatory mites are attracted by volatile signals to the fungal food sources of their collembolan prey (Hall and Hedlund 1999), and injured collembola emit warning signals to alert conspecifics (Pfander and Zettel 2004). (*7*) Roots interfere with bacterial communication by emitting quorum sensing mimic compounds (Mathesius et al. 2003). (*8*) Bacterivores, such as amoebae and nematodes regulate rhizosphere bacterial community composition (Rosenberg et al. 2009), they affect the production of bacterial metabolites (see Fig. 3) and release NH_4-N from consumed bacterial biomass (Griffiths 1994). (*9*) Grazing resistant bacteria benefit from bacterivores (Jousset et al. 2008) and stimulate exudation (see text). (*10*) Grazing-induced changes in the composition of rhizosphere bacteria lead to enhanced production of lateral roots (Mao et al. 2007), thereby favouring a positive feedback on steps (8) and (9) (Bonkowski 2004)

a battery of defences that they employ to try to repel nematodes once they have been detected within the roots. Plant-parasitic nematodes have responded to this by evolving a series of physical and biochemical adaptations that help them either avoid eliciting a host response or to reduce the toxic effects of any plant defence response. A large number of the genes that are induced by nematode infection are likely to contribute to establishing the parasitic interaction (Puthoff et al. 2003). Thus, while nematodes themselves will inject cell-wall degrading enzymes via the stylet, they also stimulate the plant to upregulate genes that encode host cell-wall degrading enzymes such as endoglucanase and polygalacturonase (Goellner et al. 2001; Vercauteren et al. 2002). Analysis of mutants and reporter-gene constructs indicate that auxin-response genes are also induced in the formation of nematode feeding sites (Goverse et al. 2000). There are remarkable similarities in the signaling pathways between the formation of nematode feeding sites, root nodules induced by *Rhizobium*, mycorrhizal infections of roots and lateral root formation (Mathesius 2003). In all cases, it is possible that the perturbance of the root auxin balance is mediated by plant flavonoids, the only plant derived auxin transport inhibitors known (Jacobs and Rubery 1988). The specificity of each interaction might be determined by temporal and spatial patterns of expression and by the induction of specific isoforms of enzyme classes. Overall, the web of interactions between internal and external signals regulating plant responses to microbial signals shows that the plant is actively orchestrating its interactions with microorganisms (Mathesius et al. 2003).

Root feeding invertebrates can have major direct effects by reducing plant performance and so facilitating ecological succession or invasion. Phytophagous nematodes have been estimated to take up as much as one quarter of the net primary production of grassland vegetation (Stanton 1988), and they affect plant quality (Troelstra et al. 2001), plant diversity, and vegetation succession (De Deyn et al. 2003a). By focusing on agricultural systems, important and beneficial functions of root feeders, such as plant-parasitic nematodes, in natural systems are frequently overlooked. Organisms that derive their energy from living roots, such as nematodes, affect rhizodeposition and thereby influence the supply of root C to rhizosphere microorganisms (Yeates et al. 1998, 1999a, b; Bardgett et al. 1999a; Bardgett and Wardle 2003; Ayres et al. 2007; Haase et al. 2007; Poll et al. 2007). The activities of root feeders affect the turnover rate of root tissue (Dawson et al. 2002; Treonis et al. 2005), enhance rhizodeposition of organic compounds (Murray et al. 1996; Yeates et al. 1998; Treonis et al. 2005) and alter root architecture (Treonis et al. 2007). This has important consequences for micro-food webs of free-living bacteria and their consumers on plant roots (see below).

At low levels of nematode infestation, losses of C from the plant have been shown to be offset by changes in the root architecture that reduce rhizodeposition (Treonis et al. 2007). N fluxes in the soil may also be affected. For example, clover-cyst nematodes have been shown to stimulate the transfer of N from clover to neighbouring grasses (Hatch and Murray 1994; Bardgett et al. 1999b; Denton et al. 1999; Yeates et al. 1999a), potentially affecting plant succession in grasslands. Thus, root herbivores in natural communities can have net positive effects on nutrient mineralization, and soil microbial community structure and functioning (Denton et al. 1999; Grayston et al. 2001; Bardgett and Wardle 2003; Treonis et al. 2005).

Root-feeding insect larvae

Plant interactions with root-feeding insect larvae are as complex as plant-nematode interactions. According to the proverb "the enemy of my enemy is my friend", roots have been shown to actively attract pathogens of insect root herbivores. This mechanism has been demonstrated for the roots of very distantly related plant species, such as the evergreen shrub *Thuja* (Cupressaceae) (van Tol et al. 2001) and the monocotyledenous *Zea mays* (Poaceae) (Rasmann et al. 2005), suggesting a wide occurrence among plants. Upon attack by insect larvae, plant roots have been shown to release (E)-β-caryophyllene, a specific volatile that attracts entomopathogenic nematodes of the genus *Heterorhabditis* (Rhabditidae) (Boff et al. 2001; Rasmann et al. 2005). The nematodes themselves exist in a further obligate symbiosis and employ bacteria of the genus *Photorhabdus* (Enterobacteriaceae) to kill the insect hosts and to protect the insect corpse against competitors, such as saprophytic microorganisms, bacteriovorous nematodes and scavenging insects. The bacterial symbionts also serve as substrate for growth and reproduction of the nematode. In turn, the *Photorhabdus* bacteria utilize the nematode as a vector for delivery into another insect hemocoel and to persist outside the insect host (Ciche et al. 2006). This example of a positive feedback interaction vividly illustrates the complexity and high degree of multitrophic coevolution between plant roots, soil fauna and microorganisms.

Thus, roots produce specific volatile signals to attract entomopathogenic nematodes in much the same way as attacked plant leaves have been shown to release volatiles to attract the parasitoids of insect leaf herbivores (Erb et al. 2008; Rasmann and

Agrawal 2008; Dicke 2009). Since plants have to integrate and coordinate the signals and activities from both the below- and aboveground parts, it is not surprising that consumers of plant roots also influence aboveground food webs and plant signaling pathways. In fact, belowground interactions do not stop at the soil surface, but may significantly affect plant performance and food webs of herbivores and their consumers aboveground (Gange and Brown 1997; Tscharntke and Hawkins 2002). Rasmann and Turlings (2007) exposed young maize plants to either root herbivores (*Diabrotica virgifera* (beetle) larvae) or leaf herbivores (*Spodoptera littoralis* larvae), or to both herbivores, in the presence of their predators. They found that the parasitic wasp *Cotesia marginiventris* and the entomopathogenic nematode *H. megidis* were strongly attracted if their respective host was feeding on a plant, but the attraction was significantly reduced if both insect herbivores occurred together on a plant (Rasmann and Turlings 2007). The emission of root volatile signals was reduced by the double infestation. Although leaf volatiles did not change, the parasitoid wasp was able to learn the differences in odour emissions and showed reduced attraction to the odour of a doubly infested plant (Rasmann and Turlings 2007). Similarly, in a semi-field experiment with mustard plants (*Brassica nigra*), Soler et al. (2007) demonstrated that root herbivory by the dipteran larvae *Delia radicum* (Anthomyiidae) affected the behaviour of *Cotesia glomerata*, a parasitoid wasp on caterpillars of the leaf herbivore *Pieris brassicae* (Lepidoptera), mediated by changes in plant volatiles. Plants exposed to root herbivory were shown to emit high levels of specific sulphur volatile compounds with known toxicity for insects, combined with low levels of several compounds, i.e. beta-farnesene, reported to act as attractants for insect herbivores. By exploiting root-induced signals to evaluate and select the most suitable host for their offspring, females of the parasitoid *C. glomerata* preferred to search for hosts and to oviposit in hosts feeding on plants with no root herbivory. Root herbivory on neighbouring plants significantly affected search efficiency of aboveground parasitoids (Soler et al. 2007). Biocontrol activity of parasitoids on aboveground herbivores can, therefore, be influenced by belowground herbivores through changes in the composition of plant volatiles. These studies shed light on the importance of volatile signals in orchestrating above- and belowground plant defences and clearly demonstrate how root herbivory can influence aboveground tritrophic signalling and parasitoid feed-back (van Dam et al. 2003; Bezemer and Van Dam 2005). A review on plant responses to below- and aboveground herbivory is given by van Dam et al. (2003).

Direct plant defence compounds can also be induced by root herbivory. Bezemer et al. (2003) demonstrated that the relative growth rate and food consumption of the herbivorous caterpillar *Spodoptera exigua* was reduced by more than 50%, on cotton plants exposed to previous root damage by larvae of the beetle *Agriotes lineatus*, even though plant growth and foliar nitrogen levels were not affected by root herbivory. Exposure to root herbivores resulted in an increase in terpenoid levels in both roots and foliage, demonstrating that root herbivores may change the level and distribution of plant secondary chemistry and thus direct plant defences aboveground (Bezemer et al. 2003). A meta-analysis (Kaplan et al. 2008) indicated that root feeders generally induce strong responses in roots as well as in shoots, whereas leaf feeders tend to induce responses only in the aboveground parts of the plants.

Root herbivores have been shown to affect plant performance both directly and indirectly by their influence on bottom-up and top-down control of aboveground invertebrate herbivores Bezemer et al. 2005) with important consequences for plant community composition (Brown and Gange 1989; De Deyn et al. 2003b; Schädler et al. 2004; van Ruijven et al. 2005). The relationship between plants, their belowground herbivores, and the soil microbial community is likely to be dynamic, depending on plant growth stage, the degree of herbivory and the life-cycle of the herbivore. An overview of the interactions between above- and below-ground plant associated organisms and their biodiversity is given by Bardgett et al. (1998); Wardle (2002); Bardgett and Wardle (2003); De Deyn and Van der Putten (2005).

Rhizodeposition and interactions of rhizosphere fauna with microorganisms on plant roots

In contrast to conventional wisdom, recent evidence suggest that a major part of the soil animal food web strongly relies on the C-inputs from plant roots and less so on the carbon and nutrient inputs via leaf litter

(Albers et al. 2006; Larsen et al. 2007; Pollierer et al. 2007), but see Elfstrand et al. (2008). Therefore special attention must be given to the consumers of microorganisms in the rhizosphere, because they are at the base of the soil food web channeling the energy to the higher trophic levels via two distinct routes, the fungal and the bacterial energy channel (Moore and Hunt 1988) and determine the rates of nutrient cycling and the availability of mineral nutrients to plants (Clarholm 1985; Kuikman et al. 1990; Ekelund and Rønn 1994; Laakso and Setälä 1999; Bonkowski 2004).

Interactions of fauna with fungi in the plant rhizosphere

A considerable amount of microbial biomass in soil is contained within the extensive hyphal network of soil fungi. In arable and forest ecosystems fungal hyphae may gain a length of up to 400 and 2,000 m per gram of soil, respectively (Christensen 1989). Not surprisingly fungal feeders are found among all soil animal taxa and have been shown to play a significant role in the release of nutrients to plants (Beare et al. 1995; Bardgett and Chan 1999; Chen and Ferris 1999; Bonkowski et al. 2000a; Gange 2000). Rhizosphere fauna may even serve as a source of nutrients for mycorrhiza (Fig. 2) (Klironomos and Hart 2001; Wilkinson 2008). Some of the most important root symbionts and pathogens are fungi, and apart from the liberation of nutrients soil fauna plays an important role in shaping plant-fungus interactions.

Unfortunately, we know little about fungal defence strategies against fungivores which are probably as complex as bacteria-predator relationships (Kampichler et al. 2004; Scheu and Folger 2004; Scheu and Simmerling 2004; Harold et al. 2005; Bretherton et al. 2006; Tordoff et al. 2006, 2008; Wood et al. 2006).

In multiple choice feeding experiments, fungivores have shown surprisingly similar feeding preferences over a broad range of animal taxa with plant pathogens being among the most preferred fungal diet (Bonkowski et al. 2000c), and mycorrhizal fungi being less attractive (Thimm and Larink 1995; Klironomos and Kendrick 1996; Gange 2000; Sabatini and Innocenti 2001; Bracht Jørgensen et al. 2005). In correspondence with these findings collembola were important for shifting competition between arbuscular mycorrhizal and saprophytic fungi in the rhizosphere of the invasive grass *Cynodon dactylon* (Fig. 2) (Tiunov and Scheu 2005) and significant reductions of fungal plant pathogens have been reported in both laboratory and field studies for protozoa (Tapilskaja 1967; Chakraborty 1983), collembola (Sabatini and Innocenti 2001; Shiraishi et al. 2003) and earthworms (Stephens and Davoren 1997; Clapperton et al. 2001).

There is increasing evidence that fungivores and higher trophic levels in the food web depend in large part on root-derived carbon. This is convincingly exemplified for the best studied group of soil fungivores, the Collembola (Larsen et al. 2007; Ostle et al. 2007). By tracing ^{13}C-signatures of grasses in grassland soil Jonas et al. (2007) have shown that depending on soil type and plant species, saprophytic fungi made up between 40–80%, while arbuscular mycorrhizal fungi constituted up to 60% of the collembolan diet, respectively. Similarly, by investigating the stable isotope composition of soil fauna under a C4-plant (maize) growing in an arable field with C3-plant derived organic matter Albers et al. (2006) demonstrated that 40–50% collembolan body carbon was root-derived within a growing season.

It is an important fact that the effects of fungivores on plant performance are strongly density dependent (Harris and Boerner 1990; Klironomos and Ursic 1998; Bakonyi et al. 2002). The fungivores themselves seem to be regulated mainly by resource levels. Steinaker and Wilson (2008) made a detailed field study on the relationships between roots, mycorrhiza and collembola by using a non-invasive minirhizotron camera technique. Collembola correlated well with root production over the whole growing season in grasslands and forests, suggesting a general dependence of fungivores on plant belowground allocation. Strongly negative exponential relationships between both root and mycorrhizal growth with collembola over short sampling intervals indicated that consumers were driving rhizosphere resource levels at short time scales. By severing the hyphal mycorrhizal network from plant roots collembola have been shown to significantly reduce plant carbon allocation belowground and may impair mycorrhizal function (Johnson et al. 2005), but this may not ultimately result in negative effects on plant growth (Setälä 1995). In fact Steinaker and Wilson (2008) demonstrated that annual root and mycorrhizal production were at a maximum at intermediate collembolan densities with 300–700 and 600–1,500 ind. m^{-2}, in grassland and forest soils, respectively. Predators at higher trophic levels have been shown to regulate population densities of fungivores, thus channelling root-derived energy higher up the food chain. The predatory mite *Hypoaspis aculeifer* which is attracted to the fungal food sources of their collembolan prey by volatiles of the fungus *Alternaria alternata* (Hall and Hedlund 1999) was shown to regulate population densities of collembola to levels which resulted in maximum microbial respiration (Hedlund and Sjögren Öhrn 2000, Fig. 2). In fact, (Pollierer et al. 2007) demonstrated in a stable isotope labelling study of forest trees that root-derived C was quickly incorporated in high amounts into virtually all decomposer species. These studies suggest that the fungal energy channel in fact consists of two distinct routes: a fast energy pathway linked to plant roots and a slower pathway linked to decomposing organic matter.

Root-derived carbon rapidly enters the soil food web (Fitter et al. 2005; Leake et al. 2006) and since coupling of the fungal and bacterial energy channels via top predators is thought to confer the extraordinary stability of soil food webs (Rooney et al. 2006), the food webs connected with both, mycorrhizal fungi and rhizosphere bacteria seem crucial for soil food web functioning.

Effects of collembola on root growth

Soil zoologists have recently begun to pay attention to fauna-induced changes in root architecture. Endlweber and Scheu (2006) investigated the effect of the collembola *Protaphorura fimata* on growth and competition between *Cirsium arvense* (creeping thistle) and *Epilobium adnatum* (willow herb) in a laboratory experiment. Although Collembola did not affect total root biomass they influenced root morphology of both plant species. Roots grew longer and thinner and had more root tips in presence of Collembola. Comparable effects were shown in a subsequent experiment studying competition between clover (*Trifolium repens*) and the grass *Lolium perenne* (Endlweber and Scheu 2007). The authors hypothesized that changes in root morphology in presence of collembola were due to collembola-mediated changes in nutrient availability and distribution.

Interactions between bacterivores and bacteria in the rhizosphere

In the rhizosphere of plants, pulses of root-derived carbon (exudates) fuel bacterial growth and activity on and around roots (Kuzyakov et al. 2000; Paterson 2003). Protozoa and bacterivorous nematodes are the major consumers of bacterial production in the rhizosphere, forming the basis of the heterotrophic eukaryotic food web that channels the energy flow via the bacterial energy channel to higher trophic levels in soil.

Although bacterivores are usually small, their abundance, biomass and in particular their turnover in soil is high, suggesting a significant impact on bacterial turnover in the rhizosphere (Venette and Ferris 1998; Christensen et al. 2007). In a comparison of Dutch farming systems, the average biomass and annual production rates of bacterivores under winter wheat (0–25 cm soil depth) were estimated as 16 kg C ha^{-1} and 105 kg C ha^{-1} yr-1, respectively for protozoa and 0.33 kg C ha^{-1} and 11.6 kg C ha^{-1} yr^{-1} for nematodes (Bouwman and Zwart 1994). Consequently, bacteria in the rhizosphere are strongly top-down regulated via grazing by protozoa and nematodes (Wardle and Yeates 1993). From a gross nutrient perspective, the interactions between plants, bacteria and protozoan grazers in the rhizosphere have been described to form a loop: plant exudates stimulate bacterial growth and through grazing on bacteria protozoa liberate nutrients which in turn stimulate plant growth ("microbial loop in soil", Clarholm 1985). In fact, bacterivores have been well known for their plant growth-promoting properties (Ingham et al. 1985; Clarholm 1994; Ekelund and Rønn 1994; Griffiths 1994; Zwart et al. 1994). Recent investigations indicate that the nutrients released from microbial biomass are only a small part of an intimate, but indirect symbiosis of bacterivores with plant roots (Fig. 2). To understand these complex multitrophic interactions between plant roots, bacteria and bacterivore soil fauna we will focus on mechanisms of selective grazing of bacterivores, bacterial defence and resulting changes in microbial community composition with subsequent feed-backs on root growth.

Grazing of bacterivores is not random

Although the interactions between bacteria and protozoa form one of the oldest and most highly evolved predator-prey systems on earth, predation is a factor rarely included for understanding microbial community structure. In fact, there exists a great diversity of feeding modes of protozoa and nematodes on bacteria, suggesting high competition and distinct niche partitioning of bacterivores (Weisse 2002; Bjørnlund et al. 2006; Blanc et al. 2006) and bacteria evolved sophisticated physical and chemical defence strategies to escape consumption by bacterivores which in their diversity are comparable to the strategies of plants to escape herbivore grazers aboveground (Bonkowski 2004; Huber et al. 2004; Matz and Kjelleberg 2005; Pernthaler 2005; Young 2006; Montagnes et al. 2007).

In general, size-selection has been shown to play an important role for predation on bacteria, resulting in distinct shifts in the size and morphology of grazed suspended planktonic cells (Pernthaler 2005; Young 2006). Bacteria in the rhizosphere, however, are organized in surface associated biofilms (Rudrappa et al. 2008). These biofilms are regulated in a population density dependent manner by quorum sensing (Shapiro 1998; Matz et al. 2004a) where chemical defence is probably most important.

It is well known that both, protozoa and nematodes show distinct feeding preferences and different growth rates according to bacterial prey quality (Weekers et al. 1993; Venette and Ferris 1998; Arndt et al. 2003; Newsham et al. 2004; Blanc et al. 2006; Pickup et al. 2007), but only recently have studies begun to reveal the different roles of bacterial toxin production in defence of predators (Köthe et al. 2003; Matz et al. 2004b; Matz and Kjelleberg 2005; Jousset et al. 2006). Species-specific differences in bacterial consumption depend on the general feeding mode and size of the buccal cavity, the numbers of bacteria ingested, and the ability of bacterivores to digest the prey. For example, high food selectivity appears imperative for small flagellates which successively ingest only single bacteria, (Boenigk and Arndt 2002). Some nematodes have been shown to sense bacterial quorum sensing signals and learn to avoid well-defended colonies (Beale et al. 2006; Köthe et al. 2003), while some species of bacteria-feeding nematodes exhibit stronger food preferences (e.g. monhysterids) than others (e.g. *Panagrolaimus*) (De Mesel et al. 2004). In certain nematode taxa a sclerotized cuticular lining of the terminal bulb serves to grind food particles (Munn and Munn 2002). Due to this mechanical comminution, nematodes seem able to thrive on bacteria undigestible for protozoa (Bjørnlund et al. 2006). However, non-selective feeding bears a cost. In a study using antibiotic-producing *Pseudomonas fluorescens* and non-toxic mutants in rhizosphere systems, Jousset et al. (2009) found that the amoeba *Acanthamoeba castellanii* was able to keep up high population densities even at high densities of toxin-producing bacteria by selectively preying on the mutants. The nematode *C. elegans* in contrast showed high consumption rates of all bacterial strains, even at high densities of toxin-producing bacteria, which produced strongly negative effects on nematode reproduction (Jousset et al. 2009). In more diverse bacterial communities, nematode predators may even benefit from *P. fluorescens* in the rhizosphere. Germinating pea seedlings had a nematicidal effect on *C. elegans* which was reversed by inoculation with *P. fluorescens* strains, suggesting that the nematicidal compounds were metabolised by the introduced bacteria (Brimecombe et al. 2000).

Grazing-induced shifts in rhizosphere bacterial community structure and feed backs on root architecture

Predation is a major factor influencing rhizosphere bacterial community structure and function (Blanc et al. 2006; Murase et al. 2006). Microbivores enhance microbial turnover, carbon transfer and nutrient recycling in soils out of proportion to their own biomass with significant feedbacks on root growth and plant performance. Predators will affect the fitness of bacteria at the individual level by selective feeding on non-toxic cells, and at the group level by preferentially consuming bacteria from populations containing few toxin producing bacteria.

More detailed studies have now shed light on the complex mechanisms by which certain bacteria thrive under grazing and have identified some of the key traits.

The formation of biofilms is a quorum-sensing regulated key trait in the defence against bacterivores and it is interesting to note that plant roots have been shown to interfere with bacterial communication in the rhizosphere (Bauer and Mathesius 2004). A well described example is the formation of biofilm by *P. aeruginosa* in response to protozoan grazing. Within the first 3 days, the bacteria form microcolonies that are physically protected from flagellate predators. At a later state, a mature biofilm forms and quorum sensing regulated toxins are released. These toxins are possibly targeting amoebae that often arrive to the scene later than flagellates in the course of a natural decomposer succession (Matz et al. 2004a; Weitere et al. 2005). This pattern indicates that protection against grazers for individual strains of bacteria may have developed to match the most likely succession of bacterial feeders. The experimental results suggested that the competitive success of a bacterial strain depended on its ability to cope with the prevailing bacterial predator.

In the rhizosphere, fluorescent pseudomonads are important and common root colonizers which increase their competitiveness by producing a broad array of secondary metabolites which inhibit competitors and repel predators (Haas and Keel 2003; Jousset et al. 2006, 2008). These toxins often also inhibit plant pathogens, making pseudomonads potent biological control agents in agricultural systems (Haas and Defago 2005; Siddiqui et al. 2005). *P. fluorescens* produces the phenolic antifungal metabolite 2,4-diacetylphloroglucinol (DAPG) (Keel et al. 1992; Maurhofer et al. 2004), which Jousset et al. (2006) identified as being also the most potent toxin conferring grazing resistance against protozoan predators. Recently we have shown that *P. fluorescens* Q2-87 responded to the addition of *A. castellanii* by increasing DAPG production (Fox, Bonkowski and Phillips, unpublished) (Fig. 3). A sterile-filtered rinse of an overnight culture of amoebae in mineral water also stimulated *P. fluorescens* Q2-87 to significantly increase its toxin production. These results demonstrate that *P. fluorescens* Q2-87 was able to sense the presence of the amoebae and up-regulate DAPG production in anticipation of the predators (Fig. 4).

Toxic exoproducts not only protect bacteria by repelling grazers, but result in prey-switching of predators towards more palatable prey (Jezbera et al. 2006; Liu et al. 2006). For example, Jousset et al. (2008) introduced *gfp*-tagged *P. fluorescens* CHA0 or an isogenic bacterial mutant defective in secondary metabolite production to a complex bacterial community in the rhizosphere of rice. Comparing the competitive outcomes between *P. fluorescens* CHA0, a biocontrol strain with antifungal activity, and its mutants in absence and presence of the amoeba *Acanthamoeba castellanii* clearly demonstrated that

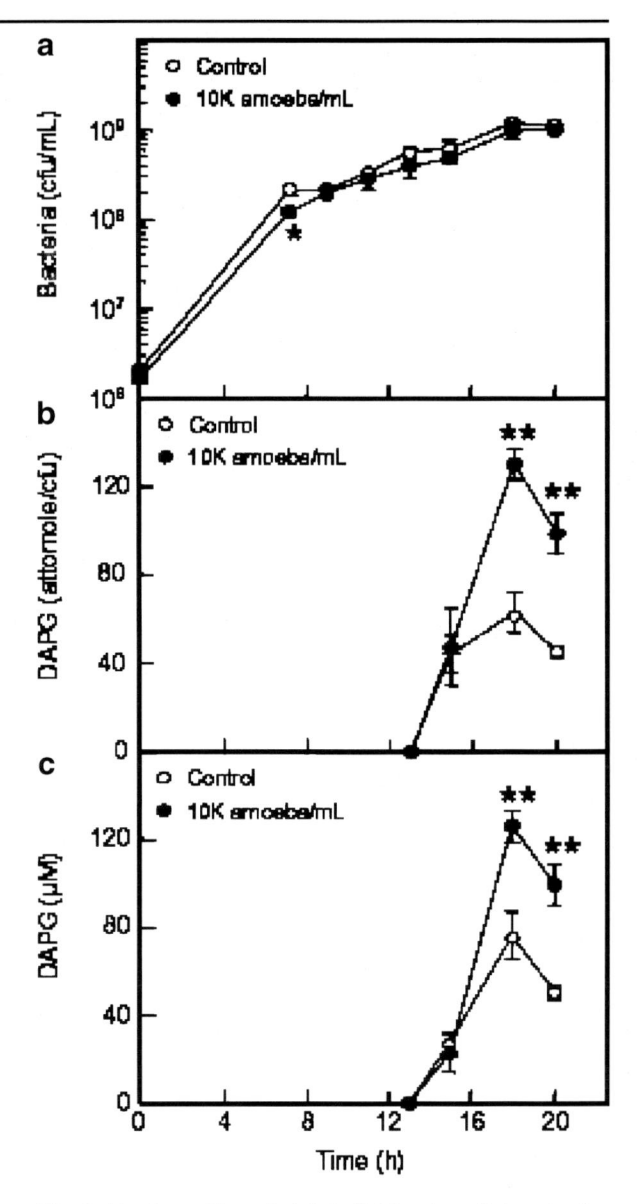

Fig. 3 Numbers of bacteria (cfu / ml) (**a**) and parallel production of the metabolite 2,4-diacetylphloroglucinol (DAPG) (both amount per bacterial cell (**b**) and total amount produced (**c**)) by the bacterial strain *Pseudomonas fluorescens* Q2-87 during 20 h of growth (Control) or in response to the addition of 1×10^4 *Acanthamoeba castellanii* ml^{-1} (10 K amoebae/ml). Note that numbers of bacteria generally did not significantly change in the presence of amoebae. Asterisks indicate significant differences between bacteria and bacteria + amoeba treatments (*$P<0.05$; **$P<0.01$) (Phillips, Fox and Bonkowski, unpublished)

the biocontrol strain *P. fluorescens* CHA0 was not only rejected by the protozoan predator but doubled in numbers because their bacterial competitors were preferentially consumed. In addition, growth-limiting

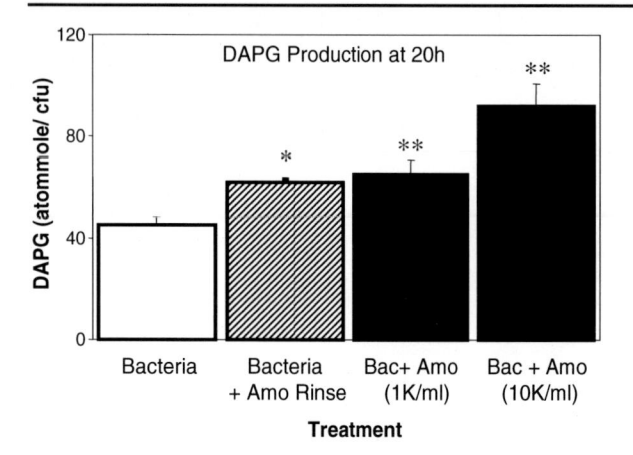

Fig. 4 Production of the metabolite 2,4-diacetylphloroglucinol (DAPG) by the bacterial strain *Pseudomonas fluorescens* Q2-87 after 20 h growth in treatments with (**a**) bacteria only (Bacteria), (**b**) in response to a sterile-filtered supernatant (rinse) of an axenic culture of *Acanthamoeba castellanii* kept for 24 h in mineral water (Bacteria + Amo Rinse), (**c**) in response to the addition of 1×10^3 *A. castellanii* ml^{-1} and (**d**) in response to the addition of 1×10^4 *A. castellanii* ml^{-1}. Asterisks indicate significant differences between bacteria and amoeba treatments (*$P<0.05$; **$P<0.01$) (Phillips, Fox and Bonkowski, unpublished)

nutrients released from the protozoa will further benefit growth of the toxin producing strain (Griffiths 1994). Thus grazing-resistant bacteria will gain a threefold benefit by avoiding predation losses, elimination of competitors and by having more resources at their disposal. These results reveal a basic principle how specific rhizosphere bacteria consistently gain a competitive edge over less defended bacterial taxa in the presence of bacterivores.

Microorganisms in soil are generally limited by the availability of low-molecular weight carbon compounds and rhizodeposition plays a crucial role as energy source in the coupling of plant and microbial productivity (Paterson 2003). With this in mind let us consider that *P. fluorescence* has also been noted to stimulate root branching (Beyeler et al. 1999; De Leij et al. 2002). Since root tips are the major sites of root exudation, grazing-resistant bacteria might have evolved mechanisms to gain carbon from plant roots and nutrients via their interaction with bacterivores at the same time.

This view corresponds well with contemporary theory on rhizosphere interactions. In their conceptual review Phillips et al. (2003) introduced an evolutionary view on rhizosphere interactions and assumed that during the colonization of land, plants would have benefited from preexisting associations between soil microorganisms and their consumers. Many rhizosphere bacteria are known to manipulate plant exudation and root branching by the release of hormones (Spaepen et al. 2007) and other signal molecules (Phillips et al. 1999, 2004; Dakora and Phillips 2002; Joseph and Phillips 2003) but rhizosphere bacteria are inextricably linked with their consumers. Therefore interactions with the microbial rhizosphere food web have to be considered to gain a deeper mechanistic understanding of the evolutionary forces shaping microbial rhizosphere interactions. Plant roots can be viewed as a source and receptor of molecular signals for mutualistic organisms (microbes and their consumers) and there are genetic 'control points' at which interactions between organisms (including the plant) determine the outcome of the interaction (Phillips et al. 2003). Such a 'control point' is the signalling between exudates—bacterial grazers—bacteria—plant root growth as has been postulated by Bonkowski (2004).

Bonkowski and Brandt (2002) demonstrated strong growth-stimulating effects of amoebae on the root system of garden cress (*Lepidium sativum*). The numbers and length of first order lateral roots increased four- and fivefold, respectively in treatments with amoebae. A concomitant proportional increase in auxin-producing bacteria led Bonkowski and Brandt (2002) to suggest that specific plant growth-promoting bacteria were favoured by amoebae and stimulated root growth which allowed more nutrients to be absorbed, but would also increase exudation rates, thereby further stimulating bacteria-bacterivore interactions. Thus a mutual feedback exists between plant roots, bacteria and bacterivores. Several studies with grasses, cereals, forbs and tree seedlings have confirmed a strongly stimulating effect of amoebae on the number and length of lateral roots (Jentschke et al. 1995; Kreuzer et al. 2006; Herdler et al. 2008; Somasundaram et al. 2008). The increase in laterals plays a crucial role in plant development because they form the scaffold for the architecture of the branched root system (Malamy and Benfey 1997) and enhance root uptake of nutrients released by protozoa (Bonkowski et al. 2000b; Somasundaram et al. 2008). In line with this concept, it has been found that the presence of bacterial-feeding nematodes increased the total amount of organic-C exuded by

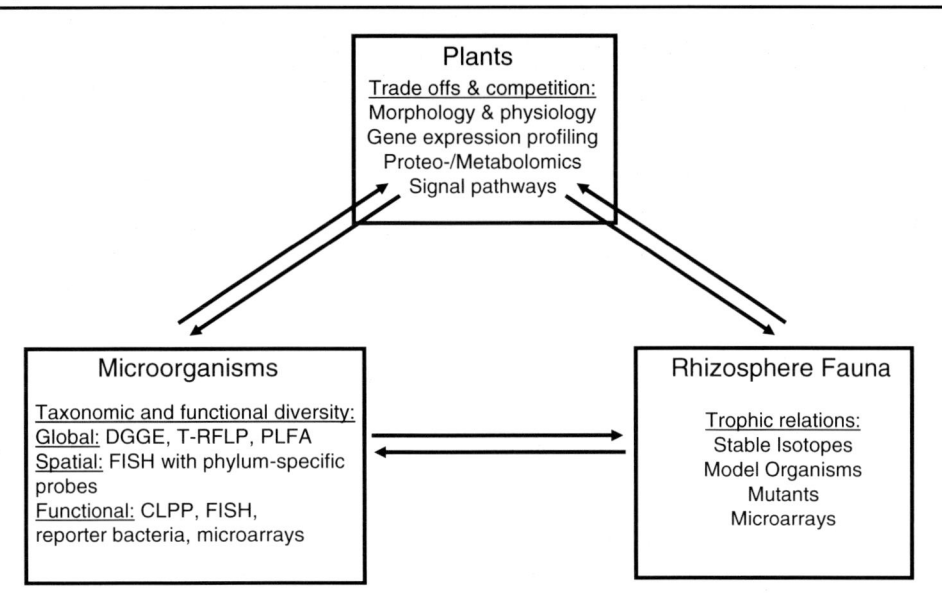

Fig. 5 A perspective on methods in "Rhizosphere Molecular Ecology" to detect molecular control points in rhizosphere food webs. In plants, apart from mapping changes in (root) morphology and physiology, changes in the regulation of gene expression are crucial to uncover the specific pathways involved in plant-microbial-faunal interactions. Subsequently mutants silenced in specific pathways are being used to verify the functional significance of the mechanisms involved. Proteomics and metabolomics in parallel with stable isotope labelling provide means to check for the transcription and expression of specific plant-derived signaling cascades. A broad spectrum of methods allows the analysis of changes in microbial biofilms on roots at different levels of complexity. Denaturing Gradient Gel Electrophoresis (DGGE), Terminal Restriction Fragment Length Polymorphism (T-RFLP) analysis or Phospholipid Fatty Acid (PLFA) profiling are some methods often used for a global assessment of microbial diversity. Fluorescence in situ Hybridization (FISH) with phylum-specific probes gives quantitative data on the spatial arrangement of micoorganisms on roots, whereas Community Level Physiological Profiles (CLPP), FISH or microarrays on functional genes or reporter bacteria are sensitive methods to monitor functional changes in microbial consortia along roots. Stable isotope studies are increasingly used to quantify trophic interactions with the soil fauna, but model organisms (e.g. the bacterial-feeding nematode *Caenorhabditis elegans*) already allow the testing of more subtle interactions (e.g. attraction or stress responses) of rhizosphere fauna to plant- and microbial signals (metabolites and volatiles) by using mutants (e.g. knock out or *gfp*-reporter) or measuring gene expression (e.g. microarrays, quantitative real-time PCR)

roots of *Brassica napus* almost 3-fold over bacteria only controls (Sundin et al. 1990). The positive feedback effects of bacterial feeding nematodes on bacterial biomass and activity varies with the nematode species and population size. *Cruznema tripartum*, for example, increased bacterial biomass and activity approximately 4-fold while *Acrobeloides bodenheimeri* was responsible for only a 2-fold increase (Fu et al. 2005). A future goal therefore must lie in the identification of bacterivore taxa most important for plant growth promotion.

In soil with greater populations of bacterial-feeding nematodes, either the native soil population or added *C. elegans*, tomato plants developed a more highly branched root system with longer and thinner roots accompanied by an increase of soil auxin content, such as indole-acetic acid (IAA) and an altered microbial community structure. Bacterial-feeding nematodes may have affected plant growth by stimulating hormone production through grazing-induced changes to the soil microbial community (Mao et al. 2006, 2007), suggesting corresponding mechanisms of nematodes and protozoa (Bonkowski and Brandt 2002). The role of IAA in microbial and microbial-plant signaling have been reviewed by Spaepen et al. (2007) but the inclusion of bacterial grazers adds another level of interaction. The feedback between bacterial-feeding fauna and IAA producing bacteria and root growth maybe strongest with young plants, as the interaction with the microbial community seems to diminish with plant age (Vestergård et al. 2007).

We suggest that significant progress will be made in understanding the molecular and genetic basis of the interactions involving signaling compounds in the rhizosphere and how microfauna mediate and even control these outcomes (Fig. 5). However, not all plant cultivars are similarly responsive to bacterivores, Somasundaram et al. (2008) found a general strong growth increase of rice in the presence of *Acanthamoebae*, but some rice cultivars did not respond at all to protozoa. In particular upland and lowland Japonica rice cultivars differed strongly in their response to the presence of amoebae, suggesting that during rice breeding some essential genes involved in the signalling between rhizosphere bacteria or bacterivores with plant roots were lost. Similar findings have been made with maize (R. Koller, personal communication).

Moreover, plants exert species specific effects on the composition of root colonizing bacteria (Chanway et al. 1991; Stephan et al. 2000; Wieland et al. 2001), and rhizosphere populations of bacterial-feeding protozoa and nematodes seem partly dependent on the plant species (Henderson and Katznelson 1961; Geltzer 1963; Griffiths 1990; Brimecombe et al. 2000). Venette et al. (1997), for example, observed significant increases in bacterial-feeding nematode abundance in the rhizosphere of *Crotalaria juncea* (sun hemp) and *Vicia villosa* (vetch) but not *Tagetes patula* (marigold), *Eragrostis curvula* (love grass) or *Sesanum indicum* (sesame) compared with bulk soil in a pot experiment.

In view of these findings, it seems crucial to narrow down specific plant traits, and ultimately genetic rhizospere control points (*sensu* Phillips et al. 2003) interacting with the rhizosphere microbial food web for plant breeding to improve the management of nutrient availability in soils (Phillips and Streit 1998; Rengel and Marschner 2005; Joshi et al. 2007).

Soil fauna as vectors of rhizosphere microorganisms

Since microorganisms are not very mobile in soil, it has been suggested that the soil fauna plays important roles as vectors of microbial symbionts in rhizosphere.

Nematodes seem particularly important in spreading bacteria around the rhizosphere. Nematodes can defaecate 30–60% of ingested bacteria in a viable form (Chantanao and Jensen 1969) and can harbour and protect bacteria from adverse environmental conditions (Caldwell et al. 2003). They are also able to carry bacteria and fungal spores externally, adhering to their cuticular mucilage. It is well known that on an agar plate, for example, one can follow the trails of where nematodes have been from the bacterial colonies that subsequently grow there (Young et al. 1996). A recent finding that legume roots, by emitting specific volatile signals, recruit bacterivorous nematodes for their inoculation with rhizobia is again testimony of the so far unnoticed, but fundamental role of signal exchange between roots and soil fauna. Horiuchi et al. (2005) showed the transfer of *Sinorhizobium meliloti* by *C. elegans* to the roots of the legume *Medicago truncatula* in response to the plant-released volatile dimethylsulfide that attracted the nematode to the roots. Bacterial-feeding nematodes therefore, may significantly foster the initiation of the N-fixing symbiosis in legumes, but they have also been repeatedly shown to spread plant growth-promoting bacteria. For example in mushroom cultures, *C. elegans* has been shown to feed selectively on a biocontrol species of *Pseudomonas* in preference to plant pathogenic *Pseudomonas* species and so reduced the incidence of bacterial blotch by spreading the antagonistic species (Grewal 1991). Rhizosphere colonisation of seed applied biocontrol agents was substantially increased by bacterial-feeding nematodes (Knox et al. 2003, 2004) and the bacterial-feeding *Diplogaster iheritieri* preferred to feed on growth promoting rhizobacteria in laboratory tests and it was suggested that it could move viable cells around plant roots and enhance plant growth in the field (Kimpinski and Sturz 1996).

Microarthropods have been shown to disseminate viable spores of ectomycorrhizal and arbuscular mycorrhizal fungi (Lilleskov and Bruns 2005; Seres et al. 2007), and they are potential vectors of entomopathogenic and saprophytic fungi (Dromph 2003; Renker et al. 2005). But Gormsen et al. (2004) showed that earthworms and collembola did not affect the spread of AM fungi over a distance greater than 20 cm. Although the dispersal range of microbes by soil fauna might be quite limited, this distance seems not without significance in respect to the tiny size of microorganisms and microbial competition within the tight network of roots in dense stands of established vegetation. Similarly Rantalainen et al. (2004) demonstrated in a laboratory study an important role of enchytraeids in spreading saprophytic fungi between habitat patches through corridors, but in a successive field study it was shown that most fungi had quite good dispersal abilities via hyphal growth independent of soil faunal activity (Rantalainen et al. 2005). At present, evidence suggests nematodes as being important vectors of bacteria, whereas fungivores appear to have more important roles as consumers than as vectors of fungi in the rhizosphere.

Rhizosphere macrofauna: new functional roles of earthworms

Due to their important function in the acceleration of decomposition by physicochemical processes, the role of earthworms is usually considered important in long-term processes such as decomposition of litter materials (Scheu 1993; Brussaard 1998; Schulman and Tiunov 1999). It would be incorrect, however, to assume that earthworms solely affect plant performance by enhanced liberation of nutrients. For example, humus compounds released from earthworm worked soils have been shown to exert hormone-like effects (Nardi et al. 1994; Muscolo et al. 1999; Zandonadi et al. 2007) and recent studies provide convincing evidence that earthworms can induce subtle host-mediated changes that determine the disposition of plants to herbivore attack (Scheu et al. 1999; Wurst and Jones 2003).

For example, Blouin et al. (2005) found that presence of earthworms strongly increased the tolerance of rice to root-feeding nematodes. Although earthworms had no direct effect on nematode population size, the detrimental effects of nematodes on root biomass and photosynthesis of rice disappeared in the presence of earthworms. Since the expression of three stress-responsive genes in leaves (coding for lipoxygenase, phospholipase D and cysteine protease) was modulated in presence of earthworm activity, Blouin et al. (2005) convincingly demonstrated that earthworms triggered the induced defense of rice against root parasitic nematodes.

Similarly, Wurst et al. (2003) investigated the effects of *A. caliginosa* and *O. tyrtaeum* on aphid performance on plants of different functional groups (the grass *Lolium perenne*, the forb *Plantago lanceolata*, the legume *Trifolium repens*) that differed in root morphology and N allocation strategies. Earthworm activity generally enhanced nitrogen mobilization from litter and from soil. However, the earthworm-mediated increase in plant nitrogen uptake differed between plant species. Earthworms enhanced N uptake from litter and soil in all plant species but shoot and root growth only in *L. perenne* and *P. lanceolata*. Earthworms increased aboveground biomass and contents of total nitrogen and ^{15}N in both *L. perenne* and *P. lanceolata*, and root growth in the grass. Due to the increase in plant nitrogen content one would expect positive effects of earthworms on aphid reproduction (Dixon 1985). But reproduction of *M. persicae* was reduced on *P. lanceolata* in presence of earthworms (Wurst et al. 2003), most likely as a result of earthworm effects on plant defense compounds. Wurst et al. (2004) subsequently confirmed that earthworms and organic matter distribution strongly affected the contents of phytosterols in *P. lanceolata*. Phytosterols serve as precursors of moulting hormones in the diet of herbivorous insects, including aphids (Campell and Nes 1983). Phytosterols and iridoid gylcosides were positively correlated with plant nitrogen content, suggesting that the production of defense compounds might be indirectly driven by increased N availability as a result from earthworm activity.

These results demonstrate that plant vigour and susceptibility to insect herbivores are driven by a complexity of interactions with soil macrofauna, reaching far beyond the standard view on nutrient liberation (Bonkowski et al. 2000a; Scheu and Setälä 2002).

Conclusions and directions for future research

This review highlights the importance of the many recent findings on significant indirect interactions between soil fauna, rhizosphere microorganisms and plant roots. Even "direct" feeding relationships between plant roots and herbivores are not as simple and straight forward as seen in the past. The exchange of chemical signals which formerly was only considered important between roots and closely associated microbial symbionts now appears common between all rhizosphere players on all levels of interaction. The arms race between plants and herbivores belowground seems as highly coevolved as plant interactions with the aboveground food web, and both are intricately connected. Bacteria and bacterivores mutually cooperate with significant effects on root architecture, but there is considerable uncertainty to which degree plants are in control of these processes. Animal behaviour has been intensively studied for decades. Now it becomes apparent that plant roots "behave" in much the same sense (Dudley and File 2007). Upon specific key signals, roots have been shown to emit signal compounds to communicate with soil organisms, including fauna. The signals are used to attract entomopathogenic nematodes against root herbivores, or bacterivores for root inoculation with beneficial rhizosphere bacteria, or to disturb communication among potential harmful soil bacteria, but also to manipulate plant belowground allocation. Undoubtedly more such examples will be discovered soon, and the challenge will be to identify the genetic control points which determine these interactions and to determine how the interactions between multiple plant symbionts are orchestrated (Fig. 5).

In view of the increasing efforts to increase crop resistance to agricultural pests by transformation of crops with foreign genes, the study of natural plant defence systems has a practical application. However, lack of consideration by scientists and plant breeders of root ecology has already led to the loss of genes important in root-fauna communication (Rasmann et al. 2005; Somasundaram et al. 2008). In the light of these results, it seems timely to consider the study of rhizosphere ecology as a multidisciplinary task to improve plant breeding efforts.

Acknowledgements We are very grateful to Prof. Dr. Donald Phillips and Dr. Tama Fox, Plant Sciences Department, University of California, Davis, USA, for their collaborative support for MB and for providing the data on DAPG production by pseudomonads for this review.

References

Albers D, Schaefer M, Scheu S (2006) Incorporation of plant carbon into the soil animal food web of an arable system. Ecology 87:235–245. doi:10.1890/04-1728

Arndt H, Schmidt-Denter K, Auer B, Weitere M (2003) Protozoans and biofilms. In: Krumbein WE, Paterson DM, Zavarzin GA (eds) Fossil and recent biofilms. Kluwer Academic, Dordrecht, pp 173–189

Ayres E, Dromph KM, Cook R, Ostle N, Bardgett RD (2007) The influence of below-ground herbivory and defoliation of a legume on nitrogen transfer to neighbouring plants. Funct Ecol 21:256–263. doi:10.1111/j.1365-2435.2006.01227.x

Bais H, Park S, Weir T, Callaway R, Vivanco J (2004) How plants communicate using the underground information superhighway. Trends Plant Sci 9:26–32. doi:10.1016/j.tplants.2003.11.008

Bais H, Weir T, Perry L, Gilroy S, Vivanco J (2006) The role of root exudates in rhizosphere interactions with plants and other organisms. Annu Rev Plant Biol 57:233–266. doi:10.1146/annurev.arplant.57.032905.105159

Bakonyi G, Posta K, Kiss I, Fábián M, Nagy P, Nosek JN (2002) Density-dependent regulation of arbuscular mycorrhiza by collembola. Soil Biol Biochem 34:661–664. doi:10.1016/S0038-0717(01)00228-0

Bardgett RD, Chan KF (1999) Experimental evidence that soil fauna enhance nutrient mineralization and plant nutrient uptake in montane grassland ecosystems. Soil Biol Biochem 31:1007–1014. doi:10.1016/S0038-0717(99)00014-0

Bardgett RD, Wardle DA (2003) Herbivore-mediated linkages between aboveground and belowground communities. Ecology 84:2258–2268. doi:10.1890/02-0274

Bardgett R, Wardle D, Yeates G (1998) Linking above-ground and below-ground interactions: how plant responses to foliar herbivory influence soil organisms. Soil Biol Biochem 30:1867–1878. doi:10.1016/S0038-0717(98)00069-8

Bardgett R, Cook R, Yeates G, Denton C (1999a) The influence of nematodes on below-ground processes in grassland ecosystems. Plant Soil 212:23–33. doi:10.1023/A:1004642218792

Bardgett R, Denton C, Cook R (1999b) Below-ground herbivory promotes soil nutrient transfer and root growth in grassland. Ecol Lett 2:357–360. doi:10.1046/j.1461-0248.1999.00001.x

Bauer W, Mathesius U (2004) Plant responses to bacterial quorum sensing signals. Curr Opin Plant Biol 7:429–433. doi:10.1016/j.pbi.2004.05.008

Beale E, Li G, Tan M-W, Rumbaugh KP (2006) *Caenorhabditis elegans* senses bacterial autoinducers. Appl Environ Microbiol 72:5135–5137

Beare M, Coleman D, Crossley D, Hendrix P, Odum E (1995) A hierarchical approach to evaluating the significance of soil biodiversity to biogeochemical cycling. Plant Soil 170:5–22. doi:10.1007/BF02183051

Beyeler M, Keel C, Michaux P, Haas D (1999) Enhanced production of indole-3-acetic acid by a genetically modified strain of *Pseudomonas fluorescens* CHA0 affects root growth of cucumber, but does not improve protection of the plant against *Phytium* root rot. FEMS Microbiol Ecol 28:225–233. doi:10.1111/j.1574-6941.1999.tb00578.x

Bezemer TM, Van Dam NM (2005) Linking aboveground and belowground interactions via induced plant defenses. Trends Ecol Evol 20:617–624. doi:10.1016/j.tree.2005.08.006

Bezemer TM, Wagenaar R, Dam NMV, Wäckers FL (2003) Interactions between above- and belowground insect herbivores as mediated by the plant defense system. Oikos 101:555–562. doi:10.1034/j.1600-0706.2003.12424.x

Bezemer TM, De Deyn GB, Bossinga TM, Van Dam NM, Harvey JA, Van Der Putten WH (2005) Soil community composition drives aboveground plant-herbivore-parasitoid interactions. Ecol Lett 8:652–661. doi:10.1111/j.1461-0248.2005.00762.x

Bjørnlund L, Mørk S, Vestergard M, Rønn R (2006) Trophic interactions between rhizosphere bacteria and bacterial feeders influenced by phosphate and aphids in barley. Biol Fertil Soils 43:1–11. doi:10.1007/s00374-005-0052-7

Blanc C, Sy M, Djigal D, Brauman A, Normand P, Villenave C (2006) Nutrition on bacteria by bacterial-feeding nematodes and consequences on the structure of soil bacterial community. Eur J Soil Biol 42, Suppl 1:S70–S78

Blouin M, Zuily-Fodil Y, Pham-Thi AT, Laffray D, Reversat G, Pando A, Tondoh J, Lavelle P (2005) Belowground organism activities affect plant aboveground phenotype, inducing plant tolerance to parasites. Ecol Lett 8:202–208

Boenigk J, Arndt H (2002) Bacterivory by heterotrophic flagellates: community structure and feeding strategies. Antonie Leeuwenhoek 81:465–480

Boff MIC, Zoon FC, Smits PH (2001) Orientation of *Heterorhabditis megidis* to insect hosts and plant roots in a Y-tube sand olfactometer. Entomol Exp Appl 98:329–337. doi:10.1023/A:1018907812376

Bonkowski M (2004) Protozoa and plant growth: the microbial loop in soil revisited. New Phytol 162:617–631. doi:10.1111/j.1469-8137.2004.01066.x

Bonkowski M, Brandt F (2002) Do soil protozoa enhance plant growth by hormonal effects? Soil Biol Biochem 34:1709–1715. doi:10.1016/S0038-0717(02)00157-8

Bonkowski M, Cheng W, Griffiths BS, Alphei J, Scheu S (2000a) Microbial-faunal interactions in the rhizosphere and effects on plant growth. Eur J Soil Biol 36:135–147. doi:10.1016/S1164-5563(00)01059-1

Bonkowski M, Griffiths B, Scrimgeour C (2000b) Substrate heterogeneity and microfauna in soil organic 'hotspots' as determinants of nitrogen capture and growth of ryegrass. Appl Soil Ecol 14:37–53. doi:10.1016/S0929-1393(99)00047-5

Bonkowski M, Griffiths BS, Ritz K (2000c) Food preferences of earthworms for soil fungi. Pedobiologia (Jena) 44:666–676. doi:10.1078/S0031-4056(04)70080-3

Bouwman LA, Zwart KB (1994) The ecology of bacterivorous protozoans and nematodes in arable soil. Agric Ecosyst Environ 51:145–160

Bracht Jørgensen H, Johansson T, Canbäck B, Hedlund K, Tunlid A (2005) Selective foraging of fungi by collembolans in soil. Biol Lett 1:243–246. doi:10.1098/rsbl.2004.0286

Bretherton S, Tordoff GM, Jones TH, Boddy L (2006) Compensatory growth of Phanerochaete velutina mycelial systems grazed by Folsomia candida (Collembola). FEMS Microbiol Ecol 58:33–40. doi:10.1111/j.1574-6941.2006.00149.x

Brimecombe M, De Leij F, Lynch J (2000) Effect of introduced *Pseudomonas fluorescens* strains on soil nematode and protozoan populations in the rhizosphere of wheat and pea. Microb Ecol 38:387–397. doi:10.1007/s002489901004

Brown V, Gange A (1989) Differential effects of abobe- and below-ground insect herbivory during early plant succession. Oikos 54:67–76. doi:10.2307/3565898

Brussaard L (1998) Soil fauna, guilds, functional groups and ecosystem processes. Appl Soil Ecol 9:123–135

Caldwell KN, Anderson GL, Williams PL, Beuchat LR (2003) Attraction of a free-living nematode, *Caenorhabditis elegans*, to foodborne pathogenic bacteria and its potential as a vector of *Salmonella poona* for preharvest contamination of cantaloupe. J Food Prot 66:1964–1971

Campell BC, Nes WD (1983) A reappraisal of sterol biosynthesis and metabolism in aphids. J Insect Physiol 29:149–156. doi:10.1016/0022-1910(83)90138-5

Chakraborty S (1983) Population dynamics of amobae in soils suppressive and non-suppressive to wheat take-all. Soil Biol Biochem 15:661–664. doi:10.1016/0038-0717(83)90029-9

Chantanao A, Jensen HJ (1969) Saprozoic nematodes as carriers and disseminators of plant pathogenic bacteria. J Nematol 1:216–218

Chanway C, Turkington R, Holl F (1991) Ecological implications of specificity between plants and rhizosphere microorganisms. Adv Ecol Res 21:121–169. doi:10.1016/S0065-2504(08)60098-7

Chen J, Ferris H (1999) The effects of nematode grazing on nitrogen mineralization during fungal decomposition of organic matter. Soil Biol Biochem 31:1265–1279. doi:10.1016/S0038-0717(99)00042-5

Christensen M (1989) A view of fungal ecology. Mycologia 81:1–19. doi:10.2307/3759446

Christensen S, Bjørnlund L, Vestergard M (2007) Decomposer biomass in the rhizosphere to assess rhizodeposition. Oikos 116:65–74

Ciche TA, Darby C, Ehlers R-U, Forst S, Goodrich-Blair H (2006) Dangerous liaisons: the symbiosis of entomopathogenic nematodes and bacteria. Biol Control 38:22–46. doi:10.1016/j.biocontrol.2005.11.016

Clapperton MJ, Lee NO, Binet F, Conner RL (2001) Earthworms indirectly reduce the effects of take-all (Gaeumannomyces graminis var. tritici) on soft white spring wheat (Triticum aestivum cv. Fielder). Soil Biol Biochem 33:1531–1538. doi:10.1016/S0038-0717(01)00071-2

Clarholm M (1985) Interactions of bacteria, protozoa and plants leading to mineralization of soil nitrogen. Soil Biol Biochem 17:181–187. doi:10.1016/0038-0717(85)90113-0

Clarholm M (1994) The microbial loop in soil. In: Ritz K, Dighton J, Giller KE (eds) Beyond the biomass. Wiley-Sayce, Chichester, pp 221–230

Dakora FD, Phillips DA (2002) Root exudates as mediators of mineral acquisition in low-nutrient environments. Plant Soil 245:35–47. doi:10.1023/A:1020809400075

Davis E, Hussey R, Baum T, Bakker J, Schots A (2000) Nematode parasitism genes. Annu Rev Phytopathol 38:365–396. doi:10.1146/annurev.phyto.38.1.365

Dawson LA, Grayston SJ, Murray PJ, Pratt SM (2002) Root feeding behaviour of Tipula paludosa (Meig.) (Diptera : Tipulidae) on Loliumn perenne (L.) and Trifolium repens (L.). Soil Biol Biochem 34:609–615. doi:10.1016/S0038-0717(01)00217-6

De Deyn GB, Van der Putten WH (2005) Linking aboveground and belowground diversity. Trends Ecol Evol 20:625–633. doi:10.1016/j.tree.2005.08.009

De Deyn G, Raaijmakers C, Zoomer H, Berg M, de Ruiter P, Verhoef H, Bezemer T, van der Putten W (2003a) Soil invertebrate fauna enhances grassland succession and diversity. Nature 422:711–713. doi:10.1038/nature01548

De Deyn GB, Raaijmakers CE, Zoomer HR, Berg MP, De Ruiter PC, Verhoef HA, Bezemer TM, Van der Putten WH (2003b) Soil invertebrate fauna enhances grassland succession and diversity. Nature 422:711–713. doi:10.1038/nature01548

De Leij FAAM, Dixon-Hardy JE, Lynch JM (2002) Effect of 2, 4-diacetylphloroglucinol-producing and non-producing strains of Pseudomonas fluorescens on root development of pea seedlings in three different soil types and its effect on nodulation by Rhizobium. Biol Fertil Soils 35:114–121. doi:10.1007/s00374-002-0448-6

De Mesel I, Derycke S, Moens T, Van Der Gucht K, Vincx M, Swings J (2004) Top-down impact of bacterivorous nematodes on the bacterial community structure: a microcosm study. Environ Microbiol 6:733–744. doi:10.1111/j.1462-2920.2004.00610.x

Denton CS, Bardgett RD, Cook R, Hobbs PJ (1999) Low amounts of root herbivory positively influence the rhizosphere microbial community in a temperate grassland soil. Soil Biol Biochem 31:155–165. doi:10.1016/S0038-0717(98)00118-7

Dicke M (2009) Behavioural and community ecology of plants that cry for help. Plant Cell Environ . doi:10.1111/j.1365-3040.2008.01913.x

Dixon AFG (1985) Aphid ecology. Blackie, Glasgow London, p 157

Dromph KM (2003) Collembolans as vectors of entomopathogenic fungi. Pedobiologia (Jena) 47:245–256. doi:10.1078/0031-4056-00188

Dudley SA, File AL (2007) Kin recognition in an annual plant. Biol Lett 3:435–438. doi:10.1098/rsbl.2007.0232

Ekelund F, Rønn R (1994) Notes on protozoa in agricultural soil with emphasis on heterotrophic flagellates and naked amoebae and their ecology. FEMS Microbiol Rev 15:321–353. doi:10.1111/j.1574-6976.1994.tb00144.x

Elfstrand S, Lagerlöf J, Hedlund K, Mårtensson A (2008) Carbon routes from decomposing plant residues and living roots into soil food webs assessed with 13C labelling. Soil Biol Biochem 40:2530–2539. doi:10.1016/j.soilbio.2008.06.013

Endlweber K, Scheu S (2006) Effects of Collembola on root properties of two competing ruderal plant species. Soil Biol Biochem 38:2025–2031

Endlweber K, Scheu S (2007) Interactions between mycorrhizal fungi and Collembola: effects on root structure of competing plant species. Biol Fertil Soils 43:741–749. doi:10.1007/s00374-006-0157-7

Erb M, Ton J, Degenhardt J, Turlings TCJ (2008) Interactions between arthropod-induced aboveground and belowground defenses in plants. Plant Physiol 146:867–874

Fitter AH, Gilligan CA, Hollingworth K, Kleczkowski A, Twyman RM, Pitchford JW (2005) Biodiversity and ecosystem function in soil. Funct Ecol 19:369–377. doi:10.1111/j.0269-8463.2005.00969.x

Fu SL, Ferris H, Brown D, Plant R (2005) Does the positive feedback effect of nematodes on the biomass and activity of their bacteria prey vary with nematode species and population size? Soil Biol Biochem 37:1979–1987

Gange A (2000) Arbuscular mycorrhizal fungi, Collembola and plant growth. Trends Ecol Evol 15:369–372. doi:10.1016/S0169-5347(00)01940-6

Gange A, Brown V (1997) Multitrophic interactions in terrestrial systems. Blackwell, Oxford

Geltzer JG (1963) On the behaviour of soil amoebae in the rhizospheres of plants. Pedobiologia (Jena) 2:249–251

Gheysen G, Jones J (2006) Molecular aspects of plant-nematode interactions. In: Perry R, Moens M (eds) Plant Nematology. CABI, pp 234–254

Goellner M, Wang X, Davis EL (2001) Endo-1, 4-glucanase expression in compatible plant nematode interactions. Plant Cell 13:2241–2255

Gormsen D, Olsson PA, Hedlund K (2004) The influence of collembolans and earthworms on AM fungal mycelium. Appl Soil Ecol 27:211–220

Goverse A, Overmars H, Engelbertink J, Schots A, Bakker J, Helder J (2000) Both induction and morphogenesis of cyst nematode feeding cells are mediated by auxin. Mol Plant Microbe Interact 13:1121–1129. doi:10.1094/MPMI.2000.13.10.1121

Grayston SJ, Dawson LA, Treonis AM, Murray PJ, Ross J, Reid EJ, MacDougall R (2001) Impact of root herbivory by insect larvae on soil microbial communities. Eur J Soil Biol 37:277–280. doi:10.1016/S1164-5563(01)01098-6

Grewal PS (1991) Effects of Caenorhabditis elegans(Nematoda: Rhabditidae) on the spread of the bacterium Pseudomonas tolaasii in mushrooms (Agaricus bisporus). Ann Appl Biol 118:47–55. doi:10.1111/j.1744-7348.1991.tb06084.x

Griffiths BS (1990) A comparison of microbial-feeding nematodes and protozoa in the rhizosphere of different plants. Biol Fertil Soils 9:83–88. doi:10.1007/BF00335867

Griffiths BS (1994) Soil nutrient flow. In: Darbyshire J (ed) Soil protozoa. CAB International, Wallingford, pp 65–91

Haas D, Defago G (2005) Biological control of soil-borne pathogens by fluorescent pseudomonads. Nat Rev Microbiol 3:307–319. doi:10.1038/nrmicro1129

Haas D, Keel C (2003) Regulation of antibiotic production in root-colonizing Pseudomonas spp. and relevance for biological control of plant disease. Annu Rev Phytopathol 41:117–153. doi:10.1146/annurev.phyto.41.052002.095656

Haase S, Ruess L, Neumann G, Marhan S, Kandeler E (2007) Low-level herbivory by root-knot nematodes (Meloidogyne incognita) modifies root hair morphology and rhizodeposition in host plants (Hordeum vulgare). Plant Soil 301:151–164. doi:10.1007/s11104-007-9431-1

Hall M, Hedlund K (1999) A soil mite uses fungal cues in search for its collembolan prey. Pedobiologia (Jena) 43:11–17

Harold S, Tordoff GM, Jones TH, Boddy L (2005) Mycelial responses of Hypholoma fasciculare to collembola grazing: effect of inoculum age, nutrient status and resource quality. Mycol Res 109:927–935. doi:10.1017/S095375620500331X

Harris KK, Boerner REJ (1990) Effects of belowground grazing by collembola on growth, mycorrhizal infection, and P uptake of Geranium robertianum. Plant Soil 129:203–210

Hatch D, Murray P (1994) Transfer of nitrogen from damaged roots of white clover (Trifolium repens L.) to closely associated roots of intact perennial ryegrass (Lolium perenne L.). Plant Soil 166:181–185. doi:10.1007/BF00008331

Hedlund K, Sjögren Öhrn M (2000) Tritrophic interactions in a soil community enhance decomposition rates. Oikos 88:585–591

Henderson V, Katznelson H (1961) The effect of plant roots on the nematode population of the soil. Can J Microbiol 7:163–167

Herdler S, Kreuzer K, Scheu S, Bonkowskia M (2008) Interactions between arbuscular mycorrhizal fungi (Glomus intraradices, Glomeromycota) and amoebae (Acanthamoeba castellanii, Protozoa) in the rhizosphere of rice (Oryza sativa). Soil Biol Biochem 40:660–668. doi:10.1016/j.soilbio.2007.09.026

Horiuchi J-I, Prithiviraj B, Bais H, Kimball B, Vivanco J (2005) Soil nematodes mediate positive interactions between legume plants and rhizobium bacteria. Planta 222:848–857. doi:10.1007/s00425-005-0025-y

Huber B, Feldmann F, Köthe M, Vandamme P, Wopperer J, Riedel K, Eberl L (2004) Identification of a novel virulence factor in Burkholderia cenocepacia H111 required for efficient slow killing of Caenorhabditis elegans. Infect Immun 72:7220–7230. doi:10.1128/IAI.72.12.7220-7230.2004

Huber-Sannwald E, Pyke DA, Caldwell MM (1997) Perception of neighbouring plants by rhizomes and roots: morphological manifestations of a clonal plant. Can J Bot 75: 2146–2157

Ingham RE, Trofymow JA, Ingham ER, Coleman DC (1985) Interactions of bacteria, fungi, and their nematode grazers: Effects on nutrient cycling and plant growth. Ecol Monographs 55:119–140

Jacobs M, Rubery PH (1988) Naturally-occuring auxin transport regulators. Science 241:346–349. doi:10.1126/science.241.4863.346

Jentschke G, Bonkowski M, Godbold DL, Scheu S (1995) Soil protozoa and forest tree growth: non-nutritional effects and interaction with mycorrhizae. Biol Fertil Soils 20:263–269. doi:10.1007/BF00336088

Jezbera J, Hornak K, Simek K (2006) Prey selectivity of bacterivorous protists in different size fractions of reservoir water amended with nutrients. Environ Microbiol 8:1330–1339. doi:10.1111/j.1462-2920.2006.01026.x

Johnson D, Krsek M, Wellington EMH, Stott AW, Cole L, Bardgett RD, Read DJ, Leake JR (2005) Soil invertebrates disrupt carbon flow through fungal networks. Science 309:1047. doi:10.1126/science.1114769

Jonas JL, Wilson GWT, White PM, Joern A (2007) Consumption of mycorrhizal and saprophytic fungi by Collembola in grassland soils. Soil Biol Biochem 39:2594–2602

Joseph C, Phillips D (2003) Metabolites from soil bacteria affect plant water relations. Plant Physiol Biochem 41:189–192. doi:10.1016/S0981-9428(02)00021-9

Joshi A, Chand R, Arun B, Singh R, Ortiz R (2007) Breeding crops for reduced-tillage management in the intensive, rice-wheat systems of South Asia. Euphytica 153:135–151. doi:10.1007/s10681-006-9249-6

Jousset A, Lara E, Wall LG, Valverde C (2006) Secondary metabolites help biocontrol strain Pseudomonas fluorescens CHA0 to escape protozoan grazing. Appl Environ Microbiol 72:7083–7090. doi:10.1128/AEM.00557-06

Jousset A, Scheu S, Bonkowski M (2008) Secondary metabolite production facilitates establishment of rhizobacteria by reducing both protozoan predation and the competitive effects of indigenous bacteria. Funct Ecol 22:714–719

Jousset A, Péchy-Tarr M, Rochat L, Keel C, Scheu S, Bonkowski M (2009) Cheating and predation determine the toxin production by the biocontrol bacterium *Pseudomonas fluorescens* CHA0. (submitted)

Kampichler C, Rolschewski J, Donnelly DP, Boddy L (2004) Collembolan grazing affects the growth strategy of the cord-forming fungus Hypholoma fasciculare. Soil Biol Biochem 36:591–599. doi:10.1016/j.soilbio.2003.12.004

Kaplan I, Halitschke R, Kessler A, Sardanelli S, Denno RF (2008) Constitutive and induced defenses to herbivory in above- and belowground plant tissues. Ecology 89:392–406. doi:10.1890/07-0471.1

Keel C, Schnider U, Maurhofer M, Voisard C, Laville J, Burger U, Wirthner P, Haas D, Defago G (1992) Suppression of root diseases by Pseudomonas fluorescens CHA0: importance of the bacterial secondary metabolite 2, 4-diacetylphloroglucinol. Mol Plant Microbe Interact 5:4–13

Kimpinski J, Sturz A (1996) Population growth of a rhabditid nematode on plant growth promoting bacteria from potato tubers and rhizosphere soil. J Nematol 28:682–686

Klironomos JN, Hart MM (2001) Animal nitrogen swap for plant carbon. Nature 410:651–652. doi:10.1038/35070643

Klironomos JN, Kendrick WB (1996) Palatability of microfungi to soil arthropods in relation to the functioning of arbuscular mycorrhizae. Biol Fertil Soils 21:43–52. doi:10.1007/BF00335992

Klironomos JN, Ursic M (1998) Density-dependent grazing on the extraradical hyphal network of the arbuscular mycorrhizal fungus, Glomus intraradices, by the collembolan, Folsomia candida. Biol Fertil Soils 26:250–253. doi:10.1007/s003740050375

Knox OGG, Killham K, Mullins CE, Wilson MJ (2003) Nematode-enhanced microbial colonization of the wheat rhizosphere. FEMS Microbiol Lett 225:227–233. doi:10.1016/S0378-1097(03)00517-2

Knox OGG, Killham K, Artz RRE, Mullins C, Wilson M (2004) Effect of nematodes on rhizosphere colonization by seed-applied bacteria. Appl Environ Microbiol 70:4666–4671. doi:10.1128/AEM.70.8.4666-4671.2004

Köthe M, Antl M, Huber B, Stoecker K, Ebrecht D, Steinmetz I, Eberl L (2003) Killing of *Caenorhabditis elegans* by *Burkholderia cepacia* is controlled by the cep quorum-sensing system. Cell Microbiol 5:343–351. doi:10.1046/j.1462-5822.2003.00280.x

Kreuzer K, Adamczyk J, Iijima M, Wagner M, Scheu S, Bonkowski M (2006) Grazing of a common species of soil protozoa (*Acanthamoeba castellanii*) affects rhizosphere bacterial community composition and root architecture of rice (*Oryza sativa* L.). Soil Biol Biochem 38:1665–1672. doi:10.1016/j.soilbio.2005.11.027

Kuikman PJ, Jansen AG, van Veen JA, Zehnder AJB (1990) Protozoan predation and the turnover of soil organic carbon and nitrogen in the presence of plants. Biol Fertil Soils 10:22–28

Kuzyakov Y, Friedel J, Stahr K (2000) Review of mechanisms and quantification of priming effects. Soil Biol Biochem 32:185–1498

Laakso J, Setälä H (1999) Sensitivity of primary production to changes in the architecture of belowground food webs. Oikos 87:57–64. doi:10.2307/3546996

Larsen T, Gorissen A, Krogh P, Ventura M, Magid J (2007) Assimilation dynamics of soil carbon and nitrogen by wheat roots and Collembola. Plant Soil 295:253–264. doi:10.1007/s11104-007-9280-y

Leake JR, Ostle NJ, Rangel-Castro JI, Johnson D (2006) Carbon fluxes from plants through soil organisms determined by field 13CO2 pulse-labelling in an upland grassland. Appl Soil Ecol 33:152–175. doi:10.1016/j.apsoil.2006.03.001

Lilleskov EA, Bruns TD (2005) Spore dispersal of a resupinate ectomycorrhizal fungus, Tomentella sublilacina, via soil food webs. Mycologia 97:762–769. doi:10.3852/mycologia.97.4.762

Liu XY, Shi M, Liao YH, Gao Y, Zhang ZK, Wen DH, Wu WZ, An CC (2006) Feeding characteristics of an amoeba (Lobosea: Naegleria) grazing upon cyanobacteria: food selection, ingestion and digestion progress. Microb Ecol 51:315–325. doi:10.1007/s00248-006-9031-2

Malamy J, Benfey P (1997) Lateral root formation in *Arabidopsis thaliana*. Plant Physiol 114:277

Mao X, Hu F, Griffiths B, Li H (2006) Bacterial-feeding nematodes enhance root growth of tomato seedlings. Soil Biol Biochem 38:1615–1622

Mao X, Hu F, Griffiths B, Chen X, Liu M, Li H (2007) Do bacterial-feeding nematodes stimulate root proliferation through hormonal effects? Soil Biol Biochem 39:1816–1819. doi:10.1016/j.soilbio.2007.01.027

Mathesius U (2003) Conservation and divergence of signalling pathways between roots and soil microbes—the Rhizobium-legume symbiosis compared to the development of lateral roots, mycorrhizal interactions and nematode-induced galls. Plant Soil 255:105–119. doi:10.1023/A:1026139026780

Mathesius U, Mulders S, Gao M, Teplitski M, Caetano-Anollés G, Rolfe B, Bauer W (2003) Extensive and specific responses of a Eukaryote to bacterial quorum-sensing signals. Proc Natl Acad Sci USA 100:1444–1449. doi:10.1073/pnas.262672599

Matz C, Kjelleberg S (2005) Off the hook—how bacteria survive protozoan grazing. Trends Microbiol 13:302–307. doi:10.1016/j.tim.2005.05.009

Matz C, Bergfeld T, Rice SA, Kjelleberg S (2004a) Microcolonies, quorum sensing and cytotoxicity determine the survival of *Pseudomonas aeruginosa* biofilms exposed to protozoan grazing. Environ Microbiol 6:218–226. doi:10.1111/j.1462-2920.2004.00556.x

Matz C, Deines P, Boenigk J, Arndt H, Eberl L, Kjelleberg S, Jürgens K (2004b) Impact of violacein-producing bacteria on survival and feeding of bacterivorous nanoflagellates. Appl Environ Microbiol 70:1593–1599. doi:10.1128/AEM.70.3.1593-1599.2004

Maurhofer M, Baehler E, Notz R, Martinez V, Keel C (2004) Cross talk between 2, 4-diacetylphloroglucinol-producing biocontrol pseudomonads on wheat roots. Appl Environ Microbiol 70:1990–1998. doi:10.1128/AEM.70.4.1990-1998.2004

McKenzie Bird D (2004) Signaling between nematodes and plants. Curr Opin Plant Biol 7:372–376. doi:10.1016/j.pbi.2004.05.005

Montagnes DJS, Barbosa AB, Boenigk J, Davidson K, Jurgens K, Macek M, Parry JD, Roberts EC, Simek K (2007) Selective feeding behaviour of key free-living protists: avenues for continued study. In 10th Symposium on Aquatic Microbial Ecology (SAME 10). pp 83–98. Inter-Research, Faro, PORTUGAL.

Moore JC, Hunt WH (1988) Resource compartmentation and the stability of real ecosystems. Nature 333:261–263. doi:10.1038/333261a0

Munn E, Munn P (2002) Feeding and digestion. In: Lee D (ed) The biology of nematodes. Taylor & Francis, Singapore, pp 211–232

Murase J, Noll M, Frenzel P (2006) Impact of protists on the activity and structure of the bacterial community in a rice field soil. Appl Environ Microbiol 72:5436–5444. doi:10.1128/AEM.00207-06

Murray PJ, Hatch DJ, Cliquet JB (1996) Impact of insect root herbivory on the growth and nitrogen and carbon contents of white clover (*Trifolium repens*) seedlings. Can J Bot 74:1591–1595. doi:10.1139/b96-192

Muscolo A, Bovalo F, Gionfriddo F, Nardi S (1999) Earthworm humic matter produces auxin-like effects on *Daucus carota* cell growth and nitrate metabolism. Soil Biol Biochem 31:1303–1311. doi:10.1016/S0038-0717(99)00049-8

Nardi S, Panuccio MR, Abenavoli MR, Muscolo A (1994) Auxin-like effect of humic substances extracted from faeces of *Allolobophora caliginosa* and *A. rosea*. Soil Biol Biochem 26:1341–1346. doi:10.1016/0038-0717(94)90215-1

Newsham KK, Rolf J, Pearce DA, Strachan RJ (2004) Differing preferences of Antarctic soil nematodes for microbial prey. Europ J Soil Biol 40:1–8

Ostle N, Briones MJI, Ineson P, Cole L, Staddon P, Sleep D (2007) Isotopic detection of recent photosynthate carbon flow into grassland rhizosphere fauna. Soil Biol Biochem 39:768–777. doi:10.1016/j.soilbio.2006.09.025

Paterson E (2003) Importance of rhizodeposition in the coupling of plant and microbial productivity. Europ J Soil Sci 54:741–750

Pernthaler J (2005) Predation on prokaryotes in the water column and its ecological implications. Nat Rev Microbiol 3:537–546. doi:10.1038/nrmicro1180

Pfander I, Zettel J (2004) Chemical communication in *Ceratophysella sigillata* (Collembola: Hypogastruridae): intraspecific reaction to alarm substances. Pedobiologia (Jena) 48:575–580. doi:10.1016/j.pedobi.2004.06.002

Phillips DA, Streit W (1998) Modifying rhizosphere microbial communities to enhance nutrient availability in cropping systems. Field Crops Res 56:217–221. doi:10.1016/S0378-4290(97)00133-0

Phillips D, Joseph C, Yang G, Martinez-Romero E, Sanborn J, Volpin H (1999) Identification of lumichrome as a Sinorhizobium enhancer of alfalfa root respiration and shoot growth. Proc Natl Acad Sci USA 96:12275–12280. doi:10.1073/pnas.96.22.12275

Phillips D, Ferris H, Cook D, Strong D (2003) Molecular control points in rhizosphere food webs. Ecology 84:816–826

Phillips D, Fox T, King M, Bhuvaneswari T, Teuber L (2004) Microbial products trigger amino acid exudation from plant roots. Plant Physiol 136:2887–2894. doi:10.1104/pp.104.044222

Pickup ZL, Pickup R, Parry JD (2007) Effects of bacterial prey species and their concentration on growth of the amoebae Acanthamoeba castellanii and Hartmannella vermiformis. Appl Environ Microbiol 73:2631–2634

Poll J, Marhan S, Haase S, Hallmann J, Kandeler E, Ruess L (2007) Low amounts of herbivory by root-knot nematodes affect microbial community dynamics and carbon allocation in the rhizosphere. FEMS Microbiol Ecol 62:268–279. doi:10.1111/j.1574-6941.2007.00383.x

Pollierer M, Langel R, Körner C, Maraun M, Scheu S (2007) The underestimated importance of belowground carbon input for forest soil animal food webs. Ecol Lett 10:729–736. doi:10.1111/j.1461-0248.2007.01064.x

Popeijus H, Overmars H, Jones J, Blok V, Goverse A, Helder J, Schots A, Bakker J, Smant G (2000) Enzymology—Degradation of plant cell walls by a nematode. Nature 406:36–37. doi:10.1038/35017641

Puthoff D, Nettleson D, Rodermel S, Baum T (2003) *Arabidopsis* gene expression changes during cyst nematode parasitism revealed by statistical analyses of microarray expression profile. Plant J 33:911–921. doi:10.1046/j.1365-313X.2003.01677.x

Rantalainen ML, Fritze H, Haimi J, Kiikkila O, Pennanen T, Setala H (2004) Do enchytraeid worms and habitat corridors facilitate the colonisation of habitat patches by soil microbes? Biol Fertil Soils 39:200–208

Rantalainen M-L, Fritze H, Haimi J, Pennanen T, Setälä H (2005) Species richness and food web structure of soil decomposer community as affected by the size of habitat fragment and habitat corridors. Glob Change Biol 11:1614–1627. doi:10.1111/j.1365-2486.2005.000999.x

Rasmann S, Agrawal AA (2008) In defense of roots: a research agenda for studying plant resistance to belowground herbivory. Plant Physiol 146:875–880

Rasmann S, Turlings TCJ (2007) Simultaneous feeding by aboveground and belowground herbivores attenuates plant-mediated attraction of their respective natural enemies. Ecol Lett 10:926–936. doi:10.1111/j.1461-0248.2007.01084.x

Rasmann S, Köllner TG, Degenhardt J, Hiltpold I, Toepfer S, Kuhlmann U, Gershenzon J, Turlings TCJ (2005) Recruitment of entomopathogenic nematodes by insect-damaged maize roots. Nature 434:732–737. doi:10.1038/nature03451

Rengel Z, Marschner P (2005) Nutrient availability and management in the rhizosphere: exploiting genotypic differences. New Phytol 168:305–312. doi:10.1111/j.1469-8137.2005.01558.x

Renker C, Otto P, Schneider K, Zimdars B, Maraun M, Buscot F (2005) Oribatid mites as potential vectors for soil microfungi: study of mite-associated fungal species. Microb Ecol 50:518–528. doi:10.1007/s00248-005-5017-8

Riga E (2004) Orientation behavior. In: Gaugler R, Bilgrami AL (eds) Nematode behaviour. CABI, Wallingford, pp 63–90

Robinson AF (2003) Nematode behaviour and migrations through soil and host tissue. In: Zhongxiao X, Chen SY, Dickson DW (eds) Nematology advances and perspectives. Volume 1, Nematode morphology, physiology, and ecology. CABI, Wallingford, pp 330–405

Rodger S, Bengough AG, Griffiths BS, Stubbs V, Young IM (2003) Does the presence of detached root border cells of *Zea mays* alter the activity of the pathogenic nematode *Meloidogyne incognita*?. Phytopathology 93:1111–1114. doi:10.1094/PHYTO.2003.93.9.1111

Rooney N, McCann K, Gellner G, Moore JC (2006) Structural asymmetry and the stability of diverse food webs. Nature 442:265–269. doi:10.1038/nature04887

Rosenberg K, Bertaux J, Krome K, Hartmann A, Scheu S, Bonkowski M (2009) Soil amoebae rapidly change bacterial community composition in the rhizosphere of *Arabidopsis thaliana*. ISME J, The ISME Journal advance online publication 26 February 2009. doi:10.1038/ismej.2009.11

Rudrappa T, Biedrzycki ML, Bais HP (2008) Causes and consequences of plant-associated biofilms. FEMS Microbiol Ecol 64:153–166. doi:10.1111/j.1574-6941.2008.00465.x

Sabatini MA, Innocenti G (2001) Effects of Collembola on plant-pathogenic fungus interactions in simple experimental systems. Biol Fertil Soils 33:62–66. doi:10.1007/s003740000290

Schädler M, Jung G, Brandl R, Auge H (2004) Secondary succession is influenced by belowground insect herbivory on a productive site. Oecologia 138:242–252. doi:10.1007/s00442-003-1425-y

Scheu S (1993) Cellulose and lignin decomposition in soils from different ecosystems on limestone as affected by earthworm processing. Pedobiologia 37:167–177

Scheu S, Folger M (2004) Single and mixed diets in Collembola: effects on reproduction and stable isotope fractionation. Funct Ecol 18:94–102. doi:10.1046/j.0269-8463.2004.00807.x

Scheu S, Setälä H (2002) Multitrophic interactions in decomposer food-webs. In: Tscharntke T, Hawkins BA (eds) Multitrophic level interactions. Cambridge University Press, Cambridge, pp 223–264

Scheu S, Simmerling F (2004) Growth and reproduction of fungal feeding Collembola as affected by fungal species, melanin and mixed diets. Oecologia 139:347–353. doi:10.1007/s00442-004-1513-7

Scheu S, Theenhaus A, Jones TH (1999) Links between the detritivore and the herbivore system: effects of earthworms and collembola on plant growth and aphid development. Oecologia 119:541–551. doi:10.1007/s004420050817

Schulman O, Tiunov A (1999) Leaf litter fragmentation by the earthworm Lumbricus terrestris L. Pedobiologia 43:453–458

Seres A, Bakonyi G, Posta K (2007) Collembola (Insecta) disperse the arbuscular mycorrhizal fungi in the soil: pot experiment. Pol J Ecol 55:395–399

Setälä H (1995) Growth of birch and pine seedlings in relation to grazing by soil fauna on ectomycorrhizal fungi. Ecology 76:1844–1851. doi:10.2307/1940716

Shapiro J (1998) Thinking about bacterial populations as multicellular organisms. Annu Rev Microbiol 52:81–104. doi:10.1146/annurev.micro.52.1.81

Shiraishi H, Enami Y, Okano S (2003) Folsomia hidakana (Collembola) prevents damping-off disease in cabbage and Chinese cabbage by Rhizoctonia solani. Pedobiologia (Jena) 47:33–38. doi:10.1078/0031-4056-00167

Siddiqui IA, Haas D, Heeb S (2005) Extracellular protease of *Pseudomonas fluorescens* CHA0, a biocontrol factor with activity against the root-knot nematode *Meloidogyne incognita*. Appl Environ Microbiol 71:5646–5649. doi:10.1128/AEM.71.9.5646-5649.2005

Soler R, Bezemer TM, Cortesero AM, Van Der Putten WH, Vet LEM, Harvey JA (2007) Impact of foliar herbivory on the development of a root-feeding insect and its parasitoid. Oecologia 152:257–264. doi:10.1007/s00442-006-0649-z

Somasundaram S, Bonkowski M, Iijima M (2008) Functional role of mucilage-border cells: a complex facilitating protozoan effects on plant growth. Plant Prod Sci 11:344–351. doi:10.1626/pps.11.344

Spaepen S, Vanderleyden J, Remans R (2007) Indole-3-acetic acid in microbial and microorganism-plant signaling. FEMS Microbiol Rev 31:425–448. doi:10.1111/j.1574-6976.2007.00072.x

Stanton NL (1988) The underground in grasslands. Annu Rev Ecol Syst 19:573–589. doi:10.1146/annurev.es.19.110188.003041

Steinaker DF, Wilson SD (2008) Scale and density dependent relationships among roots, mycorrhizal fungi and collembola in grassland and forest. Oikos 117:703–710

Stephan A, Meyer A, Schmid B (2000) Plant diversity positively affects soil bacterial diversity in experimental grassland ecosystems. J Ecol 88:988–998. doi:10.1046/j.1365-2745.2000.00510.x

Stephens P, Davoren C (1997) Influence of the earthworms Aporrectodea trapezoides and *A. rosea* on the disease severity of Rhizoctonia solani on subterranean clover and ryegrass. Soil Biol Biochem 29:511–516. doi:10.1016/S0038-0717(96)00108-3

Sundin P, Valeur A, Olsson S, Odham G (1990) Interactions between bacteria-feeding nematodes and bacteria in the rape rhizosphere: effects on root exudation and distribution of bacteria. FEMS Microbiol Ecol 73:13–22

Tapilskaja N (1967) Amoeba albida Nägler und ihre Beziehungen zu dem Pilz Verticillium dahliae Kleb, dem Erreger der Welkekrankheit von Baumwollpflanzen. Pedobiologia (Jena) 7:156–165

Thimm T, Larink O (1995) Grazing preferences of some collembola for endomycorrhizal fungi. Biol Fertil Soils 19:266–268. doi:10.1007/BF00336171

Tiunov A, Scheu S (2005) Arbuscular mycorrhiza and Collembola interact in affecting community composition of saprotrophic microfungi. Oecologia 142:636–642. doi:10.1007/s00442-004-1758-1

Tordoff GM, Boddy L, Jones TH (2006) Grazing by Folsomia candida (Collembola) differentially affects mycelial morphology of the cord-forming basidiomycetes Hypholoma fasciculare, Phanerochaete uelutina and Resinicium bicolor. Mycol Res 110:335–345. doi:10.1016/j.mycres.2005.11.012

Tordoff GM, Boddy L, Jones TH (2008) Species-specific impacts of collembola grazing on fungal foraging ecology. Soil Biol Biochem 40:434–442. doi:10.1016/j.soil bio.2007.09.006

Treonis AM, Grayston SJ, Murray PJ, Dawson LA (2005) Effects of root feeding, cranefly larvae on soil microorganisms and the composition of rhizosphere solutions collected from grassland plants. Appl Soil Ecol 28:203–215. doi:10.1016/j.apsoil.2004.08.004

Treonis AM, Cook R, Dawson L, Grayston SJ, Mizen T (2007) Effects of a plant parasitic nematode (*Heterodera trifolii*) on clover roots and soil microbial communities. Biol Fertil Soils 43:541–548. doi:10.1007/s00374-006-0133-2

Troelstra S, Wagenaar R, Smant W, Paters B (2001) Interpretation of bioassays in the study of interactions between soil organisms and plants: involvement of nutrient factors. New Phytol 150:697–706. doi:10.1046/j.1469-8137.2001.00133.x

Tscharntke T, Hawkins B (2002) Multitrophic level interactions. Princeton University Press, New Jersey

van Dam NM, Harvey JA, Wäckers FL, Bezemer TM, Van Der Putten WH, Vet LEM (2003) Interactions between aboveground and belowground induced responses against phytophages. Basic Appl Ecol 4:63–77

van Ruijven J, De Deyn G, Raaijmakers CE, Berendse F, van der Putten W (2005) Interactions between spatially separated herbivores indirectly alter plant diversity. Ecol Lett 8:30–37. doi:10.1111/j.1461-0248.2004.00688.x

van Tol R, van der Sommen A, Boff M, van Bezooijen J, Sabelis M, Smits P (2001) Plants protect their roots by alerting the enemies of grubs. Ecol Lett 4:292–294. doi:10.1046/j.1461-0248.2001.00227.x

Venette R, Ferris H (1998) Influence of bacterial type and density on population growth of bacterial-feeding nematodes. Soil Biol Biochem 30:949–960

Venette R, Mostafa F, Ferris H (1997) Trophic interactions between bacterial-feeding nematodes in plant rhizospheres and the nematophagous fungus *Hirsutella rhossiliensis* to suppress *Heterodera schachtii*. Plant Soil 191:213–223

Vercauteren I, Engler JD, De Groodt R, Gheysen G (2002) An Arabidopsis thaliana pectin acetylesterase gene is upregulated in nematode feeding sites induced by root-knot and cyst nematodes. Mol Plant Microbe Interact 15:404–407. doi:10.1094/MPMI.2002.15.4.404

Veronico P, Jones J, Di Vito M, De Giorgi C (2001) Horizontal transfer of a bacterial gene involved in polyglutamate biosynthesis to the plant-parasitic nematode Meloidogyne artiellia. FEBS Lett 508:470–474. doi:10.1016/S0014-5793(01)03132-5

Vestergård M, Bjørnlund L, Henry F, Ronn R (2007) Decreasing prevalence of rhizosphere IAA producing and seedling root growth promoting bacteria with barley development irrespective of protozoan grazing regime. Plant Soil 295:115–125. doi:10.1007/s11104-007-9267-8

Wardle D (2002) Communities and ecosystems: Linking the aboveground and belowground components. Princeton University Press, New Jersey

Wardle DA, Yeates GW (1993) The dual importance of competition and predation as regulatory forces in terrestrial ecosystems: evidence from decomposer food-webs. Oecologia 93:303–306

Weekers PHH, Bodelier PLE, Wijen JPH, Vogels GD (1993) Effects of grazing by the free-living soil amobae *Acanthamoeba castellanii, Acanthamoeba polyphaga*, and *Hartmannella vermiformis* on various bacteria. Appl Environ Microbiol 59:2317–2319

Weisse T (2002) The significance of inter- and intraspecific variation in bacterivorous and herbivorous protists. Antonie Leeuwenhoek 81:327–341

Weitere M, Bergfeld T, Rice SA, Matz G, Kjelleberg S (2005) Grazing resistance of Pseudomonas aeruginosa biofilms depends on type of protective mechanism, developmental stage and protozoan feeding mode. Environ Microbiol 7:1593–1601. doi:10.1111/j.1462-2920.2005.00851.x

Wieland G, Neumann R, Backhaus H (2001) Variation of microbial communities in soil, rhizosphere, and rhizoplane in response to crop species, soil type, and crop development. Appl Environ Microbiol 67:5849–5854. doi:10.1128/AEM.67.12.5849-5854.2001

Wilkinson DM (2008) Testate amoebae and nutrient cycling: peering into the black box of soil ecology. Trends Ecol Evol 23:596–599. doi:10.1016/j.tree.2008.07.006

Williamson VM, Gleason CA (2003) Plant–nematode interactions. Curr Opin Plant Biol 6:327–333. doi:10.1016/S1369-5266(03)00059-1

Wood J, Tordoff GM, Jones TH, Boddy L (2006) Reorganization of mycelial networks of Phanerochaete velutina in response to new woody resources and collembola (Folsomia candida) grazing. Mycol Res 110:985–993. doi:10.1016/j.mycres.2006.05.013

Wurst S, Jones H (2003) Indirect effects of earthworms (*Aporrectodea caliginosa*) on an above-ground tritrophic interaction. Pedobiologia (Jena) 47:91–97. doi:10.1078/0031-4056-00173

Wurst S, Langel R, Reineking A, Bonkowski M, Scheu S (2003) Effects of earthworms and organic litter distribution on plant performance and aphid reproduction. Oecologia 137:90–96

Wurst S, Dugassa-Gobena D, Langel R, Bonkowski M, Scheu S (2004) Combined effects of earthworms and vesicular-arbuscular mycorrhizas on plant and aphid performance. New Phytol 163:169–176

Wyss U (2002) Feeding behaviour of plant parasitic nematodes. In: Lee DL (ed) The biology of nematodes. Taylor & Francis, London, pp 462–513

Yeates GW, Saggar S, Denton CS, Mercer CF (1998) Impact of clover cyst nematode (Heterodera trifolii) infection on soil microbial activity in the rhizosphere of white clover (Trifolium repens)—A pulse-labelling experiment. Nematologica 44:81–90

Yeates G, Bardgett R, Mercer C, Saggar S, Feltham C (1999a) Increase in 14C-carbon translocation to the soil microbial biomass when five species of pant parasitic nematodes infect roots of white clover. Nematology 1:295–300. doi:10.1163/156854199508298

Yeates GW, Saggar S, Hedley CB, Mercer CF (1999b) Increase in 14C-carbon translocation to the soil microbial biomass when five species of plant-parasitic nematodes infect roots of white clover. Nematology 1:295–300. doi:10.1163/156854199508298

Young IM, Griffiths BG, Robertson WM (1996) Continuous foraging by bacterial-feeding nematodes. Nematologica 42:378–382

Young KD (2006) The selective value of bacterial shape. Microbiol Mol Biol Rev 70:660–703. doi:10.1128/MMBR.00001-06

Zandonadi D, Canellas L, Façanha A (2007) Indolacetic and humic acids induce lateral root development through a concerted plasmalemma and tonoplast H + pumps activation. Planta 225:1583–1595. doi:10.1007/s00425-006-0454-2

Zwart KB, Kuikman PJ, Van Veen JA (1994) Rhizosphere protozoa: Their significance in nutrient dynamics. In: Darbyshire J (ed) Soil protozoa. CAB International, Wallingford, pp 93–121

REVIEW ARTICLE

Plant-driven selection of microbes

Anton Hartmann · Michael Schmid ·
Diederik van Tuinen · Gabriele Berg

Received: 1 July 2008 / Accepted: 20 October 2008 / Published online: 15 November 2008
© Springer Science + Business Media B.V. 2008

Responsible editor: Philippe Lemanceau.

A. Hartmann (✉) · M. Schmid
Department Microbe-Plant Interactions,
Helmholtz Zentrum München,
German Research Center for Environmental Health (GmbH),
Ingolstaedter Landstrasse 1,
D-85764 Neuherberg, Germany
e-mail: anton.hartmann@helmholtz-muenchen.de

D. v. Tuinen
UMR INRA Université de Bourgogne,
Plante-Microbe-Environnement CMSE-INRA,
17 rue Sully, BP 86510, Dijon Cedex F-21065, France

G. Berg
Institute for Environmental Biotechnology,
Graz University of Technology,
Petersgasse 12,
A-8010 Graz, Austria

Abstract The rhizodeposition of plants dramatically influence the surrounding soil and its microflora. Root exudates have pronounced selective and promoting effects on specific microbial populations which are able to respond with chemotaxis and fast growth responses, such that only a rather small subset of the whole soil microbial diversity is finally colonizing roots successfully. The exudates carbon compounds provide readily available nutrient and energy sources for heterotrophic organisms but also contribute e.g. complexing agents, such as carboxylates, phenols or siderophores for the mobilization and acquisition of rather insoluble minerals. Root exudation can also quite dramatically alter the pH- and redox-milieu in the rhizosphere. In addition, not only specific stimulatory compounds, but also antimicrobials have considerable discriminatory effect on the rhizosphere microflora. In the "biased rhizosphere" concept, specific root associated microbial populations are favored based on modification of the root exudation profile. Rhizosphere microbes may exert specific plant growth promoting or biocontrol effects, which could be of great advantage for the plant host. Since most of the plant roots have symbiotic fungi, either arbuscular or ectomycorrhizal fungi, the impact of plants towards the rhizosphere extends also to the mycorrhizosphere. The selective effect of the roots towards the selection of microbes also extends towards the root associated and symbiotic fungi. While microbes are known to colonize plant roots endophytically, also mycorrhiza are now known to harbor closely associated bacterial populations even within their hyphae.

The general part of the manuscript is followed by the more detailed presentation of specific examples for the selection and interaction of roots and microbes, such as in the rhizosphere of strawberry, potato and oilseed rape, where the soil-borne plant pathogen *Verticillium dahliae* can cause high yield losses; the potential of biocontrol by specific constituents of the rhizosphere microbial community is demonstrated. Furthermore, plant cultivar specificity of microbial communities is described in different potato lines including the case of transgenic lines. Finally, also the specific selective effect of different *Medicago* species on the selection of several arbuscular mycorrhizal taxa is presented.

Keywords Root exudation · Rhizodeposition · Microbial diversity · Rhizosphere bacteria · Mycorrhizal fungi · Arbuscular mycorrhiza · Ectomycorrhiza · Antimicrobials · Signalling compounds · Plant growth promotion · Biological control · "Biased rhizosphere concept"

Introduction

The rhizosphere, an unique environment for microbial colonization

According to the general view of the rhizosphere, it includes plant roots and the surrounding soil. This is a wide and wise definition, already coined more than hundred years ago by Lorenz Hiltner, as documented in detail by Hartmann et al. (2008). In the rhizosphere, a biologically and chemically highly diverse, complex and dynamic interaction occurs between plant roots, soil (micro) biota and the physicochemical conditions of the soil. The autotrophic plant partner is providing substrate and energy flow into the rhizosphere and gets in return essentials for its development and growth: nutrients, minerals and water. Heterotrophic soil biota usually are limited in the supply of carbon and energy and thus a complex sequence of responses are initiated, which in due course also influence the plants. Soil biota (bacteria, fungi, micro-fauna and the plant root) are themselves embedded in food webs and thus interactions with consumers or predators in the microbial as well as micro- and mesofaunal world are important to understand rhizosphere processes.

From the viewpoint of the plant, the rhizosphere is characterized by the investment of the plant into an effective development of the root architecture and the return of mineral nutrients and water from the soil. In the root system, sloughing off of root cells (in particular at the root tip), root death (root hair cells and epidermis cells in older root parts) and the exudation of carbon compounds are processes which support soil biota and select a specific rhizosphere community according to their composition (see below). Already in the initial phases of the evolution of terrestrial plants, the necessity and opportunity appeared to integrate the abilities of soil microbes to explore the soil for nutrients and water into the development of plants. Vice versa a high number of soil microbes attained properties enabling them to interact more efficiently with roots and withstand the quite challenging conditions of rhizosphere life (see below). This process can be regarded as an ongoing process of micro-evolution in low-nutrient environments, which are quite common in natural ecosystems (Schloter et al. 2000). For example, the interaction with soil fungi lead to the development of mycorrhiza which explore the soil for phosphate, nitrogen and other nutrients and micronutrients much beyond the physical expansion of the root system. The bacteria with their impressive metabolic versatility and originality - expanding much beyond the oxygen-dependent spheres - were also taken on board to gain e. g. better access and even independence from the often most limiting nitrogen supply. The size and dynamic of the plants investment into the rhizosphere is dependent from the aboveground part of the plant and differs by ecosystem type, plant species and growth stage of the plant. In grasslands, the ratio of shoot to root (S:R) development is roughly 1:1 and is much different to forests, where far more photosynthate is allocated into the aboveground parts. Far more carbon is accumulating in the rhizospheres of e. g. arctic tundra, where the turnover is very slow and the range of S:R can be found as low as 0.1 (Farrar et al. 2003; Moore et al. 2007).

From the viewpoint of soil microbiota which are mostly heterotrophic organisms depending on exogenous supply of carbon substrates as energy as well as nutrient sources for growth and development, roots are providing almost all, what is lacking in soil. While in pathogens or root grazers complete utilization and destruction is one extreme version of this interaction with plant roots, the balance of utilization of the resource and coexistence or even symbiosis is reached in many other rhizosphere colonizing microbiota. Surely, the plant is restricting or directing the development of the attracted organisms in a way to keep control of these guests by excreting quite selective mixtures of substances which provide selective conditions for rhizosphere organisms. Furthermore, the rhizosphere is a quite heavily populated microhabitat which is characterized by competition and even predation among the inhabitants. Therefore, soil organisms do experience the rhizosphere environment as micro-habitat of great opportunities but also of big challenges. Rhizospheres can only be successfully colonized with the appropriate tools of efficient substrate acquisition, resistance mechanisms as well as competitive traits. Thus, evolution shaped soil biota to fit into these specific niche conditions which are also characterized by specificities based on the diversity of plants and soil environments. Furthermore, the colonization of the interior of plant roots by microbial endophytes appears as most attractive goal, because there plant nutrient resources can be explored even more effectively without the tough competition with the high number of other microbes colonizing the root surface and environment (Rosenblueth and Martinez-Romero 2006; Schulz et al. 2006). However, in this case the efficient interaction with the plant host gets even more important.

In this review, we follow the original hypothesis of Lorenz Hiltner in 1904, that plant roots set the stage for the development of a unique rhizosphere microbial community and we extend this by including mycorrhizal fungi and their associated bacteria as well as root endophytic microbes. The present state of the art based on modern and molecular methods is presented in four major chapters: (i) Plant traits shaping the conditions for microbial colonization, (ii) Microbial responses to specific rhizosphere conditions, (iii) The mycorrhizosphere and its specific traits and interactions with other rhizosphere constitutents and (iv) Selected examples for the specific selection and interaction of roots and microbes. The influence of the soil on the composition of rhizosphere microbial communities is not treated in this chapter, because it is obvious, that the rhizosphere microflora is recruited in most part from the given soil microflora (which is certainly different in different soils).

Plant traits shaping the conditions for microbial colonization

Soluble carbon compounds and other rhizodepositions

Plants exude a variety of organic compounds (e.g. carbohydrates, carboxylic acids, phenolics, amino acids) as well as inorganic ions (protons and other ions) into the rhizosphere to change the chemistry and biology of the root microenvironment. This exudation is plant specific and generally accepted to reflect the evolution and/or specific physiological adaptation to particular soil habitat conditions (Crowley and Rengel 1999). In order to withstand challenges like deficiencies of various macro- and micro-nutrients, like iron (Marschner and Römheld 1994) zinc, manganese or phosphate, plants have different strategies to cope with (see below) (Rengel 1999). The type of root exudates is crucial for the ecosystem distribution and niche-specificity of certain plants (Dakora and Phillips 2002). For example, a so called "calcifuge" plant does not tolerate alkaline (basic) soil. The word is derived from the Latin 'to flee from chalk'. These plants are also described as ericaceous, as the prototypical calcifuge is the genus *Erica* (heaths). It is not the presence of carbonate or hydroxide ions per se that these plants cannot tolerate, but the fact that under alkaline conditions, iron becomes less soluble (see below). There are many horticultural plants which are calcifuges, most of which require an 'ericaceous' compost with a low pH, composed principally of Sphagnum moss peat. These include heathers, Camellias, Rhododendrons, Azaleas, and most carnivorous plants. In contrast, "calcicole" plants can cope very well with alkaline soil conditions.

While so called "calcicole" plants exude mostly di- and tricarboxylic acids, the "calcifuge" plants exude mostly monocarboxylic acids. While the former plants efficiently complex phosphate and ferric iron ions, the latter complex only poorly phosphate and iron from alkaline soils. The capacity of plants to respond to environmental conditions of nutrient deprivation can be separated into three major processes: (1) the signaling, (2) a powerful biosynthetic capacity, and (3) specific membrane transporter processes to foster the transfer of effective organic compounds into the rhizosphere.

Roots products and rhizodepositions may probably consist of every type of plant compound, except specific compounds involved in photosynthesis. Most root products are regular plant compounds which became available as substrate of rhizosphere colonizing microbes, including specific compounds typical of the secondary metabolism of each plant species. This is especially true, when root hairs are decaying or the rhizodermis is partially degraded in older root parts. Root products with a certain biological activity may also be actively secreted. These secretions can be classified as allelochemicals, phytotoxins, phytoalexins, phytohormons and ectoenzymes—they are discussed in more detail below.

The amounts of exudates (also as proportions of photosynthate) vary considerably in different plants, plant growth cycle and root segments and the methodological and ecological implications were reviewed extensively (Jones et al. 2004; van Veen et al. 2007). To be able to realistically evaluate the relevance of certain excreted compounds and their individual fluxes (excretion and uptake), the exudation / uptake rate per unit root length per hour may be important to know and to be specified for a particular root segment (e.g. root cap or elongation zone). The types and amounts of root products excreted or secreted into the root environment is in detail reviewed by Uren (2007). In general, the fate of the more or less diffusible exudates or secretions in the rhizosphere is determined by their physicochemical properties and nutrient quality. The rhizosphere can be regarded as a gradient system, where free diffusible compounds spread into the root surroundings (Fig. 1). The longer the distance of diffusion is the more binding to soil mineral or humic compounds as well as microbial degradation is going to occur. Thus, the immediate surface of the root, the rhizoplane, is certainly exposed to the highest rates and concentrations of beneficial and harmful exudation / secretion products. With distance and time, the organic carbon compounds are progressively metabolized to carbon dioxide or into recalcitrant or colloidal bound forms of humic carbon compounds. Therefore, the molecular scenario a root creates is most complex and variable also due to the soil conditions. The selective effects on the rhizosphere microflora is to be expected very complex and only at the right set of conditions, an targeted effect on the behavior of a certain microbial subpopulation, e.g. an introduced inoculum, which is supposed to get established and interactive with the plant root is possible.

pH- and redox-modulating factors

Both protons and electrons are secreted as C-compounds in the form of undissociated acids or compounds with reducing abilities. In addition, plasma membrane enzymatic processes are the main sites of proton or electron transport processes (Bérczi and Møller 2000; Yan et al. 2002). The origin and adaptation to changing environmental conditions of root mediated pH-changes have been recently been reviewed by Hinsinger et al. (2003). The reducing power is a long observed property of plant roots and was demonstrated in several different approaches, such as the reduction of insoluble manganese oxide by roots (Uren 1981). Although other biogenic acids can affect soil acidification and weathering dissolution, root uptake of nutrient ions, organic acid production, redox-cycling of electron-deficient metals and the carbonic acid system are major contributors to

Fig. 1 Stimulatory and inhibitory factors on rhizosphere microbial communities and the different quality of feedback effects of microbes and plants

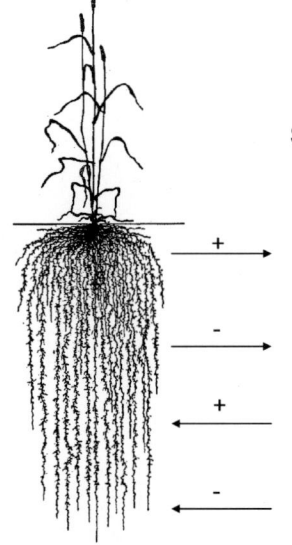

Selective interactions in the rhizosphere

Stimulatory factors:
Carbon substances, vitamines, Complexingagents, specific substrates, hydrogen

Inhibitory factors:
Volatile and soluble antimicrobials, QS-inhibitors

Stimulatory feedback:
Solubilisation of minerals, enlarge drooting volume stimulation of pathogen resistance; growth regulators; biological control of pathogens; degradation of inhibitors

Inhibitory feedback:
Competition for nutrients, phytotoxins, allelochemicals

rhizosphere acid production and soil formation (Richter et al. 2007). Rhizosphere acidification, which may be as much as two pH units, can result from the "excess cation uptake" by plant roots leaving behind protons in the rhizosphere soil, since the overall plant nutrient uptake process is regarded as electro-neutral. Since oxygen is very actively consumed in the rhizosphere due to high rates of microbial decomposition and root respiration, steep redox gradients can develop between the root environment and the surrounding bulk soil. In contrast, roots are providing the rhizosphere with oxygen in waterlogged soils and sediments. As a consequence, iron-oxidizing bacteria precipitate Fe plaque as oxidized coatings at root surfaces (Uren 2007). Finally, carbonic acid weathering involves all three phases of the soil system: CO_2 in the gas phase, carbonic acid and associated ions in the liquid phase and mineral surfaces and structures in the solid phase. Since partial pressures of CO_2 increases with soil depth, B- and C-horizons are subject to this type of acidification process. The detailed estimation of the sources and sinks of CO_2 in the rhizosphere was recently reviewed by Kuzyakov (2005).

Complexing agents: siderophores, phenols and carboxylates

Together with the excretion of protons and carboxylates, siderophores and phenols play significant roles in mineral weathering. Roots are most active in the excretion of these compounds to overcome nutrient limitation and join in soil microbes to overcome these limitations too. Thus, rhizosphere conditions assist microbes and microbes assist roots in this most important interaction with soil minerals and nutrient mineral solubilization.

Iron deficiency

Under conditions of iron deficiency, plants employ two basic mechanisms. Strategy I involves the stimulation of proton extrusion, enhanced exudation of reductants and chelators (carboxylates and phenolics) and an enhanced activity of plasma-membrane-bound Fe-(III)-reductase (Marschner 1991; Marschner and Römheld 1994; Rengel 1999). In strategy II, which is restricted to *Gramineae*, the biosynthesis and excretion of phytosiderophores like mugineic acid is specifically induced under iron-limited conditions. The amount of phytosiderophore release is different between plant species. Relative efficiency of phytosiderophore exudation decreases in the order barley, maize, sorghum, and correlates with Fe-deficiency tolerance. Interestingly, also Zn-deficiency leads to increased phytosiderophore exudation. In maize, the particular Fe-deficiency-sensitive yellow-stripe mutant (*ys1*) shows a comparable rate of phytosiderophore release, but it has a defect in the uptake system for the phytosiderophore (von Wiren et al. 1994). It could be demonstrated that e. g. Alice maize has a high affinity iron uptake component (von Wiren et al. 1995). In the rhizosphere, also microorganisms excrete siderophores under iron limited conditions which prevail in aerobic soils, when iron is present in mostly insoluble Fe-(III)-oxide / hydroxide complexes. It could be demonstrated, that plants profit from the Fe-(III)-solubilizing activity of microbes as well as microbes can take advantage from phytosiderophores after ligand exchange or using heterologous siderophores (Marschner and Crowley 1998). Since the acquisition of the highly insoluble ferric iron complexes is an essential component of rhizosphere competence and fitness, it is usually well developed in rhizosphere microorganisms. However, important differences in the iron uptake capabilities occur at the strain level in bacteria (Hartmann 1988) and this trait appears essential to be regarded in selecting successful rhizosphere colonizing plant growth promoting or biocontrol inoculant strains (Moenne-Loccoz et al. 1996).

Phosphate deficiency

In many plants, phosphate deficiency enhances the production and exudation of phenolic compounds (Chishaki and Horiguchi 1997) (see Fig. 1). Phenolic biosynthesis can be regarded as a metabolic bypass reaction or response of P-starved cells to solubilize inorganic P_i from soil. Since certain phenolic compounds have antibiotic properties, this component of exudates could be relevant in counteracting infectious root pathogens, but may also prevent fast microbial breakdown of exudates. On the other hand, root exudation of metal-chelating carboxylates (e.g. citrate, malate, malonate, and oxalate) in sufficient amounts to mediate P mobilization in soils can comprise up to 25% of the assimilated carbon (Dinkelaker et al. 1995). This is comparable to the carbon investment for mycorrhizal associations. Cluster rooted plant species (Neumann and Martinoia 2002), characterized by the highest P-deficiency induced carboxylate (mostly citrate) exudation are mostly nonmycorrhizal plants and cluster roots are regarded as alternate strategy for nutrient acquisition (Skene 2000).

Exudation of antimicrobials

Most plant species are resistant to most potential pathogens. However, it is not known in a comprehensive manner, why most plant-microbe interactions do not lead to disease, although many resistance mechanisms are known and the disease resistance is certainly a multi-factorial response (Thordal-Christensen 2003). A very widely found mechanism of local defense is the generation of reactive oxygen species and the subsequent stress on the colonizing microbes or neighboring roots. Thus, antioxidant enzymes, like catalase, laccase, superoxide dismutase (SOD) and the glutathione system are important in order to face this challenge in the root environment. However, ROS and also activated nitrogen species (like NO) are known as part of the innate immunity system of plants posing selective pressure on the rhizosphere microflora (Apel and Hirt 2004; Zeidler et al. 2004). When analyzing the response in gene expression of root colonizing *Pseudomonas putida* (Matilla et al. 2007) it appeared that also bacterial antioxidant enzymes are important to face the challenge of the root environment.

While in the case of leave pathogens, several specific resistance mechanisms have been developed by plants, the investment into the antagonistic potential within the rhizosphere microbial community appears to be an important part of biological control of pathogens in roots (Cook et al. 1995). In addition, the exudation of specific root derived antimicrobial metabolites is certainly a major mechanism, as has been shown for many plant species (Bais et al. 2006). Plant secondary metabolites such as butanoic acid, cinnamic acid, o- and p-coumaric acid, vanillic acid or p-hydroxybenzamide were shown to occur in the rhizosphere of Arabidopsis plants when challenged with the Gram-negative bacterial pathogen *Pseudomonas syringae* pv. tomato (Bais et al. 2005). Most interestingly, bacteria can effectively modify this antimicrobial plant response, because the strains which are partly resistant to these compounds are able to block the exudation of antimicrobials using a mechanism based on the type III secretory system (Bais et al. 2005). Therefore, overcoming host response is not only due to a resistance towards exuded antimicrobial compounds but in a second step also in the blockage of further production of antimicrobials. Most interesting, volatile substances with antimicrobial activity have also been identified (Ryu et al. 2004) which could play an important role in selecting the plant associated microflora.

Another most interesting group of plant derived compounds are quorum sensing inhibitors. Many Gram-negative rhizosphere bacteria, including pathogens, use auto-inducer signaling compounds in order to coordinate their activity when colonizing the rhizoplane (Fuqua et al. 2001; Gantner et al. 2006). This density dependent response is called "quorum sensing" (QS), since it becomes effective only in high density cell populations. More recently it was suggested, that the auto-inducers, e.g. of the N-acylhomoserine lactone type, have even a more general relevance in sensing the quality of the micro-environment for its diffusion characteristics and to acquire the important information whether it is economically sound to initiate the expensive biosynthesis of exoenzymes, termed efficiency sensing (Hense et al. 2007). Thus, auto-inducer signaling is very common in bacteria, and plants are facing these compounds regularly. The interference with bacterial signaling by excreting QS-mimic or -inhibitory substances would have the advantage to disturb the coordination of the bacterial attack - especially in the case of pathogens. It has been indeed found in several plants, that QS-inhibitors are produced (Bauer and Mathesius 2004; Degrassi et al. 2007; Hentzer et al. 2002; Rasmussen et al. 2005; Teplitski et al. 2000). On the other hand, it was demonstrated that mostly leguminous plants are able to degrade bacterial N-acyl homoserine lactone (AHL) autoinducers by excreting lactonases or AHL-hydrolases (d'Angelo-Picard et al. 2005; Delalande et al. 2005). In other plants it was shown (Götz et al. 2007; von Rad et al. 2008), that the short side chain C4-, C6- and C8-AHLs are taken up by plants and they could be found also in the shoots (Götz et al. 2007; von Rad et al. 2008). A quite wide range of responses of plants were shown to be triggered by the exposition to bacterial AHL-compounds. In tomato plants, C6-and C8-AHLs induced a systemic resistance response (Schuhegger et al. 2006) with stimulated induction of the PR1-protein and chitinase. On the other hand, *Arabidopsis thaliana* responded by altered phytohormone levels (increased auxin/cytokinin-ratio) in roots and stimulated root growth (von Rad et al. 2008). Thus, plant roots are responding to the presence of bacteria and their signaling compounds and their selective effect on the rhizosphere microflora has to be regarded as a dynamic one which is modulated by the "history" of the rhizosphere colonization events.

Exudation of specific stimulatory compounds

In contrast to inhibitory compounds which are present in root exudates of plants under particular conditions, many root exudates, like sugar, organic acids or amino acids, stimulate a positive chemotactic response of bacteria (Somers et al. 2004) (see Fig. 1). Thus, a flagellum-driven chemotaxis towards roots and their exudates is an important trait for root colonization in many root-bacteria interactions (de Weert et al. 2002). Certain compounds have even specific sites of exudation. It has been shown that tryptophan, the precursor of bacterial indole acetic acid (IAA) production, is mainly exuded near the root tip in *Avena barbata* (Jaeger et al. 1999). Since rhizobacterial IAA-production by root associated bacteria is a major mechanism of plant growth promotion this has important implication for the development of the root system under the influence of rhizosphere microbes. The *ipdC*-gene, coding for indole pyruvate decarboxylase, in the major route of IAA-biosynthesis of the plant growth promoting rhizobacterium *Azospirillum brasilense* was shown to be induced in rhizoplane colonizing bacteria (Rothballer et al. 2005). On the other hand, a high diversity of rhizosphere bacteria have recently been characterized to be able to use indole acetic acid as carbon source for growth (Leveau and Gerards 2008).

Since exudation is a major driving force for microbial root colonization, plant root exudation could be specifically engineered to selectively stimulate specific microbial colonization. Oger at al. (1997) has demonstrated that genetically engineered plants which produce opines have an altered rhizosphere community. In transgenic *Lotus* plants producing two opines, mannopine and nopaline, specific microbial rhizosphere communities were stimulated by the modified root exudation. Opine-utilizing microbes represented a large community in the rhizosphere of opine-producing *Lotus* plants and opine utilizes were found to belong to the Gram-positive as well as Gram-negative bacteria (Oger et al. 1997). The natural example of opine production is provided by the genetic colonization of plant roots by *Agrobacterium tumefaciens* and its T_i plasmid in order to divert assimilate flow from the plant specifically to support the growth of the bacterium which is equipped with opine degrading enzymes. This strategy, termed "biased rhizosphere", could be successfully used to engineer "artificial symbioses" (O'Connell et al. 1996) and was successfully applied to engineer e. g. transgenic tobacco plants capable of releasing substantial amounts of opine compounds in the rhizosphere (Oger et al. 2004).

A different quality of specific stimulatory rhizosphere effect can be found in leguminous roots, where nitrogen fixing nodules release substantial amounts of molecular hydrogen (Dong et al. 2003). The generation of hydrogen is an unavoidable side reaction in the nitrogenase reaction (Simpson and Burris 1984). Since rhizobia and legumes harbor hydrogen uptake activities to different degrees, an escape of hydrogen is present in certain leguminous species. It could be shown that this hydrogen release from nodules has a substantial effect on both the general microbial activity and the bacterial numbers as well as on the activity level of the bacteria (Stein et al. 2005). In addition, a shift in the bacterial community composition was shown which was most pronounced in the group of Betaproteobacteria and *Cytophaga/Flavobacteria*. Interestingly, beyond a certain threshold level of hydrogen flux, an onset of carbon dioxide fixation was observed to occur in these communities (Dong et al. 2003; Stein et al. 2005), which was also indicated by the demonstration of mRNA of bacterial ribulose bisphosphate carboxylase (*cbbL*) in these communities (Rohe and Hartmann, unpublished results). Thus, under these specific rhizosphere conditions of hydrogen releasing legumes, carbon dioxide fixation occurs in a wide variety of soil and rhizosphere bacteria harboring the *cbbL*-genes (Selesi et al. 2005).

Shaping specific habitat conditions by physicochemical forces

The rhizosphere is the critical interface between biota and geologic environments. Growing roots and their mycorrhizal hyphae follow pores and channels that are usually not less than their own diameter. When perennial roots (e.g. tree roots) grow, they expand in volume radially and exert enormous pressures on the surrounding soil (Richter et al. 2007). Even unweathered rocks are susceptible to these physical effects of roots. Mechanical weathering is stimulated by these root forces, accelerating chemical weathering by increasing minerals surface area that is contacted by microbes, organic compounds, electrons and protons. In B-horizons, root growth results in a significant increase in bulk densities of soils, creating reduced porosity, hydraulic conductivity and aeration. Thus biogeochemical processes and the activity of soil microbes may well be affected by this phenomenon, which is quite less studied. A higher degree of permanently water-filled pores my favor anaerobic microbial processes which may lead to increased denitrification (possibly NO and N_2O production) or other microaerobic or anaerobic microbial processes. This steep spatial redox gradients and dynamics may have considerable consequences for the microbial degradation of complex xenobiotic organic substances, because aerobic and anaerobic conditions may foster very different metabolic potentials in different microbial clades leading to more complete removal of xenobiotic compounds.

Microbial responses to specific rhizosphere conditions

The unique rhizosphere conditions are the basis for the sustainable fertility of soils providing soil biota specific support for their continuous activity in nutrient cycling and provision of nutrients for plants and they are fundamentally important for pedogenetic processes. Due to the high diversity of chemical influences in the rhizosphere of different plants, roots drive specific selections of microbes out of the almost indefinite pool of soil microbial diversity. In addition to the selection of pre-existing diversity, also microevolution towards most properly adapted microbial life forms (Schloter et al. 2000) is supported in the well nourished rhizosphere environment, because e. g. of the possibility of genetic exchange employing horizontal gene transfer (van Elsas et al. 2003) and/or creation and selection of spontaneous mutants, transconjugants and transformants with improved properties for specific selective rhizosphere conditions. Thus, engineering root exudation towards the production of two novel carbon compounds leads to the selection of distinct microbial populations in the rhizosphere (Oger et al. 2004; Savka et al. 2002). This concept of "artificial symbiosis" provides an interesting concept for engineering specific microbe-plant interactions. The impact of this modification of root exudation has been studied extensively as model system of the impact of engineered plants on soil ecosystem (Bruinsma et al. 2003; Kowalchuk et al. 2003).

There are many independent evidences using microbiological and molecular techniques that roots stimulate selectively soil microbial communities creating unique rhizosphere communities (Duineveld et al. 1998; Marschner et al. 2001; Rengel and Marschner 2005) (see also below for specific examples). However, only with the availability of the Stable Isotope Probing (SIP) technique (Radajewski et al. 2000) the utilization and fate of plant carbon substrates in the rhizosphere by associated microbes and the dynamic feature of this process in the microbial food web could be investigated in detail (Prosser et al. 2006). There are two major rather independent approaches concerning the molecular markers for carbon assimilation: the analysis of (i) the phospholipids fatty acids (PLFA) (Butler et al. 2004; Butler et al. 2003; Paterson et al. 2007; Treonis et al. 2004) and (ii) the ribosomal genes or ribosomal RNA (Lu and Conrad 2005; Lu et al. 2006; Prosser et al. 2006; Rangel-Castro et al. 2005) (see separate chapter).

Microbial traits important for the performance in the rhizosphere

Microorganisms living in the rhizosphere can have a neutral, pathogenic or beneficial interaction with their host plant (Raaijmakers et al. 2008; Whipps 2001). This reaction depends on the balance of the plant-microbe interaction. Interestingly, the colonization strategy of microbes is highly similar independent of their effect on host. Steps of colonization include recognition, adherence, invasion (only endophytes and pathogens), colonization and growth, and several strategies to establish interaction. The importance to recognize and adhere to plant roots for all plant-associated microorganisms is underlined in many studies. Factors that contribute to recognition include the ability to sense and use root exudates composed of small organic molecules like carbonic acids, amino acids or sugars etc. Chemotaxis especially to plant root exudates is an important trait for colonization of the rhizosphere: this was shown for pathogenic and symbiotic plant-associated bacteria, e.g. *Ralstonia solanacearum* (Yao and Allen 2006) as well as *Rhizobium leguminosarum* (Miller et al. 2007). Interestingly, chemotaxis experiments of cyanobacteria with host plants like *Gunnera* and *Blasia* and non-host plants like *Arabidopsis* showed the possibility to attract cyanobacteria may be widespread in plants. Using comparative transcriptome analysis of *Pseudomonas aeruginosa*, root exudates of sugar beet altered gene expression of genes involved in chemotaxis (Mark et al. 2005). An early step in the establishment of a plant-bacterium interaction is attachment of cells to plant roots, in which for example fimbriae and cell-surface proteins are involved. For the colonization of *Pseudomonas* of plant roots, flagella, pili, O-antigen of lipopolysaccharides (LPS), the growth rate and the ability to grow on root exudates are important (Lugtenberg and Dekkers 1999). Attachment also is an initial step for the formation of microbial biofilms on plant roots as reviewed by Rodríguez-Navarro et al. (2007). The same authors explain mechanisms of attachment of rhizobia on legumes: the first phase of attachment, which is a weak, reversible, and unspecific binding of plant lectins, a Ca^{2+}-binding bacterial protein (rhicadhesin), and bacterial surface polysaccharide and a second attachment step, which requires the synthesis of bacterial cellulose fibrils that cause a tight and irreversible binding of the bacteria to the roots. In *Agrobacterium*, cyclic glucans, capsular polysaccharide, and cellulose fibrils also appear to be involved in the attachment of to plant cells while in *Azospirillum brasilense* the attachment to cereals roots also can be divided into two different steps (Rodriguez-Navarro et al. 2007). Not only bacteria but also fungi attach to the root surface. Fungal adhesion to plants is a key step for establishment of interaction (Tucker and Talbot 2001). Some plant-microbe interaction have evolved complex signal exchange mechanisms that allow a specific bacterial species to induce its host plant to form invasion structures through which the bacteria can enter the plant root, e. g. *Sinorhizobium* (Jones et al. 2007). For the grass endophyte *Azoarcus*, the putative type IV pilus retraction protein PilT is not essential for the bacterial colonization of the plant surface, but twitching motility is necessary for invasion of and establishment inside the plant (Bohm et al. 2007). Plant-associated bacteria used quorum-sensing signals and two-component regulatory systems to coordinate, in a cell density-dependent manner or in response to changing environmental conditions, the expression of important factors for host colonization and invasion (Soto et al. 2006). The success of invasion and survival within the host also requires that bacteria overcome plant defense responses triggered after microbial recognition, a process in which surface polysaccharides, antioxidant systems, ethylene biosynthesis inhibitors and virulence genes are involved (Soto et al. 2006).

There are microbial species, which can colonize only a few or single plant species. This fact is well-known for the beneficial interactions, e.g. *Rhizobium* - legumes, as well as for many plant pathogens. On the other side, some microbial species such as *Pseudomonas* and *Trichoderma* occur ubiquitous and more or less on each plant species. However, it was shown for *Pseudomonas fluorescens/putida* group as well as for *Serratia* that plant specific genotypes exists (Berg et al. 2006; Berg et al. 2002). Theoretically, the composition of microbes, which colonize the rhizosphere, can be a result of a positive or negative selection procedure or both. However, little is known about this important issue, and there are only a few examples for both ways. *Stenotrophomonas maltophilia* is a member of the rhizobacterial populations of cruciferous plants, which produce particularly high levels of sulphur-containing compounds, f. e. amino acids like methionine (Debette and Blondeau 1980). *Stenotrophomonas maltophilia* requires methionine (Ikemoto et al. 1980). These results can base on a positive relationship between both partners. Due to the fact that root exudates are highly plant species specific the use of specific compounds can explain plant specificity of microbial communities. Interestingly, flavonoids, a diverse class of polyphenolic compounds secreted by plants, often serve as signals in plant-microbe interactions (Shaw et al. 2006). On the other hand, plants produce and secrete a variety of secondary metabolites, which can be toxic to microorganisms. Those plants, which are known for their high production of toxins, e. g. walnut (*Juglans regia* L.) are colonized by specific microbial population which can degrade or detoxify metabolites via specific hydrolases (Rettenmaier and Lingens 1985). Another strategy to survive despite the occurrence of toxins are efflux pumps, which pump toxic components outside the body. In addition, production of toxins can be very effective in maintaining microbial diversity (Czárán et al. 2002). A global analysis of *Pseudomonas putida* gene expression during their interaction with maize roots showed the importance of two selective forces of *Pseudomonas* cells to colonize the rhizosphere: stress adaptation and availability of particular nutrients (Matilla et al. 2007). More in detail, genes involved in nutrition (amino acid uptake and metabolism of aromatic compounds) and adaptation (induction of efflux pumps and enzymes for glutathione metabolism) were preferentially expressed in the rhizosphere.

Many plant-associated bacteria, especially root endophytic bacteria, intimately interact with plant metabolism. Fascinating examples are endophytic methylobacteria, which use C1 bodies from the plant for their energy production (Zabetakis 1997). The chemical compound hydroxypropanol is given back to the plant and works as precursor of the flavor compounds mesifuran und 2,5-dimethyl-4-hydroxy-2H-furan (DMHF). The latter posses additional antifungal activity and can be responsible for pathogens defense. These bacteria show also a strong plant growth promoting effect. Interestingly, a recent report provided evidence that hormone-producing methylobacteria are essential for germination and development of protonema of *Funaria hygrometrica* (Hornschuh et al. 2002).

The mycorhizosphere: specific traits and interactions with other rhizosphere constituents

Mycorrhizal symbioses - mutualistic root-fungus associations - are present in almost all land plants and are essential biological constituents of the rhizosphere. Mycorrhizae are grouped in two main categories: endomycorrhizae such as arbuscular (AM), ericoid and orchid mycorrhiza and ectomycorrhiza (EM). The arbuscular mycorrhizal symbiosis represents the most widespread and ancient plant symbiosis. From molecular data and information from fossils in the Devonian, a period during which plant started to colonize land, a close interaction between plant roots and soil born fungi has been established about 450 millions years ago (Remy et al. 1994). These fungi have recently been grouped, on the basis of molecular data, in a new phylum the *Glomeromycota* (Schüßler et al. 2001). In the ectomycorrhizae thousands of *Asco-* and *Basidiomycota* species are known as being mycorrhizal. Nowadays, more than 80% of surveyed plant species and 92% of plant families, present in most ecosystems are mycorrhizal (Wang and Qiu 2006), and mycorrhizal fungi are, on a biomass basis, the largest fungal group in soils (Olsson et al. 1999). In a single cubic centimeter of soil, the mycorrhizal fungal network can represent up to 20 meters (Pearson et al. 1993). For these reasons, the rhizosphere concept has been extended to include the fungal component of this symbiosis, resulting in the mycorrhizosphere (Rambelli 1973). This term includes the physical zone influenced by the root, but also by the mycorrhizal fungus mycelium or hyphosphere. Mycorrhizae are regarded at least as tripartite symbioses since they commonly interact with bacteria and other soil organisms producing beneficial effects on plant nutrition and health as well as on soil structure and stability (Frey-Klett et al. 2007). The fungal and bacterial communities associated with roots vary at different developmental stages, e.g. in *Medicago truncatula* (Mougel et al. 2006). The diversity of arbuscular mycorrhizal communities and their relation with bacteria on grass roots and in grassland ecosystems is documented (Oehl et al. 2003; Singh et al. 2008). The mycorrhizal fungus through the mycelium network increases by several orders of magnitude the soil volume which can be explored. This is achieved by the extension of the mycelium network, but also as the size of the fungal hyphae, which are thinner than the roots, and therefore can enter soil particles in a more efficient than the roots. The strongly reduced mobility of P_i and the rapid uptake of P_i by the plant root generates P_i depleted zones around the plant root hairs followed by a decline in the Pi uptake by the plant (Marschner and Dell 1994; Roose and Fowler 2004). In non-mycorrhizal roots, the P_i depletion zone is closely related to the extension of the root system, whereas in a mycorrhizal root system the P_i depletion zone exceeds greatly the root cylinder (Harley and Smith 1983). The fine fungal mycelia can explore soil particles and then translocate P_i from the soil to root cells, improving the phosphate nutrition. In AM-symbioses, the transfer of the phosphate from the fungus to the plant occurs mainly at the level of the arbuscule, symbiotic organ formed by the fungal hyphae penetrating cortex cells, and forming a arbuscule like structure (Smith and Read 1997). In EM-symbioses the fungal hyphae forming the Hartig net surrounding root cortical cells are the structures of nutrient exchange from root to the fungus. These structures are essential for active symbioses and the development of extraradical mycelium (Smith and

Read 1997). In return, the plant transfers carbon to the fungus. The amount of photosynthates transferred from the plant to the fungus can be has high as 20% (Johnson et al. 2002). This carbon is essential for these fungi, which relies on the plant for their growth, but part of it is transferred through the fungal mycelium to the soil and the atmosphere. Johnson et al. (2002) showed that under field conditions a large amount of carbon, provided under the form of CO_2 was released with a peak 9 – 14 h after labeling by the mycorrhizal fungus. The colonization of roots by AM-fungi was also shown to decrease root exudation (Jones et al. 2004). It has also been shown, that mycorrhization also has a qualitative effect of plant exudation which affects the associated microflora and the soil adjacent to the roots (Jones et al. 2004) resulting in changing the bacterial community composition in the rhizosphere (Marschner and Baumann 2003).

The extension of mycorrhizal fungi into the soil and thus their effect on rhizosphere processes is very dependent on the fungal taxa. Hart and Reader (2002) showed that fungi belonging to the *Glomus* or *Acaulospora* genus had a mean extraradical hyphal length of 1 m to 2 m per cm^{-3}, whereas the fungi belonging to the *Gigaspora* or *Scutellospora* genus had an average hyphal length in the soil of 6 to 9 m per cm^{-3}. The colonization rate was also very dependent on the fungal taxa. *Glomus* species colonized the first plant roots in their conditions after 1 week, whereas *Gigaspora* or the *Scutellopsora* colonized the first roots after 4 to 6 weeks. It has been shown (van Tuinen et al. 1998), that colonization efficiency of *Glomus* and *Gigaspora* fungi, were not identical if these fungi were inoculated alone or in a mixed community. *Gig. rosea* colonized pea or onion roots more efficiently when other mycorrhizal fungi, mainly from the genus *Glomus* where present, suggesting a synergetic behavior of these fungi. These important differences in the colonization pattern could be a reason for the predominant present of *Glomus* species in field collected mycorrhizal roots. (Hempel et al. 2007; Mathimaran et al. 2005; Pivato et al. 2007; Turnau et al. 2001; Vandenkoornhuyse et al. 2003) The differences in colonization speed and mycelium extension observed between the different arbuscular mycorrhizal fungi emphasizes the dynamics of the mycorrhizosphere in time and space. Among the ectomycorrhizae, four different "exploration types" with respect to their ecologically important contact area with the soil substrate were characterized: contact, short, medium and long distance types (Agerer 2001). The long distance type reaches out far into the soil and is also able to bridge different root systems. The mycorrhizal fungal community in a soil is evolving over the season, the mycorrhizosphere impacts also the soil components and the other microorganisms in a temporal way.

In the interaction of mycorrhizal fungi and roots specific signaling factors are known. Root exudates of *Lotus japonicum* contain a "branching factor" which was recently identified as a strigolactone, 5-deoxystrigol (Akiyama et al. 2005). At very low concentrations, this sesquiterpene induces an extensive branching Gigaspora at very low concentrations. On the other hand, the AM fungal partners release diffusible molecules (socalled Myc factor) perceived by the host root in the absence of direct physical contact. In addition, fungal auxins play a key role in ectomycorrhizal development, root morphology and branching of ectomycorrhizae. Events in the early development and ethylene production by *P. microcarpus* are presumably triggered by the production of indole acetic acid. The influences of the mycorrhizal fungi on soil structure and quality are due to the physical presence of the fungal mycelium, and also to the secretion of glomalin by hyphae of *Glomeromycota*. The extensive extrametrical mycelium of the EM-fungi is ideally placed for nutrient acquisition in the top 10 cm of soils, where most of the local nutrient pools are present.

Although little detailed information is available on the direct impact and interaction of bacteria on mycorrhizal fungi, it has been shown that the germination of mycorrhizal spores, can be affected by the presence of some bacteria (Daniels and Trappe 1980; Mayo et al. 1986; Mosse 1959). In a similar way, the establishment of an active symbiosis has an impact on the rhizospheric bacteria population. It has been shown that mycorrhizal symbiosis does not have a significant influence on the number of cultivable bacteria in the mycorrhizosphere (Andrade et al. 1997; Mansfeld-Giese et al. 2002), but on the contrary has a qualitative effect. When plant roots were colonized by a mycorrhizal fungus, *Paenibacillus* spp. were preferentially isolated from the mycorrhizosphere when compared to non-mycorrhizal plant roots (Artursson et al. 2005; Mansfeld-Giese et al. 2002). This suggests a close relation between mycorrhizal fungi and soil bacteria. By using root organ cultures

Toljander and collaborators (Toljander et al. 2007), showed that mycorrhizal fungi through their exudates had a direct impact on the soil bacterial community. This observation could explain the differential influence different taxa of arbuscular mycorrhizal fungi of on associated bacteria (Rillig et al. 2006). Some of the bacteria associated with arbuscular mycorrhiza fungi, can improve the mycorrhizal colonization (Barea et al. 1998; Budi et al. 1999), improve root branching (Gamalero et al. 2002), or present antifungal properties (Budi et al. 1999). From the mycorhizosphere of sorghum, a bacterial strain, *Paenibacillus* sp B2, has been isolated (Budi et al. 1999). This bacteria beside stimulating mycorrhizal colonization (Budi et al. 1999) produces a molecule with biopesticide properties. This molecule which has been characterized, has a structure with some similarities to polymyxin B (Selim et al. 2005), has a very broad inhibitory spectrum against positive or negative bacteria, but also fungi such as *Fusarium accuminatum* or *F. solani* (Selim et al. 2005). Nevertheless this molecule is compatible with the growth of mycorrhizal fungi (Budi et al. 1999). The influences of mycorrhizal fungi on the bacterial diversity, and through them on some soil characteristics such as soil aggregation (Rillig et al. 2005), reveals some of the links between the plant, the mycorrhizal fungus and soil bacteria at the diversity and functional level.

Concerning ectomycorrhiza, mycorrhizal helper bacteria are also known (Frey-Klett et al. 2007; Garbaye 1994). The mycorrhizal mantle and the emanating hyphae are densely colonized by a biofilm-like structure of diverse bacterial community. Using the direct fluorescence *in situ* hybridization (FISH) approach, Mogge et al. (2000) could demonstrate that Betaproteobacteria commonly are found to colonize the *Laccaria subdulcis* / beech mycorrhizosphere although they could not be cultured from this mycorrhiza. In addition, *Acidobacteria* were also shown to be very frequent colonizers of different ectomycorrhizas (C. Kellermann, R. Agerer, A. Hartmann, unpublished results) by 16S rDNA clone bank and FISH-studies, although their cultivation in pure culture was not possible up to date. Depending on the mycorrhizal type and environmental conditions, more than 10.000 bacteria per mm^2 could be counted. While most of the bacteria are usually found to colonize the surface of mycorrhizal fungi, some bacteria were also found to enter the fungal hyphae, such as *Paenibacillus* sp. 101 in cultures of the ectomycorrhizal fungus *Laccaria bicolour* S238N (Bertaux et al. 2003). In contrast, in the natural environment the fungus was colonized intracellularly by Alphaproteobacteria mostly (Bertaux et al. 2005). Also a *Streptomyces* strain *(*GB 4−2), belonging to the *Actinomycetales* (Gram-positive organisms with high DNA G+C DNA content) was characterized as effective colonizer of the ectomycorrhizosphere of Norway spruce (Lehr et al. 2007). Most interestingly, this bacterium caused a systemic response in the spruce plants resulting in inhibition of the phytopathogenic fungus *Heterobasidium abietinum*. This bacterium also increased the general photosynthetic yield and peroxidase activity in the needles leading to decreased infection by *Botrytis cinerea*. Another mycorrhizal helper bacterium, *Streptomyces* AcH 505, from the mycorrhizosphere of fly agaric produces the antibiotic auxofuran, which causes changes of microbial communities in the mycorrhizosphere and additionally stimulates plant growth (Riedlinger et al. 2006). Thus the interaction of mycorrhiza with bacteria can alter the function of plants and rhizosphere communities considerably.

Selected examples for the specific selection and interaction of roots and microbes

After the separate discussion of plant traits shaping microbial communities, of microbial responses towards rhizosphere conditions and the mycorrhizosphere, this chapter finally presents selected case studies, demonstrating the interaction of plant and microbial activities in the rhizosphere to result in plant specificity of rhizosphere microbial communities. This plant specificity has e. g. great relevance for the development of biological control of phytopathogens (case study of the biological control of the soil-borne plant pathogen *Verticillium*), the influence of roots of genetically engineered plants on the rhizosphere microflora (case study of transgenic modified potato lines) and the selection of mycorrhizal microdiversity (case study of arbuscular mycorrhiza in different *Medicago* lines).

Plant specificity of microbial communities in the rhizosphere of Verticillium

Although several studies using cultivation-dependent techniques found indications for plant specificity in the rhizosphere (Germida et al. 1998; Grayston et al. 1998;

Kremer et al. 1990; Miller et al. 1989), Smalla et al. (2001) showed for the first time that roots of each model plant species are colonized by its own bacterial communities using cultivation-independent methods. Three phylogenetically different and economically important crops - strawberry (*Fragaria x ananassa* (Duchense) Decaisne and Naudin [*Rosaceae*]; potato (*Solanum tuberosum* L. [*Solanaceae*]; and oilseed rape *Brassica napus* L. [*Brassicaceae*] - were analyzed in this study. All species belong to the broad host range of the soil-borne fungal pathogen *Verticillium dahliae* Kleb., which cause high yield losses world-wide (Tjamos et al. 2000). Besides the analysis of whole bacterial community structures in rhizospheres, the functional group of antagonists was another indicator to analyze differences and similarities between the three plants. Antagonists are naturally occurring micro-organisms that express traits that enable them to interfere with pathogen growth, survival, and infection. They form the antagonistic potential against plant pathogens interacting by diverse mechanisms with pathogens and host plants (Cook et al. 1995).

It was possible to differentiate plant species on the basis of the rhizosphere microbial communities using denaturing gradient gel electrophoresis (DGGE) in a randomized field trial (Smalla et al. 2001). The DGGE fingerprints showed plant-dependent shifts in the relative abundance of bacterial populations in the rhizosphere which became more pronounced in the second year. Interestingly, all rhizospheres showed some bands in common but also specific bands, e.g. *Nocardia* populations were identified as strawberry-specific bands. The proportion and composition of bacterial antagonists of potato, oilseed rape and strawberry towards *V. dahliae* was also shown to be influenced by the plant species (Berg et al. 2002). Furthermore, plant specific genotypes of 34 *Pseudomonas putida* A isolates were observed, suggesting that these bacteria were specifically enriched in each rhizosphere. When the field experiment with the three *Verticillium* host plants was performed at different sites (Rostock, Berlin, Braunschweig in Germany) plant specificity of rhizosphere communities was detected again while also different bulk soil community fingerprints were revealed for each sampling site (Costa et al. 2006a). Universal and group-specific (Alphaproteobacteria, Betaproteobacteria and *Actinobacteria*) primers were used to PCR-amplify 16S rRNA gene fragments of bacterial and 18S rRNA amplificates for the fungal communities prior to DGGE analysis. The plant species was the determinant factor in shaping similar Actinobacterial communities in the strawberry rhizosphere from different sites in both years. The rhizosphere effect on the antagonistic bacterial community was demonstrated by an enhanced proportion of antagonistic isolates, by enrichment of specific ARDRA types, species and genotypes as evidenced by BOX-PCR, and by a reduced diversity of antagonistic bacteria in the rhizosphere in comparison to bulk soil (Berg et al. 2006). Since bacteria of the genus *Pseudomonas* are prominent root-associated bacteria (Haas and Défago 2005) they were investigated by culture dependent and independent techniques. Based on the data of this field trial, the factors sampling site, plant species and year-to-year variation were shown to significantly influence the community structure of *Pseudomonas* in rhizosphere soils (Costa et al. 2006b). The composition of *Pseudomonas* 16S rRNA gene fragments in the rhizosphere differed from that in the adjacent bulk soil and the rhizosphere effect tended to be plant-specific. The clone sequences of most dominant bands analyzed belonged to the *Pseudomonas fluorescens* lineage and showed closest similarity to culturable *Pseudomonas* known for displaying antagonistic properties. In addition, *Pseudomonas*-specific *gac*A fingerprints of total-community rhizosphere DNA were surprisingly diverse, plant-specific and differed markedly from those of the corresponding bulk soils (Costa et al. 2007). By combining multiple culture-dependent and independent surveys, a group of *Pseudomonas* isolates antagonistic towards *V. dahliae* was shown to be genotypically conserved, to carry the *phl*D biosynthetic locus (involved in the biosynthesis of 2,4-diacetylphloroglucinol – 2,4-DAPG) (Costa et al. 2007). This group of *Pseudomonas* isolates corresponded to a highly frequent *Pseudomonas* population in the rhizosphere of field-grown strawberries planted at three sites in Germany which have different land use histories. It belongs to the *Pseudomonas fluorescens* phylogenetic lineage and showed closest relatedness to *P. fluorescens* strain F113 (97% *gac*A gene sequence identity in 492-bp sequences), a biocontrol agent and 2,4-DAPG producer. Partial *gac*A gene sequences derived from isolates, clones of the strawberry rhizosphere and DGGE bands retrieved in this study represent previously unknown *Pseudomonas gacA* gene clusters as revealed by phylogenetic analysis (Costa et al. 2007).

Concerning fungal rhizosphere communities, a plant-specific composition could also be detected, but not in all cases (Costa et al. 2006a). Higher heterogeneity of DGGE profiles within soil and rhizosphere replicates was observed for the fungi than for bacteria. A high proportion of fungi antagonistic towards the pathogen *V. dahliae* were found for bulk and rhizosphere soil at all sites (Berg et al. 2005). A plant- and site-dependent specificity of the composition of antagonistic morphotypes and their genotypic diversity was found. *Trichoderma* strains displayed high diversity in all soils, and a high degree of plant specificity could be shown by BOX-PCR fingerprints. The diversity of rhizosphere-associated antagonists was lower than that of antagonists in bulk soil, suggesting that some fungi were specifically enriched in each rhizosphere. In Fig. 2, a model for the rhizosphere effect of bacterial and fungal antagonists was developed. Altogether, the rhizosphere effect for fungal antagonists was lower pronounced than for bacteria. Altogether, these results obtained in these six-year field studies proofed a clear influence of plant species on the structure and function of rhizosphere bacterial communities. Although there are several contrasting reports in the literature indicating plant or soil type as dominant factor (Girvan et al. 2003; Grayston et al. 1998; Nunan et al. 2005) in more or less all studies the influence of plant species is clearly visible [rev. in (Berg and Smalla 2008; Garbeva et al. 2004)]. Extend of plant specificity is influenced by the selected plant species, applied methods as well as by the experimental design. Exemplarily, Fig. 3 corroborates the plant specific selection of rhizosphere microbial communities since it was clearly demonstrated that different herbaceous plants select a very different bacterial community from the very same soil (Dohrmann and Tebbe 2005).

Cultivar specificity of microbial communities in the rhizosphere of different potatoes (*Solanum tuberosum* L.) including transgenic lines

Rhizosphere microbial communities are not only influenced at the plant species level, but also at the cultivar level (Germida and Siciliano 2001). Genetically modified plants with altered root exudates are interesting model systems to study cultivar-specific effects, which were found in several studies for root-associated bacterial as well as fungal communities (Mansouri et al. 2002; Oger et al. 1997; Oger et al. 2004; Oliver et al. 2008). T4 lysozyme potatoes are a well-studied model system to investigate the potential risk of pathogen resistant plants (Düring et al. 1993). For example, in a 6-year field release of T4 lysozyme producing potatoes cv. Désirée the changes in structure and function of plant-associated bacterial populations were monitored by a polyphasic approach. However, in both microenvironments (rhizosphere and geocaulosphere) no statistically significant

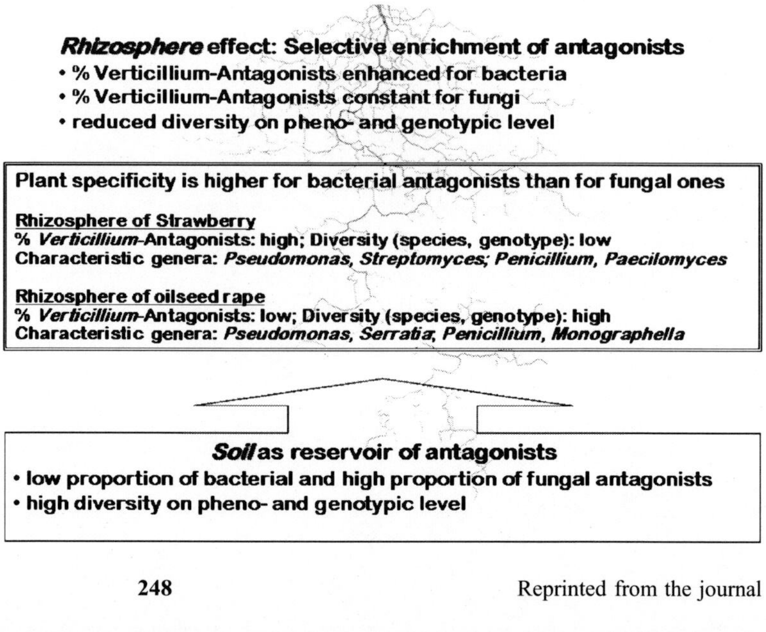

Fig. 2 Selective enrichments of antagonistic populations in the rhizosphere according to results obtained in a six-year field trial with Verticillium host plants (Berg et al. 2002, 2005, 2006)

***Rhizosphere* effect: Selective enrichment of antagonists**
- % Verticillium-Antagonists enhanced for bacteria
- % Verticillium-Antagonists constant for fungi
- reduced diversity on pheno- and genotypic level

Plant specificity is higher for bacterial antagonists than for fungal ones

Rhizosphere of Strawberry
% *Verticillium*-Antagonists: high; Diversity (species, genotype): low
Characteristic genera: *Pseudomonas, Streptomyces; Penicillium, Paecilomyces*

Rhizosphere of oilseed rape
% *Verticillium*-Antagonists: low; Diversity (species, genotype): high
Characteristic genera: *Pseudomonas, Serratia; Penicillium, Monographella*

***Soil* as reservoir of antagonists**
- low proportion of bacterial and high proportion of fungal antagonists
- high diversity on pheno- and genotypic level

Fig. 3 Plant driven selection of bacterial communities in the rhizosphere of different herbaceous plants. Six different herbaceous plants were cultivated in the same soil and the rhizosphere bacterial communities were analyzed using the SSCP-profiling technique (according to Dohrmann and Tebbe (2005))

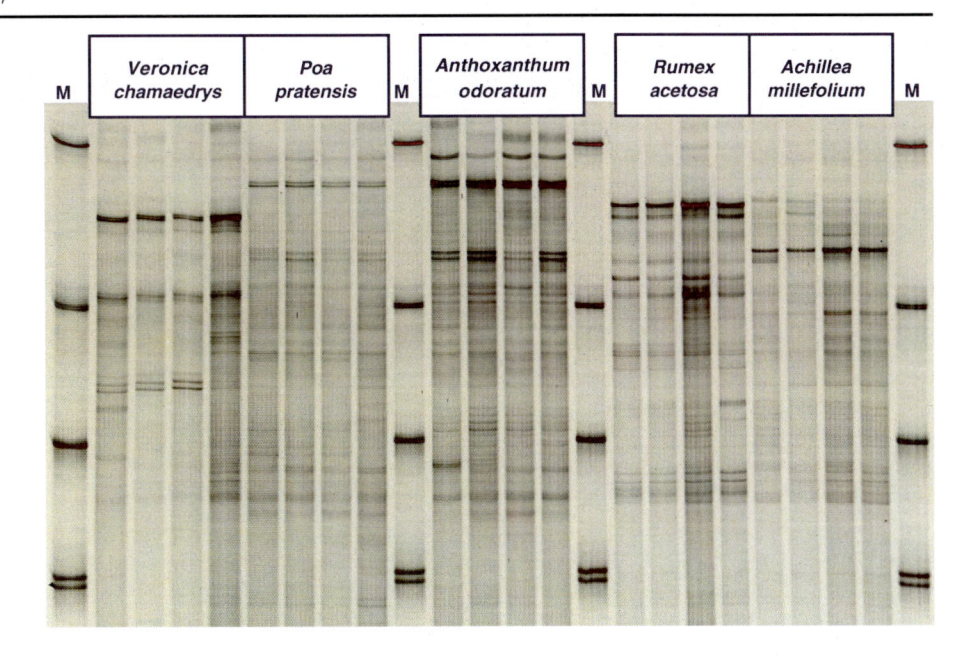

differences between transgenic and non transgenic plants were found in the following parameters: (i) the abundances of bacteria, (ii) the percentage of auxin-producing bacteria, (iii) the percentage of antagonistic bacteria, and (iv) on the diversity of bacterial antagonists on genotypic and phenotypic level (Lottmann and Berg 2001; Lottmann et al. 1999; Lottmann et al. 2000). In an additional approach, rifampicin resistant mutants of two antagonistic plant-associated bacteria were used for seed tuber inoculation of transgenic and non-transgenic potato lines. During flowering of plants, significantly more colony counts of the lysozyme tolerant *Pseudomonas putida* QC14-3-8 were recovered from the transgenic T4 lysozyme plant than from the non-transgenic control and the parental line. Furthermore, using a root hair - *Bacillus subtilis* model, roots from potato lines expressing the T4 lysozyme gene always showed significantly (1.5- to 3.5-fold) higher killing (Ahrenholtz et al. 2000). However, using cultivation-independent methods the influence of environmental factors on potato associated bacteria was much higher than of the transgene (Heuer et al. 2002). Altogether, an influence of transgenic T4 lyzozyme on bacterial strains and community was to be seen in special experiments. In field trials, the influence was negligible compared to other environmental factors.

To assess potential effects of T4 lysozyme on culturable plant-associated fungi in the rhizosphere, the abundances of colony forming units, the percentage of antagonistic fungi and their diversity were investigated (Berg, unpublished results). The results from this study suggest that transgenic plants producing T4 lysozyme did not affect the fungal communities and the abundance and composition of fungi with antagonistic activity. Furthermore, the composition and relative abundance of endophytic fungi in roots of T4-lysozyme producing potatoes and the parental line were assessed by classical isolation from root segments and cultivation-independent techniques to test the hypothesis that endophytic fungi are affected by T4-lysozyme (Götz et al. 2006). Fungi were isolated from the majority of root segments of both lines and at least 63 morphological groups were obtained with *Verticillium dahliae, Cylindrocarpon destructans, Colletotrichum coccodes* and *Plectosporium tabacinum* as the most frequently isolated species. Dominant bands in the fungal fingerprints obtained by denaturing gradient gel electrophoresis analysis of 18S rRNA gene fragments amplified from total community DNA corresponded to the electrophoretic mobility of the 18S rRNA gene fragments of the three most abundant fungal isolates, *V. dahliae, C. destructans* and *Col. coccodes*, but not to *P. tabacinum*. The assignment of the bands to these isolates was confirmed for *V. dahliae* and *Col. coccodes* by sequencing of clones. *Verticillium dahliae* was the most abundant endophytic fungus in the roots of healthy potato plants. Differences in the relative abundance of endophytic fungi colonizing the roots of T4-lysozyme producing potatoes and the parental line could be detected by both methods.

These results obtained for T4-lysozyme potato were confirmed in greenhouse experiments (Rasche et al. 2006) and in other studies analyzing transgenic plants (Bruinsma et al. 2003). There was only a minor, and in comparison to other environmental factors, negligible influence on the structure and function of microbial communities to be seen. In another approach comparing zeaxanthine producing potatoes with different potato cultivars, the effect of cultivar was much higher than of the transgene (Schloter, pers. communication).

Impact of *Medicago* species on arbuscular mycorrhizal fungi

A relationship of the diversity of arbuscular mycorrhizal fungi and plant diversity has been documented several times under field conditions (Gollotte et al. 2004; Husband et al. 2002; Oehl et al. 2005; van der Heijden et al. 1998; Vandenkoornhuyse et al. 2003; Wolfe et al. 2007). These knowledge has been obtained by analyzing the root-associated mycorrhizal community on the basis of spore counting (Husband et al. 2002; Oehl et al. 2005; Vandenkoornhuyse et al. 2003; Wolfe et al. 2007), PCR-RFLP analysis of the small ribosomal sub-unit (Husband et al. 2002; Scheublin et al. 2004; Vandenkoornhuyse et al. 2003) or by sequencing of the large ribosomal subunit of the fungi (Gollotte et al. 2004). More recently, Real-Time PCR has been used to quantify mycorrhizal fungi in plant roots inoculated with a single (Filion et al. 2003) or two fungal isolates (Alkan et al. 2004; Alkan et al. 2006). This method opens the possibility of quantifying the fungal ribosomal operon in the soil or plant roots, by a very refined and sensitive method, and has been used to estimate the fungal community in the roots, by monitoring several arbuscular mycorrhizal taxa, in closely related *Medicago* species (Pivato et al. 2007). As it is known that the plant used to trap arbuscular mycorrhizal has a great influence on the species detected (Jansa et al. 2002), a preliminary experiment was carried out to identify the indigenous arbuscular mycorrhizal fungi present in the soil. This was performed on a low fertility soil from the Mediterranean basin, corresponding to the naturally growing zone of annual medics. The arbuscular mycorrhizal fungi were identified by sequencing the large ribosomal subunit (LSU), directly amplifying from soil extracted DNA. By using *Glomeromycota* specific primers, targeting the 5' end of the LSU (Gollotte et al. 2004; van Tuinen et al. 1998), Pivato and co-workers (2007) were enabled to group, after a phylogenetic analysis, the 246 obtained sequences in 12 *Glomus* species, or OTU, belonging to the *Glomus* A group (Schwarzott et al. 2001; van Tuinen et al. 1998). From these 12 OTU, only two could be identified on the basis of their homology with well-identified *Glomus* species, whose sequences are available in the databases, namely *G. mosseae* and *G. intraradices*. No *Acaulosporaceae* nor *Gigasporaceae* were detected, although the *Glomeromycota* specific primers used were also able to positively amplify the former taxa (Gollotte et al. 2004). Primers specifically amplifying 4 of the 12 identified *Glomus* OTU, were then designed, and used to quantify by Real-Time PCR, the presence of the corresponding OTU in bare soil or in roots of 4 closely related *Medicago* species, namely *Medicago laciniata* L., *M. murex* Wild., *M. polymorpha* L. and *M. truncatula* Gaertn, grown for 34 days in the same soil. For each OTU specific primer the primer binding site was fully conserved within the sequences of the corresponding OTU, this was important as it is well known that different ribosomal operons are harbored in the same single *Glomeromycota* spore. This approach enabled to show that the amount of the four selected OTU varied between bare soil and the plants, but also between the *Medicago* species. No statistical differences were found in the roots of *M. laciniata* and *M. murex*, whereas statistical differences were observed between *M. polymorpha* and *M. truncatula* for 2 of the selected OTU. Interestingly one of the OTU was present in the same amount in all 4 root systems. This approach demonstrated a subtle modification in the arbuscular mycorrhizal fungi community composition between closely related plants grown in their native soil environment. These measurements were performed at a single time point, and it is possible that the community structure varies over time. Thus, this technique offers the possibility to accurately monitor the selective impact of plants on arbuscular mycorrhizal fungi associated with their roots or their mycorhizospheres.

Summary and conclusions

Ample evidences exist which clearly demonstrate the selection of microbes by roots of plants. Due to many usually overlapping and interfering mechanisms, the roots provide a specific microhabitat for the proliferation of a specific subset of soil microbes. Usually, new interactions amongst colonizing microbes arise in the quite densely colonized rhizosphere. For example, specific mycorrhiza-bacteria interactions lead to a stimulation of symbiosis and other microbe-microbe interactions result in the biological control of phytopathogenic microbes by plant beneficial root colonizers. Vice versa, the plant is profiting manifold from microbial activities in the rhizosphere and is additionally influenced by root colonizing microbes through signaling pathways, leading to potentially improved plant fitness. It is probably not the mere utilization of available carbon sources which selects the rhizosphere community but rather the presence of selective and inhibitory interactions which creates the bias into root-associated microbial populations.

To get deeper insight into key rhizosphere processes and the major steering factors involved, a combination of stable isotope probing (through e. g. ^{13}C-CO_2 labeling of plant assimilates) with molecular biological techniques of community characterization will certainly gain even more importance in future. Since m-RNA can be retrieved from soil with more confidence now, studies on functional gene expression of specific bacterial genes and even rhizosphere transcriptome analysis are feasible. These studies should be combined with metagenome, proteome and metabolome studies to get a complete picture. However, to be able to cope with the complexity and amount of data, bioinformatics and mathematical modeling will have to be included in these endeavors. Finally, on site field studies of rhizosphere research should be performed more intensively to be able to proof the experience of more or less laboratory model studies to the field and forest setting. Climatic or environmental simulation chambers will help to reduce the unpredictable climatic factors to well designed factorial analyses while keeping the complexity of a realistic ecosystem setting. This will allow studying the influence of extreme climatic conditions on plant performance and how plants influence rhizosphere communities and processes under these conditions.

Rhizosphere driven selection of microbes has high potential to improve the development and health of plants. Although some promising products are in applications, this route needs to be much more developed and applied for the sake of sustainable agriculture, silviculture and horticulture. In future, both the plant side and the microbial side should be included in concerted biotechnological development and breeding programs.

References

Agerer R (2001) Exploration types of ectomycorrhizal mycelial systems: A proposal to classify mycorrhizal mycelial systems with respect to their ecologically important contact area with the substrate. Mycorrhiza 11:107–114. doi:10.1007/s005720100108

Ahrenholtz I, Harms K, de Vries J, Wackernagel W (2000) Increased killing of *Bacillus subtilis* on the hair roots of transgenic T4 lysozyme-producing potatoes. Appl Environ Microbiol 66:1862–1865. doi:10.1128/AEM.66.5.1862-1865.2000

Akiyama K, Matsuzaki K-i, Hayashi H (2005) Plant sesquiterpenes induce hyphal branching in arbuscular mycorrhizal fungi. Nature 435:824–827. doi:10.1038/nature03608

Alkan N, Gadkar V, Coburn J, Yarden O, Kapulnik Y (2004) Quantification of the arbuscular mycorrhizal fungus *Glomus intraradices* in host tissue using real-time polymerase chain reaction. New Phytol 161:877–885. doi:10.1046/j.1469-8137.2004.00975.x

Alkan N, Gadkar V, Yarden O, Kapulnik Y (2006) Analysis of quantitative interactions between two species of arbuscular mycorrhizal fungi, *Glomus mosseae* and *G. intraradices*, by Real-Time PCR. Appl Environ Microbiol 72:4192–4199. doi:10.1128/AEM.02889-05

Andrade G, Mihara KL, Linderman RG, Bethlenfalvay GJ (1997) Bacteria from the rhizosphere and hyphoshere soils of different arbuscular-mycorrhizal fungi. Plant Soil 192:71–79. doi:10.1023/A:1004249629643

Apel K, Hirt H (2004) Reactive oxygen species: Metabolism, oxidative stress, and signal transduction. Annu Rev Plant Biol 55:373–399. doi:10.1146/annurev.arplant.55.031903.141701

Artursson V, Finlay RD, Jansson JK (2005) Combined bromodeoxyuridine immunocapture and terminal-restriction fragment length polymorphism analysis highlights differences in the active soil bacterial metagenome due to *Glomus mosseae* inoculation or plant species. Environ Microbio l7:1952–1966. doi:10.1111/j.1462-2920.2005.00868.x

Bais HP, Prithiviraj B, Jha AK, Ausubel FM, Vivanco JM (2005) Mediation of pathogen resistance by exudation of antimicrobials from roots. Nature 434:217–221. doi:10.1038/nature03356

Bais HP, Weir TL, Perry LG, Gilroy S, Vivanco JM (2006) The role of root exudates in rhizosphere interactions with plants and other organisms. Annu Rev Plant Biol 57:233–266. doi:10.1146/annurev.arplant.57.032905.105159

Barea JM, Andrade G, Bianciotto V, Dowling D, Lohrke S, Bonfante P, O'Gara F, Azcon-Aguilar C (1998) Impact on arbuscular mycorrhiza formation of *Pseudomonas* strains used as inoculants for biocontrol of soil-borne fungal plant pathogens. Appl Environ Microbiol 64:2304–2307

Bauer WD, Mathesius U (2004) Plant responses to bacterial quorum sensing signals. Curr Opin Plant Biol 7:429–433. doi:10.1016/j.pbi.2004.05.008

Bérczi A, Møller IM (2000) Redox enzymes in the plant plasma membrane and their possible roles. Plant Cell Environ 23:1287–1302. doi:10.1046/j.1365-3040.2000.00644.x

Berg G, Smalla K (2008) Plant species versus soil type: which factors influence the structure and function of the microbial communities in the rhizosphere? FEMS Microbiol Ecol (submitted)

Berg G, Roskot N, Steidle A, Eberl L, Zock A, Smalla K (2002) Plant-dependent genotypic and phenotypic diversity of antagonistic rhizobacteria isolated from different *Verticillium* host plants. Appl Environ Microbiol 68:3328–3338. doi:10.1128/AEM.68.7.3328-3338.2002

Berg G, Zachow C, Lottmann J, Gotz M, Costa R, Smalla K (2005) Impact of plant species and site on rhizosphere-associated fungi antagonistic to *Verticillium dahliae* Kleb. Appl Environ Microbiol 71:4203–4213. doi:10.1128/AEM.71.8.4203-4213.2005

Berg G, Opelt K, Zachow C, Lottmann J, Gotz M, Costa R, Smalla K (2006) The rhizosphere effect on bacteria antagonistic towards the pathogenic fungus *Verticillium* differs depending on plant species and site. FEMS Microbiol Ecol 56:250–261. doi:10.1111/j.1574-6941.2005.00025.x

Bertaux J, Schmid M, Prevost-Boure NC, Churin JL, Hartmann A, Garbaye J, Frey-Klett P (2003) In situ identification of intracellular bacteria related to *Paenibacillus* spp. in the mycelium of the ectomycorrhizal fungus *Laccaria bicolor* S238N. Appl Environ Microbiol 69:4243–4248. doi:10.1128/AEM.69.7.4243-4248.2003

Bertaux J, Schmid M, Hutzler P, Hartmann A, Garbaye J, Frey-Klett P (2005) Occurrence and distribution of endobacteria in the plant-associated mycelium of the ectomycorrhizal fungus *Laccaria bicolor* S238N. Environ Microbiol 7:1786–1795. doi:10.1111/j.1462-2920.2005.00867.x

Bohm M, Hurek T, Reinhold-Hurek B (2007) Twitching motility is essential for endophytic rice colonization by the N_2-fixing endophyte *Azoarcus* sp. Strain BH72. Mol Plant Microbe Interact 20:526–533. doi:10.1094/MPMI-20-5-0526

Bruinsma M, Kowalchuk GA, van Veen JA (2003) Effects of genetically modified plants on microbial communities and processes in soil. Biol Fertil Soils 37:329–337

Budi SW, van Tuinen D, Martinotti G, Gianinazzi S (1999) Isolation from the *Sorghum bicolor* mycorrhizosphere of a bacterium compatible with arbuscular mycorrhiza development and antagonistic towards soilborne fungal pathogens. Appl Environ Microbiol 65:5148–5150

Butler JL, Williams MA, Bottomley PJ, Myrold DD (2003) Microbial community dynamics associated with rhizosphere carbon flow. Appl Environ Microbiol 69:6793–6800. doi:10.1128/AEM.69.11.6793-6800.2003

Butler JL, Bottomley PJ, Griffith SM, Myrold DD (2004) Distribution and turnover of recently fixed photosynthate in ryegrass rhizospheres. Soil Biol Biochem 36:371–382. doi:10.1016/j.soilbio.2003.10.011

Chishaki N, Horiguchi T (1997) Responses of secondary metabolism in plants to nutrient deficiency. Soil Sci Plant Nutr 43:987–991

Cook RJ, Thomashow LS, Weller DM, Fujimoto D, Mazzola M, Bangera G, Kim D (1995) Molecular mechanisms of defense by rhizobacteria against root disease. Proc Natl Acad Sci U S A 92:4197–4201. doi:10.1073/pnas.92.10.4197

Costa R, Gotz M, Mrotzek N, Lottmann J, Berg G, Smalla K (2006a) Effects of site and plant species on rhizosphere community structure as revealed by molecular analysis of microbial guilds. FEMS Microbiol Ecol 56:236–249. doi:10.1111/j.1574-6941.2005.00026.x

Costa R, Salles JF, Berg G, Smalla K (2006b) Cultivation-independent analysis of *Pseudomonas* species in soil and in the rhizosphere of field-grown *Verticillium dahliae* host plants. Environ Microbiol 8:2136–2149. doi:10.1111/j.1462-2920.2006.01096.x

Costa R, Gomes NCM, Krogerrecklenfort E, Opelt K, Berg G, Smalla K (2007) *Pseudomonas* community structure and antagonistic potential in the rhizosphere: insights gained by combining phylogenetic and functional gene-based analyses. Environ Microbiol 9:2260–2273. doi:10.1111/j.1462-2920.2007.01340.x

Crowley DE, Rengel Z (1999) Biology and chemistry of rhizosphere influencing nutrient availability. In: Rengel Z (ed) Mineral nutrition of crops: Fundamental mechanisms and implications. The Haworth Press, New York, pp 1–40

Czárán TL, Hoekstra RF, Pagie L (2002) Chemical warfare between microbes promotes biodiversity. Proc Natl Acad Sci U S A 99:786–790. doi:10.1073/pnas.012399899

Dakora FD, Phillips DA (2002) Root exudates as mediators of mineral acquisition in low-nutrient environments. Plant Soil 13:35–47

d'Angelo-Picard C, Faure D, Penot I, Dessaux Y (2005) Diversity of N-acyl homoserine lactone-producing and -degrading bacteria in soil and tobacco rhizosphere. Environ Microbiol 17:1796–1808

Daniels BA, Trappe JM (1980) Factors affecting spore germination of the vesicular-arbusuclar mycorrhizal fungus *Glomus epigaeus*. Mycologia 72:457–471

Debette J, Blondeau R (1980) Présence de *Pseudomonas maltophilia* dans la rhizosphère de quelque plantes cultivée. Can J Microbiol 26:460–463

Degrassi G, Devescovi G, Solis R, Steindler L, Venturi V (2007) *Oryza sativa* rice plants contain molecules that activate different quorum-sensing N-acyl homoserine lactone biosensors and are sensitive to the specific AiiA lactonase. FEMS Microbiol Lett 269:213–220

Delalande L, Faure D, Raffoux A, Uroz S, D'Angelo-Picard C, Elasri M, Carlier A, Berruyer R, Petit A, Williams P, Dessaux Y (2005) N-hexanoyl-l-homoserine lactone, a mediator of bacterial quorum-sensing regulation, exhibits plant-dependent stability and may be inactivated by germinating *Lotus corniculatus* seedlings. FEMS Microbiol Ecol 52:13–20

de Weert S, Vermeiren H, Mulders IHM, Kuiper I, Hendrickx N, Bloemberg GV, Vanderleyden J, De Mot R, Lugtenberg BJJ (2002) Flagella-driven chemotaxis towards exudate components is an important trait for tomato root colonization by *Pseudomonas fluorescens* Mol Plant Microbe Interact 15:1173–1180

Dinkelaker B, Hengeler C, Marschner H (1995) Distribution and function of proteoid roots and other root clusters. Bot Acta 108:183–200

Dohrmann AB, Tebbe CC (2005) Effect of elevated tropospheric ozone on the structure of bacterial communities inhabiting the rhizosphere of herbaceous plants native to Germany. Appl Environ Microbiol 71:7750–7758

Dong Z, Wu L, Kettlewell B, Caldwell CD, Layzell DB (2003) Hydrogen fertilization of soils - is this a benefit of legumes in rotation? Plant Cell Environ 261:875–1879

Duineveld BM, Rosado AS, van Elsas JD, van Veen JA (1998) Analysis of the dynamics of bacterial communities in the rhizosphere of the Chrysanthemum via denaturing gradient gel electrophoresis and substrate utilization patterns. Appl Environ Microbiol 64:4950–4957

Düring K, Porsch P, Fladung M, Lörz H (1993) Transgenic potato plants resistant to the phytopathogenic bacterium Erwinia carotovora Plant J 3:587–598

Farrar J, Hawes M, Jones D, Lindow S (2003) How roots control the flux of carbon to the rhizosphere. Ecology 84:827–837

Filion M, St-Arnaud M, Jabaji-Hare SH (2003) Direct quantification of fungal DNA from soil substrate using real-time PCR. J Microbiol Meth 53:67–76

Frey-Klett P, Garbaye J, Tarkka M (2007) The mycorrhiza helper bacteria revisited. New Phytol 176:22–36

Fuqua C, Parsek MR, Greenberg EP (2001) Regulation of gene expression by cell-to-cell communication: Acyl-homoserine lactone quorum sensing. Annu Rev Genet 35:439–468

Gamalero E, Martinotti MG, Trotta A, Lemanceau P, Berta G (2002) Morphogenetic modifications induced by Pseudomonas fluorescens A6RI and Glomus mosseae BEG12in the root system of tomato differ according to the plant growth conditions. New Phytol 155:293–300

Gantner S, Schmid M, Duerr C, Schuhegger R, Steidle A, Hutzler P, Langebartels C, Eberl L, Hartmann A, Dazzo FB (2006) In situ quantitation of the spatial scale of calling distances and population density-independent N-acylhomoserine lactone-mediated communication by rhizobacteria colonized on plant roots. FEMS Microbiol Ecol 56:188–194

Garbaye J (1994) Helper bacteria: A new dimension to the mycorrhizal symbiosis. New Phytol 128:197–210

Garbeva P, van Veen JA, van Elsas JD (2004) Microbial diversity in soil: Selection of Microbial Populations by Plant and Soil Type and Implications for Disease Suppressiveness. Annu Rev Phytopathol 42:243–270

Germida JJ, Siciliano SD (2001)Taxonomic diversity of bacteria associated with the roots of modern, recent and ancient wheat cultivars. Biol Fert Soils 33:410–415

Germida JJ, Siciliano SD, Renato de Freitas J, Seib AM (1998) Diversity of root-associated bacteria associated with field-grown canola (Brassica napus L.) and wheat (Triticum aestivum L.). FEMS Microbiol Ecol 26:43–50

Girvan MS, Bullimore J, Pretty JN, Osborn AM, Ball AS (2003) Soil type is the primary determinant of the composition of the total and active bacterial communities in arable soils. Appl Environ Microbiol 69:1800–1809

Gollotte A, van Tuinen D, Atkinson D (2004) Diversity of arbuscular mycorrhizal fungi colonising roots of the grass species Agrostis capillaris and Lolium perenne in a field experiment. Mycorrhiza 14:111–117

Götz C, Fekete A, Gebefuegi I, Forczek S, Fuksová K, Li X, Englmann M, Gryndler M, Hartmann A, Matucha M, Schmitt-Kopplin P, Schröder P (2007) Uptake, degradation and chiral discrimination of N-acyl-D/L -homoserine lactones by barley (Hordeum vulgare) and yam bean (Pachyrhizus erosus) plants. Anal Bioanal Chem 389:1447–1457

Götz M, Nirenberg H, Krause S, Wolters H, Draeger S, Buchner A, Lottmann J, Berg G, Smalla K (2006) Fungal endophytes in potato roots studied by traditional isolation and cultivation-independent DNA-based methods. FEMS Microbiol Ecol 58:404–413

Grayston SJ, Wang S, Campbell CD, Edwards AC (1998) Selective influence of plant species on microbial diversity in the rhizosphere. Soil Biol Biochem 30:369–378

Haas D, Défago G (2005) Biological control of soil-borne pathogens by fluorescent pseudomonads. Nat Rev Microbiol 3:307–319

Harley JL, Smith SE (1983) Mycorrhizal Symbioses. Academic Press Inc., London, New York, pp 483

Hart MM, Reader RJ (2002) Taxonomic basis for variation in the colonization strategy of arbuscular mycorrhizal fungi. New Phytol 153:335–344

Hartmann A (1988) Ecophysiological aspects of growth and nitrogen fixation in Azospirillum spp. Plant Soil 110:225–238

Hartmann A, Rothballer M, Schmid M (2008) Lorenz Hiltner, a pioneer in rhizosphere microbial ecology and soil bacteriology research. Plant Soil 312:7–14

Hempel S, Renker C, Buscot F (2007) Differences in the species composition of arbuscular mycorrhizal fungi in spore, root and soil communities in a grassland ecosystem. Environ Microbiol 9:1930–1938

Hense BA, Kuttler C, Müller J, Rothballer M, Hartmann A, Kreft J-U (2007) Does efficiency sensing unify diffusion and quorum sensing? Nat Rev Microbiol 5:230–239

Hentzer M, Riedel K, Rasmussen TB, Heydorn A, Andersen JB, Parsek MR, Rice SA, Eberl L, Molin S, Hoiby N, Kjelleberg S, Givskov M (2002) Inhibition of quorum sensing in Pseudomonas aeruginosa biofilm bacteria by a halogenated furanone compound. Microbiol 148:87–102

Heuer H, Kroppenstedt RM, Lottmann J, Berg G, Smalla K (2002) Effects of T4 lysozyme release from transgenic potato roots on bacterial rhizosphere communities are negligible relative to natural factors. Appl Environ Microbiol 68:1325–1335

Hinsinger P, Plassard C, Tang C, Jaillard B (2003) Origins of root-mediated pH changes in the rhizosphere and their responses to environmental constraints: A review. Plant Soil 248:43–59

Hornschuh M, Grotha R, Kutschera U (2002) Epiphytic bacteria associated with the bryophyte Funaria hygrometrica: Effect of Methylobacterium strains on protonema development. Plant Biol 4:682–682

Husband R, Herre EA, Young JPW (2002) Temporal variation in the arbuscular mycorrhizal communities colonising seedlings in a tropical forest. FEMS Microbiol Ecol 42:131–136

Ikemoto S, Suzuki K, Kaneko T, Komagata K (1980) Characterization of strains of Pseudomonas maltophilia which do not require methionine. Int J Syst Bacteriol 30:437–447

Jaeger CHIII, Lindow SE, Miller W, Clark E, Firestone MK (1999) Mapping of sugar and amino acid availability in soil around roots with bacterial sensors of sucrose and tryptophan. Appl Environ Microbiol 65:2685–2690

Jansa J, Mozafar A, Anken T, Ruh R, Sanders I, Frossard E (2002) Diversity and structure of AMF communities as affected by tillage in a temperate soil. Mycorrhiza 12: 225–234

Johnson D, Leake JR, Ostle N, Ineson P, Read DJ (2002) In situ $^{13}CO_2$ pulse-labelling of upland grassland demonstrates a rapid pathway of carbon flux from arbuscular mycorrhizal mycelia to the soil. New Phytol 153:327–334

Jones DL, Hodge A, Kuzyakov Y (2004) Plant and mycorrhizal regulation of rhizodeposition. New Phytol 163:459–480

Jones KM, Kobayashi H, Davies BW, Taga ME, Walker GC (2007) How rhizobial symbionts invade plants: the *Sinorhizobium-Medicago* model. Nat Rev Microbiol 5:619–633

Kowalchuk GA, Bruinsma M, van Veen JA (2003) Assessing responses of soil microorganisms to GM plants. Trends Ecol Evol 18:403–410

Kremer RJ, Begonia MFT, Stanley L, Lanham ET (1990) Characterization of rhizobacteria associated with weed seedlings. Appl Environ Microbiol 56:1649–1655

Kuzyakov Y, Bol R (2005) Three sources of CO_2 efflux from soil partitioned by ^{13}C natural abundance in an incubation study. Rapid Commun Mass Spectrom 19:1417–1423

Lehr NA, Schrey SD, Bauer R, Hampp R, Tarkka MT (2007) Suppression of plant defence response by a mycorrhiza helper bacterium. New Phytol 174:892–903

Leveau JH, Gerards S (2008) Discovery of a bacterial gene cluster for catabolism of the plant hormone indole 3-acetic acid. FEMS Microbiol Ecol 65:238–250

Lottmann J, Berg G (2001) Phenotypic and genotypic characterization of antagonistic bacteria associated with roots of transgenic and non-transgenic potato plants. Microbiol Res 156:75–82

Lottmann J, Heuer H, Smalla K, Berg G (1999) Influence of transgenic T4-lysozyme-producing potato plants on potentially beneficial plant-associated bacteria. FEMS Microbiol Ecol 29:365–377

Lottmann J, Heuer H, Vries J, Mahn A, During K, Wackernagel W, Smalla K, Berg G (2000) Establishment of introduced antagonistic bacteria in the rhizosphere of transgenic potatoes and their effect on the bacterial community. FEMS Microbiol Ecol 33:41–49

Lu Y, Conrad R (2005) In situ stable isotope probing of methanogenic *Archaea* in the rice rhizosphere. Science 309:1088–1090

Lu Y, Rosencrantz D, Liesack W, Conrad R (2006) Structure and activity of bacterial community inhabiting rice roots and the rhizosphere. Environ Microbiol 8(8):1351–1360

Lugtenberg BJJ, Dekkers LC (1999) What makes *Pseudomonas* bacteria rhizosphere competent? Environ Microbiol 1: 9–13

Mansfeld-Giese K, Larsen J, Bodker L (2002) Bacterial populations associated with mycelium of the arbuscular mycorrhizal fungus *Glomus intraradices*. FEMS Microbiol Ecol 41:133–140

Mansouri H, Petit A, Oger P, Dessaux Y (2002) Engineered rhizosphere: the trophic bias generated by opine-producing plants is independent of the opine type, the soil origin, and the plant species. Appl Environ Microbiol 68:2562–2566

MarkG L, Dow JM, Kiely PD, Higgins H, Haynes J, Baysse C, Abbas A, Foley T, Franks A, Morrissey J, O, Gara F (2005) Transcriptome profiling of bacterial responses to root exudates identifies genes involved in microbe-plant interactions. Proc Natl Acad Sci U S A 102:17454–17459

Marschner H (1991) Root-induced changes in the availability of micronutrients in the rhizosphere. In: Waise lY, Eshel A, Kakafi U (eds) Plant Roots: The Hidden Half, Marcel Dekker, New York, U S A, p. 503

Marschner H, Dell B (1994) Nutrient uptake in mycorrhizal symbiosis. Plant Soil 159:89–102

Marschner H, Römheld V (1994) Strategies of plants for acquisition of iron. Plant Soil 165:261–274

Marschner P, Crowley DE (1998) Phytosiderophores decrease iron stress and pyoverdine production of *Pseudomonas fluorescens* PF-5 (PVD-INAZ). Soil Biol Biochem 30: 1275–1280

Marschner P, Baumann K (2003) Changes in bacterial community structure induced by mycorrhizal colonisation in split-root maize. Plant Soil 251:279–289

Marschner P, Yang CH, Lieberei R, Crowley DE (2001) Soil and plant specific effects on bacterial community composition in the rhizosphere. Soil Biol Biochem 33:1437–1445

Mathimaran N, Ruh R, Vullioud P, Frossard E, Jansa J (2005) *Glomus intraradices* dominates arbuscular mycorrhizal communities in a heavy textured agricultural soil. Mycorrhiza 16:61–66

Matilla M, Espinosa-Urgel M, Rodriguez-Herva J, Ramos J, Ramos-Gonzalez M (2007) Genomic analysis reveals the major driving forces of bacterial life in the rhizosphere. Genome Biol 8:R179

Mayo K, Davies RE, Motta J (1986) Stimulation of germination of spores of *Glomus versiforme* by spore associated bacteria. Mycologia 78:426–431

Miller HJ, Henken G, van Veen JA (1989) Variation and composition of bacterial populations in the rhizospheres of maize, wheat and grass cultivars. Can J Microbiol 35: 656–660

Miller LD, Yost CK, Hynes MF, Alexandre G (2007) The major chemotaxis gene cluster of *Rhizobium leguminosarum* bv. *viciae* is essential for competitive nodulation. Mol Microbiol 63:348–362

Moenne-Loccoz Y, McHugh B, Stephens PM, McConnell FI, Glennon JD, Dowling DN, O'Gara F (1996) Rhizosphere competence of fluorescent *Pseudomonas* spB24 genetically modified to utilise additional ferric siderophores. FEMS Microbiol Ecol 19:215–225

Mogge B, Loferer C, Agerer R, Hutzler P, Hartmann A (2000) Bacterial community structure and colonization patterns of *Fagus sylvatica* L. ectomycorrhizospheres as determined by fluorescence *in situ* hybridization (FISH) and confocal laser scanning microscopy (CLSM). Mycorrhiza 9:272–278

Moore JC, McCann K, de Ruiter PC (2007) Soil rhizosphere food webs, their stability, and implications for soil processes in ecosystems. In: Cardon ZG, Whitbeck JL (eds) The rhizosphere: An ecological perspective. Academic Press Inc., London, New York, pp 101–125

Mosse B (1959) The regular germination of resting spores and some observations on the growth requirements of an

Endogone sp. causing vesicular-arbuscular mycorrhiza. Trans Br Mycol Soc 42:273–286

Mougel C, Offre P, Ranjard L, Corberand T, Gamalero E, Robin C, Lemanceau P (2006) Dynamic of the genetic structure of bacterial and fungal communities at different developmental stages of *Medicago truncatula* Gaertn. cv. Jemalong line J5. New Phytol 170:165–175

Neumann G, Martinoia E (2002) Cluster roots - an underground adaptation for survival in extreme environments. Trends Plant Sci 7:162–167

Nunan N, Daniell TJ, Singh BK, Papert A, McNicol JW, Prosser JI (2005) Links between Plant and Rhizoplane Bacterial Communities in Grassland Soils, Characterized Using Molecular Techniques. Appl Environ Microbiol 71:6784–6792

O'Connell KP, Goodman RM, Handelsman J (1996) Engineering the rhizosphere: expressing a bias. Trends Biotechnol 14:83–88

Oehl F, Sieverding E, Ineichen K, Mader P, Boller T, Wiemken A (2003) Impact of land use intensity on the species diversity of arbuscular mycorrhizal fungi in agroecosystems of central europe. Appl Environ Microbiol 69:2816–2824

Oehl F, Sieverding E, Ineichen K, RisE-A, Boller T, Wiemken A (2005) Community structure of arbuscular mycorrhizal fungi at different soil depths in extensively and intensively managed agroecosystems. New Phytol 165:273–283

Oger P, Petit A, Dessaux Y (1997) Genetically engineered plants producing opines alter their biological environment. Nat Biotechnol 15:369–372

Oger PM, Mansouri H, Nesme X, Dessaux Y (2004) Engineering root exudation of *Lotus* toward the production of two novel carbon compounds leads to the selection of distinct microbial populations in the rhizosphere. Microb Ecol 47:96–103

Oliver KL, Hamelin RC, Hintz WE (2008) Effects of transgenic hybrid aspen over-expressing P 1 olyphenol oxidase on rhizosphere diversity. Appl Environ Microbiol. doi:10.1128/AEM.02836-02807

Olsson PA, Thingstrup I, Jakobsen I, Baath F (1999) Estimation of the biomass of arbuscular mycorrhizal fungi in a linseed field. Soil Biol Biochem 31:1879–1887

Paterson E, Gebbing T, Abel C, Sim A, Telfer G (2007) Rhizodeposition shapes rhizosphere microbial community structure in organic soil. New Phytol 173:600–610

Pearson JN, Abbott LK, Jasper DA (1993) Mediation of competition between two colonizing VA mycorrhizal fungi by host plants. New Phytol 123:93–98

Pivato B, Mazurier S, Lemanceau P, Siblot S, Berta G, Mougel C, van Tuinen D (2007) *Medicago* species affect the community composition of arbuscular mycorrhizal fungi associated with roots. New Phytol 176:197–210

Prosser JI, Rangel-Castro JI, Killham K (2006) Studying plant-microbe interactions using stable isotope technologies. Curr Opin Biotechnol 17:98–102

Raaijmakers JM, Paulitz CT, Steinberg C, Alabouvette C, Moenne-Loccoz Y (2008) The rhizosphere: a playground and battlefield for soilborne pathogens and beneficial microorganisms. Plant Soil. doi:10.1007/s11104-11008-19568-11106

Radajewski S, Ineson P, Parekh NR, Murrell JC (2000) Stable-isotope probing as a tool in microbial ecology. Nature 403:646–649

Rambelli A (1973) The Rhizosphere of mycorrhizae. In: Mg L, Koslowski TT (eds) Ectomycorrhizae. Academic Press, New York, pp 299–343

Rangel-Castro JI, Killham K, Ostle N, Nicol GW, Anderson IC, Scrimgeour CM, Ineson P, Meharg A, Prosser JI (2005) Stable isotope probing analysis of the influence of liming on root exudate utilization by soil microorganisms. Environ Microbiol 7:828–838

Rasche F, Hodl V, Poll C, Kandeler E, Gerzabek MH, van Elsas JD, Sessitsch A (2006) Rhizosphere bacteria affected by transgenic potatoes with antibacterial activities compared with the effects of soil, wild-type potatoes, vegetation stage and pathogen exposure. FEMS Microbiol Ecol 56:219–235

Rasmussen TB, Bjarnsholt T, Skindersoe ME, Hentzer M, Kristoffersen P, Kote M, Nielsen J, Eberl L, Givskov M (2005) Screening for quorum-sensing inhibitors (QSI) by use of a novel genetic system, the QSI selector. J Bacteriol 187:1799–1814

Remy W, Taylor TN, Hass H, Kerp H (1994) Four hundred-million-year-old vesicular arbuscular mycorrhizae. Proc Natl Acad Sci U S A 91:11841–11843

Rengel Z (1999) Physiological mechanisms underlying differential nutrient efficiency of crop genotypes. In: Rengel Z (ed) Mineral nutrition of crops: Mechanisms and implications, The Haworth Press, New York, U S A, pp 227–265

Rengel Z, Marschner P (2005) Nutrient availability and management in the rhizosphere: exploiting genotypic differences. New Phytol 168:305–312

Rettenmaier H, Lingens F (1985) Purification and some properties of two isofunctional juglone hydroxylases from *Pseudomonas putida* J1. Biol Chem Hoppe Seyler 366 (7):637–646

Richter DD, OhN-H, Fimmen R, Jackson J (2007) The rhizosphere and soil formation. In: Cardon ZG, Whitbeck JL (eds) The rhizosphere: An ecological perspective. Elsevier Academic Press, Burlington, U S A, pp 179–200

Riedlinger J, Schrey SD, Tarkka MT, Hampp R, Kapur M, Fiedler H-P (2006) Auxofuran, a novel metabolite that stimulates the growth of fly agaric, is produced by the mycorrhiza helper bacterium *Streptomyces* strain AcH 505. Appl Environ Microbiol 72:3550–3557

Rillig MC, Lutgen ER, Ramsey PW, Klironomos JN, Gannon JE (2005) Microbiota accompagning different arbuscular mycorrhizal fungal isolates influence soil aggregation. Pedobiologia 49:251–259

Rillig MC, Mummey DL, Ramsey PW, Klironomos JN, Gannon JE (2006) Phylogeny of arbuscular mycorrhizal fungi predicts community composition of symbiosis-associated bacteria. FEMS Microbiol Ecol 57:389–395

Rodriguez-Navarro DN, Dardanelli MS, Ruiz-Sainz JE (2007) Attachment of bacteria to the roots of higher plants. FEMS Microbiol Lett 272:127–136

Roose T, Fowler AC (2004) A mathematical model for water and nutrient uptake by plant root systems. J Theor Biol 228:173–184

Rosenblueth M, Martinez-Romero E (2006) Bacterial endophytes and their interactions with hosts. Mol Plant Microbe Interact 19:827–837

Rothballer M, Schmid M, Fekete A, Hartmann A (2005) Comparative in situ analysis of *ipd*C-gfpmut3 promoter fusions of *Azospirillum brasilense* strains Sp7 and Sp245. Environ Microbio l7:1839–1846

Ryu C-M, Farag MA, Hu C-H, Reddy MS, Kloepper JW, Pare PW (2004) Bacterial volatiles induce systemic resistance in *Arabidopsis*. Plant Physiol 134:1017–1026

Savka MA, Dessaux Y, Oger P, Rossbach S (2002) Engineering bacterial competitiveness and persistence in the phytosphere. Mol Plant Microbe Interact 15:866–874

Scheublin TR, Ridgway KP, Young JPW, van der Heijden MGA (2004) Nonlegumes, legumes, and root nodules harbor different arbuscular mycorrhizal fungal communities. Appl Environ Microbiol 70:6240–6246

Schloter M, Lebuhn M, Heulin T, Hartmann A (2000) Ecology and evolution of bacterial microdiversity. FEMS Microbiol Rev 24:647–660

Schuhegger R, Ihring A, Gantner S, Bahnweg G, Knappe C, Vogg G, Hutzler P, Schmid M, van Breusegem F, Eberl L, Hartmann A, Langebartels C (2006) Induction of systemic resistance in tomato by N-acylhomoserine lactone-producing rhizosphere bacteria. Plant Cell and Environment 29: 909–918

Schulz B, Boyle C, Sieber N (2006) Microbial root endophytes. Springer VerlagBerlin, Heidelberg, New York

Schüßler A, Schwarzott D, Walker C (2001) A new fungal phylum, the *Glomeromycota*: Phylogeny and evolution. Mycol Res 105:1413–1421

Schwarzott D, Walker C, Schuler A (2001)*Glomus*, the largest genus of the arbuscular mycorrhizal fungi (*Glomales*), is non monophyletic. Mol Phylogenet Evol 21:190–197

Selesi D, Schmid M, Hartmann A (2005) Diversity of green-like and red-like ribulose−1,5-bisphosphate carboxylase/oxygenase large-subunit genes (*cbb*L) in differently managed agricultural soils. Appl Environ Microbiol 71:175–184

Selim S, Negrel J, Govaerts C, Gianinazzi S, van Tuinen D 92005) Isolation and partial characterization of antagonistic peptides produced by *Paenibacillus* spstrain B2 isolated from the *Sorghum* mycorrhizosphere. Appl Environ Microbiol 71:6501–6507

Shaw LJ, Morris P, Hooker JE (2006) Perception and modification of plant flavonoid signals by rhizosphere microorganisms. Environ Microbiol 8:1867–1880

Simpson FB, Burris RH (1984) A nitrogen pressure of 50 atmospheres does not prevent evolution of hydrogen by nitrogenase. Science 224:1095–1097

Singh BK, Nunan N, Ridgway KP, McNicol J, Young JPW, Daniell TJ, Prosser JI, Millard P (2008) Relationship between assemblages of mycorrhizal fungi and bacteria on grass roots. Environ Microbiol 10:534–541

Skene KR (2000) Pattern formation in cluster roots: Some developmental and evolutionary considerations. Ann Bot 85:901–908

Smalla K, Wieland G, Buchner A, Zock A, Parzy J, Kaiser S, Roskot N, Heuer H, Berg G (2001)Bulk and rhizosphere soil bacterial communities studied by denaturing gradient gel electrophoresis: Plant-dependent enrichment and seasonal shifts revealed. Appl Environ Microbiol 67:4742–4751

Smith SE, Read DJ (1997) Mycorrhizal Symbiosis. Academic Press, London

Somers E, Vanderleyden J, Srinivasan M (2004)Rhizosphere bacterial signalling: A love parade beneath our feet. Crit Rev Microbiol 304:205–240

Soto MJ, Sanjuan J, Olivares J (2006) Rhizobia and plant-pathogenic bacteria: common infection weapons. Microbiol 152:3167–3174

Stein S, Selesi D, Schilling R, Pattis I, Schmid M, Hartmann A (2005) Microbial activity and bacterial composition of H_2-treated soils with net CO_2 fixation. Soil Biol Biochem 37:1938–1945

Teplitski M, Robinson JB, Bauer WD (2000) Plants secrete substances that mimic bacterial N-Acyl Homoserine Lactone signal activities and affect population density-dependent behaviors in associated bacteria. Mol Plant Microbe Interact 13:637–648

Thordal-Christensen H (2003) Fresh insights into processes of nonhost resistance. Curr Opin Plant Biol 6:351–357

Tjamos EC, Rowe RC, Heale JB, Fravel DR (2000) Advances in *Verticillium* research and disease managementAPS Press. The American Phytopathological Society, Minnesota, USA, 357

Toljander JF, Lindahl BD, Paul LR, Elfstrand M, Finlay RD (2007) Influence of arbuscular mycorrhizal mycelial exudates on soil bacterial growth and community structure. FEMS Microbiol Ecol 61:295–304

Treonis AM, Ostle NJ, Stott AW, Primrose R, Grayston SJ, Ineson P (2004) Identification of groups of metabolically-active rhizosphere microorganisms by stable isotope probing of PLFAs. Soil Biol Biochem 36:533–537

Tucker SL, Talbot NJ (2001) Surface attachment and pre-penetration stage development by plant pathogenic fungi. Annu Rev Phytopathol 39:385–417

Turnau K, Ryszka P, Gianinazzi-Pearson V, van Tuinen D (2001)Identification of arbuscular mycorrhizal fungi in soils and roots of plants colonizing zinc wastes in southern Poland. Mycorrhiza 10:169–174

Uren NC (1981) Chemical reduction of an insoluble higher oxide of manganese by plant roots. J Plant Nutr Soil Sci 4:65–71

Uren NC (2007) Types, amounts and possible functions of compounds released into the rhizosphere by soil-grown plants. In: Pinto RZ, Varanini PN (eds) The Rhizosphere: Biochemistry and organic substances at the soil-plant interface. CRC Press, Boca Raton, Florida, USA, pp 1–21

van der Heijden MGA, Klironomos JN, Ursic M, Moutoglis P, Streitwolf-Engel R, Boller T, Wiemken A, Sanders IR (1998) Mycorrhizal fungal diversity determines plant biodiversity, ecosystem variability and productivity. Nature 396:69–72

van Elsas JD, Turner S, Bailey MJ (2003) Horizontal gene transfer in the phytosphere. New Phytol 157:525–537

van Tuinen D, Jacquot E, Zhao B, Gollotte A, Gianinazzi-Pearson V (1998) Characterization of root colonization profiles by a microcosm community of arbuscular mycorrhizal fungi using 25S rDNA-targeted nested PCR. Mol Ecol 7:879–887

van Veen JA, Morgan JAW, Whipps JM (2007) Methodological approaches to the study of carbon flow and the associated microbial population dynamics in the rhizospherePintoRZ, VaraniniPNThe Rhizosphere: Biochemistry and organic substances at the soil-plant interface. CRC Press, Boca Raton, Florida, USA, 371–399

Vandenkoornhuyse P, Ridgway KP, Watson IJ, Fitter AH, Young JPW (2003) Co-existing grass species have distinctive arbuscular mycorrhizal communities. Mol Ecol 12:3085–3095

von Rad U, Klein I, Dobrev PI, Kottova J, Zazimalova E, Fekete A, Hartmann A, Schmitt-Kopplin P, Durner J (2008) Response of *Arabidopsis thaliana* to N-hexanoyl-DL-homoserinelactone, a bacterial quorum sensing molecule produced in the rhizosphere. Planta. doi:10.1007/s00425-008-0811-4

von Wiren N, Marschner H, Römheld V (1995) Uptake kinetics of iron-phytosiderophores in two maize genotypes differing in iron efficiency. Physiol Plant 93:611–616

von Wiren N, Mori S, Marschner H, Römheld V (1994) Iron inefficiency in maize mutant ys1 (*Zea mays* Lcv Yellow-Stripe) is caused by a defect in uptake of iron phytosiderophores. Plant Physiol 106:71–77

Wang B, Qiu YL (2006) Phylogenetic distribution and evolution of mycorrhizas in land plants. Mycorrhiza 16:299

Whipps JM (2001) Microbial interactions and biocontrol in the rhizosphere. J Exp Bot 52:487–511

Wolfe B, Mummey D, Rillig M, Klironomos J (2007) Small-scale spatial heterogeneity of arbuscular mycorrhizal fungal abundance and community composition in a wetland plant community. Mycorrhiza 17:175–183

Yan F, Zhu Y, Muller C, Zorb C, Schubert S (2002) Adaptation of H^+-pumping and plasma membrane H^+ ATPase activity in proteoid roots of white Lupin under phosphate deficiency. Plant Physiol 129:50–63

Yao J, Allen C (2006) Chemotaxis is required for virulence and competitive fitness of the bacterial wilt pathogen *Ralstonia solanacearum.* J Bacteriol 188:3697–3708

Zabetakis I (1997) Enhancement of flavour biosynthesis from strawberry (*Fragaria ananassa*) callus cultures by *Methylobacterium* species. Plant Cell Tissue Organ Cult 50:179–183

Zeidler D, Zahringer U, Gerber I, Dubery I, Hartung T, BorsW, Hutzler P, Durner J (2004) From The Cover: Innate immunity in *Arabidopsis thaliana*: Lipopolysaccharides activate nitric oxide synthase (NOS) and induce defense genes. Proc Natl Acad Sci U S A 101:15811–15816

REVIEW ARTICLE

Rhizosphere microbiota interfers with plant-plant interactions

A. Sanon · Z. N. Andrianjaka · Y. Prin · R. Bally ·
J. Thioulouse · G. Comte · R. Duponnois

Received: 29 December 2008 / Accepted: 21 April 2009 / Published online: 9 May 2009
© Springer Science + Business Media B.V. 2009

Abstract Diversity, structure and productivity of above-ground compartment of terrestrial ecosystems have been generally considered as the main drivers of the relationships between diversity and ecosystem functioning. More recently it has been suggested that plant population dynamics may be linked with the development of the below-ground community. The biologically active soil zone where root-root and root-microbe communications occur is named "Rhizosphere" where root exudates play active roles in regulating rhizosphere interactions. Root exudation can regulate the soil microbial community, withstand herbivory, facilitate beneficial symbioses, modify the chemical and physical soil properties and inhibit the growth of competing plant species. In this review, we explore the current knowledge assessing the importance of root exudates in plant interactions, in communications between parasitic plants and their hosts and how some soil microbial components could regulate plant species coexistence and change

Responsible Editor: Phillipe Lemanceau.

A. Sanon · R. Duponnois (✉)
IRD, Laboratoire Commun de Microbiologie IRD/ISRA/UCAD,
Centre de Recherche de Bel Air,
BP 1386 Dakar, Sénégal
e-mail: Robin.Duponnois@ird.fr

A. Sanon
Département de Biologie Végétale,
Université Cheick Anta Diop,
BP 5000 Dakar, Sénégal

Z. N. Andrianjaka · R. Bally
Centre National de la Recherche Scientifiques (CNRS),
Laboratoire d'Ecologie Microbienne,
UMR 5557, Université Claude Bernard Lyon 1- 69622,
Villeurbanne Cedex, France

Z. N. Andrianjaka · G. Comte
Centre d'Etude des Substances Naturelles (CESN),
Laboratoire d'Ecologie Microbienne,
UMR 5557, Université Claude Bernard Lyon 1- Bât. Forel,
69622 Villeurbanne Cedex, France

Y. Prin
CIRAD, UMR 113 CIRAD/INRA/IRD/SUP-AGRO/UM2,
Laboratoire des Symbioses Tropicales
et Méditerranéennes (LSTM),
TA10/J, Campus International de Baillarguet,
Montpellier, France

J. Thioulouse
Laboratoire de Biométrie et Biologie Evolutive,
Université de Lyon, F-69000, Lyon ; Université Lyon 1 ;
CNRS, UMR5558,
F-69622 Villeurbanne, France

R. Duponnois
IRD, UMR 113 CIRAD/INRA/IRD/SUP-AGRO/UM2,
Laboratoire des Symbioses Tropicales
et Méditerranéennes (LSTM),
TA10/J, Campus International de Baillarguet,
Montpellier, France

relationships between plants. This review will be focussed on several well documented biological processes regulating plant-plant communications such as exotic plant species invasions, negative root-root communication (allelopathy) and parasitic plant / host plant interactions and how some soil microbial components can interfere with signal traffic between roots. The reported data show that the overall effect of one plant to another results from multiple interacting mechanisms where soil microbiota can be considered as a key component.

Keywords Allelopathy · Rhizosphere · Plant invasions · Plant-soil feedbacks · Parasitic plant

Introduction

Plant biodiversity and species composition are regulated and maintained in terrestrial ecosystems by different biological processes such as competition between neighbouring plants (Aarsen 1990; Grace and Tilman 1990), spatial and temporal resource partitioning (Ricklefs 1977; Tilman 1982), disturbance creating new patches for plant colonization (Grubb 1977) and interactions with other organisms in the ecosystems (Bever et al. 1997). Diversity, structure and productivity of above-ground compartment of terrestrial ecosystems have been generally considered as the main drivers of the relationships between diversity and ecosystem functioning. There is also extensive knowledge on how abiotic and biotic soil factors interact with vegetation (Wardle 2002). For instance it is well known that at local scales the composition and activity of microbial communities are mainly subjected to plant factors such as species composition and formation age (Priha et al. 1999; Grayston et al. 2001) as well as environmental factors such as soil type, nutrient status, pH and moisture (Stotzky 1997). Recent studies have reported that local interactions between plants and microbial communities strongly influence both plant and soil community composition and ecosystem processes (Bever 2003). The biologically active soil zone where root-root and root-microbe communications occur is named "Rhizosphere" (Hiltner 1904). The rhizosphere is a densely populated area where root exudates play active roles in regulating rhizosphere interactions. Root exudation can regulate the soil microbial community, withstand herbivory, facilitate beneficial symbioses, modify the chemical and physical soil properties and inhibit the growth of competing plant species (Bais et al. 2004). In terrestrial ecosystems, most of plant species are commonly associated with arbuscular mycorrhizal (AM) fungi that are considered as a key component of the microbial populations influencing plant growth and uptake of nutrients (Johansson et al. 2004). AM symbiosis generally increases root exudation (Graham et al. 1981), modifies carbohydrate metabolism of the host plant (Shachar-Hill et al. 1995) and influences rhizosphere microbial communities (Marschner and Timonen 2005). In addition, mycorrhizal fungi themselves can exude substances that have a selective effect on soil microbiota (Andrade et al. 1998; Marschner and Timonen 2005; Offre et al. 2007). Root function and microbial equilibrium changes in the rhizosphere following AM symbiosis establishment, lead to a new microbial compartment influenced by both the roots and the mycorrhizal fungus that is commonly named "mycorrhizosphere" (Linderman 1988). It also includes the more specific term "hyphosphere" which only referres to the zone surrounding individual fungal hyphae (Johansson et al. 2004). AM fungi and mycorrhizosphere microbial communities significantly act on soil bio-functioning and plant coexistence (van der Heijden et al. 1998; Kisa et al. 2007) and reciprocally plant genotypes affect the structure of the AM fungal community (Pivato et al. 2007).

Understanding the biological factors that govern the abundance and diversity of plant species remains one of the major goals in plant ecology. For instance, the opportunities of changes in ecosystem processes induced by invader plants might also be a feature of both their invasibility and the susceptibility of the recipient community to invasion as such mechanims would have important ramifications for the management of invasions and restoration of native communities. Invaded communities often differ from native communities in organismal composition and may have altered ecosystem functions compared with native communities, including the rate and dynamics of biogeochemical processes (Vitousek and Walker 1989; Belnap and Phillips 2001; Evans et al. 2001; Ehrenfeld 2003; Wolfe and Klironomos 2005) and the suitability of habitat for over organisms (Roberts and Anderson 2001; Duda et al. 2003; Levine et al. 2003; Stinson et al. 2006; Chen et al. 2007).

In this review, we explore the current knowledge assessing the importance of root exudates in plant interactions, in communications between parasitic plants and their hosts and how some soil microbial components known to act on plant root exudation (i.e. Arbuscular mycorrhizal fungi, AMF) could regulate plant species coexistence and change relationships between plants. This review will be focussed on several well documented biological processes regulating plant-plant communications such as exotic plant species invasions, negative root-root communication (allelopathy) and parasitic plant / host plant interactions and on the biological processes from which some soil microbial components can interfere with signal traffic between roots.

Invasive exotic plant species and soil microbiota

Exotic-species invasions are among the most important global-scale problems facing natural ecosystems. Recent reviews of the extent of the homogenization of the world biota have shown that it is not only islands and disturbed sites that are affected but mainland areas and minimally disturbed ecosystems are also often invaded, even dominated, by newly established species originating from distant places. A general definition has been proposed by Shine et al. (2000): *"an invasive species is considered as an alien species that becomes established in natural or semi-natural ecosystems or habitat and is "an agent of change and threatens native biological diversity"*. It is now well recognized that plant exotic invasions induce habitat destruction and the endangerment and extinction of native species (Vitousek et al. 1997; Wilcove et al. 1998; Simberloff 2003) and thus, drastically threatened the global biological diversity (Pimentel et al. 2000; Cabin et al. 2002; CBD 2006; Meiners 2007). Likewise, Vitousek et al. (1997) noted that for managers of parks and reserves, exotic species are *"ongoing threat to the persistence of native assemblages because they can consume native species, infect them with diseases to which they have no resistance, outcompete them, or alter ecosystem functions, making it difficult and expensive to return the ecosystem to its prior, often more desirable condition"*. Moreover, several well documented studies have shown that the species composition of communities can have far-reaching effects on ecosystem processes: changes in overall species richness, in the type of species present ('functional groups'), or in the presence of a 'keystone' species change food-web architecture, leading to changes in standing stocks and flows of energy and nutrients. It has been also suggested that the species composition disturbances following exotic plant invasions should alter ecosystem processes, particularly their functioning and stability (Vitousek and Walker 1989; D'Antonio and Mahall 1991; D'Antonio and Vitousek 1992; Couto and Betters 1995; Hutchinson and Vankat 1997; Hamilton et al. 1999; Belnap and Phillips 2001; Hierro and Callaway 2003; Chen et al. 2007).

Invasive species also present an economic problem, costing the United States alone as much as U.S. $137 million annually in loss ecosystem services control measures, and public health (Wilcove et al. 1998; Pimentel et al. 2000). Indeed, managers of many reserves estimated they spend a significant amount of their annual operating budget on control of non-indigenous species. For example, at Hawaii Volcanoes National Park, 80% of their annual budget has been spent in controlling exotic species activities (D'Antonio and Meyerson 2002).

Exotic plants may become aggressive invaders outside their home ranges for a number of reasons, including release from native, specialized antagonist (Mitchell and Power 2003), higher relative performance in a new site (Thébaud and Simberloff 2001), direct chemical (allelopathic) interference with native plant performance (Callaway and Ridenour 2004), and variability in the responses and resistance of native systems to invasion (Hobbs and Huenneke 1992; Levine and D'Antonio 1999).

Although soil organisms play important roles in regulating ecosystem-level processes (Wardle et al. 2004; Wolfe and Klironomos 2005) and contain most of the terrestrial ecosystems biodiversity (Torsvik et al. 1990; Vandenkoornhuyse et al. 2002), the effects of plant invasions have been mainly studied on aboveground flora and fauna (Levine et al. 2003). However, the composition and functioning of soil microbiota are closely linked with aboveground composition (Wardle et al. 2004) and exotic plant species can directly or indirectly disrupt these links after their invasion (Duda et al. 2003 ; Stinson et al. 2006 ; Kisa et al. 2007).

The large ecological and economic impacts of invasive plant species on terrestrial ecosystems and agrosytems has lead to a great interest in order to elucidate the biological mechanisms that regulate the interactions between exotic plant species and soil microbiota (Pimentel et al. 2000). In this manuscript, we review some of the recent scientific advances on exotic plant species *vs* soil microbiota interactions with an emphasis on mycorrhizal fungi. These microsymbiots, widely widespreaded in soil, form a key component of sustainable soil-plant interactions (Bethlenfalvay 1992; van der Heijden et al. 1998; Johansson et al. 2004) and indeed, might play a crucial role in plant invasion processes.

Effects of exotic plant invasions on soil microbiota

The soil immediately surrounding plant roots constitutes a particular physical, biochemical, and ecological environment that has been named "rhizosphere". The rhizosphere is to a large extent controlled by the root system itself through chemicals exuded/secreted into the surrounding soil. Through the exudation of a wide variety of compounds, roots may regulate the functionalities and the structure of soil microbial communities in their immediate vicinity, encourage beneficial microbial symbioses, change the chemical and physical properties of the soil, and inhibit the development of competing plant species (Nardi et al. 2000; Bais et al. 2002).

Native plant influence soil communities but invasive plant-mediated modifications may be more pronounced, or may introduce novel biological mechanisms in the native community environment. In addition, more than one mechanism may involve in exotic plants invasion processes exacerbating their effects on soil communities (Wolfe and Klironomos 2005).

Main mechanisms involved in soil community alteration

Plants supply resources for soil communities by providing organic matter through leaf-litter inputs, through the release of root exudates, or through other ways of deposition of organic compounds into the soil environment (Grayston et al. 1996). Plants develop diverse ways of supplying these resources to the soil, and as a result, specific soil communities form under different plant species (Bever et al. 1996; Westover et al. 1997; Wolfe and Klironomos 2005; Pivato et al. 2007) and under plant communities that differ in composition and abundance (Zak et al. 2003; Johnson et al. 2004). Hence it is well established that distinct microbial communities, in their structure as well as their function, might develop under different plants species (Roberts and Anderson 2001; Duda et al. 2003; Kourtev et al. 2003).

As an exotic plant species invades a community, it can alter links between native aboveground communities and belowground communities, including the timing, quality, quantity, and spatial structure of plant-derived soil inputs (Wolfe and Klironomos 2005). Kourtev et al. (2003) clearly showed that exotic invasive plants might induce drastic modifications in soil communities. Their results indicated that the structure and functional diversity of soil microbial communities (established by PhosphoLipid Fatty Acids (PLFA) profiles and substrate-induced respiration patterns) are strongly affected by the invasion processes. Moreover, these shifts are accompanied by alterations in soil chemical properties (soil pH and nitrogen content, nitrogen mineralization processes) as microbial communities are the drivers of main biogechemical cycles. In addition, Ehrenfeld (2003) suggested that plant invasions-mediated shifts in soil salinity, moisture, pH, carbon and nitrogen content, are also susceptible to significantly modify belowground microbial communities.

Furthermore, Kourtev et al. (2002, 2003) investigated the effects of two exotic understory species, Japanese barberry (*Berberis thunbergii*) and Japanese stilt grass (*Microstegium vimineum*), on soil biota in northeastern hardwood forest of North America. In field conditions, the structure of the microbial community, determined by PLFA profiles, was different under these two invaders compared with the soil under native plant species. In barberry soils, there was an overall decrease in fungal abundance, indicating conversion to a community dominated by bacteria. In stilt grass soils, one of the most pronounced structural changes was an increase in the abundance of arbuscular mycorrhizal fungi (AMF). Using molecular biological tools (i.e. Terminal Restriction Fragment Length Polymorphism T-RFLP), some authors indicated a significant reduction of AMF diversity in soil samples collected in *Centaurea maculosa* Lam (spotted knapweed)-dominated areas compared to uninvaded ones. In addition, extraradical hyphal lengths exhibited a significant reduction in *C. maculosa*—versus native grass-dominated sites

(Mummey and Rillig 2006). Belnap and Phillips (2001) also found that most aspects of the vegetative and soil food-web communities changed after the introduction of *Bromus tectorum*, an invasive annual grass in western United States: a decrease in the overall soil biological diversity, fewer fungi and unvertebrate abundance, and higher number of active bacteria, resulted from carbon and nitrogen availability-mediated by *B. tectorum* organic compound inputs.

Plants release secondary compounds into the soil environment from their roots as exudates, and if the compounds released by an exotic plant are newly represented in soil environment, they may alter the structure and function of soil community (Wolfe and Klironomos 2005). Allelochemicals resealed from plant roots have been widely used as an explanation of the success of exotic plants in the context of plant-plant interactions (Hierro and Callaway 2003), but evidence of allelochemicals altering the interactions between native plants and soil communities has only recently been established (Bais et al. 2002 ; Wolfe and Klironomos, 2005). Bais et al. (2002) investigated the allelopathic capacities of the noxious weed spotted knapweed (*Centaura maculosa*) and they showed that this plant species exuded (±)-catechin in their rhizosphere; (−)-catechin enantiomer was phytotoxic whilst (+)-catechin had antibacterial activity against root-infesting pathogens, which (−)-catechin did not show. It suggests that the exudation of racemic catechin had a biological significance in giving different properties that are beneficial for plant growth and survival. Diffuse knapweed (*Centaurea diffusa*) is an Eurasian knapweed species that has invaded many natural ecosystems in western North America. This plant species could release the chemical 8-hydroxyquinoline from its roots. This chemical compound has been demonstrated to be an antimicrobial agent (Vivanco et al. 2004) and diffuse knapweed could cause shifts in the composition of the soil microbial community probably through the release of these allelochemicals (Callaway et al. 2004). Garlic mustard (*Alliaria petiolata*), another exotic species in North America and native from Europe, is a member of the Brassicaceae, a family of plants in which many species produce glucosinolates (Wolfe and Klironomos 2005). These compounds are deposited into the soil, through root exudation or litter production, and they may cause changes in soil microbial communities (Vaughn and Berhow 1999). The dominance of garlic mustard in North American forests has been shown to cause significant alterations in AMF communities (Roberts and Anderson 2001; Mummey and Rillig 2006; Stinson et al. 2006). Stinson et al. (2006) hightlighted that the antifungal activity of this invasive plant suppressed native plant growth. This exotic plant species acted by disrupting mutuaslistic associations between native canopy tree seedlings and belowground AMF. Their results elucidated an indirect mechanism by which invasive plants can impact native flora, and explain how this plant successfully invades relatively undisturbed forest habitat.

Other traits of exotic plants, such as novel nutrient acquisition strategies, could also have implications on the structure and function of soil communities (Wolfe and Klironomos 2005). A widely recognized example of an exotic plant altering the attributes of an ecosystem is the invasion of firetree (*Myrica faya*) in Hawaii. In this area, this plant species and its nitrogen-fixing root symbionts (*Frankia* spp.) have invaded nitrogen-limited communities, altering nitrogen cycling and native plant community composition (Vitousek and Walker 1989).

Some exotic plants cause increases in litter production that can lead to increases in fire intensity and frequency (D'Antonio and Vitousek 1992) and changes in fire regime could indirectly alter soil communities (Boerner and Brinkman 2003). Furthermore, as invasive plants affected soil physical properties and erosion processes, they could alter soil microbial communities through soil microorganism habitat perturbation (Rillig et al. 2002).

From these different studies, it remains apparent that belowground effects of exotic plant invasions can be highly variable. Studies have documented negative, neutral, or positive effects on soil composition and functionalities, depending on the exotic plant species, the community or ecosystem invaded, the methods used to assess changes in structure or functions, and the temporal and spatial scales considered (Wolfe and Klironomos 2005). In addition to variation in the effects of exotic invasion on soil communities across plant species and systems, the authors indicate that it is also interesting to note that different taxonomic groups within a soil community may not respond similarly to the presence of an invasive exotic. These complexities, sole or together, may thus limit researchers's ability to predict the effects of exotic plant species invasions on soil microbial communities.

Soil microbial communities in exotic plant invasion processes

Role of specific components of the soil community on exotic plant

To discuss exotic plant invasions in the context of aboveground vs belowground relationships, it is crucial to consider not only how exotic plants could affect soil microbiota but, conversely, how the structure and function of soil communities might play a role in exotic plant invasions (Wolfe and Klironomos 2005). Indeed, soil microbiota and their feedback effect on plant growth and survival can strongly influence the relative abundance of plant species within a community.

A main mechanism by which soil biota influence the invasion of plants into native plants communities is through direct effects (either positive or negative) of specific soil organisms on plant growth (Wolfe and Klironomos 2005). The view that biotic resistance determines invasion success or failure has been introduced by Chapman (1931) with the concept of "environmental resistance". It describes the forces that oppose the establishment of species in a new location. This concept is principally based on biological factors (complexe of native predators, pathogens, parasites, competitors, mutualists, etc) and it could be resumed as a "biotic resistance" (Simberloff 1974; Simberloff and Von Holle 1999). Several well-illustrative examples of how soil organisms, particulary mycorrhizal fungi, can play a major role in the establishment and dominance of an invading plant have already been described. The first situation is the facilitation of the invasion of pine (*Pinus* spp.) by ectomycorrhizal fungi in parts of the Southern Hemisphere. Most members of the genus *Pinus* symbiotically grow with ectomycorrhizal fungi. Nevertheless there were no or few ectomycorrhizal fungal symbionts of pine native to many regions of the Southern Hemisphere. Then, with the introduction of suitable fungal symbionts with introduced trees, exotic pine species have been able to invade many plant communities in the regions (Richardson et al. 1994). Similarly, Fumanal et al. (2006) also proposed that, with regards to the invasive status of the common ragweed *Ambrosia artemisiifolia* L., increases in growth and development resulting from AMF colonization might be the main factor facilitating the spread of this plant species. Moreover, others studies (Marler et al. 1999; Callaway et al. 2001, 2003; Zabinski et al. 2002; Carey et al. 2004) indicated that the presence of AMF is a major factor to facilitate the invasiveness of the spotted knapweed, *Cautaurea maculosa* Lam. The specificity of AMF-facilitated *C. maculosa* competitiveness could be due to a number of potentially interacting factors, including alteration of AMF functionalities resulting from differential host responses to AMF species and/or alteration of AMF community composition comprising mycelial networks (Mummey and Rillig 2006).

On the other hand, Stinson et al. (2006) found that the mechanism by which garlic mustard, *Alliaria petiolata* (Brassicaceae), invaded the mesic temperate forests in North America result from a disruption in belowground mutualisms, notably mutualistic associations between native plants and AM fungi. As this plant species is non-mycorrhizal, thus the presence of AM fungi propagules in the soil should promote plant coexistence by decreasing the competitive abilities of this dominant non mycotrophic plant (Zobel and Moora 1995; Moora and Zobel 1996), hinding *A. petiolata* invasion. The authors showed that garlic mustard inhibited AM formation in native tree species through phytochemical inhibition, by reducing germination rates of native AM spores (Stinson et al. 2006) and this remains the major strategy leading to *A. petiolata* invasion.

These two kinds of results (positive or negative relationships between the presence of AMF and the invasive plant species) may not necessarily be in conflict and the outcome may be highly dependent to the invasive plant mycorrhizal dependency in the local dominance hierarchy (Urcelay and Diaz 2003). In this context, some authors argued that if an otherwise less competitive plant species is infected by more AMF than is a highly competitive plant species, then AMF should promote plant coexistence by increasing the ability of less competitive species to access nutrients (Zobel and Moora 1995; Moora and Zobel 1996). Alternatively, if a highly competitive plant species is also more infected by AMF, then AMF would simply reinforce competitive dominance by that species (West 1996).

Feedbacks between exotic plants and soil biota

Although it is useful to understand the effects of specific soil biota to predict the relative importance of different soil organisms in the invasion process, knowledge of the net effect of the soil community is more useful to understand the role of soil microbiota in the invasion process in field conditions (Wolfe and Klironomos 2005). Plants have different abilities to influence their abundance by changing the structure of their associated soil microbiota that is an important regulator of plant community structure (Klironomos 2002).

Soil microbes have profound negative or beneficial effects on plant growth and survival through pathogenic effects, root-fungus mutualisms and by driving the nutrient cycles on which plants depend (Newsham et al. 1994; Packer and Clay 2000; Mitchell and Power 2003). These effects, and the reciprocal effects of plants on soil microbes, contribute to two contrasting dynamic feedback interactions between plants and root soil microbiota (Bever et al. 1997). As a plant grows in a local soil community, it may modify the composition of the soil organisms by altering abiotic or biotic components of the soil environment. These modifications can lead to changes into the effects of soil microbiota on plant growth (either positive, negative, or neutral), leading to feedbacks between plants and soil biota (Bever et al. 1997; Bever 1994, 2003; Wolfe and Klironomos 2005). Positive feedbacks occur when a plant species promote microbes near their roots that have beneficial effects on the same plant species, such as mycorrhizal fungi and nitrogen fixers. Positive feedbacks are thought to lead to a loss of local community diversity (Bever et al. 1997; Bever 2003). Negative feedbacks occur when plant species accumulate pathogenic microbes in their rhizospheres, creating conditions that are increasingly hostile to the plants that favor the pathogens (Klironomos 2002; Bever 1994). Negative feedbacks are thought to enhance community diversity by increasing species turnover rates (Bever et al. 1997; Bever 2003).

Within a plant community, the feedback between plants and the soil microbiota can explain the relative abundance of plant species, with the most abundant species having positive or neutral feedbacks with soil microbiota and the least abundant species having negative feedbacks (Klironomos 2002; Wolfe and Klironomos 2005). Wolfe and Klironomos (2005) cultured five of North America's most notorious exotic invaders in soil that had been cultivated with each of the five species. A positive growth effect was observed compared with plant growth in soil that had been cultured by a different species. They argued that changes in the soil microbiota resulting from the presence of these plants, would not result in negative growth effects on the same plant. But, when five rare native species were treated in the same way, a negative growth effect was observed when growing in their own soil compared with the growth of these plants in the soil of others species, suggesting that the plants accumulated pathogens in their local soil community. Hence, they concluded that exotic plants, and in some cases widespread native plants, could be abundant within native communities because they do not experience the same negative feedback with soil biota as do rare native species (Klironomos 2002).

These initial feedback studies suggested that exotic plant may escape the negative effects of soil pathogens in their novel ranges (Wolfe and Klironomos 2005), supporting the enemy-release hypothesis that has been demonstrated for some exotic plants with aboveground antagonists, such as herbivores (Reader 1998; Maron and Vila 2001) and fungal and viral pathogens (van der Putten et al. 1993; Mitchell and Power 2003). Several recent studies have followed up these previous works by comparing the soil feedbacks of exotic plants in their native and exotic ranges. Spotted knapweed (*Centaurea maculosa*), a major exotic plant that dominates many grasslands of western North America, is native from Europe. Callaway et al. (2004) indicated that *C. maculosa* was able to modify soil microbiota in invaded soils to its advantage, thus favouring its invasion process. In contrast, *C. maculosa* is inhibited by a negative feedback in its native soil, propably due to the accumulation of pathogens and potentially also due to adaptation of inhibitory microbial populations to antimicrobial compounds produced by the spotted knapweed. Otherwise, if plants and pathogens co-evolved locally, it would be expected that feedback between a plant species and soil microbes from its native range will be negative, and that exotic invaders may escape more pathogens than they acquire in their new habitat (Mitchell and Power 2003). In addition and in contrast to the host-specific tendency of pathogenic microbes, many mycorhizal fungi tend to infect a broad range of hosts (Eom and Hartnett 2000). Hence it was possible for mycotrophic invaders to use the native AMF in their introduction area. Therefore, the feedback of soil microbiota from the invaded range of an exotic weed to the weed itself is likely to be neutral or positive because of the potential for the invader to accumulate mutualistic fungi in the absence of host-specific soil pathogens (Callaway et al. 2004). These results indicated that soil organisms and their feedback effects on plants could strongly drive plant species relative abundance within a community.

Positive feeback responses were also found between black cherry (*Prunus serotina*) and soil microbiota in its introduced range (Europe) and negative responses to soil microbiota in its home range (North America) (Reinhart et al. 2003).

Overall, these combined findings support the hypothesis that feedback between plants and soil communities may strongly determine the ability of a plant to establish, invade and persist in a local habitat (Klironomos 2002). Thus, feedback could be an important mechanism for coexistence and/or invasion, and the regulation of plant biodiversity in communities. Beyond this evident effect of feedback mechanisms in invasion processes, conflicting studies on the release of exotic plants from negative soil feedbacks in invasive ranges make it difficult to generalize how important this mechanism may be in explaining the success of invasive plant (Wolfe and Klironomos 2005). These authors cited the example of European beachgrass (*Ammophila arenaria*) which was introduced into California in the 1800s. In its native range (Europe), this plant is an early-successional dune species that is replaced by others species as it accumulates soil organisms that negatively affect its growth (van der Putten et al. 1993). In California, soil microbiota was found to have similar negative effects on this species (Beckstead and Parker 2003), suggesting that European beachgrass did not escape the negative effects of soil biota in its invasive range (Wolfe and Klironomos 2005). However, escape from negative feedback from soil microbiota was observed in populations in South Africa, another region where European beachgrass has invaded plant communities (Knevel et al. 2004). Theses discrepancies suggested that soil communities will not have the same magnitude or direction (positive versus negative) effect on the invasion of all exotic species in all novel ranges (Levine et al. 2004; Wolfe and Klironomos 2005).

Soil microbiota and ecological restoration following plant invasion

A major goal of restoration practitioners is to return a habitat to more desirable conditions involving a particular species composition, community structure, and/or set of ecosystem functions (Noss 1990). They are several reasons why both 'natural' and direct human disturbances are known to promote invasive exotic species in plant communities (Huenneke et al. 1990; Hobbs and Huenneke 1992; Hughes and Vitousek 1993; Maron and Connors 1996; Lozon and MacIsaac 1997; Baskin and Baskin 1998; Tardiff and Stanford 1998; D'Antonio et al. 1999), and an understanding of these processes may provide insight into management options.

After the establishment of an invasive species that begins to dominate in an area, land managers try to stop the spread of the invasive plant, to remove plants that have estbalished, and to restore attributes of the pre-invaded community. Traditionally, these restoration approaches have been 'aboveground-focused', only considering the components of the community that can be easily seen and monitored over time as the restoration progresses (Wolfe and Klironomos 2005). A variety of processes have been used to remove exotic species from reserves or restoration sites. These most commonly include hand or mechanical removal, herbicides, fire, planned disturbance, biological control, or some combinations of the above techniques (Masters and Nissen 1998; D'Antonio and Meyerson 2002) but soil microbiota has generally been ignored (Wolfe and Klironomos 2005).

More recently, the opportunity of using microbial control has been investigated. In this context, selective manipulation of soil fertility through soil microorganisms may be used for control of some undesired species (D'Antonio and Meyerson 2002). Although application to natural areas may be difficult, this approach may potentially be useful in a restoration project where the particular nutrient requirements of an invader are known. Where high N-demanding exotic species are present, several investigators have suggested the addition of sawdust or a carbon 'cocktail' to decrease soil-available N (Wilson and Gerry 1995; Reever-Morghan and Seastedt 1999; D'Antonio & Meyerson 2002). The underlying reasoning behind this cultural approach is that labile C will stimulate microbial population growth and increased microbial populations will then immobilize soil N. The resulting lower soil N will differentially affect the faster growing more N-demanding plant species, decreasing their competitive advantage over native species for at least a brief window of time (D'Antonio and Meyerson 2002). Likewise, it has been demonstrated that the spread of the annual meadow grass, *Poa annua* L., was controlled by the bacterial species *Xanthomonas campestris* pv. *poae* (Zhou and Neal 1995; Imaizumi et al. 1997, 1998). Futhermore, others authors successfully used AMF to reduce the growth and the invasion of golf putting greens by annual meadow grass (Gange et al. 1999) ; this weed grass is generally considered to be undesirable in putting greens because its shallow root system makes it particularly susceptible to abiotic stress, especially water availability (Adams and Gibbs 1994) and this is important because water use is an expensive and often controversial aspect of golf course management (Kneebone et al. 1992). It has been previously reported that the abundance of *P. annua* in one golf course was negatively related to the amount of AMF in the soil (Gange 1994). In others respects, when the abundance of AMF was very low then *P. annua* was common, and *vice versa*. Accordingly, in a manipulative experiment, Gange et al. (1999) added mycorrhizal inoculum to a golf green soil and observed that mycorrhizal inoculation could eventually decrease the abundance of *P. annua* contrary to the abundance of *Agrostis stolonifera*, one of the most widely sown and desirable species in golf greens.

Parasitic plant—host plant interactions

Intra and interspecific plant chemical signaling is of great importance in biology and more particularly in non-beneficial underground interactions (Hirsch et al. 2003). One example of plant-plant underground communication is the recognition by the parasitic plants (*Orobranche* spp. and *Striga* spp.) of chemical signals exuded by the roots of susceptible plant species. Broomrapes (*Orobranche* spp.) and witchweeds (*Striga* spp.) can heavily infest crops with a large negative impact on agriculture in many countries. Since the life cycle of *Striga* and *Orobranche* spp. are essentially similar, the knowledge of rhizosphere mediated parasitic host plant interaction will be reviewed on the *Striga* genus.

Striga spp. belong to the hemiparasites whereas *Orobranche* spp. are holoparasites (both *Scrophulariacea*) which largely depend on a host plant to obtain their nutrients and water (Parker and Riches 1993). More than thirty species has been described

and eighty percent have been found in Africa. Nine species are indigenous to Africa and three to Australia (Musselman 1987). The majority of *Striga* species does not impact agriculture production, but those which parasitize crops are extremely harmful. The main species that have an important economical impact is *Striga hermonthica* (Del.) Benth, *S. asiatica* (L.) Kuntze and *S. gesneroïdes* Benth (Berner et al. 1995). *Striga hermonthica*, causes extensive damage in subsaharan dry areas, particularly West Africa (Olivier 1995). In sub-Saharan area, this obligate root hemiparasite can cause important yield losses in cereals such as maize (*Zea mays* L.), sorghum (*Sorghum bicolor* L.) and millet (*Pennisetum typhoides* L.) (Parker 1991). *Striga* is considered as the largest biological constraint on food production in sub-Saharan Africa and crop losses resulting from *Striga* infestation are estimated to more than 7 billions US$ (Lenné 2000).

Botanical description and physiological aspects

Striga plant can reach a height of 80 cm and is characterized by green, rigid and rough stems and by bright irregular flowers. The seeds are light and tiny (250 μm x 150 μm), and 40000 to 100000 seeds per plant are produced according to the species. They are covered with a hard and brownish integument with typical ornamentations. These characteristics allow an easy dissemination and a seed protection in the soil for many years. Seeds can keep their germination potential for more than 15 years in the soil until suitable environmental conditions are reached (Sallé and Raynal-Roques 1989). *Striga* is well adapted to climatic conditions encountered in semi-arid tropical regions (Salle and Aber 1986). The seeds require a period of after-ripening (4 to 6 months) and need a period of pre-treatment or conditioning (2–4 weeks) before they have the potential to germinate. The seeds need to be water imbibed and stocked at 30°C in the dark. Then they become responsive to the stimulant signals secreted by host plant roots (Worsham 1987). These stimulants induce seed germination within 3 to 6 mm around the host plant root. *Striga* radicles grow towards the host plant roots by chemotropism and some papilla develop at their tip facilitating their attachment (Sallé and Raynal-Roques 1989). At that time, a factor inducing the development of the haustorium is excreted by the host plant roots to allow their attachment. Once xylem connection established between both plant partners, the parasitic plant develop an underground vegetative form and draws its nutritive elements from its host (Fig. 1) (Joel et al. 1995).

The parasitic plant emerges after 4 or 5 weeks. The strongest morphological and physiological perturbations of the host plant are expressed at this stage. The *Striga* plant develops chlorophyllian structures but remains dependent of the host root for its mineral and water nutrition. The parasitic plant reaches to its maturity, flowers and produces mature seeds (Fig. 1).

Chemical communications regulating *Striga* development

In many of the steps of the life cycle of *Striga* development (germination, attachment, penetration and nutritional demand by the parasite), chemical communication occur between the host plant and the parasite. It begins by the exudation of secondary metabolites from the roots of the host that induce the germination of the seeds of the parasite. Then other host-derived secondary metabolites regulate the plant-plant interactions. It has been suggested that the *Striga* radicule is oriented towards the host root by the concentration gradient of the germination stimulant or other host plant secondary metabolites (Dube and Olivier 2001). Hausterium formation that allows the attachment of *Striga* radicule to the host plant root and the host-parasite xylem connection is initiated by host-derived metabolites, more particularly phenolic compounds (Hirsch et al. 2003). *Striga* seeds need phenolic compound (2,6 -dimethyl-p-benzoquinone) as signal for haustorium induction (Kim et al. 1998). Finally, hydrolytic enzymes produced by the parasite facilitate the penetration of intrusive cells into the host root xylem (Losner-Goshen et al. 1998) and the transition from vegetative to flowering stage can be induced by phenolic compounds (Albrecht et al. 1999).

Germination stimulants

As described above, parasitic weed seeds require a chemical signal to initiate germination. Hence the germination stimulants play a crucial in the life cycle of parasitic plants and are generally considered as an important target to ensure the control of parasitic plants.

It has been demonstrated that germination stimulants were mainly synthesized and exuded from the apex of the root. The first naturally occurring germination stimulant identified was "Strigol" (Fig. 2), an unstable tetracyclic sesquiterpene isolated from non host root exudates (*Gossypium hirsutum* L.) (Cook et al. 1966).

Fig. 1 Biological cycle of *Striga* spp

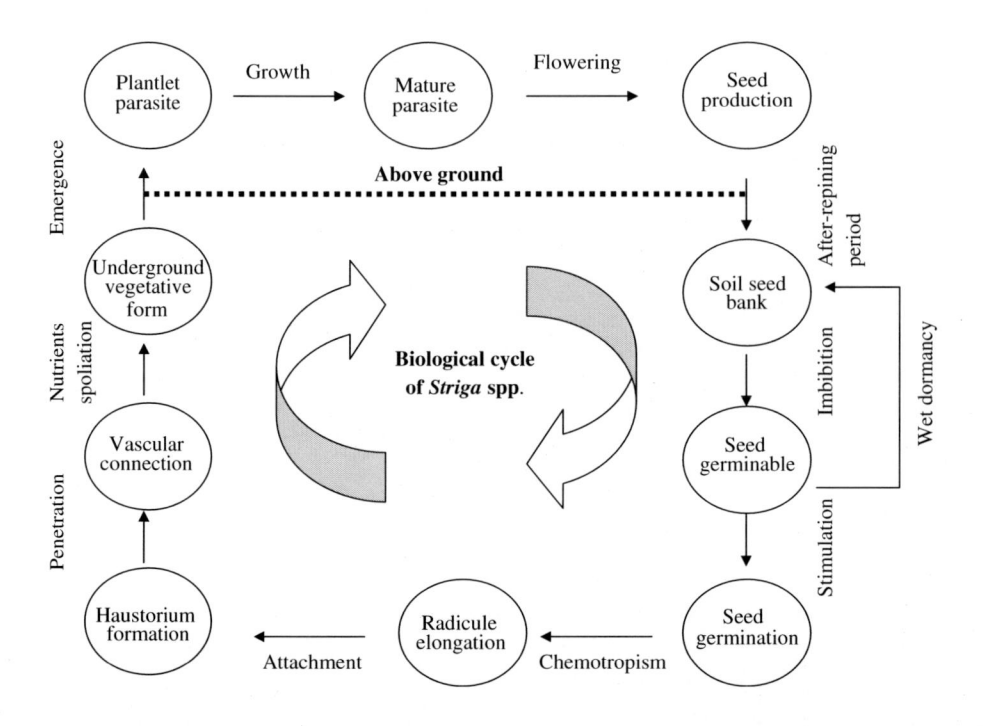

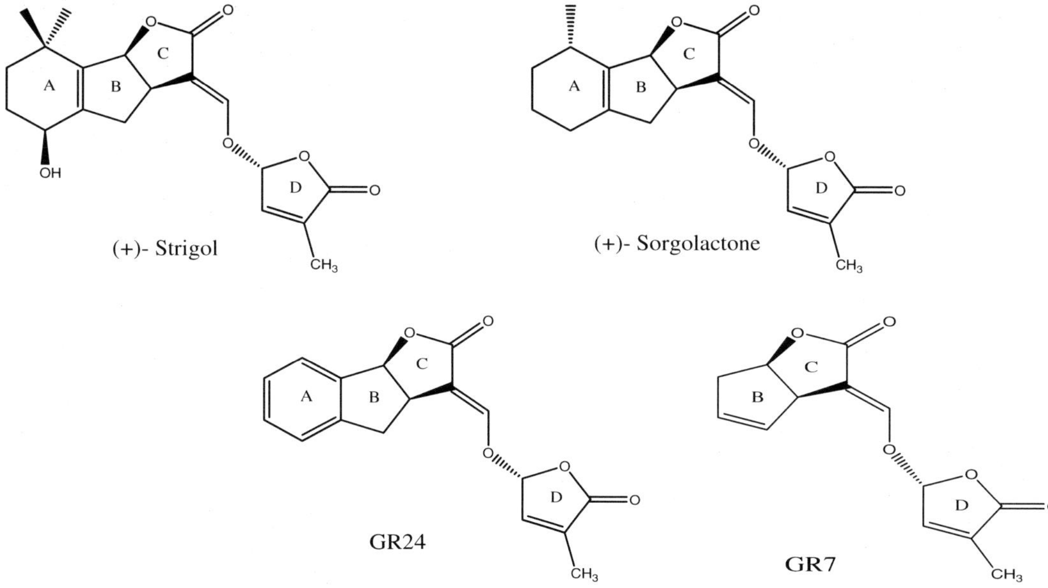

Fig. 2 Chemical structure of natural [(+)- Strigol and (+)- Sorgolactone] and synthetic (GR24 and GR7) germination stimulants of *Striga* seeds

In 1992 "Sorgolactone" (Fig. 2), an analogous of "Strigol", was isolated from sorghum (*Sorghum bicolor* L. Moench) roots (Hauck et al. 1992).

During the ten last years, several synthetic compounds analogous of Strigol were artificially synthesized and the most potent and active, currently used, compounds are GR7 (Gerry Roseberry 7, Fig. 2) and GR24 (Wigchert et al. 1999). These molecules are unstable, widely distributed in the plant kingdom and active at very small concentrations. For example, seedlings of cotton produce about 14 pg of strigol per plant per day (Sato et al. 2005).

Factors affecting *Striga* development

Low nitrogen availability in the soil promotes *S. hermonthica* infection (Farina et al. 1985). Generally, nitrogen fertilizer application reduces the crop damage caused by these parasitic plants (Khan et al. 2002). It has been established that reduced forms of nitrogen, exogenously applied, affected the biosynthesis and/or leakage of germination stimulants from the host root (Raju et al. 1990) and had a negative influence on the attachment and development of *Striga* seedlings (Cechin and Press 1993). With *Orobanche spp.*, nitrogen in ammonium form inhibits the elongation of seedling radicles by half (Westwood and Foy 1999).

In addition, some amino acids cause severe physiological disorders in germinating seeds (inhibiting germination, germ tube elongation). Among the amino acids tested, methionine was able to inhibit almost totally the germination of *Orobanche* seeds by reducing strongly the number of *Orobanche* plantlets (Vurro et al. 2006).

Nitrogen metabolism in *Striga* is characterized by a very high level of asparagine accumulation. This amino acid seems to be a storage form of the supra-optimal nitrogen amounts derived from xylem sap and the important nitrogen uptake appears to be one of the main factors responsible of the drastic effect of *Striga* on its host plant (Pageau et al. 2000).

The use of *Azospirillum* spp. to biologically control parasitic plant development

Importance of plant growth promoting rhizobacteria (PGPR) Azospirillum

This Gram negative diazotrophic bacterium (α-subclass of proteobacteria) able to fix atmospheric nitrogen, lives in close association with plant roots and forms associative symbioses. *Azospirillum* are isolated from the rhizosphere of many grasses and cereals all over the world, in tropical as well as in temperate climates.

Agronomic applications of these beneficial effects have been recorded in many studies (from 1997 to 2003) listed by Bashan et al. (2004). In many cases, inoculation reduced the utilization of chemical fertilizers from 20 to 50%, particularly nitrogen fertilization. In many developing countries, PGPR inoculation has improved the cost / benefit ratio of crops. The main effect of these inoculations was recorded on the morphological modifications of the host root system which increased in volume, weight and surface (Bashan and Levanony 1990). They also improved other plant growth parameters such as height, total biomass and nitrogen level in the stem and the seeds of the host plant (Jacoud 1997). Indeed, *Azospirillum* is mainly known by its capacity to produce and to exude plant hormones, polyamines and amino acids (Thuler et al. 2003). Among the hormones, auxins like IAA (indole acetic acid) and gibberellins play the most important roles.

The bacterium *Azospirillum* can also protect its host against phytopathogenic stresses *via* an antagonism phenomenon or by changing the host plant susceptibility, for example, by induced resistance (Dobbelaere et al. 2003). This protection was not limited to deleterious microorganisms and soil-borne pathogens, but also to the parasitic plants such as *Striga*. Several authors have demonstrated that some strains of *Azospirillum* can inhibit the germination of *Striga* seeds (Miché et al. 2000; Dadon et al. 2004). Therefore, the plant growth-promoting effect of *Azospirillum* in fields infested by *Striga*, could result from a direct effect on the plant growth but also from an indirect effect (*Azospirillum* antagonistic effect against *Striga* development). Many studies focused on the molecular basis of plant growth promotion and biocontrol by *Azospirillum* (Dobbelaere et al. 2003; Bashan et al. 2004) but none so far on the potential control by PGPR bacterium of parasitic plants development.

Potential effects of Azospirillum metabolites on Striga seed germination and infestation

Many hypotheses have been suggested on the potential role of substances secreted by *Azospirillum* strains against *Striga* infestation (Fig. 3). Indeed, it was already shown that *Azospirillum*, through chemical signals, could inhibit *Striga* invasion by reducing its germination rate and by blocking its radicle elongation (Miché et al. 2000).

Phytohormones and Striga germination

The influence of auxins on root elongation is well known, as well as the production of this hormone by *Azospirillum*. Al-Menoufi et al. (1986) have demonstrated that auxins have not a direct role in seed germination of the *Orobanche* parasitic plant. If the application of auxin alone does not stimulate germination, the application of the germination stimulant (GR24) strongly increases the seed germination rate (Al-Menoufi et al. 1986). The same effect of root elongation inhibition could also occur in the case of germinated *Striga* seeds.

Ethylene is involved in seed germination (Babiker et al. 2000). It was reported that germination induced by the germination stimulants, natural or synthetic, depends on the endogenous synthesis of ethylene (Logan and Stewart 1991). These stimulants induce the biosynthesis of ethylene that is a mediator of a biochemical cascade leading to seed germination (Babiker et al. 2000). Conversely, the addition of exogenic ethylene reduced the germination of the seeds in the presence of GR24, in particular with *Orobanche* spp. (Zehhar et al. 2002).

Ethylene production by *Azospirillum* spp is related to the phytohormones released into the medium and on the presence or absence of ACC (1-aminocyclopropane-1-carboxylate) desaminase. Indeed, it has been reported that a synergistic action of auxin and cytokinine on the stimulation of ethylene production (Babiker et al. 1994).

Striga germination and aminoethoxyvinyl glycine (AVG)

It has been showed that *Striga* germination was inhibited by an inhibitor of ethylene biosynthesis, the Aminoethoxyvinyl Glycine (AVG) known to inhibit the 1-amino-cyclopropane-1-carboxylate (ACC) synthase (Adams and Yang 1979). Babiker et al. (2000) showed that AVG inhibits the induction of ethylene carried out by GR24. Sugimoto et al. (2003) mentioned the inhibition of *Striga* germination by the inhibitor of the ethylene biosynthesis "AVG". Zehhar et al. (2002) confirmed the same significant decrease of germination when AVG and GR24 were applied simultaneously. These results led to the hypothesis that the AVG or similar compound acting in the same way could be produced by *Azospirillum* to protect its

Fig. 3 Compounds secreted by *Azospirillum spp.* potentially antagonistic against *Striga spp.* 1-amino-cyclopropane-1-carboxylate (ACC), Acyl-homoserine lactone (AHLs), Aminoethoxyvinyl Glycine (AVG), Diacetylphloroglucinol (DAPG), 2,6-dimethyl-p-benzoquinone (2,6-DMBQ), Indole acetic acid (IAA), Phenylacetic acid (PAA)

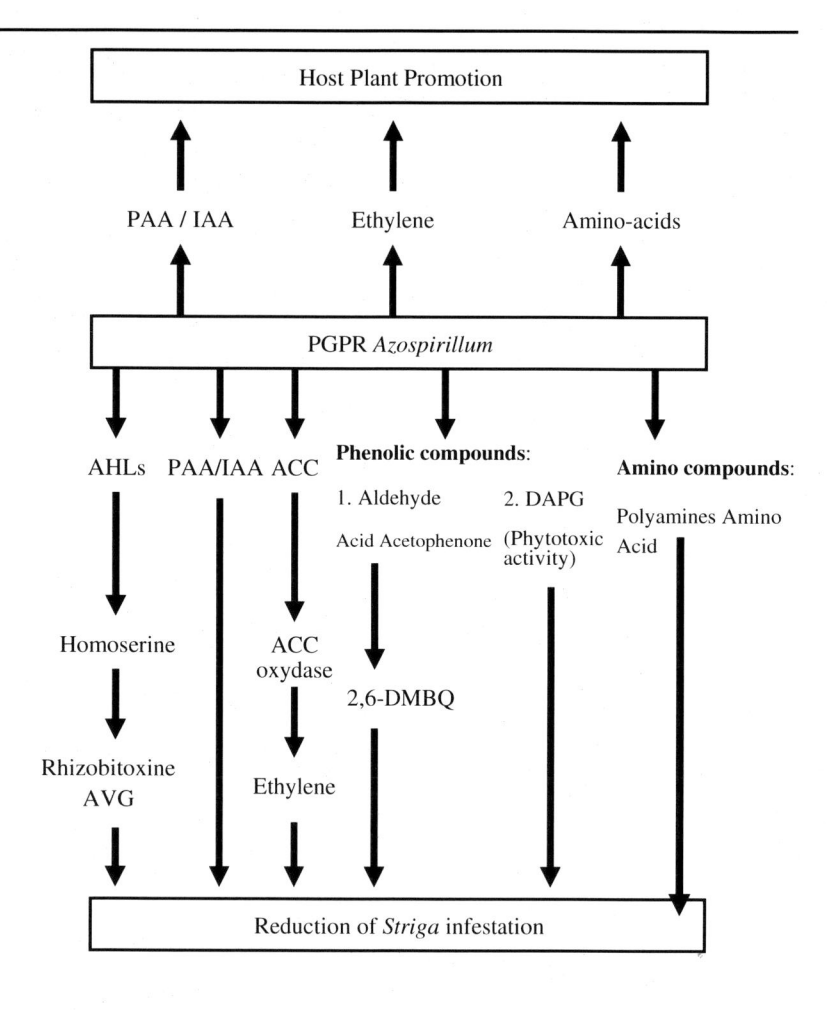

host against *Striga*. Nevertheless, the production of AVG-type compounds has not yet described in *Azospirillum* species.

Allelopathy and soil microbiota

Plant interactions result from the product of complex interactions based on combinations of specific mechanisms (Chapin et al. 1994). Allelopthy effect is generally considered as a competition by interference (Mahall and Callaway 1992) but its relative contribution to the total negative effect of one species on another remains poorly evaluated (Nilsson 1994). Allelopathy refers to the harmful effects of one plant on another plant by the release of chemicals from plant parts by leaching, root exudates, volatilization, residue decomposition and other biological processes in both natural and agricultural systems. In the present review, the role of root secretions in interactions among plants will be explored as well the effects of rhizosphere microbes on such biological relationships. Here we will particularly review our own work performed on *Eucalyptus camaldulensis* and on the effects of AM fungi against the allelopathy effect of this exotic tree species on native plant cover in West Africa.

Negative root-root communication

Root synthetises and accumulates a great diversity of micro- and macromolecular metabolites that are secreted into the rhizosphere as root exudates (Bais et al. 2004). Although recognition of the importance of root exudates in plant interactions has increased with recent studies of inter- and intraspecific root communication (Mahall and Callaway 1991), there were a few studies that had separated the effects of competition and allelopathy in plant interferences.

Root-mediated allelopathy has been suggested since the early 1800s by De Candolle (1832). This author suspected that plants could exude some chemical compounds that are detrimental for other plant species. Other studies have reported that these deleterious substances to plant growth were excreted into the soil by growing roots and they have found that "carbon black" soil amendment decreases soil toxicity (Schreiner and Reed 1908). The use of "carbon black" (charcoal powder) led to a decrease in the ability of *Larrea* roots to inhibit the elongation of neighboring roots (Mahall and Callaway 1991). More recently, the relative importance of allelopathy and resource competition in plant-plant interactions has been clearly assessed (Ridenour and Callaway 2001). Using "carbon black" to manipulate the effects of root exudates of *Centaurea maculosa* (noxious weed in western North America) on the development of native brunchgrass *Festuca idahoensis*, they found that the decrease of *Festuca* growth resulting from *Centaurea* was modulated by the "carbon black" soil amendment. They concluded that allelopathy was of great importance in the mechanisms of interference between both plant species. However these results have been obtained under controlled conditions and the mechanisms could be more complex in natural conditions. In the field, the secretion of root allelochemical compounds depends on many factors such as plant densities, root distributions, root densities and microbial activity.

Interferences between rhizosphere microbes and root-mediated allelopathy

Eucalyptus camaldulensis, one of the most widely planted eucalypts in the world, is extremely damaging ecologically to many native plant species. The annual vegetation adjacent to naturalized stands of *E. camaldulensis* is severely inhibited and annual herbs rarely survive to maturity when *Eucalyptus* litter accumulates. In addition, it is generally observed that the introduction of this tree species leads to a depletion of soil nutrients, acidification and to an excessive water utilization (Couto and Betters 1995). A study was conducted in controlled conditions to evaluate the impacts of *E. camaldulensis* on bacterial community structure and functional diversity and to determine the effects of arbuscular mycorrhizal inoculation on this exotic plant species effect in a sahelian soil (Kisa et al. 2007). The results showed that this plant species clearly modified the soil bacterial community. Both microbial community structure and microbial functions were significantly affected. These changes were accompanied by disturbances in the composition of the herbaceous plant species layer and mycorrhizal soil infectivity (reduction of the total number of mycorrhizal spores and of the mycelial network). However the negative impact of this exotic tree species was significantly moderated when it was inoculated with an efficient arbuscular mycorrhizal (AM) fungus. Beside a significant promoting effect on *E. camaldulensis* tree growth, the inoculation of *Glomus intraradices* (an AM fungus species) tended to return the soil to its initial conditions with a similar bacterial community structure and soil mycorrhizal potential. In addition AM inoculation has increased the development of herbaceous plant species under AM inoculated *E. camaldulensis* plants. The well-developed mycelial network measured under inoculated *E. camaldulensis* trees could explain this positive effect by equalizing the distribution of soil resources among competitively dominant and sub-dominant species (Wirsel 2004). But it has been also reported that soil microbes can act against allelochemical mediators, inactivating or metabolizing toxic compounds (Renne et al. 2004). In particular, it has been suggested that AM fungi could protect seedlings from allelopathy (Renne et al. 2004).

More recently, this biological property of AM symbiosis has been studied with an invasive plant species, *Amaranthus viridis*, in Senegal (Sanon et al., unpublished data). The experiment was conducted in Senegal at two sites: (i) one invaded by *A. viridis* and the other covered by other plant species but without *A. viridis*. Additionally, five sahelian *Acacia* species were grown in (1) soil disinfected or not collected from both sites, (2) un-invaded soil exposed to *A. viridis* plant aqueous extract, (3) soil collected from invaded and un-invaded sites and inoculated or not with the AM fungus *Glomus intraradices*. The results showed that the invasion of *A. viridis* increased soil nutrient availability, bacterial abundance and microbial activities. In contrast symbiotic microorganisms (AM fungi, Rhizobia) development and *Acacia* species growth were severely reduced in the *A. viridis* invaded soil. However, the inoculation of *G. intraradices* was highly beneficial to the growth and nodulation of *Acacia* species irrespective to the soil origin. Hence this negative impact of the invasive plant species was modified when *Acacia* species were inoculated by an efficient AM fungus. These results highlight the role of AM symbiosis in interacting with root-root communication and modifying plant coexistence.

Conclusion

All these results show that microbial soil communities are of great importance the biological processes driven plant co-existence. Hence it also shows that specific management regimes could be used to favour the development of target soil communities that are compatible with the development of desired aboveground communities. This approach will probably not became pratical until the technical challenges and costs of current soil microbial community analyses are reduced and until we obtain a better understanding of what measures of microbial community structure and function can serve as reliable and meaningful indicators (Harris 2003; Wolfe and Klironomos 2005). It shows that the overall effect of one plant to another result from multiple interacting mechanisms where soil microbiota can be considered as a key component. It highlights the need to consider soil microbiota in future management practices in order to maintain plant diversity in terrestrial ecosystems and improve the productivity in agrosystems.

References

Aarsen WL (1990) Ecological combining ability and competitive combining in plants toward a general evolutionary theory of coexistence in systems of competition. Am Nat 122:707–731. doi:10.1086/284167

Adams DO, Yang SF (1979) Ethylene biosynthesis: identification of 1-aminocyclopropane-1-carboxilc acid as an intermediate in the conversion of methionine to ethylene. Proc Natl Acad Sci USA 76:170–174. doi:10.1073/pnas.76.1.170

Adams WA, Gibbs RJ (1994) Natural turf for sport and emenity: science and practice. CAB International, Wallingford, UK

Albrecht H, Yoder JI, Phillips DA (1999) Flavonoids promote haustoria formation in the root parasite Tryphysaria versicolor. Plant Physiol 119:585–591. doi:10.1104/pp.119.2.585

Al-Menoufi MOA, Mostafa AK, Zaitoun FMF (1986) Studies on *Orobanche* spp. Seed germination as affected by auxins and kinetin. Alexandria Sci Exch 7(3):277–292

Andrade G, Mihara KL, Linderman RG, Bethlenfalvay GJ (1998) Soil aggregation status and rhizobacteria in the mycorrhizosphere. Plant Soil 202:89–96. doi:10.1023/A:1004301423150

Babiker ACT, Cai T, Ejeta G, Butler LG, Woodson WR (1994) Enhancement of biosynthesis and germination with thidiazuron and some selected auxins in *Striga asiatica* seeds. Physiol Plant 91:529–536. doi:10.1111/j.1399-3054.1994.tb02984.x

Babiker ACT, Ma Y, Sugimoto Y, Inanaga S (2000) Conditioning period, CO_2 and GR24 influence ethylene biosynthesis and germination of *Striga hermonthica*. Physiol Plant 109:75–80. doi:10.1034/j.1399-3054.2000.100111.x

Bais HP, Walker TS, Stermitz FR, Hufbauer RA, Vivanco JM (2002) Enantiomeric-dependent phytotoxic and antimicrobial activity of (±)-catechin. A rhizosecreted racemic mixture from spotted knapweed. Plant Physiol 128:1173–1179. doi:10.1104/pp.011019

Bais HP, Park S-W, Weir TL, Callaway RM, Vivanco JM (2004) How plants communicate using the underground information superhighway. Trends Plant Sci 9:26–32. doi:10.1016/j.tplants.2003.11.008

Bashan Y, Levanony H (1990) Current status of *Azospirillum* inoculation technology: *Azospirillum* as a challenge of agriculture. Can J Microbiol 36:591–608

Bashan Y, Holguin G, De-Bashan LE (2004) Azospirillum-plant relationships: physiological, molécular, agricultural and environmental advances (1997–2003). Can J Microbiol 50:521–577. doi:10.1139/w04-035

Baskin CC, Baskin JM (1998) Seeds: ecology, biogeography and evolution of dormancy and germination. Academic Press, New York, p 700

Beckstead J, Parker IM (2003) Invasiveness of *Ammophila arenaria*: Realease from spoil-born pathogens? Ecology 84:2824–2831. doi:10.1890/02-0517

Belnap J, Phillips SL (2001) Soil biota in an un-grazed grassland: response to annual grass (*Bromus tectorum*) invasion. Ecol Appl 11:1261–1275. doi:10.1890/1051-0761(2001)011[1261:SBIAUG]2.0.CO;2

Berner DK, Kling JG, Singh BB (1995) *Striga* research and control: a perspective from Africa. Plant Dis 79:23–35

Bethlenfalvay GJ (1992) Mycorrhizae and crop productivity. In: Bethlenfalvay GJ, Linderman RG (eds) Mycorrhizae in sustainable agriculture. American Society of Agronomy pp 1–27

Bever JD (1994) Feedback between plants and their soil communities in an old field community. Ecology 75:1965–1977. doi:10.2307/1941601

Bever JD (2003) Soil community feedback nd the coexistence of competitors: conceptual frameworks and empirical tests. New Phytol 157:465–473. doi:10.1046/j.1469-8137.2003.00714.x

Bever JD, Morton JB, Antonovics J, Schultz PA (1996) Host-dependent sporulation and species diversity of arbuscular mycorrhizal fungi in a mown grassland. J Ecol 3:205–214

Bever JD, Westover KM, Antonovics J (1997) Incorporating the soil community into plant population dynamics: The utility of the feedback approach. J Ecol 85:561–573. doi:10.2307/2960528

Boerner REJ, Brinkman JA (2003) Fire frequency and soil enzyme activity in southern Ohio oak-hickory forests. Appl Soil Ecol 23:137–146. doi:10.1016/S0929-1393(03)00022-2

Cabin RJ, Weller SG, Lorence DH, Cordell S, Hadway LJ, Montgomery R, Goo D, Urakami A (2002) Effects of light, alien grass and native species additions on Hawaiian dry forest restoration. Ecol Appl 12:1595–1610. doi:10.1890/1051-0761(2002)012[1595:EOLAGA]2.0.CO;2

Callaway RM, Ridenour WM (2004) Novel weapons: Invasive success and the evolution of increased competitive ability. Front Ecol Environ 2:436–443

Callaway RM, Newingham B, Zabinski CA, Mahall BE (2001) Compensatory growth and competitive ability of an invasive weed are enhanced by soil fungi and native neighbors. Ecol Lett 4:429–433. doi:10.1046/j.1461-0248.2001.00251.x

Callaway RM, Mahall BE, Wicks C, Pankey J, Zabinski C (2003) Soil fungi and the effects of an invasive forb on grasses: neighbor identity matters. Ecology 84:129–135. doi:10.1890/0012-9658(2003)084[0129:SFATEO]2.0.CO;2

Callaway RM, Thelen GC, Rodriguez A, Holben WE (2004) Soil biota and exotic plant invasion. Nature 427:731–733. doi:10.1038/nature02322

Carey EV, Marler MJ, Callaway RM (2004) Mycorrhizae transfer carbon from a native grass to an invasive weed: evidence from stable isotopes and physiology. Plant Ecol 172:133–141. doi:10.1023/B:VEGE.0000026031.14086.f1

CBD (2006) Invasive Alien Species. Convention on Biological Diversity. Web site: http://www.biodiv.org/programmes/cross-cutting/alien/

Cechin I, Press MC (1993) Nitrogen relations of the sorghum-*Striga hermonthica* host parasite association: germination, attachment and early growth. New Phytol 124:681–687. doi:10.1111/j.1469-8137.1993.tb03858.x

Chapin FS III, Walker LR, Fastie CL, Sharman LC (1994) Mechanisms of primary succession following deglaciation at Glacier Bay, Alaska. Ecol Monogr 64:149–175. doi:10.2307/2937039

Chapman RN (1931) Animal Ecology. McGraw-Hill, New York

Chen H, Li B, Fang C, Chen J, Wu J (2007) Exotic plant influences soil nematode communities through litter input. Soil Biol Biochem 39:1782–1793. doi:10.1016/j.soilbio.2007.02.011

Cook CE, Whichard LP, Turner B, Wall ME, Egley GH (1966) Germination of witchweed (*S. lutea* Lour.): isolation and properties of a potent stimulant. Science 154:1189–1190. doi:10.1126/science.154.3753.1189

Couto L, Betters DR (1995) Short rotation Eucalypt plantations in Brazil: social and environmental issues. OKA Ridge National laboratory, Tennessee, USA

D'Antonio CM, Mahall BE (1991) Root profiles and competition between the invasive, exotic perennial, *Carpobortus edulis*, and two native shrub species in California coastal scrub. Am J Bot 78:885–894. doi:10.2307/2445167

D'Antonio CM, Vitousek PM (1992) Biological invasions by exotic grasses, the grass/fire cycle and global change. Annu Rev Ecol Evol Syst 23:63–87

D'Antonio C, Meyerson LA (2002) Exotic plant species as problems and solutions in ecological restoration: a synthesis. Restor Ecol 10:703–713. doi:10.1046/j.1526-100X.2002.01051.x

D'Antonio CM, Dudley TL, Mack MC (1999) Disturbance and biological invasions: direct effects and feedbacks. In: Walker L (ed) Ecosystems of disturbed ground. Elsevier, Amsterdam, pp 413–452

Dadon T, Nun NB, Mayer RM (2004) A factor from *Azospirillum brasilense* inhibits germination and radicle growth of *Orobanche* aegyptiaca. Isr J Plant Sci 52:83–86. doi:10.1560/Q3BA-8BJW-W7GH-XHPX

De Candolle AP (1832) Physiologie végétale. Bechet Jeune, Paris

Dobbelaere S, Vanderleyden J, Okon Y (2003) Plant growth promoting effects of diazotrophs in the rhizosphere. Crit Rev Plant Sci 22:107–149. doi:10.1080/713610853

Dube MP, Olivier A (2001) *Striga gesnerioides* and its host, cowpea: interaction and methods of control. Can J Bot 79 (10):1225–1240. doi:10.1139/cjb-79-10-1225

Duda JJ, Freeman DC, Emlen JM, Belnap J, Kitchen SG, Zak JC, Sobek E, Tracy M, Montante J (2003) Differences in native soil ecology associated with invasion of the exotic annual chenopod, *Halogeton glomeratus*. Biol Fertil Soils 38:72–77. doi:10.1007/s00374-003-0638-x

Ehrenfeld JG (2003) Effects of exotic plant invasions on soil nutrient cycling processes. Ecosystems (N Y, Print) 6:503–523. doi:10.1007/s10021-002-0151-3

Eom A, Hartnett DC (2000) Host plant species effects on arbuscular mycorhizal fungi communities in tallgrass prairie. Oecologia 122:435–444. doi:10.1007/s004420050050

Evans RD, Rimer R, Sperry L, Belnap J (2001) Exotic plant invasion alters nitrogen dynamics in a arid grassland. Ecol Appl 11:1301–1310. doi:10.1890/1051-0761(2001)011[1301:EPIAND]2.0.CO;2

Farina MPW, Thomas PEL, Channon P (1985) Nitrogen, phosphorus and potassium effects on the incidence of Striga asiatica (L.) Kuntze in maize. Weed Res 25:443–447. doi:10.1111/j.1365-3180.1985.tb00667.x

Fumanal B, Plenchette C, Chauvel B, Bretagnolle F (2006) Which role can arbuscular mycorrhizal fungi play in the facilitation of *Ambrosia artemisiifolia* L. invasion in France? Mycorrhiza 17:25–35. doi:10.1007/s00572-006-0078-1

Gange AC (1994) Subterranean insects and fungi: hidden costs and benefits to the greenkeeper. Cochran AJ, Ferrally MR (Eds) Science and Golf II: Proccedings of the World Scientific Congress of Golf. E. and F.N. Spon, London, UK. pp 461–466

Gange AC, Lindsay DE, Ellis LS (1999) Can arbuscular mycorrhizal fungi be used to control the undesirable grass *Poa annua* on golf courses? J Appl Ecol 36:909–919. doi:10.1046/j.1365-2664.1999.00456.x

Grace JD, Tilman D (1990) Perspectives on plant competition. Academic, San Diego

Graham JH, Leonard RT, Menge JA (1981) Membrane mediated decrease of root-exudation responsible for phosphorus inhibition of vesicular-arbuscular mycorrhiza formation. Plant Physiol 68:548–552. doi:10.1104/pp.68.3.548

Grayston SJ, Vaughan D, Jones D (1996) Rhizosphere carbon flow in trees, in comparison with annual plants: the importance of root exudation and its impact on microbial activity and nutrient availability. Appl Soil Ecol 5:29–56. doi:10.1016/S0929-1393(96)00126-6

Grayston SJ, Griffith GS, Mawdsley JL, Campbell CD, Bardgett RD (2001) Accounting for variability in soil microbial communities of temperate upland grassland ecosystems. Soil Biol Biochem 33:533–551. doi:10.1016/S0038-0717(00)00194-2

Grubb P (1977) The maintenance of species richness in plant communities: the importance of the regeneration niche. Biol Rev Camb Philos Soc 52:107–145. doi:10.1111/j.1469-185X.1977.tb01347.x

Hamilton JG, Holzapfel C, Mahall BE (1999) Coexistence and interference between native perennial grass and non-native annual in California. Oecologia 121:518–526. doi:10.1007/s004420050958

Harris JA (2003) Measurements of the soil microbial community for estimating the success of restoration. Eur J Soil Sci 54:801–808. doi:10.1046/j.1351-0754.2003.0559.x

Hauck C, Müller S, Schildknecht H (1992) A germination stimulant for parasitic flowering plants from *Sorghum bicolor*, a genuine host plant. J Plant Physiol 139:474–478

Hierro JL, Callaway RM (2003) Allelopathy and exotic plant invasion. Plant Soil 256:29–39. doi:10.1023/A:1026208327014

Hiltner L (1904) Uber neuere Erfahrungen und Probleme auf dem Gebiet der Bodenbakteriologie unter besonderer Beruksichtigung der Grundungung und Brache (On recent insights and problems in the area of soil bacteriology under special consideration of the use of green manure and fallowing). Arb Dtsch Landwirt Ges 98:59–78

Hirsch AM, Dietz Bauer W, Bird DM, Cullimore J, Tyler B, Yoder JI (2003) Molecular signals and receptors: controlling rhizosphere interactions between plants and other organisms. Ecolog 84:858–868. doi:10.1890/0012-9658(2003)084[0858:MSARCR]2.0.CO;2

Hobbs RJ, Huenneke LF (1992) Disturbance, diversity, and invasion: Implications for conservation. Conserv Biol 6:324–337. doi:10.1046/j.1523-1739.1992.06030324.x

Huenneke LF, Hamburg SP, Koide R, Mooney HA, Vitousek PM (1990) Effects of soil resources on plant invasions and community structure in Californian serpentine grassland. Ecology 71:478–491. doi:10.2307/1940302

Hughes RF, Vitousek PM (1993) Barriers to shrub re-establishment following fire in the seasonal submontane zone of Hawai. Oecologia 93:557–563. doi:10.1007/BF00328965

Hutchinson TF, Vankat JL (1997) Invasibility and effects of Amur honeysuckle in southwestern Ohio forest. Conserv Biol 11:1117–1124. doi:10.1046/j.1523-1739.1997.96001.x

Imaizumi S, Nishino T, Miyabe K, Fujimori T, Yamada M (1997) Biological control of annual bluegrass (*Poa annua* L.) with Japenese isolate of *Xanthomonas campestris* pv. *poae* (JT-P482). Biol Control 8:7–14. doi:10.1006/bcon.1996.0475

Imaizumi S, Teteno A, Fujimori T (1998) Effect of bacterial concentration of *Xanthomonas campestris* pv. *poae* (JT-P482) on the control of annual bluegrass (*Poa annua* L.). J Pest Sci 23:141–144

Jacoud C (1997) Induction précoce de l'effet«PGPR»par *Azospirillum lipoferum* CRT1 chez le mais. Thèse pour l'obtention du diplôme de doctorat, UCB Lyon 1

Joel DM, Steffens JC, Mathews DE (1995) Germination of weedy root parasites. Seed Development and Germination, In

Johansson JF, Paul LR, Finlay RD (2004) Microbial interactions in the mycorrhizosphere and their significance for sustainable agriculture. FEMS Microbiol Ecol 48:1–13. doi:10.1016/j.femsec.2003.11.012

Johnson D, Vandenkoornhuyse PJ, Leake JR, Gilbert L, Booth RE, Grime JP, Young JPW (2004) Plant communities affect arbuscular mycorrhizal fungal diversity and community composition in grassland microcosms. New Phytol 161:503–515. doi:10.1046/j.1469-8137.2003.00938.x

Khan ZR, Hassanali A, Overholt W, Khamis TM, Hooper AM, Pickett JA, Wadhams LJ, Woodcock CM (2002) Control of witchweed *Striga hermonthica* by intercropping with *Desmodium* spp., and the mechanism defined as allelopathic. J Chem Ecol 28:1871–1884. doi:10.1023/A:1020525521180

Kim D, Koiz R, Boone L, Keyes WJ, Lynn DG (1998) On becoming a parasite evaluating the role of wall oxidases in parasitic plant development. Chem Biol 5:103–117. doi:10.1016/S1074-5521(98)90144-2

Kisa M, Sanon A, Thioulouse J, Assigbetse K, Sylla S, Spichiger R, Dieng L, Berthelin J, Prin Y, Galiana A, Lepage M, Duponnois R (2007) Arbuscular mycorrhizal symbiosis counterbalance the negative influence of the exotic tree species *Eucalyptus camaldulensis* on the structure and functioning of soil microbial communities in a Sahelian soil. FEMS Microbiol Ecol 62:32–44. doi:10.1111/j.1574-6941.2007.00363.x

Klironomos JN (2002) Feedback with soil biota contributes to plant rarity and invasiveness in communities. Nature 417:67–70. doi:10.1038/417067a

Kneebone WR, Kopec DM, Mancino CF (1992) Water requirements and irrigation. Waddington DV, Carrow RN, Shearman RC (eds) Turfgrass. American Society of Agronomy, Madison, MI, pp 441–472

Knevel I, Lans T, Menting FBJ, Hertling UM, van der Putten WH (2004) Release from native root herbivores and biotic resistance by soil pathogens in a new habitat both affect the alien *Ammophila arenaria* in South Africa. Oecologia 141:502–510. doi:10.1007/s00442-004-1662-8

Kourtev PS, Ehrenfeld JG, Häggblom M (2002) Exotic plant species alter the microbial community structure and function in the soil. Ecology 83:3152–3166

Kourtev PS, Ehrenfeld JG, Häggblom M (2003) Experimental analysis of the effect of exotic and native plant species on the structure and function of soil microbial communities. Soil Biol Biochem 35:895–905. doi:10.1016/S0038-0717(03)00120-2

Lenné J (2000) Pests and poverty: the continuing need for crop protection research. Outlook Agric 29:235–250

Levine JM, D'Antonio CM (1999) Elton revisited: a review of evidence linking diversity and invasibility. Oikos 87:15–26. doi:10.2307/3546992

Levine JM, Vilà M, D'Antonio CM, Dukes JS, Grigulis K, Lavorel S (2003) Mechanisms underlying the impacts of exotic plant invasion. P Roy B-Biol Sci 270:775–781

Levine JM, Adler PB, Yelenik SG (2004) A meta-analysis of biotic resistance to exotic plant invasions. Ecol Lett 7:975–989. doi:10.1111/j.1461-0248.2004.00657.x

Linderman RG (1988) Mycorrhizal interactions with the rhizosphere microflora: the mycorrhizosphere effect. Phytopathology 78:366–371

Logan DC, Stewart GR (1991) Germination of the seeds of parasitic angiosperms. Seed Sci Res 2:179–190

Losner-Goshen D, Portnoy VH, Mayer AM, Joel DM (1998) Pectolytic activity by the haustorium of the parasitic plant *Orobanche* L. (*Orobanchaceae*) in host roots. Ann Bot (Lond) 81:319–326. doi:10.1006/anbo.1997.0563

Lozon JD, MacIsaac HJ (1997) Biological invasions: are they dependent on disturbance? Environ Rev 5:131–144. doi:10.1139/er-5-2-131

Mahall BE, Callaway RM (1991) Root communication among desert shrubs. Proc Natl Acad Sci USA 88:874–876. doi:10.1073/pnas.88.3.874

Mahall BE, Callaway RM (1992) Root communications mechanisms and intracommunity distributions of two Mojave Desert Shrubs. Ecology 73:2145–2151. doi:10.2307/1941462

Marler MJ, Zabinski CA, Callaway RM (1999) Mycorrhizae indirectly enhance competitive effects of an invasive forb on a native bunchgrass. Ecology 80:1180–1186

Maron J, Connors P (1996) A native nitrogen fixing plant facilitates weed invasion. Oecologia 105:305–312. doi:10.1007/BF00328732

Maron JL, Vila M (2001) Do herbivores affect plant invasion? Evidence for the natural enemies and biotic resistance hypotheses. Oikos 95:363–373. doi:10.1034/j.1600-0706.2001.950301.x

Marschner P, Timonen S (2005) Interactions between plant species and mycorrhizal colonization on the bacterial community composition in the rhizosphere. Appl Soil Ecol 28:23–36. doi:10.1016/j.apsoil.2004.06.007

Masters RA, Nissen SJ (1998) Revegetating leafy spurge (Euphorbia esula)-infested rangeland with native tallgrass. Weed Technol 12:381–390

Meiners SJ (2007) Apparent competition: an impact of exotic shrub invasion on tree regeneration. Biol Invasions 9:849–855. doi:10.1007/s10530-006-9086-5

Miché L, Bouillant ML, Rohr R, Sallé G, Bally R (2000) Physiological and cytological studies on the inhibition of *Striga* seed germination by the plant growth promoting bacterium *Azospirillum brasilense*. Eur J Plant Pathol 106:347–351. doi:10.1023/A:1008734609069

Mitchell CE, Power AG (2003) Release of invasive plants from viral and fungal pathogens. Nature 421:625–627. doi:10.1038/nature01317

Moora M, Zobel M (1996) Effect of arbuscular mycorrhiza and inter- and intraspecific competition of two grassland species. Oecologia 108:79–84. doi:10.1007/BF00333217

Mummey DL, Rillig MC (2006) The invasive plant species *Centaurea maculosa* alters arbuscular mycorrhizal fungal communities in the field. Plant Soil 288:81–90. doi:10.1007/s11104-006-9091-6

Musselman L (1987) Taxonomy of witchweeds. Parasitic Weeds in Agriculture Vol. 1

Nardi S, Concheri G, Pizzeghello D, Sturaro A, Rella R, Parvoli G (2000) Soil organic matter mobilization by root exudates. Chemosphere 41:653–658. doi:10.1016/S0045-6535(99)00488-9

Newsham KK, Fitter AH, Watkinson AR (1994) Root pathogenic and arbuscular mycorrhizal fungi determine fecundity of asymptomatic plants in the field. J Ecol 82:805–814. doi:10.2307/2261445

Nilsson MC (1994) Separation of allelopathy and resource competition by the boreal dwarf shrub *Empetrum hermaphrodium*. Oecologia 98:1–7. doi:10.1007/BF00326083

Noss R (1990) Indicators for monitoring biodiversity: a hierarchical approach. Conserv Biol 4:355–364. doi:10.1111/j.1523-1739.1990.tb00309.x

Offre P, Pivato B, Siblot S, Gamalero E, Corberand T, Lemanceau P, Mougel C (2007) Identification of Bacterial Groups Preferentially Associated with Mycorrhizal Roots of *Medicago truncatula*. Appl Environ Microbiol 73:913–921. doi:10.1128/AEM.02042-06

Olivier A (1995) Le *Striga*, mauvaises herbes parasites des céréals africaines: biologie et méthode de lutte. Agronomie 15:517–525. doi:10.1051/agro:19950901

Parker C (1991) Protection of crops against parasitic weeds. Crop Prot 10:6–22. doi:10.1016/0261-2194(91)90019-N

Parker C, Riches CR (1993) *Orobanche* species: the broomrape. In: Parker C, Riches CR (eds) Parasitic weeds of the world: biology and control. Wallingford, CAB International, pp 111–163

Packer A, Clay K (2000) Soil pathogens and spatial patterns of seedlings mortality in a temperate tree. Nature 440:278–281. doi:10.1038/35005072

Pageau K, Rousset A, Simier P, Delavault P, Zehhar N, Fer A (2000) Special features of mechanisms controlling germination and carbon and nitrogen metabolism in two parasitic angiosperms: *Striga hermonthica* and *Orobanche ramosa*. Comptes rendus de l'Academie d'Agriculture de France 86(8):69–84

Pimentel DL, Lach L, Zuniga R, Morrison D (2000) Environmental and economic costs of nonindigenous species in the United States. Bioscience 50:53–64. doi:10.1641/0006-3568(2000)050[0053:EAECON]2.3.CO;2

Pivato B, Mazurier S, Lemanceau P, Siblot S, Berta G, Mougel C, van Tuinen D (2007) Medicago species affect the community of arbuscular mycorrhizal fungi associated with roots. New Phytol 176:197–210. doi:10.1111/j.1469-8137.2007.02151.x

Priha O, Grayston SJ, Pennanen T, Smolander A (1999) Microbial activities related to C and N cycling and microbial community structure in the rhizosphere of Pinus sylvestris, Picea abies and Betula pendula seedlings in an organic and mineral soil. FEMS Microbiol Ecol 30:187–199. doi:10.1111/j.1574-6941.1999.tb00647.x

Raju PS, Osman MA, Soman P, Peacock JM (1990) Effects of N, P and K on *Striga asiatica* (L.) Kuntze seed germinatio and infestation of sorghum. Weed Res 30:139–144. doi:10.1111/j.1365-3180.1990.tb01697.x

Reader RJ (1998) Relationship between species relative abundance and plant traits for an infertile habitat. Plant Ecol 134:43–51. doi:10.1023/A:1009700100343

Reever-Morghan K, Seastedt TR (1999) Effects of soil nitrogen reduction on nonnative plants in restored grasslands. Restor Ecol 7:51–55. doi:10.1046/j.1526-100X.1999.07106.x

Reinhart KO, Packer A, van der Putten WH, Clay K (2003) Plant-soil biota interactions and spatial distribution of black cherry in its native and invasive ranges. Ecol Lett 6:1046–1050. doi:10.1046/j.1461-0248.2003.00539.x

Renne IJ, Rios BG, Fehmi JS, Tracy BF (2004) Low allelopathic potential of an invasive forage grass on native grassland plant: a cause for encouragements? Basic Appl Ecol 5:261–269. doi:10.1016/j.baae.2003.11.001

Richardson DM, Williams PA, Hobbs RJ (1994) Pine invasions in the Southern Hemisphere-determinants of spread and invadability. J Biogeogr 21:511–527. doi:10.2307/2845655

Ricklefs RW (1977) Environmental heterogeneity and plant species diversity: a hypothesis. Am Nat 111:376–381. doi:10.1086/283169

Ridenour WM, Callaway RM (2001) The relative importance of allelopathy in interference: the effects of an invasive weed on a native bunchgrass. Oecologia 126:444–450. doi:10.1007/s004420000533

Rillig MC, Wright SF, Eviner VT (2002) The role of arbuscular Mycorrhizal fungi and glomalin in soil aggregation: comparing effects of five plant species. Plant Soil 238:325–333. doi:10.1023/A:1014483303813

Roberts KJ, Anderson RC (2001) Effect of garlic mustard [*Alliaria petiolata* (Beib. Cavara and Grande)] extracts on plants and arbuscular mycorrhizal (AM) fungi. Am Midl Nat 146:146–152. doi:10.1674/0003-0031(2001)146[0146:EOGMAP]2.0.CO;2

Salle G, Aber M (1986) Les phanérogames parasites : biologie et stratégies de lutte. Bull Soc Bot France. Lettre bot 3:235–263

Sallé G, Raynal-Roques A (1989) Le Striga. Recherche 206:44–51

Sato D, Awad AA, Takeuchi Y, Yoneyama K (2005) Confirmation and quantification of strigolactones, germination stimulants for root parasitic plants *Striga* and *Orobanche*, produced by cotton. Biosci Biotechnol Biochem 69:98–102. doi:10.1271/bbb.69.98

Schreiner O, Reed HS (1908) The toxic action of certain organic plant constituents. Bot Gaz 34:73–102. doi:10.1086/329469

Shachar-Hill Y, Reffer PE, Douds D, Osman SF, Doner LW, Ratcliffe RG (1995) Partitioning of intermediary carbon metabolism in vesicular-arbuscular mycorrhizal leek. Plant Physiol 108:7–15

Shine C, Williams N, Gundling L (2000) A guide to designing legal and institutional frameworks on alien invasive species. Vol. 40. IUCN, Switzerland

Simberloff D (1974) Equilibrium theory of island biogeography and ecology. Annu Rev Ecol Evol Syst 5:161–182

Simberloff D, Von Holle B (1999) Positive interactions of nonindigenous species: invasional meltdown. Biol Invasions 1:21–32. doi:10.1023/A:1010086329619

Simberloff D (2003) Confronting introduced species: a form of xenophobia? Biol Invasions 5:179–192. doi:10.1023/A:1026164419010

Stinson KA, Campbell SA, Powell JR, Wolfe BE, Callaway RM, Thelen GC, Hallett SG, Prati D, Klironomos JN (2006) Invasive plant suppresses the growth of native tree seedling by disrupting belowground mutualisms. PLoS Biol 4:727–731. doi:10.1371/journal.pbio.0040140

Stotzky G (1997) Soil as an environment for microbial life. In: Van Elsas JD, Trevors JT, Wellington EMH (eds) Modern soil microbiology. Dekker, New York, pp 1–20

Sugimoto Y, Ali MA, Yabuta S, Kinoshita H, Inanaga S, Itai A (2003) Germination strategy of *Striga hermonthica* involves regulation of ethylene biosynthesis. Physiol Plant 119:137–145. doi:10.1034/j.1399-3054.2003.00162.x

Tardiff S, Stanford JA (1998) Grizzly bear digging: effects on subalpine meadow plants in relation to mineral nitrogen availability. Ecology 79:2219–2228

Thébaud C, Simberloff D (2001) Are plants really larger in their introduced ranges? Am Midl Nat 157:231–236

Thuler DS, Floh EIS, Handro W, Barbosa HR (2003) Plant growth regulators and amino acids released by *Azospirillum* sp in chemically defined media. Lett Appl Microbiol 37:174–178. doi:10.1046/j.1472-765X.2003.01373.x

Tilman D (1982) Resource competition and community structure. Monographs in population biology, vol 17. Princeton University Press, Princeton

Torsvik V, Goksoyr J, Daae FL (1990) High diversity in DNA of soil bacteria. Appl Environ Microbiol 56:782–787

Urcelay C, Diaz S (2003) The mycorrhizal dependence of surbordinates determines the effect of arbuscular mycorrhizal fungi on plant diversity. Ecol Lett 6:388–391. doi:10.1046/j.1461-0248.2003.00444.x

van der Heijden MGA, Klironomos JN, Ursic M, Moutoglis P, Streitwolf-Engel R, Boller T, Wiemken A, Sanders IR (1998) Mycorrhizal fungal diversity determines plant biodiversity, ecosystem variability and productivity. Nature 396:69–72. doi:10.1038/23932

van der Putten WH, Van Dijk C, Peters BAM (1993) Plant-specific soil-borne diseases contribute to succession in foredune vegetation. Nature 362:53–56. doi:10.1038/362053a0

Vandenkoornhuyse P, Husband R, Daniell TJ, Watson IJ, Duck JM, Fitter AH, Young JPW (2002) Arbuscular mycorrhizal community composition associated with two plant species in a grassland ecosystem. Mol Ecol 11:1555–1564. doi:10.1046/j.1365-294X.2002.01538.x

Vaughn SF, Berhow MA (1999) Allelochemicals isolated from tissues of the invasive weed garlic mustard (*Alliaria petiolata*). J Chem Ecol 25:2495–2504. doi:10.1023/A:1020874124645

Vitousek PM, D'Antonio CM, Loope LL, Rejmanek M, Westbrooks R (1997) Introduced species: a significant component of human-caused global change. N Z J Ecol 21:1–16

Vitousek PM, Walker LR (1989) Biological invasion by *Myrica faya* in Hawaii-plant demography, nitrogen fixation, ecosystem effects. Ecol Monogr 59:247–265. doi:10.2307/1942601

Vivanco JM, Bais HP, Stermitz FR, Thelen GC, Callaway RM (2004) Biogeographical variation in community response to root allelochemistry: Novel weapons and exotic invasion. Ecol Lett 7:285–292. doi:10.1111/j.1461-0248.2004.00576.x

Vurro M, Boari A, Pilgeram AL, Sands DC (2006) Exogenous acids inhibit seed germination and tubercle formation by *Orobanche ramosa* (boomrape): Potential application for management of parasitic weeds. Biol Control 36:258–265. doi:10.1016/j.biocontrol.2005.09.017

Wardle DA (2002) Communities and Ecosystems: Linking the Aboveground and Belowground Components. Princeton University Press, Princeton, NJ, p 392

Wardle DA, Bardgett RD, Klironomos JN, Setala H, van der Putten WH, Wall DH (2004) Ecological linkages between aboveground and belowground biota. Science 304:1629–1633. doi:10.1126/science.1094875

West HM (1996) Influence or arbuscular mycorrhizal infection on competition between *Holcus lanatus* and *Dactylis glomerata*. J Ecol 84:429–438. doi:10.2307/2261204

Westover K, Kennedy A, Kelley S (1997) Patterns of rhizosphere microbial community structure associated with co-occurring plant species. J Ecol 85:863–873. doi:10.2307/2960607

Westwood JH, Foy CL (1999) Influence of nitrogen on germination and early development of boomrape (*Orobanche sp.*). Weed Sci 47:2–7

Wigchert SCM, Kuiper E, Boelhouwer GJ, Nefkens GHL, Vekleij JAC, Zwanenburg B (1999) Dose-response of seeds of parasitic weeds and *Orobanche* towards the synthetic germination stimulants GR24 and Njimegen-1. J Agric Food Chem 47:1705–1710. doi:10.1021/jf981006z

Wilcove DS, Rothstein D, Bubow J, Phillips A, Losos E (1998) Quantifying threats to imperilled species in the United states. Bioscience 48:607–615. doi:10.2307/1313420

Wilson SD, Gerry AK (1995) Strategies for mixed-grass prairie restoration: herbicide, tilling, and nirogen manipulation. Restor Ecol 3:290–298. doi:10.1111/j.1526-100X.1995.tb00096.x

Wirsel SGR (2004) Homogenous stands of wetland grass harbour dicerse consortia of arbuscular mycorrhizal fungi. FEMS Microbiol Ecol 48:129–138. doi:10.1016/j.femsec.2004.01.006

Wolfe BE, Klironomos JN (2005) Breaking new ground: soil communities and exotic plant invasion. Bioscience 55:477–487. doi:10.1641/0006-3568(2005)055[0477:BNGSCA]2.0.CO;2

Worsham AD (1987) Germination of witchweed seeds. In: Musselman LJ (ed) Parasitic Weeds in Agriculture Vol. 1: Striga. CRC, Boca Raton, Florida, pp 45–61

Zabinski CA, Quinn L, Callaway RM (2002) Phosphorus uptake, not carbon transfer, explains arbuscular mycorrhizal enhancement of *Centaurea maculosa* in the presence of native grassland species. Funct Ecol 16:758–765. doi:10.1046/j.1365-2435.2002.00676.x

Zak DR, Holmes WE, White DC, Peacock AD, Tilman D (2003) Plant diversity, soil microbial communities, and ecosystem function: Are there any links? Ecology 84:2042–2050. doi:10.1890/02-0433

Zehhar N, Ingouff M, Bouya D, Fer A (2002) Possible involvement of gibberellins and ethylene in *Orobanche ramosa* germination. Weed Res 42:464–469. doi:10.1046/j.1365-3180.2002.00306.x

Zhou T, Neal JC (1995) Annual bluegrass (*Poa annua*) control with *Xanthomonas campestris* pv. *poae* in New York State. Weed Technol 9:173–177

Zobel M, Moora M (1995) Interspecific competition and arbuscular mycorrhiza: importance for the coexistence of two calcareous grassland species. Folia Geobot 30:223–230. doi:10.1007/BF02812100

REVIEW ARTICLE

Molecular communication in the rhizosphere

Denis Faure · Danny Vereecke ·
Johan H. J. Leveau

Received: 7 October 2008 / Accepted: 13 November 2008 / Published online: 6 December 2008
© Springer Science + Business Media B.V. 2008

Abstract This paper will exemplify molecular communications in the rhizosphere, especially between plants and bacteria, and between bacteria and bacteria. More specifically, we describe signalling pathways that allow bacteria to sense a wide diversity of plant signals, plants to respond to bacterial infection, and bacteria to coordinate gene expression at population and community level. Thereafter, we focus on mechanisms evolved by bacteria and plants to disturb bacterial signalling, and by bacteria to modulate hormonal signalling in plants. Finally, the dynamics of signal exchange and its biological significance we elaborate on the cases of *Rhizobium* symbiosis and *Agrobacterium* pathogenesis.

Keywords Rhizosphere · Signal · Rhizobium · Agrobacterium · Quorum-sensing · Plant hormones · Plantbacteria interactions

Responsible Editor: Yves Dessaux.

D. Faure (✉)
Institut des Sciences du Végétal,
Centre national de la Recherche Scientifique,
Avenue de la Terrasse,
91198 Gif-sur-Yvette, France
e-mail: faure@isv.cnrs-gif.fr

D. Vereecke
Department of Plant Systems Biology,
Flanders Institute for Biotechnology (VIB),
Ghent University,
Technologie park 927,
9052 Zwijnaarde, Belgium
e-mail: Danny.Vereecke@psb.ugent.be

J. H. J. Leveau
Department of Plant Pathology,
University of California at Davis,
One Shields Avenue, 476 Hutchison Hall,
Davis, CA 95616, USA
e-mail: jleveau@ucdavis.edu

Introduction

This paper will exemplify molecular communications in the rhizosphere, especially between plants and bacteria, and between bacteria and bacteria. More specifically, we describe signalling pathways that allow bacteria to sense a wide diversity of plant signals, plants to respond to bacterial infection, and bacteria to coordinate gene expression at population and community level. Thereafter, we focus on mechanisms evolved by bacteria and plants to disturb bacterial signalling, and by bacteria to modulate hormonal signalling in plants. Finally, the dynamics of signal exchange and its biological significance we elaborate on the cases of *Rhizobium* symbiosis and *Agrobacterium* pathogenesis. For a complete overview of communication in the rhizosphere, we recommend other papers that illustrate plant-plant interactions, and that give additional insights about nitrogen-fixing microorganisms, plant-driven selection of microbes, plant growth promoting microorganisms, and plant pathogens.

Communication in the rhizosphere: mechanisms and functions

How bacteria sense plant signals

The bulk soil is generally a very poor, nutrient-diluted and therefore hostile environment in which nutrient bioavailability is often hampered by the soil biochemistry. Within this nutritional desert, the presence of plant roots provides the means for the formation of true oases with flourishing microbial populations because all roots have the ability to actively secrete low- and high-molecular-weight molecules into the rhizosphere. Root exudation is largely mediated by the root hairs, but also the root cap and apical epidermal cells make a significant contribution. These actively secreted compounds are composed of excretions—waste products from the plants' internal metabolic processes without any identified function— and secretions—a mixture of compounds that facilitate external processes like lubrication or nutrient acquisition. Moreover, root growth is accompanied by sloughing-off of living cells, senescence, cell wounding and leakage from plant cells which represent more passive release mechanisms of diverse components that nonetheless are very important for the provision of carbon in the soil. The compounds released by these processes are termed mucilages and exudates, respectively. Finally, the microbial community actively participates in defining the composition of the rhizosphere by degrading and secreting complex organics compounds, and by lysing plant cells. These types of molecules are part of mucilages and lysates, respectively (Bertin et al. 2003; Somers et al. 2004). The whole of these root-associated components accumulating in the rhizosphere is termed rhizodeposit and it has a large impact on plant growth and soil ecology. Rhizodeposition is a dynamic process that is developmentally regulated and varies with the plant species and cultivar; it is also altered upon biotic and abiotic stress. Moreover, the microbial community influences the composition of the exudates to its advantage (Yang and Crowley 2000; Paterson et al. 2006; Shaw et al. 2006; Yoneyama et al. 2007).

From the above it is clear that root exudates are complex molecular mixtures and in Table 1 the diversity of molecules identified in rhizodeposits is illustrated. Generally, rhizodeposition is involved in primary and secondary plant metabolic processes, nutrient and water acquisition, plant defence and stimulatory or inhibitory interactions with other soil organisms (Bertin et al. 2003). However, depending on their relative abundance, the different components of rhizodeposits also affect the soil microorganisms. It is not difficult to envision that many of these compounds are chemoattractants and welcome nutrients for the microbes living in or nearby the rhizosphere (Somers et al. 2004; Brencic and Winans 2005). Whereas many micro-organisms can only utilise rather general plant metabolites, some bacteria have the capacity to catabolize certain plant secondary metabolites providing a selective advantage to colonize the rhizosphere of specific plants (Savka et al. 2002). Examples of such nutritional mediators are glycosides and aryl-glycosides (Faure et al. 1999, 2001), calystegin (Tepfer et al. 1988; Guntli et al. 1999), certain flavonoids (Hartig et al. 1991), proline (Jiménez-Zurdo et al. 1997), 1-aminocyclopropane-1-carboxylic acid (Penrose and Glick 2001), and homoserine and betaines (Boivin et al. 1990; Goldmann et al. 1991). Another well described effect of often unidentified components of rhizodeposits is the activation of bacterial gene expression culminating in more or less intimate interactions with the producing plant host (Stachel et al. 1985; Koch et al. 2002; Brencic et al. 2005; Brencic and Winans 2005; Cooper 2007; Reddy et al. 2007; Franks et al. 2008; Johnston et al. 2008). Recent genome-wide studies have shown that root exudates modulate the expression of a significant number of bacterial genes of which the function in rhizosphere colonisation and competitiveness had not been anticipated (Mark et al. 2005; Matilla et al. 2007; Yuan et al. 2008b). Moreover, plants have been shown to secrete components that interfere with quorum sensing (Teplitski et al. 2000; Dunn and Handelsman 2002; Gao et al. 2003), a cell-cell signalling mechanism in bacteria that is very important in group-coordinated processes that impact interactions with Eukaryotes (von Bodman et al. 2003; Waters and Bassler 2005). In order to trigger these diverse molecular, physiological and behaviour responses, soil bacteria first have to sense the presence of the root exudates via one- and two-component signal perception systems.

A widespread mechanism by which bacteria sense their environment and respond accordingly is the two-component system which is typically comprised of a usually membrane-bound sensor histidine protein

Table 1 Organic compounds and enzymes released by plants in root exudates and their function in the rhizosphere

Class of compounds	Components	Functions
Sugars	arabinose, desoxyribose, fructose, galactose, glucose, maltose, oligosaccharides, raffinose, rhamnose, ribose, sucrose, xylose, mannitol, complex polysaccharides	lubrication; protection of plants against toxins; chemoattractants; microbial growth stimulation
Amino acids and amides	all 20 proteinogenic amino acids, γ-aminobutyric acid, cystathionine, cystine, homoserine, mugenic acid, ornithine, phytosiderophores, betaine, stachydrine	inhibit nematodes and root growth; microbial growth stimulation; chemoattractants, osmoprotectants; iron scavengers
Aliphatic acids	acetic, acetonic, aconitic, aldonic, butyric, citric, erythronic, formic, fumaric, gluconic, glutaric, glycolic, isocitric, lactic, maleic, malic, malonic, oxalic, oxaloacetic, oxaloglutaric, piscidic, propionic, pyruvic, shikimic, succinic, tartaric, tetronic, valeric acid	plant growth regulation; chemoattractants; microbial growth stimulation
Aromatic acids	p-hydroxybenzoic, caffeic, p-coumeric, ferulic, gallic, gentisic, protocatechuic, sinapic, syringic acid	plant growth regulation; chemoattractants
Phenolics	flavanol, flavones, flavanones, anthocyanins, isoflavonoids, acetosyringone	plant growth regulation; allelopathic interactions; plant defence; phytoalexins; chemoattractants; initiate legume-rhizobia, arbuscular mycorrhizal and actinorhizal interactions; microbial growth stimulation; stimulate bacterial xenobiotic degradation
Fatty acids	linoleic, linolenic, oleic, palmitic, stearic acid	plant growth regulation
Vitamins	p-aminobenzoic acid, biotin, choline, n-methionylnicotinic acid, niacin, panthothenate, pyridoxine, riboflavin, thiamine	microbial growth stimulation
Sterols	campestrol, cholesterol, sitosterol, stigmasterol	plant growth regulation
Enzymes and proteins	amylase, invertase, phosphatase, polygalacturonase, protease, hydrolase, lectin	plant defence; Nod factor degradation
Hormones	auxin, ethylene and its precursor 1-aminocyclopropaan-1-carboxylic acid (ACC), putrescine, jasmonate, salicylic acid	plant growth regulation
Miscellaneous	unidentified acyl homoserine lactone mimics, saponin, scopoletin, reactive oxygen species, nucleotides, calystegine, trigonelline, xanthone, strigolactones	quorum quenching; plant growth regulation; plant defence; microbial attachment; microbial growth stimulation; initiate arbuscular mycorrhizal interactions

Adapted from Bertin et al. (2003) and Somers et al. (2004)

kinase and a response regulator most often mediating differential gene expression. The structural genes are frequently organised as an operon, although many orphan kinases have been detected in the available bacterial genomes, and both encoded proteins consist of at least two domains. Via an N-terminal input domain the sensor perceives a specific stimulus, which upon interaction results in a conformational change of the cytoplasmic transmitter domain resulting in autophosphorylation at a conserved histidine residue. The activated transmitter domain will then in turn activate the N-terminal receiver domain of the cognate response regulator by phosphotransfer to a conserved aspartate residue. Next, the activated response regulator will mediate a cellular response via its C-terminal effector or output domain mainly by protein-protein interactions or protein-DNA interactions (see Fig. 1). Finally, dephosphorylation of the response regulator brings the system back to the pre-stimulus state (Laub and Goulian 2007). A common variation on this prototypical two-component system is the phosphorelay in which the histidine kinase has an additional receiver-like domain. Such hybrid histidine kinases will, upon signal perception and subsequent autophosphorylation, transfer their phosphoryl group intramolecularly to an aspartate residue in their receiver domain. This phosphoryl group is then transferred to a histidine residue of a cytoplasmic histidine phosphotransferase which finally shuttles it to the aspartate residue of the terminal response regulator (see Fig. 1a; Hoch and Varughese 2001). Whereas the histidine kinase domains of the sensors and the receiver domains of the response regulators comprise paralogous gene families that share considerable sequence and structural similarity, their input and output domains vary extensively although many conserved modules have been identified (Galperin 2006; Mascher et al. 2006; Szurmant et al. 2007). Classification of the sensor histidine protein kinases based on their domain architecture, reflecting the mechanism of sensing and signal transduction, revealed three major groups: the periplasmic- or

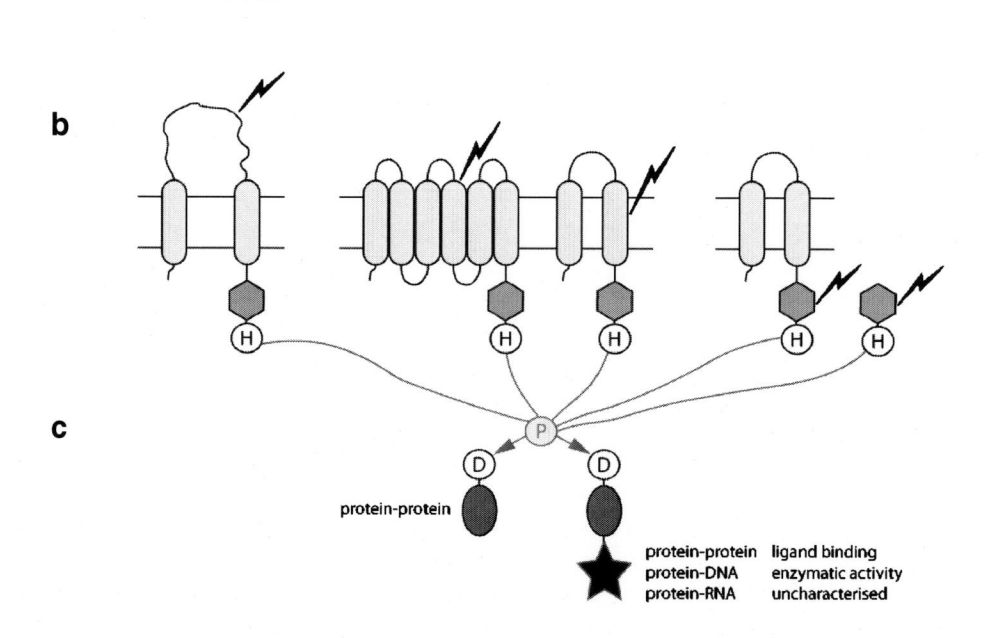

Fig. 1 Different signal perception and transduction mechanisms in bacteria. **a** Overview of the signal transduction systems; **b** Classification of the sensor histidine protein kinases: the periplasmic- or extracellular-sensing histidine kinases (*left*), histidine kinases with sensing mechanisms linked to transmembrane regions (*middle*), and cytoplasmic-sensing histidine kinases (*right*); **c** Classification of the response regulators: stand-alone receiver domains (*left*), and receiver domains combined with an output domain (*right*). H, histidine residue; D, aspartate residue; P, phosphotransfer; HPT, histidine phosphotransferase; lightening flash, incoming signal

extracellular-sensing histidine kinases, histidine kinases with sensing mechanisms linked to transmembrane regions, and cytoplasmic-sensing histidine kinases (see Fig. 1b; Mascher et al. 2006). The first class is the largest and signal detection occurs directly via binding of a small molecule to the sensor domain, indirectly through interaction with a periplasmic solute-binding protein, or via a conformational change of the input domain after a mechanical or electrochemical stimulus. The hybrid histidine kinases of phosphorelay systems belong to this class. Typically, the periplamic sensor kinases recognise solutes and nutrients, and so are part of many two-component systems involved in rhizosphere sensing. The second and smallest class of sensor kinases lack elaborate extracellular input domains and rely mainly on their transmembrane helices for perception of stimuli that are either associated with the membrane or occur within the membrane interphase. The third class groups the cytoplasmic-sensing histidine kinases which can be membrane-anchored or soluble and detect diffusible or internal stimuli (Mascher et al. 2006). A structural classification of the response regulators based on their domain architectures resulted in six major types that reflect their functionality: stand-alone receiver domains, and receiver domains combined with DNA- or RNA-binding, enzymatic, protein- or ligand-binding and uncharacterized output domains (see Fig. 1c; Galperin 2006). The transcriptional regulators with a DNA-binding output domain encompass 75% of all response regulators and typically have a important role in rhizosphere signal transduction.

Although two-component systems have been considered as the paradigm signal perception and transduction systems in prokaryotes, large scale genome analyses have recently shown that a bacterial cell contains a plethora of the much simpler one-component systems (Ulrich et al. 2005). Typically, these systems are single proteins that contain input and output domains, but lack the phosphotransfer histidine kinase and receiver domains (see Fig. 1a). Another type of one-component systems resembles fusions of classical histidine kinases with full-length response regulators and consists of single proteins with input, transmitter, receiver and output domains (see Fig. 1a; Galperin 2006). The repertoire of input and output domains in one-component systems is much more diverse than in two-component systems, with many domains unique for the one-component systems. This finding suggests that one-component systems likely perceive similar stimuli and elicit similar responses as two-component systems; their greater variability however is related to their extensive involvement in cytoplasmic sensing (Ulrich et al. 2005).

The list of one- and mainly two-component systems involved in plant recognition by rhizospheric bacteria is obviously extensive and several of them will be described in detail throughout this chapter. The following examples illustrate that almost every class of sensors and response regulators are involved in rhizosphere sensing. The periplasmic-sensing histidine kinase VirA of *Agrobacterium tumefaciens* recognises acidic pH, phenolic compounds, and monosaccharides (the latter via the periplasmic sugar-binding protein ChvE) released by wounded plant cells and its cognate DNA-binding response regulator VirG activates *vir* gene expression initiating T-DNA transfer (Mukhopadhyay et al. 2004). The GacS hybrid histidine kinase of many proteobacteria recognises environmental signals and activates the transcription factor GacA that controls for instance the biosynthesis of extracellular enzymes and secondary metabolites involved in virulence (Heeb and Haas 2001). The CbrAB system of *Pseudomonas aeruginosa* senses the intracellular carbon/nitrogen ratio via the transmembrane sensor CbrA and CbrB adjusts its catabolism by modulating expression of catabolic operons (Nishijyo et al. 2002); in *Pseudomonas putida*, a CbrAB system is involved in the degradation of IAA (Leveau and Gerards 2008). The membrane-associated cytoplasmic sensor kinase FixL and its cognate response regulator FixJ mediate O_2-controlled gene expression in root-nodulating bacteria (Gilles-Gonzalez and Gonzalez 2004). The soluble sensor CheA with its response regulator CheY controls chemotaxis in many bacteria via protein-protein interactions (Szurmant and Ordal 2004). The best described one-component system implicated in perception of the plant is likely the cytoplasmic possibly membrane-associated NodD protein of rhizobia (Brencic and Winans 2005). It is a LysR type transcription factor that perceives flavonoids and then activates transcription of the *nod* genes that encode the biosynthesis of the lipochito-oligosaccharide Nod factor (Peck et al. 2006).

Although many signal transduction systems have been described and keep on being identified via genome-wide approaches (Mascher et al. 2006; Qian et al. 2008), a lot of the mechanistic details and the identity of many of the primary stimuli remain to be uncovered. Nevertheless, from the above it is clear that the simple and exchangeable modular design of one- and two-component systems combined with extensive cross-regulation permits bacteria to perform sophisticated information processing allowing them to survive in the dynamic rhizosphere environment.

How plants sense bacteria

Plants evolved complex and diverse mechanisms to sense and respond to bacterial presence. Morphogens such as cytokinines, auxins and Nod factors, can profoundly affect plants. Plants sense Nod factors via receptor kinases of the LysM family (primary Nod factor receptor MtLYK4/NFP, and secondary Nod factor receptor or entry receptor MtLYK3/HCL), upon which a complex signal transduction cascade is triggered involving other extracellular-domain-containing receptors (Jones et al. 2007 and references therein). Plants also respond to presence of bacterial quorum-sensing signals, but the mechanism involved is still unknown (Mathesius et al. 2003; Schuhegger et al. 2006; von Rad et al. 2008). However pathogen-associated molecular patterns (PAMPs) such as lipopolysaccharides (LPS), peptidoglycan, and abundant proteins like the translational factor EF-tu and flagellin, are perceived via specific receptors, the pattern recognition receptors (PRRs). Perception of PAMPS constitutes the primary immune response of plants and is referred as PAMP-triggered immunity (PTI) (Zipfel 2008). Regulatory cascades implicating several classes of kinases activate the PTI. Until today, FLS2 and EFR are the only known PRRs in Arabidopsis; other examples of plant PRRs are very scarce (Zipfel 2008). However, the genome of *Arabidopsis* possesses numerous (hundreds) potential PRRs (Schwessinger and Zipfel 2008). The availability of new genomic resources and novel tools should enable the discovery of additional PRRs from crop species.

In the co-evolution of host-microbe interactions (Chisholm et al. 2008), pathogens acquired the capacity to suppress PTI by interfering with recognition at the plasma membrane or by secreting effector proteins into the plant cell cytosol that alter resistance signalling and PTI. Remarkably, the ability to deliver proteins directly into plant host cells is a common feature among phytopathogens. Bacterial effectors that are released into plant cells can possess enzyme activities, such as proteases and phosphatases, which are responsible for modifying host protein to enhance pathogen virulence and evade detection. Some other effectors are protein chaperones protecting the pathogen itself from these potentially detrimental enzymatic activities or keeping the effector protein unfolded prior to secretion.

In response to the delivery of pathogen effector proteins, plants acquired surveillance proteins (R proteins) to recognize and either directly or indirectly block or modify the properties of bacterial effectors. This response constitutes the secondary immune response of plants and is referred to as effector-triggered immunity (ETI). The connection between PTI and ETI is an emerging field of research. Furthermore, the role of small RNAs in immunity and that of PTI in symbiosis are valuable areas to investigate.

How bacteria sense bacteria

Bacteria have evolved sophisticated mechanisms to coordinate gene expression at population and community levels via the synthesis and perception of diffusible molecules. Because the concentration of the emitted signal in a confined environment reflects the bacterial cell number per volume unit (commonly cell density), such a regulatory pathway was termed quorum sensing (QS) (Fuqua et al. 1994). In an open environment, however, the concentration of the signal reflects both the bacterial cell number and the signal diffusion coefficient. In such open environments, the term diffusion sensing was proposed (Redfield 2002). A recent tentative to unify quorum and diffusion sensing states that the perception of a signal by a cell (efficiency sensing) is modulated by three essential factors: cell density (quorum sensing), mass-transfer properties (diffusion sensing), and spatial distribution of the cells (Hense et al. 2007). However, additional environmental factors may directly modify the synthesis rate and stability of the signals in the rhizosphere, which will be discussed in a latter paragraph.

The nature of QS signals is highly diverse (Schaefer et al. 2008; Whitehead 2001). Oligopeptides and substituted gamma-butyrolactones have been described in Gram-positive bacteria, while other substituted gamma-butyrolactones, the N-acyl-homoserine lactones (AHLs), are synthesized by a large number of Gram-negative bacteria. In this latter bacterial group, 3-hydroxypalmitic acid methyl ester (Flavier et al. 1997), 3,4-dihydroxy-2-heptylquinoline (Holden et al. 1999), and a furanosyl borate diester (Chen et al. 2002) can also act as QS signals. The most studied QS signals among rhizobacteria are AHLs (Whitehead et al. 2001). The synthesis of AHL depends upon synthases generally belonging to two classes: the LuxI and the AinS homologs. The perception of the signal relies on a sensor protein, a LuxR homolog, which is also the transcriptional regulator controlling the expression of QS-regulated genes.

The rhizosphere is potentially favorable for QS signalling, because it is a spatially structured habitat that is colonized, at a high cell density, by diverse bacterial populations. Experimental evidence supports this assertion. Ten to twenty percent of the cultivable bacteria in soil and rhizospheric environments are AHL-producing (D'Angelo-Picard et al. 2004). They are able to communicate both at the intra- and inter-species level (Steidle et al. 2001, 2002). Moreover, AHL signalling is implicated in the manifestation of plant-associated phenotypes in pathogenic, symbiotic, and biocontrol bacterial strains. The functions controlled by QS are highly diverse, including the horizontal transfer of plasmids, and the regulation of rhizospheric competence factors such as antibiotics, as well as functions that are directly implicated in plant-bacteria associations, such as virulence factors (Whitehead et al. 2001).

The AHL QS-signals show variations in the length and side chains of a core structure, and each AHL receptor can recognize a specific AHL structure. Even though some correlation exists between the genetic position of a strain or a group of strains and their AHL production patterns (D'Angelo et al. 2005), most AHL profiles are not strictly conserved at the genus or species level. Indeed, some phylogenetically distant species exhibit similar AHL profiles, supporting inter-species communication. Several explanations may account for this phenomenon. At the molecular level, the amino acid sequences of the AHL synthases are sometimes more distant within one species than between distinct species (Gray and Garey 2001). At the ecological and evolutionary levels, the presence of multiple AHL synthase homologues in species such as in Rhizobium legumi-nosarum (González and Marketon 2003) and the fact that multiple *luxI-luxR* determinants in a bacterium may be acquired independently (Gray and Garey 2001), can explain the occurrence of these complex patterns of AHLs. As an example, in the genus *Rhizobium*, some strains produce a single AHL, while others synthesize several AHLs (González and Marketon 2003). Such heterogeneity within AHL profiles may result from a selective pressure that tends to stimulate the emergence of distinct molecular languages at sub-species level, especially when related organisms share common ecological niches. An alternative explanation calls for another selective pressure that would authorize inter-species cooperation. One can not exclude the possibility that bacterial populations use distinct communication pathways to discriminate different levels of genetic proximity (clone, population and community). One of the QS signals facilitating communication at community level would be a furanosyl borate diester (AI-2) that is synthesized and recognized by a large range of Gram-positive and Gram-negative bacteria (Chen et al. 2002). The multiplicity of QS-signals, their interconnection and their modulation by environmental factors, especially the plant host, as well as spatial and temporal constraints remain to be elaborated.

How bacteria and plants interfere with bacterial signals

Bacteria and plants, as well as their genetically modified derivatives generated for research and biotechnological purposes, can produce QS-signal biomimics or QS-interfering molecules, including QS-signal modifying enzymes (Dong et al. 2007). The term quorum quenching (QQ) encompasses various natural phenomena or engineered procedures that lead to the perturbation of the expression of QS-regulated functions.

QS-biomimics were discovered in plants and in bacteria; their function is still speculative (McDouglas et al. 2007). In contrast, numerous reports evaluated QQ mechanisms, their function *in vivo*, and their potential agricultural applications (Dong et al. 2007). The three main steps of QS regulation that seem to be targeted are signal synthesis, and the much better described signal stability and sensing. For instance, the red algae *Delisea pulchra* limits bacterial colonization (fouling) of its lamina by interfering with the

QS-controlled motility and biofilm-formation ability. This process is mediated by halogenated furanones produced by the algae that bind the bacterial LuxR receptor, prevent the binding of or displace the AHL signal, and thereby accelerate the degradation of the LuxR protein (Rasmussen and Givskov 2006). Other inhibitors have been found in plants such as pea and soybean, Medicago, fruit extracts such as those from grape and strawberry, garlic, vanilla, lily and pepper, *Clematis vitalba*, *Geranium molle*, and *Tropaeolum majusi* (Rasmussen and Givskov 2006). Fungi such as *Penicillum* species also produce inhibitors of QS, identified as the lactones patulin and penicillic acid (Rasmussen et al. 2005). Interestingly, patulin naturally occurs in fruits such as apple, pear, peach, apricot, banana, pineapple, and grape, where the compound may also contribute to the inhibition of QS. The impact of these molecules on the behavior of rhizobacteria remains to be clarified. Aside from the investigations on natural inhibitors, efforts have been made to identify or design chemical compounds that may target the LuxR-like receptor(s). Most of the designs are based on actual AHL structures and analogues with either activating or inhibitory activity have been identified (Reverchon et al. 2002).

QS-signals are subject to enzymatic degradation. The AHL- lactonases catalyze a reaction that is identical to pH-mediated lactonolysis (opening the gamma-butyrolactone ring), while acylases/amidohydrolases convert AHL to homoserine lactone and a fatty acid. These enzymatic activities were observed in bacteria such as *Variovorax* (Leadbetter and Greenberg 2000) and *Bacillus* (Dong et al. 2000). Since these pioneer reports, numerous bacteria inactivating AHLs have been identified (Faure and Dessaux 2007). Some dissimilate AHL, i.e. use these substrates as growth substrates, and some do not (Leadbetter and Greenberg 2000; Uroz et al. 2003). To date, AHL inactivation has been described in α-proteobacteria (e.g. *Agrobacterium*, *Bosea*, *Sphingopyxis* and *Ochrobactrum*), β-proteobacteria (e.g. *Variovorax*, *Ralstonia*, *Comamonas*, and *Delftia*), and γ-proteobacteria (e.g. *Pseudomonas* and *Acinetobacter*). AHL inactivation also occurs in Gram-positive strains, both amongst low-G+C% strains or firmicutes such as *Bacillus* and high-G+C% strains or actinobacteria, e.g. *Rhodococcus*, *Arthrobacter*, and *Streptomyces*. *Rhodococcus erythropolis* has lactonase and acylase activies, as well as an oxidoreductase that converts 3-oxo-AHL to 3-hydroxy-AHL, which represents a different AHL-modifying activity that is not sensu stricto degrading (Uroz et al. 2005, 2008). Since the substitution at C3 is crucial for signal specificity, the oxidoreductase leads to a change in or loss of the signaling capability of the QS molecules. Aside from bacteria, AHL-degradation abilities have also been observed in animals (Chun et al. 2004) and plants (Delalande et al. 2005).

Several authors have proposed to take advantage of quenching to develop novel medical and animal therapies or novel biocontrol strategies for plant pathogens (Dong et al. 2007; Rasmussen and Givskov 2006). QQ applications therefore fall into the family of anti-virulence/anti-disease strategies. QQ-enzymes may be also used to identify the QS-regulated functions in bacteria (Smadja et al. 2004). For agricultural developments, the frequently proposed strategies imply the degradation of QS signal by plants and bacteria. They are illustrated by the following examples: (i) plants, which are genetically modified to gain the capacity to inactivate AHL because they express the AHL-lactonase AiiA of *Bacillus*, were more resistant to *Pectobacterium carotovorum* infection than the parental, wild-type plants (Dong et al. 2001); (ii) QQ bacteria were proposed as biocontrol agents to interfere with the virulence of plant pathogens (Uroz et al. 2003); (iii) chemicals that either directly interfere with QS-signalling or stimulate the growth of QQ-bacteria in the treated rhizosphere (Cirou et al. 2007). All QQ strategies were developed in vitro or under greenhouse conditions, so their efficiency in the field remains to be evaluated. However, QQ strategies may also prevent QS-regulated functions in plant benefic bacteria, such as antifungal synthesis by biocontrol strains (Molina et al. 2003).

How bacteria can interfere with plant hormones

Plant hormones control plant growth and development by acting as signal molecules. They affect the spatial and temporal expression of various phenotypes such as plant cell elongation, division, and differentiation. In addition, they play an important role in a plant's response to biotic and abiotic stresses. Several plant-associated bacteria have evolved ways to tap into these hormone signalling pathways and to manipulate plant physiology accordingly and to their own advantage.

One such way is stimulation of hormone synthesis by the plant itself. For example, the pathogenic bacterium *Pseudomonas syringae* pv *tomato* DC3000 is able to induce the biosynthesis of the hormones auxin (Schmelz et al. 2003) and abscisic acid (de Torres-Zabala et al. 2007) in *Arabidopsis thaliana*. Another intriguing example is the ability of bacterial quorum sensing molecules such as AHLs to downregulate auxin-induced genes (Mathesius et al. 2003). A different and well-known type of bacterial manipulation of plant hormone levels is the transfer, integration and expression of bacterial DNA coding for the biosynthesis of auxin and cytokinin in plant tissues, as described for *Agrobacterium tumefaciens* and *A. rhizogenes* (Francis and Spiker 2005).

Another route for exploitation of the plant hormone system is through bacterial synthesis or degradation of plant hormones (Costacurta and Vanderleyden 1995; Patten and Glick 1996; Tsavkelova et al. 2006; Spaepen et al. 2007; Glick et al. 2007). Table 2 shows examples for the five classical plant hormones (Kende and Zeevaart 1997), i.e. auxin (indole 3-acetic acid or IAA), ethylene, abscisic acid (ABA), cytokinin (zeatin) and gibberellin (gibberellic acid or GA). As is clear from the table, every one of these hormones can be synthesized and/or degraded by bacteria. Obviously, our understanding of the pathways, genes, and enzymes underlying bacterial synthesis and/or degradation is biased towards what is known about a small number of intensively studied cases. These include the synthesis of IAA (Patten and Glick 1996; Spaepen et al. 2007) and the activity of 1-aminocyclopropane-1-carboxylate (ACC) deaminase, which lowers ethylene concentrations through degradation of the ethylene precursor ACC (Glick 2005; Glick et al. 2007). Much less is known about other activities, such as the phenomenon of bacterial IAA degradation which has long been recognized but until recently (Leveau and Lindow 2005) did not receive serious attention as a means by which bacteria might affect plant physiology. Only very recently the first bacterial genes for IAA degradation were discovered in a *Pseudomonas putida* species (Leveau and Gerards 2008).

Many bacteria are capable of producing more than one type of plant hormone (Boiero et al. 2007; Karadeniz et al. 2006). Moreover, some bacteria can produce and degrade the same hormone (Leveau and Lindow 2005), produce one and degrade the precursor of another (Patten and Glick 2002), or harbor the genes for more than one biosynthetic pathway, e.g. *Pantoea agglomerans* pv *gypsophilae*, which features an IAM as well as an IPyA biosynthetic pathway for IAA (Manulis et al. 1998). This potential of even single bacterial strains to interfere differently with plant hormone levels remains one of the challenges towards better understanding, predicting, and possibly controlling plant hormone manipulation in complex plant-associated bacterial communities.

Plant signalling and physiology are affected by bacterial hormone synthesis and/or degradation in different ways, depending on the physiological role of the hormone, on the recalcitrance of plant tissue to changes in the hormone pool, and on the magnitude of the hormonal sink or source that these bacteria represent. Bacterially produced IAA may be beneficial or detrimental to plants. In *Azospirillum brasilense* (Dobbelaere et al. 1999) and *P. putida* GR12-2 (Patten and Glick 2002) it enhances root proliferation which results in greater root surface area through which more nutrients and water can be absorbed from the soil. In *P. syringae* pv *savastanoi* (Robinette and Matthysse 1990), *Erwinia chrysanthemi* (Yang et al. 2007) and *Rhodococcus fascians* (Vandeputte et al. 2005), IAA synthesis has been shown to be necessary for pathogenesis. Bacteria with ACC deaminase activity are generally considered beneficial to plants, as they promote root elongation and increase root density (Glick 2005). For cytokinins, it was suggested that bacteria are indispensable to plant growth because they would represent the only source of this type of hormone in plants (Holland 1997). This hypothesis was later rejected however with the discovery of plant genes encoding cytokinin synthesis (Sakakibara and Takei 2002).

From the bacterial perspective, there are several advantages to invest in plant hormone production or degradation. It has been suggested (Robert-Seilaniantz et al. 2007) that plant pathogens benefit from the production of phytohormones as this suppresses plant defense responses. In galls and tumours, production of IAA and cytokinin stimulates cell division, which acts as a sink for exploitable nutrients from other parts of the plant. IAA production may also locally stimulate ethylene biosynthesis, which indirectly prevents water and nutrient losses to the shoot organs above the tumor (Aloni et al. 1995). IAA production or ACC deaminase activity by plant-growth promoting rhizobacteria results in increased root density and therefore more

Table 2 Bacterial synthesis and degradation of plant hormones

Hormone	Pathway	Key enzyme(s)	Gene(s)	Representative species	Reference
IAA	Trp→IAM→IAA	Trp 2-monooxygenase, IAM hydrolase	iaaM, iaaH	Agrobacterium tumefaciens	(Inze et al. 1984)
				Pseudomonas syringae pv. savastanoi	(Yamada et al. 1985)
				Bradyrhizobium japonicum (IAM→IAA)	(Sekine et al. 1989)
	Trp→IPyA→IAAld→IAA	IPyA decarboxylase	ipdC	Pantoea agglomerans pv. gypsophilae	(Clark et al. 1993)
				Pseudomonas syringae pv. syringae	(Mazzola and White 1994)
				Enterobacter cloacae	(Koga et al. 1991)
				Azospirillum brasilense	(Costacurta et al. 1994)
				Pantoea agglomerans	(Brandl and Lindow 1996)
				Pseudomonas putida	(Patten and Glick 2002)
	Trp→TAM→IAAld→IAA	Trp decarboxylase, TAM oxidase	–	Bacillus cereus (Trp → TAM)	(Perley and Stowe 1966)
				Azospirillum brasilense (TAM → IAA)	(Hartmann et al. 1983)
	Trp→IAAld→IAA	Trp side-chain oxidase	–	Pseudomonas fluorescens	(Oberhansli et al. 1991)
	Trp→IAN→IAA	IAN nitrilase	nitA	Alcaligenes faecalis	(Kobayashi et al. 1993)
				Pseudomonas fluorescens	(Kiziak et al. 2005)
	IAA→IAA-Lys	IAA-lysine synthase	iaaL	Pseudomonas savastanoi	(Glass and Kosuge 1986)
	IAA→Cat→	catechol ortho cleavage into β-ketoadipate pathway	iac locus, catABC-pcaD	Pseudomonas putida	(Leveau and Gerards 2008)
	IAA→Ska→Ind→Sal→Cat	–	–	Pseudomonas sp.	(Proctor 1958)
	IAA→Dio→Isa→IsA→Ant	isatin amidohydrolase	–	Bradyrhizobium japonicum	(Olesen and Jochimsen 1996)
	IAA→2-FABA→Ant	–	–	unidentified	(Tsubokura et al. 1961)
C_2H_4	Met→KMBA→C_2H_4	methionine transaminase	–	Escherichia coli	(Ince and Knowles 1985)
	Glu→2-OG→C_2H_4	ethylene-forming enzyme	efe	Agrobacterium rhizogenes	(Kepczynska et al. 2003)
	ACC→C_2H_4	–	–	Pseudomonas syringae	(Nagahama et al. 1994)
				Bacillus sp.	(Bae and Kim 1997)
	ACC→2-OBA	ACC deaminase	acdS	Agrobacterium rhizogenes	(Kepczynska et al. 2003)
				Enterobacter cloacae	(Shah et al. 1998)
				Achromobacter, Azospirillum, Burkholderia, Pseudomonas, Ralstonia, Rhizobium, Kluyvera species	(Blaha et al. 2005)
	C_2H_4→CO_2	–	–	Pseudomonas sp.	(Kim 2006)
ABA	–	–	–	Bradyrhizobium japonicum	(Boiero et al. 2007)
				Azospirillum brasilense	(Perrig et al. 2007)
Z/ZR	AMP→iAMP→Z/ZR	isopentenyl transferase (cytokinin synthase)	ipt	Agrobacterium tumefaciens	(Akiyoshi et al. 1984)
			ptz	Pseudomonas savastanoi	(Powell and Morris 1986)
			etz	Rhodococcus fascians	(Crespi et al. 1992)
			fas1	Erwinia herbicola	(Lichter et al. 1995)

Hormone	Conversion	Enzyme	Gene	Organism	Reference
	tRNA→isopentenyl-tRNA→iAMP→Z/ZR	tRNA:isopentenyl transferase	miaA	Agrobacterium tumefaciens Methylobacterium sp.	(Gray et al. 1996) (Koenig et al. 2002)
	Z/ZR→GA	cytokinin oxidase/dehydrogenase	fas5	Rhodococcus fascians	(Galis et al. 2005)
		–	–	Rhizobium phaseoli, Azospirillum sp. A. diazotrophicus, H. seropedicae	(Rademacher 1994) (Bastian et al. 1998) (Gutierrez-Manero et al. 2001)
GA	GA→	–	–	Bacillus sp.	
				Unspecified	(Riviere et al. 1966)

Abbreviations: -, not known, *2-FABA* 2-formaminobenzoylacetic acid, *2-OBA* 2-oxobutyric acid, *2-OG* 2-oxoglutarate, *ABA* abscisic acid, *ACC* 1-aminocyclopropane-1-carboxylic acid, *Ant* anthranilic acid, *C₂₄* ethylene, *Cat* catechol, *Dio* dioxindole, *GA* gibberellic acid, *Glu* glutamate, *IAA* indole 3-acetic acid, *IAA-Lys* IAA-lysine, *IAAld* indole 3-acetaldehyde, *IAM* indole 3-acetamide, *iAMP* isopentenyl-AMP, *IAN* indole 3-acetonitrile, *Ind* indoxyl, *IPyA* indole 3-pyruvate, *Isa* isatin, *IsA* isatinic acid, *KMBA* 2-keto-4-methylthiobutyric acid, *Sal* salicylic acid, *Ska* skatole, *TAM* tryptamine, *Trp* Tryptophan, *Z* zeatin, *ZR* zeatin riboside

surface to colonize and greater return in root exudation. Several studies have shown that the ability to grow on or in plants is reduced in bacterial mutants unable to produce IAA (Brandl et al. 2001; Suzuki et al. 2003) or ethylene (Weingart et al. 2001), although the basis of this remains unclear.

There might also be other reasons for bacteria to produce or degrade plant hormones. Ethylene, for example, is a fungistatic (Smith 1973), the production of which might help bacteria to compete with fungi for plant-derived nutrients. Similarly, IAA has been shown to be inhibitory at high concentrations to plant-associated bacteria (Liu and Nester 2006). A less obvious reason to degrade plant hormones is that they represent sources of nutrition. For example, *P. putida* 1290 can use IAA as sole source of carbon and energy (Leveau and Lindow 2005). Given the relatively low concentrations of IAA and other hormones in the plant environment, it is doubtful that these compounds contribute greatly to bacterial biomass. However, it is noteworthy that three of the five classic hormones represent sources of nitrogen which might be of importance under conditions of nitrogen limitation. In fact, several of the degrading enzymes listed in Table 2 release readily available nitrogen from plant hormones or their precursors. For example, ACC deaminase produces ammonia, a property that has greatly facilitated the search for bacteria with ACC deaminase activity by selection for growth on ACC as sole source of nitrogen (Penrose and Glick 2003). Similarly, the transaminase enzyme involved in bacterial ethylene production from methionine releases the amino group from methionine as a source of nitrogen for growth (Ince and Knowles 1985). Several bacteria can use IAA as sole source of nitrogen (Leveau and Lindow 2005), but more than one enzymatic step is required for the release of nitrogen from the indole ring.

Degradation and utilization of plant hormones represent an extreme form of hormone inactivation, analogous to IaaL activity which conjugates and biologically inactivates IAA (Glass and Kosuge 1986). However, it is worth noting that the bacterial degradation products of some plant hormones are in turn signal molecules. For example, a *Pseudomonas* sp. from soil (Proctor 1958) was shown to convert IAA to catechol via salicylate, which is a plant hormone (Raskin 1992) involved in the plant response to pathogens. Thus, bacteria may have the potential to re-circuit certain plant signalling pathways by conversion of one hormone to another. Such bacterially induced re-circuiting may not be limited to plant signalling pathways. For example, the IAA degradation pathway described for *Bradyrhizobium japonicum* (Jensen et al. 1995) and an Alcaligenes sp. (Claus and Kutzner 1983) features isatin, which has a demonstrated signalling function in bacteria, e.g. in biofilm formation by strains of *E. coli* (Lee et al. 2007). Furthermore, there is a growing body of evidence to suggest that IAA can actually act as a signal molecule in bacteria and fungi (Spaepen et al. 2007). For example, IAA induces the expression of genes in *E. coli* related to survival under stress conditions (Bianco et al. 2006), stimulates by a positive feedback mechanism its own synthesis in *Azospirillum* species (Vande Broek et al. 1999), and provokes invasive growth in *Saccharomyces cerevisiae* (Prusty et al. 2004). Thus, the use of hormones as signalling molecules does not appear to be exclusive to plants, but may also underlie part of the communication between bacteria and other microorganism.

Integrative examples

In the rhizobia-plant interaction

The legume rhizosphere has a strong attractive power on rhizobia since abundantly secreted polycyclic aromatic compounds called flavonoids trigger chemotactic responses directing the bacteria to their compatible host (Reddy et al. 2007). Subsequently, specific flavonoids are perceived by the NodD protein, a LysR-type transcription factor, which initiates the transcription of *nod*ulation genes that encode the biosynthetic machinery for the primary bacterial signal, the Nod factor. This lipochito-oligosaccharide consists of a β-1,4-linked *N*-acetyl-glucosamine backbone with four or five residues, carries an acyl chain at the C-2 position at the non-reducing end, and can be decorated at defined positions with acetyl, sulfonyl, carbamoyl, fucosyl or arabinosyl moieties depending on the rhizobial strain (reviewed by D'Haeze and Holsters 2002, 2005). Upon perception of the Nod factors by the plant multiple signal transduction pathways are redirected culminating in the initiation of nodule formation. However, the paramount role of legume flavonoids and rhizobial Nod factors in the initiation of the rhizobium-plant interaction has masked the appreciation of other signals derived from both partners in mediating the onset of a successful interaction. Moreover, it has become increasingly clear that flavonoids play several roles (in addition to *nod* gene induction), and likewise that Nod factors are not only essential for inducing plant responses like root hair curling and cortical cell division (Cooper 2007). The complexity of the molecular dialogue between both partners of the rhizobial symbiosis will be illustrated by two examples: the interaction between Sinorhizobium meliloti and Medicago, and between Rhizobium sp. NGR234 and one of its many hosts.

Typically, the rhizodeposits of alfalfa (*Medicago sativa*) and one of the model legumes, barrel medic (*M. truncatula*), are complex and consist of flavonoids, sugars, amino acids, dicarboxylic acids, hydroxy-aromatic acids, biotin and other vitamins that trigger chemotactic responses in and support growth of their microsymbiotic partner *Sinorhizobium meliloti* (Cooper and Rao 1995; Streit et al. 1996; Heinz et al. 1999). These plant metabolites are sensed and appropriate responses generated via one- and two-component systems, but recently a downstream role for trans-acting riboregulators has been revealed (del Val et al. 2007). The nutritional advantage for the bacteria inhabiting the rhizosphere is reinforced by the secretion of a riboflavin degradation product, lumichrome, by *S. meliloti*. It is suggested that lumichrome enhances root respiration and that the root-evolved CO_2 increases net carbon accumulation improving both plant and bacterial growth, but alternative mechanisms explaining the plant growth stimulatory effect have not been ruled out (Phillips et al. 1999; Matiru and Dakora 2005). Moreover, once a functional nodule is established, bacteroids synthesize rhizopines that are secreted into the rhizosphere and can be utilised by some *S. meliloti* strains, further strengthening the nutritional relation between both partners (Galbraith et al. 1998). At high cell densities long chain acyl homoserine lactones (AHLs), quorum sensing signals secreted by *S. meliloti*, accumulate beyond a threshold level and trigger responses in the population that positively affect the efficiency of root colonisation and nodule invasion, such as the down-regulation of bacterial motility (Hoang et al. 2008) and the production of symbiotically active galactoglucan (Marketon et al. 2003). Unexpectedly it was shown that the AHLs produced by *S. meliloti* had a strong impact on the proteome of *M. truncatula*, modulating 7% of the total resolved proteins affecting diverse functions such as primary metabolism, protein processing, transcriptional regulation, host defence, hormone responses and cytoskeletal activity (Gao et al. 2003; Mathesius et al. 2003). *M. truncatula* itself produces quorum sensing mimics that can potentially modulate the bacterial behaviour in the rhizosphere (Teplitski et al. 2000), and interestingly, exposure of the roots to AHLs of *S. meliloti* altered the amounts and types of AHL mimics secreted by *M. truncatula* (Mathesius et al. 2003), illustrating a strong interplay between both partners. At this point of the interaction the bacterial population is located close to the root, sufficiently dense and not motile which allows it to colonize the root hairs. Biofilm formation represents the "natural way of life" for bacterial populations because it offers a protective environment and the possibility for co-operative behaviour (Morris and Monier 2003; Lasa 2006). Typically surface polysaccharides play an important role in biofilm maturation (Branda et al. 2005), and in *S. meliloti* cyclic β-glucans are mediating efficient root hair attachment (Dickstein et al. 1988), a first and essential step in biofilm formation. The *nod*D-like gene *syr*M is involved in controlling biosynthesis of succinoglycan, which contributes to the capacity to form highly structured biofilms (Fujishige et al. 2006). Interestingly, it was discovered that core Nod factors synthesized by the common *nod* genes *nod*ABC, and regulated by NodD1 but independent of *nod* gene-inducing plant flavonoids, are also required for biofilm formation and efficient attachment to roots. The core Nod factors apparently facilitate cell-to-cell adhesion which is thought to allow the bacteria to remain closely attached to the roots until, in response to plant inducers, a sufficient localized concentration of the host-specific signalling Nod factor is produced, required for triggering plant developmental processes that mark the onset of the symbiotic interaction (Fujishige et al. 2008). Indeed, upon perception of luteolin by NodD1 (Peck et al. 2006), or non-flavonoid inducers by NodD2 (Phillips et al. 1992; Gagnon and Ibrahim 1998), expression of both the common and the host-specific *nod* genes is activated and fully decorated Nod factors are synthesized (Lerouge et al. 1990). However, the rhizobial response to plant flavonoids goes far beyond the synthesis of host-specific Nod factors. Several genome-wide studies have identified multiple luteolin- or apigenin-induced genes that have no *nod*-box in their promotors and hence do not belong to the *nod* gene family (Barnett et al. 2004; Zhang and Cheng 2006). The function of most these genes awaits elucidation, but these results strongly suggest that the early stages of symbiosis are likely to be more complex than originally anticipated. In a last step of the rhizospheric signalling between *S. meliloti* and its legume host, the localized production of host-specific Nod factors is perceived by the plant via the LysM-type receptor kinases and the subsequent complex signal transduction cascade that culminates in early plant responses such as initiation of cortical cell division, calcium spiking and formation of colonized curled root hairs (Jones et al. 2007 and references therein). From the latter, the bacteria induce inward tip growth of the root hair and via these infection threads gain access to plant tissues, start their endophytic life phase and initiate their journey to the nodule primordium. Although beyond the scope of this chapter, clearly, during this endophytic part of the infection process many signals are exchanged, some of which are identified and known to be involved for instance in formation and progression of the infection threads (Nod factors, EPS and LPS; Jones et al. 2007, 2008), suppression of and protection against plant defence (SPS; Campbell et al. 2002; Ferguson et al. 2005; Jones et al. 2008) and activation of cortical cell division (flavonoids and cytokinins; Gonzalez-Rizzo et al. 2006; Wasson et al. 2006); many signals however remain to be discovered.

As for other legume-rhizobium examples, the signal exchange occurring at the onset of the symbiotic interaction between the promiscuous nodulator *Rhizobium* sp. NGR234 (hereafter NRG234) and one of its over 112 hosts (Pueppke and Broughton 1999) overlaps with the one described above for the *S. meliloti-Medicago* interaction. Indeed, flavonoid and non-flavonoid *nod* gene inducers (Le Strange et al. 1990), rhizopines, bacterial surface polysaccharides (Broughton et al. 2006; Staehelin et al. 2006) and Nod factors are important players in the communication between this bacterium and its host, but other signals play a role also. The Nod factors secreted by NGR234 activate flavonoid release in soybean (Schmidt et al. 1994), and the flavonoids activate transcription of 19 *nod*-box-containing promotors and 147 other genes in a *nod*-box-independent way. Whereas the functions of the latter largely remain to be discovered, the *nod*-box controlled genes encode typical pathways involved in Nod factor biosynthesis, rhizopine catabolism, SPS synthesis and modification, and nitrogen fixation, but also in transcriptional control, hopanoid synthesis, auxin (IAA) production, and type III secretion (Kobayashi et al. 2004). The presence of *nod*-boxes in the promotors of transcriptional regulators creates a complex regulatory network that allows sequential activation of gene expression. In this network, NodD1 is the key regulator of all 19 flavonoid-inducible loci including *syr*M2. SyrM2 in its turn controls the delayed flavonoid-induction of a number of loci that have SyrM binding sites in their promotors. One of these is *nod*D2 of which the gene product is required for the optimal activation of specific-*nod* boxes that control the expression of genes involved in the later stages of the symbiotic interaction. NodD2 also represses *nod*D1 expression, which results in a self-attenuation of the flavonoid-induced regulatory cascade (Kobayashi et al. 2004). Expression of hopanoid biosynthetic genes is NodD1 dependent and thus flavonoid inducible (Kobayashi et al. 2004). These lipids function as membrane reinforcers and could mediate resistance to environmental stress in the soil. However, hopanoids have been discovered in a number of nitrogen-fixing soil bacteria (Kannenberg et al. 1996), and in the actinomycete Frankia they are located in the envelope of specialised nitrogenase-containing vesicles possibly reducing oxygen diffusion and thereby protecting the nitrogenase (Rosa-Putra et al. 2001; Alloisio et al. 2007). Hence, hopanoids might function either during the rhizospheric or the endophytic phase of the symbiotic interaction. NodD1-controlled expression of the response regulator TtsI results in the activation of genes that carry a *tts*-box in their promotors and, amongst others, code for part of a type III secretion system, nodulation outer proteins (Nops) and homologs of effectors of pathogens, and the rhamnan component of LPS (Marie et al. 2004). The proteins secreted via the type III secretion system are rhizobial keys that are needed when the bacteria have entered the root hairs and, upon injection into the plant cells, they are thought to interfere with the eukaryotic cellular metabolism, altering plant defence or signalling networks permitting the continuation of nodule development (Marie et al. 2004; Skorpil et al. 2005). The rhamnose-rich LPS is likely also only implicated in the later stages of the interaction, and could be required for protection against plant defence molecules and for bacterial release from infection threads (Marie et al. 2004; Broughton et al. 2006). Auxin production is widespread amongst plant-associated bacteria including rhizobia, and it is often related to epiphytic fitness and suppression of defence (Prinsen et al. 1991; Robert-Seilaniantz et al. 2007; Spaepen et al. 2007). NGR234 synthesizes IAA via three independent pathways: the indole-3-acetamide, the tryptamine and the indole-3-pyruvic acid pathway. The latter is predominant and expression of the genes encoding this pathway is controlled by the NodD1-SyrM2-NodD2 regulatory circuit implying a function during the later stages of the interaction when a more intimate contact between both partners has been established (Theunis et al. 2004). Although no obvious nodulation phenotype was obtained upon mutation of the indole-3-pyruvic acid pathway, a putative role has been postulated in vascularisation of the nodule tissue, facilitating carbon and nitrogen exchange, or acting as a synergistic factor for other signals (Theunis et al. 2004).

From the above it is clear that the action radius of flavonoids and Nod factors has been underestimated. Moreover, the molecular dialogue between legumes and rhizobia has proven to go far beyond these two established signals. Instead a true communication network is established between both partners reflecting the complexity of setting up a successful interaction in the rhizosphere.

In the agrobacteria-plant interaction

Agrobacterium tumefaciens is a soil α-proteobacterium that can infect a broad range of dicotyledonous plants and transfers an oncogenic DNA fragment, the T-DNA, from its tumour-inducing (Ti) plasmid to the nuclear genome of plants (Gelvin 2000). This natural engineer largely contributed to the enormous advances in plant sciences. In the transformed plant tissues, the expression of T-DNA genes leads to the uncontrolled synthesis of growth regulators, auxin and cytokinins, resulting in the formation of tumours, a phenomenon known as crown gall disease. Three main steps could be proposed to describe the dynamics of the *A. tumefaciens*-plant interaction: (1) the colonization of rhizosphere and plant tissues by virulent and avirulent (free of Ti plasmid) agrobacteria; (2) the transfer of T-DNA from virulent agrobacteria to plants; (3) the emergence and development of a tumour in which avirulent bacteria may be converted into virulent ones by horizontal transfer of the Ti plasmid. In the course of their interaction, plants and agrobacteria exchange a wide variety of signals including, sugars, amino acids, phenolics, and lactones.

The number of agrobacteria increases (from 100 to 1,000 fold), as the structure of these populations varies, when the plant environment was compared to bulk soil (Sanguin et al. 2006). Agrobacteria can survive inside roots and root nodules (Wang et al. 2006), and invade the plants via vessels and apoplasm (Cubero et al. 2006). Microarray analysis of bacterial diversity revealed the predominance of agrobacteria in rhizosphere of maize (Sanguin et al. 2006). A high diversity of agrobacteria can coexist in one cubic centimetre of soil (Vogel et al. 2003). Commonly, most of the agrobacteria recovered from soil and rhizospheric samples are avirulent, lacking the Ti plasmid (Mougel et al. 2001). However, in conductive soils, virulent strains may dominate (Krimi et al. 2002). Several functions contribute to the capacity of agrobacteria to colonize the root, including motility, chemotaxis, surface characteristics and assimilation of a large spectrum of plant compounds. The genome of *A. tumefaciens* C58 is rich in ABC-genes that would participate in the sensing and transport of a large range of organic and inorganic compounds (Wood et al. 2001).

A complex machinery is required for the transfer of T-DNA to a plant cell. The *A. tumefaciens* VirB/D4 system is an archetypal Type IV secretion system composed of 11 VirB mating pair formation subunits and a VirD4 substrate receptor that form a trans-envelope secretion channel (Christie et al. 2005). The substrate of translocation is a single-stranded copy of the T-DNA that becomes integrated into the plant nuclear genome. Transfer of T-DNA operates in a few of hours (Sykes and Matthysse 1986). The transcription of the *vir* regulon is induced by specific plant-released phenolic compounds in combination with several other stimuli, such as monosaccharides, acidic pH and temperature below 30°C (Brencic and Winans 2005). The VirA-VirG two-component system and ChvE sugar binding protein are involved in the perception of these stimuli. Activation of *vir* genes and T-DNA transfer were observed in wounded and unwounded plant tissues. In unwounded transformed plant tissues, the synthesis of opines from T-DNA genes is observed even in the absence of tumour (Brencic et al. 2005), suggesting that cell division during wound healing may play a role in tumour formation.

T-DNA encodes the synthesis of the plant growth factors, cytokinines and auxin, as well as opines, which are specific growth substrates and signals for the bacteria colonizing the plant host. The cytokinine biosynthesis enzyme, which is encoded by the T-DNA, is targeted to and functions in plastids to shunt the original cytokinine pathway (Sakakibara et al. 2005). This feature illustrates that agrobacteria manipulate several compartments of the plant cells. The emergence and development of a tumour is a complex process in which overproduction of auxin and its gradual, flavonoid-dependent retention in the tissue, play an essential role (Schwalm et al. 2003). Furthermore, high vascularisation and epidermal disruption are associated with the establishment of tumours. These phenomena are linked to the redirection of the nutrient-bearing water flow and carbohydrate delivery for growth of the tumour tissues and the inhabiting bacteria (Wächter et al. 2003).

The synthesis of opines defines a specific microhabitat in the plant host. The assimilation of opines as carbon and nitrogen sources confer a selective advantage to the Ti plasmid harbouring bacteria in plant tumours, the so called opine niche. Some opines, termed conjugative opines, are required for high-rate of synthesis of 3-oxo-octanoyl-homoserine lactone (OC8HSL), a cell-to-cell signal implicated in the QS regulation of the conjugative transfer of the Ti plasmid (Piper et al. 1993). The recipient bacteria for the Ti plasmid may be Ti plasmid free agrobacteria, which represent up to 1% of the total cultivable bacteria in the rhizosphere, as well as other rhizobacteria belonging to different genera, such as *Sinorhizobium*, *Rhizobium*, and *Phyllobacterium* (Teyssier-Cuvelle et al. 1999, 2004). The Ti plasmid confers to these non-*Agrobacterium* hosts the capacity to assimilate opine and, in some instances, to induce tumours on the plant hosts; it also remains transferable to other bacteria (Teyssier-Cuvelle et al. 2004). These data strongly suggest that the Agrobacterium populations may not be unique reservoirs for the maintenance and propagation of the Ti plasmid in the rhizosphere. In addition to conjugation, Ti plasmid copy-number (Li and Farrand 2000) and severity of tumour symptoms are also subjected to QS regulation (Pappas and Winans 2003; Chevrot et al. 2006). Even though the mechanism that places emergence of tumours under QS regulation remains unknown, anti-virulence strategies targeting QS, termed quorum-quenching, have been proposed to decrease the *Agrobacterium*-induced symptoms on plants (Molina et al. 2003; Chevrot et al. 2006).

In *A. tumefaciens* C58-induced tumours, the conjugative opines, agrocinopines A and B, tightly control the synthesis of the OC8HSL signal at the transcriptional level. The AccR-mediated transcriptional repression of the *arc* (agrocinopine catabolism) operon (*orfA-orfB-splA-traR-mcpA*) of the Ti plasmid is released in the presence of agrocinopines A and B (Beck von Bodman et al. 1992; Piper et al. 1999). The *traR* gene of the *arc* operon encodes the transcriptional regulator TraR that binds OC8HSL and permits the expression of the OC8HSL synthase encoded by the *traI* gene. This latter gene belongs to the *trb* operon, located on the Ti plasmid. The TraR/OC8HSL complex also activates the expression of the *tra* and *rep* operons that are required for conjugative transfer and copy-number amplification, respectively, of the Ti plasmid. However, TraR activity is modulated at the post-translational level by TraM, which directly interacts with TraR (Luo et al. 2000) and thereby prevents the interaction between the TraR/OC8HSL complex and target DNA-sequences of QS-regulated promoters. In the presence of conjugative opines, the antagonistic effect of TraM would be compensated by the high synthesis rate of TraR.

The enzymatic inactivation of OC8HSL by lactonases AttM (Zhang et al. 2002) and AiiB (Carlier et al. 2003) also participates in the fine tuning of QS-controlled functions in *A. tumefaciens* C58. The expression of the lactonase AttM is regulated at the transcriptional level by plant signals, such as gamma-aminobutyrate (GABA) and its by-products such as gamma-hydroxybutyrate (GHB) and succinic semialdehyde (SSA) (Carlier et al. 2004; Chevrot et al. 2006). In wounded tissues and in *A. tumefaciens*-induced plant tumours GABA accumulates to high levels (Deeken et al. 2006). Noticeably, increasing evidences would suggest that GABA plays a key-role in interactions between plants and other organisms, including bacteria, fungi and insects (Shelp et al. 2006). In *A. tumefaciens*, the lactonase-encoding gene *attM* is part of the *attKLM* operon, the expression of which is controlled by the transcription factor AttJ (Zhang et al. 2002). In the presence of SSA and GHB, the repressing activity of AttJ is altered and the *attKLM* operon is expressed (Chai et al. 2007). Although GABA and gamma-butyrolactone (GBL) do not directly alter the repressing activity of AttJ, the expression of *attKLM* is also observed in the presence of these compounds. It is assumed that GABA and GBL are converted to SSA and GHB by *A. tumefaciens* and/or the plant host. In addition to the implication of AttM in the GBL-ring cleavage of OC8HSL, the *attKLM* operon encodes a complete degradation pathway of GBL into succinate, with GHB and SSA as intermediates (Carlier et al. 2004; Chai et al. 2007).

Plants recognize agrobacteria as invaders, and induce plant defense genes; in parallel agrobacteria have developed strategies to avoid plant defenses, including phenolics and reactive oxygen species (Kalogeraki et al. 1999; Citovsky et al. 2007; Saenkham et al. 2007). Noticeabely, benefic bacteria are also able to induce and avoid some chemical plant-defenses (examples in Faure et al. 1995, 1996; Dombrecht et al. 2005; Madhaiyan et al. 2006). *A. thaliana* detects different *A. tumefaciens* effectors, such as a conserved domain of flagellin and the transcriptional factor EF-Tu of A. tumefaciens. Specific receptors belonging to the Leu-rich repeat transmembrane receptor (LRR) family are implicated in perception of these effectors, such as EFR for EF-Tu and FLS2 for flagellin in *A. thaliana* (Chinchilla et al. 2006; Zipfel et al. 2006). Remarkably, Nicotiana benthamiana, a plant unable to perceive EF-Tu, acquires EF-Tu binding sites and responsiveness upon transient expression of the EFR receptor of A. thaliana. The LRR receptor kinase activates the mitogen-activated protein kinases (MAPK) to activate the immune response. One of the phosphorylated targets of MAPK3 is the transcription factor VIP1 that relocalizes from the cytoplasm to the nucleus and regulates the expression of the *PR1* pathogenesis-related gene. *A. tumefaciens* uses a Trojan horse strategy by hijacking VIP1 to import the VirE2 protein (associated with the T-DNA) into the nucleus (Djamei et al. 2007). Finally, two recent studies described the essential role of salicylic acid (SA) and auxin (IAA) in the control of virulence. IAA inhibits the expression of *vir* genes and the growth of *A. tumefaciens* (Liu and Nester 2006). This feature suggests a retro-control of T-DNA transfer by a product encoded by the T-DNA; therefore IAA avoids the cost—for plant and bacteria— of an additional transformation. However, SA, which accumulates upon bacterial infection, also shuts down the expression of the *vir* regulon (Yuan et al. 2007). Recently, multidisciplinary approaches are taken to give an integrative view of the fascinating A. tumefaciens-plant host interaction (Deeken et al. 2006; Yuan et al. 2008a).

Conclusions and perspectives

A multiplicity of signals controls the responses of plants and their associated organisms in the rhizosphere. The deciphering of the interconnections between all these signals is a future challenge that will be supported by global and fine analytic tools including transcriptomics, proteomics and metabolomics. Moreover, the analysis of temporal and spatial factors in these processes will give more precise insights into the dynamics of the interactions in the rhizosphere. Finally, in addition to model organisms, approaches such as metagenomics (Leveau 2007; Riaz et al. 2008), will take into account the diversity of organisms that communicate in the rhizosphere and the mechanisms implicated in this communication.

References

Akiyoshi DE, Klee H, Amasino RM, Nester EW, Gordon MP (1984) T-DNA of *Agrobacterium tumefaciens* encodes an enzyme of cytokinin biosynthesis. Proc Natl Acad Sci USA 81:5994–5998 doi:10.1073/pnas.81.19.5994

Alloisio N, Felix S, Marechal J et al (2007) *Frankia alni* proteome under nitrogen-fixing and nitrogen-depleted conditions. Physiol Plant 130:440–453 doi:10.1111/j.1399-3054.2007.00859.x

Aloni R, Pradel KS, Ullrich CI (1995) The 3-dimensional structure of vascular tissues in *Agrobacterium tumefaciens*-induced crown galls and in the host stems of *Ricinus communis* L. Planta 196:597–605 doi:10.1007/BF00203661

Bae M, Kim MY (1997) A new alkalophilic bacterium producing ethylene. J Microbiol Biotechnol 7:212–214

Barnett MJ, Tolman CJ, Fisher RF et al (2004) A dual-genome symbiosis chip for coordinate study of signal exchange and development in a prokaryote-host interaction. Proc Natl Acad Sci USA 101:16636–16641 doi:10.1073/pnas.0407269101

Bastian F, Cohen A, Piccoli P, Luna V, Baraldi R, Bottini R (1998) Production of indole-3-acetic acid and gibberellins A(1) and A(3) by *Acetobacter diazotrophicus* and *Herbaspirillum seropedicae* in chemically-defined culture media. Plant Growth Regul 24:7–11 doi:10.1023/A:1005964031159

Beck von Bodman S, Hayman GT, Farrand SK (1992) Opine catabolism and conjugal transfer of the nopaline Ti plasmid pTiC58 are coordinately regulated by a single repressor. Proc Natl Acad Sci USA 89:643–647 doi:10.1073/pnas.89.2.643

Bertin C, Yang X, Weston L (2003) The role of root exudates and allelochemicals in the rhizosphere. Plant Soil 256:67–83 doi:10.1023/A:1026290508166

Bianco C, Imperlini E, Calogero R, Senatore B, Amoresano A, Carpentieri A et al (2006) Indole-3-acetic acid improves *Escherichia coli*'s defences to stress. Arch Microbiol 185:373–382 doi:10.1007/s00203-006-0103-y

Blaha D, Prigent-Combaret C, Mirza MS, Moënne-Loccoz Y (2005) Phylogeny of the 1-aminocyclopropane-1-carboxylic acid deaminase-encoding gene *acdS* in phytobeneficial and pathogenic Proteobacteria and relation with strain biogeography. FEMS Microbiol Ecol 56:455–470 doi:10.1111/j.1574-6941.2006.00082.x

Boiero L, Perrig D, Masciarelli O, Penna C, Cassan F, Luna V (2007) Phytohormone production by three strains of *Bradyrhizobium japonicum* and possible physiological and technological implications. Appl Microbiol Biotechnol 74:874–880 doi:10.1007/s00253-006-0731-9

Boivin C, Camut S, Malpica CA et al (1990) *Rhizobium meliloti* genes encoding catabolism of trigonelline are induced under symbiotic conditions. Plant Cell 2:1157–1170

Branda SS, Vik Å, Friedman L et al (2005) Biofilms: the matrix revisited. Trends Microbiol 13:20–26 doi:10.1016/j.tim.2004.11.006

Brandl MT, Lindow SE (1996) Cloning and characterization of a locus encoding an indolepyruvate decarboxylase involved in indole-3-acetic acid synthesis in *Erwinia herbicola*. Appl Environ Microbiol 62:4121–4128

Brandl MT, Quinones B, Lindow SE (2001) Heterogeneous transcription of an indoleacetic acid biosynthetic gene in *Erwinia herbicola* on plant surfaces. Proc Natl Acad Sci USA 98:3454–3459 doi:10.1073/pnas.061014498

Brencic A, Winans SC (2005) Detection of and response to signals involved in host-microbe interactions by plant-associated bacteria. Microbiol Mol Biol Rev 69:155–194 doi:10.1128/MMBR.69.1.155-194.2005

Brencic A, Angert ER, Winans SC (2005) Unwounded plants elicit *Agrobacterium* vir gene induction and T-DNA transfer: transformed plant cells produce opines yet are tumour free. Mol Microbiol 57:1522–1531 doi:10.1111/j.1365-2958.2005.04763.x

Broughton WJ, Hanin M, Relić B et al (2006) Flavonoid-inducible modifications of rhamnan O antigens are necessary for *Rhizobium* sp. strain NGR234-legume symbiosis. J Bacteriol 188:3654–3663 doi:10.1128/JB.188.10.3654-3663.2006

Campbell GRO, Reuhs BL, Walker GC (2002) Chronic intracellular infection of alfalfa nodules by *Sinorhizobium meliloti* requires correct lipopolysaccharide core. Proc Natl Acad Sci USA 99:3938–3943 doi:10.1073/pnas.062425699

Carlier A, Uroz S, Smadja B, Fray R, Latour X, Dessaux Y, Faure D (2003) The Ti plasmid of *Agrobacterium tumefaciens* harbors an *attM*-paralogous gene, *aiiB*, also encoding N-Acyl homoserine lactonase activity. Appl

Environ Microbiol 69:4989–4993 doi:10.1128/AEM.69.8.4989-4993.2003

Carlier A, Chevrot R, Dessaux Y, Faure D (2004) The assimilation of gamma-butyrolactone in *Agrobacterium tumefaciens* C58 interferes with the accumulation of the N-acyl-homoserine lactone signal. Mol Plant Microbe Interact 17:951–957 doi:10.1094/MPMI.2004.17.9.951

Chai Y, Tsai CS, Cho H, Winans SC (2007) Reconstitution of the biochemical activities of the AttJ repressor and the AttK, AttL, and AttM catabolic enzymes of *Agrobacterium tumefaciens*. J Bacteriol 189:3674–3679 doi:10.1128/JB.01274-06

Chen X, Schauder S, Potier N, Van Dorsselaer A, Pelczer I, Bassler BL, Hughson FM (2002) Structural identification of a bacterial quorum-sensing signal containing boron. Nature 415:545–549 doi:10.1038/415545a

Chevrot R, Rosen R, Haudecoeur E, Cirou A, Shelp BJ, Ron E, Faure D (2006) GABA controls the level of quorum-sensing signal in *Agrobacterium tumefaciens*. Proc Natl Acad Sci USA 103:7460–7464 doi:10.1073/pnas.0600313103

Chinchilla D, Bauer Z, Regenass M, Boller T, Felix G (2006) The Arabidopsis receptor kinase FLS2 binds flg22 and determines the specificity of flagellin perception. Plant Cell 18:465–476 doi:10.1105/tpc.105.036574

Chisholm ST, Coaker G, Day B, Staskawisz BJ (2008) Host-microbe interactions: shaping the evolution of the plant immune response. Cell 124:803–814 doi:10.1016/j.cell.2006.02.008

Christie PJ, Atmakuri K, Krishnamoorthy V, Jakubowski S, Cascales E (2005) Biogenesis, architecture, and function of bacterial type IV secretion systems. Annu Rev Microbiol 59:451–485 doi:10.1146/annurev.micro.58.030603.123630

Chun CK, Ozer EA, Welsh MJ, Zabner J, Greenberg EP (2004) Inactivation of a *Pseudomonas aeruginosa* quorum-sensing signal by human airway epithelia. Proc Natl Acad Sci USA 101:3587–3590 doi:10.1073/pnas.0308750101

Cirou A, Diallo S, Kurt C, Latour X, Faure D (2007) Growth promotion of quorum-quenching bacteria in the rhizosphere of *Solanum tuberosum*. Environ Microbiol 9:1511–1522 doi:10.1111/j.1462-2920.2007.01270.x

Citovsky V, Kozlovsky SV, Lacroix B, Zaltsman A, Dafny-Yelin M, Vyas S, Tovkach A, Tzfira T (2007) Biological systems of the host cell involved in *Agrobacterium* infection. Cell Microbiol 9:9–20 doi:10.1111/j.1462-5822.2006.00830.x

Clark E, Manulis S, Ophir Y, Barash I, Gafni Y (1993) Cloning and characterization of *iaaM* and *iaaH* from *Erwinia herbicola* pathovar *gypsophilae*. Phytopathol 83:234–240 doi:10.1094/Phyto-83-234

Claus G, Kutzner HJ (1983) Degradation of indole by *Alcaligenes* spec. Syst Appl Microbiol 4:169–180

Cooper JE (2007) Early interactions between legumes and rhizobia: disclosing complexity in a molecular dialogue. J Appl Microbiol 103:1355–1365 doi:10.1111/j.1365-2672.2007.03366.x

Cooper JE, Rao JR (1995) Flavonoid metabolism by rhizobia—Mechanisms and products. Symbiosis 19:91–98

Costacurta A, Vanderleyden J (1995) Synthesis of phytohormones by plant-associated bacteria. Crit Rev Microbiol 21:1–18 doi:10.3109/10408419509113531

Costacurta A, Keijers V, Vanderleyden J (1994) Molecular cloning and sequence analysis of an *Azospirillum brasilense* indole-3-pyruvate decarboxylase gene. Mol Gen Genet 243:463–472

Crespi M, Messens E, Caplan AB, Vanmontagu M, Desomer J (1992) Fasciation induction by the phytopathogen *Rhodococcus fascians* depends upon a linear plasmid encoding a cytokinin synthase gene. EMBO J 11:795–804

Cubero J, Lastra B, Salcedo CI, Piquer J, López MM (2006) Systemic movement of *Agrobacterium tumefaciens* in several plant species. J Appl Microbiol 101:412–421 doi:10.1111/j.1365-2672.2006.02938.x

D'Angelo-Picard C, Faure D, Carlier A, Uroz S, Raffoux A, Fray R, Dessaux Y (2004) Bacterial populations in the rhizosphere of tobacco plants producing the quorum-sensing signals hexanoyl-homoserine lactone and 3-oxo-hexanoyl-homoserine lactone. FEMS Microbiol Ecol 51:19–29 doi:10.1016/j.femsec.2004.07.008

D'Angelo-Picard C, Faure D, Penot I, Dessaux Y (2005) Diversity of N-acyl homoserine lactone-producing and -degrading bacteria in soil and tobacco rhizosphere. Environ Microbiol 7:1796–1808 doi:10.1111/j.1462-2920.2005.00886.x

Deeken R, Engelmann JC, Efetova M, Czirjak T, Müller T, Kaiser WM, Tietz O, Krischke M, Mueller MJ, Palme K, Dandekar T, Hedrich R (2006) An integrated view of gene expression and solute profiles of *Arabidopsis* tumors: a genome-wide approach. Plant Cell 18:3617–3634 doi:10.1105/tpc.106.044743

Delalande L, Faure D, Raffoux A, Uroz S, D'Angelo C, Elasri M, Carlier A, Berruyer R, Petit A, Williams P, Dessaux Y (2005) N-Hexanoyl-L-homoserine lactone, a mediator of bacterial quorum-sensing regulation, exhibits a plant-dependent stability in the rhizosphere and may be inactivated by germinating *Lotus corniculatus* seedlings. FEMS Microbiol Ecol 52:13–20 doi:10.1016/j.femsec.2004.10.005

del Val C, Rivas E, Torres-Quesada O et al (2007) Identification of differentially expressed small non-coding RNAs in the legume endosymbiont *Sinorhizobium meliloti* by comparative genomics. Mol Microbiol 66:1080–1091 doi:10.1111/j.1365-2958.2007.05978.x

de Torres-Zabala M, Truman W, Bennett MH, Lafforgue G, Mansfield JW, Egea PR et al (2007) *Pseudomonas syringae* pv. *tomato* hijacks the *Arabidopsis* abscisic acid signalling pathway to cause disease. EMBO J 26:1434–1443 doi:10.1038/sj.emboj.7601575

D'Haeze W, Holsters M (2002) Nod factor structures, responses, and perception during initiation of nodule development. Glycobiology 12:79R–105R doi:10.1093/glycob/12.6.79R

D'Haeze W, Holsters M (2005) Surface polysaccharides enable bacteria to evade plant immunity. Trends Microbiol 12:555–561 doi:10.1016/j.tim.2004.10.009

Dickstein R, Bisseling T, Reinhold V et al (1988) Expression of nodule-specific genes in alfalfa root-nodules blocked blocked at an early stage of development. Genes Dev 2:677–687 doi:10.1101/gad.2.6.677

Djamei A, Pitzschke A, Nakagami H, Rajh I, Hirt H (2007) Trojan horse strategy in *Agrobacterium* transformation: abusing MAPK defense signaling. Science 318:453–456 doi:10.1126/science.1148110

Dobbelaere S, Croonenborghs A, Thys A, Vande Broek A, Vanderleyden J (1999) Phytostimulatory effect of *Azospirillum brasilense* wild type and mutant strains altered in IAA production on wheat. Plant Soil 212:155–164 doi:10.1023/A:1004658000815

Dombrecht B, Heusdens C, Beullens S, Verreth C, Mulkers E, Proost P, Vanderleyden J, Michiels J (2005) Defence of *Rhizobium etli* bacteroids against oxidative stress involves a complexly regulated atypical 2-Cys peroxiredoxin. Mol Microbiol 55:1207–1221 doi:10.1111/j.1365-2958.2005.04457.x

Dong YH, Xu JL, Li XZ, Zhang LH (2000) AiiA, an enzyme that inactivates the acylhomoserine lactone quorum-sensing signal and attenuates the virulence of *Erwinia carotovora*. Proc Natl Acad Sci USA 97:3526–3531 doi:10.1073/pnas.060023897

Dong YH, Wang LH, Xu JL, Zhang HB, Zhang XF, Zhang LH (2001) Quenching quorum-sensing-dependent bacterial infection by an N-acyl homoserine lactonase. Nature 411:813–817 doi:10.1038/35081101

Dong YH, Wang LY, Zhang LH (2007) Quorum-quenching microbial infections: mechanisms and implications. Philos Trans R Soc Lond B Biol Sci 362:1201–1211 doi:10.1098/rstb.2007.2045

Dunn AK, Handelsman J (2002) Towards understanding of microbial communities through analysis of communication networks. Antonie Van Leeuwenhoek 81:565–574 doi:10.1023/A:1020565807627

Faure D, Dessaux Y (2007) Quorum sensing as a target for developing biocontrol strategies towards the plant pathogen *Pectobacterium*. Eur J Plant Pathol 119:353–365 doi:10.1007/s10658-007-9149-1

Faure D, Bouillant M, Bally R (1995) Comparative Study of Substrates and Inhibitors of *Azospirillum lipoferum* and *Pyricularia oryzae* Laccases. Appl Environ Microbiol 61:1144–1146

Faure D, Bouillant ML, Jacoud C, Bally R (1996) Phenolic derivatives related to lignin metabolism as substrates for *Azospirillum* laccase activity. Phytochem 42:357–359 doi:10.1016/0031-9422(95)00869-1

Faure D, Desair J, Keijers V, Bekri MA, Proost P, Henrissat B, Vanderleyden J (1999) Growth of *Azospirillum irakense* KBC1 on the aryl beta-glucoside salicin requires either salA or salB. J Bacteriol 181(10):3003–3009

Faure D, Henrissat B, Ptacek D, Bekri MA, Vanderleyden J (2001) The celA gene, encoding a glycosyl hydrolase family 3 beta-glucosidase in *Azospirillum irakense*, is required for optimal growth on cellobiosides. Appl Environ Microbiol 67:2380–2383 doi:10.1128/AEM.67.5.2380-2383.2001

Ferguson GP, Datta A, Carlson RW et al (2005) Importance of unusually modified lipid A in *Sinorhizobium* stress resistance and legume symbiosis. Mol Microbiol 56:68–80 doi:10.1111/j.1365-2958.2005.04536.x

Flavier AB, Clough SJ, Schell MA, Denny TP (1997) Identification of 3-hydroxypalmitic acid methyl ester as a novel autoregulator controlling virulence in *Ralstonia solanacearum*. Mol Microbiol 26:251–259 doi:10.1046/j.1365-2958.1997.5661945.x

Francis KE, Spiker S (2005) Identification of *Arabidopsis thaliana* transformants without selection reveals a high occurrence of silenced T-DNA integrations. Plant J 41:464–477

Franks A, Mark-Byrne GL, Dow JM et al (2008) A putative RNA-binding protein has a role in virulence in *Ralstonia solanacearum*. Mol Plant Pathol 9:67–72

Fujishige NA, Kapadia NN, De Hoff PL et al (2006) Investigations of *Rhizobium* biofilm formation. FEMS Microbiol Ecol 56:195–205 doi:10.1111/j.1574-6941.2005.00044.x

Fujishige NA, Lum MR, De Hoff PL et al (2008) *Rhizobium* common *nod* genes are required for biofilm formation. Mol Microbiol 67:504–515

Fuqua WC, Winans SC, Greenberg EP (1994) Quorum sensing in bacteria: the LuxR/LuxI family of cell density-responsive transcriptional regulators. J Bacteriol 176:269–275

Gagnon H, Ibrahim RK (1998) Aldonic acids: a novel family of nod gene inducers of *Mesorhizobium loti*, *Rhizobium lupini* and *Sinorhizobium meliloti*. Mol Plant Microbe Interact 11:988–998 doi:10.1094/MPMI.1998.11.10.988

Galbraith MP, Feng SF, Borneman J et al (1998) A functional myo-inositol catabolism pathway is essential for rhizopine utilization by *Sinorhizobium meliloti*. Microbiology 144:2915–2924

Galis I, Bilyeu K, Wood G, Jameson PE (2005) *Rhodococcus fascians*: shoot proliferation without elevated cytokinins? Plant Growth Regul 46:109–115 doi:10.1007/s10725-005-7752-8

Galperin M (2006) Structural classification of bacterial response regulators: diversity of output domains and domain combinations. J Bacteriol 188:4169–4182 doi:10.1128/JB.01887-05

Gao MS, Teplitski M, Robinson JB et al (2003) Production of substances by *Medicago truncatula* that affect bacterial quorum sensing. Mol Plant Microbe Interact 16:827–834 doi:10.1094/MPMI.2003.16.9.827

Gelvin SB (2000) *Agrobacterium* and plant genes involved in T-DNA transfer and integration. Annu Rev Plant Physiol Plant Mol Biol 51:223–256 doi:10.1146/annurev.arplant.51.1.223

Gilles-Gonzalez MA, Gonzalez G (2004) Signal transduction by heme-containing PAS-domain proteins. J Appl Physiol 96:774–783 doi:10.1152/japplphysiol.00941.2003

Glass NL, Kosuge T (1986) Cloning of the gene for indoleacetic acid-lysine synthetase from *Pseudomonas syringae* subsp *savastanoi*. J Bacteriol 166:598–603

Glick BR (2005) Modulation of plant ethylene levels by the bacterial enzyme ACC deaminase. FEMS Microbiol Lett 251:1–7 doi:10.1016/j.femsle.2005.07.030

Glick BR, Cheng Z, Czarny J, Duan J (2007) Promotion of plant growth by ACC deaminase-containing soil bacteria. Eur J Plant Pathol 119:329–339 doi:10.1007/s10658-007-9162-4

Goldmann A, Boivin C, Fleury V et al (1991) Betaine use by rhizospere bacteria: genes essential for trigonelline, stachydrine, and carnitine catabolism in *Rhizobium meliloti* are located on pSym in the symbiotic region. Mol Plant Microbe Interact 4:571–578

González JE, Marketon MM (2003) Quorum sensing in nitrogen-fixing rhizobia. Microbiol Mol Biol Rev 67:574–592 doi:10.1128/MMBR.67.4.574-592.2003

Gonzalez-Rizzo S, Crespi M, Frugier F (2006) The *Medicago truncatula* CRE1 cytokinin receptor regulates lateral root development and early symbiotic interaction with *Sino-*

rhizobium meliloti. Plant Cell 18:2680–2693 doi:10.1105/tpc.106.043778

Gray KM, Garey JR (2001) The evolution of bacterial LuxI and LuxR quorum sensing regulators. Microbiology 147:2379–2387

Gray J, Gelvin SB, Meilan R, Morris RO (1996) Transfer RNA is the source of extracellular isopentenyladenine in a Ti-plasmidless strain of *Agrobacterium tumefaciens*. Plant Physiol 110:431–438

Guntli D, Heeb M, Moenne-Loccoz Y et al (1999) Contribution of calystegine catabolic plasmid to competitive colonization of the rhizosphere of calystegine-producing plants by *Sinorhizobium meliloti* Rm41. Mol Ecol 8:855–863 doi:10.1046/j.1365-294X.1999.00640.x

Gutierrez-Manero FJ, Ramos-Solano B, Probanza A, Mehouachi J, Tadeo FR, Talon M (2001) The plant-growth-promoting rhizobacteria *Bacillus pumilus* and *Bacillus licheniformis* produce high amounts of physiologically active gibberellins. Physiol Plant 111:206–211 doi:10.1034/j.1399-3054.2001.1110211.x

Hartmann A, Singh M, Klingmuller W (1983) Isolation and characterization of *Azospirillum* mutants excreting high amounts of indoleacetic acid. Can J Microbiol 29:916–923

Hartwig UA, Joseph CM, Phillips DA (1991) Flavonoids released naturally from alfalfa seeds enhance growth rate of *Rhizobium meliloti*. Plant Physiol 95:797–803

Heeb S, Haas D (2001) Regulatory roles of the GacS/GacA two-component system in plant-associated and other Gram-negative bacteria. Mol Plant Microbe Interact 14:1351–1363 doi:10.1094/MPMI.2001.14.12.1351

Heinz EB, Phillips DA, Streit WR (1999) BioS, a biotin-induced, stationary-phase, and possible LysR-type regulator in *Sinorhizobium meliloti*. Mol Plant Microbe Interact 12:803–812 doi:10.1094/MPMI.1999.12.9.803

Hense BA, Kuttler C, Müller J, Rothballer M, Hartmann A, Kreft JU (2007) Does efficiency sensing unify diffusion and quorum sensing? Nat Rev Microbiol 5:230–239 doi:10.1038/nrmicro1600

Hoang HH, Gurich N, González JE (2008) Regulation of motility by the ExpR/Sin quorum-sensing system in *Sinorhizobium meliloti*. J Bacteriol 190:861–971 doi:10.1128/JB.01310-07

Hoch JA, Varughese KI (2001) Keeping signals straight in phosphorelay signal transduction. J Bacteriol 183:4941–4949 doi:10.1128/JB.183.17.4941-4949.2001

Holden MT, Ram Chhabra S, de Nys R, Stead P, Bainton NJ et al (1999) Quorum-sensing cross talk: isolation and chemical characterization of cyclic dipeptides from *Pseudomonas aeruginosa* and other gram-negative bacteria. Mol Microbiol 33:1254–1266 doi:10.1046/j.1365-2958.1999.01577.x

Holland MA (1997) Occam's razor applied to hormonology—are cytokinins produced by plants? Plant Physiol 115:865–868

Ince JE, Knowles CJ (1985) Ethylene formation by cultures of *Escherichia coli*. Arch Microbiol 141:209–213 doi:10.1007/BF00408060

Inze D, Follin A, Vanlijsebettens M, Simoens C, Genetello C, Vanmontagu M et al (1984) Genetic analysis of the individual T-DNA genes of *Agrobacterium tumefaciens*—further evidence that 2 genes are involved in indole-3-acetic-acid synthesis. Mol Gen Genet 194:265–274 doi:10.1007/BF00383526

Jensen JB, Egsgaard H, Vanonckelen H, Jochimsen BU (1995) Catabolism of indole-3-acetic acid and 4-chloroindole-3-acetic and 5-chloroindole-3-acetic acid in *Bradyrhizobium japonicum*. J Bacteriol 177:5762–5766

Jiménez-Zurdo JI, García-Rodríguez FM, Toro N (1997) The *Rhizobium meliloti putA* gene: its role in the establishment of the symbiotic interaction with alfalfa. Mol Microbiol 23:85–93 doi:10.1046/j.1365-2958.1997.1861555.x

Johnson EG, Joshi MV, Gibson DM et al (2007) Cello-oligosaccharides released from host plants induce pathogenicity in scab-causing *Streptomyces* species. Physiol Mol Plant Pathol 7:18–25 doi:10.1016/j.pmpp.2007.09.003

Jones K, Kobayashi H, Davies BW et al (2007) How rhizobial symbionts invade plants: the *Sinorhizobium-Medicago* model. Nat Rev Microbiol 5:619–633 doi:10.1038/nrmicro1705

Jones K, Sharopova N, Lohar DP et al (2008) Differential response of the plant *Medicago truncatula* to its symbiont *Sinorhizobium meliloti* or an exopolysaccharide-deficient mutant. Proc Natl Acad Sci USA 105:704–709 doi:10.1073/pnas.0709338105

Kalogeraki VS, Zhu J, Eberhard A, Madsen EL, Winans SC (1999) The phenolic vir gene inducer ferulic acid is O-demethylated by the VirH2 protein of an *Agrobacterium tumefaciens* Ti plasmid. Mol Microbiol 34:512–522 doi:10.1046/j.1365-2958.1999.01617.x

Kannenberg EL, Perzl M, Muller P et al (1996) Hopanoid lipids in *Bradyrhizobium* and other plant-associated bacteria and cloning of the *Bradyrhizobium japonicum* squalene-hopene cyclase gene. Plant Soil 186:107–112 doi:10.1007/BF00035063

Karadeniz A, Topcuoglu SF, Inan S (2006) Auxin, gibberellin, cytokinin and abscisic acid production in some bacteria. World J Microbiol Biotechnol 22:1061–1064 doi:10.1007/s11274-005-4561-1

Kende H, Zeevaart JAD (1997) The five "classical" plant hormones. Plant Cell 9:1197–1210 doi:10.1105/tpc.9.7.1197

Kepczynska E, Zielinska S, Kepczynski J (2003) Ethylene production by *Agrobacterium rhizogenes* strains in vitro and in vivo. Plant Growth Regul 39:13–17 doi:10.1023/A:1021897203840

Kim JG (2006) Assessment of ethylene removal with *Pseudomonas* strains. J Hazard Mater 131:131–136 doi:10.1016/j.jhazmat.2005.09.019

Kiziak C, Conradt D, Stolz A, Mattes R, Klein J (2005) Nitrilase from *Pseudomonas fluorescens* EBC191: cloning and heterologous expression of the gene and biochemical characterization of the recombinant enzyme. Microbiol 151:3639–3648 doi:10.1099/mic.0.28246-0

Kobayashi M, Izui H, Nagasawa T, Yamada H (1993) Nitrilase in biosynthesis of the plant hormone indole-3-acetic-acid from indole-3-acetonitrile—cloning of the *Alcaligenes* gene and site-directed mutagenesis of cysteine residues. Proc Natl Acad Sci USA 90:247–251 doi:10.1073/pnas.90.1.247

Kobayashi H, Naciri-Graven Y, Broughton WJ et al (2004) Flavonoids induce temporal shifts in gene-expression of *nod-*

box controlled loci in *Rhizobium* sp. NGR234. Mol Microbiol 51:335–347 doi:10.1046/j.1365-2958.2003.03841.x

Koch B, Nielsen TH, Sorensen D et al (2002) Lipopeptide production in *Pseudomonas* sp strain DSS73 is regulated by components of sugar beet seed exudate via the Gac two-component regulatory system. Appl Environ Microbiol 68:4509–4516 doi:10.1128/AEM.68.9.4509-4516.2002

Koenig RL, Morris RO, Polacco JC (2002) tRNA is the source of low-level trans-zeatin production in *Methylobacterium* spp. J Bacteriol 184:1832–1842 doi:10.1128/JB.184.7.1832-1842.2002

Koga J, Adachi T, Hidaka H (1991) Molecular cloning of the gene for indolepyruvate decarboxylase from *Enterobacter cloacae*. Mol Gen Genet 226:10–16 doi:10.1007/BF00273581

Krimi Z, Petit A, Mougel C, Dessaux Y, Nesme X (2002) Seasonal fluctuations and long-term persistence of pathogenic populations of *Agrobacterium* spp. in soils. Appl Environ Microbiol 68:3358–3365 doi:10.1128/AEM.68.7.3358-3365.2002

Lasa I (2006) Towards the identification of the common features of bacterial biofilm development. Int Microbiol 9:21–28

Laub MT, Goulian M (2007) Specificity in two-component signal transduction pathways. Annu Rev Genet 41:121–145 doi:10.1146/annurev.genet.41.042007.170548

Leadbetter JR, Greenberg EP (2000) Metabolism of acyl-homoserine lactone quorum-sensing signals by *Variovorax paradoxus*. J Bacteriol 182:6921–6926 doi:10.1128/JB.182.24.6921-6926.2000

Lee J, Bansal T, Jayaraman A, Bentley WE, Wood TK (2007) Enterohemorrhagic *Escherichia coli* biofilms are inhibited by 7-hydroxyindole and stimulated by isatin. Appl Environ Microbiol 73:4100–4109 doi:10.1128/AEM.00360-07

Lerouge P, Roche P, Faucher C et al (1990) Symbiotic host-specificity of *Rhizobium meliloti* is determined by a sulphated and acylated glucosamine oligosaccharide signal. Nature 344:781–784 doi:10.1038/344781a0

Le Strange KK, Bender GL, Djordjevic MA et al (1990) The *Rhizobium* strain NGR234 nodD1 gene product responds to activation by the simple phenolic compounds vanillin and isovanillin present in wheat seedling extracts. Mol Plant Microbe Interact 3:214–220

Leveau JHJ (2007) The magic and menace of metagenomics: prospects for the study of plant growth-promoting rhizobacteria. Eur J Plant Pathol 119:279–300 doi:10.1007/s10658-007-9186-9

Leveau JHJ, Gerards S (2008) Discovery of a bacterial gene cluster for catabolism of the plant hormone indole 3-acetic acid. FEMS Microbiol Ecol 65:238–250 doi:10.1111/j.1574-6941.2008.00436.x

Leveau JHJ, Lindow SE (2005) Utilization of the plant hormone indole-3-acetic acid for growth by *Pseudomonas putida* strain 1290. Appl Environ Microbiol 71:2365–2371 doi:10.1128/AEM.71.5.2365-2371.2005

Li PL, Farrand SK (2000) The replicator of the nopaline-type Ti plasmid pTiC58 is a member of the *repABC* family and is influenced by the TraR-dependent quorum-sensing regulatory system. J Bacteriol 182:179–188

Lichter A, Manulis S, Sagee O, Gafni Y, Gray J, Meilan R et al (1995) Production of cytokinins by *Erwinia herbicola* pv *gypsophilae* and isolation of a locus conferring cytokinin biosynthesis. Mol Plant Microbe Interact 8:114–121

Liu P, Nester EW (2006) Indoleacetic acid, a product of transferred DNA, inhibits vir gene expression and growth of *Agrobacterium tumefaciens* C58. Proc Natl Acad Sci USA 103:4658–4662 doi:10.1073/pnas.0600366103

Luo ZQ, Qin Y, Farrand SK (2000) The antiactivator TraM interferes with the autoinducer-dependent binding of TraR to DNA by interacting with the C-terminal region of the quorum-sensing activator. J Biol Chem 275:7713–7722 doi:10.1074/jbc.275.11.7713

Madhaiyan M, Suresh Reddy BV, Anandham R, Senthilkumar M, Poonguzhali S, Sundaram SP (2006) Plant growth-promoting *Methylobacterium* induces defense responses in groundnut (*Arachis hypogaea* L.) compared with rot pathogens. Curr Microbiol 53:270–276

Manulis S, Haviv-Chesner A, Brandl MT, Lindow SE, Barash I (1998) Differential involvement of indole-3-acetic acid biosynthetic pathways in pathogenicity and epiphytic fitness of *Erwinia herbicola* pv *gypsophilae*. Mol Plant Microbe Interact 11:634–642 doi:10.1094/MPMI.1998.11.7.634

Marie C, Deakin WJ, Ojanen-Reuhs T et al (2004) TtsI, a key regulator of *Rhizobium* species NGR234 is required for type III-dependent protein secretion and synthesis of rhamnose-rich polysaccharides. Mol Plant Microbe Interact 17:958–966 doi:10.1094/MPMI.2004.17.9.958

Mark GL, Dow M, Kiely PD et al (2005) Transcriptome profiling of bacterial responses to root exudates identifies genes involved in microbe-plant interactions. Proc Natl Acad Sci USA 102:17454–17459 doi:10.1073/pnas.0506407102

Marketon MM, Glenn SA, Eberhard A et al (2003) Quorum sensing controls exoplysaccharide production in *Sinorhizobium meliloti*. J Bacteriol 185:325–331 doi:10.1128/JB.185.1.325-331.2003

Mascher T, Helmann JD, Unden G (2006) Stimulus perception in bacterial signal-transducing histidine kinases. Microbiol Mol Biol Rev 70:910–938 doi:10.1128/MMBR.00020-06

Mathesius U, Mulders S, Gao M et al (2003) Extensive and specific responses of a eukaryote to bacterial quorum-sensing signals. Proc Natl Acad Sci USA 100:1444–1449 doi:10.1073/pnas.262672599

Matilla MA, Espinosa-Urgel M, Rodríguez-Herva JJ et al (2007) Genomic analysis reveals the major driving forces of bacterial life in the rhizosphere. Genome Biol 8:R179 doi:10.1186/gb-2007-8-9-r179

Matiru VN, Dakora FD (2005) The rhizosphere signal molecule lumichrome alters seedling development in both legumes and cereals. New Phytol 166:439–444 doi:10.1111/j.1469-8137.2005.01344.x

Mazzola M, White FF (1994) A mutation in the indole-3-acetic-acid biosynthesis pathway of *Pseudomonas syringae* pv *syringae* affects growth in *Phaseolus vulgaris* and syringomycin production. J Bacteriol 176:1374–1382

McDougald D, Rice SA, Kjelleberg S (2007) Bacterial quorum sensing and interference by naturally occurring biomimics. Anal Bioanal Chem 387:445–453 doi:10.1007/s00216-006-0761-2

Molina L, Constantinescu F, Michel L, Reimmann C, Duffy B, Défago G (2003) Degradation of pathogen quorum-sensing molecules by soil bacteria: a preventive and curative bilogical control mechanism. FEMS Microbiol Ecol 1522:1–11

Morris CE, Monier J-M (2003) The ecological significance of biofilm formation by plant-associated bacteria. Annu Rev Phytopathol 41:429–453 doi:10.1146/annurev.phyto.41.022103.134521

Mougel C, Cournoyer B, Nesme X (2001) Novel tellurite-amended media and specific chromosomal and Ti plasmid probes for direct analysis of soil populations of *Agrobacterium* biovars 1 and 2. Appl Environ Microbiol 67:65–74 doi:10.1128/AEM.67.1.65-74.2001

Mukhopadhyay A, Gao R, Lynn DG (2004) Integrating input from multiple signals: The VirA/VirG two-component system of *Agrobacterium tumefaciens*. ChemBioChem 5:1535–1542 doi:10.1002/cbic.200300828

Nagahama K, Yoshino K, Matsuoka M, Sato M, Tanase S, Ogawa T et al (1994) Ethylene production by strains of the plant-pathogenic bacterium *Pseudomonas syringae* depends upon the presence of indigenous plasmids carrying homologous genes for the ethylene-forming enzyme. Microbiology 140:2309–2313

Nishijyo T, Haas D, Itoh Y (2002) The CbrA-CbrB two-component regulatory system controls the utilization of multiple carbon and nitrogen sources in *Pseudomonas aeruginosa*. Mol Microbiol 40:917–931 doi:10.1046/j.1365-2958.2001.02435.x

Oberhansli T, Defago G, Haas D (1991) Indole-3-acetic acid (IAA) synthesis in the biocontrol strain CHA0 of *Pseudomonas fluorescens*—role of tryptophan side-chain oxidase. J Gen Microbiol 137:2273–2279

Olesen MR, Jochimsen BU (1996) Identification of enzymes involved in indole-3-acetic acid degradation. Plant Soil 186:143–149 doi:10.1007/BF00035068

Pappas KM, Winans SC (2003) The RepA and RepB autorepressors and TraR play opposing roles in the regulation of a Ti plasmid *repABC* operon. Mol Microbiol 49:441–455 doi:10.1046/j.1365-2958.2003.03560.x

Paterson E, Gebbing T, Abel C et al (2006) Rhizodeposition shapes rhizosphere microbial community structure in organic soil. New Phytol 173:600–610 doi:10.1111/j.1469-8137.2006.01931.x

Patten CL, Glick BR (1996) Bacterial biosynthesis of indole-3-acetic acid. Can J Microbiol 42:207–220

Patten CL, Glick BR (2002) Role of *Pseudomonas putida* indoleacetic acid in development of the host plant root system. Appl Environ Microbiol 68:3795–3801 doi:10.1128/AEM.68.8.3795-3801.2002

Peck MC, Fisher RF, Long SR (2006) Diverse flavonoids stimulate NodD1 binding to *nod* gene promotors in *Sinorhizobium meliloti*. J Bacteriol 188:5417–5427 doi:10.1128/JB.00376-06

Penrose DM, Glick BR (2001) Levels of ACC and related compounds in exudate and extracts of canola seeds treated with ACC deaminase-containing plant growth-promoting bacteria. Can J Microbiol 47:368–372

Penrose DM, Glick BR (2003) Methods for isolating and characterizing ACC deaminase-containing plant growth-promoting rhizobacteria. Physiol Plant 118:10–15 doi:10.1034/j.1399-3054.2003.00086.x

Perley JE, Stowe BB (1966) Production of tryptamine from tryptophan by *Bacillus cereus*. Biochem J 100:169

Perrig D, Boiero ML, Masciarelli OA, Penna C, Ruiz OA, Cassan FD et al (2007) Plant-growth-promoting compounds produced by two agronomically important strains of *Azospirillum brasilense*, and implications for inoculant formulation. Appl Microbiol Biotechnol 75:1143–1150 doi:10.1007/s00253-007-0909-9

Phillips DA, Joseph CM, Maxwell CA (1992) Trigonelline and stachydrine released from alfalfa seeds activate NodD2 in *Rhizobium meliloti*. Plant Physiol 99:1526–1531

Phillips DA, Joseph CM, Yang GP et al (1999) Identification of lumichrome as a *Sinorhizobium* enhancer of alfalfa root respiration and shoot growth. Proc Natl Acad Sci USA 96:12275–12280 doi:10.1073/pnas.96.22.12275

Piper KR, Beck von Bodman S, Farrand SK (1993) Conjugation factor of *Agrobacterium tumefaciens* regulates Ti plasmid transfer by autoinduction. Nature 362:448–450 doi:10.1038/362448a0

Piper KR, Beck Von Bodman S, Hwang I, Farrand SK (1999) Hierarchical gene regulatory systems arising from fortuitous gene associations: controlling quorum sensing by the opine regulon in *Agrobacterium*. Mol Microbiol 32:1077–1089 doi:10.1046/j.1365-2958.1999.01422.x

Powell GK, Morris RO (1986) Nucleotide sequence and expression of a *Pseudomonas savastanoi* cytokinin biosynthetic gene—homology with *Agrobacterium tumefaciens tmr* and *tzs* loci. Nucleic Acids Res 14:2555–2565 doi:10.1093/nar/14.6.2555

Prinsen E, Chauvaux N, Schmidt J et al (1991) Stimulation of indole-3-acetic acid production in *Rhizobium* by flavonoids. FEBS Lett 282:53–55 doi:10.1016/0014-5793(91)80442-6

Proctor MH (1958) Bacterial dissimilation of indoleacetic acid—new route of breakdown of the indole nucleus. Nature 181:1345–1345 doi:10.1038/1811345a0

Prusty R, Grisafi P, Fink GR (2004) The plant hormone indoleacetic acid induces invasive growth in *Saccharomyces cerevisiae*. Proc Natl Acad Sci USA 101:4153–4157 doi:10.1073/pnas.0400659101

Pueppke SG, Broughton WJ (1999) *Rhizobium* sp. NGR234 and *R. fredii* USDA257 share exceptionally broad, nested host ranges. Mol Plant Microbe Interact 12:293–318 doi:10.1094/MPMI.1999.12.4.293

Qian W, Han ZJ, He CZ (2008) Two-component signal transduction systems of *Xanthomonas* spp.: a lesson from genomics. Mol Plant Microbe Interact 21:151–161 doi:10.1094/MPMI-21-2-0151

Rademacher W (1994) Gibberellin formation in microorganisms. Plant Growth Regul 15:303–314 doi:10.1007/BF00029903

Raskin I (1992) Salicylate, a new plant hormone. Plant Physiol 99:799–803

Rasmussen TB, Givskov M (2006) Quorum sensing inhibitors: a bargain of effects. Microbiology 152:895–904 doi:10.1099/mic.0.28601-0

Rasmussen TB, Skindersoe ME, Bjarnsholt T, Phipps RK, Christensen KB, Jensen PO, Andersen JB, Koch B, Larsen TO, Hentzer M, Eberl L, Hoiby N, Givskov M (2005) Identity and effects of quorum-sensing inhibitors produced by *Penicillium* species. Microbiology 151:1325–1340 doi:10.1099/mic.0.27715-0

Reddy PM, Rendón-Anaya M, de los Dolores Soto del Río M et al (2007) Flavonoids as signaling molecules and regulators of root nodule development. Dynamic Soil. Dyn Plant 1:83–94

Redfield RJ (2002) Is quorum sensing a side effect of diffusion sensing? Trends Microbiol 10:365–370 doi:10.1016/S0966-842X(02)02400-9

Reverchon S, Chantegrel B, Deshayes C, Doutheau A, Cotte-Pattat N (2002) New synthetic analogues of N-acyl homoserine lactones as agonists or antagonists of transcriptional regulators involved in bacterial quorum sensing. Bioorg Med Chem Lett 12:1153–1157 doi:10.1016/S0960-894X(02)00124-5

Riaz K, Elmerich C, Moreira D, Raffoux A, Dessaux Y, Faure D (2008) A metagenomic analysis of soil bacteria extends the diversity of quorum-quenching lactonases. Environ Microbiol 10:560–570 doi:10.1111/j.1462-2920.2007.01475.x

Riviere J, Laboureu P, Sechet M (1966) Bacterial degradation of indole-3-acetic acid and of gibberellin A3 in soil. Ann Physiol Vegetale 8:209

Robert-Seilaniantz A, Navarro L, Bari R et al (2007) Pathological hormone imbalances. Curr Opin Plant Biol 10:372–379 doi:10.1016/j.pbi.2007.06.003

Robinette D, Matthysse AG (1990) Inhibition by *Agrobacterium tumefaciens* and *Pseudomonas savastanoi* of development of the hypersensitive response elicited by *Pseudomonas syringae* pv *phaseolicola*. J Bacteriol 172:5742–5749

Rosa-Putra S, Nalin R, Domenach A-M et al (2001) Novel hopanoids from *Frankia* spp. and related soil bacteria. Eur J Biochem 268:4300–4306 doi:10.1046/j.1432-1327.2001.02348.x

Saenkham P, Eiamphungporn W, Farrand SK, Vattanaviboon P, Mongkolsuk S (2007) Multiple superoxide dismutases in *Agrobacterium tumefaciens*: functional analysis, gene regulation, and influence on tumorigenesis. J Bacteriol 189:8807–8817 doi:10.1128/JB.00960-07

Sakakibara H, Takei K (2002) Identification of cytokinin biosynthesis genes in *Arabidopsis*: A breakthrough for understanding the metabolic pathway and the regulation in higher plants. J Plant Growth Regul 21:17–23 doi:10.1007/s003440010043

Sakakibara H, Kasahara H, Ueda N, Kojima M, Takei K, Hishiyama S, Asami T, Okada K, Kamiya Y, Yamaya T, Yamaguchi S (2005) *Agrobacterium tumefaciens* increases cytokinin production in plastids by modifying the biosynthetic pathway in the host plant. Proc Natl Acad Sci USA 102:9972–9977 doi:10.1073/pnas.0500793102

Sanguin H, Remenant B, Dechesne A, Thiouleuse J, Vogel TM, Nesme X, Moënne-Loccoz Y, Grundmann GL (2006) Potential of a 16S rRNA-based taxonomic microarray for analyzing the rhizosphere effects of maize on *Agrobacterium* spp. and bacterial communities. Appl Environ Microbiol 72:4302–4312 doi:10.1128/AEM.02686-05

Savka MA, Dessaux Y, Oger P et al (2002) Engineering bacterial competitiveness and persistence in the phytosphere. Mol Plant Microbe Interact 15:866–874 doi:10.1094/MPMI.2002.15.9.866

Schaefer AL, Greenberg EP, Oliver CM, Oda Y, Huang JJ, Bittan-Banin G, Peres CM, Schmidt S, Juhaszova K, Sufrin JR, Harwood CS (2008) A new class of homoserine lactone quorum-sensing signals. Nature. doi:10.1038/nature07088

Schmelz EA, Engelberth J, Alborn HT, O'Donnell P, Sammons M, Toshima H et al (2003) Simultaneous analysis of phytohormones, phytotoxins, and volatile organic compounds in plants. Proc Natl Acad Sci USA 100:10552–10557 doi:10.1073/pnas.1633615100

Schmidt PE, Broughton WJ, Werner D (1994) Nod factors of *Bradyrhizobium japonicum* and *Rhizobium* sp. NGR234 induce flavonoid accumulation in soybean root exudate. Mol Plant Microbe Interact 7:384–390

Schuhegger R, Ihring A, Gantner S, Bahnweg G, Knappe C, Vogg G, Hutzler P, Schmid M, Van Breusegem F, Eberl L, Hartmann A, Langebartels C (2006) Induction of systemic resistance in tomato by N-acyl-L-homoserine lactone-producing rhizosphere bacteria. Plant Cell Environ 29:909–918 doi:10.1111/j.1365-3040.2005.01471.x

Schwalm K, Aloni R, Langhans M, Heller W, Stich S, Ullrich CI (2003) Flavonoid-related regulation of auxin accumulation in *Agrobacterium tumefaciens*-induced plant tumors. Planta 218:163–178 doi:10.1007/s00425-003-1104-6

Schwessinger B, Zipfel C (2008) News from the frontline: recent insights into PAMP-triggered immunity in plants. Curr Opin Plant Biol 11:1–17 doi:10.1016/j.pbi.2008.06.001

Sekine M, Watanabe K, Syono K (1989) Nucleotide sequence of a gene for indole-3-acetamide hydrolase from *Bradyrhizobium japonicum*. Nucleic Acids Res 17:6400–6400 doi:10.1093/nar/17.15.6400

Shah S, Li JP, Moffatt BA, Glick BR (1998) Isolation and characterization of ACC deaminase genes from two different plant growth-promoting rhizobacteria. Can J Microbiol 44:833–843 doi:10.1139/cjm-44-9-833

Shaw LJ, Morris P, Hooker JE (2006) Perception and modification of plant flavonoid signals by rhizosphere microorganisms. Environ Microbiol 8:1867–1880 doi:10.1111/j.1462-2920.2006.01141.x

Shelp BJ, Bown AW, Faure D (2006) Extracellular gamma-aminobutyrate mediates communication between plants and other organisms. Plant Physiol 142:1350–1352 doi:10.1104/pp.106.088955

Skorpil P, Saad MM, Boukli NM et al (2005) NopP, a phosphorylated effector of *Rhizobium* sp. NGR234, is a major determinant of nodulation of the tropical legumes *Flemingia congesta* and *Tephrosia vogelii*. Mol Microbiol 57:1304–1317 doi:10.1111/j.1365-2958.2005.04768.x

Smadja B, Latour X, Faure D, Chevalier S, Dessaux Y, Orange N (2004) Involvement of N-acylhomoserine lactones throughout the plant infection by *Erwinia carotovora* subsp. *atroseptica* (*Pectobacterium atrosepticum*). Mol Plant Microbe Interact 17:1269–1278 doi:10.1094/MPMI.2004.17.11.1269

Smith AM (1973) Ethylene as a cause of soil fungistasis. Nature 246:311–313 doi:10.1038/246311a0

Somers E, Vanderleyden J, Srinivasan M (2004) Rhizosphere bacterial signalling: a love parade beneath our feet. Crit Rev Microbiol 30:205–240 doi:10.1080/10408410490468786

Spaepen S, Vanderleyden J, Remans R (2007) Indole-3-acetic acid in microbial and microorganism-plant signalling. FEMS Microbiol Rev 31:425–448 doi:10.1111/j.1574-6976.2007.00072.x

Stachel SE, Messens E, Van Montagu M et al (1985) Identification of the signal molecules produced by wounded plant cells that activate T-DNA transfer in *Agrobacterium tumefaciens*. Nature 318:624–629 doi:10.1038/318624a0

Staehelin C, Forsberg LS, D'Haeze W et al (2006) Exo-oligosaccharides of *Rhizobium* sp strain NGR234 are required for symbiosis with various legumes. J Bacteriol 188:6168–6178 doi:10.1128/JB.00365-06

Steidle A, Sigl K, Schuhegger R, Ihring A, Schmid M, Gantner S, Stoffels M, Riedel K, Givskov M, Hartmann A, Langebartels C, Eberl L (2001) Visualization of N-acylhomoserine lactone-mediated cell—cell communication between bacteria colonizing the tomato rhizosphere. Appl Environ Microbiol 67:5761–5770 doi:10.1128/AEM.67.12.5761-5770.2001

Steidle A, Allesen-Holm M, Riedel K, Berg G, Givskov M, Molin S, Eberl L (2002) Identification and characterization of an N-acylhomoserine lactone-dependent quorum-sensing system in *Pseudomonas putida* strain IsoF. Appl Environ Microbiol 68:6371–6382 doi:10.1128/AEM.68.12.6371-6382.2002

Streit WR, Joseph CM, Phillips DA (1996) Biotin and other water-soluble vitamins are key growth factors for alfalfa root colonization by *Rhizobium meliloti* 1021. Mol Plant Microbe Interact 9:330–338

Suzuki S, He YX, Oyaizu H (2003) Indole-3-acetic acid production in *Pseudomonas fluorescens* HP72 and its association with suppression of creeping bentgrass brown patch. Curr Microbiol 47:138–143 doi:10.1007/s00284-002-3968-2

Sykes LC, Matthysse AG (1986) Time required for tumor induction by *Agrobacterium tumefaciens*. Appl Environ Microbiol 52:597–598

Szurmant H, Ordal GW (2004) Diversity in chemotaxis mechanisms among the Bacteria and Archaea. Microbiol Mol Biol Rev 68:301–319 doi:10.1128/MMBR.68.2.301-319.2004

Szurmant H, White RA, Hoch JA (2007) Sensor complexes egulating two-component signal transduction. Curr Opin Struct Biol 17:706–715 doi:10.1016/j.sbi.2007.08.019

Tepfer D, Goldmann A, Pamboukdjian N et al (1988) A plasmid of *Rhizobium meliloti* 41 encodes catabolism of two compounds from root exudate of *Calystegium sepium*. J Bacteriol 170:1153–1161

Teplitski M, Robinson JB, Bauer WD (2000) Plants secrete compounds that mimic bacterial N-acyl homoserine lactone signal activities and affect population density-dependent behaviours in associated bacteria. Mol Plant Microbe Interact 13:637–648 doi:10.1094/MPMI.2000.13.6.637

Teyssier-Cuvelle S, Mougel C, Nesme X (1999) Direct conjugal transfers of Ti plasmid to soil microflora. Mol Ecol 8:1273–1284 doi:10.1046/j.1365-294X.1999.00689.x

Teyssier-Cuvelle S, Oger P, Mougel C, Groud K, Farrand SK, Nesme X (2004) A highly selectable and highly transferable Ti plasmid to study conjugal host range and Ti plasmid dissemination in complex ecosystems. Microb Ecol 48:10–18 doi:10.1007/s00248-003-2023-6

Theunis M, Kobayashi H, Broughton WJ et al (2004) Flavonoids, NodD1, NodD2, and *nod*-box NB15 modulate expression of the y4wEFG locus that is required for indole-3-acetic acid synthesis in *Rhizobium* sp. strain NGR234. Mol Plant Microbe Interact 17:1153–1161 doi:10.1094/MPMI.2004.17.10.1153

Tsavkelova EA, Klimova SY, Cherdyntseva TA, Netrusov AI (2006) Hormones and hormone-like substances of microorganisms: a review. Appl Biochem Microbiol 42:229–235 doi:10.1134/S000368380603001X

Tsubokura S, Sakamoto Y, Ichihara K (1961) Bacterial decomposition of indoleacetic acid. J Biochem 49:38

Ulrich LE, Koonin EV, Zhulin IB (2005) One-component systems dominate signal transduction in prokaryotes. Trends Microbiol 13:52–56 doi:10.1016/j.tim.2004.12.006

Uroz S, Dangelo C, Carlier A, Faure D, Petit A, Oger P, Sicot C, Dessaux Y (2003) Novel bacteria degrading N-acyl homoserine lactones and their use as quenchers of quorum-sensing regulated functions of plant pathogenic bacteria. Microbiology 149:1981–1989 doi:10.1099/mic.0.26375-0

Uroz S, Chhabra SR, Cámara M, Williams P, Oger P, Dessaux Y (2005) N-Acylhomoserine lactone quorum-sensing molecules are modified and degraded by *Rhodococcus erythropolis* W2 by both amidolytic and novel oxidoreductase activities. Microbiol 151:3313–3322 doi:10.1099/mic.0.27961-0

Uroz S, Oger PM, Chapelle E, Adeline MT, Faure D, Dessaux Y (2008) A *Rhodococcus qsdA*-encoded enzyme defines a novel class of large-spectrum quorum-quenching lactonases. Appl Environ Microbiol 74:1357–1366 doi:10.1128/AEM.02014-07

Vande Broek A, Lambrecht M, Eggermont K, Vanderleyden J (1999) Auxins upregulate expression of the indole-3-pyruvate decarboxylase gene in *Azospirillum brasilense*. J Bacteriol 181:1338–1342

Vandeputte O, Oden S, Mol A, Vereecke D, Goethals K, El Jaziri M et al (2005) Biosynthesis of auxin by the gram-positive phytopathogen *Rhodococcus fascians* is controlled by compounds specific to infected plant tissues. Appl Environ Microbiol 71:1169–1177 doi:10.1128/AEM.71.3.1169-1177.2005

Vogel J, Normand P, Thioulouse J, Nesme X, Grundmann GL (2003) Relationship between spatial and genetic distance in *Agrobacterium* spp. in 1 cubic centimeter of soil. Appl Environ Microbiol 69:1482–1487 doi:10.1128/AEM.69.3.1482-1487.2003

von Bodman SB, Bauer WD, Coplin DL (2003) Quorum sensing in plant-pathogenic bacteria. Annu Rev Phytopathol 41:455–482 doi:10.1146/annurev.phyto.41.052002.095652

von Rad U, Klein I, Dobrev PI, Kottova J, Zazimalova E, Fekete A, Hartmann A, Schmitt-Kopplin P, Durner J (2008) Response of *Arabidopsis thaliana* to N-hexanoyl-DL: -homoserine-lactone, a bacterial quorum sensing molecule produced in the rhizosphere. Planta 229:73–85

Wächter R, Langhans M, Aloni R, Götz S, Weilmünster A, Koops A, Temguia L, Mistrik I, Pavlovkin J, Rascher U, Schwalm K, Koch KE, Ullrich CI (2003) Vascularization, high-volume solution flow, and localized roles for enzymes of sucrose metabolism during tumorigenesis by *Agrobacterium tumefaciens*. Plant Physiol 133:1024–1037 doi:10.1104/pp.103.028142

Wang LL, Wang ET, Liu J, Li Y, Chen WX (2006) Endophytic occupation of root nodules and roots of Melilotus dentatus by *Agrobacterium tumefaciens*. Microb Ecol 52:436–443 doi:10.1007/s00248-006-9116-y

Waters CM, Bassler BL (2005) Quorum sensing: communication in bacteria. Annu Rev Cell Dev Biol 21:319–346 doi:10.1146/annurev.cellbio.21.012704.131001

Weingart H, Ullrich H, Geider K, Volksch B (2001) The role of ethylene production in virulence of *Pseudomonas syringae* pvs. *glycinea* and *phaseolicola*. Phytopathology 91:511–518 doi:10.1094/PHYTO.2001.91.5.511

Whitehead NA, Barnard AML, Slater H, L SNJ, Salmond GPC (2001) Quorum-sensing in Gram-negative bacteria. FEMS Microbiol Rev 25:365–404 doi:10.1111/j.1574-6976.2001.tb00583.x

Wood DW, Setubal JC, Kaul R, Monks DE, Kitajima JP, Okura VK, Zhou Y et al (2001) The genome of the natural genetic engineer *Agrobacterium tumefaciens* C58. Science 294:2317–2323 doi:10.1126/science.1066804

Yamada T, Palm CJ, Brooks B, Kosuge T (1985) Nucleotide sequences of the *Pseudomonas savastanoi* indoleacetic acid genes show homology with *Agrobacterium tumefaciens* T-DNA. Proc Natl Acad Sci USA 82:6522–6526 doi:10.1073/pnas.82.19.6522

Yang CC, Crowley DE (2000) Rhizosphere microbial community structure in relation to root location and plant iron nutritional status. Appl Environ Microbiol 66:345–351

Yang SH, Zhang Q, Guo JH, Charkowski AO, Glick BR, Ibekwe AM et al (2007) Global effect of indole-3-acetic acid biosynthesis on multiple virulence factors of *Erwinia chrysanthemi* 3937. Appl Environ Microbiol 73:1079–1088 doi:10.1128/AEM.01770-06

Yoneyama K, Xie X, Kusumoto D et al (2007) Nitrogen deficiency as well as phosphorus deficiency in sorghum promotes the production and exudation of 5-deoxystrigol, the host recognition signal for arbuscullar mycorrhizal fungi and root parasites. Planta 227:125–132 doi:10.1007/s00425-007-0600-5

Yuan ZC, Edlind MP, Liu P et al (2007) The plant signal salicylic acid shuts down expression of the vir regulon and activates quormone-quenching genes in *Agrobacterium*. Proc Natl Acad Sci USA 104:11790–11795 doi:10.1073/pnas.0704866104

Yuan ZC, Haudecoeur E, Faure D, Kerr KF, Nester EW (2008a) Comparative transcriptome analysis of *Agrobacterium tumefaciens* in response to plant signal salicylic acid, indole-3-acetic acid and gamma-amino butyric acid reveals signalling cross-talk and *Agrobacterium*-plant co-evolution. Cell Microbiol 10:2339–2354 doi:10.1111/j.1462-5822.2008.01215.x

Yuan ZC, Liu P, Saenkham P et al (2008b) Transcriptome profiling and functional analysis of *Agrobacterium tumefaciens* reveals a general conserved response to acidic conditions (pH 5.5) and a complex acid-mediated signaling involved in *Agrobacterium*-plant interactions. J Bacteriol 190:494–507 doi:10.1128/JB.01387-07

Zhang XS, Cheng HP (2006) Identification of *Sinorhizobium meliloti* early symbiotic genes by use of a positive functional screen. Appl Environ Microbiol 72:2738–2748 doi:10.1128/AEM.72.4.2738-2748.2006

Zhang HB, Wang LH, Zhang LH (2002) Genetic control of quorum-sensing signal turnover in *Agrobacterium tumefaciens*. Proc Natl Acad Sci USA 99:4638–4643 doi:10.1073/pnas.022056699

Zipfel C (2008) Pattern-recognition receptors in plant innate immunity. Curr Opin Immunol 20:10–16 doi:10.1016/j.coi.2007.11.003

Zipfel C, Kunze G, Chinchilla D et al (2006) Perception of the bacterial PAMP EF-Tu by the receptor EFR restricts *Agrobacterium*-mediated transformation. Cell 125:749–760 doi:10.1016/j.cell.2006.03.037

REVIEW ARTICLE

Acquisition of phosphorus and nitrogen in the rhizosphere and plant growth promotion by microorganisms

Alan E. Richardson · José-Miguel Barea · Ann M. McNeill · Claire Prigent-Combaret

Received: 24 May 2008 / Accepted: 5 January 2009 / Published online: 27 February 2009
© Springer Science + Business Media B.V. 2009

Abstract The rhizosphere is a complex environment where roots interact with physical, chemical and biological properties of soil. Structural and functional characteristics of roots contribute to rhizosphere processes and both have significant influence on the capacity of roots to acquire nutrients. Roots also interact extensively with soil microorganisms which further impact on plant nutrition either directly, by influencing nutrient availability and uptake, or indirectly through plant (root) growth promotion. In this paper, features of the rhizosphere that are important for nutrient acquisition from soil are reviewed, with specific emphasis on the characteristics of roots that influence the availability and uptake of phosphorus and nitrogen. The interaction of roots with soil microorganisms, in particular with mycorrhizal fungi and non-symbiotic plant growth promoting rhizobacteria, is also considered in relation to nutrient availability and through the mechanisms that are associated with plant growth promotion.

Keywords Soil microorganisms · PGPR · Mycorrhizal fungi · Exudate · Phosphorus · Nitrogen · Uptake · Mineralization

Responsible Editor: Philippe Hinsinger.

A. E. Richardson (✉)
CSIRO Plant Industry,
PO Box 1600, Canberra 2601, Australia
e-mail: alan.richardson@csiro.au

J.-M. Barea
Departamento de Microbiología del Suelo y Sistemas Simbióticos, Estación Experimental del Zaidín, CSIC,
Prof. Albareda 1,
Granada 18008, Spain

A. M. McNeill
University of Adelaide, Soil and Land Systems,
Waite Campus,
Adelaide 5005, Australia

C. Prigent-Combaret
Université de Lyon,
Lyon, France

C. Prigent-Combaret
Université Lyon 1,
Villeurbanne, France

C. Prigent-Combaret
CNRS, UMR 5557, Ecologie Microbienne,
Villeurbanne, France

Introduction

The rhizosphere can be defined as the zone of soil around plant roots whereby soil properties are influenced by the presence and activity of the root. Changes to the physical, chemical and biological properties of rhizosphere soil has significant influence on the subsequent growth and health of plants. In terms of nutrient acquisition, both the structural and functional characteristics of roots have long been recognized as being important in determining the capacity for plants to access and mediate the availability of essential nutrients in soil and to alleviate against those that are toxic (Darrah 1993; Hinsinger 1998; Marschner 1995). Furthermore, roots interact with diverse populations of soil microorganisms which have significant implication for growth and nutrition (Curl and Truelove 1986; Bowen and Rovira 1999; Mukerji et al. 2006; Brimecombe et al. 2007). Soil nutrients are transferred towards the root surface through the rhizosphere or, in the case of roots associated with mycorrhizal fungi, through the mycorrhizosphere, prior to acquisition.

Plants modify the physico-chemical properties and biological composition of the rhizosphere through a range of mechanisms, which include acidification through proton extrusion and the release of root exudates. Along with changes to soil pH, root exudates directly influence nutrient availability or have indirect effects through interaction with soil microorganisms. An outstanding feature of the rhizosphere is that rhizodeposition and root turnover account for up to 40% of the carbon input into soil and clearly is the major driver for soil microbiological processes (Grayston et al. 1996; Jones et al. 2009). Interactions between plant roots and soil microorganisms are ubiquitous across various trophic levels and are an essential component of ecosystem function. It has become increasingly evident that root interactions with soil microorganisms are intricate and involve highly complex communities that function in very heterogeneous environments (Giri et al. 2005). Microbial interactions with roots may involve either endophytic or free living microorganisms and can be symbiotic, associative or casual in nature. Beneficial symbionts include N_2-fixing bacteria (e.g. rhizobia) in association with legumes and interaction of roots with mycorrhizal fungi, with the later being particularly important in relation to plant P uptake. Associative and free-living microorganisms may also contribute to the nutrition of plants through a variety of mechanisms including direct effects on nutrient availability (e.g. N_2-fixation by diazotrophs and P-mobilization by many microorganisms), enhancement of root growth (i.e. through plant growth promoting rhizobacteria, or PGPR), as antagonists of root pathogens (Raaijmakers et al. 2009) or as saphrophytes that decompose soil detritus and subsequently increase nutrient availability through mineralization and microbial turnover. Such processes are likely to be of greater significance for nutrient availability in the rhizosphere where there is increased supply of readily metabolizable carbon and where mobilized nutrients can be more easily captured by roots.

In this review we address the acquisition of nutrients from soil by plants with specific emphasis on the structural and functional characteristics of roots that influence the availability and uptake of P and N. In particular, the importance of soil microorganisms and their interactions with roots in relation to nutrient availability is considered, along with their associated mechanisms of plant growth promotion. The review complements previous reviews that have specifically focused on either plant-based traits or mechanistic processes associated with P and N uptake (Raghothama 1999; Vance et al. 2003; Bucher 2007, Miller and Cramer 2004; Jackson et al. 2008). Although the rhizosphere is important for the efficient uptake of a wider range of macro and micronutrients (including Fe; Lemanceau et al. 2009), the review specifically focuses on N and P which are key nutrients that limit sustainable agricultural production across much of the globe (Tilman et al. 2002).

Mechanisms of nutrient acquisition by plant roots

Efficient capture of nutrients from soil by roots is a critical issue for plants given that in many environments nutrients have poor availability and may be deficient for optimal growth. Whilst nutrient supply in soil is often augmented by the application of fertilizers, the availability of nutrients is governed by a wide range of physico-chemical parameters, environmental and seasonal factors and biological interactions. Competition for nutrient uptake across different plant species, between different roots and with microorganisms is also significant (Hodge 2004). The rate of root growth and the plasticity of root architecture along with development of the rhizosphere, through either root growth or extension of root hairs, are clearly important for effective exploration of soil and interception of nutrients (Lynch 1995). Biochemical changes in the rhizosphere and interaction with microorganisms are also significant. However, the importance of different root traits and rhizosphere-mediated processes is dependent on the nutrient in question and other factors that include plant species and soil type (Tinker and Nye 2000). For example, for nutrients present at low concentrations in soil solution and/or with poor diffusivity (e.g. P as either HPO_4^{2-} or $H_2PO_4^-$, and micronutrients, such as Fe and Zn), root growth and proliferation into new regions of soil and release of root exudates are of particular importance (Barber 1995; Darrah 1993). In contrast, nutrients present in either higher concentrations (e.g. K^+, NH_4^+), or with greater diffusion coefficients (e.g. NO_3^-, SO_4^- and Ca^{2+}), are able to move more freely toward the root through mass flow, where root distribution and architectural characteristics that facilitate water uptake are of greater relative significance (Barber 1995; Tinker and Nye, 2000; Lynch 2005). The relative significance of such factors in the acquisition and uptake of P and N is therefore considered in more detail below.

Acquisition of phosphorus by plants

Phosphorus availability and uptake

Although soils generally contain a large amount of total P only a small proportion is immediately available for plant uptake. Plants obtain P as orthophosphate anions (predominantly as HPO_4^{2-} and $H_2PO_4^{1-}$) from the soil solution. In most soils the concentration of orthophosphate in solution is low (typically 1 to 5 µM; Bieleski 1973) and must therefore be replenished from other pools of soil P to satisfy plant requirements. Orthophosphate is rapidly depleted in the immediate vicinity of plant roots, and as such a large concentration gradient occurs across the rhizosphere between bulk soil and the root surface (Gahoonia and Nielsen 1997; Tinker and Nye 2000; Fig. 1). However, for most soils the rate of diffusion of orthophosphate is insufficient to overcome 'localized' gradients, which in most cases limits the uptake of sufficient P. Evidence from both modelling and empirical studies also suggests that actual P uptake capacity at the root surface is unlikely to be limiting for plant growth (Barber 1995). This is supported by more recent studies on the expression of genes that encode for transport proteins with high affinity for uptake of orthophosphate (e.g. K_m ~3 µM), which are predominantly expressed in root hair cells on the epidermis (Mitsukawa et al. 1997; Mudge et al. 2002; Schünmann et al, 2004). Whilst expression of these genes has been shown to facilitate the P uptake capacity of cells in suspension culture (Mitsukawa et al. 1997), their over-expression in transgenic plants did not result in increased P uptake by barley (*Hordeum vulgare* L.) when grown at a range of P concentrations in either solution or soil (Rae et al. 2004). This is consistent with the view that plants are well adapted for uptake of P from the low concentrations that are

Fig. 1 Physiological and chemical processes that influence the availability and transformation of phosphorus in the rhizosphere (adapted from Richardson 1994; Richardson 2001)

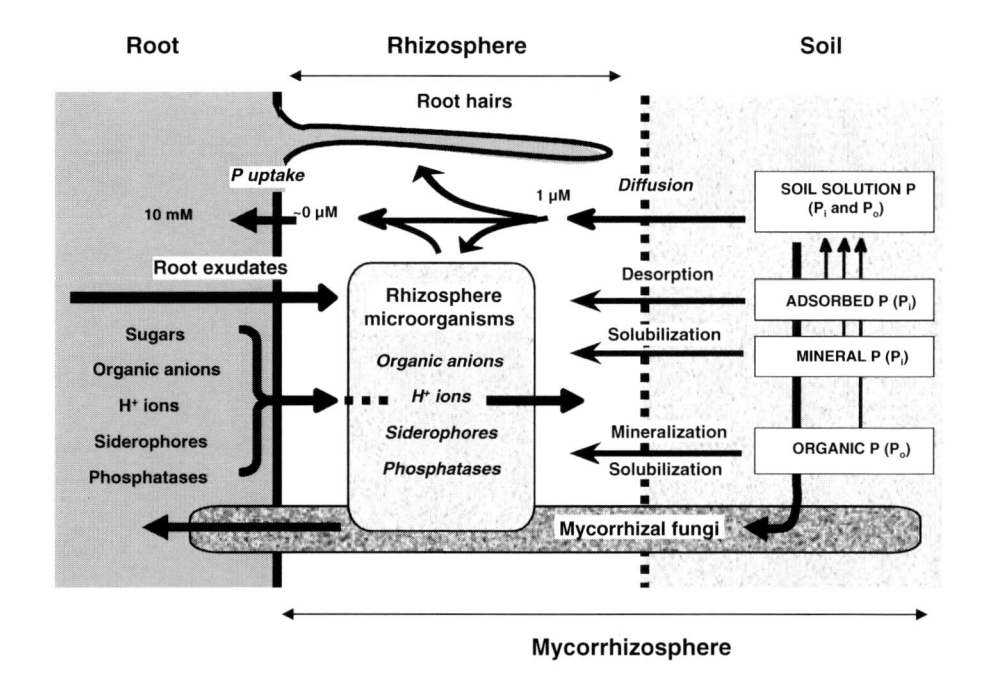

typical of soil solutions as indicated by minimum uptake concentrations (C_{lim} values) of 0.01 to 0.1 μM for different species (Jungk 2001). On this basis it is suggested that the supply of P to the root surface, and its availability as influenced by root and microbial processes (as outlined below), or the capacity of roots to exploit new regions of soil are of greater importance for P acquisition than the kinetics associated with its uptake.

The importance of root growth and architecture for the efficient capture of P is well documented and in many cases is a specific response of plants to P deficiency (reviewed by Hodge 2004; Lynch 2005; Raghothama 1999; Richardson et al. 2009; Vance et al. 2003). Characteristics of roots that facilitate soil exploration and hence P uptake include; rapid rate of root elongation and high root to shoot biomass ratio (Hill et al. 2006), increased root branching and root angle particularly in surface soils and nutrient rich regions (Lynch and Brown 2001; Manske et al. 2000; Rubio et al. 2003), high specific root length (i.e. length per unit mass or root fineness; Silberbush and Barber, 1983), the presence of root hairs (Föhse et al. 1991; Gahoonia and Nielsen 1997; Gahoonia and Nielsen 2003) and, in some species, the formation of specialized root structures such as aerenchyma (Fan et al. 2003), dauciform roots in the Cyperaceae (Shane et al. 2006) and proteoid roots (or cluster roots) in the Proteaceae and certain *Lupinus* spp. (Dinkelaker et al. 1995; Gardner et al. 1981; Roelof et al. 2001; Lambers et al. 2006).

Depletion of phosphorus in the rhizosphere

The extent of P depletion in both the rhizosphere and mycorrhizosphere has been highlighted in a number of studies. Root hairs in particular contribute to increased root volume and can constitute up to 70% of the total root surface area (Itoh and Barber 1983; Jungk 2001; Fig. 1). As such they are the major site for nutrient acquisition and can account for up to 80% of total P uptake in non-mycorrhizal plants (Föhse et al. 1991). Variation in length and density of root hairs is important particularly under conditions of low P and their contribution to P uptake has been verified through modelling studies and the use of root-hairless mutants (Gahoonia and Nielsen 2003; Ma et al. 2001). Mycorrhizal colonization of roots (as discussed below) similarly provides a significant increase in the effective volume of soil explored with associated depletion of soil P (Tarafdar and Marschner 1994 and as modelled by Schnepf et al. 2008). Interestingly, the benefit derived from mycorrhizal fungi has been shown to be inversely associated with root hair length across a range of plant species, suggesting a complementary function of these traits (Baon et al. 1994; Schweiger et al. 1995). Reduced P acquisition by a root-hairless mutant of barley at low soil P was similarly compensated for by the presence of mycorrhiza (Jakobsen et al. 2005a).

Depletion of P in the rhizosphere occurs from both inorganic and organic P which includes soluble orthophosphate and various forms of extractable P (both inorganic and organic) that are widely considered to be labile. In addition, it is evident that pools of P that are more recalcitrant to extraction (e.g. NaOH-extractable P; Fig. 2), and thus previously considered to be only of poor availability to plants, can also be depleted in the rhizosphere. Such studies have used a range of different plants species and soil types whereby roots, root hairs and mycorrhizas are separated from bulk soil using meshed-compartments in rhizobox systems (e.g. Chen et al. 2002; Gahoonia and Nielsen 1997; George et al. 2002; Morel and Hinsinger 1999; Nuruzzaman et al. 2006; Tarafdar and Jungk 1987). Such approaches are useful in identifying different processes that contribute to P depletion even if they may exaggerate rhizosphere effects. For example, Chen et al. (2002) investigated P dynamics around the roots of ryegrass (*Lolium perenne* L.) and radiata pine (*Pinus radiata* D. Don) and showed a significant depletion of various pools of P at distances of up to 2 and 5 mm from the root surface of the two species respectively, which was related to different rhizosphere properties (Fig. 2). Both species also showed a significant increase in microbial biomass around the roots and associated increases in bicarbonate-extractable organic P may be a consequence of microbial-mediated immobilization of orthophosphate within the rhizosphere (Richardson et al. 2005). Further work to elucidate the role of microorganisms in influencing P availability within the rhizosphere and the extent to which they either complement or compete with plant processes in P acquisition is required (Jakobsen et al. 2005b).

Role of root exudates in phosphorus mobilization

The availability of P in the rhizosphere is influenced significantly by changes in pH and root exudates which

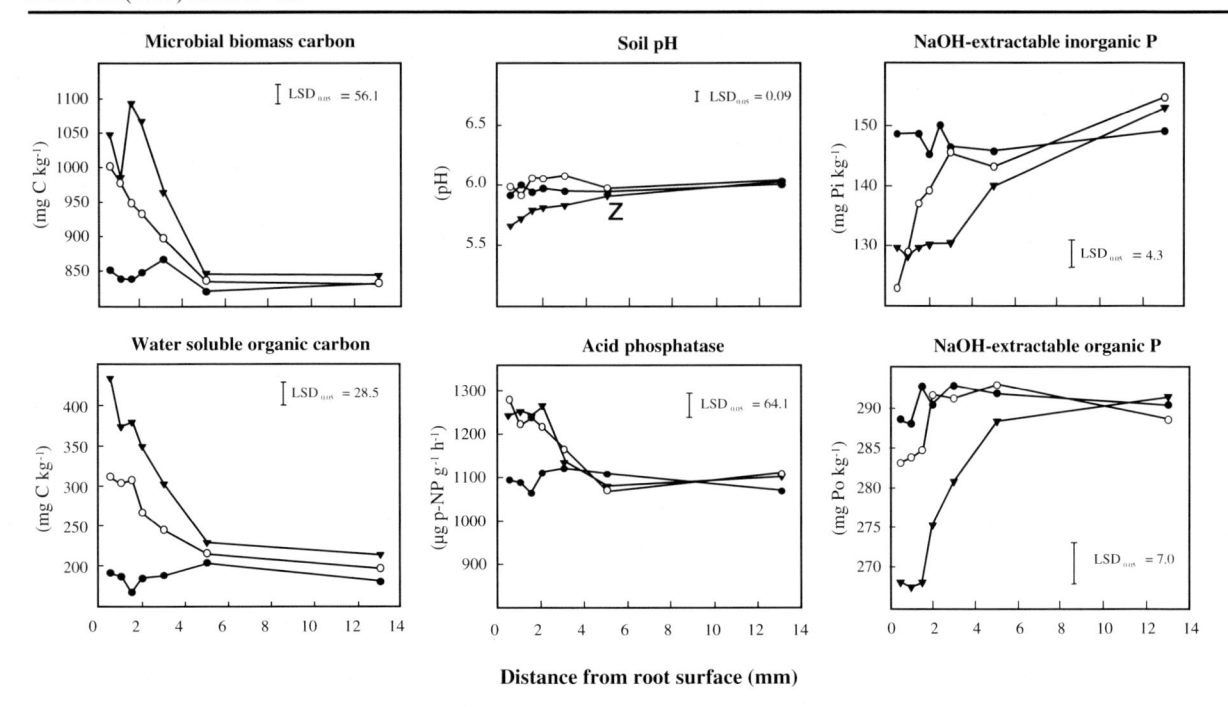

Fig. 2 Features of rhizosphere soil from perennial ryegrass (*Lolium perenne* L.; ○······○) and radiata pine (*Pinus radiata* D. Don; ▼- -▼) when grown in a rhizobox system and compared to an unplanted (control; ●—●). Shown are changes in microbial biomass C, water-soluble C, pH, acid phosphatase activity and NaOH-extractable inorganic and organic P contents of the soil at various distances from the root surface. The soil (orthic brown soil; Dystrochrept; Hurunui, New Zealand) had a total P content of 958 mg P kg^{-1} soil. For each panel the error bar (LSD $P=0.05$) shows least significant difference (data taken from Chen et al. 2002)

can either directly or indirectly affect nutrient availability and/or microbial activity (Fig. 1; Richardson 1994). Acidification of the rhizosphere in response to P deficiency has been demonstrated for a number of species (see review by Hinsinger et al. 2003) and can alter the solubility of sparingly-soluble inorganic P compounds (particularly Ca-phosphates in alkaline soils), or affect the kinetics of orthophosphate adsorption-desorption reactions in soil and the subsequent availability of orthophosphate and various micronutrients in soil solution (Hinsinger and Gilkes 1996; Gahoonia and Nielsen 1992; Hinsinger 2001; Neumann and Römheld 2007).

Organic anions are released into the rhizosphere in response to various nutritional stresses including P, Fe and micronutrient deficiency and Al toxicity (see reviews by Hocking 2001; Neumann and Römheld 2007; Ryan et al. 2001). The concentration of different organic anions is typically greater in the rhizosphere (around 10-fold) compared with that in bulk soil (Jones et al. 2003). Organic anions are commonly released from roots in association with protons which results in an acidification of the rhizosphere (Dinkelaker et al. 1989; Hoffland et al. 1989; Neumann and Römheld 2002). In addition to this change in rhizosphere pH, organic anions can also directly facilitate the mobilization of P through reduced sorption of P by alteration of the surface characteristics of soil particles, desorption of orthophosphate from adsorption sites (ligand exchange and ligand-promoted dissolution reactions), and through chelation of cations (e.g. Al and Fe in acidic soils or Ca in alkaline soils) that are commonly associated with orthophosphate in soil (Bar-Yosef 1991; Jones 1998; Jones and Darrah 1994b). Organic anions also mobilise P bound in humic-metal complexes (Gerke 1993) and have been shown to increase both the availability of organic P and its amenability to dephosphorylation by phosphatases (Hayes et al. 2000). However, the effectiveness of different organic anions in nutrient mobilization depends on various factors including; the form and amount of the particular anion released, with citrate and oxalate being more effective relative to others (e.g. malate, malonate and tartrate followed by succinate, fumarate, acetate and lactate; Bar-Yosef 1991) and interactions of the anion within the soil environment, including its effective concentration in soil solution and relative turnover rate (Jones 1998). The presence of microorganisms is of further importance because of their capacity to either rapidly metabolize different organic anions within the rhizosphere or through their own ability to release anions (Jones et al. 2003). For example, in a calcareous soil, Ström et al. (2001) showed greater stability and resistance to microbial degradation of oxalate compared with citrate and malate which resulted in greater mobilization of P within localized regions of soil with a subsequent increase in P uptake by maize (*Zea mays* L.) roots (Ström et al. 2002).

The effectiveness of organic anions in mobilizing P from soil is highlighted by studies with white lupin (*Lupinus albus* L.) which exudes significant amounts of citrate (and to some extent malate) from cluster roots that are formed in response to P deficiency (Dinlelaker et al. 1989; Gardner et al. 1983; Keerthisinghe et al. 1998; Neumann and Martinoia 2002; Vance et al. 2003). Citrate is effective in mobilizing orthophosphate from pools of soil P that are otherwise not available to plants that either do not exude, or show limited release of organic anions, such as soybean (*Glycine max* (L.) Merr.) and wheat (*Triticum aestivum* L.) (Braum and Helmke 1995; Hocking et al. 1997). In addition, and analogous to the role of root hairs, cluster roots have a relatively short life span and form on lateral roots as closely packed tertiary roots with a dense covering of root hairs. This provides a zone for both concentrated release of organic anions and high surface area for the uptake of mobilized P (Dinkelaker et al. 1995, Neumann and Martinoia 2002). Interestingly, cluster roots form on plant species that are essentially non-mycorrhizal and therefore appear to provide an important alternative strategy for plant acquisition of soil P (Shane and Lambers 2005). Indeed, the formation of cluster roots and high rates of organic anion release are reported for various native Australian species (e.g. the Proteaceae and Casuarinaceae families) which have evolved on low P soils (Roelofs et al. 2001; Shane and Lambers 2005). Increased organic anion efflux from roots in response to P-deficiency also occurs in other species including chickpea (*Cicer arietinum* L.) and pigeon pea (*Cajanus cajan* L.) and to a lesser extent in lucerne (alfalfa; *Medicago sativa* L.), canola (oil seed rape; *Brassica napus* L.) and rice (*Oryza sativa* L.) (Ae et al. 1991; Hedley et al. 1982; Hoffland et al. 1989; Lipton et al. 1987; Otani et al. 1996; Pearse et al. 2006a; Veneklaas et al. 2003; Wouterlood et al. 2004). The increase in organic anion efflux by these species in response to P deficiency however, is considerably less than for the Proteaceae or *Lupinus* spp, and in many cases the agronomic significance of organic anion release remains to be verified in soil environments, as does the role of various organic anions in mobilizing P from different forms of soil P (Pearse et al. 2006b).

Activity of phosphatases is significantly greater in the rhizosphere and is considered to be a general response of plants to mobilize P from organic forms in response to P deficiency (Richardson et al. 2005). Phosphatases are required for the hydrolysis (mineralization) of organic P, and in bulk soil microbial-mediated mineralization of organic P contributes significantly to plant availability (Frossard et al. 2000; Oehl et al. 2004). Depending on soil type and land management, organic forms of P commonly constitute around 50% of the total P in soil and is the predominant form of P found in soil solutions (Ron Vaz et al. 1993; Shand et al. 1994). Dissolved organic P is derived largely from the turnover of soil microorganisms and, relative to orthophosphate, has greater mobility in solution (Helal and Dressler 1989; Seeling and Zasoski 1993) and is therefore of critical importance to the dynamics and subsequent availability of P within the rhizosphere (Jakobsen et al. 2005a; Richardson et al. 2005).

Extracellular phosphatases released from roots have been characterized for a wide range of plant species and been shown to be effective for the *in vitro* hydrolysis of various organic P substrates (George et al. 2008; Hayes et al. 1999; Tadano et al. 1993; Tarafdar and Claassen 1988). Products of microbial turnover also contain high amounts of dissolved organic P (>80%) (primarily as nucleic acids and phospholipids), and are rapidly mineralized in soil and as such are of high availability to plants (Macklon et al. 1997). Direct hydrolysis of organic P and subsequent utilization of released orthophosphate by roots has been demonstrated in soil using both whole plant systems (e.g. McLaughlin et al. 1988) and in rhizobox studies. In the later case, depletion of various pools of extractable organic P from the rhizosphere is associated with higher activities of phosphatases around plant roots (Chen et al. 2002; Tarafdar and Claassen 1988; Fig. 2). In the study by Chen et al. (2002), greater net depletion of organic P by radiata pine (compared to ryegrass) occurred despite similar increase in phosphatase activity for both species and the development of a lesser 'root mat' at the soil interface for pine roots, which suggests the possible involvement of other mechanisms. Indeed pine roots acidified the rhizosphere to a greater extent and had higher water soluble carbon and microbial biomass (Chen et al. 2002). Higher water soluble C suggests either greater root exudation or higher turnover of microbial biomass which could result in increased P mobilization. Alternatively, greater radial depletion of P by pine roots maybe associated with the presence of longer root hairs or more likely by association with ectomycorrhizal fungi. This latter possibility is suggested by the authors and is supported by observations from more recent studies where mycorrhizal fungi have been shown to be particular effective for the capture of P by pine roots (Liu et al. 2005; Scott and Condron 2004). Casarin et al. (2004) similarly showed the importance of ectomycorrhizas for mobilization of poorly available soil P around roots of maritime pine (*Pinus piaster* Ait.) and that this benefit was from both increased soil exploration and, depending on the species of mycorrhizal fungi, due to the release of oxalate and protons. However, the relative importance of such microbial processes as compared to direct plant mechanisms and other processes remains to be fully established. In the study by Chen et al. (2002), numbers and activities of free-living and root-associated microorganisms were also enhanced significantly within the rhizosphere (e.g. as shown by increased microbial biomass; Fig. 2) and these may also contribute substantially to the mechanisms of P depletion and solubilization.

Acquisition of nitrogen by plants

Nitrogen availability and uptake

Nitrogen occurs in soil in both organic and inorganic forms and in addition to marked seasonal changes is characterised by a heterogeneous distribution within the soil. Nitrogen inputs through fixation reactions (by either symbiotic microorganisms or potentially through free-living diazotrophs, as discussed below) and transformations of N between different pools have important implications for plant growth and for the loss of N from soil systems (Jackson et al. 2008). Microbial-mediated mineralization of organic forms of N to ammonium (NH_4^+) and its subsequent nitrification to nitrate (NO_3^-) is of major significance to N availability (Fig. 3) and has influence on root behaviour and rhizosphere dynamics. Although mineral forms of N have classically been considered to dominate plant uptake (see review by Miller and Cramer 2004), there is evidence that soluble organic forms of N (e.g. low molecular weight compounds such as amino acids) may also play a significant role (Chapin et al. 1993), but few studies have quantified the relative importance of each (Leadley et al. 1997; Schimel and Bennett 2004). Of particular significance in the rhizosphere is the effect that uptake of different N forms has on soil pH in the immediate vicinity of the root and subsequent influence of this on nutrient acquisition, especially in relation to the availability of P and various micronutrients (e.g. Zn, Mn and Fe) (Marschner 1995). Changes in rhizosphere pH, caused by the influx of protons that occurs with uptake of NO_3^-, or the net release of protons for NH_4^+ uptake, can also bring about changes in the nature of substrates exuded from roots or the quantities of exudates released, and consequently may have major impact on the structure of microbial communities around the root (Bowen and Rovira 1991; Meharg and Killham 1990; Smiley and Cook 1983).

Both NO_3^- and NH_4^+ reach the root surface via a combination of mass flow and diffusion (De Willigen 1986). Nitrate is typically present in soil solution at mM concentrations and, relative to orthophosphate, is more mobile (Tinker and Nye 2000) and thus, is potentially able to move in soil by up to several mm per day (Gregory 2006). Ammonium is less mobile since it readily adsorbs to the cation exchange sites in soil and has lower rates for both mass flow and diffusion. Nevertheless, diffusion and mass flow is the major pathway for inorganic N uptake and, although it is difficult to differentiate diffusion from root interception, it is generally considered that interception of N in soil solution following root extension accounts for a small percentage only of N taken up by plants (Barber 1995; Miller and Cramer 2004).

Fig. 3 The plant-soil N cycle and pathways for N transformation mediated by physiological processes (DON = dissolved organic nitrogen; redrawn from McNeill and Unkovich, 2007)

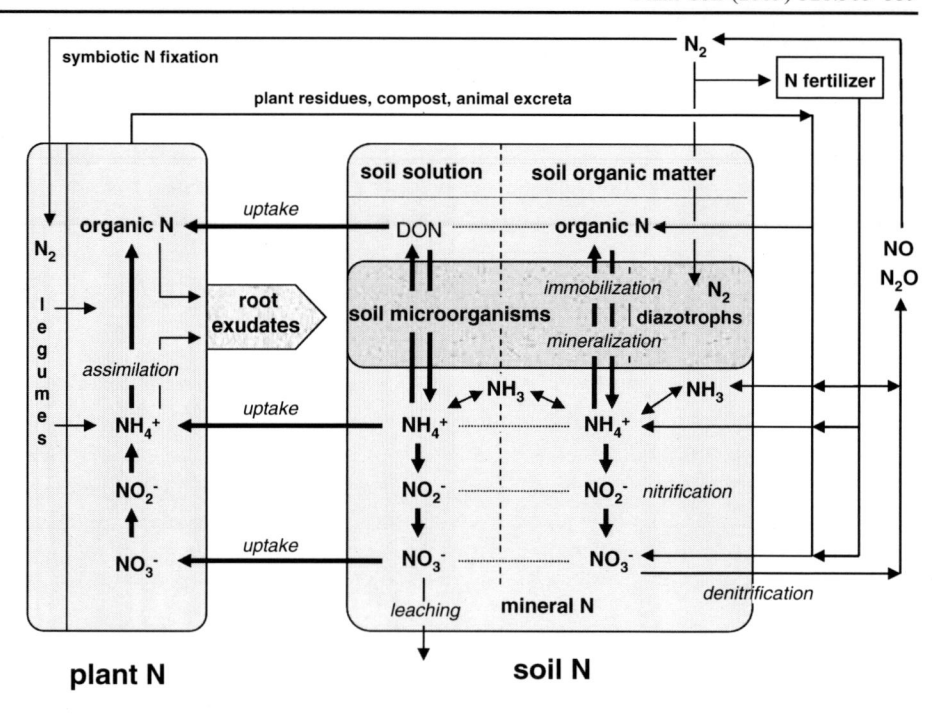

Plant uptake of NH_4^+ and NO_3^- is a function of their concentrations in soil and soil solution, root distribution, soil water content and plant growth rate. The latter is most important under conditions of liberal N supply, whereas mineral N concentration and root distribution are more critical under N limiting conditions. Whilst some plant species show a preference for either NH_4^+ or NO_3^- uptake, the significance of this for N uptake at the field level is usually less than the abovementioned factors, especially in agro-ecosystems (McNeill and Unkovich 2007). Indeed, NH_4^+ tends to dominate in many natural ecosystems for reasons which may include a suppression of microbial-mediated nitrification (reviewed by Subbarao et al. 2006). In addition and depending on soil type, non-exchangeable forms of NH_4^+ may contribute significantly to crop nutrition (Scherer and Ahrens 1996; Mengel et al. 1990), especially for lowland rice grown under flooded conditions (Keerthisinghe et al. 1985). Although many crop plants differ in their sensitivity to toxic effects of NH_4^+, the crucial factor seems to be the relative concentration of the two ions, with the optimal mix being dependent on factors such as plant species, age and soil pH (Britto and Kronzucker 2002; Badalucco and Nannipieri 2007). Furthermore, uptake of NO_3^- across the plasma membrane is more costly in terms of energy expenditure but nonetheless occurs effectively over almost the entire range of NO_3^- concentrations found in soil solutions (Forde and Clarkson 1999). However, for crop production systems the regulation of whole plant N uptake remains relatively poorly understood (Gastal and Lemaire 2002; Jackson et al. 2008), although there is evidence that the concentrations of both NO_3^- and NH_4^+ in soil solution, as well as plant N status are involved (Aslam et al. 1996; Devienne-Barret et al. 2000).

The significance of soluble organic N for plant nutrition was first highlighted in solution culture studies and has since been demonstrated for soils in a range of different ecosystems (Jones and Darrah 1994a; Schimel and Bennett 2004). Amino acids typically constitute up to half of the total soluble N in soil solution (concentrations ranging from 0.1 to 50 mM) and thus comprise a significant part of the potentially plant available N pool (Christou et al. 2005; Jones et al. 2002). Amino acids in soil solution occur as a result of either direct exudation by roots or from the breakdown of proteins and peptides from soil organic matter and microbial biomass turnover as a result of microbial-derived proteases (Jaeger et al. 1999; Owen and Jones 2001). In addition the involvement of plant-exuded proteases in digestion of proteins at the root surface has been reported along with the possibility of direct uptake of proteins by roots through endocytosis (Paungfoo-Lonhienne et al. 2008). Uptake of organic N compounds by plants may also be facilitated by association with ectomycorrhizal fungi (Chalot and Brun 1998; Nasholm et al. 1998). However, the relative importance of such mechanisms in many ecosystems remains debatable since the diffusion rates of amino acids and proteins are typically orders of magnitude lower than for NO_3^- (Kuzyakov et al. 2003), and thus mycorrhizas may offer distinct advantages, whereas microorganisms in the rhizosphere are likely to compete more strongly for these compounds (Hodge et al. 2000a). Although direct evidence that plant roots can out-compete microorganisms for N is limited to a few studies (Hu et al. 2001; Jingguo and Bakken 1997), there is some evidence from ^{15}N time-course studies where plants accumulate ^{15}N and thus benefit over the longer term which may be associated with microbial turnover (Hodge et al. 2000b; Kaye and Hart 1997; Yevdokimov and Blagodatsky 1994).

Given the heterogeneous distribution of N in soil in terms of chemical form, temporal dynamics and spatial distribution, plants adopt one or more of three main strategies to optimize their acquisition of N. Broadly these are; (i) to explore greater volume of soil and/or soil solution by extending root length and branching or increasing root surface area via changes in root diameter or root hair morphology, (ii) specific adaptive response mechanisms in order to exploit spatial and temporal 'niches' such as N-rich patches or due to the presence of particular forms of N (amino acids, NH_4^+ or NO_3^-), and (iii) influences on plant available N in the rhizosphere through plant-microbial interactions.

Root growth and morphological responses to nitrogen

The size and architecture of the root system is an important feature for ensuring adequate access to soil N, and root system size (relative to shoot growth) has generally been shown to increase when N is limiting (Chapin 1980; Ericsson 1995). However, changes to biomass alone are not necessarily indicative of the total absorptive area of a root system and morphological changes can occur without change in biomass. Although the architecture of root systems is intrinsically determined by genotype and the pattern of root branching, species specific attributes related to size and architecture are also strongly determined by external physical, chemical and biological factors (Miller and Cramer 2004). Whilst primary root growth is generally less sensitive to nutritional effects than is the growth of secondary or higher root orders (Forde and Lorenzo 2001), the diameter of first and second order laterals were significantly thicker in cereals grown at high concentrations of NO_3^- (Drew et al. 1973). Thicker roots may be more costly to produce but have greater capacity for the transport of water and nutrients and are less vulnerable to adverse edaphic conditions (Fitter 1987). Conversely, fine roots allow greater exploration of soil and plants appear to accommodate trade-off between the two by exhibiting plasticity in root diameter (and morphology) according to the environmental conditions (Forde and Lorenzo 2001). Root angle, an important component of root architecture in soil in relation to P deficiency (Rubio et al. 2003), appears to be unaffected by N deficiency.

Apart from the size and depth of root systems other attributes may also influence the capacity for efficient capture of N. Only a limited portion of the root may actually be effective in the uptake of N (Robinson 2001) and thus the spatial localization of roots is important when nutrient is distributed heterogeneously (Ho et al. 2005). Electrophysiological and molecular evidence supports a role for root hairs in the uptake and transport of both NO_3^- and NH_4^+ (Gilroy and Jones 2002) and proliferation of fine roots and root hairs in response to localized patches of N has been demonstrated (Jackson et al., 2008).

Rooting depth, which varies greatly between species, influences the capture of N by plants, particularly of NO_3^- during periods of leaching, and is clearly an important characteristic for many perennial agricultural species and tree crops (Gastal and Lemaire 2002). However, a dimorphic root system, having both shallow and deep roots to enable acquisition of mineralized N in the topsoil as well as leached N at depth, is considered to be important (Ho et al. 2005). Indeed, vigorous wheat lines with faster vertical root growth and more extensive horizontal root development have been shown to take up significantly more N (Liao et al. 2006).

Roots exhibit high plasticity as a physiological response to localized patches of organic and inorganic nutrients in soil, including proliferation in N-rich zones (Hodge et al. 1999a; Robinson and van Vuuren

1998). This proliferation essentially involves the initiation of new laterals, but may also include increases in the elongation rate of individual roots and expansion of the rhizosphere through root hairs. Roots can also enhance their physiological ion-uptake capacities in localized nutrient-rich zones. This root 'foraging' capability is considered to be an important plant response to optimize resource allocation in regard to N capture and is of particular ecological importance in situations where there is competition with neighbouring roots for limited resources (Hodge et al. 1999b; Robinson et al. 1999). However, foraging is not a fixed property but varies within species in response to different environmental conditions (Wijesinghe et al. 2001), indicating that the environmental context in which the root response is expressed is as important as the response itself. However, most studies investigating root growth in response to patches of N have largely ignored the attributes of the patch itself despite the fact that the dynamics of nutrient transformation within the patch and microbial interactions are of major importance. There is need for research to follow both together, including consideration of the complexity of interactions with other root systems, soil microorganisms and fauna, and physical/chemical interactions in the soil (Hodge 2004).

Association with microorganisms and plant growth promotion

Microbial associations with roots are complex in soil and can enhance the ability of plants to acquire nutrients from soil through a number of mechanisms. These include; i) an increase in the surface area of roots by either a direct extension of existing root systems (e.g. mycorrhizal associations) or ii) by enhancement of root growth, branching and/or root hair development (e.g. through plant growth promoting rhizobacteria), (iii) a direct contribution to nutrient availability though either N fixation (e.g. rhizobia and diazotrophs) or by stimulation and/or contribution to metabolic processes that mobilize nutrients from poorly available sources (e.g. organic anions) or, an indirect effect on nutrient availability by (iv) displacement of sorption equilibrium that results in increased net transfer of nutrients into solution and/or as the mediators of transformation of nutrients between different pools (e.g. nitrification inhibitors and microbial-mediated processes that alter the distribution of nutrients between inorganic and organic forms) or v) through the turnover of microbial biomass within rhizosphere (Gyaneshwar et al. 2002; Jakobsen et al. 2005a; Kucey et al. 1989; Richardson 2007; Tinker 1980).

In this respect, mycorrhizal fungi, rhizobia and *Frankia* microsymbionts, and plant growth promoting (PGP) microorganisms are of particular importance and as such have been studied most widely. PGP microorganisms represent a wide diversity of bacteria and fungi that typically colonize the rhizosphere and are able to stimulate plant growth through either a 'biofertilizing' (direct) effect or through mechanisms of 'biocontrol' (indirect effect; Bashan and Holguin 1997; and see Raaijmakers et al. 2009; Harman et al. 2004; Fig. 4). Biofertilizing-PGPR (considered in more detail below) specifically refers to rhizobacteria that are able to promote growth by enhancing the supply of nutrients to plants (Vessey 2003). For rhizobia and *Frankia* this involves symbiotic relationships with host legume and actinorhizal plants, respectively (see review by Franche et al. 2009), and are therefore not considered here. Rhizobiaceae (rhizobia) can also develop non-specific associative interactions with roots and promote the growth of non-legumes, and as such are also commonly considered as PGPR (Sessitsch et al. 2002). However, PGP is a complex phenomenon that often cannot be attributed to a single mechanism and, as outlined below, PGPR may typically display a combination of mechanisms (Ahmad et al. 2008; Kuklinsky-Sobral et al. 2004). In addition, the effects of individual PGPR may not occur alone but through synergistic interactions between different microorganisms.

Association with mycorrhizal fungi

Mycorrhizal symbioses are found in almost all ecosystems and can enhance plant growth through a number of processes which include improvement of plant establishment, increased nutrient uptake (particularly P and essential micronutrients such as Zn and Cu, but also N and, depending on soil pH, may enhance the uptake of K, Ca and Mg; Clark and Zeto 2000), protection against biotic and abiotic stresses and improved soil structure (Buscot 2005; Smith and Read 2008). Mycorrhizal fungi typically

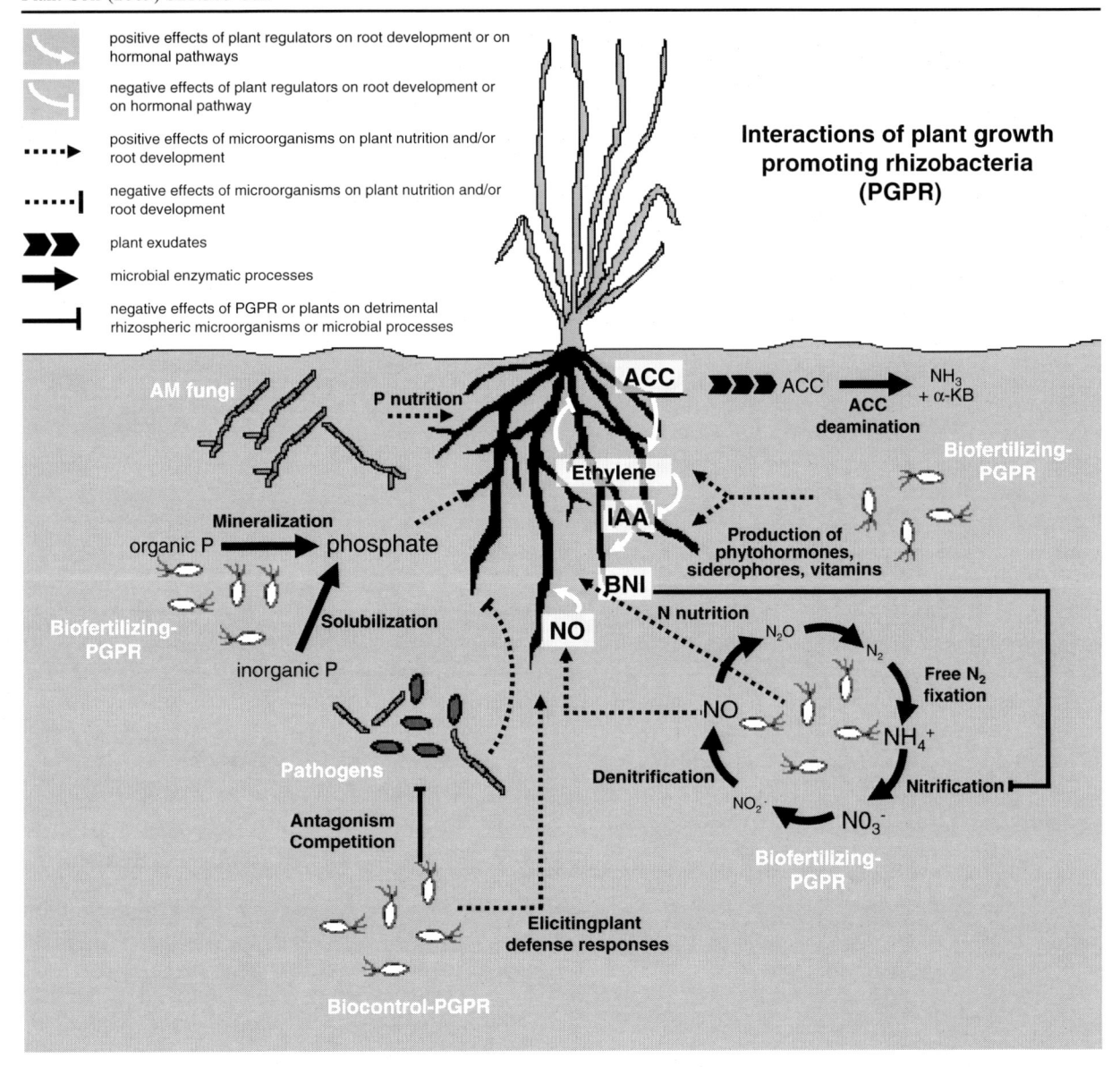

Fig. 4 Plant growth promotion mechanisms (positive and negative effects) associated with soil and rhizosphere (PGPR) microorganisms. Biofertilizing-PGPR and arbuscular mycorrhizal (AM) fungi stimulate plant nutrition by directly increasing the supply of nutrients to plants (e.g. through N fixation, P solubilization and/or mineralization, vitamin and siderophore production) or by increasing the plants access to nutrients due to enhancement of root volume. Promotion of root growth is linked to the ability of PGPR to produce phytohormones (e.g. IAA, ethylene, NO) or by direct influences on plant hormone levels (e.g. deamination of ACC precursor to plant ethylene). Biocontrol-PGPR improve plant heath by inhibiting the growth of plant pathogens or by eliciting plant defense responses

colonize the root cortex biotrophically and develop external hyphae (or extra-radical mycelia) which connect the root with the surrounding soil. All but few vascular plant species are able to associate with mycorrhizal fungi. The universality of this symbiosis implies a great diversity in the taxonomic features of the fungi and the plants involved. At least five types of mycorrhizas are recognized, the structural and functional features of which are reviewed in detail elsewhere (Smith and Read 2008; Brundrett, 2002) and are thus only considered briefly here.

Higher plants commonly form associations with ectomycorrhizas, mainly forest trees in the Fagaceae, Betulaceae, Pinaceae, *Eucalyptus,* and some woody legumes. The fungi involved are usually Basidiomycetes and Ascomycetes which colonize cortical root tissues but without intracellular penetration (Smith and Read 2008). Three other types of mycorrhizas can be grouped as endomycorrhizas, in which the fungus colonizes the root cortex intercellularly. One of these is restricted to species within the Ericaceae (ericoid mycorrhizas), the second to the Orchidaceae (orchid mycorrhizas) and the third, which is by far the most widespread (and therefore considered in most detail here), are the arbuscular mycorrhizas (AM). A fifth group, the ectendomycorrhizas, are associated with plant species in families other than Ericaceae, including the Ericaless and the Monotropaceae (arbutoid and monotropoid mycorrhizas). The majority of plant families form arbuscular associations, with the AM fungi being an obligate symbiont that is unable to complete its life cycle without colonization of a host plant. The AM fungi were formerly included in the order Glomales, Zygomycota, but are now considered as new phylum, the Glomeromycota (Redecker 2002, Schübler et al. 2001).

The establishment of mycorrhizal fungi in roots changes key aspects of plant physiology, including mineral nutrient composition in tissues, plant hormonal balance and patterns of C allocation. The fungi may also alter the chemical composition of root exudates, whilst the development of mycelium in soil can act as a C source for microbial communities and introduce physical modifications to the soil environment (Gryndler 2000). Such changes in the rhizosphere can affect microbial populations both quantitatively and qualitatively such that the rhizosphere of mycorrhizal plants (known as the mycorrhizosphere) generally has features that differ substantially from those of non-mycorrhizal plants (Barea et al. 2002; Johansson et al. 2004; Offre et al. 2007). As discussed in more detail below, a wide range of bacteria (including actinomycetes and various PGPR) associate with mycorrhizas within the mycorrhizosphere (Rillig et al. 2006; Toljander et al. 2007).

Contribution of mycorrhizas to the phosphorus nutrition of plants

It is well established that mycorrhizal fungi contribute significantly to the P nutrition of plants, particularly under low P conditions (Barea et al. 2008). This is most evident for the ectomycorrhizal fungi which are largely associated with non-agricultural plants and appear to show greater functional diversity than the AM fungi (Brundrett 2002). Whilst it is generally accepted that mycorrhizal fungi have similar access to sources of P in soil solution that are also directly available to plants (reviewed by Bolan 1991) there is some evidence to suggest that both AM and ectomycorrhizal fungi have enhanced ability to use alternative sources of P (Bolan et al. 1984; Casarin et al. 2004). For example, for AM fungi, Tawaraya et al. (2006) showed that exudates from fungal hyphae solubilized more P than root exudates alone, suggesting that the mycorrhiza contribute to increased P uptake through solubilization. The extra-radical mycelium of AM fungi have also been shown to excrete phosphatases which could potentially enhance the mineralization and utilization of organic P (Koide and Kabir 2000; Joner and Johansen 2000). However, it is generally considered that this is unlikely to be of major significance to the overall contribution of AM fungi to plant P nutrition (Joner et al. 2000, Richardson et al. 2007). Alternatively, indirect effects of mycorrhizal fungi on P mobility may occur through changes in soil microbial communities within the mycorrhizosphere (Barea et al. 2005b).

The increased efficiency of P acquisition by mycorrhizal plants is based mainly on the existence of the extra-radical mycelia which develop into soil and allow P to be accessed by the mycorrhiza from soil solution at distances up to several cm away from the root and then subsequently transferred to the plant (Jakobsen et al. 1992). High mycorrhizal hyphae density also provides considerably greater surface area for the absorption of orthophosphate by plants and, due to the smaller size of hyphae in relation to roots and root hairs and their greater length relative to root hairs, hyphae are also most effective in exploiting soil pores and nutrient patches that may not be directly accessible to roots (Jakobsen et al. 2005a). Thus, higher uptake of P by mycorrhizal plants (both ectomycorrhizas and AM) can generally be explained in term of increased hyphal exploitation of the soil as modelled by Schneph et al. (2008) and the competitive ability of the hyphae to absorb localized sources of orthophosphate and organic nutrient patches (Tibbett and Sanders 2002; Cavagnaro et al. 2005). In this context, numerous studies have shown positive correlations between fungal variables, such as hyphal length or hyphal density, with growth response variables of colonized plants such as shoot biomass, P uptake and total P content (Avio et al. 2006; Jakobsen et al. 2001). However, this cannot be taken as a general conclusion since high hyphal development does not always correlate with plant growth responses (Smith et al. 2004) and in some situations, particularly under fertilized field conditions, the presence of AM mycorrhizal fungi appears to provide little or no benefit in terms of plant P nutrition (Ryan and Angus 2003, Ryan et al. 2005).

Apart from the physical extension of root systems, mycorrhizal fungi may also acquire orthophosphate from soil solution at lower concentrations than roots, but whether this contributes significant advantage to the P nutrition of plants remains uncertain. Some studies report higher affinity for orthophosphate uptake by mycorrhizal plants (i.e. lower K_m values than for non-mycorrhizal roots; Cardoso et al. 2006). Genes encoding for the high-affinity phosphate transport in AM fungi have been identified and shown to be preferentially expressed in the extraradical mycelium (Benedetto et al. 2005; Maldonado-Mendoza et al. 2001). Mycorrhizal plants therefore have 'two' pathways for P uptake, the 'direct' pathway via the plant-soil interface through root hairs, and the 'mycorrhizal' pathway via the fungal mycelium (Smith et al. 2003). Interestingly, several studies have shown that expression of plant epidermal P transporters is reduced in roots that are colonized by AM fungi, and that under these circumstances P uptake proceeds predominantly via fungal transporters with subsequent transfer of P to plants at the arbuscular-root interface (Burleigh et al. 2002; Chiou et al. 2001; Liu et al. 1998; Rausch et al. 2001). In some cases AM colonization results in a complete inactivation of the direct P uptake pathway via root hairs with essentially all of the P in plant tissues being provided through the mycorrhizal route (Smith et al. 2004).

Contribution of mycorrhizas to nitrogen nutrition

Several studies have also shown increased N uptake from soil by roots associated with mycorrhizal fungi (Barea et al. 2005a). For example, Ames et al. (1983) first showed that mycorrhizal hyphae were able to absorb, transport and utilize NH_4^+ and, using ^{15}N-based techniques, Barea et al. (1987) demonstrated that mycorrhizal plants under field conditions had increased N uptake. Further studies using ^{15}N and compartmented rhizobox systems verified that mycorrhizal hyphae in root-free compartments were able to access ^{15}N (Tobar et al. 1994b). To investigate whether the mycorrhizal contribution to N acquisition was from pools of N that were unavailable to non mycorrhizal plants, the apparent pool size of plant available N has been determined by isotopic dilution (i.e. the A_N value of the soil; Zapata 1990). Using this approach, higher A_N values for plants inoculated with AM fungi compared to non-inoculated controls were obtained, suggesting that the AM mycelium were able to access N from forms that were otherwise less available than that for non-mycorrhizal plants (Barea et al. 1991). Indeed the presence of a functional transporter for high-affinity uptake of NH_4^+ has recently been identified in the extraradical mycelium of *Glomus* sp. (López-Pedrosa et al. 2006).

Association with free-living microorganisms

Phosphate-mobilizing microorganisms

Free-living bacteria and fungi that are able to mobilize orthophosphate from different forms of organic and inorganic P have commonly been isolated from soil and in particular from the rhizosphere of plants (Kucey et al. 1989; Barea et al. 2005b). Phosphate-solubilizing microorganisms (PSM) are characterized by their capacity to solubilize precipitated forms of P when cultured in laboratory media and include a wide range of both symbiotic and non-symbiotic organisms, such as *Pseudomonas*, *Bacillus* and *Rhizobium* spp., actinomycetes and various fungi such as *Aspergillus* and *Penicillium* spp. (see reviews by Gyaneshwar et al. 2002; Kucey et al. 1989; Rodriguez and Frago 1999; Subba-Rao 1982; Whitelaw 2000). Selection of PSM is routinely based on the solubilization of sparingly soluble Ca phosphates (typically, tri-calcium phosphate [$Ca_3(PO_4)_2$] and rock phosphates containing hydroxy- and fluor-apatites [$Ca_5(PO_4)_3OH$ and $Ca_{10}(PO_4)_6F_2$]) and Fe and Al phosphates such as strengite ($FePO_4 \cdot 2H_2O$) and variscite ($AlPO_4 \cdot 2H_2O$). The amount of P solubilized is highly dependent on the source (solubility) of the P and, for different microorganisms, is influenced to a large extent by the culture conditions. For example, fungi are commonly reported to be more effective at solubilization of Fe and Al phosphates, whereas the ability of different organisms to solubilize Ca-phosphates is influenced by the source of carbon and nitrogen in the media, by the buffering capacity of the media and the stage at which cultures are sampled (Kucey 1983; Illmer and Schinner 1995; Nahas 2007; Whitelaw et al. 1999). From various studies it is evident that change in pH of the media is particularly important for solubilization of Ca-phosphates, whereby cultures supplied with NH_4^+ are more effective than those with NO_3^- due to associated proton release and acidification of the media. Acidification is also commonly associated with the release of organic anions which have been widely reported for various microorganisms (i.e. with citrate, oxalate, lactate, and gluconate being most common). Organic anions themselves may further increase the mobilization of particular forms of poorly soluble P (e.g. Al-P and Fe-P) through chelation reactions (Whitelaw 2000).

Under controlled growth conditions various studies have demonstrated enhanced growth and P nutrition of plants inoculated with PSM which is often attributed to the P-solubilizing activity of the microorganisms involved (see reviews by Gyaneshwar et al. 2002; Kucey et al. 1989; Rodriguez and Fraga 1999; Whitelaw 2000). However, clear effect of PSM in more complex soil environments and in field conditions, have proved more difficult to demonstrate and inconsistent response of plants and performance of different microorganisms have been observed. As discussed by Richardson (2001) this may be due to a range of factors that include, insufficient knowledge for introducing and understanding the dynamics of microorganisms and their interaction with complex microbial communities in soil, the apparent lack of any specific association between partners, and poor understanding of the actual mechanisms involved, both for the microorganisms and their interaction and efficacy within different soil environments. For example, whilst *Penicillium radicum* effectively solubilized P in laboratory media and was able to promote the growth of wheat, evidence for improved P uptake in response to inoculation was only evident in glasshouse trials particularly where fertilizer P was applied (Whitelaw et al. 1997). However, growth promotion may not necessarily be directly associated with P solubilization and production of phytohormones is likely to be involved (Wakelin et al. 2006). Similarly, promotion of root growth and enhanced P nutrition of plants inoculated with *P. bilaii* has been shown to be primarily associated with increased root growth, including greater specific root length and production of longer root hairs (Gulden and Vessey 2000). Such studies highlight the difficulties in determining the actual mechanism associated with growth promotion, as stimulation of root growth also contributes to greater potential for P acquisition. It is important therefore that experiments directed at demonstrating the benefits of PSM be conducted across a range of P supplies, whereby benefits of inoculation should be negated at higher levels of applied P, and that specific measures of P acquisition (e.g. by isotopic dilution) be made to confirm P mobilization from pools of soil that are otherwise poorly available to roots.

The mineralization of organic P in soil is largely mediated by microbial processes and as such microorganisms play a significant role in maintaining plant available P. Microorganisms are able to hydrolyze a wide range of organic P substrates when grown in culture and, when added to soil, different forms of organic P have been shown to be rapidly mineralized (Adams and Pate 1992; Macklon et al. 1997; and see review by Richardson et al. 2005). Indeed benefits of microbial inoculation for the utilization of organic P by plants under controlled growth conditions have been demonstrated (Richardson et al. 2001a). In addition, the microbial biomass is important for maintaining both inorganic and organic P in soil solution (Seeling and Zososki 1993) and turnover of the biomass represents an important potential supply of P to plants (Oberson and Joner 2005). This contribution is likely to be of greater significance in the rhizosphere where there is increased amount of readily metabolizable carbon and higher density of microorganisms (Brimecombe et al. 2007; Jakobsen et al. 2005b). However, the relative importance of microbial mineralization relative to the short-term immobilization of P by microorganisms in the rhizosphere and its impact on the availability of orthophosphate to plants requires more detailed investigation.

Whilst it is evident that microbial-mediated solubilization and mineralization of inorganic and organic P are important processes whereby microorganisms are able to acquire P from soil, it has been argued that they are unlikely to mobilize sufficient P above their own requirements to meet plant demand (Tinker

1980). Indeed, few studies have unequivocally demonstrated a direct release of P by microorganisms in soil and benefits to plant nutrition are therefore often inferred. Nevertheless, the cycling of P within the microbial biomass and its subsequent release is paramount to the P cycle in soil and represents an important pathway for movement of P from various soil pools into plant-available forms and may also serve to protect orthophosphate from becoming unavailable in soil due to various physicochemical reactions (Magid et al. 1996; Oberson et al. 2001). The significance of this in the rhizosphere warrants further research.

Microbial interactions and nitrogen availability

Root exudation also has important implications for N availability. Although the chemical composition of exudates varies widely for different species and root types and is primarily comprised of C compounds, exudates can also contain significant quantities of N, which is either available to microorganisms in the rhizosphere or can be recaptured by plants (Bertin et al. 2003; Uren 2007). In addition, root exudates are the major energy supply for the soil food web and play a significant role in the turnover of soil organic matter and associated nutrients. However, as suggested by Jones et al. (2004), although root exudates have been hypothesized to be involved in the enhanced mobilization and acquisition for many nutrients in soil, there is little mechanistic evidence from soil-based systems to verify this, which is further highlighted by a recent analysis of the literature concerning rhizodeposition by maize (Amos and Walters 2006). Despite this, some studies have demonstrated enhanced N cycling in the vicinity of plant roots (Jackson et al. 2008). For example, following the application of ^{15}N labelled fertilizer, the excess of ^{15}N in the microbial biomass increased significantly in both planted and control soils at up to 8 weeks after plant emergence, but then declined in control soils only (Qian et al. 1997). Retention of ^{15}N in the microbial biomass in planted soil was attributed to the release of root-derived C from maize as estimated using ^{13}C abundance methodology. This release of exudate was suggested to promote microbial immobilization of the N and is consistent with greater microbial biomass in the rhizosphere.

Increased rates of N mineralization have similarly been demonstrated in the rhizosphere of slender wild oats (*Avena barbata* Pott ex Link), where N mineralization was 10 times higher than in bulk soil, but this was highly dependent on location along the root (Herman et al. 2006). In addition, a rhizosphere 'priming effect' has been suggested to be involved in the decomposition of native soil organic matter around roots (Cheng et al. 2003; Kuzyakov 2002). However, stimulation of mineralization may be dependent on plant species and on the C:N ratio of substrates as highlighted in a study of peas (*Pisum sativum* L.) inoculated with *Pseudomonas fluorescens*, where increased uptake of N from ^{15}N enriched organic residues occurred, whereas decreased uptake was observed for wheat (Brimecombe et al. 1999). Additional work demonstrated that microbial-microfaunal interactions in the rhizosphere were also involved in this differential response, with lower numbers of nematodes and protozoa being present in the rhizosphere of uninoculated peas which appeared to exert a nematicidal effect (Brimecombe et al. 2000). On the contrary, higher numbers of nematodes and protozoa (e.g. up to 6-fold) have generally been reported in the rhizosphere where they specifically feed on bacteria, fungi and yeasts (Zwart et al. 1994). These interactions enhance N flows in the rhizosphere both directly, via the excretion of consumed nutrients and mineralization of nutrients on death (Griffiths 1989), and indirectly via changes to the composition and activity of the microbial community (Griffiths et al. 1999). For example, bacteriophagous nematodes mineralized up to six times more N than an equivalent biomass of protozoa grazing on bacteria (Griffiths 1990). Net mineralization is due to differences in C:N ratio between the protozoan (or nematode) predator and the bacterial prey and their relatively low assimilation efficiency, whereby ~60% of ingested nutrients are typically excreted and thus potentially available for plant or microbial uptake (Bonkowski 2004). Moreover, plants grown under controlled conditions have been observed to develop more highly branched root systems in the presence of protozoa which may partly be explained as a response to NO_3^- formed from mineralization of NH_4^+ excreted by protozoa (Forde 2002). However, other work has suggested that root responses are due to a direct phytohormone effect by the presence of either protozoa or a consequence of auxin producing bacteria stimulated by the presence of the protozoa (Bonkowski 2004; Bonkowski and Brandt 2002). The consequences of soil organisms promoting a mutually beneficial relationship between plant roots and bacteria in the rhizosphere on root architecture, nutrient uptake and plant productivity is therefore of current research interest (Mantelin and Touraine 2004). For example, in an experiment using $^{15}N/^{13}C$ labelled organic nutrient sources combined with manipulation of the composition of microfaunal populations, Bonkowski et al. (2000) observed effects on ryegrass growth and concluded that microfaunal grazing increased the temporal coupling of nutrient release and plant uptake, whereas root foraging in organic nutrient-rich zones enhanced the spatial coupling of mineralization and plant uptake. More recent work has shown interaction at higher trophic levels where collembola that feed on bacteria, fungi, nematodes and protozoa further enhanced N mineralization without alteration to the microbial biomass C (Kaneda and Kaneko 2008).

At a larger scale there is increasing evidence from using *in situ* ^{15}N labelling of plant root systems of the importance of N rhizodeposition in sustaining the N cycle of agro-ecosystems (Hogh-Jensen 2006; Mayer et al. 2004; McNeill and Fillery 2008). A recent review (Wichern et al. 2008) highlights the wide variability in results and suggests there is need for more investigations on key environmental factors influencing the amounts of N released under field conditions from different species. Apart from a direct influence of the rhizosphere on N deposition, there is also need to more fully understand the role that plant roots have in interacting with soil microorganisms and influencing other parts of the soil N cycle. For example, the presence of plant-derived biological nitrification inhibitors (BNI) in the root zone can influence the conversion of NH_4^+ to NO_3^- (nitrification) and subsequently to the potential for gaseous losses of N through denitrification (Fig. 3). Indeed nitrification inhibitors have been recognized for some time in native plant ecosystems, but more recently have been shown to be effective as root exudates from brachiaria (*Brachiaria humidicola* (Rendle) Schweick a tropical grass species (Subbarao et al. 2006). In these systems there is greater retention of NH_4^+ in soil which has important implication for improving N-use efficiency by reducing potential N losses through NO_3^- leaching and/or its conversion (denitrification) to N_2O gas (via NO), which contributes significantly to greenhouse-gas emissions (Fillery 2007; Subbarao et al. 2006). However, presently it appears that BNI is poorly expressed in many agricultural crop species, although activities has been reported for sorghum (*Sorghum bicolour* (L.) Moench.), pearl millet (*Pennisetum glaucum* (L.) R.Br.) and peanut (*Arachis hypogaea* L.) (Subbarao et al. 2007a) and more recently BNI has been shown to be effective in wild rye (*Leymus racemosus* (Lam.) Tzvelev), a wild relative of wheat (Subbarao et al. 2007b).

Nitrogen fixation by diazotrophs

Diazotrophs (i.e. N_2-fixing bacteria) are classified as being either symbiotic (rhizobia and *Frankia* species) or as free-living (associative) and/or root endophytic microorganisms (Cocking 2003). Rhizobia develop symbiotic relationships with host legumes and through atmospheric N_2 fixation within nodules can provide up to 90% of the N requirements of the plant (see Franche et al. 2009; Höflich et al. 1994). Free-living N_2-fixers also have the potential for providing N to host plants but so far, the direct contribution of N-fixation by diazotrophs to the N nutrition of plants and subsequent growth promotion remains in question. Free-living diazotrophs have been identified in several genera of common rhizosphere-inhabiting microorganisms such as *Acetobacter, Azoarcus, Azospirillum, Azotobacter, Burkholeria, Enterobacter, Herbaspirillum, Gluconobacter* and *Pseudomonas* (Baldani et al. 1997; Mirza et al. 2006; Vessey 2003), with some being recognized as endophytes. Endophytic diazotrophs may have advantage over root-surface associated organisms, as they can colonize the interior of plant roots and establish themselves within niches that are more conducive to effective N_2 fixation and subsequent transfer of the fixed N to host plants (Baldani et al. 1997; Reinhold-Hurek and Hurek 1998).

Mutants deficient in nitrogenase activity (i.e. Nif mutants) have been constructed in various PGPR, including *Azospirillum brasilense, Azoarcus* sp. and *Pseudomonas putida*, and importantly, have been shown in several cases to retain their ability to promote plant growth of certain crops (Hurek et al. 1994; Lifshitz et al. 1987). This questions the relative contribution of N_2 fixation to the growth promotion effect. On the contrary, Hurek et al. (2002) showed that an endophytic strain of *Azoarcus* sp. was able to fix and transfer N when associated with kallar grass (*Leptochloa fusca* (L.) Kunth), as Nif⁻ mutants gave lower plant growth stimulation than the wild-type strain. Similar results were obtained in the case of the associative symbiosis between sugar-cane (*Saccharum officinarum* L.) and the endophytic diazotroph *Acetobacter diazotrophicus* (renamed *Gluconobacter diazotrophicus*) (Sevilla et al. 2001). Furthermore, N-balance, ^{15}N natural abundance and ^{15}N dilution studies, performed in either pot experiments or in the field, have provided clear evidence of the ability of endophytic N_2-fixing bacteria to supply significant inputs of nitrogen to some grasses and cereals (Boddey et al. 2001; Oliveira et al. 2002).

In contrast to symbiotic N_2 fixation, where there is direct transfer of N across the symbiotic interface, it is evident that root surface associated diazotrophs seem not able to readily release fixed N to the host plant and that this occurs only through microbial turnover (Lethbridge and Davidson 1983; Rao et al. 1998). This may account for inconsistent response of plants inoculated with diazotrophs and indicates that there is need for better understanding of the potential for free-living and endophytic diazotrophs to supply N to host plants. In addition to their potential for supplying plants with N, free-living diazotrophs may also promote plant growth and nutrition through various other mechanisms.

Other mechanisms of PGPR to enhance plant nutrition

Microbial production of phytohormones

Many PGPR produce phytohormones that are considered to enhance root growth and greater surface area (e.g. bigger roots, more lateral roots and root hairs) leading to an increase in explored soil volume and thus plant nutrition (Fig. 4). Such microorganisms, commonly termed 'phytostimulators', include a wide range of soil bacteria and fungi. The most common phytohormones produced by PGPR are auxins, cytokinins, giberellins and to a lesser extent ethylene, with auxins being the most well characterized (Khalid et al. 2004; Patten and Glick 1996; and see review by Arshad and Frankenberger 1998). Indeed in *Azospirillum*, auxin (IAA) production, rather than N_2 fixation, is generally considered to be the major factor responsible for the PGPR response through stimulation of root growth (Dobbelaere et al. 1999).

Indole-3-acetic acid (IAA) controls a wide variety of processes in plant development and plant growth and plays a key role in shaping plant root architecture such as regulation of lateral root initiation, root vascular tissue differentiation, polar root hair positioning, root meristem maintenance and root gravitrophism (Aloni et al. 2006; Fukaki et al. 2007). Production of IAA is widespread among rhizobacteria (Khalid et al 2004; Patten and Glick 1996; Spaepen et al. 2007), with increasing numbers of endophytic IAA-producing PGPR being reported (Tan and Zou 2001). For example, Kuklinsky-Sobral et al. (2004) screened a collection of root-associated bacteria from soybean for their ability to produce IAA and showed that it was present in 28% of isolates. More recently, the distribution of IAA biosynthetic pathways among annotated bacterial genomes suggests that 15% (from 369 analysed) contain genes necessary for synthesis of IAA (Spaepen et al. 2007). IAA production has also been observed in rhizobia and in phytopathogenic bacteria, although the amount of auxins produced by different rhizobacteria seems to differ according to their mode of interaction with plants (Kawaguchi and Syōno 1996). Several IAA biosynthetic pathways, classified according to their intermediates, exist in bacteria and for most, tryptophan has been identified as the precursor of IAA (Patten and Glick 1996; Spaepen et al. 2007). However, only few specific genes and proteins involved in IAA biosynthesis have been characterized to date, and only in a small number of PGPR (e.g. *Azospirillum brasilense*, *Enterobacter cloacae*, *Pantoea agglomerans* and *Pseudomonas putida*; Koga et al. 1991; Patten and Glick 2002; Zimmer et al. 1998). In phytopathogenic bacteria, IAA seems to be mainly produced from tryptophan via the intermediate indole-3-acetamide (IAM pathway), whilst in beneficial phytostimulatory bacteria, IAA appears to be synthesized predominantly via indole-3-pyruvic acid (IPyA pathway) (Manulis et al. 1998; Patten and Glick 1996; 2002; Zimmer et al. 1998).

In *A. brasilense*, inactivation of a key enzyme in the IPyA pathway (the *ipdC* gene, encoding an indole-3-pyruvate decarboxylase) resulted in up to 90% reduction of IAA production (Dobbelaere et al. 1999), but mutants were not completely abolished in IAA biosynthesis, suggesting some redundancy in pathways. Irrespective of this, various *ipdC* mutants displayed altered phenotypes compared to the wild-type strains in their ability to alter wheat root morphology (i.e. either no increase in root hair and lateral root formation and no decrease in root length; Malhotra and Srivastava 2008; Dobbelaere et al 1999). The impact of exogenous auxin on plant development ranges from positive to negative effects, and occurs as a function of the amount of IAA produced, the cell number of auxin-producing rhizobacteria and on the sensitivity of the host plant to changes in IAA concentration (Dobbelaere et al 1999; Spaepen et al 2008; Xie et al. 1996). For example, Remans et al. (2008) highlighted cultivar specificity in the response of plants to auxin-producing bacterial strains. In many PGPR, genes involved in IAA production are fine-regulated by stress factors that commonly occur in soil and potentially within the rhizosphere (e.g. acidic pH and osmotic stress), and in some cases have been shown to be activated by plant extracts (e.g. amino acids such as tryptophan, tyrosine and phenylalanine, auxins and flavonoids; Ona et al. 2005; Prinsen et al. 1991; Vande Broek et al. 1999; Zimmer et al. 1998).

Cytokinins stimulate plant cell division, control root meristem differentiation, inhibit primary root elongation and lateral root formation but can promote root hair development (Riefler et al. 2006; Silverman et al. 1998). Cytokinin production has been reported in various PGPR including, *Arthrobacter* spp., *Azospirillum* spp., *Pseudomonas fluorescens*, and *Paenibacillus polymyxa* (Cacciari et al. 1989; de Salamone et al. 2001; Perrig et al. 2007; Timmusk et al. 1999). However, genes involved in the biosynthesis of bacterial cytokinins have not yet been characterized in PGPR and therefore their involvement in plant growth promotion largely remains speculative.

Gibberellins enhance the development of plant tissues particularly stem tissue and promote root elongation and lateral root extension (Barlow et al. 1991; Yaxley et al. 2001). Production of gibberellins have been documented in several PGPR such as *Azospirillum* spp., *Azotobacter* spp., *Bacillus pumilus*, *B. licheniformis*, *Herbaspirillum seropedicae*, *Gluconobacter diazotrophicus* and rhizobia (Bottini et al. 2004; Gutiérrez-Mañero et al. 2001). Some *Azospirillum* strains are also able to hydrolyze, both *in vitro* (Piccoli et al. 1997) and *in vivo* (Cassán et al. 2001), glucosyl-conjugates of gibberellic acid, which correspond to reserve or transport forms of gibberellic acid produced by plants (Schneider and Schliemann 1994). This activity leads to an increase in the release of active forms of the phytohormone into the rhizosphere. However, the bacterial genetic determinants involved in this mechanism remain to be identified as does the precise role of gibberellins in plant growth promotion by PGPR.

Ethylene is a key phytohormone that can inhibit root elongation, nodulation and auxin transport, and promotes seed germination, senescence and abscission of various organs and fruit ripening (Bleecker and Kende 2000; Glick et al. 2007b). Ethylene is required for the induction of systemic resistance in plants during associative and symbiotic plant-bacteria interactions and, at higher concentrations, is involved in plant defence pathways induced in response to pathogen infection (Broekaert et al. 2006; Glick et al. 2007a). Certain PGPR such as *A. brasilense* have been shown to produce small amounts of ethylene from methionine as a precursor (Perrig et al. 2007; Thuler et al. 2003), and this ability seems to promote root hair development in tomato plants (Ribaudo et al. 2006). However, a better knowledge (i.e. characterization of bacterial biosynthesis pathway and genetic determinants involved) has to be gained in order to determine the role of the production of this plant growth regulator in the growth promoting effect of PGPR.

Some plant-associated bacteria such as *A. brasilense* (strain Sp245) are able to produce nitric oxide (NO) due to the activity of nitrite reductases (Creus et al. 2005; Pothier et al. 2007). The formation of NO is an intermediate in the denitrification pathway, during which nitrate (or nitrite) is converted to nitrogen oxides (N_2O) and to N_2 (Zumft 1997; Fig. 3). This pathway is utilized by soil bacteria to gain energy under oxygen-limited conditions that may occur in the rhizosphere (Højberg et al. 1999). Although denitrification by rhizobacteria diminishes the amount of NO_3^- available for plant nutrition, it may have positive effects on root development by means of NO production, which is a key signal molecule that controls root growth and nodulation, stimulates seed germination and is involved in plant defence responses against pathogens (Lamattina et al. 2003; Pii et al. 2007). Furthermore, NO can interact with other plant hormone signalling networks including that for IAA (Lamattina et al 2003; Fig. 4). Bacterial denitrification and production of NO by *A. brasilense* has been demonstrated on wheat roots (Creus et al.

2005; Neuer et al. 1985; Pothier et al. 2007) and NO produced during tomato root colonization stimulated the formation of lateral roots (Creus et al. 2005). Nitrous oxide production therefore is potentially another plant-beneficial trait displayed by *Azospirillum* PGPR.

Microbial enzymatic activities influencing plant hormone levels

Certain PGPR are able to stimulate plant growth by directly lowering plant ethylene levels through the action of 1-aminocyclopropane-1-carboxylic acid (ACC) deaminase (i.e. deamination of the plant ethylene precursor; Fig. 4). ACC deaminase (encoded by the *acdS* gene) catalyses the conversion of ACC, the immediate plant precursor for ethylene, into NH_3 and α-ketobutyrate and is widely distributed in soil fungi and bacteria, especially the Proteobacteria (Blaha et al. 2006; Prigent-Combaret et al. 2008). Whilst the ecological significance of ACC-deaminase in soil microorganisms is largely unknown, in plant-beneficial bacteria it may serve to diminish the amount of ACC available for production of ethylene (Glick et al. 2007a). Since ethylene inhibits growth and elongation of root, this may lead to enhanced root system development (Glick et al. 2007a). Indeed, this model has been validated by analysis of the root growth promoting effect of a *Pseudomonas putida* PGPR, where *acdS* was inactivated (Glick et al. 1994; Li et al. 2000). In the case of *Azospirillum*, complementation of AcdS⁻ strains with an *acdS* gene from *P. putida* enhanced the plant-beneficial effects of these PGPR on both tomato (*Lycopersicon esculentum* Mill.) and canola (Holguin and Glick 2001; Holguin and Glick 2003). Similar results were obtained following the introduction of the *Pseudomonas acdS* gene into AcdS⁻ *Escherichia coli*, *Agrobacterium tumefaciens* and biocontrol strains of *Pseudomonas* spp., where expression of ACC deaminase promoted root elongation, inhibition of crown gall development and improved protection against phytopathogens, respectively (Hao et al. 2007; Shah et al. 1998; Wang et al. 2000). From a number of studies it appears that the growth promotion effect of ACC deaminase in rhizobacteria is most effective in stress environments such as in flooded, heavy-metal contaminated or saline soils (Cheng et al. 2007; Farwell et al. 2007) and in response to phytopathogens (Wang et al. 2000).

Cross-talk between plant-growth promoting pathways in plants

It is evident that plant regulatory molecules (e.g. auxin, ethylene, NO, gibberellin etc) do not act alone but interact with one another in a variety of complex ways (Fu and Harberd 2003; Glick et al. 2007a; Lamattina et al. 2003). Moreover, it is clear that the PGPR effect occurs as a result of a combination of different mechanisms (additive hypothesis). For example, a model has been proposed by Glick et al. 2007a to describe cross-talk between auxin and ethylene in both PGPR and plants. In response to root exudates containing tryptophan, PGPR produce IAA that can be taken up by plant cells. Besides the direct effect of IAA on plant cell proliferation and elongation, it also induces the synthesis of ACC synthase in plants (Abel et al. 1995) and thereby the production of ethylene. A negative feedback loop, involving inhibition by ethylene of the transcription of auxin response factors, would lead *in fine* to a slow down of ACC synthase activity and decrease of ACC and ethylene biosynthesis (Glick et al. 2007a). Other close interactions between IAA and ethylene pathways have been recently reported. It appears that ethylene triggers the accumulation of auxin in the root apex (Stepanova et al. 2005) and that the transport of auxin from the apex to the elongation zone of roots is required for ethylene to inhibit root growth (Swarup et al. 2007). Overall, the molecular mechanisms by which ethylene and auxin interact to competitively regulate root development remain largely unknown and future prospects will aim to clarify their respective contribution. AcdS- and IAA-producing PGPR might promote root growth by both a lowering of plant ethylene production and ethylene-dependent signalling pathways, and through an increase in an ethylene-independent manner, of the content of IAA in roots.

Facilitating plant iron and vitamins absorption

In addition to phytohormones, PGPR may influence the growth of plant roots through the production of siderophores and vitamins. Roots of strategy II type plants (e.g. the Graminaceae; Marschner 1995; Robin et al. 2008) secrete phytosiderophores (Fe-chelators) which bind Fe^{3+} and maintain its concentration in soil solution (see Lemanceau et al. 2009). At the root surface chelated Fe^{3+} can then be directly taken up as a phytosiderophore-Fe complex (strategy II; see Lemanceau et al. 2009), whereas in non-graminaceous species (strategy I type plants; Marschner 1995; Robin et al. 2008) Fe^{3+} must first be reduced to Fe^{2+} prior to being absorbed by the plant (Lemanceau et al. 2009). Rhizobacteria (and fungi) also produce siderophores and it has been shown that plants can absorb bacterial-Fe^{3+} complexes, which includes strategy I species possibly by endocytosis (Bar-Ness et al. 1991; Vansuyt et al. 2007). The capture of these bacterial complexes by plants may play a significant role in nutrition and growth, especially in alkaline and calcareous soils where Fe availability is low (Bar-Ness et al. 1991; Masalha et al. 2000). Moreover, this mechanism is involved in biocontrol activities of PGPR and has been linked to competitive effects with phytopathogens and other detrimental rhizosphere microorganisms (Duijff et al. 1993; Longxian et al. 2005; Robin et al. 2008).

Plants under optimal growing conditions synthesize vitamins but when grown in stressed environments, vitamin-producing rhizobacteria may stimulate plant growth and yield. In particular, production of vitamins of the B group (e.g. thiamine, biotin, riboflavine, niacin) has been documented in some *Azospirillum*, *Azotobacter*, *Pseudomonas fluorescens* and *Rhizobium* strains (Marek-Kozaczuk and Skorupska 2001; Revillas et al. 2000; Rodelas et al. 1993; Sierra et al. 1999). There is evidence that exogenous supply of B-group vitamins to plants favours root development (Mozafar and Oertli 1992), but there is presently no direct evidence (e.g. using mutants) that PGPR can stimulate plant growth through this mechanism (Marek-Kozaczuk and Skorupska 2001), although further work is warranted.

Interactions between plant growth promoting microorganisms

Mycorrhizal associations

Microbial populations in the rhizosphere, including known PGPR, can either interfere with or benefit the formation and function of mycorrhizal symbioses (Gryndler 2000). A typical beneficial effect is that exerted by 'mycorrhizal-helper-bacteria' (MHB) which stimulate mycelial growth and/or enhance mycorrhizal formation (Garbaye 1994). This applies both to ectomycorrhizal fungi (Frey-Klett et al. 2005) and to AM fungi (Barea et al. 2005b; Johansson et al. 2004) and involves a range of bacterial species, commonly including *Bacillus* and *Pseudomonas*. Responses to MHB are associated with both the production of compounds that increase root cell permeability and rates of root exudation, which either stimulate AM fungal mycelia in the rhizosphere or facilitate root penetration by the fungus, and the production of phytohormones that influence AM establishment (Barea et al. 2005b). Specific rhizobacteria are also known to affect the pre-symbiotic stages of AM development, such as spore germination and rate of mycelial growth (Barea et al. 2005b). Recently, Frey-Klett et al. (2007) revisited the significance of MHB and differentiated the effects based on either AM formation or AM function, including nutrient mobilization, N_2 fixation and protection of plants against root pathogens.

Given that the external mycelium of mycorrhizas act as a link between roots and the surrounding soil, the fungus can also synergistically interact with soil microorganisms that mobilize soil P, through either solubilization or mineralization (Azcon et al. 1976; Barea 1991; Barea et al. 2005a; Kucey 1987; Tarafdar and Marschner 1995). Such interactions have been investigated with ^{32}P-based methodologies using reactive rock phosphate in a non-acidic soil (Toro et al. 1997) and in an experiment conducted by Barea et al. (2002) using various treatments that included i) AM inoculation, ii) PSB inoculation, iii) AM plus PSB dual inoculation and iv) non-inoculated controls in a soil containing natural populations of both AM fungi and PSB (Fig. 5). Soils were either un-amended (without P application) or fertilized with rock phosphate. Both rock phosphate addition and microbial inoculation improved biomass production and P accumulation in plants, with dual inoculation being the most effective (Barea et al. 2002; Fig. 5). Independent of rock phosphate addition, AM-inoculated plants showed lower specific activity for ^{32}P than compared to non-AM inoculated controls, particularly when inoculated with PSB, suggesting that the PSB were effective in releasing P from sparingly soluble sources either directly from the soil or from added rock phosphate. Other studies have similarly showed a positive interaction between ectomycorrhizal fungi and the presence of bacterial isolates that show potential for the weathering of soil minerals and its associated release of nutrients for plant uptake (Uroz et al. 2007). On such evidence it suggested that mycor-

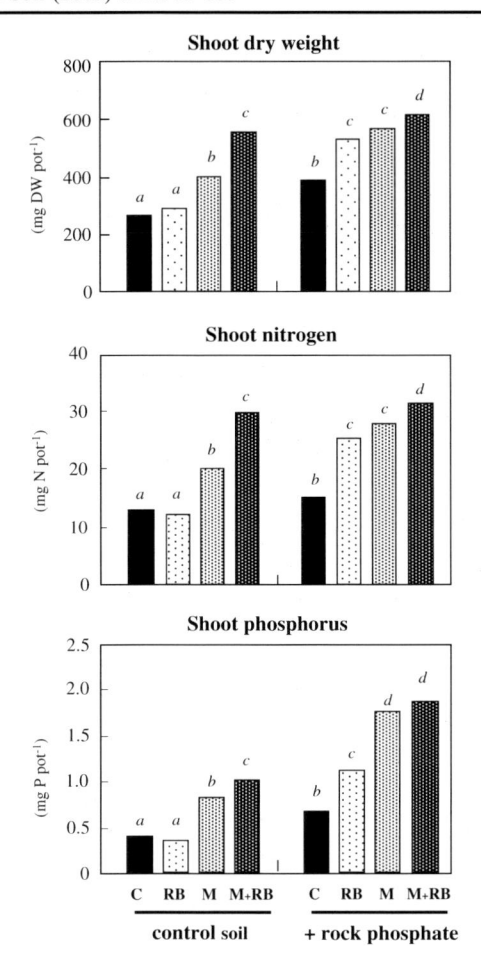

Fig. 5 Growth, N and P content of shoots of lucerne (alfalfa; *Medicago sativa* L.) inoculated with plant growth promoting rhizobacteria (PGPR). The plants were inoculated with *Rhizobium meliloti* and grown in a Cambisol soil (Granada; Spain) with either no added P (control) or supplied with rock phosphate (Riecito Venezuela; 11.4% P provided at 100 mg P kg^{-1} soil). Additionally, plants were grown as either control (C) or were further inoculated with a phosphate-solubilizing bacteria (RB; *Enterobacter* sp.), an arbuscular mycorrhizal fungi (M; *Glomus mosseae*) or with a comibation of both microorganisms (M+RB). For each panel, columns not designated with the same letter are significantly ($P<0.05$) different (data taken from Barea et al. 2002)

rhizas are highly effective for improving the capture of mobilized nutrients in soil, especially in relation to the mobilization and capture of soil P.

Microbial interactions and enhanced nitrogen fixation

Interaction between mycorrhizal fungi, PGPR and diazotrophs, including both rhizobia and associative N-fixers, has also received considerable attention. In legumes it is evident that AM can improve both nodulation and N$_2$ fixation within nodules (Barea et al. 2005a; Barea et al. 2005b). Co-inoculation of *Rhizobium* sp. with AM fungi gave greater growth promotion in lucerne (alfalfa) and pea than inoculation of either symbiont alone (Höflich et al. 1994). The physiological and biochemical basis of AM fungal x *Rhizobium* interactions in improving legume productivity suggests that the main effect of AM in enhancing nodule activity is through a generalized stimulation of host nutrition, but specific hormonal effects on root and nodule development may also occur (Barea et al. 2005a).

Several reports have demonstrated a direct PGPR effect on legume nodulation. Various PGPR including *Azotobacter vinelandii, A. brasilense, Bacillus* sp., *Pseudomonas* sp., and *Serratia* spp. increased root and shoot growth, the number and mass of nodules, N$_2$ fixation and plant N content, and grain yield in various legumes when co-inoculated with *Rhizobium* sp. (Burdman et al. 1996; Parmar and Dadarwal 1999; Sivaramaiah et al. 2007). The most commonly implicated mechanism involved the production of IAA (Molla et al. 2001). PGPR stimulation of root growth potentially provides more infection sites for nodule initiation and, through their ability to induce flavonoid production in the plant (Burdman et al. 1996), an induction in the expression of rhizobial *nod* genes. Co-inoculation of *Rhizobium* sp. with *Pseudomonas fluorescens* (strain PsIA12), which is able to promote the development of roots and protect against root pests, also has been shown to improve the benefit of rhizobia on various legumes including, lucerne, pea and broad bean (*Vicia faba* var *major* Harz.) (Höflich et al. 1994).

Interaction between AM fungi and various PGPR has also been reported and in many cases shown to be beneficial for plant growth (see review by Barea et al. 2005a). For example, increased root colonization by AM fungi was observed when coinoculated with a range of PGPR including *Azospirillum, Azotobacter croococcum, Bacillus polymyxa* and *Pseudomonas stricta* (Artursson et al. 2006; Gamalero et al. 2004; Toro et al. 1997). Barea et al. (1983) similarly showed that maize and ryegrass inoculated with *A. brasilense* and mycorrhizal fungi had comparable N and P contents as to plants grown with fertilizer. Inoculation of plants with *Azospirillum* enhanced mycorrhizal formation and conversely, *Azospirillum* establishment in the rhizosphere was also shown to be improved (Barea et al. 2005a). However, increased N acquisition by dual-inoculated plants was attributed to greater N uptake capacity by mycorrhizal infected roots, rather than a direct effect through N_2 fixation. Multi-level interactions between AM fungi, *Azospirillum* and PSB have also been reported with indication of synergistic effects when inoculated simultaneously (Muthukumar et al. 2001). However, it is important to also consider that relationships between PGPR and mycorrhizas may not always be positive (Walley and Germida 1997).

Conclusions and future directions

The acquisition of nutrients from soils is governed by root growth and its interaction with the abiotic and biotic components of soil. This interaction is manifest largely by the physical, chemical and biological properties of the rhizosphere (Hinsinger et al. 2009). Through better understanding of rhizosphere interactions and how roots associate with soil microorganisms there is opportunity for enhancing the efficiency of nutrient uptake by plants (Rengel and Marschner, 2005). This may occur through either direct plant selection, manipulation of root growth or by management of indigenous microbial communities and/or specific symbiotic and associative interactions through inoculation.

Manipulation of roots

Growth and development of roots and formation of the rhizosphere through root hairs or via the capacity to associate with mycorrhizas is of critical importance for effective soil exploration and access to nutrients. Indeed identification of germplasm and its use in breeding programs that are directed at modifications to root architecture have been successful in improving nutrient efficiency under field conditions through either direct plant selection (Gahoonia et al. 1999; Liao et al. 2006; Lynch and Brown 2001) or by intercropping of different plant species for enhanced N and P nutrition (e.g. Li et al. 2007; Knudsen et al. 2004). Further success can be expected as molecular tools become available that allow root traits to be more readily identified and manipulated, such as genome-wide analyses and the development of molecular markers for specific root traits and/or the identification of key regulatory genes including transcription factors that are involved in root development (Wissuwa et al. 2009). For example, genes associated with branching and lateral root growth, proliferation in N-rich regions and root hair development in response to nutrient deficiency have been identified, some of which are associated with hormonal responses in plants (Grierson et al. 2001; Yi et al. 2005; Zhang and Forde 2000). Ultimately such genes may be used to manipulate root growth for greater nutrient acquisition. Alternatively, it is possible that biochemical traits of roots may also be manipulated to influence the availability of nutrients in soil. For example, microbial phytase genes have been expressed in plant roots and when grown in controlled environments these plants have increased ability to access organic forms of P when supplied as inositol phosphates (George et al. 2005a; Richardson et al. 2001b). Genes that encode for the synthesis and transport of organic anions in plants have similarly been identified and, in some cases, used to enhance the release of organic anions from roots with benefits reported in terms of tolerance to Al (de la Fuente-Martínez et al. 1997; Delhaize et al. 2004) and improved access to soil P (Lopez-Bucio et al. 2000). However, the reliability of over-expressing genes for organic anion biosynthesis as a general means to improve organic anion release from roots has been questioned (Delhaize et al. 2001). In addition there is a need to demonstrate widespread applicability of such approaches for plants when grown in different soil environments where restricted performance has either been observed or where growth may be impeded by other limitations (George et al. 2005b). There is also a need to better understand the importance of spatial and temporal variations in root growth and rhizosphere function in different soil environments in relation to the uptake of specific nutrients (primarily N and P, but also for micronutrients), so that plant manipulations and breeding objectives are directed at appropriate traits for maximum benefit. Indeed the use of appropriate models have shown that small changes to root growth or rhizosphere processes can have significant effect on the acquisition of N and P by different plant species (Wissuwa 2003; Dunbabin et al. 2006). Likewise, the importance of microbial-mediated processes for nutrient mobilization in the rhizosphere and their interaction with plant mechanisms needs to be considered along with the significance of rhizosphere interactions under field conditions (Watt et al., 2006).

Interaction with microorganisms, management and development of inocula

Understanding of the interaction of roots with soil microorganisms that are associated with enhanced nutrient acquisition has also advanced in recent years. For instance, significant progress in knowledge of mycorrhizal symbioses and their role in agro-ecosystem function have been made, but further research, particularly in relation to the potential contribution of AM fungi to agricultural systems, is required. This includes better understanding of; (i) population biology and diversity of mycorrhizal fungi in soils, (ii) identifying genetic determinants involved in the compatibility and synergism between plant and fungal partners, (iii) examining ecological traits that control the beneficial effects of the fungus on plant growth and soil quality, (iv) increased knowledge of interactions within the mycorrhizosphere, particularly those involving other PGPR, including phosphate-mobilizing microorganisms, and (v) improved management systems for realizing mycorrhizal benefits for both sustainable agricultural production systems and in natural and restoration ecosystems. For AM fungi in particular, this includes improved methods for production of inocula and development of appropriate management practices (e.g. either by crop rotation, intercropping or with inclusion of various PGPR, including mycorrhiza helper bacteria) to promote the presence and function of various AM in different environments.

Effort has also been directed at understanding of the ecology and management of PGPR in soil, yet their development as inoculants (with the exception of rhizobia which have been successful over many decades) remains a considerable challenge. Whilst biocontrol products based on specific strains of *Pseudomonas* or *Bacillus* have been developed successfully, exploitation of phytostimulatory (or biofertilizer) PGPR-based inoculants remains less advanced. One exception is *Azospirillum*, which has now been commercialized widely as an inoculant and shown considerable promise in different agriculture environments. In this case it is evident that growth promotion occurs primarily through phytostimulation and enhanced root growth, rather than direct effects on N_2 fixation. A number of soil fungi, originally isolated and characterized on the basis of their capacity to solubilize inorganic soil P under laboratory conditions, have similarly been developed as commercial biofertilizer inoculants. However, again it is evident that these inoculants appear to promote plant growth through stimulation of root growth rather than directly through P-mobilization (Wakelin et al. 2006). Increased root growth subsequently allows greater soil exploration and thus increased P uptake which results in an apparent increase in P efficiency. In contrast, Wakelin et al. (2007) have recently reported the isolation of a wider range of soil fungi (predominantly *Penicillium* sp.) that show greater potential for direct P mobilization. With these few examples it is evident that growth promotion and microbial interaction with roots is a complex phenomenon and that better understanding of the various mechanisms involved and how they interact with roots is required. Indeed it is generally accepted that greater nutrient uptake by plants in response to PGPR appears to occur generally as a result of stimulated root growth as compared to direct effects on increased plant uptake of nutrients.

Paramount to further success of PGPR is a need to better understanding the ecology of microorganisms, either indigenous or introduced, within the rhizosphere both as individual organisms and through their interaction with other microorganisms (e.g. N_2-fixing rhizobacteria or endophytic diazotrophs with P-mobilizing AM fungi) and directly with host plants. Inconsistent response of PGPR in different environments and on various hosts remains as a significant impediment to their widespread development and application (Richardson 2001). Identification of traits involved in the ability of specific organisms to establish themselves in the rhizosphere (rhizosphere competence) at levels sufficient to exert effects on plant growth, to effectively compete with indigenous microorganisms or to cooperative with other beneficial organisms, and understanding of the signalling processes that occur between plants and microorganisms is required. Moreover, methods of inoculation (e.g. use of carriers, rates, and product longevity etc) and opportunity for developing microbial consortia as inocula needs to be considered. Opportunities for this will be advanced by development and application of molecular techniques and biotechnological approaches to microbial ecology. This includes non-disruptive *in situ* visualization techniques (e.g. confocal laser scanning microscopy and fluorescent tagged DNA probes, proteins, or microorganisms), functional genomics and analysis of the molecular basis of root colonization and signaling in the rhizosphere. In addition there is need for better understanding of the actual mechanisms that contribute to growth promotion and interaction with plants roots, including the possibility for genetic modification of specific microorganisms or plants.

References

Abel S, Nguyen MD, Chow W, Theologis A (1995) ACS4, a primary indoleacetic acid-responsive gene encoding 1-aminocyclopropane-1-carboxylate synthase in *Arabidopsis thaliana*: Structural characterization, expression in *Escherichia coli*, and expression characteristics in response to auxin. J Biol Chem 270:19093–19099

Adams MA, Pate JS (1992) Availability of organic and inorganic forms of phosphorus to lupins (*Lupinus* spp.). Plant Soil 145:107–113 doi:10.1007/BF00009546

Ae N, Arihara J, Okada K (1991) Phosphorus uptake mechanisms of pigeon pea grown in Alfisols and Vertisols. In: Johansen C, Lee KK, Sahrawat KL (eds) Phosphorus nutrition in grain legumes in the semi-arid tropics. ICRISAT, Patancheru, pp 91–98

Ahmad F, Ahmad I, Khan MS (2008) Screening of free-living rhizospheric bacteria for their multiple plant growth promoting activities. Microbiol Res 163:173–181

Aloni R, Aloni E, Langhans M, Ullrich CI (2006) Role of cytokinin and auxin in shaping root architecture: regulating vascular differentiation, lateral root initiation, root apical dominance and root gravitropism. Ann Bot 97:883–893

Ames RN, Reid CPP, Porter LK, Cambardella C (1983) Hyphal uptake and transport of nitrogen from two ^{15}N labelled sources by *Glomus mosseae*, a vesicular arbuscular mycorrhizal fungus. New Phytol 95:381–396

Amos B, Walters DT (2006) Maize root biomass and net rhizodeposited carbon: An analysis of the literature. Soil Sci Soc Am J 70:1489–1503

Arshad M, Frankenberger WT (1998) Plant growth regulating substances in the rhizosphere. Microbial production and functions. Adv Agron 62:46–151

Artursson V, Finlay RD, Jansson JK (2006) Interactions between arbuscular mycorrhizal fungi and bacteria and their potential for stimulating plant growth. Environ Microbiol 8:1–10

Aslam M, Travis RL, Rains DW (1996) Evidence for substrate induction of a nitrate efflux system in barley roots. Plant Physiol 112:1167–1175

Avio L, Pellegrino E, Bonari E, Giovannetti M (2006) Functional diversity of arbuscular mycorrhizal fungal isolates in relation to extraradical mycelial networks. New Phytol 172:347–357

Azcon R, Barea JM, Hayman DS (1976) Utilization of rock phosphate in alkaline soils by plants inoculated with mycorrhizal fungi and phosphate-solubilizing bacteria. Soil Biol Biochem 8:135–138

Badalucco L, Nannipieri P (2007) Nutrient transformations in the rhizosphere. In: Pinton R, Varanini Z, Nannipieri P (eds) The rhizosphere biochemistry and organic substances at the soil-plant interface. CRC Press, Boca Raton, Florida, pp 111–133

Baldani JI, Caruso L, Baldani VLD, Goi SR, Döbereiner J (1997) Recent advances in BNF with non-legume plants. Soil Biol Biochem 29:911–922

Baon JB, Smith SE, Alston AM (1994) Growth response and phosphorus uptake of rye with long and short root hairs: Interactions with mycorrhizal infection. Plant Soil 167:247–254 doi:10.1007/BF00007951

Barber SA (1995) Soil nutrient bioavailability: A mechanistic approach. 2nd edition. Wiley, New York

Barea JM (1991) Vesicular-arbuscular mycorrhizas as modifiers of soil fertility. In: Stewart BA (ed) Advances in Soil Science. Springer-Verlag, New York, pp 1–40

Barea JM, Azcón R, Azcón-Aguilar C (2005a) Interactions between mycorrhizal fungi and bacteria to improve plant nutrient cycling and soil structure. In: Buscot F, Varma A (eds) Microorganisms in soils: Roles in genesis and functions. Springer-Verlag, Berlin, pp 195–212

Barea JM, Azcón-Aguilar C, Azcón R (1987) Vesicular-arbuscular mycorrhiza improve both symbiotic N_2 fixation and N uptake from soil as assessed with a ^{15}N technique under field conditions. New Phytol 106:717–725

Barea JM, Bonis AF, Olivares J (1983) Interactions between *Azospirillum* and VA mycorrhiza and their effects on growth and nutrition of maize and ryegrass. Soil Biol Biochem 15:705–709

Barea JM, Pozo MJ, Azcón R, Azcón-Aguilar C (2005b) Microbial co-operation in the rhizosphere. J Exp Bot 56:1761–1778

Barea JM, Toro M, Orozco MO, Campos E, Azcón R (2002) The application of isotopic (^{32}P and ^{15}N) dilution techniques to evaluate the interactive effect of phosphate-solubilizing rhizobacteria, mycorrhizal fungi and *Rhizobium* to improve the agronomic efficiency of rock phosphate for legume crops. Nutr Cycl Agroecosys 63:35–42

Barea JM, Ferrol N, Azcón-Aguilar C, Azcón R (2008) Mycorrhizal symbioses. In: White PJ, Hammond JP (eds) The ecophysiology of plant-phosphorus interactions. Springer, Dordrecht, pp 143–163

Barlow PW, Brain P, Parker JS (1991) Cellular growth in roots of a gibberellin-deficient mutant of tomato (*Lycopersicon esculentum* Mill.) and its wild-type. J Exp Bot 42:339–351

Bar-Ness E, Chen Y, Hadar Y, Marschner H, Römheld V (1991) Siderophores of *Pseudomonas putida* as an iron source for dicot and monocot plants. Plant Soil 130:231–241 doi:10.1007/BF00011878

Bar-Yosef B (1991) Root excretions and their environmental effects. Influence on availability of phosphorus. In: Waisel Y, Eshel A, Kafkafi U (eds) Plant roots: The hidden half. Marcel Dekker, New York, pp 529–557

Bashan Y, Holguin G (1997) Proposal for the division of plant growth-promoting rhizobacteria into two classifications: biocontrol-PGPB (plant growth-promoting bacteria) and PGPB. Soil Biol Biochem 30:1225–1228

Benedetto A, Magurno F, Bonfante P, Lanfranco L (2005) Expression profiles of a phosphate transporter gene

(*GmosPT*) from the endomycorrhizal fungus *Glomus mosseae*. Mycorrhiza 15:620–627

Bertin C, Yang XH, Weston L (2003) The role of root exudates and allelochemicals in the rhizosphere. Plant Soil 256:67–83 doi:10.1023/A:1026290508166

Bieleski RL (1973) Phosphate pools, phosphate transport and phosphate availability. Ann Rev Plant Physiol 24:225–252

Blaha D, Prigent-Combaret C, Mirza MS, Moënne-Loccoz Y (2006) Phylogeny of the 1-aminocyclopropane-1-carboxylic acid deaminase-encoding gene *acdS* in phytobeneficial and pathogenic Proteobacteria and relation with strain biogeography. FEMS Microbiol Ecol 56:455–470

Bleecker AB, Kende H (2000) A gaseous signal molecule in plants. Annu Rev Cell Dev Biol 16:1–18

Boddey RM, Polidoro JC, Resende AS, Alves BJR, Urquiaga S (2001) Use of the ^{15}N natural abundance technique for the quantification of the contribution of N_2 fixation to grasses and cereal. Aust J Plant Physiol 28:889–895

Bolan NS (1991) A critical review on the role of mycorrhizal fungi in the uptake of phosphorus by plants. Plant Soil 134:189–207 doi:10.1007/BF00012037

Bolan NS, Robson AD, Barrow NJ, Aylmore LAG (1984) Specific activity of phosphorus in mycorrhizal and non-mycorrhizal plants in relation to the availability of phosphorus to plants. Soil Biol Biochem 16:299–304

Bonkowski M (2004) Protozoa and plant growth: The microbial loop in soil revisited. New Phytol 162:617–631

Bonkowski M, Brandt F (2002) Do soil protozoa enhance plant growth by hormonal effects? Soil Biol Biochem 34:1709–1715

Bonkowski M, Griffiths BS, Scrimgeour CM (2000) Substrate heterogeneity and microfauna in soil organic 'hotspots' as determinants of nitrogen capture and growth of ryegrass. Appl Soil Ecol 14:37–53

Bottini R, Cassán F, Piccoli P (2004) Gibberellin production by bacteria and its involvement in plant growth promotion and yield increase. Appl Microbiol Biotechnol 65:497–503

Bowen GD, Rovira AD (1991) The rhizosphere, the hidden half of the hidden half. In: Waisel Y, Eshel A, Kafkafi U (eds) Plant roots: The hidden half. Marcel Dekker, New York, pp 641–649

Bowen GD, Rovira AD (1999) The rhizosphere and its management to improve plant growth. Adv Agron 66:1–102

Braum SM, Helmke PA (1995) White lupin utilizes soil-phosphorus that is unavailable to soybean. Plant Soil 176:95–100 doi:10.1007/BF00017679

Brimecombe MJ, de Leij FAAM, Lynch JM (1999) Effect of introduced *Pseudomonas fluorescens* strains on the uptake of nitrogen by wheat from ^{15}N-enriched organic residues. World J Microbiol Biotech 15:417–423

Brimecombe MJ, de Leij FAAM, Lynch JM (2000) Effect of introduced *Pseudomonas fluorescens* strains on soil nematode and protozoan populations in the rhizosphere of wheat and pea. Microb Ecol 38:387–397

Brimecombe MJ, De Leij FAAM, Lynch JM (2007) Rhizodeposition and microbial populations. In: Pinton R, Varanini Z, Nannipieri P (eds) The rhizosphere biochemistry and organic susbstances at the soil-plant interface. CRC Press, Boca Raton, Florida, pp 73–109

Britto DT, Kronzucker HJ (2002) NH_4^+ toxicity in higher plants: A critical review. J Plant Physiol 159:567–584

Broekaert WF, Delauré SL, De Bolle MFC, Cammue BPA (2006) The role of ethylene in host-pathogen interactions. Ann Rev Phytopathol 44:393–416

Brundrett MC (2002) Coevolution of roots and mycorrhizas of land plants. New Phytol 154:275–304

Bucher M (2007) Functional biology of plant phosphate uptake at root and mycorrhiza interfaces. New Phytol 173:11–26

Burdman S, Volpin H, Kigel J, Kapulnik Y, Okon Y (1996) Promotion of *nod* gene inducers and nodulation in common bean (*Phaseolus vulgaris*) roots inoculated with *Azospirillum brasilense* Cd. Appl Environ Microbiol 62:3030–3033

Burleigh SH, Cavagnaro TR, Jakobsen I (2002) Functional diversity of arbuscular mycorrhizas extends to the expression of plant genes involved in P nutrition. J Exp Bot 53:1–9

Buscot F (2005) What are soils? In: Buscot F, Varma A (eds) Microorganisms in soils: Roles in genesis and functions. Soil Biology Vol 3. Springer-Verlag, Heidelberg, pp 3–17

Cacciari I, Lippi D, Pietrosanti T, Pietrosanti W (1989) Phytohormone-like substances produced by single and mixed diazotrophic cultures of *Azospirillum* and *Arthrobacter*. Plant Soil 115:151–153 doi:10.1007/BF02220706

Cardoso IM, Boddington CL, Janssen BH, Oenema O, Kuyper TW (2006) Differential access to phosphorus pools of an oxisol by mycorrhizal and nonmycorrhizal maize. Comm Soil Sci Plant Anal 37:1537–1551

Cassán F, Bottini R, Schneider G, Piccoli P (2001) *Azospirillum brasilense* and *Azospirillum lipoferum* hydrolyze conjugates of GA20 and metabolize the resultant aglycones to GA1 in seedlings of rice dwarf mutants. Plant Physiol 125:2053–2058

Casarin V, Plassard C, Hinsinger P, Arvieu JC (2004) Quantification of ectomycorrhizal fungal effects on the bioavailability and mobilization of soil P in the rhizosphere of *Pinus pinaster*. New Phytol 163:177–185

Cavagnaro TR, Smith FA, Smith SE, Jakobsen I (2005) Functional diversity in arbuscular mycorrhizas: exploitation of soil patches with different phosphate enrichment differs among fungal species. Plant Cell Environ 28:642–650

Chalot M, Brun A (1998) Physiology of organic nitrogen acquisition by ectomycorrhizal fungi and ectomycorrhizas. FEMS Microbiol Rev 22:21–44

Chapin FS (1980) The mineral nutrition of wild plants. Ann Rev Ecol Syst 11:233–260

Chapin FS (1993) Preferential use of organic nitrogen for growth by a nonmycorrhizal artic sedge. Nature 361:150–153

Chen CR, Condron LM, Davis MR, Sherlock RR (2002) Phosphorus dynamics in the rhizosphere of perennial ryegrass (*Lolium perenne* L.) and radiata pine (*Pinus radiata* D.Don). Soil Biol Biochem 34:487–499

Cheng W, Johnson DW, Fu S (2003) Rhizosphere effects on decomposition: Controls of plant species, phenology, and fertilization. Soil Sci Soc Am J 67:1418–1427

Cheng Z, Park E, Glick BR (2007) 1-Aminocyclopropane-1-carboxylate deaminase from *Pseudomonas putida* UW4 facilitates the growth of canola in the presence of salt. Can J Microbiol 53:912–918

Chiou TJ, Liu H, Harrison MJ (2001) The spatial expression patterns of a phosphate transporter (*MtPT1*) from *Medicago truncatula* indicate a role in phosphate transport at the root/soil interface. Plant J 25:281–293

Christou M, Avramides EJ, Roberts JP, Jones DL (2005) Dissolved organic nitrogen in contrasting agricultural ecosystems. Soil Biol Biochem 37:1560–1563

Clark RB, Zeto SK (2000) Mineral acquisition by arbuscular mycorrhizal plants. J Plant Nutr 23:867–902

Cocking EC (2003) Endophytic colonization of plant roots by nitrogen-fixing bacteria. Plant Soil 252:169–175 doi:10.1023/A:1024106605806

Creus CM, Graziano M, Casanovas EM, Pereyra MA, Simontacchi M, Puntarulo S, Barassi CA, Lamattina L (2005) Nitric oxide is involved in the *Azospirillum brasilense*-induced lateral root formation in tomato. Planta 221:297–303

Curl EA, Truelove B (1986) The rhizosphere. Advanced Series in Agricultural Sciences 15. Springer-Verlag, Heidelberg

Darrah PR (1993) The rhizosphere and plant nutrition: A quantitative approach. Plant Soil 156:1–20 doi:10.1007/BF00024980

de Freitas JR, Banerjee MR, Germida JJ (1997) Phosphate-solubilizing rhizobacteria enhance the growth and yield but not phosphorus uptake of canola (*Brassica napus* L.). Biol Fert Soil 24:358–364

de la Fuente-Martínez JM, Ramirez-Rodriguez V, Cabrera-Ponce JL, Herrera-Estrella L (1997) Aluminum tolerance in transgenic plants by alteration of citrate synthesis. Science 276:1566–1588

de Salamone IEG, Hynes RK, Nelson LM (2001) Cytokinin production by plant growth promoting rhizobacteria and selected mutants. Can J Microbiol 47:404–411

De Willigen P (1986) Supply of soil nitrogen to the plant during the growing season. In: Lambers H, Neeteson JJ, Stulen I (eds) Fundamental, ecological and agricultural aspects of nitrogen metabolism in higher plants. Martinus Nijhoff, Dordrecht, pp 417–432

Delhaize E, Hebb DM, Ryan PR (2001) Expression of a *Pseudomonas aeruginosa* citrate synthase gene is not associated with either enhanced citrate accumulation or efflux. Plant Physiol 125:2059–2067

Delhaize E, Ryan PR, Hebb DM, Yamamoto Y, Sasaki T, Matsumoto H (2004) Engineering high level aluminium tolerance in barley with the ALMT1 gene. Proc Nat Acad Sci USA 101:15249–15254

Devienne-Barret F, Justes E, Machet JM, Mary B (2000) Integrated control of nitrate uptake by crop growth rate and soil nitrate availability under field conditions. Ann Bot 86:995–1005

Dinkelaker B, Hengeler C, Marschner H (1995) Distribution and function of proteoid and other root clusters. Bot Acta 108:183–200

Dinlelaker B, Römheld V, Marschner H (1989) Citric acid exudation and precipitation of calcium citrate in the rhizosphere of white lupin (*Lupinus albus* L.). Plant Cell Environ 12:265–292

Dobbelaere S, Croonenborghs A, Thys A, Vande Broek A, Vanderleyden J (1999) Phytostimulatory effect of *Azospirillum brasilense* wild type and mutant strains altered in IAA production on wheat. Plant Soil 212:155–164

Drew MC, Saker LR, Ashley TW (1973) Nutrient supply and the growth of the seminal root system in barley I. The effect of nitrate concentration on the growth of axes and laterals. J Exp Bot 24:1189–1202

Duijff BJ, Meijer JW, Bakker PAHM, Schippers B (1993) Siderophore-mediated competition for iron and induced resistance in the suppression of fusarium wilt of carnation by fluorescent *Pseudomonas* spp. Eur J Plant Pathol 99:277–289

Dunbabin VM, McDermott S, Bengough AG (2006) Upscaling from rhizosphere to whole root system: Modelling the effects of phospholipid surfactants on water and nutrient uptake. Plant Soil 283:57–72 doi:10.1007/S11104-005-0866-y

Ericsson T (1995) Growth and shoot:root ratio of seedlings in relation to nutrient availability. Plant Soil 168/169:205–214 doi:10.1007/BF00029330

Fan MS, Zhu JM, Richards C, Brown KM, Lynch JP (2003) Physiological roles for aerenchyma in phosphorus-stressed roots. Func Plant Biol 30:493–506

Farwell AJ, Veselya S, Neroa V, Rodriguez H, McCormack K, Shah S, Dixona DG, Glick BR (2007) Tolerance of transgenic canola plants (*Brassica napus*) amended with plant growth-promoting bacteria to flooding stress at a metal-contaminated field site. Environ Pollut 147:540–545

Fillery IRP (2007) Plant-based manipulation of nitrification in soil: A new approach to managing N loss? Plant Soil 294:1–4 doi:10.1007/S11104-007-9263-z

Fitter AH (1987) An architectural approach to the comparative ecology of plant-root systems. New Phytol 106:61–77

Föhse D, Claassen N, Jungk A (1991) Phosphorus efficiency of plants II. Significance of root radius, root hairs and cation-anion balance for phosphorus influx in seven plant species. Plant Soil 132:261–272 doi:10.1007/BF00010407

Forde B (2002) Local and long range signalling pathways regulating plant response to nitrate. Ann Rev Plant Biol 53:203–224

Forde B, Clarkson DT (1999) Nitrate and ammonium nutrition of plants: Physiological and molecular perspectives. Adv Bot Res 30:1–90

Forde B, Lorenzo H (2001) The nutritional control of root development. Plant Soil 232:51–68 doi:10.1023/A:1010329902165

Franche C, Lindström K, Elmerich C (2009) Nitrogen-fixing bacteria associated with leguminous and non-leguminous plants. Plant Soil doi:10.1007/S11104-008-9833-8

Frey-Klett P, Garbaye J, Tarkka M (2007) The mycorrhiza helper bacteria revisited. New Phytol 176:22–36

Frossard E, Condron LM, Oberson A, Sinaj S, Fardeau JC (2000) Processes governing phosphorus availability in temperate soils. J Environ Qual 29:15–23

Fu X, Harberd NP (2003) Auxin promotes *Arabidopsis* root growth by modulating gibberellin response. Nature 421:740–743

Fukaki H, Okushima Y, Tasaka M (2007) Auxin-mediated lateral root formation in higher plants. Int Rev Cytol 256:111–137

Gahoonia TS, Nielsen NE (1992) The effect of root induced pH changes on the depletion of inorganic and organic phosphorus in the rhizosphere. Plant Soil 143:185–191 doi:10.1007/BF00007872

Gahoonia TS, Nielsen NE (1997) Variation in root hairs of barley cultivars doubled soil phosphorus uptake. Euphytica 98:177–182

Gahoonia TS, Nielsen NE (2003) Phosphorus (P) uptake and growth of a root hairless barley mutant (bald root barley, *brb*) and wild type in low- and high-P soils. Plant Cell Environ 26:1759–1766

Gahoonia TS, Nielsen NE, Lyshede OB (1999) Phosphorus acquisition of cereal cultivars in the field at three levels of P fertilization. Plant Soil 211:269–281 doi:10.1023/A:1004742032367

Gamalero E, Trotta A, Massa N, Copetta A, Martinotti MG, Berta G (2004) Impact of two fluorescent Pseudomonads and an arbuscular mycorrhizal fungus on tomato plant growth, root architecture and P acquisition. Mycorrhiza 14:185–192

Garbaye J (1994) Helper bacteria, a new dimension to the mycorrhizal symbiosis. New Phytol 128:197–210

Gardner WK, Barber DA, Parbery DG (1983) The acquisition of phosphorus by *Lupinus albus* L. III. The probable mechanism by which phosphorus movement in the soil/root interface is enhanced. Plant Soil 68:19–32 doi:10.1007/BF02374724

Gardner WK, Parbury DG, Barber DA (1981) Proteoid root morphology and function in *Lupinus albus*. Plant Soil 60:143–147 doi:10.1007/BF02377120

Gastal F, Lemaire G (2002) N uptake and distribution in crops: An agronomical and ecophysiological perspective. J Exp Bot 53:789–799

George TS, Gregory PJ, Hocking PJ, Richardson AE (2008) Variation in root-associated phosphatase activities in wheat contributes to the utilization of organic P substrates *in-vitro*, but does not explain differences in the P-nutrition of plants when grown in soils. Environ Experim Bot 64:239–249

George TS, Gregory PJ, Robinson JS, Buresh RJ (2002) Changes in phosphorus concentrations and pH in the rhizosphere of some agroforestry and crop species. Plant Soil 246:65–73 doi:10.1023/A:1021523515707

George TS, Richardson AE, Smith JB, Hadobas PA, Simpson J (2005b) Limitations to the potential of transgenic *Trifolium subterraneum* L. plants that exude phytase, when grown in soils with a range of organic phosphorus content. Plant Soil 278:263–274 doi:10.1007/S11104-005-8699-2

George TS, Simpson RJ, Hadobas PA, Richardson AE (2005a) Expression of a fungal phytase gene in *Nicotiana tabacum* improves phosphorus nutrition in plants grown in amended soil. Plant Biotechnol J 3:129–140

Gerke J (1993) Solubilization of Fe (III) from humic-Fe complexes, humic Fe oxide mixtures and from poorly ordered Fe-oxide by organic acids: consequences for P-adsorption. Z Pflanz Bodenkunde 156:253–257

Gilroy S, Jones DL (2002) Through form to function: Root hair development and nutrient uptake. Trends Plant Sci 5:56–60

Giri B, Giang PH, Kumari R, Prasad R, Varma A (2005) Microbial diversity in soils. In: Buscot F, Varma A (eds) Microorganisms in soils: Roles in genesis and functions. Springer-Verlag, Heidelberg, pp 195–212

Glick BR, Cheng Z, Czarny J, Duan J (2007a) Promotion of plant growth by ACC deaminase-producing soil bacteria. Eur J Plant Pathol 119:329–339

Glick BR, Jacobson CB, Schwarze MMK, Pasternak JJ (1994) 1-aminocyclopropane-1-carboxylic acid deaminase mutants of the plant growth promoting rhizobacterium *Pseudomonas putida* GR 12-2 do not stimulate canola root elongation. Can J Microbiol 40:911–915

Glick BR, Todorovic B, Czarny J, Cheng Z, Duan J, McConkey B (2007b) Promotion of plant growth by bacterial ACC deaminase. Crit Rev Plant Sci 26:227–242

Grayston SJ, Vaughan D, Jones D (1996) Rhizosphere carbon flow in trees, in comparison with annual plants: The importance of root exudation and its impact on microbial activity and nutrient availability. Appl Soil Ecol 5:29–56

Gregory PJ (2006) Plant roots: Growth, activity and interaction with soils. Blackwell Publishing, Oxford

Grierson CS, Parker JS, Kemp AC (2001) Arabidopsis genes with roles in root hair development. J Plant Nutr Soil Sci 164:131–140

Griffiths BS (1989) Enhanced nitrification in the presence of bacteriophagous protozoa. Soil Biol Biochem 21:1045–1051

Griffiths BS (1990) A comparison of microbial-feeding nematodes and protozoa in the rhizosphere of different plants. Biol Fert Soil 9:83–88

Griffiths BS, Bonkowski M, Dobson G, Caul S (1999) Changes in soil microbial community structure in the presence of microbial-feeding nematodes and protozoa. Pedobiologia 43:297–304

Gryndler M (2000) Interactions of arbuscular mycorrhizal fungi with other soil organisms. In: Kapulnik Y, Douds DD (eds) Arbuscular mycorrhizas: Physiology and function. Kluwer Academic Publishers, Dordrecht, pp 239–262

Gulden RH, Vessey JK (2000) *Penicillium bilaii* inoculation increases root hair production in field pea. Can. J. Plant Sci. 80:801–804

Gutiérrez-Mañero FJ, Ramos-Solano B, Probanza A, Mehouachi J, Tadeo FR, Talon M (2001) The plant-growth-promoting rhizobacteria *Bacillus pumilus* and *Bacillus licheniformis* produce high amounts of physiologically active gibberellins. Physiol Plantarum 111:206–211

Gyaneshwar P, Naresh Kumar G, Parekh LJ, Poole PS (2002) Role of soil microorganisms in improving P nutrition of plants. Plant Soil 245:83–93 doi:10.1023/A:1020663916259

Hao Y, Charles TC, Glick BR (2007) ACC deaminase from plant growth-promoting bacteria affects crown gall development. Can J Microbiol 53:1291–1299

Harman GE, Howell CR, Viterbo A, Chet I, Lorito M (2004) *Trichoderma* species-opportunistic avirulent plant symbionts. Nature Rev 2:43–56

Hayes JE, Richardson AE, Simpson RJ (1999) Phytase and acid phosphatase activities in roots of temperate pasture grasses and legumes. Aust J Plant Physiol 26:801–809

Hayes JE, Simpson RJ, Richardson AE (2000) The growth and phosphorus utilization of plants in sterile media when supplied with inositol hexaphosphate, glucose 1-phosphate or inorganic phosphate. Plant Soil 220:165–174 doi:10.1023/A:1004782324030

Hedley MJ, White RE, Nye PH (1982) Plant-induced changes in the rhizosphere of rape (*Brassica napus* var. Emerald) seedlings III. Changes in L value, soil phosphate fractions and phosphatase activity. New Phytol 91:45–56

Helal HM, Dressler A (1989) Mobilization and turnover of soil phosphorus in the rhizosphere. Z Pflanz Bodenkunde 152:175–180

Herman DJ, Johnson KK, Jaeger CH, Schwartz E, Firestone MK (2006) Root influence on nitrogen mineralization and nitrification in *Avena barbata* rhizosphere soil. Soil Sci Soc Am J 70:1504–1511

Hill JO, Simpson RJ, Moore AD, Chapman DF (2006) Morphology and response of roots of pasture species to phosphorus and nitrogen nutrition. Plant Soil 286:7–19 doi:10.1007/S11104-006-0014-3

Hinsinger P (1998) How do plant roots acquire mineral nutrients? Chemical processes involved in the rhizosphere. Advan Agron 64:225–265

Hinsinger P (2001) Bioavailability of inorganic P in the rhizosphere as affected by root-induced chemical changes: A review. Plant Soil 237:173–195 doi:10.1023/A:1013351617532

Hinsinger P, Bengough AG, Vetterlein D, Young IM (2009) Rhizosphere: biophysics, biogeochemistry and ecological relevance. Plant Soil doi:10.1007/S11104-008-9885-9

Hinsinger P, Gilkes RJ (1996) Mobilization of phosphate from phosphate rock and alumina-sorbed phosphate by the roots of ryegrass and clover as related to rhizosphere pH. Euro J Soil Sci 47:533–544

Hinsinger P, Plassard C, Tang CX, Jaillard B (2003) Origins of root-mediated pH changes in the rhizosphere and their responses to environmental constraints: A review. Plant Soil 248:43–59 doi:10.1023/A:1022371130939

Ho MD, Rosas JC, Brown KM, Lynch JP (2005) Root architectural tradeoffs for water and phosphorus acquisition. Funct Plant Biol 32:737–748

Hocking P (2001) Organic acids exuded from roots in phosphorus uptake and aluminium tolerance of plants in acid soils. Adv Agron 74:63–97

Hocking PJ, Keerthisinghe G, Smith FW, Randall PJ (1997) Comparison of the ability of different crop species to access poorly-available soil phosphorus. In: Ando T, Fujita K, Mae T, Matsumoto H, Mori S, Sekiya J (eds) Plant nutrition for sustainable food production and agriculture. Kluwer Academic Publishers, Dordrecht, pp 305–308

Hodge A (2004) The plastic plant: root responses to heterogeneous supplies of nutrients. New Phytol 162:9–24

Hodge A, Robinson D, Fitter AH (2000a) Are microorganisms more effective than plants at competing for nitrogen. Trends Plant Sci 5:304–308

Hodge A, Robinson D, Fitter AH (2000b) An arbuscular mycorrhizal inoculum enhances root proliferation in, but not nitrogen capture from, nutrient-rich patches in soil. New Phytol 145:575–584

Hodge A, Robinson D, Griffith BS, Fitter AH (1999a) Nitrogen capture by plants grown in N-rich organic patches of contrasting size and strength. J Exp Bot 50:1243–1252

Hodge A, Robinson D, Griffiths BS, Fitter AH (1999b) Why plants bother: Root proliferation results in increased nitrogen capture from am organic patch when two grasses compete. Plant Cell Environ 22:811–820

Hoffland E, Findenegg GR, Nelemans JA (1989) Solubilization of rock phosphorus by rape. II. Local root exudation of organic acids in response to P-starvation. Plant Soil 113:161–165 doi:10.1007/BF02280176

Höflich G, Wiehe W, Kühn G (1994) Plant growth stimulation by inoculation with symbiotic and associative rhizosphere microorganisms. Experientia 50:897–905

Hogh-Jensen H (2006) The nitrogen transfer between plants: An important but difficult flux to quantify. Plant Soil 282:1–5 doi:10.1007/S11104-005-2613-9

Højberg O, Schnider U, Winteler HV, Sørensen J, Haas D (1999) Oxygen-sensing reporter strain of *Pseudomonas fluorescens* for monitoring the distribution of low-oxygen habitats in soil. Appl Environ Microbiol 65:4085–4093

Holguin G, Glick BR (2001) Expression of the ACC deaminase gene from *Enterobacter cloacae* UW4 in *Azospirillum brasilense*. Microb Ecol 41:281–288

Holguin G, Glick BR (2003) Transformation of *Azospirillum brasilense* Cd with an ACC deaminase gene from *Enterobacter cloacae* UW4 fused to the Tet^r gene promoter improves its fitness and plant growth promoting ability. Microb Ecol 46:122–133

Hu S, Chapin FS, Firestone MK, Field CB, Chiarello NR (2001) Nitrogen limitation of microbial decomposition in a grassland under elevated CO_2. Nature 409:188–190

Hurek T, Handley LL, Reinhold-Hurek B, Piché Y (2002) *Azoarcus* grass endophytes contribute fixed nitrogen to the plant in an unculturable state. Mol Plant Microbe Interact 15:233–242

Hurek T, Reinhold-Hurek B, Van Montagu M, Kellenberger E (1994) Root colonization and systemic spreading of *Azoarcus* sp. strain BH72 in grasses. J Bacteriol 176:1913–1923

Illmer P, Schinner F (1995) Solubilization of inorganic calcium phosphates-solubilization mechanisms. Soil Biol Biochem 27:257–263

Itoh S, Barber SA (1983) Phosphorus uptake by six plant species as related to root hairs. Agron J 75:457–461

Jackson LE, Burger M, Cavagnaro TR (2008) Roots, nitrogen transformations, and ecosystem services. Annu Rev Plant Biol 59:341–363

Jaeger CH, Monson RK, Fisk MC, Schmidt SK (1999) Seasonal partitioning of nitrogen by plants and soil microorganisms in an alpine ecosystem. Ecology 80:1883–1891

Jakobsen I, Abbott LK, Robson AD (1992) External hyphae of vesicular-arbuscular mycorrhizal fungi associated with *Trifolium subterraneum* L. 1. Spread of hyphae and phosphorus inflow into roots. New Phytol 120:371–380

Jakobsen I, Chen BD, Munkvold L, Lundsgaard T, Zhu YG (2005a) Contrasting phosphate acquisition of mycorrhizal fungi with that of root hairs using the root hairless barley mutant. Plant Cell Environ 28:928–938

Jakobsen I, Gazey C, Abbott IK (2001) Phosphate transport by communities of arbuscular mycorrhizal fungi in intact soil cores. New Phytol 149:95–103

Jakobsen I, Leggett ME, Richardson AE (2005b) Rhizosphere microorganisms and plant phosphorus uptake. In: Sims JT, Sharpley AN (eds) Phosphorus, agriculture and the environment. American Society for Agronomy, Madison, pp 437–494

Jingguo W, Bakken LR (1997) Competition for nitrogen during decomposition of plant residues in soil: Effect of spatial placement of N-rich and N-poor plant residues. Soil Biol Biochem 29:153–162

Johansson JF, Paul LR, Finlay RD (2004) Microbial interactions in the mycorrhizosphere and their significance for sustainable agriculture. FEMS Microbiol Ecol 48:1–13

Joner EJ, Johansen A (2000) Phosphatase activity of external hyphae of two arbuscular mycorrhizal fungi. Mycol Res 104:81–86

Joner EJ, van Aarle IM, Vosatka M (2000) Phosphatase activity of extra-radical arbuscular mycorrhizal hyphae. Plant Soil 226:199–210 doi:10.1023/A:1026582207192

Jones DL (1998) Organic acids in the rhizosphere: A critical review. Plant Soil 205:25–44 doi:10.1023/A:1004356007312

Jones DL, Darrah PR (1994a) Amino-acid influx at the soil-root interface of *Zea mays* L. and its implications in the rhizosphere. Plant Soil 163:1–12 doi:10.1007/BF00033935

Jones DL, Darrah PR (1994b) Role of root derived organic acids in the mobilization of nutrients in the rhizosphere. Plant Soil 166:247–257 doi:10.1007/BF00008338

Jones DL, Dennis PG, Owen AG, van Hees PAW (2003) Organic acid behavior in soils—misconceptions and knowledge gaps. Plant Soil 248:31–41 doi:10.1023/A:1022304332313

Jones DL, Nguyen C, Finlay RD (2009) Carbon flow in the rhizosphere: carbon trading at the soil-root interface. Plant Soil doi:10.1007/S11104-009-9925-0

Jones DL, Owen AG, Farrar JF (2002) Simple method to enable the high resolution determination of total free amino acids in soil solutions and soil extracts. Soil Biol Biochem 34:1893–1902

Jones DL, Shannon DV, Murphy D, Farrar J (2004) Role of dissolved organic nitrogen (DON) in soil N cycling in grassland soils. Soil Biol Biochem 36:749–756

Jungk A (2001) Root hairs and the acquisition of plant nutrients from soil. J Plant Nutr Soil Sci 164:121–129

Kaneda S, Kaneko N (2008) Collembolans feeding on soil affect carbon and nitrogen mineralization by their influence on microbial and nematode activities. Biol Fert Soils 44:435–442

Kawaguchi M, Syōno K (1996) The excessive production of indole-3-acetic acid and its significance in studies of the biosynthesis of this regulator of plant growth and development. Plant Cell Physiol 37:1043–1048

Kaye JP, Hart SC (1997) Competition for nitrogen between plants and soil microorganisms. Trends Ecol Evol 12:139–143

Keerthisinghe G, DeDatta SK, Mengel K (1985) Importance of exchangeable and nonexchangeable soil NH_4^+ in nitrogen nutrition of lowland rice. Soil Sci 140:194–201

Keerthisinghe G, Hocking PJ, Ryan PR, Delhaize E (1988) Effect of phosphorus supply on the formation and function of proteoid roots of white lupin (*Lupinus albus* L.). Plant Cell Environ 21:467–478

Khalid A, Arshad M, Zahir ZA (2004) Screening plant growth-promoting rhizobacteria for improving growth and yield of wheat. J Appl Microbiol 96:473–480

Knudsen MT, Hauggaard-Nielsen H, Jørnsgård B, Jensen ES (2004) Comparison of interspecific competition and N use in pea-barley, faba bean-barley and lupin-barley intercrops grown at two temperature locations. J Agric Sci 142:617–627

Koga J, Adachi T, Hidaka H (1991) Molecular cloning of the gene for indolepyruvate decarboxylase from *Enterobacter cloacae*. Mol Gen Genet 226:10–16

Koide RT, Kabir Z (2000) Extraradical hyphae of the mycorrhizal fungus *Glomus intraradices* can hydrolyse organic phosphate. New Phytol 148:511–517

Kucey RMN (1987) Increased phosphorus uptake by wheat and field bean inoculated with phosphorus-solubilizing *Penicillium bilaji* strain and with vesicular arbuscular mycorrhizal fungi. Appl Environ Microbiol 53:2699–2703

Kucey RMN (1983) Phosphate-solubilizing bacteria and fungi in various cultivated and virgin Alberta soils. Can J Soil Sci 63:671–678

Kucey RMN, Janzen HH, Leggett ME (1989) Microbially mediated increases in plant-available phosphorus. Adv Agron 42:199–228

Kuklinsky-Sobral J, Araújo WL, Mendes R, Geraldi IO, Pizzirani-Kleiner AA, Azevedo JL (2004) Isolation and characterization of soybean-associated bacteria and their potential for plant growth promotion. Environ Microbiol 6:1244–1251

Kuzyakov Y (2002) Review: Factors affecting rhizosphere priming effects. J Plant Nutr Soil Sci 165:382–396

Kuzyakov Y, Leinweber P, Sapronov D, Eckhardt K (2003) Qualitative assessment of rhizodeposits in non-sterile soil by analytical pyrolysis. J Plant Nutr Soil Sci 166:719–723

Lamattina L, Garcia-Mata C, Graziano M, Pagnussat G (2003) Nitric oxide: The versatility of an extensive signal molecule. Ann Rev Plant Biol 54:109–136

Lambers H, Shane MW, Cramer MD, Pearse SJ, Veneklaas EJ (2006) Root structure and functioning for efficient acquisition of phosphorus: matching morphological and physiological traits. Ann Bot 98:693–713

Leadley P, Reynolds J, Chapin FS (1997) A model of nitrogen uptake by *Eriophorum vaginatum* roots in the field: Ecological implications. Ecol Monogr 67:1–22

Lemanceau P, Kraemer SM, Briat JF (2009) Iron in the rhizosphere: a case study. Plant Soil (this volume-in press)

Lethbridge G, Davidson MS (1983) Root-associated nitrogen-fixing bacteria and their role in the nitrogen nutrition of wheat estimated by ^{15}N isotope dilution. Soil Biol Biochem 15:365–374

Li J, Ovakim DH, Charles TC, Glick BR (2000) An ACC deaminase minus mutant of *Enterobacter cloacae* UW4 no longer promotes root elongation. Curr Microbiol 41:101–105

Li L, Li SM, Sun JH, Zhou LL, Bao XG, Zhang HG, Zhang FS (2007) Diversity enhances agricultural productivity via rhizosphere phosphorus facilitation on phosphorus-deficient soils. Proc Natl Acad Sci 104:11192–11196

Liao M, Palta JA, Fillery IRP (2006) Root characteristics of vigorous wheat improve early nitrogen uptake. Aust J Agric Res 57:1097–1107

Lifshitz R, Kloepper JW, Kozlowski M, Simonson C, Carlson J, Tipping EM, Zaleska I (1987) Growth promotion of canola (rapeseed) seedlings by a strain of *Pseudomonas putida* under gnotobiotic conditions. Can J Microbiol 33:390–395

Lipton DS, Blanchar RW, Blevins DG (1987) Citrate, malate, and succinate concentration in exudates from P-sufficient and P-stressed *Medicago sativa* L. seedlings. Plant Physiol 85:315–317

Liu H, Trieu AT, Blaylock LA, Harrison MJ (1998) Cloning and characterization of two phosphate transporters from *Medicago truncatula* roots: Regulation in response to phosphate and to colonization by arbuscular mycorrhizal (AM) fungi. Mol Plant-Microbe Interact 11:14–22

Liu Q, Loganathan P, Hedley MJ (2005) Influence of ectomycorrhizal hyphae on phosphate fractions and dissolution of phosphate rock in rhizosphere soils of *Pinus radiata*. J Plant Nutrit 28:1525–1540

LongXian R, MiaoLian X, Bin Z, Bakker PAHM (2005) Siderophores are the main determinants of fluorescent *Pseudomonas* strains in suppression of grey mould in *Eucalyptus urophylla*. Acta Phytopathol Sin 35:6–12

Lopez-Bucio J, de la Vega OM, Guevara-García A, Herrera-Estrella L (2000) Enhanced phosphorus uptake in transgenic tobacco plants that overproduce citrate. Nat Biotechnol 18:450–453

López-Pedrosa A, González-Guerrero M, Valderas A, Azcón-Aguilar C, Ferrol N (2006) *GintAMT1* encodes a functional high-affinity ammonium transporter that is expressed in the extraradical mycelium of *Glomus*. Fungal Genet Biology 43:102–110

Lynch J (1995) Root architecture and plant productivity. Plant Physiol 109:7–13

Lynch JP (2005) Root architecture and nutrient acquisition. In: BassiriRad H (ed) Nutrient acquisition by plants: an ecological perspective. Springer-Verlag, Berlin, pp 147–183

Lynch JP, Brown KM (2001) Topsoil foraging–an architectural adaptation of plants to low phosphorus availability. Plant Soil 237:225–237 doi:10.1023/A:10133

Ma Z, Walk TC, Marcus A, Lynch JP (2001) Morphological synergism in root hair length, density, initiation and geometry for phosphorus acquisition in *Arabidopsis thaliana*: A modeling approach. Plant Soil 236:221–235

Macklon AES, Grayston SJ, Shand CA, Sim A, Sellars S, Ord BG (1997) Uptake and transport of phosphorus by *Agrostis capillaris* seedlings from rapidly hydrolysed organic sources extracted from ^{32}P-labelled bacterial cultures. Plant Soil 190:163–167

Magid J, Tiessen H, Condron LM (1996) Dynamics of organic phosphorus in soils under natural and agricultural ecosystems. In: Piccolo A (ed) Humic substances in terrestrial ecosystems. Elsevier Science, Amsterdam, pp 429–466

Maldonado-Mendoza IE, Dewbre GR, Harrison MJ (2001) A phosphate transporter gene from the extra-radical mycelium of an arbuscular mycorrhizal fungus *Glomus intraradices* is regulated in response to phosphate in the environment. Mol Plant-Microbe Interact 14:1140–1148

Malhotra M, Srivastava S (2008) An *ipdC* gene knock-out of *Azospirillum brasilense* strain SM and its implications on indole-3-acetic acid biosynthesis and plant growth promotion. Anton Leeuw 93:425–433

Manske GGB, Ortiz-Monasterio JI, Van Grinkel M, Rajaram S, Molina E, Vick PLG (2000) Traits associated with improved P-uptake efficiency in CIMMYT's semidwarf spring bread wheat grown on an acid Andisol in Mexico. Plant Soil 221:189–204

Mantelin S, Touraine B (2004) Plant growth-promoting bacteria and nitrate availability: impacts on root development and nitrate uptake. J Exp Bot 55:27–34

Manulis S, Haviv-Chesner A, Brandl MT, Lindow SE, Barash I (1998) Differential involvement of indole-3-acetic acid biosynthetic pathways in pathogenicity and epiphytic fitness of *Erwinia herbicola* pv. *gypsophilae*. Mol Plant-Microbe Interact 11:634–642

Marek-Kozaczuk M, Skorupska A (2001) Production of B-group vitamins by plant growth-promoting *Pseudomonas fluorescens* strain 267 and the importance of vitamins in the colonization and nodulation of red clover. Biol Fert Soils 33:146–151

Marschner H (1995) Mineral nutrition of higher plants. 2nd edition. Academic Press, London

Masalha J, Kosegarten H, Elmaci Ö, Mengel K (2000) The central role of microbial activity for iron acquisition in maize and sunflower. Biol Fert Soils 30:433–439

Mayer J, Buegger F, Jensen ES, Schloter M, Hess J (2004) Turnover of grain legume N rhizodeposits and effect of rhizodeposition on the turnover of crop residues. Biol Fert Soils 39:153–164

McLaughlin MJ, Alston AM, Martin JK (1988) Phosphorus cycling in wheat-pasture rotations. II. The role of the microbial biomass in phosphorus cycling. Aust J Soil Res 26:333–342

McNeill AM, Fillery IRP (2008) Field measurement of lupin belowground nitrogen accumulation and recovery in the subsequent cereal-soil system in a semi-arid mediterranean-type climate. Plant Soil 302:297–316

McNeill AM, Unkovich M (2007) The nitrogen cycle in terrestrial ecosystems. In: Marschner P, Rengel Z (eds) Nutrient cycling in terrestrial ecosystems. Springer-Verlag, Berlin, pp 37–64

Meharg AA, Killham K (1990) The effect of soil pH on rhizosphere carbon flow of *Lolium perenne*. Plant Soil 123:1–7

Mengel K, Horn D, Tributh H (1990) Availability of interlayer ammonium as related to root vicinity and mineral type. Soil Sci 149:131–137

Miller A, Cramer M (2004) Root nitrogen acquisition and assimilation. Plant Soil 274:1–36

Mirza SM, Mehnaz S, Normand P, Prigent-Combaret C, Moënne-Loccoz Y, Bally R, Malik KA (2006) Molecular characterization and PCR detection of a nitrogen -fixing *Pseudomonas* strain promoting rice growth. Biol Fert Soils 43:163–170

Mitsukawa N, Okumura S, Shirano Y, Sato S, Kato T, Harashima S, Shibata D (1997) Overexpression of an *Arabidopsis thaliana* high-affinity phosphate transporter gene in tobacco cultured cells enhances cell growth under phosphate-limited conditions. Proc Nat Acad Sci USA 94:7098–7102

Molla AH, Shamsuddin ZH, Halimi MS, Morziah M, Puteh AB (2001) Potential for enhancement of root growth and nodulation of soybean co-inoculated with *Azospirillum* and *Bradyrhizobium* in laboratory systems. Soil Biol Biochem 33:457–463

Morel C, Hinsinger P (1999) Rood-induced modifications of the exchange of phosphate ion between soil solution and soil solid phase. Plant Soil 211:103–110

Mozafar A, Oertli JJ (1992) Uptake of a microbially-produced vitamin (B12) by soybean roots. Plant Soil 139:23–30

Mudge SR, Rae AL, Diatloff E, Smith FW (2002) Expression analysis suggests novel roles for members of the *Pht1* family of phosphate transporters in *Arabidopsis*. Plant J 31:341–353

Mukerji KG, Manoharachary C, Singh J (2006) Microbial activity in the rhizosphere. Soil Biology Vol 7. Springer, Heidelberg

Muthukumar T, Udaiyan K, Rajeshkannan V (2001) Response of neem (*Azadirachta indica* A. Juss) to indigenous arbuscular mycorrhizal fungi, phosphate-solubilizing and asymbiotic nitrogen-fixing bacteria under tropical nursery conditions. Biol Fert Soils 34:417–426

Nahas E (2007) Phosphate solubilising microorganisms: Effect of carbon, nitrogen and phosphorus sources. In: Valázquez E, Rodríguez-Barrueco C (eds) First international meeting on microbial phosphate solubilization. Developments in Plant and Soil Science, vol 102. Springer, Dordrecht, pp 111–115

Nasholm T, Ekblad A, Nordin A, Giesler R, Hogberg M, Hogberg P (1998) Boreal forest plants take up organic nitrogen. Nature 392:914–916

Neuer G, Kronenberg A, Bothe H (1985) Denitrification and nitrogen fixation by *Azospirillum*. III. Properties of a wheat-*Azospirillum* association. Arch Microbiol 141:364–370

Neumann G, Martinoia E (2002) Cluster roots–an underground adaptation for survival in extreme environments. Trends Plant Sci 7:162–167

Neumann G, Römheld V (2002) Root-induced changes in the availability of nutrients in the rhizosphere. In: Waisel Y, Eshel A, Kafkafi U (eds) Plant roots: The hidden half. Marcel Dekker, New York, pp 617–649

Neumann G, Römheld V (2007) The release of root exudates as affected by the plant's physiological status. In: Pinton R, Varanini Z, Nannipieri P (eds) The rhizosphere biochemistry and organic substances at the soil-plant interface, 2nd edition. CRC Press, New York, pp 23–72

Nuruzzaman M, Lambers H, Bolland MDA (2006) Distribution of carboxylates and acid phosphatase and depletion of different phosphorus fractions in the rhizosphere of a cereal and three grain legumes. Plant Soil 281:109–120

Oberson A, Friesen DK, Rao IM, Bühler S, Frossard E (2001) Phosphorus transformations in an oxisol under contrasting land-use systems: The role of the microbial biomass. Plant Soil 237:197–210

Oberson A, Joner EJ (2005) Microbial turnover of phosphorus in soil. In: Turner BL, Frossard E, Baldwin DS (eds) Organic phosphorus in the environment. CABI, Wallingford, UK, pp 133–164

Oehl F, Frossard E, Fliessbach A, Dubois D, Oberson A (2004) Basal organic phosphorus mineralization in soils under different farming systems. Soil Biol Biochem 36:667–675

Offre P, Pivato B, Siblot S, Gamalero E, Corberand T, Lemanceau P, Mougel C (2007) Identification of bacterial groups preferentially associated with mycorrhizal roots of *Medicago truncatula*. Appl Environ Microbiol 73:913–921

Oliveira ALM, Urquiaga S, Dobereiner J, Baldani JI (2002) The effect of inoculating endophytic N_2-fixing bacteria on micropropagated sugarcane plants. Plant Soil 242:205–215

Ona O, Van Impe J, Prinsen E, Vanderleyden J (2005) Growth and indole-3-acetic acid biosynthesis of *Azospirillum brasilense* Sp245 is environmentally controlled. FEMS Microbiol Lett 246:125–132

Otani F, Ae N, Tanaka H (1996) Uptake mechanisms of crops grown in soils with low P status. II. Significance of organic acids in root exudates of pigeon pea. Soil Sci Plant Nutr 42:533–560

Owen AG, Jones DL (2001) Competition for amino acids between wheat roots and rhizosphere microorganisms and the role of amino acids in plant N acquisition. Soil Biol Biochem 33:651–657

Parmar N, Dadarwal KR (1999) Stimulation of nitrogen fixation and induction of flavonoid-like compounds by rhizobacteria. J Appl Microbiol 86:36–44

Patten CL, Glick BR (1996) Bacterial biosynthesis of indole-3-acetic acid. Can J Microbiol 42:207–220

Patten CL, Glick BR (2002) Role of *Pseudomonas putida* indoleacetic acid in development of the host plant root system. Appl Environ Microbiol 68:3795–3801

Paungfoo-Lonhienne C, Lonhienne TGA, Rentsch D, Robinson N, Christie M, Webb RI, Gamage HK, Carroll BJ, Schenk PM, Schmidt S (2008) Plants can use protein as a nitrogen source without assistance from other organisms. Proc Natl Acad Sci 105:4524–4529

Pearse SJ, Venaklaas EJ, Cawthray G, Bolland MDA, Lambers H (2006a) Carboxylate release of wheat, canola and 11 grain legume species as affected by phosphorus status. Plant Soil 288:127–139

Pearse SJ, Venaklaas EJ, Cawthray G, Bolland MDA, Lambers H (2006b) Carboxylate composition of root exudates does not relate consistently to a crop species' ability to use phosphorus from aluminium, iron or calcium phosphate sources. New Phytol 173:181–190

Perrig D, Boiero ML, Masciarelli OA, Penna C, Ruiz OA, Cassán FD, Luna MV (2007) Plant-growth-promoting compounds produced by two agronomically important strains of *Azospirillum brasilense*, and implications for inoculant formulation. Appl Microbiol Biotechnol 75:1143–1150

Piccoli P, Lucangeli CD, Bottini R, Schneider G (1997) Hydrolysis of [17,17-$^2H_2$]gibberellin A_{20}-glucoside and [17,17-$^2H_2$]gibberellin A_{20}-glucosyl ester by *Azospirillum lipoferum* cultured in a nitrogen-free biotin-based chemically-defined medium. Plant Growth Regul 23:179–182

Pii Y, Crimi M, Cremonese G, Spena A, Pandolfini T (2007) Auxin and nitric oxide control indeterminate nodule formation. BMC Plant Biol 7:21

Pothier JF, Wisniewski-Dyé F, Weiss-Gayet M, Moënne-Loccoz Y, Prigent-Combaret C (2007) Promoter trap identification of wheat seed extract-induced genes in the plant growth-promoting rhizobacterium *Azospirillum brasilense* Sp245. Microbiol 153:3608–3622

Prigent-Combaret C, Blaha D, Pothier JF, Vial L, Poirier MA, Wisniewski-Dyé F, Moënne-Loccoz Y (2008) Physical organization and phylogenetic analysis of *acdR* as leucine-responsive regulator of the 1-aminocyclopropane-1-carboxylate deaminase gene *acdS* in phytobeneficial *Azospirillum lipoferum* 4B and other Proteobacteria. FEMS Microbiol Ecol 65:202–219

Prinsen E, Chauvaux N, Schmidt J, John M, Wieneke U, De Greef J, Schell J, Van Onckelen H (1991) Stimulation of indole-3-acetic acid production in *Rhizobium* by flavonoids. FEBS-Lett 282:53–55

Qian JH, Doran JW, Walters DT (1997) Maize plant contributions to root zone available carbon and microbial transformations of nitrogen. Soil Biol Biochem 29:1451–1462

Raaijmakers JM, Paulitz TC, Steinberg C, Alabouvette C, Moënne-Loccoz Y (2009) The rhizosphere: a playground and battlefield for soilborne pathogens and beneficial microorganisms. Plant Soil doi:10.1007/s11104-008-9568-6

Rae AL, Jarmey JM, Mudge SR, Smith FW (2004) Overexpression of a high-affinity phosphate transporter in transgenic barley plants does not enhance phosphate uptake rates. Funct Plant Biol 31:141–148

Raghothama KG (1999) Phosphate acquisition. Ann Rev Plant Physiol Plant Mol Biol 50:665–693

Rao VR, Ramakrishnan B, Adhya TK, Kanungo PK, Nayak DN (1998) Current status and future prospects of associative nitrogen fixation in rice. World J Microbiol Biotechnol 14:621–633

Rausch C, Daram P, Brunner S, Jansa J, Laloi M, Leggewie G, Amrhein N, Bucher M (2001) A phosphate transporter expressed in arbuscule-containing cells in potato. Nature 414:462–466

Redecker D (2002) New views on fungal evolution based on DNA markers and the fossil record. Res Microbiol 153:125–130

Reinhold-Hurek B, Hurek T (1998) Life in grasses: diazotrophic endophytes. Trends Microbiol 6:139–144

Remans R, Beebe S, Blair M, Manrique G, Tovar E, Rao I, Croonenborghs A, Torres-Gutierrez R, El-Howeity M, Michiels J, Vanderleyden J (2008) Physiological and genetic analysis of root responsiveness to auxin-producing plant growth-promoting bacteria in common bean (*Phaseolus vulgaris* L.). Plant Soil 302:149–161

Rengel Z, Marschner P (2005) Nutrient availability and management in the rhizosphere: Exploiting genotypic differences. New Phytol 168:305–312

Revillas JJ, Rodelas B, Pozo C, Martínez-Toledo MV, González-López J (2000) Production of B-group vitamins by two *Azotobacter* strains with phenolic compounds as sole carbon source under diazotrophic and adiazotrophic conditions. J Appl Microbiol 89:486–493

Ribaudo CM, Krumpholz EM, Cassán FD, Bottini R, Cantore ML, Curá JA (2006) *Azospirillum* sp. promotes root hair development in tomato plants through a mechanism that involves ethylene. J Plant Growth Regul 25:175–185

Richardson AE (1994) Soil microorganisms and phosphorus availablility. In: Pankhurst CE, Doube BM, Gupta VVSR, Grace PR (eds) Management of the soil biota in sustainable farming systems. CSIRO Publishing, Melbourne, pp 50–62

Richardson AE (2001) Prospects for using soil microorganisms to improve the acquisition of phosphorus by plants. Aust J Plant Physiol 28:897–906

Richardson AE (2007) Making microorganims mobilize soil phosphorus. In: Valázquez E, Rodríguez-Barrueco C (eds) First international meeting on microbial phosphate solubilization. Developments in Plant and Soil Science, vol 102. Springer, Dordrecht, pp 85–90

Richardson AE, George TS, Hens M, Simpson RJ (2005) Utilization of soil organic phosphorus by higher plants. In: Turner BL, Frossard E, Baldwin DS (eds) Organic phosphorus in the environment. CABI, Wallingford, UK, pp 165–184

Richardson AE, George TS, Jakobsen I, Simpson RJ (2007) Plant utilization of inositol phosphates. In: Turner BL, Richardson AE, Mullaney EJ (eds) Inositol phosphates: linking agriculture and the environment. CABI, Wallingford, UK, pp 242–260

Richardson AE, Hadobas PA, Hayes JE (2001a) Extracellular secretion of *Aspergillus* phytase from *Arabidopsis* roots enables plants to obtain phosphorus from phytate. Plant J 25:641–649

Richardson AE, Hadobas PA, Hayes JE, O, Hara CP, Simpson RJ (2001b) Utilization of phosphorus by pasture plants supplied with *myo*-inositol hexaphosphate is enhanced by the presence of soil microorganisms. Plant Soil 229:47–56

Richardson AE, Simpson RJ, George TS, Hocking PJ (2009) Plant mechanisms to optimize access to soil phosphorus. Crop Pasture Sci 60(2): (in press)

Riefler M, Novak O, Strnad M, Schmülling T (2006) *Arabidopsis* cytokinin receptor mutants reveal functions in shoot growth, leaf senescence, seed size, germination, root development, and cytokinin metabolism. Plant Cell 18:40–54

Rillig MC, Mummey DL, Ramsey PW, Klironomos JN, Gannon JE (2006) Phylogeny of arbuscular mycorrhizal fungi predicts community composition of symbiosis-associated bacteria. FEMS Microbiol Ecol 57:389–395

Robin A, Vansuyt G, Hinsinger P, Meyer JM, Briat JF, Lemanceau P (2008) Iron dynamics in the rhizosphere: consequences for plant health and nutrition. Adv Agron 99:183–225

Robinson D (2001) Root proliferation, nitrate inflow and their carbon costs during nitrogen capture by competing plants in patchy soil. Plant Soil 232:41–50

Robinson D, Hodge A, Griffiths BS, Fitter AH (1999) Plant root proliferation in nitrogen patches confers competitive advantage. Proc Roy Soc Lond B Biol Sci 266:431–435

Robinson D, van Vuuren MMI (1998) Responses of wild plants to nutrient patches in relation to growth rate and life form. In: Lambers H, Poorter H, van Vuuren MMI (eds) Variation in plant growth. Backhuys, Leiden, pp 237–257

Rodelas B, Salmerón V, Martinez-Toledo MV, González-López J (1993) Production of vitamins by *Azospirillum brasilense* in chemically-defined media. Plant Soil 153:97–101

Rodríguez H, Frago R (1999) Phosphate solubilizing bacteria and their role in plant growth promotion. Biotechnol Adv 17:319–339

Roelofs RFR, Rengel Z, Cawthray GR, Dixon KW, Lambers H (2001) Exudation of carboxylates in Australian Proteaceae: chemical composition. Plant Cell Environ 24:891–903

Ron-Vaz MD, Edwards AC, Shand CA, Cresser MS (1993) Phosphorus fractions in soil solution: influence of soil acidity and fertiliser additions. Plant Soil 148:175–183

Rubio G, Liao H, Yan XL, Lynch JP (2003) Topsoil foraging and its role in plant competitiveness for phosphorus in common bean. Crop Sci 43:598–607

Ryan MH, Angus JF (2003) Arbuscular mycorrhizas in wheat and field pea crops on a low P soil: increased Zn-uptake but no increase in P-uptake or yield. Plant Soil 250:225–239

Ryan MH, van Herwaarden AF, Angus JF, Kirkegaard JA (2005) Reduced growth of autumn-sown wheat in a low-P soil is associated with high colonization by arbuscular mycorrhizal fungi. Plant Soil 270:275–286

Ryan PR, Delhaize E, Jones DL (2001) Function and mechanism of organic anion exudation from plant roots. Ann Rev Plant Physiol Plant Mol Biol 52:527–560

Scherer HW, Ahrens G (1996) Depletion of non-exchangable NH_4^+-N in the soil-root interface in relation to clay mineral composition and plant species. Eur J Agron 5:1–7

Schimel JP, Bennett J (2004) Nitrogen mineralization: Challenges of a changing paradigm. Ecology 85:591–602

Schneider G, Schliemann W (1994) Gibberellin conjugates: an overview. Plant Growth Regul 15:247–260

Schnepf A, Roose T, Schweiger P (2008) Impact of growth and uptake patterns of arbuscular mycorrhizal fungi on plant phosphorus uptake—a modelling study. Plant Soil 312:85–99

Schünmann PHD, Richardson AE, Smith FW, Delhaize E (2004) Characterisation of promoter expression patterns derived from the *Pht1* phosphate transporter genes of barley (*Hordeum vulgare* L.). J Exp Bot 55:855–865

Schübler A, Schwarzott D, Walker C (2001) A new fungal phylum, the *Glomeromycota*, phylogeny and evolution. Mycol Res 105:1413–1421

Schweiger PF, Robson AD, Barrow NJ (1995) Root hair length determines beneficial effect of a *Glomus* species on shoot growth of some pasture species. New Phytol 131:247–254

Scott JT Condron LM (2004) Short term effects of radiata pine and selected pasture species on soil organic phosphorus mineralization. Plant Soil 266:153–163

Seeling B, Zasoski RJ (1993) Microbial effects in maintaining organic and inorganic solution phosphorus concentrations in a grassland topsoil. Plant Soil 148:277–284

Sessitsch A, Howieson JG, Perret X, Antoun H, Martínez-Romero E (2002) Advances in *Rhizobium* research. Crit Rev Plant Sci 21:323–378

Sevilla M, Burris RH, Gunapala N, Kennedy C (2001) Comparison of benefit to sugarcane plant growth and $^{15}N_2$ incorporation following inoculation of sterile plants with *Acetobacter diazotrophicus* wild-type and Nif mutants strains. Mol Plant-Microbe Interact 14:358–366

Shah S, Li J, Moffatt BA, Glick BR (1998) Isolation and characterization of ACC deaminase genes from two different plant growth-promoting rhizobacteria. Can J Microbiol 44:833–843

Shand CA, Macklon AES, Edwards AC, Smith S (1994) Inorganic and organic P in soil solutions from three upland soils. I. Effects of soil solution extraction conditions, soil type and season. Plant Soil 159:255–264

Shane MW, Lambers H (2005) Cluster roots: A curiosity in context. Plant Soil 274:99–123

Shane MW, Cawthray GR, Cramer MD, Kuo J, Lambers H (2006) Specialized 'dauciform' roots of Cyperaceae are structurally distinct, but functionally analogous with 'cluster' roots. Plant Cell Environ 29:1989–1999

Sierra S, Rodelas B, Martínez-Toledo MV, Pozo C, González-López J (1999) Production of B-group vitamins by two *Rhizobium* strains in chemically defined media. J Appl Microbiol 86:851–858

Silberbush M, Barber SA (1983) Sensitivity of simulated phosphorus uptake to parameters used by a mechanistic-mathematical model. Plant Soil 74:93–100

Silverman FP, Assiamah AA, Bush DS (1998) Membrane transport and cytokinin action in root hairs of *Medicago sativa*. Planta 205:23–31

Sivaramaiah N, Malik DK, Sindhu SS (2007) Improvement in symbiotic efficiency of chickpea (*Cicer arietinum*) by coinoculation of *Bacillus* strains with *Mesorhizobium* sp. Cicer. Indian J Microbiol 47:51–56

Smiley RW, Cook R (1983) Relationship between take-all of wheat and rhizosphere pH in soils fertilised with ammonium versus nitrate. Phytopath 63:822–825

Smith SE, Read DJ (2008) Mycorrhizal symbiosis. 3rd Edition. Academic Press, Elsevier, Amsterdam

Smith SE, Smith FA, Jakobsen I (2003) Mycorrhizal fungi can dominate phosphate supply to plants irrespective of growth responses. Plant Physiol 133:16–20

Smith SE, Smith FA, Jakobsen I (2004) Functional diversity in arbuscular mycorrhizal (AM) symbioses: the contribution of the mycorrhizal P uptake pathway is not correlated with mycorrhizal responses in growth or total P uptake. New Phytol 162:511–524

Spaepen S, Dobbelaere S, Croonenborghs A, Vanderleyden J (2008) Effects of *Azospirillum brasilense* indole-3-acetic acid production on inoculated wheat plants. Plant Soil 312:15–23

Spaepen S, Vanderleyden J, Remans R (2007) Indole-3-acetic acid in microbial and microorganism-plant signaling. FEMS Microbiol Rev 31:425–448

Stepanova AN, Hoyt JM, Hamilton AA, Alonso JM (2005) A link between ethylene and auxin uncovered by the characterization of two root-specific ethylene-insensitive mutants in *Arabidopsis*. Plant Cell 17:2230–2242

Ström L, Godbold DL, Owen AG, Jones DL (2001) Organic acid behaviour in a calcareous soil: sorption reactions and biodegradation rates. Soil Biol Biochem 33:2125–2133

Ström L, Godbold DL, Owen AG, Jones DL (2002) Organic acid mediated P mobilization in the rhizosphere and uptake by maize roots. Soil Biol Biochem 34:703–710

Subbarao GV, Ito O, Sahrawat KL, Berry WL, Nakahara K, Ishikawa T, Watanabe T, Suenaga K, Rondon M, Rao IM (2006) Scope and strategies for regulation of nitrification in agricultural systems: Challenges and opportunities. Crit Rev Plant Sci 25:303–335

Subbarao GV, Rondon M, Ito O, Ishikawa T, Rao IM, Nakahara K, Lascano C, Berry WL (2007a) Biological nitrification inhibition (BNI)-Is it a widespread phenomenon? Plant Soil 294:5–18

Subbarao GV, Tomohiro B, Masahiro K, Osamu I, Samejima H, Wang HY, Pearse SJ, Gopalakrishnan S, Nakahara K, Zakir Hossain AKM, Tsujimoto N, Berry WL (2007b) Can biological nitrification inhibition (BNI) genes from perennial *Leymus racemosus* (*Triticeae*) combat nitrification in wheat farming? Plant Soil 299:55–64

Subba-Rao NS (1982) Phosphate solubilization by soil microorganisms. In: Subba-Rao NS (ed) Advances in agricultural microbiology. Oxford and IBH Publishing, New Delhi, pp 295–303

Swarup R, Perry P, Hagenbeek D, Van Der Straeten D, Sandberg G, Bhalerao R, Ljung K, Bennett MJ (2007) Ethylene upregulates auxin biosynthesis in *Arabidopsis* seedlings to enhance inhibition of root cell. Plant Cell 19:2186–2196 doi:10.1105/tpc.107.052100

Tadano T, Ozawa K, Sakai H, Osaki M, Matsui H (1993) Secretion of acid phosphatase by the roots of crop plants under phosphorus-deficient conditions and some properties of the enzyme secreted by lupin roots. Plant Soil 155/156:95–98 doi:10.1007/BF00024992

Tan RX, Zou WX (2001) Endophytes: A rich source of functional metabolites. Nat Prod Rep 18:448–459 doi:10.1039/b100918o

Tarafdar JC, Claassen N (1988) Organic phosphorus compounds as a phosphorus source for higher plants through the activity of phosphatases produced by plant roots and microorganisms. Biol Fertil Soils 5:308–312 doi:10.1007/BF00262137

Tarafdar JC, Jungk A (1987) Phosphatase activity in the rhizosphere and its relation to the depletion of soil organic phosphorus. Biol Fertil Soils 3:199–204 doi:10.1007/BF00640630

Tarafdar JC, Marschner H (1994) Efficiency of VAM hyphae in utilization of organic phosphorus by wheat plants. Soil Sci Plant Nutr 40:593–600

Tarafdar JC, Marschner H (1995) Dual inoculation with *Aspergillus fumigatus* and *Glomus mosseae* enhances biomass production and nutrient uptake in wheat (*Triticum aestivum* L) supplied with organic phosphorus as Na-phytate. Plant Soil 173:97–102 doi:10.1007/BF00155522

Tawaraya K, Naito M, Wagatsuma T (2006) Solubilization of insoluble inorganic phosphate by hyphal exudates of arbuscular mycorrhizal fungi. J Plant Nutr 29:657–665 doi:10.1080/01904160600564428

Thuler DS, Floh EIS, Handro W, Barbosa HR (2003) Plant growth regulators and amino acids released by *Azospirillum* sp. in chemically defined media. Lett Appl Microbiol 37:174–178 doi:10.1046/j.1472-765X.2003.01373.x

Tibbett M, Sanders FE (2002) Ectomycorrhizal symbiosis can enhance plant nutrition through improved access to discrete organic nutrient patches of high resource quality. Ann Bot (Lond) 89:783–789 doi:10.1093/aob/mcf129

Tilman D, Cassman KG, Matson PA, Naylor R, Polasky S (2002) Agricultural sustainability and intensive production practices. Nature 418:671–677 doi:10.1038/nature01014

Timmusk S, Nicander B, Granhall U, Tillberg E (1999) Cytokinin production by *Paenibacillus polymyxa*. Soil Biol Biochem 31:1847–1852 doi:10.1016/S0038-0717(99)00113-3

Tinker PB (1980) The role of rhizosphere microorganisms in mediating phosphorus uptake by plants. In: Kwasenah FE, Sample EC, Kamprath EJ (eds) The role of phosphorus in agriculture. American Society of Agronomy, Madison, WI, pp 617–654

Tinker PB, Nye PH (2000) Solute movement in the rhizosphere. Oxford University Press, New York

Tobar RM, Azcón R, Barea JM (1994b) Improved nitrogen uptake and transport from ^{15}N-labeled nitrate by external hyphae of arbuscular mycorrhiza under water-stressed conditions. New Phytol 126:119–122 doi:10.1111/j.1469-8137.1994.tb07536.x

Toljander JF, Lindahl BD, Paul LR, Elfstrand M, Finlay RD (2007) Influence of arbuscular mycorrhizal mycelial exudates on soil bacterial growth and community structure. FEMS Microbiol Ecol 61:295–304 doi:10.1111/j.1574-6941.2007.00337.x

Toro M, Azcon R, Barea JM (1997) Improvement of arbuscular mycorrhiza development by inoculation of soil with phosphate solubilizing rhizobacteria to improve rock phosphate bioavailability (^{32}P) and nutrient cycling. Appl Environ Microbiol 63:4408–4412

Uren N (2007) Types, amounts and possible function of compounds released into the rhizosphere by soil-grown plants. In: Pinton R, Varanini Z, Nannipieri P (eds) The rhizosphere biochemistry and organic substances at the soil-plant interface. CRC Press, Boca Raton, Florida, pp 1–21

Uroz S, Calvaruso C, Turpaul MP, Pierrat JC, Mustin C, Frey-Klett P (2007) Effect of the mycorrhizosphere on the genotypic and metabolic diversity of the bacterial communities involved in mineral weathering in a forest soil. Appl Environ Microbiol 73:3019–3027 doi:10.1128/AEM.00121-07

Vansuyt G, Robin A, Briat JF, Curie C, Lemanceau P (2007) Iron acquisition from Fe-pyoverdine by *Arabidopsis thaliana*. Mol Plant Microbe Interact 20:441–447 doi:10.1094/MPMI-20-4-0441

Vance CP, Ehde-Stone C, Allan DL (2003) Phosphorus acquisition and use: Critical adaptations by plants for securing a nonrenewable resource. New Phytol 157:423–447 doi:10.1046/j.1469-8137.2003.00695.x

Vande Broek A, Lambrecht M, Eggermont K, Vanderleyden J (1999) Auxins upregulate expression of the indole-3-pyruvate decarboxylase gene in *Azospirillum brasilense*. J Bacteriol 181:1338–1342

Veneklaas EJ, Stevens J, Cawthray GR, Turner S, Grigg AM, Lambers H (2003) Chickpea and white lupin rhizosphere carboxylates vary with soil properties and enhance phosphorus uptake. Plant Soil 248:187–197 doi:10.1023/A:1022367312851

Vessey JK (2003) Plant growth promoting rhizobacteria as biofertilizers. Plant Soil 255:571–586 doi:10.1023/A:1026037216893

Wakelin S, Anstis S, Warren R, Ryder M (2006) The role of pathogen suppression on the growth promotion of wheat by *Penicillium radicum*. Aust Plant Pathol 35:253–258 doi:10.1071/AP06008

Wakelin SA, Gupta VVSR, Harvey PR, Ryder MH (2007) The effect of *Penicillium* fungi on plant growth and phosphorus mobilization in neutral to alkaline soils from southern Australia. Can J Microbiol 53:106–115 doi:10.1139/W06-109

Walley FL, Germida JJ (1997) Response of spring wheat (*Triticum aestivum*) to interactions between *Pseudomonas* species and *Glomus clarum* NT4. Biol Fertil Soils 24:365–371 doi:10.1007/s003740050259

Wang C, Knill E, Glick BR, Défago G (2000) Effect of transferring 1-aminocyclopropane-1-carboxylic acid (ACC) deaminase genes into *Pseudomonas fluorescens* strain CHA0 and its gacA derivative CHA96 on their growth-promoting and disease-suppressive capacities. Can J Microbiol 46:898–907 doi:10.1139/cjm-46-10-898

Watt M, Kirkegaard JA, Passioura JB (2006) Rhizosphere biology and crop productivity—a review. Aust J Soil Res 44:299–317 doi:10.1071/SR05142

Whitelaw M (2000) Growth promotion of plants inoculated with phosphate-solubilizing fungi. Adv Agron 69:99–151 doi:10.1016/S0065-2113(08)60948-7

Whitelaw MA, Harden TJ, Bender GL (1997) Plant growth promotion of wheat inoculated with *Penicillium radicum* sp. nov. Aust J Soil Res 35:291–300 doi:10.1071/S96040

Whitelaw MA, Harden TJ, Helayer KR (1999) Phosphate solubilization in solution culture by the soil fungus *Penicillium radicum*. Soil Biol Biochem 31:655–665 doi:10.1016/S0038-0717(98)00130-8

Wichern F, Eberhardt E, Mayer J, Joergensen RG, Muller T (2008) Nitrogen rhizodeposition in agricultural crops: Methods, estimates and future prospects. Soil Biol Biochem 40:30–48 doi:10.1016/j.soilbio.2007.08.010

Wijesinghe DK, John EA, Beurskens S, Hutchings MJ (2001) Root system size and precision in nutrient foraging: Responses to spatial patterns of nutrient supply in six herbaceous species. J Ecol 89:972–983 doi:10.1111/j.1365-2745.2001.00618.x

Wissuwa M (2003) How do plants achieve tolerance to phosphorus deficiency? Small causes with big effects. Plant Physiol 133:1947–1958 doi:10.1104/pp.103.029306

Wissuwa M, Mazzola M, Picard C (2009) Novel approaches in plant breeding for rhizosphere-related traits. Plant Soil doi:10.1007/s11104-008-9693-2

Wouterlood M, Cawthray GR, Turner S, Lambers H, Veneklaas EJ (2004) Rhizosphere carboxylate concentrations of chickpea are affected by genotype and soil type. Plant Soil 261:1–10 doi:10.1023/B:PLSO.0000035568.28893.f6

Xie H, Pasternak JJ, Glick BR (1996) Isolation and characterization of mutants of the plant growth-promoting rhizobacterium *Pseudomonas putida* GR12-2 that overproduce indoleacetic acid. Curr Microbiol 32:67–71 doi:10.1007/s002849900012

Yaxley JR, Ross JJ, Sherriff LJ, Reid JB (2001) Gibberellin biosynthesis mutations and root development in pea. Plant Physiol 125:627–633 doi:10.1104/pp.125.2.627

Yevdokimov IV, Blagodatsky SA (1994) Nitrogen immobilisation and remineralization by microorganisms and nitrogen uptake by plants: interactions and rate calculations. Geomicrobiol J 11:185–193

Yi K, Wu Z, Zhou J, Du L, Guo L, Wu Y, Wu P (2005) *OsPTF1*, a novel transcription factor involved in tolerance to phosphate starvation in rice. Plant Physiol 138:2087–2096 doi:10.1104/pp.105.063115

Zapata F (1990) Isotope techniques in soil fertility and plant nutrition studies. In: Hardarson G (ed) Use of nuclear techniques in studies of soil-plant relationships. IAEA, Vienna, pp 61–128

Zhang H, Forde B (2000) Regulation of *Arabidopsis* root development by nitrate availability. J Exp Bot 51:51–59 doi:10.1093/jexbot/51.342.51

Zimmer W, Wesche M, Timmermans LC (1998) Identification and isolation of the indole-3-pyruvate decarboxylase gene from *Azospirillum brasilense* Sp7: Sequencing and functional analysis of the gene locus. Curr Microbiol 36:327–331 doi:10.1007/s002849900317

Zumft WG (1997) Cell biology and molecular basis of denitrification. Microbiol Mol Biol Rev 61:533–616

Zwart KB, Kuikman PJ, vanVeen JA (1994) Rhizosphere protozoa: Their significance in nutrient dynamics. In: Darbyshire JF (ed) Soil Protozoa. CAB International, Wallingford, pp 91–122

The rhizosphere: a playground and battlefield for soilborne pathogens and beneficial microorganisms

Jos M. Raaijmakers · Timothy C. Paulitz ·
Christian Steinberg · Claude Alabouvette ·
Yvan Moënne-Loccoz

Received: 14 December 2007 / Accepted: 4 February 2008 / Published online: 23 February 2008
© The Author(s) 2008

Abstract The rhizosphere is a hot spot of microbial interactions as exudates released by plant roots are a main food source for microorganisms and a driving force of their population density and activities. The rhizosphere harbors many organisms that have a neutral effect on the plant, but also attracts organisms that exert deleterious or beneficial effects on the plant. Microorganisms that adversely affect plant growth and health are the pathogenic fungi, oomycetes, bacteria and nematodes. Most of the soilborne pathogens are adapted to grow and survive in the bulk soil, but the rhizosphere is the playground and infection court where the pathogen establishes a parasitic relationship with the plant. The rhizosphere is also a battlefield where the complex rhizosphere community, both microflora and microfauna, interact with pathogens and influence the outcome of pathogen infection. A wide range of microorganisms are beneficial to the plant and include nitrogen-fixing bacteria, endo- and ectomycorrhizal fungi, and plant growth-promoting bacteria and fungi. This review focuses on the population dynamics and activity of soilborne pathogens and beneficial microorganisms. Specific attention is given to mechanisms involved in the tripartite interactions between beneficial microorganisms, pathogens and the plant. We also discuss how agricultural practices affect pathogen and antagonist populations and how these practices can be adopted to promote plant growth and health.

Keywords Epidemiology of soilborne pathogens ·
Microbial interactions · Induced systemic resistance ·
Antibiotic resistance in pathogens ·
Biofumigation and organic amendment

Responsible Editor: Philippe Lemanceau

J. M. Raaijmakers (✉)
Laboratory of Phytopathology, Wageningen University,
Binnenhaven 5,
6709 PD Wageningen, the Netherlands
e-mail: jos.raaijmakers@wur.nl

T. C. Paulitz
USDA-ARS,
345 Johnson Hall,
Pullman, WA 99164-6430, USA

C. Steinberg · C. Alabouvette
INRA-Université de Bourgogne, UMR-MSE,
21065 Dijon, France

Y. Moënne-Loccoz
Université Lyon 1,
Lyon 69003, France

Y. Moënne-Loccoz
UMR CNRS 5557, Ecologie Microbienne,
Villeurbanne 69622, France

Introduction

The rhizosphere is an environment that the plant itself helps to create and where pathogenic and beneficial microorganisms constitute a major influential force on plant growth and health (Lynch 1990). Microbial groups and other agents found in the rhizosphere include bacteria, fungi, nematodes, protozoa, algae and microarthropods (Lynch 1990; Raaijmakers 2001). Many members of this community have a neutral effect on the plant, but are part of the complex food web that utilizes the large amount of carbon that is fixed by the plant and released into the rhizosphere (i.e. rhizodeposits). The microbial community in the rhizosphere also harbors members that exert deleterious or beneficial effects on the plant. Microorganisms that adversely affect plant growth and health are the pathogenic fungi, oomycetes, bacteria and nematodes, whereas microorganisms that are beneficial include nitrogen-fixing bacteria, endo- and ectomycorrhizal fungi, and plant growth-promoting rhizobacteria (PGPR) and fungi. The number and diversity of deleterious and beneficial microorganisms are related to the quantity and quality of the rhizodeposits and to the outcome of the microbial interactions that occur in the rhizosphere (Somers et al. 2004). Understanding the processes that determine the composition, dynamics, and activity of the rhizosphere microflora has attracted the interest of scientists from multiple disciplines and can be exploited for the development of new strategies to promote plant growth and health. In this review, we will focus on the epidemiology (spatial and temporal aspects) of soilborne pathogens and the economic importance of soilborne diseases. Specific attention is given to mechanisms, both offensive and defensive, involved in the interactions between soilborne pathogens and beneficial microorganisms. Also direct positive effects of rhizosphere microorganisms on the plant are addressed. Finally, we discuss the effects of agricultural practices on pathogen and antagonist populations and how these practices can be manipulated to induce soil suppressiveness and to promote plant growth and health.

Soilborne pathogens and their economic importance

In most agricultural ecosystems, soilborne plant pathogens can be a major limitation in the production of marketable yields. They are also more recalcitrant to management and control compared to pathogens that attack the above-ground portions of the plant (Bruehl 1987). Soilborne pathogens are adapted to grow and survive in the bulk soil, but the rhizosphere is the infection court where the pathogen encounters the plant and establishes a parasitic relationship. This is also where the complex rhizosphere community, both microflora and microfauna, can interact with the pathogen and influence the outcome of pathogen infection.

There are four main groups of plant pathogens (Agrios 2005), but only two of them are major players in the soil: fungi (true fungi and oomycetes) and nematodes. Only a few groups of bacteria are considered to be soilborne, probably because non-spore forming bacteria cannot survive well in soil for long periods. Bacteria also require a wound or natural opening to penetrate into the plant and cause infection. Examples are *Ralstonia solanacearum*, cause of bacterial wilt of tomato (Genin and Boucher 2004), and *Agrobacterium tumefaciens*, the well-studied causal agent of crown gall (Nester et al. 2005). Some filamentous bacteria (*Streptomyces*) can also infect plants and are better adapted to survive in the soil. Only a few viruses can infect roots. Like bacteria, they require a wound to infect the plant and are mostly transmitted by vectors. In soil, they can be transmitted by nematodes (Nepoviruses; Brown et al. 1995) or by zoosporic fungi such as *Olpidium* and *Polymyxa* (Campbell 1996).

Nematodes are complex, worm-like eukaryotic invertebrate animals and probably among the most numerous animals on the planet (Perry and Moens 2006). Most nematodes in soil are free-living, consuming bacteria, fungi, and other nematodes, but some can parasitize plants. Some feed on the outside of the root (migratory ectoparasitic), some penetrate and move in the interior of the root (migratory endoparasitic), and some set up a feeding site in the interior of the root and remain there for reproduction (sedentary endoparasites).

Fungi and oomycetes are the most important soilborne pathogens and will be the focus of this review. Fungi are eukaryotic, filamentous, multicellular, heterotrophic organisms that produce a network of hyphae called the mycelium and absorb nutrients from the surrounding substrate (Alexopoulos et al. 1996). Oomycetes have a morphology similar to fungi, but are phylogenetically more closely related to brown algae. They produce swimming spores (zoospores; Fig. 1) and contain cellulose in their cell walls as opposed to chitin in true fungi. Nevertheless,

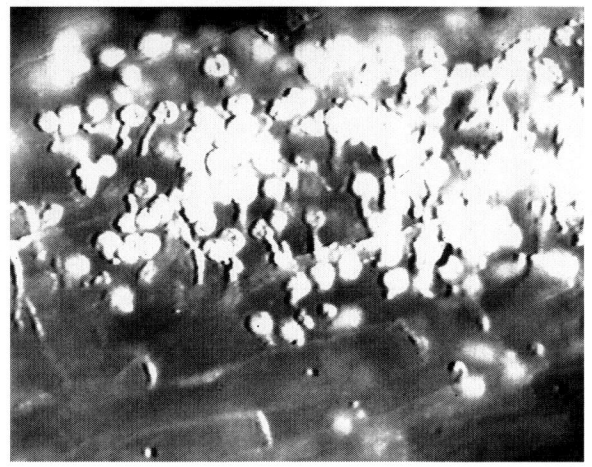

Fig. 1 Aggregation of encysted zoospores of *Pythium aphanidermatum* in the rhizoplane of roots of cucumber seedlings grown in hydroponic solution. Encysted zoospores were visualised by UV epifluorescence after staining with acridine orange and malachite green (micrograph from experiments described in Zhou and Paulitz 1993)

the mechanisms of parasitism and the diseases they cause are similar to true fungi, and therefore will be considered together in this discussion. Almost all soilborne fungi are necrotrophic, meaning they kill host tissue with enzymes and toxins in advance of the hyphae and do not require a living cell to obtain nutrients. Most of the biotrophic pathogens, such as rusts and powdery mildews, occur on the aboveground portions of the plants and require a living cell to obtain nutrients. A few root pathogens such as *Phytophthora sojae* are semibiotrophic. Surprisingly few root pathogens are biotrophic. Some examples are lower zoosporic fungi and Oomycetes, such as *Plasmodiophora brassicae* and *Plasmopara halstedii*. Most necrotrophic pathogens are generalists with a wide host range, as opposed to biotrophic pathogens with narrow host ranges that have co-evolved with the plant. Thus, there is usually no race structure within necrotrophic pathogen populations and no specific single-gene resistances in the plant.

Environmental conditions in the soil are generally not favorable for fungal growth, due to high or low temperatures (frozen ground) or extremely dry conditions. Pathogens survive in the soil as resistant propagules, such as chlamydospores, sclerotia, thick-walled conidia or hyphae, or survive in plant roots and crop residues (Bruehl 1987). When conditions are favorable and when a seed or root approaches the dormant propagule, the fungus is stimulated to germinate by root or seed exudates and chemotactically grows toward the plant. The germ tube or zoospore can attach to the surface of the root, penetrate and infect the epidermal cells of the root tips, secondary roots, and root hairs, or attack the emerging shoots and radicles of seedlings. Some fast-growing pathogens, such as *Pythium* species, can attack seeds and embryos before they emerge. Fungi penetrate through intact cell walls via cell wall-degrading enzymes and mechanical turgor pressure, and colonize the root cortex. Most soilborne fungi attack young, juvenile roots as opposed to secondary woody roots. After the roots have been killed and the fungus ramifies through the cortex, it reproduces and forms spores within the root tissue. Mycelium can continue to spread up the root, internally or externally, or can spread to other roots in close proximity. A specialized group of pathogens that cause wilt diseases (e.g. *Fusarium oxysporum*, *Verticillium dahliae*) can penetrate through the endodermis into the vascular tissue and move up the xylem to above-ground parts of the plant, impeding the flow of water (Beckman 1987).

A number of diseases and symptoms can be manifested by plants infected with fungal soilborne pathogens. However, these diseases can be difficult to diagnose, because most of the symptoms occur below ground, and the above-ground symptoms may be non-distinct or similar to those caused by abiotic factors such as drought, stress, and lack of nutrients. Soilborne pathogens can cause seed decay, damping-off (both pre- and post-emergence), and can also move into the base of the stem, causing crown rot and wilt. In perennial trees, fungi can move into the collar of the tree, girdling the tree, or inoculum can splash onto the fruit, causing decay and rot. However, the primary disease is root rot. By killing root tips, root growth on that axis is eliminated. By destroying fine feeder roots and root hairs, the ability of the plant to absorb water and nutrients is diminished. This leads to reduced plant size, stunting, drought stress and nutrient deficiencies.

Economic impacts of soilborne pathogens and root diseases

Attainable yield has been defined as the potential yield in a given environment (temperature, water) without the limitations of pests and diseases (Cook and Veseth 1991). The actual yield is that obtained after biological factors (pests, pathogens, weeds) have acted on the crop. The difference between these two yields (*attainable − actual*) is often called crop loss, but in reality it is potential yield that was never attained. But in considering the economic impact, one also has to consider the costs of control and management of diseases. How much crop losses do diseases and in particular soilborne pathogens cause? Estimating crop loss from pathogens is difficult and there are only a few well-documented studies. From 2001–2003, an average of 7% to 15% of crop loss occurred on major world crops (wheat, rice, potato, maize and soybean) due to fungi and bacteria (Oerke 2005). From 1996 to 1998, these pathogens caused an actual loss of 9.9%, but the potential loss without controls would be 14.9% (Oerke and Dehne 2004). Nematode crop losses have been estimated at 10% up to 20%, with worldwide losses exceeding $US 100 billion (Bird and Kaloshian 2003). Losses from soilborne pathogens are even more difficult to estimate, because of the difficulty of diagnosis. Some estimate that soilborne pathogens cause 50% of the crop loss in the US (Lewis and Papavizas 1991). The most accurate studies are based on replicated field plots where fumigants, fungicides and nematicides are applied and yield is compared to non-treated plots. Some studies compare resistant to susceptible varieties, but for most soil-borne pathogens there are no resistant varieties. Most studies are done with natural inoculum, and the pathogen is quantified and used as a predictor. Probably the most comprehensive studies on crop loss have been done with soilborne pathogens of cereals and serve to illustrate more realistic crop losses. Folwell et al. (1991) estimated that take-all disease caused 20% yield reduction on wheat, based on soil fumigation studies and grower surveys. Treatment with metalaxyl, a fungicide specific for oomycetes, resulted in wheat yield increases of 1–2 tons/acre, due to control of *Pythium* (Cook et al. 1980). Treatment with soil fumigation increased wheat yields by 3–36% (Cook et al. 1987). Based on detailed disease measurements, Fusarium crown rot was estimated to cause a 9.9% loss on wheat, with some losses up to 35% (Smiley et al. 2005a). With severe Rhizoctonia bare patch, close to 20% of the field can be covered with patches with essentially no yield in the patches (Cook et al. 2002). Using a combination of a nematicide (aldicarb) and tolerant and intolerant varieties, yield suppression caused by *Pratylenchus neglectus* ranged from 8% to 36% in Oregon (Smiley et al. 2005b). Aldicarb increased wheat yields 67% and 113% in soil infested with *P. thornei* (Smiley et al. 2005c). However, yield losses from combinations of pathogens and disease complexes are not well understood. For example, fungal wilt pathogens and nematodes can have synergistic interactions (Back et al. 2002).

Epidemiology of soilborne diseases: temporal and spatial aspects

Temporal spread of soilborne pathogens and diseases

Like all biological organisms, pathogens can grow, multiply and reproduce on their plant hosts, and as a consequence diseases increase over time. The increase of disease over time can be described by a disease progress curve, where disease (counts, incidence or severity) is plotted as the dependent variable measured over various times in the season. Vanderplank (1963) described two types of disease progress curves: one where inoculum multiplies many times over the season (compound interest diseases) and one where there was only one infection cycle and no increase in inoculum during the growing season (simple interest diseases). The former disease progress curve can be described by a logistic or other similar model with an S-shaped curve and an upper asymptote or plateau, because the rate slows when the density of susceptible host tissue decreases and becomes limiting. The most important part of this model is the infection rate or the slope of the line, which describes how quickly the epidemic increases and changes over time. In past literature, this type of disease progress curve was assumed to be typical of foliar diseases such as rust, which may have many cycles of infection and inoculum production (sporulation) over a single season. Soilborne diseases were assumed to be described by a monomolecular model, based on monomolecular chemical reactions of the first order, or also known as the negative exponential model. This model also has an asymptote, but the rate is more constant. This type of model has been used to describe a number of soilborne epidemics (Hao and Subbarao 2005; Stanghellini et al. 2004a). With this model, the initial inoculum plays an important role in the outcome of the epidemic, but has a minor role in a logistic model. Thus, there is a strong relationship between the inoculum in the soil and disease, described by the inoculum density-disease incidence (ID/DI) plot. The ID/DI curve can be described by an S-shaped logistic model (Baker 1978); Baker further theorized that the slope of a linearized transformation would approach 1 for a rhizosphere effect, and 0.67 for a rhizoplane effect, based on theories from physical chemistry and packing of particles. However, Gilligan and colleagues have developed a model based on probabilistic models and proposed the "pathozone" around a root (Gilligan and Bailey 1997). The pathozone is based on the probability of infection with the distance of the inoculum from the host. Essentially, this is the zone in which a propagule can germinate and successfully infect a root. Infection efficiency, or the efficiency of a propagule, declines in an exponential or sigmoidal manner as the distance from the root increases.

However, the simple interest or monomolecular models may be overly simplistic for most soilborne pathogens, although they may fit a disease such as Fusarium wilt where there is little transmission from plant to plant. These models assume there is a uniform environment over time, that the population of pathogen and host are uniformly virulent and susceptible, and that the spatial pattern of the pathogen is uniform or random. However, soil temperature may increase over the season, host roots may become more resistant over time, and pathogens are often aggregated or clustered. In most soilborne diseases, there is a primary infection step with new infection of healthy plants from a reservoir of inoculum in the soil. But once a root is infected, the pathogen can spread to adjacent roots in the same season. For example, zoosporic pathogens such as *Pythium* and *Phytophthora* can produce zoospores that can swim to and encyst on adjacent roots (Fig. 1). Mycelium of *Rhizoctonia* or *Gaeumannomyces* can grow into the soil from an infected root and can spread to adjacent healthy roots. Gilligan has produced a series of elegant models and experiments with take-all (caused by *Gaeumannomyces graminis* var. *tritici*) that demonstrated in the early part of the epidemic, the proportions of diseased roots increased monotonically to an initial plateau and then increased sigmoidally to an asymptotic level (Bailey et al. 2005; Bailey and Gilligan 1999, 2004). Their model has components for primary infection, secondary infection, and also accounts for root production and decay of inoculum. As roots grow and move through the soil, they encounter more inoculum and infected roots. The disease itself will alter the density of roots and in some cases increase root production (Bailey et al. 2006). Over time, inoculum will decay as it exhausts endogenous nutrients and propagules are parasitized and colonized by other microbes and antagonists or consumed by predators. Pathogens have a latent period (period of time from infection to production of inoculum) and an infectious period (time during which inoculum is produced), which must also be considered in models. For a more in-depth review of the temporal aspects of root disease epidemics we refer to Gilligan (1994).

Spatial aspects of soilborne pathogens and root diseases

Soilborne pathogens not only spread through time but also through space. However, the dynamics of spread and spatial patterns of soilborne pathogens are very different from foliar pathogens, which produce spores or propagules that can rapidly spread aerially by wind and rain over large distances or by water splash over small distances. Soilborne pathogens are confined within the soil, a three-dimensional matrix of mineral soil particles, pores, organic matter in various stages of decomposition, and a biological component. Thus, the spread of soilborne pathogens over time and space is more limited. Some soilborne pathogens in infected crop debris or soil can be spread by wind that blows during harvesting or cultivation. Some soilborne pathogens, such as *Sclerotinia sclerotiorum* or *Rhizoctonia solani*, produce aerial sexual spores that are ejected into the air and spread by wind. Pathogens can move above ground with irrigation water or rain run-off, which can carry soil particles into adjacent fields. The oomycetes, which produce motile swimming zoospores (Fig. 1), are especially adapted for movement in water. Both *Pythium* and *Phytophthora* have frequently been recovered from lakes, streams and irrigation ponds by using baits or molecular methods. Recent work with *Phytophthora ramorum*, an introduced pathogen and causal agent of sudden oak death, has documented movement in soil and streams in natural ecosystems (Davidson et al. 2005). However, most soilborne pathogens move and spread directly through the soil profile as mycelium. Soil texture and water (matric) potential are probably the two most important factors that determine spread, based on the size of the soil pores. For example, *Rhizoctonia* growth is restricted at high matric potentials and spreads faster in soils with high porosity with larger pores (Otten et al. 1999, 2004; Harris et al. 2003), an observation confirmed in the field in Australia (Gill et al. 2000).

How does the biology of soilborne pathogens affect the spatial patterns or distribution of plant diseases? In general, soilborne pathogens tend to be more aggregated or clustered, compared to foliar pathogens. An aggregated distribution has been demonstrated with pathogens such as *Phytophthora*, *Verticillium*, *Gaeumannomcyes*, and *Macrophomina* (Ristaino et al. 1993; Johnson et al. 2006; Gosme et al. 2007; Mihail and Alcorn 1987). These patterns are also more likely to be preserved from year to year. For example, the aggregated patterns of *Rhizoctonia oryzae* in wheat over a 30-acre field were evident from sampling the following year (Paulitz et al. 2003). The distance of spread each year is also likely to be less than that for aerial pathogens. However, long-distance movement can occur from soil attached to cultivation or harvesting equipment or by movement of nursery material across a country. A classic example was the movement of *P. ramorum* across the US in 2004 on infected potted camellias (Stokstad 2004).

How are the spatial patterns or distributions described or quantified? How does one determine whether the pathogen is clustered or randomly distributed? It is beyond the scope of this review to cover this in detail and for more information on this topic we refer to reviews by Campbell and Madden (1990), Campbell and Benson (1994) and Madden et al. (2007). Basically, disease or pathogens are sampled in regular quadrats, transects, or rows of plants, and the spatial location of each sample is recorded. The data can be mapped and the frequency distributions can be fitted to various models. For example, a Poisson distribution would indicate randomness, whereas fitting to a beta binomial distribution (Madden and Hughes 1994) would indicate a clustered pattern. Indices of dispersion can be calculated, the simplest being the ratio of the variance to the mean, which should be greater than 1 for an aggregated pattern, and equal to 1 for a random distribution. Other widely used indices, including Lloyd's Index of Patchiness and Morista, are based on the mean and variance. With intensively-mapped or quadrat data, the spatial information can be utilized to pick out patterns, such as with distance-based or nearest neighbor analyses (Madden et al. 2007). If the distribution is random, the distance between neighboring samples should not have any effect. A similar reasoning is used for spatial autocorrelation techniques and geostatistics, which have been widely used in earth sciences. The assumption is that plants or samples that are close together will be more similar than samples that are further apart. Another technique that is becoming more widely used is SADIE (Spatial analysis by distance indices; Perry 1998). This method is based on how much samples must be moved in a grid to attain a regular pattern, based on random rearrangements, called the distance to regularity. With a more clustered distribution, more rearrangements are required. One overriding factor in spatial analysis is the scale of measurement. Most studies in agriculture are done at the scale of a few meters in replicated agronomic plots. Very few studies have looked at a microscale (millimeter or micrometer) of a rhizosphere or at a mesoscale (kilometer) of a county, district, province, or country. The spatial patterns may vary with the grain or resolution of the measurement. For example, at a microscale level (square millimeter), propagules of *Macrophomina phaseolini* in the maize rhizosphere exhibited a random distribution (Olanya and Campbell 1989), but at a larger scale in the field (square meter) the pattern was aggregated (Mihail and Alcorn 1987). On the other hand, Paulitz and Rossi (2004) found that *R. solani* and *R. oryzae* showed a similar pattern of aggregation at a 30, 3, or 0.3 m scale. Large patches of *Rhizoctonia* in the field were composed of smaller patches, which themselves are composed of smaller patches.

In conclusion, although we know more about how soilborne pathogens are distributed on scales applicable to agriculture, we know little about how pathogens are arranged or interact with beneficial microorganisms at the microscale of the rhizosphere. Advances in molecular techniques, fluorescence labeling, and imaging such as confocal laser microscopy, may be useful in the future. These techniques have been used to study colonization of roots by biocontrol bacteria and fungi (Gamalero et al. 2005; Bloemberg et al. 2000; Lu et al. 2004) and infection of roots by *Fusarium* (Lagopodi et al. 2002; Bolwerk et al. 2005; Olivain et al. 2006), but these have been descriptive and have not employed spatial statistics.

Interactions between beneficial microorganisms and soilborne pathogens

The rhizosphere is the playground and infection court where soilborne pathogens establish a parasitic relationship with the plant. However, the rhizosphere is also a battlefield where the complex rhizosphere community, both microflora and microfauna, interact with soilborne pathogens and influence the outcome of pathogen infection. The growth or activity of soilborne pathogenic fungi, oomycetes, bacteria, and/ or nematodes can be inhibited by several beneficial rhizosphere microorganisms. The activity and effects of beneficial rhizosphere microorganisms on plant growth and health are well documented for bacteria belonging to the Proteobacteria (noticeably *Pseudomonas* and *Burkholderia*) and Firmicutes (*Bacillus* and related genera), and for fungi from the Deuteromycetes (e.g. *Trichoderma*, *Gliocladium* and non-pathogenic *F. oxysporum*). In the remainder of this section, these beneficial microorganisms will be referred to as biocontrol microorganisms or biocontrol agents.

Biocontrol microorganisms may adversely affect the population density, dynamics (temporal and spatial) and metabolic activities of soilborne pathogens via mainly three types of interactions, which are competition, antagonism and hyperparasitism. In the rhizosphere, competition takes place for space at the root surface (Fig. 2) and for nutrients, noticeably those released as seed or root exudates. Competitive colonisation of the rhizosphere and successful establishment in the root zone is a prerequisite for effective biocontrol, regardless of the mechanism(s) involved (Weller 1988; Raaijmakers et al. 1995). In the case of biocontrol bacteria, this is explained in part by the fact that production of several antagonistic traits and compounds is subjected to cell-density dependent regulation or quorum sensing (Pierson et al. 1998; Pierson and Pierson 2007). In addition, competition can in itself be a biocontrol mechanism, often for organic compounds necessary for reactivation of propagules and/or subsequent proliferation and root colonisation by the pathogen (Paulitz et al. 1992; Van Dijk and Nelson 2000; Fravel et al. 2003). Competition can also take place for micronutrients, especially iron, that are essential for growth and activity of the pathogen. Competition for soluble ferric iron is based on production and/or utilisation of high-affinity chelators termed siderophores (Lemanceau et al. 1992; Neilands 1995). Once complexed with iron,

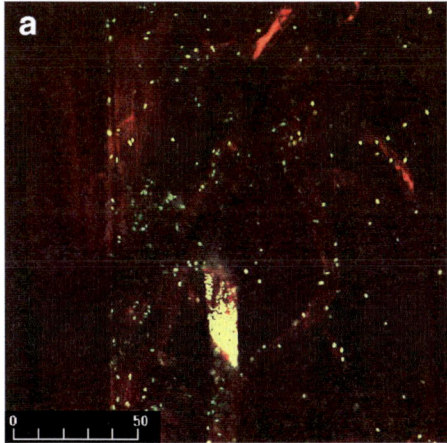

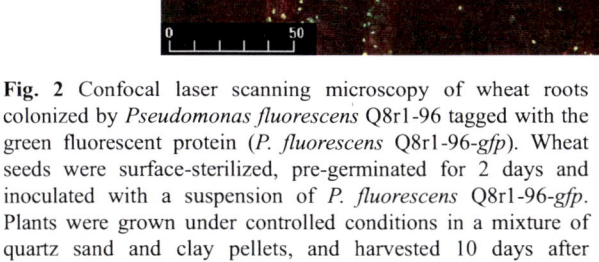

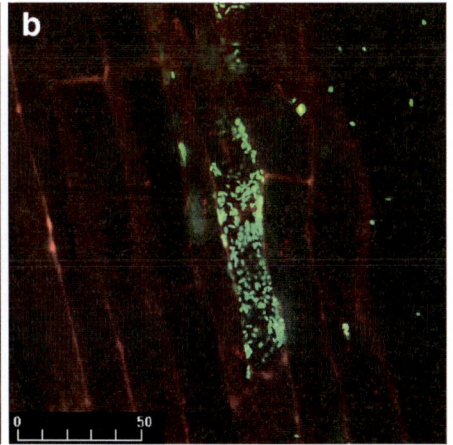

Fig. 2 Confocal laser scanning microscopy of wheat roots colonized by *Pseudomonas fluorescens* Q8r1-96 tagged with the green fluorescent protein (*P. fluorescens* Q8r1-96-*gfp*). Wheat seeds were surface-sterilized, pre-germinated for 2 days and inoculated with a suspension of *P. fluorescens* Q8r1-96-*gfp*. Plants were grown under controlled conditions in a mixture of quartz sand and clay pellets, and harvested 10 days after inoculation. Root samples were stained for 20 min with propidium-iodide. Microcolonies of *P. fluorescens* Q8r1-96-*gfp* on mature root hairs (**a**) and along the junctions of epidermal cells (**b**). Courtesy of Olga Mavrodi and Dmitri Mavrodi, Department of Plant Pathology, Washington State University, USA

siderophores are taken up via specific membrane receptors. Competition for iron as well as competition for carbon are documented as important modes of action for several biocontrol bacteria and fungi (Lemanceau et al. 1992; Alabouvette et al. 2006), with iron competition being particularly significant in calcareous soils where high pH leads to low iron solubility.

Antagonism is usually mediated by the production of secondary antimicrobial metabolites (antibiosis), lytic enzymes and/or effectors. Often, antagonistic microorganisms can produce a range of different antimicrobial secondary metabolites, e.g. 2,4-diacetylphloroglucinol (DAPG), pyrrolnitrin, pyoluteorin, phenazines, cyclic lipopeptides and hydrogen cyanide in the case of certain fluorescent pseudomonads (Raaijmakers et al. 2002, 2006; Picard and Bosco 2008; Cook et al. 1995; Weller 2007). Antimicrobial secondary metabolites are also involved in antagonistic effects of fungi such as *Trichoderma* and *Gliocladium* (Kubicek et al. 2001). The concentration at which these compounds are toxic towards pathogenic bacteria, fungi and nematodes depends on the compound and the target. In fungal pathogens, they may affect the electron transport chain (phenazines, pyrrolnitrin), metalloenzymes such as copper-containing cytochrome *c* oxidases (hydrogen cyanide), membrane integrity (biosurfactants), or cell membrane and zoospores (DAPG, biosurfactants; Haas and Défago 2005; Raaijmakers et al. 2006), but the modes of action of many known antimicrobial metabolites are still poorly understood. The role of several antimicrobial secondary metabolites in plant protection has been demonstrated by comparing wild-type strains and mutant derivatives. Results of these studies have indicated that multiple antimicrobial metabolites can play an important role in the same pathosystem (Haas and Défago 2005). For instance, both the abilities of *Pseudomonas* sp. CHA0 to produce hydrogen cyanide (Voisard et al. 1989) and DAPG (Keel et al. 1992) contribute to suppression of *Thielaviopsis basicola*-mediated black root rot of tobacco. However, population-level comparisons of biocontrol strains indicated that some of these compounds play a more significant role than others in plant protection (Sharifi-Tehrani et al. 1998; Ellis et al. 2000; Rezzonico et al. 2007). Antimicrobial secondary metabolites have received extensive research attention, in part because they are thought to contribute largely to soil disease suppressiveness. This is particularly the case of DAPG-producing pseudomonads, which are involved in soil suppressiveness in different pathosystems (Weller et al. 2002), noticeably black root rot of tobacco (Stutz et al. 1986) and take-all of wheat (Raaijmakers and Weller 1998; Raaijmakers et al. 1997, 1999).

Production of extracellular lytic enzymes is quite common among antagonistic microorganisms (Adesina et al. 2007), but it does not contribute to antagonism in all cases (Sharifi-Tehrani et al. 1998; Dunne et al. 1997). Extracellular lytic enzymes act in different ways: many of them can affect the cell wall of pathogens, and this is documented for cellulases, chitinases and proteases produced by various bacteria. Inactivation of genes involved in their biosynthesis has been used to provide evidence for their contribution in biocontrol *in planta* (Dunne et al. 1997; Kobayashi et al. 2002). Other lytic enzymes from Proteobacteria target virulence factors, such as the phytotoxin fusaric acid produced by *F. oxysporum*, thereby enabling protection of tomato plants from wilt disease (Toyoda et al. 1988).

Antagonism can also implicate effectors (not yet identified) secreted by the type III secretion system of biocontrol bacteria, leading to reduced virulence in certain pathogens (Rezzonico et al. 2005). Type III protein secretion systems, first discovered in pathogenic bacteria (Stuber et al. 2003), enable direct introduction of effectors into eukaryotic host cells. Type III secretion genes are also present in many saprophytic pseudomonads, including biocontrol strains (Preston et al. 2001; Mazurier et al. 2004; Rezzonico et al. 2004). Inactivation of the type III secretion gene *hrcV* impaired the ability of *P. fluorescens* KD to diminish polygalacturonase activity of *P. ultimum* in vitro, and reduced its biocontrol efficacy against this pathogen on cucumber (Rezzonico et al. 2005). This gene is upregulated in presence of the pathogen rather than the plant, which further suggests that the Oomycete is the target of the type III secretion system in this pseudomonad.

In addition to competition and antagonism, direct biocontrol effects on soilborne plant pathogens can result from hyperparasitism. This is mainly documented for *Trichoderma* and *Gliocladium*, and it affects various fungal pathogens, such as *Rhizoctonia*, *Sclerotinia*, *Verticillium* and *Gaeumannomyces* (Harman et al. 2004). Hyperparasitism by *Trichoderma* involves secretion of chitinases and cellulases, which release small molecules from the target pathogen and trigger chemotropism towards the latter (Zeilinger et al. 1999). Contact is followed with coiling of hyphae around the hyphae of the pathogen, further enzymatic digestion of its cell wall, and penetration by *Trichoderma* (Djonović et al. 2006; Woo et al. 2006). Cell wall damage caused by endochitinases was also shown to play an important role in the activity of *Gliocladium virens* against *Botrytis cinerea* (Di Pietro et al. 1993). Hyperparasitim enables also the Firmicute *Pasteuria penetrans* to control the plant parasitic nematode *Meloidogyne* (Duponnois et al. 1999), but the mechanisms involved are still poorly understood.

Direct positive effects of rhizosphere microorganisms on the plant

Next to the biocontrol activity of rhizosphere microorganisms, several can have a direct positive effect on plant growth and health. Often, it is one of several modes of actions by which these microorganisms can benefit plant health. First, phytostimulatory and biofertilising microbes can promote plant health by making the plant 'stronger'. Second, many rhizosphere microorganisms can induce a systemic response in the plant, resulting in the activation of plant defence mechanisms (Pieterse et al. 2003). This capacity has been identified in a wide range of bacteria (Van Loon et al. 1998; Haas and Défago 2005), including endophytes (Compant et al. 2005) as well as saprophytic (Fuchs et al. 1997), hyperparasitic (Woo et al. 2006) and arbuscular mycorrhizal fungi (Pozo et al. 2002). Induced systemic resistance (ISR) does not confer complete protection, but it does protect the plant from various types of phytopathogens (including root pathogens), without requiring direct interaction between the resistance-inducing microorganisms and the pathogen (Van Loon et al. 1998; Zehnder et al. 2001). In addition, ISR can be effective under field conditions and in commercial greenhouses (Zehnder et al. 2001; Pieterse et al. 2003).

ISR exhibits similarities but also several differences with systemic acquired resistance (SAR), which is the plant response triggered upon exposure to pathogens. ISR involves jasmonate and ethylene signals, as evidenced in experiments with specific *Arabidopsis* mutants (Pieterse et al. 2003). Unlike SAR, ISR is typically salicylate-independent, although certain plant-beneficial microorganisms can activate a salicylate-dependent pathway in the plant (De Meyer et al. 1999). Several cell surface constituents of biocontrol bacteria, i.e. lipopolysaccharides and flagella, can trigger ISR (Pieterse et al. 2003; Haas and Défago 2005). ISR can also take place following exposure of the plant to compounds produced by plant-beneficial bacteria, e.g. the volatile 2,3-butanediol (Ryu et al. 2004), the siderophore pyoverdine (Maurhofer et al. 1994), DAPG (Iavicoli et al. 2003), and cyclic lipopeptide surfactants (Ongena et al. 2007; Tran et al. 2007).

Adaptation and defense of plant pathogens to microbial antagonism

To date, microbial interactions in the rhizosphere are mostly viewed from the perspective of how beneficial microorganisms inhibit the growth or activity of pathogenic microorganisms. However, also pathogens have a diverse array of mechanisms to counteract antagonism, including active efflux and degradation of antimicrobial compounds, and interference with the regulation and biosynthesis of enzymes and antimicrobial metabolites produced by antagonistic microorganisms (reviewed in Duffy et al. 2003). Resistance development in pathogen populations to chemical control agents is a common and well-studied phenomenon. In contrast, resistance in pathogens to antimicrobial compounds produced by antagonistic microorganisms is presumed not to develop or at least relatively slowly, because antagonistic microorganisms operate in microsites in the rhizosphere where only a small fraction of the pathogen population is exposed to the antimicrobial compounds during a short period of its life cycle (Handelsman and Stabb 1996). Furthermore, in contrast to the inundative application of chemical pesticides, only minute amounts of the antimicrobial compounds are produced by the antagonistic microorganisms in the rhizosphere (Séveno et al. 2002; Duffy et al. 2003). Nevertheless, a number of studies have shown that substantial variation in sensitivity against antimicrobial compounds exists within pathogen populations and that pathogens harbour a wide range of defense mechanisms against microbial antagonism.

Variation in sensitivity of pathogen populations to antimicrobial compounds

Studies on the effects of antimicrobial compounds produced by antagonistic microorganisms on plant pathogens often consider only one single strain and one specific stage in the life cycle of the pathogen. Most pathogen populations and life cycles, however, are diverse and comprise numerous structures that allow pathogens to respond adequately to selection pressure exerted by competing microorganisms. Several studies have addressed the variation in sensitivity of pathogenic fungi, oomycetes and bacteria to several antimicrobial compounds, including agrocin 84 (Cooksey and Moore 1982; Stockwell et al. 1996), the volatile hydrogen cyanide (Mackie and Wheatley 1999), phenazines (Gurusiddaiah et al. 1986; Mazzola et al. 1995), the phenolic antibiotic DAPG (Mazzola et al. 1995; De Souza et al. 2003; Schouten and Raaijmakers 2004), gliotoxin (Jones and Hancock 1988), kanosamine (Milner et al. 1996), and the cyclic lipopeptide massetolide A (Mazzola et al. 2007).

One of the most detailed studies on the variation in sensitivity to antimicrobial compounds produced by antagonistic microorganisms was performed by Mazzola et al. (1995), who screened a total of sixty-six individual isolates of the take-all fungus *G. graminis* var. *tritici* (*Ggt*) for sensitivity to DAPG and phenazine-1-carboxylic acid (PCA). Substantial variation in sensitivity to both antimicrobials was observed among a range of Ggt isolates obtained from a single wheat field. In interactions with antagonistic *Pseudomonas* strains producing either DAPG or PCA, the antibiotic-insensitive *Ggt* isolates could not be controlled effectively in the rhizosphere of wheat plants. At least one of the PCA-insensitive *Ggt* isolates was also insensitive to DAPG, suggesting similar mechanisms of resistance to both antimicrobial compounds. Studies by Schouten and Raaijmakers (2004) showed that also among pathogenic and non-pathogenic *F. oxysporum*, substantial variation in sensitivity to DAPG exists. There was no clear relationship between DAPG insensitivity and geographical origin or formae speciales of *F. oxysporum*, suggesting that the traits responsible for DAPG insensitivity are relatively ancient, have developed independently, or are easily transferred within and between populations.

Mechanisms of resistance in plant pathogens against antimicrobial compounds

Pathogenic fungi, oomycetes and bacteria have developed a range of strategies to tolerate or resist the deleterious effects of antimicrobial compounds produced by antagonistic microorganisms (Duffy et al. 2003). One of the best studied examples in plant pathogenic bacteria is resistance of the crown gall pathogen *A. tumefaciens* to agrocin 84 produced by *Agrobacterium rhizogenes* strain K84. Agrocin 84 is believed to inhibit DNA replication and is transported into cells of *A. tumefaciens* via agrocinopine permease, a periplasmic protein encoded by genes carried on certain types of the Ti plasmid present in sensitive strains of *A. tumefaciens* (Stockwell et al. 1996). In strain K84, agrocin 84 biosynthesis and resistance genes are located on the conjugative plasmid pAgK84 (Ryder et al. 1987). Among 65 strains and isolates of *A. tumefaciens*, all of the biotype 3 strains tested were resistant to K84, whereas many of the biotype 1 and 2 strains were susceptible (Van Zyl et al. 1986). Cooksey and Moore (1982) earlier showed that in three *A. tumefaciens* strains and one *A. rhizogenes* strain, mutation rates for resistance to agrocin 84 were relatively high ranging from 2.5×10^{-3} to 4.2×10^{-4}. Conjugal transfer of plasmid pAgK84 was demonstrated in vitro and in crown gall tissue of infected plants and is regarded as one of the main mechanisms of agrocin 84 resistance in pathogenic *A. tumefaciens* strains. The fact that agrocin 84 resistant *A. tumefaciens* strains were isolated from different soils worldwide (Van Zyl et al. 1986) suggested that conjugal transfer may also have occurred in natural environments. Stockwell et al. (1996) carried out a field experiment with cherry seedlings treated with K84 and *A. tumefaciens* and showed that transconjugants were detected in four out of 13 galls and estimated that the frequency of pAgK84 transfer was approximately 10^{-4} transconjugants per recipient. A transconjugant strain retained the plasmid for up to seven months in the rhizosphere of plants grown in the field, colonized the rhizosphere of cherry plants to the same extent as its parental strain and caused crown gall disease.

Studies on resistance mechanisms of soilborne pathogenic fungi to antimicrobial compounds produced by antagonistic microorganisms is limited and fragmentary (Duffy et al. 2003). Although most of the studies in this research area have been performed with pathogenic fungi infecting aerial plant parts, several of the resistance mechanisms described below most likely also operate in soilborne fungi. Work by Levy et al. (1992) suggested that resistance in *Mycosphaerella graminicola* against phenazines produced by *Pseudomonas aeruginosa* is based, in part, on degradation of these antimicrobials and on the presence of superoxide dismutase and catalase, enzymes involved in the detoxification of oxygen radicals resulting from the oxidative stress generated by the phenazines. Also for the nematode *Caenorhabditis elegans*, Mahajan-Miklos et al. (1999) reported that a mutant with increased levels of catalase and superoxide dismutase was more resistant to fast killing by the phenazine pyocyanin. Degradation of antimicrobial compounds produced by antagonistic microorganisms was shown to be an important mechanism of DAPG tolerance in several *F. oxysporum* isolates (Schouten and Raaijmakers 2004). DAPG tolerance was correlated with the ability of the *F. oxysporum* isolates to convert this antimicrobial metabolite, via deacetylation, into the less toxic derivatives monoacetylphloroglucinol and phloroglucinol (Schouten and Raaijmakers 2004).

Among the non-degradative resistance mechanisms, membrane-bound efflux transporters not only enables target pathogens to resist exogenous toxic compounds, but also play an important role in preventing self-intoxication in antimicrobial metabolite-producing microorganisms (De Waard 1997; Stergiopoulos et al. 2002). Schoonbeek et al. (2002) demonstrated that in *B. cinerea* the efflux pump BcAtrB (*Botrytis cinerea* ABC transporter B) plays an important role in defense against phenazines produced by antagonistic *Pseudomonas* strains: several phenazines induced expression of *BcatrB* in a dose-dependent manner and *BcatrB* replacement mutants were more sensitive to phenazines and phenazine-producing *Pseudomonas* strains than their parental strain. BcATRB also confers increased tolerance of *B. cinerea* to the phytoalexin resveratrol and the phenylpyrrole fungicides fenpiclonil and fludioxinil (Schoonbeek et al. 2001). Recent studies by Schouten et al. (2008) showed that DAPG also induces expression of *BcatrB* and that *BcatrB* replacement mutants are more sensitive to DAPG. Collectively, these results indicate that plant pathogens harbor efflux transporters that confer resistance to multiple and structurally different antimicrobials produced by antagonistic microorganisms.

Interference with the biosynthesis of antimicrobial compounds in rhizosphere microorganisms

The first example of interference of a soil-borne pathogenic fungus with the biosynthesis of an antimicrobial compound in a beneficial bacterium was described by Duffy and Défago (1997). In their study, DAPG biosynthesis in fluorescent *Pseudomonas* strain CHA0 was repressed by *F. oxysporum* f.sp. *radicis-lycopersici*. Fusaric acid produced by *Fusarium* was shown to be the fungal metabolite that specifically repressed DAPG biosynthesis (Duffy and Défago 1997; Notz et al. 2002): blocking fusaric acid production in *Fusarium* by addition of zinc relieved repression of the *phlA* gene and improved the activity of strain CHA0. Subsequent studies showed that among a collection of genotypically different DAPG-producing *Pseudomonas* strains, several were relatively insensitive to fusaric acid-mediated repression of DAPG biosynthesis (Duffy et al. 2004).

Another example of pathogen-antagonist signalling was described for the interaction between mycotoxigenic *Fusarium* and mycoparasitic *Trichoderma* (Lutz et al. 2003). Their study showed that the mycotoxin deoxynivalenol produced by *Fusarium culmorum* and *Fusarium graminearum* acts as a negative signal repressing the expression of the *nag1* chitinase gene in *Trichoderma atroviridae*. Repression appeared to be specific for *nag1* since no adverse effect was observed on the expression of *ech42*, another important chitinase gene in *T. atroviridae* (Lutz et al. 2003).

For many antagonistic bacteria living in the rhizosphere, expression of a range of genes is regulated by autoinducers, such as *N*-acylhomoserine lactones (AHLs), which act as intercellular signals (reviewed in Somers et al. 2004; Zhang and Dong 2004). This phenomenon of cell to cell communication, also referred to as quorum sensing, drives the expression of several beneficial traits in rhizosphere bacteria and of a range of virulence traits in human and plant pathogenic bacteria (Pierson et al. 1998; Zhang and Dong 2004). The ability of antagonists to interfere with quorum sensing in plant pathogens provides a means to control plant diseases and to promote plant health. Conversely, soilborne plant pathogens can utilize similar strategies to interfere with quorum-regulated antibiotic biosynthesis as a defense strategy against microbial antagonism. Several strategies of quorum sensing inhibition, also refered to as quorum quenching, have been unraveled in the past decade and include repression or blockage of the production of signal molecules, inactivation of the signal molecules or interference with signal perception (reviewed in Zhang and Dong 2004; Rasmussen and Givskov 2006). To date, two types of enzymes that inactivate AHLs have been identified in a range of bacterial species and genera; these include the AHL-lactonases that hydrolyse the lactone ring to yield acyl homoserines with reduced biological activity, and the AHL-acylases that break the amide linkage of AHLs resulting in homoserine lactone and fatty acids, which do not exhibit biological activity (reviewed in Zhang and Dong 2004; Uroz et al. 2007). Work by Molina et al. (2003) elegantly demonstrated that lactonolysis of AHLs by the soil bacterium *Bacillus* sp. A24 or by the rhizosphere isolate *P. fluorescens* P3 modified with the lactonase gene *aiiA*, significantly reduced potato soft rot caused by *Pectobacterium carotovorum* and crown gall of tomato caused by *A. tumefaciens*. Other studies, including those by Uroz et al. (2003) and Jafra et al. (2006), have shown that various other bacterial species are able to degrade AHLs.

Finally, plant pathogens can also fight back without targeting pathways involved in the biosynthesis of specific antimicrobial compounds in biocontrol microorganisms. However, these mechanisms are comparatively much less documented. In the case of *P. fluorescens* strain F113, genes necessary for competitive colonisation of sugar beet roots are downregulated by signal(s) released by the oomycete pathogen *Pythium ultimum*, and thus strain F113 does not reach population densities in the rhizosphere high enough for effective biocontrol of *P. ultimum* (Fedi et al. 1997). Some of the genes targeted in *P. fluorescens* F113 include rRNA genes (Smith et al. 1999), which play a key role in cell physiology during growth.

Influence of agricultural practices on pathogen and antagonist populations

Management of the biotic and abiotic properties of a soil is an important approach to promote the activities of beneficial microorganisms in the rhizosphere and thus limiting the densities and activities of soilborne pathogens to a tolerable level (Janvier et al. 2007). Adaptation of cultural practices has been proposed as a means to decrease the soil inoculum potential or increase the level of suppressiveness to diseases (Steinberg et al. 2007). Indeed, disease suppressiveness has been obtained through crop rotation (Cook et al. 2002), intercropping (Schneider et al. 2003), residue destruction (Baird et al. 2003), organic amendments (Tilston et al. 2002), tillage management practices (Sturz et al. 1997; Pankhurst et al. 2002) and a combination of those regimes (Hagn et al. 2003; Garbeva et al. 2004). Forty years ago, the use of heat-treatment of soils by steaming was a common practice in intensive vegetable cultivation in greenhouses. Most of the pathogens are highly susceptible to heat, the lethal temperatures for pathogenic fungi being reached at 55–65°C for 15 to 30 min (Bollen 1969). With the oil crisis, soil steaming became too expensive and the growers moved to application of chemical biocides which are hazardous for man and environment. These biocides kill not only the pathogens but also most of the beneficial microorganisms, leading to an unbalanced equilibrium in soil and rhizosphere environments. Many of these biocides, except methyl bromide, are still in use and produce ephemeral results including uncontrolled side effects on both existing and forthcoming microbial communities, leading to the infernal circle of applying repeatedly the same treatments. Fortunately, less drastic techniques of pathogen eradication have been proposed that do not kill every soil microorganism, but instead modify the microbial balance in a positive direction for pathogen control and stimulation of plant growth and health (Mazzola 2004).

Solarisation

Solarisation or solar heating is a method that uses the solar energy to enhance the soil temperature to levels at which many plant pathogens will be killed or sufficiently weakened to obtain significant control of the diseases. Solarisation does not destroy all soil microorganisms, but modifies the microbial balance in favour of the beneficial microorganisms. Many studies report that the efficacy of soil solarisation is not only due to a decrease of pathogen populations, but also to an increase of the density and activity of populations of antagonistic microorganisms such as *Bacillus* spp., *Pseudomonas* spp. and *Talaromyces flavus*. Several review papers are available that describe both the technology of solar heating and mechanisms involved in the control of pests, pathogens and weeds by solarisation (DeVay 1995; Katan 1996).

Solarisation is a hydrothermal process; its effectiveness is not only related to the temperature but also to the soil moisture. Temperature maxima are obtained when the soil water content is about 70% of the field capacity in the upper layers and the soil should be moist to a depth of 60 cm. The duration of solarisation is an important factor determining the effectiveness of the treatment. The longer the mulching period, the greater the depth of effective activity and the higher the pathogen killing rates are. In Mediterranean areas, four weeks are usually required to achieve disease control. An important characteristic of soil solarisation is its broad spectrum of activity, including activity against fungi, nematodes, bacteria, weeds, arthropod pests and some unidentified agents. It should be noted, however, that not all of the pathogens present the same susceptibility to solar heating and that failures have been reported. Solarisation often results in increased yield when applied to monoculture soils where specific pathogens have not been identified. In this case, solarisation probably controls the weak pathogens or deleterious microorganisms responsible for "soil sickness". Another interesting property of solarisation is its long-term effect. Disease control and yield increase have been reported two and sometimes three years after solarisation (Gallo et al. 2007). This long term effect is probably due to both the reduction of the inoculum density and some induced level of disease suppressiveness of the soil. The efficiency of the process can be improved by combining soil solarization and organic amendments, leading to an accumulation of ammonium/ammonia in the soil which reduces the inoculum densities and may weaken the remaining inoculum, including nematodes (Ndiaye et al. 2007; Oka et al. 2007). Obviously, solarisation is effective in warm and sunny areas in the world and, in Europe, adopted in the Mediterranean area (Katan 1996).

Biofumigation or biodisinfection

A strategy better adapted to the cooler regions of the world is biological soil disinfection, which is based on plastic mulching of the soil after incorporation of fresh organic matter (Blok et al. 2000). The mechanisms involved are not fully understood yet, but two main mechanisms have been proposed to contribute to the efficacy of biological soil disinfection: the fermentation of organic matter under plastic results in (a) anaerobic conditions in soil and (b) in the production of toxic metabolites, and both processes contribute to the inactivation or destruction of pathogenic fungi. Based on the type of mechanisms involved, two definitions have been proposed by Lamers et al. (2004): biofumigation corresponds to the use of specific plant species containing identified toxic molecules, whereas biodisinfection refers to the use of high quantities of organic matter which, after soil tarping, result in anaerobic conditions mainly responsible for the destruction of pathogens.

Many species of *Brassicaceae* (*Cruciferae*) produce glucosinolates, a class of organic molecules that may represent a source of allelopathic control of various soilborne plant pathogens (Kirkegaard and Sarwar 1998). Toxicity is not attributed to glucosinolates but to products such as isothiocyanates, organic cyanides or ionic thiocyanates resulting from their enzymatic degradations achieved by a group of enzymes called myrosinases. Myrosinase and glucosinolates are separated from each other in intact plant tissues. When the *Brassicaceae* (cabbage, mustard, horseradish) are grown as an intermediate crop and subsequently buried into soil as green manure, the disruption of cellular tissues allows mixing of glucosinolates and myrosinases resulting in the rapid release of glucosinolate degradation products. The hydrolysis products have a broad biocidal activity towards nematodes, insects and fungi as well as putative phytotoxic effects. They act either as selective fungicides or as fungistatic compounds thereby limiting the development and activity of fungal populations, some of them being pathogenic on the forthcoming crop (Sarwar et al. 1998). Also other plant families, including the *Alliacae*, release toxic compounds. Degradation of garlic, onion, and leek tissues releases sulfurous volatiles such as thiosulfinates and zwiebelanes which are converted into disulfides that have biocidal activities against fungi, nematodes and arthropods (Arnault et al. 2004).

However, not all pathogens are equally susceptible to volatile compounds. For example, soil amendment with *Brassica napus* seed meal controlled root infection by *Rhizoctonia* spp. and the nematode *Pratylenchus penetrans*, but did not consistently suppress soil populations of *Pythium* spp. and control apple root infection (Mazzola et al. 2001). Mazzola (2007) suggested that the role of the isothiocyanates could be mediated by select groups of indigenous populations of microorganisms whose presence and sufficient population density are necessary to achieve disease control. A plant systemic protection was proposed to explain, at least in part, the positive relation observed between the increase in population densities and activity of *Streptomyces* spp. and disease control obtained in a soil amended with *B. napus* seed meal (Cohen and Mazzola 2006).

Crop rotation versus mono-cropping

In general, continuous cropping with a susceptible host causes the build-up of populations of specific plant pathogens resulting in increases in disease incidence and/or severity. In contrast, rotation with non-host plants or plants that are less susceptible to the pathogen will limit the build-up of pathogen populations, and in some cases may even lead to a decrease of the pathogen inoculum density. Some non-host plants are able to trigger the germination of pathogen survival structures (sclerotia, chlamydospores, oospores) and in the absence of a susceptible host, some pathogens are not able to survive saprophytically in soil. Therefore, cropping of a non-host plant will result in a decrease of the inoculum potential of the soil. Moreover, crops in a rotation scheme may also stimulate antagonistic microbial populations that adversely affect the growth or activity of the pathogens. For example, Mazzola (1999) showed that growing wheat in orchard soil prior to planting apple seedlings significantly reduced infection by a complex of pathogens including *Cylindrocarpon destructans*, *Phytophthora* and *Pythium* spp., and *R. solani*. This beneficial effect correlated with an increased population of specific antagonistic populations of fluorescent pseudomonads making the soil more suppressive towards *R. solani*.

The case of take-all decline of wheat, however, illustrates that longterm monocropping may also be beneficial to plant health. In this case, monoculture of wheat or barley results first in an increase of take-all disease which in turn stimulates antagonists of the take-all pathogen. Therefore, take-all disease of wheat can be naturally controlled by monocropping wheat or barley provided that monoculture lasts for more than 4 years (Dulout et al. 1997; Weller et al. 2002). Take-all suppressiveness was related to the development of populations of DAPG-producing fluorescent pseudomonads in the rhizosphere which adversely affect the growth and activity of the take-all pathogen (Raaijmakers and Weller 1998; Weller et al. 2002). It should be noted that the best yields following take-all decline are rarely equal to those achieved with crop rotation. Nevertheless, in several countries wheat monoculture is a common practise and preferred by growers (Cook 2003).

Residue management

Plant residues left on or near the soil surface may contribute to an increase of disease suppressiveness through the promotion of the general microbial activity. In some cases, however, the debris not only promotes the microbial activity but also helps to preserve the pathogens, preventing a decrease of the inoculum density. This is the case for *Macrophomina phaseolina* causing charcoal rot in soybean (Baird et al. 2003), *Fusarium* sp. causing root and crown rot on maize (Cotten and Munkvold 1998), and *R. solani* causing crown and root rot on sugar beet (Guillemaut 2003). Some practices used by growers to kill living plants at crop termination (e.g. foliar application of herbicide and mechanical destruction of the vines) could be counterproductive with respect to disease management. Indeed, such strategies might enhance the fungal reproduction and increase the soil inoculum as it was shown in the case of the root-infecting fungus *Monosporascus cannonballus* causing vine decline of melons. In such cases, destruction of infected roots prior to pathogen reproduction would be a method of preventing inoculum build-up in soil (Stanghellini et al. 2004b). Therefore, attention should be paid to residue management by burial through tillage practices or promotion of rapid decomposition (Toresani et al. 1998). When residues are buried, the pathogens are displaced from their niche to deeper layers in the soil and their ability to survive is severely decreased. Repeated incorporations of crop residues can affect a change in the activity of residue-borne microorganisms that in turn influence the decomposition of crop residues. Carbon released from this decomposition contributes to an increase of soil microbial activity and thereby enhances the level of general suppression. Developing disease suppressive soils by introducing organic amendments and crop residue management takes time, but the benefits accumulate across successive years leading to an improvement of soil health and structure (Bailey and Lazarovits 2003).

Soil tillage

It is difficult to assess the role of tillage on disease suppression as its evaluation is often combined with the effects of other agricultural practices such as organic amendments and green manure burial, residue management or crop rotations (Bailey and Lazarovits 2003). Therefore tillage appears as giving conflicting effects on disease suppression. Conventional tillage results in considerable disturbance of the soil but removes residue from the surface. Tillage also disrupts hyphae thereby affecting the ability of fungi such as *R. solani* to survive (Roget et al. 1996; Bailey and Lazarovits 2003). Reduced tillage can also favour pathogens by protecting the pathogen's refuge in the residue from microbial degradation, lowering soil temperature, increasing soil moisture, and leaving soil undisturbed (Bockus and Shroyer 1998). The impact of tillage practices depends on specific pathogen–soil–crop–environment interactions, with the environment being sometimes the most important factor limiting disease severity regardless of tillage or crop rotation practices (Bailey et al. 2000).

Organic amendments

Some years ago organic amendments were proposed to control soilborne diseases (Lumsden et al. 1983). Although their effects were not studied in relation to the induction of suppressiveness in soil, many papers reported that organic amendment can reduce disease incidence or severity. Hoitink (1980) developed a growth medium based on composted bark to grow rhododendron and azaleas. This substrate is suppressive towards root rots induced by several species of *Pythium* and *Phytophthora*. After heating, the compost can be colonized by a great diversity of microorganisms some being antagonistic to the pathogens. The level of disease control obtained depends on many factors such as the chemical properties of the parent material, the composting process and obviously the type of microorganisms present. This is probably why such contrasted data have been published regarding the efficacy of disease control obtained by organic amendments of soil (Termorshuizen et al. 2006). To enhance the suppressive potential of composts and thus to improve the efficacy of disease control, it has been proposed to inoculate these composts after peak heating with specific strains of antagonistic microorganisms. Although promising, this strategy has not yet been successfully applied. Composts can also mimic a non-host plant: an interesting example is provided by the incorporation of onion wastes into the soil to control Allium white rot due to *Sclerotium cepivorum*. This fungus is an obligatory parasite which can survive as dormant sclerotia in the soil for many years but can only germinate in the presence of the host plants. The stimulus for germination is the exudation of alk(en)yl cysteine sulphoxides by the roots of *Allium* species. Properly composted, onion wastes contained some sulphoxides (di-*n*-propyl disulphide) which trigger the dormant sclerotia to germinate in absence of the root. These germinated sclerotia are unable to survive without the living host, which contributes to the decrease in the primary inoculum faced by the next onion crop (Coventry et al. 2002).

To date, compost amendment has been successfully used to increase soil suppressiveness to diseases in agricultural crops, including nematode diseases (Erhart et al. 1999; Lumsden et al. 1983; Oyarzun et al. 1998; Serra-Wittling et al. 1996; Steinberg et al. 2004; Widmer et al. 2002), as well as disease suppression in horticultural crops (Cotxarrera et al. 2002; Hoitink and Boehm 1999). The mechanisms involved, however, are not fully understood yet and subject of ongoing studies.

Acknowledgements We are grateful to Drs. Olga Mavrodi and Dmitri Mavrodi, Department of Plant Pathology, Washington State University, USA, for providing the confocal laser microscopic pictures of roots colonized by beneficial bacteria. We also would like to thank the editors Drs. Philippe Lemanceau, Yves Dessaux and Philippe Hinsinger for the opportunity to contribute to this special issue on the rhizosphere.

Open Access This article is distributed under the terms of the Creative Commons Attribution Noncommercial License which permits any noncommercial use, distribution, and reproduction in any medium, provided the original author(s) and source are credited.

References

Adesina MF, Lembke A, Costa R, Speksnijder A, Smalla K (2007) Screening of bacterial isolates from various European soils for *in vitro* antagonistic activity towards *Rhizoctonia solani* and *Fusarium oxysporum*: site-dependent composition and diversity revealed. Soil Biol Biochem 39:2818–2828

Agrios GN (2005) Plant pathology, 5th edn. Elsevier, New York

Alabouvette C, Olivain C, Steinberg C (2006) Biological control of plant diseases: the European situation. Eur J Plant Pathol 114:329–341

Alexopoulos CJ, Mims CW, Blackwell M (1996) Introductory mycology, 4th edn. Wiley, New York

Arnault I, Mondy N, Diwo S, Auger J (2004) Soil behaviour of sulfur natural fumigants used as methyl bromide substitutes. Int J Environ Anal Chem 84:75–82

Back MA, Haydock PPJ, Jenkinson P (2002) Disease complexes involving plant parasitic nematodes and soilborne pathogens. Plant Pathol 51:683–697

Bailey DJ, Gilligan CA (1999) Dynamics of primary and secondary infection in take-all epidemics. Phytopathol 89:84–91

Bailey DJ, Gilligan CA (2004) Modeling and analysis of disease-induced host growth in the epidemiology of take-all. Phytopathol 94:535–540

Bailey KL, Lazarovits G (2003) Suppressing soil-borne diseases with residue management and organic amendments. Soil Tillage Res 72:169–180

Bailey KL, Gossen GB, Derksen DA, Watson PR (2000) Impact of agronomic practices and environment on diseases of wheat and lentil in southeastern Saskatchewan. Can J Plant Sci 80:917–927

Bailey DJ, Paveley N, Pillinger C, Foulkes J, Spink J, Gilligan CA (2005) Epidemiology and chemical control of take-all on seminal and adventitious roots of wheat. Phytopathol 95:62–68

Bailey DJ, Kleczkowski A, Gilligan CA (2006) An epidemiological analysis of the role of disease-induced root growth in the differential response of two cultivars of winter wheat to infection by *Gaeumannomyces graminis* var *tritici*. Phytopathol 96:510–516

Baird RE, Watson CE, Scruggs M (2003) Relative longevity of *Macrophomina phaseolina* and associated mycobiota on residual soybean roots in soil. Plant Dis 87(5):563–566

Baker R (1978) Inoculum potential. In: Horsfall JG, Cowling EB (eds) Plant disease, vol 1: How disease is managed. Academic, NY, pp 137–157

Beckman CH (1987) The nature of wilt diseases of plants. APS, St Paul, MN

Bird DM, Kaloshian I (2003) Are nematodes special? Nematodes have their say. Physiol Mol Plant Pathol 62:115–123

Bloemberg GV, Wijfjes AHM, Lamers GEM, Stuurman N, Lugtenberg BJJ (2000) Simultaneous imaging of *Pseudomonas fluorescens* WCS365 populations expressing three different autofluorescent proteins in the rhizosphere: new perspectives for studying microbial communities. Mol Plant–Microb Interact 13:1170–1176

Blok WJ, Lamers JG, Termorshuizen AJ, Bollen AJ (2000) Control of soilborne plant pathogens by incorporating fresh organic amendments followed by tarping. Phytopathol 30:253–259

Bockus WW, Shroyer JP (1998) The impact of reduced tillage on soilborne plant pathogens. Ann Rev Phytopathol 36:485–500

Bollen GJ (1969) The selective effect of heat treatment on the microflora of a greenhouse soil. Neth J Plant Pathol 75:157–163

Bolwerk A, Lagopodi AL, Lugtenberg BJJ, Bloemberg GV (2005) Visualization of interactions between a pathogenic and a beneficial *Fusarium* strain during biocontrol of tomato foot and root rot. Mol Plant–Microb Interact 18:710–721

Brown DJF, Robertson WM, Trudgill DL (1995) Transmission of viruses by plant nematodes. Ann Rev Phytopathol 33:223–249

Bruehl GW (1987) Soilborne plant pathogens. Macmillan, NY

Campbell RN (1996) Fungal transmission of plant viruses. Ann Rev Phytopathol 34:87–108

Campbell CL, Benson DM (1994) Spatial aspects of the development of root disease epidemics. In: Campbell CL, Benson DM (eds) Epidemiology and management of root diseases. Springer, Berlin, pp 195–243

Campbell CL, Madden LV (1990) Introduction to plant disease epidemiology. Wiley, NY

Cohen MF, Mazzola M (2006) Resident bacteria, nitric oxide emission and particle size modulate the effect of *Brassica napus* seed meal on disease incited by *Rhizoctonia solani* and *Pythium* spp. Plant Soil 286:75–86

Compant S, Reiter B, Sessitsch A, Nowak J, Clément C, Ait Barka E (2005) Endophytic colonization of *Vitis vinifera* L by a plant growth-promoting bacterium *Burkholderia* sp strain PsJN. Appl Environ Microbiol 71:1685–1693

Cook RJ (2003) Take-all of wheat. Phys Mol Plant Pathol 62:73–86

Cook RJ, Veseth RJ (1991) Wheat health management. APS, St Paul, MN

Cook RJ, Sitton JW, Waldher JT (1980) Evidence for *Pythium* as a pathogen of direct-drilled wheat in the Pacific Northwest. Plant Dis 64:1061–1066

Cook RJ, Sitton JW, Haglund WA (1987) Influence of soil treatments on growth and yield of wheat and implications for control of Pythium root rot. Phytopathol 77:1172–1198

Cook RJ, Thomashow LS, Weller DM, Fujimoto D, Mazzola M, Bangera G, Kim DS (1995) Molecular mechanisms of defense by rhizobacteria against root disease. Proc Natl Acad Sci U S A 92:4197–4201

Cook RJ, Schillinger WF, Christensen NW (2002) Rhizoctonia root rot and take-all of wheat in diverse direct-seed spring cropping systems. Can J Plant Pathol 24:349–358

Cooksey DA, Moore LW (1982) High frequency spontaneous mutations to Agrocin 84 resistance in *Agrobacterium tumefaciens* and *A rhizogenes*. Physiol Plant Pathol 20:129–35

Cotten TK, Munkvold GP (1998) Survival of *Fusarium moniliforme*, *F. proliferatum* and *F. subglutinans* in maize stalk residue. Phytopathol 88:550–555

Cotxarrera L, Trillas-Gay MI, Steinberg C, Alabouvette C (2002) Use of sewage sludge compost and *Trichoderma asperellum* isolates to suppress Fusarium wilt of tomato. Soil Biol Biochem 34:467–476

Coventry E, Noble R, Mead A, Whipps JM (2002) Control of allium white rot *Sclerotium cepivorum* with composted onion waste. Soil Biol Biochem 34:1037–1045

Davidson JM, Wickland AC, Patterson HA, Falk KR, Rizzo DM (2005) Transmission of *Phytophthora ramorum* in mixed-evergreen forest in California. Phytopathol 95:587–596

De Meyer G, Audenaert K, Höfte M (1999) *Pseudomonas aeruginosa* 7NSK2-induced systemic resistance in tobacco depends on in planta salicylic acid accumulation but is not associated with PR1a expression. Eur J Plant Pathol 105:513–517

De Souza JT, Arnould C, Deulvot C, Lemanceau P, Gianinazzi-Pearson V, Raaijmakers JM (2003) Effect of 2,4-diacetylphloroglucinol on *Pythium*: cellular responses and variation in sensitivity among propagules and species. Phytopathol 93:966–975

DeVay JE (1995) Solarization: an environmental-friendly technology for pest management. Arab J Plant Prot 13:56–61

De Waard MA (1997) Significance of ABC transporters in fungicide sensitivity and resistance. Pest Sci 51:271–275

Di Pietro A, Lorito M, Hayes CK, Broadway RM, Harman GE (1993) Endochitinase from *Gliocladium virens*: isolation, characterization, and synergistic antifungal activity in combination with gliotoxin. Phytopathol 83:308–313

Djonović S, Pozo MJ, Kenerley CM (2006) Tvbgn3 a b-16-glucanase from the biocontrol fungus *Trichoderma virens* is involved in mycoparasitism and control of *Pythium ultimum*. Appl Environ Microbiol 72:7661–7670

Duffy BK, Défago G (1997) Zinc improves biocontrol of *Fusarium crown* and root rot of tomato by *Pseudomonas fluorescens* and represses the production of pathogen metabolites inhibitory to bacterial antibiotic biosynthesis. Phytopathol 87:1250–1257

Duffy BK, Schouten A, Raaijmakers JM (2003) Pathogen self defense: mechanisms to counteract microbial antagonism. Ann Rev Phytopathol 41:501–538

Duffy BK, Keel C, Defago G (2004) Potential role of pathogen signaling in multitrophic plant–microbe interactions involved in disease protection. Appl Environ Microbiol 70:1836–1842

Dulout A, Lucas P, Sarniguet A, Dore T (1997) Effects of wheat volunteers and blackgrass in set-aside following a winter wheat crop on soil infectivity and soil conduciveness to take-all. Plant Soil 197:149–155

Dunne C, Crowley JJ, Moënne-Loccoz Y, Dowling DN, de Bruijn FJ, O'Gara F (1997) Biological control of *Pythium ultimum* by *Stenotrophomonas maltophilia* W81 is mediated by an extracellular proteolytic activity. Microbiol 143:3921–3931

Duponnois R, Bâ AM, Mateille T (1999) Beneficial effects of *Enterobacter cloacae* and *Pseudomonas mendocina* for biocontrol of *Meloidogyne incognita* with the endospore-forming bacterium *Pasteuria penetrans*. J Nematol 1:95–101

Ellis RJ, Timms-Wilson TM, Bailey MJ (2000) Identification of conserved traits in fluorescent pseudomonads with antifungal activity. Environ Microbiol 2:274–284

Erhart E, Burian K, Hartl W, Stich K (1999) Suppression of *Pythium ultimum* by biowaste composts in relation to compost microbial biomass activity and content of phenolic compounds. J Phytopathol 147:299–305

Fedi S, Tola E, Moënne-Loccoz Y, Dowling DN, Smith LM, O'Gara F (1997) Evidence for signaling between the phytopathogenic fungus *Pythium ultimum* and *Pseudomonas fluorescens* F113: *P. ultimum* represses the expression of genes in *P. fluorescens* F113, resulting in altered ecological fitness. Appl Environ Microbiol 63:4261–4266

Folwell RJ, Cook RJ, Heim MN, Moore DL (1991) Economic significance of take-all on winter wheat in the Pacific Northwest USA. Crop Prot 10:391–395

Fravel D, Olivain C, Alabouvette C (2003) *Fusarium oxysporum* and its biocontrol. New Phytol 157:493–502

Fuchs J-G, Moënne-Loccoz Y, Défago G (1997) Nonpathogenic *Fusarium oxysporum* strain Fo47 induces resistance to Fusarium wilt in tomato. Plant Dis 81:492–496

Gallo L, Siverio F, Rodriguez-Perez AM (2007) Thermal sensitivity of *Phytophthora cinnamomi* and long-term effectiveness of soil solarisation to control avocado root rot. Ann Appl Biol 150:65–73

Gamalero E, Lingua G, Tombolini R, Avidano L, Pivato B, Berta G (2005) Colonization of tomato root seedling by *Pseudomonas fluorescens* 92rkG5: spatio-temporal dynamics localization, organization, viability and culturability. Microb Ecol 50:289–297

Garbeva P, van Veen JA, van Elsas JD (2004) Assessment of the diversity and antagonism toward *Rhizoctonia solani* AG3 of *Pseudomonas* species in soil from different agricultural regimes. FEMS Microbiol Ecol 47:51–64

Genin S, Boucher C (2004) Lessons learned from the genome analysis of *Ralstonia solanacearum*. Ann Rev Phytopathol 42:107–134

Gill JS, Sivasithamparam K, Smettem KRJ (2000) Soil types with different texture effects development of Rhizoctonia root rot of wheat seedlings. Plant Soil 221:113–120

Gilligan CA (1994) Temporal aspects of the development of root disease epidemics. In: Campbell CL, Benson DM (eds) Epidemiology and management of root diseases. Springer, Berlin, pp 148–194

Gilligan CA, Bailey DJ (1997) Components of pathozone behaviour. New Phytol 135:475–490

Gosme M, Willocquet L, Lucas P (2007) Size, shape and intensity of aggregation of take-all disease during natural epidemics in second wheat crops. Plant Pathol 56:87–96

Guillemaut C (2003) Identification et étude de l'écologie de *Rhizoctonia solani* responsable de la maladie de pourriture brune de la betterave sucrière PhD thesis, Université Claude Bernard-Lyon I, Lyon, France

Gurusiddaiah S, Weller DM, Sarkar A, Cook RJ (1986) Characterization of an antibiotic produced by a strain of *Pseudomonas fluorescens* inhibitory to *Gaeumannomyces graminis* var *tritici* and *Pythium* spp. Antimicrob Agents Chemother 29:488–495

Haas D, Défago G (2005) Biological control of soil-borne pathogens by fluorescent pseudomonads. Nat Rev Microbiol 3:307–319

Hagn A, Pritsch K, Schloter M, Munch JC (2003) Fungal diversity in agricultural soil under different farming management systems with special reference to biocontrol strains of *Trichoderma* spp. Biol Fertil Soils 38:236–244

Handelsman J, Stabb EV (1996) Biocontrol of soilborne plant pathogens. Plant Cell 8:1855–1869

Hao JJ, Subbarao KV (2005) Comparative analysis of lettuce drop epidemics caused by *Sclerotinia* minor and *S sclerotiorum*. Plant Dis 89:717–725

Harman GE, Petzoldt R, Comis A, Chen J (2004) Interactions between *Trichoderma harzianum* strain T22 and maize inbred line Mo17 and effects of these interactions on diseases caused by *Pythium ultimum* and *Colletotrichum graminicola*. Phytopathol 94:147–153

Harris K, Young IM, Gilligan CA, Otten W, Ritz K (2003) Effect of bulk density on the spatial organization of the fungus *Rhizoctonia solani* in soil. FEMS Microbiol Ecol 44:45–56

Hoitink HAJ (1980) Composted bark a lightweight growth medium with fungicidal properties. Plant Dis 66:142–147

Hoitink HAJ, Boehm MJ (1999) Biocontrol within the context of soil microbial communities: a substrate-dependent phenomenon. Ann Rev Phytopathol 37:427–446

Iavicoli A, Boutet E, Buchala A, Métraux J-P (2003) Induced systemic resistance in *Arabidopsis thaliana* in response to root inoculation with *Pseudomonas fluorescens* CHA0. Mol Plant–Microb Interact 16:851–858

Jafra S, Przysowa J, Czajkowski R, Michta A, Garbeva P, Van der Wolf JM (2006) Detection and characterization of bacteria from the potato rhizosphere degrading N-acyl-homoserine lactone. Can J Microbiol 52:1006–1015

Janvier C, Villeneuve F, Alabouvette C, Edel-Hermann V, Mateille T, Steinberg C (2007) Soil health through soil disease suppression: which strategy from descriptors to indicators? Soil Biol Biochem 39:1–23

Johnson DA, Zhang H, Alldredge JR (2006) Spatial pattern of Verticillium wilt in commercial mint fields. Plant Dis 90:789–797

Jones RW, Hancock JG (1988) Mechanism of gliotoxin action and factors mediating gliotoxin sensitivity. J Gen Microbiol 134:2067–2075

Katan J (1996) Soil solarization: integrated control aspects. In: Hal R (ed) Principles and practice of managing soilborne plant pathogens. APS, St Paul, MN, pp 250–278

Keel C, Schnider U, Maurhofer M, Voisard C, Laville J, Burger U, Wirthner P, Haas D, Défago G (1992) Suppression of root diseases by *Pseudomonas fluorescens* CHA0: importance of the bacterial secondary metabolite 2,4-diacetylphloroglucinol. Mol Plant–Microb Interact 5:4–13

Kirkegaard JA, Sarwar M (1998) Biofumigation potential of brassicas: variation in glucosinolate profiles of diverse field-grown brassicas. Plant Soil 201:71–89

Kobayashi DY, Reedy RM, Bick JA, Oudemans PV (2002) Characterization of a chitinase gene from *Stenotrophomonas maltophilia* strain 34S1 and its involvement in biological control. Appl Environ Microbiol 68:1047–1054

Kubicek CP, Mach RL, Peterbauer CK, Lorito M (2001) *Trichoderma*: from genes to biocontrol. J Plant Pathol 83:11–23

Lagopodi AL, Ram AFJ, Lamers GEM, Punt PJ, Van den Hondel CAMJJ, Lugtenberg BJJ, Bloemberg GV (2002) Novel aspects of tomato root colonization and infection by *Fusarium oxysporum* f. sp. *radicis-lycopersici* revealed by confocal laser scanning microscopic analysis using the green fluorescent protein as a marker. Mol Plant–Microb Interact 15:172–179

Lamers J, Wanten P, Blok W (2004) Biological soil disinfestation: a safe and effective approach for controlling soilborne pests and diseases. Agroindustria 3:289–291

Lemanceau P, Bakker PAHM, de Kogel WJ, Alabouvette C, Schippers B (1992) Effect of pseudobactin 358 production by *Pseudomonas putida* WCS358 on suppression of Fusarium wilt of carnation by nonpathogenic *Fusarium oxysporum* Fo47. Appl Environ Microbiol 58:2978–2982

Levy E, Eyal Z, Chet I, Hochman A (1992) Resistance mechanisms of *Septoria tritici* to antifungal products of *Pseudomonas*. Physiol Mol Plant Pathol 40:163–171

Lewis JA, Papavizas GC (1991) Biocontrol of plant diseases: the approach for tomorrow. Crop Prot 10:95–105

Lu ZX, Tombolini R, Woo S, Zeilinger S, Lorito M, Jansson JK (2004) In vivo study of *Trichoderma*-pathogen–plant interactions using constitutive and inducible green fluorescent protein reporter systems. Appl Environ Microbiol 70:3073–3081

Lumsden RD, Lewis JA, Millner PD (1983) Effect of composted sewage sludge on several soilborne pathogens and diseases. Phytopathol 73:1543–1548

Lutz M, Feichtinger G, Défago G, Duffy B (2003) Mycotoxigenic *Fusarium* and deoxynivalenol production repress chitinase gene expression in the biocontrol agent *Trichoderma atroviridae*. Appl Environ Microbiol 69:3077–3084

Lynch J (1990) The rhizosphere. Wiley, London, UK, p 458

Mackie AE, Wheatley RE (1999) Effects and incidence of volatile organic compound interactions between soil bacterial and fungal isolates. Soil Biol Biochem 31:375–385

Madden LV, Hughes G (1994) BBD: computer software for fitting the beta-binomial distribution to disease incidence data. Plant Dis 78:536–540

Madden LV, Hughes G, Van den Bosch F (2007) The study of plant disease epidemics. APS, St Paul, MN

Mahajan-Miklos S, Tan Man W, Rahme LG, Ausubel FM (1999) Molecular mechanisms of bacterial virulence elucidated using a *Pseudomonas aeruginosa*–*Caenorhabditis elegans* pathogenesis model. Cell 96:47–56

Maurhofer M, Hase C, Meuwly P, Métraux J-P, Défago G (1994) Induction of systemic resistance of tobacco to tobacco necrosis virus by the root colonizing *Pseudomonas fluorescens* strain CHA0: influence of the *gacA* gene and of pyoverdin production. Phytopathol 89:139–146

Mazurier S, Lemunier M, Siblot S, Mougel C, Lemanceau P (2004) Distribution and diversity of type III secretion system-like genes in saprophytic and phytopathogenic fluorescent pseudomonads. FEMS Microbiol Ecol 49:455–467

Mazzola M (1999) Transformation of soil microbial community structure and *Rhizoctonia*-suppressive potential in response to apple roots. Phytopathol 89:920–927

Mazzola M (2004) Assessment and management of soil microbial community structure for disease suppression. Ann Rev Phytopathol 42:35–59

Mazzola M (2007) Manipulation of rhizosphere microbial communities to induce suppressive soils. J Nematol 39:213–220

Mazzola M, Fujimoto DK, Thomashow LS, Cook RJ (1995) Variation in sensitivity of *Gaeumannomyces graminis* to antibiotics produced by fluorescent *Pseudomonas* spp and effect on biological control of take-all of wheat. Appl Environ Microbiol 61:2554–2559

Mazzola M, Granatstein DM, Elfving DC, Mullinix K (2001) Suppression of specific apple root pathogens by *Brassica napus* seed meal amendment regardless of glucosinolate content. Phytopathol 91:673–679

Mazzola M, Zhao X, Cohen MF, Raaijmakers JM (2007) Cyclic lipopeptide surfactant production by *Pseudomonas fluorescens* SS101 is not required for suppression of complex *Pythium* spp. populations. Phytopathol 97:1348–1355

Mihail JD, Alcorn SM (1987) *Macrophomina phaseolina*: spatial patterns in a cultivated soil and sampling strategies. Phytopathol 77:1126–1131

Milner JL, Silo-Suh LA, Lee JC, He H, Clardy J, Handelsman J (1996) Production of kanosamine by *Bacillus cereus* UW85. Appl Environ Microbiol 62:3061–3065

Molina L, Constantinescu F, Michel L, Reimmann C, Duffy B, Défago G (2003) Degradation of pathogen quorum-sensing molecules by soil bacteria: a preventive and curative biological control mechanism. FEMS Microbiol Ecol 45:71–81

Ndiaye M, Termorshuizen AJ, Van Bruggen AHC (2007) Combined effects of solarization and organic amendment on charcoal rot caused by *Macrophomina phaseolina* in the Sahel. Phytoparasitica 35:392–400

Neilands JB (1995) Siderophores: structure and function of microbial iron transport compounds. J Biol Chem 270:26723–26726

Nester E, Gordon MP, Kerr A (2005) *Agrobacterium tumefaciens*: From plant pathology to biotechnology. APS, St. Paul, MN

Notz R, Maurhofer M, Dubach H, Haas D, Défago G (2002) Fusaric acid-producing strains of *Fusarium oxysporum* alter 2,4-diacetylphloroglucinol biosynthetic gene expression in *Pseudomonas fluorescens* CHA0 in vitro and in the rhizosphere of wheat. Appl Environ Microbiol 68:2229–2235

Oerke EC (2005) Crop losses to pests. J Agr Sci 144:31–43

Oerke EC, Dehne HW (2004) Safeguarding production: losses in major crops and the role of crop protection. Crop Prot 23:275–285

Oka Y, Shapira N, Fine P (2007) Control of root-knot nematodes in organic farming systems by organic amendments and soil solarization. Crop Prot 26:1556–1565

Olanya OM, Campbell CL (1989) Density and spatial pattern of propagules of *Macrophomina phaseolina* in corn rhizospheres. Phytopathol 79:1119–1123

Olivain C, Humbert C, Nahalkova J, Fatehi J, L'Haridon F, Alabouvette C (2006) Colonization of tomato root by pathogenic and nonpathogenic *Fusarium oxysporum* strains inoculated together and separately into the soil. Appl Environ Microbiol 72:1523–1531

Ongena M, Jourdan E, Adam A, Paquot M, Brans A, Joris B, Arpigny JL, Thonart P (2007) Surfactin and fengycin lipopeptides of *Bacillus subtilis* as elicitors of induced systemic resistance in plants. Environ Microbiol 9:1084–1090

Otten W, Gilligan CA, Watts CW, Dexter AR, Hall D (1999) Continuity of air-filled pores and invasion thresholds for a soilborne fungal plant pathogen *Rhizoctonia solani*. Soil Biol Biochem 31:1803–1810

Otten W, Harris K, Young IM, Ritz K, Gilligan CA (2004) Preferential spread of the pathogenic fungus *Rhizoctonia solani* through structured soil. Soil Biol Biochem 36:203–210

Oyarzun P, Gerlagh JM, Zadoks JC (1998) Factors associated with soil receptivity to some fungal root rot pathogens of peas. Appl Soil Ecol 10:151–169

Pankhurst CE, McDonald HJB, Hawke G, Kirkby CA (2002) Effect of tillage and stubble management on chemical and microbiological properties and the development of suppression towards cereal root disease in soils from two sites in NSW Australia. Soil Biol Biochem 34:833–840

Paulitz TC, Rossi R (2004) Spatial distribution of *Rhizoctonia solani* and *Rhizoctonia oryzae* at three different scales in direct-seeded wheat. Can J Plant Pathol 26:419

Paulitz TC, Anas O, Fernando DG (1992) Biological control of Pythium damping-off by seed treatment with *Pseudomonas putida*: relationship with ethanol production by pea and soybean seeds. Biocontrol Sci Technol 2:193–201

Paulitz TC, Zhang H, Cook RJ (2003) Spatial distribution of *Rhizoctonia oryzae* and Rhizoctonia root rot in direct seeded cereals. Can J Plant Pathol 25:295–303

Perry JN (1998) Measures of spatial pattern for counts. Ecol 79:1008–1017

Perry RN, Moens M (2006) Plant nematology. CABI, Cambridge, MA

Picard C, Bosco M (2008) Genotypic and phenotypic diversity in populations of plant-probiotic *Pseudomonas* spp. colonizing roots. Naturwissenschaften 95:1–16

Pierson LS III, Pierson EA (2007) Roles of diffusible signals in communication among plant-associated bacteria. Phytopathol 97:227–232

Pierson LS III, Wood DW, Pierson EA (1998) Homoserine lactone-mediated gene regulation in plant-associated bacteria. Ann Rev Phytopathol 36:207–25

Pieterse CMJ, van Pelt JA, Verhagen BWM, Ton J, van Wees SCM, Leon-Kloosterziel KM, van Loon LC (2003) Induced systemic resistance by plant growth-promoting rhizobacteria. Symbiosis 35:39–54

Pozo MJ, Cordier C, Dumas-Gaudot E, Gianinazzi S, Barea JM, Azcón-Aguilar C (2002) Localized versus systemic effect of arbuscular mycorrhizal fungi on defence responses to *Phytophthora* infection in tomato plants. J Exp Bot 53:525–534

Preston GM, Bertrand N, Rainey PB (2001) Type III secretion in plant growth-promoting *Pseudomonas fluorescens* SBW25. Mol Microbiol 41:999–1014

Raaijmakers JM (2001) Rhizosphere and rhizosphere competence. In: Maloy OC, Murray TD (eds) Encyclopedia of plant pathology. Wiley, USA, pp 859–860

Raaijmakers JM, Weller DM (1998) Natural plant protection by 2,4-diacetylphloroglucinol-producing *Pseudomonas* spp. in take-all decline soils. Mol Plant–Microb Interact 11:144–152

Raaijmakers JM, Leeman M, Van Oorschot MMP, Van der Sluis I, Schippers B, Bakker PAHM (1995) Dose–response relationships in biological control of fusarium wilt of radish by *Pseudomonas* spp. Phytopathol 85:1075–1081

Raaijmakers JM, Weller DM, Thomashow LS (1997) Frequency of antibiotic-producing *Pseudomonas* spp. in natural environments. Appl Environ Microbiol 63:881–887

Raaijmakers JM, Bonsall RF, Weller DM (1999) Effect of cell density of *Pseudomonas fluorescens* on the production of 2,4-diacetylphloroglucinol in the rhizosphere of wheat. Phytopathol 89:470–475

Raaijmakers JM, Vlami M, de Souza JT (2002) Antibiotic production by bacterial biocontrol agents. Antonie van Leeuwenhoek 81:537–547

Raaijmakers JM, de Bruijn I, de Kock MJD (2006) Cyclic lipopeptide production by plant-associated *Pseudomonas* species: diversity, activity, biosynthesis and regulation. Mol Plant–Microb Interact 19:699–710

Rasmussen TB, Givskov M (2006) Quorum sensing inhibitors: a bargain of effects. Microbiol 152:895–904

Rezzonico F, Défago G, Moënne-Loccoz Y (2004) Comparison of ATPase-encoding type III secretion system *hrcN* genes in biocontrol fluorescent pseudomonads and in phytopathogenic proteobacteria. Appl Environ Microbiol 70:5119–5131

Rezzonico F, Binder C, Défago G, Moënne-Loccoz Y (2005) The type III secretion system of biocontrol *Pseudomonas fluorescens* KD targets the phytopathogenic Chromista *Pythium ultimum* and promotes cucumber protection. Mol Plant–Microb Interact 18:991–1001

Rezzonico F, Zala M, Keel C, Duffy B, Moënne-Loccoz Y, Défago G (2007) Is the ability of biocontrol fluorescent pseudomonads to produce the antifungal metabolite 2,4-diacetylphloroglucinol really synonymous with higher plant protection? New Phytol 173:861–872

Ristaino JB, Larkin RP, Campbell CL (1993) Spatial and temporal dynamics of Phytophthora epidemics in commercial bell pepper fields. Phytopathol 83:1312–1320

Roget DK, Neate SM, Rovira AD (1996) Effect of sowing point out design and tillage practice on the incidence of rhizoctonia root rot take all and cereal cyst nematode in wheat and barley. Aust J Exp Agric 36:683–693

Ryder MH, Slota JE, Scarim A, Farrand SK (1987) Genetic analysis of agrocin 84 production and immunity in *Agrobacterium* spp. J Bacteriol 169:4184–4189

Ryu C-M, Farag MA, Hu C-H, Reddy MS, Kloepper JW, Paré PW (2004) Bacterial volatiles induce systemic resistance in Arabidopsis. Plant Physiol 134:1017–1026

Sarwar M, Kirkegaard JA, Wong PTW, Desmarchelier JM (1998) Biofumigation potential of brassicas—III. In vitro toxicity of isothiocyanates to soil-borne fungal pathogens. Plant Soil 201:103–112

Schneider O, Aubertot JN, Roger-Estrade J, Doré T (2003) Analysis and modelling of the amount of oilseed rape residues left at the soil surface after different soil tillage operations. 7th International Conference on Plant Pathology, Tours France, 3–5 December 2003

Schoonbeek H, Del Sorbo G, De Waard MA (2001) The ABC transporter BcatrB affects the sensitivity of *Botrytis cinerea* to the phytoalexin resveratrol and the fungicide fenpiclonil. Mol Plant–Microb Interact 14:562–571

Schoonbeek HJ, Raaijmakers JM, De Waard MA (2002) Fungal ABC transporters and microbial interactions in natural environments. Mol Plant–Microb Interact 15:1165–1172

Schouten A, Raaijmakers JM (2004) Defense responses of *Fusarium oxysporum* against 24-diacetylphloroglucinol a broad-spectrum antibiotic produced by antagonistic *Pseudomonas fluorescens*. Mol Plant–Microb Interact 17:1201–1211

Schouten A, Maksimova O, Cuesta-Arenas Y, van den Berg G, Raaijmakers JM (2008) Involvement of the ABC-transporter BcatrB and the laccase BcLcc2 in defense of *Botrytis cinerea* against the broad-spectrum antibiotic 2,4-diacetylphloroglucinol. Environ Microbiol. DOI 10.1111/j.1462-2920.2007.01531.x

Serra-Wittling C, Houot S, Alabouvette C (1996) Increased soil suppressiveness to fusarium wilt of flax after addition of municipal solid waste compost. Soil Biol Biochem 28:1207–1214

Séveno NA, Kallifidas D, Smalla K, van Elsas JD, Collard JM (2002) Occurrence and reservoirs of antibiotic resistance genes in the environment. Rev Med Microbiol 13:15–27

Sharifi-Tehrani A, Zala M, Natsch A, Moënne-Loccoz Y, Défago G (1998) Biocontrol of soil-borne fungal plant diseases by 2,4-diacetylphloroglucinol-producing fluorescent pseudomonads with different restriction profiles of amplified 16S rDNA. Eur J Plant Pathol 104:631–643

Smiley RW, Gourlie JA, Easley SA, Patterson LM, Whittaker RG (2005a) Crop damage estimates for crown rot of wheat and barley in the Pacific Northwest. Plant Dis 89:595–604

Smiley RW, Whittaker RG, Gourlie JA, Easley SA (2005b) Suppression of wheat growth and yield by *Pratylenchus neglectus* in the Pacific Northwest. Plant Dis 89:958–968

Smiley RW, Whittaker RG, Gourlie JA, Easley SA (2005c) *Pratylenchus thornei* associated with reduced wheat yield in Oregon. J Nematol 37:45–54

Smith LM, Tola E, de Boer P, O'Gara F (1999) Signalling by the fungus *Pythium ultimum* represses expression of two ribosomal RNA operons with key roles in the rhizosphere ecology of *Pseudomonas fluorescens* F113. Environ Microbiol 1:495–502

Somers E, Vanderleyden J, Srinivasan M (2004) Rhizosphere bacterial signalling: a love parade beneath our feet. Crit Rev Microbiol 30:205–240

Stanghellini ME, Kim DH, Waugh MM, Ferrin DM, Alcantara T, Rasmussen SL (2004a) Infection and colonization of melon roots by *Monosporascus cannonballus* in two cropping season in Arizona and California. Plant Pathol 53:54–57

Stanghellini ME, Waugh MM, Radewald KC, Kim DH, Ferrin DM, Turini T (2004b) Crop residue destruction strategies that enhance rather than inhibit reproduction of *Monosporascus cannonballus*. Plant Pathol 53:50–53

Steinberg C, Edel-Hermann V, Guillemaut C, Pérez-Piqueres A, Singh P, Alabouvette C (2004) Impact of organic amendments on soil suppressiveness to diseases. In: Sikora RA, Gowen S, Hauschild R, Kiewnick S (eds) Multitrophic interactions in soil and integrated control. IOBC wprs Bulletin/Bulletin OILB, pp 259–266

Steinberg C, Edel-Hermann V, Alabouvette C, Lemanceau P (2007) Soil suppressiveness to plant diseases. In: van Elsas JD, Jansson J, Trevors JT (eds) Modern soil microbiology. CRC, New York, pp 455–478

Stergiopoulos I, Zwiers LH, De Waard MA (2002) Secretion of natural and synthetic toxic compounds from filamentous fungi by membrane transporters of the ATP-binding cassette and major facilitator superfamily. Eur J Plant Pathol 108:719–734

Stockwell VO, Kawalek MD, Moore LW, Loper JE (1996) Transfer of pAgK84 from the biocontrol agent *Agrobacterium radiobacter* K84 to *A tumefaciens* under field conditions. Phytopathol 86:31–37

Stokstad E (2004) Nurseries may have shipped sudden oak death pathogen nationwide. Science 303:1959

Stuber K, Frey J, Burnens AP, Kuhnert P (2003) Detection of type III secretion genes as a general indicator of bacterial virulence. Mol Cell Probes 17:25–32

Stutz E, Défago G, Kern H (1986) Naturally occurring fluorescent pseudomonads involved in suppression of black root rot of tobacco. Phytopathol 76:181–185

Sturz AV, Carter MR, Johnston HW (1997) A review of plant disease pathogen interactions and microbial antagonism under conservation tillage in temperate humid agriculture. Soil Tillage Res 41:169–189

Termorshuizen AJ, van Rijn E, van der Gaag DJ, Alabouvette C, Chen Y, Lagerlöf J, Malandrakis AA, Paplomatas EJ, Rämert B, Ryckeboer J, Steinberg C, Zmora-Nahum S (2006) Suppressiveness of 18 composts against 7 soilborne plant pathogens. Soil Biol Biochem 38:2461–2477

Tilston E, Pitt LD, Groenhof AC (2002) Composted recycled organic matter suppresses soil-borne diseases of field crops. New Phytol 154:731–740

Toresani S, Gomez E, Bonel B, Bisaro V, Montico S (1998) Cellulolytic population dynamics in a vertic soil under three tillage systems in the humid pampa of Argentina. Soil Tillage Res 49:79–83

Toyoda H, Hashimoto H, Utsumi R, Kobayashi H, Ouchi S (1988) Detoxification of fusaric acid by a fusaric acid-resistant mutant of *Pseudomonas solanacearum* and its application to biological control of fusarium wilt of tomato. Phytopathol 78:1307–1311

Tran HTT, Ficke A, Asiimwe T, Hofte M, Raaijmakers JM (2007) Role of the cyclic lipopeptide surfactant massetolide A in biological control of *Phytophthora infestans* and colonization of tomato plants by *Pseudomonas fluorescens*. New Phytol 175:731–742

Uroz S, D'Angelo-Picard C, Carlier A, Elasri M, Sicot C, Petit A, Oger P, Faure D, Dessaux Y (2003) Novel bacteria degrading *N*-acylhomoserine lactones and their use as quenchers of quorum-sensing-regulated functions of plant-pathogenic bacteria. Microbiol 149:1981–1989

Uroz S, Oger P, Chhabra SR, Camara M, Williams P, Dessaux Y (2007) *N*-acyl homoserine lactones are degraded via an amidolytic activity in *Comamonas* sp. strain D1. Arch Microbiol 187:249–256

Vanderplank JE (1963) Plant diseases: epidemics and control. Academic, NY

Van Dijk K, Nelson EB (2000) Fatty acid competition as a mechanism by which *Enterobacter cloacae* suppresses *Pythium ultimum* sporangium germination and damping-off. Appl Environ Microbiol 66:5340–5347

Van Loon LC, Bakker PAHM, Pieterse CMJ (1998) Systemic resistance induced by rhizosphere bacteria. Ann Rev Phytopathol 36:453–483

Van Zyl FGH, Strijdom BW, Staphorst JL (1986) Susceptibility of *Agrobacterium tumefaciens* strains to two agrocin-producing *Agrobacterium* strains. Appl Environ Microbiol 52:234–238

Voisard C, Keel C, Haas D, Défago G (1989) Cyanide production by *Pseudomonas fluorescens* helps suppress black root of tobacco under gnotobiotic conditions. EMBO J 8:351–358

Weller DM (1988) Biological control of soilborne pathogens in the rhizosphere with bacteria. Annu Rev Phytopathol 26:379–407

Weller DM (2007) *Pseudomonas* biocontrol agents of soilborne pathogens: looking back over 30 years. Phytopathol 97:250–256

Weller DM, Raaijmakers JM, McSpadden Gardener BB, Thomashow LS (2002) Microbial populations responsible for specific soil suppressiveness to plant pathogens. Ann Rev Phytopathol 40:309–348

Widmer TL, Mitkowski NA, Abawi GS (2002) Soil organic matter and management of plant-parasitic nematodes. J Nematol 34:289–295

Woo SL, Scala F, Ruocco M, Lorito M (2006) The molecular biology of the interactions between *Trichoderma* spp. phytopathogenic fungi and plants. Phytopathol 96:181–185

Zehnder GW, Murphy JF, Sikora EJ, Kloepper JW (2001) Application of rhizobacteria for induced resistance. Eur J Plant Pathol 107:39–50

Zeilinger S, Galhaup C, Payer K, Woo SL, Mach RL, Fekete C, Lorito M, Kubicek CP (1999) Chitinase gene expression during mycoparasitic interaction of *Trichoderma harzianum* with its host. Fungal Genet Biol 26:131–140

Zhang L-H, Dong Y-H (2004) Quorum sensing and signal interference: diverse implications. Mol Microbiol 53:1563–1571

Zhou T, Paulitz TC (1993) In vitro and in vivo effects of *Pseudomonas* spp. on *Pythium aphanidermatum*: zoospore behavior in exudates and on the rhizoplane of bacteria-treated cucumber roots. Phytopathol 83:872–876

REVIEW ARTICLE

Rhizosphere engineering and management for sustainable agriculture

Peter R. Ryan · Yves Dessaux ·
Linda S. Thomashow · David M. Weller

Received: 26 November 2008 / Accepted: 9 April 2009 / Published online: 12 May 2009
© Springer Science + Business Media B.V. 2009

Abstract This paper reviews strategies for manipulating plants and their root-associated microorganisms to improve plant health and productivity. Some strategies directly target plant processes that impact on growth, while others are based on our knowledge of interactions among the components of the rhizosphere (roots, microorganisms and soil). For instance, plants can be engineered to modify the rhizosphere pH or to release compounds that improve nutrient availability, protect against biotic and abiotic stresses, or encourage the proliferation of beneficial microorganisms. Rhizobacteria that promote plant growth have been engineered to interfere with the synthesis of stress-induced hormones such as ethylene, which retards root growth, and to produce antibiotics and lytic enzymes active against soilborne root pathogens. Rhizosphere engineering also can involve the selection by plants of beneficial microbial populations. For example, some crop species or cultivars select for and support populations of antibiotic-producing strains that play a major role in soils naturally suppressive to soil-borne fungal pathogens. The fitness of root-associated bacterial communities also can be enhanced by soil amendment, a process that has allowed the selection of bacterial consortia that can interfere with bacterial pathogens. Plants also can be engineered specifically to influence their associated bacteria, as exemplified by quorum quenching strategies that suppress the virulence of pathogens of the genus *Pectobacterium*. New molecular tools and powerful biotechnological advances will continue to provide a more complete knowledge of the complex chemical and biological interactions that occur in the rhizosphere, ensuring that strategies to engineer the rhizosphere are safe, beneficial to productivity, and substantially improve the sustainability of agricultural systems.

Keywords Plant engineering · Proton efflux ·
Organic cation efflux · Microbe engineering ·
Biological control · PGPR · Plant genotype selection ·
Take-all decline · Soil amendment · Biostimulation

Responsible Editor: Philippe Lemanceau.

P. R. Ryan
CSIRO Plant Industry,
GPO Box 1600, Canberra ACT 2601, Australia

Y. Dessaux (✉)
Institut des sciences du végétal,
CNRS UPR 2355, avenue de la terrasse,
91198 Gif-sur-Yvette, France
e-mail: dessaux@isv.cnrs-gif.fr

L. S. Thomashow · D. M. Weller
USDA-ARS, Washington State University,
P.O. Box 646430, Pullman, WA 99164-6430, USA

Abbreviations
ACC   1-aminocyclopropane-1-carboxylate
ALMT Al^{3+}-activated malate transporter
DAPG 2,4-diacetylphloroglucinol
EST   expressed sequence tags

Ggt	*Gaeumannomyces graminis* var. *tritici*
MATE	multidrug and toxic compound extrusion
NAHL	N-acyl homoserine lactone
PCA	phenazine-1-carboxylic acid
PGPR	plant growth-promoting rhizobacteria
QS	quorum-sensing
QTL	quantitative trait locus(i)
TAD	take-all decline

Why engineer the rhizosphere?

The rhizosphere is the zone of soil around roots that is influenced by root activity. The intimacy of this interface between plants and their environment is essential for the acquisition of water and nutrients and for beneficial interactions with soil-borne microorganisms. Yet, this same intimacy increases the vulnerability of plants to a range of biotic and abiotic stresses. Plants have evolved a variety of strategies to modify the rhizosphere to lessen the impact of these environmental stresses, and an understanding of the processes involved will suggest ways in which the rhizosphere can be manipulated to improve plant health and productivity. Rhizosphere engineering may ultimately reduce our reliance on agrochemicals by replacing their functions with beneficial microbes, biodegradable biostimulants or transgenic plants. Some of these materials and techniques are still being developed, while others are being tested in the field. This paper reviews aspects of rhizosphere engineering and management with the view of developing novel approaches for sustainable agricultural production.

How to engineer the rhizosphere?

The physical and chemical milieu of the rhizosphere is the culmination of many competing and interacting processes that depend on the soil type and water content, the composition and biological activities of root-associated microbial communities and the physiology of the plant itself (Pinto et al. 2007). The rhizosphere can be modified over short periods of a plant's growth cycle by agronomic management (Bowen and Rovira 1999). Farmers influence the environment around the roots of their crops and pasture species every time they irrigate their fields or apply fertilizer. Ammonium-based fertilizers applied to plants tend to acidify the rhizosphere whereas nitrate-based fertilizers result in a more alkaline rhizosphere. Shifts in pH can alter soil chemistry around roots and influence the growth and composition of microbial communities. More prolonged changes in the rhizosphere that persist through the growth cycle can be generated with plant breeding and biotechnology. Selecting plants with favourable rhizosphere characteristics and introgressing these traits into elite breeding lines can lead to more permanent changes. This approach relies on being able to identify characters that are useful, heritable and easily detected: all of which are difficult when dealing with below-ground traits. Nevertheless, plant breeders have unintentionally engineered beneficial rhizosphere traits into crops by simply selecting for the best performing lines in field trials and recurrently crossing them to other breeding lines. A good example of this is the breeding for aluminium (Al^{3+}) resistance in wheat, which relies on the release of malate anions from the root apices—a rhizosphere process. The physiology of this mechanism has been understood for 15 years (Delhaize et al. 1993; Ryan et al. 1995) and the gene controlling it was isolated relatively recently (Sasaki et al. 2004), yet breeders have inadvertently been selecting for it for over a century by selecting the best performing lines on acid soils (de Sousa 1998). Microorganisms are a vital component of the rhizosphere, and the total biomass and composition of rhizosphere microbial populations markedly affects interactions between plants and the soil environment. There is considerable interest in developing methods for encouraging the proliferation of beneficial introduced or indigenous microbial populations that facilitate nutrient uptake (e.g., rhizobia and mycorrhiza), promote plant growth directly, or suppress plant pathogens. As the complexity of the rhizosphere is unraveled we can attempt to create conditions most beneficial to plant growth by amending the soil, breeding or engineering better plants, and manipulating plant/microorganism interactions. A summary of the above approaches is given in Fig. 1.

Engineering << individual >> rhizosphere components

Engineering plants

One of the main ways plants modify the rhizosphere is through the release of inorganic and organic

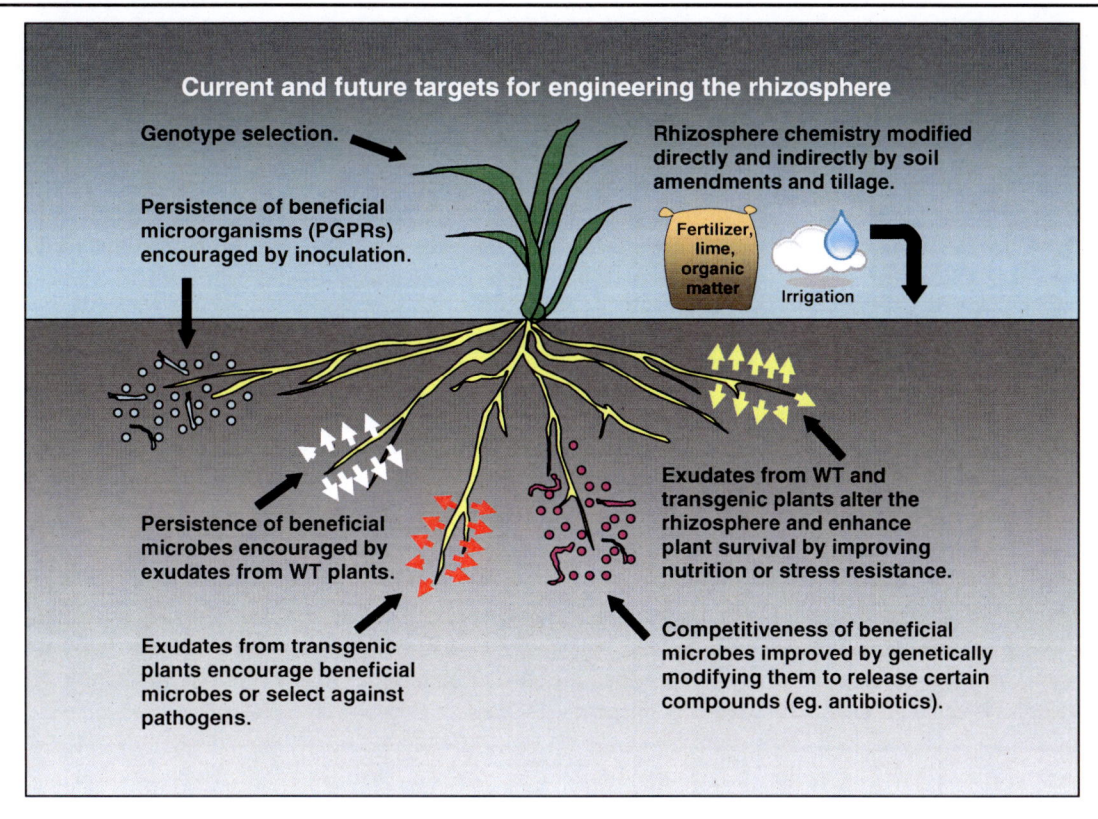

Fig. 1 Current and future targets for rhizosphere engineering The rhizosphere can be manipulated or engineered with agronomic practices, plant selection, soil inoculation or with biotechnology. Common practices such as soil tillage, fertilizer application or even irrigation can alter the chemistry of the plant-sol interface by changing aeration, root function or microbial communities. Plants with favorable roots traits that improve performance can be selected for by breeders. These traits could include exudates that increase nutrient accessibility, minimize stress or that encourage the persistence of beneficial micro-organisms. Biotechnology can be used to accentuate these useful traits or generate plants and micro-organisms with novel phenotypes that help plant survival. Transgenic plants and micro-organisms can be engineered to exude exogenous compounds that improve plant nutrition, repress pathogenic microbes and minimize the consequences of biotic or abiotic stresses

substances from their roots (rhizodeposition). These exudates can enhance nutrient acquisition, help to avoid mineral stresses or encourage the growth of favourable microorganisms. For instance, most transport processes across the plasma membrane in plant cells, including the uptake of essential nutrients, rely on the efflux of H^+ to generate a membrane potential difference and an electrochemical gradient for H^+. In addition to generating a driving force for membrane transport, H^+ efflux can contribute to nutrient acquisition by acidifying the rhizosphere (Hinsinger et al. 2003). Local acidification can increase the availability of Fe^{3+} and phosphorus, which are often fixed to the surface of minerals or present as insoluble complexes. Iron uptake in dicots and non-graminaceous monocots also involves the release of organic compounds that chelate Fe^{3+} and facilitate its reduction to Fe^{2+} prior to uptake. Grass species, by contrast, rely on the release of compounds called phytosiderophores that help mine the poorly soluble Fe^{3+} and deliver it to the root surface for uptake (Neumann and Römheld 2007). The release of organic anions such as citrate and malate, as well as phytases and phosphatases, helps some species to access poorly-soluble organic and inorganic phosphorus in a similar way (Dinkelaker et al. 1995; Richardson et al. 2001; Ryan et al. 2001; Vance et al. 2003). Organic anion efflux also protects some crop species from Al^{3+} toxicity in acid soils by chelating harmful Al^{3+} ions and preventing their damaging interactions with root apices (Ma et al. 2001; Ryan et al. 2001). Therefore the release of organic anions from roots can have an important influence on plant growth and nutrition. Since many of the genes controlling these exudates have now been identified, it is possible to manipulate conditions in the rhizosphere by modifying their expression via genetic engineering.

Plant genetic engineering is a relatively new science and several hurdles need to be crossed to achieve a successful outcome. Transgenic plants can exhibit phenotypes that are subtle, unexpected or even totally absent. The activity of exogenously-expressed proteins can be limited by metabolic feedback, substrate supply or post-translational regulation. Even when a desired trait, such as a root exudate, has been engineered successfully into a plant, the compounds might be rapidly degraded or otherwise inactivated in the soil, or the rate of exudation might simply be too small to influence the rhizosphere as predicted. The constitutive high or low expression of some genes can have pleiotropic effects on plant development, fertility or function that are detrimental to production. Also, transgenic plants can also display unexpected phenotypes that have nothing to do with the activity of the engineered protein (loss of function due to random insertion of transgene, somaclonal variation, etc.). Although many of these problems occur infrequently, they emphasise the need for careful analysis of the transgenic material and the use of suitable controls.

Once transgenic plants have been generated it is essential to analyze several independent transgenic lines and to compare them with control lines. There is no simple answer for how many transgenic and control lines should be analyzed. This will depend on how subtle or strong the phenotype is but in general, "more is better." The analysis of several independently transformed lines provides confidence that the measured phenotype is caused by transgene expression. It also is reassuring to show that the magnitude of the phenotype is correlated with the level of transgene expression in independent lines.

Wild-type plants commonly are compared with transgenic lines, but other types of controls can be used as well. When transgenic lines are developed from tissue culture, many workers include control lines that are transformed with an empty vector or non-transformed plants that have passed through tissue culture. This practice improves the chances of detecting real differences between the transgenic and control lines because it simulates more of the variability that may be introduced into transgenic lines during tissue culture. However, the preferred controls are the null segregants generated by selfing the primary transformants and analyzing the resulting lines. Null plants are excellent controls because they have experienced the same treatments as the transgenic plants except that they no longer contain the transgene. The null segregants also are useful when transgenic lines are generated without tissue culture, such as by the flower-dipping method for transforming Arabidopsis (Clough and Bent 1998).

The following case studies describe attempts to engineer the rhizosphere by manipulating the efflux of H^+ and organic anions from the roots in transgenic plants.

Case Study 1: Manipulating rhizosphere pH Proton efflux from plant cells is largely controlled by a large family of H^+-ATPase proteins that use energy from ATP catalysis to pump H^+ across the plasma membrane against a steep electrochemical gradient. The genes encoding these proteins have been isolated and studied for many years (Palgren 1991; Gaxiola et al. 2007). Therefore the objective of manipulating rhizosphere pH by over-expressing these genes would appear relatively straightforward. Yet, this has not proved to be the case and few studies have successfully increased H^+-efflux enough to change root function. For instance, when *Nicotiana tabacum* and Arabidopsis plants were transformed with cDNA encoding H^+-ATPase genes, they showed no symptoms despite significant increases in protein levels (Gévaudent et al. 2007; Young et al. 1998; Zhao et al. 2000). It became clear that activity of the H^+-ATPase protein was being regulated by an auto-inhibitory domain on its C-terminal tail. This domain suppresses enzyme activity by preventing protein phosphorylation and subsequent interactions with 14-3-3 regulatory proteins (Palgren 1991). It was only when plants were transformed with a modified H^+-ATPase protein in which the auto-inhibitory region was removed that phenotypes such as increased H^+-efflux from roots, more acidic rhizosphere (Gévaudent et al. 2007; Yang et al. 2007), or improved growth at low pH (Young et al. 1998) appeared in the transgenic lines. Constitutive expression of a modified H^+-ATPase gene, *PMA4*, in tobacco was also associated with minor developmental phenotypes and better growth on 200 mM NaCl than wild-type controls (Gévaudent et al. 2007). The improved salinity resistance was attributed to a greater capacity for Na^+/H^+ exchange across the plasma membrane.

Proton pumps also occur in subcellular compartments, and the transport of H^+ across the tonoplast membrane in plant cells is catalyzed by a vacuolar H^+-ATPase and a H^+-pyrophosphatase. Since these enzymes pump H^+ into the vacuole, and not across the plasma membrane, there was no *a priori* reason to suspect that modulating their activity would affect rhizosphere pH. Nevertheless, over-expression of the *AVP1* pyrophosphatase in Arabidopsis did induce a more acidic rhizosphere, apparently by up-regulating the activity of the plasma membrane H^+-ATPase (Yang et al. 2007). These transgenic Arabidopsis lines displayed an array of other phenotypes that appeared to be related to auxin transport (Li et al. 2005), since the changes in apoplasmic pH and the concomitant up-regulation of auxin transport proteins altered the distribution of auxin along the roots.

Case Study 2: Enhancing organic anion efflux from roots Once it was established that organic anions play a pivotal role in many rhizosphere processes, several groups began to investigate how efflux could be manipulated (see reviews by de la Fuente-Martinez and Herrera-Estrella 1999; López-Bucio et al. 2000b; Ryan et al. 2001). Two approaches have been tried to increase organic anion release from roots: (1) engineering plants with a greater capacity to synthesise organic anions, and (2) engineering plants with a greater capacity to transport organic anions out of the cell. These studies have mostly targeted malate and citrate anions because they occur in most living cells and are known to be released from roots in response to nutrient deficiency and mineral stress. In view of the large outwardly-directed electrochemical gradient for anions across the plasma membrane, some workers have argued that efflux from roots is more likely to be limited by transport across the plasma membrane than by synthesis (Ryan et al. 2001). If so, strategies aimed at enhancing transport capacity will be more successful than those attempting to modify metabolism. However both approaches rely on important assumptions as explained below.

Transgenic plants generated via the first approach typically exhibit enhanced expression of an enzyme involved in synthesis of an organic anion or reduced expression of an enzyme involved its consumption, the assumption being that these changes will increase anion accumulation in the cytosol and greater efflux from the cell. This approach will have a better chance of success if the target enzyme is rate-limiting in a pathway so that changes in protein concentration correspond to changes in flux through that pathway. Doubling or tripling the amount of an enzyme that catalyses a near-equilibrium reaction is unlikely to affect metabolism enough to generate a phenotype. Similarly, it is best to avoid enzymes whose activity is closely regulated by secondary modifications such as phosphorylation or other restrictive feedback mechanisms. Unwanted regulation may be minimized if the transgene is expressed in a sub-cellular compartment different from where it usually is located (i.e. expressing a mitochondrial gene in the cytosol) or by choosing a gene from a disparate source (expressing a bacterial gene in a plant) or a different species (expression of a carrot gene in Arabidopsis). Finally, this strategy assumes that an endogenous transport system is present for moving the organic anions across the plasma membrane and into the rhizosphere.

The second approach for increasing organic anion efflux transforms plants with genes encoding proteins which facilitate organic anion movement across the plasma membrane. This approach assumes that activity of the transport protein will not be regulated by the cell and that efflux will not be limited by the plant's capacity to synthesize organic anions.

Approach 1 Engineering metabolic pathways for greater organic anion efflux

One of the first attempts to manipulate organic anion release from plants was reported by de la Fuente et al. (1997). They increased citrate efflux from tobacco roots by transforming plants with a citrate synthase gene from *Pseudomonas aeruginosa* under the control of a constitutive promoter. The two- to three-fold increases in citrate synthase activity measured in the transgenic lines were associated with three-to ten-fold higher concentrations of citrate in the root tissue and a four-fold greater efflux of citrate from seedlings suspended in sterile water. Compared to control plants transformed with an empty vector, the transgenic plants reportedly showed enhanced resistance to Al^{3+} stress and a greater ability to acquire phosphorus from poorly soluble forms present in an alkaline soil (Lòpez-Bucio et al. 2000a).

Similarly, Koyama et al. (1999) transformed carrot (*Daucus carota*) suspension cells with a mitochondrial citrate synthase gene from Arabidopsis, and the four-fold greater release of citrate enhanced cell growth on media supplemented with AlPO$_4$. Koyama et al. (2000) later transformed Arabidopsis with a mitochondrial citrate synthase gene from carrot and although the increase in citrate release was relatively small at 60%, it was sufficient to confer a marginal increase in Al^{3+} resistance and an improved capacity to utilize sparingly soluble P compared to wild-type plants and null segregants derived from the transgenic lines. Other studies using similar approaches have reported some success in increasing the organic anion efflux from *Medicago sativa* (Tesfaye et al. 2001) and *Brassica napus* (Anoop et al. 2003).

More recently, rice and tomato plants were transformed with the Arabidopsis vacuolar H$^+$-pyrophosphatase gene *AVP1*. The transgenic lines showed approximately 50% greater citrate and malate efflux than wild-type plants when treated with AlPO$_4$. The authors argued that this, combined with greater rhizosphere acidification, was sufficient to enhance resistance to Al^{3+} stress and improve the ability to utilize insoluble phosphorus (Yang et al. 2007). The transgenic lines also exhibited larger roots, longer root hairs and greater shoot biomass, which appear to be caused by perturbations in auxin transport (Li et al. 2005). Since the transgenic plants exhibited many strong phenotypes, the claims of nutrient efficiency and Al^{3+} resistance need to be carefully reconciled with the differences in plant size under control conditions.

Other studies using this strategy have been less successful in manipulating organic anion efflux from roots (Delhaize et al. 2001; Delhaize et al. 2003). For example, attempts by independent groups to repeat the findings of de la Fuente et al. (1997) were unsuccessful despite using identical constructs and increasing the citrate synthase activity in transgenic tobacco lines to similar, or even greater levels, than those published originally (Delhaize et al. 2001; Jian Feng Ma, pers. comm.). These results demonstrate that increases in enzyme activity do not necessarily lead to anion accumulation and enhanced efflux, and suggest that metabolic or environmental factors can influence the effectiveness of this approach.

Approach 2   Engineering transport proteins for greater organic anion efflux

Until recently, there was no prospect of manipulating organic anion efflux from roots by targeting the transport step because the genes encoding the transport proteins had not been identified. Fortunately, over the last few years, the potential for engineering the rhizosphere has been revolutionized by the identification of candidate genes in two different families: the Al^{3+}-activated malate transporter (*ALMT*) and the multidrug and toxic compound extrusion (*MATE*) gene families. Members of these families encode membrane-bound proteins that facilitate the efflux of malate and citrate from plant cells. These recent developments have provided new possibilities for manipulating organic anion efflux from roots.

(i) *ALMT family*

The first gene identified to encode a transport protein that facilitates organic anion efflux from plant cells is *TaALMT1* from wheat (*Triticum aestivum*) (Sasaki et al. 2004). *TaALMT1* encodes the first member of a novel membrane protein family that functions as an anion channel to mediate malate efflux from roots (Ryan et al 1997; Zhang et al. 2008). Constitutive expression of *TaALMT1* in plant and animal cells including tobacco suspension cells, rice (*Oryza sativa*) and *Xenopus* oocytes (Sasaki et al. 2004), barley (*Hordeum vulgare*) (Delhaize et al. 2004) as well as Arabidopsis (Peter Ryan, unpublished data) all generate an Al^{3+}-activated efflux of malate anions. In some cases (tobacco suspension cells, barley and Arabidopsis), malate efflux also increased resistance to Al^{3+} stress. The finding that TaALMT1 confers the same phenotype to a diverse range of plant and animal systems demonstrates that a sustained efflux of organic anions from all these cell types and tissues will occur once a transport pathway across the membrane is provided.

TaALMT1 is a valuable tool for manipulating malate release into the rhizosphere but it has an important limitation that restricts its wider application to plant nutrition—it needs to be activated by Al^{3+}. Even when TaALMT1 is constitutively expressed in a heterologous system, malate efflux will not occur unless Al^{3+} is added to the bathing solution. The most likely explanation for this response is that TaALMT1 is a ligand-gated anion channel that requires direct interaction with Al^{3+} to shift the protein into an active configuration. While this property is ideal for conferring Al^{3+} resistance to transgenic plants, the practical application of TaALMT1 is limited to acid soils where the concentration of Al^{3+} is sufficient to activate the protein. TaALMT1 is one member of a large protein family distributed widely among plant species (Delhaize et al. 2007), of which four additional members have been characterized in detail. Three of these from Arabidopsis and *Brassica napus* are involved in Al^{3+} resistance and are similarly activated by Al^{3+} (Hoekenga et al. 2006; Ligaba et al. 2006). The fifth member of the ALMT family, also from Arabidopsis, functions differently from the others. This protein, AtALMT9, localises to the tonoplast of shoot cells and transports malate out of the cytoplasm and into the vacuole with no requirement for Al^{3+} (Kovermann et al. 2007). AtALMT9 and other proteins that function similarly to it may be more useful members of the ALMT family for manipulating malate release from roots.

(ii) *MATE family*

MATE genes are widely distributed among all kingdoms of living organisms, where they function to export a wide range of small organic compounds including secondary metabolites such as flavanoids and alkaloids, as well as exogenous antibiotics (Eckardt 2001; Omote et al. 2006). More recently, members of the *MATE* family have been shown to facilitate citrate efflux from plant cells. Recent studies have demonstrated that *MATE* genes in sorghum (*Sorghum bicolor*) and barley (*Hordeum vulgare*) control Al^{3+} resistance in these species by facilitating citrate efflux from the roots (Magalhaes et al. 2007; Furukawa et al. 2007; Wang et al. 2007). The citrate binds with the toxic Al^{3+} ions in the rhizosphere and protects the root apices much like malate efflux protects wheat from Al^{3+} stress. The heterologous expression of *SbMATE1* in Arabidopsis, and *HvMATE* in tobacco, conferred Al^{3+}-activated efflux of citrate and increased resistance to Al^{3+} stress (Magalhaes et al. 2007; Furukawa et al. 2007).

Two additional *MATE* genes, *Frd3* and *OsFRDL1* from Arabidopsis and rice (Oryza sativa), respectively, also facilitate citrate efflux from cells but not as part of an Al^{3+}-resistance mechanism. Instead, they control the release of citrate from vascular cells into the xylem, where it forms a complex with iron and moves to the shoot (Durrett et al. 2007; Yokosho et al. 2009). Expression of *Frd3* and *OsFRDL1* in *Xenopus laevis* oocytes revealed a fundamentally different mode of action from MATEs involved in Al^{3+} resistance because both facilitated citrate efflux in the absence of Al^{3+}. Over-expression of *Frd3* in Arabidopsis under the control of the CaMV 35S promoter also conferred greater Al^{3+} resistance because the plants constitutively released citrate from their roots. That FRD3 is active without exposure to Al^{3+} highlights a real advantage over the ALMT1 genes examined so far. FRD3 provides an opportunity to enhance citrate efflux into the rhizosphere over a wider range of soil types. This type of efflux engineering should take in account the metabolic and energy costs associated with both the production and exudation of organic acids anions in the rhizosphere.

This section has described how biotechnology can influence the chemical properties of the rhizosphere. The examples presented provide proof of principle that manipulation of root exudates has the potential to improve plant nutrition and protect plants from stress. It seems likely that increased exudation of organic acids by transgenic plants also will influence the composition and density of rhizosphere microbial populations capable of metabolizing these carbon sources, but the magnitude and potential consequences of increased nutrient availability have yet to be determined.

Engineering microbes

In addition to their direct influence on soil chemistry, root exudates support large microbial populations that afford a basal level of protection to roots against pathogens simply through their metabolic activity (see Section "Natural engineering of microbial populations" below). Some isolates which have additional beneficial properties are plant growth-promoting rhizobacteria (PGPR)—free living strains able to colonize roots and stimulate plant growth. Growth stimulation can be mediated directly, as through enhanced nutrient acquisition or modulation of phytohormone synthesis, or indirectly, through induction of the plant's own defense responses or antagonism of soil-borne pathogens, and several mechanisms of promotion can operate simultaneously in a single strain. Hundreds, if not thousands, of PGPR representing diverse genera have been described over the past 50 years. Despite their appeal as a "natural" means of plant protection few strains have been developed commercially. This is partly because uneconomically large doses often must be applied and performance can be inconsistent in the field. Genetically engineered strains have been valuable in identifying some of the causes of variable strain performance and they offer a means to develop PGPR that are effective at low inoculum doses and under a variety of environmental conditions.

Strategic issues for strain development To be effective, introduced PGPR must establish and maintain biologically active populations in competition with the already-adapted resident microflora. A variety of cell surface molecules contribute to the colonization process (Lugtenberg et al. 2001), as does efficient utilization of the major carbon and nitrogen sources available in root exudates (Kamilova et al. 2005, 2006), but genetic engineering of individual fitness determinants generally has not led to improved strain performance. In contrast, genes involved in growth promotion have proven effective as targets for strain improvement, either by modifying the timing or level of their expression or by transferring and expressing them in alternate hosts with other desirable attributes. Important considerations for field application of such strains include not only whether plant growth is enhanced, but also whether the engineered trait is stably maintained and expressed, its effects on the fitness of the modified strain, and the effects of the modified strain on nontarget organisms in the environment.

PGPR activity is enhanced in engineered strains The *chiA* gene, encoding a chitinase targeted to chitin in the cell walls of fungal root pathogens, was among the first to be transferred to heterologous bacteria with the goal of enhancing PGPR activity via pathogen suppression. *Escherichia coli* (which itself lacks the ability to control plant pathogens) expressing *chiA* caused rapid and extensive bursting of the hyphal tips of *Sclerotium rolfsii* and effectively reduced its ability to cause disease on bean (Shapira et al. 1989). An engineered strain of *Pseudomonas* expressing a *chiA* gene from *Serratia marcescens* more effectively controlled *Fusarium oxysporum* f. sp. *redolens* and *Gaeumannomyces graminis* var. *tritici* (Sundheim et al. 1988), and *P. fluorescens* expressing the introduced gene constitutively, either on a plasmid or integrated into the genome (Downing and Thomson 2000; Koby et al. 1994), provided improved control of fungal pathogens in the growth chamber or greenhouse.

Many rhizobacteria produce phytohormones that undoubtedly affect root growth, but efforts to engineer the rhizosphere through hormone manipulation have focused mainly on degradation of so-called "stress" ethylene, which is synthesized by plants upon exposure to stresses such as flooding, drought, salt, and the presence of metals, organic contaminants and pathogens. Ethylene is required for proper plant development, growth and survival, including the induction of systemic defense mechanisms against pathogens, but the elevated levels associated with stress can trigger damaging physiological processes that exacerbate environmental pressures and exaggerate disease symptoms (reviewed by Glick et al. 2007). Rhizobacteria that produce 1-aminocyclopropane-1-carboxylate (ACC) deaminase, which catalyzes the cleavage of ACC, the immediate precursor of ethylene, are a metabolic sink for ACC, diminishing ethylene concentrations to levels too low to initiate the physiological cascade leading to reduced growth or plant tissue damage. *P. fluorescens* CHA0 transformed with the ACC deaminase gene *acdS* from *P. putida* UW4 (formerly classified as *Enterobacter cloacae*) increased root length in canola seedlings and provided improved protection of cucumber against *Pythium,* demonstrating the involvement of ethylene in this plant-pathogen interaction (Wang et al. 2000). Canola, tomato and tobacco plants grown in the presence of ACC deaminase-producing bacteria or transformed directly with the bacterial *acdS* gene generally exhibited improved growth upon exposure to abiotic stresses in the greenhouse (Glick et al. 2007). Ethylene's central regulatory role in plant growth and survival, and the ability of PGPR to modulate stress ethylene levels, make it an attractive target for rhizosphere engineering, particularly when plant growth is limited due to abiotic stress. A more complex situation requiring empirical evaluation may prevail with pathogen-induced stresses where ethylene is needed to trigger the systemic defense cascade but stress ethylene levels can exacerbate pathogen-induced apoptosis and tissue damage.

Successful disease control by PGPR that produce antibiotics requires that such strains produce active compounds in sufficient quantity and within a window of opportunity dictated by the disease cycle of the target pathogen. This can be particularly challenging for the suppression of pre-and post-emergence damping-off diseases caused by *Pythium* and *Rhizoctonia* spp., which can attack seeds and seedlings before introduced strains are able to synthesize inhibitory levels of metabolites. Ligon et al. (2000) addressed this problem by engineering derivatives of *P. fluorescens* BL915, which synthesizes the antifungal compound pyrrolnitrin. In one derivative the regulatory gene *gacA*, which controls the synthesis of pyrrolnitrin in *Pseudomonas*, was constitutively expressed on a multicopy plasmid in BL915. This regulatory derivative produced about 2.5-fold more pyrrolnitrin than the parental strain. A second derivative in which the entire four-gene *prnABCD* operon was constitutively expressed from a plasmid produced 4-fold more pyrrolnitrin, and production was increased to levels 10-fold over those of the parental strain when both plasmids were expressed in the same cells. In greenhouse trials, the derivative strains were as protective of cucumber and impatiens in soil infested with *R. solani* as BL915 applied at 10-fold higher doses. In the field, control of *R. solani* and *P. ultimum* on cotton was better than that provided by BL915 and not significantly different from chemically-treated and healthy controls. Other experiments showed that control of *R. solani* on cotton was proportional to the amount of pyrrolnitrin synthesized in vitro (Ligon et al. 2000).

The results of Ligon et al. (2000) suggest that constitutive production or overproduction of antibiotics can improve strain effectiveness and are supported by other works in which biosynthesis operons for the antibiotics phenazine-1-carboxylic acid (PCA) and 2,4-diacetylphloroglucinol (DAPG) were transferred to heterologous strains to enhance control of damping-off diseases. In one case, the constitutively expressed seven-gene PCA operon was introduced into random sites in the genome of *P. fluorescens* SBW25. Chromosomal insertion provides genetic stability and effective gene containment, which are important in minimizing the potential for horizontal gene transfer in the rhizosphere. The PCA-producing derivatives reduced damping-off disease on pea seedlings more effectively than did wild-type SBW25 and the level of PCA produced was correlated directly with the efficacy of the bacteria. Moreover, pretreatment of the soil with the engineered strain effectively decontaminated it, reducing disease incidence (Timms-Wilson et al. 2000). In another study, engineered derivatives of *P. fluorescens* 5-2/4 expressing an integrated cassette carrying the DAPG biosynthesis operon from *P. fluorescens* Q2-87 (a B genotype strain; see Section "Take-all decline in Washington State: a case study") provided increased control of *P. ultimum* (Alsanius et al. 2002).

The PCA integrative cassette has been exploited to extend the range of diseases controlled by *P. fluorescens* Q8r1-96 (a D-genotype strain; see Section "Take-all decline in Washington State: a case study"), which produces DAPG and is highly effective against the take-all pathogen *Gaeumannomyces graminis* var. *tritici* but less effective against root rot caused by *Rhizoctonia solani*. Expression of the phenazine biosynthesis operon normally is regulated by quorum sensing, and as expected, a constitutively engineered derivative in Q8r1-96 produced more PCA, both in vitro and in the wheat rhizosphere (Huang et al. 2004), than did *P. fluorescens* 2-79, from which the operon had been cloned. Surprisingly, however, the PCA-producing derivatives also synthesized more DAPG than did the parental strain, suggesting that a PCA biosynthesis intermediate might have derepressed expression of the DAPG operon. Levels of antibiotic synthesis also differed among individual recombinants, presumably due to positional effects resulting from random insertion of the PCA cassette within the Q8r1-96 genome. As in transgenic plant lines, it is necessary to compare several independent recombinant clones with the parental and plasmid-transformed control strains. Vectors targeting introduced genes to a neutral chromosomal site (Craig 2002) also can be used. In the greenhouse, PCA-producing strains suppressed rhizoctonia root rot at inoculum doses 10-to 100-fold lower than the dose of Q8r1-96 required for comparable disease suppression (Huang et al. 2004) and in the field, wheat treated with the engineered strains consistently had yields 8–20% greater than those treated with Q8r1-96 alone (D. M. Weller et al., unpublished), supporting the feasibility of pyramiding antibiotic pathways to control mixed populations of yield-limiting soil-borne pathogens that typically exist in the field.

Recombinant strains and rhizosphere competence The ecological fitness of PGPR, whether engineered or not, is an important factor in evaluating the potential risks associated with their release into the environment. Using isogenic derivatives of *P. fluorescens* SBW25 tagged with marker genes, De Leij et al. (1998) detected no effect of metabolic burden on the bacteria in the rhizosphere of pea or wheat. The question of fitness also has been addressed in strain Q8r1-96 and its PCA-producing derivatives. Because PCA contributes to the competitiveness of *Pseudomonas* strains (Mazzola et al. 1992), it is conceivable that Q8r1-96 constitutively producing PCA would be more competitive than the wild type. On the other hand, PCA synthesis is energetically costly, and constitutive production, combined with the upregulation of DAPG synthesis in the recombinant strains, likely would create a metabolic load that would reduce the fitness of the recombinants. Indeed, this could explain why wild-type isolates producing both PCA and DAPG are seldom if ever found in nature. To distinguish between these possibilities, the persistence on wheat of Q8r1-96 and its PCA-producing derivative Z30-97 was monitored under greenhouse conditions (Bankhead et al. 2004a) and in a 3-year field study (Bankhead et al. 2004b). No consistent strain-specific differences in rhizosphere competence were observed when Q8r1-96 and Z30-97 colonized separate rhizospheres. In contrast, when the strains were co-inoculated on the same plants to provide more intense competitive pressure, Q8r1-96 displaced Z30-97 in the wheat rhizosphere (Bankhead et al. 2004a). Collectively, the data suggest that on wheat, any benefit of PCA synthesis ultimately was overridden by its metabolic cost to the engineered strain, but apparently that cost was not enough to adversely affect competitiveness against indigenous rhizosphere microorganisms, even over 3 field seasons. In related studies in which these same strains were co-inoculated on barley, pea and navy bean under controlled conditions to evaluate the effect of the plant host, wild-type Q8r1-96 outcompeted Z30-97 on barley, both strains maintained similar population densities on navy bean, and surprisingly, the engineered strain displaced the wild type on pea (S.B. Bankhead, L.S. Thomashow and D.M. Weller, unpublished). The results indicate that the crop species modulates strain competitiveness and must be considered when assessing the potential fate of and risk posed by the release of recombinant strains into the environment.

Non-target effects of wild-type and genetically engineered PGPR Studies of the non-target effects of antibiotic-producing and non-producing PGPR introduced into the rhizosphere have considered the abundance and community structure of microorganisms that are closely related or unrelated to the introduced rhizobacteria, soil enzyme activities and available nutrients, microbial indicators such as rhizobia, protozoa and nematodes, and effects on the plant (review: Winding et al. 2004). Some of the best studies to date of the population dynamics and non-target effects of recombinant rhizobacteria have been conducted with *P. putida* WCS358r, modified to produce either PCA or DAPG (Glandorf et al. 2001; Leeflang et al. 2002; Viebahn et al. 2003). This work is especially notable because it was conducted in the field, PCA was shown to be produced in the rhizosphere by the recombinant strain, and both cultivation-dependent and cultivation-independent methods were employed to quantify non-target effects. The results indicated that the wild-type and recombinant strains both had transient effects on the composition of the rhizosphere fungal and bacterial microflora of wheat, and the effects of the recombinant strains sometimes were longer-lasting. Perhaps more importantly, the impact of the recombinant strains differed from year to year and study to study. These results, which mirror those of most other studies conducted under controlled and field conditions, are not so surprising given that WCS358r and other PGPR typically establish very high population densities immediately after inoculation, and then the densities decline (sometimes precipitously) over time and distance from the inoculum source. In addition, introduced rhizobacteria do not become uniformly distributed throughout the rhizosphere or among roots of the same or different plants. Collectively, studies of the non-target effects of wild-type and transgenic rhizobacteria indicate that while the bacteria have definite impacts on non-target bacterial, fungal and protozoan populations, the effects vary from study to study and are transient.

Engineering microbial populations and plant-microbe interactions

The influence of plant species

The plant genotype profoundly influences both the quantity and composition of indigenous microorganisms and the population dynamics of introduced rhizobacteria. This occurs because rhizosphere microorganisms are dependent for their growth on substrates liberated from the root and because rhizodeposition is largely under the genetic control of the plant. Numerous studies demonstrating this concept occur in the earlier literature (Miller et al. 1989). The classic studies conducted by Neal, Atkinson and Larson (Neal et al. 1970, 1973; Atkinson et al. 1975) utilizing chromosomal substitution lines of wheat demonstrated beautifully the impact of plant genotype on rhizosphere microflora. The wheat lines S-615 and Rescue, both of which were susceptible to common root rot, harbored larger populations of rhizosphere bacteria than did the root rot-resistant line Apex. Substitution of the chromosome pair 5B from Apex for its homologue in S-615 yielded the chromosomal substitution line S-A5B that was as resistant as Apex to root rot. Furthermore, the bacterial population in the rhizosphere of S-A5B was similar to that of Apex. In addition, the resistant lines had greater percentages of bacteria antagonistic to the common root rot pathogen than did the susceptible lines.

Lemanceau et al. (1995) isolated fluorescent pseudomonads from a silty loam soil and from the rhizosphere, rhizoplane or root interior of flax and tomato grown in that soil. Isolates were characterized on the basis of enzyme activities and their ability to utilize organic substrates. Numerical analysis of these characteristics was used to cluster the isolates, and the results indicated that the plant species had a selective influence on the populations. Isolates from flax and tomato were grouped into nine clusters. Flax isolates were distributed in six clusters, three of which were specific for flax. Tomato isolates also formed six clusters, two of which were specific for tomato. Furthermore, a much greater proportion of tomato isolates than flax isolates could assimilate inositol, ribose, saccharose, trehalose, erythritol, m-hydroxybenzoate, and 5-cetogluconate.

More recently, studies targeting DAPG-producing *P. fluorescens* ($phlD^+$) strains known for their PGPR activity have demonstrated that plant species and varieties differentially enrich and support populations (Bergsma-Vlami et al. 2005; De La Fuente et al. 2006; Mazzola et al. 2004) and genotypes (see Section "Take-all decline in Washington State: a case study"; Landa et al. 2002; Landa et al. 2006; Mazzola et al. 2004; Picard et al. 2004) of this specific group of *P. fluorescens*, and that DAPG accumulation in the rhizosphere also differs among plant hosts (Bergsma-Vlami et al. 2005). For example, wheat, sugar beet and potato grown in a Dutch soil supported greater DAPG production and population densities of $phlD^+$ isolates than did lily (Bergsma-Vlami et al. 2005). Notz et al. (2001) showed that DAPG accumulation by *P. fluorescens* CHA0 is correlated with expression of the DAPG biosynthesis gene *phlA*, and expression was significantly greater in the rhizosphere of monocots (maize and wheat) than dicots (bean and cucumber). They also observed differences in gene expression on six maize cultivars. Wheat varieties also differentially support $phlD^+$ populations, genotypes of $phlD^+$ strains, and DAPG accumulation on roots. For example, Mazzola et al. (2004) showed that the population densities of $phlD^+$ isolates from wheat varieties grown in apple orchard soils differed significantly. For instance, variety 'Lewjain' supported densities $>10^5$ CFU g^{-1} root, but no $phlD^+$ isolates were detected on 'Eltan.' Okubara and Bonsall (2008) quantified DAPG on roots of wheat varieties 'Tara,' 'Buchanan,' and 'Finley' inoculated with *P. fluorescens* Q8r1-96 (D genotype) or Q2-87 (B genotype) and grown on moist filter paper. Both strains colonized roots equally, but DAPG accumulation differed significantly between strains and among wheat varieties.

Natural engineering of microbial populations

Disease-suppressive soils provide some of the best examples in which plants protect themselves against soil-borne pathogens by "naturally engineering" the composition of rhizosphere microbial populations. As defined by Baker and Cook (1974), suppressive soils are those "in which the pathogen does not establish or persist, establishes but causes little or no damage, or establishes and causes disease for awhile but thereafter the disease is less important, although the pathogen may persist in the soil." In contrast, conducive (non-suppressive) soils are soils in which disease readily occurs. "General suppression," the aforementioned ability of essentially all soils to suppress the growth or activity of soilborne pathogens to a limited extent, is due to the activity of the total microbial biomass in soil competing with the pathogen and is not transferable between soils. "Specific suppression" is superimposed over the background of general suppression, is highly effective, is due to the activity of individual or select groups of microorganisms, and is transferable between soils (Weller et al. 2002). Suppressive soils owe their activity to both general and specific suppression.

Suppressive soils are known for many soil-borne pathogens (Weller et al. 2002) and have been further categorized as "long-standing" or "induced" (Hornby 1983, 1998). Long-standing suppression is naturally associated with soil and is of unknown origin whereas induced suppression develops as a result of a cropping practice such as monoculture. Two well-known examples of suppressive soils are take-all decline of wheat (Weller et al. 2002; Hornby 1983; 1998) and scab decline of potato (Menzies 1959).

Take-all decline in Washington State: a case study Take-all, caused by *G. graminis* var. *tritici* (*Ggt*), generally develops at a soil pH of 5.5 to 8.5 and is most severe where wheat is grown under moist conditions, but it also is common under dryland conditions. As a result, take-all occurs throughout the world and is considered the most important root disease of wheat (Cook 2003; Freeman and Ward 2004). The pathogen survives saprophytically in dead roots, crowns, and tiller bases, the inoculum source for the next crop. Primary infection occurs with the growth of dark runner hyphae on the root surface. Hyaline hyphae penetrate the cortex and colonize the vascular tissue, causing characteristic black lesions. Runner hyphae continue to grow over the root surface, to other roots, and upward to the crown and stem bases. Early infection ultimately causes yellowing of lower leaves, stunting, and premature death of plants in patches. Wheat is highly susceptible to take-all but other Poaceae (barley, rye and triticale) also are infected (Cook 2003; Freeman and Ward 2004). Crop rotation and tillage are widely used as controls for take-all, but trends in modern cereal-based production systems are toward less tillage and two or three consecutive wheat crops, and these practices exacerbate the disease. No resistant commercial varieties exist and chemical controls are limited (Cook 2003; Paulitz et al. 2002). Take-all decline (TAD), the spontaneous decrease in take-all incidence and severity induced by continuous wheat or barley monoculture after a severe outbreak of the disease (Hornby 1998, Weller et al. 2002, 2007), also is widely used to manage take-all.

TAD follows a similar pattern worldwide, but local biotic and abiotic factors modulate the speed of its development and its robustness. Amongst these factors, soil physico-chemical properties such as manganese (Heckman et al. 2003) or ammonium concentrations (Sarniguet et al. 1991) may play a significant role. The number of continuous crops of wheat or barley required to initiate TAD varies considerably, but often ranges from four to six (Weller et al. 2002). It has been known for decades that TAD involves microbiological changes in the soil or rhizosphere that suppress the pathogen (Weller et al. 2007). The specific suppression associated with TAD is transferable to a conducive soil and is eliminated by soil pasteurization (60°C, 30 min) or fumigation, and by rotation with non-cereal crops (Hornby 1998; Weller et al. 2002). TAD effectively suppresses severe disease, but yearly fluctuations in its robustness are common. Several types of microorganisms have been studied as potential mechanisms of TAD suppressiveness (Hornby 1998) and beginning in the 1970s, antagonistic *Pseudomonas* spp. were implicated (Smiley 1979; Weller et al. 1988).

A series of studies of soils from irrigated and non-irrigated, and conventionally cultivated and direct-seeded continuous monoculture wheat fields throughout Washington State, U.S.A. (dating back to the late 1960s) resulted in the demonstration of a key role for DAPG-producing *P. fluorescens* in TAD (reviewed in Weller et al. 2007). For example, it was shown that DAPG producers colonized wheat roots from TAD soils at densities above the threshold (10^5 CFU g^{-1} root) (Raaijmakers and Weller 1998) required for take-all control, but were below the threshold or not detected on roots from conducive soils (Raaijmakers et al. 1997). Suppressiveness was lost when DAPG producers were eliminated from TAD soils by soil pasteurization, and adding TAD soil to conducive soil transferred suppressiveness and established populations of DAPG producers above the threshold required for disease control (Raaijmakers and Weller 1998). Introduction of DAPG producers into conducive soils rendered the soils suppressive (Raaijmakers and Weller 1998). Finally, DAPG was detected on roots of wheat grown in TAD soil but not conducive soil (Raaijmakers et al. 1999). The role of DAPG producers in TAD is not restricted to Washington State. DAPG-producing strains of *P. fluorescens* occurred above the threshold needed for take-all suppression on wheat grown in monoculture soils collected from major wheat growing regions of the U.S.A. (Landa et al. 2006; Weller et al. 2007), and De Souza et al. (2003) demonstrated that they also play a role in Dutch TAD soils.

Among worldwide collections of DAPG-producing strains of *P. fluorescens*, 22 genotypes (A-T, Pfy, Pfz) currently have been recognized, with many genotypes showing substantial endemicity while others such as A, D and F are more widely distributed. Several genotypes occur in TAD soils in Washington State, but the D genotype comprises 50–90% of all $phlD^+$ isolates. Genotype D isolates play the major role in Washington TAD soils because of their unique ability to colonize wheat and barley roots and maintain densities necessary for take-all suppression throughout the growing season and from year to year of monoculture (Raaijmakers and Weller 2001).

Of the root pathogens infecting Pacific Northwest wheat fields, *Ggt* is the most sensitive to DAPG (Allende-Molar 2006). Mazzola et al. (1995) screened isolates of *Ggt* from different countries and regions of the U.S.A., and found differential sensitivity to the antibiotic. In addition, they reported that isolates not inhibited by 3 μg ml^{-1} of DAPG were not suppressed by the DAPG-producing strain Q2-87 on wheat roots. A common question about the role of DAPG in take-all suppression is whether isolates of the pathogen with tolerance to the antibiotic become enriched in TAD fields as a result of exposure during wheat monoculture. Kwak et al. (2009) addressed this question by determining the sensitivity of *Ggt* isolated from TAD and non-TAD fields. For isolates from all fields, the 90% effective dose (ED_{90}) of DAPG ranged from 3.1 to 11.1 μg ml^{-1}. Sensitivity of isolates to the antibiotic was normally distributed in all fields and was not correlated with geographic origin or cropping history of the field, indicating that tolerance to DAPG does not develop in the take-all pathogen during monoculture. This study was the first to address the question of emergence of tolerance in a soil-borne pathogen to introduced or indigenous PGPR.

Soil amendment as a means of modifying microbial populations

Modification of the soil or rhizosphere microbial component is often the unintentional consequence of human activity. For instance, soil pollutants may drastically affect the composition of soil and plant-associated microflora (e.g. Colores et al. 2000; Siciliano et al. 2001; reviewed by Lynch 2002), and repeated cultivation of certain crop species can lead to the emergence of disease-suppressive soils (see above).

The development of new techniques in microbiology and microbial ecology has provided opportunities to modify the soil microbiota in a manner paralleling the selective "rhizosphere engineering" that occurs in Nature. For instance, the plant species *Calystegia sepium* (hedge bindweed), *Convolvulus arvensis* (morning glory), *Scopolia japonica* (scopolia) and *Morus alba* (white mulberry) all produce alkaloid calystegyns (Asano et al. 1994, 1996), and numerous calystegin-degrading bacteria (eg. *Sinorhizobium sp.*) have been identified in the rhizosphere of these plants (Tepfer et al. 1988). The microflora around the roots of some legume species is similarly influenced by root exudates. *Leucanea sp.* and *Mimosa sp.* plants produce the amino acid mimosine, which is toxic to many bacteria (Hammond 1995). Nitrogen-fixing *Rhizobiaceae* that inhabit the nodules on these plants have acquired the ability to degrade mimosine, a feature not found in non-colonizing *Rhizobiaceae* (Soedarjo et al. 1994). This example of mimosine "engineering" appears advantageous for both the plant and symbionts: the plants select beneficial, well-adapted nitrogen-fixing symbionts, and the symbionts are able to out-compete other microbes in the root environment (Fox and Borthakur 2001; Soedarjo and Borthakur 1996).

Several examples of microbial engineering have involved the application of chemical and biological amendments to the soil. Devliegher et al. (1995) selected bacterial communities able to degrade Igepal and di-octyl sulfosuccinate from soils amended with these molecules. Amongst the selected bacteria, a detergent-habituated PGPR *Pseudomonas* strain survived better in amended bulk and rhizosphere soils than in control soils. Amendments have also been used to favor plant growth. Though the exact mechanism was not fully understood, a cocktail of Triton X100, EDTA and a strain of *Sinorhizobium sp.* altered the soil microflora enough to improve the growth of *Brassica juncea* plants (Di Gregorio et al. 2006). In earlier series of studies, a strain of *P. fluorescens* was engineered to utilize salicylic acid as a source of carbon (Colbert et al. 1993a, b). Growth of this strain was enhanced in salicylate-amended bulk and rhizosphere soils but the magnitude of the response was sensitive to inoculum level, field site, and soil depth. When dusted onto leaves, salicylate stimulated the epiphytic growth of similarly engineered pseudomonads (Wilson and Lindow 1995) including a strain with biological control activity against bacterial speck disease of tomato (Ji and Wilson 2003).

Bacterial strains capable of interfering with the pathogenicity of *Pectobacterium carotovorum* (formerly *Erwinia carotovora*; Whitehead et al. 2001; Barnard et al. 2007; Faure and Dessaux 2007) were selected by an amendment/enrichment-based approach. The virulence of *Pectobacterium* depends upon the production and sensing of quorum sensing (QS) N-acyl-homoserine-lactone (NAHL) molecules, the concentration of which parallels cell density. When a certain cell density (quorum) of the pathogen is reached, a threshold NAHL concentration is detected by a sensor protein (reviewed in Whitehead et al. 2001; Miller and Bassler 2001; Williams et al. 2007), which activates the synthesis of maceration enzymes such as cellulase, pectate lyase, etc. (review Faure and Dessaux 2007). Bacteria that naturally degrade NAHL have the potential to break this cycle, quenching the pathogenicity of *Pectobacterium*. Several candidates, including promising strains of *Rhodococcus erythropolis*, have been selected from soils amended with NAHL or NAHL analogues (Leadbetter and Greenberg 2000; Dong et al. 2002; Uroz et al. 2003; D'Angelo-Picard et al. 2005). Another strategy developed by Cirou et al. (2007) identified gamma-caprolactone and 4-heptanolide as potent growth stimulators of bacterial communities able to degrade NAHL and quench *Pectobacterium* pathogenicity in potato tuber assays. This ecological engineering approach is both elegant and environmentally friendly because the NAHL analogues, which are used as flavoring agents by the food industry, are degraded by the same bacterial community selected to degrade NAHL.

Plant genetic engineering as a tool to shape plant-associated microbial populations

The strategies described above exploit either natural processes or exogenous soil amendments to manipulate microbial populations. The following section describes alternative approaches which engineer plants to produce compounds that either modulate the growth of defined bacterial rhizosphere populations or modify their biochemical functions.

The opine model Numerous studies have generated transgenic plants that release xenotopic compounds into the rhizosphere. Xenotopic compounds are those that are not naturally present in a particular environment. The compounds targeted in these studies, the opines, are low molecular weight amino acid derivatives that occur in tumors induced by *Agrobacterium tumefaciens* (see Dessaux et al. 1998). Opines can be synthesized and released into the rhizosphere by plants transformed with one to three biosynthesis genes (Savka et al. 1996).

In one such experiment, bacteria colonizing the rhizosphere of *Lotus corniculatus* plants engineered to produce opines were compared with those colonizing near-isogenic wild-type plants (Oger et al. 1997). The population densities of culturable bacteria, agrobacteria, pseudomonads, sporulating, and thermotolerant bacteria were identical in the rhizospheres of the transgenic and control plants. However, the density of the bacterial community able to degrade opines was two to three orders of magnitude greater in the rhizosphere of the opine-producing plants than in that of wild-type plants. Furthermore, the size of this community was related to the magnitude of opine production by the transgenic plants (Oger et al. 2004). More subtle modifications also were noted. For instance, while the total density of all pseudomonads was unaltered by opine production, the proportion of those capable of degrading opines was greater in the rhizosphere of opine-producing plants than in that of wild-type plants. However, this shift in community composition depended upon the type of opines produced (Oger et al. 1997). When opine-producing plants were removed from the soil and replaced by wild-type plants, the population density of the opine-degrading microorganisms declined over time but remained higher than that in control experiments involving only wild-type plants (Oger et al. 2000). The opine-induced bias was observed in two different soils (a clay-rich and a sandy-loamy soil), and for three different plant species (*Lotus corniculatus*, *L. japonicus*, and *Solanum nigrum*), indicating that it was not specific for a single soil type or plant system (Mansouri et al. 2002).

These results support three important conclusions. Firstly they demonstrate that plant exudates directly affect the composition of the rhizosphere microflora. Secondly, they indicate that the rhizosphere "bias" can be extended beyond the period of opine production, an interesting feature in terms of ecological engineering. Thirdly, they provide a precautionary observation that the composition of the rhizosphere microbial community is influenced by root exudation that occurred previously, as well as that in progress at the time of the analysis.

Other studies have investigated whether the opine approach could be used to favor the multiplication of single bacterial strains. Growth of an epiphytic strain of *Pseudomonas syringae* capable of degrading mannityl opines (Wilson et al. 1995) showed a modest 2-to 3-fold increase on the leaves of opine-producing as compared with control plants. In similar studies, growth of a rhizosphere strain of *Pseudomonas* sp. engineered to utilize certain opines was moderately enhanced on the roots of opine-producing plants (Savka and Farrand 1997). Several reasons may account for the limited bacterial growth stimulation observed in these experiments. Firstly, the strains investigated were already well adapted to either the leaf or the root environment. Second, only a single strain was investigated in each study, and competition with other organisms was limited (on the leaf surface) or non-existant (in sterilized soil). In spite of these issues, the opine model is a promising strategy for favoring the growth of a bacterial strain in the plant environment, mindful that factors other than favorable carbon allocation are likely to contribute to strain fitness (Savka et al. 2002).

The opine system from *Agrobacterium* is not the only one that can be used to select for specific plant-bacteria interactions. Other studies have attempted to engineer interactions based on the synthesis (by plants) and the degradation (by bacteria) of rhizopine (i.e., 3-O-methyl-*scyllo*-inosamine), a compound found in nitrogen-fixing nodules of legumes (Murphy et al. 1987). To this end, genes involved in the synthesis and degradation of rhizopine have been isolated from *Sinorhizobium meliloti* (Saint et al. 1993; Murphy et al. 1993). Three rhizopine biosynthesis genes were transformed into *Arabidopsis* and while all were expressed, no rhizopine was detected in any of the transgenic plants (McSpadden-Gardener et al. 1998). The four genes involved in rhizopine degradation also were cloned and tentatively expressed in various bacteria. These genes enabled strains of the *Rhizobiaceae*, but not other bacterial hosts including members of the alpha-proteobacteria, to degrade rhizopine (Rossbach et al. 1994; Rossbach, personal communication). Engineering plant-microbe interactions based on rhizopine metabolism appears to be more complex than anticipated. Nevertheless, since rhizopine-degrading strains are favoured in a rhizopine-rich environment (Gordon et al. 1996), the strategy is worth pursuing, especially because rhizopine could favor the growth of beneficial nitrogen-fixing bacteria.

The quorum quenching model As indicated above, damage caused by the plant pathogen *Pectobacterium* can be reduced by microorganisms that interfere with the QS NAHL signal essential to the pathogen's life cycle. Some workers have taken this concept a step further by modifying plants to degrade or alternatively, to synthesize NAHL signals. In one approach, plants were transformed with a bacterial lactonase gene, *aiiA,* enabling them to degrade NAHL molecules. The *aiiA* gene from *Bacillus sp.* (Dong et al. 2000, 2002) was transferred to potato (*Solanum tuberosum*) to generate transgenic lines that indeed inactivated NAHL (Dong et al. 2001). These lines were more tolerant—not to say resistant—to *Pectobacterium*, as no maceration symptoms were visible on tubers inoculated with the pathogen. Tobacco (*Nicotiana tabacum*) is a non-host species for *Pectobacterium*, so the pathogen only induces a localized hypersensitive reaction in response to the QS-dependent production of the toxic peptide harpin. Transgenic tobacco lines expressing *aiiA* did not exhibit this hypersensitive reaction when inoculated with the pathogen. Therefore, expression of AiiA by these plant species completely abolished the pathogenicity of *Pectobacterium* (Dong et al. 2001).

Other studies have modified plants to constitutively synthesize and release NAHLs so that damaging maceration enzymes are released prematurely from *Pectobacterium*. This benefits plants by enabling them to defend themselves more efficiently against limited numbers of pathogenic bacteria. NAHL synthase genes used in this way include *yenI* from *Yersinia enterocolitica* (Fray et al. 1999) and *expI* from *Pectobacterium* (Mae et al. 2001). Tobacco plants transformed with *yenI* targeted for expression in the chloroplast were able to complement the biocontrol ability of *Pseudomonas aureofaciens* strains defective in NAHL synthesis, indicating that crosstalk indeed occurred between the plant and bacteria (Fray et al. 1999). Reduced hypersensitivity to *Pectobacterium* was seen on non-host tobacco plants producing NAHL (Mae et al. 2001), but increased virulence was reported upon inoculation of potato, a host plant, with the pathogen (Toth et al. 2004).

Collectively, these studies suggest that while quorum quenching strategies incorporating genetically modified plants are conceivable (reviews: Fray 2002; Zhang and Dong 2004), strategies relying on signal degradation appear to be more efficient at present than those involving signal overproduction. Interestingly, the impact of signal overproduction on microbial populations and QS pathways appears limited or non-existent (D'Angelo-Picard et al. 2004). Whether this is true for the signal degradation approach remains to be determined, especially considering that several bacterial functions beneficial to plants (e.g. production of antifungal compounds, root colonisation) rely on QS regulation (e.g. Chin-A-Woeng et al. 2001; Maddula et al. 2006).

Future research orientations

Research to date has shown that the rhizosphere can be engineered through appropriate selection of crop species and varieties, by the introduction of microorganisms or soil amendments, and by genetic modification of plant and microbial biological activities. The emergence of molecular techniques now allows the direct manipulation of genes that influence rhizosphere functions, and continuing advances in biotechnology ensure more progress for the future. High throughput and "omics" techniques further make it possible to screen and analyse large and complex microbial communities in the soil. Genomics has given rise to metagenomics, an approach that will benefit from the remarkable development of mass sequencing procedures and which will enable us to explore the microbial diversity of the rhizosphere more rapidly and in greater detail.

While progress in a number of avenues has been encouraging, our ability to reliably and predictably engineer the rhizosphere remains a challenge. One major scientific obstacle impedes further progress: a detailed understanding of the complex chemical and biological interactions that occur in this zone. The complexity of rhizosphere chemistry and biology continues to present a multitude of "black boxes" to our understanding. Fundamental issues concerning microbial abundance and diversity in the soil remain unresolved. The role of predation and the phenomenon of resilience are poorly understood. The complex relationships between the structure of microbial communities and their function make attempts to predict and manipulate their ecology very difficult, and will certainly remain for several more years the touch stone of rhizosphere ecology.

Societal lack of acceptance of the emerging technology remains a stumbling block as well. Though relatively limited in Canada, China, and the USA, the controversy surrounding genetically modified organisms is strong in Europe and parts of South America, even amongst members of the scientific community. Indeed, trials of genetically-modified organisms, including plants that could benefit the environment, are extremely unlikely in Italy, Germany and France in the near future. The situation might improve if more restrictive regulations for the use of agrochemicals are implemented. However, a movement towards banning of agrochemicals exists. This should encourage the development of more ecologically friendly alternatives, such as non polluting amendments, novel natural biocontrol agents, and —possibly—genetically modified options. The demands of an ever-increasing world population conjugated to a risk of reduction of arable surface (for instance in fertile river deltas, or low lands) will only make these needs more pressing. It is imperative that scientists continue their work so that a more receptive public of the future can benefit from safe, sustainable and environmentally sound agricultural practices.

References

Allende-Molar R (2006) Role of 2, 4-diacetylphloroglucinol-producing Pseudomonas fluorescens in suppression of take-all and Pythium root rots of wheat. Department of Plant Pathology. Washington State University, Pullman, WA USA

Alsanius BW, Hultberg M, Englund JE (2002) Effect of lacZY-marking of the 2, 4-diaceyl-phloroglucinol producing Pseudomonas fluorescens strain 5–2/4 on its physiological performance and root colonization ability. Microbial Res 157:39–45

Anoop VM, Basu U, McCammon MT, McAlister-Henn L, Taylor GJ (2003) Modulation of citrate metabolism alters aluminum tolerance in yeast and transgenic canola overexpressing a mitochondrial citrate synthase. Plant Physiol 132:2205–2217

Asano N, Oseki K, Tomioka E, Kizu H, Matsui K (1994) N-containing sugars from *Morus alba* and their glycosidase inhibitory activities. Carbohydr Res 259:243–255

Asano N, Kato A, Kizu H, Matsui K, Watson AA, Nash RJ (1996) Calystegine B4, a novel trehalase inhibitor from *Scopolia japonica*. Carbohydr Res 293:195–204

Atkinson TG, Neal JL Jr, Larson RI (1975) Genetic control of the rhizosphere microflora of wheat. In: Bruehl GW (ed) Biology and Control of Soil-Borne Plant Pathogens. Am Phytopathol Soc, Saint-Paul, pp 116–122

Baker KF, Cook RJ (1974) Biological control of plant pathogens. Am Phytopathol Soc, Saint Paul, MN, p 433

Bankhead SB, Landa BB, Lutton E, Weller DM, Gardener BBM (2004a) Minimal changes in rhizobacterial population structure following root colonization by wild type and transgenic biocontrol strains. FEMS Microbiol Ecol 49:307–318

Bankhead SB, Schroeder K, Son MY, Thomashow LS, Weller DM (2004) Population dynamics and movement of transgenic *Pseudomonas fluorescens* in the rhizosphere of field-grown wheat. In: Hartmann A, Schmid M, Wenzel W, Hinsinger Ph (eds) Rhizosphere 2004: Perspectives and Challenges. A Tribute to Lorenz Hiltner. GSF Forschungszentrum, p 165

Barnard AM, Bowden SD, Burr T, Coulthurst SJ, Monson RE, Salmond GP (2007) Quorum sensing, virulence and secondary metabolite production in plant soft-rotting bacteria. Philos Trans R Soc Lond B Biol Sci 362:1165–1183

Bergsma-Vlami M, Prins ME, Raaijmakers JM (2005) Influence of plant species on population dynamics, genotypic diversity and antibiotic production in the rhizosphere by indigenous *Pseudomonas* spp. FEMS Microbiol Ecol 52:59–69

Bowen GD, Rovira AD (1999) The rhizosphere and its management to improve plant growth. Adv Agron 66:1–102

Chin-A-Woeng TF, van den Broek D, de Voer G, van der Drift KM, Tuinman S, Thomas-Oates JE, Lugtenberg BJ, Bloemberg GV (2001) Phenazine-1-carboxamide production in the biocontrol strain *Pseudomonas chlororaphis* PCL1391 is regulated by multiple factors secreted into the growth medium. Mol Plant Microbe Interact 14:969–979

Cirou A, Diallo S, Kurt C, Latour X, Faure D (2007) Growth promotion of quorum-quenching bacteria in the rhizosphere of *Solanum tuberosum*. Environ Microbiol 9:1511–1522

Clough SJ, Bent AF (1998) Floral dip: a simplified method for *Agrobacterium*-mediated transformation of *Arabidopsis thaliana*. Plant J 16:735–743

Colbert SF, Hendson M, Ferri M, Schroth MN (1993a) Enhanced growth and activity of a biocontrol bacterium genetically engineered to utilize salicylate. Appl Environ Microbiol 59:2071–2076

Colbert SF, Schroth MN, Weinhold AR, Hendson M (1993b) Enhancement of population densities of *Pseudomonas putida* ppg7 in agricultural ecosystems by selective feeding with the carbon source salicylate. Appl Environ Microbiol 59:2064–2070

Colores GM, Macur RE, Ward DM, Inskeep WP (2000) Molecular analysis of surfactant-driven microbial population shifts in hydrocarbon-contaminated soil. Appl Environ Microbiol 66:2959–2964

Cook RJ (2003) Take-all of wheat. Physiol Mol Plant Path 62:73–86

Craig N (2002) Tn7. In: Craig N, Craigie R, Gellert M, Lambowitz A (eds) Mobile DNA II. ASM, Washington, DC, pp 423–456

D'Angelo-Picard C, Faure D, Carlier A, Uroz S, Raffoux A, Fray R, Dessaux Y (2004) Bacterial populations in the rhizosphere of tobacco plants producing the quorum-sensing signals hexanoyl-homoserine lactone and 3-oxo-hexanoyl-homoserine lactone. FEMS Microbiol Ecol 51:19–29

D'Angelo-Picard C, Faure D, Penot I, Dessaux Y (2005) Diversity of N-acyl homoserine lactone-producing and -degrading bacteria in soil and tobacco rhizosphere. Environ Microbiol 7:1796–1808

de la Fuente JM, Ramírez-Rodríguez V, Cabrera-Ponce JL, Herrera-Estrella L (1997) Aluminum tolerance in transgenic plants by alteration of citrate synthesis. Science 276:1566–1568

de la Fuente-Martinez JM, Herrera-Estrella L (1999) Advances in the understanding of aluminum toxicity and the development of aluminum-tolerant transgenic plants. Adv Agron 66:103–120

De La Fuente L, Landa BB, Weller DM (2006) Host crop affects rhizosphere colonization and competitiveness of 2, 4-diacetylphloroglucinol-producing *Pseudomonas fluorescens*. Phytopathology 96:751–762

De Leij FAAM, Thomas CE, Bailey MJ, Whipps JM, Lynch JM (1998) Effect of insertion site and metabolic load on the environmental fitness of a genetically modified *Pseudomonas fluorescens* isolate. Appl Environ Microbiol 64:2634–2638

de Sousa CNA (1998) Classification of Brazilian wheat cultivars for aluminium toxicity in acid soil. Plant Breed 117:217–221

de Souza JT, Weller DM, Raaijmakers JM (2003) Frequency, diversity, and activity of 2, 4-diacetylphloroglucinol-producing fluorescent Pseudomonas spp. in Dutch take-all decline soils. Phytopathology 93:4–63

Delhaize E, Gruber BD, Ryan PR (2007) The roles of organic anion permeases in aluminium tolerance and mineral nutrition. FEBS Lett 581:2255–2262

Delhaize E, Hebb DM, Ryan PR (2001) Expression of a *Pseudomonas aeruginosa* citrate synthase gene in tobacco is not associated with either enhanced citrate accumulation or efflux. Plant Physiol 125:2059–2067

Delhaize E, Ryan PR, Hebb DM, Yamamoto Y, Sasaki T, Matsumoto II (2004) Engineering high-level aluminum tolerance in barley with the ALMT1 gene. Proc Natl Acad Sci USA 101:15249–15254

Delhaize E, Ryan PR, Hocking PJ, Richardson AE (2003) Effects of altered citrate synthase and isocitrate dehydrogenase expression on internal citrate concentrations of tobacco (*Nicotiana tabacum* L.). Plant Soil 248V:137–144

Delhaize E, Ryan PR, Randall PJ (1993) Aluminum tolerance in wheat (*Triticum aestivum* L.) II. Aluminum stimulated excretion of malic acid from root apices. Plant Physiol 103:695–702

Dessaux Y, Petit A, Farrand SK, Murphy PM (1998) Opines and opine-like molecules in plant-Rhizobiaceae interactions. In: Kondorosi A, Spaink H, Hooykaas P (eds) The Rhizobiaceae. Kluwer Academic, Dordrecht, pp 173–197

Devliegher W, Arif M, Verstraete W (1995) Survival and plant growth promotion of detergent-adapted *Pseudomonas fluorescens* ANP15 and *Pseudomonas aeruginosa* 7NSK2. Appl Environ Microbiol 61:3865–3871

Di Gregorio S, Barbafieri M, Lampis S, Sanangelantoni ΛM, Tassi E, Vallini G (2006) Combined application of Triton X-100 and *Sinorhizobium sp.* Pb002 inoculum for the improvement of lead phytoextraction by *Brassica juncea* in EDTA amended soil. Chemosphere 63:293–299

Dinkelaker B, Hengeler C, Marschner H (1995) Distribution and function of proteoid roots and other root clusters. Bot Acta 108:183–200

Dong YH, Xu JL, Li XZ, Zhang LH (2000) AiiA, an enzyme that inactivates the acylhomoserine lactone quorum-sensing signal and attenuates the virulence of *Erwinia carotovora*. Proc Natl Acad Sci USA 97:3526–3531

Dong YH, Wang LH, Xu JL, Zhang HB, Zhang XF, Zhang LH (2001) Quenching quorum-sensing-dependent bacterial infection by an N-acyl homoserine lactonase. Nature 411:813–817

Dong YH, Gusti AR, Zhang Q, Xu JL, Zhang LH (2002) Identification of quorum-quenching N-acyl homoserine lactonases from *Bacillus* species. Appl Environ Microbiol 68:1754–1759

Downing K, Thomson JA (2000) Introduction of the *Serratia marcescens chiA* gene into an endophytic *Pseudomonas fluorescens* for the biocontrol of phytopathogenic fungi. Can J Microbiol 46:363–369

Durrett TP, Gassmann W, Rogers EE (2007) The FRD3-mediated efflux of citrate into the root vasculature is necessary for efficient iron translocation. Plant Physiol 144:197–205

Eckardt NA (2001) Move it on out with MATEs. Plant Cell 13:1477–1480

Faure D, Dessaux Y (2007) Novel biocontrol strategies directed at *Pectobacterium carotovorum*. Eur J Plant Pathol 119:353–365

Fox PM, Borthakur D (2001) Selection of several classes of mimosine-degradation-defective Tn3Hogus-insertion mutants of *Rhizobium sp.* strain TAL1145 on the basis of mimosine-inducible GUS activity. Can J Microbiol 47:488–494

Fray RG (2002) Altering plant-microbe interaction through artificially manipulating bacterial quorum sensing. Ann Bot (Lond) 89:245–253

Fray RG, Throup JP, Daykin M, Wallace A, Williams P, Stewart GS, Grierson D (1999) Plants genetically modified to produce N-acylhomoserine lactones communicate with bacteria. Nat Biotechnol 17:1017–1020

Freeman J, Ward E (2004) *Gaeumannomyces graminis*, the take-all fungus and its relatives. Mol Plant Pathol 5:235–252

Furukawa J, Yamaji N, Wang H, Mitani N, Murata Y, Sato K, Katsuhara M, Takeda K, Ma JF (2007) An aluminum-activated citrate transporter in barley. Plant Cell Physiol 48:1081–1091

Gaxiola RA, Palmgren MG, Schumacher K (2007) Plant proton pumps. FEBS Lett 581:2204–2214

Gévaudant F, Duby G, von Stedingk E, Zhao RM, Morsomme P, Boutry M (2007) Expression of a constitutively activated plasma membrane H+-ATPase alters plant development and increases salt tolerance. Plant Physiol 144:1763–1776

Glandorf DCM, Verheggen P, Jansen T, Jorritsma JW, Smit E, Leeflang P, Wernars K, Thomashow LS, Laureijs E, Thomas-Oates JE, Bakker PAHM, Van Loon LC (2001) Effect of genetically modified *Pseudomonas putida* WCS358r on the fungal rhizosphere microflora of field-grown wheat. Appl Environ Microbiol 67:3371–3378

Glick BR, Cheng Z, Czarny J, Duan J (2007) Promotion of plant growth by ACC deaminase-producing soil bacteria. Eur J Plant Pathol 119:329–339

Gordon DM, Ryder MH, Heinrich K, Murphy PJ (1996) An experimental test of the rhizopine concept in *Rhizobium meliloti*. Appl Environ Microbiol 62:3991–3996

Hammond AC (1995) *Leucaena* toxicosis and its control in ruminants. J Anim Sci 73:1487–1492

Heckman JR, Clarke BB, Murphy JA (2003) Optimizing manganese fertilization for the suppression of take-all patch disease on creeping bentgrass. Crop Science 43:1395–1398

Hinsinger P, Plassard C, Tang CX, Jaillard B (2003) Origins of root-mediated pH changes in the rhizosphere and their responses to environmental constraints: A review. Plant Soil 248:43–59

Hoekenga OA, Maron LG, Cançado GMA, Piñeros MA, Shaff J, Kobayashi Y, Ryan PR, Dong B, Delhaize E, Sasaki T, Matsumoto M, Koyama H, Kochian LV (2006) *AtALMT1*, which encodes a malate transporter, is identified as one of several genes critical for aluminium tolerance in Arabidopsis. Proc Natl Acad Sci USA 103:9738–9743

Hornby D (1983) Suppressive soils. Ann Rev Phytopathol 21:65–85

Hornby D (1998) Take-all of cereals: a regional perspective. CAB International, Wallingford

Huang ZY, Bonsall RF, Mavrodi DV, Weller DM, Thomashow LS (2004) Transformation of *Pseudomonas fluorescens* with genes for biosynthesis of phenazine-1-carboxylic acid improves biocontrol of rhizoctonia root rot and in situ antibiotic production. FEMS Microbiol. Ecol. 49:243–251

Ji P, Wilson M (2003) Enhancement of population size of a biological control agent and efficacy in control of bacterial speck of tomato through salicylate and ammonium sulfate amendments. Appl Environ Microbiol 69:1290–1294

Kamilova F, Kravchenko LV, Shaposhnikov AI, Azarova T, Makarova N, Lugtenberg B (2006) Organic acids, sugars, and L-tryptophane in exudates of vegetables growing on stonewool and their effects on activities of rhizosphere bacteria. Mol Plant-Microbe Interact 19:250–256

Kamilova F, Validov S, Azarova T, Mulders I, Lugtenberg B (2005) Enrichment for enhanced competitive root tip colonizers selects for a new class of biocontrol bacteria. Environ Microbiol 7:1809–1817

Koby S, Schickler H, Chet I, Oppenheim AB (1994) The chitinase encoding Tn7-based *chiA* gene endows *Pseudomonas fluorescens* with the capacity to control plant-pathogens in soil. Gene 147:81–83

Koyama H, Kawamura A, Kihara T, Hara T, Takita E, Shibata D (2000) Overexpression of mitochondrial citrate synthase in *Arabidopsis thaliana* improved growth on a phosphorus-limited soil. Plant Cell Physiol 41:1030–1037

Koyama H, Takita E, Kawamura A, Hara T, Shibata D (1999) Over expression of mitochondrial citrate synthase gene improves the growth of carrot cells in Al-phosphate medium. Plant Cell Physiol 40:482–488

Kovermann P, Meyer S, Hortensteiner S, Picco C, Scholz-Starke J, Ravera S, Lee Y, Martinoia E (2007) The arabidopsis vacuolar malate channel is a member of the ALMT family. Plant J 52:1169–1180

Kwak YS, Bakker PAHM, Glandorf DM, Rice JT, Paulitz TC, Weller DM (2009) Diversity, virulence and 2,4-diacetylphloroglucinol sensitivity of *Gaeumannomyces graminis* var. *tritici* isolates from Washington State. Phytopathol (in press)

Landa BB, Mavrodi OV, Raaijmakers JM, Gardener BBM, Thomashow LS, Weller DM (2002) Differential ability of genotypes of 2, 4-diacetylphloroglucinol-producing Pseudomonas fluorescens strains to colonize the roots of pea plants. Appl Environ Microbiol 68:3226–3237

Landa BB, Mavrodi OV, Schroeder KL, Allende-Molar R, Weller DM (2006) Enrichment and genotypic diversity of *phlD*-containing fluorescent *Pseudomonas* spp. in two soils after a century of wheat and flax monoculture. FEMS Microbiol Ecol 55:351–368

Leadbetter JR, Greenberg EP (2000) Metabolism of acyl-homoserine lactone quorum-sensing signals by *Variovorax paradoxus*. J Bacteriol 182:6921–6926

Leeflang P, Smit E, Glandorf DCM, van Hannen EJ, Wernars K (2002) Effects of *Pseudomonas putida* WCS358r and its genetically modified phenazine producing derivative on the *Fusarium* population in a field experiment, as determined by 18S rDNA analysis. Soil Biol Biochem 34:1021–1025

Lemanceau P, Corberand T, Gardan L, Latour X, Laguerre G, Boeufgras J, Alabouvette C (1995) Effect of two plant species, flax (*Linum usitatissinum* L.) and tomato (*Lycopersicon esculentum* Mill.), on the diversity of soilborne populations of fluorescent pseudomonads. Appl Environ Microbiol 61:1004–1012

Li JS, Yang HB, Peer WA, Richter G, Blakeslee J, Bandyopadhyay A, Titapiwantakun B, Undurraga S, Khodakovskaya M, Richards EL, Krizek B, Murphy AS, Gilroy S, Gaxiola R (2005) *Arabidopsis* H^+-PPase AVP1 regulates auxin-mediated organ development. Science 310:121–125

Ligaba A, Katsuhara M, Ryan PR, Shibasaka M, Matsumoto H (2006) The *BnALMT1* and *BnALMT2* genes from *Brassica napus* L. encode aluminum-activated malate transporters that enhance the aluminum resistance of plant cells. Plant Physiol 142:1294–1303

Ligon JM, Hill DS, Hammer PE, Torkewitz NR, Hofmann D, Kempf HJ, van Pee KH (2000) Natural products with antifungal activity from *Pseudomonas* biocontrol bacteria. Pest Manag Sci 56:688–695

López-Bucio J, Martínez de la Vega O, Guevara-García A, Herrera-Estrella L (2000a) Enhanced phosphorus uptake in transgenic tobacco plants that overproduce citrate. Nature Biotech 18:450–453

López-Bucio J, Nieto-Jacobo MF, Ramirez-Rodriguez V, Herrera-Estrella L (2000b) Organic acid metabolism in plants: from adaptive physiology to transgenic varieties for cultivation in extreme soils. Plant Sci 160:1–13

Lugtenberg BJJ, Dekkers L, Bloemberg GV (2001) Molecular determinants of rhizosphere colonization by *Pseudomonas*. Annu Rev Phytopathol 39:461–490

Lynch JM (2002) Resilience of the rhizosphere to anthropogenic disturbance. Biodegradation 13:21–27

Ma JF, Ryan PR, Delhaize E (2001) Aluminium tolerance in plants and the complexing role of organic acids. Trends Plant Sci 6:273–278

Maddula VS, Zhang Z, Pierson EA, Pierson LS 3rd (2006) Quorum sensing and phenazines are involved in biofilm formation by Pseudomonas chlororaphis (aureofaciens) strain 30–84. Microb Ecol 52:289–301

Mae A, Montesano M, Koiv V, Palva ET (2001) Transgenic plants producing the bacterial pheromone N-acyl-homoserine lactone exhibit enhanced resistance to the bacterial phytopathogen *Erwinia carotovora*. Mol Plant–Microbe Interact 14:1035–1042

Magalhaes JV, Liu J, Guimaraes CT, Lana UGP, Alves VMC, Wang YH, Schaffert RE, Hoekenga OA, Pineros MA, Shaff JE, Klein PE, Carneiro NP, Coelho CM, Trick HN, Kochian LV (2007) A gene in the multidrug and toxic compound extrusion (MATE) family confers aluminum tolerance in sorghum. Nat Genet 39:1156–1161

Mansouri H, Petit A, Dessaux Y (2002) Engineered rhizosphere: the trophic bias generated by opine-producing plants is independent of the opine-type, the soil origin and the plant species. Appl Environ Microbiol 68:2562–2566

Mazzola M, Cook RJ, Pierson LS, Thomashow LS, Weller DM (1992) Contribution of phenazine antibiotic biosynthesis to the ecological competence of fluorescent pseudomonads in soil habitats. Appl Environ Microbiol 58:2616–2624

Mazzola M, Fujimoto DK, Thomashow LS, Cook RJ (1995) Variation in sensitivity of *Gaeumannomyces graminis* to antibiotics produced by fluorescent *Pseudomonas* spp. and effect on biological control of take-all of wheat. Appl Environ Microbiol 61:2554–2559

Mazzola M, Funnell DL, Raaijmakers JM (2004) Wheat cultivar-specific selection of 2, 4-diacetylphloroglucinol-producing fluorescent Pseudomonas species from resident soil populations. Microbiol Ecol 48:338–348

McSpadden-Gardener BB, de Bruijn FJ (1998) Detection and isolation of novel rhizopine-catabolizing bacteria from the environment. Appl Environ Microbiol 64:4944–4949

Menzies JD (1959) Occurrence and transfer of a biological factor in soil that suppresses potato scab. Phytopathology 49:648–652

Miller MB, Bassler BL (2001) Quorum sensing in bacteria. Annu Rev Microbiol 55:165–99

Miller HJ, Henken G, van Veen JA (1989) Variation and composition of bacterial populations in the rhizosphere of maize, wheat, and grass cultivars. Can J Microbiol 35:656–660

Murphy PJ, Heycke N, Banfalvi Z, Tate ME, de Bruijn F, Kondorosi A, Tempé J, Schell J (1987) Genes for the catabolism and synthesis of an opine-like compound in *Rhizobium meliloti* are closely linked and on the Sym plasmid. Proc Natl Acad Sci USA 84:493–497

Murphy PJ, Trenz SP, Grzemski W, De Bruijn FJ, Schell J (1993) *Rhizobium meliloti* rhizopine *mos* locus is a mosaic structure facilitating its symbiotic regulation. J Bacteriol 175:5193–5204

Neal JL Jr, Atkinson TG, Larson RI (1970) Changes in the rhizosphere microflora of spring wheat induced by disomic substitution of a chromosome. Can J Microbiol 16:153–158

Neal JL Jr, Larson RI, Atkinson TG (1973) Changes in rhizosphere populations of selected physiological groups

of bacteria related to substitution of specific pairs of chromosomes in spring wheat. Plant Soil 39:209–212
Neumann G, Römheld V (2007) The release of root exudates as affected by the root physiological status. In: Pinton R, Varanini Z, Nannipieri P (eds) The Rhizosphere: Biochemistry and organic substances at the soil-plant interface. CRC, Boca Raton, pp 23–72
Notz R, Maurhofer M, Schnider-Keel U, Duffy B, Haas D, Défago G (2001) Biotic factors affecting expression of the 2, 4-diacetylphloroglucinol biosynthesis gene phlA in Pseudomonas fluorescens biocontrol strain CHA0 in the rhizosphere. Phytopathology 91:873–81
Oger P, Petit A, Dessaux Y (1997) Transgenic plants producing opines alter their biological environment. Nature/Biotechnology 15:369–372
Oger P, Mansouri H, Dessaux Y (2000) Effect of crop rotation and soil cover on the alteration of the soil microflora generated by the culture of transgenic plants producing opines. Mol Ecol 9:881–890
Oger P, Mansouri H, Nesme X, Dessaux Y (2004) Engineering root exudation of Lotus towards the production of two novel carbon compounds leads to the selection of distinct microbial populations in the rhizosphere. Microbiol Ecol 47:96–103
Okubara PA, Bonsall RF (2008) Accumulation of Pseudomonas-derived 2, 4-diacetylphloroglucinol on wheat seedling roots is influenced by host cultivar. Biol Control 46:322–331
Omote H, Hiasa M, Matsumoto T, Otsuka M, Moriyama Y (2006) The MATE proteins as fundamental transporters of metabolic and xenobiotic organic cations. Trends Pharmacol Sci 27:587–593
Palmgren MG (1991) Regulation of plant plasma-membrane H+-ATPase activity. Physiol Plant 83:314–323
Paulitz TC, Smiley RW, Cook RJ (2002) Insights into the prevalence and management of soilborne cereal pathogens under direct seeding in the Pacific Northwest, USA. Can J Plant Path 24:416–428
Picard C, Frascaroli E, Bosco M (2004) Frequency and biodiversity of 2, 4-diacetylphloroglucinol-producing rhizobacteria are differentially affected by the genotype of two maize inbred lines and their hybrid. FEMS Microbiol Ecol 49:207–215
Pinto R, Varanini Z, Nannipieri P (2007) The Rhizosphere: Biochemistry and organic substances at the soil-plant interface. CRC, Boca Raton
Raaijmakers JM, Weller DM (1998) Natural plant protection by 2, 4-diacetylphloroglucinol-producing Pseudomonas spp. in take-all decline soils. Mol Plant-Microbe Interact 11:144–152
Raaijmakers JM, Weller DM (2001) Exploiting genotypic diversity of 2, 4-diacetylphloroglucinol-producing Pseudomonas spp.: characterization of superior root-colonizing P. fluorescens strain Q8r1–96. Appl Environ Microbiol 67:2545–2554
Raaijmakers JM, Bonsall RF, Weller DM (1999) Effect of population density of Pseudomonas fluorescens on production of 2, 4-diacetylphloroglucinol in the rhizosphere of wheat. Phytopathology 89:470–475
Raaijmakers JM, Weller DM, Thomashow LS (1997) Frequency of antibiotic-producing Pseudomonas spp. in natural environments. Appl Environ Microbiol 63:881–887
Richardson AE, Hadobas PA, Hayes JE (2001) Extracellular secretion of Aspergillus phytase from Arabidopsis roots enables plants to obtain phosphorus from phytate. Plant J 25:641–649
Rossbach S, McSpadden B, Kulpa D, de Bruijn FJ (1994) Rhizopine synthesis and catabolism genes for the creation of "biased rhizospheres" and a marker system to detect (genetically modified) microorganisms in the soil. In: Levin M, Grim C, Angle JS (eds) Biotechnology risk assessment. University of Maryland Biotechnology Insitute, College Park, pp 223–244
Ryan PR, Delhaize E, Jones DL (2001) Function and mechanism of organic anion exudation from plant roots. Ann Rev Plant Physiol Plant Mol Bio l 52:527–560
Ryan PR, Delhaize E, Randall PJ (1995) Characterisation of Al-stimulated efflux of malate from the apices of Al-tolerant wheat roots. Planta 196:103–111
Ryan PR, Skerrett M, Flindlay G, Delhaize E, Tyerman SD (1997) Aluminium activates an anion channel in the apical cells of wheat roots. Proc Natl Acad Sci USA 94:6547–6552
Saint CP, Wexler M, Murphy PJ, Tempé J, Tate ME, Murphy PJ (1993) Characterization of genes for synthesis and catabolism of a new rhizopine induced in nodules by Rhizobium meliloti Rm220–3: extension of the rhizopine concept. J Bacteriol 175:5205–5215
Sarniguet A, Lucas P, Lucas M (1991) Relationships between take-all, soil conduciveness to the disease, populations of fluorescent pseudomonads and nitrogen fertilizers. Plant Soil 145:17–21
Sasaki T, Yamamoto Y, Ezaki B, Katsuhara M, Ahn SJ, Ryan PR, Delhaize E, Matsumoto H (2004) A wheat gene encoding an aluminum-activated malate transporter. Plant J 37:645–653
Savka MA, Black RC, Binns AN, Farrand SK (1996) Translocation and exudation of tumor metabolites in crown galled plants. Mol Plant Microbe Interact. 9:310–313
Savka MA, Farrand SK (1997) Modification of rhizobacterial populations by engineering bacterium utilization of a novel plant-produced resource. Nat Biotechnol 15:363–368
Savka MA, Dessaux Y, Oger P, Rossbach S (2002) Engineering bacterial competitiveness and persistence in the phytosphere. Mol Plant Microbe Interact 15:866–874
Shapira R, Ordentlich A, Chet I, Oppenheim AB (1989) Control of plant diseases by chitinase expressed from cloned DNA in Escherichia coli. Phytopathology 79:1246–1249
Siciliano SD, Fortin N, Mihoc A, Wisse G, Labelle S, Beaumier D, Ouellette D, Roy R, Whyte LG, Banks MK, Schwab P, Lee K, Greer CW (2001) Selection of specific endophytic bacterial genotypes by plants in response to soil contamination. Appl Environ Microbiol 67:2469–2475
Smiley RW (1979) Wheat-rhizoplane pseudomonads as antagonists of Gaeumannomyces graminis. Soil Biol Biochem 11:371–376
Soedarjo M, Hemscheidt TK, Borthakur D (1994) Mimosine, a toxin present in leguminous trees (Leucaena spp.), induces a mimosine-degrading enzyme activity in some Rhizobium strains. Appl Environ Microbiol 60:4268–4272
Soedarjo M, Borthakur D (1996) Mimosine produced by the tree legume Leucaena provides growth advantages to

some *Rhizobium* strains that utilize it as a source of carbon and nitrogen. Plant and Soil 186:87–92

Sundheim L, Poplawsky AR, Ellingboe AH (1988) Molecular cloning of two chitinase genes from *Serratia marcescens* and their expression in *Pseudomonas* species. Physiol Molec Plant Pathol 33:483–491

Tepfer D, Goldmann A, Pamboukdjian N, Maille M, Lepingle A, Chevalier D, Dénarié J, Rosenberg C (1988) A plasmid of *Rhizobium meliloti* 41 encodes catabolism of two compounds from root exudate of *Calystegium sepium*. J Bacteriol 170:1153–1161

Tesfaye M, Temple SJ, Allan DL, Vance CP, Samac DA (2001) Overexpression of malate dehydrogenase in transgenic alfalfa enhances organic acid synthesis and confers tolerance to aluminum. Plant Physiol 127:1836–1844

Timms-Wilson TM, Ellis RJ, Renwick A, Rhodes DJ, Mavrodi DV, Weller DM, Thomashow LS, Bailey MJ (2000) Chromosomal insertion of phenazine-1-carboxylic acid biosynthetic pathway enhances efficacy of damping-off disease control by *Pseudomonas fluorescens*. Mol Plant-Microbe Interact 13:1293–1300

Toth IK, Newton JA, Hyman LJ, Lees AK, Daydin M, Ortori C, Williams P, Fray RG (2004) Potato plants genetically modified to produce N-acylhimoserine lactones increase susceptibility to soft rot *Erwiniae*. Mol Plant-Microbe Interact 17:880–887

Uroz S, D'Angelo-Picard C, Carlier A, Elasri M, Sicot C, Petit A, Oger P, Faure D, Dessaux Y (2003) Novel bacteria degrading N-acylhomoserine lactones and their use as quenchers of quorum-sensing-regulated functions of plant-pathogenic bacteria. Microbiology 149:1981–1989

Vance C, Uhde-Stone C, Allan DL (2003) Phosphorus acquisition and use: critical adaptations by plants for securing a non-renewable resource. New Phytol 157:423–447

Viebahn M, Glandorf DCM, Ouwens TWM, Smit E, Leeflang P, Wernars K, Thomashow LS, Van Loon LC, Bakker PAHM (2003) Repeated introduction of genetically modified *Pseudomonas putida* WCS358r without intensified effects on the indigenous microflora of field-grown wheat. Appl Environ Microbiol 69:3110–3118

Wang CX, Knill E, Glick BR, Defago G (2000) Effect of transferring 1-aminocyclopropane-1-carboxylic acid (ACC) deaminase genes into *Pseudomonas fluorescens* strain CHA0 and its gacA derivative CHA96 on their growth-promoting and disease-suppressive capacities. Can J Microbiol 46:898–907

Wang J, Raman H, Zhou M, Ryan PR, Delhaize E, Hebb DM, Coombes N, Mendham N (2007) High-resolution mapping of Alp, the aluminium tolerance locus in barley (Hordeum vulgare L.), identifies a candidate gene controlling tolerance. Theor Appl Genet 115:265–276

Weller DM, Howie WJ, Cook RJ (1988) Relationship between in vitro inhibition of *Gaeumannomyces graminis* var. *tritici* and suppression of take-all of wheat by fluorescent pseudomonads. Phytopathology 78:1094–1100

Weller DM, Landa BB, Mavrodi OV, Schroeder KL, De La Fuente L, Blouin Bankhead S, Allende Molar R, Bonsall RF, Mavrodi DV, Thomashow LS (2007) Role of 2, 4-diacetylphloroglucinol-producing fluorescent Pseudomonas spp. in the defense of plant roots. Plant Biol 9:4–20

Weller DM, Raaijmakers JM, Gardener BBM, Thomashow LS (2002) Microbial populations responsible for specific soil suppressiveness to plant pathogens. Ann Rev Phytopathol 40:309–48

Whitehead NA, Barnard AM, Slater H, Simpson NJ, Salmond GP (2001) Quorum-sensing in Gram-negative bacteria. FEMS Microbiol Rev 25:365–404

Williams P, Winzer K, Chan WC, Cámara M (2007) Look who's talking: communication and quorum sensing in the bacterial world. Philos Trans R Soc Lond B Biol Sci. 362:1119–1134

Wilson M, Lindow SE (1995) Enhanced epiphytic coexistence of near-isogenic salicylate-catabolizing and non-salicylate-catabolizing *Pseudomonas putida* strains after exogenous salicylate application. Appl Environ Microbiol 61:1073–1076

Wilson M, Savka MA, Hwang I, Farrand SK, Lindow SE (1995) Altered epiphytic colonization of mannityl opine-producing transgenic tobacco plants by a mannityl opine-catabolizing strain of *Pseudomonas syringae*. Appl Environ Microbiol 61:2151–2158

Winding A, Binnerup SJ, Pritchard H (2004) Non-target effects of bacterial biological control agents suppressing root pathogenic fungi. FEMS Microbiol Ecol 47:129–141

Yang H, Knapp J, Koirala P, Rajagopal D, Peer WA, Silbart LK, Murphy A, Gaxiola RA (2007) Enhanced phosphorus nutrition in monocots and dicots over-expressing a phosphorus-responsive type IH$^+$-pyrophosphatase. Plant Biotech J 5:735–745

Yokosho K, Yamaji N, Ueno D, Mitani N, Ma JF (2009) *OsFRDL1* is a citrate transporter required for efficient translocation of iron in rice. Plant Physiol 149:297–305

Young JC, DeWitt ND, Sussman MR (1998) A transgene encoding a plasma membrane H+-ATPase that confers acid resistance in *Arabidopsis* thaliana seedlings. Genetics 149:501–507

Zhang LH, Dong YH (2004) Quorum sensing and signal interference: diverse implications. Mol Microbiol 53:1563–1571

Zhang W-H, Ryan PR, Delhaize E, Sasaki T, Yamamoto Y, Sullivan W, Tyerman SD (2008) Electrophysiological characterisation of the TaALMT1 protein in transfected tobacco (*Nicotiana tabacum* L.) cells. Plant J (submitted)

Zhao RM, Dielen V, Kinet JM, Boutry M (2000) Cosuppression of a plasma membrane H$^+$-ATPase isoform impairs sucrose translocation, stomatal opening, plant growth, and male fertility. Plant Cell 12:535–546

REVIEW ARTICLE

Rhizosphere processes and management in plant-assisted bioremediation (phytoremediation) of soils

Walter W. Wenzel

Received: 20 December 2007 / Accepted: 9 June 2008 / Published online: 11 July 2008
© Springer Science + Business Media B.V. 2008

Abstract Plant-assisted bioremediation or phytoremediation holds promise for in situ treatment of polluted soils. Enhancement of phytoremediation processes requires a sound understanding of the complex interactions in the rhizosphere. Evaluation of the current literature suggests that pollutant bioavailability in the rhizosphere of phytoremediation crops is decisive for designing phytoremediation technologies with improved, predictable remedial success. For phytoextraction, emphasis should be put on improved characterisation of the bioavailable metal pools and the kinetics of resupply from less available fractions to support decision making on the applicability of this technology to a given site. Limited pollutant bioavailability may be overcome by the design of plant–microbial consortia that are capable of mobilising metals/metalloids by modification of rhizosphere pH (e.g. by using *Alnus* sp. as co-cropping component) and ligand exudation, or enhancing bioavailability of organic pollutants by the release of biosurfactants. Apart from limited pollutant bioavailability, the lack of competitiveness of inoculated microbial strains (in particular degraders) in field conditions appears to be another major obstacle. Selecting/engineering of plant–microbial pairs where the competitiveness of the microbial partner is enhanced through a "nutritional bias" caused by exudates exclusively or primarily available to this partner (as known from the "opine concept") may open new horizons for rhizodegradation of organically polluted soils. The complexity and heterogeneity of multiply polluted "real world" soils will require the design of integrated approaches of rhizosphere management, e.g. by combining co-cropping of phytoextraction and rhizodegradation crops, inoculation of microorganisms and soil management. An improved understanding of the rhizosphere will help to translate the results of simplified bench scale and pot experiments to the full complexity and heterogeneity of field applications.

Keywords Bioremediation · Phytoremediation · Rhizosphere · Rhizosphere manipulation · Organic pollutants · Metals

Introduction

Bioremediation, i.e. the use of living organisms to manage or remediate polluted soils, is an emerging technology. It is defined as the elimination, attenuation or transformation of polluting or contaminating substances by the use of biological processes. Initially, bioremediation employed microorganisms

Responsible Editor: Philippe Hinsinger.

W. W. Wenzel (✉)
Universität für Bodenkultur Wien,
Wien, Austria
e-mail: walter.wenzel@boku.ac.at

to degrade organic pollutants, but since the use of green plants was proposed for in situ soil remediation (Baker et al. 1991; Chaney 1983; Salt et al. 1995), phytoremediation has become an attractive topic of research and development. Plant-assisted bioremediation, or phytoremediation, is commonly defined as the use of green or higher terrestrial plants for treating chemically or radioactively polluted soils. The following fundamental processes/technologies are distinguished (Salt et al. 1995; Wenzel et al. 1999):

- Phytostabilisation (and immobilisation) is a containment process using plants—often in combination with soil additives to assist plant installation—to mechanically stabilising the site and reducing pollutant transfer to other ecosystem compartments and the food chain;
- Phytoextraction is a removal process taking advantage of the unusual ability of some plants to (hyper-)accumulate metals/metalloids in their shoots;
- Phytovolatilisation/rhizovolatilisation are removal processes employing metabolic capabilities of plants and associated rhizosphere microorganisms to transform pollutants into volatile compounds that are released to the atmosphere;
- Phytodegradation/rhizodegradation refer to the use of metabolic capabilities of plants and rhizosphere microorganims to degrade organic pollutants.

Phytoremediation is generally considered as an environmentally friendly, gentle management option for polluted soil as it uses solar-driven biological processes to treat the pollutant. Phytoremediation appears attractive because in contrast to most other remediation technologies, it is not invasive and, in principle, delivers intact, biologically active soil. However, the in situ applicability of phytoremediation is constrained to the root zone, i.e., down to a few decimetres or metres from the soil surface, and often by the relatively long time required to achieve the remedial target.

Enhancement of the phytoremediation process is therefore a primary goal of current research. Selection, traditional breeding and genetic engineering of plants focus on increasing pollutant tolerance, root and shoot biomass, root architecture and morphology, pollutant uptake properties, degradation capabilities for organic pollutants etc. Other approaches are directed to the management of microbial consortia: the selection and engineering of microorganisms with capabilities for pollutant degradation, beneficial effects on the phytoremediation crops, or modifying effects on pollutant bioavailability. Additional strategies include proper management of the soil, e.g. via fertilisation or chelant addition to increase pollutant bioavailability, and of the phytoremediation crops, e.g. via optimisation of coppicing, harvest cycles, development of mixed cropping systems etc.

Many—if not all—enhancement strategies will have some impact on the rhizosphere, either deliberately or unintentionally.

Justification and objectives of this review

The general subject of phytoremediation and particular phytoremediation processes such as phytoextraction of metals have been reviewed in numerous journal articles and book chapters. Whereas many of these articles contain some section on rhizosphere aspects of phytoremediation, relatively few focus on the role of the rhizosphere and its manipulation during phytoremediation of inorganic or organic pollutants.

The earliest review specifically addressing rhizosphere action in bioremediation of organic compounds (Anderson et al. 1993), closely followed by that of Anderson and Coats (1994), summarised pioneer work, spread the concept of "rhizoremediation", and stimulated further research activities. Plant–bacterial interactions and their role in phytodegradation of organic pollutants were reviewed by Siciliano and Germida (1998). More recently, Kuiper at al. (2004) and Newman and Reynolds (2004) published reviews on rhizodegradation of organic pollutants, and Dzantor (2007) addressed the state of rhizosphere "engineering" for rhizodegradation of xenobiotic contaminants. Meharg and Cairney (2000) provide an overview on the potential role of ectomycorrhizal associations in rhizosphere remediation of persistent organic pollutants.

The role of rhizosphere processes in the phytoremediation of inorganic pollutants, in particular metals/metalloids, is much less investigated and only a few specific reviews are available on this topic (McGrath et al. 2001; Fitz and Wenzel 2002; Wenzel et al. 2004).

As concluded by Anderson et al. (1993), "further understanding of critical factors influencing the plant–microbe–toxicant interaction in soils will permit more rapid realisation" of plant-assisted bioremediation. The present paper aims to contribute towards this goal by examining the current concepts and published data on the role of rhizosphere processes and major controls that may be used for their management in phytoremediation of inorganic and organic soil pollutants. Attention is particularly given to the limitations and challenges associated with the rhizosphere, including system-inherent complications for its predictability and manageability such as the complexity and heterogeneity of this microenvironment (Hinsinger et al. 2005), and those arising from experimental approaches, including scaling problems and simplifications of experimental systems.

To this end, the first main section provides information on rhizosphere processes generally relevant to the establishment of vital, functioning phytoremediation crops irrespective which particular technology (e.g. phytoextraction or rhizodegradation) is considered. As pollutant bioavailability is a major target and/or obstacle in all phytoremediation technologies, its controls in the rhizosphere during phytoremediation are discussed for inorganic and organic pollutants in a subsequent chapter. This is followed by overviews on the rhizosphere controls of metal/metalloid volatilisation and the biodegradation of organic pollutants. The concluding section summarises the obstacles and perspectives of rhizosphere manipulation and management for each of the main phytoremediation processes and ends up with exploring the feasibility of integrated/combined approaches.

Rhizosphere interactions important for the establishment of functioning plant–microbial phytoremediation consortia in adverse, toxic environments

The efficiency of phytoremediation relies on the establishment of vital plants with sufficient shoot and root biomass growth, active root proliferation and/or root activities that can support a flourishing microbial consortium assisting phytoremediation in the rhizosphere. In turn, a healthy microbial consortium can benefit the plant. Rhizosphere controls on general vitality and functioning of plant–microbial consortia are compiled in Table 1, including beneficial as well as adverse effects. As most aspects of resource (water, nutrients) availability/use efficiency and plant health are not specific to phytoremediation crops, the reader is referred for details to other sources (e.g. Brix et al. 1996; Curl and Truelove 1986; Dziejowski et al. 1997; Jones et al. 2004; Marschner 1995; Reddy et al. 1989; Uren 2001). Here, I want to highlight two aspects of particular relevance in the context of phytoremediation of infertile, polluted soils: (1) the role of rhizosphere interactions in resource competition and (2) alleviation of pollutant toxicity.

Apart from many beneficial interactions (Table 1), plants and microbes also compete for resources, including nutrients and water (Kaye and Hart 1997). In soils with low available nutrient levels as commonly found at polluted sites, resource competition, especially for nitrogen and phosphorus, may become a limiting factor of microbial growth and biodegradation (Joner et al. 2006; Moorehead et al. 1998; Unterbrunner et al. 2007). This problem can be even more severe in situations where water supply is also limited, because nutrient transport into the rhizosphere via mass flow and diffusion is restricted by low soil moisture contents (Smith et al. 1998; Unterbrunner et al. 2007). There are also reports of reduced microbial activity in the vicinity of ectomycorrhizal mycelia (Olsson et al. 1996), possibly indicating resource competition and/or antagonistic interaction. Note that while excessive supply of nutrients can stimulate heterotrophic and pollutant-adapted bacteria, this may not necessarily be reflected by enhanced rates of phytoremediation. This was shown for hydrocarbon degradation in a crude-oil-contaminated soil where no effect or even inhibition of pollutant degradation was observed following nutrient addition (Chaîneau et al. 2005).

The second aspect to be highlighted relates to rhizosphere controls of pollutant toxicity. Plants and microorganism can become adapted to toxic pollutant concentrations. Phyto-/rhizoremediation processes that remove pollutants (phytoextraction, -degradation, -volatilisation) also contribute to alleviation of toxicity by decreasing the pollutant concentration in the rhizosphere. Resistance of plants and beneficial rhizosphere microorganisms (e.g. plant-growth-promoting rhizobacteria, PGPR) against (multiple) pollutants is a prerequisite for their use in any phytoremediation technology (Belimov et al. 2005; Burd et al. 2000).

Table 1 Rhizosphere controls of the performance and vitality of plants and microorganisms in phytoremediation systems

		Plant activity			Microbial activity		
	Soil properties involved	Beneficial to		Impeding microorganism	Beneficial to		Impeding plants
		Plant	Microorganisms		Plant	Microorganisms	
Physical environment (microbial habitat, niche, structural support)	Soil porosity and microsites		Proliferation, architecture, morphology and density of roots			Rhizosphere competence	
Resource availability and use efficiency							
Oxygen	Diffusivity, porosity	Aerenchyma, creation of soil pores by proliferating roots			(Mycorrhiza)		
Water		Proliferation, architecture, morphology and density of roots; root products (exudates, enzymes)		Competition for resources	N_2-fixation, mycorrhiza, PGPR (e.g. P of Fe solubilising bacteria), microbial enzymes and exudates		Competition for resources
Mineral nutrients	Nutrient content and solubility						
Organic compounds	Soil organic matter content, turnover and composition		Root products (exudates, enzymes), root debris		Microbial exudates and lysates		
Plant health							
General (biomass production, root growth pattern, competitiveness					PGPR (producing phytohormones)		Phytotoxins
Stress					ACC deaminase		
Plant deseases		Allelopathy		Allelopathy	Predators of pathogens, biocontrol, PGPR		Pathogens
Resistance against toxic pollutants in soil		Degradative enzymes, exudates (e.g. organic acids) alleviating toxicity, adaptation	Degradative enzymes, exudates		Degradative enzymes, ACC deaminase, protection by mycorrhiza, PGPR	Degradative enzymes, adaptation	

Beneficial interactions between phytoremediation crops and bacteria have been demonstrated to alleviate metal toxicity (and nutrient deficiency). Inoculants of cadmium-resistant, rhizosphere-competent bacterial strains that had been isolated from metal-polluted soil substantially improved root elongation, root and shoot biomass production of *Brassica napus* grown on cadmium-polluted soil (Sheng and Xia 2006). Similarly, growth and biomass production of *Brassica juncea* grown on Pb–Zn mine tailings was improved substantially upon inoculation with a PGPR consortium consisting of N_2-fixing *Azotobacter chroococcum* HKN5, P-solubilising *Bacillus megaterium* HKP-1, and K-solubilising *Bacillus mucilaginosus* HKK-1 (Wu et al. 2006b). The PGPR *Bacillus subtilis* strain SJ-101 capable of producing the phytohormone indole acetic acid and solubilising inorganic phosphates stimulated *Brassica juncea* growth on nickel polluted soil (Zaidi et al. 2006). Root growth and proliferation in polluted soils has also been shown to increase in the presence of ACC (1-aminocyclopropane-1-carboxylate) deaminase-producing bacteria (Arshad et al. 2007). High ethylene concentrations produced by plant roots in response to toxicity and other stresses inhibit root growth and proliferation. ACC deaminase regulates ethylene concentrations in plants via metabolisation of the ethylene precursor ACC into a-ketobutyric acid and ammonia (Arshad et al. 2007; Glick 2005). Indigenous ACC deaminase producing bacteria were found in the rhizosphere of the nickel hyperaccumulator *Thlaspi goesingense* (Idris et al. 2004), indicating their potential role in metal resistance of this hyperaccumulator species. Recent work has demonstrated that metal-resistant PGPR inoculates containing ACC-deaminase-producing bacteria protected *Brassica napus* and *B. campestris* against metal toxicity and stimulated plant growth (Burd et al. 1998; Belimov et al. 2001).

Ectomycorrhizal associations can display considerable resistance against toxicity in soil polluted with metals (Leyval et al. 1997; Meharg and Cairney 2000) and organic compounds such as *m*-toluate (Sarand et al. 1999), petroleum (Sarand et al. 1998), or polycyclic aromatic hydrocarbons (Leyval and Binet 1998). Densely packed mycorrhizal sheaths and phenolic inter-hyphal material can protect plant roots from direct contact with the pollutant (Ashford et al. 1988), while the large surface and cation exchange capacity of extramatrical mycelia may reduce bioavailable pollutant concentrations, at least for some metals, through their substantial adsorption capacities (Colpaert and Asche, 1993; Marschner et al. 1998). The structure of the fungal sheath and the density and surface area of the mycelium are likely to be important characteristics determining the efficiency of an ectomycorrhizal association to withstand metal toxicity and to protect the host plant from pollutant contact (Hartley et al. 1997). Beneficial effects on growth and biomass production of *Salix* x *dasyclados* inoculated with an ectomycorrhizal fungus were observed in a phytoextraction study conducted by Baum et al. (2006). In addition to their protective role, mycorrhizae may contribute to the resistance of plant–microbial associations through enhanced degradation of organic pollutants in the mycorrhizosphere (Meharg and Cairney 2000), thus lowering the bioavailable concentration of the pollutant in soil.

Wang et al. (2005) found increased biomass production of *Elsholtzia splendens* inoculated with a single arbuscular mycorrhizal strain (*Glomus claedonium* 90036) or a mix of five different arbuscular mycorrhizal fungi. Beneficial effects of a consortium of arbuscular mycorrhizae and a *Penicillium* fungus on biomass production of *Elsholtzia splendens* was recently confirmed in the field (Wang et al. 2007b) but where inoculated plants were grown on sterilised soil (Wang et al. 2007a), increased biomass production in inoculated treatments may simply reflect inhibited growth in the absence of soil microbes. However, while there is some evidence of beneficial effects of mycorrhizal inoculation on plant growth in polluted soils, this is not always the case. For example, in a study where the arbuscular mycorrhizal fungus *Glomus mossae* was inoculated to *Cannabis sativa* grown on metal-contaminated soil, plant growth was decreased and the extent of the decrease was related to the degree of mycorrhization (Citterio et al. 2005).

Rhizosphere controls of pollutant bioavailability during phytoremediation

Bioavailability refers to the fraction of a chemical that can be taken up or transformed by living organisms (Semple et al. 2003) from the surrounding bio-influenced zone where organism-mediated (bio)chemical changes occur (Harmsen et al. 2005). Pollutant bioavailability is understood as the result of many interacting factors associated with soil properties, pollutant characteristics and effects of plant roots and the associated microbial community. The efficiency of any phytoremediation system depends on the bioavailability of the targeted pollutant and root-microbial modifications of their solubility and chemical speciation in the rhizosphere.

Plant controls on metal/metalloid bioavailability in the rhizosphere

Plants can control metal/metalloid bioavailability in their rhizosphere via uptake mechanisms, properties of their root system, and root activities as largely documented in non-accumulating plants (Hinsinger and Courchesne 2008; McLaughlin et al. 1998). These bioavailability controls are discussed below with emphasis on the specific root/rhizosphere traits of hyperaccumulator plants for which there is much less published reports than in other plants.

The unique ability of hyperaccumulator plants to take up excessive amounts of metals or metalloids has been related to high-affinity transport systems across the plasma membranes of root tissues (Lasat et al. 1996; Lombi et al. 2001; Pence et al. 2000). Another unique trait shown for the Cd and Zn hyperaccumulator *Thlaspi caerulescens* refers to preferential proliferation of roots into metal-rich patches of soil (Schwartz et al. 1999, 2003; Whiting et al. 2000). Hyperaccumulator plants (*T. caerulescens* and *T. goesingense*) have also been shown to develop a dense root system with a large proportion of fine roots, which may also contribute to enhanced uptake of metals (Keller et al. 2003; Himmelbauer et al. 2005).

Plant roots and root debris can decrease the bioavailability of cationic metals/metalloids via adsorption onto root surfaces or root-derived biomolecules (e.g., Breckle and Kahle 1992; Cathala and Salsac 1975). Moreover, uptake of metals/metalloids can reduce the bioavailability of inorganic pollutants to microorganisms. Another important trait of plant roots relates to the typical avoidance (Keller et al. 2003) or—in the case of metal hyperaccumulator *Thlaspi caerulescens*—active fetching of metal-polluted soil patches (Schwartz et al. 2003).

Plant root activities that potentially increase metal/metalloid solubility and may change speciation include acidification/alkalinisation, modification of the redox potential, exudation of metal chelants and organic ligands (in particular low molecular organic acids and phytosiderophores) that compete with anionic species (e.g. arsenate) for binding sites (Jones et al. 2004; Fitz and Wenzel 2002; Wenzel et al. 2003a). The mobilising effect of root exudates has been demonstrated *in vitro* (Mench and Martin 1991) and recently in resin buffered nutrient solutions and soil experiments (Degryse et al. 2007; Shenker et al. 2001). However, increased mobility was not necessarily associated with increased uptake in plants (Shenker et al. 2001).

In anaerobic soils, wetland plants such as cattail (*Typha latifolia*) and common reed (*Phragmites australis*) can release oxygen into the rhizosphere (Brix et al. 1996) and increase the redox potential (Flessa and Fischer 1992). This induces the formation of ferric iron plaque and subsequent sorption and immobilisation of metals (e.g. zinc; Doyle and Otte 1997) and metalloids (e.g. arsenic; Blute et al. 2004).

As metal/metalloid (hyper-)accumulator plants represent the main model for the development and investigation of phytoextraction technology, the question has arisen as to whether specific rhizosphere traits exist in these plants in terms of metal mobilisation. While it was shown that the initial soil pH can considerably affect metal uptake in (hyper-)accumulator plants (Brown et al. 1994; Wieshammer et al. 2007), and metal uptake in hyperaccumulator crops might be optimised by adjusting pH using appropriate soil amendments (Wang et al. 2006), hyperaccumulation could not be related to changes in rhizosphere pH which were found to be typically small (McGrath et al. 2001; Wenzel et al. 2003a).

In hydroponic experiments, root exudates of the nickel hyperaccumulator *Thlaspi goesingense* (Salt et al. 2000) and cadmium/zinc hyperaccumulator *T. caerulescens* (Zhao et al. 2001) could not explain the excessive metal uptake compared to a closely related nonaccumulator. However, in rhizobox and field studies, sustained or even enhanced metal/metalloid concentrations measured in soil solutions from depleted rhizospheres of *T. goesingense* and arsenic hyperaccumulator *Pteris vittata* were associated with increased levels of dissolved organic carbon (Wenzel et al. 2003a; Fitz et al. 2003).

Microbial controls on metal/metalloid bioavailability in the rhizosphere

Microorganisms may either increase or decrease metal/metalloid solubility. The various processes by which microorganisms in soil may alter metal/metalloid mobility are depicted in Fig. 1 and briefly discussed below.

Microorganisms can increase solubility and change speciation of metals/metalloids through the production of organic ligands via microbial decomposition of soil organic matter, and exudation of metabolites (e.g. organic acids) and microbial siderophores that can complex cationic metals or desorb anionic species (e.g. arsenate) by ligand exchange (Gadd 2004). Depending on the surface charge of soil minerals and below metal-specific pH values, siderophores produced by microbes (and plants) may also immobilise cationic metals such as cadmium, copper, or zinc. This can be explained by differential surface charge of metal–siderophore complexes resulting either in attraction to (immobilisation) or repulsion from (mobilisation) charged soil minerals (Neubauer et al 2002). The complexity of interactions between organic ligands and metals was also demonstrated by modelling of copper solubility and transport in the root zone in the absence and presence of organic ligands, showing that ligands do not necessarily increase the solubility and bioavailability of metals (Seuntjens et al. 2004). Here, pH was also identified as a major control of mobilisation versus immobilisation.

Enhanced (co-)dissolution of metal/metalloid compounds or desorption triggered by ion competition between the metal and the proton may arise from heterotrophic proton efflux via plasma membrane H^+-ATPases or from dissociation of carboxylic acid accumulated from carbon dioxide respiration (Gadd 2004). The associated changes in pH can also affect speciation of metals/metalloids in solution. Proton-induced solubilisation of metals has been demonstrated for numerous microorganisms including ericoid mycorrhizal and ectomycorrhizal fungi (Fomina et al. 2005, Martino et al. 2003)

Microbially mediated reduction and oxidation processes can also modify the solubility of metals and metalloids. Specialised anaerobic bacteria can use iron (III) as terminal electron acceptor, dissimilatory metal-reducing bacteria can use various metals/metalloids such as chromium (VI), iron (III), manganese (IV), mercury (II), selenium (VI), and uranium (VI) (Gadd 2004). Microbial reduction results in mobilisation of iron and manganese, but can immobilise elements such as uranium, chromium (Gadd 2004), and selenium (Di Gregorio et al. 2006). Metals/metalloids, even if their solubility is not directly affected by changes of the redox potential, can be subject to

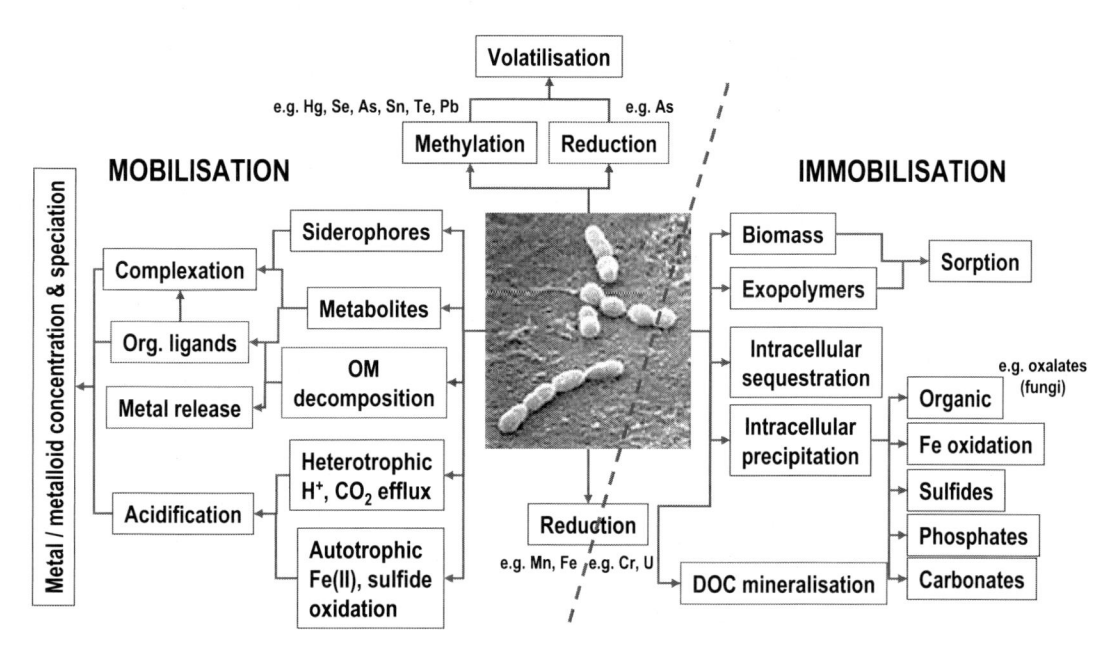

Fig. 1 Processes involved in microbial mobilisation and immobilisation of metals and metalloids in soil

co-dissolution if adsorbed to or occluded in sesquioxides (McBride 1989).

Microorganisms can immobilise metals/metalloids in various other ways (Gadd 2004). They can take up the elements and accumulate them in their biomass via intracellular sequestration (Berthelsen et al. 2000) or precipitation, or adsorb them onto cell walls (Fein et al. 2001; Leyval and Joner 2000; Zaidi and Musarrat 2004) and exopolymers released into their surroundings (He et al. 2000). The high sorption capacity of the *Bacillus subtilis* strain SJ-101 for nickel (Zaidi and Musarrat 2004) was shown to protect *Brassica juncea* against nickel toxicity when inoculated with this strain (Zaidi et al. 2006). Mineralization of dissolved metal–organic complexes may be another cause of microbially mediated immobilisation.

Microbial controls of metal/metalloid bioavailability in the rhizosphere of (hyper-)accumulator plants have been little investigated in contrast to microorganisms indigenously associated with non-accumulator plants. Idris et al. (2004) characterised the indigenous bacterial communities associated with the nickel hyperaccumulator *Thlaspi goesingense* using cultivation and cultivation-independent techniques. They showed that the majority of bacterial strains had been able to produce siderophores, indicating that—provided the metal–siderophore complex—soil interaction is repulsive (Neubauer et al. 2002)—solubilisation by rhizobacteria may contribute to the sustained nickel concentrations in a nickel-depleted rhizosphere as reported by Wenzel et al. (2003a). Similar results were obtained in studies of the indigenous microbial communities in the rhizospheres of the nickel accumulator *Alyssum murale* (Abou-Shanab et al. 2003) and the zinc accumulator *T. caerulescens* (Lodewyckx et al. 2002), indicating the potential of microbial inoculation to enhance metal uptake in phytoextraction crops. In fact, inoculation of several rhizobacteria strains obtained from the rhizosphere of *A. murale* grown on a serpentine site proved to enhance nickel extractability from soil and to increase its uptake in *A. murale* by up to 40% compared to non-inoculated controls (Abou-Shanab et al. 2006). Similarly, rhizosphere bacteria increased concentrations of Zn in the hyperaccumulator species *Thlaspi caerulescens* (Whiting et al. 2001) and in the non-hyperaccumulator *Brassica juncea* (De Souza et al. 1999b).

Mycorrhizae can efficiently explore the soil volume and, due to their small diameter, microsites that are not accessible for plant roots. They can further modify pollutant bioavailablity in several ways, including competition with roots and other microorganisms for water and pollutant uptake, protection of roots from direct interaction with the pollutant via formation of the ectomycorrhizal sheath, and impeded pollutant transport through increased soil hydrophobicity (Meharg and Cairney 2000). This is equally valid for the bioavailability of organic pollutants discussed in the subsequent section.

Substantially increased arsenic concentrations in fronds of the arsenic-hyperaccumulating fern *Pityrogramma calomelanos* were observed in pot and field experiments when amended with an uncharacterised rhizobacterial inoculum obtained from arsenic-contaminated soil, whereas no increased uptake was found with rhizofungal inoculum (Jankong et al. 2007). Wieshammer et al. (personal communication) found foliar Zn and Cd concentrations in the metal-accumulating willow *Salix caprea* to be increased by 85% and 51%, respectively, in treatments inoculated with a combination of five metal-resistant bacterial strains isolated from a metal-contaminated soil during the first vegetation period. Similar results were obtained after the second vegetation period, demonstrating sustained effects of the microbial inoculants in this experiment. However, no or only short-term effects of the inoculate were observed for two other willows, *S. rubens* and *S. fragilis*, indicating that the effect of the bacterial strains on metal uptake is dependent on the willow species used. These findings may be explained by the fact that the site from which the inoculum was obtained was primarily vegetated by *S. caprea*.

A metal-accumulating willow species inoculated with the ectomycorrhizal fungus *Cadophora finlandica* and a native mix of microbes was grown on a metal-contaminated but sterilised soil from which the microbes had been isolated. However, no significant increase in metal uptake in the willows was observed (Dos Santos Utmazian et al. 2007). Sell et al. (2005) inoculated *Salix viminalis* and *Populus canadensis* with different ectomycorrhizal fungi and found increased uptake and translocation of cadmium from a sterilised soil only for the association of *P. canadensis* with *Paxillus involutus*. In a similar experiment, *Salix dasyclados* was inoculated with two different *Paxillus involutus* strains, only one of which increased the extractability of Cd from the experimental soils (Baum et al. 2006).

In some cases, arbuscular mycorrhizal fungi have been shown to increase uptake of metals (Liao et al. 2003; Whitfield et al. 2004; Citterio et al. 2005) and arsenic (Liu et al. 2005; Leung et al. 2006) in plants but other studies showed no effect (Trotta et al. 2006; Wu et al. 2007) or decreased concentrations in plant tissues (Weissenhorn et al. 1995). The contrasting results are difficult to evaluate and may be partly due to different experimental settings, e.g. greenhouse (Liu et al. 2005; Leung et al. 2006) versus field studies (Trotta et al. 2006; Wu et al. 2007) as in the case of arsenic uptake in *Pteris vittata* inoculated with arbuscular mycorhizal fungi.

Phytoremediation practices may benefit from microbial processes but in turn may also influence the composition and functioning of the microbial consortia in the rhizosphere of phytoremediation crops. It could be shown that choice of plant species and soil amendments had substantial impacts on the indigenous community of arbuscular mycorrhizal fungi associated with the hyperaccumulator *Thlaspi caerulescens* and the nonaccumulators *Silene vulgaris* and *Zea mays* grown on a metal-contaminated landfill site (Pawlowska et al. 2000).

Bioavailability controls of organic pollutants in the rhizosphere

Major factors and processes controlling the bioavailability of organic pollutants in the rhizosphere are summarised in Fig. 2.

The bioavailability of organic pollutants has been related to the octanol–water partitioning coefficient K_{ow} (Burken and Schnoor 1998; Simonich and Hites 1995), with an optimal uptake between log K_{ow} 1 and 3 (Bromilow and Chamberlain 1995). Other important pollutant properties controlling their fate in the environment include the vapour pressure and the

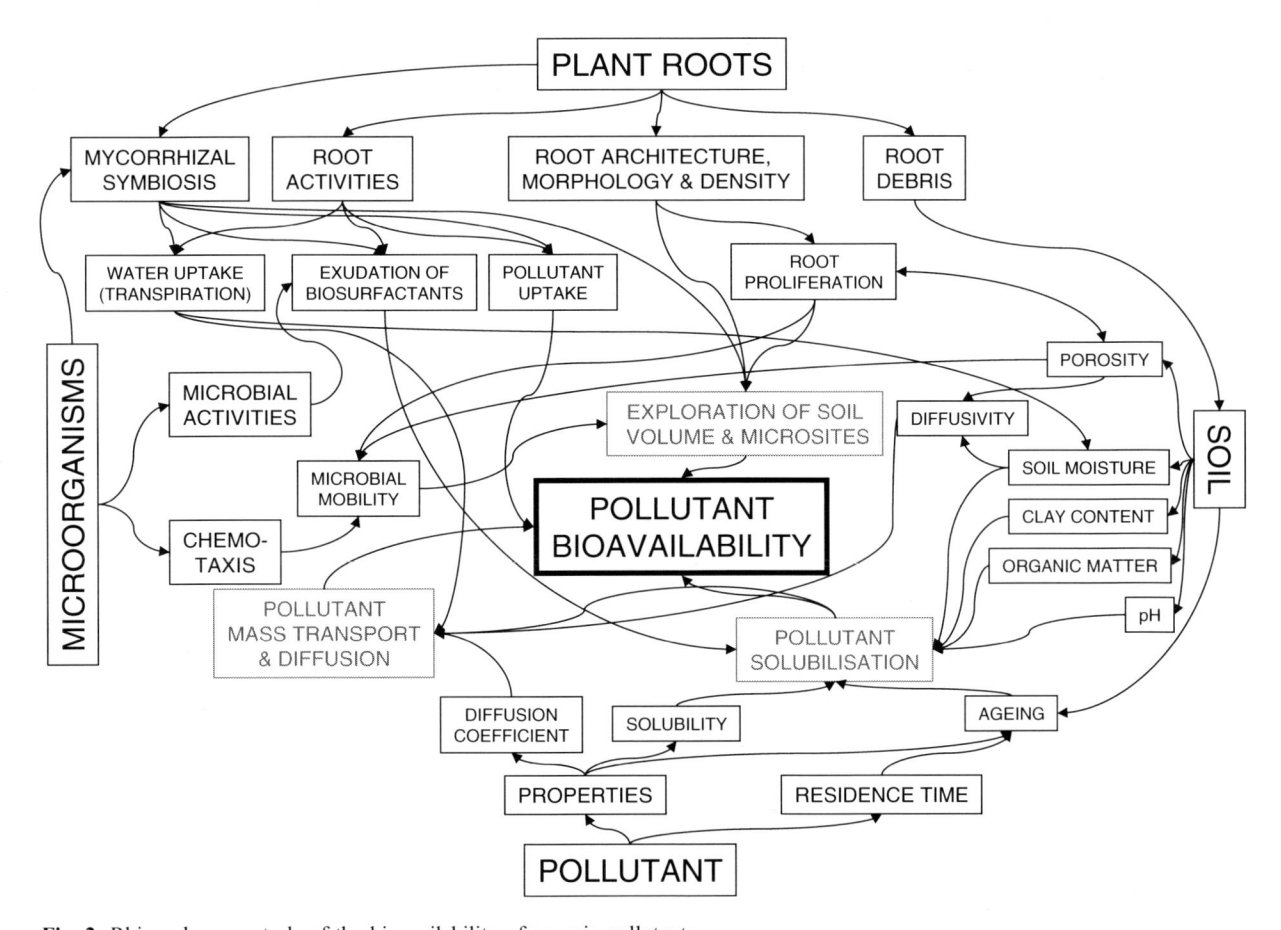

Fig. 2 Rhizosphere controls of the bioavailability of organic pollutants

Henry constant. The vapour pressure indicates whether or not a pollutant is easily volatilised in dry soil conditions, the Henry constant provides a better measure of the volatilisation potential in wet and flooded soil. As the residence time in soil of highly volatile compounds such as chloroethene will be short, they are not a primary target of rhizodegradation.

The solubility of a pollutant is further modified by soil properties. Organic matter quality and content, clay content, mineral composition, type of mineral surface, pH and redox potential are known as important controls of organic pollutant solubility (Luthy et al. 1997; Reid et al. 2000), with hydrophobic, nonpolar organic matter being of particular importance for binding organic pollutants. Binding of organic pollutants to the soil matrix is known to progress as the contact time increases, rendering pollutants less bioavailable. This phenomenon is known as "ageing" and is attributed to sorption onto minerals and organic matter in soil, and subsequent interparticle diffusion in minerals and entrapment within humic complexes, nano- and micropores (Luthy et al. 1997; Reid et al. 2000; Semple et al. 2003).

Apart from the absorption capability of the organisms (biology), the bioavailability of a pollutant in soil not only depends on its solubility (chemistry), but also its diffusion and mass transport (physics) towards the sites and niches where degrader populations are abundant (Semple et al. 2003). Soils are heterogeneous environments (Hinsinger et al. 2005) where microsites and niches hosting microorganisms with degradation capabilities are often separated from micropores containing the pollutant. Frequently, a substantial proportion of the pollutant is not accessible for the degrader community. Therefore, interconnected soil factors such as porosity, water content and diffusivity controlling the transport of water and solutes (Young and Crawford 2004) are important controls of pollutant bioavailability.

It is well established that bioavailability is one of the most limiting factors in bioremediation of persistent organic pollutants in soil (Mohan et al. 2006; Reid et al. 2000). In bioreactor systems this problem is often addressed by agitation and mixing, and/or by the addition of surfactants (Mohan et al. 2006). But how can roots and microbial associates in their rhizosphere improve pollutant bioavailability?

Our current knowledge of the processes involved yields a complex picture of root-microbe-mediated modifications of pollutant bioavailability (Fig. 2). Roots create pores which can improve connectivity and diffusivity of the soil (Young and Crawford 2004) thus facilitating mass flow and diffusion of water and pollutants. Roots induce transpiration-driven mass flow towards the rhizosphere, delivering dissolved pollutant compounds to sites of (typically) increased microbial activity (Ferro et al. 1994). Roots can also serve as a carrier of (degrader) microorganisms through the soil, thus extending the contact between degraders and the pollutant (Gilbertson et al. 2007). Some microorganisms may also actively fetch the pollutant via chemotaxis (Valenzuela et al. 2006).

The exudation of biosurfactants, i.e. small molecules with a hydrophilic head and a hydrophobic tail, by roots (Read et al. 2003) and, perhaps more importantly, associated microorganisms (Bento et al. 2005) may mobilise hydrophobic pollutants from soil particle surfaces, enabling their transport to sites of high degradation activity. Bacterial biosurfactants such as rhamnolipids have also been used in soil washing, but the amount of crude oil removed from aged (weathered) contaminated soil was low (Urum et al. 2004). Soil properties such as texture and clay mineralogy have been shown to influence desorption of hydrophobic organic pollutants by surfactants (Lee et al. 2002).

Sloughed-off cells, mucilage and root debris can bind organic pollutants and decrease their bioavailability. Roots can also take up, i.e., phytoextract hydrophilic pollutants such as pentachlorophenol (Simonich and Hites 1995; Ferro et al. 1994) or trinitrotoluene (Schnoor et al. 1995) and therefore compete with degrader microorganisms for the substrate. The efficiency of root proliferation in enhancing rhizodegradation may also be limited by avoidance of pollution hot spots (Kechavarzi et al. 2007).

Rhizosphere controls of metal/metalloid volatilisation

Soil microorganisms are known to convert some metals and metalloids (i.e. arsenic, boron, antimony, selenium, tin, tellurium, lead, mercury) to their volatile species. These species are generally represented by methyl and hydride derivatives and also the elemental form in the case of mercury (Meyer et al. 2007). This microbial conversion is usually

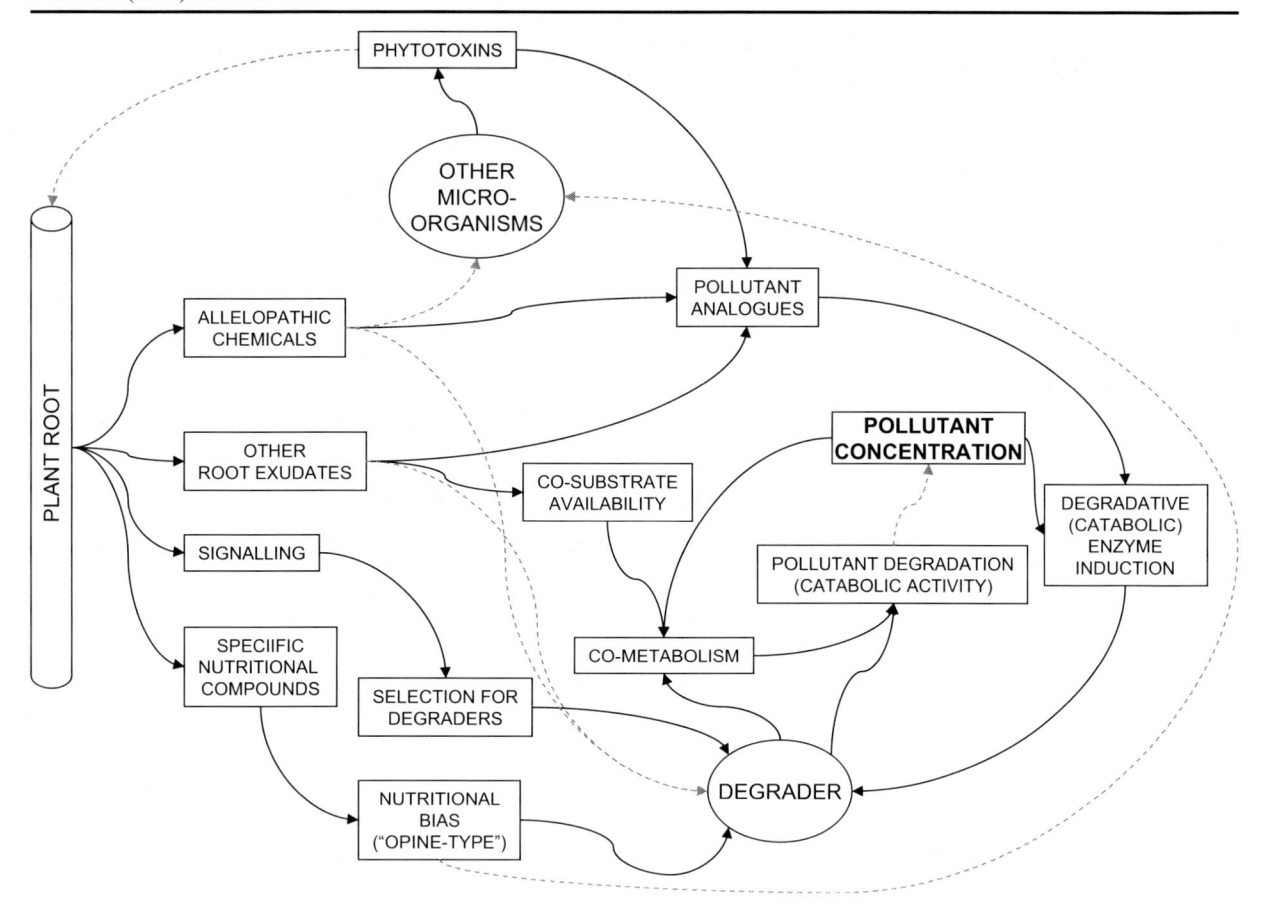

Fig. 3 Plant–degrader interactions potentially involved in rhizodegradation (*solid-line arrows* indicate positive, *dashed-line arrows* negative influence on the targeted process or component)

considered as a detoxification mechanism by which the microorganisms decrease the toxicity of the surrounding microenvironment.

Phytovolatilisation of selenium (e.g. Terry and Zayed 1998; Frankenberger and Karlson 1994) is carried out by both plants and microorganisms and involves reduction of selenate to selenite (and/or selenide). The reduced inorganic selenium is assimilated into organic forms such as selenomethionine and selenocystein which are then methylated to nonvolatile dimethylselenium compounds and finally converted to volatile dimethylselenide (DMSe) and dimethyldiselenide (DMDSe; Zhang and Frankenberger 2000). Both bacteria and fungi are able to volatilise selenium with bacterial methylation being dominant under anaerobic conditions. Whereas a large proportion of the literature deals with the combined volatilisation potential of the plant–microbial system, only a few studies (Azaizeh et al. 1997, 2003) have highlighted the importance of rhizosphere processes on selenium volatilisation, e.g.

for bulrush (*Scirpus robustus*) and common reed (*Phragmites australis*). Rhizosphere bacteria seem to be also responsible for enhancing plant-mediated selenium volatilisation and the accumulation of selenium in plants (De Souza et al. 1999a).

In the case of mercury, microbial rhizovolatilisation is achieved by the degradation of methylmercury and reduction of Hg^{2+} to elemental Hg (Hg^0; Barkay et al. 2003). In bacteria this process is linked to the *mer* operon which enhances mercury resistance. This operon consists of a battery of genes including *merA*, encoding for mercuric reductase, and *merB*, encoding for organomercurial lyase, which is responsible for the breakdown of organic mercury to Hg^{2+} (Heaton et al. 1998). These mercury resistance genes are ubiquitous in soils as they are generally carried by mobile genetic elements such as transposons and plasmids (Schneiker et al. 2001). The few studies available on the role of rhizosphere in mercury volatilisation found either enhanced (Moreno et al. 2005) or decreased

(Johnson et al. 2003) Hg^0 volatilisation in planted as compared to non-planted soil. The contrasting results may reflect the plant-specifity of stimulation of microbial strains capable of volatilising mercury.

Fungi, yeast and bacteria can volatilise arsenic as a detoxification mechanism. Arsenic can be volatilised in the form of arsine and its methylated species mono-, di- and tri-methylarsine (Frankenberger and Arshad 2002). However, the overall volatilisation rate for arsenic has been reported to be typically low (Gao and Burau 1997; Prohaska et al. 1999; Turpeinen et al. 2002), except in an incubation experiment conducted by Edvantoro et al. (2004). Yet, the role of rhizosphere processes in arsenic volatilisation has not been explored.

Rhizosphere controls of biodegradation of organic compounds

The susceptibility of organic pollutants to bacterial metabolism and enzymatic attack differs widely and is related to the molecular structure (Sabljic and Piver 1992). As a consequence of the co-evolution of natural organic compounds and soil microorganisms, xenobiotics that structurally resemble naturally occurring compounds are more likely to be susceptible to biodegradation (Semple et al. 2003).

Among other activities, plant roots can release degradative enzymes into the rhizosphere (Schnoor et al. 1995). Reports are available on the degradation of nitroaromatic compounds (e.g. trinitrotoluene) by plant-derived nitroreductases and laccases at the laboratory scale (Boyajian and Carreira 1997) and in field tests (Wolfe et al. 1993). Other plant-derived enzymes with the potential to contribute to the degradation of organic pollutants in the rhizosphere include dehalogenase involved in dehalogenising chlorinated solvents such as hexachloroethane and trichloroethylene, peroxidases degrading phenols, and phosphatases cleaving phosphate groups from large organophosphate pesticides (Susarla et al. 2002). Our knowledge of the relative importance and efficiency of plant extracellular enzymes in the presence of degrading microorganisms is still very limited, but the measured half-life of these enzymes suggest that they may actively degrade organic pollutants for days following their release from plant tissues (Schnoor et al. 1995).

Apart from the direct release of degradative enzymes, plants are able to stimulate the activities of microbial degrader organisms/communities. Plant–degrader interactions that are thought to be most relevant for the success of rhizodegradation are depicted in Fig. 3 and described in the following paragraphs.

Mechanisms of nonspecific stimulation potentially involved in rhizodegradation include exudates that serve as analogues or co-metabolites of organic pollutants (Siciliano and Germida 1998). Pollutant analogues include various known root exudates and allelopathic chemicals excreted by plants in response to pathogen attack in the rhizosphere. Such compounds may well be the evolutionary cause of the development of detoxifying microbial communities in the rhizosphere (Siciliano and Germida 1998). Similarly, one might speculate that phytotoxins produced by microorganisms in the rhizosphere could serve as pollutant analogues.

Allelopathic chemicals released by plant roots can induce catabolic enzymes in degrader organisms and thus induce/enhance rhizodegradation of structurally similar pollutants. Allelopathic compounds of interest for rhizodegradation include flavonoids, as well as other chemicals, e.g. salicin, hirsutin, 2 (3H)-benzoxazolinone or cyanide (Dzantor 2007; Siciliano and Germida 1998). Other pollutant analogues may be derived from unspecific exudation of compounds such as acetylene, biphenyl, p-coumaric acid, morin or palmatic acid (Siciliano and Germida 1998). For instance, it has been demonstrated that polychlorinated biphenyl (PCB) degradation and growth of PCB-degrading bacteria was enhanced by the flavonoids apigenin and naringin (Fletcher and Hedge 1995).

Plant roots exude compounds that can serve as co-metabolites in microbial pollutant degradation (Hedge and Fletcher 1996). This is important especially where microorganisms cannot utilize the pollutant as a sole carbon source as for instance in the aerobic degradation of trichloroethylene (Hyman et al. 1995). Enhanced degradation of the polycyclic aromatic hydrocarbon benzo[a]pyrene by the rhizobacterium *Sphingomonas yanoikuyae* JAR02 was demonstrated *in vitro* in the presence of root extracts or exudates obtained from several plant species, including mulberry (*Morus alba*) and hybrid willow (*Salix alba x matsudana*; Rentz et al. 2005).

In turn, co-metabolism may be associated with a shift in the composition of the microbial consortium (Hubert et al. 2005). Root-exuded compounds may also selectively support specific microbial strains relative to others. This, for instance, was shown for the growth of different PCB-degrading strains *Alcaligenes eutrophus* H850, *Corynebacterium* sp. MB1 and *Pseudomonas putida* LB400 on various model plant compounds such as catechin, morin and naringin (Donnelly et al. 1994). As root exudation patterns differ substantially among plant species (Fletcher and Hedge 1995; Jones et al. 2004), this and similar findings suggest that rhizodegradation efficiency may benefit from the selection of appropriate plant–degrader pairs (Siciliano and Germida 1998).

Another mechanism by which plant roots, in theory, could select for microbial strains or populations with degradation capabilities would be the microbial-induced root exudation of compounds that can only be utilised by selected organisms (Siciliano and Germida 1998; Dzantor 2007). One such relationship is known for *Agrobacterium tumefaciens* which causes crown gall disease. Here, infection of the root induces a tumour (gall formation) that excretes opines, unusual carbon substrates that initially may have been utilised exclusively by the inducing crown gall bacterium (Moore et al. 1997). However, due to the genetic exchanges occurring in the rhizosphere other rhizobacteria also evolved the ability to utilise opine compounds (Moore et al. 1997; Dzantor 2007). The "opine concept" may provide guidance to identify or create similar but perhaps less selective plant–microbial interactions in rhizodegradation systems.

As yet there is no convincing evidence for direct stimulation of degrader microorganisms by plant roots through signalling. Increased abundance and activities of degrader populations in the rhizosphere (rhizodegradation systems; e.g., Johrdal et al. 1997; Ryslava et al. 2003) could not be separated from other ecological interactions such as the effect of contamination (Siciliano and Germida 1998). Kamath et al. (2004) found induction of *nahG*, one of the genes responsible for naphthalene dioxygenase transcription, in bioluminescent *Pseudomonas fluorescens* when potentially root-derived compounds such as salicylate were offered, but no induction in the presence of real root extracts derived from various plants (e.g. mulberry, *Morus rubra* and hybrid willow, *Salix alba x matsudana*). The expression of *nahG* was even decreased at the individual cell level as root extract concentrations were increased in the presence of naphthalene. However, they found significantly enhanced microbial growth and overall *nahG* expression associated with enhanced biodegradation of naphthalene and concluded that plant-promoted proliferation of competent genotypes had compensated for the inhibitory effect of root-derived compounds (Kamath et al. 2004).

Obstacles and perspectives for rhizosphere manipulation/management in phytoremediation

Phytostabilisation/phytoimmobilisation

Pollutant toxicity, adverse soil conditions (e.g. low organic matter content, poor soil structure etc.), water stress and nutrient deficiency are typical problems challenging the establishment of vegetation on contaminated sites (Tordoff et al. 2000; Bradshaw and Johnson 1992). Apart from the selection of pollutant-tolerant plants, rhizosphere processes and their proper management may be crucial for the success of phytostabilisation.

To ameliorate nutrient limitations, advantage can be taken of rhizosphere processes associated with co-cropping of legumes, inoculation with phosphorus- and iron-solubilising bacteria, and inoculation with pollutant-resistant mycorrhizal fungi. As discussed above, inoculation with metal-resistant PGPR can support the establishment and improve vitality of the phytostabilisation crops, and detoxification mechanisms in the rhizosphere may be enhanced by inoculation with microbial associates. Some plants and microorganisms are able to precipitate metal compounds in the rhizosphere. This was shown for lead pyromorphite (Cotter-Howells et al. 1994, 1999) and may provide an effective means to reduce metal toxicity as well as metal mobility (phytoimmobilisation; Cotter-Howells and Caporn 1996).

The main challenge for the design of phytostabilisation systems relates to combining different approaches to ameliorate multiple constraints (i.e., nutrient and water deficiency, toxicity due to mixed contamination) and to control their efficiency in field conditions. A recent article by Roy et al. (2007) reviews the combined use of alders, frankiae and mycorrhizae for the remediation of contaminated ecosystems. This system is thought to improve plant nutrition through both the actinorhizal symbiosis (frankiae) and the mycorrhizal symbioses, and to protect the plant from toxicity through the mechanisms discussed above for metals and organic pollutants.

Phytoextraction

In spite of an "explosion" of literature addressing phytoextraction of metals and metalloids during the past decade, there is still only limited evidence for satisfactory extraction rates (Van Nevel et al. 2007). This lack of success is largely related to the small biomass of most true hyperaccumulator plants or to metal accumulation by high-biomass (crop) plants being too low. However there are also further obstacles. For example while contaminant mixtures appear to be the rule rather than the exception at polluted sites, metal tolerance, as well as efficient metal accumulation by a given plant species, is typically restricted to one or few elements.

Moreover, high metal/metalloid uptake rates in plants as required for phytoextraction can only be achieved if the metal/metalloid activity in the rhizosphere soil solution is sustained by rapid re-supply from the solid phase (Fitz et al. 2003; Lehto et al. 2006). The rate of re-supply, or in other words, the response time of the soil to the depletion of the element in soil solution (Lehto et al. 2006) is likely to be related to metal speciation in the solid phase but further work is required to characterise this system.

To improve our understanding of soil factors controlling the phytoextraction process and its predictability, it is useful to distinguish between different combinations of K_d and soil response time (Fig. 4) as suggested by Lehto et al. (2006). At small K_d and low metal concentration or activity in the soil solution, both the labile and soluble pool will be rapidly depleted but plant uptake will be rather low. At low K_d but high metal activity in soil solution, uptake by the plant would be initially high but decline rapidly as the labile pool is depleted.

Conversely in soils with high K_d values, depletion will be generally slow because of the large labile pool. At low solution metal activity there will be a rather low, but sustained, rate of uptake, whereas at high solution activities, uptake will be sustained at a high rate. As long as only a small proportion of the labile pool is depleted, the kinetics of re-supply are not important for uptake, but a rapid rate of resupply from the fixed to the labile pool will further delay depletion of the latter.

Under conditions where resupply is not limiting to uptake, hyperaccumulators and non-accumulators are likely to forage from the same pool. Indeed, based on isotopic dilution methods, i.e. the determination of E and L values, most studies appear to support this hypothesis (Gerard et al. 2000; Hutchinson et al. 2000; Massoura et al. 2004; Schwartz et al. 2003). Hutchinson et al. (2000) inferred that the hyperaccumulator plant *Thlaspi caerulescens* was accessing Cd from the same labile (=isotopically exchangeable) metal pool as non-accumulator *Lepidium heterophyllum* and concluded that *T. caerulescens* did not mobilise Cd from non-labile pools via root exudates or alteration of rhizosphere pH. Similar conclusions were also reported from experiments using high-metal soils or model soil compounds spiked with high amounts of metals (Gérard et al. 2000; Hammer et al. 2006; Schwartz et al. 2003). However further work is required to definitively establish whether hyperaccumulators and nonaccumulators obtain metal from the same pool under conditions where re-supply is limiting to uptake, and where there is also a large non-labile source of metal in the soil. Under such conditions, rhizosphere processes and their manipulation may well form an essential aspect of the phytoextraction exercise. For example, mobilisation by microbial exudation could prevent the soil response time from becoming the limiting factor.

To overcome the long times required for phytoextraction, bioavailable contaminant stripping (BCS) has been proposed as alternative to the removal of contaminants to target values based on the total metal content (Fitz et al. 2003; Hamon and McLaughlin 1999). BCS aims at reducing the labile (bioavailable) fraction below an ecotoxicological target to keep metal leaching and transfer into biota below critical limits. Again, the metal removal rate achieved during phytoextraction will depend on the combination of solute activity, buffer power and soil response time. In addition, the definition of the target value for the remaining labile pool after termination of phytoextraction needs careful attention in the light of soil and rhizosphere processes that will determine the labile and soluble metal fraction according to subsequent

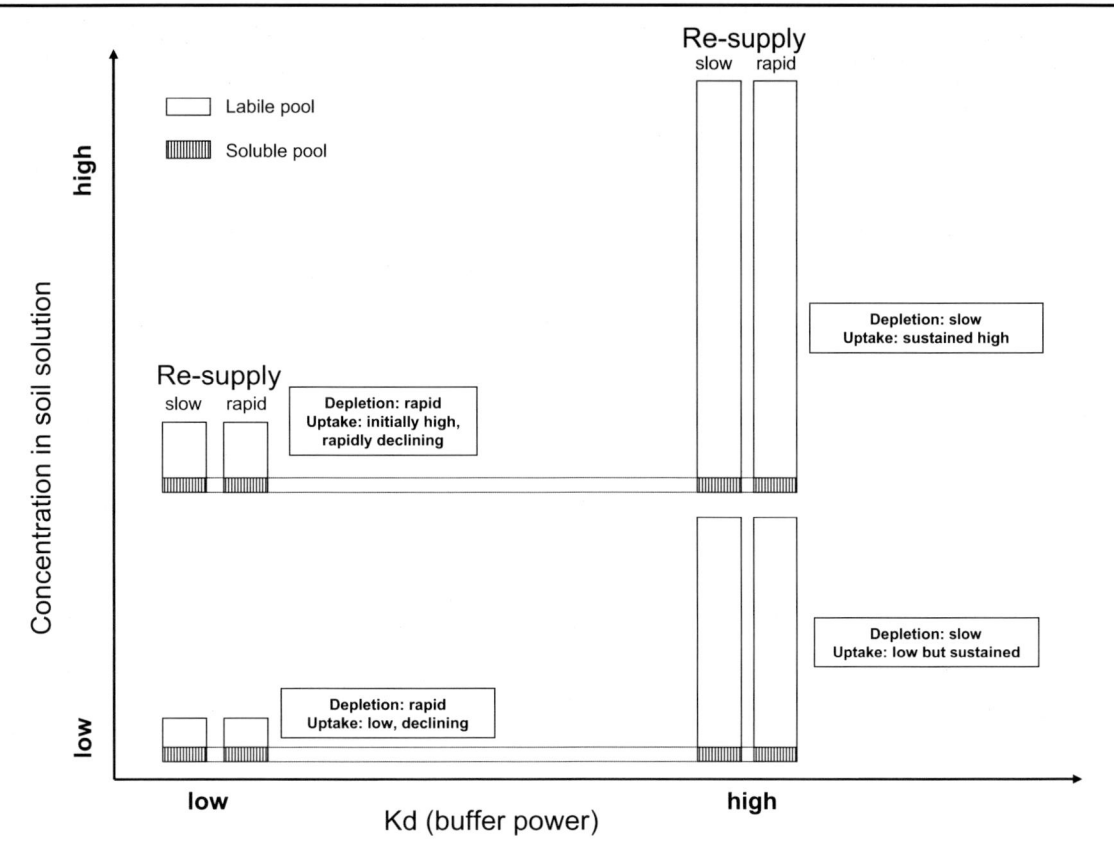

Fig. 4 Schematic representation of possible combinations of soil buffer power (Kd), soil solution activities, size of labile pool and soil response time (Tc) and their expected consequences on the depletion of labile and soluble metal pools and plant uptake

land use, e.g. food, fodder, bio-energy or fibre crops (Fitz et al. 2003). BCS may be particularly attractive where the labile metal fraction is relatively small compared to the total metal content. In this case, BCS will save time because of the smaller amount of metal to be removed and because (rate-limited) removal from fixed pools during the final stages of phytoextraction is avoided. However, predicting the rate of replenishment is not trivial and requires additional experimental and modelling work before becoming applicable.

One of the problems expected in upscaling bench and greenhouse experiments to the field relates to the heterogeneity of pollutant distribution in soil in comparison to the homogenous distribution of metals in soils from most pot studies. Plant roots typically avoid growing into zones or patches of polluted soil material when this is possible. The unique ability of hyperaccumulator *Thlaspi caerulescens* roots to preferentially proliferate into contaminated patches (e.g.

Schwartz et al. 2003) remains to be demonstrated for other metal hyperaccumulator plants as well as for high biomass accumulator species (e.g. *Salix* sp.). Apart from the field study of Keller et al. (2003) little information is available on root distribution pattern of phytoremediation crops in field conditions. Selection, breeding and genetic engineering of metal accumulator plants that preferentially forage from polluted zones may be a promising approach to improve phytoextraction efficiency in field conditions.

Soil/rhizosphere management to enhance metal availability in the rhizosphere of phytoextraction crops has been tested in numerous experiments. The most-studied approach is chelant-assisted phytoextraction using EDTA and other artificial chelants. In a recent review, Nowack et al. (2006) provide convincing evidence that at the level of chelant addition that is required for substantially improved metal accumulation in the plant, only a small proportion of the mobilised metal complexes can be taken up. In consequence, leaching of metals as has previously been shown in bench scale (e.g. Sun et al. 2001), pot and field lysimeter experiments (e.g. Wenzel et al. 2003b) is unavoidable and—along with the high costs of the chelants—therefore restricts the use of this technique to sites that are not connected to groundwater (Nowack et al. 2006). Other researchers have employed acidifying amendments such as elemental sulphur (Kayser et al. 2000; Wang et al. 2006) and ammonium fertilisation along with nitrification inhibitors (Puschenreiter et al. 2001; Zaccheo et al. 2006) to enhance metal mobility in the rhizosphere of phytoextraction crops. These studies show that optimisation of rhizosphere pH may considerably improve phytoextraction efficiency in pot experiments. However the efficacy of this practise in field applications, where metals and pH are distributed heterogeneously, as well as the potential for metal leaching associated with soil acidification remains to be demonstrated.

Co-cropping of different plant species has been proposed as a strategy to increase metal bioavailability (Gove et al. 2002) and to better explore the soil volume and address the heterogeneous distribution of pollutants in field soils (Keller et al. 2003), for instance by combining deep rooting metal accumulating willows with small hyperaccumulator species that can efficiently explore the uppermost soil horizons. Co-cropping Cd/Zn accumulator *Salix caprea* and hyperaccumulator *Arabidopsis halleri* in an outdoor lysimeter experiment did not meet expectations, probably because of competition for uptake of water, nutrients and pollutant metals (Wieshammer et al. 2007). In the field, where roots are not confined to a restricted volume of soil, competition may not be the main obstacle however, as observed in the study of Wieshammer et al. (2007) sustained establishment of the small, herbaceous hyperaccumulator and weed control in the understory of trees may become the major challenge. Co-cropping of metal accumulators with alder trees (*Alnus* sp.) may offer an interesting alternative to chemical mobilisation of metals in phytoextraction crops (Roy et al. 2007). Alder species are associated with N_2-fixing actinorhizal symbionts (frankiae). Nitrogen fixation has been shown to result in substantial acidification in alder rhizospheres (Van Miegroet and Cole 1984) because nitrogen uptake relying on N_2-fixation rather than anionic nitrate results in enhanced proton exudation to maintain the cation–anion balance (Hinsinger et al. 2003). These indigenous processes could be used to increase metal bioavailability to co-cropped metal accumulators and to improve nitrogen nutrition. The gradual production of protons in the alder rhizosphere may limit unwarranted metal leaching compared to addition of acidifying amendments or chelants and also reduce costs and management efforts required (Roy et al. 2007). However, to our knowledge this concept has not yet been tested.

Another interesting approach to enhance pollutant tolerance, plant performance and accumulation of metals at root surfaces using recombinant rhizobacterium *Pseudomonas putida* expressing a metal-binding peptide (E20) was recently demonstrated in hydroponic culture by Wu et al. (2006a). Inoculation to sunflower resulted in marked decrease in cadmium phytotoxicity and 40% increase in cadmium accumulation in the plant root. They suggest that this approach may be extended to mixed metal–organic-polluted soil.

Rhizovolatilisation

Rhizovolatilisation of inorganic contaminants differs significantly from other remediation techniques in respect to the fact that it releases the contaminants in the atmosphere. In the case of selenium, the volatile methylated species are less toxic than inorganic forms (Wilber 1980). The concern related to volatilisation of contaminants is significant especially for elements such as mercury and arsenic, which are not essential and can form extremely toxic volatile compounds (Buchet and Lauwerys 1981). Elemental mercury is far less toxic than methylmercury and its half-life in the atmosphere is in the order of years which should enable a substantial dilution into the large atmospheric pool (Meagher et al. 2000). For this reason initial research efforts focussed on the genetic manipulation of plants with the *merA* and *merB* genes to enhance Hg^{2+} uptake and volatilisation of Hg^0 (Heaton et al. 1998; Rugh et al. 1998). However, more recently an approach based on the accumulation of Hg^{2+} in the plant shoot rather than volatilisation of elemental mercury has been proposed as an alternative strategy (Meagher and Heaton 2005).

In view of the concern discussed above, rhizovolatilisation of inorganic contaminants needs to be further investigated before the potentials and limitations of this approach can be accurately determined. Meyer et al. (2007) pointed out that despite the ecological and toxicological importance of volatilisation processes, little is known in terms of the composition of the metalloid-volatilising microflora, its capacity for derivatisation and the environmental factors controlling conversion to volatile compounds. The role of rhizosphere processes in volatilisation of inorganic contaminants is, with the exception of some studies regarding selenium, largely unexplored and needs to be addressed before the potential of this approach can be properly evaluated.

Rhizodegradation

In the past two decades, a large number of publications on rhizodegradation of various organic toxicants using different plants and/or microbial inoculants have been published (see review of this literature in Anderson et al. 1993; Wenzel et al. 1999; Susarla et al. 2002; Kuiper et al. 2004; Newman and Reynolds 2004; Mohan et al. 2006; Dzantor 2007). Here, we analysed twenty three recent publications to evaluate reasons for success or failure of rhizodegradation efficiencies. Field-contaminated soils that have undergone prolonged periods of ageing (e.g. Phillips et al. 2006; Ryslava et al. 2003; Liste and Prutz 2006; Liste and Felgentreu 2006; Olson et al. 2007; Muratova et al. 2003; Demnerova et al. 2005) generally appear to be much less responsive to rhizodegradation than freshly spiked soil (e.g. Dams et al. 2007; Kaimi et al. 2007; Lin et al. 2006; Chiapusio et al. 2007; Kim et al. 2006; He et al. 2005, 2007; Gunderson et al. 2007; Child et al. 2007). Similarly, studies using spiked soils that had been aged for at least several weeks before the experiment started were less successful than those using freshly spiked substrate in terms of the "plant effect" (i.e., the percentage of additional degradation in the presence of a plant relative to an unplanted control at termination of the experiment) on pollutant degradation (e.g. Singer et al. 2003; Unterbrunner et al. 2007). The highest additional effects of plant treatments were obtained with graminaceous species (Dams et al. 2007; Chiapusio et al. 2007; He et al. 2005, 2007) on spiked soil material. I conclude that low bioavailability is a main cause of failure of rhizodegradation in field-contaminated and aged spiked soils. This has important implications for the applicability of rhizodegradation as well as for the evaluation of data obtained on freshly or only shortly aged, spiked soil material. Other strategies to enhance rhizodegradation (e.g. inoculation of degrader strains) are likely to fail where low bioavailability is the main constraint.

Six (Dams et al. 2007; Singer et al. 2003; Van Dillewijn et al. 2007; Johnson et al. 2004; Mehmannavaz et al. 2002; Child et al. 2007) out of the twenty three studies compared the effect of plant inoculation on pollutant degradation. In four studies, biodegradation in the presence of inoculate was slightly (Singer et al. 2003) to substantially (Dams et al. 2007; Johnson et al. 2004; Child et al. 2007) enhanced. However, microbial inoculation inhibited rhizodegradation relative to the noninoculated control in two studies (van Dillewijn et al. 2007; Mehmannavaz et al. 2002). Interestingly, microbial treatments appeared to be successful at the bench and pot experiment scale (Dams et al. 2007; Singer et al. 2003; Johnson et al. 2004; Child et al. 2007) but failed when applied to long-term contaminated soil on field experiments (Van Dilleweijn et al. 2007; Siciliano et al. 2003). This indicates again the importance of the experimental scale and of bioavailability.

In view of the still disappointing and controversial results of traditional inoculation (see also Dzantor 2007), enhanced rhizodegradation requires more sophisticated approaches. Enhanced degradation capabilities of inoculated microorganisms may be obtained by induction of a nutritional bias towards the inoculated strains. This will require the selection, breeding and engineering of plants that exude specific carbon substrates such as opines or flavonoids that can be preferentially used by the microbial degrader strains/populations either inoculated or natively present in the polluted soil. Profiling root exudation in terms of chemical composition and quantity and investigation of utilisation pattern by microbial strains/consortia competent to degrade organic pollutants will be a prerequisite for this purpose. Only recently, Narasimhan et al. (2003) identified root exudate compounds (phenylpropanoids) that created a nutritional bias in favor of enhanced PCB degradation.

In long-term field contaminated soil, enhancement of bioavailability appears to be the key of successful biodegradation. Selection and engineering of plants and microbial strains that can modify solubility and transport of organic pollutants through exudation of biosurfactants holds promise (e.g. Wang et al. 2007c) but the applicability of this approach has yet not been fully demonstrated.

Recent attempts to genetically engineer plant–microbial systems to enhance rhizodegradation include gene cloning of plants containing bacterial enzymes for the degradation of organic pollutants such as PCBs (Francova et al. 2003) and of recombinant, root-colonising bacteria (e.g. *Pseudomonas fluorescens*) expressing degradative enzymes (e.g. ortho-monooxygenase for toluene degradation; Yee et al. 1998). Similar work was presented by Saiki et al. (2003) for the degradation of dioxine-like compounds by a recombinant *Rhizobium tropici* strain expressing 1,9a-dioxygenase. These developments are still largely restricted to the bench scale and require careful investigation of ecological consequences and consideration of public concerns before they can be applied to field conditions. The development of biological containment systems for the control of death and survival of recombinant bacteria (Ronchel and Ramos 2001) may provide a tool to overcome such limitations, and recent work of Aguirre de Cárcer et al. (2007) indicates that genetic changes on native bacteria induced by genetically modified microorganisms may be limited to the rhizosphere. However, further work is needed to solicitate these findings.

Integrated approaches

The complexity and heterogeneity of sites often polluted with multiple metals, metalloids and organic compounds requires the design of integrated phytoremediation systems that combine different processes and approaches. Co-cropping different species may enhance the overall capabilities of a phytoremediation system to explore the contaminated soil volume, address different pollutants, and support differential microbial consortia in their rhizospheres. Shared rhizospheres may be designed to optimise the nutritional status, e.g. by combining plants that support N_2-fixing and P-solubilising microorganisms, or—as common reed (*Phragmites australis*) — supply oxygen via aerenchyma in submerged conditions. Co-cropping could be also used to modify the bioavailability of pollutants, e.g. by combining *Alnus* sp. with metal-accumulating willows (Roy et al. 2007) or to combine metal phytoextraction crops (e.g. willows) with plants that support rhizodegradation of organic pollutants (e.g. grasses). Engineering rhizobacteria capable of heavy metal accumulation and enhanced degradation of organic pollutants such as trichloroethylene offers further opportunities to address multiply contaminated sites (Lee et al. 2006).

While some phyto-/rhizoremediation technologies are being used commercially, it is obvious that the complexity of interactions in the plant–microbe–soil-pollutant system requires substantial further research efforts to improve our understanding of the rhizosphere processes involved. Emphasis should be put on evaluating results obtained in simplified bench and pot experiments to heterogeneous, multiply polluted field sites and the functioning of phyto-/rhizoremediation systems under various ecological conditions.

Acknowledgements The author is grateful to Dr. Rebecca Hamon for critically reading and her suggestions to shorten and improve the manuscript and to Dr. Enzo Lombi for his contributions to the sections on rhizovolatilisation.

References

Abou-Shanab RA, Angle JS, Delorme TA, Chaney RL, van Berkum P, Moawad H et al (2003) Rhizobacterial effects on nickel extraction from soil and uptake by *Alyssum murale*. New Phytol 158:219–224

Abou-Shanab RAI, Angle JS, Chaney RL (2006) Bacterial inoculants affecting nickel uptake by *Alyssum murale* from low, moderate and high Ni soils. Soil Biol Biochem 38:2882–2889

Aguirre de Cárcer D, Martin M, Mackova M, Macek T, Karlson U, Rivilla R (2007) The introduction of genetically modified microorganisms designed for rhizoremediation induces changes on native bacteria in the rhizosphere but not in the surrounding soil. ISME J 1:215–223

Anderson TA, Coats JE (1994) Bioremediation through rhizosphere technology. ACS Symp Ser:563. Am Chem Soc, Washington, DC

Anderson TA, Guthrie EA, Walton BT (1993) Bioremediation in the rhizosphere. Plant roots and associated microbes clean contaminated soil. Environ Sci Technol 27:2630–2636

Arshad M, Saleem M, Hussain S (2007) Perspectives of bacterial ACC deaminase in phytoremediation. Trends Biotechnol 25:356–362

Ashford AE, Peterson CA, Carpenter JL, Cairney JWG, Allaway WG (1988) Structure and permeability of the fungal sheath in the *Pisonia* mycorrhiza. Protoplasma 147:149–161

Azaizeh H, Gowthaman S, Terry N (1997) Microbial selenium volatilisation in rhizosphere and bulk soils from a constructed wetland. J Environ Qual 26:666–672

Azaizeh HA, Salhani N, Sebesvari Z, Emons H (2003) The potential of rhizosphere microbes isolated from a constructed wetland to biomethylate selenium. J Environ Qual 32:55–62

Baker AJM, Reeves RD, McGrath SP (1991) In situ decontamination of heavy metal polluted soils. Using crops of metal-accumulating plants: a feasibility study. In: Hinchee RE, Olfenbuttel RF (eds) In situ bioreclamation. Butterworth-Heinemann, Stoneham, MA, p 539

Barkay T, Miller SM, Summers AO (2003) Bacterial mercury resistance from atoms to ecosystems. FEMS Microbiol Rev 27:355–384

Baum C, Hrynkiewicz K, Leinweber P, Meißner R (2006) Heavy-metal mobilization and uptake by mycorrhizal and nonmycorrhizal willows (*Salix* x *dasyclados*). J Plant Nutr Soil Sci 169:516–522

Belimov AA, Safronova VI, Sergeyeva TA, Egorova TN, Matveyeva VA, Tsyganov VE et al (2001) Characterization of plant growth-prompting rhizobacteria isolated from polluted soils and containing 1-aminocyclopropane-1-carboxylate deaminase. Can J Microbiol 47:642–652

Belimov AA, Hontzeas N, Safronova VI, Demchinskaya SV, Piluzza G, Bullitta S et al (2005) Cadmium-tolerant plant growth-promoting bacteria associated with the roots of Indian mustard (*Brassica juncea* L. Czern.). Soil Biol Biochem 37:241–250

Bento FM, de Oliveira Camargo FA, Okeke BC, Frankenberger WT (2005) Diversity of biosurfactant producing microorganisms isolated from soils contaminated with diesel soil. Microbiol Res 160:249–255

Berthelsen BO, Lamble GM, MacDowell AA, Nicholson DG (2000) Analysis of metal speciation and distribution in symbiotic fungi (ectomycorrhiza) studied by micro X-ray absorption spectroscopy and X-ray fluorescence. In: Gobran GR, Wenzel WW, Lombi E (eds) Trace elements in the rhizosphere.. CRC, Boca Raton, USA, pp 149–164

Blute NK, Brabander DJ, Hemond HF, Sutton SR, Newville MG, Rivers ML (2004) Arsenic sequestration by ferric iron plaque on cattail roots. Environ Sci Technol 38:6074–6077

Boyajian GE, Carreira LH (1997) Phytoremediation: a clean transition from laboratory to marketplace. Nat Biotechnol 15:127–128

Bradshaw AD, Johnson MS (1992) Revegetation of metalliferous mine waste: the range of practical techniques used in Western Europe. In: Minerals, metals and the environment. Institute of Mining and Mettalurgy, London, 491 p

Breckle S-W, Kahle H (1992) Effect of toxic heavy metals (Cd, Pb) on growth and mineral nutrition of beech (*Fagus sylvatica* L.). Vegetatio 101:43–53

Brix H, Sorrell BK, Schierup H-H (1996) Gas fluxes by in situ convective flow in *Phragmites australis*. Aquat Bot 54:151–163

Bromilow RH, Chamberlain K (1995) Principles governing uptake and transport of chemicals. In: Trapp S, McFarlane JC (eds) Plant contaminations, modeling and simulation of organic chemical processes. CRC, Boca Raton, FL, USA, pp 37–68

Brown SL, Chaney RL, Angle JS, Baker AJM (1994) Phytoremediation potential of Thlaspi caerulescens and bladder campion for zinc- and cadmium-contaminated soil. J Environ Qual 23:1151-1157

Buchet JP, Lauwerys R (1981) Evaluation of exposure to inorganic arsenic in man. Analytical techniques for heavy metals in biological fluids. Elsevier, Amsterdam, pp 75–89

Burd GI, Dixon DC, Click BR (1998) A plant growth promoting bacterium that decreases nickel toxicity in seedlings. Appl Environ Microbiol 64:3663–3668

Burd GI, Dixon DG, Glick BR (2000) Plant growth-promoting bacteria that decrease heavy metal toxicity in plants. Can J Microbiol 46:237–245

Burken JB, Schnoor JL (1998) Predictive relationships for uptake of organic contaminants by hybrid poplar trees. Environ Sci Technol 32:3379–3385

Cathala N, Salsac L (1975) Absorption du cuivre par les racines de mais (*Zea mays* L.) et de tournesol (*Helianthus annuus* L.). Plant Soil 42:65–83

Chaîneau CH, Rougeux G, Yéprémian C, Oudot J (2005) Effects of nutrient concentration on the biodegradation of crude oil and associated microbial populations in the soil. Soil Biol Biochem 37:1490–1497

Chaney RL (1983) Plant uptake of organic waste constituents. In: Parr et al (ed) Land treatment of hazardous wastes. Noyes Data, Park Ridge, NJ, pp 50–76

Chiapusio G, Pujol S, Toussaint ML, Badot PM, Binet P (2007) Phenanthrene toxicity and dissipation in the rhizosphere of grassland plants (*Lolium perenne* L. and *Trifolium pratense* L.) in three spiked soils. Plant Soil 294:103–112

Child R, Miller CD, Liang Y, Sims RC, Anderson AJ (2007) Pyrene mineralization by *Mycobacterium* sp. strain KMS in a barley rhizosphere. J Environ Qual 36:1260–1265

Citterio S, Prato N, Fumagalli P, Aina R, Massa N, Santagostino A et al (2005) The arbuscular mycorrhizal fungus *Glomus mosseae* induces growth and metal accumulation changes in *Cannabis sativa* L. Chemosphere 59:21–29

Colpaert JV, Asche JA (1993) The effects of cadmium on ectomycorrhizal *Pinus sylvestris* L. New Phytol 123:325–333

Cotter-Howells JD, Champness PE, Charnock JM, Pattrick RAD (1994) Identification of pyromorphite in mine-waste contaminated soils by ATEM and EXAFS. Eur J Soil Sci 45:393–402

Cotter-Howells J, Caporn S (1996) remediation of contaminated land by formation of heavy metal phosphates. Appl Geochem 11:335–342

Cotter-Howells JD, Champness PE, Charnock JM (1999) Mineralogy of lead-phosphorus grains in the roots of *Agrostis capillaris* L. by ATEM and EXAFS. Min Mag (Lond) 63:777–789

Curl EA, Truelove B (1986) The rhizophere. Advanced series in agricultural science 15. Springer, Berlin, Germany

Dams RI, Paton GI, Killham K (2007) Rhizoremediation of pentachlorphenol by *Sphinobium chlorophenolicum* ATCC 39723. Chemosphere 68:864–870

Degryse F, Verma VK, Smolders E (2007) Mobilization of Cu and Zn by root exudates of dicotyledonous plants in resin buffered solutions and in soil. Plant Soil 306:69–84

Demnerova K, Mackova M, Spevakova V, Beranova K, Kochankova L, Lovecka P et al (2005) Two approaches to biological decontamination of groundwater and soil polluted by aromatics—characterization of microbial populations. Int Microbiol 8:205–211

De Souza MP, Chu D, Zhao M, Zayed AM, Ruzin SE, Schichnes D et al (1999a) Rhizosphere bacteria enhance selenium accumulation and volatilisation by Indian Mustard. Plant Physiol 119:563–573

De Souza MP, Huang CPA, Chee N, Terry N (1999b) Rhizosphere bacteria enhance the accumulation of selenium and mercury in wetland plants. Planta 209:259–263

Di Gregorio S, Lampis S, Malorgio F, Petruzzelli G, Pezzarossa B, Vallini G (2006) *Brassica juncea* can improve selenite and selenate abatement in selenium contaminated soils through the aid of its rhizosperic bacterial population. Plant Soil 285:233–244

Donnelly PK, Hedge RS, Fletcher JS (1994) Growth of PCB-degrading bacteria on compounds from photosynthetic plants. Chemosphere 28:981–988

Dos Santos Utmazian MN, Schweiger P, Sommer P, Gorfer M, Strauss J, Wenzel WW (2007) Influence of *Cadophora finlandica* and other microbial treatments on cadmium and zinc uptake in willows grown on polluted soil. Plant Soil Environ 53:158–166

Doyle MO, Otte ML (1997) Organism-induced accumulation of iron. zinc and arsenic in wetland soils. Environ Pollut 96:1–11

Dzantor EK (2007) Phytoremediation: the state of rhizosphere "engineering" for accelerated rhizodegradation of xenobiotic contaminants. J Chem Technol Biotechnol 82:228–232

Dziejowski JE, Rimmer A, Steenhuis TS (1997) Preferential movement of oxygen in soils. Soil Sci Soc Am J 6:1607–1610

Edvantoro BB, Naidu R, Megharaj M, Merrington G, Singleton I (2004) Microbial formation of volatile arsenic in cattle dip site soils contaminated with arsenic and DDT. Appl Soil Ecol 25:207–217

Fein JB, Martin AM, Wightman PG (2001) Metal adsorption onto bacterial surfaces: development of a predictive approach. Geochim Cosmochim Acta 65:4267–4273

Ferro AM, Sims RC, Bugbee B (1994) Hycrest crested wheatgrass accelerates the degradation of pentachlorophenol in soil. J Environ Qual 23:272–279

Fitz WJ, Wenzel WW (2002) Arsenic transformations in the soil–rhizosphere–plant system: fundamentals and potential application to phytoremediation. J Biotechnol 99:259–278

Fitz WJ, Wenzel WW, Zhang H, Nurmi J, Štipek K, Fischerova Z et al (2003) Rhizosphere characteristics of the arsenic hyperaccumulator *Pteris vittata* L. and monitoring of phytoremoval efficiency. Environ Sci Technol 37:5008–5014

Flessa H, Fischer WR (1992) Plant-induced changes in the redox potential of rice rhizospheres. Plant Soil 143:55–60

Fletcher JS, Hedge RS (1995) Release of phenols by perennial plant roots and their potential importance in bioremediation. Chemosphere 31:3009–3016

Fomina MA, Alexander IJ, Copaert JV, Gadd GM (2005) Solubilization of toxic metal minerals and metal tolerance of mycorrhizal fungi. Soil Biol Biochem 37:851–866

Francova K, Sura M, Macek T, Szekeres M, Bancos S, Demnerova K et al (2003) Preparation of plants containing bacterial enzyme for the degradation of polychlorinated biphenyls. Fresenius Environ Bull 12:309–313

Frankenberger WT, Karlson U (1994) Soil-management factors affecting volatilization of selenium from dewatered sediments. Geomicrobiol J 12:265–278

Frankenberger WT, Arshad M (2002) Volatilisation of arsenic. In: Frankenberger WT (ed) Environmental chemistry of arsenic. Marcel Dekker, New York, pp 363–380

Gadd GM (2004) Microbial influence on metal mobility and application to bioremediation. Geoderma 122:109–119

Gao S, Burau RG (1997) Environmental factors affecting rates of arsine evolution from mineralization of arsenicals in soil. J Environ Qual 26:753–763

Gerard E, Echevarria G, Sterckeman T, Morel J-L (2000) Cadmium availability to three plants species varying in cadmium accumulation pattern. J Environ Qual 29:1117–1123

Gilbertson AW, Fitch MW, Burken JG, Wood TK (2007) Transport and survival of GFP-tagged root-colonizing microbes: implications for rhizodegradation. Eur J Soil Biol 43:224–232

Glick BR (2005) Modulation of plant ethylene levels by the bacterial enzyme ACC deaminase. FEMS Microbiol Lett 251:1–7

Gove B, Hutchinson JJ, Young SD, Craigon J, McGrath SP (2002) Uptake of metals by plants sharing a rhizosphere with the hyperaccumulator *Thlaspi caerulescens*. Int J Phytorem 4:267–281

Gunderson JJ, Knight JD, Van Rees KCJ (2007) Impact of ectomycorrhizal colonization of hybrid poplar on the remediation of diesel-contaminated soil. J Environ Qual 36:927–934

Hammer D, Keller C, McLaughlin MJ, Hamon RE (2006) Fixation of metals in soil constituents and potential remobilization by hyperaccumulating and non-hyperaccumulating plants: results from an isotopic dilution study. Environ Pollut 143:407–415

Hamon RE, McLaughlin MJ (1999) Use of the hyperaccumulator *Thlaspi caerulescens* for bioavailable contaminant striping. In: Wenzel WW et al (eds) Proc 5th International Conference on the Biogeochemistry of Trace Elements. Vienna, Austria, pp 908–909

Harmsen J, Rulkens W, Eijsackers H (2005) Bioavailability: concept for understanding or tool for predicting. Land Contam Recl 13:161–171

Hartley J, Cairney JWG, Meharg AA (1997) Do ectomycorrhizal fungi exhibit adaptive tolerance to potentially toxic metals in the environment. Plant Soil 189:303–319

He LM, Neu MP, Vanderberg LA (2000) Bacillus lichenformis γ-glutamyl exopolymer: physicochemical characterization and U(VI) interaction. Environ Sci Technol 34:1694–1701

He Y, Xu J, Tang C, Wu Y (2005) Facilitation of pentachlorophenol degradation in the rhizosphere of ryegrass (*Lolium perenne* L.). Soil Biol Biochem 37:2017–2024

He Y, Xu J, Ma Z, Wng H, Wu Y (2007) Profiling of PLFA: implications for nonlinear spatial gradient of PCP degradation in the vicinity of *Lolium perenne* L. roots. Soil Biol Biochem 39:1121–1129

Heaton ACP, Rugh CL, Wang NJ, Meagher RB (1998) Phytoremediation of mercury- and methylmercury-polluted soils using genetically engineered plants. J Soil Contam 7:497–509

Hedge RS, Fletcher JS (1996) Influence of plant growth stage and season on the release of root phenolics by mulberry as related to the development of phytoremediation technology. Chemosphere 32:2471–2479

Himmelbauer M, Puschenreiter M, Schnepf A, Loiskandl W, Wenzel WW (2005) Root morphology of *Thlaspi goesingense* Halacsy grown on a serpentine soil. J Plant Nutr Soil Sci 168:138–144

Hinsinger P, Courchesne F (2008) Biogeochemistry of metals and metalloids at the soil–root interface. In: Violante A,

Huang PM, Gadd GM (eds) Biophysic-chemical processes of heavy metals and metalloids in soil environments. Wiley, Hoboken, USA, pp 267–311

Hinsinger P, Gobran GR, Gregory PJ, Wenzel WW (2005) Rhizosphere geometry and heterogeneity arising from root-mediated physical and chemical processes. New Phytol 168:293–303

Hinsinger P, Plassard C, Tang C, Jaillard B (2003) Origins of root-induced pH changes in the rhizosphere and their responses to environmental constraints: A review. Plant Soil 248:43-59

Hubert C, Shen Y, Voordouw G (2005) Changes in soil microbial community composition induced by cometabolism of toluene and trichloroethylene. Biodegradation 16:11–22

Hutchinson JJ, Young SD, McGrath SP, West HM, Black CR, Baker AJM (2000) Determining uptake of "non-labile" soil cadmium by *Thlaspi caerulescens* using isotopic dilution techniques. New Phytol 146:453–460

Hyman MR, Russell SA, Ely RL, Williamson KJ, Arp DJ (1995) Inhibition, inactivation, and recovery of ammonia-oxidizing activity in co-metabolism of trichloroethylene by *Nitrosomonas europaea*. Appl Environ Microbiol 61:1480–1487

Idris R, Trifinova R, Puschenreiter M, Wenzel WW, Sessitsch A (2004) Bacterial communities associated with flowering plants of the Ni hyperaccumulator *Thlaspi goesingense*. Appl Environ Microbiol 70:2667–2677

Jankong P, Visoottiviseth P, Khokiattiwong S (2007) Enhanced phytoremediation of arsenic contaminated land. Chemosphere 68:1906-1912

Johnson DW, Benesch JA, Gustin MS, Schorran DS, Lindberg SE, Coleman JS (2003) Experimental evidence against diffusion control of Hg evasion from soils. Sci Total Environ 304:175–184

Johnson DL, Maguire KL, Anderson DR, McGrath SP (2004) Enhanced dissipation of chrysene in planted soil: the impact of rhizobial inoculum. Soil Biol Biochem 36:33–38

Johrdal JL, Foster L, Schnoor JL, Alvarez PJJ (1997) Effect of hybrid poplar trees on microbial populations important to hazardous waste bioremediation. Environ Toxicol Chem 16:1318–3121

Joner EJ, leyval C, Colpaert JV (2006) Ectomycorrhizas impede phytoremediation of polycyclic aromatic hydrocarbons (PAHs) both within and beyond the rhizosphere. Environ Poll 142:34-38

Jones DL, Hodge A, Kuzyakov Y (2004) Plant and mycorrhizal regulation of rhizodeposition. New Phytol 163:459–480

Kaimi E, Mukaidani T, Tamaki M (2007) Effect of rhizodegradation in diesel-contaminated soil under different soil condition. Plant Prod Sci 10:105–111

Kamath R, Schnoor JL, Alvarez PJJ (2004) Effect of root-derived substrates on the expression of nah-lux genes in Pseudomonas fluorescens HK44: implications for PAH biodegradation in the rhizophsere. Environ Sci Technol 38:1740–1745

Kaye JP, Hart SC (1997) Competition for nitrogen between plants and soil microorganisms. Trends Ecol Evol 12:139–143

Kayser A, Wenger K, Keller A, Attinger W, Felix HR, Gupta SK et al (2000) Enhancement of phytoextraction of Zn, Cd and Cu from calcareous soil: the use of NTA and sulphur amendments. Environ Sci Technol 34:1778–1783

Kim J, Kang S-H, Min K-A, Cho-K-S, Lee I-S (2006) Rhizosphere microbial activity during phytoremediation of diesel-contaminated soil. J Environ Sci Health Part A 41:2503-2516

Kechavarzi C, Pettersson K, Leeds-Harrisson P, Ritchie L, Ledin S (2007) Root establishment of perennial ryegrass (*L. perenne*) in diesel contaminated subsurface soil layers. Environ Pollut 145:68–74

Keller C, Hammer D, Kayser A, Richner W, Brodbeck M, Sennhauser M (2003) Root development and heavy metal phytoextraction efficiency: comparison of different plant species in the field. Plant Soil 249:67–81

Kuiper I, Lagendijk EL, Bloemberg GV, Lugtenberg BJJ (2004) Rhizoremediation: a beneficial plant–microbe interaction. Mol Plant Microbe Interact 17:6–15

Lasat MM, Baker AJM, Kochian LV (1996) Physiological Characterisation of root Zn^{2+} absorption and translocation to shoots in Zn hyperaccumulator and nonaccumulator species of *Thlaspi*. Plant Physiol 112:1715–1722

Lee D-H, Cody RD, Kim D-J, Choi S (2002) Effect of soil texture on surfactant-based remediation of hydrophobic organic-contaminated soil. Environ Int 27:681–688

Lee W, Wood TK, Chen W (2006) Engineering TCE-degrading rhizobacteria for heavy metal accumulation and enhanced TCE degradation. Biotechnol Bioeng 95:399–403

Lehto NJ, Davison W, Zhang H, Tych W (2006) Theoretical comparison of how soil processes affect uptake of metals by diffusive gradients in thinfilms and plants. J Environ Qual 35:1903–1913

Leung HM, Ye ZH, Wong MH (2006) Interactions of mycorrhizal fungi with *Pteris vittata* (As hyperaccumulator) in As-contaminated soils. Environ Pollut 139:1–8

Leyval C, Binet P (1998) Effect of poyaromatic hydrocarbons in soil on arbuscular mycorrhizal plants. J Environ Qual 27:402–407

Leyval C, Joner EJ (2000) Bioavailability of metals in the mycorhizosphere. In: Gobran GR, Wenzel WW, Lombi E (eds) Trace elements in the rhizosphere. CRC, Boca Raton, USA, pp 165–185

Leyval C, Turnau K, Haselwandter K (1997) Effect of heavy metal pollution on mycorrhizal colonization and function: physiological, ecological and applied aspects. Mycorrhiza 7:139–153

Liao JP, Lin XG, Cao ZH, Shi YQ, Wong MH (2003) Interactions between arbuscular mycorrhizae and heavy metals under sand culture experiment. Chemosphere 50:847–853

Lin Q, Wang Z, Ma S, Chen Y (2006) Evaluation of dissipation mechanisms by *Lolium perenne* L, and *Raphanus sativus* for pentachlorophenol (PCP) in copper co-contaminated soil. Sci Total Environ 368:814–822

Liste H-H, Felgentreu D (2006) Crop growth, culturable bacteria, and degradation of petrol hydrocarbons (PHCs) in a long-term contaminated field soil. Appl Soil Ecol 31:43–52

Liste H-H, Prutz I (2006) Plant performance, dioxygenase-expressing rhizosphere bacteria, and biodegradation of weathered hydrocarbons in contaminated soil. Chemosphere 62:1411–1420

Liu Y, Zhu YG, Chen BD, Christie P, Li XL (2005) Influence of the arbuscular mycorrhizal fungus *Glomus mosseae* on uptake of arsenate by the As hyperaccumulator fern *Pteris vittata* L. Mycorrhiza 15:187–192

Lodewyckx C, Mergeay M, Vangronsveld J, Clijsters H, van der Lelie D (2002) Isolation, characterization, and identification of bacteria associated with the zinc hyperaccumulator *Thlaspi caerulescens* subsp. *calaminaria*. Int J Phytoem 4:101–115

Lombi E, Zhao FJ, McGrath SP, Young SD, Sacchi GA (2001) Physiological evidence for a high-affinity cadmium transporter highly expressed in a *Thlaspi caerulescens* ecotype. New Phytol 149:53–60

Luthy RG, Aiken GR, Brusseau ML, Cunnningham SD, Gschwend PM, Pignatello JJ et al (1997) Sequestration of hydrophobic organic contaminants by geosorbents. Environ Sci Technol 31:3341–3347

Marschner H (1995) Mineral nutrition of higher plants, 2nd edn. Academic, San Diego, CA

Marschner P, Jentschke G, Godbold DL (1998) Cation exchange capacity and lead soption in ectomycorrhizal fungi. Plant Soil 205:93–98

Martino E, Perotto S, Parsons R, Gadd GM (2003) Solubilization of insoluble inorganic zinc compounds by ericoid mycorrhizal fungi derived from heavy metal polluted sites. Soil Biol Biochem 35:133–141

Massoura ST, Echevarria G, Leclerc-Cessac E, Morel JL (2004) Response of excluder, indicator and hyperaccumulator plants to nickel availability in soils. Aust J Soil Res 42:933–938

McBride NM (1989) Reactions controlling heavy metal solubility in soils. Adv Soil Sci 10:1–56

McGrath SP, Zhao FJ, Lombi E (2001) Plant and rhizosphere processes involved in phytoremediation of metal-contaminated soils. Plant Soil 232:207–214

McLaughlin MJ, Smolders E, Merckx R (1998) Soil–root interface: physicochemical processes. In: Huang PM, Adriano DC, Logan TJ, Checkai RT (eds) Soil chemistry and ecosystem health. Soil Science Society of America, Madison, Wisconsin, USA, pp 233–277 Special Publication n°52

Meagher RE, Heaton ACP (2005) Strategies for the engineered phytoremediation of toxic element pollution: mercury and arsenic. J Ind Microbiol Microtechnol 32:502–513

Meagher RB, Rugh CL, Kandasamy MK, Gragson G, Wang NJ (2000) Engineered phytoremediation of mercury pollution in soil and water using bacterial genes. In: Terry N, Bañuelos G (eds) Phytoremediation of contaminated soil and water. Lewis , Boca Raton, USA, pp 201–220

Meharg AA, Cairney JWG (2000) Extomycorrhizas—extending the capabilities of rhizosphere remediation. Soil Biol Biochem 32:1475–1484

Mehmannavaz R, Prasher SO, Ahmad D (2002) Rhizospheric effects of alfalfa on biotransformation of polychlorinated bipheyls in a contaminated soil augmented with *Sinorhizobium meliloti*. Process Biochem 37:955–963

Mench M, Martin E (1991) Mobilization of cadmium and other metals from two soils by root exudates of *Zea mays* L., *Nicotiana tabacum* L. and *Nicotiana rustica* L. Plant Soil 132:187–196

Meyer J, Schmidt A, Michalke K, Hensel R (2007) Volatilisation of metals and metalloids by the microbial population of an alluvial soil. Syst Appl Microbiol 30:229–238

Mohan SV, Kisa T, Ohkuma T, Kanaly RA, Shimizu Y (2006) Bioremediation technologies for treatment of PAH-contaminated soil and strategies to enhance process efficiency. Rev Environ Sci Biotechnol 5:347–374

Moore LW, Chilton WS, Canfield ML (1997) Diversity of opines and opine-catabolizing bacteria isolated from naturally occurring crown gall tumors. Appl Environ Microbiol 63:201–207

Moorehead DL, Westerfield MM, Zak JC (1998) Plants retard litter decay in a nutrient-limited soil: a case of exploitative competition. Oecologia 113:530–536

Moreno FN, Anderson CWN, Stewart RB, Rosinson BH, Nomura R, Ghomshei M et al (2005) Effect of thioligands on plant Hg accumulation and volatilization from mercury-contaminated mine tailings. Plant Soil 275:233–246

Muratova AY, Turkovskaya OV, Hübner T, Kuschk P (2003) Studies of the efficacy of alfalfa and reed in the phytoremediation of hydrocarbon-polluted soil. Appl Biochem Microbiol 39:599–605

Narasimhan K, Basheer C, Bajic VB, Swarup S (2003) Enhancement of plant–microbe interactions using a rhizosphere metabolomics-driven approach and its application in the removal of polychlorinated biphenyls. Plant Physiol 132:146–153

Neubauer U, Furrer G, Schulin R (2002) Heavy metal sorption on soil minerals affected by the siderophore desferrioxamine B: the role of Fe(III) (hydr)oxides and dissolved Fe(III). Eur J Soil Sci 53:45–55

Newman LA, Reynolds CM (2004) Phytodegradation of organic compounds. Curr Opin Biotechnol 15:225–230

Nowack B, Schulin R, Robinson BH (2006) Critical assessment of chelant-enhanced metal phytoextraction. Environ Sci Technol 17:5225–5232

Olson PE, Castro A, Joern M, DuTeau NM, Pilon-Smits EAH, Reardon KF (2007) Comparison of plant families in a greenhouse phytoremediation study on an aged polycyclic aromatic hydrocarbon-contaminated soil. J Environ Qual 36:1461–1469

Olsson PA, Chalot M, Baath E, Finlay RD, Söderström B (1996) Ectomycorrhizal mycelia reduce bacterial activity in a sandy soil. FEMS Microbiol Ecol 21:77–86

Pawlowska TE, Chaney RL, Chin M, Charvat I (2000) Effects of metal phytoextraction practices on the indigenous community of arbuscular mycorrhizal fungi at a metal-contaminated landfill. Appl Environ Microbiol 66:2526–2530

Pence NS, Larsen PB, Ebbs SD, Letham DLD, Lasat MM, Garvin DF et al (2000) The molecular physiology of heavy metal transporter in the Zn/Cd hyperaccumulator *Thlaspi caerulescens*. Proc Natl Acad Sci USA 97:4956–4960

Phillips LA, Greer CW, Germida JJ (2006) Culture-based and culture-independent assessment of the impact of mixed and single plant treatments on rhizosphere microbial communities in hydrocarbon contaminated flare-pit soil. Soil Biol Biochem 38:2823–2833

Prohaska T, Pfeffer M, Yulipan M, Stingeder G, Mentler A, Wenzel WW (1999) Speciation of arsenic of liquid and gaseous emission from soil in a microcosmos experiment by liquid and gas chromatography with inductively coupled plasma mass spectrometer (ICP–MS) detection. Fresenius J Anal Chem 364:467–470

Puschenreiter M, Stöger G, Lombi E, Horak O, Wenzel WW (2001) Phytoextraction of heavy metal contaminated soils with *Thlaspi goesingense* and *Amaranthus hybridus*:

rhizosphere manipulation using EDTA and ammonium sulphate. J Plant Nutr Soil Sci 164:615–621

Read DB, Bengough AG, Gregory PJ, Crawford JW, Robinson D, Scrimgeour CM et al (2003) Plant roots release phospholipids surfactants that modify the physical and chemical properties of soil. New Phytol 157:315–326

Reddy KR, Patrick WH Jr, Lindau CW (1989) Nitrification—denitrification at the plant root–sediment interface in wetlands. Limnol Oceanogr 34:1004–1013

Reid BJ, Jones KC, Semple KT (2000) Bioavailability of persistent pollutants in soils and sediments—a perspective on mechanisms, consequences and assessment. Environ Pollut 108:103–112

Rentz JA, Alvarez PJJ, Schnoor JL (2005) Benzo[a]pyrene co-metabolism in the presence of plant root extracts and exudates. Implications for phytoremediation. Environ Pollut 136:477–484

Ronchel MC, Ramos JL (2001) Dual system to reinforce biological containment of recombinant bacteria designed for rhizoremediation. Appl Environ Microbiol 67:2649–2656

Roy S, Khasa DP, Greer CW (2007) Combining alders, frankiae, and mycorrhizae for the revegetation and remediation of contaminated ecosystems. Can J Bot 85:237–251

Rugh CL, Senecoff JF, Meagher RB, Merkle SA (1998) Development of transgenic yellow-poplar for mercury phytoremediation. Nat Biotechnol 33:616–621

Ryslava E, Krejcik Z, Macek T, Novakova H, Demnerova K, Mackova M (2003) Study of PCB degradation in real contaminated soil. Fresenium Environ Bull 12:296–301

Sabljic A, Piver WT (1992) Quantitative modelling of environmental fate and impact of commercial chemicals. Environ Toxicol Chem 11:961–972

Saiki Y, Habe H, Yuuki T, Ikeda M, Yoshida T, Nojiri H et al (2003) Rhizoremediation of dioxine-like compounds by a recombinant *Rhizobium tropici* strain expressing carbazole 1,9a-dioxigenase constitutively. Biosci Biotechnol Biochem 67:1144–1148

Salt DE, Blaylock M, Kumar NPBA, Dushenkov V, Ensley BD, Chet I et al (1995) Phytoremediation: a novel strategy for the removal of toxic metals from the environment using plants. Biotechnology 13:468–474

Salt DE, Kato N, Krämer U, Smith RD, Raskin I (2000) The role of root exudates in nickel hyperaccumulation and tolerance in accumulator and nonaccumulator species of *Thlaspi*. In: Terry N, Banuelos G (eds) Phytoremediation of contaminated soil and water. Lewis, Boca Raton, pp 189–200

Sarand I, Timonen S, Nurmiaho-Lassila E-L, Koivila T, Haahtela K, Romantschuk M et al (1998) Microbial biofilms and catabolic plasmid harbouring degradative fluorescent pseudomonads in Scots pine ectomycorrhizospheres developed on petroleum contaminated soil. FEMS Microbiol Ecol 27:115–126

Sarand I, Timonen S, Koivula T, Peltola R, Haahtela K, Sen R et al (1999) Tolerance and biodegradation of *m*-toluate by Scots pine, a mycorrhizal fungus and fluorescent pseudomonads individually and under associative conditions. J Appl Microbiol 86:817–826

Schneiker S, Keller M, Droege M, Lanka E, Puller A, Selbitschka W (2001) The genetic organization and evolution of the broad host range mercury resistance plasmid pSB102 isolated from a microbial population residing in the rhizosphere alfalfa. Nucleic Acids Res 29:5169–5181

Schnoor JL, Licht LA, McCutcheon SC, Wolfe NL, Carreira LH (1995) Phytoremediation of organic and nutrient contaminants. Environ Sci Technol 29:318–323

Schwartz C, Morel JL, Saumier S, Whiting SN, Baker AJM (1999) Root development of the zinc-hyperaccumulator plant *Thlaspi caerulescens* as affected by metal origin, content and localization in soil. Plant Soil 208:103–115

Schwartz C, Echevarria G, Morel JL (2003) Phytoextraction of cadmium with *Thlaspi caerulescens*. Plant Soil 249:27–35

Sell J, Kayser A, Schulin R, Brunner I (2005) Contribution of ectomycorrhizal fungi to cadmium uptake of poplars and willows from a heavily polluted soil. Plant Soil 277:245-253

Semple KT, Morriss AWJ, Paton GI (2003) Bioavailability of hydrophobic contaminants in soils: fundamental concepts and techniques for analysis. Eur J Soil Sci 54:809–818

Seuntjens P, Nowack B, Schulin R (2004) Root-zone modelling of heavy metal uptake and leaching in the presence of organic ligands. Plant Soil 265:61–73

Sheng X-F, Xia J-J (2006) Improvement of rape (*Brassica napus*) plant growth and cadmium uptake by cadmium-resistant bacteria. Chemosphere 64:1036–1042

Shenker M, Fan TWM, Crowley DE (2001) Phytosiderophores influence on cadmium mobilization and uptake by wheat and barley plants. J Environ Qual 30:2091–2098

Siciliano SD, Germida JJ (1998) Mechanisms of phytoremediation: biochemical and ecological interactions between plants and bacteria. Environ Rev 6:65–79

Siciliano SD, Germida JJ, Banks K, Greer CW (2003) Changes in microbial community composition and function during a polyaromatic hydrocarbon phytoremediation field trial. Appl Environ Microbiol 69:483–489

Singer CA, Smith D, Jury WA, Hathuc K, Crowley DE (2003) Impact of the plant rhizosphere and augmentation on remediation of polychlorinated biphenyl contaminated soil. Environ Toxicol Chem 22:1998–2004

Simonich SL, Hites RA (1995) organic pollutant accumulation in vegetation. Environ Sci Technol 29:2905–2914

Smith VH, Graham DW, Cleland DD (1998) Application of resource-ratio theory to hydrocarbon biodegradation. Environ Sci Technol 32:3386–3395

Sun B, Zhao FJ, Lombi E, McGrath SP (2001) Leaching of heavy metals from contaminated soils using EDTA. Environ Pollut 113:111–120

Susarla S, Medina VF, McCutcheon SC (2002) Phytoremediation: an ecological solution to organic chemical contamination. Ecol Engineer 18:647–658

Terry N, Zayed AM (1998) Phytoremediation of selenium. In: Frankenberger WT, Engberg RA (eds) Environmental chemistry of selenium. Marcel Dekker, New York, pp 633–657

Tordoff GM, Baker AJM, Willis AJ (2000) Current approaches to the revegetation and reclamation of metalliferous mine wastes. Chemosphere 41:219–228

Trotta A, Falaschi P, Cornara L, Minganti V, Fusconi A, Drava G et al (2006) Arbuscular mycorrhizae increase the arsenic translocation factor in the As hyperaccumulating fern *Pteris vittata* L. Chemosphere 65:74–81

Turpeinen R, Pantsar-Kallio M, Kairesalo T (2002) Role of microbes in controlling the speciation of arsenic and production of arsines in contaminated soils. Sci Total Environ 285:133–145

Unterbrunner R, Wieshammer G, Hollender U, Felderer B, Wieshammer-Zivkovic M, Puschenreiter M et al (2007) Plant and fertiliser effects on rhizodegradation of crude oil in two soils with different nutrient status. Plant Soil 300:117–126

Uren NC (2001) Types, amounts, and possible functions of compounds released into the rhizosphere by soil-grown plants. In: Pinton R, Varanini Z, Nannipieri P (eds) The rhizophere. Marcel Dekker, New York, USA, pp 19–40

Urum K, Pekdemir T, Copur M (2004) Surfactants treatment of crude oil contaminated soil. J Colloid Interface Sci 276:456–464

Valenzuela L, Chi A, Beard S, Orell A, Guiliani N, Shabanowitz J et al (2006) Genomics, metagenomics and proteomics in biomining microorganisms. Biotechnol Adv 24:197–211

Van Dillewijn P, Caballero A, Paz JA, Gonzales-Perez MM, Oliva JM, Ramos JL (2007) Bioremediation of 2,4,6-trinitrotoluene under field conditions. Environ Sci Technol 41:1378–1383

Van Miegroet H, Cole DW (1984) The impact of nitrification on soil acidification and cation leaching in red alder ecosystems. J Environ Qual 13:586–590

Van Nevel L, Mertens J, Oorts K, Verheyen K (2007) Phytoextraction of metals from soils: how far from practice. Environ Pollut 150:31–40

Wang F, Lin X, Yin R (2005) Heavy metal uptake by arbuscular mycorrizas of *Elsholtzia splendens* and the potential for phytoremediation of contaminated soil. Plant Soil 269:225–232

Wang AS, Angle JS, Chaney RL, Delorme TA, Reeves RD (2006) Soil pH effects on uptake of Cd and Zn by *Thlaspi caerulescens*. Plant Soil 281:325–337

Wang FY, Lin XG, Yin R (2007a) Inoculation with arbuscular mycorrhizal fungus *Acaulospora mellea* decreases Cu phytoextraction by maize from Cu-contaminated soil. Pedobiologia (Jena) 51:99–109

Wang FY, Lin XG, Yin R (2007b) Role of microbial inoculation and chitosan in phytoextraction of Cu, Zn, Pb, and Cd by *Elsholtzia splendens*—a field case. Environ Pollut 147:248–255

Wang Q, Fang X, Bai B, Liang X, Shuler PJ, Goddard WA III et al (2007c) Engineering bacteria for production of rhamnolipid as an agent for enhanced oil recovery. Biotechnol Bioeng 98:842–853

Wenzel WW, Adriano DC, Salt D, Smith R (1999) Phytoremediation: a plant–microbe-based remediation system. In: Adriano DC, Bollag J-M, Frankenberger WT Jr, Sims RC (eds) Agronomy Monograph 37, Madison, USA, pp 457–508

Wenzel WW, Bunkowski M, Puschenreiter M, Horak O (2003a) Rhizosphere characteristics of indigenously growing nickel hyperaccumulator and tolerant plants on serpentine soil. Environ Pollut 123:131–138

Wenzel WW, Unterbrunner R, Sommer P, Sacco P (2003b) Chelate assisted phytoextraction using canola (*Brassica napus* L.) in outdoors pot and lysimeter experiments. Plant Soil 249:83–96

Wenzel WW, Lombi E, Adriano DC (2004) Root and rhizosphere processes in metal hyperaccumulation and phytoremediation technology. In: Prasad MNV (ed) Heavy metals in plants: from biomolecules to ecosystems. Springer, Berlin, pp 313–344

Whitfield L, Richards AJ, Rimmer DL (2004) Effects of mycorrhizal colonization on *Thymus polytrichus* from heavy-metal-contaminated sites in northern England. Mycorrhiza 14:47–54

Wieshammer G, Unterbrunner R, Bañares García T, Zivkovic MF, Puschenreiter M, Wenzel WW (2007) Phytoextraction of Cd and Zn from agricultural soils by *Salix* ssp. and intercropping of *Salix caprea* and *Arabidopsis halleri*. Plant Soil 298:255–264

Wilber CG (1980) Toxicology of selenium: a review. Clin Toxicol 17:171–230

Whiting SN, Leake JR, McGrath SP, Baker AJM (2000) Positive responses to Zn and Cd by roots of the Zn and Cd hyperaccumulator *Thlaspi caerulescens*. New Phytol 145:199–210

Whiting SN, De Souza M, Terry N (2001) Rhizosphere bacteria mobilize Zn for hyperaccumulator *Thlaspi caerulescens*. Environ Sci Technol 35:3144–3150

Wolfe NL, Ou T-Y, Carreira L (1993) Biochemical remediation of TNT contaminated soils. Tech. Rep. prepared for the U.S. Army Corps Eng. U.S. Army Eng. Waterways Exp. Stn., Vicksburg, MS, USA

Wu CH, Wood TK, Mulchandani A, Chen W (2006a) Engineering of plant–microbe symbiosis for rhizoremediation of heavy metals. Appl Environ Microbiol 72:1129–1134

Wu SC, Cheung KC, Luo YM, Wong MH (2006b) Effects of inoculation of plant growth-promoting rhizobacteria on metal uptake by *Brassica juncea*. Environ Pollut 140:124–135

Wu FY, Ye ZH, Wu SC, Wong MH (2007) Metal accumulation and arbuscular mycorrhizal status in metallicolous and nonmetallicolous populations of *Pteris vittata* L. and *Sedum alfredii* Hance. Planta 226:1363–1378

Yee DC, Maynard JA, Wood TK (1998) Rhizoremediation of trichloroethylene by e recombinant, root-colonizing *Pseudomonas fluorescens* strain expressing toluene *ortho*-monooxygenese constitutively. Appl Environ Microbiol 64:112–118

Young IM, Crawford JW (2004) Interactions and self-organization in the soil–microbe complex. Science 304:1634–1637

Zaccheo P, Crippa L, Di Muzio Pasta V (2006) Ammonium nutrition as a strategy for cadmium mobilisation in the rhizosphere of sunflower. Plant Soil 283:43–56

Zaidi S, Musarrat J (2004) Characterisation and nickel sorption kinetics of a new metal hyper-accumulator *Bacillus* sp. J Environ Sci Health A 39:681–691

Zaidi S, Usmani S, Singh BR, Musarrat J (2006) Significance of *Bacillus subtilis* strain SJ-101 as a bioinoculant for concurrent plant growth promotion and nickel accumulation in Brassica juncea. Chemosphere 64:991–997

Zhang Y, Frankenberger WT (2000) Formation of dimethylselenonium compounds in soil. Environ Sci Technol 34:776–783

Zhao FJ, Hamon R, McLaughlin MJ (2001) Root exudates of the hyperaccumulator *Thlaspi caerulescens* do not enhance metal mobilization. New Phytol 151:613–620

REVIEW ARTICLE

Novel approaches in plant breeding for rhizosphere-related traits

Matthias Wissuwa · Mark Mazzola · Christine Picard

Received: 6 December 2007 / Accepted: 11 June 2008 / Published online: 4 July 2008
© Springer Science + Business Media B.V. 2008

Abstract Selection of modern varieties has typically been performed in standardized, high fertility systems with a primary focus on yield. This could have contributed to the loss of plant genes associated with efficient nutrient acquisition strategies and adaptation to soil-related biotic and abiotic stresses if such adaptive strategies incurred a cost to the plant that compromised yield. Furthermore, beneficial interactions between plants and associated soil organisms may have been made obsolete by the provision of nutrients in high quantity and in readily plant available forms. A review of evidence from studies comparing older traditional varieties to modern high yielding varieties indeed showed that this has been the case. Given the necessity to use scarce and increasingly costly fertilizer inputs more efficiently while also raising productivity on poorer soils, it will be crucial to reintroduce desirable rhizosphere-related traits into elite cultivars. Traits that offer possibilities for improving nutrient acquisition capacity, plant–microbe interactions and tolerance to abiotic and biotic soil stresses in modern varieties were reviewed. Despite the considerable effort devoted to the identification of suitable donors and of genetic factors associated with these beneficial traits, progress in developing improved varieties has been slow and has so far largely been confined to modifications of traditional breeding procedures. Modern molecular tools have only very recently started to play a rather small role. The few successful cases reviewed in this paper have shown that novel breeding approaches using molecular tools do work in principle. When successful, they involved close collaboration between breeders and scientists conducting basic research, and confirmation of phenotypes in field tests as a 'reality check'. We concluded that for novel molecular approaches to make a significant contribution to breeding for rhizosphere related traits it will be essential to narrow the gap between basic sciences and applied breeding through more interdisciplinary research that addresses rather than avoids the complexity of plant–soil interactions.

Keywords Beneficial rhizobacteria · Mycorrhizal fungi · Nutrient acquisition · Nutrient deficiency · Nutrient toxicity · Soilborne disease tolerance

Responsible Editor: Philippe Hinsinger.

M. Wissuwa (✉)
Crop Production and Environment Division,
Japan International Research Center for Agricultural Sciences (JIRCAS),
1-1 Ohwashi,
Tsukuba, Ibaraki 305-8686, Japan
e-mail: wissuwa@affrc.go.jp

M. Mazzola
USDA-ARS Tree Fruit Research Laboratory,
Wenatchee, WA, USA

C. Picard
Dipartimento di Scienze e Tecnologie Agroambientali,
Area di Microbiologia, Alma Mater Studiorum,
Università di Bologna,
Viale Fanin 42,
40127 Bologna, Italy

Introduction: plant breeding and the rhizosphere—a historical perspective

Agriculture in the twentieth century was characterized by impressive productivity gains. During the past 4 decades global food production has more than doubled and outpaced population growth (Evenson and Gollin 2003). Most of that increase has been due to a sharp rise in crop yields as a result of a bundle of "Green Revolution" technologies, namely improved varieties, increased fertilizer and pesticide use, and improvements in water supply through irrigation. The contribution of improved varieties to yield increases varied between crops and regions with estimates typically ranging from 30–50% (Evenson and Gollin 2003; Khush 1999; Tilman et al. 2002).

During the next four decades global population is projected to increase by 30–50%, which will require, at minimum, an equally large increase in food production to avert widespread malnutrition (Tilman et al. 2002). This increase in production will most likely be dependent upon a diminishing area of highly productive land, not only as a result of agricultural land being diverted to other uses but also due to the recent land-use change from food production to production of renewable energy resources. The need to increase yields further therefore seems to be a necessity. However, in many of the most productive environments crops have reached a yield ceiling (Duvick and Cassman 1999) and unless new means are found to overcome this barrier much of the additional food production will have to come from less favorable environments where nutrient deficiencies and toxicities are a significant obstacle.

Thus, the main present day challenges in agriculture can be summarized as (1) continued expansion of production to feed a growing population, (2) to do so in a more sustainable manner that reduces environmental impact, (3) to improve living standards for rural poor in developing countries, and (4) provide staple food crops with higher micronutrient content. Meeting these challenges will require new approaches and in that context the often overlooked role of rhizosphere traits in breeding programs will be the topic of this review.

The global shift towards sustainable agricultural production systems that do not compromise environmental or human health will necessitate reduced reliance on fertilizer and pesticide inputs in regions of high crop productivity. Maintaining high yield levels under sustainable systems will require gains to be made in nutrient efficiency traits. Attempts to achieve this goal have primarily focused on maintaining high yields with reduced nitrogen (N) inputs, either through selection of genotypes with higher N use efficiency (Bänziger et al. 1997; Presterl et al. 2002) or through enhancing microbial conversion of atmospheric N_2 to a plant usable N form. However, the realization that phosphorus (P) fertilizers are a finite resource that will be depleted towards the end of this century (Runge-Metzger 1995) is going to make improvements in P efficiency a necessity. Attention should therefore focus on improving the abilities of crop plants to access soil-bound P, and other nutrients, that have accumulated as a result of continued fertilization (Vance et al. 2003). A focus on sustainable agriculture will also dictate reduced pesticide application. One opportunity that has received minimal attention in this realm is the management and enhancement of native rhizosphere associated microbial communities that are antagonistic toward pathogens. Breeding efforts are in their infancy but this is an exciting new field as recent studies have demonstrated the genotype specific nature of interactions with these saprophytic microorganisms.

If plant breeding is to play a crucial role in providing varieties that are highly productive with reduced fertilizer and pesticide inputs, it will be important to examine past effects of breeding on rhizosphere function prior to developing novel approaches for plant breeding. In this paper we will first review some of the historic developments with regard to the effect of plant breeding on rhizosphere-related traits and then proceed to identify potentially useful traits for assimilation into breeding programs. Additional related processes such as root growth, architecture and development affect the development of the rhizosphere over space and time (Hinsinger et al. 2005). As such, the following discussion will extend beyond the traditional definition of rhizosphere traits to include specific elements which have definitive impacts on the structure and function of the rhizosphere. Our discussion will conclude with an outline of possible breeding strategies to incorporate these traits into modern varieties. This will include the areas of plant nutrient acquisition, nutrient supply via soil microorganisms, and tolerance to soil-related biotic and abiotic stresses.

Have modern varieties lost genes associated with efficient direct and symbiotic nutrient acquisition?

Modern cultivar selection has typically been performed in standardized, high fertility soil conditions with a primary focus on yield. What consequence would this have on rhizosphere related traits? Under intensified agriculture, it is possible that crop selection undermined the capacity of some crops to access soil nutrients existing in forms not readily plant available. Furthermore, under such conditions, rhizosphere microbial communities are faced with an environment that differs substantially from the one in which plant–microbial interactions originally evolved (Drinkwater and Snapp 2007). Benefits incurred through interactions between plants and beneficial soil microorganisms may have been made obsolete by the excess provision of nutrients in readily plant available forms. Studies comparing allelic diversity within landraces and modern varieties tended to conclude that allele richness declined in modern varieties (Fu et al. 2005; Grau Nersting et al. 2006). It is likely that alleles contributing to beneficial plant–microbe interactions and associated with efficient nutrient acquisition strategies were among the alleles lost if such strategies incurred a cost to the plant and therefore compromised yield (Lambers et al. 2006).

Does that automatically imply that modern cultivars are nutrient inefficient? That depends largely on how nutrient efficiency is defined. If the definition is based on grain yield produced per amount of P applied or P contained in shoots, modern wheat (*Triticum aestivum* L.) cultivars tend to be equal to or more efficient compared to traditional varieties (Batten 1992). However, this is predominantly a function of their high harvest index and not some specific nutrient efficiency trait. Efficiency rankings typically reverse if P uptake or root efficiency (RE), the amount of P taken up per unit root size, is used to evaluate different germplasm and this was particularly apparent when no or very little P fertilizer was applied (Batten 1992; Wissuwa and Ae 2001).

In a screen of 30 diverse rice (*Oryza sativa* L.) genotypes on an Andosol containing a large pool of P fixed in forms of low plant availability, Wissuwa and Ae (2001) detected large genotypic differences in capacity to access P from this pool of soil-bound P (Fig. 1). Traditional varieties did not automatically show high P uptake ability but all genotypes capable of maintaining high relative P uptake were traditional varieties. It was interesting to note that the three genotypes with the highest root efficiency were all traditional upland cultivars from Japan that presumably were selected on similar soils. These results suggest that some traditional varieties possess specific adaptations to soils with low P availability. Similar conclusions were reached in other studies comparing the mycorrhizal competence of several wheat cultivars: those developed prior to 1950 were found to be more reliant on mycorrhizal symbiosis than modern

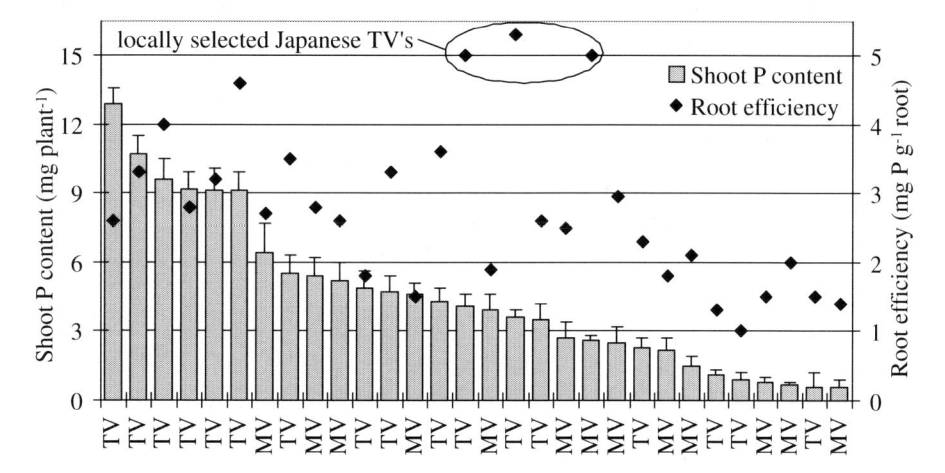

Fig. 1 Shoot P content and root efficiency (RE) of 30 rice genotypes of diverse origin and plant type. Depending on growth habit genotypes were classified as either (semidwarf) modern varieties (*MV*) or as tall traditional varieties (*TV*). Plants were grown under upland conditions in a highly P fixing soil (Andosol) that had never received P fertilizer. *Bars* represent standard errors of means (*n*=5; adapted from Wissuwa and Ae 2001; a more detailed description of genotypes used is provided in that paper)

wheat cultivars (Hetrick et al. 1992, 1993). Furthermore, landraces of mycorrhizal wheat grown in low-P soils produced a higher yield than modern varieties grown under the same conditions (Fgle et al. 1989). This is in agreement with Zhu et al. (2001) who found that mycorrhizal responsiveness of modern wheat cultivars, measured in terms of shoot P, was generally lower than that of older cultivars. Such an observation has also been reported for plant associations with beneficial microorganisms other than mycorrhizal fungi. For example, root endophytes such as *Azoarcus* spp. or *Neotyphodium* and *Acremonium* preferentially colonized wild species and older varieties over modern cultivars of rice (Engelhard et al. 2000) and wheat (Marshall et al. 1999), respectively.

The reintroduction of genes regulating such adaptive traits into the genepool of modern varieties may represent the most promising means to improve their direct or symbiotic nutrient acquisition capacity (Ismail et al. 2007; Bosco et al. 2006).

Past efforts to breed for tolerance to abiotic and biotic stresses

Host plant resistance is an economically sustainable and environmentally sound measure for the management of abiotic and biotic plant stresses. Acid soils, and their associated toxicity (aluminum (Al) and manganese) and deficiency (P) syndromes (Kochian et al. 2004; Rao et al. 1993) are major constraints to crop production (von Uexküll and Mutert 1995). Thus, development of tolerant cultivars has received considerable attention, particularly since the 1980s (Magnavaca and Bahia Filho 1993; Rao et al. 1993; Sarkarung 1986). Local varieties with high aluminum tolerance existed for several crops but were of limited value due to various factors including low yield potential, poor grain quality and susceptibility to diseases and drought (Hede et al. 2001; Rao et al. 1993). Modern varieties, on the other hand, were not adapted to the acid soil syndrome, which meant the typical Green Revolution strategy in cultivar development could not be followed where soil acidity posed severe growth restrictions. Instead localized selection based on crosses between adapted genotypes and high-yielding cultivars seems to have been very common (Sarkarung 1986). Subsequent selection of adapted germplasm was typically done without liming and at low levels of N and P inputs (Rao et al. 1993; Sarkarung 1986). This strategy has produced improved cultivars of rice, wheat and maize (*Zea mays* L.) that combine tolerance to the acid soil syndrome with higher yield potential and tolerance to several important diseases (Hede et al. 2001; Magnavaca and Bahia Filho 1993; Rao et al. 1993).

Breeding for plant resistance towards foliar pathogens has been essential to the continued productivity of certain crops such as wheat (Line and Chen 1995) and rice (Wang et al. 2005; Zhai et al. 2002). However, with a few exceptions there has been an absence of similar programs addressing resistance or tolerance toward pathogens inciting diseases of the plant root system. Interestingly, the capacity for such efforts to yield benefit were inadvertently demonstrated, such as in the instance of tree fruit rootstock breeding programs focused on horticultural traits (Khanizadeh and Granger 1998), which unintentionally produced material with tolerance toward specific soilborne pathogens (Browne and Mircetich 1993). Mechanisms other than host resistance likely contribute to plant tolerance towards soilborne pathogens. It has been suggested that plants have evolved strategies of stimulating and supporting specific groups of antagonistic microorganisms in the rhizosphere (Cook 2006; Smith et al. 1999). Differential capacity of plant genotypes to support specific groups of resident as well as introduced microbial antagonists has been repeatedly reported (Berg et al. 2002; Larkin et al. 1993; Mazzola and Gu 2002; Mazzola et al. 2004). More importantly, a genetic basis for specialization between a microbial antagonist and a plant host has been demonstrated (Smith et al. 1997, 1999). A more comprehensive understanding of the impact of plant genotype on the non-symbiotic microbial community resident to the rhizosphere may have significant benefit to plant selection programs focused on development of disease tolerance/resistance.

Plant breeding for rhizosphere-related traits

The desire for more efficient utilization of resources in high input agriculture and enhanced adaptation to low fertility and other biotic and abiotic soil stresses in developing countries have already precipitated several changes in the conduct of plant breeding. Two key strategies currently being pursued are (1) broadening the genetic base of modern varieties through the reintroduction of genes from suitable donors such as traditional varieties or wild relatives (Ismail et al. 2007), and (2) selection of cultivars in systems employing no or reduced fertilizer and pesticide inputs (Bänziger et al. 1997). The problem remains that heterogeneous environments make it difficult to consistently apply high selection pressure to identify few superior genotypes across environments. Increasingly this is offset by highly decentralized selection in on-farm trials in unfavorable environments, which may involve farmer participation to select locally adapted and accepted material (Ceccarelli and Grando 2007; Dawson et al. 2007). In doing so breeders will inadvertently be selecting for beneficial rhizosphere related genes such as host genes affecting plant–microbe associations, or genes improving nutrient acquisition.

Useful traits of direct nutrient acquisition capacity and tolerance to abiotic soil stresses

The capacity of any genotype to acquire essential nutrients from the soil relies essentially on the two factors, (1) nutrient interception, which is dependent on root size and architecture, and (2) the acquisition process as affected by efficient uptake of available nutrients and solubilization or mineralization of less plant-available nutrients (Hinsinger et al. 2005; Ismail et al. 2007; Lambers et al. 2006). Plant–microbe interactions may play a role in both processes but will be discussed in following sections. Here the focus shall be on processes directly affected by plant adaptations.

Where low nutrient availability limits overall plant growth an increase in the root–shoot ratio has typically been observed as an adaptive response. This does not mean, however, that root growth is not reduced in absolute terms. In fact it has been shown for rice that the ability to maintain root growth is more closely linked to tolerance to P or zinc (Zn) deficiency than the non-stress root growth potential (Wissuwa and Ae 2001; Ismail et al. 2007). To understand what enables tolerant genotypes to maintain such high relative root growth rates will be a key factor in designing crops with improved nutrient acquisition capability (Lynch 2007). Through a modeling approach Wissuwa (2003) showed that small genotypic differences in factors improving P uptake per root biomass may provide enough extra P to maintain root growth with large subsequent effects on P uptake through improved root interception. From a physiological perspective, a focus should be placed on traits leading to enhanced root exudation of compounds capable of solubilizing soil-bound nutrients such as P (Lambers et al. 2006), Zn (Hoffland et al. 2006) or iron (Fe; Ishimaru et al. 2006). Enhancing the excretion of phosphatases or phytases to access organic P forms in soils appears less promising because of complex sorption/desorption characteristics in soil (George et al. 2004).

Several root structure related traits may also offer possibilities for improving nutrient acquisition. These include enhancing density or length of root hairs (Gahoonia and Nielsen 2004), modifications in root architecture to preferentially concentrate root biomass in soil strata containing the maximum amount of a limiting nutrient (Rubio et al. 2003), and ways to maintain new root development with minimal expenditure of additional resources either through more rapid root turnover (Lambers et al. 2006) or through etoilation (Lynch 2007).

Often nutrient deficiencies are due to interactions of multiple soil-related stresses. The acid soil syndrome is a good example as P deficiency is exacerbated by root growth reductions due to Al toxicity (Kochian et al. 2004). Similarly Zn deficiency can be aggravated by the negative effect high bicarbonate concentrations, typically encountered in alkaline/sodic soils, have on root growth in intolerant genotypes (Ismail et al. 2007). Therefore, in attempts to enhance nutrient acquisition capacity, one must take into account that interactions between several stress factors may occur in problem soils and that improvements in single traits may not be sufficient to improve genotype performance.

In selecting for improved nutrient efficiency under high-input systems, it is not clear whether nutrient acquisition or internal nutrient use efficiency plays a dominant role and therefore what traits would be beneficial. Maize lines with higher N efficiency have successfully been selected but their improved efficiency was apparently not due to better N uptake or any other rhizosphere trait (Paponov et al. 2005). In rice internal N use efficiency was also the more stable and therefore more selectable trait but genotypic differences in N uptake were also reported (Tirol-Padre et al. 1996). A more clear argument for the importance of rhizosphere traits in the uptake of nutrients in sustainable but highly productive agricultural systems can probably be made for nutrients such as P and Zn that are easily sorbed onto soil surfaces and hence show low mobility in soils (Singh et al. 2005). For both nutrients, traits discussed above for deficient conditions should also be beneficial under mild deficiency.

Useful traits for better nutrition via beneficial plant–microbe interactions

Harnessing the potential of beneficial plant–microbe interactions to manage nutrition (and diseases) in agro-ecosystems can be considered as an alternative, or a supplementary means, for reducing the use of chemicals in agriculture (Rengel and Marschner 2005; Picard and Bosco 2008). Managing these beneficial plant–microbe interactions can be achieved through various agricultural practices, such as the selection of the appropriate host plant species or cultivar, the degree and type of fertilization and crop rotation or soil tillage (Mazzola and Gu 2002; Oehl et al. 2004; Sarniguet al. 1992), and in some cases by inoculations. Concerning the plant-genotype selection approach, the first step should be aimed at evaluating genetic variability of the crop plant. This phase should be devoted to the screening of plant genotypes available for selection (genetic resources) and characterization of beneficial microbial populations in the rhizosphere of such genotypes (Tanksley and McCouch 1997). These evaluations should be performed in low-input environments in order to reveal optimal plant–microbial interactions in limiting environments.

Host variation in responsiveness to beneficial microorganisms generally has been expressed as microbial root-colonization density and diversity, as well as effective plant growth stimulation. A major challenge in screening for enhanced microbial root-colonization is the design of effective procedures that provide a reproducible classification in the field environment. Level of root colonization exhibits dramatic variation related to plant age (Picard et al. 2000, 2004; Roesh et al. 2006), thus genotype rank can vary significantly depending upon plant age at the time of analysis. For example, the frequency of beneficial root-colonizing microorganisms was found to be very low in the initial stage of plant growth, greatly increased at the flowering stage, and then decreased with physiological maturation, at least for maize (Picard et al. 2000, 2004, Roesh et al. 2006) and for *Achillea ageratum* L plants (Picard and Bosco 2003). Interestingly, it is also at the flowering stage that greater differences in root colonization are observed between maize cultivars (Picard and Bosco 2005, 2006; Picard et al. 2004, 2008). Therefore, for maize and *Achillea ageratum* L, it is recommended that screening of plant-genotypes for the ability to support rhizosphere colonization by beneficial microorganisms be conducted at the flowering stage.

Independent of the beneficial association studied, it has been established that host genotype has a substantial impact in determining the extent of microbial colonization. For example, diversity in mycorrhizal responsiveness, defined in terms of mycorrhizal dependency (Tawaraya 2003) was observed among wheat genotypes by Hetrick et al. (1992, 1995, 1996). Interestingly, it was reported that diversity in capacity of wheat to sustain root colonization by mycorrhizal fungi was associated with yield responses, varying from zero to positive or negative values (Xavier and Germida 1998). Furthermore, mycorrhizal dependency is often negatively correlated with root morphological traits, such as root length, root dry weight, root hair length and density of root hairs, traits known to improve the ability of the non-mycorrhizal plant to acquire P directly from soil (reviewed in Tawaraya 2003).

Concerning atmospheric N_2 fixation, root nitrogenase activity and quantification of plant N derived from the atmosphere (Ndfa) have also been taken into consideration, both for rhizobial–legume symbiosis and plant associations with free-living microorganisms. Specifically for legumes, number of nodules was a quantitative trait measured to assess host reaction. Significant genotypic variation in the responsiveness of legume cultivars to *Rhizobium* has been reported. A range of bean (*Phaseolus vulgaris* L.; Hungria and Phillips 1993), soybean [*Glycine max* (L.) Merr.; Cregan 1989] and Lucerne (*Medicago sativa* L.; Hungria and Phillips 1993) genotypes differed in relative nodulation. Furthermore, high variability in N_2 fixation was observed among crop legume genotypes, varying from 0% to 97% (in percentage of crop N derived from N_2 fixation; see Herridge and Rose 2000). Differential capacity to support associative N_2 fixation has been clearly established among cereal species as well as among genotypes within cereal species (Jagnow 1990). Maximal nitrogenase activity was reported to be dependent upon maize genotype

(Ela et al. 1982; Neyra and Dobereiner 1977). Furthermore, Picard et al. (2008) recently gave clear evidence that maize genotype influences the size of microbial communities involved in N_2 fixation, as well as the diversity of the mycorrhizal fungi colonizing population. App et al. (1986) indicated for the first time that rice lines differentially influence beneficial N_2 fixers. By comparing 69 rice lines from diverse backgrounds (indica, japonica, and javonica; traditional and improved) and from different maturity groups (early, medium, and long), Shrestha and Ladha (1996) demonstrated that Ndfa differed significantly amongst the various lines, ranging from 1.3 to 20%. Those with high Ndfa were mostly traditional varieties, but some improved lines also had high associative N_2 fixation. Furthermore, appropriate combinations of *Acetobacter diazotrophicus* and rice genotype are needed for the relationship to be effective: some plants grew for up to 12 months on N-free medium obtaining N from the associative fixation (Rolfe et al. 1997).

Interestingly, these cultivar-dependent variations in interactions with beneficial microorganisms seem to have resulted from evolution over generations (Engelhard et al. 2000). In most cases, the capacity of a plant-genotype to positively interact with beneficial microorganisms appears to be an inherited trait (Rengel 2002; Smith and Goodman 1999).

Useful traits for a better barrier to pathogen infection of plant root

Although soil type has a considerable influence on the structure of microbial populations (Dalmastri et al. 1999; Latour et al. 1996; Marschner et al. 2001), plant species and genotype also are significant factors determining composition of microbial communities resident to soils and the rhizosphere (Berg et al. 2002; Dalmastri et al. 1999; Lemanceau et al. 1995; Marschner et al. 2001; Mazzola and Gu 2002; Miethling et al. 2000). As microbial populations indigenous to the rhizosphere can be considered an initial barrier to pathogen infection of plant roots, active manipulation of this community may be an effective means to suppress soilborne plant pathogens. The evolving view is that a degree of host specificity exists in interactions between plants and microbial antagonists (da Mota et al. 2002; Mazzola and Gu 2002; Notz et al. 2001; Smith et al. 1999), both those introduced as potential biological control agents and those that are resident to the soil ecosystem. Although the vast majority of studies have focused on the saprophytic bacterial community as the source of antagonists, there is also evidence to indicate that plant species differentially support root colonization by fungi, such as *Trichoderma*, *Penicillium*, and non-pathogenic *Fusarium* spp., with potential to suppress plant pathogens (Berg et al. 2005; Larkin et al. 1996; Rengel and Marschner 2005). Thus, utilization of crop genotypes with an elevated capacity to select for specific functional microbial genotypes would appear to be a viable means to enhance crop disease tolerance/resistance and ultimately productivity. However, as acceptance of this phenomenon as a valuable parameter to crop improvement is still considered novel, there exists a lack of breeding effort in this realm.

Although the body of work consists primarily of studies conducted in controlled environments, plant genotype-dependent selection of specific resident soil microorganisms having a functional role in disease suppression has been reported. The vast majority of work addressing the impact of plant genotype on selection of microbial antagonists has focused on fluorescent *Pseudomonas* spp. strains producing the antibiotic 2,4-diacetylphloroglucinol (2,4-DAPG) which has activity toward numerous soilborne plant pathogens (Keel et al. 1992), including take-all of wheat. Wheat genotypes were found to differentially support populations of an introduced 2,4-DAPG-producing strain (Mazzola et al. 2004) and varied in the capacity to selectively enhance resident populations of these bacteria when soils were cropped to successive wheat plantings (Fig. 2). Such a finding is of significance as development of take-all suppressive soils in response to continuous wheat monoculture results from an enrichment in populations of 2,4-DAPG-producing *Pseudomonas fluorescens* to a density of 10^5 CFU/g of root, the threshold required to suppress the causal pathogen, *Gaeumannomyces graminis* var. *tritici* (Raaijmakers et al. 1999). Likewise, qualitative attributes of the 2,4-DAPG-producing bacterial genotype recovered from the rhizosphere also varied with plant genotype (Mazzola et al. 2004) and age (Picard et al. 2000). Individual wheat cultivars selected for distinct dominant 2,4-DAPG genotypes from the same soil microbial community and the genetic diversity of the population

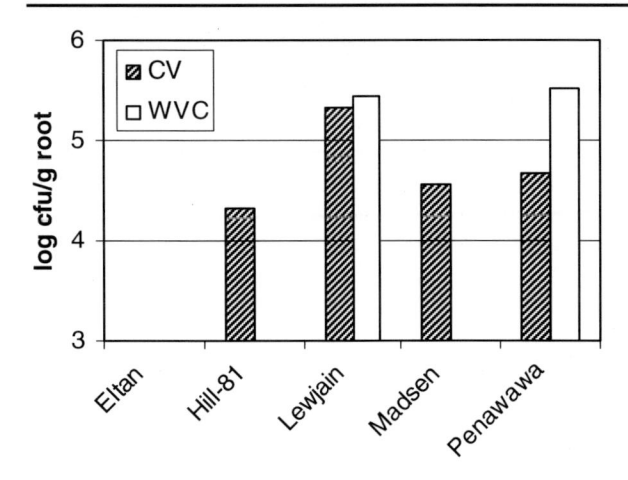

Fig. 2 Recovery of 2,4-DAPG-producing fluorescent pseudomonads from the rhizosphere of five wheat genotypes grown in Columbia View (*CV*) and Wenatchee Valley College (*WVC*) orchard soils. The limit of detection was 10^4 cfu g^{-1} root and indigenous populations of these bacteria were not recovered from the rhizosphere of cv. Eltan when grown in either soil, nor the rhizosphere of cv Hill-81 or Madsen grown in WVC soil. For a given soil, *values designated with the same letter* are not significantly ($P>0.05$) different (Mazzola et al. 2004)

recovered from the maize rhizosphere was related to plant age. Certain 2,4-DAPG-producing genotypes are known to possess superior disease control potential in part due to superior colonization abilities (Raaijmakers and Weller 2001).

Plants may also have direct impact on regulation of microbial genetic elements directly conferring activity of introduced biological control agents, thereby affecting disease control efficacy. Significant differences in expression of the 2,4-DAPG biosynthetic gene *phlA* were detected in the rhizosphere of different plant species with greater expression in the rhizosphere of monocots than in the rhizosphere of dicot species (Notz et al. 2001). Plant genotype effect on *phlA* expression was also observed within species, with significant differences detected among six maize cultivars. Similarly, Okubara (2006) determined that wheat cultivars differed in the ability to sustain root populations of these bacteria and to accumulate 2,4-DAPG in the rhizosphere.

Perhaps the most persuasive finding that could motivate the breeding of plant materials specifically for the capacity to modulate resident soil microbial populations in a manner that yields effective disease control is the observation that soil suppressiveness is induced in a plant genotype dependent manner. Larkin et al. (1993) documented the development of *Fusarium* wilt suppressive soils in response to repeated cultivation of soil with watermelon [*Citrullus lanatus* (Thunb.) Matsum & Nakai]. This response was only observed when cultivars resistant to *Fusarium oxysporum* f. sp. *niveum* were employed, and was associated with increases in specific populations of non-pathogenic *Fusarium oxysporum*, with many of the isolates possessing the capacity to induce host systemic resistance (Larkin et al. 1996). Soil suppressiveness toward Rhizoctonia root rot in response to repeated cultivation of wheat was also found to occur in a wheat genotype dependent manner (Mazzola and Gu 2002). Capacity of a wheat genotype to induce disease suppression was associated with enhancing rhizosphere populations of specific fluorescent pseudomonad genotypes demonstrating antagonistic activity toward *Rhizoctonia solani* AG-5 and AG-8 (Gu and Mazzola 2003). Cultivation of wheat genotypes that did not modify the fluorescent pseudomonad population in this manner did not elicit a disease suppressive soil.

Rhizosphere-related breeding: current approaches

Several approaches are being followed to introduce beneficial rhizosphere traits into modern varieties. At present, most involve modifications in conventional breeding strategies but with the rapid advances being made in molecular techniques they will increasingly rely on alternative breeding strategies such as marker assisted selection or genetic modifications. This section will discuss the recent changes in breeding approaches.

Changes in conventional approaches—indirect selection

As a result of difficulty in assessing rhizosphere related traits in breeding programs that typically evaluate thousands of segregating progeny, selection for beneficial rhizosphere traits is typically carried out indirectly, e.g. by selecting for yield in an environment where a particular rhizosphere related trait should confer an advantage. To assure that such indirect selection actually exposes benefits of rhizosphere traits the following strategies have been employed:

1. Choice of donors to complement modern varieties in traits that are so far missing

2. Subsequent selection conducted in environments that should favor progeny with introgression of beneficial traits
3. Participatory plant breeding

Choice of donors to complement modern varieties

The selection of donors with a highly favorable phenotype for a trait of interest has been the most commonly used strategy in improving plant adaptation to edaphic factors in the past and can therefore not be called a novel approach. However, as our understanding of the complexity of plant–soil interactions increases, criteria for the selection of donors continue to evolve and include novel traits such as root shallowness as measured by root growth angle in bean or soybean seedlings (Lynch 2007), root hair length (Gahoonia and Nielsen 2004) or adventitious root number (Ochoa et al. 2006), all traits associated with improved P uptake.

The identification of germplasm with resistance to pathogens has had a prominent role in the historical control of plant pathogens. The development and use of resistant cultivars has been effective in controlling *Fusarium* wilt of numerous crop plants caused by the ubiquitous and specialized fungal pathogen *Fusarium oxysporum* (El Mohtar et al. 2007; Herman and Perl-Treves 2007).

Although the effectiveness of such a strategy for the management of less specialized root pathogens, such as *Pythium* and *Rhizoctonia* spp. has been questioned (Cook et al. 1995; Cook 2006), certain reports suggest that some useful measure of tolerance is attainable. However, the sources are commonly wild plant populations and will require transfer to commercially relevant plant genotypes. Novel sources of resistance to *Rhizoctonia solani* were identified in a screening of wild *Beta* germplasm (Luterbacher et al. 2005) and polygenic resistance to Rhizoctonia root rot was identified in sugar beet (*Beta vulgaris* L.; Panella 1998). Evaluation of 16 apple (*Malus domestica* Borkh.) genotypes, representing commercially available rootstocks as well as developmental material from an apple-rootstock breeding program (Fazio and Mazzola 2005) documented wide variation in susceptibility to root infection by *Rhizoctonia solani* AG-5, with infection rates ranging from 6 to 65%. Furthermore, the wild apple *Malus sieversii* is highly resistant to this fungal pathogen, and crosses between this resistant wild material and Geneva dwarfing elite rootstocks has yielded resistant genotypes (Fazio et al. 2006). Differential susceptibility to indigenous populations of *Pythium* spp. was also detected, with Geneva rootstocks in general less susceptible to infection than those of the Malling series. Resistance to pre-emergence damping off incited by *Pythium* spp. has been reported for several crops including soybean (Kirkpatrick et al. 2006) and subterranean clover (*Trifolium subterraneum*; You et al. 2005). However, in screening of a subterranean clover breeding line, there was no association between resistance to pre-emergence damping-off incited by *Pythium irregulare* and root rot caused by this same pathogen. Among 50 lines tested, a single line exhibited resistance to root rot incited by *P. irregulare*.

Selection in environments that favor progeny with introgression of beneficial traits

Following the introgression of beneficial traits from donors into a segregating breeding population it is crucial to conduct selection among progeny in environments that expose advantages of such introgression. This represents a deviation from the classical breeding approach so successfully employed in the development of the highly productive Green Revolution varieties. The advantage of selection in high-input environments has led breeders to assume that a carry-over effect of high yield potential into unfavorable environments can be realized. However, experimental evidence suggests that this is not necessarily the case and that unfavorable environments are frequently below the cross-over point at which genotypic rankings in response to environment are reversed (Ceccarelli and Grando 2007). Rather than reducing the impact of environment on genotype performance through fertilization, soil amendments or disease control, genotype x environment effects therefore need to be exposed and selected for under low input conditions and with the presence of biotic and abiotic soil stresses.

One of the first breeding programs targeting low input environments was the upland rice breeding program at CIAT, Colombia. Initiated in 1982, emphasis was first placed on identifying donor genotypes among a broad collection of landrace and improved upland varieties in screening experiments conducted on acid soils with low fertility (Rao et al. 1993). Subsequent 'decentralized' selection in segregating populations was done under a range of soil fertilities to identify lines that showed both improved yield potential in low-fertility conditions and high yield potential in more favorable environments. As a result several new rice cultivars have been released, emphasizing benefits of decentralized selection over a range of target environments. A similar yet more targeted approach, as donors were selected for their favorable expression of root anatomical traits, is currently being followed in developing bean, maize and soybean cultivars with adaptation to low P availability (Lynch 2007; Yan et al. 2006).

A similar approach was utilized in the selection of sugarcane (*Saccharum officinarum* L.) varieties in Brazil in which breeding programs used both local and introduced material and were conducted in the absence of large N inputs. This method has led to Brazilian sugarcane varieties that are able to incorporate up to 70% of their total N from associative bacterial N_2 fixation (Andrews et al. 2003; Baldani et al. 2002). This may be the reason why today the best materials have little demand for N fertilizer, and an effective association has developed between endophytic N_2-fixing bacteria and the plant (Baldani et al. 2002).

Breeding programs for enhancing N_2 fixation in legumes have had particular application to cropping systems encountered in unfavorable environments of developing countries. One of these successful legume breeding programs is that of Bliss, which over a period of 13 years made substantial progress in selecting for yield of common bean in low N soils. This work resulted in the release of five high N_2 fixing lines well adapted to the South America (Bliss 1993). A second example is the soybean breeding program realized in Africa by the International Institute of Tropical Agriculture (IITA). In this case, soybean genotypes nodulated by indigenous rhizobia and demonstrating an acceptable yield were successfully selected among the locally used cultivars (Nangju 1980).

Selection in most breeding programs for maize, based on the phenomenon of heterosis, has resulted in superior modern maize hybrids that produce yields that range from 150% to over 300% of their parental mean (Hallauer and Miranda 1988; Shull 1908). Interestingly, by comparing populations of plant-beneficial microorganisms isolated from maize hybrids and from their respective parental lines, it was shown that microbial root colonization could be related to heterosis (Picard and Bosco 2005, 2006; Picard et al. 2004). Repeated field experiments demonstrated that maize hybrids are more abundantly colonized by beneficial bacterial strains with greater genetic differences and more effective in disease suppression, root stimulation and associative N_2 fixation, than those supported by their respective parental lines. Similar results were also obtained for mycorrhizal symbiosis (Picard et al. 2008). These findings suggest that, in this particular instance, the conventional breeding programs for maize based on heterozygosis have inadvertently selected for maize genotypes able to support large and genetically diverse populations of efficient plant beneficial microorganisms.

Establishing environments suitable for the effective evaluation of host resistance to soilborne plant pathogens is not a trivial undertaking. Disease severity is a product of numerous interactions among the resident soil biota in addition to the general nutritional status of the plant. For diseases of complex disease etiology, attending to one biological issue is likely to elevate the impact of another component as often the contributors are direct competitors for a specific niche and once released from this competition may result in elevated damage levels (Mazzola 1998; Pieczarka and Abawi 1978). While addressing all such issues in the design of a specific environment for screening plant germplasm is not likely to be practical, addressing certain basic parameters will improve the validity of screening programs. Much like the limitations concerning the selection of cultivars in systems possessing unlimited nutrient availability, screening of crop material in the absence of any potential limits placed upon it through the activity of soil microbes can have unforeseen negative outcomes. Such may have been the case with regard to certain strawberry (*Fragaria* Duch.) breeding programs where selections were concluded based upon growth in fumigated soils, which yielded varieties that are highly susceptible to a wealth of soilborne pathogens. While the highly methyl bromide fumigation-dependent strawberry material from the California breeding system initially yielded no obvious genetic diversity for developing cultivars adapted to sublethal effects of organisms in non-fumigated soils (Shaw and Larson 1996), tolerance to black root rot, incited by a complex of non-lethal pathogens, was detected among three genotypes released from the breeding program in Nova Scotia, Canada (Particka and Hancock 2005).

While field level evaluations of host resistance are critical to the screening program, a common hurdle to the conduct of these trials is the absence of suitable disease pressure. A typical manner to circumvent this difficulty is to augment the target population by infestation with laboratory raised inoculum, which usually emanates from one or a few strains of the pathogen. In certain instances, this may be essential where resistance to a specific race or *formae specialis* of the pathogen is the subject of the breeding program. However, this has the possibility of generating a further limitation to the process by restricting plant exposure to the genetic diversity which exists in any pathogen population. For many soilborne pathogens, a straightforward means of enhancing indigenous inoculum is through cultivation of a susceptible host prior to establishment of the screening trial (Gutierrez and Cramer 2005). In actuality, some combination of controlled and field assessment is likely as a limit to the amount of material screened is often necessitated by field trial expenses.

Participatory plant breeding

Marginal environments are typically characterized by the concurrence of several stresses. When rainfall patterns vary from year to year, risk-averse farmers typically do not apply fertilizers or other costly inputs. Thus environmental stresses tend to be heterogeneous and site or region specific. Under such conditions the "on-station" approach of developing varieties with broad adaptation has had limited success (Ceccarelli 1994). A second limitation of the centralized approach to cultivar development is that attributes other than grain yield may determine whether cultivars are accepted locally (Morris and Bellon 2004). Such secondary attributes may involve specific grain quality traits or an emphasis on straw yield and suitability as animal feed.

To develop locally adapted and accepted varieties several participatory plant breeding strategies have been developed that involve farmers at various stages of the selection process (Ceccarelli and Grando 2007; Dawson et al. 2007). One such participatory plant breeding approach in which farmers evaluated progeny from several crosses in their own fields, yielded release of two rice varieties that showed particular advantages over a check variety under low input/low yield conditions while being inferior to the check in more productive environments (Virk et al. 2003). A slightly different approach is to let breeders conduct selection within their segregating material but to disseminate near-finished lines for farmer participatory variety evaluation, possibly conducted at farmer fields with little or no fertilizer inputs. A perceived advantage of this approach is testing in a large number of locations and rapid dissemination of farmer preferred seed at respective locations. Bänziger and Diallo (2001) summarized results from maize evaluation trials in low N/low rainfall environments and concluded that this participatory evaluation approach was very successful not only in developing adapted cultivars but importantly also in bringing them into farmer's fields.

Direct selection for rhizosphere related traits

Direct selection for rhizosphere traits in segregating populations remains rare because of the potential high cost of evaluating hundreds if not thousands of lines. However, this additional cost can be justified if the heritability realized in direct evaluations is far greater than in indirect screens under field conditions. Yan et al. (2006) and Lynch (2007) suggested that rapid screens available for certain root morphology/architecture traits may actually be more cost-efficient than selection for yield in fields with high spatial soil heterogeneity. Root hair length and density is one example of a highly heritable trait (Gahoonia and Nielson 2004; Yan et al. 2004) that is currently being used successfully in selections among segregating populations of bean and soybean (J. P. Lynch, pers. comm.). Another example is the glasshouse screening method developed at CIAT to evaluate Al tolerance and root growth vigor in vegetative propagules of the forage grass *Brachiaria* (Ishitani et al. 2004).

However, it may be concluded that direct selection for rhizosphere traits remains an exception, either because few suitable traits have been identified to date or because the expression of such traits are prone to variation depending on growth stages or environmental conditions. Much could be gained if phenotypic evaluations are replaced by selection for molecular markers tightly linked with the trait of interest. Considerable effort has therefore been invested in mapping quantitative trait loci (QTLs) associated with rhizosphere traits.

Targeted introgression of rhizosphere traits via marker assisted selection (MAS)

Using tightly linked markers to indirectly select for a trait of interest should be ideally suited to transfer important rhizosphere traits to modern varieties because of the difficulties in evaluating and selecting for rhizosphere traits directly. Numerous studies have been conducted to map QTLs associated with rhizosphere traits, however, very few of these QTLs have been used in practical breeding programs. In reviewing factors responsible for the apparent lack of success in turning mapped QTLs into tools for plant breeding, Wissuwa (2005) concluded that the bottleneck does not lie in a scarcity of identified QTLs but possibly in a lack of relevant QTLs identified by screening procedures that convince breeders of their usefulness in target environments. One challenge therefore is to identify or at least confirm benefits of QTLs in target environments. Given that the spatial variability in field trials is inherently large for nutrient availability/toxicity traits, accurate phenotyping will represent an obstacle in mapping nutritional traits that should not be underestimated.

Furthermore, not all QTLs identified can automatically be considered useful in breeding. Frequently one parent used in mapping populations was chosen because it showed an unfavorable phenotype that contrasted well with a second parent that may not have been outstanding with regard to the trait studied. A locus capable of improving a rather poorly performing parent is not necessarily suited for improving breeding material that already shows the favorable phenotype to some degree. Breeders are more inclined to include novel alleles with large effects in their programs and these have in the past been detected in mapping populations that included a donor showing a very favorable phenotype as one of the parents (Mackill 2006).

For the practical application of a QTL in marker assisted breeding, QTL positions need to be known with higher precision than typically achieved in initial QTL mapping experiments (Yano et al. 2000). The resources required to achieve this may be several-fold higher compared to the initial QTL mapping experiment. One of the few traits for which mapping has progressed to a stage where marker assisted back-crossing is feasible is tolerance to P deficiency as conferred by the *Pup1* locus in rice (Wissuwa et al. 2002). The strategy followed by the International Rice Research Institute (IRRI) and collaborators is outlined in Fig. 3. Key factors for the advances made with *Pup1* were the choice of a highly tolerant donor genotype (Kasalath) in QTL mapping, subsequent confirmation of the QTL effect in a near isogenic line (NIL) background, reliance on phenotyping in the field, precision mapping based on secondary mapping populations derived from a backcross of the tolerant NIL to intolerant recurrent parent 'Nipponbare', and identification of suitable recipient varieties. The aim at present is to simultaneously transfer tolerance to P deficiency with higher rice blast resistance to upland varieties having high yield potential under non-stress conditions (Ismail et al. 2007). Once the gene at the *Pup1* locus is cloned, efforts will be directed towards identifying additional and possibly even stronger alleles at this locus by screening additional gene bank accessions for novel polymorphisms within the locus (allele mining).

Probably the most extensively and successfully mapped rhizosphere related trait is Al tolerance. Part of the success with Al tolerance is due to the simplicity of phenotyping for this particular trait and

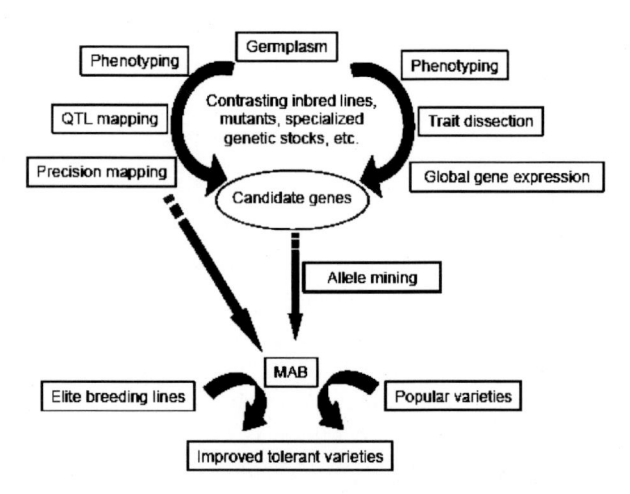

Fig. 3 Outline of a strategy for germplasm improvement for problem soils. Proper germplasm and genetic stalks were used for mapping traits associated with tolerance to salinity, P and Zn deficiency. The two major QTLs (*Saltol* and *Pup1*) were fine-mapped and QTL and gene-specific markers were identified and are being used in their introgression into elite breeding lines and popular varieties using marker assisted backcrossing (*MAB*). Genes in the two regions were also annotated, short-listed based on converging evidences from positional, functional and expression analysis and are being further validated (our unpublished data). Figure reproduced from Ismail et al. (2007) with kind permission of Springer Science and Business Media

the close correlation between nutrient solution screens and actual field tolerance. Furthermore, Al tolerance in several crop species [wheat, barley (*Hordeum vulgare* L.), sorghum (*Sorghum bicolor* (L.)) Moench] is in large part due to a single gene, subsequently found to be coding an aluminum-induced efflux transporter for malate (Sasaki et al. 2004) or citrate (Magalhaes et al. 2007; Wang et al. 2007). Based on these efforts marker assisted breeding is currently being practiced in sorghum breeding programs in Brazil (J. V. Magalhaes, pers. comm.).

Molecular biology, combined with Mendelian and quantitative genetics in quantitative trait locus (QTL) mapping and marker-assisted selection are more and more used for revealing the inheritance of root association with beneficial microorganisms. For example, it has been shown that interactions of maize with mycorrhizal fungi, or of rice with associative N_2 fixers, were associated with two and four QTLs, respectively (Kaeppler et al. 2000; Wu et al. 1995). Another interesting example is the case of the single nucleotide amplified polymorphism (SNAP) markers identified by Kim et al. (2005) for supernodulation in soybean. However, marker-assisted selection (MAS) for enhanced root-beneficial microbe association typically does not form a part of routine cultivar improvement programs, and incorporating selection criteria for such associations, such as nodule mass, N_2 fixation and responsiveness to mycorrhizal fungi, into breeding schemes while attending to other breeding objectives remains a challenge. If MAS for beneficial root-microbe interactions could be integrated into breeding programs that are already practicing MAS for other traits, this would avoid the necessity of additional phenotypic selection methodologies purely for determination of beneficial microbial-root associations.

Genetic determinants of resistance to *Fusarium oxysporum* for a specific host to the same formae specialis can vary, and in melon (*Cucumis melo* L.) has been reported to be monogenic and dominant, as well as conferred through polygenic effects depending upon the physiological race of the causal fungus, *F. oxysporum* f. sp. *melonis*, and the cultivar examined (Herman and Perl-Treves 2007). Similar differences in the genetic governance of wilt resistance has been described in chickpea (*Cicer arietinum* L.), with monogenic resistance described toward certain races of *F. oxysporum* f. sp. *ciceris* and oligogenic resistance toward others (Sharma and Muehlbauer 2007). Analysis of a recombinant inbred line population of melon possessing polygenic resistance to *F. oxysporum* f. sp. *melonis* resulted in the resolution of nine qualitative trait loci (Perchepied et al. 2005), and QTL for Fusarium wilt resistance have been identified in a number of other plant hosts (Iruela et al. 2007).

In tomato (*Solanum lycopersicum* var. *lycopersicum* L.), the *I-2* resistance gene for resistance to *F. oxysporum* f. sp. *lycopersici* Race 2 was mapped (Segal et al. 1992; Ori et al. 1997) and sequenced. Primers specific to the *I-2* gene were designed (El Mohtar et al. 2007) and used to screen a series of tomato varieties, with the expected amplicon only detected in varieties known to possess the *I-2* gene. When used to screen an uncharacterized F2 seedling population, the *I-2* gene was detected in 72 seedlings, 69 of 72 seedlings showed resistance to the pathogen, and among those lacking the resistance gene based on the PCR assay, 26 of 29 exhibited susceptibility. This method could be effectively employed in breeding programs to detect the presence of the *I-2* gene.

Introduction of new traits and genes through genetic modifications

Rapid developments in the field of plant molecular biology have accelerated the rate of gene discovery and promise to provide scientists and breeders with new genes of potential importance for the improvement of plant adaptation to unfavorable soil environments. Initial attempts to improve nutrient acquisition through overexpression of candidate genes have, however, yielded mixed results.

Transgenic *Trifolium subterraneum* transformed with a phytase gene from *Aspergillus niger* improved P uptake of plants supplied with phytate and grown in agar under sterile conditions (Richardson et al. 2001), but this positive effect was not seen in soil grown plants, possibly because phytate was bound to soil particles and therefore not available (George et al. 2005). Delhaize et al. (2001) overexpressed a citrate synthase gene from *Pseudomonas aeruginosa* in tobacco (*Nicotiana tabacum* L.) and detected greatly enhanced citrate synthase protein levels in the transgenic plants. However, neither citrate accumulation nor efflux was enhanced and consequently no positive effect on tolerance to Al toxicity was observed in transgenic lines. However, expressing a malate efflux transporter gene from wheat in barley did significantly improve aluminum tolerance in the transgenic barley lines (Delhaize et al. 2004).

A more robust approach may therefore be the transfer of genes known to confer tolerance to edaphic stresses from highly tolerant plant species to other species lacking certain components of tolerance mechanisms. The work of Nishizawa and coworkers has eloquently shown merits of this approach. Rice has evolved under flooded conditions characterized by high concentrations of Fe^{2+} in the soil solution. The adaptation to such high Fe^{2+} environments may have selected for an Fe uptake system predominantly relying on direct uptake of Fe^{2+} via the Fe^{2+} transporter OsIRT1 without a need to first reduce Fe^{3+} to Fe^{2+} (through Fe^{3+} chelate reductase) as would be required for other Strategy I species that rely on the IRT transport pathway and that evolved on non-flooded soils that predominantly contain the oxidized Fe^{3+} form. When grown in such aerobic soils rice is prone to suffer from Fe deficiency because of low Fe^{3+} chelate reductase activity and because of low phytosiderophore secretion compared to other graminaceous strategy II species like barley or wheat (Ishimaru et al. 2007). Nishizawa and coworkers followed two independent approaches to engineer rice plants with high tolerance to Fe deficiency. The first approach involved strengthening the Strategy II system by expressing two barley nicotianamine aminotransferase genes in rice. Transgenic rice plants secreted much larger amounts of phytosiderophores resulting in better performance in calcareous soils (Takahashi et al. 2001). The second approach was to reconstitute the first step of the Strategy I uptake system by introducing a Fe^{3+} chelate reductase gene from yeast fused to the Fe-regulated OsIRT1 promoter (Ishimaru et al. 2007). Transgenic rice plants had eight times higher grain yield compared to non-transformed plants when grown on calcareous soil. First field trials on calcareous soils confirmed experimental results even though yield benefits were lower than seen in pot experiments (Suzuki et al. 2008).

Future transgenic approaches will increasingly rely on using alleles from tolerant genotypes for transformation within a given species. At present this approach is limited by the low number of useful naturally occurring alleles identified. However, with sequence data available for an increasing number of relevant crop species and concerted international efforts directed towards allele mining (Generation Challenge Program 2007), progress will be made in the coming years. The identification of a major Al tolerance gene in sorghum through positional cloning of the Alt_{SB} locus is one of the most promising cases being actively pursued at the moment (Magalhaes et al. 2007). Several allelic variants of the Alt_{SB} gene have been identified from different donors and efforts are under way to identify the most effective allele, both through transgenic approaches and through marker assisted backcrossing (Caniato et al. 2007).

A similar strategy based on the identification of superior alleles via QTL mapping has led to the identification of several putative candidate genes at the Pup1 locus in rice (Ismail et al. 2007). NILs carrying the tolerance allele at the Pup1 locus showed threefold higher P uptake and grain yield in field trials conducted on a highly P deficient Andosol. Currently transgenic plants are produced for several candidate genes in an attempt to clone and characterize the responsible gene. A different approach based on screening for genes induced by P starvation led to the molecular cloning of the OsPTF1 gene in rice (Yi et al. 2005). OsPTF1 is a transcription factor that enhances root growth under P deficiency. Transgenic rice plants overexpressing the OsPTF1 gene showed 20% higher P uptake and biomass accumulation when grown in P deficient soil.

The development of transgenic resistance through molecular breeding programs may be central to the generation of plant genotypes suitably resistant to non-specific root pathogens. Attempts to confer resistance to *Rhizoctonia solani* have focused on the transfer of hydrolytic enzymes including chitinases (Broglie et al. 1991; Sareena et al. 2006) and glucanases capable of degrading cell walls of invading fungal pathogens to susceptible crop plants. While a modest level of resistance to foliar blight diseases caused by *R. solani* has been obtained (Datta et al. 2001), similar effort yielded minimal improvement in resistance to root rot disease incited by this pathogen. Transfer of the thaumatin gene from *Thaumatococcus daniellii* into tobacco delayed disease development incited by *R. solani* and *Pythium aphanidermatum* (Rajam et al. 2007). However, in the vast majority of instances trials assessing the conference of resistance have been conducted using extremely artificial systems or have considered disease development only in the seedling stage of plant development (damping-off) and have not assessed impacts on root rot diseases incited by these pathogens.

RNA interference technology is a phenomenon of gene silencing at the mRNA level that is a powerful research tool used in functional genomics. RNAi is being used to create plants having novel traits, including resistance to plant pathogens and parasites. In this realm initial efforts to develop resistance to diseases focused particularly on those caused by viruses. Recently, certain efforts have examined the utility of RNAi for the control of plant root parasitic nematodes, and most specifically the obligate sedentary endoparasites belonging to the genera *Globodera*, *Heterodera* and *Meloidogyne* (Lilley et al. 2007). Studies demonstrated that an RNAi like response could be achieved when artificially exposed to dsRNA (Urwin et al. 2002). Fewer nematodes were recovered from plants inoculated with pre-parasitic juvenile forms that had ingested dsRNA *in vitro* than plants inoculated with non-treated control nematodes. Subsequent studies in transgenic Arabidopsis [*Arabidopsis thaliana* (L.) Heynh.] and tobacco demonstrated that *Meloidogyne* could acquire plant derived dsRNA through root feeding activities (Fairbairn et al. 2007; Huang et al. 2006). In vitro ingestion of dsRNA of the parasitism gene *16D10*, which encodes for a parasitism peptide secreted by second stage juveniles, silenced the target gene in J2 juveniles of *M. incognita*. Reproduction of four *Meloidogyne* species was dramatically reduced on transgenic Arabidopsis expressing *16D10* dsRNA. The breadth of efficacy realized in *16D10* dsRNA expressing plants across species of *Meloidogyne* has not been realized in response to known resistance genes.

One difficulty associated with the transgenic approach in terms of making the leap from experimental verification to practical breeding is the lack of acceptance by a public skeptical of gene transfers, particularly if this involves different species. In this regard the transfer of useful alleles within a species should be less problematic and from this standpoint alone, more effort should be directed towards the identification of such useful alleles. A different approach that might be more acceptable is the generation of new allelic variation within species through mutagenesis. One example is that of Picard et al. (2007), who selected from 10 mutant lines belonging to a tomato (cv *Red Setter*) ethylmethane sulphonate (EMS) mutant collection, those possessing both good mycorrhizal competence and high productivity in order to obtain tomato genotypes adapted to low-input cropping systems. It is important to note that EMS mutations did not induce loss in fruit quality, since no relevant differences were observed between mutants and the isogenic line. In contrast, differences in the arbuscular mycorrhizal fungi (AMF) infection dynamics were observed between lines. Some differences were also found in the diversity of root-colonizing AMF taxa and in fruit quality. These differences in values for mycorrhization level, root infection dynamics, AMF population structure and fruit quality have been used to classify these mutants. Interestingly, results of this classification by these four characters tended to correlate. One tomato mutant was especially mycorrhizal-competent, showed a highly structured AMF population, and was highly productive. This mutant is the most promising culture for low-input cropping systems. Another example is the 15 EMS-induced mutants obtained by Carrol et al. (1985) from the parental Bragg soybean, which formed up to 40 times the number of nodules as the parent and continued acetylene reduction (N_2 fixation) activity.

Outlook: opportunities and limitations

It is obvious that there is a strong demand for plant breeding to produce new cultivars with improved yield potentials for application in low-input less favorable crop production environments. From the review of recent literature, it appears that it may be difficult to select cultivars for lower-input agriculture from the elite cultivars currently used in conventional agriculture. It has been broadly recognized that desirable traits will need to be reintroduced into elite cultivars and considerable effort has been devoted to the identification of beneficial traits and suitable donors for that purpose. Yet progress has been slow and has so far largely been confined to modifications of traditional breeding procedures such as decentralized selection under low input conditions. This is in stark contrast to the explosion of knowledge generated in plant sciences during the past decade. Based on the review of potentially useful traits in this paper and by others in this volume it is obvious that there is no shortage of ideas. How to turn these ideas into better varieties remains a critical and often unresolved question.

Several successful cases have shown that concepts do work in principle. Insights into plant adaptation to nutrient scarcity have identified traits that are being used in direct selection, for example root morphology in beans and soybeans (J. P. Lynch, pers. comm.) and symbiotic N_2 fixation in common bean (Bliss 1993). QTL mapping can lead to identification of new and useful alleles, such as those at the Alt_{SB} or Pup1 loci that are being used in MAS. Transgenic approaches did show their efficacy in improving tolerance even under field conditions (e.g. improving tolerance of Fe deficiency in rice; Suzuki et al. 2008).

However, given the large number of traits studied and that numerous QTLs have been mapped, it is surprising how few examples do find their way into practical breeding. Several factors, both political and scientific, may be responsible for this obvious gap between basic research and applied breeding. The success of the Green Revolution has up to very recently resulted in a food surplus and that has precipitated a shift in focus away from food security, particularly in highly developed countries. Consequently, investments in public research were predominantly channeled into basic research dominated by model systems such as *Arabidopsis thaliana*. Breeding on the other hand has increasingly been done by the private sector with a natural focus on high value seeds but not on poor people's varieties or crops. With funding and esteem being progressively more associated with basic research, training at universities has also shifted away from breeding and that may have further widened the gap between basic research and applied sciences. These economic and political changes in combination with the entirely new opportunities offered by ever-evolving molecular tools may be responsible for the shift in research focus away from complex systems such as plant–soil interactions to single gene/traits under highly artificial conditions.

Yet in order to bring novel approaches generated in basic science to applied plant breeding it will be crucial to narrow the gap between the two disciplines. In certain instances this has been recognized by public funding sources that have moved away from supporting model system research to specifically focus on research systems directly applicable to crop plants. The few success stories (breeding for enhanced Al tolerance, improved P acquisition) involved close collaboration between breeders and scientists conducting basic research and, by periodically testing material under field conditions, they did not avoid the complexity of plant–soil interactions. While stating the need for more interdisciplinary and field-based research may appear mundane, it seems obvious that the lack of such approaches has resulted in a significant impediment to making further progress through plant breeding, especially with regard to materials of value to low-input "sustainable" systems. We therefore want to put forward a few points that we believe are key to making progress in this regard:

- Phenotypes need to be confirmed under realistic field conditions in order to establish that traits warrant attention in breeding programs. Such a 'reality check' is also of benefit for basic research because it assures that mechanisms studied or genes identified are practically relevant.
- Rapid screening protocols that correlate closely to field performance are essential for applied breeding and considerable effort should be directed towards their development.
- It is important to realize that allelic variation, as detected in QTL mapping studies, is not equal to useful allelic variation. Without confirmation of QTL effects in NIL backgrounds, QTL mapping will remain a purely academic exercise. Only strong alleles that can significantly improve a phenotype will find their way into breeding programs.
- Similarly, genes identified in unsystematic reverse genetic screens may be of little practical use. More effort should be directed towards identification of superior phenotypes and their underlying genes.
- Finally a more holistic view will be needed that takes into account the trade-offs or synergistic effects existing for many traits.

Employing such a holistic view will eventually enable us to redesign plants by pyramiding of genes or QTLs in order to combine traits or genes to enhance processes that need simultaneous activity of several components. An ideal new variety would show enhanced root interactions with beneficial microorganisms (mycorrhizal fungi and rhizobacteria), soilborne disease resistance and/or nutrient acquisition capability and tolerance to abiotic soil stresses. For the case of improved capability to access soil-bound P, plants of the future would present high affinity with mycorrhizal fungi (Bosco et al. 2006; Rengel 2002), or would show enhanced carboxylate exudation to release inorganic as well as organic P sources into the soil solution (Gahoonia and Nielson 2004); they should furthermore have higher phytase/phosphatase excretion rates to break down the organic P forms and this would be combined with increased root hair length to facilitate exudation and uptake at larger distances from the root surface. All of these traits have so far been studied in isolation and with limited success in terms of producing the desired phenotype (Delhaize et al. 2001; Richardson et al. 2001). The more difficult task of combining them in a single genotype awaits realization.

With further advances in molecular techniques and in our understanding of which genes govern highly complex traits ever more possibilities for engineering crops will arise in the future. Turning cereals into efficient N_2 fixing crops or reducing N losses through enhancing the release of natural nitrification inhibitors by crops (Subbarao et al. 2007) may be feasible one day. However, in the meantime non-renewable resources are being diminished and poverty in rural areas of developing countries remains extremely high. Improved varieties are therefore needed urgently and more effort to combat present problems with present technological means is warranted. Further emphasis on bridging the gap between breeding and basic sciences is therefore needed from policy makers, science managers, funding agencies and, ultimately from the scientists involved. It is encouraging to see that this is successfully done in research organizations of Brazil, China, Australia, and in the research centers belonging to the Consultative Group on International Agricultural Research (CGIAR); however, much could be gained if such a shift in emphasis were realized in some of the most developed countries as well.

References

Andrews M, James EK, Cummings SP, Zavalin AA, Vinogradova LV, McKenzie BA (2003) Use of nitrogen fixing bacteria inoculants as a substitute for nitrogen fertiliser for dryland graminaceous crops: Progress made, mechanisms of action and future potential. Symbiosis 35:209–229

App A, Watanabe I, Ventura TS, Bravo M, Jurey CD (1986) The effect of cultivated and wild rice varieties on the nitrogen balance of flooded soil. Soil Sci 141:448–452

Baldani JL, Reis VM, Baldani VLD, Dobereiner J (2002) A brief story of nitrogen fixation in sugarcane—reasons for success in Brazil. Funct Plant Biol 29:417–423

Bänziger M, Diallo AO (2001) Progress in developing drought and N stress tolerant maize cultivars for Eastern and Southern Africa. Seventh Eastern and Southern Africa Regional Maize Conference, 11–15 February 2001, pp 189–194

Bänziger M, Betran FJ, Lafitte HR (1997) Efficiency of high-nitrogen selection environments for improving maize for low-nitrogen target environments. Crop Sci 37:1103–1109

Batten GD (1992) A review of phosphorus efficiency in wheat. Plant Soil 146:163–168

Berg G, Roskot N, Steidle A, Eberl L, Zock A, Smalla K (2002) Plant-dependent genotypic and phenotypic diversity of antagonistic rhizobacteria isolated from different *Verticillium* host plants. Appl Environ Microbiol 68:3328–3338

Berg G, Zachow C, Lottmann J, Götz M, Costa R, Smalla K (2005) Impact of plant species and site on rhizosphere-associated fungi antagonistic to *Verticillium dahliae* Kleb. Appl Environ Microbiol 71:4203–4213

Bliss FA (1993) Breeding common bean for improved biological nitrogen-fixation. Plant Soil 152:71–79

Bosco M, Baruffa E, Picard C (2006) Organic breeding should select for plant genotypes able to efficiently exploit indigenous Probiotic Rhizobacteria. In: Andreasen CB, Elsgaard L, Sondegaard SL, Hansen G (eds) Organic farming and European rural development. Proceedings of the European Joint Organic Congress, pp 376–377

Broglie KE, Chet I, Holliday M, Cressman R, Biddle O, Knowlton S et al (1991) Transgenic plants with enhanced resistance to the fungal pathogen *Rhizoctonia solani*. Science 254:1194–1197

Browne GT, Mircetich SM (1993) Relative resistance of thirteen apple rootstocks to three species of *Phytophthora*. Phytopathology 83:744–749

Caniato F, Guimarães C, Schaffert R, Alves V, Kochian L, Borém A et al (2007) Genetic diversity for aluminum tolerance in sorghum. Theor Appl Genet 114:863–876

Carroll BJ, McNeil DL, Gresshoff PM (1985) Isolation and properties of soybean [Glycine max (L.) Merr.] mutants that nodulate in the presence of high nitrate concentrations. Proc Natl Acad Sci U S A 82:4162–4166

Ceccarelli S (1994) Specific adaptation and breeding for marginal conditions. Euphytica 77:205–219

Ceccarelli S, Grando S (2007) Decentralized-participatory plant breeding: an example of demand driven research. Euphytica 155:349–360

Cook RJ (2006) Toward cropping systems that enhance productivity and sustainability. Proc Natl Acad Sci U S A 103:18389–18394

Cook RJ, Thomashow LS, Weller DM, Fujimoto DK, Mazzola M, Bangera G et al (1995) Molecular mechanisms of defense by rhizobacteria against root disease. Proc Natl Acad Sci U S A 92:4197–4201

Cregan PB (1989) Host plant effects on nodulation and competitiveness of the *Bradyrhizobium japonicum* serotype strains constituting serocluster 123. Appl Environ Microbiol 55:2532–2536

da Mota FF, N'obrega A, Marriel IE, Paiva E, Seldin L (2002) Genetic diversity of *Paenibacillus polymyxa* populations isolated from the rhizosphere of four cultivars of maize (*Zea mays*) planted in Cerrado soil. Appl Soil Ecol 20: 119–132

Dalmastri C, Chiarini L, Cantale C, Bevivino A, Tabacchioni S (1999) Soil type and maize cultivar affect the genetic diversity of maize root-associated *Burkholderia cepacia* populations. Microb Ecol 38:273–284

Datta K, Tu J, Oliva N, Ona I, Velazahan R, Wew TW et al (2001) Enhanced resistance to sheath blight by constitutive expression of infection-related rice chitinase in transgenic elite indica rice cultivars. Plant Sci 160:405–414

Dawson J, Murphy K, Jones S (2007) Decentralized selection and participatory approaches in plant breeding for low-input systems. Euphytica (in press) doi:10.1007/s10681-007-9533-0

Delhaize E, Hebb DM, Ryan PR (2001) Expression of a *Pseudomonas aeruginosa* citrate synthase gene in tobacco is not associated with either enhanced citrate accumulation of efflux. Plant Physiol 125:2059–2067

Delhaize E, Ryan PR, Hebb DM, Yamamota Y, Sasaki T, Matsumoto H (2004) Engineering high-level aluminum tolerance in barley with the ALMT1 gene. Proc Natl Acad Sci U S A 101:15249–15254

Drinkwater LE, Snapp SS (2007) Understanding and managing the rhizosphere in agroecosystems. In: ZG Cardon, JL Whitbeck (eds) The rhizosphere—an ecological perspective. Elsevier, pp 155–178

Duvick DN, Cassman KG (1999) Post-green revolution trends in yield potential of temperate maize in the North-Central United States. Crop Sci 39:1622–1630

Egle K, Manse G, Roemer W, Vlek PLG (1989) Improved phosphorus efficiency of three new wheat genotypes from CIMMYT in comparison with an older Mexican variety. J Plant Nutr Soil Sci 162:353–358

El Mohtar CA, Atamian HS, Dagher RB, Abou-Jawdah Y, Salus MS, Maxwell DP (2007) Marker-assisted selection of tomato genotypes with the *I-2* gene for resistance to *Fusarium oxysporum* f. sp. *lycopersici* race 2. Plant Dis 91:758–762

Ela SW, Anderson MA, Brill WJ (1982) Screening and selection of maize to enhance associative bacterial nitrogen fixation. Plant Physiol 70:1564–1567

Engelhard M, Hurek T, Reinhold-Hurek B (2000) Preferential occurrence of diazotrophic endophytes, *Azoarcus* spp., in wild rice species and land races of *Oryza sativa* in comparison with modern races. Environ Microbiol 2:131–141

Evenson RE, Gollin D (2003) Assessing the impact of the Green Revolution, 1960 to 2000. Science 300:758–762

Fairbairn DJ, Cavallaro AS, Bernard M, Mahalinga-Iyer J, Graham MW, Botella JR (2007) Host-delivered RNAi: an effective strategy to silence genes in plant parasitic nematodes. Planta 226:1525–1533

Fazio G, Mazzola M (2005) Target traits for the development of marker assisted selection of apple rootstocks—prospects and benefits. Acta Hortic 663:823–827

Fazio G, Robinson T, Aldwinckle H, Mazzola M, Leinfelder M, Parra R (2006) Traits of the next wave of geneva apple rootstocks. Compact Fruit Tree 38:7–11

Fu Y-B, Peterson GW, Richards KW, Somers D, DePauw RM, Clarke JM (2005) Allelic reduction and genetic shift in the Canadian hard red spring wheat germplasm released from 1845 to 2004. Theor Appl Genet 110:1505–1516

Gahoonia TS, Nielsen NE (2004) Barley genotypes with long root hairs sustain high grain yields in low-P field. Plant Soil 262:55–62

Generation Challenge Programme (2007) Project mid-year and final reports: Competitive and commissioned project. Generation Challenge Programme, Texcoco, Mexico, pp 42–46

George TS, Richardson AE, Hadobas PA, Simpson RJ (2004) Characterisation of transgenic *Trifolium subterraneum* L. which expresses *phyA* and releases extracellular phytase: growth and P nutrition in laboratory media and soil. Plant Cell Environ 27:1351–1361

George TS, Richardson AE, Smith JB, Hadobas PA, Simpson RJ (2005) Limitations to the Potential of transgenic *Trifolium subterraneum* L. plants that exude phytase when grown in soils with a range of organic P content. Plant Soil 278:263–274

Grau Nersting L, Bode Andersen S, von Bothmer R, Gullord M, Bagger Jørgensen R (2006) Morphological and molecular diversity of Nordic oat through one hundred years of breeding. Euphytica 150:327–337

Gu Y-H, Mazzola M (2003) Modification of fluorescent pseudomonad community and control of apple replant disease induced in a wheat cultivar-specific manner. Appl Soil Ecol 24:57–72

Gutierrez JA, Cramer CS (2005) Screening short-day onion cultivars for resistance to fusarium basal rot. HortScience 40:157–160

Hallauer AR, Miranda Fo JB (1988) Quantitative genetics in plant breeding. Iowa State Univ. Press, Ames, pp 337–348

Hede A, Skovmand B, Lopez-Cesati J (2001) Acid soils and aluminum toxicity. In: Reynolds M (ed) Application of physiology in wheat breeding. CIMMYT, Mexico, pp 172–182

Herman R, Perl-Treves R (2007) Characterization and inheritance of a new source of resistance to *Fusarium oxysporum* f. sp. *melonis* race 1.2 in *Cucumis melo*. Plant Dis 91:1180–1186

Herridge D, Rose I (2000) Breeding for enhanced nitrogen fixation in crop legumes. Field Crops Res 65:229–248

Hetrick BAD, Wilson GWT, Cox TS (1992) Mycorrhizal dependence of modern wheat-varieties, landraces, and ancestors. Can J Bot 70:2032–2040

Hetrick BAD, Wilson GWT, Cox TS (1993) Mycorrhizal dependence of modern wheat cultivars and ancestors—a synthesis. Can J Bot 71:512–518

Hetrick BAD, Wilson GWT, Gill BS, Cox TS (1995) Chromosome location of mycorrhizal responsive genes in wheat. Can J Bot 73:891–897

Hetrick BAD, Wilson GWT, Todd TC (1996) Mycorrhizal response in wheat cultivars: Relationship to phosphorus. Can J Bot 74:19–25

Hinsinger P, Gobran GR, Gregory PJ, Wenzel WW (2005) Rhizosphere geometry and heterogeneity arising from root-mediated physical and chemical processes. New Phytol 168:293–303

Hoffland E, Wei C, Wissuwa M (2006) Organic anion exudation by lowland rice (*Oryza sativa* L.) at Zinc and Phosphorus deficiency. Plant Soil 283:155–162

Huang G, Allen R, Davis EL, Baum TJ, Hussey RS (2006) Engineering broad root-knot resistance in transgenic plants by RNAi silencing of a conserved and essential root-knot nematode parasitism gene. Proc Natl Acad Sci U S A 103:14302–14306

Hungria M, Phillips DA (1993) Effects of a seed color mutation on rhizobial nod-gene-inducing flavanoids and nodulation in common bean. Mol Plant Microbe Interact 6:418–422

Iruela M, Castro P, Rubio J, Cubero JI, Jacinto C, Millán T et al (2007) Validation of a QTL for resistance to ascochyta blight linked to resistance to fusarium wilt race 5 in chickpea (*Cicer arietinum* L.). Eur J Plant Pathol 119:29–37

Ishimaru Y, Suzuki M, Tsukamoto T, Suzuki K, Nakazono M, Kobayashi T et al (2006) Rice plants take up iron as an Fe^{3+}-phytosiderophore and as Fe^{2+}. Plant J 45:335–346

Ishimaru Y, Kim S, Tsukamoto T, Oki H, Kobayashi T, Watanabe S et al (2007) Mutational reconstructed ferric chelate reductase confers enhanced tolerance in rice to iron deficiency in calcareous soil. Proc Natl Acad Sci U S A 104:7373–7378

Ishitani M, Rao I, Wenzl P, Beebe S, Tohme J (2004) Integration of genomics approach with traditional breeding towards improving abiotic stress adaptation: drought and aluminum toxicity as case studies. Field Crops Res 90:35–45

Ismail AM, Heuer S, Thomson JT, Wissuwa M (2007) Genetic and genomic approaches to develop rice germplasm for problem soils. Plant Mol Biol 65:547–570

Jagnow G (1990) Differences between cereal crop cultivars in root-associated nitrogen-fixation. Possible causes of variable yield response to seed inoculation. Plant Soil 123:255–259

Kaeppler SM, Parke JL, Mueller SM, Senior L, Stuber C, Tracy WF (2000) Variation among maize inbred lines and detection of quantitative trait loci for growth at low phosphorus and responsiveness to arbuscular mycorrhizal fungi. Crop Sci 40:358–364

Keel C, Schnider U, Maurhofer M, Voisard C, Laville J, Burger P et al (1992) Suppression of root diseases by *Pseudomonas fluorescens* CHA0: importance of the secondary metabolite 2,4-diacetylphlorglucinol. Mol Plant Microbe Interact 5:4–13

Khanizadeh S, Granger R (1998) An overview on history, progress, present and future objectives of the Quebec apple cultivar and rootstock breeding program. Acta Hortic 513:477–482

Khush GS (1999) Green revolution: preparing for the 21st century. Genome 42:646–655

Kim MY, Van K, Lestari P, Moon JK, Lee SH (2005) SNP identification and SNAP marker development for a GmNARK gene controlling supernodulation in soybean. Theor Appl Genet 110:1003–1010

Kirkpatrick MT, Rothrock CS, Rupe JC, Gbur EE (2006) The effect of *Pythium ultimum* and soil flooding on two soybean cultivars. Plant Dis 90:597–602

Kochian LV, Hoekenga OA, Pineros MA (2004) How do plants tolerate acid soils? Mechanisms of aluminum tolerance and phosphorus efficiency. Annu Rev Plant Biol 55:459–493

Lambers H, Shane MW, Cramer MD, Pearse SJ, Veneklaas EJ (2006) Root structure and functioning for efficient acquisition of Phosphorus: Matching morphological and physiological traits. Ann Bot (Lond) 98:693–713

Larkin RP, Hopkins DL, Martin FN (1993) Effect of successive watermelon plantings on *Fusarium oxysporum* and other microorganisms suppressive and conducive to Fusarium wilt of watermelon. Phytopathology 83:1097–1105

Larkin RP, Hopkins DL, Martin FN (1996) Suppression of *Fusarium* wilt of watermelon by nonpathogenic *Fusarium oxysporum* and other microorganisms recovered from a disease suppressive soil. Phytopathology 86:812–819

Latour X, Corberand T, Laguerre G, Allard F, Lemanceau P (1996) The composition of fluorescent pseudomonad populations associated with roots is influenced by plant and soil type. Appl Environ Microbiol 62:2449–2456

Lemanceau P, Corberand T, Gardan L, Latour X, Laguerre G, Boeufgras J-M et al (1995) Effect of two plant species, flax (*Linum usitatissinum* L.) and tomato *(Lycopersicon esculentum* Mill.), on diversity of soilborne populations of fluorescent pseudomonads. Appl Environ Microbiol 61:1004–1012

Lilley CJ, Bakhetia M, Charlton WL, Urwin PE (2007) Recent progress in the development of RNA interference for plant parasitic nematodes. Mol Plant Pathol 8:701–711

Line RF, Chen X (1995) Successes in breeding for and managing durable resistance to wheat rusts. Plant Dis 79:1254–1255

Luterbacher MC, Asher MJC, Beyer W, Mandolino G, Scholtern OE, Frese L et al (2005) Sources of resistance to diseases of sugar beet in related *Beta* germplasm: II. Soil-borne diseases. Euphytica 141:49–63

Lynch JP (2007) Roots of the Second Green Revolution. Aust J Bot 55:493–512

Mackill DJ (2006) Breeding for resistance to abiotic stresses in rice: the value of quantitative trait loci. In: Lamkey KR, Lee M (eds) Plant breeding: the Arnel R Hallauer International Symposium. Blackwell, Ames, IA, pp 201–212

Magalhaes JV, Liu J, Guimaraes CT, Lana UGP, Alves VMC, Wang Y-H et al (2007) A gene in the multidrug and toxic compound extrusion (MATE) family confers aluminum tolerance in sorghum. Nat Genet 39:1156–1161

Magnavaca R, Bahia Filho AFC (1993) Success in maize acid soil tolerance. In: Maranvill JW (ed) Adaptation of plants to soil stress. Intsormil Publ. No. 94-2. University of Nebraska, Lincoln, pp 209–220

Marschner P, Yang C-H, Lieberei R, Crowley DE (2001) Soil and plant specific effects on bacterial community composition in the rhizosphere. Soil Biol Biochem 33:1437–1445

Marshall D, Tunali B, Nelson LR (1999) Occurrence of fungal endophytes in species of wild *Triticum*. Crop Sci 39:1507–1512

Mazzola M (1998) Elucidatin of the microbial complex having a causal role in development of apple replant disease in Washington. Phytopathology 88:930–938

Mazzola M, Gu Y-H (2002) Wheat genotype-specific induction of soil microbial communities suppressive to *Rhizoctonia solani* AG 5 and AG 8. Phytopathology 92:1300 1307

Mazzola M, Funnell DL, Raaijmakers JM (2004) Wheat cultivar-specific selection of 2,4-diacetylphloroglucinol-producing fluorescent *Pseudomonas* species from resident soil populations. Microb Ecol 48:338–348

Miethling R, Wieland G, Backhaus H, Tebbe CC (2000) Variation of microbial rhizosphere communities in response to crop species, soil origin, and inoculation with *Sinorhizobium meliloti* L33. Microb Ecol 41:43–56

Morris M, Bellon M (2004) Participatory plant breeding research: opportunities and challenges for the international crop improvement system. Euphytica 136:21–35

Nangju D (1980) Soybean response to indigenous rhizobia as influenced by cultivar origin. Agron J 72:403–406

Neyra CA, Dobereiner J (1977) Nitrogen fixation in grasses. Adv Agron 29:1–38

Notz R, Maurhofer M, Schnider-Keel U, Duffy B, Haas D, Défago G (2001) Biotic factors affecting expression of the 2,4-diacetylphloroglucinol biosynthesis gene *phlA* in *Pseudomonas fluorescens* biocontrol strain CHA0 in the rhizosphere. Phytopathology 91:873–881

Ochoa IE, Blair MW, Lynch JP (2006) QTL analysis of adventitious root formation in common bean under contrasting phosphorus availability. Crop Sci 46:1609–1621

Oehl F, Sieverding E, Mäder P, Dubois D, Ineichen K, Boller T et al (2004) Impact of long-term conventional and organic farming on the diversity of arbuscular mycorrhizal fungi. Oecol 138:574–583

Okubara PA (2006) Molecular responses of wheat to soilborne fungal pathogens. Kado Lab Science Colloquium, July 22, Davis, CA. Abstract no. 7, pp 17–19

Ori N, Eshed Y, Paran I, Presting G, Aviv D, Tanksley S et al (1997) The *I2C* family from the wilt disease resistance locus *I2* belongs to the nucleotide binding leucine-rich repeat superfamily of plant resistance genes. Plant Cell 9:521–532

Panella L (1998) Screening and utilizing *Beta* genetic resources with resistance to *Rhizoctonia* root rot and *Cercospora* leaf spot in a sugar beet breeding programme. In: L Frese, L Panella, HM Srivastava, W Lange (eds.) 4th International Beta Genetic resources Workshop & World Beta Network Conference, Izmir, Turkey. IPGRI, Rome

Paponov IA, Sambo P, Schulte auf'm Erley G, Presterl T, Geiger HH, Engels C (2005) Kernel set in maize genotypes differing in nitrogen use efficiency in response to resource availability around flowering. Plant Soil 272:101–110

Particka CA, Hancock JF (2005) Field evaluation of strawberry genotypes for tolerance to black root rot on fumigated and nonfumigated soil. J Am Soc Hortic Sci 130:688–693

Perchepied L, Dogimont C, Pitrat M (2005) Strain-specific and recessive QTLs involved in the control of partial resistance to *Fusarium oxysporum* f. sp. *melonis* race 1.2 in a recombinant inbred line population of melon. Theor Appl Genet 111:431–438

Picard C, Bosco M (2003) Soil antimony pollution and plant growth stage affect the biodiversity of auxin-producing bacteria isolated from the rhizosphere of *Achillea ageratum* L. FEMS Microbiol Ecol 46:73–80

Picard C, Bosco M (2005) Maize heterosis affects the structure and dynamics of indigenous rhizospheric auxins-producing *Pseudomonas* populations. FEMS Microbiol Ecol 53:349–357

Picard C, Bosco M (2006) Heterozygosis drives maize hybrids to select elite 2,4-diacethylphloroglucinol-producing *Pseudomonas* strains among resident soil populations. FEMS Microbiol Ecol 58:193–204

Picard C, Bosco M (2008) Genotypic and phenotypic diversity in populations of plant-probiotic *Pseudomonas* spp. colonizing roots. Naturwissenschaften 95:1–16

Picard C, Di Cello F, Ventura M, Fani R, Guckert A (2000) Frequency and biodiversity of 2,4-diacetylphloroglucinol-producing bacteria isolated from the maize rhizosphere at different stages of plant growth. Appl Environ Microbiol 66:948–955

Picard C, Frascaroli E, Bosco M (2004) Frequency and biodiversity of 2,4-diacetylphloroglucinol-producing rhizobacteria are differentially affected by the genotype of two maize inbred lines and their hybrid. FEMS Microbiol Ecol 49:207–215

Picard C, Carriero F, Petrozza A, Zamariola L, Baruffa E, Bosco M (2007) Selecting tomato (*Solanum lycopersicon* L.) lines for mycorrhizal competence: a pre-requisite for breeding the plants for the future. International Congress Rhizosphere 2. 26–31 August 2007, Montpellier, France

Picard C, Baruffa E, Bosco M (2008) Enrichment and diversity of plant-probiotic microorganisms in the rhizosphere of hybrid maize during four growth cycles. Soil Biol Biochem 40:106–115

Pieczarka DJ, Abawi GS (1978) Effects of interaction between *Fusarium*, *Pythium*, and *Rhizoctonia* on severity of bean root rot. Phytopathology 68:403–408

Presterl T, Groh S, Landbeck M, Seitz G, Schmidt W, Geiger HH (2002) Nitrogen uptake and utilization efficiency of European maize hybrids developed under conditions of low and high nitrogen input. Plant Breed 121:480–486

Raaijmakers JM, Weller DM (2001) Exploiting genotypic diversity of 2,4-diacetylphloroglucinol-producing *Pseudomonas* spp.: characterization of superior root colonizing *P. fluorescens* strain Q8r1 M-96. Appl Environ Microbiol 63:881–887

Raaijmakers JM, Bonsall RF, Weller DM (1999) Effect of population density of *Pseudomonas fluorescens* on production of 2,4-diacetylphloroglucinol in the rhizosphere of wheat. Phytopathology 89:470–475

Rajam MV, Chandola N, Saiprasad, Goud P, Singh D, Kashyap V, Choudhary ML, Sihachakr D (2007) Thaumatin gene confers resistance to fungal pathogens as well as tolerance to abiotic stress in transgenic tobacco plants. Biol Plant 51:135–141

Rao IM, Zeigler RS, Vera R, Sarkarung S (1993) Selection and breeding for acid-soil tolerance in crops. Bioscience 43:454–465

Rengel Z (2002) Breeding for better symbiosis. Plant Soil 245:147–162

Rengel Z, Marschner P (2005) Nutrient availability and management in the rhizosphere: exploiting genotypic differences. New Phytol 168:305–312

Richardson AE, Hadobas PA, Hayes JE (2001) Extracellular secretion of *Aspergillus* phytase from *Arabidopsis* roots enables plants to obtain phosphorus from phytate. Plant J 25:641–649

Roesh LFW, Olivares FL, Pereira Passaglia LM, Selbach PA, Saccol de Sa EL, Oliveira de Camargo FA (2006) Characterization of diazotrophic bacteria associated with maize: effect of plant genotype, ontogeny and nitrogen-supply. World J Microbiol Biotechnol 22:967–974

Rolfe BG, Djordjevic MA, Weinman JJ, Mathesius U, Pittock C, Gartner E et al (1997) Root morphogenesis in legumes and cereals and the effect of bacterial inoculation on root development. Plant Soil 194:131–144

Rubio G, Liao H, Yan X, Lynch JP (2003) Topsoil foraging and its role in plant competitiveness for phosphorus in common bean. Crop Sci 43:598–607

Runge-Metzger A (1995) Closing the cycle: obstacles to efficient P management for improved global food security.

In: Tiessen H (ed) Phosphorus in the global environment: transfers, cycles and management. John Wiley and Sons, New York, pp 27–42

Sareena S, Poovannan K, Kumar KK, Raja JAJ, Samiyappan R, Sudhakar D et al (2006) Biochemical responses in transgenic rice plants expressing a defence gene deployed against the sheath blight pathogen, *Rhizoctonia solani*. Curr Sci 91:1529–1532

Sarkarung S (1986) Screening upland rice for aluminum tolerance and blast resistance. In: Progress in Upland Rice Research. International Rice Research Institute, Manila, pp 271–281

Sarniguet A, Lucas P, Lucas M (1992) Relationship between take-all, soil conduciveness to the disease, populations of fluorescent pseudomonads and nitrogen fertilizers. Plant Soil 145:17–27

Sasaki T, Yamamoto Y, Ezaki B, Katsuhara M, Ahn SJ, Ryan PR et al (2004) A wheat gene encoding an aluminum-activated malate transporter. Plant J 37:645–653

Segal G, Sarfatti M, Schaffer MA, Zamier D, Fluhr R (1992) Correlation of genetic and physical structure in the region surrounding the *I-2 Fusarium oxysporum* locus in tomato. Mol Gen Genet 231:179–185

Sharma KD, Muehlbauer FJ (2007) Fusarium wilt of chickpea: physiological specialization, genetics of resistance and resistance gene tagging. Euphytica 157:1–14

Shaw DV, Larson KD (1996) Relative performance of strawberry cultivars from California and other North American sources in fumigated and nonfumigated soils. J Am Soc Hortic Sci 121:764–767

Shull GH (1908) The composition of a field of maize. Rep Am Breeders Assoc 4:296–301

Shrestha RK, Ladha JK (1996) Genotypic variation in promotion of rice dinitrogen fixation as determined by nitrogen-15 dilution. Soil Sci Soc Am J 60:1815–1821

Singh B, Natesan SKA, Singh RK, Usha K (2005) Improving zinc efficiency of cereals under zinc deficiency. Curr Sci 88:36–84

Smith KP, Goodman RM (1999) Host variation for interactions with beneficial plant-associated microbes. Annu Rev Phytopathol 37:473–491

Smith KP, Handelsman J, Goodman RM (1997) Modeling dose–response relationships in biological control: partitioning host responses to the pathogen and biocontrol agent. Phytopathology 87:720–729

Smith KP, Handelsman J, Goodman RM (1999) Genetic basis in plants for interactions with disease-suppressive bacteria. Proc Natl Acad Sci U S A 96:4786 4790

Subbarao G, Rondon M, Ito O, Ishikawa T, Rao I, Nakahara K et al (2007) Biological nitrification inhibition (BNI) is it a widespread phenomenon. Plant Soil 294:5–18

Suzuki M, Morikawa KC, Nakanishi H, Takahashi M, Saigusa M, Mori S et al (2008) Transgenic rice lines that include barley genes have increased tolerance to low iron availability in a calcareous paddy soil. Soil Sci Plant Nutr 54:77–85

Takahashi M, Nakanishi H, Kawasaki S, Nishizawa NK, Mori S (2001) Enhanced tolerance of rice to low iron availability in alkaline soils using barley nicotianamine aminotransferase genes. Nat Biotechnol 19:466–469

Tanksley SD, McCouch R (1997) Seed banks and molecular maps: unlocking genetic potential from the wild. Science 277:1063–1066

Tawaraya K (2003) Arbuscular mycorrhizal dependency of different plant species and cultivars. Soil Sci Plant Nutr 9:655–668

Tilman D, Cassman KG, Matson PA, Naylor R, Polasky S (2002) Agricultural sustainability and intensive production practices. Nature 418:671–677

Tirol-Padre A, Ladha JK, Singh U, Laureles E, Punzalan G, Akita S (1996) Grain yield performance of rice genotypes at suboptimal levels of soil N as affected by N uptake and utilization efficiency. Field Crops Res 46:127–143

Urwin PE, Lilley CJ, Atkinson H (2002) Ingestion of double-stranded RNA by preparasitic juvenile cyst nematodes leads to RNA interference. Mol Plant Microbe Interact 15:747–752

Vance CP, Uhde-Stone C, Allan DL (2003) Phosphorus acquisition and use: critical adaptations by plants for securing a nonrenewable resource. New Phytol 157:423–447

Virk DS, Singh DN, Prasad SC, Gangwar JS, Witcombe JR (2003) Collaborative and consultative participatory plant breeding of rice for the rainfed uplands of eastern India. Euphytica 132:95–108

von Uexküll HR, Mutert E (1995) Global extent, development and economic impact of acid soils. Plant Soil 171:1–15

Wang Y, Xue Y, Li J (2005) Towards molecular breeding and improvement of rice in China. Trends Plant Sci 10:610–614

Wang J, Raman H, Zhou M, Ryan P, Delhaize E, Hebb D et al (2007) High-resolution mapping of the Alp locus and identification of a candidate gene HvMATE controlling aluminium tolerance in barley (*Hordeum vulgare* L.). Theor Appl Genet 115:265–276

Wissuwa M (2003) How do plants achieve tolerance to phosphorus deficiency? Small causes with big effects. Plant Physiol 133:1947–1958

Wissuwa M (2005) Mapping nutritional traits in crop plants. In: Broadley MR, White PJ (eds) Plant nutritional genomics. Blackwell, Oxford, UK, pp 220–241

Wissuwa M, Ae N (2001) Genotypic variation for tolerance to phosphorus deficiency in rice and the potential for its exploitation in rice improvement. Plant Breed 120:43–48

Wissuwa M, Wegner J, Ae N, Yano M (2002) Substitution mapping of *Pup1*: a major QTL increasing phosphorus uptake of rice from a phosphorus-deficient soil. Theor Appl Genet 105:890–897

Wu P, Zhang G, Ladha JK, McCouch SR, Huang N (1995) Molecular-marker facilitated investigation on the ability to stimulated N_2 fixation in the rhizosphere by irrigated rice plants. Theor Appl Genet 91:1177–1183

Xavier LJC, Germida JJ (1998) Response of spring wheat cultivars to *Glomus clarum* NT4 in a P-deficient soil containing arbuscular mycorrhizal fungi. Can J Soil Sci 78:481–484

Yan X, Liao H, Beebe SE, Blair MW, Lynch JP (2004) QTL mapping of root hair and acid exudation traits and their relationship to phosphorus uptake in common bean. Plant Soil 265:17–29

Yan X, Wu P, Ling H, Xu G, Xu F, Zhang Q (2006) Plant nutriomics in China: an Overview. Ann Bot (Lond) 98:473–482

Yano M, Katayose Y, Ashikari M, Yamanouchi U, Monna L, Fuse T et al (2000) *Hd1*, a major photoperiod sensitivity quantitative trait locus in rice, is closely related to the

Arabidopsis flowering time gene *CONSTANS*. Plant Cell 12:2473–2483

Yi K, Wu Z, Zhou J, Du L, Guo L, Wu Y et al (2005) OsPTF1, a novel transcription factor involved in tolerance to phosphate starvation in rice. Plant Physiol 138:2087–2096

You MP, Barbetti MJ, Nichols PGH (2005) New sources of resistance identified in *Trifolium subterraneum* breeding lines and cultivars to root rot caused by *Fusarium avenaceaum* and *Pythium irregulare* and their relationship to seedling survival. Australas Plant Pathol 34:237–244

Zhai W, Wang W, Zhou Y, Li X, Zhang Q, Wang G et al (2002) Breeding bacterial blight-resistant hybrid rice with the cloned bacterial blight resistance gene Xa21. Mol Breed 8:285–293

Zhu YG, Smith SE, Barritt AR, Smith FA (2001) Phosphorus (P) efficiencies and mycorrhizal responsiveness of old and modern wheat cultivars. Plant Soil 237:249–255

REVIEW ARTICLE

Strategies and methods for studying the rhizosphere—the plant science toolbox

Günter Neumann · Timothy S. George · Claude Plassard

Received: 3 July 2008 / Accepted: 26 February 2009 / Published online: 3 April 2009
© Springer Science + Business Media B.V. 2009

Abstract This review summarizes and discusses methodological approaches for studies on the impact of plant roots on the surrounding rhizosphere and for elucidation of the related mechanisms, covering a range from simple model experiments up to the field scale. A section on rhizosphere sampling describes tools and culture systems employed for analysis of root growth, root morphology, vitality testing and for monitoring of root activity with respect to nutrient uptake, water, ion and carbon flows in the rhizosphere. The second section on rhizosphere probing covers techniques to detect physicochemical changes in the rhizosphere as a consequence of root activity. This comprises compartment systems to obtain rhizosphere samples, visualisation techniques, reporter gene approaches and remote sensing technologies for monitoring the conditions in the rhizosphere. Approaches for the experimental manipulation of the rhizosphere by use of molecular and genetic methods as tools to study rhizosphere processes are discussed in a third section. Finally it is concluded that in spite of a wide array of methodological approaches developed in the recent past for studying processes and interactions in the rhizosphere mainly under simplified conditions in model experiments, there is still an obvious lack of methods to test the relevance of these findings under real field conditions or even on the scale of ecosystems. This also limits reliable data input and validation in current rhizosphere modelling approaches. Possible interactions between different environmental factors or plant-microbial interactions (e.g. mycorrhizae) are frequently not considered in model experiments. Moreover, most of the available knowledge arises from investigations with a very limited number of plant species, mainly crops and studies considering also intraspecific genotypic differences or the variability within wild plant species are just emerging.

Keywords Genotypic variation · Imaging · Ion uptake · Nutrient acquisition · Rhizosphere management · Root exudates · Root growth

Responsible Editor: Philippe Hinsinger.

G. Neumann (✉)
Institute of Plant Nutrition (330), Hohenheim University,
Stuttgart 70593, Germany
e-mail: gd.neumann@t-online.de

T. S. George
Scottish Crop Research Institute (SCRI),
Invergowrie,
Dundee DD2 5DA Scotland, United Kingdom

C. Plassard
UMR Eco&Sols (Ecologie Fonctionnelle
& Biogéochimie des Sols INRA-IRD-SupAgro),
2 Place Pierre Viala,
Montpellier Cedex 1 F-34060, France

Introduction

The activity of plant roots has an impact on the physicochemical conditions as well as on the biological activity in the surrounding rhizosphere compartment, and vice versa. These processes are determining nutrient availability, cycling of nutrients and solubility of toxic elements for plants and microorganisms, thereby creating the rhizosphere as a unique microecosystem, which can exhibit completely different properties compared with the bulk soil, not directly influenced by the activity of roots. Therefore, agricultural production, development of strategies for plant stress management, ecosystem research, soil science and soil microbiology strongly depend on the understanding of rhizosphere processes. Major challenges of plant sciences in rhizosphere research comprise: (i) the in situ detection and quantification of root distribution and turnover under natural soil conditions, (ii) monitoring of root activity, reflected by root-induced physicochemical changes in the rhizosphere, (iii) the characterization of the underlying regulatory mechanisms at the physiological and molecular level and (iv) knowledge transformation into modelling approaches and strategies of rhizosphere management.

This article summarizes the related methodological approaches, including tools for monitoring of root growth and morphology, culture systems employed for rhizosphere sampling, and the detection and quantification of root activity with respect to nutrient uptake, ion and carbon flows in the rhizosphere. A section on rhizosphere probing includes techniques to detect and to quantify physicochemical changes in the rhizosphere as a consequence of root activity, and the use of plant-physiological responses, reporter gene strategies or remote sensing technologies as indicators for rhizosphere processes. The third section focuses on perspectives of molecular and genetic approaches for the experimental manipulation of the conditions in the rhizosphere.

Although the frame of this review does not allow detailed method descriptions, the advantages, limitations, scales of application and perspectives for the various methodological approaches, are discussed. For more details the reader is referred to the respective method collections and methodological monographs (e.g. Smit et al. 2000a; Luster and Finlay (eds) 2006).

Rhizosphere sampling

Due to their capacity to be simultaneously a source for organic C and a sink for mineral nutrients and water, the roots of plants play a key role in the soil surrounding them by creating the rhizosphere. Unlike airborne organs of plants, roots are not directly accessible to observation and the first step to study the impact of plant effects in its rhizosphere is to use experimental systems enabling us to detect and to analyse, qualitatively and quantitatively, growth and morphology of the roots and root-induced changes determining rhizosphere processes and the extension of the rhizosphere.

Root growth and root morphology

Assessing root growth in the field

Depending on the objective of the study, several methods exist to assess root growth under field conditions. One possibility is to use excavation-related methods, comprising: (i) excavation or uprooting of root systems; (ii) excavation of undisturbed cores or blocks of soil with or without insertion of pinboard matrices to hold the root system in the original position during subsequent washing-out, manual soil removal or soil removal by air pressure; and (iii) installation of in-growth cores, filled with root-free soil and subsequently colonised by the roots growing through the inserted mesh bags (Makkonen and Helmisaari 1999; Oliveira et al. 2000; Polomski and Kuhn 2002; Danjon and Reubens 2008). Each of these methods gives access to root biomass, root length distribution and root morphology description, partly also to root geometry and topology on one hand and to the soil either adhering to roots (rhizosphere) or not adhering to roots (bulk soil) on the other hand. Excavation at different time points during the culture period (for annual species) or the year (for perennial plants) can give the temporal variation of root growth and its consequences on mineral composition of the rhizosphere and soil solution. Recently, root excavation of whole trees has been used by Turpault et al. (2005) to study the influence of mature Douglas fir roots on the solid phase of the rizosphere and its solution chemistry. The same group also used soil sampling to follow the temporal variations of rhizosphere and bulk soil chemistry in another Douglas fir stand (Turpault et al. 2007). As major limitations, excavation-based techniques are generally destructive and time-consuming, the sampling process is associated with losses and damage particularly of fine root structures and is frequently restricted to analysis of the upper soil layers (Danjon and Reubens 2008).

Manual, semiautomatic and automatic techniques have been described to record the 3D orientation of plant roots, as well as data on root geometry and root morphology in excavated or uprooted root systems or in subsamples, such as undisturbed soil blocks or cores. These techniques are usually based on manual recording of the XYZ coordinates of plant roots, using a frame of movable rulers after step by step removal of the surrounding soil. Also 3D-scanning devices have been employed for this purpose. The described techniques have provided the data base for the development of root growth models, which then in turn can be employed to obtain entire information on root systems, using the data originating from partial samplings (Danjon and Reubens 2008).

Direct monitoring of root growth and short-term root growth dynamics is possible using the trench wall technique or root windows (Polomski and Kuhn 2002). The principle of these techniques is based on recording of root images in situ in a non-destructive way along soil profiles cut with steel blades beneath the growing plants of interest (Fig. 1), to create an observation plane (van Noordwijk et al. 2000; Smit et al. 2000b; Polomski and Kuhn 2002). Alternatively, transparent minirhizotron tubes containing video imaging facilities are inserted into undisturbed soil profiles after removing cylindrical soil cores with hand augers or hydraulic coring machines (Majdi 1996; Smit et al. 2000b; Maniero 2006; Thorup-Kristensen 2006). Analyses of the root images with appropriate software can produce quantitative data about root growth rate, root morphology and root turnover at the observation interface. Particularly the root window approach additionally offers the opportunity for measurements of spatial and temporal alterations of nutrient uptake, rhizosphere chemistry and rhizosphere microbial populations (Smit et al. 2000b). The main limitation of these techniques is that the window used for root observations is static and represents a limited, two-dimensional area which does not provide information on the total extension of root systems. Although non-destructive sampling along single roots is possible, the experimental setup does not necessarily represent a completely undisturbed system and soil mechanical impedance, temperature, moisture distribution, solute concentrations and redox conditions along the two-dimensional observation planes may differ from the conditions in undisturbed soils. As a general restriction, soil coring, as well as root windows and minirhizotrone-based methods are difficult to apply on stony soils (Majdi 1996; Thorup-Kristensen 2006).

Particularly for ecological studies, frequently working with mixed stands of different plant species, it is also important to identify the origin of root samples. This has been performed using anatomical and morphological identification keys (Cutler et al. 1987). More recently, molecular methods have been developed, using restriction fragment length polymorphisms (RFLP) of plastid DNA for species identification e.g. of tree root samples (Bobowski et al. 1999; Brunner et al. 2001).

Assessing root growth in the laboratory

Studies under laboratory conditions offer the opportunity to investigate root growth characteristics in controlled environments. Root growth and morphology can be easily monitored without background interference in hydroponics or in aeroponic culture systems, where nutrients are supplied to the root system as a mist of nutrient solution (Bucher 2006). However, root growth patterns and root morphology can be substantially altered in the absence of solid substrates. In pot experiments with soil culture, a more complete excavation of root systems is possible than in field experiments but root growth characteristics can be affected by the disturbed soil structure, differences in root zone temperature, soil water distribution, redox potential and limited rooting volume in pots.

Non-destructive measurements of root growth under laboratory conditions are carried out in rhizoboxes or in rhizotrone facilities (Polomski and Kuhn 2002). The basic rhizobox system is made from PVC boxes equipped with transparent root observation windows (Fig. 2). Depending on plant species and developmental stage studied, different sizes of rhizoboxes are available, containing soil amounts ranging from several hundred grams up to 80–100 kg (Neumann 2006a). Root development along the observation windows can be followed with time by taking images at regular intervals. However, direct digital analysis of photographic root images in soil culture is still frequently complicated by an insufficient ability of the digitalization software packages for background elimination. Therefore, most frequently root growth is still recorded by repeated drawings of the root system on transparent foils with subsequent digitalization. Image analysis

Fig. 1 Installation of root observation windows in a potato field. **a)** cutting a soil profile by insertion of a steel plate; **b)** replacement of the steel plate by a plexiglass observation window; **c)** fixation and isolation of the observation window; **d)** Root window in the field 3 weeks after installation with development of new roots along the observation plane

using appropriate software will generate quantitative data about dynamics of root growth and root architecture, according to the treatments applied. Morphological characteristics of single roots, such as diameter of fine roots, length and density of root hairs have been documented non-destructively by video image analysis (Haase et al. 2007a). Rhizobox studies are usually carried out with young plants or annual plant species. As an example, Boukcim et al. (2001, 2006) characterized root growth and architecture of young plants of *Cedrus atlantica* in rhizoboxes as a function of nitrogen nutrition and ectomycorrhizal inoculation. However, in some cases rhizobox studies have been carried out even with older trees (Dinkelaker et al. 1997; Aviani et al. 2006).

For long-term studies, rhizotrone facilities have been employed, representing subterranean root observation chambers, where the roots are growing along observation windows. Although not representing an undisturbed system, this approach combines controlled laboratory conditions with more field-orientated long-term observations (Polomski and Kuhn 2002).

Rhizobox systems can be modified by creating several compartments using nylon nets, that are unpermeable for roots but can be crossed by the fungal hyphae associated with the roots (Fig. 2) e.g. for studies on effects of mycorrhizal colonization on nutrient acquisition (e.g. Hawkins and George 1999). An interesting approach is the use of split-root systems to follow the effect of root symbioses on the development of roots, with inoculated and non-inoculated compartments containing roots of a common root system. This approach was employed for studying the effect of rhizobia inoculation (Hacin et al. 1997) but could be also used to follow the effect of other root symbioses, such as associations with

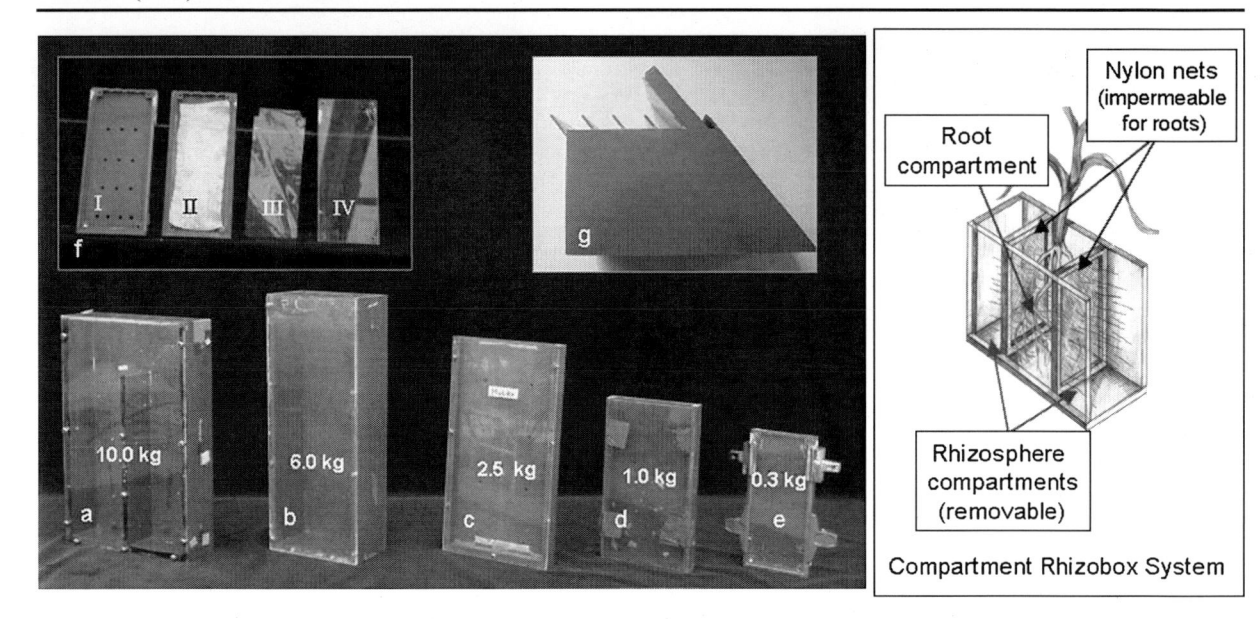

Fig. 2 Rhizobox systems: different sizes (**a–e**) and parts (**f**) of PVC-rhizoboxes consisting of a corpus with irrigation holes (f-I); a viscose fleece for moisture distribution (f-II); transparent plastic foil for soil-covering (f-III) and a Perspex front lid (f-IV) mounted with screws (**a–c**), adhesive tape (**d**) or clamps (**e**). Rhizobox holder for fixing five rhizoboxes during the culture period (**g**). Right: compartment rhizobox system with removable rhizosphere compartments separated by nylon nets from a central root compartment (adapted from Neumann 2006a)

arbuscular- or ectomycorrhizal fungi. Nevertheless, split root systems do not necessarily allow independent observations in the different root compartments, since manipulations in one compartment may influence the other compartment via systemic feed back loops. Systems with complete or partial partitioning of different soil compartments have been employed for intercropping studies both, under laboratory conditions or in the field to investigate synergistic or competitive interactions with respect to acquisition of nutrients (Li et al. 2003, 2004). A special, transparent rhizobox design was used to study root proliferation induced by nutrient-rich patches in soils (Hodge et al. 1999).

Although, rhizobox systems allow repeated and non-destructive measurements of root development and rhizosphere processes, it should be kept in mind that the system still suffers from the general disadvantages of pot experiments associated with the disturbed soil structure, altered root-zone temperatures and the limited rooting volume. Moreover, rhizosphere processes occurring in soils at a three dimensional level are reduced to the two dimensions of the observation windows.

A comparatively novel approach for non-destructive monitoring of root growth is the use of computer-aided tomography techniques, based on X-ray computed tomography, gamma-ray computer-assisted tomography or magnetic resonance (NMR) imaging. Sophisticated technical equipment and restrictions to homogenous soil densities, to non-swelling soils or to the absence of clay minerals and paramagnetic elements (Fe, Mn, Cu) are limitations for a more general use of these techniques in root growth analysis (Asseng et al. 2000). Also geophysical techniques, such as Ground Penetrating Radar (GPR) and 3D laser scanning methods have been occasionally employed for root measurements. However, particularly the limited resolution for overlapping objects make these techniques currently applicable only for small root systems with limited complexity (reviewed by Danjon and Reubens 2008).

Assessing vitality of roots, root biomass and necro-mass

For studies on root turn-over in soils, it is essential distinguish active, living roots from damaged and dead roots. Sequential soil coring or the use of ingrowth cores (see section: "Assessing root growth in the field") are destructive methods, allowing an assessment of root vitality in root samples washed out from the cores (Majdi 1996) by visual scoring (colour, consistency, mycorrhizal colonisation) (Helmisaari and Makkonen 2006) or by physiological indicators, such as polarographic respiration measurements or tetrazolium-based vital staining methods. (Comass and Eissenstat 2004; Richter et al. 2007). Colourless tetrazolium salts, such as triphenyltetrazolium chloride (TTC) are taken up into intact tissues and reduced by respiratory activity to coloured formazan derivatives. However, Richter et al. (2007) demonstrated that TTC reduction can occur also in roots containing high levels of reducing compounds, such as phenolics, independent of the vitality status. Damaged roots have been identified by intracellular callose formation in the root tips as physiological stress indicator e.g. for aluminium (Al) toxicity (Hirano et al. 2006), by measuring the Ca/Al molar ratio in fine roots, which declines under conditions of Al toxicity and soil acidity (Vanguelova et al. 2007) or by assessing the carbohydrate status of fine roots.

Non-destructive techniques are based on root observations using minirhizotrons (Majdi 1996; see section: "Assessing root growth in the field"). Although these techniques allow continuous observations of root development, vitality scoring is mainly restricted to visual evaluation and to monitoring of root growth. Recently, a novel approach combined multispectral imaging (visible and near-infrared spectroscopy) with the rhizobox technique, offering perspectives for non-destructive characterisation of root vitality and developmental stage and the differentiation from other components, such as leaf litter and bulk soil (Nakaji et al. 2008).

Fluxes of ions, organic compounds and water in the rhizosphere

Ion fluxes

Undoubtly, one of the driving forces determining changes in chemistry and of biological processes in the rhizosphere are the ion fluxes occurring along the root during the root activity. In hydroponic culture, ion fluxes are frequently estimated at the whole plant level by measuring concentration changes in the solution bathing the roots (Engels et al. 2000). However, measuring uptake particularly of cationic micronutrients, such as iron (Fe), zinc (Zn) and manganese (Mn) by depletion studies frequently leads to overestimations due to strong adsorption in the root apoplast (Strasser et al. 1999) and more reliable estimations require removal of adsorbed micronutrients by root washings with metal-chelators (Bienfait et al. 1985). Alternatively, short-term studies can be conducted with isotope tracers, measuring the appearance of the isotope in xylem sap and in the shoot tissue.

Estimations of the in situ root activity for uptake of a specific nutrient averaged over the whole system in pot or field experiments can be calculated by a combination of growth analysis and determination of nutrient concentrations in plant tissues incremented between sequential harvest dates (Engels et al. 2000). However, ion fluxes are not uniformly distributed along single roots, and root activity can vary greatly within different root zones and different types of roots. This is of particular importance when working with highly heterogenous root systems, such as roots of woody plants. Localisation of active root zones for mineral uptake and proton fluxes has been successfully performed along the roots of herbaceous and woody species using self-referencing, ion-selective microelectrodes. The main advantage of the microelectrode ion flux estimation (MIFE) methodology is the possibility to measure simultaneously several ion fluxes, such as H^+, K^+, NH_4^+ and NO_3^- using successively different experimental conditions, without disturbing the plant. One of the main limitations of this methodology is that it is not possible to use high external ion concentrations, particularly when the root uptake is low, measurements are mainly restricted to hydroponics or sand culture systems, while applications in soil culture are usually impossible. This method was first used in herbaceous species (Newman et al. 1987; Henriksen et al. 1992; Newman 2001) and later in woody species, such as Eucalypt (Garnett et al. 2001) and maritime pine (Plassard et al. 2002). The results obtained showed that uptake or release of ions varied greatly along the roots, with the subapical zone being most active. In addition, this methodology made it possible to quantify the impact of mycorrhizal symbiosis on nutrient uptake: in maritime pine, NO_3^- net fluxes into ectomycorrhizal roots formed with the Basidiomycetes *Hebeloma cylindrosposporum* and *Rhizopogon roseolus*, were decreased or increased, respectively compared with the fluxes measured into the non-mycorrhizal short roots (Plassard et al. 2002). Other experiments showed that *R. roseolus* significantly modified the kinetics of nitrate uptake into ectomycorrhizal roots, suppressing the induction step of nitrate uptake observed in non-mycorrhizal roots (Gobert and Plassard 2002). So far, the MIFE method is restricted to measurements of mineral ion fluxes. However, the development of enzyme-coupled electrochemical biosensors may offer the opportunity for the construction of microelectrodes with the ability to detect also fluxes of organic molecules, such as sucrose, glucose, or organic acid anions, released as exudates by plant roots and microorganisms (Marshall Porterfield 2002).

Measurements of spatial variations of proton or electron fluxes along roots of soil-grown plants in rhizoboxes or along root windows in the field are possible using antimony pH-electrodes (Häussling et al. 1985) or platinum redox-electrodes (Fischer and Schaller 1980; Fiedler and Fischer 1994) with a spatial resolution of approximately 1–3 mm. Recently, the fluorometric measurement of rhizosphere pH by so-called "imaging optodes" has been reported (Blossfeld and Gansert 2007). The technique is based on pH-sensitive indicator dyes, which are permanently fixed in a polymer proton permeable matrix, drawn out to a sensor foil of constant thickness (10 μM). The foil is applied to the root surface of plants grown in rhizoboxes equipped with a root observation window, which is subsequently closed with a transparent lid. Continuous fluorescence measurements of the undisturbed system are taken from outside using a mobile polymer optical fibre system to apply the light for fluorescence excitation and to monitor fluorescence emission from the detector foil. Similar techniques are available for measuring O_2 concentrations and imaging optodes with potential perspectives for rhizosphere applications have been developed also for the detection of minerals such as NH_4^+, various heavy metals or enzyme activities (Stromberg 2006, 2008).

Other approaches to measure spatial variations of nutrient uptake in plant roots comprise: (i) the injection of tracers into the soil in various depths and distances to the plant with subsequent measurements of tracer depletion in the soil or its accumulation in the plant; (ii) application of tracers in agar blocks onto the surface of distinct root zones of plants grown in nutrient solution, in rhizobox soil culture (Fig. 2) or in the field with root observation windows (Fig. 1) and subsequent measurement of tracer depletion in the agar blocks or the accumulation in the plant tissue; (iii) use of root containers with compartments filled with tracer solutions, separating different zones along the root and sealed with silicone grease. Nutrient uptake is calculated from tracer depletion in the root bathing solution or accumulation in the plant tissue. (iv) Measuring uptake of tracers form root bathing solutions into isolated root segments (Engels et al. 2000). Problems to be considered with the use of these techniques arise from unspecific losses of tracers by leaching, adsorption in the substrates and in the root apoplast, chemical precipitation or biological processes (sequestration, mineralisation). A risk of root injury with impact on nutrient uptake characteristics is associated with the techniques involving physical manipulations of isolated roots such as compartmented root containers and isolated root segments (Engels et al. 2000).

Methods, application range and limitations for studying ion fluxes in the rhizosphere are summarized in Table 1.

Rhizodeposition and root exudates

A substantial proportion of carbon fixed during photosynthesis by higher plants (20–60%) can be translocated belowground (Grayston et al. 1996; Kuzyakov and Domanski 2000). Up to 70% of this carbon fraction in perennial and up to 40% in annual plants, enter the soil as organic rhizodeposition corresponding to 800–4,500 kg carbon $ha^{-1}y^{-1}$ (Kuzyakov and Domanski 2000; Lynch and Whipps 1990). This is associated with a concomitant input between 4 and 70% of the assimilated nitrogen. (Wichern et al. 2008). Shoot isotope labelling techniques ($^{14}CO_2$, $^{13}CO_2$, ^{15}N) are frequently employed to quantify the fluxes of organic compounds out of the roots (Cheng and Gershenson 2007), referred as rhizodeposition. While the term "rhizodeposition" comprises all organic compounds released both, from healthy roots and also from damaged and senescent root structures during root turn-over, root exudates are defined as organic compounds released from undamaged roots (Neumann and Römheld 2007).

Pulse-chase studies with short isotope labelling pulses, ranging between <1 h up to several days, are usually employed to trace the short-term fate and transfer speed of photosynthetically fixed carbon into the rhizosphere or into the associated microflora. However, the limited observation time prevents the use of the data for carbon budgets, which is possible only by long-term labelling, frequently conducted

Table 1 Overview on methods for determination of nutrient uptake and efflux by plant roots

Method	Measurements	Application range	Limitations
Methods considering the whole root system			
Net uptake according to Williams (1948)	Sequential analysis of biomass production and tissue concentrations of mineral nutrients	Hydroponics, greenhouse studies, field experiments	Only net uptake considered, destructive method. Difficulties in determination of root biomass and nutrient concentrations in soil grown plants No differentiation between effects of soil nutrient availability and root activity.
Analysis of the growth medium	Measuring changes of nutrient concentrations in the root incubation medium	Hydroponic culture	Not applicable for plants in soil culture, potential side effects by adsorption, sequestration or precipitation of nutrients in the incubation medium
Localized sampling			
Compartment containers	Localized tracer application into root compartments. Measuring tracer depletion in the container or appearance in the plant	Hydroponic culture	Not applicable for plants in soil culture, potential side effects by adsorption, sequestration or precipitation of nutrients in the incubation medium, risk of root damage during installation.
Tracer injection	Tracer injection into distinct soil zones. Measuring tracer depletion in the substrate or appearance in the plant.	Soil culture, field experiments	Potential side effects by adsorption, sequestration or precipitation of nutrients in the incubation medium, risk of root damage during installation.
Agar block technique	Tracer application to agar blocks applied onto the root surface. Measuring tracer depletion in the agar blocks or appearance in the plant.	Hydroponics Soil culture (Rhizoboxes, root windows in the field)	Potential side effects by adsorption, sequestration or precipitation of nutrients in the agar blocks, risk of root damage during installation.
Root segments	Measuring short-term uptake of nutrients into isolated root segments	Hydroponics, soil-grown plants	Artificial system. Risk of artefacts due to root injury
Microelectrodes	Measuring ion activities at the root surface with ion-selective micro-electrodes	Hydroponics (soil culture: rhizoboxes, root windows for pH)	For mineral nutrients not applicable in soil culture and in presence of high external nutrient concentrations. risk of root damage during installation.

For reviews see: Engels et al 2000; Marshall Porterfield 2002)

over time periods of several weeks or months. A major disadvantage of long-term labelling is the high requirement of expensive isotopes and the limited suitability for field studies. Alternatively, differences in the natural abundance of isotopes between root-derived and soil-derived materials can be employed for determination of rhizosphere carbon budgets. However, using this approach, only large differences can be detected (Cheng and Gershenson 2007).

Most isotope studies have been conducted to quantify the release of organic compounds over the whole root system. Similar approaches with non-labelled plants used root washings in hydroponic culture, percolation of solid substrates or extraction of the soil adhering to plant roots (for review, see Neumann and Römheld 2007). However, in many cases root exudation is not homogenously distributed along the roots. Particularly adaptive responses of root exudation involved in nutrient mobilization, attraction of microbial symbionts or detoxification of toxic elements are frequently restricted to special root structures, such as sub-apical root zones, the root-hair zone or cluster roots. The localized and concentrated release enables an accumulation of root exudates in the rhizosphere above the threshold levels required for the specific functions (Neumann and Römheld 2007). Therefore, an understanding of rhizosphere processes frequently requires methodological approaches to detect spatial variations in rhizosphere chemistry along single roots.

In hydroponics, compartment containers filled with root washing solution or even isolated root segments (Hoffland et al. 1989; Keerthisinghe et al. 1998; Kania et al. 2003; Delhaize and Ryan 2006) have been used to collect root exudates from distinct root zones. In soil culture, localized sampling techniques (Fig. 3) comprise the use of sorption media such as glass-fiber,- or paper filters, membrane filters or foils and small gaze bags with ion exchange resins, placed onto the surface of the respective root zones of plants grown in rhizoboxes (Kamh et al. 1999; Neumann 2006b; Haase et al. 2007a, b). Also the insertion of micro-suction cups close to the roots (Fig. 3) has been reported (Dessureault-Rompré et al. 2006; Shen and Hoffland 2007). The major problems of these techniques arise from limited and variable recovery of exudate compounds due to rapid microbial degradation in the soil solution, selective and rapid adsorption of certain compounds (e.g. carboxylates, phenolics, proteins) at the soil matrix and at the root surface (Jones 1998; Neumann and Römheld 2007). Sterile culture is frequently not possible for soil-grown plants with longer culture periods and the presence of microorganisms can alter the patterns and the intensity of root exudation. Therefore, short collection periods of 1–2 h (Neumann 2006b) or the use of sorption media, selectively binding certain exudate compounds from the soil solution (ion exchange resins, charcoal, membrane filters) are frequently employed to minimize microbial degradation and unspecific adsorption (Kamh et al. 1999; Haase et al. 2007b). Also the soil structure and water content are determinants of exudate recovery and it is difficult to differentiate exudates from other rhizosphere products in the soil solution. At least some studies used isotope labelling to identify also the localized release of exudates in different root zones (Johnson et al. 1996; Schilling et al. 1998; Haase et al. 2007b). The use of specific transgenic bacteria with reporter genes indicating the presence of specific compounds e.g. by colour reactions or light emission may offer new tools to study the localized release of selected exudate compounds at a micro-scale of spatial resolution (Jaeger et al. 1999; Darwent et al. 2003).

Most of the investigations have been conducted as model experiments under laboratory or greenhouse

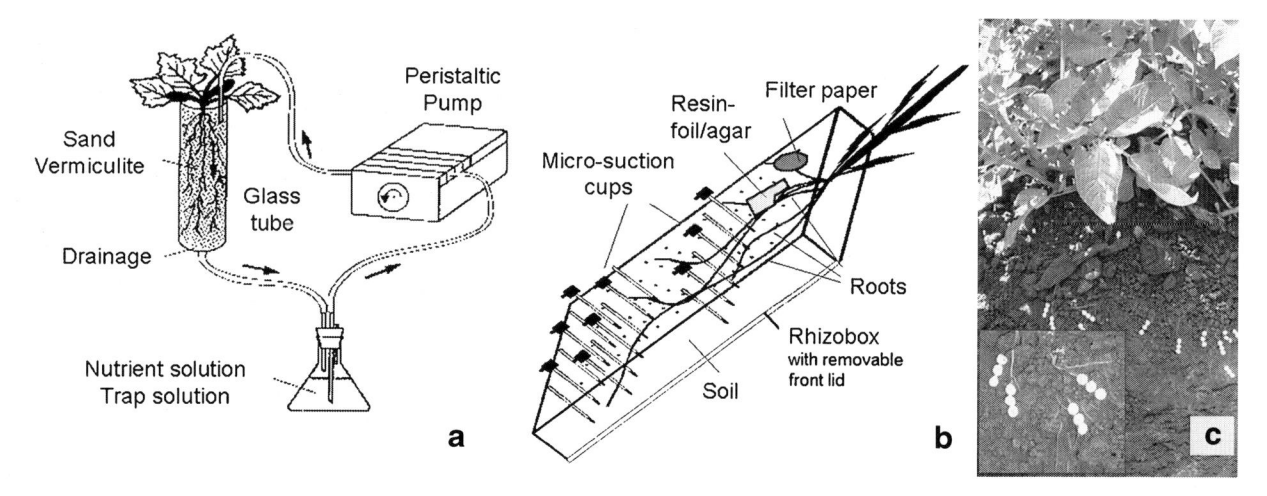

Fig. 3 Techniques for collection of root exudates and rhizosphere soil solution from plants grown in solid substrates. **a** Percolation of the culture vessel with trap solution. **b** Rhizobox system with removable front lid. Insertion of micro-suction cups close to the roots or application of sorption media onto the root surface. **c** Root window in the field. Application of sorption media onto the root surface (adapted from Neumann and Römheld 2007)

conditions and field studies on spatial variation of root exudation along single plant roots are rare. Techniques, application ranges and limitations for collection of root exudates and rhizodeposits are summarized in Table 2.

Exudate sampling usually results in small sample volumes (localized sampling techniques) and/or highly diluted exudate solutions (root washing techniques), requiring sample pre-concentration and sensitive analytical facilities. HPLC, HPLC-MS, capillary electrophoresis, GC-MS but also enzymatic methods are the major analytical approaches employed for exudate analysis (Neumann and Römheld 2007). For profiling the wide array of different compounds which can be present in root exudates, silyl-derivatization coupled with GC-MS, ^{13}C-NMR and pyrolysis field ionisation mass spectrometry (PyFIMS) have been employed (Van Veen et al. 2007). However, due to the highly advanced technological requirements, long analysis times (PyFIMS) and derivatisation steps requiring non-aqueous samples and yielding multiple peaks for some compounds, these techniques are not widely used for routine analysis.

Table 2 Overview on methods for collection of organic compounds released from plant roots

Method	Measurements	Application range	Limitations
Methods considering the whole root system			
Root washings in trap solutions	Incubation of root systems in trap solutions for limited time periods (best 2–6 h)	Hydroponics,	Should not be employed for excavated roots in soil culture, diluted sampling solution, risk of precipitation during concentration, risk of root damage during handling.
Percolation techniques	Percolation of trap solutions through the culture vessels of intact plants (after pre-washing steps to remove rhizosphere products accumulated prior to collection period.	Hydroponics with solid substrates (soil culture)	Potential side effects by adsorption or precipitation of rhizodeposits in the substrate, diluted sampling solution.
Rhizosphere extraction	Collection and extraction of soil adhering to plant roots or separated by compartment rhizobox systems	Soil culture	Risk of contamination with organic compounds from damaged roots and microorganisms. Potential side effects by adsorption of rhizodeposits in the substrate.
Localized sampling			
Compartment containers	Localized application of trap solution into root compartments.	Hydroponics	Not applicable for soil-grown plants, risk of root damage during installation, diluted sampling solution
Root segments	Short term release of rhizodeposits from isolated root segments into trap solutions	Hydroponics Soil culture	Artificial system. Risk of artefacts due to root injury
Micro-suction cups	Collection of rhizosphere soil solution with micro-capillaries equipped with ceramic filter tips	Soil culture (rhizoboxes, root windows in the field)	Potential side effects by adsorption of rhizodeposits in the substrate, small sample volumes, recovery dependent on soil moisture levels
Sorption media	Application of sorption media (membrane filters, paper discs, ion exchange membranes, agarose gels) onto the root surface.	Hydroponics Soil culture (rhizoboxes, root windows in the field)	Potential side effects by competitive adsorption of rhizodeposits in the substrate, small sample volumes, recovery dependent on soil moisture levels, soil structure, collection time.
Reporter genes	Inoculation with bacteria with expression of reporter genes detecting the presence specific compounds.	Hydroponics Soil culture (rhizoboxes, root windows in the field).	Restricted to the detection of specific compounds. Potential side effects by competitive adsorption of rhizodeposits in the substrate.

For reviews see: Cheng and Gershenson 2007; Neumann and Römheld 2007; Luster and Finlay 2006

Water flux

Time domain reflectometry (TDR), tensiometry and thermocouple psychrometry are standard techniques to assay the influence of plant growth on soil moisture at scales of 0.01–1 m (Fernandez et al. 2000; Cardon and Gage 2006). At smaller scales, relevant to study moisture conditions in the rhizosphere (0.1–10 mm), micro-tensiometry and miniaturized TDR sensors have been employed (Vetterlein et al 1993; Göttlein et al. 1996; Vetterlein and Jahn 2004). For micro-scale measurements (0.001–1 mm), recently genetically engineered microbes have been developed that report soil moisture by increased expression of green fluorescent protein (GFP) as the soil dries, using fluorescence detection by epifluorescence microscopy (Cardon and Herron 2005; Cardon and Gage 2006).

Also various non-invasive techniques for imaging of root systems, such as magnetic resonance imaging (NMRI) (Macfall et al. 1990), X-ray computed tomography (CT) (Hainsworth and Aylmore 1989), ground penetrating radar (GPR) and electrical resistivity tomography (ERT) have the potential to report soil moisture conditions along plant root systems in situ. Garrigues et al. (2006) reported a method to demonstrate water uptake along plant roots based on 2D light transmission imaging with plants cultivated in transparent rhizoboxes, filled with a 4 mm layer of a translucent sand-clay mixture. However, the spatial resolution of these techniques (approx. 0.01 m–1 m) is not yet sufficient for direct rhizosphere measurements (Fernandez et al. 2000; Luster et al. 2009), frequently requires expensive instrumentation and is restricted to certain growth substrates to avoid interferences.

Water uptake into roots and their hydraulic conductivity has been assessed by use of pressure probes fixed to decapitated plants for detecting root pressure as a driving force for water flow in the xylem. Changes in pressure over time can be converted into water flow rates and calculations of root water conductivity are possible (Fernandez et al. 2000; Knipfer and Steudle 2008).The root pressure probe allows measurements with whole root systems over time periods of up to 10 days (Steudle1994), while cell pressure probes are employed to measure the hydraulic conductivity of single roots and isolated root segments for up to 1–2 days (Fernandez et al 2000). In xylem pressure probes, micro-probes are used for sampling of single xylem vessels (Zimmermann et al. 1994). For calculations of sap flow rates also various techniques have been employed, based on tracing temperature changes after application of heat pulses to stems and roots in some distance to the location of a heat sensor, with a wide range of applications particularly in woody plants (Smith and Allen 1996; Fernandez et al. 2000).

Probing the rhizosphere

Apart from analysing ion fluxes or release of organic compounds by plant roots, understanding of rhizosphere processes implies also the estimation of the actual impact of these root effects on the soil conditions and vice-versa.

Compartment systems

To address this question, appropriate devices have been designed (see Fig. 2), most of them being derived from the setup of Kuchenbuch and Jung (1982). Plants are generally grown in a two-step procedure. The first step is a pre-culture period, where the plants are grown from seeds on the surface of a polyamide net impermeable for roots or root hairs, in hydroponics or soil culture until a dense, planar mat of roots is obtained on the surface of the net. The second step is a test culture period, where the plants are transferred on top of a soil (or any other solid substrate to be tested, such as pure minerals). Depending on its thickness, the soil can be either considered as whole rhizosphere (ie when the layer is 1 to 3 mm thick, e.g. Chaignon and Hinsinger 2003) or used after cutting it with a razor blade or a freeze microtome parallel to the root mat, in order to measure rhizosphere gradients (e.g. Hinsinger and Gilkes 1996). The effect of plant mineral nutrition can be easily assessed by varying the composition of the nutrient solution supplied to the soil/substrate layer and the root mat compartment. Measuring nutrient depletion in the rhizosphere layer, it is also possible to assess the availability of nutrients in different soils and the potential of plants for acquisition of these nutrients. For nutrient depletion studies with sparingly soluble nutrients, such as P, Fe and micronutrients it is important to consider the buffering capacity of the soil matrix for these minerals, e.g. by employing sequential soil extraction protocols with solvents of increasing extraction potential. Of course, plant effects on rhizosphere microbial populations could also be analysed. For practical reasons, the methodology is mainly limited to the study of fast growing species, rapidly forming mats of active roots. Moreover, it should be kept in mind, that the high density of roots in the root mat can lead to unrealisticly high levels of root exudate accumulation and associated chemical changes in the adjacent rhizosphere compartment and thus to an overestimation of rhizosphere effects. Care has to be taken also for the selection of membrane materials separating the rooting compartment from rhizosphere compartments since some materials can exhibit selective adsorption of certain exudate compounds, e.g. adsorption of phenolics and proteins by nylon membranes (Neumann 2006b).

Visualisation techniques

Methods enabling the localisation of active root zones are of great interest to measure the effect of plant roots on the physicochemical conditions in the rhizosphere with respect to pH and redox conditions or nutrient availability. This can be achieved by applying gels (agar, agarose, polyacrylamide) containing indicator reagents onto the surface of root systems previously grown in nutrient solution or in rhizoboxes. This includes dyes that reveal Al complexation (Dinkelaker et al. 1993) or Fe^{3+} and Mn reduction (Dinkelaker et al. 1993; Engels et al. 2000). Quantitative data could be obtained from digital image analysis taken at regular time intervals, for monitoring proton fluxes using videodensitometry of dye indicator in agar gels (Ruiz and Arvieu 1990; Jaillard et al. 1996; Plassard et al. 1999). Examples for visualisation techniques are summarized in Fig. 4.

Various techniques, such as scanning and transmission electron microscopy, X-ray tomography, microparticle-induced X-ray emission (μPIXE) or nano-scale secondary ion mass spectrometry (Nano-SIMS) offer perspectives to study the microscale element distribution in the rhizosphere and also in plant roots, fungal hyphae or even in single bacteria (Strasser et al. 1999; Frey and Turnau 2006; Hermann et al. 2007; for review see also Luster et al. 2009). Sample preparation is frequently a critical step for this type of microscopical approaches and requires rapid dehydration and cryo-conservation of the sample to avoid artefacts. These emerging techniques can provide a novel level of resolution for rhizosphere studies. However, the restricted availability of the required technological equipment still limits routine applications and measurements are frequently only possible in cooperation projects.

On a macro-scale, various remote sensing approaches have been developed for probing nutrient availability in the rhizosphere under field conditions as a key technology for crop nutrient management in precision agriculture. Major plant-based sensor approaches are currently being employed and tested for nitrogen (N) management and comprise (i) hand-held chlorophyll meters, (ii) ground-based spectral radiometers including mobile tractor-mounted sensor systems, and (iii) aerial imaging (Schmidhalter et al. 2003; Kitchen 2007). Variations in colour and size of the plant canopy are easily detectable by remote sensing technologies. Since the leaf chlorophyll content is closely related with the plant N status, chlorosis scoring by remote sensing has been claimed as suitable information for the detection of N deficiencies in the field, to be used for an adapted nitrogen fertilization management, considering spatial heterogeneities of N availability (Baret and Fourty 1997; Blackmer and Schepers 1996). However, variations in the spectral reflectance of the canopy can be affected by more than one factor, such as plant disease, P, S, micronutrient or water availability, bearing the risk of an erroneous N fertilisation (Lilienthal and Schnug 2007). The introduction of improved remote sensor technologies including high resolution and hyperspectral imaging of several 100 wavelengths per image pixel (Lilienthal and Schnug 2007), fluorescence imaging (Liew et al. 2008) or the combination with reporter gene strategies may offer perspectives for the development of indicator parameters for a more specific and even earlier detection of plant nutritional disorders on a field scale (Liew et al. 2008), which is currently possible only by destructive analytical tests.

Reporter gene strategies

A novel approach is the combination of rhizobox studies with plants able to probe the concentrations of minerals. Such an approach could be used with plant species easily transformable such as *Arabidopsis, Medicago truncatula, Nicatiana tabacum* or *Oriza sativa*. As an example, promoters of different phosphate transporter genes from barley were fused to β-glucuronidase and green fluorescent protein (GFP)

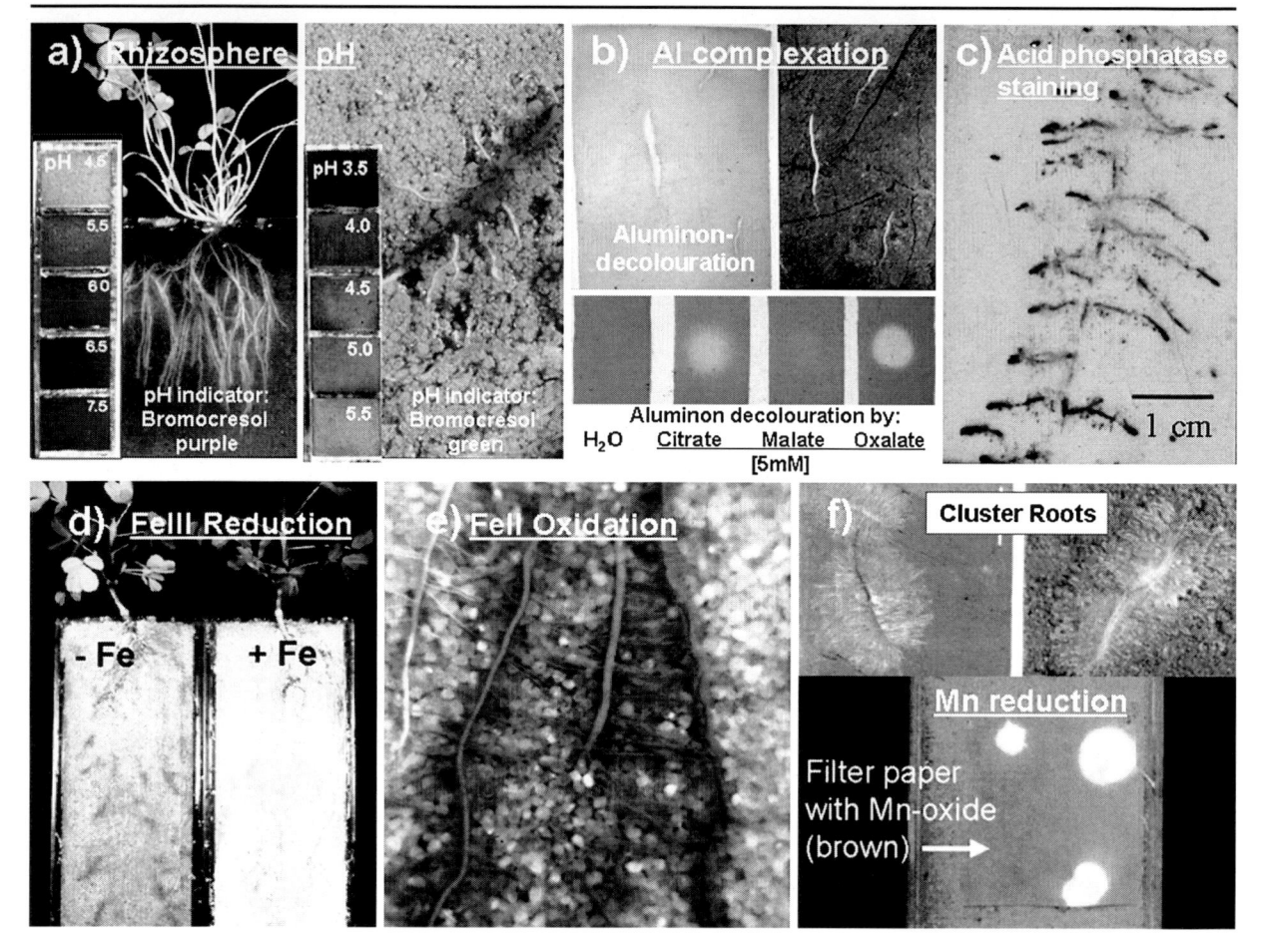

Fig. 4 Visualisation of root-induced chemical changes in the rhizosphere by use of indicator media applied to the root surface of plants grown in rhizobox systems. **a** Root-induced alterations of rhizosphere pH by infiltration with agar containing pH indicators **b** Aluminium complexation: application agar gels containing red Al-aluminon complexes. Competetive Al complexation with root exudate compounds leads to decolouration. **c** Acid phosphatase activity in the rhizosphere detected by application of filter papers soaked with naphtylphosphate-fast red TR as artificial substrate. **d** Fe III reduction by application of indicator gels containing Fe III and the Fe II chelator BPDS (bathophenantroline-disulfonic acid). Root induced Fe III reduction leads to formation of a red FeII-BPDS complex. **e** FeII oxidation by formation of brown FeII-oxide; **f** Manganese reduction indicated by decolouration of filter paper impregnated with brown Mn-oxide.(adapted from Neumann 2006b)

reporter genes were used to transform rice plants (Schünmann et al. 2004). After regeneration, the plants expressed a higher level of the reporter gene when grown in P-starved conditions. Such a system could be used to reveal complex rhizosphere conditions, especially for P availability which is difficult to measure directly. Currently, the approach is mainly conducted with easy-transformable model plants but it will be applicable also for other plant species as soon as transformation protocols are available.

Alternatively, plants could be inoculated with rhizobacteria tagged with the *lux*-luminescence gene for light emission to report the actual conditions in the rhizosphere. The bioluminescence produced by short-term carbon-starved *lux*-marked *Pseudomonas fluorescens* depends on the source and concentration of carbon (Yeomans et al. 1999). Therefore these bacteria have the potential to both quantify and specify the carbohydrates released by the roots (Killham and Yeomans 2001). In theory, many other genes could be tagged with *lux*-reporting systems (Killham and Yeomans 2001; Cardon and Gage 2006). As an example, luminescence from *lux*-tagged *P. fluorescens* indicating P starvation was found in the rhizosphere of barley roots (Kragelund et al. 1997). *Lux*-tagged *Nitrosomonas europaea* has been employed to detect and to quantify natural nitrification inhibitors released from plant roots in root exudates (Iizumi et al. 1998; Subbarao et al. 2006; Hossain et al. 2008).As it is easier to manipulate bacteria than plants, these systems may offer a broad perspective to probe the specific conditions occurring in the rhizosphere, as a result of numerous exchanges between the roots and their environment. Limitations and challenges for application of these approaches arise from observations that the expression of the reporter genes is sometimes not strictly limited to the experimentally defined target inducers but can be additionally influenced by other conditions or compounds, limiting the specificity of these systems (Cardon and Gage 2006).

Experimental strategies to elucidate rhizosphere mechanisms by manipulating rhizosphere traits

It is evident that plants have a wide range of mechanisms and physiological traits that manipulate the rhizosphere. Genotypic variation within plants, either natural, bred or heterologously expressed, in traits could be used to elucidate rhizosphere mechanisms more clearly. Manipulating the rhizosphere for experimental purposes using plants will however, require; (i) identification of the plant trait or gene that affects acquisition; (ii) identification of sufficient genetic variation in these traits within crop germplasm or heterologous expression of the genes controlling the trait. If the findings are to be applied to agricultural systems it will also require (iii) introgression of the trait into elite germplasm (see Wissuwa et al. 2009; Vance et al. 2003). Numerous techniques are now available for cloning rhizosphere specific genes and approaches for validating and deploying these genes in elite germplasm are highlighted in Fig. 5.

With the plethora of techniques at our disposal to identify the variation between plant genotypes and to create isogenic (transgenics, mutants, RNAi, smiRNA) or near isogenic lines (RIL's, mapping populations) of plants to manipulate the rhizosphere, many studies have been performed. However there have been few successful attempts to take this line of research to its ultimate conclusion with few studies leading to an increase in the efficiency of rhizosphere processes through genes whose expression manipulates the rhizosphere. The comparison of the impact of isogenic

Fig. 5 A functional genomics approach to study abiotic stress and rhizosphere processes, and deployment of these traits into crop gerplasm. 2 DGE, two-dimensional gel electrophoresis; EST, expression sequence tag; GPR, ground penetrating radar; MALDI-TOF, matrix-assisted laser desorption/ionization-time of flight; MPSS, massively parallel signature sequencing; QTL, quantitative trait loci; RNAi, RNA interference; SAGE, serial analysis of gene expression; smiRNA, synthetic micro RNA. Adapted from Vij and Tyagi (2007)

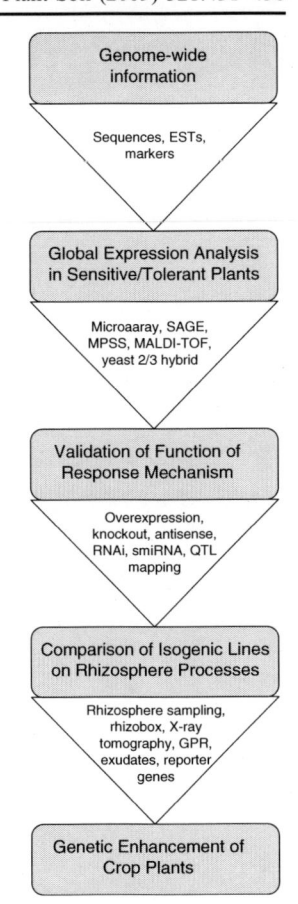

and near isogenic lines of plants on their rhizosphere has, however, given us an appreciation for the complexity of interactions between such traits and their variable efficacy in a range of typical agro-environments (Wissuwa 2003; George et al. 2005c). Irrespective of these complications, a number of promising approaches for beneficially manipulating the rhizosphere of plants have been identified, including the identification of key morphological and physiological traits associated with nutrient and water uptake efficiency. A range of studies which have successfully impacted the physical, chemical, biochemical and biological environment of the rhizosphere are highlighted in Table 3.

The availability of the genome sequences of certain important plant species including *Arabidopsis thaliana*, *Medicago truncatula*, *Lotus japonicus*, and *Populus* and a range of crops (rice, potato, barley, tomato, and maize) has enabled the use of strategies, such as genome wide expression profiling to identify genes controlling mechanisms associated with the rhizosphere (Vij and Tyagi 2007). This,

Table 3 Approaches aimed at manipulating the rhizosphere through plant genotypes (adapted from Richards et al. 2007)

Target limitation	Species	Rhizosphere environment	Mechanism	Approach	Reference
Acid soils	Wheat	Chemical	Release of organic acids to chelate Al^{3+}	Recombinant inbred line screening Transgenics	Ryan et al. 1995 Delhaize et al. 2004
P-deficit	Barley	Physical	Longer root hairs	Mutant screening	Gahoonia et al. 1999
Mn-deficiency	Wheat	Biological	Promote Mn-reducing bacteria	Comparison of cultivars	Rengel et al. 1998
P-deficit	Subterranean clover	Biochemical	Release phytase to mineralise inositol phosphates	Transgenics	George et al. 2004
P-deficit	Potato	Biochemical	Release phytase to mineralise inositol phosphates	Transgenics	Zimmermann et al. 2003
Fe-deficit	Rice	Chemical	Release phytosiderophores to chelate Fe	Transgenics	Takahashi et al. 2001
N-deficit	Rice	Biological	Promotion of NH_4^+ oxidising bacteria	Comparison of cultivars	Briones et al. 2002
P-deficit	Rice	Physical	Not known-possibly root characteristics	Recombinant inbred line screening	Wissuwa and Ae 2001a
N-deficit	Non-legumes	Biological	Nodulation without nitrogen fixing bacteria	Transgenics	Gleason et al. 2006; Tirichine et al. 2006
Salinity	Wheat (Durum)	Chemical	Enhanced K^+/Na^+ discrimination	Comparison of species	Munns et al. 2000
Dryland salinity	Wheat	Chemical	Boron tolerance	Doubled haploid population screening	Jefferies et al. 2000

followed by verification of gene function by the analysis of mutants and transgenics may lead to crop varieties which can manage the rhizosphere predictably and directly. A range of functional genomic studies on abiotic stress have been performed, but only a small proportion of them have identified genes that impact directly on the rhizosphere (Table 4). Even fewer of these studies have identified genes with a known function that have been successfully deployed to enhance our understanding of the

Table 4 Examples of functional genomic studies performed on crop species where genes identified may have an impact on the rhizosphere (adapted from Vij and Tyagi 2007).

Species	Stress	Reference
Barley (Hordeum vulgare)	Drought and salinity	Ozturk et al. 2002
	Fe-deficit	Negishi et al. 2002
Rice (Oryza sativa)	Salt stress	Kawasaki et al. 2001
	Drought and salinity	Rabbani et al. 2003
	Salt stress	Chao et al. 2005
	Drought	Hazen et al. 2005
	Drought	Lan et al. 2005
	Drought	Markandeya et al. 2005
	Salt stress	Walia et al. 2005
	N-deficit	Lian et al. 2006
Potato (Solanum tuberosum)	Salt stress	Rensink et al. 2005
Sorghum (Sorghum bicolor)	Drought and salt stress	Buchanan et al. 2005
	Drought	Pratt et al. 2005
Wheat (Triticum aestivum)	Salt stress	Kawaura et al. 2006
Maize (Zea mays)	Water stress	Yu and Setter 2003
White lupin (Lupinus albus)	P-deficit	Uhde-Stone et al. 2003

rhizosphere. One example for the successful application of this approach is the transgenic overexpression of a transcription factor identified in a genomic study of barley, which has resulted in an increase of our understanding of plant tolerance to drought, salinity and low temperature stress in rice (Oh et al. 2007).

Use of plant populations to understand the genetic control of rhizosphere processes

A common finding in many screens for processes which take place in the rhizosphere is that traditional varieties or landraces have traits and efficiencies that do not exist in elite germplasm (Manske et al. 2000; Wissuwa and Ae 2001a). This is often attributed to the impact of breeding programmes being performed under agronomically favourable conditions such as nutrient and water sufficiency (Rengel and Marschner 2005; Buso and Bliss 1988). The use of landraces has been successful in targeting rhizosphere processes to improve the P-deficiency tolerance of upland rice on volcanic soils in the Phillipinnes (Wissuwa and Ae 2001a). Near isogenic lines (NIL's) to a P-inefficient rice variety (Nipponbare), made in a cross with a P-efficient landrace (Kasalath), took up more P than the inefficient parental line under P-deficient conditions. Similarly, there are key differences between species which can be a source of genetic information. For example, in comparisons made between hexaploid bread wheat (*Triticum aestivum*) and tetraploid durum wheat (*Triticum durum*), key differences in tolerance to excess aluminium and salinity (excess Na) in the rhizosphere were evident. These tolerances were further associated with the telomeric region of one of the chromosomes present in bread wheat and absent in durum wheat (Munns et al. 2000). Such associations and their impacts on the rhizosphere can be elucidated by the comparison of populations of chromosome substitution lines.

Another possibility to elucidate the genetic control of specific rhizosphere traits by genotype screening is to use association mapping populations which allow the screening of a range of crop cultivars for specific traits and association of these with chromosome maps annotated with several thousand SNP's (single nucleotide polymorphisms). This allows the identification of QTLs and specific markers for genes. As these populations are generally found in elite gerplasm, this approach has the advantage of easy deployment of traits into relevant high yielding cultivars, although if the trait of interest has been "bred out" of the elite cultivars, then the amount of relevant variation available may be inadequate. Use of mapping populations has elucidated potential QTLs for unknown N- and P-use efficiency mechanisms in wheat (Liao et al 2004, 2008) and these mechanisms are likely to impact the rhizosphere. Comparison of lines with and without the relevant QTLs or markers elucidated by these approaches will allow a greater understanding of the way in which plants manipulate the rhizosphere at a genetic scale.

Other populations that can be used for screening for the impact of plant traits on rhizosphere processes include mutant populations. Such populations exist not only in *Arabidopsis* but also in crop plants including wheat, barley and rice. The genomes of these populations are saturated with mutations throughout the cDNA, with mutations either "knocking-out" genes or up- or down-regulating genes downstream of the mutation. Such an approach produces many thousand individual mutants and subsequent phenotypic screens to identify a specific trait of interest can allow very rapid elucidation of the gene controlling the expression of that trait. It is possible to use such an approach to identify traits which are useful for manipulating the rhizosphere. In contrast to the forward genetic approach, a reverse genetic approach can be used in which a mutation in a gene with a putative role in the rhizosphere can be compared with a wild-type control with respect to the rhizosphere process of interest. Examples comprise, the use of mutants to estimate the contribution of ammonium or urea transporters to nitrogen uptake (Yuan et al. 2007; Kojima et al. 2006), while the contribution of root hairs to nutrient acquisition was investigated by use of a barley mutant without root hairs (Gahoonia et al. 2004).

Genes identified using these various population screening approaches can then be validated for their rhizosphere specific mechanisms by overexpression through transgenic approaches, coupling the gene with specific promotors, or by monitoring their loss of function after down-regulating the gene of interest by RNAi (RNA-interference) or smiRNA (synthetic micro RNA) technologies (Miki and Shimamoto 2004; Alvarez et al. 2006).

Genotypic variation in mechanisms apparent in the rhizosphere

A number of studies have investigated genotypic variation in mechanisms which take place, completely or partially, in the rhizosphere of cereals, particularly maize, wheat and rice (Fageria et al. 1988; Fageria and Baligar 1997a, b; Ciarelli et al. 1998; Manske et al. 2000; Wissuwa and Ae 2001a, b; Osborne and Rengel 2002; Wang et al. 2005; Zhu et al. 2005; Liao et al. 2004; 2008). Most of these studies have been targeted to nutrient efficiencies, most notably that of N, P and K with most progress being made for P (Rengel and Marschner 2005). Frequently the main focus was nutrient-use efficiency (e.g. yield per amount of element taken up or yield per fertiliser applied) and may therefore not address the mechanisms apparent in the rhizosphere directly. Other studies have, however, targeted below-ground traits including root architecture in common bean and maize (Lynch and Brown 2001; Zhu et al. 2005), root hair formation in barley (Gahoonia and Nielsen 1996; Gahoonia et al. 1999); rhizosphere acidification in common bean (Yan et al. 2004) and extracellular enzyme activity in wheat (George et al. 2008). Almost exclusively, all such studies have found significant genotypic variation between cultivars of the crop species tested. Suggesting that meaningful variation in germplasm will allow the use of plant genotypes to experimentally challenge the rhizosphere.

Although some traits evident in the rhizosphere are thought to be highly heritable and simply controlled by one or two genes (e.g. Al toxicity tolerance in wheat (Delhaize et al. 1993) or heavy metal tolerance in *Thlaspi caerulescens* (Eapen and D'Souza 2005)), it is important to consider that with a multi-mechanistic tolerance, such as that to P-deficiency, markers or genes identified may only be applicable to specific soil conditions. For example, George et al. (2008) demonstrated that a phenotype of enhanced activity of root-associated phosphatase was effective in promoting plant growth in-vitro and in some P-deficient soils but was ineffective in other P-limited soils. This complexity due to gene-environment interactions is highlighted when attempts are made to assign nutrient efficiency traits to a single QTL, and many studies conclude that the trait is complex and not attributed to a single mechanism or QTL in isolation (Wissuwa and Ae 2001b; Liao et al. 2008).

Systems where single genes have been used experimentally to influence rhizosphere traits

The approach outlined in Fig. 5 has been used with varying success to identify, validate and deploy genes which impact the rhizosphere. Two contrasting examples of this are given by the identification, cloning and expression of genes involved in the production of extracellular organic anions and extracellular phosphatases.

Use of plant lines to investigate the role of extracellular organic anions

Given the importance of organic anions in the availability of a range of nutrients (P, Fe, Mn, Zn, K), it is possible that genetic variation in the capacity to exude organic anions can be exploited to beneficially manipulate the rhizosphere.

There are large differences between species in the amount and composition of organic acids exuded to the rhizosphere. Typically crop species such as wheat produce few extracellular organic anions, while *Lupinus albus* and other lupin species produce high amounts. Crops such as *Brassica napus* and bean species tend to be intermediate (Jones 1998; Pearse et al. 2006). There is also evidence for intraspecific variation in production of extracellular organic anions for example in landraces of *Lupinus albus* (Pearse et al. 2008). Likewise, differences in the ability of cultivars of pigeon pea to exude organic anions have been identified (Subbarao et al. 1997; Ishikawa et al. 2002). In addition, there is considerable variation in the composition of root exudates among various genotypes of wheat (Delhaize et al. 1993; Cieslinski et al. 1998), rice (Kirk et al. 1999) and maize (Gaume et al. 2001) under a range of deficiencies and toxicities. Therefore, comparison of the dynamics of nutrient availability and other rhizosphere processes with the growth of plant genotypes with known organic acid signatures can be used to elucidate rhizosphere mechanisms.

Heterologous expression of genes for enzymes involved in organic anion synthesis has also been used in an attempt to improve both the tolerance of plants to Al-toxicity conditions and conditions of P-deficit. Over-expression of a bacterial citrate synthase (CS) gene in tobacco has been reported to increase citrate efflux from roots of transgenic lines (de la Fuente-Martínez et al. 1997) allowing tolerance of toxic levels of Al and access to P from Ca-P found in the rhizosphere (Lopez-Bucio et al. 2000). However, repeating the work using similar gene constructs, and in some cases the same transgenic tobacco lines, did not confirm these results (Delhaize et al. 2001). Moreover, tobacco plants transformed to over-express tobacco CS and to down-regulate isocitrate dehydrogenase showed no increase in citrate efflux (Delhaize et al. 2003). Notwithstanding this, over-expression of a plant gene for a mitochondrial CS in *Arabidopsis thaliana* did enhance citrate efflux, with associated small improvements in P acquisition (Koyama et al. 2000). Other approaches have attempted to increase organic-anion exudation by over-expressing key genes of the tricarboxylic acid (TCA) cycle such as, phosphoenol pyruvate carboxylase (PEPC) and malate dehydrogenase (MDH) (Raghothama 1999). Over-expression of MDH in lucerne increased the efflux of organic anions from roots, but over-expression of PEPC did not (Tesfaye et al. 2001). Regardless of the approach, it is worth noting that the increase in organic-anion efflux achieved for transgenic *A. thaliana* or lucerne was not sufficient to allow us to understand fully the mechanistic significance of organic anions in the rhizosphere.

A more successful approach addressed the genes that encode channels involved in the transport of organic anions from roots to the rhizosphere. Citrate-permeable channels in the plasma membrane of cluster roots of Lupinus albus have been identified (Zhang et al. 2004), and a gene encoding a malate channel that is activated by the presence of toxic levels of aluminium has been identified and cloned from wheat (Sasaki et al. 2004). The expression of this latter gene in barley was able to mitigate the sensitivity of this barley genotype to Al-toxicity in the rhizosphere (Delhaize et al. 2004). This approach has not only completely explained the rhizospheric mechanism of malate induced Al-toxicity tolerance (Ryan et al. 1995), the gene identified is now being used as a "perfect" marker for breeding programmes with aims to identify cultivars of wheat with tolerance to acid soil conditions (Raman et al. 2005).

Use of plant lines to investigate the role of extracellular phosphatase

A number of studies have investigated extracellular phosphatase activities of plant roots and identified significant genotypic variation across and within different species (Tadano et al. 1993; Li et al. 1997; Asmar et al. 1995; Gaume et al. 2001; George et al. 2008), which may be useful in selection of genotypes for greater utilisation of soil organic P (Asmar et al. 1995; Rengel and Marschner 2005). In recent years, research using plants to influence the rhizosphere has focussed on improving the ability of plants to acquire P directly from the most common form of organic P in soils, inositol phosphates. This has culminated in the heterologous expression of microbial phytase genes (from both bacteria and fungi) in plants (Richardson et al. 2001; Zimmermann et al. 2003; Lung et al. 2005; Xiao et al. 2005) and comparison of the impact of these plants on rhizosphere processes with those of isogenic non-expressing controls. It was demonstrated that transgenic plants which release extracellular phytase to the rhizosphere had a novel ability to mineralise and utilise P from inositol phosphates and, when grown under controlled conditions, showed enhanced growth and P nutrition (George et al. 2004; Richardson et al. 2001; Mudge et al. 2003; Zimmermann et al. 2003). However, when grown in a range of soils, these plants show only a limited capacity to access additional P (George et al. 2004; George et al. 2005b). The use of the transgenic plants compared to their isogenic controls has allowed the identification of factors in the rhizosphere, which limit the capacity of the plants to access inositol phosphates when grown in soil. These include, immobilization of phytase protein by sorption on soil components (George et al. 2005a), poor solubility of inositol phosphate in soils (George et al. 2005b), and compensatory effects due to the presence of soil microorganisms (Richardson et al. 2007). Without the use of isogenic lines produced by transgenic approaches our understanding of phytase dynamics in the rhizosphere would not have been elucidated as quickly or clearly as it has. This approach has also highlighted the complexity inherent in attempting to improve multi-mechanistic tolerance traits by single gene approaches.

It is therefore clear that much potential exists to understand the genetic control of rhizosphere properties by comparing traits in the rhizosphere of isogenic or near-isogenic plants. At the same time it is also possible to elucidate rhizosphere mechanisms by comparing the dynamics of processes in rhizospheres of plants with vastly different phenotypes, either by comparing species or cultivars within species. As our knowledge of the range and number of plant genes involved in the control of rhizosphere processes expands through whole-genome approaches there will be opportunity to develop more experimental material which will allow the manipulation of many more specific rhizosphere traits.

Concluding remarks

During the last two decades, enormous progress has been achieved in terms of (i) rhizosphere sampling, considering gradients and spatial variations of root-induced changes in rhizosphere chemistry; (ii) miniaturisation of sampling techniques; (iii) introduction of biosensors, macro-, and micro-scale imaging techniques for detection of rhizosphere pH, redox changes, enzyme activities, element composition and nutrient availability in the rhizosphere; (iv) analytical tools for detection and quantification of rhizosphere compounds mainly based on coupling of various chromatographic and spectroscopic methods; (v) elucidation of physiological and molecular mechanisms involved in the regulation of chemical changes in the rhizosphere and (vi) the characterization of the rhizosphere microflora and of plant-microbial interactions.

However, most of the present knowledge still originates from model experiments with plants grown in rhizobox or rhizotron systems or even in hydroponic culture. Although, the number of studies in soil culture systems considerably increased during the last decade, investigations under real field conditions are rare. Many analytical tools developed and successfully applied for rhizosphere studies in model experiments are not easily applicable under field conditions, requiring robust routine methods for processing of large sample numbers to account for heterogenity and variability on field sites. Another limiting factor arises from the sampling process itself: even rhizobox-, or root-window approaches employed for in situ sampling techniques along soil-grown roots with minimal disturbance, cannot be necessarily considered as identical with natural growth conditions.

The lack of knowledge concerning the composition, pool sizes, fluxes, origin and binding forms of rhizosphere compounds under field conditions limits the integration of these processes into rhizosphere models and model validation (for a review on rhizosphere models see Luster et al. 2009). For example, first attempts to consider the effects of root exudation of carboxylates on phosphorus availability in the rhizosphere in models for nutrient uptake are still based on exudation data obtained from artificial culture systems (Hoffland 1992; Geelhoed et al. 1999).

The majority of the current investigations were carried out only with a limited number of model plants, comprising mainly crop species. Studies at the ecosystem level, considering wild plant species, genotypic differences and plant communities are just emerging and may unravel yet unidentified rhizosphere processes involved in nutrient cycling, and plant-microbial interactions.

Another aspect, frequently overlooked in model experiments, is the possibility of synergistic or antagonistic effects of simultaneous chemical and biological processes in the rhizosphere under field conditions: e.g., studies on root-induced responses to nutrient limitations should consider the fact that plants under natural growth conditions are usually exposed simultaneously to a wide array of additional stress factors with differential impact on the expression of rhizosphere responses (Neumann and Römheld 2007). Despite of the widespread occurrence of mycorrhizal associations in higher plants, also the potential impacts of mycorrhizae on the various rhizosphere processes are frequently not considered in simple model experiments.

Moreover, information gained in controlled laboratory experiments usually relates to the micro-scale by analysing single roots or root systems. In contrast, many types of models aim at quantifying processes on larger scales up to plant communities or to the ecosystem level. Upscaling from the micro-scale observed under laboratory conditions to larger scales in the field is very difficult and poses a major challenge for future research.

It is clear that much potential exists to manipulate rhizosphere properties at a genetic scale but it is unlikely to be universally successful until such attempts are supported by an understanding of the control of the mechanisms imposed by the complex soil environment in which they need to be effective.

As the knowledge on range and number of plant genes involved in the control of resource capture expands through whole-genome approaches, there will be opportunity to develop more resource efficient plants through the manipulation of specific traits. This will be achieved through selection and marker-assisted plant breeding for specific rhizosphere traits or by direct manipulation of plants by gene technologies. However, at least in Europe, public acceptance of efforts to investigate and to use transgenic organisms under field conditions is limited and a proper consideration of bio-safety issues is indispensable.

References

Alvarez JP, Pekkera I, Goldshmidt A, Blum E, Amsellem Z, Eshed Y (2006) Endogenous and synthetic MicroRNAs stimulate simultaneous, efficient, and localized regulation of multiple targets in diverse species. Plant Cell 18:1134–1151. doi:10.1105/tpc.105.040725

Asmar F, Gahoonia TS, Nielsen NE (1995) Barley genotypes differ in activity of soluble extracellular phosphatase and depletion of organic phosphorus in the rhizosphere soil. Plant Soil 172:117–122. doi:10.1007/BF00020865

Asseng S, Aylmore LAG, MacFall JS, Hopmanns JW, Gregory PJ (2000) Computer-assisted tomography and magnetic resonance imaging. In: Smit AL, Bengough AG, Engels C, Van Noordwijk M, Pellerin S, Van de Geijn SC (eds) Root methods. A handbook. Springer, Heidelberg, Germany, pp 343–364

Aviani I, Laor Y, Raviv M (2006) Limitations and potential of in situ rhizobox sampling for assessing microbial activity in fruit tree rhizosphere. Plant Soil 279:327–332. doi:10.1007/s11104-005-2189-4

Baret F, Fourty T (1997) Radiometric estimates of nitrogen status of leaves and canopies. In: Lemaire G (ed) Diagnosis of the nitrogen status in crops. Springer, Heidelberg, pp 201–227

Bienfait HF, van den Briel W, Mesland-Mul NT (1985) Free space iron pools in roots: generation and mobilization. Plant Physiol 78:596–600. doi:10.1104/pp. 78.3.596

Blackmer TM, Schepers JS (1996) Aerial photography to detect nitrogen stress in corn. J Plant Physiol 148:440–444

Blossfeld S, Gansert D (2007) A novel non-invasive optical method for quantitative visualization of pH dynamics in the rhizosphere of plants. Plant Cell Environ 30:176–186. doi:10.1111/j.1365-3040.2006.01616.x

Bobowski BR, Hole D, Wolfs PG, Bryant L (1999) Identification of roots of woody species using polymerase chain reaction (PCR) and restriction fragment length polymorphism (RFLP) analysis. Mol Ecol 8:485–491. doi:10.1046/j.1365-294X.1999.00603.x

Boukcim H, Pages L, Plassard C, Mousain D (2001) Root system architecture and receptivity to mycorrhizal infection in seedlings of *Cedrus atlantica* as affected by nitrogen source and concentration. Tree Physiol 21:109–115

Boukcim H, Pages L, Mousain D (2006) Local NO_3^- or NH_4^+ supply modifies the root system architecture of *Cedrus atlantica* seedlings grown in a split-root device. J Plant Physiol 163:1293–1304. doi:10.1016/j.jplph.2005.08.011

Briones AM, Okabe S, Umemiya Y, Ramsing NW, Reichardt-Okuyama H (2002) Influence of different cultivars on populations of ammonia-oxidising bacteria in the root environment of rice. Appl Environ Microbiol 68:3067–3075. doi:10.1128/AEM.68.6.3067-3075.2002

Brunner I, Brodbeck S, Büchler U, Sperisen C (2001) Molecular identification of fine roots of trees from the alps: reliable and fast DNA extraction and PCR-RFLP analyses of plastid DNA. Mol Ecol 10:2079–2087. doi:10.1046/j.1365-294X.2001.01325.x

Buchanan CD, Lim S, Salzman RA, Kagiampakis I, Morishige DT (2005) Sorghum bicolor's transcriptome response to dehydration, high salinity and ABA. Plant Mol Biol 58:699–720. doi:10.1007/s11103-005-7876-2

Bucher M (2006) Aeroponic culture. In: Luster J, Finlay R (eds) Handbook of methods used in rhizosphere research. Swiss Federal Research Institute WSL, Birmersdorf, pp 119–120

Buso GSC, Bliss FA (1988) Variability among lettuce cultivars grown at two levels of available phosphorus. Plant Soil 111:67–73. doi:10.1007/BF02182038

Cardon ZG, Gage DJ (2006) Resource exchange in the rhizosphere: molecular tools and the microbial perspective. Annu Rev Ecol Evol Syst 37:459–488. doi:10.1146/annurev.ecolsys.37.091305.110207

Cardon ZG, Herron PM (2005) Sweeping water, oozing carbon: Long distance transport and patterns of rhizosphere resource exchange. In: Holbrook NM, Zwieniecki MA (eds) Vascular transport in plants. San Diego Academic, USA, pp 257–76

Chaignon V, Hinsinger P (2003) A biotest for evaluating copper availability to plants in a contaminated soil. J Environ Qual 32:824–833

Chao DY, Luo YH, Shi M, Luo D, Lin HX (2005) Salt responsive genes in rice revealed by cDNA microarray analysis. Cell Res 15:796–810. doi:10.1038/sj.cr.7290349

Cheng W, Gershenson A (2007) Carbon fluxes in the rhizosphere. In: Cardon CG, Whitbeck JL (eds) The rhizosphere. An ecological perspective. Elsevier Academic, Burlington, USA, pp 31–56

Ciarelli DM, Furlani AMC, Dechen AR, Lima M (1998) Genetic variation among maize genotypes for phosphorus uptake and phosphorus-use efficiency in nutrient solution. J Plant Nutr 21:2219–2229. doi:10.1080/01904169809365556

Cieslinski G, Rees KCJ, van Szmigielska AM, Krishnamurti GSR, Huang PM (1998) Low molecular weight organic acids in rhizosphere soils of durum wheat and their effect on cadmium bioaccumulation. Plant Soil 203:109–117. doi:10.1023/A:1004325817420

Comas LH, Eissenstat DM (2004) Linking fine root traits to maximum potential growth rate among 11 mature temperate tree species. Funct Ecol 18:388–397. doi:10.1111/j.0269-8463.2004.00835.x

Cutler DF, Rudall PJ, Gasson PE, Gale RMO (1987) Root identification manual of trees and shrubs. A guide to the anatomy of trees and shrubs hardy in Britain and Northern Europe. Chapman & Hall, London

Danjon F, Reubens B (2008) Assessing and analyzing 3D architecture of woody root systems, a review of methods and applications in tree and soil stability, resource acquisition and allocation. Plant Soil 303:1–34. doi:10.1007/s11104-007-9470-7

Darwent MJ, Paterson E, James A, McDonald S, Deri Tomos A (2003) Biosensor reporting of root exudation from *Hordeum vulgare* in relation to shoot nitrate concentration. J Exp Bot 54:325–334. doi:10.1093/jxb/54.381.325

de la Fuente-Martínez JM, Ramirez-Rodriguez V, Cabrera-Ponce JL, Herrera-Estrella L (1997) Aluminum tolerance in transgenic plants by alteration of citrate synthesis. Science 276:1566–1588. doi:10.1126/science.276.5318.1566

Delhaize E, Ryan PR (2006) Aluminium-induced malate exudation. In: Luster J, Finlay R (eds) Handbook of methods used in rhizosphere research. Swiss Federal Research Institute WSL, Birmersdorf, pp 285–286

Delhaize E, Ryan PR, Randall PJ (1993) Aluminium tolerance in wheat (*Triticum aestivum* L.) 2. Aluminium stimulated excretion of malic acid from root apices. Plant Physiol 103:695–702

Delhaize E, Hebb DM, Ryan PR (2001) Expression of a *Pseudomonas aeruginosa* citrate synthase gene is not associated with either enhanced citrate accumulation or efflux. Plant Physiol 125:2059–2067. doi:10.1104/pp.125.4.2059

Delhaize E, Ryan PR, Hocking PJ, Richardson AE (2003) Effects of altered citrate synthase and isocitrate dehydrogenase expression on internal citrate concentrations in tobacco (*Nicotiana tabacum* L.). Plant Soil 248:137–144. doi:10.1023/A:1022352914101

Delhaize E, Ryan PR, Hebb DM, Yamamoto Y, Sasaki T, Matsumoto H (2004) Engineering high level aluminum tolerance in barley with the ALMT1 gene. Proc Natl Acad Sci USA 101:15249–15254. doi:10.1073/pnas.0406258101

Dessureault-Rompré J, Nowack B, Schulin R, Luster J (2006) Modified microsuction cup/rhizobox approach for the in-situ detection of organic acids in rhizosphere soil solution. Plant Soil 286:99–107. doi:10.1007/s11104-006-9029-z

Dinkelaker B, Hahn G, Römheld V, Wolf GA (1993) Non-destructive methods for demonstrating chemical changes in the rhizosphere I. Description of methods. Plant Soil 155/156:67–74. doi:10.1007/BF00024985

Dinkelaker B, Hengeler C, Neumann G, Eltrop L, Marschner H (1997) Root exudates and mobilization of nutrients. In: Rennenberg H, Eschrich W, Ziegler H (eds) Trees contributions to modern tree physiology. Backhuys, Leiden, The Netherlands, pp 441–452

Eapen S, D'Souza SF (2005) Prospects of genetic engineering of plants for phytoremediation of toxic metals. Biotechnol Adv 23:97–114. doi:10.1016/j.biotechadv.2004.10.001

Engels C, Neumann G, Gahoonia T, George E, Schenk M (2000) Assessment of the ability of roots for nutrient acquisition. In: Smit AL, Bengough AG, Engels C, Van Noordwijk M, Pellerin S, Van de Geijn SC (eds) Root methods. A handbook. Springer, Heidelberg, Germany, pp 403–459

Fageria NK, Baligar VC (1997a) Phosphorus use efficiency by corn genotypes. J Plant Nutr 20:1267–1277. doi:10.1080/01904169709365334

Fageria NK, Baligar VC (1997b) Upland rice genotypes evaluation for phosphorus use efficiency. J Plant Nutr 20:499–509. doi:10.1080/01904169709365270

Fageria NK, Wright RJ, Baligar VC (1988) Rice cultivar evaluation for phosphorus use efficiency. Plant Soil 111:105–109. doi:10.1007/BF02182043

Fernandez JE, Clothier BE, van Noordwijk M (2000) Water uptake. In: Smit AL, Bengough AG, Engels C, Van Noordwijk M, Pellerin S, Van de Geijn SC (eds) Root Methods. A Handbook. Springer, Heidelberg, Germany, pp 461–507

Fiedler S, Fischer WR (1994) Automatic device for longtime measurements of redox potentials under field condition. Z Pflanzenernahr Bodenk 157:305–308. doi:10.1002/jpln.19941570410

Fischer WR, Schaller G (1980) Ein Elektrodensystem zur Messung des Redoxpotentials im Kontaktbereich Boden/Wurzel. Z Pflanzenernaehr Bodenk 143:344–348. doi:10.1002/jpln.19801430312

Frey B, Turnau K (2006) Elemental analysis of roots and fungi. In: Luster J, Finlay R (eds) Handbook of methods used in rhizosphere research. Swiss Federal Research Institute WSL, Birmensdorf, pp 200–213

Gahoonia TS, Nielsen NE (1996) Variation in acquisition of soil phosphorus among wheat and barley genotypes. Plant Soil 178:223–230. doi:10.1007/BF00011587

Gahoonia TS, Nielsen NE, Lyshede OB (1999) Phosphorus acquisition of cereal cultivars in the field at three levels of P fertilization. Plant Soil 211:269–281. doi:10.1023/A:1004742032367

Gahoonia TS, Nielsen NE, Joshi PA, Jahoor A (2004) A root hairless barley mutant for elucidating genetic of root hairs and phosphorus uptake. Plant Soil 235:211–219. doi:10.1023/A:1011993322286

Garnett TP, Shabala SN, Smethurst PJ, Newman IA (2001) Simultaneous measurements of ammonium, nitrate and proton fluxes along the length of eucalypt roots. Plant Soil 236:55–62. doi:10.1023/A:1011951413917

Garrigues E, Doussan C, Pierret A (2006) Water uptake by plant roots: I—Formation and propagation of a water extraction front in mature root systems as evidenced by 2D light transmission imaging. Plant Soil 283:83–98. doi:10.1007/s11104-004-7903-0

Gaume A, Machler F, Deleon C, Narro L, Frossard E (2001) Low-P tolerance by maize (*Zea mays* L.) genotypes: significance of root growth, and organic acids and acid phosphatase root exudation. Plant Soil 228:253–264. doi:10.1023/A:1004824019289

Geelhoed JS, van Riemsdijk WH, Findenegg GR (1999) Simulation of the effect of citrate exudation from roots on the plant availability of phosphate adsorbed on goethite. Eur J Soil Sci 50:379–390. doi:10.1046/j.1365-2389.1999.00251.x

George TS, Richardson AE, Hadobas PA, Simpson RJ (2004) Characterisation of transgenic *Trifolium subterraneum* L. which expresses phyA and releases extracellular phytase: growth and phosphorus nutrition in laboratory media and soil. Plant Cell Environ 27:1351–1361. doi:10.1111/j.1365-3040.2004.01225.x

George TS, Richardson AE, Simpson RJ (2005a) Behaviour of plant-derived extracellular phytase upon addition to soil.

Soil Biol Biochem 37:977–988. doi:10.1016/j.soilbio.2004. 10.016

George TS, Simpson RJ, Hadobas PA, Richardson AE (2005b) Expression of a fungal phytase gene in *Nicotiana tabacum* improves phosphorus nutrition in plants grown in amended soil. Plant Biotechnol J 3:129–140. doi:10.1111/j.1467-7652.2004.00116.x

George TS, Richardson AE, Smith JB, Hadobas PA, Simpson J (2005c) Limitations to the potential of transgenic *Trifolium subterraneum* L. plants that exude phytase, when grown in soils with a range of organic phosphorus content. Plant Soil 278:263–274. doi:10.1007/s11104-005-8699-2

George TS, Gregory PJ, Hocking PJ, Richardson AE (2008) Variation in root-associated phosphatase activities in wheat contributes to the utilisation of organic P substrates in vitro, but does not effectively predict P-nutrition in different soils. Environ Exp Bot 64:239–249. doi:10.1016/j.envexpbot.2008.05.002

Gleason C, Chaudhuri S, Yang T, Muñoz A, Poovaiah BW, Oldroyd GED (2006) Nodulation independent of rhizobia induced by a calcium-activated kinase lacking autoinhibition. Nature 441:1149–1152. doi:10.1038/nature04812

Gobert A, Plassard C (2002) Differential NO_3^- dependent patterns of NO_3^- uptake in *Pinus pinaster, Rhizopogon roseolus* and their ectomycorrhizal association. New Phytol 154:509–516. doi:10.1046/j.1469-8137.2002.00378.x

Göttlein A, Hell U, Blasek R (1996) A system for microscale tensiometry and lysimetry. Geoderma 69:147–156. doi:10.1016/0016-7061(95)00059-3

Grayston SJ, Vaughan D, Jones D (1996) Rhizosphere carbon flow in trees, in comparison with annual plants: the importance of root exudation and its impact on microbial activity and nutrient availability. Appl Soil Ecol 5:29–56. doi:10.1016/S0929-1393(96)00126-6

Haase S, Ruess L, Neumann G, Marhan S, Kandeler E (2007a) Low-level herbivory by root-knot nematodes (*Meloidogyne incognita*) modifies root hair morphology and rhizodeposition in host plants (*Hordeum vulgare*). Plant Soil 301:151–164. doi:10.1007/s11104-007-9431-1

Haase S, Neumann G, Kania A, Kuzyakov Y, Römheld V, Kandeler E (2007b) Atmospheric CO_2 and the N-nutritional status modifies nodulation, nodule-carbon supply and root exudation of *Phaseolus vulgaris* L. Soil Biol Biochem 39:2208–2221. doi:10.1016/j.soilbio.2007. 03.014

Hacin JI, Ben Bohlool B, Singleton PW (1997) Partitioning of ^{14}C-labelled photosynthate to developing nodules and roots of soybean (Glycine max). New Phytol 137:257–265. doi:10.1046/j.1469-8137.1997.00812.x

Hainsworth JM, Aylmore LAG (1989) Non-uniform soil water extraction by plant roots. Plant Soil 113:121–124. doi:10. 1007/BF02181929

Häussling M, Leisen E, Marschner H, Römheld V (1985) An improved method for non-destructive measurements of the pH at the root-soil interface (rhizosphere). J Plant Physiol 117:371–375

Hawkins HJ, George E (1999) Effect of plant nitrogen status on the contribution of arbuscular mycorrhizal hyphae to plant nitrogen uptake. Physiol Plant 105:694–700. doi:10.1034/j.1399-3054.1999.105414.x

Hazan SP, Pathan MS, Sanchez A, Baxter I, Dunn M, Estes B, Chang HS, Zhu T, Kreps JA, Nguyen HT (2005) Expression profiling of rice segregating for drought tolerance QTLs using a rice genome array. Funct Integr Genomics 5:104–116. doi:10.1007/s10142-004-0126-x

Helmisaari HS, Makkonen K (2006) Root biomass and necromass. In: Finlay R, Luster J (eds) Handbook of methods used in rhizosphere research. Swiss Federal Research Institute WSL, Birmensdorf, pp 153–164

Henriksen GH, Raman DR, Walker LP, Spanwick RM (1992) Measurement of net fluxes of ammonium and nitrate at the surface of barley roots using ion-selective microelectrodes. II-Patterns of uptake along the root axis and evaluation of the microelectrode flux estimation technique. Plant Physiol 99:734–747. doi:10.1104/pp. 99.2.734

Herrmann AM, Ritz K, Nunan N, Clode PL, Pett-Ridge J, Kilburn MR, Murphy DV, O'Donnell AG, Stockdale EA (2007) Nanoscale secondary ion mass spectrometry—a new analytical tool in biogeochemistry and soil ecology. UCRL-JRNL-225506, Lawrence Livermore National Laboratory, CA, USA

Hinsinger P, Gilkes RJ (1996) Mobilization of phosphate from phosphate rock and alumina-sorbed phosphate by the roots of ryegrass and clover as related to rhizosphere pH. Eur J Soil Sci 47:533–544. doi:10.1111/j.1365-2389.1996.tb01 853.x

Hirano Y, Walthert L, Brunner I (2006) Callose in root apices of European chestnut seedlings: a physiological indicator of aluminum stress. Tree Physiol 26:431–440

Hodge A, Robinson D, Griffiths B, Fitter AH (1999) Why plants bother: root proliferation results in increased nitrogen capture from an organic patch when two grasses compete. Plant Cell Environ 22:811–820. doi:10.1046/j.1365-3040.1999.00454.x

Hoffland E (1992) Quantitative evaluation of the role of organic acid exudation in the mobilization of rock phosphate by rape. Plant Soil 140:279–289. doi:10.1007/BF00010605

Hoffland E, Findenegg GR, Nelemans JA (1989) Solubilization of rock phosphorus by rape II. Local root exudation of organic acids in response to P-starvation. Plant Soil 113:161–165. doi:10.1007/BF02280176

Hossain AK, Zakir M, Subbarao GV, Pearse SJ, Gopalakrishnan S, Ito O, Ishikawa T, Kawano N, Nakahara K, Yoshihashi T, Ono H, Yoshida M (2008) Detection, isolation and characterization of a root-exuded compound methyl 3-(4-hydroxyphenyl) propionate, responsible for biological nitrification inhibition by sorghum (Sorghum bicolor). New Phytol 180:442451

Iizumi Z, Mizumoto M, Nakamura K (1998) A bioluminescence assay using Nitrosomonas europaea for rapid and sensitive detection of nitrification inhibitors. Appl Environ Microbiol 64:3656–3662

Ishikawa S, Adu-Gyamfi JJ, Nakamura T, Yoshihara T, Watanabe T, Wagatsuma T (2002) Genotypic variability in phosphorus solubilising activity of root exudates by pigeon pea grown in low-nutrient environments. Plant Soil 245:71–81. doi:10.1023/A:1020659227650

Jaeger CHIII, Lindow SE, Miller W, Clark E, Firestone MK (1999) Mapping of sugar and amino acid availability in soil around roots with bacterial sensors of sucrose and tryptophan. Appl Environ Microbiol 65:2685–2690

Jaillard B, Ruiz L, Arvieu JC (1996) pH mapping in transparent gel using color indicator videodensitometry. Plant Soil 183:1–11. doi:10.1007/BF02185568

Jefferies SP, Pallotta MA, Paull JG, Katakousis A, Kretchmer JM, Manning S, Islam AKMR, Langridge P, Chalmers KJ

(2000) Mapping and validation of chromosome regions conferring boron toxicity tolerance in wheat (*Triticum aestivum*). Theor Appl Genet 101:767–777. doi:10.1007/s001220051542

Johnson F, Allan DL, Vance CP, Weiblen G (1996) Root carbon dioxide fixation by phosphorus-deficient *Lupinus albus* (Contribution to organic acid exudation by proteoid roots). Plant Physiol 112:19–30

Jones DL (1998) Organic acids in the rhizosphere—a critical review. Plant Soil 205:25–44

Kamh M, Horst WJ, Amer F, Mostafa H, Maier P (1999) Mobilization of soil and fertilizer phosphate by cover crops. Plant Soil 211:19–27

Kania A, Langlade N, Martinoia E, Neumann G (2003) Phosphorus deficiency-induced modifications in citrate catabolism and in cytosolic pH as related to citrate exudation in cluster roots of white lupin. Plant Soil 248:117–127

Kawasaki S, Borchert C, Deyholos M, Wang H, Brazile S, Kawai K, Galbraith D, Bohnert HJ (2001) Gene expression profiles during the initial phase of salt stress in rice. Plant Cell 13:889–905

Kawaura K, Mochida K, Yamazaki Y, Ogihara Y (2006) Transcriptome analysis of salinity stress responses in common wheat using a 22 k oligo-DNA microarray. Funct Integr Genomics 6:132–142

Keerthisinghe G, Hooking PJ, Ryan PR, Delhaize E (1998) Effect of phosphorus supply on the formation and function of proteoid roots of white lupin (*Lupinus albus* L.). Plant Cell Environ 21:467–478

Killham K, Yeomans C (2001) Rhizosphere carbon flow measurement and implications: from isotopes to reporter genes. Plant Soil 232:91–96

Kirk GJD, Santos EE, Findenegg GR (1999) Phosphate solubilization by organic anion secretion from rice (*Oryza sativa* L.) growing in aerobic soil. Plant Soil 211:11–18

Kitchen NR (2007) Incorporating nutrient sensing technology in production agriculture. Fluid Fertilizer Forum, February 18–20, 2007, Scottsdale. AZ, pp 35–40

Knipfer T, Steudle E (2008) Root hydraulic conductivity measured by pressure clamp is substantially affected by internal unstirred layers. J Exp Bot 59:2071–2084

Kojima S, Bohner A, von Wirén N (2006) Molecular mechanisms of urea transport in plants. J Membr Biol 212:83–91

Koyama H, Kanamura A, Kihara T, Hara T, Takita E, Shibata D (2000) Overexpression of mitochondrial citrate synthase in *Arabidopsis thaliana* improved growth on a phosphorus limited soil. Plant Cell Physiol 41:1030–1037

Kragelund L, Hosond C, Nybroe O (1997) Distribution of metabolic activity and phosphate starvation response of *lux*-tagged *Pseudomonas fluorescens* reporter bacteria in the barley rhizosphere. Appl Environ Microbiol 63:4920–4928

Kuchenbuch R, Jung A (1982) A method for determining concentration profiles at the root-soil interface by thin slicing rhizospheric soil. Plant Soil 68:391–394

Kuzyakov Y, Domanski G (2000) Carbon input by plants into the soil. Review. Journal of Plant Nutr Soil Sci 163:421–431

Lan L, Li M, Lai Y, Xu W, Kong Z, Ying K, Han B, Xue Y (2005) Microarray analysis reveals similarities and variations in genetic programs controlling pollination/fertilization and stress responses in rice (*Oryza sativa* L.). Plant Mol Biol 59:151–164

Li M, Osaki M, Rao IM, Tadano T (1997) Secretion of phytase from the roots of several plant species under phosphorus-deficient conditions. Plant Soil 195:161–169

Li L, Zhang F, Li X, Christie P, Sun J, Yang S, Tang C (2003) Interspecific facilitation of nutrient uptake by intercropped maize and faba bean. Nutrient Cycling in Agroecosystems 65:61–71

Li SM, Li L, Zhang FS, Tang C (2004) Acid phosphatase role in Chickpea/Maize intercropping. Annals of Botany 94:297–202

Lian X, Wang S, Zhang J, Feng Q, Zhang L, Fan D, Li X, Yuan D, Han B, Zhang Q (2006) Expression profiles of 10,422 genes at early stages of low nitrogen stressing rice assayed using a cDNA microarray. Plant Mol Biol 60:617–631

Liao MT, Fillery IRP, Palta JA (2004) Early vigorous growth is a major factor influencing nitrogen uptake in wheat. Funct Plant Biol 31:121–129

Liao M, Hocking PJ, Dong B, Delhaize E, Richardson AE, Ryan PR (2008) Variation in early phosphorus-uptake efficiency among wheat genotypes grown on two contrasting Australian soils. Aust J Agric Res 59:157–166

Liew OW, Ching P, Chong J, Li B, Asundi AK (2008) Signature optical cues: emerging technologies for monitoring plant health. Sensors 8:3205–3239

Lilienthal H, Schnug E (2007) New issues for remote sensing in agriculture—a critical overview. Dahlia Greidinger Symposium—advanced technologies for monitoring nutrient and water availability to plants, March 2007. Haifa, Israel, pp 87–104

Lopez-Buccio J, de la Vega OM, Guevara-Garcia A, Herrera-Estrella L (2000) Enhanced phosphorus uptake in transgenic tobacco plants that overproduce citrate. Nature Biotech 18:450–453

Lung SC, Chan WL, Yip W, Wang L, Yeung EC, Lim BL (2005) Secretion of beta-propeller phytase from tobacco and *Arabidopsis* roots enhances phosphorus utilisation. Plant Sci 169:341–349

Luster J, Finlay R (eds) (2006) Handbook of methods used in rhizosphere research. Swiss Federal Institute for Forest, Snow, and Landscape Research, Birmensdorf, Switzerland, online at www.rhizo.at/handbook

Luster J, Göttlein A, Nowack B, Sarret G (2009) Sampling, defining, characterising and modeling the rhizosphere—the soil science tool box. Plant Soil (This volume), in press. doi:10.1007/s11104-008-9781-3

Lynch JP, Brown KM (2001) Topsoil foraging—an architectural adaptation of plants to low phosphorus availability. Plant Soil 237:225–237

Lynch JM, Whipps JM (1990) Substrate flow in the rhizosphere. Plant Soil 129:1–10

Macfall JS, Johnson GA, Kramer PJ (1990) Observation of a water-depletion region surrounding loblolly pine roots by magnetic resonance imaging. Proc Natl Acad Sci USA 87:1203–1207

Majdi H (1996) Root sampling methods-applications and limitations of the minirhizotron technique. Plant and Soil 185:255–258

Makkonen K, Helmisaari HS (1999) Assessing fine-root biomass and production in a Scots pine stand—comparison of soil core and ingrowth core methods. Plant Soil 210:43–50

Maniero R (2006) Fine root dynamics. In: Luster J, Finlay R (eds) Handbook of methods used in rhizosphere research. Swiss Federal Research Institute WSL, Birmensdorf, pp 165–166

Manske GGB, Ortiz-Monasterio JI, Van Grinkel M, Rajaram S, Molina E, Vlek PLG (2000) Traits associated with improved P-uptake efficiency in CIMMYT's semidwarf spring bread wheat grown on an acid andisol in Mexico. Plant Soil 221:189–204

Markandeya G, Babu PR, Lachagari VBR, Feltus FA, Paterson AH, Reddy AR (2005) Functional genomics of drought-stress response in rice: transcript mapping of annotated unigenes of an *indica* rice (*Oryza sativa* L. cv. Nagina 22). Curr Sci 89:496–514

Marshall Porterfield D (2002) Use of microsensors for studying the physiological activity of plant roots. In: Waisel Y, Eshel A, Kafkafi U (eds) Plant roots the hidden half, 3rd edn. Marcel Dekker, New York, USA, pp 333–347

Miki D, Shimamoto K (2004) Simple RNAi vectors for stable and transient suppression of gene function in rice. Plant Cell Physiol 45:490–495

Mudge SR, Smith FW, Richardson AE (2003) Root-specific and phosphate-regulated expression of phytase under the control of a phosphate transporter promoter enables *Arabidopsis* to grow on phytate as a sole phosphorus source. Plant Sci 165:871–878

Munns R, Hare RA, James RA, Rebetzke GJ (2000) Genetic variation for improving the salt tolerance of durum wheat. Aust J Agric Res 51:69–74

Nakaji T, Noguchi K, Oguma H (2008) Classification of rhizosphere components using visible-near infrared spectral images. Plant Soil 310:245–261

Negishi T, Nakanishi H, Yazaki J, Kishimoto N, Fujii F, Shimbo K et al (2002) cDNA microarray analysis of gene expression during Fe-deficiency stress in barley suggests that polar transport of veiscles is implicated in phytosiderophore secretion in Fe-deficient barley roots. Plant J 30:83–94

Neumann G (2006a) Construction and setup of rhizoboxes. In: Luster J, Finlay R (eds) Handbook of methods used in rhizosphere research. Swiss Federal Research Institute WSL, Birmensdorf, pp 143–144

Neumann G (2006b) Collection of root exudates and rhizosphere soil solution from soil-grown plants. In: Luster J, Finlay R (eds) Handbook of methods used in rhizosphere research. Swiss Federal Research Institute WSL, Birmensdorf, pp 317–318

Neumann G, Römheld V (2007) The release of root exudates as affected by the plant physiological status. In: Pinton R, Varanini Z, Nannipieri Z (eds) The rhizosphere: biochemistry and organic substances at the soil-plant interface, 2nd edn. CRC, Boca Raton, Florida, USA, pp 23–72

Newman IA (2001) Ion transport in roots: measurement of fluxes using ion-selective microelectrodes to characterize transporter function. Plant Cell Environm 24:1–14

Newman IA, Kochian LV, Grusak MA, Lucas WJ (1987) Fluxes of H^+ and K^+ in corn roots: characterization and stoichiometries using ion-selective microelectrodes. Plant Physiol 84:1177–1184

Oh SJ, Kwon CW, Choi DW, Song SI, Kim JK (2007) Expression of barley HvCBF4 enhances tolerance to abiotic stress in transgenic rice. Plant Biotechnol J 5:646–656

Oliveira MG, van Noordwijk M, Gaze SR, Brouwer G, Bona S, Mosca G, Hairiah K (2000) Auger sampling, ingrowth cores and pinboard methods. In: Smit AL, Bengough AG, Engels C, Van Noordwijk M, Pellerin S, Van de Geijn SC (eds) Root methods. A handbook. Springer, Heidelberg, Germany, pp 175–210

Osborne LD, Rengel Z (2002) Screening cereals for genotypic variation in the efficiency of phosphorus uptake and utilization. Aust J Agric Res 53:295–303

Ozturk ZN, Talame V, Deyholos M, Michalowski CB, Galbraith DW, Gozukirmizi N, Tuberosa R, Bohnert HJ (2002) Monitoring large-scale changes in transcript abundance in drought-and salt-stressed barley. Plant Mol Biol 48:551–573

Pearse SJ, Venaklaas EJ, Cawthray G, Bolland MDA, Lambers H (2006) Carboxylate composition of root exudates does not relate consistently to a crop species' ability to use phosphorus from aluminium, iron or calcium phosphate sources. New Phytol 173:181–190

Pearse SJ, Venaklaas EJ, Cawthray G, Bolland MDA, Lambers H (2008) *Lupinus albus* landraces from different ecogeographical origins vary in their ability to use phosphorus from sparingly soluble forms. Austr J Agr Res 59:626–623

Plassard C, Meslem M, Souche G, Jaillard B (1999) Localization and quantification of net fluxes of H^+ along maize roots by combined use of pH-indicator dye videodensitometry and H^+-selective microelectrodes. Plant Soil 211:29–39

Plassard C, Guérin-Laguette A, Véry AA, Casarin V, Thibaud TB (2002) Local measurements of nitrate and potassium fluxes along roots of maritime pine. Effects of ectomycorrhizal symbiosis. Plant Cell Environm 25:75–84

Polomski J, Kuhn N (2002) Root research methods. In: Waisel Y, Eshel A, Kafkafi U (eds) Plant roots the hidden half, 3rd edn. Marcel Dekker, New York, USA, pp 295–321

Pratt LH, Liang C, Shah M, Sun F, Wang H, Reid SP et al (2005) Sorghum expressed sequence tags identify signature genes for drought, pathogenesis, and skotomorphogenesis from a milestone set of 16 801 unique transcripts. Plant Physiol 139:869–884

Rabbani MA, Maruyama K, Abe H, Khan MA, Katsura K, Ito Y, Yoshiwara K, Seki M, Shinozaki K, Yamaguchi-Shinozaki K (2003) Monitoring expression profiles of rice genes under cold, drought and high salinity stresses and abscisic acid application using cDNA microarray and RNA gel blot analyses. Plant Physiol 133:1755–1767

Raghothama KG (1999) Phosphate acquisition. Ann Rev Plant Physiol Plant Mol Biol 50:665–693

Raman H, Zhang K, Cakir M, Appels R, Garvin DF, Maron LG, Kochian LV, Moroni JS, Raman R, Imtiaz M, Drake-Brockman F, Waters I, Martin P, Sasaki T, Yamamoto Y, Matsumoto H, Hebb DM, Delhaize E, Ryan PR (2005) Molecular characterisation and mapping of ALMT1, the aluminium tolerance gene in bread wheat (*Triticum aestivum* L.). Genome 48:781–791

Rengel Z, Marschner P (2005) Nutrient availability and management in the rhizosphere: exploiting genotypic differences. New Phytol 168:305–312

Rengel Z, Ross G, Hirsch P (1998) Plant genotype and micronutrient status influence colonization of wheat roots by soil bacteria. J Plant Nutr 21:99–113

Rensink WA, Lobst S, Hart A, Stegalkina S, Liu J, Buell CR (2005) Gene expression profiling of potato responses to cold heat and salt stress. Funct Integr Genomics 5:201–207

Richards RA, Watt M, Rebetzke GJ (2007) Physiological traits and cereal germplasm for sustainable agricultural systems. Euphytica 154:409–425

Richardson AE, Hadobas PA, Hayes JE (2001) Extracellular secretion of *Aspergillus* phytase from *Arabidopsis* roots enables plants to obtain phosphorus from phytate. Plant J 25:641–649

Richardson AE, George TS, Jakobsen I, Simpson RJ (2007) Plant utilization of inositol phosphates. In: Turner BL, Richardson AE, Mullaney EJ (eds) Inositol phosphates: linking agriculture and the environment. CAB International, Wallingford, pp 242–260

Richter AK, Frossard E, Brunner I (2007) Polyphenols in woody roots of Norway spruceand European beech reduce TTC. Tree Physiol 27:155–160

Ruiz L, Arvieu JC (1990) Measurements of pH gradients in the rhizosphere. Symbiosis 9:71–75

Ryan PR, Delhaize E, Randall PJ (1995) Malate efflux from root apices and tolerance to aluminium are highly correlated in wheat. Aust J Plant Phys 22:531–536

Sasaki T, Yamomoto Y, Ezaki B, Katsuhara M, Ahn SJ, Ryan PR, Delhaize E, Matsumoyo H (2004) A wheat gene encoding an aluminium-activated malate transporter. Plant J 37:645–653

Schilling G, Gransee A, Deubel A, Lezovic G, Ruppel S (1998) Phosphorus availability, root exudates, and microbial activity in the rhizosphere. Z Pflanzenernähr Bodenk 161:465–478

Schmidhalter U, Jungert S, Bredemeier C, Gutser R, Manhart R, Mistele B et al (2003) Field-scale validation of a tractor-based multispectral crop scanner to determine biomass and nitrogen uptake of winter wheat. In: Stafford J, Werner A (eds) Precision agriculture: Papers from the 4th European Conference on Precision Agriculture, Berlin, pp 615–619

Schünmann PHD, Richardson AE, Smith FW, Delhaize E (2004) Characterization of promoter expression patterns derived from the Pht1 phosphate transporter genes of barley (*Hordeum vulgare* L.). J Exp Bot 55:855–865

Shen J, Hoffland E (2007) In situ sampling of small volumes of soil solution using modified micro cups. Plant Soil 292:161–169

Smit AL, Bengough AG, Engels C, Van Noordwijk M, Pellerin S, Van de Geijn SC (2000a) Root methods. A handbook. Springer, Heidelberg, Germany

Smit AL, George E, Groenwold J (2000b) Root observations and measurements at (transparent) interfaces with soil. In: Smit AL, Bengough AG, Engels C, Van Noordwijk M, Pellerin S, Van de Geijn SC (eds) Root methods. A handbook. Springer, Heidelberg, Germany, pp 236–271

Smith DM, Allen SJ (1996) Measurement of sap flow in plant stems. J Exp Bot 47:1833–1844

Steudle E (1994) Water transport across roots. Plant Soil 167:79–90

Strasser O, Köhl K, Römheld V (1999) Overestimation of apoplastic Fe in roots of soil grown plants. Plant Soil 210:179–187

Stromberg N (2006) Imaging optodes. Dept. of Chemistry Analytical Chemistry, Goteborg University, Sweden

Stromberg N (2008) Determination of ammonium turnover and flow patterns close to roots using imaging optodes. Environm. Sci Technol 42:1630–1637

Subbarao GV, Ae N, Otani T (1997) Genotypic variation in the iron- and aluminium-phosphate solubilising activity of pigeon pea root exudates under P deficient conditions. Soil Sci Plant Nutr 43:295–305

Subbarao GV, Ishikawa T, Ito O, Nakahara K, Wang HY, Berry WL (2006) A bioluminescence assay to detect nitrification inhibitors released from plant roots: a case study with *Brachiaria humidicola*. Plant Soil 288:101–112

Tadano T, Ozawa K, Sakai H, Osaki M, Matsui H (1993) Secretion of acid phosphatase by the roots of crop plants under phosphorus-deficient conditions and some properties of the enzyme secreted by lupin roots. Plant Soil 155 (156):95–98

Takahashi M, Nakanishi H, Kawasaki NK, Mori S (2001) Enhanced tolerance of rice to low iron availability in alkaline soils using barley nicotinamine aminotransferase genes. Nat Biotechnol 19:466–469

Tesfaye M, Temple SJ, Allan DL, Vance CP, Samac DA (2001) Overexpression of malate dehydrogenase in transgenic alfalfa enhances organic acid synthesis and confers tolerance to aluminium. Plant Physiol 127:1836–1844

Thorup-Kristensen K (2006) Root density and rooting depth, root turnover, short term root growth responses. In: Luster J, Finlay R (eds) Handbook of methods used in rhizosphere research. Swiss Federal Research Institute WSL, Birmensdorf, pp 177–178

Tirichine L, Imaizumi-Anraku H, Yoshida S, Murakami Y, Madsen LH, Miwa H, Nakagawa T, Sandal N, Albrektsen AS, Kawaguchi M, Downie A, Sato S, Tabata S, Kouchi H, Parniske M, Kawasaki S, Stougaard J (2006) Deregulation of a Ca^{2+}/calmodulin-dependent kinase leads to spontaneous nodule development. Nature 441:1153–1156

Turpault MP, Utérano C, Boudot JP, Ranger J (2005) Influence of mature Douglas fir roots on the solid phase of the rhizosphere and its solution chemistry. Plant Soil 275:327–336

Turpault MP, Gobran GR, Bonnaud P (2007) Temporal variations of rhizosphere and bulk soil chemistry in a Douglas fir stand. Geoderma 137:490–496

Uhde-Stone C, Zinn KE, Ramirez-Yáñez M, Li A, Vance CP, Allan DL (2003) Nylon filter arrays reveal differential gene expression in proteoid roots of white lupin in response to P deficiency. Plant Physiol 131:1064–1079

Van Noordwijk M, Brouwer G, Meijboom F, Oliveira MG, Bengough AG (2000) Trench profile techniques and core break methods. In: Smit AL, Bengough AG, Engels C, Van Noordwijk M, Pellerin S, Van de Geijn SC (eds) Root methods. A handbook. Springer, Heidelberg, Germany, pp 211–233

Van Veen JA, Kay E, Vogel TM, Simonet P (2007) Methodological approaches to the study of carbon flow and the associated microbial population dynamics in the rhizosphere. In: Pinton R, Varanini Z, Nannipieri Z (eds) The rhizosphere: biochemistry and organic substances at the soil-plant interface, 2nd edn. CRC, Boca Raton, Florida, USA, pp 371–399

Vance CP, Uhde-Stone C, Allan DL (2003) Phosphorus acquisition and use: critical adaptations by plants for securing a nonrenewable resource. New Phytol 157:423–447

Vanguelova E, Hirano Y, Eldhuset TD, Sas-Paszt L, Bakker MR, Püttsepp U et al (2007) Tree fine root Ca/Al molar ratio—indicator of Al and acidity stress. Plant Biosyst 141:460–480

Vetterlein D, Jahn R (2004) Combination of micro suction cups and time-domain reflectometry to measure osmotic potential gradients between bulk soil and rhizosphere at high resolution in time and space. Eur J Soil Sci 55:497–504

Vetterlein D, Marschner H, Horn R (1993) Microtensiometer technique to study hydraulic lift in a sandy soil planted with pearl millet (*Pennisetum americanum* L. Leeke). Plant Soil 149:275–282

Vij S, Tyagi AK (2007) Emerging trends in the functional genomics of the abiotic stress response in crop plants. Plant Biotechnol J 5:361–380

Walia H, Wilson C, Condamine P, Liu X, Ismail AM, Zeng L, Wanamaker SI, Mandal J, Xu J, Cui X, Close TJ (2005) Comparative transcriptional profiling of two contrasting rice genotypes under salinity stress during the vegetative growth stage. Plant Physiol 139:822–835

Wang QR, Li JY, Li ZS, Christie P (2005) Screening Chinese wheat germplasm for phosphorus efficiency in calcareous soils. J Plant Nutr 28:489–505

Wichern F, Eberhardt E, Mayer J, Joergensen RG, Müller T (2008) Nitrogen rhizodeposition in agricultural crops: methods, estimates and future prospects. Soil Biol Biochem 40:30–48

Williams RF (1948) The effects of phosphorus supply on the rates of intake of phosphorus and nitrogen and upon certain aspects of phosphorus metabolism in graminaceous plants. Aust J Sci Res B1:333–361

Wissuwa M (2003) How do plants achieve tolerance to phosphorus deficiency? Small causes with big effects. Plant Physiol 133:1947–1958

Wissuwa M, Ae N (2001a) Further characterization of two QTLs that increase phosphorus uptake of rice (*Oryza sativa* L.) under phosphorus deficiency. Plant Soil 237:275–286

Wissuwa M, Ae N (2001b) Genotypic variation for tolerance to phosphorus deficiency in rice and the potential for its exploitation in rice improvement. Plant Breeding 120:43–48

Wissuwa M, Mazzola M, Picard C (2009) Novel approaches in plant breeding for rhizosphere-related traits. Plant Soil (This volume), in press. doi:10.1007/s11104-008-9693-2

Xiao K, Harrison MJ, Wang ZY (2005) Transgenic expression of a novel *Medicago truncatula* phytase gene results in improved acquisition of organic phosphorus by *Arabidopsis*. Planta 222:27–36

Yan X, Liao H, Beebe SE, Blair MW, Lynch JP (2004) QTL mapping of root hair and acid exudation traits and their relationship to phosphorus uptake in common bean. Plant Soil 265:17–29

Yeomans C, Porteous F, Paterson E, Meharg AA, Killham K (1999) Assessment of *lux*-marked *Pseudomonas fluorescent* for reporting on organic compounds. FEMS Microbiol Letters 176:79–83

Yu LX, Setter TL (2003) Comparative transcriptional profiling of placenta and endosperm in developing maize kernels in response to water deficit. Plant Physiol 131:568–582

Yuan L, Loque D, Kojima S, Rauch S, Ishiyama K, Inoue E, Takahashi H, von Wiren N (2007) The organization of high-affinity ammonium uptake in *Arabidopsis* roots depends on the spatial arrangement and biochemical properties of AMT1-type transporters. Plant Cell 19:2636–2652

Zhang WH, Ryan PR, Tyerman SD (2004) Citrate-permeable channels in the plasma membrane of cluster roots from white lupin. Plant Physiol 136:3771–3783

Zhu J, Kaeppler SM, Lynch JP (2005) Mapping of QTL controlling root hair length in maize (*Zea mays* L.) under phosphorus deficiency. Plant Soil 270:299–310

Zimmermann U, Meinzer FC, Benker R, Zhu JJ, Schneider H, Goldstein G, Kuchenbrod E, Haase A (1994) Xylem water transport: ist the available evidence consistent with the cohesion theory? Plant Cell Environ 17:1169–1181

Zimmermann P, Zardi G, Lehmann M, Zeder C, Amrhein N, Frossard E, Bucher M (2003) Engineering the root-soil interface via targeted expression of a synthetic phytase gene in trichoblasts. Plant Biotech J 1:353–360

REVIEW ARTICLE

Sampling, defining, characterising and modeling the rhizosphere—the soil science tool box

Jörg Luster · Axel Göttlein · Bernd Nowack · Géraldine Sarret

Received: 16 December 2007 / Accepted: 8 September 2008 / Published online: 21 October 2008
© Springer Science + Business Media B.V. 2008

Abstract We review methods and models that help to assess how root activity changes soil properties and affects the fluxes of matter in the soil. Subsections discuss (1) experimental systems including plant treatments in artificial media, studying the interaction of model root and microbial exudates with soil constituents, and microcosms to distinguish between soil compartments differing in root influence, (2) the sampling and characterization of rhizosphere soil and solution, focusing on the separation of soil at different distances from roots and the spatially resolved sampling of soil solution, (3) cutting-edge methodologies to study chemical effects in soil, including the estimation of bioavailable element or ion contents (biosensors, diffusive gradients in thin-films), studying the ultrastructure of soil components, localizing elements and determining their chemical form (microscopy, diffractometry, spectroscopy), tracing the compartmentalization of substances in soils (isotope probing, autoradiography), and imaging gradients in-situ with micro electrodes or gels or filter papers containing dye indicators, (4) spectroscopic and geophysical methods to study the plants influence on the distribution of water in soils, and (5) the modeling of rhizosphere processes. Macroscopic models with a rudimentary depiction of rhizosphere processes are used to predict water or nutrient requirements by crops and forests, to estimate biogeochemical element cycles, to calculate soil water transport on a profile scale, or to simulate the development of root systems. Microscopic or explanatory models are based on mechanistic or empirical relations that describe processes on a single root or root system scale and/or chemical reactions in soil solution. We conclude that in general we have the tools at hand to assess individual processes on the microscale under rather artificial conditions. Microscopic, spectroscopic and tracer methods to look at processes in small "aliquots" of naturally structured soil seem to step out of their infancy and have become promising tools to better understand the complex

Responsible Editor: Philippe Hinsinger.

J. Luster (✉)
Swiss Federal Institute for Forest, Snow,
and Landscape Research WSL,
Zürcherstrasse 111,
CH-8903 Birmensdorf, Switzerland
e-mail: joerg.luster@wsl.ch

A. Göttlein
Center of Life and Food Sciences Weihenstephan,
Division of Forest Nutrition and Water Resources,
TU München,
85354 Freising, Germany

B. Nowack
Empa-Swiss Federal Laboratories
for Materials Testing and Research,
CH-9014 St. Gallen, Switzerland

G. Sarret
Environmental Geochemistry Group,
LGIT, Université J. Fourier and CNRS,
38041 Grenoble Cedex 9, France

interactions between plant roots, soil and microorganisms. On the field scale, while there are promising first results on using non-invasive geophysical methods to assess the plant's influence on soil moisture, there are no such tools in the pipeline to assess the spatial heterogeneity of chemical properties and processes in the field. Here, macroscopic models have to be used, or model results on the microscopic level have to be scaled up to the whole plant or plot scale. Upscaling is recognized as a major challenge.

Keywords Geophysics · Imaging · Isotope probing · Microcosms · Soil solution · Spectroscopy

Introduction

There are two basic questions involved with this part of rhizosphere research. (1) How are physical and chemical soil properties and related functional parameters (e.g. structural stability, availability of water, nutrients or toxic substances) affected by root growth, root physiological processes involved in nutrient acquisition and uptake and related root–microbe interactions, and how far do these effects extend from the root (Hinsinger et al. 2005)? (2) How do these root-related processes affect the fluxes of water, elements and ions in the soil, and thus biogeochemical cycles? On principle all methods for the analysis and modeling of the properties of the respective soil phases apply and can be looked up in standard textbooks such as Weaver et al. (1994; biochemical and isotopic methods), Sparks (1996; chemical methods), Dane and Topp (2002; physical methods), Pansu and Gautheyrou (2006; mineralogical and chemical methods) and Nollet (2007; water analysis with implications for soil solution analysis). The critical issue, which is the red-line of this chapter, is to separate, define or identify the rhizosphere. In a first section, the various degrees of simplifying real soil and experimental systems to study the interaction of model root and microbial exudates with soil constituents are discussed. Laboratory and field systems are presented that allow a distinction of soil compartments in terms of root influence, that facilitate the sampling of rhizosphere soil or soil solution, or that enable the in-situ analysis of the root's influence on soil properties. In the second section, methods to separate rhizosphere from bulk soil and to sample rhizosphere solution and gas are presented together with a brief overview of analytical methods for their characterization. Soil biological methods are described by Sørensen et al. (2008, Strategies and methods- the microbial ecology toolbox, submitted). The third section is devoted to cutting-edge methodologies to study chemical effects in soils. This includes techniques to assess bioavailable contents, to trace the compartmentalization of organic carbon, and to map the distribution of elements and species in-situ. In the fourth section, the prospects of spectroscopic and geophysical methods to image non-invasively the plant influence on soil moisture distribution in the laboratory and field are discussed. Modeling, the topic of the fifth section, is an important tool to understand and predict plant influence on soil properties, and vice versa, how to manage the soil to fulfill plant water and nutrient requirements. In addition, models are useful to estimate how plant activity affects terrestrial element cycles, and vice versa, how plants react to climatic changes. Scaling model results up from the single-root level to the whole-plant, plot or catchment level is one of the most demanding current research issues. In a sixth and last section we discuss this and other challenges ahead. An alternative treatment of aspects dealt with in this paper can be found in Luster and Finlay (2006).

Experimental systems

Field soil is a complex three-phase system with varying degrees of spatial and temporal heterogeneity of physical and chemical properties. Soil fauna, microorganisms and growing plant roots increase this heterogeneity. In particular, growing plant roots add spatial gradients in two directions (Fig. 1). Along the growth direction, root segments differ in their functionality in terms of uptake (water, nutrients) or exudation, causing a variability of root-induced changes in the properties of the surrounding soil. This root influence decreases with increasing distance from the root surface leading to gradients from the rhizosphere to the bulk soil. In addition, there is a temporal variation in root influence due to diurnal, seasonal or age related changes in the physiological activity of root segments. Dead parts of the root system first become local sources of organic matter, and after their degradation macropores can be created which can have a strong impact on the soils transport

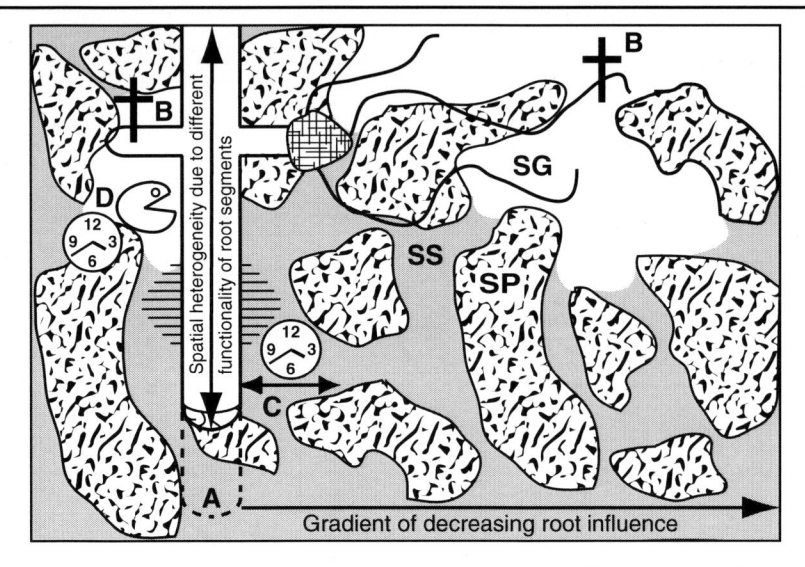

Fig. 1 Rhizosphere as three-phase system with soil solid phase (*SP*), soil solution (*SS*), and soil gas phase (*SG*); spatial heterogeneity along and perpendicular to root growth added by a developing root system is emphasised and is overlaid by temporal variability: root growth (*A*), turnover of roots and fungal hyphae (*B*), diurnal or seasonal changes in the activity of roots (exudation, uptake; *C*), or associated organisms (*D*)

properties. The goal of rhizosphere research being to assess these plant influences, minimising the heterogeneity of the soil itself is an important consideration. The degree of simplification in terms of substrate properties and/or system geometry must be adequate for the problem and allow a correct interpretation of the data.

Artificial substrates

The nature of artificial growth media relates to the fact that root activity generally needs water as medium. They either contain no solid phase at all (hydroponics) or employ a solid phase with low chemical reactivity suspended in or irrigated with nutrient or treatment solution. Artificial solid substrates are often easier to sterilize than soil material. Sterilization of soils can alter their chemical and physical properties (Wolf and Skipper 1994) and it is difficult to maintain sterility during longer experiments. As such artificial substrates are excellent tools to study plant physiological reactions (Neumann et al. 2008, Strategies and methods-the plant science toolbox, submitted), but also potential plant effects on soil solution can be investigated.

In hydroponic culture the composition of root exudates can be studied without adsorption losses to a solid phase, whereas the effect of mechanical impedance experienced by roots growing in soil on exudation is neglected (Neumann and Römheld 2001). The in- or efflux of ions from root segments can be measured in hydroponics using micro electrodes (Plassard et al. 2002), or in gelatinized solutions by visualizing gradients with dye indicators and quantification with videodensitometry (Plassard et al. 1999). In order to add mechanical impedance to growing roots, while maintaining the advantage of controlled soil solution composition, glass beads (Hodge et al. 1996) or sand mixtures (Tang and Young 1982) have been used as growth media for the collection of root exudates. The chemical inertness of these media, however, is limited (Sandnes and Eldhuset 2003). Volcanic glasses like perlite or clays like vermiculite are excellent preculture media, but are of limited use to assess root exudation or chemical gradients around roots (Heim et al. 2003).

Testing root influence on specific soil materials

An effective way of investigating the influence of root activity on the structure or reactivity of soil components like clay minerals or oxides is to study their interaction with isolated root exudates or model compounds (e.g., carboxylates, siderophores) in the absence of plants (Ochs et al. 1993; Reichard et al. 2005). Data on sorption of organic compounds by soil materials can give clues about their migration potential in soils (Jones and Brassington 1998). The compilation of Martell and Smith (1974–1989) provides thermodynamic data on equilibria between exudates as ligands and dissolved metal ions. The behavior of carboxylate anions in soils was reviewed by Jones (1998), that of phytosiderophores by Kraemer et al. (2006). An elegant way to test the effect of individual compounds on the bioavailability of nutrients was presented by Ström et al. (2002). They grew maize seedlings in "rhizotubes", added a solution with carboxylate anions to a ^{33}P labeled patch of soil, and measured the ^{33}P uptake.

Alternatively, minerals can be mixed into an inert substrate and the effect of a growing root system with or without microbial inoculation on weathering can be assessed (Leyval and Berthelin 1991). The spatial extent of root exudation on weathering can be studied effectively using root mat systems as described below (Hinsinger and Gilkes 1997).

Laboratory systems to assess gradients in soil

When studying root influence on soil, simplifications with respect to soil structure and root system geometry are usually involved, and/or compartments with a high root density separated from root-free soil. Depending on the system, destructive methods for the collection of rhizosphere soil can be applied, rhizosphere soil solution can be sampled, or gradients can be assessed by non-invasive tools. There is no unambiguous nomenclature for such systems. For example, rhizotrones and rhizo-boxes are often used for similar types of flat growth systems in which plants form quasi 2D root systems. In the following we will use the term "microcosm" and differentiate between types by the way how roots interact with the soil and how rhizosphere is defined.

Microcosms in which roots are in direct contact with soil

Pot and column studies belong into this category. Differences between bulk and rhizosphere soil can be assessed by separating rhizosphere from bulk soil by shaking or washing (Liu et al. 2004), by resin impregnation followed by microscopic or spectroscopic inspection of thin sections, or by non-invasive 3D tomography (Pierret et al. 2003). Both repacked soil (aggregate structure destroyed) and soil monoliths can be studied.

Flat boxes, in which quasi 2D root systems are formed in a narrow slit filled with soil come in various dimensions. The so-called "Hohenheim" box is inclined to force the root system to develop preferentially along the lower cover plate (Dinkelaker and Marschner 1992). This type of microcosms is usually filled with repacked soil or artificial substrates, which may be arranged in zones of different properties (Hodge et al. 1999). Often the boxes are at least partly transparent to allow the visual observation of root development. Rhizosphere gradients can be assessed by sampling the soil in different distances from the root. More importantly, such microcosms are ideal for the application of non-invasive methods for in-situ characterization of gradients. Soil solution can be sampled in defined distances from given root segments as described below. The advantage of having roots in direct contact with soil is contrasted by the difficulties of detecting small effects by individual roots.

Microcosms in which membranes are used to separate compartments or root mats

Membranes, usually made of poly-amide, are used to separate microcosms into different compartments. Membranes with a mesh size of 20–30 μm can be penetrated by fungal hyphae and root hairs, but not roots. Membranes with a mesh size of 0.45 μm allow exchange of soil solution and gases but neither hyphae nor roots can penetrate.

Compartment systems are devices, in which membranes are used to separate "root zone", "fungal hyphae zone" and root/hyphae free soil. Often the properties of the different compartments are compared as a whole. If root density in the root compartment is large, rhizosphere gradients may be observed in an adjacent soil compartment (Corgié et al. 2003; Vetterlein and Jahn 2004).

In other systems dense root mats are formed which are in contact with the soil via the membrane (Fig. 2). The root mat itself can be in contact with soil or an artificial substrate (Gahoonia and Nielsen 1991), or it is formed in an air-filled compartment (Wenzel et al. 2001). Such systems are ideal for assessing chemical rhizosphere gradients by sampling the soil or the soil solution in the root-free compartment in defined distances from the membrane. The root mat approach has the advantage of amplifying the root influence, and thus to enable the detection also of otherwise small effects. However, the results may not be representative for field conditions with less dense root systems. Also,

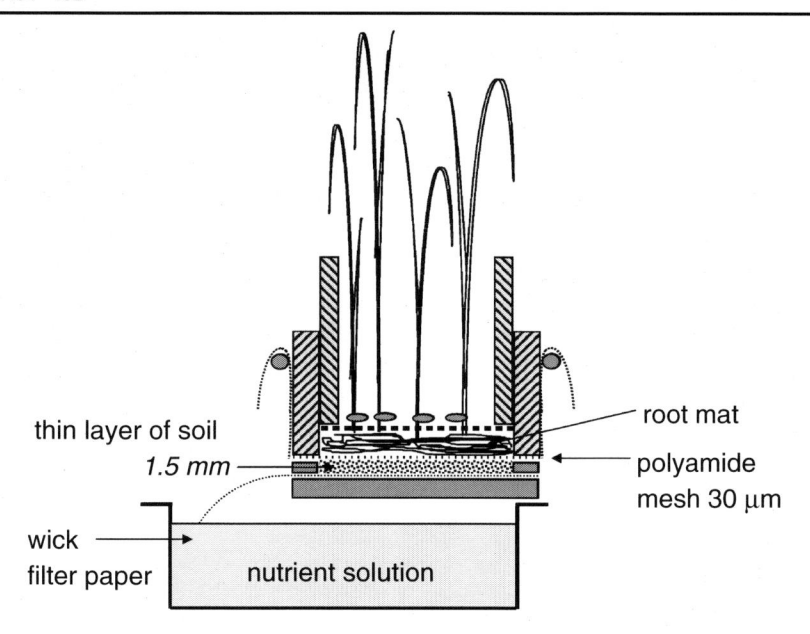

Fig. 2 Example of a root mat type microcosm. It is composed of a lower part containing a thin soil layer (1–3 mm thick; or, alternatively, a soil cylinder of greater height if aiming at studying rhizosphere gradients), and of an upper part containing the root mat, separated by a polyamide membrane. For pregrowth, the upper part is immersed in aerated nutrient solution (adapted from Guivarch et al. 1999, Fig. 1; with kind permission from Springer Science+Business Media); for further explanations see Chaignon and Hinsinger (2003)

the exchange of water and ions between root and soil can be affected by the membrane (Fitz et al. 2006).

Field systems

Lysimeters are large 3D, usually cylindrical, and often weighable structures to study water, element and ion fluxes in larger soil volumes under field conditions (not to be confused with tension or tension-free lysimeters which are soil solution collection devices). Lysimeters either contain a soil monolith or are refilled with loose soil material. While refilled lysimeters allow to establish experimental setups with several treatments under the same soil conditions (Luster et al. 2008), monolith lysimeters provide a controlled access to naturally structured soil (Bergström and Stenström 1998). Rhizosphere in a microscopic sense cannot be studied unless coupled to observation tools such as mini-rhizotrons (Majdi 1996). However, plant effects on soil can be studied by comparing planted and plant-free lysimeters.

There are several designs of root windows described in the literature (Polomski and Kuhn 2002). The most common type consists of glass- or plexiglass plates pressed onto a soil profile and can be combined with sampling and observation methods similar to microcosms of the "flat box" type (Dieffenbach and Matzner 2000).

Sampling and characterization of rhizosphere soil and soil solution

Dependent on soil texture and structure, plant species and observed parameter, root induced changes of most soil properties can be observed up to a distance of a few micrometers to about 7 mm from the surface of an active root segment or a root mat (Jungk and Claassen 1997; Jones et al. 2003). Sampling procedures for rhizosphere soil and solution have to cope with this demand for spatial resolution.

However, rhizosphere effects may also reach beyond this range when considering highly mobile compounds like water or CO_2 (Gregory 2006, Hinsinger et al. 2005) or when including the effects of fungal hyphae extending from mycorrhizal root segments ("mycor-rhizosphere", e.g. Agerer 2001).

Sampling rhizosphere soil

For the separation of rhizosphere soil from so-called bulk soil several procedures based on shaking or washing-off soil particles adhering to roots have been proposed. First, the root system, together with adhering soil is carefully removed from the soil. Then Naim (1965) obtained rhizosphere soil by shaking the root system for 5 min in water. Turpault (2006) defined bulk soil, rhizosphere soil (detaches spontaneously when drying the root system) and rhizosphere interface (falls off when shaking the dried root system). Others define the soil falling off when shaking the root system as bulk soil and only the soil that is removed by subsequent brushing as rhizosphere soil (Yanai et al. 2003). Because soil texture and actual soil moisture strongly influence the amount of soil adhering to the root system, results from different experiments should be compared with caution.

Slicing techniques require root mat type microcosms. Gahoonia and Nielsen (1991) sliced the frozen soil with a microtome in different distances to the root mat. Because freezing the soil may alter its chemical properties, Fitz et al. (2003a) developed a device that allows thin-slicing without freezing.

Characterization of rhizosphere soil

For the characterization of separated rhizosphere soil in principle all soil analytical methods published in text books (see "Introduction") or recommended by organizations such as Deutsches Institut für Normung (www.din.de), US Environmental Protection Agency (www.epa.gov) or United Nations Economic Commission for Europe (www.unece.org) may be used.

There are two major groups of methods for chemical soil properties. The first deals with the total analysis of the soil solid phase, which is generally of little interest to rhizosphere research. The exception is total C and N analysis which is well applicable because of the small amounts of sample required by modern elemental analyzers. The second group comprises a large variety of extraction procedures to characterize different fractions of soil bound molecules or ions. Extractions for organic compounds (root and microbial exudates, contaminants) usually aim at complete recovery. Volatile organic compounds with a boiling point <200°C are purged from a heated soil suspension in water or methanol by an inert gas and trapped on suitable sorbents, while less volatile compounds are extracted using suitable solvents and applying different techniques (Sawhney 1996). By contrast, extractants for elements, inorganic ions and inorganic or organometallic compounds are often chosen to obtain a bioavailable fraction. An overview of commonly used extractants for this purpose is given in Table 1. Note that fractions are defined mainly operationally, and thus results obtained with

Table 1 Common extractants for elements and ions grouped approximately in decreasing order of plant availability as compiled from standard method collections

Phytoavailability of extracted species	N	P	K, Ca, Mg	Fe, Al	Trace metals
↑	H_2O	H_2O	H_2O	H_2O	H_2O
	Hot H_2O; NH_4^+, NO_3^- in salt extracts (KCl, $CaCl_2$..)	Ca-lactate; NH_4-lactate; Citrate	NH_4Cl^a; $BaCl_2^a$	NH_4Cl^a; $BaCl_2^a$	$NaNO_3$; NH_4Cl^a; $BaCl_2^a$; NH_4-acetate
		Ca acetate/lactate; $NaHCO_3$; NH_4F/HCl	HNO_3; HCl	EDTA; NH_4-oxalate	NH_4–EDTA; NH_4-oxalate
	H_2SO_4		HCl/HNO_3	Na-dithionite; HCl/HNO_3	HNO_3; HCl/HNO_3

For most extractants there are several slightly different protocols in terms of extractant concentration, extraction time, etc. Also, there can be large differences in the extractive power of a given extractant depending on soil properties such as pH or soil organic matter content (e.g. some extractants can only be used either for calcareous or acidic soils)

[a] Methods to determine exchangeable cation contents; from the sum of all major cations the cation exchange capacity of the soil can be calculated

different methods may not be easily compared. Nevertheless, depending on extractant, element and plant species there may be good correlations between extractable element concentration and plant uptake (citations in Sparks 1996 or Pansu and Gautheyrou 2006). A comprehensive characterization of soil-bound elements can be achieved by sequential extractions. There are protocols defining several fractions for organic nitrogen and carbon (Stevenson 1996; VonLützow et al. 2007), phosphorus (Psenner et al. 1988; Kuo 1996) and trace metals (Tessier et al. 1979; Zeien and Brümmer 1989). Since extraction methods have been developed without sample volume restrictions, the often limited sample amount may hamper their application in rhizosphere research, depending on analyte content in the soil and on the sensitivity of the analytical method. Generally extracts can be analysed by commonly available analytical equipment such as potentiometry, molecular absorption spectrometry, gas and liquid chromatography, atomic absorption spectrometry (AAS) or inductively-coupled plasma optical emission spectrometry (ICP-OES). Only the detection of less-abundant analytes asks for more specialised equipment involving mass-spectrometric detection. Because the availability of standard reference materials for extractable contents in soils is limited (www.nist.gov/srm; www.erm-crm.org), most extraction methods require the use of internal references and the traceability of instrument calibration to certified standards.

Isotopic exchange is another method for determining bioavailable contents applicable to ions of a few elements with radioactive isotopes (PO_4^{3-}, SO_4^{2-}, K^+, Zn^{2+}, Cd^{2+}; Frossard and Sinaj 1997). A small amount of isotopic tracer is added to a soil suspension and the dilution of the label by homoionic exchange with the non-labeled ions at the soil solid phase is characterized. Either so-called *E*-values (contents in the soil solid phase that are exchanged within a defined incubation time), or kinetic parameters of the exchange are determined.

Collection of soil solution

Göttlein et.al. (1996) presented a system for the microscale collection of soil solution based on micro suction cups made of ceramic capillaries with an outer diameter of 1 mm. Their system was used successfully to detect gradients in the rhizosphere (Göttlein et al. 1999). Matrices of micro suction cups placed in front of a developing root system allowed to monitor the changes in soil solution chemistry when the root system passed through (Fig. 3; Dieffenbach et al. 1997). This micro suction cup system was slightly modified by Dessureault-Rompré et al. (2006) to allow for localized collection of carboxylate anions and by Shen and Hoffland (2007) who introduced polyethersulfone as porous cup material. Puschenreiter et al. (2005a) presented a suction cup with a different geometry based on a nylon membrane (diameter 3 mm) suitable for sampling soil solution in a defined distance to root mats. Sampling soil solution with micro suction cups faces the same problems and restrictions as with ordinary suction cups, just on a smaller scale. Firstly, sampling is influenced by the contact with the soil matrix, and by texture and actual moisture of the soil. Secondly, analytes may be sorbed by or released from the sampling system (Rais et al. 2006), which asks for thorough testing of a particular system for a given problem. Nevertheless, the method has been applied successfully to assess rhizosphere gradients for major inorganic cations and anions (Wang et al. 2001), organic acid anions (Dessureault-Rompré et al. 2006) and trace metals (Shen and Hoffland 2007).

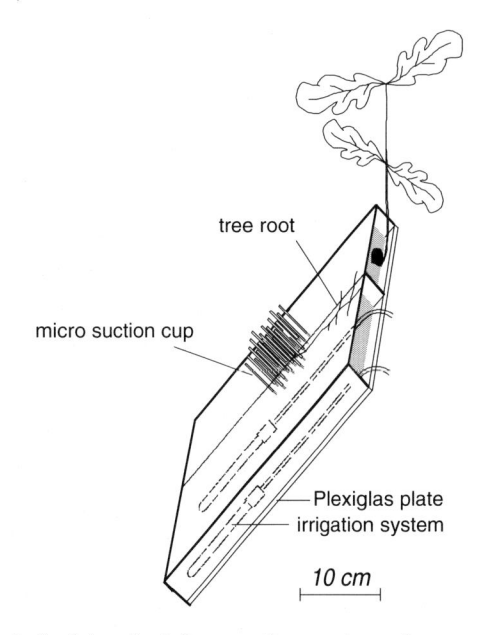

Fig. 3 Studying the influence of a growing oak root on soil solution chemistry using a micro suction cup array installed in a "Hohenheim" type microcosm (adapted from Göttlein et al. 1999; with kind permission from Springer Science+Business Media)

Alternatively, soil solution can be trapped by the application of filter papers, cellulose acetate filters or blotting membranes onto roots exposed in flat rhizoboxes, a method which has been used mainly for the collection of root exudates or root-secretory enzymes (Neumann 2006).

Analysis of small volumes of aqueous solution

The miniaturization of sampling devices also minimizes the sample volume available for analysis. In principle all common analytical methods like ICP-OES, AAS, HPLC (high performance liquid chromatography), IC (ion chromatography), or colorimetry (manual or automatic as in flow-injection and auto analyzers) can be used, because except for flame AAS and standard ICP applications the sample amount needed for the measurement itself is not very high. The main task in adapting analytical methods to small sample volumes often is to optimize the autosampling system (Table 2). There are techniques available that significantly reduce the sample consumption of ICP-OES (Mermet and Todoli 2004) or ICP-MS (Prabhu et al. 1993; Lofthouse et al. 1997), which is normally in the range of several milliliters. Capillary electrophoresis (CE) offers the possibility to analyze samples as small as one droplet. Göttlein and Blasek (1996) optimized CE for the analysis of major cations and anions in soil solutions. Because CE is a true ion-analytical method it offers the possibility to detect the potentially phytotoxic Al^{3+} ion, which is of particular interest for studies of acidic soils (Göttlein 1998). Combining the analysis of labile species by CE or miniaturized voltammetric systems (Tercier-Waeber et al. 2002) with total analysis by graphite furnace AAS or micro-injection ICP methods (Göttlein 2006) allows metal speciation in rhizosphere solutions (Dessureault-Rompré et al. 2008). ISFET-sensors enable pH measurements in one to two droplets (Göttlein and Blasek 1996), and afterwards the sample can be used for other analyses, because the sensors do not contaminate the sample like standard pH electrodes. Dissolved organic carbon (DOC) in small sample volumes can be measured using TC analyzers with a direct sample injection option, or, taking the UV absorption as an indirect measure, using an HPLC system with a UV-detector but without separation column (Göttlein and Blasek 1996). Employing the microanalytical methods described above, a comprehensive characterization of soil solution including metal speciation is possible with a sample volume of about 250 µl. If only pH measurement and CE analysis of cations and anions are done, 30 to 50 µl are sufficient. Very small liquid sample volumes may also be analyzed by scanning electron microscopy coupled with energy-dispersive X-ray analysis, however after sophisticated sample preparation (Bächmann and Steigerwald 1993).

Table 2 Techniques for analyzing main parameters of aqueous solutions and their applicability to rhizosphere research

Technique (analytes)	Availability, costs	Suitability for/adaptation to rhizosphere research (limited sample amount)
Potentiometry (pH)	Common, low	ISFET instead of glass electrodes
Flow injection analysis (NH4)	Common, low	Autosampler and sample loop limiting
Voltammetry (labile metal cations)	Special, low	Micro-sensors necessary, however sample demand still in ml-range
TC/TN analyser (DOC, CO_3, N_{tot})	Common, intermediate	Autosampler and sample injection limiting; direct injection option reduces sample demand to 50 µl
Ion chromatography (inorganic anions, organic acids, NH4)	Common, intermediate	Autosampler and sample loop limiting; microbore systems allow reduction of sample demand to the sub-µl-range
HPLC (organic acids, sugars, etc.)	Common, intermediate	As for ion chromatography
Flame AAS (total metal conc.)	Common, intermediate	Hardly possible because of high sample demand
Graphite furnace AAS (total metal conc.)	Special, intermediate	Suitable, sample demand of 20 to 50 µl for single element analysis
Capillary electrophoresis (inorganic anions, organic acids, free metal cations, NH4)	Special, intermediate	With a demand of 20 nL suitable for the analysis of minimal sample amounts
ICP-OES	Common, expensive	Special nebulizers for lowering sample demand to about 100 µl for multielement analysis
ICP-MS	Special, expensive	As for ICP-OES

Since for small solution samples the risk of contamination or adsorption losses is particularly high, the proper preconditioning and cleaning of all devices and containers that the sample comes in contact with are pivotal to reliable results (for recommended methods see Nollet 2007). Furthermore, evaporation losses during sampling should be minimised (Göttlein et al. 1996). Some natural water standard reference materials (www.nist.gov/srm; www.erm-crm.org) can be used for total analysis. For speciation, quality assurance must rely on internal references.

Sampling and analysis of soil gases

Measuring the total efflux of CO_2 in-situ from a given, usually circular surface area of soil using infrared gas analysers is a well established and routinely used method. The contribution of rhizosphere respiration has been estimated either by comparing total soil respiration with respiration measured after terminating autotrophic respiration by detopping of plants (Andersen and Scagel 1997), girdling (Ekberg et al. 2007) or trenching (Sulzmann et al. 2005), or by applying suitable modeling to the soil respiration data (Raich and Mora 2005). Alternatively, rhizosphere respiration can be assessed by coupling ^{13}C labeling of the plant shoots with sampling of the soil CO_2 efflux and analysing its $\partial^{13}C$ using isotope-ratio mass spectrometry (Yevdokimov et al. 2007).

Membrane probes allow the diffusive sampling of soil gases like CO_2, N_2O, CH_4 or H_2 at various soil depths in the field or in microcosms (Rothfuss and Conrad 1994; Yu and DeLaune 2006), and are sometimes coupled with on-line analysis (Panikov et al. 2007). It should be tested whether gradients in the partial pressure of gases from the rhizosphere to the bulk soil can be assessed with this technique. The oxygen concentration in soil can be measured with microelectrodes in high spatial resolution (Rappoldt 1995).

Cutting-edge methods for studying plant effects on rhizosphere soil

In-situ assessment of soil solution

In-situ measurements of chemical variables in the rhizosphere involve both the characterization of the solid and the solution phase. Impregnating rooted soil "profiles" in microcosms with dye indicators dissolved in agarose gel has been used for assessing root induced changes in pH (Fig. 4) and the exudation of

Fig. 4 Effect of soil-buffering capacity ($CaCO_3$ content) on the extension of root-induced rhizosphere acidification of chickpea *(Cicer arietinum* L.) seedlings 12 DAS, detected in "Hohenheim" type microcosms by soil impregnation with pH-indicator (bromocresol purple) agar (from Römheld 1986; courtesy of the International Potash Institute, Switzerland)

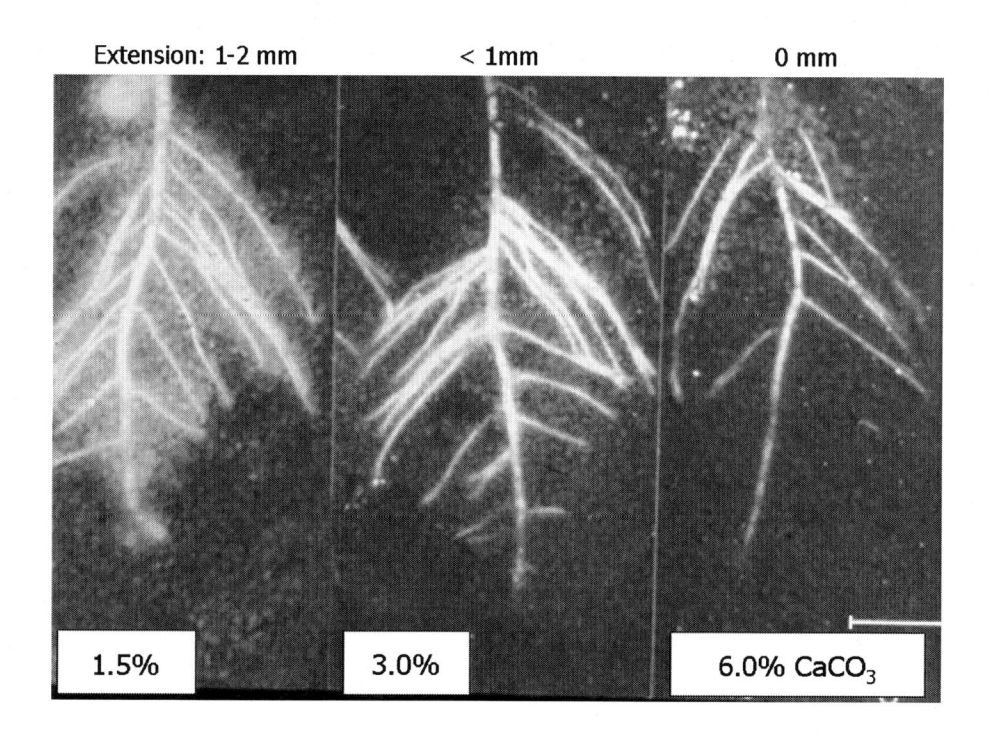

aluminum complexing ligands or Fe(III) reducing agents (Engels et al. 2000; Neumann 2006). Root-induced Mn reduction and the excretion of acid phosphatases can be detected by applying specially impregnated filter papers to the rooted soil "profiles" (Dinkelaker and Marschner 1992; Dinkelaker et al. 1993). While such staining methods can be used to monitor pH changes in the rhizosphere with time in artificial systems composed of agarose gel (Plassard et al. 1999), they can hardly be used for a continuous monitoring in real soil. Recently, a novel non-invasive method was presented by Blossfeld and Gansert (2007) for the visualisation of rhizosphere pH dynamics in waterlogged soils using a pH-sensitive fluorescent indicator dye in a proton permeable polymer matrix (pH planar optode). However, the applicability of this method to non-saturated soils has still to be proven. In aerated soils, antimony microelectrodes allow high resolution monitoring of root induced changes of pH in the rhizosphere (Häussling et al. 1985; Fischer et al. 1989; Zhang and Pang 1999). Measuring soil redox potential with Pt microelectrodes dates back to Lemon and Erickson (1952) and has seen improvements to date (Hui and Tian 1998; VanBochove et al. 2002; Cornu et al. 2007). In particular, they were used in microcosms to monitor redox gradients in the rhizosphere of rice in order to study the formation of iron plaque on roots (Bravin et al. 2008). Except for a single application for the Na^+ ion by Hamza and Aylmore (1991) selective electrodes have not been applied to other chemical parameters due to the lack of suitable electrodes that can be operated reliably in soil.

The DGT-technique (diffusive gradients in thin-films, Zhang et al. 1998) has been developed to evaluate the phytoavailable pool of metals and phosphorus. A DGT device consists of a gel-embedded resin layer acting as a sink for the species of interest, overlaid by another gel layer and a filter through which the molecules or ions have to diffuse to reach the resin. Element and ion contents in soil extracted by DGT correlate well with contents in plants (Zhang et al. 2001). Up to now, DGT devices have been applied mostly to moist pastes of separated soil samples. However, they are particularly promising tools for direct application to the surface of rooted soil "profiles" in rhizoboxes (Fitz et al. 2003b; Nowack et al. 2004). Spatially resolved maps of DGT extractable species can be obtained by slicing the resin gel prior to analysis (Zhang et al. 2001) or by measuring the metal in the resin gel by laser ablation ICP-MS (Warnken et al. 2004).

Biosensors

Whole-cell bacterial biosensors are constructed by insertion of a gene coding for an autofluorescent protein, the most common one being the *lux* gene for the green fluorescent protein (GFP; Killham and Yeomans 2001). Three types have been developed, differing by the physiological process the expression of bioluminescence is related to. Firstly, in non-specific biosensors, bioluminescence is related to the basal metabolism. They can be used to detect C rhizodeposition (strains with a broad range of substrates should be chosen to account for all exudates) and rhizosphere bacterial colonization. In semi-specific biosensors, luminescence is linked to a generic process such as oxidative stress. In specific biosensors, lighting reports on the expression of a specific pathway such as the utilisation of a particular exudate compound, the degradation of or resistance to a given contaminant. A number of biosensors have been developed to estimate the bioavailability of organic and inorganic contaminants (Hansen and Sørensen 2001). While the simplicity and rapidity of the measurement, and the possibility to monitor *in situ* various substances over time make biosensors attractive, their application to real-world environmental samples is still a challenge (Rodriguez-Mozaz et al. 2006). They cannot be applied directly to soils because soil particles absorb part of the emitted light, and some soil constituents are autofluorescent. Usually, either the biosensor is inoculated and then extracted from the soil before analysis, or the biosensor is applied to a solution after an extraction stage. Several parameters should be considered carefully during the analysis such as the colonization of the medium, the survival of the organisms over time, and possible matrix effects due to the presence of organic matter, other contaminants, etc. The distribution of compounds can be visualised by combining biosensors with imaging by a CCD camera, as shown for root exudates in sand microcosms (Paterson et al. 2006). In most cases, the measured signals are used to compare different conditions, but not to determine the actual concentration of a compound.

Characterization of ultrastructure and element mapping using microscopic, diffractometric and spectroscopic techniques

This subsection is restricted to studies of the soil solid phase, while the characterization of roots is addressed in Neumann et al. (2008, Strategies and methods- the plant science toolbox, submitted). Standard techniques for two-dimensional element mapping are scanning electron microscopy (SEM) and transmission EM (TEM) coupled with energy dispersive X-ray microanalysis (EDX). Energy filtered TEM (EFTEM) offers a higher resolution and better detection limit (about 10 nm and 1–10 μg g^{-1}, respectively). Other tools for two-dimensional element mapping include synchrotron-based micro X-ray fluorescence (μSXRF), micro-particle induced X-ray emission (μPIXE), secondary ion mass spectrometry (SIMS) and laser ablation (LA)-ICP-MS. SIMS and LA-ICP-MS have been coupled with stable isotope probing (SIP) to image the distribution of C isotopes in the soil at a sub-micrometer (nanoSIMS) and sub-millimeter (LA-ICP-MS) resolution (Bruneau et al. 2002; DeRito et al. 2005). Three-dimensional images of soil porosity can be obtained non-invasively by X-ray computed tomography (CT; Mooney et al. 2006a), a method also used to study root architecture in-situ (Hodge et al. 2008, Plant roots: growth and architecture, submitted). Alternatively, Moran et al. (2000) used X-ray absorption and phase contrast imaging to study the relation between roots and soil structure, and Mooney et al. (2006b) investigated the relation between the structure of a mineral landfill cap and root penetration by polarising microscopy.

The various microscopic techniques listed above can be used on any growth system (artificial, microcosm or field soil) after appropriate sample preparation. This sample preparation is a critical step for rhizosphere samples because they contain living and hydrated components. Classical procedures involving dehydration, chemical fixation, resin embedding and staining are progressively replaced by cryo fixation. The latter enables the measurement of hydrated samples with techniques such as SEM, TEM, μXRF and μPIXE, thus limiting possible artefacts related to dehydration and keeping the systems in a more natural state (Fomina et al. 2005). Environmental SEM (ESEM) also enables observation and analysis of hydrated root and soil samples with minimal perturbation (e.g. Cabala and Teper 2007), however at a limited resolution.

Despite recent advances in data acquisition time each analysis by a microscopic technique implies a compromise between resolution and size of the sample. Therefore, the representativeness of the samples should be evaluated, possibly by upscaling from high resolution to coarser observation scales.

Mineral weathering and formation of secondary minerals have been studied intensively by EM techniques, particularly by SEM-EDX (Gadd 2007) and TEM-EDX (Hinsinger et al. 1993). Observing the size and shape of minerals and estimating their composition allow to predict the nature of the minerals present. X-ray diffraction (XRD) allows a direct identification of minerals. Standard powder diffractometers are limited by the amount of sample required (1 g), but recent instruments require only a few tens of mg. Using EM and XRD, various precipitates and products of mineral weathering were detected in the vicinity of fungi and roots (Hinsinger et al. 1993; April and Keller 2005; Gadd 2007). However, the weak sensitivity of XRD for minor phases remains a major limitation. It can be partly overcome by micro-XRD (μXRD) using laboratory or synchrotron X-ray sources, or by separation prior to XRD analysis. Furthermore, XRD on oriented clays, which requires only a few mg of particles, is suited to trace changes in clay mineralogy occurring in the rhizosphere, as shown in artificial substrates (Hinsinger et al. 1993) and in soils (Kodama et al. 1994). Recently, Barré et al. (2007) proposed a more quantitative approach for studying changes in the composition of the clay fraction in the rhizosphere.

The local chemical environment of metals can be assessed by X-ray absorption spectroscopy (XAS), including X-ray absorption near edge structure (XANES, also called NEXAFS for near-edge X-ray absorption fine structure) and extended X-ray absorption fine structure (EXAFS) spectroscopy. Major advantages of these techniques include element specificity, sensitivity to amorphous and weakly crystalline species, and detection limits for soil samples of 10 to 100 mg kg^{-1} for XANES and of 100 to 300 mg kg^{-1} for EXAFS, depending on target element and matrix. Bulk XAS provides information on major metal species. This technique was combined with μXRF (Voegelin et al. 2007) and X-ray fluorescence microtomography (Hansel et al. 2001; Blute et al. 2004) to study the distribution and speciation of heavy metals in the root plaque of plants growing in flooded environments.

These studies revealed a heterogeneous composition of Fe(III) and Fe(II) phases with associated trace element species including As(V) and Zn(II), whereas Pb(II) was complexed by organic functional groups possibly belonging to bacterial biofilms. Micro-XAS (μXAS), generally combined with bulk XAS and μXRF, provides information on the chemical form of metals with a lateral resolution of a few square micrometers to a few hundreds of square nanometers (Manceau et al. 2002). These tools were used to study the impacts of remediation treatments on metal speciation in contaminated substrates (Fig. 5; Nachtegaal et al. 2005; Panfili et al. 2005, Manceau et al. 2008). Micro XRD, available as additional tool on some spectrometers, allows the simultaneous identification of crystalline metal bearing phases (Lanson et al. 2008). These tools can be applied to any growth system (artificial, microcosm or field soil) after homogenizing and grinding (for bulk XAS), or after resin impregnation followed by thin sectioning (for μXRF/μXAS/μXRD). A major limitation of these synchrotron-based techniques (and of state-of-the art microscopic facilities in general) is their restricted access due to the small number of beamlines and microscopes worldwide.

The speciation of light elements including carbon, nitrogen, sulfur and phosphorus can be studied by bulk XANES and by scanning transmission X-ray microscopy (STXM, including μXRF and μXANES) using soft X-rays (Myneni 2002). The X-ray spot sizes are generally <1 μm and can be as small as few tens nm. Most STXM spectrometers allow the study of wet systems. These techniques have been used to study soil colloids (Schumacher et al. 2005) and bacterial biomineralization (Benzerara et al. 2004) at the single-particle and single-cell scale, respectively. Electron energy loss spectrometry (EELS) is a more exotic technique for speciating elements. Main advantages are the coupling with TEM imaging and the very good lateral resolution of around 10 nm (Watteau and Villemin 2001).

^{13}C, ^{31}P, ^{15}N and ^{1}H solid and liquid state nuclear magnetic resonance (NMR) spectroscopies are classical tools for the characterization of molecular structures and functional groups in soil organic matter (SOM) and for the identification of low molecular weight molecules (Fan et al. 1997). Advanced techniques such as high-resolution magic-angle spinning and 2D NMR open new possibilities (Kelleher et al. 2006). The large sample size required for solid state NMR (0.5 to 1 g of isolated SOM compared to a few tens of mg for liquid state NMR), limits its use for rhizosphere applications. Fourier transformed infrared (FTIR) spectroscopy is another classical tool for the characterization of molecular structures in SOM. Attenuated total reflectance (ATR)-FTIR allows the study of wet systems, and FTIR microscopy enables 2D mapping with a resolution of a few micrometers (Raab and Vogel 2004). Electron paramagnetic resonance (EPR) has been used to quantify free radicals in organic molecules, and to

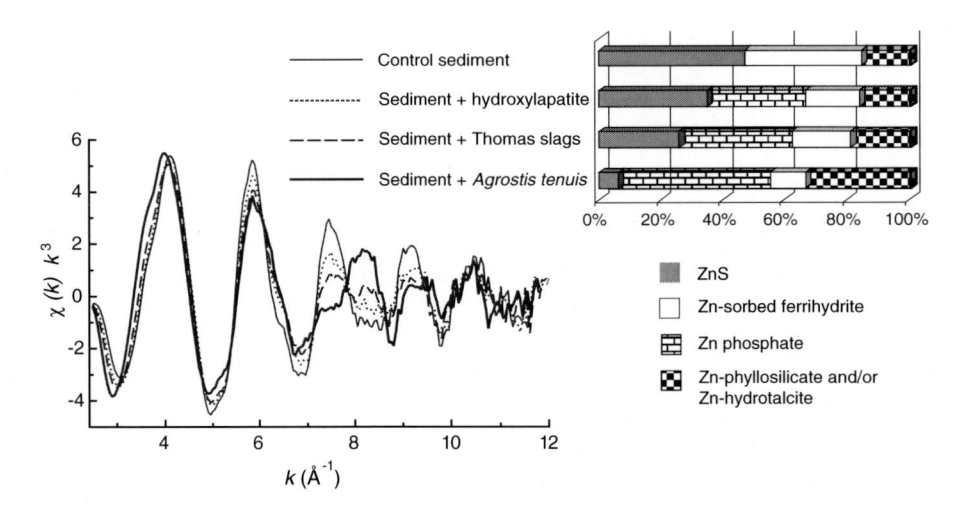

Fig. 5 Zn K-edge bulk EXAFS spectra of a Zn-contaminated sediment (control), treated with mineral amendments and planted with *Agrostis tenuis*, and distribution of Zn species determined from the analysis of these data and μEXAFS spectra. The amendments induce a significant oxidation of ZnS and the formation of secondary species. These effects are strongly enhanced in the presence of *A. tenuis*, with an almost complete removal of ZnS [adapted from Panfili et al. 2005; Copyright Elsevier (2005)]

study the interaction of paramagnetic metals with SOM in terms of oxidation state, ligand types and coordination geometry (Senesi 1996). For EPR, the same sample size restrictions apply as for solid state NMR.

Labelling with and tracing/imaging of stable and radioactive isotopes

Carbon fluxes in the rhizosphere can be assessed by $^{14}CO_2$ or $^{13}CO_2$ pulse-labelling the atmosphere of a plant soil system, and measuring the radioactivity or the $\partial^{13}C$ value in the compartment of interest (soil, isolated DOC, microbial biomass, roots, etc.) by liquid scintillation or isotope ratio mass spectrometry (IRMS), respectively (Killham and Yeomans 2001; Rangel Castro et al. 2005). Gas chromatography may be coupled with IRMS in order to probe a specific molecule or family of molecules (Derrien et al. 2005). A more exotic method is the labelling with ^{11}C (Minchin and McNaughton 1984).

Laterally resolved information on the distribution of an isotope can be obtained in different ways. Gradients around roots can be determined using microcosms of the root mat type and analyzing slices of soil at various distances from the root mat (Kuzyakov et al. 2003). Microcosms of the "Hohenheim" type allowed to assess the equilibration of stable isotope labels for Mg, K and Ca between rhizosphere soil and solution (Göttlein et al. 2005). Autoradiography on flat microcosms provides non-invasive 2D imaging of the distribution of radioactive isotopes. Images were classically obtained on films or photographic emulsions, then on phosphor storage screens, and more recently by electronic autoradiography (Fig. 6; Rosling et al. 2004). Apart from following C fluxes, this versatile method can be used to characterize the spatial distribution and its change over time of added radioactive P (Hendriks et al. 1981; Hübel and Beck 1993; Lindahl et al. 2001), SO_4^{2-} (Jungk and Claassen 1997) or Zn and Cd (Whiting et al. 2000).

The use of stable isotope probing (SIP) to assess microbial activity in the rhizosphere is treated by Sørensen et al. (2008, Strategies and methods- the microbial ecology toolbox, submitted).

Mapping the plants influence on soil moisture

Using micro-tensiometers and small time-domain reflectometry sensors installed in rhizoboxes and compartment systems, one-dimensional rhizosphere gradients in soil moisture and differences between root and root-free compartments could be shown (Göttlein et al. 1996; Vetterlein and Jahn 2004). Recently, microorganisms have been genetically altered to indicate changes in soil moisture by varying the expression of the green fluorescent protein as detected by epifluorescence microscopy (Cardon and Gage 2006).

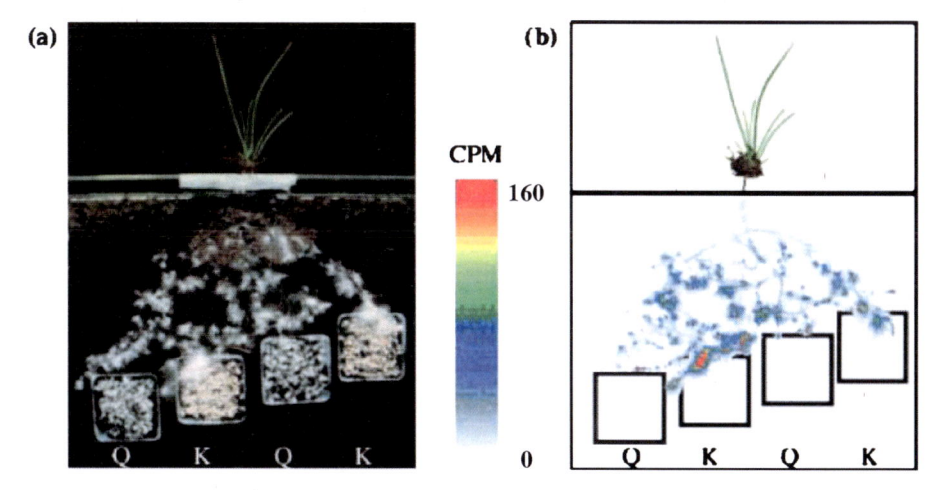

Fig. 6 Peat microcosm containing *Pinus sylvestris* seedlings colonised by *Hebeloma crustuliniforme* and pure mineral patches of either K feldspar (*K*) or quartz (*Q*). Fifteen weeks after introducing mineral patches at the growing mycelial front (**a**), the shoots were pulse labelled with $^{14}CO_2$. Greater amounts of labelled carbon are allocated to root tips and mycelia associated with patches of K feldspar compared to patches of quartz (**b**). *CPM* counts per minute. (adapted from Rosling et al. 2004; with kind permission from the New Phytologist Trust)

Some of the methods to image root systems in microcosms are sensitive also to differences in substrate moisture and can therefore be used to assess the plants influence on soil moisture distribution. Light transmission imaging (Garrigues et al. 2006) is a rather inexpensive method with which large quasi 2D microcosms (e.g. 1,000×500×4 mm) can be studied at a resolution of ≥1500 μm. With magnetic resonance imaging (MRI; Chudek and Hunter 1997; Herrmann et al. 2002), which depends on the accessibility to a medical imager or an NMR spectrometer with a suitable accessory, 3D images can be obtained from boxes (up to 70×70×20 mm) or cylinders (diameters up to 60 mm and heights up to 200 mm) at a resolution between 10 and several hundred micrometers. Considering the high spatial resolution, these methods are able to assess plant effects on soil moisture on the scale of a single-root. However, their applicability to real soil is limited by inherent incompatibilities. Light-transmission is restricted to translucent sand with addition of small amounts of clay and MRI to soils with low iron contents. By contrast, X-ray computed tomography allows to map root effects on structure and moisture distribution in real soils at a resolution of 100 μm to 1 mm for typically cylindrical samples with a diameter of a few cm (Hamza and Aylmore 1992; Gregory and Hinsinger 1999). The sensitivity to soil water content, however, is comparatively weak. Recently, Oswald et al. (2008) demonstrated the high sensitivity of Neutron radiography to differences in soil water content and could show variable water uptake by different parts of root systems growing in flat microcosms (170×150×13 mm) made of aluminum at a spatial resolution of ≥100 μm. Although the contrast is highest in quartz sand, the method can also be applied to natural soil (Menon et al. 2007).

Electrical resistivity tomography (ERT) and ground penetrating radar (GPR) are non-invasive geophysical methods increasingly used in hydrological studies of the vadose zone. ERT is a comparatively inexpensive method exploiting the spatial variability in the electrical conductivity of the soil (Benderitter and Schott 1999). Among other applications the method can be used to monitor changes in soil water content in the field indirectly via inverse modelling of resistivity and the use of petrophysical relationships. Large stone contents make application of ERT difficult and spatial resolution for true non-invasive surface applications decreases strongly with soil depth. GPR velocity tomography can be used for the same purpose, because the water content influences the soils permittivity to radar waves (Annan 2005). The method, however, is ineffective in soils with clay. A few studies have made the attempt to use ERT and/or GPR tomography to examine spatial variability or temporal changes in soil moisture content caused by plant water uptake on the scale of the whole root system (Fig. 7; Michot et al. 2003; AlHagrey 2007). Theoretically, depending on the electrode spacing or the antenna frequency, the spatial resolution of ERT and GPR can be increased to the cm range. However, feasability and applicability to map root–soil water interactions in the field on a smaller scale than the whole root system remain to be shown.

Rhizosphere modeling

The nature of concentration gradients in the soil caused by plant activity depends mainly on two sets of factors that modeling needs to take into account. These are (1) physical and biological factors such as

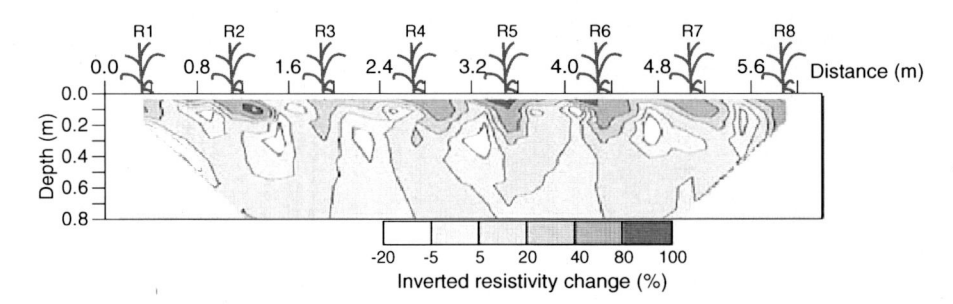

Fig. 7 Changes in soil moisture in a profile during drying shown as difference between the inverted electrical resistivity at about 8 days after irrigation and immediately after irrigation. Root zones of corn rows (*R1* to *R8*) show as dark zones that dry out quickly (adapted from Michot et al. 2003; Reproduced/modified by permission of American Geophysical Union)

geometry, morphology and symbiotic status of the root system, rates of growth, uptake and exudation by roots, and diffusion properties of the soil around roots, and (2) chemical factors such as the distribution and speciation of chemical elements in the soil.

There are two main approaches to model rhizosphere processes. The first category of models follows a macroscopic, empirical approach and operates on a whole plant or even field scale. Here the root system is treated as a single unit without considering the effect of individual roots. The second category deals with a single root or a root system and follows a microscopic approach. Table 3 gives an overview of the categories and the scales discussed in this chapter.

Macroscopic models

Macroscopic models are descriptive and explanatory and help to understand the dynamic and complex interactions occurring adjacent to roots (Darrah et al. 2006). These models can have several layers of complexity, ranging from simple single-root models to sophisticated whole-root system models.

Crop/forest models Although many models predicting the flow of nutrients between soil and plants have been developed, few of these deal in detail with root processes. Such models often use a simplified approximation of rhizosphere processes and verification is at scales larger than the individual plant. Such models have been used intensively as a tool to analyze the performance of cropping systems under variable climate (Wang and Smith 2004) or forest growth affected by different environmental variables (Pinjuv et al. 2006). They typically involve many subprocesses and satisfactory verification does not guarantee that the rhizosphere subprocesses have been modeled accurately (Darrah et al. 2006). Root water uptake is normally treated in a highly simplified submodel, usually with the root system acting as a zero-sink for nutrients, with uptake controlled by soil water potential and transpiration rate or by diffusion flux rate (Darrah 1993). These models can be used to investigate the relative impact of integrated rhizosphere processes on plant and crop scales. They normally incorporate numerical schemes for deducing nutrient concentrations at root surfaces from bulk soil parameters, but do not represent the rhizosphere as a volume of soil with properties different from the bulk soil (Dunbabin et al. 2006). Some models also incorporate the influence of exudation or microorganisms on uptake (Siegel et al. 2003).

Biogeochemical ecosystem models These models are used to identify the governing parameters in ecosystems in order to understand element or nutrient cycles or to predict ecosystem dynamics. Examples include the DNDC model which simulates soil carbon and nitrogen biogeochemistry (Li et al. 1994). A plant growth submodel is used to calculate root respiration, N uptake and plant growth and these processes are linked to climate and soil status. Biogeochemical

Table 3 Approaches and scales in rhizosphere modeling

Model type		Model scale	Main model targets	Examples
Macroscopic (empirical)		Agricultural field/forest	Plant yield, forest growth	Pinjuv et al. (2006); Siegel et al. (2003); Cosby et al. (1985)
		Ecosystem	Element and nutrient cycles	Li et al. (1994)
		Soil profile	Water transport	Somma et al. (1998)
		Whole root system	Root growth	Diggle (1988); Doussan et al. (2006); Dunbabin et al. (2002); Lynch et al. (1997)
Microscopic (explanatory)	Semi-empirical	Single root	Root processes	Nye and Tinker (1977); Barber (1995); Kirk (1999); Roose et al. (2001)
		Root system	Root system development	Roose and Fowler (2004a, b)
	Molecular	Soil solution	Speciation in solution	Calba et al. (2004); Puschenreiter et al. (2005b)
		Single root	Integration of chemical reactions	Geelhoed et al. (1999); Nowack et al. (2006)
		Soil profile	Integration of all mechanisms	Seuntjens et al. (2004)

models pay more attention to soil processes than crop models. Complexation, cation exchange, precipitation, and adsorption can be included in various degrees of complexity (Cosby et al. 1985; Alewell and Manderscheid 1998).

Soil profile scale Soil physical models describing water transport in soils also include a root water uptake term, usually a pressure head dependent sink term that is introduced into the soil water balance (Hopmans and Bristow 2002). There has been a tendency to describe the root water uptake analogous to Darcy's equation, assuming that the rate of uptake is proportional to soil hydraulic conductivity and the difference between the total pressure head at the root–soil interface and the corresponding pressure head in the soil. This approach is useful to understand the root water extraction process, but it is difficult to use for the interpretation of field data. Water transport models have been extended to include solute uptake. In one example a three-dimensional solute transport model including passive and active nutrient uptake by roots has been linked to a three-dimensional transient model for soil water flow and root growth (Somma et al. 1998).

Whole root system scale Several root architecture models are available that simulate the growth of whole root systems at high spatial resolution to generate two or three-dimensional representations of root systems, e.g. ROOTMAP (Diggle 1988), SimRoot (Lynch et al. 1997) or Root Typ (Pagès et al. 2004). An example of a modeled root system is shown in Fig. 8a. Doussan et al. (2006) extended a

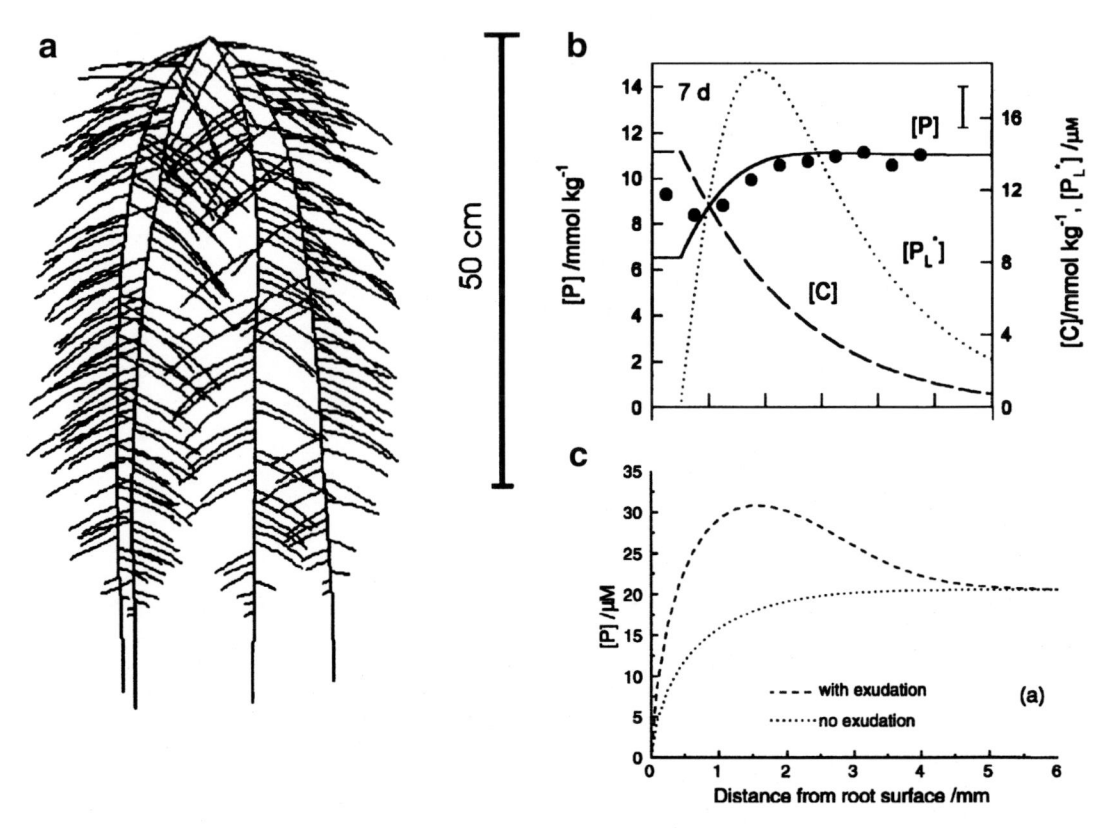

Fig. 8 Examples of different rhizosphere models. **a** Macroscopic model, whole root system scale: modeled root system of *Lupinus albus* (from Doussan et al. 2006; with kind permission from Springer Science+Business Media). **b** Microscopic, mechanistic single root model of citrate exudation and its influence on phosphate solubilization (*dots* experimental, *black line* modeled P in soil, *dotted line* P in solution, *dashed line* citrate in soil; from Kirk 1999; with kind permission from Blackwell Publishing). **c** Microscopic single root model, molecular scale: influence of citrate on phosphate mobilization (P in solution in the absence and presence of citrate exudation; from Geelhoed et al. 1999; with kind permission from Blackwell Publishing)

whole root-system model to include water transport in soils with full coupling of water transport in the root system and the influence of aging on the hydraulic conductivity of root segments and thus on water uptake. The linking of such models to the underlying biology is not yet strongly advanced (Darrah et al. 2006). However, several models have been developed that take into account interactions between root systems, water and nutrients in the environment (Dunbabin et al. 2002). Wu et al. (2007) recently presented a dynamic simulation model that is multi-dimensional, operates on a field scale, is weather driven and models C and N cycling between plants, soil and microbes.

Microscopic models

Microscopic models, also called explanatory models, help to understand the complex and dynamic interactions in the rhizosphere and are based as far as possible on mechanistic relations derived from the laws of chemistry and physics and empirical relations (Kirk 2002). These models can be divided into two subgroups, the molecular and the semi-empirical models. The molecular models are based on the description of chemical processes by a suite of single reactions, e.g. speciation in solution or surface complexation. The semi-empirical models use a more simplified description of molecular processes, e.g. a buffer power to describe adsorption, desorption or precipitation/dissolution.

Semi-empirical models on the single root scale Semi-empirical root models simulate the uptake of nutrients by an isolated root segment. The classical rhizosphere model is that of Nye and Tinker (1977) and Barber (1995). It supposes a cylindrical root surrounded by an infinite amount of soil, with convection and diffusion of nutrients through the soil and uptake through Michaelis–Menten type kinetics at the root surface. The non-linearity of the model requires a numerical solution but recently an analytical solution of the equations was obtained (Roose et al. 2001). This model has also been extended to describe P or metal uptake in microcosms of the root mat type (Kirk 1999; Puschenreiter et al. 2005b). Most of these models are based on a rather simplified description of soil chemistry and the effects of plant roots. The actions exerted by roots on their rhizosphere are generally limited to element uptake, and the chemical interactions between dissolved elements and the soil are reduced to a buffer power or Freundlich adsorption isotherm (Barber 1995; Kirk 1999). Figure 8b shows as an example the influence of citrate exudation on phosphate solubilization. The effect of exudation has been incorporated into the basic modeling concept, and conditional models parameterized for different soils have been formulated, e.g. to model the effect of organic acid exudation on phosphate mobilization (Gerke et al. 2000a, b). The application of certain rhizosphere models requires to write a new computer program or to change existing software. Schnepf et al. (2002) have shown that pde-solvers are useful in rhizosphere modeling because they make it easy to create, reproduce or link models from the known constituting equations.

Semi-empirical models on the root system scale An upscaling of single root models to the whole root system allows to predict plant uptake by integrating the flux on a unit segment basis over the total root length. The approach of Roose et al. (2001) allowed the direct incorporation of root branching structures and whole roots into plant uptake models, based on a mechanistic description of root uptake and soil processes (Roose and Fowler 2004a, b).

Molecular soil solution models In hydrogeochemistry, sophisticated computational tools have been developed to describe acid–base and redox reactions, complexation, ion exchange, adsorption and desorption, dissolution and precipitation of chemical species in soil environments using thermodynamic and kinetic relationships. Examples are PHREEQC (Parkhurst and Appelo 1999), ECOSAT (Keizer and VanRiemsdijk 1995) and ORCHESTRA (Meeussen 2003). Additionally there are computer codes that are specialized in modeling three-dimensional transport in variably saturated media that include geo-chemical modeling, e.g. MIN3P (Mayer et al. 2002). Applications of some of these models to rhizosphere research is described in the forthcoming paragraphs.

In some of the semi-empirical models mentioned above, soil solution speciation was included as input parameter. Calba et al. (2004) modeled the effect of protons, solid phase dissolution and adsorption on aluminum speciation in the rhizosphere, and Puschenreiter et al. (2005b) considered Ni speciation in soil solution when looking at Ni uptake by a hyperaccumulator. Zhao et al. (2007) used speciation modeling to elucidate the effect of plant roots on metal mobilization and speciation in soils. However, in these last two examples speciation was considered static and not to be affected by root activity. In particular the feedback loops between exudation, soil and element uptake are not considered implicitly in single root models, although many authors have demonstrated their importance in the plant availability of mineral elements (Parker and Pedler 1997).

Molecular models at the single root scale The full coupling of single-root models with speciation calculations is still in its infancy. An example of the inclusion of solution and surface speciation into rhizosphere models is the modeling of the effect of citrate exudation on phosphate uptake (Geelhoed et al. 1999). The model calculations showed that citrate exudation from roots increases the plant availability of sorbed phosphate (Fig. 8c). Recently a simple rhizosphere model was described in which the uptake into a single root was linked to three geochemical computational tools (ORCHESTRA, MIN3P, and PHREEQC; Nowack et al. 2006). The first step in this approach was an accuracy analysis of the different solution strategies by comparing the numerical results to the analytical solution of solute uptake by a single cylindrical root. All models were able to reproduce the concentration profiles as well as the uptake flux. The strength of this new approach is that it can also be used to investigate more complex and coupled biogeochemical processes in the rhizosphere. This was shown exemplarily with simulations involving both exudation and the simultaneous uptake of solute and water.

Molecular models at the soil profile scale The coupling of root uptake, speciation modeling and water transport in soils is even less advanced than on the single root scale. In order to describe metal uptake in the presence of ligands, Seuntjens et al. (2004) developed a model coupling processes under steady-state flow conditions with rhizosphere processes and speciation modeling. The simulations showed that exudation of ligands does not necessarily increase the solubility and bioavailability of metals, but that bioavailability may actually be reduced by formation of ternary surface complexes or reduction of the free metal concentration. The model can be easily extended to include further processes.

Challenges ahead

Our review on current methodology to study the effects of root and microbial activity on soil properties in the rhizosphere has shown that—although there is a need for improvements in certain aspects as outlined below—in general we have the tools at hand to assess individual processes on the microscale under rather artificial conditions. This is true mainly for looking at soil chemical properties and processes, while due to still large methodological limitations our understanding of the biophysics of the rhizosphere is comparatively limited (Gregory and Hinsinger 1999), despite major recent advances (Pierret et al. 2007; Hinsinger et al., 2008, Rhizosphere: biophysics, biogeochemistry and ecological relevance, in preparation). Microscopic, spectroscopic and tracer methods to look at individual and coupled chemical processes in small "aliquots" of naturally structured soil seem to step out of their infancy and have become promising tools to better understand the complex interactions between roots, soil and microorganisms. On the field scale, however, while there are promising first results on using non-invasive geophysical methods to assess the plant's influence on soil moisture, there are no tools in the pipeline to assess the spatial heterogeneity of chemical properties and processes in the field. For the time being, the use of macroscopic models or the upscaling of model results from the single root to the whole plant or plot scale is the only solution to this problem. However, upscaling itself is a major issue as outlined below. An optimal feedback between different developments requires a good communication between the various disciplines involved in rhizosphere research, in particular between experimental and modeling works. Both, early incorporation of new insights gained experimentally at the micro scale into explanatory models and involving models in experimental design could accelerate progress.

Methodological improvements for investigations at the micro scale

While most studies on root and microbial exudation limit their analysis to more abundant substances like sugars, carboxylates, amino acids and siderophores, the fate and role of many compounds like sterols or lactones that are exuded for signalling or as allelochemicals (Bertin et al. 2003) still need to be evaluated. Coupling of advanced chromatographic or electrophoretic separation methods with mass spectrometry allows to identify such compounds, e.g. in extracts of bacterial isolates (Frommberger et al. 2004). However, they cannot be detected in real soil solution with current methodologies.

Another challenge is to identify the source of a particular compound measured in soil solution, i.e. whether is has been exuded by plant roots, fungal hyphae or bacteria, or is the product of SOM degradation. Further advancements in compound specific isotopic analysis are needed in order to be able to trace ^{13}C labels to individual compounds. Currently, isotopic ratios can be determined for total DOC in small volumes of soil solution (Glaser 2005), while for individual compounds, even for more abundant ones, this will require drastic improvements in the detection limit of the coupled chromatography-IRMS instrumentation.

Considering the large potential of biosensors to assess the spatial heterogeneity of bioavailable molecules or ions, their in-situ application to microcosms containing real soil would be highly desirable. The difficulty to discriminate between the signals from biosensors and autofluorescent soil components must be overcome, and good correction factors for the reabsorption of the biosensor signal by soil particles must be determined. Furthermore, the development of multi-reporter gene biosensors, or the combined use of several biosensors in a given system, might help to control the influence of external factors (nutrient conditions, competition, inhibition factors, etc.), and thus to get more quantitative results in soils.

There have been great efforts to use microscopic and spectroscopic methods to assess the properties of soil and their components on the microscopic and molecular scale. The techniques are slowly getting sufficiently spatially resolved to separate components that are intimately associated. Apart from improving the capabilities of the instruments (flux and size of the incident beam, efficiency of detector systems) to get better sensitivity and resolution, efforts should focus on limiting the perturbation of the systems, e.g. by preserving their hydrated state, and better assessing or controlling the radiation damages by X-ray, electron or particle beams. Another challenge is to link the molecular- and microscopic-scale information obtained by these techniques to information obtained at higher scale.

Upscaling

On the microscale, plant physiology and soil microbiology have developed a detailed understanding of plant water and nutrient uptake, root respiration, root release of organic carbon and interactions between roots and soil microorganisms. However, there is a lack of understanding as to how the multiple complex interactions in the rhizosphere affect ecosystem functions on the macroscale (soil profile, plot, catchment). There is an urgent need to improve the mechanistic bases of models aimed at crop growth, forest production or biogeochemical element cycling by including rhizosphere processes. Closing the gaps between the different scales, or in other words making explanatory or predictive models on the macro scale more process-based, is a major challenge in biogeochemical research. At present, most of the available upscaling approaches for soil water processes ignore the effects of vegetation or use an extremely simplified approach. There is a need to develop upscaling approaches that explicitly account for the effects of growing plants under field conditions (Vereecken et al. 2007). A step into this direction is BIOCHEM-ORCHESTRA, a modeling tool that integrates eco-toxicological transfer functions with speciation and transport modeling (Vink and Meeussen 2007). The plant module, however, is still very simple and uses only empirical parameters such as the relevant rooting zone and a time-dependent uptake behavior. Root architecture models such as Root Typ (Pagès et al. 2004) have a great potential to be linked with other model approaches and could thus contribute significantly to the integration at higher scales.

On the opposite end of the scale spectrum, there is an urgent need for new modeling approaches that combine the molecular description of chemical processes in soils with pore-scale transport and root uptake. Up to now, molecular scale analytical tools and modeling approaches have developed rather independently. The coupling of 3-dimensional root growth modeling, root uptake, speciation modeling and water transport in soils presents challenges both on the computational and on the conceptual level. An example of a first step into this direction is the modeling of the effects of phospholipid surfactants on nutrient and water uptake by whole root systems (Dunbabin et al. 2006).

One key problem in the upscaling of rhizosphere processes is to assess correctly the distribution of active root segments in the soil. Non-invasive methods like X-ray computed tomography and MRI can, under certain conditions, produce well-resolved 3D images of the root system, but they are restricted to small laboratory systems. First results have demonstrated the potential of ERT and GPR to provide coarse images of root systems non-invasively and in-situ in the field via their imprint on soil moisture distribution. With GPR reflection it was even possible to resolve larger single roots in a silty sand (AlHagrey 2007). This warrants further exploration of geophysical methods in terms of delineating response from roots and soil structural heterogeneities, of improving spatial resolution (ERT), and of application to soils with higher clay contents (GPR).

References

Agerer R (2001) Exploration types of ectomycorrhizae—a proposal to classify ectomycorrhizal mycelial systems according to their patterns of differentiation and putative ecological importance. Mycorrhiza 11:107–114 doi:10.1007/s005720100108

Alewell C, Manderscheid B (1998) Use of objective criteria for the assessment of biogeochemical ecosystem models. Ecol Modell 107:213–224 doi:10.1016/S0304-3800(97)00218-4

AlHagrey SA (2007) Geophysical imaging of root-zone, trunk, and moisture heterogeneity. J Exp Bot 58:839–854 doi:10.1093/jxb/erl237

Andersen CP, Scagel CF (1997) Nutrient availability alters belowground respiration of ozone-exposed ponderosa pine. Tree Physiol 17:377–387

Annan AP (2005) GPR methods for hydrogeological studies. In: Rubin Y, Hubbard SS (eds) Hydrogeophysics. Springer, Dordrecht, pp 185–213

April R, Keller D (2005) Mineralogy of the rhizosphere in forest soils of the eastern United States. Biogeochemistry 9:1–18 doi:10.1007/BF00002714

Bächmann K, Steigerwald K (1993) The use of SEM for multielement analysis in small volumes and low concentration. Fresenius J Anal Chem 346:410–413 doi:10.1007/BF00325852

Barber SA (1995) Soil nutrient bioavailability: a mechanistic approach, 2nd edn. Wiley, New York

Barré P, Velde B, Catel N, Abbadie L (2007) Soil–plant potassium transfer: impact of plant activity on clay minerals as seen from X-ray diffraction. Plant Soil 292:137–146 doi:10.1007/s11104-007-9208-6

Benderitter Y, Schott JJ (1999) Short time variation of the resistivity in an unsaturated soil: the relationship with rainfall. Eur J Environ Eng Geophys 4:37–49

Benzerara K, Yoon T, Tyliszak T, Constantz A, Sportmann A, Brown G (2004) Scanning transmission X-ray microscopy study of microbial calcification. Geobiology 2:249–259 doi:10.1111/j.1472-4677.2004.00039.x

Bergström L, Stenström J (1998) Environmental fate of chemicals in soil. Ambio 27:16–23

Bertin C, Yang X, Weston LA (2003) The role of root exudates and allelochemicals in the rhizosphere. Plant Soil 256:67–83 doi:10.1023/A:1026290508166

Blossfeld S, Gansert D (2007) A novel non-invasive optical method for quantitative visualization of pH dynamics in the rhizosphere of plants. Plant Cell Environ 30:176–186 doi:10.1111/j.1365-3040.2006.01616.x

Blute NK, Brabander DJ, Hemond HF, Sutton SR, Newville MG, Rivers ML (2004) Arsenic sequestration by ferric iron plaque on cattail roots. Environ Sci Technol 38:6074–6077 doi:10.1021/es049448g

Bravin MN, Travassac F, Le Floch M, Hinsinger P, Garnier JM (2008) Oxygen input controls the spatial and temporal dynamics of arsenic at the surface of a flooded paddy soil and in the rhizosphere of lowland rice (*Oryza sativa* L.): a microcosm study. Plant Soil doi:10.1007/s11104-007-9532-x

Bruneau PMC, Ostle N, Davidson DA, Grieve IC, Fallick AE (2002) Determination of rhizosphere C-13 pulse signals in soil thin sections by laser ablation isotope ratio mass spectrometry. Rapid Commun Mass Spectrom 16:2190–2194 doi:10.1002/rcm.740

Cabala J, Teper L (2007) Metalliferous constituents of rhizosphere soils contaminated by Zn-Pb mining in Southern Poland. Water Air Soil Pollut 178:351–362 doi:10.1007/s11270-006-9203-1

Calba H, Firdaus, Cazevieille P, Thée C, Poss R, Jaillard B (2004) The dynamics of protons, aluminum, and calcium in the rhizosphere of maize cultivated in tropical acid soils: experimental study and modelling. Plant Soil 260:33–46

Cardon ZG, Gage DJ (2006) Resource exchange in the rhizosphere: molecular tools and the microbial perspective. Annu Rev Ecol Evol Syst 37:459–488 doi:10.1146/annurev.ecolsys.37.091305.110207

Chaignon V, Hinsinger P (2003) A biotest for evaluating copper bioavailability to plants in a contaminated soil. J Environ Qual 32:824–833

Chudek JA, Hunter G (1997) Magnetic resonance imaging of plants. Prog Nucl Magn Reson Spectrosc 31:43–62 doi:10.1016/S0079-6565(97)00005-8

Corgié S, Joner E, Leyval C (2003) Rhizospheric degradation of phenanthrene is a function of proximity to roots. Plant Soil 257:143–150 doi:10.1023/A:1026278424871

Cornu JY, Staunton S, Hinsinger P (2007) Copper concentration in plants and in the rhizosphere as influenced by the iron status of tomato (*Lycopersicum esculentum* L.). Plant Soil 292:63–77 doi:10.1007/s11104-007-9202-z

Cosby BJ, Hornberger GM, Galloway JN, Wright RF (1985) Modeling the effects of acid deposition: assessment of a lumped parameter model of soil water and streamwater chemistry. Water Resour Res 21:51–63 doi:10.1029/WR021i001p00051

Dane JH, Topp GC (2002) Methods of soil analysis. Part 4. Physical methods. SSSA Book Series 5, Soil Science Society of America, Madison Wisconsin, p 1692

Darrah PR (1993) The rhizosphere and plant nutrition: a quantitative approach. Plant Soil 155/156:1–20 doi:10.1007/BF00024980

Darrah PR, Jones DL, Kirk GJD, Roose T (2006) Modeling the rhizosphere: a review of methods for 'upscaling' to the whole-plant scale. Eur J Soil Sci 57:13–25 doi:10.1111/j.1365-2389.2006.00786.x

DeRito CM, Pumphrey GM, Madsen EL (2005) Use of field-based stable isotope probing to identify adapted populations and track carbon flow through a phenol-degrading soil microbial community. Appl Environ Microbiol 71:7858–7865 doi:10.1128/AEM.71.12.7858-7865.2005

Derrien D, Marol C, Balesdent J (2005) The dynamics of neutral sugars in the rhizosphere of wheat. An approach by 13C pulse-labelling and GC/C/IRMS. Plant Soil 267:243–253 doi:10.1007/s11104-005-5348-8

Dessureault-Rompré J, Nowack B, Schulin R, Luster J (2006) Modified micro suction cup/ rhizobox approach for the in-situ detection of organic acids in rhizosphere soil solution. Plant Soil 286:99–107 doi:10.1007/s11104-006-9029-z

Dessureault-Rompré J, Nowack B, Schulin R, Tercier-Waeber M-L, Luster J (2008) Metal solubility and speciation in the rhizosphere of *Lupinus albus* cluster roots. Environ Sci Technol doi:10.1021/es800167g

Dieffenbach A, Matzner E (2000) In situ soil solution chemistry in the rhizosphere of mature Norway spruce (*Picea abies* [L.] Karst) trees. Plant Soil 222:149–161 doi:10.1023/A:1004755404412

Dieffenbach A, Göttlein A, Matzner E (1997) In-situ investigation of soil solution chemistry in an acid soil as influenced by growing roots of Norway spruce (*Picea abies* [L.] Karst.). Plant Soil 192:57–61 doi:10.1023/A:1004283508101

Diggle AJ (1988) Rootmap—a model in 3-dimensional coordinates of the growth and structure of fibrous root systems. Plant Soil 105:169–178 doi:10.1007/BF02376780

Dinkelaker B, Marschner H (1992) In vivo demonstration of acid phosphatase activity in the rhizosphere of soil-grown plants. Plant Soil 144:199–205 doi:10.1007/BF00012876

Dinkelaker B, Hahn G, Römheld V, Wolf GA, Marschner H (1993) Non-destructive methods for demonstrating chemical changes in the rhizosphere I. Description of methods. Plant Soil 155:67–70 doi:10.1007/BF00024985

Doussan C, Pierret A, Garrigues E, Pagès L (2006) Water uptake by plant roots: II—modeling of water transfer in the soil root-system with explicit account of flow within the root system—comparsion with experiments. Plant Soil 283:99–117 doi:10.1007/s11104-004-7904-z

Dunbabin VM, Diggle AJ, Rengel Z, VanHugten R (2002) Modelling the interactions between water and nutrient uptake and root growth. Plant Soil 239:19–38 doi:10.1023/A:1014939512104

Dunbabin VM, McDermott S, Bengough AG (2006) Upscaling from rhizosphere to whole root system: modelling the effects of phospholipid surfactants on water and nutrient uptake. Plant Soil 283:57–72 doi:10.1007/s11104-005-0866-y

Ekberg A, Buchmann N, Gleixner G (2007) Rhizospheric influence on soil respiration and decomposition in a temperate Norway spruce stand. Soil Biol Biochem 39:2103–2110 doi:10.1016/j.soilbio.2007.03.024

Engels C, Neumann G, Gahoonia T, George E, Schenk M (2000) Assessment of the ability of roots for nutrient acquisition. In: Smit AL, Bengough AG, Engels C, Van Noordwijk M, Pellerin S, Van de Geijn SC (eds) Root methods. A handbook. Springer, Heidelberg, pp 403–459

Fan TWM, Lane AN, Pedler J, Crowley D, Higashi RM (1997) Comprehensice analysis of organic ligands in whole root exudates using nuclear magnetic resonance and gas chromatography-mass spectrometry. Anal Biochem 251:57–68 doi:10.1006/abio.1997.2235

Fischer WR, Flessa H, Schaller G (1989) pH values and redox potentials in microsites of the rhizosphere. Z Pflanzenernähr Bodenk 152:191–195 doi:10.1002/jpln.19891520209

Fitz WJ, Wenzel WW, Wieshammer G, Istenic B (2003a) Microtome sectioning causes artefacts in rhizobox experiments. Plant Soil 256:455–462 doi:10.1023/A:1026173613947

Fitz WJ, Wenzel WW, Zhang H, Nurmi J, Stipek K, Fischerova Z et al (2003b) Rhizosphere characteristics of the arsenic hyperaccumulator *Pteris vittata* L. and monitoring of phytoremoval efficiency. Environ Sci Technol 37:5008–5014 doi:10.1021/es0300214

Fitz WJ, Puschenreiter M, Wenzel WW (2006) Growth systems. In: Luster J, Finlay R (eds) Handbook of methods used in rhizosphere research. Swiss Federal Research Institute WSL, Birmensdorf, pp 9–15

Fomina M, Hillier S, Charnock JM, Melville K, Alexander IJ, Gadd GM (2005) Role of oxalic acid overexcretion in transformations of toxic metal minerals by *Beauveria caledonica*. Appl Environ Microbiol 71:371–381 doi:10.1128/AEM.71.1.371-381.2005

Frommberger M, Schmitt-Kopplin P, Ping G, Frisch H, Schmid M, Zhang Y et al (2004) A simple and robust set-up for on-column sample preconcentration–nano-liquid chromatography–electrospray ionization mass spectrometry for the analysis of N-acylhomoserine lactones. Anal Bioanal Chem 378:1014–1020 doi:10.1007/s00216-003-2400-5

Frossard E, Sinaj S (1997) The isotope exchange kinetic technique: a method to describe the availability of inorganic nutrients. Applications to K, P, S and Zn. Isotopes Environ Health Stud 33:61–77 doi:10.1080/10256019708036332

Gadd GM (2007) Geomycology: biogeochemical transformations of rocks, minerals, metals and radionuclides by fungi, bioweathering and bioremediation. Mycol Res 111:3–49 doi:10.1016/j.mycres.2006.12.001

Gahoonia TS, Nielsen NE (1991) A method to study rhizosphere processes in thin soil layers of different proximity to roots. Plant Soil 135:143–146 doi:10.1007/BF00014787

Garrigues E, Doussan C, Pierret A (2006) Water uptake by plant roots: I—formation and propagation of a water extraction front in mature root systems as evidenced by 2D light transmission imaging. Plant Soil 283:83–98 doi:10.1007/s11104-004-7903-0

Geelhoed JS, Van Riemsdijk WH, Findenegg GR (1999) Simulation of the effect of citrate exudation from roots on the plant availability of phosphate adsorbed on goethite. Eur J Soil Sci 50:379–390 doi:10.1046/j.1365-2389.1999.00251.x

Gerke J, Beissner L, Römer W (2000a) The quantitative effect of chemical phosphate mobilization by carboxylate anions in P uptake by a single root. I. The basic concept and determination of soil parameters. J Plant Nutr Soil Sci 163:207–212 doi:10.1002/(SICI)1522-2624(200004) 163:2<207::AID-JPLN207>3.0.CO;2-P

Gerke J, Römer W, Beissner L (2000b) The quantitative effect of chemical phosphate mobilization by carboxylate anions in P uptake by a single root. II. The importance of soil and plant parameters for uptake of mobilized P. J Plant Nutr Soil Sci 163:213–219 doi:10.1002/(SICI)1522-2624(200004) 163:2<213::AID-JPLN213>3.0.CO;2-0

Glaser B (2005) Compound-specific stable isotope ($\partial 13C$) analysis in soil science. J Plant Nutr Soil Sci 168:633–648 doi:10.1002/jpln.200521794

Göttlein A (1998) Measurement of free Al^{3+} in soil solutions by capillary electrophoresis. Eur J Soil Sci 49:107–112 doi:10.1046/j.1365-2389.1998.00133.x

Göttlein A (2006) Metal speciation in micro samples of soil solution by Capillary electrophoresis (CE) and ICP-OES with microinjection. In: Luster J, Finlay R (eds) Handbook of methods used in rhizosphere research. Swiss Federal Research Institute WSL, Birmensdorf, p 251

Göttlein A, Blasek R (1996) Analysis of small volumes of soil solution by capillary electrophoresis. Soil Sci 161:705–715 doi:10.1097/00010694-199610000-00007

Göttlein A, Hell U, Blasek R (1996) A system for microscale tensiometry and lysimetry. Geoderma 69:147–156 doi:10.1016/0016-7061(95)00059-3

Göttlein A, Heim A, Matzner E (1999) Mobilization of aluminium in the rhizosphere soil solution of growing tree roots in an acidic soil. Plant Soil 211:41–49 doi:10.1023/A:1004332916188

Göttlein A, Heim A, Kuhn AJ, Schröder WH (2005) In-situ application of stable isotope tracers in the rhizosphere of an oak seedling. Eur J For Res 124:83–86 doi:10.1007/s10342-005-0060-z

Gregory PJ (2006) Roots, rhizosphere, and soil: the route to a better understanding of soil science? Eur J Soil Sci 57:2–12 doi:10.1111/j.1365-2389.2005.00778.x

Gregory PJ, Hinsinger P (1999) New approaches to studying chemical and physical changes in the rhizosphere: an overview. Plant Soil 211:1–9 doi:10.1023/A:1004547401951

Guivarch A, Hinsinger P, Staunton S (1999) Root uptake and distribution of radiocaesium from contaminated soils and the enhancement of Cs adsorption in the rhizosphere. Plant Soil 211:131–138 doi:10.1023/A:1004465302449

Hamza M, Aylmore LAG (1991) Liquid ion-exchanger microelectrodes used to study soil solute concentrations near plant-roots. Soil Sci Soc Am J 55:954–958

Hamza M, Aylmore LAG (1992) Soil solute concentration and water uptake by single lupin and radish plant roots. I. Water extraction and solute accumulation. Plant Soil 145:187–196 doi:10.1007/BF00010347

Hansel CM, Fendorf S, Sutton S, Newville M (2001) Characterization of Fe plaque and associated metals on the roots of mine-waste impacted aquatic plants. Environ Sci Technol 35:3863–3868 doi:10.1021/es0105459

Hansen LH, Sørensen SJ (2001) The use of whole-cell biosensors to detect and quantify compounds or conditions affecting biological systems. Microb Ecol 42:483–494 doi:10.1007/s00248-001-0025-9

Häussling M, Leisen E, Marschner H, Römheld V (1985) An improved method for non-destructive measurement of pH at the root–soil interface (Rhizosphere). J Plant Physiol 117:371–375

Heim A, Brunner I, Frossard E, Luster J (2003) Aluminum Effects on *Picea abies* at low solution concentrations. Soil Sci Soc Am J 67:895–898

Hendriks L, Claassen N, Jungk A (1981) Phosphatverarmung des wurzelnahen Bodens und Phosphataufnahme von Mais und Raps. J Plant Nutr Soil Sci 144:486–499 doi:10.1002/jpln.19811440507

Herrmann K-H, Pohlmeier A, Gembris D, Vereecken H (2002) Three-dimensional imaging of pore water diffusion and motion in porous media by nuclear magnetic resonance imaging. J Hydrol (Amst) 267:244–257 doi:10.1016/S0022-1694(02)00154-3

Hinsinger P, Gilkes RJ (1997) Dissolution of phosphate rock in the rhizosphere of five plant species grown in an acid, P-fixing mineral substrate. Geoderma 75:231–249 doi:10.1016/S0016-7061(96)00094-8

Hinsinger P, Elsass F, Jaillard B, Robert M (1993) Root-induced irreversible transformation of a trioctahedral mica in the rhizosphere of rape. Eur J Soil Sci 44:535–545 doi:10.1111/j.1365-2389.1993.tb00475.x

Hinsinger P, Gobran GR, Gregory PJ, Wenzel WW (2005) Rhizosphere geometry and heterogeneity arising from root-mediated physical and chemical processes. New Phytol 168:293–303 doi:10.1111/j.1469-8137.2005.01512.x

Hodge A, Grayston SJ, Ord BG (1996) A novel method for characterization and quantification of plant root exudates. Plant Soil 184:97–104 doi:10.1007/BF00029278

Hodge A, Robinson D, Griffiths BS, Fitter AH (1999) Why plants bother: root proliferation results in increased nitrogen capture from an organic patch when two grasses compete. Plant Cell Environ 22:811–820 doi:10.1046/j.1365-3040.1999.00454.x

Hopmans JW, Bristow KL (2002) Current capabilities and future needs of root water and nutrient uptake modeling. Adv Agron 77:103–183 doi:10.1016/S0065-2113(02)77014-4

Hübel F, Beck E (1993) In-situ determination of the P-relations around the primary root of maize with respect to inorganic and phytate-P. Plant Soil 157:1–9

Hui P, Tian CZ (1998) Fabrication of redox potential microelectrodes for studies in vegetated soils or biofilm systems. Environ Sci Technol 32:3646–3652 doi:10.1021/es980024u

Jones DL (1998) Organic acids in the rhizosphere—a critical review. Plant Soil 205:25–44 doi:10.1023/A:1004356007312

Jones DL, Brassington DS (1998) Sorption of organic acids in acid soils and its implications in the rhizosphere. Eur J Soil Sci 49:447–455 doi:10.1046/j.1365-2389.1998.4930447.x

Jones DL, Dennis PG, Owen G, Hees PW (2003) Organic acid behavior in soil-misconceptions and knowledge gaps. Plant Soil 248:31–41 doi:10.1023/A:1022304332313

Jungk A, Claassen N (1997) Ion diffusion in the soil–root system. Adv Agron 61:53–110 doi:10.1016/S0065-2113(08)60662-8

Keizer MG, van Riemsdijk WH (1995) ECOSAT, a computer program for the calculation of chemical speciation and transport in soil–water systems. Wageningen Agricultural University, Wageningen

Kelleher BP, Simpson MJ, Simpson AJ (2006) Assessing the fate and transformation of plant residues in the terrestrial environment using HR-MAS NMR spectroscopy. Geochim Cosmochim Acta 70:4080–4094 doi:10.1016/j.gca.2006.06.012

Killham K, Yeomans C (2001) Rhizosphere carbon flow measurement and implications: from isotopes to reporter genes. Plant Soil 232:91–96 doi:10.1023/A:1010386019912

Kirk GJD (1999) A model of phosphate solubilization by organic anion excretion from plant roots. Eur J Soil Sci 50:369–378 doi:10.1046/j.1365-2389.1999.00239.x

Kirk GJD (2002) Use of modeling to understand nutrient acquisition by plants. Plant Soil 247:123–130 doi:10.1023/A:1021115809702

Kodama H, Nelson S, Yang F, Kohyama N (1994) Mineralogy of rhizospheric and non-rhizospheric soils in corn fields. Clays Clay Miner 42:755–763 doi:10.1346/CCMN.1994.0420612

Kraemer SM, Crowley DE, Kretzschmar R (2006) Geochemical aspects of phytosiderophore-promoted iron acquisition by plants. Adv Agron 91:1–46 doi:10.1016/S0065-2113(06)91001-3

Kuo S (1996) Phosphorus. In: Sparks DL (ed) Methods of soil analysis. Part 3. Chemical Methods. Soil Science Society of America, Madison, pp 869–919

Kuzyakov Y, Raskatov AV, Kaupenjohann M (2003) Turnover and distribution of root exudates of Zea mays. Plant Soil 254:317–327 doi:10.1023/A:1025515708093

Lanson B, Marcus MA, Fakra S, Panfili F, Geoffroy N, Manceau A (2008) Formation of Zn-Ca phyllomanganate nanoparticles in grass roots. Geochim Cosmochim Acta 72:2478-2490 doi:10.1016/j.gca.2008.02.022

Lemon ER, Erickson AE (1952) The measurement of oxygen diffusion in the soil with a platinum microelectrode. Soil Sci Soc Am J 16:160–163

Leyval C, Berthelin H (1991) Weathering of mica by roots and rhizospheric microorganisms of pine. Soil Sci Soc Am J 55:1009–1016

Li CS, Frolking S, Harriss R (1994) Modeling carbon biogeochemistry in agricultural soils. Global Biogeochem Cycles 8:237–254 doi:10.1029/94GB00767

Lindahl B, Finlay RD, Olsson S (2001) Simultaneous bidirectional translocation of ^{32}P and ^{33}P between wood blocks connected by mycelial cords of Hypholoma fasciculare. New Phytol 150:189–194 doi:10.1046/j.1469-8137.2001.00074.x

Liu Q, Loganathan P, Hedley MJ, Skinner MF (2004) The mobilisation and fate of soil and rock phosphate in the rhizosphere of ectomycorrhizal Pinus radiata seedlings in an Allophanic soil. Plant Soil 264:219–229 doi:10.1023/B:PLSO.0000047758.77661.57

Lofthouse SD, Greenway GM, Stephen SC (1997) Microconcentric nebuliser for the analysis of small sample volumes by inductively coupled plasma mass spectrometry. J Anal At Spectrom 12:1373–1376 doi:10.1039/a705047j

Luster J, Finlay R (eds) (2006) Handbook of methods used in rhizosphere research. Swiss Federal Research Institute WSL, Birmensdorf. 536 pp.; online at www.rhizo.at/handbook

Luster J, Menon M, Hermle S, Schulin R, Goerg-Günthardt MS, Nowack B (2008) Initial changes in refilled lysimeters built with metal polluted topsoil and acidic or calcareous subsoils as indicated by changes in drainage water composition. Water Air Soil Pollut Focus 8:163–176 doi:10.1007/s11267-007-9169-z

Lynch JP, Nielsen KL, Davis RD, Jablokow AG (1997) SimRoot: modelling and visualization of root systems. Plant Soil 188:139–151 doi:10.1023/A:1004276724310

Majdi H (1996) Root sampling methods—applications and limitations of minirhizotron technique. Plant Soil 185:225–258

Manceau A, Marcus MA, Tamura N (2002) Quantitative speciation of heavy metals in soils and sediments by synchrotron X-ray techniques. In: Fenter P, Rivers M, Sturchio N, Sutton S (eds) applications of synchrotron radiation in low-temperature geochemistry and environmental science. Reviews in Mineralogy and Geochemistry, Mineralogical Society of America, Washington D.C, pp 341–428

Manceau A, Nagy KL, Marcus MA, Lanson M, Geoffroy N, Jacquet T et al (2008) Formation of metallic copper nanoparticles at the soil−root interface. Environ Sci Technol 42:1766–1772 doi:10.1021/es072017o

Martell AE, Smith RM (1974–1989) Critical stability constants, vol. 1 to 6. Plenum, New York

Mayer KU, Frind EO, Blowes DW (2002) Multicomponent reactive transport modeling in variably saturated porous media using a generalized formulation for kinetically controlled reactions. Water Resour Res 38:1174–1194 doi:10.1029/2001WR000862

Meeussen JCL (2003) ORCHESTRA: an object-oriented framework for implementing chemical equilibrium models. Environ Sci Technol 37:1175–1182 doi:10.1021/es025597s

Menon M, Robinson B, Oswald SE, Kaestner A, Abbaspour KC, Lehmann E et al (2007) Visualisation of root growth in heterogeneously contaminated soil using neutron radiography. Eur J Soil Sci 58:802–810 doi:10.1111/j.1365-2389.2006.00870.x

Mermet JM, Todolí JL (2004) Towards total-consumption pneumatic liquid micro-sample-introduction systems in ICP spectrochemistry. Anal Bioanal Chem 378:57–59 doi:10.1007/s00216-003-2368-1

Michot D, Benderitter Y, Dorigny A, Nicoullaud B, King D, Tabbagh A (2003) Spatial and temporal monitoring of soil water content with an irrigated corn crop cover using surface electrical resistivity tomography. Water Resour Res 39:1138 doi:10.1029/2002WR001581

Minchin PEH, McNaughton GS (1984) Exudation of recently fixed carbon by non-sterile roots. J Exp Bot 35:74–82 doi:10.1093/jxb/35.1.74

Mooney SJ, Morris C, Berry PM (2006a) Visualization and quantification of the effects of cereal root lodging on three-dimensional soil macrostructure using X-ray computed tomography. Soil Sci 171:706–718 doi:10.1097/01.ss.0000228041.03142.d3

Mooney SJ, Foot K, Hutchings TR, Moffat AJ (2006b) Micromorphological investigations into root penetration in a landfill mineral cap, Hertfordshire, UK. Waste Manag 27:1225–1232 doi:10.1016/j.wasman.2006.07.012

Moran CJ, Pierret A, Stevenson AW (2000) X-ray absorption and phase contrast imaging to study the interplay between plant roots and soil structure. Plant Soil 223:99–115 doi:10.1023/A:1004835813094

Myneni SCB (2002) Soft X-ray spectroscopy and spectromicroscopy studies of organic molecules in the environment. In: Fenter P, Rivers M, Sturchio N, Sutton S (eds) Reviews in Mineralogy and geochemistry. Applications of synchrotron radiation in low-temperature geochemistry and environmental science. Mineralogical Society of America, pp 485–579

Nachtegaal M, Marcus MA, Sonke JE, Vangronsveld J, Livi KJT, Van der Lelie D et al (2005) Effects of in situ remediation on the speciation and bioavailability of zinc in a smelter contaminated soil. Geochim Cosmochim Acta 69:4649 doi:10.1016/j.gca.2005.05.019

Naim MS (1965) Development of rhizosphere and rhizoplane microflora of *Artistida coerulescens* in the Lybian desert. Arch Mikrobiol 50:321–325 doi:10.1007/BF00509573

Neumann G (2006) Root exudates and organic composition of plant roots. In: Luster J, Finlay R (eds) Handbook of methods used in rhizosphere research. Swiss Federal Research Institute WSL, Birmensdorf, pp 52–61

Neumann G, Römheld V (2001) The release of root exudates as affected by the plant's physiological status. In: Pinton R, Varanini Z, Nannipieri P (eds) The rhizosphere: biochemistry and organic substances at the soil-plant interface. Marcel Dekker, New York, pp 41–93

Nollet LML (ed) (2007) Handbook of water analysis, 2nd Ed. CRC Press, Boca Raton. 769 pp

Nowack B, Köhler S, Schulin R (2004) Use of diffusive gradients in thin films (DGT) in undisturbed field soils. Environ Sci Technol 38:1133–1138 doi:10.1021/es034867j

Nowack B, Mayer KU, Oswald SE, VanBeinum W, Appelo CAJ, Jacques D et al (2006) Verification and intercomparison of reactive transport codes to describe root-uptake. Plant Soil 285:305–321 doi:10.1007/s11104-006-9017-3

Nye PH, Tinker PB (1977) Solute movement in the soil–root system. Blackwell, Oxford, p 342

Ochs M, Brunner I, Stumm W, Cosovic B (1993) Effects of root exudates and humic substances on weathering kinetics. Water Air Soil Pollut 68:213–229 doi:10.1007/BF00479404

Oswald SE, Menon M, Carminati A, Vontobel P, Lehmann E, Schulin R (2008) Quantitative imaging of infiltration, root growth, and root water uptake via neutron radiography. Vadose Zone J 7:1035–1047 doi:10.2136/vzj2007.0156

Pagès L, Vercambre G, Drouet JL, Lecompte F, Collet C, LeBot J (2004) Root Typ: a generic model to depict and analyse the root system architecture. Plant Soil 258:103–119 doi:10.1023/B:PLSO.0000016540.47134.03

Panfili F, Manceau A, Sarret G, Spadini L, Kirpichtchikova T, Bert V et al (2005) The effect of phytostabilization on Zn speciation in a dredged contaminated sediment using scanning electron microscopy, X-ray fluorescence, EXAFS spectroscopy and principal components analysis. Geochim Cosmochim Acta 69:2265–2284 doi:10.1016/j.gca.2004.10.017

Panikov NS, Mastepanov MA, Christensen TR (2007) Membrane probe array: technique development and observation of CO_2 and CH_4 diurnal oscillations in peat profile. Soil Biol Biochem 39:1712–1723 doi:10.1016/j.soilbio.2007.01.034

Pansu M, Gautheyrou J (2006) Handbook of soil analysis: mineralogical, organic and inorganic methods. Springer, Berlin, p 993

Parker DR, Pedler JF (1997) Reevaluating the free-ion activity model of trace metal availability to higher plants. Plant Soil 196:223–228 doi:10.1023/A:1004249923989

Parkhurst DL, Appelo CAJ (1999) User's guide to PHREEQC (version 2). A computer program for speciation, batch-reaction, one-dimensional transport, and inverse geochemical calculations. Water resources investigations Report 99-4259. US Geological Survey. 312 pp

Paterson E, Sim A, Standing D, Dorward M, McDonald AJS (2006) Root exudation from *Hordeum vulgare* in response to localized nitrate supply. J Exp Bot 57:2413–2420 doi:10.1093/jxb/erj214

Pierret A, Doussan C, Garrigues E, McKirby J (2003) Observing plant roots in their environment: current imaging options and specific contribution of two-dimensional approaches. Agronomie 23:471–479 doi:10.1051/agro:2003019

Pierret A, Doussan C, Capowiez Y, Bastardie F, Pagès L (2007) Root functional architecture: a framework for modeling the interplay between roots and soil. Vadose Zone J 6:269–281 doi:10.2136/vzj2006.0067

Pinjuv G, Mason EG, Watt M (2006) Quantitative validation and comparison of a range of forest growth model types. For Ecol Manage 236:37–46 doi:10.1016/j.foreco.2006.06.025

Plassard C, Meslem M, Souche G, Jaillard B (1999) Localization and quantification of net fluxes of H^+ along maize roots by combined use of pH-indicator dye videodensitometry and H^+-selective microelectrodes. Plant Soil 211:29–39 doi:10.1023/A:1004560208777

Plassard C, Guérin-Laguette A, Véry AA, Casarin V, Thibaud JB (2002) Local measurements of nitrate and potassium fluxes along roots of maritime pine. Plant Cell Environ 25:75–84 doi:10.1046/j.0016-8025.2001.00810.x

Polomski J, Kuhn N (2002) Root research methods. In: Waisel Y, Eshel A, Kafkafi U (eds) Plant roots—the hidden half, 3rd edn. Marcel Dekker, New York, pp 295–321

Prabhu RK, Vijayalahshimi S, Mahalingam TR, Viswanathan KS, Methews CK (1993) Laser vaporization inductively-coupled plasma-mass spectrometry—a technique for the analysis of small volumes of solutions. J Anal At Spectrom 8:565–569 doi:10.1039/ja9930800565

Psenner R, Boström B, Dinka M, Pettersson K, Pucsko R, Sager M (1988) Fractionation of phosphorus in suspended matter and sediment. Arch Hydrobiol Beih 30:99–103

Puschenreiter M, Wenzel WW, Wieshammer G, Fitz WJ, Wieczorek S, Kanitsar K et al (2005a) Novel micro-suction-cup design for sampling soil solution at defined distances from roots. J Plant Nutr Soil Sci 168:386–391 doi:10.1002/jpln.200421681

Puschenreiter M, Schnepf A, Millan IM, Fitz WJ, Horak O, Klepp J et al (2005b) Changes of Ni biogeochemistry in the rhizosphere of the hyperaccumulator *Thlaspi goesingense*. Plant Soil 271:205–218 doi:10.1007/s11104-004-2387-5

Raab TK, Vogel JP (2004) Ecological and agricultural applications of synchrotron IR microscopy. Infrared Phys.Technol. 45:393–402 doi:10.1016/j.infrared.2004.01.008

Raich JW, Mora G (2005) Estimating root plus rhizosphere contributions to soil respiration in annual croplands. Soil Sci Soc Am J 69:634–639 doi:10.2136/sssaj2004.0257

Rais D, Nowack B, Schulin R, Luster J (2006) Sorption of trace metals by different standard and micro suction cups used as soil water samplers as influenced by dissolved organic carbon. J Environ Qual 35:50–60 doi:10.2134/jeq2005.0040

Rangel Castro JI, Killham K, Ostle N, Nicol GW, Anderson IC, Scrimgeour CM et al (2005) Stable isotope probing analysis of the influence of liming on root exudate utilization by soil microorganisms. Environ Microbiol 7:828–838 doi:10.1111/j.1462-2920.2005.00756.x

Rappoldt C (1995) Measuring the millimeter scale oxygen diffusivity in soil using microelectrodes. Eur J Soil Sci 46:169–177 doi:10.1111/j.1365-2389.1995.tb01824.x

Reichard PU, Kraemer SM, Frazier SW, Kretzschmar R (2005) Goethite dissolution in the presence of phytosiderophores: rates, mechanisms, and the synergistic effect of oxalate. Plant Soil 276:115–132 doi:10.1007/s11104-005-3504-9

Rodriguez-Mozaz S, Lopez de Alda M, Barceló D (2006) Biosensors as useful tools for environmental analysis and monitoring. Anal Bioanal Chem 386:1025–1041 doi:10.1007/s00216-006-0574-3

Römheld V (1986) pH changes in the rhizosphere of various crop plants in relation to the supply of plant nutrients. In: Potash Review 12, Internat. Potash Institute, Bern Switzerland, Subject 6, 55th Suite

Roose T, Fowler AC (2004a) A mathematical model for water and nutrient uptake by plant root systems. J Theor Biol 228:173–184 doi:10.1016/j.jtbi.2003.12.013

Roose T, Fowler AC (2004b) A model for water uptake by plant roots. J Theor Biol 228:155–171 doi:10.1016/j.jtbi.2003.12.012

Roose T, Fowler AC, Darrah PR (2001) A mathematical model of plant nutrient uptake. J Math Biol 42:347–360 doi:10.1007/s002850000075

Rosling A, Lindahl B, Finlay RD (2004) Carbon allocation to ectomycorrhizal roots and mycelium colonising different mineral substrates. New Phytol 162:795–802 doi:10.1111/j.1469-8137.2004.01080.x

Rothfuss F, Conrad R (1994) Development of a gas diffusion probe for the determination of methane concentrations and diffusion characteristics in flooded paddy soil. FEMS Microbiol Ecol 14:307–318 doi:10.1111/j.1574-6941.1994.tb00116.x

Sandnes A, Eldhuset TD (2003) Soda glass beads as growth medium in plant cultivation experiments. J Plant Nutr Soil Sci 166:660–661 doi:10.1002/jpln.200320308

Sawhney BL (1996) Extraction of organic chemicals. In: Sparks DL (ed) Methods of soil analysis. Part 3. chemical methods. Soil Science Society of America, Madison, pp 1071–1084

Schnepf A, Schrefl T, Wenzel WW (2002) The suitability of pde-solvers in rhizosphere modeling, exemplified by three mechanistic rhizosphere models. J Plant Nutr Soil Sci 165:713–718 doi:10.1002/jpln.200290008

Schumacher M, Christl I, Scheinost AC, Jacobsen C, Kretzschmar R (2005) Chemical heterogeneity of organic soil colloids investigated by scanning transmission X-ray microscopy and C-1s NEXAFS microspectroscopy. Environ Sci Technol 39:9094–9100 doi:10.1021/es050099f

Senesi N (1996) Electron spin (or paramagnetic) resonance spectroscopy. In: Sparks DL (ed) Methods of soil analysis. Part 3. Chemical methods. Soil Science Society of America, Madison, pp 323–356

Seuntjens P, Nowack B, Schulin R (2004) Root-zone modeling of heavy metal uptake and leaching in the presence of organic ligands. Plant Soil 265:61–73 doi:10.1007/s11104-005-8470-8

Shen J, Hoffland E (2007) In situ sampling of small volumes of soil solution using modified micro-cups. Plant Soil 292:161–169 doi:10.1007/s11104-007-9212-x

Siegel LS, Alshawabkeh AN, Palmer CD, Hamilton MA (2003) Modeling cesium partitioning in the rhizosphere: a focus on the role of root exudates. Soil Sediment Contam 12:47–68 doi:10.1080/713610960

Somma F, Hopmans JW, Clausnitzer V (1998) Transient three-dimensional modeling of soil water and solute transport with simultaneous root growth, root water and nutrient uptake. Plant Soil 202:281–293 doi:10.1023/A:1004378602378

Sparks DL (ed) (1996) Methods of soil analysis. Part 3. Chemical Methods. SSSA Book Series 5, Soil Science Society of America, Madison. 1358 pp

Stevenson FJ (1996) Nitrogen-organic forms. In: Sparks DL (ed) Methods of soil analysis., Part 3. Chemical methods. Soil Science Society of America, Madison, pp 1185–1200

Ström L, Owen AG, Godbold DL, Jones DL (2002) Organic acid mediated P mobilization in the rhizosphere and uptake by maize roots. Soil Biol Biochem 34:703–710 doi:10.1016/S0038-0717(01)00235-8

Sulzmann EW, Brant JB, Bowden RD, Lajtha K (2005) Contribution of aboveground litter, belowground litter, and rhizosphere respiration to total soil CO_2 efflux in an old growth coniferous forest. Biogeochemistry 73:231–256 doi:10.1007/s10533-004-7314-6

Tang CS, Young CC (1982) Collection and identification of allelopathic compounds from the undisturbed root-system of bigalta limpograss (Hemarthria altissima). Plant Physiol 69:155–160

Tercier-Waeber M-L, Buffle J, Koudelka-Hep M, Graziottin F (2002) Submersible voltammetric probes for in-situ real-time trace element monitoring in natural aquatic systems. In: Taillefert M, Rozan TF (eds) Environmental electrochemistry: analysis of trace element biogeochemistry. ACS Series No. 811, Washington D.C, pp 16–39

Tessier A, Campbell PGC, Bisson M (1979) Sequential extraction procedure for the speciation of particulate trace metals. Anal Chem 51:844–851 doi:10.1021/ac50043a017

Turpault MP (2006) Sampling of rhizosphere soil for physicochemical and mineralogical analyses by physical separation based on drying and shaking. In: Luster J, Finlay R (eds) Handbook of methods used in rhizosphere research. Swiss Federal Research Institute WSL, Birmensdorf, pp 196–197

VanBochove E, Beauchemin S, Theriault G (2002) Continuous multiple measurement of soil redox potential using platinum microelectrodes. Soil Sci Soc Am J 66:1813–1820

Vereecken H, Kasteel R, Vanderborght J, Harter T (2007) Upscaling hydraulic properties and soil water flow processes in heterogeneous soils: a review. Vadose Zone J 6:1–28 doi:10.2136/vzj2006.0055

Vetterlein D, Jahn R (2004) Combination of micro suction cups and time-domain reflectometry to measure osmotic potential gradients between bulk soil and rhizosphere at high resolution in time and space. Eur J Soil Sci 55:497–504 doi:10.1111/j.1365-2389.2004.00612.x

Vink JPM, Meeussen JCL (2007) BIOCHEM-ORCHESTRA: a tool for evaluating chemical speciation and ecotoxicological impacts of heavy metals on river flood plain systems. Environ Pollut 148:833–841 doi:10.1016/j.envpol.2007.01.041

Voegelin A, Weber F-A, Kretzschmar R (2007) Distribution and speciation of arsenic around roots in a contaminated riparian floodplain soil: Micro-XRF element mapping and EXAFS spectroscopy. Geochim Cosmochim Acta 71:5804–5820 doi:10.1016/j.gca.2007.05.030

VonLützow M, Kögel-Knabner I, Ekschmitt K, Flessa H, Guggenberger G, Matzner E et al (2007) SOM fractionation methods: relevance to functional pools and to stabilization mechanisms. Soil Biol Biochem 39:2183–2207 doi:10.1016/j.soilbio.2007.03.007

Wang E, Smith CJ (2004) Modeling the growth and water uptake function of plant root systems: a review. Aust J Agric Res 55:501–523 doi:10.1071/AR03201

Wang Z, Göttlein A, Bartonek G (2001) Effects of growing roots of Norway spruce (*Picea abies* [L.] Karst.) and European beech (*Fagus sylvatica* L.) on rhizosphere soil solution chemistry. J Plant Nutr Soil Sci 164:35–41 doi:10.1002/1522-2624(200102)164:1<35::AID-JPLN35>3.0.CO;2-M

Warnken K, Zhang H, Davison W (2004) Analysis of polyacrylamide gels for trace metals using diffusive gradients in thin films and laser ablation inductively coupled plasma mass spectrometry. Anal Chem 76:6077–6084 doi:10.1021/ac0400358

Watteau F, Villemin G (2001) Ultrastructural study of the biogeochemical cycle of silicon in the soil and litter of a temperate forest. Eur J Soil Sci 52:385–396 doi:10.1046/j.1365-2389.2001.00391.x

Weaver RW, Angle S, Bottomley P (eds) (1994) Methods of soil analysis. Part 2. Microbiological and biochemical properties. SSSA Book Series 5, Soil Science Society of America, Madison. 1121 pp

Wenzel WW, Wieshammer G, Fitz WJ, Puschenreiter M (2001) Novel rhizobox design to assess rhizosphere characteristics at high spatial resolution. Plant Soil 237:37–45 doi:10.1023/A:1013395122730

Whiting SN, Leake JR, McGrath SP, Baker AJM (2000) Positive responses to Zn and Cd by roots of the Zn and Cd hyperaccumulator *Thlaspi caerulescens*. New Phytol 145:199–210 doi:10.1046/j.1469-8137.2000.00570.x

Wolf DC, Skipper HD (1994) Soil sterilization. In: Weaver RW, Angle S, Bottomley P (eds) Methods of soil analysis, Part 2; microbiological and biochemical properties. Soil Science Society of America, Madison, pp 41–51

Wu L, McGechan MB, McRoberts N, Baddeley JA, Watson CA (2007) SPACSYS: integration of a 3D root architecture component to carbon, nitrogen and water cycling-model description. Ecol Modell 200:343–359 doi:10.1016/j.ecolmodel.2006.08.010

Yanai RD, Majdi H, Park BB (2003) Measured and modelled differences in nutrient concentrations between rhizosphere and bulk soil in a Norway spruce stand. Plant Soil 257:133–142 doi:10.1023/A:1026257508033

Yevdokimov IV, Ruser R, Buegger F, Marx M, Munch JC (2007) Carbon turnover in the rhizosphere under continuous plant labeling with (CO_2)-C-13: partitioning of root, microbial, and rhizomicrobial respiration. Eurasian Soil Sci 40:969–977 doi:10.1134/S1064229307090074

Yu KW, DeLaune RD (2006) A modified soil diffusion chamber for gas profile analysis. Soil Sci Soc Am J 70:1237–1241 doi:10.2136/sssaj2005.0332N

Zeien H, Brümmer GW (1989) Chemische Extraktion zur Bestimmung von Schwermetallbindungsformen in Böden. Mitteilgn Dtsch Bodenkundl Ges 59:505–510

Zhang TC, Pang H (1999) Applications of microelectrode techniques to measure pH and oxidation–reduction potential in rhizosphere soil. Environ Sci Technol 33:1293–1299 doi:10.1021/es981070x

Zhang H, Davison W, Knight B, McGrath S (1998) In situ measurement of solution concentrations and fluxes of trace metals in soils using DGT. Environ Sci Technol 32:704–710 doi:10.1021/es9704388

Zhang H, Zhao FJ, Sun B, Davison W, McGrath SP (2001) A new method to measure effective soil solution concentration predicts copper availability to plants. Environ Sci Technol 35:2602–2607 doi:10.1021/es000268q

Zhao LYL, Schulin R, Nowack B (2007) The effects of plants on the mobilization of Cu and Zn in soil columns. Environ Sci Technol 41:2770–2775 doi:10.1021/es062032d

REVIEW ARTICLE

Molecular tools in rhizosphere microbiology—from single-cell to whole-community analysis

Jan Sørensen · Mette Haubjerg Nicolaisen · Eliora Ron · Pascal Simonet

Received: 11 July 2008 / Accepted: 24 February 2009 / Published online: 17 March 2009
© Springer Science + Business Media B.V. 2009

Abstract It is the aim of this chapter to present an overview of new, molecular tools that have been developed over recent years to study individual, single cells and composite, complex communities of microorganisms in the rhizosphere. We have carefully focused on culture-independent assays and selected methodologies that have already been or will soon be applicable for rhizosphere microbiology. Emphasis is placed on rhizosphere bacteria and the review first describes a number of the new methodologies developed for detection and localization of specific bacterial populations using modern electron and fluorescence microscopy combined with specific tagging techniques. First half of the chapter further comprises a thorough treatise of the recent development of reporter gene technology, i.e. using specific reporter bacteria to detect microscale distributions of rhizosphere compounds such as nutrients, metals and organic exudates or contaminants. Second half of the chapter devoted to microbial community analysis contains a thorough treatise of nucleotide- and PCR-based technologies to study composition and diversity of indigenous bacteria in the natural rhizosphere. Also included are the most recent developments of functional gene and gene expression analyses in the rhizosphere based on specific mRNA transcript or transcriptome analysis, proteome analysis and construction of metagenomic libraries.

Keywords Metagenome · Microarray · Microscopy · Reporter · Proteome · Transcriptome

Responsible editor: Yves Dessaux.

J. Sørensen (✉) · M. Haubjerg Nicolaisen
Section of Genetics and Microbiology,
Department of Agriculture and Ecology,
University of Copenhagen,
Thorvaldsensvej 40, DK-1871 Frederiksberg C,
Copenhagen, Denmark
e-mail: jan@life.ku.dk

E. Ron
Department of Molecular Microbiology and Biotechnology,
Tel Aviv University,
Tel Aviv 69978, Israel

P. Simonet
Environmental Microbial Genomics Group,
Microsystems and Microbiology, Laboratoire Ampère,
UMR CNRS 5005, Ecole Centrale de Lyon,
36 avenue Guy de Collongue,
69134 Ecully cedex, France

Introduction

It is the aim of this chapter to present an overview of new, molecular tools that have been developed over recent years to study individual, single cells and composite, complex communities of microorganisms in the rhizosphere. We have carefully focused on culture-independent assays and selected methodologies that have already been or will soon be applicable for rhizosphere microbiology. Emphasis is placed on rhizosphere bacteria and the review first describes a number of the new methodologies developed for detection and localization of specific bacterial populations using modern electron and fluorescence microscopy combined with specific tagging techniques. First half of the chapter further comprises a thorough treatise of the recent development of reporter gene technology, i.e. using specific reporter bacteria to detect microscale distributions of rhizosphere compounds such as nutrients, metals and organic exudates or contaminants. Second half of the chapter devoted to microbial community analysis contains a thorough treatise of nucleotide- and PCR-based technologies to study composition and diversity of indigenous bacteria in the natural rhizosphere. Also included are the most recent developments of functional gene and gene expression analyses in the rhizosphere based on specific mRNA transcript or transcriptome analysis, proteome analysis and construction of metagenomic libraries.

Single-cell studies in the rhizosphere

Direct microscopy

Direct microscopy may be defined by a complete or near absence of physical sample disruption which is most important for specimen preparation. The particular difficulty of maintaining soil structure under microscopy has called for several solutions. The approaches of resin-imbedding and thin-sectioning, development of root-growth chambers (rhizosphere microcosms), and gentle root-washing techniques assuring a minimum of change for microbial activity at the root surface *per se* (rhizoplane), have all been introduced as alternatives to direct observations in the native rhizosphere environment. Microscopy can be used in direct visualization using e.g. scanning electron microscopy (SEM) or combined with staining of the indigenous microbial community. In addition, microscopy can be used to monitor specific microorganisms, e.g. growth activity and survival of bacterial reporter strains. In this section we will give a broad overview of these three major approaches to study rhizosphere microbiology.

Traditional and environmental SEM

Conventional SEM has typically required extensive sample preparation, including fixation and dehydration before the specimen could eventually be mounted for microscopy. Most suitable object for study is here the rhizoplane, thus excluding the loosely associated rhizosphere soil surrounding the root. Alternatively, the porous matrix of the rhizosphere soil (or bulk soil) may actually be accessible by SEM, provided the preparation includes imbedding in a resin, followed by thin-sectioning of the specimen (Postma and Altemüller 1990). Classical studies in the 1950′es were pioneering for direct observations of single microbial cells on root surfaces (Rovira 1956). Hence, conventional SEM technique dates back 50 years and the technique has over the years been used extensively to visualize microbial root-surface colonization. Fukui et al. (1994) used conventional SEM technique to study bacterial colonization of sugar beet seeds, and observed different distribution patterns for *Bacillus subtilis* and *Pseudomonas putida* strains, respectively. Similarly, Dandurand et al. (1997) obtained detailed information on distributions of a *Pseudomonas fluorescens* biocontrol strain on young pea root surfaces. This approach gave evidence of a spatial pattern for root colonization including cell aggregate formation along the entire root. Finally, Fakhouri et al. (2001) studied the ultra-structure of root colonization, including antagonistic interactions between *Pseudomonas fluorescens* and pathogenic *Fusarium oxysporum* on tomato roots.

Sample preparation for conventional SEM involves dehydration, but the so-called Environmental Scanning Electron Microscopy (ESEM) operates slightly above the saturation vapor pressure of water in the specimen chamber. Under such conditions, water remains a liquid and hydrated biological specimen may be observed without prior preparation. The technical problem of maintaining water vapor pressure during observation and thus avoiding evaporation and dehydration as outlined by DeLeo et al. (1997) needs careful attention. The technique has been used to study mineral composition in the rhizosphere (Cabala and Teper 2007). Although the technique has not yet been applied in rhizosphere microbiology, it may soon become useful for high-resolution studies of microorganisms in undistorted rhizosphere samples.

Molecular stains

Direct detection of single cells in a complex environment as the rhizosphere has required development of high-resolution technologies primarily based on unique molecular staining and tagging systems for bacteria combined with advanced fluorescence microscopy. The molecular staining techniques combined with fluorescence microscopy is often specific enough to localize complete genera or functional groups of the indigenous microorganisms in rhizosphere samples, as described below.

General cell stains

As for SEM, conventional epifluorescence microscopy (EFM) in rhizosphere samples has always been hampered by the sample destruction necessary to prepare a thin specimen for staining and observation. Hence, sample mounting in resin and subsequent thin-sectioning to provide structurally intact specimen is also important in fluorescence microscopy (Eickhorst and Tippkötter 2008). Where mechanical distortion must be avoided or kept at a minimum, an alternative is the use of long-distance objectives; such objectives may in some cases provide adequate depth of the working range for "thick" rhizosphere samples (Thrane et al. 2000).

Applications of direct epifluorescence microscopy of soil and rhizosphere specimen include DNA staining with acridine orange (AO) in thin-sectioned soil samples as reported by DeLeo et al. (1997). For direct EFM of a root specimen, the AO may be useful to stain the total microbial population. One such application documented a particularly high density of bacteria near root sections presumed to release specific exudates (sucrose or tryptophan) at the root tip or older root segments, respectively (Jaeger et al. 1999). Since the hydrophobic AO compound has a strong adsorption affinity to soil humic material, unspecific binding to the soil matrix may actually be exploited as a counterstain (Anguish and Ghiorse 1997). AO counterstaining for observation of fungal zoospores (*Pythium aphanidermatum*) on cucumber roots has also been used successfully (Zhou and Paulitz 1993). An alternative, the UV-excitable DNA stain DAPI may also stain unspecifically in some soils. Where this is a problem, a useful DNA stain is the green fluorescent SYBR Green II (Weinbauer et al. 1998). Finally, a useful stain is also Fluorescent Brightener (FB) 28 (sometimes referred to as Calcofluor White M2R), applied for wall staining of fungi in root-soil microcosms (Thrane et al. 1999) and imbedded thin-sections of soil (Postma and Altemüller 1990; Eickhorst and Tippkötter 2008). A more recent Fluorescent Brigthener Agent (FBA) 220 may actually improve staining of the soil thin-sections since crystallization problems can be avoided (Harris et al. 2002). Today these general strains are rarely used alone, but rather to gather additional information in combination with specific cell stains.

Specific cell stains

Confocal laser scanning microscopy (CLMS) introduced the possibility of 3-D reconstructions to obtain high resolution information on the structural and spatial composition of microbial communities in environmental samples. CLSM applications to study microbial populations in rhizosphere samples have become numerous during the last decade or so, as seen in the following. Modern fluorescence microscopy has thus provided an excellent insight on spatial and temporal colonization patterns for a large number of specific bacterial inoculants, i.e. from the early binding of inoculant cells to a seed and to their firm establishment within roots (endophytic bacteria), in the rhizoplane, or in the rhizosphere including the mucigel polymer matrix or root-adhering soil.

The first applications of CLSM to study bacterial root colonization were based on strain-specific fluorescent antibody (FA) staining to follow the inoculants (Schloter et al. 1993; Hansen et al. 1997). Co-inoculation experiments demonstrated that *Azospirillum brasilense* strains colonizing wheat roots were mutually competitive (Kirchhof et al. 1997), while *Pseudomonas fluorescens* strains colonizing barley roots were not (Hansen et al. 1997).

Unlike bulk soil, the rhizosphere habitat has high microbial activity due to a prevalence of root metabolites. As a consequence, fluorescence probing of rhizosphere bacteria with rRNA-targeting oligonucleotides may result in higher hybridization signals assuming that cellular rRNA contents correlate with growth activity (Assmus et al. 1997). Hence, the rapid development of novel fluorescence in situ hybridization (FISH) probes, staining technologies and CLSM application has resulted in numerous studies of root colonization, as exemplified here. Taxonomic probes for a selected strain or group of organisms can be applied, sometimes even physiological probes for specific cellular activity. Important early observations were the 3-D patterns of active sub-populations of both inoculant and indigenous bacteria on root surfaces. The early works by Assmus et al. (1995, 1997) using combinations of strain-specific monoclonal antibody (*Azospirillum brasilense* Wa3), species-specific FISH probe (*A. brasilense*), group-specific FISH probe (α-Proteobacteria) and general FISH probe (domain Bacteria) on wheat roots first demonstrated the potential of combining probes of different specificity. More recently, Kutter et al. (2005) used FISH to demonstrate different colonization patterns by the pathogens *Salmonella enterica* and *Listeria* spp. on barley roots. Watt et al. (2006) used FISH for quantitative studies of bacteria colonizing wheat roots; *Pseudomonas* and filamentous bacteria were found to comprise 10% and 4%, respectively, of the total rhizosphere community. Other Gram-negative bacteria monitored recently by FISH have been *Rhizobium* sp. on rape (Santaella et al. 2008) and *Methylobacterium suomiense* on rice and tomato (Poonguzhali et al. 2008). A Gram-positive group under study has been *Paenibacillus* on both *Arabidopsis* (Timmusk et al. 2005) and maize (von der Weid et al. 2005). Finally, Kreuzer et al. (2006) used FISH to show that root architecture was affected by interactions between plant root exudation, bacteria and their predators.

Recent advances to further develop the FISH technology have been to target mRNA (rather than rRNA) monitoring expression of toluene monooxygenase gene (*tom*) in *P. putida* in wheat rhizoplane samples (Wu et al. 2008a). At present, the first application of FISH to target cellular DNA monitoring occurrence of nitrite reductase gene (*nirK*) in indigenous denitrifier populations in environmental samples has been reported (Pratscher et al. 2009). This development is very promising for studying both occurrence (e.g. phylogeny, density, colonization pattern) and metabolic activity (gene expression) of rhizosphere bacteria. A major challenge to achieve FISH recordings in undisturbed soil and rhizosphere samples has been met by the development of resin embedding and thin-sectioning of FISH-stained soil samples (Eickhorst and Tippkötter 2008). Another major challenge in quantitative studies using fluorescence microscopy in rhizosphere samples is development of reliable quantitative and automated image analysis tools for 2D images and CLSM-3D image stacks (Dazzo et al. 2007). Recently, Daim et al. (2006) reported a novel image analysis program offering functions to quantify microbial populations and evaluate new FISH probes.

Reporter bacteria

An important approach to track single bacteria in the rhizosphere is insertion of fluorescence marker genes in specific bacteria under study. Suitable designs of plasmid or transposon vectors for insertion, driving promoters and a panel of color variants have allowed for bioluminescence (e.g. *lux* gene) or Green Fluorescence Protein (e.g. *gfp* gene) tagging in several bacteria and microfungi from soil and rhizosphere. The early paper by Gage et al. (1996) described the construction of *Rhizobium meliloti* GFP mutants and showed detailed CLSM images of their growth and behaviour during the early stages of infection and nodulation in living alfalfa roots. A number of papers subsequently reported on detailed root colonization patterns of fluorescent *Pseudomonas* sp. biocontrol strains; CLSM was used to follow constitutive GFP mutants of *P. chlororaphis* on barley seeds (Tombolini et al. 1999), *P. fluorescens* on roots of both barley (Normander et al. 1999), tomato (Götz et al. 2006), avocado (Pliego et al. 2008) and olive (Prieto and Mercado-Blanco 2008). To specifically monitor growth-active *Pseudomonas* cells, Ramos et al. (2000) followed the colonization pattern of an unstable GFP mutant of *P. putida* on young barley roots.

Noteworthy are also the GFP constructs made in both Oomycota such as *Phytophthora* spp. (Bottin et al. 1999; van West et al. 1999) and in more advanced soil microfungi, e.g. *Trichoderma harzianum* (Bae and Knudson 2000) and *Fusarium oxysporum* (Lagopodi et al. 2002).

Another approach for detection of single cells in the rhizosphere is the use of reporter genes in specific bacteria, also referred to as reporter bacteria, whole-cell biosensors, bacterial bioreporters or monitor strains. Such cells are equipped with reporter genes that encode a product, which is easily assayed in fluorescence microscopy and related to metabolic activity or specific gene expression of the host cell.

Reporter bacteria respond to the bioavailable fraction of compounds in their surroundings. A tight definition of bioavailability would then be: the fraction of the total pool of a compound that is biologically available to an organism. The bioavailable compound, defined as above, is here assessed using one, specific reporter strain. However, other microorganisms may have alternate mechanisms regulating their gene expression in response to the compound, or have different permeability properties than the reporter strain. This is an obvious limitation to the reporter technique as assessments made by a specific reporter strain may not be valid for all indigenous microorganisms, even for taxonomic groups closely related to the reporter strain.

Non-specific (general), semi-specific and specific reporters

Reporter bacteria can be classified into three groups. The first group represents the non-specific reporters, which carry e.g. *lux* reporter genes under control of a constitutive promoter. Frequently these reporters are tagged with a complete *luxCDABE* cassette encoding the luciferase as well as genes involved in production of substrate for the enzyme. These bacteria will emit constant light when supplied with oxygen and energy. As such a *lux*-tagged reporter is dependent on the intracellular energy from oxic respiration, reduced bioluminescence is a sensitive, but non-specific assay format for testing environmental stress conditions, e.g. toxicity of soil pollutants. However, specific cell location can not be obtained with non-specific *lux*-tagged reporters. To track a specific bacterial inoculant in the rhizosphere, the Green Fluorescent Protein (GFP) technology in combination with fluorescence microscopy has had greatest importance for plant-microbe interaction studies. The principle of a non-specific reporter strain is illustrated in Fig. 1. Non-specific reporter constructs can be plasmid-borne as well as chromosomal. Non-specific reporter strains with constitutive expression of both *gfp* and *lux* under the same promoter may offer simultaneous detection of cell localization and metabolic activity in soil (Unge et al. 1999). The second group of semi-specific reporters is based on expression of reporter genes in response to stress-full stimuli even when the actual stress factor is unknown. Finally, the third group of specific reporters may respond to presence of specific compounds or elements (e.g. Jaeger et al. 1999) or to

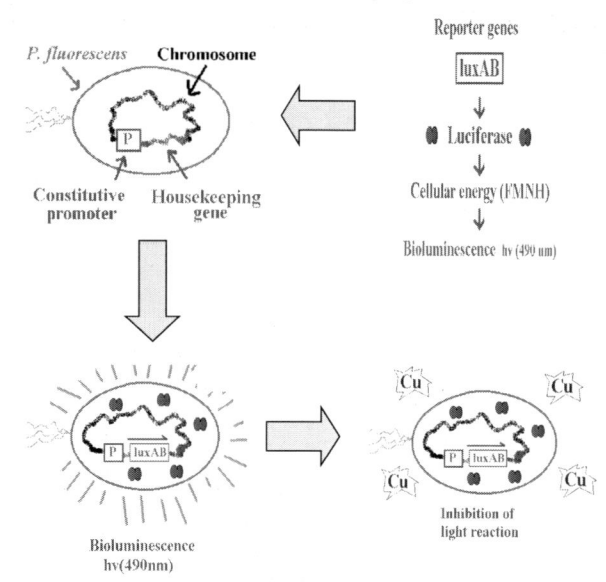

Fig. 1 In a non-specific bacterial reporter, the *luxAB* genes encoding bacterial luciferase are inserted behind a constitutive promoter located on the chromosome (or a plasmid) of e.g. *Pseudomonas fluorescens*. Bioluminescence depends on cellular energy, but also on oxygen and aldehyde substrate (not shown). Toxic compounds or elements such as Cu inhibit generation of cellular energy and thereby the light reaction

their absence (e.g. Koch et al. 2001). Figure 2 illustrates the principle of a specific reporter.

Non-specific (general) reporters were used relatively early to monitor metabolic activity of bacterial inoculants in samples from soil and rhizosphere. (Meikle et al. 1995) reported a loss of metabolic activity over time in drying soils. With a comparable approach, Kragelund et al. (1997) found that metabolic activity in a root-colonizing *P. fluorescens* strain varied along the root and was different in the rhizoplane compared to the surrounding rhizosphere and bulk soil.

Semi-specific stress reporters may be used where bacterial physiology under adverse conditions (e.g. exposure to high temperature or osmolarity, drought, or presence of reactive oxygen species) is controlled by global regulatory circuits (regulons) of gene expression. Typically, these reporters will respond to a broad range of conditions and can be useful to characterize a soil environment in terms of general cytotoxicity. Due to the extended knowledge on such systems in Enteric bacteria, *Escherichia coli* or *Salmonella typhimurium* have been used in much of

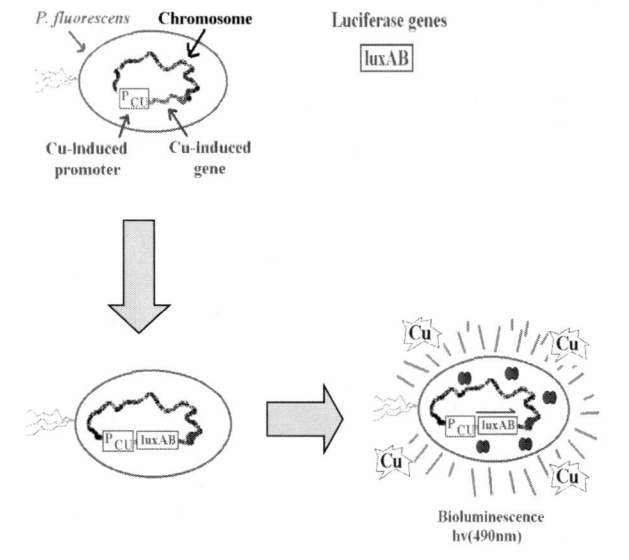

Fig. 2 In a specific bacterial reporter, the *luxAB* genes are inserted behind an environmentally induced promoter located on the chromosome (or a plasmid) of e.g. *Pseudomonas fluorescens*. Expression of bioluminescence, e.g. by Cu induction, occurs at concentrations below the level that inhibits the light reaction

the early work, e.g. linking the *lux* reporter system to "general stress" promoters such as the heat shock promoters *dnaK* or *grpE* (van Dyk et al. 1995). A more recent contribution by Park et al. (2002) demonstrated the use of a *dnaK-luxCDABE* construct in *Pseudomonas* sp. (strain DJ-12) to detect aromatic compounds (biphenyl and 4-chlorobiphenyl) related to PCB occurrence and degradation in the environment (see also below).

Specific reporters of bacterial growth have been constructed in *Pseudomonas* bacteria by fusing bioluminescence or *gfp* reporter genes to the promoter regions of operons supporting macromolecular synthesis of ribosomal RNA. An early example is the work of Marschner and Crowley (1996), who used a ribosomal promoter-driven *lux* reporter (emitting bioluminescence during growth, when rDNA genes are highly expressed); growth of a *P. fluorescens* strain was found to be higher in natural rhizosphere (pepper) than in bulk soil. Ramos et al. (2000), using a ribosomal promoter-driven *gfp* reporter (unstable GFP variant) in a *P. putida* strain, found that bacterial growth was detectable in rhizosphere (barley), primarily at the root tip.

Studies of the rhizosphere environment using reporter organisms

Carbon, nitrogen, phosphorous and oxygen availability

Responses to specific nutrient limitation in soil and rhizosphere represent some of the first applications of reporter bacteria to study specific gene expression in this environment. The approach has been dominated by intensive studies of *Pseudomonas* sp. strains, most certainly due to the importance in degradation and nutrient cycling by this organism *per se*, but also due to the potential exploitation of the organism in agricultural biotechnology, e.g. for plant protection, plant growth promotion and bioremediation.

Root exudates have long been considered to be the major C source supporting growth of root-colonizing bacteria in the rhizosphere of young plants. An advancement has been the use of non-specific *lux*-tagged reporters to detect the actual C-source composition and availability in soil and rhizosphere samples through changes in metabolic activity. Hence, inoculant *Pseudomonas* sp. (shortly pre-starved for C) responded to both source and concentration of C; wheat root exudates gave a response comparable to that of reducing sugar monomer (glucose), rather than that of common amino acid (glutamate) or carboxylic acid (succinate) components in root exudate (Yeomans et al. 1999). The cells were later shown to be capable of discriminating composition of root exudates from plants grown with or without herbicide treatment (Porteous et al. 2000). Koch et al. (2001) and van Overbeek et al. (1997) both used specific *lacZ*-based C-limitation reporter systems in *P. fluorescens* strains to demonstrate C limitation in bulk soil, but not in rhizosphere. Paterson et al. (2006) and Puglisi et al. (2008) used a *lux*-marked *P. fluorescens* biosensor (strain 10,586 pUCD607) to monitor C exudation from roots and C availability in thin slices of soil, respectively. While C may not generally be limiting in the rhizosphere, the composition and availability of specific organic components may still be important for actual C status of *Pseudomonas* sp.

The significance of N and P limitation in soil and rhizosphere has been addressed by studies including bioluminescent *Pseudomonas* reporter strains (Jensen and Nybroe 1999; Koch et al. 2001; Kragelund et al. 1997; Standing et al. 2003). In bulk soil neither N nor

P limitation could be observed in agreement with the above observations of C limitation in this habitat (Jensen and Nybroe 1999; Kragelund et al. 1997). However, soil amendment with barley straw changed the life conditions for the inoculated *Pseudomonas* strain, encountering N limitation when C-rich polymers from the barley residues were degraded (Jensen and Nybroe 1999; Koch et al. 2001). The rhizosphere (barley) demonstrated significant N limitation, whereas P limitation was not observed (Jensen and Nybroe 1999; Kragelund et al. 1997). This work was the first identification of a major nutrient limitation by N of potential significance for growth and activity of *Pseudomonas* sp. in natural rhizosphere. DeAngelis et al. (2005) made a nitrate reporter (nitrate-regulated promoter of *narG* in *E. coli* fused to GFP genes) in *Enterobacter cloacae* and demonstrated lower nitrate abundance in wild oat rhizosphere compared to bulk soil. An important improvement has further been the development of double reporters, which may address changes in C and N availabilities in one reporter strain (Koch et al. 2001). The concomitant application of several reporters addressing C, N and P availabilities in the same samples (Standing et al. 2003) may be useful to reveal the nutrient conditions affecting growth and survival of *Pseudomonas* spp. in rhizosphere environments.

The N reporter strain used by Jensen and Nybroe (1999) reacted towards limitation by both NH_4^+ and common amino acids (e.g. glutamate), and further work should address if specific, reduced N components in exudates may regulate *Pseudomonas* spp. growth in the rhizosphere. Specific reporter bacteria responding to individual amino acids show great promise for identification of such growth-limiting compounds. For example, induction of a lysine-responsive *P. putida* reporter was demonstrated in rhizosphere (corn), but not in bulk soil (Espinosa-Urgel and Ramos 2001).

The study by Jaeger et al. (1999) illustrates another advantage by reporter studies, namely that precise information on the spatial distribution of C and N compounds can be obtained. These authors made a *Erwinia herbicola* tryptophan-reporter strain with a fusion between the *aatI* gene encoding a tryptophane aminotransferase and a *inaZ* ice nucleation reporter. In the rhizosphere of an annual grass (*Avena barbata*), the reporter showed significant induction in older root segments with lateral root formation, but not at the root tip (Jaeger et al. 1999). In the same set-up, however, bioavailable sucrose was most abundant at the root tip as demonstrated by another *E. herbicola* reporter strain. Finally, a *P. fluorescens* reporter strain was used by Kuiper et al. (2001) to show that uptake regulation of putrescine, a common polyamine in tomato root exudate, was important for growth rate and thus competitive colonization ability in the rhizosphere.

Oxygen availability in the soil environment is of fundamental importance to expression of several distinguishing traits in *Pseudomonas* spp., notably denitrification but also a number of redox-regulated traits like fluorescent siderophore and HCN production. High consumption and limited supply rates of oxygen may be expected in rhizosphere, organic aggregates (hot spots), or highly compacted soil. In the first attempt to determine oxygen availability by reporter strains, Højberg et al. (1999) found induction of a low-oxygen-sensitive *lacZ*-based *P. fluorescens* reporter strain in wetted (85% WHC) but not in unwetted (60% WHC) rhizosphere (barley) and in compacted bulk soil. The work demonstrated that common water and texture conditions easily promoted low-oxygen and thus denitrifying conditions in both rhizosphere and bulk soil. More work based on reporter strains is needed to elucidate the role of redox-regulated phenotypes in soil. Hence Ghiglione et al. (2000), using a nitrate reductase-deficient mutant of a denitrifying *P. fluorescens* strain, demonstrated that this function may confer a selective advantage in the rhizosphere (corn).

Pollutant aromatics and their degradation

Several specific reporters containing bioluminescence or *gfp* fusions in the functional degradation genes have been presented. A first example to illustrate the progress being made to detect occurrence and degradation of specific chlorinated aromatics, involves the herbicide 2,4-dichlorophenoxyacetic acid (2,4-D) and the degrading soil bacterium, *Cupriavidus necator* JMP134 (formerly *R. eutropha* JMP134). This strain, harboring the degradation plasmid pJP4, has become the model of 2,4-D degradation; two modules of *tfd* genes, $R_{regulatory}$-$D_{II}C_{II}E_{II}F_{II}$ and $T_{regulatory}$-$C_{I}D_{I}E_{I}F_{I}$, are both involved in the degradation steps from ring cleavage of 2,4-dichlorocatechol to formation of a product entering the TCA cycle. Hay et al. (2000) constructed a reporter in the organism containing a chromosomal insert of the *tfdR-tfdD$_{II}$* (including promoter) sequence linked to promoterless *lux*; the reporter responded sensitively and linearly (up to 100 μM) to both 2,4-D and the first degradation intermediate 2,4-dichlorophenol (2,4-DCP). Füchslin et al. (2003) presented an alternative reporter containing a chromosomal insert of the *tfdC$_I$* (including promoter) linked to *gfp*; the reporter system was supported by the indigenous regulator from *tfdT* located on pJP4. Such a reporter construct appears promising for application in rhizosphere environments, e.g. for single-cell studies using advanced microscopy.

A second example of pollutant compounds being assessed with reporter bacteria in soil and rhizosphere degradation studies is that concerning the polychlorinated biphenyls (PCB). This group of much-attended pollutants is degraded by several soil bacteria, including typical degrading genera such as *Ralstonia*, *Burkholderia* and *Pseudomonas*. The initial degradation step is mediated by a biphenyl dioxygenase, encoded by the *bphA* gene and Layton et al. (1998) constructed an *R. eutropha* ENV307 strain harboring the reporter plasmid pUTK60 with an insert of *orf0$_{regulatory}$-bphA$_1$* (including promoter) linked to promoterless *lux*. The detection limit was approx. 1 μM for both monochlorinated (2-CB, 3-CB and 4CB) and polychlorinated biphenyls (Arachlor 1,242 mixture), but the linear range appeared narrow (approx. 1–10 μM). By comparison, Brazil et al. (1995) constructed a PCB reporter in *P. fluorescens* F113 by chromosomal insertion of *orf0-bphA$_1$* linked to promoterless *lacZ* (beta-galactosidase). This strain was a strong root colonizer showing potential for bioremediation of polluted soils; however, since the organism did not naturally contain biphenyl-degrading genes, a complete array of *bph* genes was inserted independently of the reporter genes. The authors demonstrated that the reporter strain had beta-galactosidase activity and thus expression of *bph* genes for at least 5 days after inoculation on sugar beet seeds. Boldt et al. (2004) modified the PCB-degrading *P. fluorescens* F113 strain to report on *bph* activity by expression of stable or unstable *gfp* (Andersen et al. 1998); further GFP reporter constructs in F113 also sensing chlorobenzoic acid derivates from PCB degradation was recently reported by Liu et al. (2007). Using confocal laser scanning microscopy (CLSM), Boldt et al. (2004) could discern single F113 cells showing *bph* activity (and thus biphenyl/PCB degradation) on roots developing from inoculated alfalfa seeds in PCB-amended soil extract medium.

Cell-cell interactions in the rhizosphere

Some of the most fascinating applications of reporter bacteria have addressed the complicated plant-bacterial, fungal-bacterial or even bacterial cell-cell interactions in soil systems, as examplified in the following. The first well-known example is that of plant signals (flavonoid compounds) controlling early stages of the legume-*Rhizobium* symbiosis via activation of the bacterial *nod* genes; an early *Rhizobium* reporter based on a *nodC-lacZ* fusion was constructed by Bolanos Vasquez and Warner (1997) to study activation by six different flavonoids from host plants (bean). Further, the identification of new, specific environmental signals regulating bacterial growth and activity in the rhizosphere has become feasible with reporter techniques that can sort out activated gene promoters under in vivo conditions. A number of different reports (Rainey 1999; Timms-Wilson et al. 2000; Allaway et al. 2001; Marco et al. 2003) have described such reporter systems to identify specific rhizosphere-activated promoters in both *Rhizobium* sp. and *Pseudomonas* sp. Briefly, a promoterless reporter is here fused into a host strain and the activated host cells are subsequently recovered to identify the promoter control by gene sequencing at the site of reporter insertion. Representing an example, a *Rhizobium* reporter with promoterless *gfp* captured both rhizosphere-activated promoters controlling syntheses of thiamine and cyclic glucan synthesis and surface growth-activated promoters controlling methionine synthesis or putrescine uptake (Allaway et al. 2001). Another recent approach with promoter-less *gfp* in *Rhizobium* demonstrated 29 rhizosphere-induced loci encoding proteins involved in environmental sensing, control of gene expression, metabolic reactions and membrane transport (Barr et al. 2008).

There has also been research using *Pseudomonas* reporter bacteria to study gene-regulating signals of importance for interaction with plant-pathogenic microfungi, including both Oomycota and "true" microfungi. A major interest is here to study the molecular mechanisms of hyphal colonization and antagonism by *Pseudomonas* bacteria in plant-protecting biological control. Using a *P. fluorescens* F113 reporter strain, Smith et al. (1999) found that *Pythium ultimum* (Oomycota) released a molecular signal, which down-regulated *rrn* promoters (ribosomal RNA synthesis) in the reporter and thus controlled the cellular growth rate. In another study, Lee and Cooksey (2000) found that hyphal colonization of *Phytophthora parasitica* by a *P. putida* reporter strain led to activated promoters controlling ABC transporter proteins. The recent report of de Werra et al. (2008) demonstrates that GFP-based reporter fusions to the *phlA* and *prnA* genes essential for production of antifungal compounds in *P. fluorescens* CHA0 can be constructed to monitor biocontrol gene expression of this strain in the rhizosphere.

Finally, a role of bacterial cell-cell communication including cell density-dependent gene regulation (quorum sensing) in a large number of bacteria has also been indicated in rhizosphere systems. Steidle et al. (2001) constructed AHL-sensitive reporters with GFP in *P. putida* and were able to demonstrate that the indigenous bacterial community colonizing tomato roots in natural soil produces AHL molecules. A fraction (approx. 40%) of *Pseudomonas* spp. colonizing plant roots was reported by Elasri et al. (2001) to produce N-acyl-L-homoserine lactone (AHL) molecules, serving as bacterial cell-cell communication signals; interestingly, the AHL production appeared to be more common among plant-associated than among soilborne *Pseudomonas* spp. The role of AHL communication in the rhizosphere still needs to be elucidated for a comprehensive understanding; a recent addition to the list of possibly AHL-regulated functions was that of rhizosphere N mineralization (DeAngelis et al. 2008). These authors reported that quorum sensing was coupled to extracellular chitinase or protease expression in a large number of rhizosphere bacteria, including *α-Proteobacteria*.

Given the current excitement of possibilities arising with the reporter technology, it is also timely to warn against the pitfalls sometimes neglected and the needs for testing and standardization. Meeting such a standard, a thorough testing of *lux* reporter systems has been conducted by Maier and co-workers (Neilson et al. 1999; Dorn et al. 2003). With particular attention to soil studies, reporters are used in their most difficult environment, extremely heterogenous at the bacterial scale. Leveau and Lindow (2002) discussed the use of reporter bacteria in heterogenous samples; it is clear that reporter bacteria actually offer a unique opportunity to study this heterogeneity in soil in great detail. In the future, more studies should be made to compare results using different scales of investigation; there are unique opportunities in comparing single-cell studies based on sensitive, fluorescent reporters such as *gfp* in combination with advanced CLSM microscopy with larger samples such as soil extracts or slurries, and with whole plant-soil detection studies based on direct bioluminescence recorded under a sensitive camera. On-line monitoring using immobilized reporter cells on optic fibers (Heitzer et al. 1994) should be tested further; a recent application of fiber optics to monitor reporter bacteria in porous media was reported by Yolcubal et al. (2000).

Community studies in the rhizosphere

Molecular analysis of microbial communities

Despite improvements in cultivation techniques to study the microbial community composition in complex environmental samples (Janssen et al. 2002; Joseph et al. 2003), the organisms in culture represent only a minor fraction of the microorganisms occurring in situ as estimated by e.g. DNA reassociation (Torsvik et al. 1990). Culturing of microorganisms is hampered by difficulties in reproducing natural, ecological niches in the laboratory media. Furthermore, symbiotic relationships might be of crucial importance for community function and are often impossible to maintain in the cultivation process. Hence, there is a need for cultivation-independent techniques to accompany the cultivation-dependent methods to obtain a deeper insight into the structure and function of indigenous microbial communities.

Cultivation-independent molecular approaches to study the indigenous microbial community were adopted routinely in microbial ecology in the 1990's when the Polymerase Chain Reaction (PCR) and other DNA-based characterization methods became available. The rapid interest for these methods requiring the DNA (or RNA) to be directly extracted from the environment was related to their capacity to overcome biases in isolation and in vitro cultivation. The new discipline opened new possibilities in the search for knowledge within the "black box" of soil microbiology. Despite the introduction of other biases (Wintzingerode et al. 1997), the new DNA- and PCR-based approaches have provided completely new insight into the life of microorganisms in their natural environment. In the present overview we aim to discuss advantages and disadvantages of the DNA/RNA based methodology most often used and/or promising for studies of the indigenous rhizosphere community, including examples of knowledge obtained. Furthermore, we discuss the proteomics approach not yet fully functional, but highly promising, for application in complex environments as the rhizosphere.

Extraction of nucleic acids from complex systems

Nucleic acid extraction from environmental samples is the basis for a range of methods used in molecular microbial ecology. For complex environments such as bulk soil and rhizosphere two approaches have been developed for extracting nucleic acids. The first is direct extraction of the nucleic acids after in situ cell lysis which is then followed by DNA purification (Ogram et al. 1987). In the second approach the cell fraction is first separated from soil particles before the cells are lysed and nucleic acids purified (Holben et al. 1988; Courtois et al. 2001). Both approaches have advantages and disadvantages related to DNA yields, DNA purity for molecular purposes, and the ever-questioned representation of the entire microbial diversity.

The direct DNA extraction method

The direct DNA extraction method assumes complete in situ lysis of all microorganisms present in the sample. The disruption of the microbial cell wall must lead to the release of all nucleic acids from all bacteria, theoretically independently of the cell wall sensitivity to lysis treatments, the location of bacteria in microstructures and their interactions with soil particles. Three types of cell lysis (or membrane disruption) are used alone or in combination: (i) physical, (ii) chemical and (iii) enzymatic disruption. Physical treatments including freezing-thawing or freezing-boiling, bead-mill homogenization, bead beating, mortar-mill grinding, grinding under liquid nitrogen, ultrasonication and microwave thermal shock have shown efficiencies for disruption of soil structures and tend to give the best representation of the whole bacterial community. Vegetative forms, small cells and spores are efficiently disrupted but the physical treatments may also result in significant DNA shearing. Chemical lysis is based on the use of detergents (the most common detergent is sodium dodecyl sulfate (SDS) which dissolves the hydrophobic material of cell membranes) that have often been used in combination with heat treatment and with chelating agents such as EDTA, Chelex 100 and a variety of Tris buffer or sodium phosphate buffers. However, in the search of the best compromise between DNA quantity and purity, protocols have been developed in which cetyltrimethyl-ammonium bromide (CTAB) and polyvinylpolypyrrolidone (PVPP) have been added to remove at least partially the humic compounds. Many protocols have been developed in which enzymes such as lysozyme, achromopeptidase and proteinase K are used in combination with a chemical treatment.

Purification of the extracted DNA solution is the second step of the direct DNA extraction methods in which the aim is to get the best compromise between purification efficiency and DNA loss. Combination of organic solvent extraction, precipitation, CsCl gradient centrifugation and hydroxyapatite chromatography have been used but are now replaced by less time-consuming protocols that are also more adaptable to the processing of a higher number of samples used individually or in combination. These include electrophoresis of DNA solutions on agarose gels or gel filtration on G150, Sepharose 2B, 4B and 6B, Biogel P100 and P200. Other commercial purification products such as Wizard DNA clean-up system (Promega) and Centricon™ 50 and Microcon™ 100 concentrators (Amicon), Elutip™ D column (Schleicher and Schuell), silica-based DNA binding SpinBind Columns (FMC BioProducts) and Tip-100 and Tip-500 columns (Qiagen), have also been used (Robe et al. 2003). Several miniaturized extraction/purification kits have been specifically developed and are now on the market including those provided by QBiogene and MO BIO Laboratories Inc. and these kits have largely replaced the methods presented above. The direct DNA extraction method generally provides the highest DNA yields but some applications, such as construction of metagenomic libraries, require the extraction of large DNA fragments that are rarely obtained by direct extraction methods. In most studies, direct extraction have not provided DNA fragments larger than 20 kb although recovery of large DNA fragments (40–90 kb) was reported using lysozyme-SDS-based methods (Krsek and Wellington, 1999).

The bacteria extraction method

An alternative approach to the direct DNA extraction method was developed for the recovery of highly purified and large bacterial DNAs essential for preparing metagenomic DNA to detect complete gene clusters and biosynthetic pathways. The bacteria extraction method initially described by Faegri et al. (1977) and Torsvik and Goksoyr (1978) is based on the initial separation of bacteria from the soil particles prior to cell lysis and DNA purification with gentle treatments to preserve DNA integrity. Protocols have gradually been improved by optimizing each of the following sequential steps: dispersion of soil particles, separation of the cells from soil particles by centrifugation and/or buoyant density, lysis of extracted cells and DNA purification.

Approaches combining centrifugation-based cell separation from soil particles and in-plug lysis and pulsed field gel electrophoresis (PFGE) after bacteria are embedded in the agarose plugs are now recognized as the most efficient method yielding DNA fragments more than 300 kbp in size and providing adequate purity for further molecular cloning procedures (Robe et al. 2003). Alternative methods combining gentle lysis treatments with CsCl density purification have recovered very pure, 100 kbp long DNA fragments. This size however remains limiting for functional genomic approaches where the exploration of gene clusters and biosynthetic pathways through cosmid and bacterial artificial chromosome (BAC) cloning requires DNA greater than 200 kb.

Compared to the direct DNA extraction methods, the bacteria extraction methods offer a lower DNA yield due to the fact that only 25–50% of the total indigenous bacterial community is recovered in the bacterial fraction while direct extraction may recover more than 60% of the total bacterial DNA. Although time-consuming, cell fractionation-based methods are preferred in the cases where subsequent analyses are focusing on the prokaryotic community DNA (with the exclusion of any extracellular and eukaryotic DNA) and when highly purified DNA solutions and high molecular weight DNA are required.

RNA extraction and purification

During the last decade, focus has shifted from merely looking at diversity and community structure towards linking community structure to community function. One way has been to look at gene expression during different environmental fluctuations. Studies on environmental RNA have mainly focused on 16S rRNA, and only slowly are studies on mRNA emerging in environmental research. As mentioned previously 16S rRNA has the advantage of being found in many copies, and has been the easiest gene transcript to detect. The ubiquitous presence of RNases and the very short half-life of RNA (especially mRNA) have made studies based on RNA from environmental samples very difficult. For these reasons, the bacterial extraction procedure described above is normally not applied when mRNA is the target nucleic acid. Alternatively, quick-freezing samples in liquid nitrogen, direct extraction, pre-treatment of all solutions with the RNase inhibitor DEPC and baking all glassware as well as keeping all samples on ice during the extraction procedure, is advantageous. Several methods for extraction RNA from the complex soil and rhizosphere environment have been published (Borneman and Triplett 1997; Hurt et al. 2001; Griffiths et al. 2000; Bürgmann et al. 2003). Based on these methods, investigations on the diazotrophic community in the rhizosphere of different rice cultivars (Knauth et al. 2005) and *Spartina alterniflora* (Brown et al. 2003) by detecting *nifH* mRNA, as well as investigations on the denitrifying community detecting the *nirK* and *nirS* gene transcripts (Sharma et al. 2005), have been performed. All studies have been qualitative using different fingerprinting techniques (see below) to describe the active populations. Only few studies have succeeded in obtaining quantitative data of gene transcripts in soil (Han and Semrau 2004; Jacobsen and Holben 2007). Recently, high-resolution studies on transformation dynamics of the functional gene *tfdA* (Nicolaisen et al. 2008; Bælum et al. 2008) in natural soil have been made. The extraction protocol used co-extracts of both DNA and RNA, which has the advantage that it offers information on both the community structure in general and on the active subpopulation. The protocol also offers the possibility of relating the transcript formation over time to the actual population density thus obtaining an "activity per DNA unit". In nongrowing systems, this can help overcome sample-to-sample differences due to differences in extraction efficiency between samples. This anticipation however relies on the assumption that the extraction efficiency within one sample for both DNA and RNA is at a constant ratio. In the study by Nicolaisen et al (2008) this was shown to be the case for *tfdA* mRNA and DNA, respectively, but should be tested for each gene investigated. Despite the potential in analysing mRNA in the rhizosphere community, there are still some limitations due to low recovery of mRNA compared to DNA in environmental samples (Meckenstock et al. 1998; Nicolaisen et al. 2008). However, the potential for rhizosphere research detecting mRNA directly in the samples hopefully will make further optimization of the RNA extraction protocols a main goal for researchers.

PCR based analysis and comparative studies of diversity and function

Molecular markers

The basis of molecular microbial ecology is the molecular markers. Molecular markers can be genes or gene transcripts that can be identified in a complex pool of nucleic acids providing information on the group of organisms harbouring these genes. Information based on DNA and RNA can answer questions related to the population structure of a specific environment such as: Is a particular gene present in this population? What is the phylogenetic composition of this community? Are particular genes expressed? How is the community composition changed after perturbation of the environment? What are the spatial or temporal differences in a particular habitat?

The choice of molecular marker is very dependent on the questions asked. Different levels of phylogenetic resolution or different functional groups of organisms can be detected based on the choice of a marker. Information based on the small subunit (SSU) rRNA (16S rRNA of prokayotes or 18S rRNA of eukaryotes) reveals the phylogenetic relationship between the organisms from where the DNA or RNA arose (Woese et al. 1990). Over the last decade, sequencing of SSU rRNA from uncultured organisms has led to the development of databases (e.g. Genbank: http://www.ncbi.nlm.nih.gov/Genbank/index.html or Ribosomal Database Project: http://rdp.cme.msu.edu/) and it has been verified that only a small fraction of the soil microorganism diversity is known. Furthermore, new genera have been proposed solely on the basis of SSU rRNA sequences form environmental samples, thus having no representatives in culture (Hugenholtz 2002; Fieseler et al. 2004). An alternative to 16S rRNA as a phylogenetic marker of bacteria is the housekeeping gene *rpoB*, coding for the RNA polymerase beta-subunit. This gene is only found in one copy in all bacteria investigated (Dahllöf et al. 2000), giving some advantages over the often multicopy SSU rRNA gene when using genetic fingerprinting for diversity indexes. In addition, the *rpoB* gene has thus been shown to contain enough phylogenetic information for diversity studies (Mollet et al. 1997; Case et al. 2007). The *rpoB* approach has been used to study *Paenibacillus* spp. in the rhizosphere of sorghum under different nitrogen regimes (Coelho et al. 2007); however, the advantage of 16S rRNA for phylogenetic analysis, still lies in the huge ribosomal database available.

Another approach is to focus on a functional group within the microbial community independent of the phylogenetic relationship. Many functional genes have been/are prone to horizontal gene transfer, and hence, functional groups of organisms can in general not be identified using phylogenetic markers like 16S rRNA and *rpoB*, as functions can be found in phylogeneticly distant groups of organisms. Functional genes encoding enzymes central for specific metabolic processes found in the rhizosphere have been investigated in this environment e.g. *amoA* genes from ammonia oxidizing bacteria and archaea (Chen et al. 2008), *nir* genes from denitrifying bacteria and *nif* genes from nitrogen fixing bacteria (Babić et al. 2008).

While the use of DNA as a molecular marker reveals information on the presence of organisms or the potential function of a community, it gives no information on activity at the time of sampling. In recent years, RNA has thus been more often targeted for information on the active fraction of the population, as transcript formation is believed to follow metabolic activity. The SSU rRNA has the advantage of being present in large amounts in the cells, and thereby being the easiest of the RNA species to detect in complex samples. However, microorganisms living in the soil environment are often starved, and starved cells have been shown to maintain their ribosomes longer (Wagner et al. 2003). Hence, using 16S rRNA to describe the active populations in a soil/rhizosphere environment might be questioned. An alternative approach is to study mRNA of functional genes, which is believed to have a much shorter half-live than 16S rRNA. In a recent study it was found that a specific mRNA (*tfdA* involved in the degradation of the herbicide MCPA) was indeed present only during active degradation of the compound (Nicolaisen et al. 2008). Although more information is needed before a generalization on the link between presence of specific mRNA and function of the community can be made, direct mRNA analysis is definitely a promising approach for research on activity of microorganisms in the rhizosphere. Difficulties in obtaining high quality and quantity of mRNA from environmental samples, however, still limit the research activities in this field.

Analysing specific subpopulations based on the pool of nucleic acids directly is often not possible as most detection systems are not sensitive enough to detect the often very low amount of a specific gene in the sample. By PCR, fragments of specific marker genes are amplified to reach a concentration that can be detected by several downstream applications discussed below. The application can be divided into qualitative (describing diversity overall or active), or quantitative (describing abundance of gene copies or gene transcripts) in the environment. A challenge in using PCR is to find a good target gene (molecular marker) and to develop primers that selectively amplify this gene. First, the target gene should contain both conserved and variable regions; second, the primer should target the gene from all members of the clade investigated, and it should not target any gene from organisms outside this clade. This is not trivial as only a minor fraction of soil microorganisms have been isolated and their genes investigated, and we do not know how well the available databases represent the target clade. Consequently, PCR can introduce bias in the analysis of the microbial community due to lack of primer specificity, differential amplification of diverging target genes, etc. (Wintzingerode et al. 1997). However, to date the PCR assay has, despite its limitations, enhanced our knowledge on the soil and rhizosphere community dramatically, and will probably do so for many years to come.

For studies on RNA, a conversion of RNA into complementary DNA (cDNA) is necessary prior to PCR amplification. This is done in the reverse transcription (RT) assay and the joint protocol for reverse transcription and subsequent PCR amplification is referred to as RT-PCR. The reverse transcription can be performed with specific primers targeting a specific transcript or it can be performed using random primers converting all RNA to cDNA. Using specific primers might be advantageous when a very low transcript number is expected, whereas the use of random primers opens the possibility of analysing multiple transcripts in a single sample. The classical approaches of RNA detection like Northern blotting and the RNase protection assay do not require PCR amplification prior to analysis. However, the high sensitivity and the ability to analyse a higher number of samples in a shorter time makes the RT-PCR based detection of RNA from environmental samples the most promising, despite the biases introduced.

Clone libraries and fingerprinting

Sequencing of clone libraries based on PCR-amplified genes obtained from environmental samples offers the highest phylogenetic resolution and has led to the recognition of the impressive diversity of prokaryotes. Most obtained sequences are deposited in databases like the Ribosomal Database Project (RDP) and GenBank, and can be used for comparative studies on prokaryotic diversity in the environment and to infer phylogenetic relationships between the organisms detected. Cloning followed by sequencing of specific conserved genes is of great importance in research on indigenous populations.

In some studies it is not necessarily the exact genetic composition that is important, but rather the tracking of changes of community structure or changes in the active part of the community in relation to spatial or temporal variations, chemical gradients or specific perturbation of the environment. Fingerprinting techniques based on PCR amplicons have a high potential for screening multiple samples for differences in the genetic diversity of a group of genes. Fingerprinting techniques reveal information on the genetic diversity of the sample, but only rarely indicate phylogenetic relationship of the detected fragments. Applying a polyphasic approach combining fingerprinting with sequencing (with or without a cloning step in between) of selected fragments can improve the data analysis and add information on community structure.

Denaturing gradient gel electrophoresis (DGGE), terminal restriction fragment length polymorphism (T-RFLP) and single-strand conformation polymorphism (SSCP) of DNA and RNA are currently the fingerprinting techniques most often applied in rhizosphere studies, and are selected for further discussion here. The three techniques are based on three different separation technologies. DGGE is based on different melting behaviour of double-stranded DNA due to sequential differences in a denaturing gradient during electrophoresis (Muyzer et al. 1993). T-RFLP separates fragments based on length of the terminal fragments obtained due to differences in restriction endonuclease sites (as in RFLP and ARDRA) (Liu et al. 1997). Finally, SSCP separates fragments based on different mobility of single stranded DNA in non-denaturing gels (Schwieger and Tebbe 1998). It has been shown that the results obtained based on the three methods reveal the same clustering of the microbial members when used on the same soil sample, and the three fingerprinting methods seem equally suited for analysing differences in community patterns due to physico-chemical and biological differences between the sites of investigation (Smalla et al. 2007). Despite their equal performance, the three methods have several strengths and weaknesses when compared. Advantages of the DGGE and SSCP approaches are the possibility of isolating specific genetic elements for subsequent sequencing, whereas this is not possible when T-RFLP is used, omitting the opportunity to fully identify peaks of interest. On the other hand T-RFLP is suitable for high-throughput analyses, despite the need for a restriction digest step. Even though it is not possible to link the obtained peaks to a specific organism, the output format of the T-RFLP as an electropherogram makes comparative analysis at a higher taxonomic level possible using *in silico* digestions of database sequences, a feature included in RDP for the 16S rRNA gene. Gel-to-gel comparison is much more difficult in the DGGE approach as the handmade gradient tends to differ slightly between gels. The major drawback of SSCP is the high rate of re-annealing of single-stranded DNA during electrophoresis (Nocker et al. 2007).

All three methods have been successfully applied in rhizosphere research. SSCP has been used to show that the crenarchaeal consortia associated with the rhizosphere of a range of terrestrial plants were different from the crenarchaeal consortia found in the bulk soil (Sliwinski and Goodman 2004), and that the rhizosphere community is determined both by the plant species and the soil type (Miethling et al. 2003). DGGE is the most extensively used fingerprinting method to study structural diversity in the rhizosphere and has been used to investigate a diverse range of scientific questions related to rhizosphere microbiology e.g. dynamics of methanogenic archaeal communities in Japanese paddy soils (Watanabe et al. 2007), community structure of *Pseudomonas* spp.in relation to the antagonistic potential in the rhizosphere (Costa et al. 2007) and effects of elevated CO_2 concentrations on the structural diversity of microorganisms in a grassland (Drissner et al. 2007). Fingerprinting methods are increasingly used to study differences in profiles at the DNA and RNA level, respectively, in order to gain knowledge on the link between function and diversity of the community. Most studies on this subject have used rRNA as the marker gene, but nitrogenase (*nifH*) genes and gene transcripts were investigated using T-RFLP in a study on the influence of different rice cultivars on expression of *nifH* genes in the rhizosphere (Knauth et al. 2005). T-RFLP has also been used in more "classic" studies on the bacterial composition in soil and rhizosphere in arable field sites (Ulrich and Becker 2006). It should be noted that fingerprinting techniques based on PCR amplification do not generally provide reliable measures of diversity parameters like evenness and richness, partly due to the problem of obtaining equal amplification efficiency of all fragments in the PCR reaction and the fact that distantly related taxa can contribute to the same signal in the analysis.

Quantitative PCR

The study of rhizosphere microbiology is not only a matter of who is there, but also: What is the structure of the microbial community e.g. the abundance of different community members? What is the proportion of cells of a certain function in the habitat, and even more interesting: What are the expression patterns for selected genes involved in specific processes? Most studies based on PCR are still qualitative, and only recently has quantitative PCR (and RT-PCR) based methods been a routine in molecular microbial ecology. Traditional PCR rely on end-point detection of the amplicons. Due to chemical and physical properties included in the PCR reaction, the end-point detection of the product is not quantitative. However, several quantitative PCR methods have been introduced and used in rhizosphere research including MPN-PCR (Rosado et al. 1996), competitive PCR (Mauchline et al. 2002), and most recently real-time PCR (Mavrodi et al. 2007). Real-time PCR is presented here as the most promising method to quantify genes from environmental samples.

By real-time PCR, amplicon formation is monitored in real time using fluorescence techniques, where fluorescence detected is proportional to amplicon formation. This enables the detection of product over the full amplification curve, and inhibition of the reaction e.g. by inhibitory substances co-extracted with the nucleic acids, can easily be identified. Another major advantage of real-time PCR is the applicability over a wide range of initial DNA concentration of target gene extracted from the environment (Heid et al. 1996).

Several detection formats have been developed for the real-time PCR assay (Wilhelm and Pingoud 2003), SYBR green detection and TaqMan probes being the most commonly used in molecular microbial ecology. SYBR green detection is not sequence-specific as SYBR green binds all dsDNA in the minor groove. Upon binding to dsDNA, fluorescence enhances about 100-fold in comparison with unbound SYBR green, and it is therefore well suited for detection of product formation. Being a non-specific dye, accurate quantification of nucleic acids is dependent on high specificity of the PCR reaction *per se*. Formation of primer-dimers is a problem as these might be detected using the SYBR green detection format; hence, optimization of the PCR reaction prior to real-time detection to avoid such bias is crucial for a reliable quantification.

The TaqMan probe format relies on the specific binding of a hybridization probe to the target sequence and the cleavage of the probe by the endonuclease activity of the Taq DNA polymerase (Heid et al. 1996; Wilhelm and Pingoud 2003). The probe is double-labelled, with a quencher-dye in the 3′-end of the probe and a reporter-dye in the 5′-end of the probe. When the probe is intact, the emission spectrum is quenched by the quencher fluorophore, and no fluorescence is detected. Upon cleavage by the Taq DNA polymerase, the reporter-dye is separated from the quencher-dye, and signal can be detected. Only probes bound to single-stranded DNA during the annealing step will be cleaved during the elongation of the primer by the polymerase, and hence the fluorescent signal is proportional with the product formation.

The limitations in the use of real-time PCR are the same as encountered for the normal PCR setup regarding primer specificity, amplification efficiency, etc. However, the technical improvements in the real-time PCR technology facilitate the identification of potential problems, e.g. inhibition due to co-extracted impurities, low amplification efficiency, etc. Due to the robustness and the easy and rapid assay of real-time PCR quantification, this method is a valuable tool in soil and rhizosphere microbiology.

DNA- and RNA-based Stable Isotope Probing (SIP)

Among other applications related to the direct recovery of nucleic acids from the environment are those which can provide a direct access to relate function of a microbial community and identification of the bacteria that account for it. The establishment of such a relationship still remaining a great challenge in microbial ecology.

Stable-isotope probing (SIP) was introduced to microbial ecology by Radajewski et al. (2000) and has been used to characterize growing microorganisms in environmental samples or to determine those which have the genetic potential of metabolizing a labeled substrate. For instance, the DNA-stable isotope probing (DNA-SIP) technique that combines isotopic ^{13}C tracer incorporation into the DNA or RNA and molecular approaches as described above can help to identify soil and rhizosphere bacterial populations that are actively involved in the carbon cycle. The principle of these techniques is to provide soil bacteria with ^{13}C-labelled material, e.g. cellulose produced by *Acetobacter xylinus* as reported by El Zahar Haichar et al. (2007) and photosynthates released by plants grown under artificial atmospheres (Ostle et al. 2003) before extracting total DNA from the soil and separating the ^{13}C-labelled (heavy) and unlabelled (light) DNA fractions by ultracentrifugation.

The structure of active bacterial communities can be analysed by any downstream method e.g. those described above, with the possibility to identify the bacteria that are responsible for the rapid transfer of photosynthate C inputs to atmospheric CO_2 (Ostle et al. 2003). Other applications include identification of microorganisms capable of metabolizing labeled substrates such as polycyclic aromatic hydrocarbons (PAH) (Singleton et al. 2007), pentachlorophenol (PCP) (Mahmood et al. 2005) and phenol (Manefield et al. 2002). A further development of the procedure was introduced by Lueders et al. (2004), using fractionation of the centrifugation gradient. This fractionation allows for quantitative evaluation of isopycnic DNA or rRNA throughout the gradient, and thereby a finer resolution, which is important for interpretation of environmental SIP results. Among other, new technological developments is labelling of the microbial DNA with $H_2^{18}O$, which allows a distinction of newly grown cells incorporating the label from those whose DNA remained without the label, indicating that the latter survived but did not divide (Schwartz 2007).

Metagenomic DNA libraries

As it is now clear that culturable bacteria represent only a small fraction of bacteria present in soil, it is important to obtain information about the majority of the bacteria — the unculturable. Modern high-throughput technology has made it possible to not only look for specific genes but look at the full genomic information present in a soil sample, called the metagenome (Rondon et al. 2000). With such bacteria extraction methods clones containing soil DNA inserts ranging in size from 40 kb to 50 kb can be routinely obtained. Even without robot facilities metagenomic libraries of more than 150,000 clones can be prepared and handled for molecular, biological or chemical screenings. Although representing an equivalent of 1,500 *E. coli* genomes such libraries are only a partial representation of the initial genetic diversity of the bulk or rhizospheric soil samples tested when at least two millions of such clones would be necessary to statistically consider that all the initial genomes have been included at least once in the library (Ginolhac et al. 2004).

These limitations will be overcome in the next future with the development of libraries with an increased number of clones that will be handled and screened with the help of adapted robots. However, exploitation of these initial metagenomic libraries already demonstrates all their potential to recover, to study and to exploit the untapped microbial diversity. For example, the molecular screening of the 150,000 clones containing soil derived metagenomic DNA library mentioned above detected 139 inserts with genes encoding polyketide synthases genes. The bioinformatics analysis of 44 clones randomly chosen among the 139 demonstrated that only two protein sequences were identical whereas nucleic sequences were not redundant and that the similarity level with all the existing genes in databases was not higher than 67% (Ginolhac et al. 2004).

Several screening strategies of metagenomic DNA libraries were developed with direct recombinant expression as a first alternative to detect metabolic activities (Fig. 3). The usual option consists in using the high transformation efficiency of *E. coli* for creating metagenomic libraries although this is certainly not the best heterologous host to express genes from soil bacteria (and even if expression efficiency can be artificially increased) (Courtois et al. 2003). For instance high throughput anti-infective assays can be run by spotting the *E. coli* recombinant clones on bacteria- or fungi-seeded agar plates subsequently analyzed to detect growth inhibition halos around the positive clones (Lamprecht et al. 2007) while other bioassays are achieved by preparing an extract from each *E. coli* recombinant clone to analyze its potential activity versus a set of chemical reference compounds and of negative controls.

A second alternative relies on a molecular screening of the library requiring the spotting of colonies on high density membranes or chips before hybridizing membranes with conserved domains of targeted genes, the main difficulty consisting in designing the probe(s) encompassing the unknown genetic diversity (Demanèche et al. 2009). Using such a molecular screening strategy, recombinant clones containing new polyketide synthase genes were detected in soil metagenomic DNA libraries (Jarrin 2005) confirming the interest of the metagenomic approach to retrieve genes from unknown bacteria to study them subsequently by conventional approaches.

The metagenomics alternative has already demonstrated its utility not only to better understand the

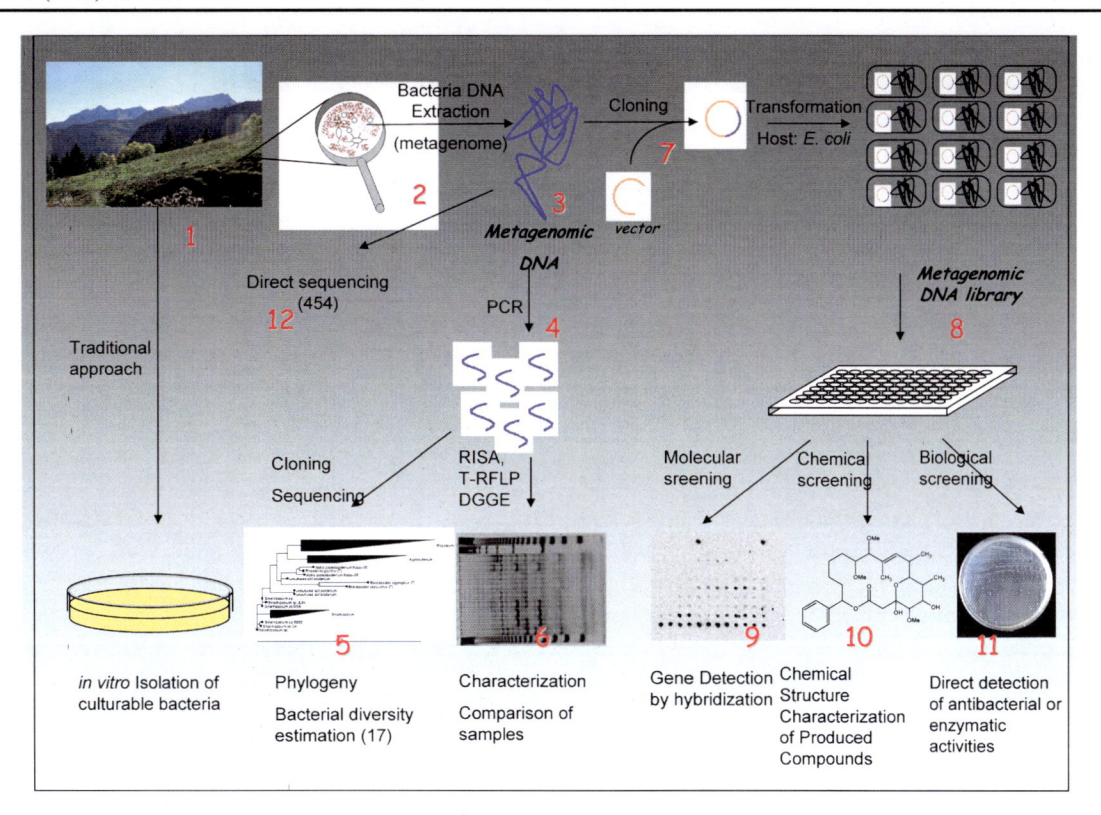

Fig. 3 Schematic representation of the metagenomic approach. A sample collected in the environment including soil (1) is submitted to treatments to extract and purify bacterial DNA (2). Several possibilities exist to deal with the extracted metagenomic DNA (3). One of the most common approaches is to use PCR amplification (4) as a fist step before products are cloned and sequences. Bacterial diversity can thus be estimated with the help of phylogenetic trees constructed from 16S rDNA or other gene sequences (5). Another approach is based on the use of fingerprinting techniques such as T-RFLP, DGGE and SSCP that produce fragments to characterize bacterial diversity in the analyzed sample (6). A second approach involves the direct cloning of extracted DNA (7) in a domesticated host such as *Escherichia coli* to produce metagenomic DNA libraries (8). These libraries can be screened according to 3 methods, including a molecular screening with clone DNA transferred to membranes for hybridization (9), or a chemical screening by analyzing the culture supernatant of metagenomic clones to detect compounds specifically produced by insert DNA (10). A last method is to spread the clones on a selective medium to detect those for which insert gene expression permits clone growth (11). Screening by these two last methods is based on expression of insert genes and transfer of recombinant plasmids, fosmids or BACs in alternative bacterial hosts can increase the recovery of genes of interest. The last approach which is called to become very popular in the next future is to sequence directly the metagenomic DNA with the help of new sequencing technologies such as 454

composition of yet unidentified bacterial communities and how ecosystems can function (Rondon et al. 2000; Beja et al. 2000; Venter et al. 2004; Rusch et al. 2007) but also in rendering the bacterial functions encoding genes from uncultivable bacteria available for an industrial exploitation (Voget et al. 2003). The main limitations for applying routinely metagenomics concepts to the rhizospheric environment are only technical. The construction of metagenomic DNA library is sometimes compromised by the relatively low amount of (sensu stricto) rhizosphere soil that can be recovered mainly when plants are grown under green house conditions. However, metagenomic DNA libraries have been already successfully constructed from the rhizosphere metagenome of plants adapted to acid mine drainage and their screening permitted to detect novel nickel resistance genes (Mirete et al. 2007)

Advances like the 454 pyrosequencing technology has opened new perspectives in sequencing the metagenome of environmental DNA (Demanèche et al. 2009). By pyrosequencing, adapters are added to each DNA fragment obtained from the environmental sample, and subsequently these fragments are added to a bead—one fragment per bead. This technology has dropped the time and costs constraints of DNA sequencing, but has the disadvantage that it currently only read 100 bases per run. This is particularly problematic when analysing environmental samples.

DNA microarrays and whole-community studies of diversity and function

The use of extraction-purification kits that permit several soil samples to be processed simultaneously has contributed to the success of the DNA-based approaches to study bacteria in their natural environment. However, the huge biodiversity and functional capabilities of bacteria in these samples cannot be monitored without the corresponding high-throughput technologies requested to analyze DNA solutions. DNA microarrays can fulfil these requirements by hybridizing in a single step soil amplified and labelled DNA targets to thousands of DNA probes targeting genes of interest immobilized on solid surfaces, the hybridization signal of each probe being subsequently and simultaneously recorded with a detector (Wagner et al. 2007; Huyghe et al. 2008).

Two main microarray systems have been developed depending on the target genes, the so-called Phylo-Chips (or phylogenetic oligonucleotide arrays) referring to the detection of the ubiquitous ribosomal genes to detect and identify theoretically any microorganism in the soil sample (Desantis et al. 2007); and the functional gene microarrays for detection and analysis of specific protein-encoding (functional) genes (Wagner et al. 2007; Ward et al. 2007; He et al. 2007). Targeting 16S or 23S rRNA genes, Phylo-Chips have begun to be used for the detection and identification of microbial strains, species, genera or higher taxa (depending on the design of the probe) (Militon et al. 2007) including the rhizosphere environments (Sanguin et al. 2008). For instance, the high-throughput analysis potential of Phylo-Chips detected a significant rhizosphere effect when a 170 probe microarray was used to compare the maize rhizosphere and the bulk soil (Sanguin et al. 2006). This study showed that taxonomic groups such as *Sphingomonas* spp., *Rhizobiaceae*, and *Actinobacteria* were identified in both rhizosphere and bulk soil with strong hybridization signals, indicating no specific habitat preference for these groups. In contrast, *Agrobacterium* spp. targeting probes yielded stronger hybridization signals with rhizosphere amplified DNA compared to the bulk soil DNA, suggesting that this group should be considered a rhizospheric group, while *Acidobacteria*, *Bacteroidetes*, *Verrucomicrobia*, and *Planctomycetes* related probes indicated that these bacteria should not be considered as rhizospheric.

However, questions arose rapidly about specificity and sensitivity of these technologies considering the difficulty to define hybridization conditions leading all probes to hybridize only to their fully matched target sequence. Only experimental verifications with reference DNA solutions can help to determine probes yielding specific signals that could be subsequently considered for the reliable interpretation of complex hybridization patterns with environmental DNA. Other controls include cloning and sequencing of PCR products from the complex DNA mixture to relate hybridization signals to DNA sequences actually retrieved in the soil DNA extract. Specificity problems for some probes do not necessarily mean a failure of the technology if each target organism is detected by more than one probe, since reliable signals by alternative matching probes may compensate for variable results.

Experimental verifications that were carried out in the maize rhizosphere study of Sanguin et al. (2006) showed that the clones that had a perfect match with corresponding probes yielded the expected positive signal in the hybridization experiment confirming the reliability of the hybridization for most of the hierarchically nested probes tested. In addition, comparison of experimental and theoretical hybridizations revealed 0.9 % false positives and 0.8 % false negatives, this specificity level being considered as sufficient for meaningful analysis of environmental samples (Wagner et al. 2007). However, much effort and care must be invested to maximize specificity when extension of the current probe set to a much wider taxonomic range is going to increase dramatically the potential of the Phylo-Chips for a systematic and more exhaustive exploration of bacterial diversity. Such controls are going to require the use of extensive collections of reference nucleic acids representing target and suitable non-target sequences for all probes on the array.

Microarray technologies were also developed to detect bacterial gene families that encode key enzymes involved in the ecosystem functioning

(Wagner et al. 2007; Ward et al. 2007; He et al. 2007; Wu et al. 2008b). Most of the potentials and limitations described for the Phylo-Chips can also apply for these functional applications. In addition, the design of functional probes suffers from the lack of sequences in the data bases compared to the ribosomal genes. Depending on the targets, specific difficulties can also be encountered to amplify variable regions when the flanking regions are not sufficiently conserved to design universal PCR primers. Finally, the functional microarrays do not benefit from the same hierarchical nesting strategy as for ribosomal genes to design probes as a result of the degeneracy of the genetic code (Loy and Bodrossy 2006). However, these functional arrays were successfully applied to analyze diversity of methanotrophs (Bodrossy et al. 2003; Stralis-Pavese et al. 2004) nitrogen fixers (Tiquia et al. 2004; Jenkins et al. 2004), sulfate reducers (Loy et al. 2004), as well as ammonia oxidizers (Adamczyk et al. 2003) in various environments and could be similarly applied to rhizosphere samples.

Transcriptome analysis

In addition to study the composition and genetic capabilities of the bacterial community in their natural environment, DNA microarrays can also be used to determe the transcript level of all the genes in a given gene pool (Schena et al. 1995; Hoheisel 2006). For this purpose the total transcript of a sample is obtained by total mRNA extraction. This mRNA is then converted into DNA (cDNA) by reverse transcription, labelled (usually with a fluorescent dye) and hybridized to microarrays containing the DNA genes whose expression is to be analyzed. The level of the signal is proportional to the level of transcript. The commercial arrays usually contain suitable control probes designed to hybridize with RNA spike-ins, which are mixed with the experiment sample during preparation. The degree of hybridization between the spike-ins and the control probes is used to normalize the hybridization measurements for the target probes.

There are two types of commercially available microarrays. One type contains the open reading frames (ORF) of the relevant genome, synthesized by PCR and spotted on a solid microarray (or macroarray) matrix. The other type is the oligonucleotide microarrays, in which the probes are short sequences designed to match parts of the sequence of known or predicted ORFs. These microarrays are produced by printing short oligonucleotide sequences designed to represent a single gene or family of gene splice-variants by synthesizing this sequence directly onto the array surface instead of depositing intact sequences. The size of the sequences varies, according to the producer (i.e., 60-mer probes such as the Agilent design or 25-mer probes such as produced by Affymetrix). Although the longer probes are more specific to individual target genes, shorter probes may be spotted in higher density across the array and the cost of their production is lower. Microarrays based on oligonucleotides enable the quantitation of transcripts from non-coding regions of the DNA. This is important for the analysis of small RNAs (Gottesman et al. 2006) and regulatory regions of the transcripts. For specific purposes, or when commercial microarrays are not available, it is possible to produce microarrays "in house". In these microarrays (spotted microarrays) the probes are oligonucleotides, cDNA or small fragments of PCR products that correspond to the expected transcripts. These arrays may be easily customized for each experiment, and are especially useful for expression profiling of environmental samples. However, it is clear that the in-house spotted microarrays do not provide the same level of sensitivity as the commercial arrays.

The DNA arrays are designed to give estimations of the absolute levels of gene expression. The expression level of each gene can be compared to other genes or to a reference gene. In order to determine changes in transcriptions resulting from physiological or environmental conditions it is necessary to compare transcriptomes—two separate single-dye hybridizations. Alternatively, it is possible to use differential labelling (Shalon 1998; Tang et al. 2007) in two-color microarrays, which may give a more appropriate quantitation of the differential expression, and are more cost-effective.

Two-Color microarrays are hybridized with cDNA prepared from two samples to be compared (e.g. bacteria at two temperatures) that are labeled with two different fluorophores (Shalon 1998), often one which has a fluorescence emission wavelength of 570 nm (corresponding to the green part of the light spectrum), and one with a fluorescence emission wavelength of 670 nm (corresponding to the red part of the light spectrum). The two differentially labelled cDNA samples are mixed and hybridized to a single microarray that is then scanned in a microarray scanner to visualize fluorescence of the two fluorophores (Fig. 4). The relative intensities of each fluorophore are analyzed to identify changes in gene expression (Tang et al. 2007). These microarrays provide data on the transcription level of the whole genome under two conditions, as well as a comparison of the two.

The transcription analysis by microarrays is sufficiently sensitive to detect transcription of most genes, including genes that are not translated into proteins. However, transcriptomics only reports on changes in the level of transcription—it misses information concerning post-transcriptional regulation (due to differences in transcript stability or transcript translation) as well as post-translational modifications. The two main drawbacks in the study of the transcriptome are that it depends on the availability of suitable microarrays, and these are available only for very few bacteria and that there are large variations in the results depending on the specific microarray used as well as on the methods for cDNA preparation and bioinformatics analysis (Bammler 2005; Clarke and Zhu 2006).

Proteome analysis

Gene expression can also be analysed at the level of translation—the final level of gene expression. Most of the experiments to analyze the composition of proteomes use high resolution two-dimensional polyacrylamide gel electrophoresis (2D gels) coupled to identification of proteins by mass spectrometry.

Two-dimensional gel electrophoresis makes it possible to resolve complex mixtures of cellular proteins. In this method proteins are separated by their isoelectric point and molecular mass. Proteins are extracted from microorganisms, tissues or other samples and are separated by their isoelectric point, on a gradient of pH (1st dimension). They are then separated by electrophoresis on SDS-polyacrylamide (SDS-PAGE) according to their molecular mass (O'Farrell 1975; Neidhardt et al. 1983; Neidhardt and van Bogelen 2000). The result of this procedure is a gel with proteins spread out on its surface. These proteins are then detected by staining (usually silver or Coomassie stains). The staining provides data on approximate protein amounts, which is adequate for most purposes.

Early on this method suffered from many technical problems, mainly due to the low reproducibility of the carrier ampholite-based electrophoresis. However, several recent developments made 2D-analyses the method of choice for most proteomic analyses. These advances include: 1) the introduction of IPG (immobilized pH gradient) gels (Gorg et al. 1988) for the separation of proteins in the first dimension, making the 2D gels a highly reproducible protein-separation method; 2) availability of mass spectromety technologies that—based on genome databases—enable the identification of the majority of the proteins; and 3) development of a large number of image analysis methods for the quantitative comparison of 2D-gel spot volumes. These types of software enable the

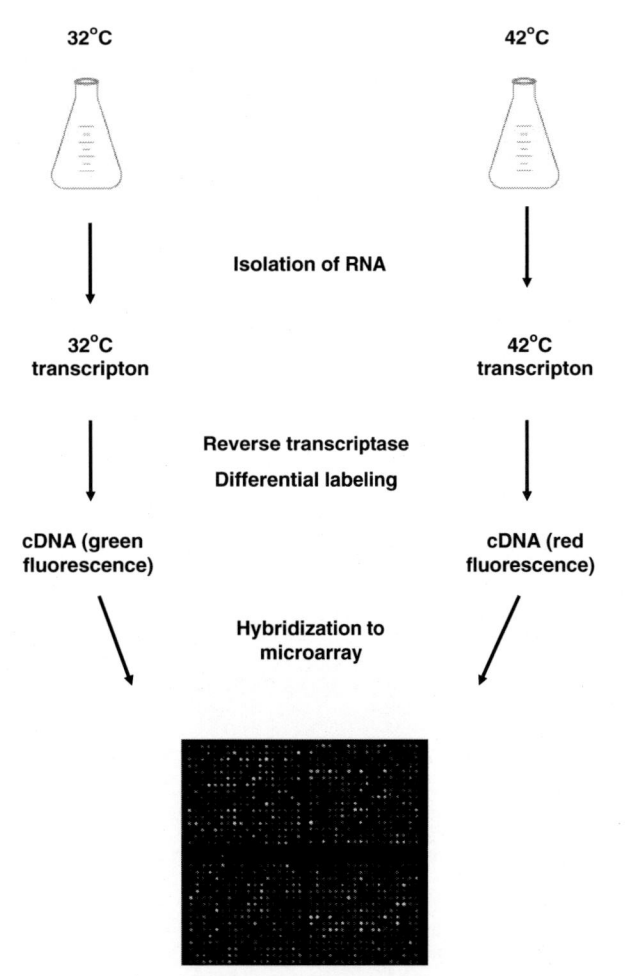

Fig. 4 Two-color microarray approach comparing transcriptomes in cultures grown at 32°C and 42°C, respectively

rapid comparative analysis of large sets of 2D gels. In addition, visualization methods, such as dual-channel imaging (Bernhardt et al. 1999), allow the rapid detection of regulatory networks which are induced by environmental changes, by combining autoradiography and a silver-stained electropherogram (see below).

Because of the above, two-dimensional gel electrophoresis has become a very powerful tool to resolve proteins, which are the final stage of gene expression (Volker and Hecker 2005). However, this method is incomplete because the routine analysis does not enable the analysis of several groups of proteins: membrane proteins, highly alkaline proteins, and rare proteins. Many of the membrane proteins are hydrophobic and are lost because the extraction and the electrophoresis of the first dimension are carried out in the absence of detergents. The second group of proteins with low compatibility for studies in 2D gels is the group of alkaline proteins. These are usually not separated because in the first dimension the pH gradient used is usually 4-7. It is possible to separate the alkaline proteins (with lower resolution) on wide pH-gradient gels (Ohlmeier et al. 2000). The third group of proteins that is not easily studied on 2D gel is the rare proteins. Because the first-dimension gels have a limited capacity, the rare proteins are present in the final gels in concentrations that may be below the detection limits, of even the very sensitive radio-labeling methods.

It should be noted that there exist alternative technologies, such as multi-dimensional-HPLC coupled mass spectrometry and protein-chip techniques, for proteomic analysis without the use of gel electrophoresis. However, these methods have not yet been extensively used in the study of gene expression in environmental samples (Wolff et al. 2007). Comparison of two samples, and analysis of proteome expression induced by environmental and physiological changes, is performed by overlaying two gels and quantifying the difference between them. The analysis can be performed by one of several image analysis programs.

More accurate results are obtained by running the two samples on one gel. This sort of experiment is often used to determine the effect of an environmental change on the composition of the proteome. The method has often been used to map regulatory networks that are induced by changes in temperature, pH, exposure to plant roots, or other environmental conditions (Rosen et al. 2002; Rosen and Ron 2002; Rosen et al. 2003; Tam et al. 2006; Wolff et al. 2007). For this type of experiment, the culture is metabolically labelled (with L-$[^{35}S]$methionine or $[^{14}C]$amino acid mixture) when exposed to the stress. The protein synthesis pattern is obtained by autoradiography can be directly compared with the protein level pattern. Because the total proteins and the newly synthesized proteins (i.e., stress-induced proteins) are on the same gel there is no need for matching several gels and proteins that belong to different stimulons or regulons can be identified (Gottesman et al. 2006). The radioactive proteome and the total proteome (obtained

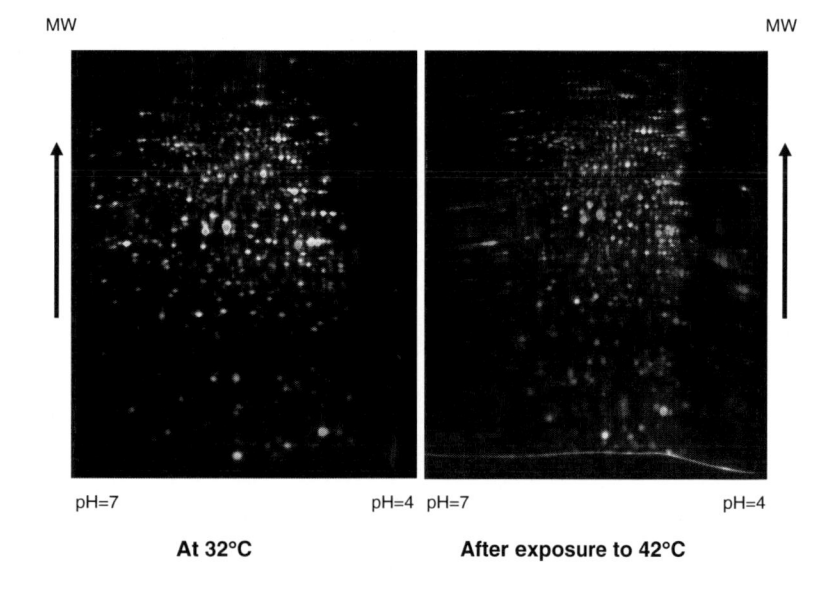

Fig. 5 Proteomic approach analyzing a temperature up-shift. A bacterial culture growing at 32°C was incubated 20 min in the presence of ^{35}Smethionine at 32°C (left side) or 42°C (right side). The proteomes were analyzed by two-dimensional electrophoresis. Each gel was autoradiographed and the image was computer-stained in red. The gel was then silver-stained and the image was computer-stained in green. The overlays of radioactive and silver-stained images are shown

by staining) can be differentially stained in the computer, and the relative gene expression is obtained by analysis of the colors. Figure 5 shows the vegetative (computer stained in green color) and heat shock (computer stained in red color) proteomes of *Agrobacterium tumefaciens*, as compared to the control in which the label was added to unstressed cultures. Using this method it is possible to define the protein signatures of microorganisms in response to a variety of environmental conditions and to identify environmentally induced regulons.

Proteomic analysis provides information about the final levels of cellular proteins, thus measuring the level of the end product of gene expression. In addition, it provides information about post-translational modifications. It has been previously assumed that post-translational modification plays only a minor role in prokaryotes. However, recent data indicate that post-translational modifications appear to be more common than expected. As an example, when a theoretical proteome was compared to the actual proteome of *A. tumefaciens*, there were many proteins that deviated from the expected (theoretical) molecular weight (MW) or pI as measured by their vertical and horizontal migration distances, respectively (Rosen et al 2004). These proteins were clearly subjected to post-translational modifications, which changed their pI and/or molecular weight. Additional support for post-translational modifications comes from the identification of multiple spots of the same gene products.

Proteomic studies can be carried out and provide solid data even on organisms whose genomes are not yet sequenced. When carried out on organisms with known genome sequences, these data can be translatable into identified proteins by the use of mass spectrometry. Because proteomics determines the level of the end product the results are not affected by parameters such as transcript stability. Moreover, proteomics enables the detection of post-translational modifications. The main drawbacks in using proteomics are the difficulties in detecting membrane-bound proteins, alkaline proteins and proteins that are unstable or present in low concentrations. Future challenges for proteomics thus include the development of efficient technologies for the study of membrane proteins and the improvement of automatic software-based analysis to overcome problems with incompletely separated (overlapping) spots and weak spots.

Another newly developed approach is "proteogenomics"—which combines metagenomics data with mass spectrometry based proteomics. It uses proteomics approaches to verify coding regions of metagenomic sequences and quantify their activity. This technology is based on the ability to directly measure peptides arising from proteolysis of expressed proteins by high-throughput liquid chromatography-tandem mass spectrometry-based proteomics. This new technology has not yet been extensively used in rhizosphere studies, but is expected to provide powerful tools in this area of research (Wilmes et al. 2008a; Wilmes et al. 2008b).

Concluding remarks

The new discipline of molecular microbial ecology was often seen as the "panacea", without considering that other limitations than those related to cultivation, could strongly bias results on the actual extent of bacterial diversity and functions. However, when researchers are aware of these biases and limitations, molecular microbial ecology still provides new insight not previously obtained by the cultivation-dependent approaches. In addition, better cultivation-dependent approaches are developing fast, and today research moves towards a polyphasic methods approach including both lines of research in order to gain most knowledge with the least bias. Furthermore, in reaching the goal of understanding the complex microbial communities in the rhizosphere, the systematic identification and quantitation of all metabolites in the rhizosphere, termed metabolomics, can be of major importance as an additional approach to include in the studies (Narasimhan et al. 2003).

Great advances in rhizosphere research has been made over the last 20 years or so, employing the major breakthroughs of advanced fluorescence microscopy for single-cell studies and molecular analysis of community structure and function. The progress is reflected well in several literature reviews including some of the present author (Sørensen 1997; Sørensen et al. 2001; Sørensen and Nybroe 2007; Sørensen and Sessitsch 2007). Major advances are to be expected also in the near future. The present availability of a large number of whole genome sequences constitutes a leap forward in the understanding of microbial communities in the rhizosphere, as well as well as the genetic composition of their individual constituents. The availability of sequences together with bioinformatics tools will enable us to translate the sequences into functions. Thus, by comparison with databank of gene sequences whose function has already been determined by genetic and physiological analyses, it is possible to make reasonable assumptions about the function of a large fraction of the genes in a newly-sequenced organism, or of genes obtained by culture-free studies. The next step is to understand the systems biology of these functions in an organism or in a community. Such functional genomic analyses can also map regulatory networks, identify the members of these networks and help determine the conditions for modulating the expression of these networks and of their individual constituents. Computational management is being more and more crucial with the increasing amount of data obtained by proteomics, metagenomic and metabolomics, and this field should be of highest priority for being able to use the information produced these years.

Acknowledgement Figures 1 and 2 and certain text fractions related to the specific treatise of reporter bacteria in this review were adopted from Sørensen and Nybroe (2007) with permission from the publisher (Springer Verlag, Berlin Heidelberg). This work was supported in part by the Center for Environmental and Agricultural Microbiology (CREAM) granted to Jan Sørensen.

References

Adamczyk J, Hesselsoe M, Iversen N, Horn M, Lehner A, Nielsen PH, Schloter M, Roslev P, Wagner M (2003) The isotope array, a new tool that employs substrate-mediated labeling of rRNA for determination of microbial community structure and function. Appl Environ Microbiol 69:6875–6887. doi:10.1128/AEM.69.11.6875-6887.2003

Allaway D, Schofield NA, Leonard ME, Gilardoni L, Finan TM, Poole PS (2001) Use of differential fluorescence induction and optical trapping to isolate environmentally induced genes. Environ Microbiol 3:397–406. doi:10.1046/j.1462-2920.2001.00205.x

Andersen JB, Sternberg C, Poulsen JK, Bjørn SP, Givskov M, Molin S (1998) New unstable variants of Green Fluorescent Protein for studies of transient gene expression in bacteria. Appl Environ Microbiol 64:2240–2246

Anguish LJ, Ghiorse WC (1997) Computer-assisted laser scanning and video microscopy for analysis of *Cryptosporidium parvum* oocysts in soil, sediment, and feces. Appl Environ Microbiol 63:724–733

Assmus B, Hutzler P, Kirchhof G, Amann RI, Lawrence JR, Hartmann A (1995) In situ localization of *Azospirillum brasilense* in the rhizosphere of wheat with fluorescently labelled, ribosomal-RNA targeted oligonucleotide probes and scanning confocal laser microscopy. Appl Environ Microbiol 61:1013–1019

Assmus B, Schloter M, Kirchhof G, Hutzler P, Hartmann A (1997) Improved *in situ* tracking of rhizosphere bacteria using dual staining with fluorescence-labelled antibodies and rRNA-targeted oligonucleotides. Microb Ecol 33:32–40. doi:10.1007/s002489900005

Babić KH, Schauss K, Hai B, Sikora S, Redzepović S, Radl V, Schloter M (2008) Influence of different *Sinorhizobium meliloti* inocula on abundance of genes involved in nitrogen transformations in the rhizosphere of alfalfa (*Medicago sativa* L.) Environ Microbiol 10: 2922–2930

Bae YS, Knudsen GR (2000) Cotransformation of *Trichoderma harzianum* with beta-glucuronidase and green fluorescent protein genes provides a useful tool for monitoring fungal growth and activity in natural soils. Appl Environ Microbiol 66:810–815. doi:10.1128/AEM.66.2.810-815.2000

Bælum J, Nicolaisen MH, Holben WE, Strobel BW, Sørensen J, Jacobsen CS (2008) Direct analysis of *tfdA* gene expression by indigenous bacteria in phenoxy acid amended agricultural soil. ISME J 2:677–687. doi:10.1038/ismej.2008.21

Bammler T (2005) Standardizing global gene expression analysis between laboratories and across platforms. Nat Methods 2:351–356. doi:10.1038/nmeth0605-477a

Barr M, East AK, Leonard M, Mauchline TH, Poole PS (2008) In vivo expression technology (IVET) selection of genes of *Rhizobium leguminosarum* biovar *viciae* A34 expressed in the rhizosphere. FEMS Microbiol Lett 282:219–227. doi:10.1111/j.1574-6968.2008.01131.x

Beja O, Suzuki MT, Koonin EV, Aravind L, Hadd A, Nguyen LP, Villacorta R, Amjadi M (2000) Construction and analysis of bacterial artificial chromosome libraries from a marine microbial assemblage. Environ Microbiol 2:516–529. doi:10.1046/j.1462-2920.2000.00133.x

Bernhardt J, Buttner K, Scharf C, Hecker M (1999) Dual channel imaging of two-dimensional electropherograms in *Bacillus subtilis*. Electrophoresis 20:2225–2240. doi:10.1002/(SICI) 1522-2683(19990801) 20:11<2225:: AID-ELPS2225>3.0.CO;2-8

Bodrossy L, Stralis-Pavese N, Murrell JC, Radajewski S, Weilharter A, Sessitsch A (2003) Development and validation of a diagnostic microbial microarray for methanotrophs. Environ Microbiol 5:566–582. doi:10.1046/j.1462-2920.2003.00450.x

Bolanos Vasquez MC, Warner D (1997) Effects of *Rhizobium tropici, R. etli*, and *R. leguminosarum* bv *phaseoli* on *nod* gene-inducing flavonoids in root exudates of *Phaseolus vulgaris*. Mol Plant Microbe Interact 10:339–346. doi:10.1094/MPMI.1997.10.3.339

Boldt T, Sørensen J, Karlson U, Molin S, Ramos C (2004) Combined use of different Gfp reporters for monitoring single-cell activities of a genetically modified PCB degrader in the rhizosphere of alfalfa. FEMS Microbiol Ecol 48:139–148. doi:10.1016/j.femsec.2004.01.002

Borneman J, Triplett EW (1997) Rapid and direct method for extraction of RNA from soil. Soil Biol Biochem 29:1621–1624. doi:10.1016/S0038-0717(97) 00084-9

Bottin A, Larche L, Villalba F, Gaulin E, Esquerre-Tugaye MT, Rickauer M (1999) Green fluorescent protein (GFP) as gene expression reporter and vital marker for studying development and microbe-plant interaction in the tobacco pathogen *Phytophthora parasitica* var. *nicotianae*. FEMS Microbiol Lett 176:51–56. doi:10.1111/j.1574-6968.1999. tb13641.x

Brazil GM, Kenefick L, Callanan M, Haro A, De Lorenzo V, Dowling DN, O'Gara F (1995) Construction of a rhizosphere pseudomonad with potential to degrade polychlorinated biphenyls and detection of *bph* gene expression in the rhizosphere. Appl Environ Microbiol 61:1946–1952

Brown MM, Friez MJ, Lovell CR (2003) Expression of *nifH* genes by diazotrophic bacteria in the rhizosphere of short form *Spartina alterniflora*. FEMS Microbiol Ecol 43:411–417. doi:10.1111/j.1574-6941.2003.tb01081.x

Burgmann H, Widmer F, Sigler WV, Zeyer J (2003) mRNA extraction and reverse transcription-PCR protocol for detection of *nifH* gene expression by *Azotobacter vinelandii* in soil. Appl Environ Microbiol 69:1928–1935. doi:10.1128/AEM.69.4.1928-1935.2003

Cabala J, Teper L (2007) Metalliferous constituents of rhizosphere soils contaminated by Zn-Pb mining in Southern Poland. Water Air Soil Pollut 178:351–362. doi:10.1007/s11270-006-9203-1

Case RJ, Boucher Y, Dahllöf I, Holmstrom C, Doolittle WF, Kjelleberg S (2007) Use of 16S rRNA and *rpoB* genes as molecular markers for microbial ecology studies. Appl Environ Microbiol 73:278–288. doi:10.1128/AEM.01177-06

Chen XP, Zhu YG, Xia Y, Shen JP, He JZ (2008) Ammonia-oxidizing archaea: important players in paddy rhizosphere soil? Environ Microbiol 10:1978–1987. doi:10.1111/j.1462-2920.2008.01613.x

Clarke JD, Zhu T (2006) Microarray analysis of the transcriptome as a stepping stone towards understanding biological systems: practical considerations and perspectives. Plant J 45:630–650. doi:10.1111/j.1365-313X.2006.02668.x

Coelho MRR, Da Mota FF, Carneiro NP, Marriel IE, Paiva E, Rosado AS, Seldin L (2007) Diversity of *Paenibacillus* spp. in the rhizosphere of four sorghum (*Sorghum bicolor*) cultivars sown with two contrasting levels of nitrogen fertilizer assessed by *rpoB*-based PCR-DGGE and sequencing analysis. J Microbiol Biotechnol 17:753–760

Costa R, Gomes NCM, Krogerrecklenfort E, Opelt K, Berg G, Smalla K (2007) *Pseudomonas* community structure and antagonistic potential in the rhizosphere: insights gained by combining phylogenetic and functional gene-based analyses. Environ Microbiol 9:2260–2273. doi:10.1111/j.1462-2920.2007.01340.x

Courtois S, Frostegard A, Goransson P, Depret G, Jeannin P, Simonet P (2001) Quantification of bacterial subgroups in soil: comparison of DNA extracted directly from soil or from cells previously released by density gradient centrifugation. Environ Microbiol 3:431–439. doi:10.1046/j.1462-2920.2001.00208.x

Courtois S, Cappellano CM, Ball M, Francou FX, Normand P, Helynck G, Martinez A, Kolvek SJ, Hopke J, Osburne MS, August PR, Nalin R, Guerineau M, Jeannin P, Simonet P, Pernodet JL (2003) Recombinant environmental libraries provide access to microbial diversity for drug discovery from natural products. Appl Environ Microbiol 69:49–55. doi:10.1128/AEM.69.1.49-55.2003

Dahllöf I, Baillie H, Kjelleberg S (2000) *rpoB*-based microbial community analysis avoids limitations inherent in 16S rRNA gene intraspecies heterogeneity. Appl Environ Microbiol 66:3376–3380. doi:10.1128/AEM.66.8.3376-3380.2000

Daime H, Lücker S, Wagner M (2006) *daime*, a novel image analysis program for microbial ecology and biofilm research. Environ Microbiol 8:200–213. doi:10.1111/j.1462-2920.2005.00880.x

Dandurand LM, Schotzko DJ, Knudsen GR (1997) Spatial patterns of rhizoplane populations of *Pseudomonas fluorescens*. Appl Environ Microbiol 63:3211–3217

Dazzo FB, Schmid M, Hartmann A (2007) Immunofluorescence microscopy and fluorescence in situ hybridization combined with CMEIAS and other image analysis tools for soil- and plant-associated microbial autecology. In: Hurst C et al (eds) Manual of environmental microbiology. ASM Press, Washington DC, pp 712–733

De Werra P, Baehler E, Huser A, Keel C, Maurhofer M (2008) Detection of palnt-modulated alterations in antifungal gene expression in *Pseudomonas fluorescens* CHA0 on roots by flow cytometry. Appl Environ Microbiol 74:1339–1349. doi:10.1128/AEM.02126-07

DeAngelis KM, Ji P, Firestone MK, Lindow SE (2005) Two novel bacterial biosensors for detection of nitrate availability in the rhizosphere. Appl Environ Microbiol 71:8537–8547. doi:10.1128/AEM.71.12.8537-8547.2005

DeAngelis KM, Lindow SE, Firestone MK (2008) Bacterial quorum sensing and nitrogen cycling in rhizosphere soil. FEMS Microbiol Ecol 66:197–207. doi:10.1111/j.1574-6941.2008.00550.x

DeLeo PC, Baveye P, Ghiorse WC (1997) Use of confocal laser scanning microscopy on soil thin-sections for improved characterization of microbial growth in unconsolidated soils and aquifer materials. J Microbiol Methods 30:193–203. doi:10.1016/S0167-7012(97) 00065-1

Demaneche S, David MM, Navarro E, Simonet P, Vogel TM (2009) Evaluation of functional gene enrichment in a soil metagenomic clone library. J Microbiol Methods 76:105–107. doi:10.1016/j.mimet.2008.09.009

Desantis TZ, Brodie EL, Moberg JP, Zubieta IX, Piceno YM, Andersen GL (2007) High-density universal 16S rRNA microarray analysis reveals broader diversity than typical clone library when sampling the environment. Microb Ecol 53:371–383. doi:10.1007/s00248-006-9134-9

Dorn JG, Frye RJ, Maier RM (2003) Effect of temperature, pH and initial cell number on *luxCDABE* and *nah* gene expression during naphthalene and salicylate catabolism in the reporter bacterium *Pseudomonas putida* RB1353. Appl Environ Microbiol 69:2209–2216. doi:10.1128/AEM.69.4.2209-2216.2003

Drissner D, Blum H, Tscherko D, Kandeler E (2007) Nine years of enriched CO_2 changes the function and structural diversity of soil microorganisms in a grassland. Eur J Soil Sci 58:260–269. doi:10.1111/j.1365-2389.2006.00838.x

Eickhorst T, Tippkötter R (2008) Detection of microorganisms in undisturbed soil by combining fluorescence in situ hybridization (FISH) and micropedological methods. Soil Biol Biochem 40:1284–1293. doi:10.1016/j.soilbio.2007.06.019

El Zahar Haichar F, Achouak W, Christen R, Heulin T, Marol C, Marais MF, Mougel C, Ranjard L, Balesdent J, Berge O

(2007) Identification of cellulolytic bacteria in soil by stable isotope probing. Environ Microbiol 9:625–634. doi:10.1111/j.1462-2920.2006.01182.x

Elasri M, Delorme S, Lemanceau P, Stewart G, Laue B, Glickmann E, Oger PM, Dessaux Y (2001) Acyl-homoserine lactone production is more common among plant-associated *Pseudomonas* spp. than among soilborne *Pseudomonas* spp. Appl Environ Microbiol 67:1198–1209. doi:10.1128/AEM.67.3.1198-1209.2001

Espinosa-Urgel M, Ramos JL (2001) Expression of a *Pseudomonas putida* involved in lysine metabolism is induced in the rhizosphere. Appl Environ Microbiol 67:5219–5224. doi:10.1128/AEM.67.11.5219-5224.2001

Faegri A, Torsvik VL, Goksoyr J (1977) Bacterial and fungal activities in soil: Separation of bacteria and fungi by a rapid fractionated centrifugation technique. Soil Biol Biochem 9:105–112. doi:10.1016/0038-0717(77) 90045-1

Fakhouri W, Kang Z, Buchenauer H (2001) Ultrastructural studies on the mode of action of fluorescent pseudomonads alone and in combination with acibenzolar-S-methyl effective against *Fusarium oxysporum* f. sp lycopersici in tomato plants. Z Pflanzenk 108:513–529

Fieseler L, Horn M, Wagner M, Hentschel U (2004) Discovery of the novel candidate phylum "Poribacteria" in marine sponges. Appl Environ Microbiol 70:3724–3732. doi:10.1128/AEM.70.6.3724-3732.2004

Füchslin HP, Rüegg I, van der Meer JR, Egli T (2003) Effect of integration of a GFP reporter gene on fitness of Ralstonia eutropha during growth with 2, 4-dichlorophenoxyacetic acid. Environ Microbiol 5:878–887. doi:10.1046/j.1462-2920.2003.00479.x

Fukui R, Poinar EI, Bauer PH, Schroth MN, Hendson M, Wang XL, Hancock JG (1994) Spatial colonization patterns and interaction of bacteria on inoculated sugar-beet seed. Phytopathol 84:1338–1345. doi:10.1094/Phyto-84-1338

Gage DJ, Bobo T, Long SR (1996) Use of green fluorescent protein to visualize the early events of symbiosis between *Rhizobium meliloti* and alfalfa (*Medicago sativa*). J Bacteriol 178:7159–7166

Ghiglione JL, Gourbiere F, Potier P, Phillippot L, Lensi R (2000) Role of respiratory nitrate reductase in ability of *Pseudomonas fluorescens* YT101 to colonize the rhizosphere of maize. Appl Environ Microbiol 66:4012–4016. doi:10.1128/AEM.66.9.4012-4016.2000

Ginolhac A, Jarrin C, Gillet B, Robe P, Pujic P, Tuphile K, Bertrand H, Vogel TM, Perriere G, Simonet P, Nalin R (2004) Phylogenetic analysis of polyketide synthase I domains from soil metagenomic libraries allows selection of promising clones. Appl Environ Microbiol 70:5522–5527. doi:10.1128/AEM.70.9.5522-5527.2004

Gorg A, Postel W, Domscheit A, Gunther S (1988) Two-dimensional electrophoresis with immobilized pH gradients of leaf proteins from barley (*Hordeum vulgare*): method, reproducibility and genetic aspects. Electrophoresis 9:681–692. doi:10.1002/elps.1150091103

Gottesman S, McCullen CA, Guillier M, Vanderpool CK, Majdalani N, Benhammou J, Thompson KM, FitzGerald PC, Sowa NA, FitzGerald DJ (2006) Small RNA regulators and the bacterial response to stress. Cold Spring Harb Symp Quant Biol 71:1–11. doi:10.1101/sqb.2006.71.016

Götz M, Gomes NCM, Dratwinski A, Costa R, Berg G, Peixoto R, Mendonca-Hagler L, Smalla K (2006) Survival of *gfp*-tagged antagonistic bacteria in the rhizosphere of tomato plants and their effects on the indigenous bacterial community. FEMS Microbiol Ecol 56:207–218. doi:10.1111/j.1574-6941.2006.00093.x

Griffiths RI, Whiteley AS, O'Donnell A, Bailey MJ (2000) Rapid method for coextraction of DNA and RNA from natural environments for analysis of ribosomal DNA- and rRNA-based microbial community composition. Appl Environ Microbiol 66:5488–5491. doi:10.1128/AEM.66.12.5488-5491.2000

Han JI, Semrau JD (2004) Quantification of gene expression in methanotrophs by competitive reverse transcription-polymerase chain reaction. Environ Microbiol 6:388–399. doi:10.1111/j.1462-2920.2004.00572.x

Hansen M, Kragelund L, Nybroe O, Sørensen J (1997) Early colonization of barley roots by *Pseudomonas fluorescens* studied by immunofluorescence technique and confocal laser scanning microscopy. FEMS Microbiol Ecol 23:353–360. doi:10.1111/j.1574-6941.1997.tb00416.x

Harris K, Crabb D, Young IM, Weaver H, Gilligan CA, Otten W, Ritz K (2002) In situ visualisation of fungi in soil thin sections: problems with crystallisation of the fluorochrome FB28 (Calcofluor M2R) and improved staining by SCRI Renaissance 2200. Mycol Res 106:293–297. doi:10.1017/S0953756202005749

Hay AG, Rice JF, Applegate BM, Bright NG, Sayler GS (2000) A bioluminescent whole-cell reporter for detection of 2, 4-dichloroophenoxyacetic acid and 2, 4-dichlorophenol in soil. Appl Environ Microbiol 66:4589–4594. doi:10.1128/AEM.66.10.4589-4594.2000

He Z, Gentry TJ, Schadt CW, Wu L, Liebich J, Chong SC, Huang Z, Wu W, Gu B, Jardine P, Criddle C, Zhou J (2007) GeoChip: a comprehensive microarray for investigating biogeochemical, ecological and environmental processes. ISME J 1:67–77. doi:10.1038/ismej.2007.2

Heid CA, Stevens J, Livak KJ, Williams PM (1996) Real time quantitative PCR. Genome Res 6:986–994. doi:10.1101/gr.6.10.986

Heitzer A, Malachowsky K, Thonnard JE, Bienkowski PR, White D, Sayler GS (1994) Optical biosensor for environmental on-line monitoring of naphthalene and salicylate bioavailability with an immobilized bioluminescent catabolic reporter bacterium. Appl Environ Microbiol 60:1487–1494

Hoheisel JD (2006) Microarray technology: beyond transcript profiling and genotype analysis. Nat Rev Genet 7:200–210. doi:10.1038/nrg1809

Højberg O, Schnider U, Winteler HV, Sørensen J, Haas D (1999) Oxygen-sensing reporter strain of *Pseudomonas fluorescens* for monitoring the distribution of low-oxygen habitats in soil. Appl Environ Microbiol 65:4085–4093

Holben WE, Jansson JK, Chelm BK, Tiedje JM (1988) DNA probe method for the detection of specific microorganisms in the soil bacterial community. Appl Environ Microbiol 54:703–711

Hugenholtz P 2002 Exploring prokaryotic diversity in the genomic era. Genome Biol. 3:reviews0003.1-reviews0003.8.

Hurt RA, Qiu XY, Wu LY, Roh Y, Palumbo AV, Tiedje JM, Zhou JH (2001) Simultaneous recovery of RNA and DNA from soils and sediments. Appl Environ Microbiol 67:4495–4503. doi:10.1128/AEM.67.10.4495-4503.2001

Huyghe A, Francois P, Charbonnier Y, Tangomo-Bento M, Bonetti EJ, Paster BJ, Bolivar I, Baratti-Mayer D, Pittet D, Schrenzel J (2008) Novel microarray design strategy to study complex bacterial communities. Appl Environ Microbiol 74:1876–1885. doi:10.1128/AEM.01722-07

Jacobsen CS, Holben WE (2007) Quantification of mRNA in *Salmonella* sp. seeded soil and chicken manure using magnetic capture hybridization RT-PCR. J Microbiol Methods 69:315–321. doi:10.1016/j.mimet.2007.02.001

Jaeger CH, Lindow SE, Miller W, Clark W, Firestone MK (1999) Mapping of sugar and amino acid availability in soil around roots with bacterial sensors of sucrose and tryptophan. Appl Environ Microbiol 65:2685–2690

Janssen PH, Yates PS, Grinton BE, Taylor PM, Sait M (2002) Improved culturability of soil bacteria and isolation in pure culture of novel members of the divisions Acidobacteria, Actinobacteria, Proteobacteria, and Verrucomicrobia. Appl Environ Microbiol 68:2391–2396. doi:10.1128/AEM.68.5.2391-2396.2002

Jarrin C (2005) Approches metagenomiques pour l'analyse des diversites genetiques et chimiques de metabolites secondaires, etude des PKS de type I d'origine bacterienne. Thesis, University Claude Bernard, Lyon 1, France.

Jenkins BD, Steward GF, Short SM, Ward BB, Zehr JP (2004) Fingerprinting diazotroph communities in the Chesapeake Bay by using a DNA macroarray. Appl Environ Microbiol 70:1767–1776. doi:10.1128/AEM.70.3.1767-1776.2004

Jensen LE, Nybroe O (1999) Nitrogen availability to *Pseudomonas fluorescens* DF57 is limited during decomposition of barley straw in bulk soil and in the barley rhizosphere. Appl Environ Microbiol 65:4320–4328

Kirchhof G, Schloter M, Assmus B, Hartmann A (1997) Molecular microbial ecology approaches applied to diazotrophs associated with non-legumes. Soil Biol Biochem 29:853–862. doi:10.1016/S0038-0717(96) 00233-7

Knauth S, Hurek T, Brar D, Reinhold-Hurek B (2005) Influence of different *Oryza* cultivars on expression of *nifH* gene pools in roots of rice. Environ Microbiol 7:1725–1733. doi:10.1111/j.1462-2920.2005.00841.x

Koch B, Worm J, Jensen LE, Højberg O, Nybroe O (2001) Carbon limitation induces σ^S-dependent gene expression in *Pseudomonas fluorescens* in soil. Appl Environ Microbiol 67:3363–3370. doi:10.1128/AEM.67.8.3363-3370.2001

Kragelund L, Hosbond C, Nybroe O (1997) Distribution of metabolic activity and phosphate starvation response of *lux*-tagged *Pseudomonas fluorescens* reporter bacteria in the barley rhizosphere. Appl Environ Microbiol 63:4920–4928

Kreuzer K, Adamczyk J, Iijima M, Wagner M, Scheu S, Bonkowski M (2006) grazing of a common species of soil protozoa (*Acanthamoeba castellani*) affects bacterial community composition and root architecture of rice (*Oryza sativa* L.). Soil Biol Biochem 38:1665–1672. doi:10.1016/j.soilbio.2005.11.027

Krsek M, Wellington EM (1999) Comparison of different methods for the isolation and purification of total community DNA from soil. J Microbiol Methods 39:1–16. doi:10.1016/S0167-7012(99) 00093-7

Kuiper I, Bloemberg GV, Noreen S, Thomas-Oates JE, Lugtenberg BJJ (2001) Increased uptake of putrescine in the rhizosphere inhibits competitive root colonization by *Pseudomonas fluorescens* strain WCS365. Mol Plant Microbe Interact 14:1096–1104. doi:10.1094/MPMI.2001.14.9.1096

Kutter S, Hartmann A, Schmid M (2005) Colonization of barley (*Hordeum vulgare*) with *Salmonella enterica* and *Listeria* spp. FEMS Microbiol Ecol 56:262–271. doi:10.1111/j.1574-6941.2005.00053.x

Lagopodi AL, Ram AFJ, Lamers GEM, Punt PJ, Van den Hondel CAMJ, Lugtenberg BJJ, Bloemberg GV (2002) Novel aspects of tomato root colonization and infection by *Fusarium oxysporum f. sp radicis-lycopersici* revealed by confocal laser scanning microscopic analysis using the green fluorescent protein as a marker. Mol Plant Microbe Interact 15:172–179. doi:10.1094/MPMI.2002.15.2.172

Lamprecht MR, Sabatini DM, Carpenter AE (2007) CellProfiler: free, versatile software for automated biological image analysis. Biotechniques 42:71–75. doi:10.2144/000112257

Layton AC, Muccini M, Ghosh MM, Sayler GS (1998) Construction of a bioluminescent reporter strain to detect polychlorinated biphenyls. Appl Environ Microbiol 64:5023–5026

Lee SW, Cooksey DA (2000) Genes expressed in *Pseudomonas putida* during colonization of a plant-pathogenic fungus. Appl Environ Microbiol 66:2764–2772. doi:10.1128/AEM.66.7.2764-2772.2000

Leveau JHJ, Lindow SE (2002) Bioreporters in microbial ecology. Curr Opin Microbiol 5:259–265. doi:10.1016/S1369-5274(02) 00321-1

Liu WT, Marsh TL, Cheng H, Forney LJ (1997) Characterization of microbial diversity by determining terminal restriction fragment length polymorphisms of genes encoding 16S rRNA. Appl Environ Microbiol 63:4516–4522

Liu XM, Germaine KJ, Ryan D, Dowling DN (2007) Development of a Gfp-based biosensor for detecting the bioavailability and biodegradation of polychlorinated biphenyls (PCBs). J. Environ. Eng. Landsc Manage 15:261–268

Loy A, Bodrossy L (2006) Highly parallel microbial diagnostics using oligonucleotide microarrays. Clin Chim Acta 363:106–119. doi:10.1016/j.cccn.2005.05.041

Loy A, Kusel K, Lehner A, Drake HL, Wagner M (2004) Microarray and functional gene analyses of sulfate-reducing prokaryotes in low-sulfate, acidic fens reveal cooccurrence of recognized genera and novel lineages. Appl Environ Microbiol 70:6998–7009. doi:10.1128/AEM.70.12.6998-7009.2004

Lüeders T, Manefield M, Friedrich MW (2004) Enhanced sensitivity of DNA- and rRNA-based stable isotope probing by fractionation and quantitative analysis of isopycnic centrifugation gradients. Environ Microbiol 6:73–78. doi:10.1046/j.1462-2920.2003.00536.x

Mahmood S, Paton GI, Prosser JI (2005) Cultivation-independent in situ molecular analysis of bacteria involved in degradation of pentachlorophenol in soil. Environ Microbiol 7:1349–1360. doi:10.1111/j.1462-2920.2005.00822.x

Manefield M, Whiteley AS, Griffiths RI, Bailey MJ (2002) RNA stable isotope probing, a novel means of linking microbial community function to phylogeny. Appl Environ Microbiol 68:5367–5373

Marco ML, Legac J, Lindow SE (2003) Conditional survival as a selection strategy to identify plant-inducible genes of *Pseudomonas syringae*. Appl Environ Microbiol 69:5793–5801. doi:10.1128/AEM.69.10.5793-5801.2003

Marschner P, Crowley DE (1996) Physiological activity of a bioluminescent *Pseudomonas fluorescens* (strain 2-79) in the rhizosphere of mycorrhizal and non-mycorrhizal pepper (*Capsicum annuum* L.). Soil Biol Biochem 28:869–876. doi:10.1016/0038-0717(96)00072-7

Mauchline TH, Kerry BR, Hirsch PR (2002) Quantification in soil and the rhizosphere of the nematophagous fungus *Verticillium chlamydospotium* by competitive PCR and comparison with selective plating. Appl Environ Microbiol 68:1846–1853. doi:10.1128/AEM.68.4.1846-1853.2002

Mavrodi OV, Mavrodi DV, Thomashow LS, Weller DM (2007) Quantification of 2, 4-diacetylphloroglucinol-producing Pseudomonas fluorescens strains in the plant rhizosphere by real-time PCR. Appl Environ Microbiol 73:5531–5538. doi:10.1128/AEM.00925-07

Meckenstock R, Steinle P, van der Meer JR, Snozzi M (1998) Quantification of bacterial mRNA involved in degradation of 1,2,4-trichlorobenzene by *Pseudomonas* sp. strain P51 from liquid culture and from river sediment by reverse transcriptase PCR (RT/PCR). FEMS Microbiol Lett 167:123–129. doi:10.1111/j.1574-6968.1998.tb13217.x

Meikle A, Amin-Hanjani S, Glover LA, Killham K, Prosser JI (1995) Matric potential and the survival and activity of a *Pseudomonas fluorescens* inoculum in soil. Soil Biol Biochem 27:881–892. doi:10.1016/0038-0717(95) 00020-F

Miethling R, Ahrends K, Tebbe CC (2003) Structural differences in the rhizosphere communities of legumes are not equally reflected in community-level physiological profiles. Soil Biol Biochem 35:1405–1410. doi:10.1016/S0038-0717(03) 00221-9

Militon C, Rimour S, Missaoui M, Biderre C, Barra V, Hill D, Mone A, Gagne G, Meier H, Peyretaillade E, Peyret P (2007) PhylArray: phylogenetic probe design algorithm for microarray. Bioinformatics 23:2550–2557. doi:10.1093/bioinformatics/btm392

Mirete S, de Figueras CG, Gonzalez-Pastor JE (2007) Novel nickel resistance genes from the rhizosphere metagenome of plants adapted to acid mine drainage. Appl Environ Microbiol 73:6001–6011. doi:10.1128/AEM.00048-07

Mollet C, Drancourt M, Raoult D (1997) *rpoB* sequence analysis as a novel basis for bacterial identification. Mol Microbiol 26:1005–1011. doi:10.1046/j.1365-2958.1997.6382009.x

Muyzer G, De Waal EC, Uitterlinden AG (1993) Profiling of complex microbial populations by denaturing gradient gel electroforesis analysis of polymerase chain reaction-amplified genes Coding for 16S rRNA. Appl Environ Microbiol 59:695–700

Narasimhan K, Basheer C, Bajic VB, Swarup S (2003) Enhancement of plant-microbe interactions using a rhizosphere metabolomics-driven approach and its application in the removal of polychlorinated biphenyls. Plant Physiol 132:146–153. doi:10.1104/pp. 102.016295

Neidhardt FC, van Bogelen RA (2000) Proteomic analysis of bacterial stress responses. In: Storz G, Hennecke H (eds) bacterial stress responses. ASM Press, Washington DC, pp 445–452

Neidhardt FC, van Bogelen RA, Lau ET (1983) Molecular cloning and expression of a gene that controls the high-temperature regulon of *Escherichia coli*. J Bacteriol 153:597–603

Neilson JW, Pierce SA, Maier RM (1999) Factors influencing expression of *luxCDABE* and *nah* genes in *Pseudomonas putida* RB1353 (NAH7, pUTK9) in dynamic systems. Appl Environ Microbiol 65:3473–3482

Nicolaisen MH, Bælum J, Jacobsen CS, Sørensen J (2008) Transcription dynamics of the functional *tfdA* gene during MCPA herbicide degradation by *Cupriavidus necator* AEO106 (pRO101) in agricultural soil. Environ Microbiol 10:571–579. doi:10.1111/j.1462-2920.2007.01476.x

Nocker A, Burr M, Camper AK (2007) Genotypic microbial community profiling: A critical technical review. Microb Ecol 54:276–289. doi:10.1007/s00248-006-9199-5

Normander B, Hendriksen NB, Nybroe O (1999) Green fluorescent protein-marked *Pseudomonas fluorescens*: Localization, viability, and activity in the natural barley rhizosphere. Appl Environ Microbiol 65:4646–4651

O'Farrell PH (1975) High resolution two-dimensional electrophoresis of proteins. J Biol Chem 250:4007–4021

Ogram A, Sayler GS, Barkay T (1987) The extraction and purification of microbial DNA from sediments. J Microbiol Methods 7:57–66. doi:10.1016/0167-7012(87) 90025-X

Ohlmeier S, Scharf C, Hecker M (2000) Alkaline proteins of *Bacillus subtilis*: first steps towards a two-dimensional alkaline master gel. Electrophoresis 21:3701–3709. doi:10.1002/1522-2683(200011) 21:17<3701::AID-ELPS3701>3.0.CO;2-5

Ostle N, Whiteley AS, Bailey MJ, Sleep D, Ineson P, Manefield M (2003) Active microbial RNA turnover in a grassland soil estimated using a (CO_2)-C-13 spike. Soil Biol Biochem 35:877–885. doi:10.1016/S0038-0717(03) 00117-2

Park SH, Lee DH, Oh KH, Lee K, Kim CK (2002) Detection of aromatic pollutants by bacterial biosensors bearing gene fusions constructed with the *dnaK* promoter of *Pseudomonas* sp DJ-12. J Microbiol Biotechnol 12:417–422

Paterson E, Sim A, Standing D, Dorward M, McDonald AJS (2006) Root exudation from *Hordeum vulgare* in response to localized nitrate supply. J Exp Bot 57:2413–2420. doi:10.1093/jxb/erj214

Pliego C, de Weert S, Lamers G, de Vicente A, Bloemberg G, Cazoria FM, Ramos C (2008) Two similar enhanced root-colonizing *Pseudomonas* strains fifer largely in the colonization strategies of avocado roots and *Rosellinia necatrix* hyphae. Environ Microbiol 10:3295–3304. doi:10.1111/j.1462-2920.2008.01721.x

Poonguzhali S, Madhaiyan M, Yim W-J, Kim K-A, Sa T-M (2008) Colonization pattern of plant root surfaces visualized by use of green-fluorescent-marked strain of *Methylobacterium suomiense* and its persistence in rhizosphere. Appl Microbiol Biotechnol 78:1033–1043. doi:10.1007/s00253-008-1398-1

Porteous F, Killham K, Meharg A (2000) Use of a *lux*-marked rhizobacterium as a biosensor to assess changes in rhizosphere C flow due to pollutant stress. Chemosphere 41:1549–1554. doi:10.1016/S0045-6535(00) 00072-2

Postma J, Altemüller HJ (1990) Bacteria in thin soil sections stained with the fluorescent brightener Calcofluor White

M2R. Soil Biol Biochem 22:89–96. doi:10.1016/0038-0717(90) 90065-8

Pratscher J, Stichternot C, Fichtl K, Schleifer K-H, Braker G (2009) Application of recognition of individual genes-fluorescence in situ hybridization (RING-FISH) to detect nitrite reductase genes (*nirK*) of denitrifiers in pure cultures and environmental samples. Appl Environ Microbiol 75:802–810. doi:10.1128/AEM.01992-08

Prieto P, Mercado-Blanco J (2008) Endophytic colonization of olive roots by the biocontrol strain *Pseudomonas fluorescens* PICF7. FEMS Microbiol Ecol 64:297–306. doi:10.1111/j.1574-6941.2008.00450.x

Puglisi E, Fragoulis G, Del Re AAM, Spaccini R, Piccolo A, Gigliotti G, Said-Pullicino D, Trevisan M (2008) Carbon deposition in soil rhizosphere following amendments with compost and its soluble fractions, as evaluated by combined soil-plant rhizobox and reporter gene systems. Chemosphere 73:1292–1299. doi:10.1016/j.chemosphere.2008.07.008

Radajewski S, Ineson P, Parekh NR, Murrell JC (2000) Stable-isotope probing as a tool in microbial ecology. Nature 403:646–649. doi:10.1038/35001054

Rainey PB (1999) Adaptation of *Pseudomonas fluorescens* to the plant rhizosphere. Environ Microbiol 1:243–257. doi:10.1046/j.1462-2920.1999.00040.x

Ramos C, Mølbak L, Molin S (2000) Bacterial activity in the rhizosphere analysed at the single-cell level by monitoring ribosome contents and synthesis rates. Appl Environ Microbiol 66:801–809. doi:10.1128/AEM.66.2.801-809.2000

Robe P, Nalin R, Capellano C, Vogel TA, Simonet P (2003) Extraction of DNA from soil. Eur J Soil Biol 39:183–190. doi:10.1016/S1164-5563(03) 00033-5

Rondon MR, August PR, Bettermann AD, Brady SF, Grossman TH, Liles MR, Loiacono KA, Lynch BA, MacNeil IA, Minor C, Tiong CL, Gilman M, Osburne MS, Clardy J, Handelsman J, Goodman RM (2000) Cloning the soil metagenome: a strategy for accessing the genetic and functional diversity of uncultured microorganisms. Appl Environ Microbiol 66:2541–2547. doi:10.1128/AEM.66.6.2541-2547.2000

Rosado AS, Seldin L, Wolters AC, van Elsas JD (1996) Quantitative 16S rDNA-targeted polymerase chain reaction and oligonucleotide hybridization for the detection of *Paenibacillus azotofixans* in soil and the wheat rhizosphere. FEMS Microbiol Ecol 19:153–164. doi:10.1111/j.1574-6941.1996.tb00208.x

Rosen R, Ron EZ (2002) Proteome analysis in the study of the bacterial heat-shock response. Mass Spectrom Rev 21:244–265. doi:10.1002/mas.10031

Rosen R, Buttner K, Becher D, Nakahigashi K, Yura T, Hecker M, Ron EZ (2002) Heat shock proteome of *Agrobacterium tumefaciens*: evidence for new control systems. J Bacteriol 184:1772–1778. doi:10.1128/JB.184.6.1772-1778.2002

Rosen R, Matthysse AG, Becher D, Biran D, Yura T, Hecker M, Ron EZ (2003) Proteome analysis of plant-induced proteins of *Agrobacterium tumefaciens*. FEMS Microbiol Ecol 44:355–360. doi:10.1016/S0168-6496(03) 00077-1

Rosen R, Sacher A, Shechter N, Becher D, Buttner K, Biran D, Hecker M, Ron EZ (2004) Two-dimensional reference map of *Agrobacterium tumefaciens* proteins. Proteomics 4:1061–1073. doi:10.1002/pmic.200300640

Rovira AD (1956) A study on the developemnt of the root surface microflora during the initial stages of plant growth. J Appl Microbiol 19:72–79. doi:10.1111/j.1365-2672.1956.tb00048.x

Rusch DB, Halpern AL, Sutton G, Heidelberg KB, Williamson S, Yooseph S, Wu D, Eisen JA, Hoffman JM, Remington K, Beeson K, Tran B, Smith H, Baden-Tillson H, Stewart C, Thorpe J, Freeman J, Andrews-Pfannkoch C, Venter JE, Li K, Kravitz S, Heidelberg JF, Utterback T, Rogers YH, Falcon LI, Souza V, Bonilla-Rosso G, Eguiarte LE, Karl DM, Sathyendranath S, Platt T, Bermingham E, Gallardo V, Tamayo-Castillo G, Ferrari MR, Strausberg RL, Nealson K, Friedman R, Frazier M, Venter JC (2007) The *Sorcerer II* Global Ocean Sampling Expedition: Northwest Atlantic through Eastern Tropical Pacific. PLoS Biol 5:398–431. doi:10.1371/journal.pbio.0050077

Sanguin H, Remenant B, Dechesne A, Thioulouse J, Vogel TM, Nesme X, Moenne-Loccoz Y, Grundmann GL (2006) Potential of a 16S rRNA-based taxonomic microarray for analyzing the rhizosphere effects of maize on *Agrobacterium* spp. and bacterial communities. Appl Environ Microbiol 72:4302–4312. doi:10.1128/AEM.02686-05

Sanguin H, Kroneisen L, Gazengel K, Kyselkova M, Remenant B, Prigent-Combaret C, Grundmann GL, Sarniguet A, Moenne-Loccoz Y (2008) Development of a 16S rRNA microarray approach for the monitoring of rhizosphere *Pseudomonas* populations associated with the decline of take-all disease of wheat. Soil Biol Biochem 40:1028–1039. doi:10.1016/j.soilbio.2007.11.023

Santaella C, Schue M, Berge O, Heulin T, Achouak W (2008) The exopolyscaccharide of *Rhizobium* sp. YAS34 is not necessary for biofilm formation on *Arabidopsis thaliana* and *Brassica napus* roots but contributes to root colonization. Environ Microbiol 10:2150–2163. doi:10.1111/j.1462-2920.2008.01650.x

Schena M, Shalon D, Davis RW, Brown PO (1995) Quantitative monitoring of gene expression patterns with a complementary DNA microarray. Science 270:467–470. doi:10.1126/science.270.5235.467

Schloter M, Borlinghaus R, Bode W, Hartmann A (1993) Direct identification and localization of *Azospirillum* in the rhizosphere of wheat using fluorescence-labelled monoclonal antibodies and confocal laser scanning microscopy. J Microscopy-Oxford 171:173–177

Schwartz E (2007) Characterization of growing microorganisms in soil by stable isotope probing with $H_2^{18}O$. Appl Environ Microbiol 73:2541–2546. doi:10.1128/AEM.02021-06

Schwieger F, Tebbe CC (1998) A new approach to utilize PCR-single-strand-conformation polymorphism for 16s rRNA gene-based microbial community analysis. Appl Environ Microbiol 64:4870–4876

Shalon D (1998) Gene expression micro-arrays: a new tool for genomic research. Pathol Biol 46:107–109

Sharma S, Aneja MK, Mayer J, Munch JC, Schloter M (2005) Diversity of transcripts of nitrite reductase genes (*nirK* and *nirS*) in rhizospheres of grain legumes. Appl Environ Microbiol 71:2001–2007. doi:10.1128/AEM.71.4.2001-2007.2005

Singleton DR, Hunt M, Powell SN, Frontera-Suau R, Aitken MD (2007) Stable-isotope probing with multiple growth substrates to determine substrate specificity of uncultivated bacteria. J Microbiol Methods 69:180–187. doi:10.1016/j.mimet.2006.12.019

Sliwinski MK, Goodman RM (2004) Comparison of crenarchaeal consortia inhabiting the rhizosphere of diverse terrestrial plants with those in bulk soil in native environments. Appl Environ Microbiol 70:1821–1826. doi:10.1128/AEM.70.3.1821-1826.2004

Smalla K, Oros-Sichler M, Milling A, Heuer H, Baumgarte S, Becker R, Neuber G, Kropf S, Ulrich A, Tebbe CC (2007) Bacterial diversity of soils assessed by DGGE, T-RFLP and SSCP fingerprints of PCR-amplified 16S rRNA gene fragments: Do the different methods provide similar results? J Microbiol Methods 69:470–479. doi:10.1016/j.mimet.2007.02.014

Smith LM, Tola E, de Boer P, O'Gara F (1999) Signalling by the fungus Pythium ultimum represses expression of two ribosomal RNA operons with key roles in the rhizosphere ecology of Pseudomonas fluorescens F113. Environ Microbiol 1:495–502. doi:10.1046/j.1462-2920.1999.00067.x

Sørensen J (1997) The rhizosphere as a habitat for soil microorganisms. In: van Elsas JD, Trevors JT, Wellington EMH (eds) Modern Soil Microbiology. Marcel Dekker Inc., New York, pp 21–45

Sørensen J, Nybroe O (2007) Reporter genes in bacterial inoculants can monitor life conditions and functions in soil. In: Nannipieri P, Smalla K (eds) Nucleic Acids and Proteins in Soil. Springer Verlag, Berlin Heidelberg, pp 375–395

Sørensen J, Sessitsch A (2007) Plant-associated bacteria—Lifestyles and molecular interactions. In: van Elsas JD, Jansson JK, Trevors JT (eds) Modern Soil Microbiology. CRC Press, Boca Raton London New York, pp 211–236

Sørensen J, Jensen LE, Nybroe O (2001) Soil and rhizosphere as habitats for Pseudomonas inoculants: new knowledge on distribution, activity and physiological state derived from micro-scale and single-cell studies. Plant Soil 232:97–108. doi:10.1023/A:1010338103982

Standing D, Meharg AA, Killham K (2003) A tripartite microbial reporter gene system for real-time assays of soil nutrient status. FEMS Microbiol Lett 220:35–39. doi:10.1016/S0378-1097(03) 00057-0

Steidle A, Sigl K, Schuhegger R, Ihring A, Schmid M, Gantner S, Stoffels M, Riedel K, Givskov M, Hartmann A, Langebartels C, Eberl L (2001) Visualization of N-acylhomoserine lactone-mediated cell-cell communication between bacteria colonizing the tomato rhizosphere. Appl Environ Microbiol 67:5761–5770. doi:10.1128/AEM.67.12.5761-5770.2001

Stralis-Pavese N, Sessitsch A, Weilharter A, Reichenauer T, Riesing J, Csontos J, Murrell JC, Bodrossy L (2004) Optimization of diagnostic microarray for application in analysing landfill methanotroph communities under different plant covers. Environ Microbiol 6:347–363. doi:10.1111/j.1462-2920.2004.00582.x

Tam LT, Antelmann H, Eymann C, Albrecht D, Bernhardt J, Hecker M (2006) Proteome signatures for stress and starvation in Bacillus subtilis as revealed by a 2-D gel image color coding approach. Proteomics 6:4565–4585

Tang T, François N, Glatigny A, Agier N, Mucchielli MH, Aggerbeck L, Delacroix H (2007) Expression ratio evaluation in two-colour microarray experiments is significantly improved by correcting image misalignment. Bioinformatics 23:2686–2691. doi:10.1093/bioinformatics/btm399

Thrane C, Olsson S, Nielsen TH, Sørensen J (1999) Vital fluorescent stains for detection of stress in Pythium ultimum and Rhizoctonia solani challenged with viscosinamide from Pseudomonas fluorescens DR54. FEMS Microbiol Ecol 30:11–23. doi:10.1111/j.1574-6941.1999.tb00631.x

Thrane C, Nielsen TH, Nielsen MN, Sørensen J, Olsson S (2000) Viscosinamide-producing Pseudomonas fluorescens DR54 exerts a biocontrol effect on Pythium ultimum in sugar beet rhizosphere. FEMS Microbiol Ecol 33:139–146. doi:10.1111/j.1574-6941.2000.tb00736.x

Timms-Wilson TM, Ellis RJ, Bailey MJ (2000) Immunocapture differential display method (IDDM) for the detection of environmentally induced promoters in rhizobacteria. J Microbiol Methods 41:77–84. doi:10.1016/S0167-7012(00) 00139-1

Timmusk S, Grantcharov N, Wagner EGH (2005) Paenibacillus polymyxa invades plant roots and forms biofilms. Appl Environ Microbiol 71:7292–7300. doi:10.1128/AEM.71.11.7292-7300.2005

Tiquia SM, Wu L, Chong SC, Passovets S, Xu D, Xu Y, Zhou J (2004) Evaluation of 50-mer oligonucleotide arrays for detecting microbial populations in environmental samples. Biotechniques 36:664–675

Tombolini R, van der Gaag DJ, Gerhardson B, Jansson JK (1999) Colonization pattern of the biocontrol strain Pseudomonas chlororaphis MA 342 on barley seeds visualized by using green fluorescent protein. Appl Environ Microbiol 65:3674–3680

Torsvik VL, Goksoyr J (1978) Determination of bacterial DNA in soil. Soil Biol Biochem 10:7–12. doi:10.1016/0038-0717(78) 90003-2

Torsvik V, Goksøyr J, Daae FL (1990) High Diversity in DNA of soil bacteria. Appl Environ Microbiol 56:782–787

Ulrich A, Becker R (2006) Soil parent material is a key determinant of the bacterial community structure in arable soils. FEMS Microbiol Ecol 56:430–443. doi:10.1111/j.1574-6941.2006.00085.x

Unge A, Tombolini R, Mølbak L, Jansson JK (1999) Simultaneous monitoring of cell number and metabolic activity of specific bacterial populations with a dual gfp-luxAB marker system. Appl Environ Microbiol 65:813–821

van Dyk TK, Smulski DR, Reed TR, Belkin S, Vollmer AC, LaRossa R (1995) Responses to toxicants of an Escherichia coli strain carrying a uspA':lux genetic fusion and an E. coli strain carrying a grpE':lux fusion are similar. Appl Environ Microbiol 61:4124–4127

van Overbeek LS, van Elsas JA, van Veen JD (1997) Pseudomonas fluorescens Tn5-B20 mutant RA92 responds to carbon limitation in soil. FEMS Microbiol Ecol 24:57–71. doi:10.1016/S0168-6496(97) 00045-7

van West P, Reid B, Campbell TA, Sandrock RW, Fry WE, Kamoun S, Gow NAR (1999) Green fluorescent protein (GFP) as a reporter gene for the plant pathogenic oomycete Phytophthora palmivora. FEMS Microbiol Lett 178:71–80. doi:10.1016/S0378-1097(99) 00320-1

Venter JC, Remington K, Heidelberg JF, Halpern AL, Rusch D, Eisen JA, Wu D, Paulsen I, Nelson KE, Nelson W, Fouts DE, Levy S, Knap AH, Lomas MW, Nealson K,

White O, Peterson J, Hoffman J, Parsons R, Baden-Tillson H, Pfannkoch C, Rogers YH, Smith HO (2004) Environmental Genome Shotgun Sequencing of the Sargasso Sea. Science 304:66–74. doi:10.1126/science.1093857

Voget S, Leggewie C, Uesbeck A, Raasch C, Jaeger KE, Streit WR (2003) Prospecting for novel biocatalysts in a soil metagenome. Appl Environ Microbiol 69:6235–6242. doi:10.1128/AEM.69.10.6235-6242.2003

Volker U, Hecker M (2005) From genomics via proteomics to cellular physiology of the Gram-positive model organism *Bacillus subtilis*. Cell Microbiol 7:1077–1085. doi:10.1111/j.1462-5822.2005.00555.x

von der Weid I, Artursson V, Seldin L, Jansson J (2005) Antifungal and root colonization patterns of GFP-tagged *Paenibacillus brasiliensis* PB177. World J Microbiol Biotechnol 12:1591–1597. doi:10.1007/s11274-005-8123-3

Wagner M, Horn M, Daims H (2003) Fluorescence in situ hybridisation for the identification and characterisation of prokaryotes. Curr Opin Microbiol 6:302–309. doi:10.1016/S1369-5274(03) 00054-7

Wagner M, Smidt H, Loy A, Zhou J (2007) Unravelling microbial communities with DNA-microarrays: challenges and future directions. Microb Ecol 53:498–506. doi:10.1007/s00248-006-9197-7

Ward BB, Eveillard D, Kirshtein JD, Nelson JD, Voytek MA, Jackson GA (2007) Ammonia-oxidizing bacterial community composition in estuarine and oceanic environments assessed using a functional gene microarray. Environ Microbiol 9:2522–2538. doi:10.1111/j.1462-2920.2007.01371.x

Watanabe T, Kimura M, Asakawa S (2007) Dynamics of methanogenic archaeal communities based on rRNA analysis and their relation to methanogenic activity in Japanese paddy field soils. Soil Biol Biochem 39:2877–2887. doi:10.1016/j.soilbio.2007.05.030

Watt M, Hugenholtz P, White R, Vinall K (2006) Numbers and locations of native bacteria on field-grown wheat roots quantified by fluorescence in situ hybridization (FISH). Environ Microbiol 8:871–884. doi:10.1111/j.1462-2920.2005.00973.x

Weinbauer MG, Beckmann C, Höfle MG (1998) Utility of green fluorescent nucleic acid dyes and aluminum oxide membrane filters for rapid epifluorescence enumeration of soil and sediment bacteria. Appl Environ Microbiol 64:5000–5003

Wilhelm J, Pingoud A (2003) Real-time polymerase chain reaction. ChemBioChem 4:1120–1128. doi:10.1002/cbic.200300662

Wilmes P, Andersson AF, Lefsrud MG, Wexler M, Shah M, Zhang B, Hettich RL, Bond PL, VerBerkmoes NC, Banfield JF (2008a) Community proteogenomics highlights microbial strain-variant protein expression within activated sludge performing enhanced biological phosphorus removal. ISME J 2:853–864. doi:10.1038/ismej.2008.38

Wilmes P, Remis JP, Hwang M, Auer M, Thelen MP, Banfield JF (2008b) Natural acidophilic biofilm communities reflect distinct organismal and functional organization. ISME J (in press).

Wintzingerode F, Gobel UB, Stackebrandt E (1997) Determination of microbial diversity in environmental samples: pitfalls of PCR-based rRNA analysis. FEMS Microbiol Rev 21:213–229. doi:10.1111/j.1574-6976.1997.tb00351.x

Woese CR, Kandler O, Wheelis ML (1990) Towards a natural system of organisms—Proposal for the domains Archaea, Bacteria, and Eucarya. Proc Natl Acad Sci USA 87:4576–4579. doi:10.1073/pnas.87.12.4576

Wolff S, Antelmann H, Albrecht D, Becher D, Bernhardt J, Bron S, Buttner K, van Dijl JM, Eymann C, Otto A, le Tam T, Hecker M (2007) Towards the entire proteome of the model bacterium *Bacillus subtilis* by gel-based and gel-free approaches. J. Chromatogr. B 849:129–140

Wu CH, Hwang Y-C, Lee W, Mulchandani TK, Yates MV, Chen W (2008a) Detection of recombinant *Pseudomonas putida* in the wheat rhizosphere by fluorescence in situ hybridization targeting mRNA and rRNA. Appl Microbiol Biotechnol 79:511–518. doi:10.1007/s00253-008-1438-x

Wu L, Kellogg L, Devol AH, Tiedje JM, Zhou J (2008b) Microarray-based characterization of microbial community functional structure and heterogeneity in marine sediments from the Gulf of Mexico. Appl Environ Microbiol 74:4516–4529. doi:10.1128/AEM.02751-07

Yeomans C, Porteous F, Paterson E, Meharg AA, Killham K (1999) Assessment of *lux*-marked *Pseudomonas fluorescens* for reporting on organic carbon compounds. FEMS Microbiol Lett 176:79–83. doi:10.1111/j.1574-6968.1999.tb13645.x

Yolcubal I, Piatt JJ, Pierce SA, Brusseau ML, Maier RM (2000) Fiber optic detection of in situ *lux* reporter gene activity in porous media: system design and performance. Anal Chim Acta 422:121–130. doi:10.1016/S0003-2670(00)01072-2

Zhou T, Paulitz TC (1993) In-Vitro and In-Vivo effects of *Pseudomonas* spp. on *Pythium-Aphanidermatum*—Zoospore behavior in exudates and on the rhizoplane of bacteria-treated cucumber roots. Phytopathol 83:872–876. doi:10.1094/Phyto-83-872

REVIEW ARTICLE

Iron dynamics in the rhizosphere as a case study for analyzing interactions between soils, plants and microbes

Philippe Lemanceau · Petra Bauer · Stephan Kraemer · Jean-François Briat

Received: 12 February 2009 / Accepted: 18 May 2009 / Published online: 11 June 2009
© Springer Science + Business Media B.V. 2009

Responsible Editor: Philippe Hinsinger.

P. Lemanceau (✉)
INRA, Université de Bourgogne, UMR1229
'Microbiologie du Sol et de l'Environnement', CMSE,
17 rue Sully, BV 86510,
21034 Dijon cedex, France
e-mail: philippe.lemanceau@dijon.inra.fr

P. Bauer
Institute for Plant Genetics and Crop Plant Research (IPK),
Corrensstrasse 3, 06466 Gatersleben, Germany

S. Kraemer
University of Vienna,
Department of Environmental Geosciences,
Althanstraße 14, 1090 Vienna, Austria

J.-F. Briat
CNRS, Université Montpellier II, SupAgro, INRA,
UMR5004 'Biochimie et Physiologie Moléculaire des Plantes',
Place Pierre Viala, 34060 Montpellier cedex I, France

Abstract Iron is an essential element for plants and microbes. However, in most cultivated soils, the concentration of iron available for these living organisms is very low because its solubility is controlled by stable hydroxides, oxyhydroxides and oxides. In the rhizosphere, there is a high demand of iron because of the iron uptake by plants, and microorganisms which density and activity are promoted by the release of root exudates. Plants and microbes have evolved active strategies of iron uptake. Iron incorporation by these organisms lead to complex interactions ranging from competition to mutualism. These complex interactions are under the control of physico-chemical properties of the soils in which they occur, and reciprocally iron uptake strategies of plants and microbes impact these soil properties. These iron-mediated interactions between soils, plants and microbes impact the plant growth and health and their analysis, together with that of the resulting iron dynamics, is of a major agronomic interest. Analysis of the complex interactions soils, plants and microbes represent also a unique opportunity to progress in our knowledge of the rhizosphere ecology. This progression requires merging complementary expertises and study strategies in soil science, plant biology and microbiology. This review provides information on (i) iron status in soil and rhizosphere, iron uptake by plants and microbes, and on (ii) the corresponding study strategies. Finally, illustrations of how integration of these approaches allows gaining knowledge in the complex interactions occurring in the rhizosphere are given.

Keywords Iron · Siderophores · Phytosiderophores · Reductases · Plant nutrition · Plant health · Competition

Introduction

Because of its electronic structure, iron is capable of reversible changes in oxidation state over a wide range of redox potential. In cultivated soils which are mainly oxic environments, iron is mostly in the Fe(III) redox state. The iron speciation in soil solution includes inorganic hydrolysis species and a range of inorganic and organic complexes. Iron is the fourth most abundant element of the earth crust. However in cultivated soils at pH values compatible with plant growth, the solubility of iron is controlled at extremely low levels by stable hydroxides, oxyhydroxides and oxides (Lindsay 1979; Marschner 1995).

Iron is essential for plants and microorganisms due to its involvement in major metabolic processes such as reduction of ribonucleotides and molecular nitrogen, and the energy-yielding electron transfer reactions of respiration and photosynthesis (Guerinot and Yi 1994). Iron is a central element of the photosynthetic electron transfer chain, and therefore plays an essential role in plant growth and ultimately in crop yield. Beside the yield, the plant iron uptake and homeostasis impact also the plant iron content, and therefore the quality of edible parts of plants. Increase of this content in order to enrich the amount of bioavailable iron in the diet is a major challenge of world agriculture because diet of humans the most affected by iron deficiency is mainly composed of plant products, poor in iron (Briat and Gaymard 2007). The absolute biological requirement for iron contrasts with the concentrations of Fe(III) species generally far below those required for optimal growth of microbes (10^{-5} to 10^{-7}) and plants (10^{-4} to 10^{-9}) (Loper and Buyer 1991; Guerinot and Yi 1994).

In this context, it is a major challenge for microorganisms and plants to acquire Fe(III) which is essential for their metabolism and growth. To meet this challenge, plants and microbes have evolved active strategies of uptake which are based on a range of chemical processes. Basically, these strategies rely on (i) acidification of soil solution mediated based on the excretion of protons or organic acids, (ii) chelation of Fe(III) by ligands including siderophores with very high affinity for Fe^{3+}, and (iii) reduction of Fe^{3+} to Fe^{2+} by reductases and reducing compounds. The efficacy of these active iron uptake strategies differ among organisms, leading to complex competitive and synergistic interactions among microbes, plants, and between plants and microbes. The chemical properties of the soil in which they occur have a strong effect on these interactions. In return the iron-uptake strategies impact the soil properties and the iron status. Thus, multiple interactions between soils, plants and microorganisms are driving a complex iron cycle in the rhizosphere.

Interactions between soils-plants-microbes in relation with iron dynamics also impact plant health as illustrated by the natural suppressiveness of some soils to a major class of soilborne diseases consisting in the plant wilting caused by pathogenic *Fusarium oxysporum* (Steinberg et al. 2007). These suppressive soils are characterized by (i) their low concentration of Fe(III) in solution due to their high pH and $CaCO_3$ content and (ii) their high microbial biomass leading to a specially strong competition for iron (Lemanceau et al. 1988a; Scher and Baker 1982). In these soils, the iron uptake of the pathogenic fungi is not efficient enough to meet their iron requirement and consequently their saprophytic growth is reduced and the frequency of root infections is decreased.

The analysis of these complex interactions and of the resulting iron dynamics is of major agronomic interest because they impact significantly the plant nutrition, growth and health. Science strives to provide solutions to this intricate problem, furthering our understanding of interactions between prokaryote and eukaryote organisms, in relation with their abiotic habitat (Robin et al. 2008). The aim of this review is to illustrate these interactions and show how complementary expertises in soil sciences, plant biology and microbiology contribute to unravel their complexity in the rhizosphere.

Iron status in soil and rhizosphere

Summary of our knowledge

As indicated above, iron is an ubiquitous element in soils and sediments, usually found in the range of crustal abundance around 3.5 % (Taylor 1964; Blume et al. 2002). It can be depleted in reducing environments due to its mobility in the divalent redox state and it can be enriched where iron rich reducing waters traverse redox gradients towards more oxidizing conditions (Cornell and Schwertmann 2003). At any rate, the availability of iron and the iron status of plants is rarely a function of the total iron content of soils. An important factor controlling iron availability is the solubility (Lindsay 1979) and dissolution rates (Kraemer et al. 2006) of pedogenic Fe(III)oxides,

-oxyhydroxides and -hydroxides (in this manuscript collectively called 'iron oxides' unless a specific mineral is addressed). Incidentally, both the solubility and dissolution rates of soil iron oxides are strongly pH dependent with minima around the neutral to slightly alkaline pH range which is characteristic for oxic calcareous soils (Lindsay 1979; Cornell and Schwertmann 2003). These important properties of pedogenic iron bearing minerals are largely responsible for the strong sensitivity of plant iron acquisition to soil pH and the presence of carbonates as pH buffering phases. However, the reactivity of pedogenic iron bearing minerals is also influenced by redox processes and by the formation of iron complexes. All three factors, pH, local redox potential and the concentration of complexing agents, can be influenced or even regulated in the rhizosphere by plant root and microbial activity in order to increase the iron availability.

Acidification of the rhizosphere seems to be the obvious response to iron limitation in calcareous soils (Table 1). Carbonate dissolution and concomitant pH buffering of the soil solution is fast compared to iron oxide dissolution, and therefore dissolution kinetics seems to work against this strategy. In other words, the acidification of the rhizosphere has to overcome, the pH buffering capacity of carbonates within the rhizosphere. However, due to the usually high total concentration of pedogenic iron oxides relative to the demand, it can be sufficient to mobilize iron from small soil volumes, for example in cluster roots or in biofilms.

Organic ligands with a sufficient affinity for iron to bind it against the high stability of iron oxides may also contribute to increasing iron availabilities (Kraemer 2004). These are ubiquitous in the rhizosphere and include natural organic matter including humic and fulvic acids as well as a range of lysates and exudates from plant cells and microorganisms. Regulated exudation of organic ligands by plants and microorganisms is a known response to iron limitation (Marschner 1995). These exudates include metabolites such as the organic acid citrate or iron specific organic ligands with a high affinity for iron binding, the so called (phyto-)siderophores. Phytosiderophores are exuded by graminaceous plants as a response to iron limitation (Takagi 1976). They promote iron oxide dissolution by a ligand controlled dissolution mechanism (Reichard et al. 2007a), similarly to microbial siderophores (Reichard et al. 2007b). The dissolution process is subject to various influences causing synergistic or inhibitory effects including soil pH (Kraemer et al. 1999), the presence of organic acids (Reichard et al. 2005) or biosurfactants (Carrasco et al. 2008).

Finally, the enzymatic reduction of iron increases the kinetic lability of iron potentially increasing its availability (Marschner and Römheld 1994). Considering the fast re-oxidation kinetics of iron in the

Table 1 Biological strategies for the mobilization of iron under iron limiting conditions

Biological activity	Response	
	Soil solution	Iron oxides
Rhizosphere acidification	- Shift of Fe speciation	- Increase of solubility
		- Acceleration of dissolution rates by proton promoted dissolution
Release of organic acids	- Complexation	- Acceleration of dissolution rates by ligand promoted dissolution
	- Ligand exchange	- Small increase of solubility
Release of siderophores	- Strong complexation	- Increase of solubility
	- Ligand exchange	- Acceleration of dissolution rates by ligand promoted dissolution
Release of reductants	- Release of Fe from ligands reduction followed by rapid reoxidation	- Acceleration of dissolution rates by reductive dissolution
Reductases at cell surfaces	- Release of Fe from ligands followed by uptake	- Acceleration of dissolution rates by reductive dissolution if direct contact between reductases and minerals is provided or if electron shuttles are available

circum-neutral pH range and the high redox potential of oxic soils, such a strategy is also bound to function locally or even on the molecular scale accelerating reaction mechanisms.

Clearly, the decisive factor for the level of plant iron nutrition is the attainment of a critical concentration of bioavailable iron in the soil solution adjacent to the root surface (Simeoni et al. 1987). As discussed above, slow kinetics and/or thermodynamic properties of solubility controlling phases may limit iron uptake. To discuss these processes it is important to keep in mind the fundamental difference between conceptional approaches based on kinetic or thermodynamic limitations to iron bioavailability. A thermodynamic description of the soil system assumes that the soil solution and all soluble iron species are in equilibrium with each other and with a large (relative to the iron requirement) pool of pedogenic iron oxides. The fact that soils usually include various pedogenic iron oxide phases that are clearly not in equilibrium with each other (Cornell and Schwertmann 2003) immediately suggests the limits of such a model. However, a reasonable assumption can be proposed that the most kinetically labile and least thermodynamically stable pedogenic iron oxide is the phase that controls the solubility of iron. If iron is distributed homogenously and transport control is not important, an equilibrium model predicts that the concentration of Fe^{3+} and of iron hydrolysis species is constant and only depends on pH. Biological uptake would have no effect and the presence of iron complexes of biogenic organic ligands would only serve as additional soluble iron pools. In such a system, a free ion activity model would not predict competition among iron acquiring organisms. On the contrary, processes such as acidification or iron reduction could potentially increase iron availability for other organisms leading to synergistic effects. The free ion activity model does not predict uptake of and competition for iron complexes. In the presence of a small and finite pool of iron specific organic ligands such as siderophores, competition among organisms possessing uptake systems for the corresponding iron complexes could occur due to depletion of the ligands.

Competition among soil organisms for iron could also imply that the soluble iron concentrations are under kinetic control due to slow iron oxide dissolution or ligand exchange rates. In such a system, biological uptake leads to decreasing concentrations of the total soluble iron pool. Moreover, exudation of siderophores or other ligands would lead to a shift of the iron speciation including a decreasing pool of iron hydrolysis species. Depending on the capability of an organism to take up iron complexes, this could lead to decreasing or increasing iron availability and organisms with uptake systems for a dominant organic iron complex would have a competitive advantage. Under kinetic control of iron concentrations, acidification, reduction and exudation of organic ligands would have an additional function in iron acquisition. They would serve to increase dissolution rates by proton promoted-, reductive-, and ligand controlled dissolution mechanisms.

It is difficult to estimate the importance of iron oxide dissolution kinetics on the iron dynamics of the rhizosphere. It hinges on the relative rates of dissolution, transport and uptake and only if the dissolution is the slowest step of all these rates it may be rate controlling. Unfortunately, factors such as pH, redox environment and the presence of organic ligands that influence dissolution rates also influence iron oxide solubility (Cornell and Schwertmann 2003) making it difficult to discriminate effects of shifting equilibria from effects of shifting reaction rates.

Strategies for studying iron status in soils and rhizospheres

Considering the different iron mobilization strategies used by plants and microorganisms and the range of intensities with which they are employed, the difficulty to define measures of iron availability as a soil property becomes apparent. Nevertheless, methods of determining soil iron availability have been developed (Table 2). The most common methods use extractants to mobilize iron from soil material during a specified time and under defined conditions (Chao and Zhou 1983). Extractants include chelating agents such as DTPA (Soltanpour and Schwab 1977; Lindsay and Norvell 1978) ammonium oxalate (Schwertmann 1964), citrate/ascorbate (Reyes and Torrent 1997) or reducing agents such as dydroxylamine (De Santiago et al. 2008). While these extractions targeted only an available iron pool, other methods attempt to detect and quantify several distinct soil iron pools using sequential extractions (Borggaard 1988; Heron et al. 1994). These sequential extraction procedures were developed not only for the investigation of solid state

Table 2 Summarized strategies and methodologies for studying the iron status in soils and rhizospheres

	Strategies & methodologies	References
Bioavailable Iron		
	DTPA extraction	Lindsay and Norvell (1978); Soltanpour and Schwab (1977)
	Ammonium oxalate extraction	Schwertmann (1964)
	Citrate/ascorbate	Reyes and Torrent (1997)
	Hydroxylamine	De Santiago et al. (2008)
Solid phase iron speciation		
Sequential extractions	Sequential application of extractants that mobilize iron from operationally defined iron pools, including, e.g.: organically bound, amorphous, crystalline, residual	Borggaard (1988); Heron et al. (1994); Wiederhold et al. (2007a, b)
X-ray diffraction	Crystalline iron bearing phases	La Force and Fendorf (2000)
X-ray spectroscopy	Dominant crystalline and amorphous iron bearing phases	La Force and Fendorf (2000)
Soil color	Crystalline and amorphous iron bearing phases	Scheinost and Schwertmann (1999)
Mössbauer spectroscopy	Crystalline and amorphous iron bearing phases	Parfitt and Childs (1988)
Magnetic measurements	Magnetite, maghemite other iron oxides	Hanesch et al. (2006)

iron pools but also for the speciation of other nutrients or pollutants (Tessier et al. 1979; Gleyzes et al. 2002). The iron pools extracted in the sequential extraction steps are usually equated with poorly-crystalline iron oxides, crystalline iron oxides, silicate-bound iron etc. Differences in iron isotope compositions of such pools have been observed and interpreted with respect to the iron cycling in soils (Wiederhold et al. 2007a, b) and plant uptake (Kiczka et al. 2007). A fundamental problem of single or sequential extraction procedures is the operational definition of the iron pools. In contrast, spectroscopic methods promise to detect soil iron species directly. In order to validate sequential extractions, La Force and Fendorf (2000) observed the effect of sequential extraction steps on concentrations of iron bearing phases in soil using x-ray diffraction, scanning electron microscopy, and x-ray absorption near edge structure (XANES) spectroscopy. This study showed that the extraction steps preferentially leached distinct iron phases, but also that the selectivity and efficiency for removing the target phase is limited. The authors caution that a sequential extraction scheme should be evaluated for any given soil using independent methods. Such methods include x-ray diffraction and X-ray spectroscopy as used by La Force and Fendorf but also other methods. Scheinost and Schwertmann (1999) used diffuse visible reflectance spectroscopy to identify goethite, akaganeite, hematite, maghemites, jarosites, and lepidocrocites. However, ferrihydrite as one of the most labile and presumably bioavailable iron bearing phase in soils was not reliably identified. Similarly, magnetic measurements can be used to identify iron bearing minerals (Hanesch et al. 2006), but the measurements are usually dominated by ferrimagnetic minerals such a maghemite or magnetite and important phases controlling iron bioavailability may not be detected. Mössbauer spectroscopy has been successfully used to identify a range of iron bearing phases in soils and has been compared to iron pools as determined to sequential extractions (Parfitt and Childs 1988). An advantage of Mössbauer spectroscopy is that it can identify x-ray amorphous phases such as ferrihydrite at low concentrations (Cornell and Schwertmann 2003).

The most common methods report iron availability defined in terms of content (mol kg^{-1}) (Harmsen 2007). However, root uptake of iron from soil solution constantly perturbs the chemical equilibrium between the soil solution and the iron pool that serves as iron source for biological uptake. If the slow release of iron from these sources limits iron uptake, then iron availability is kinetically controlled. Under these circumstances, the capacity of the iron pool (i.e. quantity of 'available' iron) as measured by the methods discussed above may be less important than the investigation of iron release kinetics and the flux of iron to the root. To our knowledge, few studies exist where kinetics of metal release from soils has been compared to root uptake. Labanowski et al. (2008) studied the release rates of metal ions (Cd, Cu, Pb, Zn) from soils in the presence of EDTA and citrate as extractants. They defined metal pools purely based on their kinetic lability and found that kinetically labile metal pools (with citrate as extractant) are related to short term plant uptake whereas labile metal pools (with EDTA as extractant) are related to long term mobilization of the metals. A different approach is the introduction of an infinite sink for metal ions that does not directly interact with the pools that resupply metal ions to the soil solution. For example, Lehto et al. (2006) have evaluated the use of chelating resins that are covered by a thin diffusive gel layer, a technique called Diffusive Gradients in Thin-films (DGT), to mimic the kinetics of micro-nutrient fluxes to roots. This modelling study addressed the relative importance of pool size and release kinetics in order to assess conditions where kinetic methods would be appropriate to study bioavailability. However, the results were not calibrated against measured plant uptake. Clearly, an infinite sink does not capture the effects of the root induced changes of rhizosphere chemistry. However, in some cases it may be an alternative to the use of an extractant that increases release rates and metal solubilities unrealistically.

The use of organisms as reporters of iron availability integrates the thermodynamic and kinetic factors influencing iron availability. Moreover, the use of microorganisms as biosensors introduces a high level of spatial resolution. This approach will be further developed in the section 'Iron uptake by microorganisms'.

Iron uptake by plants

Summary of our knowledge

Iron enters the plant via the root from where it is distributed inside the plant. Generally, iron is present in concentrations of about 10 - 500 µg Fe/g dry weight in plant tissues. Iron homeostasis is a tightly controlled process whereby control of the iron transporters is crucial. Due to its easier handling, most of our knowledge has been gained on the soil iron uptake processes taking place after germination

(detailed below). Fe mobilisation processes during germination are now being discovered (covered briefly below). Knowledge about transporters involved in iron transport in other parts inside the plant are summarized in recent reviews (Colangelo and Guerinot 2006; Kim and Guerinot 2007).

As a strategy for restricting excessive uptake of Fe, wetland species have evolved mechanisms for oxidizing ferrous Fe (Fe^{2+}) in the rhizosphere. Plants, living under aerobic soil conditions, have developed two phylogenetically distinct strategies to cope with the extremely low availability of soluble Fe compounds summarized in Marschner and Römheld (1994). Dicots and nongraminaceous monocots employ a Fe acquisition mechanism termed Strategy I based on the reduction of Fe in the rhizosphere. Under Fe-deficient conditions, such plants exhibit enhanced proton extrusion in the rhizosphere, increased Fe^{3+} reduction capacity at the root surface, followed by an uptake of Fe^{2+} via a ferrous transporter on the root plasma membrane (Chaney et al. 1972). As a result, plants elevate the Fe availability in the rhizosphere and enhance its uptake. In response to Fe deficiency, graminaceous monocots release high-affinity Fe-chelating substances from the mugineic acid family, called phytosiderophores. These substances solubilize Fe^{3+} and the resulting Fe^{3+}-phytosiderophore complexes are taken up by the root cells via a specific plasma membrane transport system without reduction of the ferric ion. This mechanism is termed Strategy II (Römheld and Marschner 1986) and it might resemble the microbial siderophore strategy.

Among the achievements of the past years is the identification of strategy I and strategy II genes that now serve as a skeleton for studying Fe uptake responses (Fig. 1).

Strategy I

FRO genes: The characterization of *Arabidopsis frd1* (= *fro2*) mutants which do not show induction of Fe^{3+}-chelate reductase under Fe-deficient conditions, confirms that Fe must be reduced prior to its transport and that Fe^{3+} reduction can be uncoupled from proton release (Yi and Guerinot 1996). FRO proteins are membrane-bound ferric chelate reductases that belong to the flavocytochrome b family, their topology was recently investigated (Robinson et al. 1999;

Strategy I

[Figure: Schematic diagram showing Strategy I with H⁺/AHA, Fe(III)/FRO, Fe(II)/IRT and FIT/FER bHLH-TF regulating Fe acquisition genes under −Fe conditions]

Strategy II

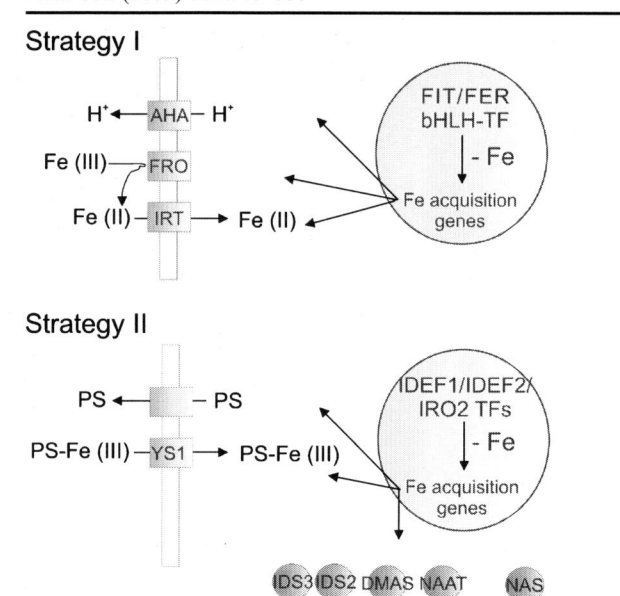

[Figure: Schematic diagram showing Strategy II with PS efflux, PS-Fe(III)/YS1 uptake, IDEF1/IDEF2/IRO2 TFs regulating Fe acquisition genes, and biosynthetic pathway SAM → NA → (NAS, NAAT, DMAS, IDS2, IDS3) → PS]

Fig. 1 Schematic representation of Fe acquisition mechanisms followed by dicot and nongraminaceous monocot (strategy I) and gramineceous monocot (strategy II) plants

Schagerlöf et al. 2006). The *Arabidopsis FRO2* is upregulated in root epidermis cells, the sites of Fe uptake, under Fe-deficiency conditions (Robinson et al. 1999). Iron deficiency-inducible and root forms of FRO2-like genes were also identified from other plant species, for example pea (*Pisum sativum* L.) and tomato (*Lycopersicon esculentum* Mill.) (Waters et al. 2002; Li et al. 2004).

Arabidopsis IRT1 encodes the founding member of a class of eukaryotic membrane-bound metal ion transporters (Eide et al. 1996), referred to as the ZIP (ZRT, IRT-LIKE TRANSPORTERS) family (Guerinot 2000). Four histidine-glycine repeats constitute potential metal-binding sites (Eng et al. 1998). In plants, IRT1 also mediates transport of Mn, Zn, Cd and Co (Vert et al. 2002). IRT1 protein was found present in the root epidermal membrane and induced under Fe-deficiency (Eide et al. 1996; Vert et al. 2002). IRT1 homologs have also been characterized in pea, tomato and other plants (Cohen et al. 1998; Eckhardt et al. 2001).

The *AHA* gene member mediating the iron mobilization response has not yet been discovered. Several *AHA* genes were described that are induced by iron deficiency and therefore could potentially mediate this function (Colangelo and Guerinot 2004; Santi et al. 2005).

Interestingly, *FRO2* and *IRT1* are mostly co-regulated in *Arabidopsis* (Vert et al. 2003) and dependent on the action of the basic helix-loop-helix transcription factor FIT (in *Arabidopsis*) (Colangelo and Guerinot 2004; Jakoby et al. 2004; Yuan et al. 2005). The *FRO2/IRT1/FIT* system is conserved. In tomato a similar system was discovered first (*FRO1/IRT1/FER*; Eckhardt et al. 2001; Ling et al. 2002; Bereczky et al. 2003; Bauer et al. 2004; Li et al. 2004). All three components *FRO2/IRT1/FIT* are essential for plant growth, and loss of a single of the three functions leads to iron deficiency and leaf chlorosis in *Arabidopsis* (for example Robinson et al. 1999; Henriques et al. 2002; Varotto et al. 2002; Vert et al. 2002; Colangelo and Guerinot 2004; Jakoby et al. 2004). *FIT/FER* genes are specifically expressed in roots. Additionally, posttranscriptional regulation controls Fe acquisition proteins FRO2, IRT1 FER/FIT (Connolly et al. 2002; Connolly et al. 2003; Jakoby et al. 2004; Brumbarova and Bauer 2005).

Interestingly, iron uptake into the root seems to start after the seedling has established (ca. 3 days after imbibition and growth under optimal conditions in *Arabidopsis*). During germination, plants rely on the mobilization of iron stored in their seed storage organs. *Arabidopsis* has emerged as an interesting model in this respect because VIT1 was isolated as a transporter for delivery of iron into storage vacuoles of vascular cells in the embryo whereas NRAMP3/NRAMP4 divalent metal transporters remobilize this vacuolar iron upon germination (Lanquar et al. 2005; Kim et al. 2006).

Strategy II

It was found that phytosiderophores are structurally related to nicotianamine. Later, it became evident that nicotianamine is an intermediate in the biosynthesis of the mugineic acid family of phytosiderophores. The precursor was found to be methionine (Mori and Nishizawa 1987). Methionine is converted to S-adenosylmethionine (SAM) by SAM synthetase (Shojima et al. 1989). Nicotianamine synthase produces nicotianamine from SAM and NA is the direct precursor for phytosiderophores (Herbik et al. 1999; Higuchi et al. 1999; Higuchi et al. 2001; Inoue et al. 2003; Mizuno et al. 2003). The subsequent critical enzymes in this specific pathway are nicotianamine aminotransferase (NAAT)

(Takahashi et al. 2001), deoxymugineic acid synthase (DMAS) (Bashir et al. 2006), IDS2 and IDS3 (IRON DEFICIENCY SPECIFIC) dioxygenases involved in the production of 3-epihydroxy-2′-deoxymugineic acid and 3-epihydroxy-mugineic acid (Nakanishi et al. 2000; Kobayashi et al. 2001). The production of phytosiderophores is increased in response to Fe deficiency, and tolerance to Fe deficiency is correlated to the quantity and the kind of phytosiderophores secreted (Negishi et al. 2002).

The uptake of Fe^{3+}-PS complexes in Strategy II plants occurs through a specialized transporter discovered by investigating the yellow stripe 1 (ys1) mutant of maize, which is unable to respond to Fe deficiency due to a defect in the uptake of Fe^{3+}-phytosiderophore complexes. The ZmYS1 gene encodes a plasma membrane protein from the OLIGOPEPTIDE TRANSPORTER (OPT) family (Curie et al. 2001).

Recently, regulators of the strategy II system were isolated, such as the basic helix-loop-helix protein OsIRO2 (Ogo et al. 2006, 2007), IDEF1 (Kobayashi et al. 2007) and IDEF2 (Ogo et al. 2008).

Rice induces the genes for iron uptake upon iron deficiency already during the first 3 days of germination (Nozoye et al. 2007) suggesting that iron uptake genes might be induced earlier in grasses than in dicots.

Rice is also able to take up iron via the iron reduction-based mechanism (Ishimaru et al. 2006). Moreover, transgenic ferric reductase renders rice plants more efficient on iron-limiting soils than regular rice plants (Ishimaru et al. 2007). However it remains unclear whether this property is linked with cultivation of rice on wet soils where Fe^{2+} is present in higher quantities than on regular soils.

Strategy and methods for studying iron transport in plants (Table 3)

Coupling heterologous expression systems and reverse genetic approaches to clone and characterize plant iron transporters

Iron uptake by plant roots has been widely documented through physiological studies up to the mid 90's (Marschner 1995). However, it is only the last decade that molecular approaches have been used to characterize the actors involved in iron uptake by plants, and to start to decipher the regulatory mechanisms which regulate their synthesis and activity.

The use of heterologous expression systems, coupled to reverse genetic approaches, turned out to be the winning choice to achieve the molecular cloning and characterization of the two major plant iron transporters, namely *Arabidopsis thaliana* IRT1

Table 3 Summarized strategies and methodologies for studying iron transport in plants

Methods	Procedures	References
Physiological	Use of radioactive iron (^{55}Fe, ^{59}Fe)	Brown et al. (1965)
	Measurement of Fe(III)-chelate reductase activity	Yi and Guerinot (1996)
Molecular	Heterologous expression systems	
	- Yeast	Eide et al. (1996)
	- *Xenopus* oocytes	Schaaf et al. (2004)
	Heterologous nucleotide probes	Robinson et al. (1999)
Genetic	EMS mutant library screening (Arabidopsis)	Yi and Guerinot (1996)
	Reverse genetic	
	- Arabidopsis	Vert et al. (2002)
	- Maize	Curie et al. (2001)
	- Tomato	Ling et al. (2002)
Cellular	Biophysic for iron imaging	
	- Transmission electron microscopy coupled with EDX	Lanquar et al. (2005)
	- Synchrotron X-ray fluorescence microtomography	Kim et al. (2006)
	- Positron emission tomography of ^{52}Fe	Ishimaru et al. (2006)

(Eide et al 1996; Henriques et al. 2002; Varotto et al. 2002; Vert et al. 2002) and *Zea mays* YS1 (Curie et al. 2001; Schaaf et al. 2004). In the case of IRT1, functional complementation of the yeast fet3fet4 mutant strain, affected both in low and high affinity iron transport, represented the first cloning of a non graminaceous plant iron transporter. However, such a strategy, in any case enabled to establish the in planta function of IRT1. Five years later, three independent laboratories (Henriques et al. 2002; Varotto et al. 2002; Vert et al. 2002) isolated T-DNA insertion mutants in the IRT1 gene. These null irt1 mutants were unable to take up iron in response to iron deficiency, demonstrating that the IRT1 transport activity measured in yeast was indeed responsible for root iron uptake.

The molecular cloning and characterization of ZmYS1, from a methodological point of view, is the illustration of the same strategy than the one used for IRT1, but set up in the inverse way. A maize Ac transposon insertion line was isolated and found to map in the same genetic locus than the one of the original maize ys1 mutant (Curie et al. 2001). From a physiological point of view, this mutant was described in detail much earlier than the molecular cloning of the corresponding gene. It was shown to be unable to take up iron from the environment by its roots (Von Wirén et al. 1994). Therefore, the maize YS1 gene could have coded either for an iron transporter or for a protein regulating the transporter expression or activity. The use of the Ac flanking regions enabled to clone the YS1 gene and to observe it codes a protein having a topology consistent with a membrane location. However, the ultimate demonstration that it has indeed a transport activity came from its expression in two heterologous systems: the yeast fet3fet4 mutant as for IRT1 (Curie et al. 2001) or *Xenopus* oocytes (Schaaf et al. 2004). This latter heterologous expression system enabled to measure directly the transport activity of YS1 through electrophysiological methods.

The use of such a strategy based on reverse genetics coupled to yeast or *Xenopus* heterologous expression is presently widely used to characterize *Arabidopsis* and rice iron transporters involved in the allocation of this metal to the various organs and tissues of a plant (Di Donato et al. 2004; Koike et al. 2004; Schaaf et al. 2005; Kim et al. 2006; Gendre et al. 2007).

Expression studies of iron transporters as a tool to document their physiological function

Although expression study of iron transporters, once they are cloned, does not allow a direct understanding of their physiological function in planta, it often gives information to set up phenotyping experiments aiming to compare wild type and mutant genotypes.

For example, expression studies of IRT1 by northern and western blots, and in situ hybridization, coupled with GFP-imaging of the subcellular localization of IRT1 evidenced its localization at the plasma membrane of root epidermal cells, consistent with a role in iron uptake that was established by heterologous expression and phenotyping null mutants (Vert et al. 2002).

Tracing and imaging iron in plants

A major methodological bottleneck in studying iron transport in plants concerns the difficulty to trace and image this metal. Although it is easy to determine the iron concentration of a given tissue by using spectrometric methods (Le Jean et al. 2005), this information does not reflect the iron status of a plant. This status not only depends of the quantity of iron but is deeply related to (i) its redox state (ferrous versus ferric), (ii) its chelation to a huge range of small organic molecules and (iii) its compartimentation between various cell types and within intracellular compartments. In addition, these parameters are highly dynamic, and would require to be measured kinetically. Radiolabelled iron (^{55}Fe or ^{59}Fe) has been used in the past to trace it and even to image it (Brown et al. 1965). However, more recently, iron imaging has been performed by using sophisticated biophysical methods, enabling to couple reverse genetic and analytical methods, leading to the discovery of in planta functions of transporters. For example, transmission electron microscopy coupled with EDX analysis has been recently used to demonstrate that the *Arabidopsis* NRAMP3 and NRAMP4 transporters are involved in iron remobilisation from the vacuoles at the germination step (Lanquar et al. 2005). This remobilized iron was previously loaded into the seed vacuoles through the VIT1 transporter. The biological function of this vacuolar transporter in *Arabidopsis* seeds has been precised by the use of synchrotron X-ray fluorescence microtomography to directly visualize iron (Kim et al. 2006). Concerning dynamic imaging of iron, the use of Positron Emission Tomography of ^{52}Fe, has enabled to follow in real time the translocation of iron from root to shoot in rice plants (Ishimaru et al. 2006).

It would be very useful in the future to develop iron-imaging methods applicable to plants, and easy to use routinely. Such methods could be based on the development of fluorescent sensors (Kikkeri et al. 2007) or chelators, as reported for bacterial siderophores, for example (Palanche et al. 1999).

Iron uptake by microorganisms

Summary of our knowledge

Microbes (bacteria and fungi) have evolved active strategies of iron uptake. Among them, the major one relies on the synthesis of low-molecular weight (generally less than 1,000 Daltons) molecules called siderophores ('iron carrier' in Greek) showing a high affinity for ferric iron (Guerinot 1994). Ferric siderophores are transported into cells via specific Fe-siderophore membrane receptors (Neilands 1981). Bacterial siderophores and membranes receptors are only synthesized under iron stress conditions and are down-regulated by a dimeric protein, the ferric uptake regulator (Fur), which acts as transcriptional repressor of iron-regulated promoters through its Fe^{2+}-dependent DNA binding activity, as first described in *E. coli* (Hantke 1981). Despite the high diversity among siderophores, with more than 500 so far characterized (Boukhalfa and Crumbliss 2002), all of them form six-coordinate octahedral complex with Fe(III) (Guerinot 1994). They are classified according to the functional groups acting as ligands: catecholates, hydroxamates, hydroxypyridonates, hydroxy- or amino-carboxylates (Bossier et al. 1988; Winkelmann 1991, 2002, 2007), some siderophores include mixed functional compounds (Mossialos and Amoutzias 2007). The affinity for iron differ upon the type of siderophore and with K values ranging between 10^{23}–10^{25} for carboxylates, 10^{29}–10^{32} for trihydroxamates, and up to 10^{52} for the catecholate siderophore, enterobactin (Drechsel and Jung 1998). Fungal siderophores belong to the hydroxamate group (Winkelmann 2001).

Among bacteria, a lot of attention has been dedicated to the siderophore-mediated iron uptake of fluorescent pseudomonads which is going to be more specifically developed as a study case. This bacterial group of Gram- bacteria belongs to the genus *Pseudomonas sensu stricto* and includes several species which are either human-animal pathogenic (*P. aeruginosa*), phytopathogenic (*P. cichorii, P. marginalis, P. syringae, P. tolaasii, P. viridiflava*) or non-pathogenic (*P. aureofaciens, P. chlororaphis, P. fluorescens, P. putida*). All these species share the ability to fluoresce under UV light when grown under iron stress conditions (Bossis et al. 2000). This fluorescence results from the synthesis of pyoverdines which are the major class of siderophores produced by fluorescent pseudomonads and show a high affinity for Fe(III) (Fepyoverdine, $K=10^{32}$) (Meyer and Abdallah 1978). The active iron uptake of the fluorescent pseudomnads also relies on the synthesis of protein membrane receptors that are usually specific (Hohnadel and Meyer 1988). Pyoverdines consist of three parts: a conserved dihydroxiquinoline chromophore, responsible for their fluorescence, a peptidic chain including six to 12 amino acids, a small dicarboxylic acid (or its monoamide) connected amidically to the NH_2-group of the chromophore. Pyoverdines present one catechol group included in the chromophore and two hydroxamate groups included in the peptide chain (Budzikiewicz 1997). Pyoverdines have a molecular mass ranging between 1,000 and 1,800 daltons (1,764 for the biggest one reported so far by Meyer et al. 2008). More than 20 genes are required for the pyoverdine synthesis (Visca et al. 2007). The chromophore and the peptide chain are synthesized by non-ribosomal peptide synthases (NRPS) (Mossialos et al. 2002). First steps of formation of the chromophore are performed by a conserved NRPS (*pvdL*), whereas the synthesis of the peptide chain is achieved by other non-conserved NRPS explaining why the peptide chains differ among fluorescent pseudomonads. Indeed, there is a high diversity among these siderophores with more than 100 different pyoverdines being characterized so far (Budzikiewicz 2004; Meyer et al. 2008). Once synthesized, pyoverdines are transported out of the cell and sequester Fe^{3+} from the environment, the resulting complexes—Fepyoverdines—are selectively recognized by and bound to TonB-dependent receptors on the outer membrane and then transported into the cell (Faraldo-Gomez and Sansom 2003). The selective recognition of the pyoverdine relies on the diversity of their peptide chain (Meyer et al. 1987). However there are exceptions and strains have been reported to use siderophores produced by other fluorescent pseudomonads (Mirleau et al. 2000) and even by other microbial groups (Poole 2004). Because of the toxicity of iron at high concentration and the cost for the cell represented by the synthesis of pyoverdines and related protein membrane receptors, siderophore-mediated iron uptake is only expressed when required under iron deficient conditions through regulation processes. Expression of siderophore synthesis and uptake genes is down regulated by a Fur protein which represses the transcription of genes encoding extracytoplasmic sigma factors which are required for the transcription of pyoverdine genes (Visca et al. 2007). Pyoverdine synthesis was shown to be also regulated when the bacterial density is high and corresponds to a significant demand in iron by the phenomenon of Quorum Sensing through the production of acyl homoserines lactones (AHLs) (Stintzi et al. 1998; Whiteley et al. 1999). Fluorescent pseudomonads may also produce additional siderophores such as pyochelin, pseudomonine, quinolobactin/thioquinolobactin and pyridine-2,6-dithiocarboxylic acid (PDTC) (Cornelis and Matthijs 2007). These siderophores usually show a lower affinity for Fe^{3+} than pyoverdine and their synthesis is repressed by this major siderophore. Further information on the iron uptake by fluorescent pseudomonads can be found in the recent reviews of Visca et al. (2007) and Cornelis et al. (2008).

Strategies for studying siderophore-mediated iron uptake and microbial interactions (Table 4)

Siderophore detection and characterization

The ability of a microorganism to produce siderophores may be assessed in vitro using the chrome azurol S (CAS) assay (Schwyn and Neilands 1987). Siderophores produced by the tested microorganisms induce a discoloration of the blue color of the CAS medium. Because siderophore synthesis is submitted to complex regulation, the synthesis ability assessed in vitro does not mean that it is actually produced in situ. However, the high density of fluorescent pseudomonads and frequency of AHL producers, and low iron bioavailability in the rhizosphere are expected to be in favour of the siderophore synthesis in this environment. Hydroxamate siderophores have indeed been directly detected in soils with concentrations in soil solution ranging from 10^{-7} to 10^{-8} M (Powell et al. 1980). However, the most often reported strategy for demonstrating siderophore synthesis relies on indirect methods. Pyoverdine synthesis was shown to occur in the rhizosphere by the use of (i) monoclonal antibodies raised against Fepyoverdine (Buyer et al. 1990) and of (ii) the ice-nucleation reporter gene *inaZ* (Loper and Lindow 1994; Duijff et al. 1999). The reporter gene is under the control of the promoters of iron-regulated genes (Loper and Lindow 1997). Such constructs were made in *P. fluorescens* Pf-5 (Loper and Lindow 1994) and in *P. putida* WCS358 (Duijff et al. 1994a) by fusing a promoter-less ice nucleation activity gene (*inaZ*) to an iron-regulated promoter regulating the production of fluorescent siderophores.

Different methods have been proposed to characterize pyoverdines. The so-called siderotyping is based on analytical and biological methods. As an example of analytical methods, isolectrofocusing (IEF) is based on the physico-chemical properties of the molecules. Indeed, molecules able to complex Fe^{3+} have electric charges which make them to migrate when submitted to electric field allowing the characterization of their respective isoelectric pH values (pHi or pI values). Isoelectrofocusing can be followed by UV illumination for pyoverdine detection or more generally, for all siderophores, by an overlay of the gel with the CAS medium (Koedam et al. 1994). This characterization can be refined on the basis of the usual specificity of the incorporation of Fepyoverdines by biological methods consisting of cross-feeding (Cornelis et al. 1989) and ^{59}Fe-siderophore-mediated uptake experiments (Meyer et al. 1997). Promotion of bacterial growth in iron stress conditions or increased ^{59}Fe bacterial content in the presence of a pyoverdine chelated with Fe, labelled or not, means that the bacterial strain is able to incorporate the tested pyoverdine and therefore the structure of this pyoverdine corresponds to its own. However, as indicated here above, there are exceptions with bacterial strains being able to incorporate so-called heterologous or xenopyoverdines produced by other strains, these exceptions being ascribed to the ability of multivalent pyoverdines to recognize several receptors or to the presence of multiple receptors for different pyoverdines in a single strain (Mirleau et al. 2000; Meyer et al. 2002; Ghysels et al. 2004). Siderotyping has been

Table 4 Summarized strategies and methodologies for studying siderophore-mediated iron uptake and microbial interactions

	Strategies & methodologies	References
Siderophores		
Detection	In vitro: CAS assay	Schwyn and Neilands (1987)
	In soils and rhizospheres:	
	monoclonal antibodies	Buyer et al. (1990)
	inaZ reporter gene	Loper and Lindow (1994); Duijff et al. (1994a, 1999)
	In plants:	
	polyclonal antibodies ^{15}N labeling	Vansuyt et al. (2007)
Characterization		
Analytical methods	EF	Koedam et al. (1994)
	HPLC/ES-MSI	Kilz et al. (1999)
Biological methods	Cross-feeding	Cornelis et al. (1989)
	^{59}Fesiderophore-mediated uptake	Meyer et al. (1997)
Iron bioavailability		
Characterization	Susceptibility of indigenous microbes to 8HQ	Lemanceau et al. (1988b); Robin et al. (2006a)
	inaZ reporter gene	Loper and Lindow (1994); Loper and Henkels (1997); Marschner and Crowley (1998)
Modification	EDDHA / FeEDTA	Scher and Baker (1982); Lemanceau et al. (1988a)
	Purified pyoverdine	Kloepper et al. (1980); Misaghi et al. (1982); Meyer et al. (1987); Lemanceau et al. (1993); Robin et al. (2007)
	WT pseudomonad / pvd- mutant	Bakker et al. (1986, 1987); Becker and Cook (1988); Loper (1988); Lemanceau et al. (1992); Duijff et al. (1993); Maurhofer et al. (1994); Buysens et al. (1996); Duijff et al. (1999); For review see Lemanceau et al. (2007)
	WT plant / OV line	Robin et al. (2006a, b, 2007)
Impact on the microbial interactions	Competitiveness/Adaptation	Raaijmakers et al. (1995); Lemanceau et al. (1988b); Mirleau et al. (2000, 2001);
	Antagonism	Kloepper et al. (1980); Misaghi et al. (1982); Bakker et al. (1986, 1987); Becker and Cook (1988); Loper (1988); Meyer et al. (1987); Lemanceau et al. (1992, 1993); Duijff et al. (1993); Maurhofer et al. (1994); Buysens et al. (1996); Duijff et al. (1999); Robin et al. (2007); For review see Lemanceau et al. (2007)

proposed has a taxonomic tool for fluorescent and nonfluorescent pseudomonads (Meyer et al. 2002). Characterization of the pyoverdine structure and identification of their mass may finally be fulfilled by High Performance Liquid Chromatography (HPLC) coupled with electrospray mass spectrometry (Kilz et al. 1999).

As indicated here above, several siderophores may be produced by the same organism and especially the synthesis of pyoverdine may mask the ability of fluorescent pseudomonads to synthesize others. The strategy described by Cornelis and Matthijs (2007) consists in testing the remaining ability of a pyoverdine-minus mutant to still discolor CAS medium indicating their ability to produce at least an additional siderophore (Mirleau et al. 2000).

Analysis of microbial interactions

As stressed in the introduction, iron is essential for most microorganisms. However, its bioavailability is low in cultivated soils and in the rhizosphere in such way that there is a strong competition among microorganisms in soils and even more in the rhizosphere.

The analysis of these interactions requires the characterization and, if possible, the quantification of the iron fraction which is bioavailable. Bioavailable iron can be defined as the portion of total iron that can be easily assimilated by living organisms, according to the general definition of bioavailability given by Harmsen et al. (2005). None of the chemical methodologies of iron quantification described in the section 'Iron status in soil and rhizosphere' can replace the final application of living organisms to test their ability to use iron in the soil environment. Iron stress conditions for microorganisms can be evaluated in soil indirectly by characterizing their susceptibility to iron starvation. This susceptibility can be assessed by determining the minimal concentrations of a strong iron chelator (8-hydroxyquinoline) inhibiting the growth in vitro of bacterial isolates originating from different soil conditions. This strategy highlighted the lower susceptibility to iron starvation of populations originating from the rhizosphere comparatively to bulk soil populations (Lemanceau et al. 1988b), indicating the lower availability of iron in the rhizosphere. Another strategy relies on so-called iron biosensors based on the use of the ice-nucleation reporter gene *inaZ* described above (Loper and Lindow 1997). Expression of ice nucleation activity from this construct is inversely related to the iron concentration (Fig. 2).

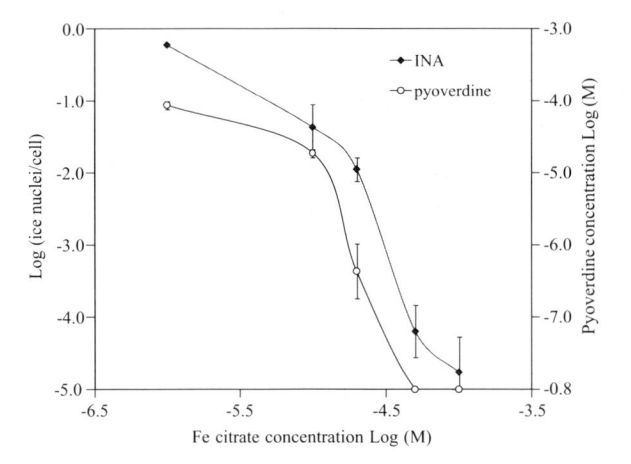

Fig. 2 Relationship between ice nucleation activity (INA) and pyoverdine production by *P. putida* WCS358*pvd-inaZ* grown in vitro with increasing concentrations of ferric citrate. Vertical bars indicate standard errors. (Reprinted from Duijff et al. 1999, Microbial antagonism at the root level is involved in the suppression of fusarium wilt by the combination of non-pathogneic *Fusarium oxysporum* Fo47 and *Pseudomonas putida* WCS358, Phytopathology 89:1073–1079)

Use of this *inaZ* reporter gene in Pf-5 construct indicated that the bacterial cells were mildly iron-stressed in the rhizosphere (Loper and Lindow 1994) and that the ice nucleation activities were similar in different root zones (Marschner and Crowley 1997). Ice nucleation activity was shown to decline with time, indicating that iron bioavailability increased during plant growth (Loper and Henkels 1997). However, the bioavailability of a nutrient or a toxic substance varies between various living organisms according to their different acquisition pathways/capabilities. These organism-dependent variations necessarily stress the limits of the use of biosensors which are based on the utilization of only one given organism for assessing the bioavailability of soil iron because it would only be relevant to those which are close enough to the species used as biosensor.

The general strategy followed to demonstrate the major role of pyoverdine-mediated iron competition in microbial interactions consisted in modifying the bioavailability of iron and looking at the resulting effects of these modifications. Iron biovailability was so far (i) decreased by introducing synthetic (8HQ, see above, and EDDHA) or not (purified pyoverdine) ligands with a high affinity for iron, (ii) increased by introducing iron chelate (FeEDTA) showing a low stability, and (iii) modified by comparing the effect of wild-type strains of fluorescent pseudomonads and their mutants impaired in their ability to produce pyoverdines.

FeEDDHA has a significantly higher stability constant ($K=10^{33.9}$) (Lindsay 1979) than Fefusarinine ($K=10^{29}$), the Fe-chelate formed by *Fusarium* siderophores (Emery 1965; Lemanceau et al. 1986). Therefore, EDDHA addition decreases iron availability to *F. oxysporum* and was shown to reduce germination of chlamydospores and germ tube length in vitro and in soil (Scher and Baker 1982). In contrast, FeEDTA has a significantly lower stability constant ($K=10^{25}$) (Lindsay 1979) than Fefusarinine and therefore increases iron availability to *F. oxysporum*. Consequently, while the reduced concentration of iron available for *F. oxysporum* by EDDHA addition led to a decreased disease severity (Scher and Baker 1982; Lemanceau et al. 1988a), the increased concentration of iron available for *F. oxysporum* by FeEDTA addition led to an increased disease severity. This strategy enabled Scher and Baker (1982) and

Lemanceau et al. (1988a) to ascribe, at least partly, the natural soil suppressiveness to fusarium wilts to the strong iron competition in these soils. The low availability of iron in naturally suppressive soils, resulting from their physico-chemical properties and from their high biomass, favors microbial populations with a more efficient siderophore-mediated iron uptake such as fluorescent pseudomonads, which express their competitive ability against pathogenic *F. oxysporum* even more strongly; consequently natural soil suppressiveness to fusarium wilts has been ascribed at least partly to these bacterial populations (Scher and Baker 1980; Lemanceau et al. 1988b).

Demonstration of the pyoverdine-mediated antagonism against pathogenic microbes was stimulated by the early report of Kloepper et al. (1980) indicating that addition of *Pseudomonas* sp. B10 or its pyoverdine to soils conducive to fusarium wilts and to take-all rendered them suppressive. They further showed that supplementation of the soils with iron overcame the positive effects of the bacterial inoculation or pyoverdine addition. Possible involvement of pyoverdine in microbial antagonism was then supported by a series of observations: (i) in vitro antagonism by some fluorescent pseudomonads against specific pathogens only occurring when pyoverdines were produced (Kloepper et al. 1980; Misaghi et al. 1982), (ii) positive correlation between the intensity of siderophore synthesis in vitro by different pseudomonads isolates and their ability to reduce chlamydospore germination of pathogenic *F. oxysporum* in soil (Sneh et al. 1984; Elad and Baker 1985), (iii) in vitro antagonistic activity of purified pyoverdine against *Pythium* (Meyer et al. 1987) and *F. oxysporum* (Lemanceau et al. 1993), and (iv) reduced chlamydospore germination of pathogenic *F. oxysporum* in soil upon pyoverdine introduction (Elad and Baker 1985). The implication of iron competition was demonstrated by showing that the antagonism achieved by the purified pyoverdines was suppressed when these siderophores were chelated with iron and therefore unable to scavenge iron from the medium (Meyer et al. 1987; Lemanceau et al. 1993). Finally, involvement of pyoverdines in the antagonism against plant pathogens (*F. oxysporum*, *Pythium* spp.) and so-called deleterious microorganisms (fluorescent pseudomonads, *Pythium* spp.) was fulfilled by the use of mutants impaired in their ability to synthesize such siderophore (Bakker et al. 1986, 1987; Becker and Cook 1988; Loper 1988; Duijff et al. 1993; Buysens et al. 1996; Leeman et al. 1996; De Boer et al. 2003). Pyoverdine-minus mutants are usually obtained by screening a mutant library obtained by random mutagenesis and selecting those which do not fluoresce anymore under iron stress conditions. Single transposon insertions are then selected and sequenced in order to localize their insertion site within the genome (Mirleau et al. 2000). As an example of the strategy based on the use of pvd- mutant, Lemanceau et al. (1992) showed that *P. putida* WCS358, but not its pvd- mutant, was able to improve the control of fusarium wilt determined by non-pathogenic *F. oxysporum* Fo47. Furthermore, pyoverdine synthesis by wild-type strain was shown indirectly by the use of the *inaZ* reporter gene. The pyoverdine-mediated improvement of the control by *P. putida* WCS358 was related to a reduced saprophytic density and activity of the pathogenic *F. oxysporum* as assessed by β-glucoronidase reporter gene in *gusA*-marked derivative of the pathogen (Duijff et al. 1999). In addition to their antagonistic activity during the saprophytic phase of the life-cycle of the pathogens, siderophores and other iron-regulated metabolites were shown to be involved in the induction of defense reactions during the parasitic phase of their life cycle (for review see Höfte and Bakker 2007). As an example, *P. fluorescens* CHA0 was shown to induce systemic resistance of tobacco against TNV, whereas its pvd- mutant was less efficient than the wild-type. These observations suggest that pyoverdine plays a role in induced systemic resistance (ISR) by CHA0 (Maurhofer et al. 1994).

Pyoverdine-mediated antagonism contributes to give a competitive advantage to fluorescent pseudomonads. The corresponding demonstrations were performed at the population and strain levels. Population studies consisted in comparing the susceptibility to iron stress of fluorescent pseudomonads isolated from rhizosphere and from bulk soils (Lemanceau et al. 1988b). The Minimal Inhibitory Concentration (MIC) of 8-8HQ, a strong iron chelator (Geels et al. 1985), was determined for each isolate. The rhizosphere isolates were significantly more represented in the classes with high MIC values than the soil isolates. It was further shown that there is a gradient between the bulk soil, rhizosphere, rhizoplane and root tissues, the MIC values of pseudomonad populations increasing when isolated closer to the roots (Robin et al. 2007). Altogether, these data indicate that populations of fluorescent pseudomonads from rhizosphere are less susceptible to iron starvation than those from bulk soil suggesting that (i) they benefit from a more efficient iron-uptake than those from bulk soil, and that (ii) this efficient iron-uptake might be implicated in their rhizosphere competence (Lemanceau et al. 1988b). The above hypothesis was tested by evaluating the possible implication of pyoverdine in the rhizosphere competence of a model strain (*P. fluorescens* C7R12) using a pyoverdine minus mutant (pvd-). The survival kinetics of the wild-type strain and of the pvd- mutant in the rhizosphere were compared in competition, in the absence (gnotobiotic conditions) and in the presence of the indigenous microorganisms (Mirleau et al. 2000; Mirleau et al. 2001). In the absence of the indigenous microorganisms, bacterial competition was favourable to the pyoverdine producer wild-type, whereas in non-gnotobiotic conditions the survival of both strains was similar. In the latter conditions, the fitness of the pvd- mutant in competition with the wild-type strain was assumed to be related to its ability to take-up, as the wild-type strain, pyoverdines from foreign origin (Mirleau et al. 2000). Altogether, these results suggest that pyoverdine-mediated iron uptake is involved in the rhizosphere competence of *P. fluorescens* C7R12. The role of siderophore receptor in rhizosphere competence was supported by experiments performed with a mutant of *P. fluorescens* WCS374 that harboured siderophore receptor PupA from *P. putida* WCS358 and was then able to utilize Fepyoverdine from WCS358. This ability was shown to give the WCS374 mutant a competitive advantage over the corresponding wild-type strain when co-inoculated with WCS358 (Raaijmakers et al. 1995).

Integrated approaches for studying the interactions between soils, plants and microbes in the rhizosphere

The previous sections report strategies and methodologies for characterizing (i) the iron status in soils and rhizospheres, (ii) the iron uptake by microbes and plants, and (iii) the iron-mediated interactions. Integration of the corresponding strategies represents a unique opportunity to expand our knowledge and to better understand the complex interactions in the rhizosphere between soils, plants and microbes in relation with the iron dynamics. The plant-microbe interactions can be represented by a feed-back loop, this loop being under the control of the soil physico-chemical properties which impact the iron bioavailability. Reciprocally these interactions also affect the soil iron status (Robin et al. 2008).

The analysis of such a feed-back loop was made recently possible by following an innovative strategy, which did not consist anymore in modifying the rhizosphere iron availability by introducing ligands but rather by monitoring this availability through the plant iron uptake. This was made possible by the use of a transgenic tobacco (strategy I) line overexpressing ferritin (ferritin overexpressor, OV) leading to an activation of the plant iron uptake mechanisms. Consequently, an increase in the plant iron content occurs (Van Wuytswinkel et al. 1999) leading to reduce the iron bioavailability in the rhizosphere of the transgenic line compared to the wild-type tobacco plants (Robin et al. 2006a). Comparison of the populations of pseudomonads associated with the wild-type and with the OV transgenic line showed that in these populations differed in the two rhizospheres (Robin et al. 2006b, 2007). Measurements of the 8HQ CMI for the isolates from the two rhizosphere indicated that those from the iron stressed rhizosphere were less susceptible to iron starvation (Robin et al. 2007), they belonged to different siderotypes, and they had a different genetic background as assessed by RAPD-PCR (Robin et al. 2007). Interestingly, these siderophores exhibited a stronger antagonism against a phytopathogenic Oomycete than those produced by populations associated with the tobacco wild-type rhizosphere. Despite this strong competition ability, these siderophores do not compete with the host-plant and even more promote its iron nutrition (Lemanceau et al. unpublished data). Indeed, plants have evolved a strategy to take up iron from Fe-pyoverdine which is not related to the mechanisms described so far in strategy I plants (Vansuyt et al. 2007). The presence of Fe-pyoverdine in the plant was observed and quantified by an enzyme-linked immunosorbent assay (ELISA) thanks to an antibody raised against the pyoverdine and by isotope measurements from ^{15}N labelled pyoverdine (Vansuyt et al. 2007). Current observations suggest that this incorporation could be ascribed to endocytosis of Fe-pyoverdine (Avoscan et al. unpublished data). Taken together these data clearly show that the decrease of the iron availability in the rhizosphere due to iron overaccumulation by the OV tobacco line leads to the selection of specific bacterial populations which are adapted to this iron depletion through the production of specific siderophores. In return, these populations improve both (i) the plant health via the decrease of phytopathogen saprophytic growth through pyoverdine-mediated iron competition, and (ii) the plant growth by promoting its iron nutrition. Therefore, the cost for the plant represented by the release in the rhizosphere of a major part of their photosynthetates as rhisodeposits is balanced by the benefit for the plant gained with the promotion of its health and growth. Furthermore, fluorescence kinetics of these siderophores showed their high ability to extract iron from soil iron oxides (ferrihydrite, goethite) as assessed by their fluorescence decrease indicating the siderophore chelation with iron (Vansuyt et al. Unpublished data).

Evidences for reciprocal interactions between iron nutrition of strategy II plants and associated microorganisms have also been reported. Thanks to the construct made in *P. fluorescens* Pf-5 with the reporter gene *pvd-inaZ* (Loper and Lindow 1994), phytosiderophores were indeed shown to repress pyoverdine synthesis, suggesting that these phytosiderophores could be a possible source of iron for the bacteria (Marschner and Crowley 1998). Similar observations were made by Jurkevitch et al. (1993) when measuring incorporation of ^{55}Fe from ^{55}Fe-mugeneic acid by a *P. putida* strain.

For strategy II plants, only indirect mechanisms may account for improved plant nutrition by pyoverdines due to their significantly higher affinity for iron compared to phytosiderophores (Meyer and Abdallah 1978; Sugiura et al. 1981). Duijff et al. (1994b) have proposed that possible degradation of Fe-pyoverdines by microbes could lead to the release of iron that would then be available for phytosiderophores. Possible interactions between phytosiderophores and microorganisms in relation with iron depend on their concentration, location and release kinetics. The diurnal production cycle of phytosiderophores results in pulses (Crowley and Gries 1994; Takagi et al. 1984; Reichman and Parker 2007), during which their concentration in the rhizosphere might be higher than that of microbial siderophores, therefore affecting the ligand exchange in favour to phytosiderophores (Jurkevitch et al. 1993; Yehuda et al. 1996). However, one should be cautious with this hypothesis because of the thermodynamic and kinetic constraints involved in such ligand-exchange processes, which even if they occurred in the rhizosphere, would be too slow to contribute significantly to plant iron nutrition. Besides these temporal aspects, the chelation efficiency of iron by phytosiderophores also relies on their spatial location. Their release occurs at greater rate in apical root zones (Marschner et al. 1987) corresponding to where the rhizodeposition is maximal (Nguyen 2003). Despite this observation, interactions between phytosiderophores and microorganisms are likely to be quite low because the rhizosphere effect on microorganisms mostly occur at some distance from root apex (Gamalero et al. 2002; Yang and Crowley 2000) due to the longer time required for root elongation than for microbial growth upon root exudation. However, in plants producing only low amounts of phytosiderophores (*Sorghum bicolour* L. Moench, *Zea mays* L.), their degradation by microorganisms could negatively affect plant iron content, whereas such negative effects were not recorded in plant (*Hordeum vulgare* L.) producing higher amounts of phytosiderophores (Von Wirén et al. 1993, 1995). According to the general model of Darrah (1991), secretion of phytosiderophores in massive amounts in restricted spatial and temporal windows would represent an efficient strategy for plants to minimize their microbial degradation in the rhizosphere, compared to the secretion of similar amounts all along the root length and all along the day.

Concluding remarks

Iron is an essential element for organisms which densities are high in the rhizosphere. However, in cultivated soils, iron is mostly complexed as hydroxides, oxyhydroxides and oxides and its solubility is low. Therefore, plants and microorganisms have evolved active strategies of iron uptake. These strategies interact together and result in complex relations between plants and microorganisms in the rhizosphere. These relations are under the control of the soil physico-chemical properties. They may either involve competition or mutualism and are driven by co-evolution processes.

An example of competition commonly illustrated is that achieved by fluorescent pseudomonads against some eukaryotic plant pathogens. This competition induces a decrease of the pathogen growth in the rhizosphere and leads to the health promotion of the host-plant which supports these bacteria via the root exudates. The mutualistic interactions between the host-plant and associated fluorescent pseudomonads involve also the promotion of the plant iron nutrition by the bacterial siderophore. On the other way round, phytosiderophores seem to enhance iron nutrition by fluorescent pseudomonads. Among microbial populations, incorporation of heterologous siderophores by microorganisms may give them a competitive advantage over those which do not show this ability.

Illustrations of the complex interactions given in the present manuscript show the added value of bringing together expertises in soil science, plant biology and microbiology. Significant progresses have been made in the methodologies to characterize the iron status in soil, the iron-uptake mechanisms by plants and microbes, and in the interactions between soils, plants and microorganisms. Further studies are now required for analyzing the cost of the different types of interactions in regard to their benefice for the organisms involved. From an ecological point of view, one may indeed expect that the stability of these complex interactions is regulated by the ratio between the cost and the benefice for each type of organism.

Analysis of these complex interactions between soils, plants and microbes in relation with iron dynamic represents a unique opportunity to progress in our knowledge of the rhizosphere ecology. These progresses are expected to provide information and tools enabling us to develop strategies to improve plant iron nutrition and health with decreases in the application of chemical inputs by making attempt to monitor these complex interactions.

References

Bakker PAHM, Lamers JG, Bakker AW, Marugg JD, Weisbeek PJ, Schippers B (1986) The role of siderophores in potato tuber yield increase by *Pseudomonas putida* in a short rotation of potato. Neth J Plant Pathol 92:249–256. doi:10.1007/BF01977588

Bakker PAHM, Bakker AW, Marugg JD, Weisbeek PJ, Schippers B (1987) Bioassay for studying the role of siderophores in potato growth stimulation by *Pseudomonas* spp. in short potato rotations. Soil Biol Biochem 4:443–449. doi:10.1016/0038-0717(87)90036-8

Bashir K, Inoue H, Nagasaka S, Takahashi M, Nakanishi H, Mori S, Nishizawa NK (2006) Cloning and characterization of deoxymugineic acid synthase genes from graminaceous plants. J Biol Chem 281:32395–32402. doi:10.1074/jbc.M604133200

Bauer P, Thiel T, Klatte M, Bereczky Z, Brumbarova T, Hell R, Grosse I (2004) Analysis of sequence, map position, and gene expression reveals conserved essential genes for iron uptake in *Arabidopsis* and tomato. Plant Physiol 136:4169–4183. doi:10.1104/pp.104.047233

Becker JO, Cook RJ (1988) Role of siderophores in suppression of *Pythium* species and production of increased-growth response of wheat by fluorescent pseudomonads. Phytopathology 78:778–782. doi:10.1094/Phyto-78-778

Bereczky Z, Wang HY, Schubert V, Ganal M, Bauer P (2003) Differential regulation of *Nramp* and *irt* metal transporter genes in wild type and iron uptake mutants of tomato. J Biol Chem 278:24697–24704. doi:10.1074/jbc.M301365200

Blume H-P, Bruemmer GW, Schwertmann U, Horn R, Koegel-Knabner I, Stahr K, Auerswald K, Beyer L, Hartmann A, Litz N, Scheinost A, Stanjek H et al (2002) Shaffer/Schachtschabel/Lehrbuch der Bodenkunde, 15th edn. Heidelberg, Spektrum Akademischer Verlag

Borggaard OK (1988) Phase identification by selective dissolution techniques. In: Stucki JW, Goodman BA, Schwertmann U (eds) Iron in soils and clay minerals. D Reidel, Dordrecht, pp 83–97

Bossier P, Höfte M, Verstraete W (1988) Ecological significance of siderophores in soil. Adv Microb Ecol 10:384–414

Bossis E, Lemanceau P, Latour X, Gardan L (2000) Taxonomy of *Pseudomonas fluorescens* and *Pseudomonas putida*: current status and need for revision. Agronomie 20:51–63. doi:10.1051/agro:2000112

Boukhalfa H, Crumbliss AL (2002) Chemical aspects of siderophore mediated iron transport. Biometals 15:325–339. doi:10.1023/A:1020218608266

Briat JF, Gaymard F (2007) Iron nutrition and interactions in plants—Preface. Plant Physiol Biochem 45:259. doi:10.1016/j.plaphy.2007.03.005

Brown AL, Yamaguchi S, Leal-Diaz J (1965) Evidence for translocation of iron in plants. Plant Physiol 40:35–38. doi:10.1104/pp.40.1.35

Brumbarova T, Bauer P (2005) Iron-mediated control of the basic helix-loop-helix protein FER, a regulator of iron uptake in tomato. Plant Physiol 137:1018–1026. doi:10.1104/pp.104.054270

Budzikiewicz H (1997) Siderophores of fluorescent pseudomonads. Z Naturforsch 52c:713–720

Budzikiewicz H (2004) Siderophores of the *Pseudomonadaceae* sensu stricto (fluorescent and non fluorescent *Pseudomonas* spp.). In: Herz W, Falk H, Kirby GW (eds) Progress in the chemistry of organic natural products. Vienna, Springer, pp 81–237

Buyer JS, Sikora LJ, Kratzke MG (1990) Monoclonal antibodies to ferric pseudobactin, the siderophore of plant growth-promoting *Pseudomonas putida* B10. Appl Environ Microbiol 56:419–424

Buysens S, Heungens K, Poppe J, Höfte M (1996) Involvement of pyochelin and pyoverdin in suppression of *Pythium*-induced damping-off of tomato by *Pseudomonas aeruginosa* 7NSK2. Appl Environ Microbiol 62:865–871

Carrasco N, Kretzschmar R, Pesch ML, Kraemer SM (2008) Effects of anionic surfactants on ligand-promoted dissolution of iron and aluminum hydroxides. J Colloid Interface Sci 321:279–287. doi:10.1016/j.jcis.2008.02.011

Chaney RL, Brown JC, Tiffin LO (1972) Obligatory reduction of ferric chelates in iron uptake by soybeans. Plant Physiol 50:208–213. doi:10.1104/pp.50.2.208

Chao TT, Zhou L (1983) Extraction techniques for selective dissolution of amorphous iron-oxides from soils and sediments. Soil Sci Soc Am J 47:225–232

Cohen CK, Fox TC, Garvin DF, Kochian LV (1998) The role of iron-deficiency stress responses in stimulating heavy-metal transport in plants. Plant Physiol 116:1063–1072. doi:10.1104/pp.116.3.1063

Colangelo EP, Guerinot ML (2004) The essential basic helix-loop-helix protein FIT1 is required for the iron deficiency response. Plant Cell 16:3400–3412. doi:10.1105/tpc.104.024315

Colangelo EP, Guerinot ML (2006) Put the metal to the petal: metal uptake and transport throughout plants. Curr Opin Plant Biol 9:322–330. doi:10.1016/j.pbi.2006.03.015

Connolly EL, Fett JP, Guerinot ML (2002) Expression of the IRT1 metal transporter is controlled by metals at the levels of transcript and protein accumulation. Plant Cell 14:1347–1357. doi:10.1105/tpc.001263

Connolly EL, Campbell NH, Grotz N, Prichard CL, Guerinot ML (2003) Overexpression of the FRO2 ferric chelate reductase confers tolerance to growth on low iron and uncovers posttranscriptional control. Plant Physiol 133:1102–1110. doi:10.1104/pp.103.025122

Cornelis P, Matthijs S (2007) *Pseudomonas* siderophores and their biological significance. In: Varma A, Chincholkar S (eds) Microbial siderophores. Springer-Verlag, Berlin Heildeberg, pp 193–203

Cornelis P, Hohnadel D, Meyer JM (1989) Evidence for different pyoverdine-mediated iron uptake systems among *Pseudomonas aeruginosa* strains. Infect Immun 57:3491–3497

Cornelis P, Baysse C, Matthijs S (2008) Iron uptake in *Pseudomonas*. In: Cornelis P (ed) *Pseudomonas*, genomics and molecular biology. Caister Academic, Norfolk, pp 213–235

Cornell RM, Schwertmann U (2003) The iron oxides, 2nd edn. Wiley–VCH, Weinheim

Crowley DE, Gries D (1994) Modelling of iron availability in the plant rhizosphere. In: Manthey JA, Crowley DE, Luster DG (eds) Biochemistry of metal micronutrients in the rhizosphere. CRC, Boca Raton, Florida, pp 199–220

Curie C, Panaviene Z, Loulergue C, Dellaporta SL, Briat JF, Walker EL (2001) Maize *yellow stripe1* encodes a membrane protein directly involved in Fe(III) uptake. Nature 409:346–349. doi:10.1038/35053080

Darrah PR (1991) Models of the rhizosphere. I. Microbial population dynamics around a root releasing soluble and insoluble carbon. Plant Soil 133:187–199. doi:10.1007/BF00009191

De Boer M, Bom P, Kindt F, Keurentjes JJB, Von der Sluis I, Van Loon LC, Bakker PAHM (2003) Control of Fusarium wilt of radish by combining *Pseudomonas putida* strains that have different disease-suppressive mechanisms. Phytopathology 93:626–632. doi:10.1094/PHYTO.2003.93.5.626

De Santiago A, Diaz I, del Campillo MD, Torrent J, Delgado A (2008) Predicting the incidence of iron deficiency chlorosis from hydroxylamine-extractable iron in soil. Soil Sci Soc Am J 72:1493–1499. doi:10.2136/sssaj2007.0366

Di Donato RJ, LA Jr R, Sanderson T, Eisley RB, Walker EL (2004) Arabidopsis *Yellow Stripe-Like2* (*YSL2*): a metal-regulated gene encoding a plasma membrane transporter of nicotianamine-metal complexes. Plant J 39:403–414. doi:10.1111/j.1365-313X.2004.02128.x

Drechsel H, Jung J (1998) Peptide siderophores. J Pept Sci 4:147–181. doi:10.1002/(SICI) 1099-1387(199805) 4:3<147::AID-PSC136>3.0.CO;2-C

Duijff BJ, Meijer JW, Bakker PAHM, Schippers B (1993) Siderophore-mediated competition for iron and induced resistance in the suppression of Fusarium wilt of carnation by fluorescent *Pseudomonas* spp. Neth J Plant Pathol 99:277–289. doi:10.1007/BF01974309

Duijff BJ, Bakker PAHM, Schippers B (1994a) Suppression of Fusarium wilt of carnation by *Pseudomonas putida* WCS358 at different levels of disease incidence and iron availability. Biocontrol Sci Technol 4:279–288. doi:10.1080/09583159409355336

Duijff BJ, De Kogel WJ, Bakker PAHM, Schippers B (1994b) Influence of pseudobactin 358 on the iron nutrition of barley. Soil Biol Biochem 26:1681–1994. doi:10.1016/0038-0717(94) 90321-2

Duijff BJ, Recorbet G, Bakker PAHM, Loper JE, Lemanceau P (1999) Microbial antagonism at the root level is involved in the suppression of Fusarium wilt by the combination of nonpathogenic *Fusarium oxysporum* Fo47 and *Pseudomonas putida* WCS358. Phytopathology 89:1073–1079. doi:10.1094/PHYTO.1999.89.11.1073

Eckhardt U, Marques AM, Buckhout TJ (2001) Two iron-regulated cation transporters from tomato complement metal uptake-deficient yeast mutants. Plant Mol Biol 45:437–448. doi:10.1023/A:1010620012803

Eide D, Broderius M, Fett J, Guerinot ML (1996) A novel iron-regulated metal transporter from plants identified by functional expression in yeast. Proc Natl Acad Sci USA 93:5624–5628. doi:10.1073/pnas.93.11.5624

Elad Y, Baker R (1985) The role of competition for iron and carbon in suppression of chlamydospore germination of *Fusarium* spp. by *Pseudomonas* spp. Phytopathology 75:1053–1059. doi:10.1094/Phyto-75-1053

Emery T (1965) Isolation, characterization, and properties of fusarinine, a hydroxamic derivative of ornithine. Biochem 4:1410–1417. doi:10.1021/bi00883a028

Eng BH, Guerinot ML, Eide D, Saier MH (1998) Sequence analyses and phylogenetic characterization of the ZIP family of metal ion transport proteins. J Membr Biol 166:1–7. doi:10.1007/s002329900442

Faraldo-Gomez JD, Sansom MS (2003) Acquisition of siderophores in Gram-negative bacteria. Nat Rev Mol Cell Biol 4:105–116. doi:10.1038/nrm1015

Gamalero E, Martinotti MG, Trotta A, Lemanceau P, Berta G (2002) Morphogenetic modifications induced by *Pseudomonas fluorescens* A6RI and *Glomus mosseae* BEG12 in the root system of tomato differ according to the plant growth conditions. New Phytol 155:293–300. doi:10.1046/j.1469-8137.2002.00460.x

Geels FP, Schmidt EDL, Schippers B (1985) The use of 8-hydroxyquinoline for the isolation and prequantification of plant growth-stimulating rhizosphere pseudomonads. Biol Fertil Soils 1:167–173. doi:10.1007/BF00257633

Gendre D, Czernic P, Conejero G, Pianelli K, Briat JF, Lebrun M, Mari S (2007) TcYSL3, a member of the YSL gene family from the hyper-accumulator *Thlaspi caerulescens*, encodes a nicotianamine-Ni/Fe transporter. Plant J 49:1–15. doi:10.1111/j.1365-313X.2006.02937.x

Ghysels B, Min Dieu BT, Beatson SA, Pirnay JP, Oschner UA, Vasil ML, Cornelis P (2004) FpvB, an alternative type I ferripyoverdine receptor of *Pseudomonas aeruginosa*. Microbiology 150:1671–1680. doi:10.1099/mic.0.27035-0

Gleyzes C, Tellier S, Astruc M (2002) Fractionation studies of trace elements in contaminated soils and sediments: a review of sequential extraction procedures. Trends Analyt Chem 21:451–467. doi:10.1016/S0165-9936(02) 00603-9

Guerinot ML (1994) Microbial iron transport. Annu Rev Microbiol 48:743–772. doi:10.1146/annurev.mi.48.100194.003523

Guerinot ML (2000) The ZIP family of metal transporters. Biochim Biophys Acta-Biomembr 1465:190–198. doi:10.1016/S0005-2736(00)00138-3

Guerinot ML, Yi Y (1994) Iron: nutritious, noxious, and not readily available. Plant Physiol 104:815–820

Hanesch M, Stanjek H, Petersen N (2006) Thermomagnetic measurements of soil iron minerals: the role of organic carbon. Geophys J Int 165:53–61. doi:10.1111/j.1365-246X.2006.02933.x

Hantke K (1981) Regulation of ferric iron transport in *Escherichia coli* K12: isolation of a constitutive mutant. Mol Gen Genet 182:288–292. doi:10.1007/BF00269672

Harmsen J (2007) Measuring bioavailability: From a scientific approach to standard methods. J Environ Qual 36:1420–1428. doi:10.2134/jeq2006.0492

Harmsen J, Rulkens W, Eijsackers H (2005) Bioavailability, concept for understanding or tool for predicting? Land Contam Reclam 13:161–171

Henriques R, Jásik J, Klein M, Martinoia E, Feller U, Schell J, Pais MS, Koncz C (2002) Knock-out of *Arabidopsis* metal transporter gene *IRT1* results in iron deficiency accompanied by cell differentiation defects. Plant Mol Biol 50:587–597. doi:10.1023/A:1019942200164

Herbik A, Koch G, Mock HP, Dushkov D, Czihal A, Thielmann J, Stephan UW, Baumlein H (1999) Isolation, characterization and cDNA cloning of nicotianamine synthase from barley—A key enzyme for iron homeostasis in plants. Eur J Biochem 265:231–239. doi:10.1046/j.1432-1327.1999.00717.x

Heron G, Crouzet C, Bourg ACM, Christensen TH (1994) Speciation of Fe(II) and Fe(III) in contaminated aquifer sediments using chemical extraction techniques. Environ Sci Technol 28:1698–1705. doi:10.1021/es00058a023

Higuchi K, Suzuki K, Nakanishi H, Yamaguchi H, Nishizawa NK, Mori S (1999) Cloning of nicotianamine synthase genes, novel genes involved in the biosynthesis of phytosiderophores. Plant Physiol 119:471–479. doi:10.1104/pp.119.2.471

Higuchi K, Watanabe S, Takahashi M, Kawasaki S, Nakanishi H, Nishizawa NK, Mori S (2001) Nicotianamine synthase gene expression differs in barley and rice under Fe-deficient conditions. Plant J 25:159–167. doi:10.1046/j.1365-313x.2001.00951.x

Höfte M, Bakker PAHM (2007) Competition for iron and induced systemic resistance by siderophores of plant growth promoting rhizobacteria. In: Varma A, Chincholkar S (eds) Microbial siderophores. Springer-Verlag, Berlin Heildeberg, pp 121–133

Hohnadel D, Meyer JM (1988) Specificity of pyoverdine-mediated iron uptake among fluorescent *Pseudomonas* strains. J Bacteriol 170:4865–4873

Inoue H, Higuchi K, Takahashi M, Nakanishi H, Mori S, Nishizawa NK (2003) Three rice nicotianamine synthase genes, *OsNAS1, OsNAS2*, and *OsNAS3* are expressed in cells involved in long-distance transport of iron and differentially regulated by iron. Plant J 36:366–381. doi:10.1046/j.1365-313X.2003.01878.x

Ishimaru Y, Suzuki M, Tsukamoto T, Suzuki K, Nakazono M, Kobayashi T, Wada Y, Watanabe S, Matsuhashi S, Takahashi M, Nakanishi H, Mori S, Nishizawa NK (2006) Rice plants take up iron as an Fe^{3+}-phytosiderophore and as Fe^{2+}. Plant J 45:335–346. doi:10.1111/j.1365-313X.2005.02624.x

Ishimaru Y, Kim S, Tsukamoto T, Oki H, Kobayashi T, Watanabe S, Matsuhashi S, Takahashi M, Nakanishi H, Mori S, Nishizawa NK (2007) From the cover: mutational reconstructed ferric chelate reductase confers enhanced tolerance in rice to iron deficiency in calcareous soil. Proc Natl Acad Sci USA 104:7373–7378. doi:10.1073/pnas.0610555104

Jakoby M, Wang HY, Reidt W, Weisshaar B, Bauer P (2004) *FRU* (*BHLH029*) is required for induction of iron mobilization genes in *Arabidopsis thaliana*. FEBS Lett 577:528–534. doi:10.1016/j.febslet.2004.10.062

Jurkevitch E, Hadar Y, Chen Y, Chino M, Mori S (1993) Indirect utilization of the phytosiderophore mugineic acid as an iron source to rhizosphere fluorescent *Pseudomonas*. Biometals 6:119–123. doi:10.1007/BF00140113

Kiczka M, Wiederhold JG, Kraemer SM, Bourdon B, Kretzschmar R (2007) The impact of Fe isotope fractionation by plants on the isotopic signature of soils. Geochim Cosmochim Acta 71:A482

Kikkeri R, Traboulsi H, Humbert N, Gumienna-Kontecka E, Arad-Yellin R, Melman G, Elhabiri M, Albrecht-Gary AM, Shanzer A (2007) Toward iron sensors: bioinspired tripods based on fluorescent phenol-oxazoline coordination sites. Inorg Chem 46:2485–2497. doi:10.1021/ic061952u

Kilz S, Lenz C, Fuchs R, Budzikiewicz H (1999) A fast screening method for the identification of siderophores from fluorescent *Pseudomonas* spp. by liquid chromatography/electrospray mass spectrometry. J Mass Spectrom 34:281–290. doi:10.1002/(SICI)1096-9888(199904) 34:4<281::AID-JMS750>3.0.CO;2-M

Kim SA, Guerinot ML (2007) Mining iron: iron uptake and transport in plants. FEBS Lett 581:2273–2280. doi:10.1016/j.febslet.2007.04.043

Kim SA, Punshon T, Lanzirotti A, Li L, Alonso JM, Ecker JR, Kaplan J, Guerinot ML (2006) Localization of iron in *Arabidopsis* seed requires the vacuolar membrane transporter VIT1. Science 314:1295–1298. doi:10.1126/science.1132563

Kloepper JW, Leong J, Teintze M, Schroth MN (1980) *Pseudomonas* siderophores: a mechanism explaining disease-suppressive soils. Curr Microbiol 4:317–320. doi:10.1007/BF02602840

Kobayashi T, Nakanishi H, Takahashi M, Kawasaki S, Nishizawa NK, Mori S (2001) In vivo evidence that *Ids3* from *Hordeum vulgare* encodes a dioxygenase that converts 2'-deoxymugineic acid to mugineic acid in transgenic rice. Planta 212:864–871. doi:10.1007/s004250000453

Kobayashi T, Ogo Y, Itai RN, Nakanishi H, Takahashi M, Mori S, Nishizawa NK (2007) The transcription factor IDEF1 regulates the response to and tolerance of iron deficiency in plants. Proc Natl Acad Sci USA 104:19150–19155. doi:10.1073/pnas.0707010104

Koedam N, Wittouck E, Gaballa A, Gillis A, Höfte M, Cornellis P (1994) Detection and differentiation of microbial siderophores by isoelectric focusing and chrome azurol S overlay. Biometals 7:287–291. doi:10.1007/BF00144123

Koike S, Inoue H, Mizuno D, Takahashi M, Nakanishi H, Mori S, Nishizawa NK (2004) OsYSL2 is a rice metal-nicotianamine transporter that is regulated by iron and expressed in the phloem. Plant J 39:415–424. doi:10.1111/j.1365-313X.2004.02146.x

Kraemer SM (2004) Iron oxide dissolution and solubility in the presence of siderophores. Aquat Sci 66:3–18. doi:10.1007/s00027-003-0690-5

Kraemer SM, Cheah SF, Zapf R, Xu JD, Raymond KN, Sposito G (1999) Effect of hydroxamate siderophores on Fe release and Pb(II) adsorption by goethite. Geochim Cosmochim Acta 63:3003–3008. doi:10.1016/S0016-7037(99)00227-6

Kraemer SM, Crowley D, Kretzschmar R (2006) Geochemical aspects of phytosiderophore-promoted iron acquisition by plants. Adv Agron 91:1–46. doi:10.1016/S0065-2113(06)91001-3

Labanowski J, Monna F, Bermond A, Cambier P, Fernandez C, Lamy I, van Oort F (2008) Kinetic extractions to assess mobilization of Zn, Pb, Cu, and Cd in a metal-contaminated soil: EDTA vs. citrate. Environ Pollut 3:693–701. doi:10.1016/j.envpol.2007.06.054

La Force MJ, Fendorf S (2000) Solid-phase iron characterization during common selective sequential extractions. Soil Sci Soc Am J 64:1608–1615

Lanquar V, Lelievre F, Bolte S, Hames C, Alcon C, Neumann D, Vansuyt G, Curie C, Schroeder A, Kramer U, Barbier-Brygoo H, Thomine S (2005) Mobilization of vacuolar iron by AtNRAMP3 and AtNRAMP4 is essential for seed germination on low iron. EMBO J 24:4041–4051. doi:10.1038/sj.emboj.7600864

Lehto NH, Davisn W, Zhang H, Tych W (2006) Analysis of micro-nutrient behaviour in the rhizosphere using a DGT parameterised dynamic plant uptake model. Plant Soil 282:227–238. doi:10.1007/s11104-005-5848-6

Le Jean M, Schikora A, Mari S, Briat JF, Curie C (2005) A loss-of-function mutation in *AtYSL1* reveals its role in iron and nicotianamine seed loading. Plant J 44:769–782. doi:10.1111/j.1365-313X.2005.02569.x

Leeman M, Den Ouden FM, Van Pelt JA, Dirkx FPM, Steijl H, Bakker PAHM, Schippers B (1996) Iron availability affects induction of systemic resistance to fusarium wilt of radish by *Pseudomonas fluorescens*. Phytopathology 86:149–155. doi:10.1094/Phyto-86-149

Lemanceau P, Alabouvette C, Meyer JM (1986) Production of fusarinine and iron assimilation by pathogenic and non-pathogenic *Fusarium*. In: Swinburne TR (ed) Iron, siderophores and plant diseases. Plenum, New York and London, pp 251–259

Lemanceau P, Alabouvette C, Couteaudier Y (1988a) Recherches sur la résistance des sols aux maladies. XIV. Modification du niveau de réceptivité d'un sol résistant et d'un sol sensible aux fusarioses vasculaires en réponse à des apports de fer ou de glucose. Agronomie 8:155–162. doi:10.1051/agro:19880209

Lemanceau P, Samson R, Alabouvette C (1988b) Recherches sur la résistance des sols aux maladies. XV. Comparaison des populations de *Pseudomonas* fluorescents dans un sol résistant et un sol sensible aux fusarioses vasculaires. Agronomie 8:243–249. doi:10.1051/agro:19880310

Lemanceau P, Bakker PAHM, De Kogel WJ, Alabouvette C, Schippers B (1992) Effect of pseudobactin 358 production by *Pseudomonas putida* WCS358 on suppression of fusarium wilt of carnations by nonpathogenic *Fusarium oxysporum* Fo47. Appl Environ Microbiol 58:2978–2982

Lemanceau P, Bakker PAHM, De Kogel WJ, Alabouvette C, Schippers B (1993) Antagonistic effect of nonpathogenic *Fusarium oxysporum* Fo47 and pseudobactin 358 upon pathogenic *Fusarium oxysporum* f. sp. *dianthi*. Appl Environ Microbiol 59:74–82

Lemanceau P, Robin A, Mazurier S, Vansuyt G (2007) Implication of pyoverdines in the interactions of fluorescent pseudomonads with soil microflora and plant in the rhizosphere. In: Varma A, Chincholkar SB (eds) Microbial siderophores. Soil biology, vol. 12. Springer-Verlag, Berlin Heidelberg, pp 165–192

Li L, Cheng X, Ling HQ (2004) Isolation and characterization of Fe(III)-chelate reductase gene *LeFRO1* in tomato. Plant Mol Biol 54:125–136. doi:10.1023/B:PLAN.0000028790.75090.ab

Lindsay WL (1979) Chemical equilibria in soils. Wiley, Chuchester

Lindsay WL, Norvell WA (1978) Development of a DTPA soil test for zinc, iron, manganese and copper. Soil Sci Soc Am J 42:421–428

Ling HQ, Bauer P, Bereczky Z, Keller B, Ganal M (2002) The tomato *fer* gene encoding a bHLH protein controls iron-uptake responses in roots. Proc Natl Acad Sci USA 99:13938–13943. doi:10.1073/pnas.212448699

Loper JE (1988) Role of fluorescent siderophore production in biological control of *Pythium ultimum* by a *Pseudomonas fluorescens* strain. Phytopathology 78:166–172. doi:10.1094/Phyto-78-166

Loper JE, Buyer JS (1991) Siderophores in microbial interactions on plant surfaces. Mol Plant Microbe Interact 4:5–13

Loper JE, Lindow SE (1994) A biological sensor for iron available to bacteria in their habitats on plant surfaces. Appl Environ Microbiol 60:1934–1941

Loper JE, Henkels MD (1997) Availability of iron to *Pseudomonas fluorescens* in rhizopshere and bulk soil evaluated with an ice nucleation reporter gene. Appl Environ Microbiol 63:99–105

Loper JE, Lindow SE (1997) Reporter gene systems useful in evaluating in situ gene expression by soil-and plant-associated bacteria. In: Hurst CJ (ed) Manual of environmental microbiology. America Society for Microbiology, Washington, pp 482–492

Marschner H (1995) Mineral nutrition of higher plants, 2nd edn. Academic, London

Marschner H, Römheld V (1994) Strategies of plants for acquisition of iron. Plant Soil 165:261–274. doi:10.1007/BF00008069

Marschner P, Crowley DE (1997) Iron stress and pyoverdin production by a fluorescent pseudomonad in the rhizosphere of white lupine (Lupinus albus L.) and barley (Hordeum vulgare L.). Appl Environ Microbiol 63:227–281

Marschner P, Crowley DE (1998) Phytosiderophores decrease iron stress and pyoverdine production of Pseudomonas fluorescens Pf-5 (pvd-inaZ). Soil Biol Biochem 30:1275–1280. doi:10.1016/S0038-0717(98)00039-X

Marschner H, Römheld V, Kissel M (1987) Localization of phytosiderophore release and iron uptake along intact barley roots. Physiol Plant 71:157–162. doi:10.1111/j.1399-3054.1987.tb02861.x

Maurhofer M, Hase C, Meuwly P, Metraux JP, Défago G (1994) Induction of systemic resistance of tobacco to tobacco necrosis virus by the root-colonizing Pseudomonas fluorescens strain CHA0: influence of the gacA gene and of pyoverdine production. Phytopathology 84:139–146. doi:10.1094/Phyto-84-139

Meyer JM, Abdallah MA (1978) The fluorescent pigment of Pseudomonas fluorescens: biosynthesis, purification and physico-chemical properties. J Gen Microbiol 107:319–328

Meyer JM, Hallé F, Hohnadel D, Lemanceau P, Ratefiarivelo H (1987) Siderophores of Pseudomonas—Biological properties. In: Winkelmann G, Van der Helm D, Neilands JB (eds) Iron transport in microbes, plants and animals. VCH, Weinheim, pp 189–205

Meyer JM, Stintzi A, De Vos D, Cornelis P, Tappe R, Taraz K, Budzikiewicz H (1997) Use of siderophores to type pseudomonads: the three Pseudomonas aeruginosa pyoverdine systems. Microbiology 143:35–43

Meyer JM, Geoffroy VA, Baida N, Gardan L, Izard D, Lemanceau P, Achouak W, Palleroni NJ (2002) Siderophore typing, a powerful tool for the taxonomy of fluorescent and non-fluorescent Pseudomonas. Appl Environ Microbiol 68:2745–2453. doi:10.1128/AEM.68.6.2745-2753.2002

Meyer JM, Gruffaz C, Raharinosy V, Bezverbnaya I, Schäfer M, Budzikiewicz H (2008) Siderotyping of fluorescent Pseudomonas: molecular mass determination by mass spectrometry as a powerful pyoverdine siderotyping method. Biometals 21:259–271. doi:10.1007/s10534-007-9115-6

Mirleau P, Delorme S, Philippot L, Meyer JM, Mazurier S, Lemanceau P (2000) Fitness in soil and rhizosphere of Pseudomonas fluorescens C7R12 compared with a C7R12 mutant affected in pyoverdine synthesis and uptake. FEMS Microbiol Ecol 34:35–44. doi:10.1111/j.1574-6941.2000.tb00752.x

Mirleau P, Philippot L, Corberand T, Lemanceau P (2001) Involvement of nitrate reductase and pyoverdine in competitiveness of Pseudomonas fluorescens strain C7R12 in soil. Appl Environ Microbiol 67:2627–2635. doi:10.1128/AEM.67.6.2627-2635.2001

Misaghi IJ, Stowell LJ, Grogan RG, Spearman LC (1982) Fungistatic activity of water-soluble fluorescent pigments of fluorescent pseudomonads. Phytopathology 72:33–36. doi:10.1094/Phyto-72-33

Mizuno D, Higuchi K, Sakamoto T, Nakanishi H, Mori S, Nishizawa NK (2003) Three nicotianamine synthase genes isolated from maize are differentially regulated by iron nutritional status. Plant Physiol 132:1989–1997. doi:10.1104/pp.102.019869

Mori S, Nishizawa NK (1987) Methionine as a dominant precursor of phytosiderophores in Gramineae plants. Plant Cell Physiol 28:1081–1092

Mossialos D, Amoutzias GD (2007) Siderophores in fluorescent pseudomonads: new tricks from an old dog. Future Microbiol 2:387–395. doi:10.2217/17460913.2.4.387

Mossialos D, Ochsner U, Baysse C, Chablain P, Pirnay JP, Koedam N, Budzikiewicz H, Fernandez DU, Schafer M, Ravel J, Cornelis P (2002) Identification of new, conserved, nonribosomal peptide synthetases from fluorescent pseudomonads involved in the biosynthesis of the siderophore pyoverdine. Mol Microbiol 45:1673–1685. doi:10.1046/j.1365-2958.2002.03120.x

Nakanishi H, Yamaguchi H, Sasakuma T, Nishizawa NK, Mori S (2000) Two dioxygenase genes, Ids3 and Ids2, from Hordeum vulgare are involved in the biosynthesis of mugineic acid family phytosiderophores. Plant Mol Biol 44:199–207. doi:10.1023/A:1006491521586

Negishi T, Nakanishi H, Yazaki J, Kishimoto N, Fujii F, Shimbo K, Yamamoto K, Sakata K, Sasaki T, Kikuchi S, Mori S, Nishizawa NK (2002) cDNA microarray analysis of gene expression during Fe- deficiency stress in barley suggests that polar transport of vesicles is implicated in phytosiderophore secretion in Fe- deficient barley roots. Plant J 30:83–94. doi:10.1046/j.1365-313X.2002.01270.x

Neilands JB (1981) Iron absorption and transport in microorganisms. Annu Rev Nutr 1:27–46. doi:10.1146/annurev.nu.01.070181.000331

Nguyen C (2003) Rhizodeposition of organic C by plants: mechanisms and controls. Agronomie 23:375–396. doi:10.1051/agro:2003011

Nozoye T, Inoue H, Takahashi M, Ishimaru Y, Nakanishi H, Mori S, Nishizawa NK (2007) The expression of iron homeostasis-related genes during rice germination. Plant Mol Biol 64:35–47. doi:10.1007/s11103-007-9132-4

Ogo Y, Itai RN, Nakanishi H, Inoue H, Kobayashi T, Suzuki M, Takahashi M, Mori S, Nishizawa NK (2006) Isolation and characterization of IRO2, a novel iron-regulated bHLH transcription factor in graminaceous plants. J Exp Bot 57:2867–2878. doi:10.1093/jxb/erl054

Ogo Y, Itai RN, Nakanishi H, Kobayashi T, Takahashi M, Mori S, Nishizawa NK (2007) The rice bHLH protein OsIRO2 is an essential regulator of the genes involved in Fe uptake under Fe-deficient conditions. Plant J 51:366–377. doi:10.1111/j.1365-313X.2007.03149.x

Ogo Y, Kobayashi T, Nakanishi IR, Nakanishi H, Kakei Y, Takahashi M, Toki S, Mori S, Nishizawa NK (2008) A novel NAC transcription factor, IDEF2, that recognizes the iron deficiency-responsive element 2 regulates the genes involved in iron homeostasis in plants. 283:13407–13417

Palanche T, Marmolle F, Abdallah MA, Shanzer A, Albrecht-Gary AM (1999) Fluorescent siderophore-based chemosensors: iron(III) quantitative determinations. J Biol Inorg Chem 4:188–198. doi:10.1007/s007750050304

Parfitt RL, Childs CW (1988) Estimation of forms of Fe and Al—a review and analysis of contrasting soils by dissolution and Mössbauer methods. Aust J Soil Res 26:121–144. doi:10.1071/SR9880121

Poole K (2004) *Pseudomonas*. In: Crosa JH, Mey AR, Payne SM (eds) Iron transport in bacteria. American Society of Microbiology Press, Washington, pp 293–310

Powell PE, Cline GR, Reid CPP, Szaniszlo PJ (1980) Occurrence of hydroxamate siderophore iron chelators in soils. Nature 287:833–834. doi:10.1038/287833a0

Raaijmakers JM, Van der Sluis L, Koster M, Bakker PAHM, Weisbeek PJ, Schippers B (1995) Utilisation of heterologous siderophores and rhizosphere competence of fluorescent *Pseudomonas* spp. Can J Microbiol 41:126–135

Reichard PU, Kraemer SM, Frazier SW, Kretzschmar R (2005) Goethite dissolution in the presence of phytosiderophores: Rates, mechanisms, and the synergistic effect of oxalate. Plant Soil 276:115–132. doi:10.1007/s11104-005-3504-9

Reichard PU, Kretzschmar R, Kraemer SM (2007a) Rate laws of steady-state and non-steady-state ligand-controlled dissolution of goethite. Colloids Surf A Physicochem Eng Asp 306:22–28. doi:10.1016/j.colsurfa.2007.03.001

Reichard PU, Kretzschmar R, Kraemer SM (2007b) Dissolution mechanisms of goethite in the presence of siderophores and organic acids. Geochim Cosmochim Acta 71:5635–5650. doi:10.1016/j.gca.2006.12.022

Reichman SM, Parker DR (2007) Probing the effect of light and temperature on diurnal rhythms of phytosiderophore release in wheat. New Phytol 174:101–108. doi:10.1111/j.1469-8137.2007.01990.x

Reyes I, Torrent J (1997) Citrate-ascorbate as a highly selective extractant for poorly crystalline iron oxides. Soil Sci Soc Am J 61:1647–1654

Robin A, Vansuyt G, Corberand T, Briat JF, Lemanceau P (2006a) The soil type affects both the differential accumulation of iron between wild type and ferritin over-expressor tobacco plants and the sensitivity of their rhizosphere bacterioflora to iron stress. Plant Soil 283:73–81. doi:10.1007/s11104-005-9437-5

Robin A, Mougel C, Siblot S, Vansuyt G, Mazurier S, Lemanceau P (2006b) Effect of ferritin over-expression in tobacco on the structure of bacterial and pseudomonad communities associated with the roots. FEMS Microbiol Ecol 58:492–502. doi:10.1111/j.1574-6941.2006.00174.x

Robin A, Mazurier S, Mougel C, Vansuyt G, Corberand T, Meyer JM, Lemanceau P (2007) Diversity of root-associated fluorescent pseudomonads as affected by ferritin over-expression in tobacco. Environ Microbiol 9:1724–1737. doi:10.1111/j.1462-2920.2007.01290.x

Robin A, Vansuyt G, Hinsinger P, Meyer JM, Briat JF, Lemanceau P (2008) Iron dynamics in the rhizosphere: consequences for plant health and nutrition. Adv Agron 99:183–225. doi:10.1016/S0065-2113(08)00404-5

Robinson NJ, Procter CM, Connolly EL, Guerinot ML (1999) A ferric-chelate reductase for iron uptake from soils. Nature 397:694–697. doi:10.1038/17800

Römheld V, Marschner H (1986) Evidence for a specific uptake system for iron phytosiderophores in roots of grasses. Plant Physiol 80:175–180. doi:10.1104/pp.80.1.175

Santi S, Cesco S, Varanini Z, Pinton R (2005) Two plasma membrane H(+)-ATPase genes are differentially expressed in iron-deficient cucumber plants. Plant Physiol Biochem 43:287–292. doi:10.1016/j.plaphy.2005.02.007

Schaaf G, Ludewig U, Erenoglu BE, Mori S, Kitahara T, Von Wirén N (2004) ZmYS1 functions as a proton-coupled symporter for phytosiderophore- and nicotianamine-chelated metals. J Biol Chem 279:9091–9096. doi:10.1074/jbc.M311799200

Schaaf G, Schikora A, Häberle J, Vert G, Ludewig U, Briat JF, Curie C, Von Wirén N (2005) A putative function of the *Arabidopsis* Fe-phytosiderophore transporter homolog AtYSL2 in Fe and Zn homeostasis. Plant Cell Physiol 46:762–774. doi:10.1093/pcp/pci081

Schagerlöf U, Wilson G, Hebert H, Al-Karadaghi S, Hägerhäll C (2006) Transmembrane topology of FRO2, a ferric chelate reductase from *Arabidopsis thaliana*. Plant Mol Biol 62:215–221. doi:10.1007/s11103-006-9015-0

Scheinost AC, Schwertmann U (1999) Color identification of iron oxides and hydroxysulfates: Use and limitations. Soil Sci Soc Am J 63:1463–1471

Scher FM, Baker R (1980) Mechanism of biological control in a Fusarium-suppressive soil. Phytopathology 70:412–417. doi:10.1094/Phyto-70-412

Scher FM, Baker R (1982) Effect of *Pseudomonas putida* and a synthetic iron chelator on induction of soil suppressiveness to Fusarium wilt pathogens. Phytopathology 72:1567–1573. doi:10.1094/Phyto-72-1567

Schwertmann U (1964) Differenzierung der eisenoxide des bodens durch extraktion mit ammoniumoxalat-lösung. Z Pflanzenernaehr Bodenkd 105:194–202. doi:10.1002/jpln.3591050303

Schwyn B, Neilands JB (1987) Universal chemical assay for the detection and determination of siderophores. Anal Biochem 106:47–56. doi:10.1016/0003-2697(87)90612-9

Shojima S, Nishizawa NK, Mori S (1989) Establishment of a cell free system for the biosynthesis of nicotianamine. Plant Cell Physiol 30:673–677

Simeoni LA, Lindsay WL, Baker R (1987) Critical iron level associated with biological control of Fusarium wilt. Phytopathology 77:1057–1061. doi:10.1094/Phyto-77-1057

Sneh B, Dupler M, Elad Y, Baker R (1984) Chlamydospore germination of *Fusarium oxysporum f.sp. cucumerinum* as affected by fluorescent and lytic bacteria from a Fusarium-suppressive soil. Phytopathology 74:1115–1124. doi:10.1094/Phyto-74-1115

Soltanpour PN, Schwab AP (1977) A new soil test for simultaneous extraction of macro- and micro-nutrients in alkaline soils. Commun Soil Sci Plant Anal 8:195–207. doi:10.1080/00103627709366714

Steinberg C, Edel-Hermann V, Alabouvette C, Lemanceau P (2007) Soil suppressiveness to plant diseases. In: Van Elsas D, Jansson JK, Trevors JT (eds) Modern Soil Microbiology, 2nd edn. CRC, Taylor & Francis Group, London, pp 455–477

Stintzi A, Evans K, Meyer JM, Poole K (1998) Quorum sensing and siderophore biosynthesis in *Pseudomonas aeruginosa*: lasR/lasI mutants exhibit reduced pyoverdine synthesis. FEMS Microbiol Lett 166:341–345. doi:10.1111/j.1574-6968.1998.tb13910.x

Sugiura Y, Tanaka H, Mino Y, Ishida T, Ota N, Nomoto K, Yosioka H, Takemoto T (1981) Structure, properties, and transport mechanism of iron(III) complex of mugineic acid, a possible phytosiderophore. J Am Chem Soc 103:6979–6982. doi:10.1021/ja00413a043

Takagi SI (1976) Naturally occurring iron–chelating compounds in oat–root and rice–root washings.1. Activity measurement and preliminary characterization. Soil Sci Plant Nutr 22:423–433

Takagi SI, Nomoto K, Takemoto T (1984) Physiological aspect of mugineic acid, a possible phytosiderophore of graminaceous plants. J Plant Nutr 7:469–477. doi:10.1080/01904168409363213

Takahashi M, Nakanishi H, Kawasaki S, Nishizawa NK, Mori S (2001) Enhanced tolerance of rice to low iron availability in alkaline soils using barley nicotianamine aminotransferase genes. Nat Biotechnol 19:466–469. doi:10.1038/88143

Taylor SR (1964) Abundance of chemical elements in the continental crust—A new table. Geochim Cosmochim Acta 28:1273–1285. doi:10.1016/0016-7037(64) 90129-2

Tessier A, Campbell PGC, Bisson M (1979) Sequential extraction procedure for the speciation of particulate trace metals. Anal Chem 51:844–851. doi:10.1021/ac50043a017

Vansuyt G, Robin A, Briat JF, Curie C, Lemanceau P (2007) Iron acquisition from Fe-pyoverdine by *Arabidopsis thaliana*. Mol Plant Microbe Interact 4:441–447. doi:10.1094/MPMI-20-4-0441

Varotto C, Maiwald D, Pesaresi P, Jahns P, Salamini F, Leister D (2002) The metal ion transporter IRT1 is necessary for iron homeostasis and efficient photosynthesis in *Arabidopsis thaliana*. Plant J 31:589–599. doi:10.1046/j.1365-313X.2002.01381.x

Vert G, Grotz N, Dedaldechamp F, Gaymard F, Guerinot ML, Briat JF, Curie C (2002) IRT1, an *Arabidopsis* transporter essential for iron uptake from the soil and for plant growth. Plant Cell 14:1223–1233. doi:10.1105/tpc.001388

Vert GA, Briat JF, Curie C (2003) Dual regulation of the *Arabidopsis* high-affinity root iron uptake system by local and long-distance signals. Plant Physiol 132:796–804. doi:10.1104/pp.102.016089

Visca P, Imperi F, Lamont IL (2007) Pyoverdine siderophores: from biogenesis to biosignificance. Trends Microbiol 15:22–30. doi:10.1016/j.tim.2006.11.004

Van Wuytswinkel O, Vansuyt G, Grignon N, Fourcroy P, Briat JF (1999) Iron homeostasis alteration in transgenic tobacco overexpressing ferritin. Plant J 17:93–97. doi:10.1046/j.1365-313X.1999.00349.x

Von Wirén N, Römheld V, Morel JL, Guckert A, Marschner H (1993) Influence of microorganisms on iron acquisition in maize. Soil Biol Biochem 25:371–376. doi:10.1016/0038-0717(93) 90136-Y

Von Wirén N, Mori S, Marschner H, Römheld V (1994) Iron inefficiency in maize mutant *ys1* (*Zea mays* L. cv Yellow-Stripe) is caused by a defect in uptake of iron phytosiderophores. Plant Physiol 106:71–77

Von Wirén N, Römheld V, Shioiri T, Marschner H (1995) Competition between micro-organisms and roots of barley and sorghum for iron accumulated in the root apoplasm. New Phytol 130:511–521. doi:10.1111/j.1469-8137.1995.tb04328.x

Waters BM, Blevins DG, Eide DJ (2002) Characterization of FRO1, a pea ferric-chelate reductase involved in root iron acquisition. Plant Physiol 129:85–94. doi:10.1104/pp.010829

Whiteley M, Kimberely ML, Greenberg EP (1999) Identification of genes controlled by quorum sensing in *Pseudomonas aeruginosa*. Proc Natl Acad Sci USA 96:13604–13609. doi:10.1073/pnas.96.24.13904

Wiederhold JG, Teutsch N, Kraemer SM, Halliday AN, Kretzschmar R (2007a) Iron isotope fractionation in oxic soils by mineral weathering and podzolization. Geochim Cosmochim Acta 71:5821–5833. doi:10.1016/j.gca.2007.07.023

Wiederhold JG, Teutsch N, Kraemer SM, Halliday AN, Kretzschmar R (2007b) Iron isotope fractionation during pedogenesis in redoximorphic soils. Soil Sci Soc Am J 71:1840–1850. doi:10.2136/sssaj2006.0379

Winkelmann G (1991) Handbook of microbial iron chelate. CRC, Bocca Raton, Florida

Winkelmann G (2001) Siderophore transport in fungi. In: Winkelmann G (ed) Microbial transport systems. Wiley-VCH, Weinheim, pp 463–480

Winkelmann G (2002) Microbial siderophore-mediated transport. Biochem Soc Trans 30:691–696. doi:10.1042/BST0300691

Winkelmann G (2007) Ecology of siderophores with special reference to the fungi. Biometals 20:379–392. doi:10.1007/s10534-006-9076-1

Yang CH, Crowley DE (2000) Rhizosphere microbial community structure in relation to root location and plant iron nutritional status. Appl Environ Microbiol 87:345–351. doi:10.1128/AEM.66.1.345-351.2000

Yehuda Z, Shenker M, Römheld V, Marschner H, Hadar Y, Chen Y (1996) The role of ligand exchange in the uptake of iron from microbial siderophores by gramineous plants. Plant Physiol 112:1273–1280

Yi Y, Guerinot ML (1996) Genetic evidence that induction of root Fe(III) chelate reductase activity is necessary for iron uptake under iron deficiency. Plant J 10:835–844. doi:10.1046/j.1365-313X.1996.10050835.x

Yuan YX, Zhang J, Wang DW, Ling HQ (2005) AtbHLH29 of *Arabidopsis thaliana* is a functional ortholog of tomato FER involved in controlling iron acquisition in strategy I plants. Cell Res 15:613–621. doi:10.1038/sj.cr.7290331